Encyclopedia of
E–Collaboration

Ned Kock
Texas A&M International University, USA

INFORMATION SCIENCE REFERENCE
Hershey · New York

Acquisitions Editor:	Kristin Klinger
Development Editor:	Kristin Roth
Senior Managing Editor:	Jennifer Neidig
Managing Editor:	Sara Reed
Copy Editor:	Killian Piraro and Becky Shore
Typesetter:	Jeff Ash, Jennifer Neidig, Sara Reed
Cover Design:	Lisa Tosheff
Printed at:	Yurchak Printing Inc.

Published in the United States of America by
Information Science Reference (an imprint of IGI Global)
701 E. Chocolate Avenue, Suite 200
Hershey PA 17033
Tel: 717-533-8845
Fax: 717-533-8661
E-mail: cust@igi-global.com
Web site: http://www.igi-global.com/reference

and in the United Kingdom by
Information Science Reference (an imprint of IGI Global)
3 Henrietta Street
Covent Garden
London WC2E 8LU
Tel: 44 20 7240 0856
Fax: 44 20 7379 0609
Web site: http://www.eurospanonline.com

Library of Congress Cataloging-in-Publication Data

Encyclopedia of e-collaboration / Ned Kock, editor.
   p. cm.
  Summary: "This encyclopedia provides the most comprehensive compilation of information on the design and implementation of e-collaboration technologies, their behavioral impact on individuals and groups, and theoretical considerations on links between the use of e-collaboration technology and behavioral patterns. It delivers indispensable content to libraries and researchers looking to develop programs of investigation into the use of e-collaboration"--Provided by publisher.
  Includes bibliographical references and index.
  ISBN 978-1-59904-000-4 (hardcover) -- ISBN 978-1-59904-001-1 (ebook)
  1. Project management. 2. Virtual work teams. 3. Human-computer interaction. 4. Information technology. I. Kock, Ned F., 1964-
  HD69.P75.E53 2007
  658.4'0220285--dc22
                         2007022233

British Cataloguing in Publication Data
A Cataloguing in Publication record for this book is available from the British Library.

Editorial Advisory Board

List of Contributors

Aiken, Milam / *University of Mississippi, USA* 457, 706

Allen, Mike / *UW–Milwaukee, USA* 499

Antunes, Pedro / *University of Lisboa, Portugal* 133

Augenbroe, Godfried / *Georgia Institute of Technology, USA* 589

Barkhi, Reza / *Virginia Tech, USA and American University of Sharjah, UAE* 377

Barthès, Jean-Paul / *UTC – Université de Technologie de Compiègne, France* 36

Batenburg, Ronald / *Utrecht University, The Netherlands* 14

Beenkens, Fernao H.C. / *Delft University of Technology, The Netherlands* 265

Bennani, Az-Eddine / *Université de Technologie de Compiègne, Reims Management School, France* 505

Blanchard, Anita L. / *University of North Carolina – Charlotte, USA* 126

Blashki, Kathy / *Deakin University, Australia* 246

Blecker, Thorsten / *Hamburg University of Technology (TUHH), Germany* 479

Boff, Elisa / *Caxias do Sul University (UCS), Brazil and Federal University of Rio Grande do Sul (UFRGS), Brazil* 74

Borges, Marcos R. S. / *IM/DCC & NCE – Universidade Federal do Estado do Rio de Janeiro, Brazil* 81, 718

Bosin, Andrea / *Università degli Studi di Cagliari, Italy* 554

Bostrom, Robert / *University of Georgia, USA* 191

Brezillon, Patrick / *University Paris 6, France* 114

Briggs, Robert O. / *University of Nebraska at Omaha, USA and University of Alaska Fairbanks, USA* 60, 139, 631

Canal, Gerome / *LORIA – Nancy Université, France* 42

Capistrano Costa, Inaldo /*Learning and Interaction Laboratory (LAI) – Aeronautical Institute of Technology (ITA), Brazil* 645

Carmona, Jesus / *Texas A&M International University, USA* 102

Cassivi, Luc / *University of Quebec – Montreal, Canada* 216

Chiasson, Mike / *Lancaster University Management School, UK* 521

Chilton, Michael / *Kansas State University, USA* 693

Cho, Kwangsu / *University of Missouri, Columbia, USA* 226

Chua, Alton / *Nanyang Technological University, Singapore* 437

Chung, T. Rachel / *University of Pittsburgh, USA* 226

Cox, Sharon A. / *Birmingham City University, UK* 279, 527

Cyr, Dianne / *Simon Fraser University, Canada* 240

Czirkos, Zoltán / *Budapest University of Technology and Economics, Hungary* 172

Damjanovic, Violeta / *Salzburg Research, Austria* 29

Daneshgar, Farhad / *University of New South Wales, Australia* 42

de Jesus Viana Sá, Eveline / *Federal Center for Technological Education of Maranhão (CEFET-MA), Brazil and Learning and Interaction Laboratory (LAI) – Aeronautical Institute of Technology (ITA), Brazil* 645

de Souza, Jano M. / *Federal University of Rio de Janeiro, Brazil* 36

de Vreede, Gert-Jan / *University of Nebraska at Omaha, USA and Delft University of Technology, The Netherlands* 60, 139, 272, 631

de Vries, Henk / *AdvancedCollaboration.nl, The Netherlands* 68

Del Aguila-Obra, Ana Rosa / *University of Málaga, Spain* 618

DeLuca, Dorrie / *University of Delaware, USA* 463, 699

Dennis, Alan R. / *Indiana University, USA* 330

Dessì, Nicoletta / *Università degli Studi di Cagliari, Italy* 554

Dew, Peter M. / *The University of Leeds, UK* 512

Dias de Figueiredo, António / *University of Coimbra, Portugal* 596

Dias, Jorge / *University of Coimbra, Portugal* 561

Diaz, Alicia / *UNLP, Argentina* 42

Dorn, Christoph / *Vienna University of Technology, Austria* 389

Dow, Kevin E. / *Kent State University, USA* 146

Du, Jianxia / *Mississippi State University, USA* 370

DuBois, Cathy L. Z. / *Kent State University, USA* 259

Dustdar, Schahram / *University of Leicester, UK* 389

Ertl, Bernhard / *Universität der Bundeswehr München, Germany* 233

Fedorowicz, Jane / *Bentley College, USA* 319

Fernandez, Amyris / *Ibmec Educacional, Brazil* 120

Ferreira Teixeira, Jeane Silva / *Federal Center for Technological Education of Maranhão (CEFET-MA), Brazil and Learning and Interaction Laboratory (LAI) – Aeronautical Institute of Technology (ITA), Brazil* 645

Feuerlicht, George / *University of Technology, Sydney, Australia* 355

Flach, John M. / *Wright State University, USA* 673

Frößler, Frank / *University College Dublin, Ireland* 487

Fugini, Maria Grazia / *Politecnico di Milano, Italy* 554

Fuks, Hugo / *Catholic University of Rio de Janeiro, Brazil* 637

Gardner, Susan / *California State University, USA* 686

Garrido-Moreno, Aurora / *University of Málaga, Spain* 667

Garza, Vanessa / *Texas A&M International University, USA* 186

Gasson, Susan / *Drexel University, USA* 699

Gerosa, Marco A. / *University of Vila Vehla, Brazil* 637

Ghosh, Anupam / *ICFAI Institute for Management Teachers, India* 319

Giuliani, Giovanni / *HP Italiana SRL, Italy* 389

Gombotz, Robert / *Vienna University of Technology, Austria* 389

Greenberg, Penelope Sue / *Widener University, USA* 146

Greenberg, Ralph H. / *Temple University, USA* 146

Gupta, Jatinder N.D. / *University of Alabama – Huntsville, USA* 655

Gupta, Saurabh / *University of North Florida, USA* 191

Haq, A. Noorul / *National Institute of Technology, India* 584

Hartman, Jackie L. / *Colorado State University, USA* 204

Hassell, Lewis / *Drexel University, USA* 577

Hayes, Niall / *Lancaster University Management School, UK* 521

Hicks, Richard C. / *Texas A&M International University, USA* 286, 292

Hooker, Robert / *Florida State University, USA* 324

Hosszú, Gábor / *Budapest University of Technology and Economics, Hungary* 107, 172, 624

Hrastinski, Stefan / *Jönköping International Business School, Sweden* 349

Hribernik, Karl A. / *Bremen Institute of Industrial Technology and Applied Work Science (BIBA), Germany* 308, 612

Janssen, Marijn / *Delft University of Technology, The Netherlands* 68

José Luzón, María / *University of Zaragoza, Spain* 1

Kalaian, Sema A. / *Eastern Michigan University, USA* 450

Kannan, G. / *National Institute of Technology, India* 584

Keogh, Kathleen / *The University of Ballarat, Australia* 493

Khazanchi, Deepak / *University of Nebraska at Omaha, USA* 472

Klobas, Jane / *Bocconi University, Italy and University of Western Australia, Australia* 712

Knolmayer, Gerhard F. / *University of Bern, Switzerland* 253

Kock, Ned / *Texas A&M International University, USA* 48, 314, 602, 699

Kolfschoten, Gwendolyn L. / *Delft University of Technology, The Netherlands* 60, 139, 631

Kong, Eu-Jin / *Universitas 21 Global, Singapore* 437

Kovács, Ferenc / *Budapest University of Technology and Economics, Hungary* 107, 172, 624

Kumar, P. Sasi / *National Institute of Technology, India* 584

Lam, Wing / *Universitas 21 Global, Singapore* 437

Lau, Lydia M. S. / *The University of Leeds, UK* 512, 547

Lewis, Carmen / *Florida State University, USA* 324

Liao, Qinyu / *University of Texas at Brownsville and Texas Southmost College, USA* 383, 680

Liberati, Diego / *Italian National Research Council, Italy* 554

Liebhart, Ursula / *Alpen-Adria-University of Klagenfurt, Austria* 479

Lim, John / *National University of Singapore, Singapore* 301

Lucena, Carlos J. P. / *Catholic University of Rio de Janeiro, Brazil* 637

Luo, Xin / *Virginia State University, USA* 383, 680

Lyons, Joseph B. / *Air Force Research Laboratory, USA* 7

Macaulay, Linda / *The University of Manchester, UK* 534, 569

Macedo Prudêncio Lopes, Tatiane / *Learning and Interaction Laboratory (LAI) – Aeronautical Institute of Technology (ITA), Brazil* 645

Mametjanov, Azamat / *University of Nebraska at Omaha, USA* 472

Maria Santoro, Flávia / *DIA – Universidade Federal do Estado do Rio de Janeiro, Brazil* 81, 718

Martz, Ben / *Northern Kentucky University, USA* 686

Masucci, Michele / *Temple University, USA* 153, 405

Mathiyalakan, Sathasivam / *Winston-Salem State University, USA* 337

McHaney, Roger / *Kansas State University, USA* 693

McKinney, Vicki R. / *University of Arkansas, USA* 499

Mejía, Ricardo / *Ecole Supérieure des Technologies Industrielles Avancées, France* 589

Mescioglu, Ibrahim / *Texas A&M International University, USA* 286

Molina, Arturo / *Centro de Innovación en Diseño y Tecnología Tecnológico de Monterrey, Mexico* 589

Moreira Silveira, D'Ilton / *Learning and Interaction Laboratory (LAI) – Aeronautical Institute of Technology (ITA), Brazil* 645

Munkvold, Bjørn Erik / *University of Agder, Norway* 411

Nagasundaram, Murli / *Boise State University, USA* 198

Nelson, W. Todd / *Air Force Research Laboratory, USA* 673

Nichol, Sophie / *Deakin University, Australia* 246

Nilsson, Michael / *Luleå University of Technology, Sweden* 308, 612

Ning, Ke / *National University of Ireland, Ireland* 389

Nolan, Terry / *Auckland University of Technology, New Zealand* 534

Ocker, Rosalie J. / *The Pennsylvania State University, USA* 363

Ortiz, Jaime / *Texas A&M International University, USA* 178

Padilla-Meléndez, Antonio / *University of Málaga, Spain* 618, 667

Panteli, Niki / *University of Bath, UK* 398

Panyasorn, Jessada / *University of Bath, UK* 398

Pardede, Raymond / *Budapest University of Technology and Economics, Hungary* 624

Parente de Oliveira, José Maria / *Aeronautical Institute of Technology (ITA), Brazil* 645

Pashnyak, Tatyana G. / *Florida State University, USA* 54, 210

Pate, George / *Mississippi State University, USA* 370

Paz Dennen, Vanessa / *Florida State University, USA* 54, 210

Peñaranda, Nicolás / *Centro de Innovación en Diseño y Tecnología Tecnológico*
de Monterrey, México 589

Pena-Sanchez, Rolando / *Texas A&M International University, USA* 286, 292

Peray, Sébastien / *European Microsoft Innovation Center, Germany* 389

Perkins, John S. / *Newman College of Higher Education, UK* 279, 527

Pes, Barbara / *Università degli Studi di Cagliari, Italy* 554

Pimentel, Mariano / *Federal University of the State of Rio de Janeiro, Brazil* 637

Powell, Philip / *University of Bath, UK* 398

Ramos de Oliveira, Alessandro / *Learning and Interaction Laboratory (LAI) –*
Aeronautical Institute of Technology (ITA), Brazil 645

Raposo, Alberto / *Catholic University of Rio de Janeiro, Brazil* 637

Reiff-Marganiec, Stephan / *University of Leicester, UK* 389

Reilly, Richard R. / *Stevens Institute of Technology, USA* 21, 660

Richly, Gábor / *Budapest University of Technology and Economics, Hungary* 107

Riemer, Kai / *Muenster University, Germany* 487

Rocha, Rui / *University of Coimbra, Portugal* 561

Rouse, Anne C. / *Deakin University, Australia* 424

Ryan, Michael R. / *Stevens Institute of Technology, USA* 21

Santos, Neide / *IME/DICC – Universidade do Estado do Rio de Janeiro, Brazil* 81, 718

Schall, Daniel / *Vienna University of Technology, Austria* 389

Schoonover, Terrance / *University of Nebraska at Omaha, USA* 343

Schwartz, Daniel H. / *Air Force Research Laboratory, USA* 7, 673

Seetharaman, Priya / *Indian Institute of Management Calcutta, India* 417

Senthil, R. / *National Institute of Technology, India* 584

Sharma, Sushil K. / *Ball State University, USA* 655

Shea, Vincent J. / *Kent State University, USA* 159

Smith, Hugh / *Florida State University, USA* 324

Sobel Lojeski, Karen / *Virtual Distance International, USA* 660

Sonenberg, Liz / *The University of Melbourne, Australia* 493

Stevens, Ken / *Memorial University of Newfoundland, Canada* 444

Stokes, Charlene K. / *Air Force Research Laboratory, USA* 7, 673

Swindler, Stephanie D. / *Air Force Research Laboratory, USA* 7

Switzer, Jamie S. / *Colorado State University, USA* 204

Tan, Yin Leng / *The University of Manchester, UK* 569

Tarmizi, Halbana / *University of Nebraska at Omaha, USA* 272

Thoben, Klaus-Dieter / *Bremen Institute of Industrial Technology and Applied Work*
Science (BIBA), Germany 308, 612

Tilly, Marcel / *European Microsoft Innovation Center, Germany* 389

Torkia, Eric / *Technology Partnerz, Ltd., Canada* 216

Torres Fernandes, Clovis /*Learning and Interaction Laboratory (LAI) – Aeronautical Institute of Technology (ITA), Brazil* 645

Tyran, Craig K. / *Western Washington University, USA*.....................................540

Tyran, Kristi M. Lewis / *Western Washington University, USA*..........................540

Vaidya, Sanjiv D. / *Indian Institute of Management Calcutta, India* 417

van Baalen, Peter J. / *RSM Erasmus University, The Netherlands* 87

van Duinkerken, Wilco / *Utrecht University, The Netherlands*................................14

van Fenema, Paul C. / *Netherlands Defense Academy, The Netherlands* 87

van Grinsven, Jürgen H. M. / *AdvancedCollaboration.nl, The Netherlands* 68

Verburg, Robert M. / *Delft University of Technology, The Netherlands*................265

Versendaal, Johan / *Utrecht University, The Netherlands*.....................................14

Vicari, Rosa Maria / *Federal University of Rio Grande do Sul (UFRGS), Brazil* 74

Vivacqua, Adriana S. / *Federal University of Rio de Janeiro, Brazil* 36

Vorisek, Jiri / *University of Economics Prague, Czech Republic* 355

Waldrup, Bobby E. / *University of North Florida, USA*....................................164

Wang, Zhen / *National University of Singapore, Singapore* 301

Wasko, Molly / *Florida State University, USA*...324

Williams, Michael L. / *Pepperdine University, USA* 330

Worrell, James / *Florida State University, USA* 324

Yoon, Tom / *Florida State University, USA* 324

Zakaria, Norhayati / *Universiti Utara Malaysia, Malaysia* 430

Zhao, Fang / *RMIT University, Australia* 606

Zhao, Wenbing / *Cleveland State University, USA* 95

Zhong, Yingqin / *National University of Singapore, Singapore* 301

Zigurs, Ilze / *University of Nebraska at Omaha, USA*................................343, 472

Contents

Academic Weblogs as Tools for E-Collaboration Among Researchers / *María José Luzón*..........................1

Adaptive Workforce as the Foundation for E-Collaboration, An / *Charlene K. Stokes,*
Joseph B. Lyons, Daniel H. Schwartz, and Stephanie D. Swindler 7

Added Value of E-Procurement for Buyer-Supplier Interaction, The / *Wilco van Duinkerken,*
Ronald Batenburg, and Johan Versendaal..............................14

Ambassadorial Leadership and E-Collaborative Teams / *Richard R. Reilly and Michael R. Ryan*...............21

Ambient Intelligent Prototype for Collaboration, An / *Violeta Damjanovic* 29

Awareness Approaches of E-Collaboration Technology / *Adriana S. Vivacqua, Jano M. de Souza,*
and Jean-Paul Barthès..............................36

Awareness Framework for Divergent Knowledge Communities, An / *Farhad Daneshgar,*
Gerome Canal, and Alicia Diaz..............................42

Basic Definition of E-Collaboration and its Underlying Concepts, A / *Ned Kock* 48

Blogging Technology and its Support for E-Collaboration / *Vanessa Paz Dennen and*
Tatyana G. Pashnyak..............................54

Collaboration Engineering for Designing Self-Directed Group Efforts / *Gert-Jan de Vreede,*
Robert O. Briggs, and Gwendolyn L. Kolfschoten..............................60

Collaboration Methods and Tools for Operational Risk Management / *Jürgen H. M. van Grinsven,*
Marijn Janssen, and Henk de Vries..............................68

Collaborative Editor for Medical Learning Environments, A / *Elisa Boff and Rosa Maria Vicari*...............74

Collaborative Writing in E-Learning Environments / *Neide Santos, Flávia Maria Santoro, and*
Marcos R. S. Borges..............................81

Collective Meaning in E-Collaborating Groups / *Paul C. van Fenema and Peter J. van Baalen*...............87

Concurrency Control in Real-Time E-Collaboration Systems / *Wenbing Zhao*..............................95

Consequences of IM on Presence Awareness and Interruptions / *Jesus Carmona* 102

Content-Based Searching in Group Communication Systems / *Gábor Richly, Gábor Hosszú, and Ferenc Kovács* 107

Context-Based Explanations for E-Collaboration / *Patrick Brezillon* 114

Cultural Influences on Virtual Teamwork Collaboration / *Amyris Fernandez* 120

Definition, Antecedents, and Outcomes of Successful Virtual Communities / *Anita L. Blanchard* 126

Design Framework for Mobile Collaboration, A / *Pedro Antunes* 133

Design Patterns for Facilitation in E-Collaboration / *Gwendolyn L. Kolfschoten, Robert O. Briggs, and Gert-Jan de Vreede* 139

Developing Synergies between E-Collaboration and Participant Budgeting Research / *Kevin E. Dow, Ralph H. Greenberg, and Penelope Sue Greenberg* 146

Digital Divide and E-Health for E-Collaboration Research / *Michele Masucci* 153

E-Collaboration and the Financial Auditor / *Vincent J. Shea* 159

E-Collaboration as a Tool in the Investigation of Occupational Fraud / *Bobby E. Waldrup* 164

E-Collaboration Enhanced Host Security / *Zoltán Czirkos, Gábor Hosszú, and Ferenc Kovács* 172

E-Collaboration for Internationalizing U.S. Higher Education Institutions / *Jaime Ortiz* 178

E-Collaboration Overview of Behavior and its Relationship with Evolutionary Factors, An / *Vanessa Garza* 186

E-Collaboration Technologies Impact on Learning / *Saurabh Gupta and Robert Bostrom* 191

E-Collaboration Through Blogging / *Murli Nagasundaram* 198

E-Collaboration Using Group Decision Support Systems in Virtual Meetings / *Jamie S. Switzer and Jackie L. Hartman* 204

E-Collaboration within Blogging Communities of Practice / *Vanessa Paz Dennen and Tatyana G. Pashnyak* 210

E-Collaboration: A Dynamic Enterprise Model / *Eric Torkia and Luc Cassivi* 216

E-Collaboration-Based Knowledge Refinement as a Key Success Factor for Knowledge Repository Systems / *T. Rachel Chung and Kwangsu Cho* 226

E-Collaborative Knowledge Construction / *Bernhard Ertl* 233

Enhancing E-Collaboration Through Culturally Appropriate User Interfaces / *Dianne Cyr* 240

Enhancing Electronic Learning for Generation Y Games Geeks / *Sophie Nichol and Kathy Blashki* 246

E-Scheduling / *Gerhard F. Knolmayer* 253

Evolving Gender Communication Issues in E-Collaboration / *Cathy L. Z. DuBois* 259

Extending TAM to Measure the Adoption of E-Collaboration in Healthcare Arenas / *Fernao H.C. Beenkens and Robert M. Verburg* 265

Facilitation of Technology-Supported Communities of Practice / *Halbana Tarmizi and Gert-Jan de Vreede* 272

Factors for Effective E-Collaboration in the Supply Chain / *Sharon A. Cox and John S. Perkins* 279

Faculty Perceptions of Traditional and Electronic Communications Channels / *Rolando Pena-Sanchez, Ibrahim Mescioglu, and Richard C. Hicks* 286

Faculty Preferences for Communications Channels / *Rolando Pena-Sanchez and Richard C. Hicks* 292

Gender Differences and Cultural Orientation in E-Collaboration / *Yingqin Zhong, Zhen Wang, and John Lim* 301

Generic Definition of Collaborative Working Environments, A / *Karl A. Hribernik, Klaus-Dieter Thoben, and Michael Nilsson* 308

Global Funding of E-Collaboration Research / *Ned Kock* 314

Governance Mechanisms for E-Collaboration / *Anupam Ghosh and Jane Fedorowicz* 319

Governing E-Collaboration in E-Lance Networks / *Robert Hooker, Carmen Lewis, Hugh Smith, James Worrell, and Tom Yoon* 324

Group Size Effects in Electronic Brainstorming / *Alan R. Dennis and Michael L. Williams* 330

GSS Research for E-Collaboration / *Sathasivam Mathiyalakan* 337

Human and Technology Leadership Roles in Virtual Teams / *Ilze Zigurs and Terrance Schoonover* 343

IM Support for Informal Synchronous E-Collaboration / *Stefan Hrastinski* 349

Impact of Collaborative Delivery of Enterprise ICT Services / *Jiri Vorisek and George Feuerlicht* 355

Impact of Personality on Virtual Team Creativity and Quality, The / *Rosalie J. Ocker* 363

Implementing Varied Discussion Forums in E-Collaborative Learning Environments / *Jianxia Du and George Pate* 370

Induced Cooperation in E-Collaboration / *Reza Barkhi* 377

Instant Messaging as an E-Collaboration Tool / *Qinyu Liao and Xin Luo* 383

Interaction and Context in Service-Oriented E-Collaboration Environments / *Christoph Dorn, Schahram Dustdar, Giovanni Giuliani, Robert Gombotz, Ke Ning, Sébastien Peray, Stephan Reiff-Marganiec, Daniel Schall, and Marcel Tilly* 389

Interaction Model in Groupware Use for Knowledge Management / *Jessada Panyasorn, Niki Panteli, and Philip Powell* 398

Interrelationships between Web-GIS and E-Collaboration Research / *Michele Masucci* 405

Levels of Adoption in Organizational Implementation of E-Collaboration Technologies / *Bjørn Erik Munkvold* 411

Macro-Level Approach to Understanding Use of E-Collaboration Technologies, A / *Sanjiv D. Vaidya and Priya Seetharaman* 417

Managing E-Collaboration Risks in Business Process Outsourcing / *Anne C. Rouse* 424

Managing Intercultural Communication Differences in E-Collaboration / *Norhayati Zakaria* 430

Managing Online Discussion Forums for Collaborative Learning / *Wing Lam, Eu-Jin Kong, and Alton Chua* 437

Matrix for E-Collaboration in Rural Canadian Schools, A / *Ken Stevens* 444

Multilevel Modeling Methods for E-Collaboration Data / *Sema A. Kalaian* 450

Multilingual Collaboration in Electronic Meetings / *Milam Aiken* 457

New Model and Theory of Asynchronous Creativity, A / *Dorrie DeLuca* 463

Practice and Promise of Virtual Project Management, The / *Ilze Zigurs, Deepak Khazanchi, and Azamat Mametjanov* 472

Prerequisites for the Implementation of E-Collaboration / *Thorsten Blecker and Ursula Liebhart* 479

Presence-Based Real-Time Communication / *Frank Frößler and Kai Riemer* 487

Prospects for E-Collaboration with Artificial Partners / *Kathleen Keogh and Liz Sonenberg* 493

Psychological Contracts' Influence on E-Collaboration / *Vicki R. McKinney and Mike Allen* 499

Reconsidering IT Impact Assessment in E-Collaboration / *Az-Eddine Bennani* 505

Reflection on E-Collaboration Infrastructure for Research Communities, A / *Lydia M. S. Lau and Peter M. Dew* 512

Research Agenda for Identity Work and E-Collaboration, A / *Niall Hayes and Mike Chiasson* 521

Role of E-Collaboration Systems in Knowledge Management, The / *Sharon A. Cox and John S. Perkins* 527

Role of Individual Trust in E-Collaboration, The / *Terry Nolan and Linda Macaulay* 534

Role of Leadership in Virtual Teams, The / *Kristi M. Lewis Tyran and Craig K. Tyran* 540

Scenarios for E-Collaboration are Only Part of the Story / *Lydia M. S. Lau* 547

Setting the Framework of E-Collaboration for E-Science / *Andrea Bosin, Nicoletta Dessì, Maria Grazia Fugini, Diego Liberati, and Barbara Pes* 554

Sharing Information Efficiently in Cooperative Multi-Robot Systems / *Rui Rocha and Jorge Dias* 561

Small Business Collaboration Through Electronic Marketplaces / *Yin Leng Tan and Linda A. Macaulay* 569

Speech Act Theory and Communication Modeling / *Lewis Hassell* 577

Support of E-Collaboration Technologies for a Blood Bank, The / *P. Sasi Kumar, R. Senthil, G. Kannan, and A. Noorul Haq* 584

Supporting Collaborative Processes in Virtual Organizations / *Ricardo Mejía, Nicolás Peñaranda, Arturo Molina, and Godfried Augenbroe* 589

Sustainability of E-Collaboration / *António Dias de Figueiredo* 596

Task Constraints as Determinants of E-Collaboration Technology Usefulness / *Ned Kock* 602

Technological Challenges in E-Collaboration and E-Business / *Fang Zhao* 606

Technological Challenges to the Research and Development of Collaborative Working Environments / *Karl A. Hribernik, Klaus-Dieter Thoben, and Michael Nilsson* 612

Telework in the Context of E-Collaboration / *Antonio Padilla-Meléndez and Ana Rosa Del Aguila-Obra* 618

Thematic-Based Group Communication / *Raymond Pardede, Gábor Hosszú, and Ferenc Kovács* 624

Thinklets for E-Collaboration / *Robert O. Briggs, Gert-Jan de Vreede and Gwendolyn L. Kolfschoten* 631

3C Collaboration Model, The / *Hugo Fuks, Alberto Raposo, Marco A. Gerosa, Mariano Pimentel, and Carlos J. P. Lucena* 637

Towards a Collaborative Educational Game Model / *Jeane Silva Ferreira Teixeira, Eveline de Jesus Viana Sá, Tatiane Macedo Prudêncio Lopes, Inaldo Capistrano Costa, D'Ilton Moreira Silveira, Alessandro Ramos de Oliveira, Clovis Torres Fernandes, and José Maria Parente de Oliveira* 645

Understanding Adverse Effects of E-Commerce / *Sushil K. Sharma and Jatinder N.D. Gupta* 655

Understanding Effective E-Collaboration Through Virtual Distance / *Karen Sobel Lojeski and Richard R. Reilly* 660

Use of E-Collaboration Technologies Among Students of Management / *Antonio Padilla-Meléndez and Aurora Garrido-Moreno* 667

Use-Centered Strategy for Designing E-Collaboration Systems, A / *Daniel H. Schwartz, John M. Flach, W. Todd Nelson, and Charlene K. Stokes* 673

Using IM to Improve E-Collaboration in Organizations / *Xin Luo and Qinyu Liao* 680

Using the Web for Contract Negotiations / *Ben Martz and Susan Gardner* 686

Videoconferencing as an E-Collaboration Tool / *Michael Chilton and Roger McHaney* 693

Virtual Teams Adapt to Simple E-Collaboration Technologies / *Dorrie DeLuca, Susan Gasson, and Ned Kock* 699

Voice-Based Group Support Systems / *Milam Aiken* 706

Wikis as Tools for Collaboration / *Jane Klobas* 712

Workflow Systems in E-Learning Environments / *Neide Santos, Flávia Maria Santoro, and Marcos R.S. Borges* 718

Foreword

What is e-collaboration? Although this term means many things to many people, Ned Kock broadly defines it as "collaboration among individuals engaged in a common task using electronic technologies." E-collaboration is not limited to computer-mediated communication (also known as CMC), or computer-supported cooperative work (known as CSCW), because other electronic technologies exist that are not (strictly speaking) computers and that can be used to support collaboration among individuals engaged in a common task.

This encyclopedia reflects the broad definition adopted by Ned Kock and will help expand the boundaries of e-collaboration. The multidimensional, interrelated e-collaboration boundaries include theoretical, technical, and use boundaries. Without overstating the obvious, e-collaboration research is very challenging. Not only does one deal with humans, one deals with understanding and facilitating the joint outcomes of groups of people. Existing theories guide further exploration but they can limit the scope and even direct attention towards paths that prove to be dead ends.

Technological enhancements are the core of e-collaboration—they make up the "e" in e-collaboration, so to speak. However, technological enhancements, in and of themselves, are not necessarily relevant to expanding e-collaboration boundaries. Functionally driven technology, without integrating with other boundaries; that is, "Build it because you can and they will use it," can waste resources and limit usefulness. For example, it is not necessarily an expansion of an e-collaboration technical boundary if one just substitutes a wireless connection for a physical connection. Some issues related to wireless technology and e-collaboration might be: Does wireless enable or limit collaboration in certain ways? Do users compensatorily adapt in overcoming limitations in unexpected ways? Does bandwidth affect collaboration technology directions? What are ways to provide collaboration support despite current limitations?

The act of using any system changes expectations of what can and should be supported. Some questions related to use boundaries include: What are the levels of use of e-collaboration? Are organizations still stuck at the lowest level of use? And, if yes, why?

What are the experiences of those who employ e-collaboration technology? What works and what are the lessons learned in implementing e-collaboration technology? How can I use e-collaboration in my situation?

Although no book can provide all the answers to these questions, Ned Kock used his skill and vision to compile a broad spectrum of work that begins to address the theoretical, technical, and use boundaries of e-collaboration. The encyclopedia provides a compendium useful to those who are just entering the field and those who have extensive knowledge and experience. I know of none more qualified than Ned Kock to tackle such an ambitious project as to compile an encyclopedia of e-collaboration.

John Teofil Paul Nosek, PhD
Professor, Computer & Information Sciences, Temple University
Associate Editor, International Journal of e-Collaboration
Senior Editor, Information Systems Journal

Preface

One sign that a research topic has become quite important is the publication of a dedicated encyclopedia full of articles addressing the topic. That is also a clear sign that there is a community of researchers whose work gravitates around the topic in question, which in the case of this volume is the topic of e-collaboration. The publication of this *Encyclopedia of E-Collaboration* is a landmark event in the path toward making e-collaboration an established field of inquiry—a field that can sustain itself and serve as a reference for other fields of inquiry. And, as the reader will certainly notice from the several applied articles contain in this volume, its publication is also a landmark event in the path toward making e-collaboration an established field of industry practice.

E-collaboration has been defined in many ways in the past, and the number of definitions has grown recently. This situation has been intensified by the emergence of an e-collaboration tools industry, with major players like Microsoft Corporation and IBM offering e-collaboration products and services.

Nevertheless, e-collaboration can be broadly defined as collaboration among individuals engaged in a common task using electronic technologies. This definition is broad enough to fit most people's interpretation of what e-collaboration is (and is not), and not to conflict with narrower definitions developed in specific technology utilization contexts.

Based on the definition above we can say safely that e-collaboration is not limited to computer-mediated communication (also known as CMC). We can also say that e-collaboration is not limited to computer-supported cooperative work (known as CSCW). Other electronic technologies exist that are not (strictly speaking) computers and that can be used to support collaboration among individuals engaged in a common task. One example is the telephone, which was one of the main targets of research conducted in the 1970s that led to the development of influential theories in the field of e-collaboration research.

It also follows as a corollary from the above definition that e-collaboration may take place without any computer-mediated communication or computer-supported collaborative work. For example, let us consider the scattered members of an army platoon, using rudimentary electronic devices to indicate their location and transmit basic information to each other while performing a joint reconnaissance task of a certain geographic area. Those platoon members are in fact engaging in e-collaboration, according to the definition of the term presented above.

That is not to say that most instances of e-collaboration will not involve computers. In fact, the opposite is the case. This is reflected in how hardware and software vendors regularly discuss related technologies. Contemporary e-collaboration technology vendors often define e-collaboration with an emphasis on technological support for electronic meetings over a private network or the Internet. Among those vendors are established companies such as Microsoft Corporation and IBM, as well as newer players such as Google and LivePerson. E-collaboration technology support is often presented by these vendors as enabling electronic meetings that incorporate many elements of face-to-face communication but that can be conducted in a geographically dispersed fashion.

Another modern trend in connection with how e-collaboration is perceived is seen in information technology (IT) publications aimed at IT managers and professionals. Those publications, which include *CIO Magazine* and *Computerworld,* often present e-collaboration technologies as tools to support electronic commerce and supply chain transactions involving two or more organizations. This view is more limited yet perfectly compatible with our adopted definition of e-collaboration.

Strictly speaking, e-collaboration could have begun as early as the mid-1800s, with the invention of the telegraph by Samuel F. B. Morse. However, that invention was probably too cumbersome to be consistently used to support the work of individuals engaged in common tasks. Even the invention of the telephone in the 1870s, and its wildfire-like diffusion in the coming years, was not enough to usher in the e-collaboration age.

In fact, e-collaboration did not become a reality with the emergence of the first commercial computers after World War II, either. Those computers were large, expensive, and generally referred to as "mainframes." At that time, organizations were very centralized, which inhibited collaborative work. Moreover, mainframes were then seen as too expensive to be used to support communication and collaboration among groups of individuals. The relatively high cost of mainframes, especially when compared with the cost of labor at the time, restricted their use to very specialized tasks conducted by expert technicians. Mainframe use was not distributed. It was highly centralized.

Arguably, one of the first and most successful e-collaboration tools, a version of e-mail, was in fact a spin-off of a large, wide area computer-networking project called ARPANET, sponsored by the U.S. Department of Defense. The project was conducted in the late 1960s. ARPANET's inventors had not envisioned it as an infrastructure to enable group communication or collaboration. At the time of its initial development, ARPANET was seen primarily as a means for researchers and computer scientists to share expensive mainframe resources.

Yet, between the early 1970s and 1980s, e-mail was discovered and used by thousands of those researchers and computer scientists. While its developers did not see it as much more than a toy system, e-mail quickly became an essential e-collaboration technology.

As the ARPANET grew, so did the use of e-mail. At the same time, new computer chip manufacturing techniques enabled the development of large-scale integrate circuits, with much lower cost and physical space demands than the circuitry used up until then in mainframe computers. This, in turn, led to the development of personal computers that were smaller, less expensive, and often more powerful (in terms of processing power) than many of the early mainframes. Soon these personal computers were connected to local area networks (LANs) through LAN operating systems, whose market was initially dominated by Novell Corporation, with its NetWare operating system.

The wide area network infrastructure created by the ARPANET, together with the development of personal computers and LANs, provided the environment in which early e-collaboration technologies flourished in the 1980s. Some of those technologies, such as Information Lens and The Coordinator, extended the functionality of early e-mail systems.

Other e-collaboration technologies, which later became known as group decision support systems (or GDSSs), were aimed at improving the efficiency of same room, same place group meetings through features such as anonymous and simultaneous idea generation and voting. Examples of early GDSSs are GroupSystems, Teamfocus, and MeetingWorks.

Still other e-collaboration technologies, such as Lotus Notes and Domino, allowed users to create asynchronous e-collaboration spaces. These latter e-collaboration technologies have often been referred to as e-collaboration systems development suites. They were in many ways similar to some of the e-learning environments that fueled the growth of distance education (e.g., Blackboard and WebCT).

The early 1990s saw what once was the ARPANET evolve into today's ubiquitous Internet; a worldwide network of computers made up of many LANs, interacting through the same general communication protocol (i.e., TCP/IP). This, in turn, provided the infrastructure necessary for the emergence of the Web, which is made up of millions of platform-dependent Web servers providing users access to static and dynamic content through platform-independent Web browsers.

Today's e-collaboration technologies are either browser-based (i.e., run on Web browsers) or non-browser-based. The latter are usually Internet-based tools enabling proprietary client software to interact with other clients either directly (peer-to-peer e-collaboration tools) or through servers (client-server e-collaboration tools). Examples of widely used browser-based e-collaboration tools are WebEx and eRoom, as well as many e-learning tools like Blackboard and WebCT. Examples of widely used non-browser-based e-collaboration tools are Groove (peer-to-peer e-collaboration), MSN Messenger, and ICQ (client-server e-collaboration).

A recent search on ABI/Inform containing the term e-collaboration suggested that the earliest articles on the topic dated back to the early to mid-1990s. (ABI/Inform is a widely used database of business and technology articles.) Yet research on topics related to e-collaboration has a long history, arguably dating back to the late 1970s. That research was conducted under different banners, some of which reflect distinctly different research traditions.

Among the main e-collaboration research traditions is that of computer-supported cooperative work (or CSCW). It dates back to the 1970s, and its first dedicated conference (called CSCW Conference) took place in the early 1980s. CSCW research has traditionally involved the search for technological solutions to e-collaboration problems, such as that of increasing social awareness of collaborators through the use of "avatars." These are visual and often metaphorical representation of a user (e.g., a unicorn). The CSCW Conference has been regularly held since its first installment, and is considered the principal meeting point for CSCW researchers.

Another main e-collaboration research tradition, of a more behavioral nature than CSCW research, has been the one targeting the family of technologies known as group decision support systems (GDSSs), and their effects on group behavior. While there is no single conference dedicated to it, GDSS research has grown over the years to become one of the main areas of research in the broader field of information systems. That research has usually focused on the match between GDSS tools and group tasks, particularly decision making tasks conducted by groups of individuals meeting at the same time and in the same room. The communication among the individuals is usually mediated by computers running GDSS software.

CSCW and GDSS research can be characterized as distinct lines of research, which, notwithstanding a tendency to benefit from multidisciplinary contributions, have their own separate and somewhat independent traditions. As with most areas of research where the scope is relatively limited, CSCW and GDSS also have distinct communities of scholars associated with them, and, among those, key contributors that are widely perceived as prominent researchers in those areas. Several of those researchers contributed articles to this encyclopedia.

The advent of the Internet, and particularly of the Web, caught many CSCW and GDSS researchers by surprise, in the sense that it brought in researchers from many other disciplines into the realm of e-collaboration research. Among those disciplines are marketing, accounting, economics, human resources management, clinical psychology, and education, just to name a few. This has led to two separate and opposing trends.

One of the trends has been the development of many subcommunities dedicated to a particular issue in connection with e-collaboration research. Examples are asynchronous learning networks and virtual social networking; the latter having experienced tremendous growth since 2005 with the emergence of social networking technologies (e.g., blogs and wikis). Unfortunately, it seems that many of those subcommunities have been unable to (or are still trying) to identify a small set of key issues that would characterize them as legitimate and to some extent independent communities of inquiry.

The other trend is that of integrating separate communities of inquiry (including the CSCW and GDSS communities) through the identification of broad issues likely to be relevant for e-collaboration research as a whole, and the creation of publication outlets aimed at bringing together scholars of different e-collaboration research traditions. Examples of broad issues that have been presented as relevant for e-collaboration researchers in general are compensatory adaptation and collaborative sense making. Examples of publication outlets aimed at bringing together scholars of different e-collaboration research traditions are the journal *IEEE Transactions on Professional Communication*, published by the prestigious Institute of Electrical and Electronics Engineers (IEEE); and the *International Journal of e-Collaboration*, whose inaugural issue was published by IGI Global in early 2005.

The Encyclopedia of E-Collaboration is one of the newest additions to the existing publication outlets aimed at bringing together scholars of different e-collaboration research traditions. What makes it a unique outlet is that it is by far the most comprehensive compilation of short articles addressing issues in connection with e-collaboration technologies and their impact on users. More than 100 articles have been contributed to this volume of the Encyclopedia by nearly 200 authors from all over the world. Each article provides a focused discussion of a topic related to e-collaboration, as well as seven or more terms and definitions related to e-collaboration. The *Encyclopedia of E-Collaboration* is expected to be a "living document" that will grow over time with the addition of new volumes.

The range of topics covered in this Encyclopedia is certainly broad and representative of the state-of-the-art discussion of conceptual, theoretical, and applied e-collaboration issues. If one looks at the broad literature on e-collaboration, as well as its impact in academic and industry circles, it becomes clear that this Encyclopedia brings together the best in terms of thinking in the field. The authors of the articles in this volume are among the most accomplished and influential e-collaboration researchers in world. I thank them for being contributors to this book, and am honored to have been able to serve as the editor of this volume.

The blend of conceptual, theoretical and applied articles found here makes me confident that the *Encyclopedia of E-Collaboration* will serve both academics and practitioners very well. I hope that this volume will stimulate further research on e-collaboration issues by academics and doctoral students and help practitioners take full advantage of the increasing new forms of work organization and social interaction modes enabled by e-collaboration technologies.

Ned Kock
Division of International Business and Technology Studies
Texas A&M International University, USA

Acknowledgment

I strongly believe that no book project can be completed successfully by an author or editor without the support of a dedicated editorial team. This is especially true of a project of the magnitude of a volume with over 100 articles, such as the *Encyclopedia of E-Collaboration*. I would like to thank the team at IGI Global for their excellent editorial support for this Encyclopedia project. Special thanks go to Mehdi Khosrow-Pour for his editorial leadership, and for convincing me over a nice lunch a few years ago to work with IGI Global's team in this project. Many thanks are also due to Jan Travers, Kristin Roth, Michelle Potter, and Corrina Chandler.

Parts of the preface, as well as of the articles that I wrote for this encyclopedia, have been published before as sections of journal articles authored or co-authored by me. The preface contains revised text from an article in which I introduced, together with my friend John Nosek, the Special Issue on Expanding the Boundaries of E-Collaboration. That special issue was published in 2005 in the journal *IEEE Transactions on Professional Communication*. The articles that I wrote for this encyclopedia contain revised text from editorial essays that I have written for the *International Journal of e-Collaboration*. I thank the publishers of those journals for permission to use text from those articles.

Several years ago, my wife and I had been discussing the idea of living in a part of the United States that had a strong Hispanic, or Latin, influence. Both of us have been raised in Brazil, a country that shares many cultural characteristics with other Latin American countries. Among those is an optimistic view of life, a preference for warm weather, and an almost fanatical love for soccer. Southern California and Florida fit the bill nicely; and Southern Texas had a particular appeal to us. Having lived mostly in the Northeast, we have always seen Texas as a somewhat unique state in the United States. So, I joined Texas A&M International University, located in the city of Laredo, near the U.S. border with Mexico. We moved to Laredo with our four kids and have been having a great time since.

My impression is that my output in terms of scholarship has gone up since I moved to Laredo, which certainly goes against the theory that warm places make people lazy. I have been joking lately that since I came to Texas I have been working toward the goal of becoming the Stephen King of information systems!

But seriously, one of the key reasons for my increased productivity is the high recognition and support given by the University's administration to faculty scholarship. Special thanks in that respect should be given to Ray Keck, the University's long-term president; Dan Jones, provost; and Jacky So, dean of the College of Business and Economics.

Another very important reason for my productivity is the support and encouragement of a wonderful group of colleagues with whom I have been sharing the third floor of Pellegrino Hall on the university's beautiful campus. Those colleagues make up the Division of International Business and Technology Studies, which I have had the pleasure to serve since 2006 in the capacity of founding chair. My special thanks go to Cindy Martinez, for being a trusted and very competent assistant; and my faculty colleagues Jacques Verville, Jackie Mayfield, Milton Mayfield, Ananda Mukherji, and Pedro Hurtado for the leadership roles that they have been playing in the Division.

Lastly, and most importantly, I would like to thank my family for their love and support. This book is dedicated to them.

Ned Kock

About the Editor

Ned Kock is associate professor and founding chair of the Division of International Business and Technology Studies at Texas A&M International University. He holds degrees in electronics engineering (BEE), computer science (MS), and management information systems (PhD). Dr. Kock has authored and edited several books, including the best-selling *Systems Analysis and Design Fundamentals: A Business Process Redesign Approach*. Dr. Kock has published his research in a number of high-impact journals including *Communications of the ACM, Decision Support Systems, European Journal of Information Systems, IEEE Transactions, Information & Management, MIS Quarterly*, and *Organization Science*. He is the founding editor-in-chief of the *International Journal of e-Collaboration*, associate editor of the *Journal of Systems and Information Technology*, and associate editor for information systems of the journal *IEEE Transactions on Professional Communication*. His research interests include action research, ethical and legal issues in technology research and management, e-collaboration, and business process improvement.

Academic Weblogs as Tools for E–Collaboration Among Researchers

María José Luzón
University of Zaragoza, Spain

INTRODUCTION

Although scientific research has always been a social activity, in recent years the adoption of Internet-based communication tools by researchers (e.g., e-mail, electronic discussion boards, electronic mailing lists, videoconferencing, weblogs) has led to profound changes in social interaction and collaboration among them. Research suggests that Internet technologies can improve and increase communication among noncollocated researchers, increase the size of work groups, increase equality of access to information by helping to integrate disadvantaged and less established researchers, help to coordinate work more efficiently, help to exchange documents and information quickly (Carley & Wendt, 1991; Nentwich, 2003). There is abundant research on new forms of group work originated from the use of computer technologies. Carley and Wendt (1991) use the term *extended research group* to refer to very large, cohesive, and highly cooperative research groups that, even being geographically dispersed, are coordinated under the supervision of a single director. The term *collaboratory* is also used to refer to similar groups (Finholt, 2002). Although there is much research on how Internet technologies are used by unified and cohesive work groups to collaborate (e.g., Moon & Sproull, 2002; Walsh & Maloney, 2002), less attention has been paid to how the Internet facilitates collaboration among researchers outside these highly cohesive groups. Weblogs (blogs) can become a useful tool for this type of collaboration and for the creation of virtual groups. Weblogs are frequently updated Web pages, consisting of many relatively short postings, organized in reverse chronological order, which tend to include the date, and a comment button so that readers can answer (Herring, Scheidt, Bonus, & Wright, 2004). They enable users to communicate with a worldwide nonrestricted community of people in similar fields, which leads to several forms of collaboration. The purpose of this article is to present a brief overview of the uses of weblogs as tools for research e-collaboration.

Defining the concept of "research e-collaboration" precisely is extremely difficult. Here we assume that members of a virtual community engage in research e-collaboration when they use e-collaborating technologies in order to share information and discuss issues which contribute to advancing knowledge in a specific area.

BACKGROUND

The term *Weblog* was coined by Jorn Barger in 1997 to refer to personal Web sites that offer frequently updated information, with commentary and links. Blood (2002) classifies blogs into two "styles": the filter type, which includes links pointing to other sites and comment on the information on those sites, and the personal-journal type, with more emphasis on personal self-expressive writing. There are many other types of blogs described in the literature, defined on the basis of different criteria; for example, knowledge blogs (k-blog), community blogs, meta-blogs.

The capabilities of blogs make them helpful tools for communication between members of a community or organisation. Some types of weblogs have originated as an answer to the communicative needs of specific communities; for example, knowledge blogs, weblogs for personal knowledge publishing. Kelleher and Miller (2006) describe "knowledge blogs" as "the online equivalent of professional journals" used by authors to document new knowledge in their disciplines. A related concept is that of "personal knowledge publishing," defined by Paquet (2002) as "an activity where a knowledge worker or researcher makes his observations, ideas, insights, interrogations, and reactions to others' writing publicly in the form of a weblog." Many corporate and academic blogs make use of capabilities that afford collaboration: they enable scholars to communicate with a wide community, fostering peer review and public discussion with researchers from different disciplines. These weblogs have a precedent in what

Harnard (1990) terms "scholarly skywriting": using multiple e-mail and topic threaded Web archives (e.g., electronic discussion) to post information that anybody can see and add their own comments to.

There are many types of academic blogs (blogs from journal editors, individual scholars' blogs, research groups' blogs, PhD blogs), each of them used for different purposes. For instance, while the main purpose of the weblogs implemented by universities is discussion, weblogs by PhD students are mainly used to comment on the day's progress and on the process of PhD writing, and blogs from journal editors are usually filter blogs, which provide links to articles or which comment on news related to the journal topic.

The uses of weblogs in research have been discussed in several papers and blog posts (Aïmeur, Brassard,, & Paquet, 2003; Efimova, 2004; Efimova & de Moor, 2005; Farmer, 2003; Mortensen & Walker, 2002; Paquet, 2002). These researchers depict blogs as facilitating scientific enquiry in two ways: (1) they help to access and manage content, through features such as archives, RSS (an automated system that enables bloggers to syndicate their content to other blogs), searchable databases, post categories; and (2) they are tools for collaboration, through communication and network features. These features include hyperlinks, comments, trackbacks (records of the Web address of the blogs that have linked to a blog posting), RSS, or blogrolls (a list of blogs the author usually reads, and that, therefore, deal with similar topics). But the most important "ingredient" for collaboration is the bloggers' perception of blogs as a means to point to, comment on, and circulate information and material from other blogs.

As a tool for collaborative research, blogs have several uses: (1) supporting community forming; (2) helping to find other competent people with relevant work; (3) facilitating connections between researchers: blogs make it easier for researchers to develop, maintain and activate connections with others; (4) making it possible to obtain speedy feedback on ideas; (5) fostering diversity and allowing for radical new ideas to be acknowledged, circulated and discussed; (6) fostering communication and collaboration between researchers from different fields; (7) supporting the initiation and development of conversations. In weblogs, ideas can be generated and developed through discussion with others.

E-COLLABORATING THROUGH ACADEMIC WEBLOGS

We used a corpus of 100 academic English-language weblogs, collected between January and March, 2006, in order to analyse how they are used as tools for e-collaboration and how the elements of weblogs contribute to this purpose. Many academic weblogs are not intended as collaborative tools and thus do not include social software. In other cases, although comments are allowed, interactivity is very limited, and the weblogs are not really used as conversation tools. Therefore, for the purpose of this research, we have selected only active weblogs that include and make use of the comment feature. Drawing on previous research on blog analysis (e.g., Herring et al., 2004) and on our initial inspection of the blogs of the corpus, we identified the functional properties that make interaction and collaboration possible (e.g., comments, links). Then, we analysed the function of the 15 most recent entries in each blog (e.g., ask for feedback), the function of the comments to these entries (e.g., disagree with a post), and the types of links included in the postings and comments (e.g., to other blogs, to news sites, to Web sites by others, to the blogger's own site).

The analysis showed that academic weblogs can be used to support the following forms of collaboration (or collaboration stages):

1.   *Finding like-minded peers and establishing contact and collaboration with them, thus helping to create a virtual community of people interested in the same topic.*

Weblogs share features of personal and public genres. As personal genres, they are used by researchers to keep records of their thoughts, ideas or impressions while carrying out research activities. They also have an open nature, since anybody can access and comment on the author's notes, and they are in fact intended to be read. By revealing the research a researcher is currently involved in and, at the same time, providing tools for immediate feedback on that research, weblogs facilitate collaboration. In addition, since weblogs are open to the public, not restricted to peers working in the same field, they facilitate contact among scholars from different disciplines and invite interdisciplinary knowledge construction (Aïmeur et al., 2003).

Blogrolling lists enable readers of a weblog to find peers with relevant work. These lists do not only point to related work, but also function as signs of personal recommendation, which help others to find relevant contacts faster (Efimova, 2004). Links and comments are also useful to expand a researcher's community. Other researchers read the postings in the research blog and leave comments, usually including links to their own research. Authors of weblogs can also include links to articles on similar work carried out by other researchers, thus enabling the readers to get into contact with them.

Many researchers use the comment feature to explain that they are working on areas similar to the ones dealt with in the original posting and to suggest the possibility of collaboration or contact, in a more or less direct way (e.g., "I would like to keep in touch about your findings"). Community blogs are sometimes used by bloggers to summarize briefly their research and invite collaboration and contact with others.

a. I am in the middle of a research project that looks at (...). I have a research log in which I make occasional observations, related and sometimes unrelated, to the ongoing work. Interested? Let me know and I will give you the URL.

The entry above got several comments of people eager to exchange ideas; for example:

b. I'd love to exchange ideas (...). How can we establish contact?

2. *Asking for and receiving advice/feedback from others.*

Weblogs enable researchers to post an idea or theory and have it circulated and discussed, thus helping the author revise and refine it. Authors who want to use weblogs for this purpose embed elements that facilitate hyperlinked conversations into them, such as links, comment software, and trackback features.

Very frequently, authors use weblogs to explicitly request feedback, comments, suggestions, and so forth on a paper, presentation, or idea. Comment software facilitates the discussion between the researchers asking for feedback and anybody wanting to contribute: Other bloggers state their agreement and disagreement with the author and with other commentators, ask questions, offer new perspectives, and so forth.

a. I was asked by the editors of the New Palgrave Dictionary of Economics to contribute a short article on the analysis of variance (...) Here's the article. Any comments (...) would be appreciated.

The entry above got eight comments suggesting ways to improve the paper; for example:

*b. * Missing from the draft is a discussion of ANOVA versus regression (...)*

* Here is one question you might address: When is ANOVA preferred to linear regression? (...)

Comments may also include links to other blogs with similar topics and thus help the original author with other people's perspectives. The weblog becomes a forum for discussion of the paper prior to publication, similar to that of a conference. When a blogger requests and gets feedback on a paper, he/she usually acknowledges the help from the others and shows the result of the revision in an update to the original entry. The result is a more consensual paper, which makes use of knowledge from other researchers.

Posting articles for peers to comment on in a weblog has several advantages. In the first place, the author can get immediate and public feedback, thus accelerating the development of new knowledge and involving many people in the review process: Anybody can contribute, even people unknown by the author, nonestablished researchers who are not usually involved in peer-reviewed publishing, and researchers from different cultural backgrounds. Besides, since bloggers usually comment on other's ideas on their own blogs, feedback may come from others than just immediate readers. Authors can post any idea, even very innovative or radical ideas which would not be accepted in traditional publishing, and, as Paquet (2002) points out, they can use weblogs to let others know not only about positive results, but also about negative results, blind alleys and problems. Peers can help by providing their ideas, suggestions, and so forth, and this way knowledge construction becomes a communal activity.

In many cases, researchers do not ask for feedback, but for references, information or resources on a topic, or help to solve a problem, hence using blogs as a tool to get information on a topic from competent peers they may not know personally.

3. *Discussing topics or theories in which several researchers are interested, thus contributing to developing or articulating an idea and to reinforcing the links between the members of the virtual community.*

Efimova and de Moor (2005) show how weblogs can serve as conversation tools which support the exchange of multiple perspectives and fast and meaningful reactions and facilitate the joint development of ideas. They define a new form of conversation enabled through the use of weblogs, "distributed weblog conversation": conversations across many weblogs, where several authors link their ideas to construct a collective piece of knowledge.

The joint articulation of ideas by several researchers relies on elements such as links, comments and trackback features. Links may be to text but also to audio or video documents presenting or illustrating an idea, to simulations, to other Web pages or to other posts. Scholars use links to present and discuss an idea by someone else, or to articulate and summarize a previous discussion, before contributing their own idea and presenting it for others to comment on. Very frequently, when presenting others' arguments on a topic, scholars include quotes accompanied by links to the original post, that way allowing the reader to follow the conversation in the original blog.

The comment feature facilitates discussion based on the posting. Bloggers write their comments, get answers back from the author of the original post, comment on other bloggers' comments, and so forth, thus engaging in a conversation that may contribute to the development of innovative ideas. Comments are used for many different purposes: to show agreement or disagreement with the blogger, to ask for clarification on a specific point that the commentator does not understand or does not agree with, to request further information, to strengthen a point, to provide links to one's own blog, and so forth. This back-and-forth may lead to collaborative revision and refinement, and even permuting, of an idea or argument, e.g.:

a. * *In general, I really side with your post. Over at Anthropology.net, there have been a couple posts and discussions on (...). However, this post of yours, much more eloquently advocates breaking the boundaries between the West and non-West (...)*

* *Thanks, X. I'll have a look at the posts you link to (...)*

Since discussion of an idea may be distributed across several weblogs, it is necessary to have mechanisms to keep track of the discussion that one's posting may have sparked. One such mechanism is trackbacks. Trackback is a special feature of the program Movable Type, which notifies the blogger when another weblog has linked to his/her weblog. Other users can view the trackbacks and thus follow the discussion in other weblogs.

FUTURE TRENDS

Weblogs are a genre whose capabilities offer a great potential for collaboration among researchers. However, there are still few academics who engage in blogging, and in many cases academic blogs are used for purposes other than collaboration. Several reasons have been suggested: the risk of sharing information and having ideas stolen or attacked before the research is published, the fear of damaging credibility, the time that blogging takes away from more traditional research activities (Lawley, 2002; Mortensen & Walker, 2002).

Since the weblog is quite a recent genre, there is little research on how weblogs are used for enquiry and knowledge creation. Their open nature makes weblogs appropriate for nonestablished researchers, who can that way collaborate with peers they don't know, and researchers who seek collaboration with a worldwide non-discipline-restricted community of researchers. There are many different types of weblogs written by academics and researchers. Further research is needed to determine which of these weblogs are more involved in collaboration, which form of collaboration takes place in the different academic weblogs, and how bloggers use language to engage in collaboration.

CONCLUSION

In this article, we start from the assumption that, since scientific knowledge is socially constructed, scientific knowledge creation is collaborative in nature. Thus, rather than considering collaboration as restricted to the work by unified and cohesive groups involved in a project under the supervision of a coordinator, we see it as any sharing of information and exchange of perspectives which contribute to advancing knowledge in a specific area.

Weblogs are a genre whose capabilities allow the exchange of any type of information and support conversations with a rhizomatic structure, thus being an appropriate tool for collaboration and for the sharing of ideas among several researchers in different places around the world. Weblogs can represent and distribute knowledge in different forms; that is, text, multimedia, simulations. They have a modularized format, relying highly on hypertext: weblogs consist of different postings, linked together to postings in other weblogs, on which they comment or which comment on them, making it possible to construct a web of knowledge on a topic. Links and comments allow connecting different pieces of related research into a coherent whole, thus facilitating the development of cumulative knowledge.

The use of CMC tools (and specifically of weblogs) does not only lead to an increase in collaboration, but also to a change in collaboration patterns. Weblogs enable almost instantaneous communication and feedback among a worldwide community, thus allowing for a much more interactive generation and articulation of ideas and theories and for the constant revision and update of theories. The informal and open nature of weblogs facilitates collaboration among anybody interested in a topic and, hence, fosters interdisciplinary collaboration.

REFERENCES

Aïmeur, E., Brassard, G., & Paquet, S. (2003). Using personal knowledge publishing to facilitate sharing across communities. In M. Gurstein (Ed.), *Proceedings of the Third International Workshop on (Virtual) Community Informatics: Electronic Support for Communities*. Retrieved February 13, 2006, from: http://is.njit.edu/vci-www2003/pkp-sharing-across-comms.pdf

Blood, R. (2002). *The Web log handbook: Practical advice on creating and maintaining your blog*. Cambridge, MA: Perseus Publishing.

Carley, K., & Wendt, K. (1991). Electronic mail and scientific communication: A study of the SOAR extended research group. *Knowledge: Creation, Diffusion, Utilisation, 12*(4), 406-440.

Efimova, L. (2004). Discovering the iceberg of knowledge work. In *Proceedings of the Fifth European Conf. on Organisational Knowledge, Learning, and Capabilities (OKLC 2004)*. Retrieved February 20, 2006, from https://doc.telin.nl/dscgi/ds.py/Get/File-34786

Efimova, L., & de Moor, A. (2005). Beyond personal webpublishing: An exploratory study of conversational blogging practices. *Proceedings of the 38th Hawaii International Conference on System Sciences (HICSS-38) (pp.107a)*. Los Alamitos, CA: IEEE Press.

Farmer, J. (2003). *Personal and collaborative publishing (PCP): Facilitating the advancement of online communication and expression within a university*. Retrieved March 13, 2006, from http://radio.weblogs.com/0120501/myimages/030829PCP.pdf

Finholt, T. A. (2002). Collaboratories. *Annual Review of Information Science and Technology, 36*, 73-108.

Harnard, S. (1990). Scholarly skywriting and the prepublication continuum of scientific inquiry. *Psychological Science, 1*, 342-343.

Herring, S. C., Scheidt, L. A., Bonus, S., & Wright, E. (2004). Bridging the gap: A genre analysis of weblogs. *Proceedings of the 37th Hawai'i International Conference on System Sciences (HICSS-37) (pp. 101-111)*. Los Alamitos, CA: IEEE Computer Society Press.

Kelleher, T., & Miller, B. M. (2006). Organizational blogs and the human voice: Relational strategies and relational outcomes. *Journal of Computer-Mediated Communication, 11*(2). Retrieved March 20, 2006, from http://jcmc.indiana.edu/vol11/issue2/kelleher.html

Lawley, L. (2002) Thoughts on academic blogging. Retrieved March 20, 2006, from http://many.corante.com/archives/2004/04/01/thoughts_on_academic_blogging_msr_breakout_session_notes.php

Mortensen, T., & Walker, J. (2002). Blogging thoughts: Personal publication as an online research tool. In A. Morrison (Ed.), *Researching ICTs in context* (249-279). Oslo, Norway: InterMedia Report.

Moon, Y., & Sproull, L. (2002). Essence of distributed work: The case of the Linux Kernel. In P. Hinds & S. Kiesle (Eds.), *Distributed work* (pp. 381-404). Cambridge, MA: MIT Press.

Nentwich, M., (2003). *Cyberscience: Research in the age of the Internet*. Vienna: Austrian Academy of Sciences Press.

A

Paquet, S. (2002). Personal knowledge publishing and its uses in research. Retrieved March 12, 2006, from http://radio.weblogs.com/0110772/stories/2002/10/03/personalKnowledgePublishingAndItsUsesInResearch.html

Walsh, J. P., & Maloney, N. G. (2002). Computer network use, collaboration structures and productivity. In P. Hinds & S. Kiesler (Eds.), *Distributed work* (pp. 433-458). Cambridge, MA: MIT Press.

KEY TERMS

Blogroll: A list of links to Web pages the author of a blog finds interesting.

Comment: A link to a window where readers can leave their comments or read others' comments and responses from the author.

Personal Knowledge Publishing: The use of Weblogs by knowledge workers or researchers to make their observations, ideas, insights and reactions to others' writing public.

Research E-Collaboration: The use of e-collaborating technologies in order to share information and discuss issues which contribute to advancing knowledge in a specific area.

Scholarly Skywriting: Using multiple e-mail and topic threaded Web archives (e.g., electronic discussion) to post information that anybody can see and add their own comments to.

Trackback: A link to notify the blogger that his/her post has been referred to in another blog.

Weblog: A frequently updated Web page, consisting of many relatively short postings, organized in reverse chronological order, which tend to include the date and a comment button so that readers can answer.

An Adaptive Workforce as the Foundation for E-Collaboration

Charlene K. Stokes
Air Force Research Laboratory, USA

Joseph B. Lyons
Air Force Research Laboratory, USA

Daniel H. Schwartz
Air Force Research Laboratory, USA

Stephanie D. Swindler
Air Force Research Laboratory, USA

INTRODUCTION

E-collaboration technologies have transformed the "world of work" as we know it today. These technologies are undeniably the predominant factor facilitating the globalization of business, and they have transformed the fundamentals of interpersonal interaction within and across organizations. Given the tremendous changes being imposed by e-collaboration technologies, we must consider the subsequent changes being elicited at the individual (or human) level. In other words, how are the users adapting not only to the technologies themselves, but to the new world of work the technologies have created?

The changes emanated from technology have been so immense that they have shifted the business world off its traditional axis. Technology has served to fundamentally transform business processes. One of the largest areas affected by technology has been the very core of business: the communication and collaboration practices of organizations. When an area so fundamental to business has been altered drastically, we must consider how this transformation has permeated throughout all areas of the business world. Moreover, understanding the breadth of influence technology has on the nature of work and adapting all levels of business accordingly will allow us to extract the benefits and avoid the hazards associated with technology and the change it has enabled. Unfortunately, such understanding and coordination is a daunting goal. We propose that establishing an adaptive workforce is an essential first step to achieving this goal.

In support of our proposition, the following article will begin with examples of diverse areas of business that have been impacted by e-collaboration and illustrate how adaptability provides the underlying theme uniting the changes that are occurring. Then, focusing on individual adaptability, we will present a relevant performance model to be implemented in organizations. Based on this performance model, we will illustrate how organizations can begin to establish an adaptive workforce that will serve as the foundation for effective e-collaboration.

BACKGROUND

A large majority of collaboration efforts in organizations today are conducted via electronic technology (e.g., video conferencing, Web-based chat tools, e-mail, group decision support systems, etc.). Such technologies are collectively referred to as e-collaboration technologies, and e-collaboration is collaboration among individuals engaged in a common task using electronic technologies (Kock & Nosek, 2005). Many organizations have implemented e-collaboration technologies as part of their standard business practices but have overlooked the impact these technologies can have on the users and on the nature of work itself. For example, many organizations fail to see the changes that occur in collaboration when switching from face-to-face to e-collaborative modalities. They assume similar efforts and results will occur and the collaboration is simply conducted via an alternative medium. However, there is an abundance of research indicating a substantial affect on collaboration efforts depending on the medium adopted (e.g., Becker-Beck & Borg, 2005; Jarvenpaa & Leidner, 1999; Kock, 2001; Ritter, Lyons, & Swindler, 2006; Straus, 1997).

There are a variety of effects on both perceptions and performance that are associated with the implementation of e-collaboration systems. These effects can be negative, neutral, or positive. For example, the use of e-collaboration can be neutral if the same level of use and similar results occur as with face-to-face collaborations. On the dark side, the use of e-collaboration can negatively affect users' perceptions and ultimately their performance if users are uncomfortable with the medium and avoid its use. Furthermore, researchers (e.g., Ritter et al., 2006) have identified specific performance barriers inherent to e-collaboration technologies, and there must be a concerted effort to address these barriers if e-collaboration is to be effective. However, e-collaborative technologies have the potential to increase productivity in organizations. In order to attain the positive effects associated with the technologies, organizations must anticipate the system-wide influence (e.g., at the organizational, technology, and human levels) the technologies will have.

THE PROLIFERATING IMPACT OF E-COLLABORATION

Adopting a dynamic systems view (Ashby, 1947), we see that nothing in an organization occurs in a vacuum. Innumerable interactions and reciprocal relations characterize all that we do. At the individual level for example, we cannot understand the full impact of technologies if we consider only the direct influence the technologies have on individuals and ignore the indirect influence. Individuals are impacted indirectly through technology-enabled transformations at higher levels of work (e.g., globalization). Essentially, we must consider both the bottom-up and top-down effect of these technologies. Indeed, the impact of communication technologies can reach far beyond the original intent of the designers or of those implementing the technologies (Cameron & Webster, 2005). A systems perspective helps us understand this impact by acknowledging the interrelatedness of levels in an organization. To illustrate, we will discuss three of the paramount, and interrelated, areas of change that have occurred in

connection with e-collaboration technologies, focusing on the impact at the individual level. These three areas of change clearly do not provide an exhaustive list but serve as an overarching descriptive framework.

1. **Globalization:** E-collaboration technologies have enabled an unparalleled degree of connectivity among businesses. This connectivity is a primary contributor to the increased globalization of business (Cheng, Love, Standing, & Gharavi, 2006). As the boundaries of business stretch across continents, the reliance on e-collaboration technologies proliferates to the point of necessity. Individuals within the globalized business world must be flexible and adaptable to changing markets and brutal competition. Again, this creates a reliance on e-collaboration technologies as they are vital to sustain a competitive advantage within the global marketplace (Cheng et al., 2006; Hesketh & Neal, 1999). The individuals employed by these global companies must learn to collaborate with people of different cultures, and do so via e-collaboration technologies in a distributed and highly dynamic environment. Moreover, individuals interacting across the globe are often faced with multilingual challenges and cultural clashes (Sutton, Pierce, Burke, & Salas, 2006).

2. **Interpersonal interaction:** The *type* of connectivity (e.g., computer mediated) afforded by e-collaboration technologies has transformed the rudiments of interpersonal interaction. Researchers have found a plethora of interpersonal processes and outcomes affected (positively and negatively) by the use of various forms of e-collaboration: conflict and affect management, motivation and confidence building (Maruping & Agarwal, 2004); degraded positive collective efficacy, reduction of self-awareness and feelings of anonymity (Cuevas, Fiore, Salas, & Bowers, 2004); equality of influence across status and expertise (Dubrovsky, Kiesler, & Sethna, 1991); delays in formation of interpersonal trust (Jarvenpaa & Leidner, 1999) as well as team cohesion (Straus, 1997); information loss (Becker-Beck & Borg, 2005). See Wainfan and Davis (2004) for an extended review of factors affected by mediated-communication. Many of the aforementioned factors are interrelated and all are based on interpersonal interaction. However, beyond noting the

obvious surface level influences on interpersonal interaction during collaboration, Kock (2004) has identified a possible explanation at the biological level for why many of these factors are affected by mediated communication. Kock's (2004) psychobiological model posits that our biological communication apparatus has evolved in a manner consistent with face-to-face communication and all its richness (e.g., facial cues, body language, etc.). Therefore, we encounter difficulties and obstacles, as described above, in the absence of this rich information. In an effort to adapt to the difficulties associated with this new form of communication, we compensate for the lack of richness via alternative mechanisms or means (e.g., increased cognitive effort resulting in increased quality of contributions; Kock, 2001). However, it is important to note that not all individuals are equally capable of such compensation, which, in turn, creates novel challenges for organizations to ensure they have the right workers for this modern era of e-collaboration.

3.   **Knowledge work:** Organizations today are moving toward "knowledge work" and "knowledge workers" (Haeckel, 1999). Knowledge work is best characterized by the intangible products of work (e.g., decision making, information, ideas, know-how, etc.), which has lead to a trend in project-based teamwork where members of distributed expertise work together collaboratively to solve a problem (Ployhart & Bliese, 2006). Obviously, such work is inherently linked to and dependent on e-collaboration technologies. E-collaboration technologies not only extend the way knowledge and information is able to be used and disseminated (Hesketh & Neal, 1999), but the technologies often serve as the repositories for the codified knowledge being produced (Gray, 2001). However, as with the previous two areas of change, the shift to knowledge work enabled by collaborative technologies has altered fundamental business processes, which in turn impact individuals. The hierarchical control of information associated with traditional organizations has been flattened, creating a wider band of available knowledge and information and, in turn, empowering the individual employee (Kasper-Fuehrer & Ashkanasy, 2001). Although employees are offered an empowered position with knowledge

work, they may not take advantage of it. That is, participating in knowledge work using collaborative technologies has increased the importance of individual skill and internal motivation, while at the same time depersonalizing ones' knowledge contribution (Gray, 2001). Furthermore, because knowledge work is easily transformed and transmitted, organizations and individuals must contend with rapid changes and consistent unpredictability (Haeckel, 1999). Thus, the newly empowered individuals must learn to act autonomously and respond quickly.

It is obvious from the above examples that e-collaboration technologies can impact (directly or indirectly) every aspect of an organization. There are influences on the breadth of an organization's boundaries, influences on the very nature of our collaborations and interactions, influences on the type of work conducted, and influences on the value of personal attributes associated with such work. Given the overwhelming impact and nested influences of e-collaboration technologies, it is difficult to see all the connections and make all the appropriate and corresponding changes in order to achieve the full potential of e-collaboration technologies. What is needed is the identification of an underlying theme that unites all the areas transformed by e-collaboration technologies. *Adaptability* appears to be that underlying theme. With every change in work associated with e-collaboration, increased flexibility and adaptability (organizational, individual, or technology) continues to emerge as the central issue. With adaptability identified as the central issue, researchers and organizations can proceed with a unified framework and formulate a coherent path to integration with e-collaboration technologies and the new world of work the technologies have helped to create.

ADAPTABILITY

Several sources have created the need for increased adaptability in numerous aspects of business, but technological changes such as e-collaboration are among the most pervasive of all recent changes and have altered the basic nature of work (Pulakos, Dorsey, & White, 2006). The need for increased flexibility and adaptability epitomizes each of the three examples described above. Not only must the organization and the e-collaboration technologies used be flexible and adaptable, but the people must be adaptable. The unique human ability to respond creatively to new situations is an integral part of an adaptive system (Haeckle, 1999). Although technological innovations to cope with changes in work abound, the implications at the human level have been neglected. However, as humans are at the core of any business, they are an obvious starting point, and changes at the human level will facilitate changes at all other levels. We must ensure that we are not simply retrofitting organizations with technology overlays without a commensurate human-centered effort to improve the performance of the workers utilizing the e-collaboration technologies. As the foundations of work itself are being altered due to the use of e-collaboration technologies, the demands placed on individuals are transforming, and in turn, the value for particular dimensions of performance are shifting (Ilgen & Pulakos, 1999; Pulakos, Arad, Donovan, & Plamondon, 2000). If the new dimensions of performance are identified, organizations will be able to select and train employees to operate more effectively in environments reliant on e-collaboration, which characterize most work settings today.

ADAPTIVE PERFORMANCE

Pulakos et al. (2000) have developed a model of job performance—the adaptive job performance model—that captures performance dimensions more pertinent to the information age. The adaptive job performance (AJP) model provides a theoretical framework for understanding adaptive behavior in jobs. As opposed to approaching adaptability as a vague notion, Pulakos et al.'s validated AJP model eliminates the elusiveness of the concept by clearly identifying adaptive performance behaviors and predictors of such behavior, which can then serve as the basis for selection and training. The behaviorally based dimensions of AJP identified in the Pulakos et al. model were derived from an extensive literature review on adaptability and content analyses on a large number of critical incidences from 21 different jobs. This effort revealed eight relevant dimensions of adaptive performance: handling work stress; solving problems creatively; handling emergencies or crisis situations; dealing with uncertain and unpredictable work situations; learning technologies, work tasks, and procedures; demonstrating interpersonal adaptability;

demonstrating cultural adaptability; demonstrating physically oriented adaptability.

As opposed to assessing absolute performance, the intent of the AJP model is to assess performance in terms of coping ability and responsiveness to changing demands (Hesketh & Neal, 1999). Such aspects of performance are imperative for the trend towards knowledge work and project-based teamwork, both of which are heavily intermingled with e-collaboration technologies. Moreover, the importance of adapting to the technologies themselves has become such an integral aspect of work today that it qualifies as its own dimension of AJP. The dimension of *learning new work tasks and technologies* is defined as demonstrating enthusiasm and effort for learning new technologies and approaches for conducting work, which is an obvious aspect of performance important for successful e-collaboration. Beyond simply *using* the e-collaboration technology, individuals must develop new approaches to collaboration that fit with the technology.

As mentioned previously, Kock's (2001) compensatory adaptation theory is a prime example of how individuals develop new approaches when using e-collaboration technologies. Kock (2001, 2004) has repeatedly found that individuals using e-collaboration technologies (e.g., e-mail) adapt their behavior to compensate for the lack of richness of information (e.g., facial cues) in the technologies. Kock (2001) has suggested that this adaptive behavior is likely attributable to general cognitive patterns. As we move away from the instinctive schemas associated with face-to-face communication, we develop (compensate with) learned schemas that support communication via technologies (Kock, 2004). However, there will be individual differences such that not all individuals will be equally as capable or as successful at developing learned schemas for e-collaboration.

The general cognitive patterns that Kock (2004) alluded to as influencing adaptive behavior in the e-collaboration domain (i.e., the technology dimension of AJP) likely influence adaptive behavior in general. It is general adaptive behavior as captured by the AJP model that serves as the keystone to, not only successful e-collaboration, but successful business today. If we want to select and train individuals to be more successful with e-collaboration technologies, we also must consider the adaptive behavior required to surmount the additional changes in work associated with the use of e-collaboration technologies. For example, the global business world requires employees that are capable of e-collaborating with individuals of different cultural (ethnic and organizational) backgrounds. The AJP dimension of *demonstrating cultural adaptability* clearly addresses this area of adaptive behavior.

FUTURE TRENDS

We are calling for a broad perspective and a proactive stance in regard to adaptive behavior in organizations, especially organizations heavily reliant on e-collaboration technologies where adaptability is paramount. As opposed to being concerned with adaptive behavior only in relation to interactions with the technologies, organizations should attempt to identify and facilitate the adaptive behavior of individuals. Fortunately, there is a growing body of research (e.g., Griffin & Hesketh, 2003; Kozlowski et al., 2001; Ployhart & Bliese, 2006; Pulakos et al., 2006; Pulakos et al., 2002; Stokes & Faas, 2006) seeking to identify the characteristics (personal and situational) associated with adaptive behavior. With the AJP model serving as a validated criterion measure, several predictors of adaptive behavior have been identified (e.g., cognitive ability, self-efficacy, cognitive flexibility, need for structure, and openness). These predictors could, in turn, be used in a selection-test battery to identify adaptive workers. Following selection, employees could participate in a training program focused on the development of *learned* schemas useful for interacting via e-collaboration technologies. Moreover, a structured training program would serve to develop shared schemas among group members, and shared schemas have been found to facilitate e-collaboration efforts by reducing cognitive load (Cuevas et al., 2004; Kock, 2004).

CONCLUSION

We believe that embracing and developing an adaptive workforce will serve as the foundation for a resolution to the many concerns and issues associated with e-collaboration technologies and the corresponding changes in work that have resulted from their use. As opposed to attacking the issues in a piecemeal fashion, adopting a systems perspective and acknowledging adaptive behavior as an underlying theme allows us to see the connections between the seemingly disparate

issues surrounding e-collaboration. We are not suggesting that adaptive behavior is a panacea, but we are suggesting that with a solid foundation in place (i.e., an adaptive workforce), the issues that remain will be less daunting.

REFERENCES

Ashby, W. R. (1947). Principles of the self-organizing dynamic system. *Journal of General Psychology, 37,* 125-128.

Becker-Beck, U., Wintermantel, M., & Borg, A. (2005). Principles of regulating interaction in teams practicing face-to-face communication versus teams practicing computer-mediated communication. *Small Group Research, 36,* 499-536.

Cameron, A. F., & Webster, J. (2005). Unintended consequences of emerging communication technologies: Instant messaging in the workplace. *Computers in Human Behavior, 21,* 85-103.

Cheng, E. W. L., Love, P. E. D., Standing, C., & Gharavi, H. (2006). Intention to e-collaborate: propagation of research propositions. *Industrial Management & Data Systems, 106*(1), 139-152.

Cuevas, H. M., Fiore, S. M., Salas, E., & Bowers, C. A. (2004). Virtual teams as sociotechnical systems. In S. H. Godar & S. P. Ferris (Eds.), *Virtual and collaborative teams: Process, technologies, and practice* (pp. 1-18). London: Idea Group.

Dubrovsky, V. J., Kiesler, S., & Sethna, B. N. (1991). The equalization phenomenon: Status effects in computer-mediated and face-to-face decision-making groups. *Human-Computer Interaction, 6,* 119-146.

Gray, P. H. (2001). The impact of knowledge repositories on power and control in the workplace. *Information Technology and People, 14*(4), 368-384.

Griffin, B., & Hesketh, B. (2003). Adaptable behaviours for successful work and career adjustment. *Australian Journal of Psychology, 55*(2), 65-73.

Haeckel, S. H. (1999). *Adaptive enterprise: Creating and leading sense and respond organizations.* Cambridge, MA: Harvard Business School.

A

Hesketh, B., & Neal, A. (1999). Technology and performance. In D. R. Ilgen, & E. D. Pulakos (Eds.), *The changing nature of performance: Implications for staffing, motivation, and development* (pp. 21-51). San Francisco: Jossey-Boss.

Ilgen, D. R., & Pulakos, E. D. (Eds.). (1999). *The changing nature of performance: Implications for staffing, motivation, and development.* San Francisco: Jossey-Bass.

Jarvenpaa, S., & Leidner, D. (1999). Communication and trust in global virtual teams. *Organizational Science, 10,* 791-815.

Kasper-Fuehrer, E. C., & Ashkanasy, N. M. (2001). Communicating trustworthiness and building trust in interorganizational virtual organizations. *Journal of Management, 27,* 235-254.

Kock, N. (2001). Compensatory adaptation to lean medium: An action research investigation of electronic communication in process improvement groups. *IEEE Transactions on Professional Communication, 44*(4), 267-285.

Kock, N. (2004). The psychobiological model: Toward a new theory of computer-mediated communication based on Darwinian evolution. *Organizational Science, 15*(3), 327-348.

Kock, N., & Nosek, J. (2005). Expanding the boundaries of e-collaboration. *IEEE Transactions on Professional Communication, 48*(1), 1-9.

Kozlowski, S. W. J., Gully, S. M., Brown, K. G., Salas, E., Smith, E. M., & Nason, E. R. (2001). Effects of training goals and goal orientation traits on multidimensional training outcomes and performance adaptability. *Organizational Behavior and Human Decision Processes, 85*(1), 1-31.

Maruping, L. M., & Agarwal, R. (2004). Managing team interpersonal processes through technology: A task-technology fit perspective. *Journal of Applied Psychology, 89*(6), 975-990.

Ployhart, R. E., & Bliese, P. D. (2006). Individual adaptability (I-Adapt) theory: Conceptualizing the antecedents, consequences, and measurement of individual differences in adaptability. In C. S. Burke, L. G. Pierce, & E. Salas (Eds.), *Understanding adaptability: A prerequisite for effective performance within com-*

plex environments. advances in human performance and cognitive engineering research (Vol. 6, pp. 3-40) Oxford, UK: Elsevier.

Pulakos, E. D., Arad, S., Donovan, M. A., & Plamondon, K. E. (2000). Adaptability in the workplace: Development of a taxonomy of adaptive performance. *Journal of Applied Psychology, 85*(4), 612-624.

Pulakos, E. D., Dorsey, D. W., & White, S. S. (2006). Adaptability in the workplace: Selecting an adaptive workforce. In C. S. Burke, L. G. Pierce, & E. Salas (Eds.), *Understanding adaptability: A prerequisite for effective performance within complex environments: Vol. 6. Advances in human performance and cognitive engineering research* (pp. 41-72). Oxford, UK: Elsevier.

Pulakos, E. D., Schmitt, N., Dorsey, D. W., Arad, S., Hedge, J. W., & Borman, W. C. (2002). Predicting adaptive performance: Further tests of a model of adaptability. *Human Performance, 15*(4), 299-323.

Ritter, J., Lyons, J. B, Swindler, & S. D. (2007). *Large-scale coordination: Developing a framework to evaluate socio-technical and collaborative issues.* Manuscript submitted for publication.

Stokes, C. K., & Faas, P. (2006). *Adaptive performance: Implications for military logistics.* In B. McQuay & W. Smari (Eds.), *Proceedings of the International Symposium on Collaborative Technologies and Systems* (pp. 248-255). Los Alamitos, CA: IEEE Computer Society Press.

Straus, S. (1997). Technology, group process, and group outcomes: Testing the connections in computer-mediated and face-to-face groups. *Human Computer Interaction, 12,* 227-266.

Sutton, J. L., Pierce, L. G., Burke, S., & Salas, E. (2006). Cultural adaptability. In C. S. Burke, L. G. Pierce, & E. Salas (Eds.), *Understanding adaptability: A prerequisite for effective performance within complex environments: Vol. 6. Advances in human performance and cognitive engineering research* (pp. 143-173). Oxford, UK: Elsevier.

Wainfan, W., & Davis, P. K. (2004). *Challenges in virtual collaboration: Videoconferencing, audioconferencing, and computer-mediated communications.* Santa Monica, CA: RAND Corporation.

A

KEY TERMS

Adaptive Performance: Altering behavior to meet new demands created by the novel and often ill-defined problems resulting from changing and uncertain work situations.

Compensatory Adaptation Theory: A theory developed by Kock (2001) that accounts for the adaptive behavior displayed by individuals engaged in computer-mediated communication, whereby individuals, applying increased cognitive effort, adapt their communication behavior (consciously and unconsciously) in order to compensate for the obstacles posed by computer mediation.

Dynamic Systems View: A dynamic systems view is based on systems theory, which emphasizes the importance of interdependence of relations.

E-Collaboration: Collaboration among individuals engaged in a common task using electronic technologies.

Knowledge Worker: One who works primarily with information or one who develops and uses knowledge in the workplace.

Psychobiological Model: A model developed by Kock (2001) that posits several propositions stating that there is a positive link between the naturalness of a communication medium and the cognitive effort required to communicate via the medium. This link is counterbalanced by the degree of schema similarity among members and the level of learned schemas (cognitive adaptation) for interacting via the medium.

Schema: A mental structure that represents some aspect of the world and assists in interacting with the world.

The Added Value of E-Procurement for Buyer-Supplier Interaction

Wilco van Duinkerken
Utrecht University, The Netherlands

Ronald Batenburg
Utrecht University, The Netherlands

Johan Versendaal
Utrecht University, The Netherlands

INTRODUCTION

Although back in 1980s Porter (1985), Kraljic (1983), Speckman (1981), and others already identified the strategic aspects of procurement, most companies have unnoted the potential of the procurement business function and e-procurement specifically until the late 1990s. The primary IT-interests of the management board were the internal processes (enterprise resource planning) as well as sales and marketing (customer relationship management). Lately however, Business-to-Business (B2B) processes and the e-collaboration between organizations have received increasing attention. (e.g., Braun & Winter, 2005; Cavalla, 2003; Versendaal & Brinkkemper, 2003; Kauffman, 2004).

Procurement is a specific area of B2B-collaboration that covers both a company's internal as well as its B2B-processes. Procurement is broader than "purchasing." Purchasing is about selecting a supplier, negotiating a price, and following up the order, while procurement also encompasses strategic sourcing, inventory control, storekeeping, disposal, and even collaborative engineering (Weele & Rietveld, 2000; Moe, 2004; Batenburg & Rutten, 2003).

Improvements in the procurement business function can lead to benefits in terms of cost savings, product quality increases and faster delivery of goods in many different ways (Chen, Paulraj, & Lado, 2004; Prahinski & Benton, 2004; Versendaal & Brinkkemper, 2003; O'Toole & Donaldson, 2002). Suppliers gain insight into the inventory levels of their buyers directly influence the production process of their suppliers (Womack, Jones, & Roos, 1991) and contract manufacturers insource production processes of their clients (e.g., Hagel, 2002).

E-procurement allows companies to use the Internet for procuring goods, as well as handling value added services like transportation, warehousing, payment, quality validation, and documentation (Johnson & Wang, 2002). With e-collaboration and e-procurement tools that support the planning, decision-making and coordination of procurement (Johnson & Wang, 2002), boundaries between buyers and suppliers can disappear.

Although there are many possible benefits anecdotal evidence shows that many initiatives and IT-implementations in the procurement domain do not deliver the suspected benefits, see for example Adamson (2001) and Pan, Pan, and Flynn (2004).

This paper searches an answer, based on empirical evidence, to the question: "Does the implementation of e-procurement (IT for procurement) positively affect the performance of buyer-supplier interaction?"

BACKGROUND

Some research has been conducted on the relation between procurement initiatives and procurement performance. Narasimhan and Das (2001) for example show that there is a correlation between the alignment of purchasing practices with a company's objectives and procurement performance. Chen et al. (2004) identify a relation between strategic purchasing, matured supply chain management and the company's financial performance. Subramaniam and Shaw (2004) found that the value of e-procurement systems in general varies depending on the process characteristics of the supported procurement function.

A

None of these papers explicitly mention e-procurement, or IT-related investments in the procurement domain, in relation to the performance of buyer-supplier interaction. Several fields of research however provide theories to answer the research questions, most notably: IT-metrics, strategic management, and business maturity.

Ever since IT-metrics have emerged as a specific field, scholars are searching for models and techniques to measure if organizations have invested the "right IT, aimed at the right processes and at the right moment" (Venkatraman, 1989). Brynjolfsson particularly contributed to this field as he defined the productivity paradox in 1993, and provided empirical evidence that IT indeed adds (tangible and intangible) benefits to the firm, provided that the right conditions are in place (Brynjolfsson & Hitt, 1995, 2000).

The capability maturity model (CMM) (Paulk et al., 1995) has become an established IS/IT maturity model. It was designed to measure, monitor and evaluate the professional development and engineering of software and related domains such as IT-governance, project management, and people management (Peppard & Ward, 1999) In the field of procurement, several authors have investigated ways for structured maturity development. (Cavinato, 1999; Kraljic, 1983; Van Weele & Rietveld, 2000) Van Weele and Rietveld defined a framework out of twelve other frameworks. The framework describes six "maturity" levels of procurement within an organization. These levels are described in the next section.

In order to help companies to choose and evaluate their e-procurement strategy this article presents an empirically tested E-procurement maturity framework

derived from Weele & Rietveld's maturity levels. The framework is called the e-procurement maturity framework (E-PMF).

THE E-PROCUREMENT MATURITY FRAMEWORK

The e-procurement maturity framework is based on Weele & Rietveld's (2000) procurement maturity model and relates e-procurement maturity to buyer-supplier interaction performance. The maturity level a company is at indicates how explicitly defined a specific process is and how strict it is managed, measured and controlled (Paulk, 1993). The six stages of maturity defined in Weele & Rietveld's procurement maturity model are:

- **Transactional orientation:** No procurement strategy, procurement is just acting on purchasing requests from the rest of the organization
- **Commercial orientation:** Mainly cost-oriented purchasing
- **Purchasing coordination:** Basic sourcing and purchasing optimization is in place within the procurement department
- **Internal integration:** The procurement department is considered as a strategic internally integrated part of the overall organization
- **External integration:** Suppliers are considered valuable integrated resources for the organization
- **Value chain integration:** The procurement department is contributing to the effectiveness of the entire consumer supply chain

Figure 1. The e-procurement maturity framework

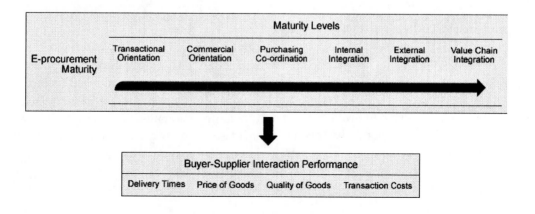

Buyer-supplier interaction performance is divided into four parts: delivery times, price of goods, quality of goods, and transaction costs. These four parts are complementary and jointly define the buyers-supplier performance (O'Toole, 2002; Humphreys, 2004). Figure 1 provides a schematic sketch of the complete e-procurement maturity framework.

THE RESEARCH MODEL AND HYPOTHESIS

Referring to our research question and the framework, as presented in Figure 1, we propose the following hypothesis:

The buyer-supplier performance of an organization is positively correlated with the e-procurement maturity, independent of the size of the company and the e-procurement budget.

In order to test this hypothesis three steps are taken. First, a correlation analysis is used to estimate the overall strength of the relationship between e-procurement maturity and the performance of buyer-supplier interaction. Next, a partial correlation analysis is conducted to perform a first direct test of the hypothesis. Then the relationship between e-procurement maturity and procurement maturity is tested in a more convincing manner, controlling for the size of the company and the annual procurement budget. Because company size and budget are likely to determine both e-procurement maturity and performance, these two volume indicators are expected to blur the relationship. Hence, controlling for size and budget should provide more insight in the net relationship between e-procurement and procurement performance. Finally, a multivariate regression analysis is executed to estimate the direct effect of e-procurement on performance. This can be considered as the second and final test of our hypothesis.

DATA

The survey was conducted in two parts with a group discussion about e-procurement in between. The first part contained questions about the company in general. The second part existed of questions related to procurement maturity and the perceived performance of the company's procurement process.

MEASURING E-PROCUREMENT MATURITY

E-procurement maturity was measured using 10 questions about the use and application of procurement and related information systems within the organization. The questions were ordered and presented to the respondent according to IT/IS complexity or maturity, starting about low complexity systems and ending with advanced e-procurement systems. Two examples of the questions/statements used are "In 'your organization,' an information system is used to track suppliers' performance for the 'spend category' of focus" and "Your organization has direct access to most databases of your main suppliers of 'spend category'-items." The spend-category was the category of items the respondent was responsible for within his company.

Procurement managers were asked to rate the extent to which they agreed with each item using a 5-point Likert scale (code "1" indicates "Strongly disagree," "5" indicates "Strongly agree"). The companies average (e-procurement) maturity level was calculated by adding all the Likert scores and subtracting 10 points from this sum.

The 10 questions or items contribute to a reliable scale supported by a Chronbach's Alpha score of 0.89. All but five of the possible intercorrelations between the 10 questions correlated significantly at 0.05 level. The companies are normally distributed over the six different maturity levels, with "Purchasing coordination" as the mean.

MEASURING BUYER-SUPPLIER INTERACTION PERFORMANCE

Buyer-supplier interaction performance was measured within the questionnaire through eight questions about the (perceived) procurement performance of the organization. Four questions were posed to the respondent to obtain an estimation of the procurement performance increase of the organization over the last two years. Four additional questions were posed to measure the extent to which the respondent's company outperforms

Table 1.

Sector	Number of Employees			Total
	< 50	50-250	>250	
IT, Tele-communications and B2B-services	3	1	7	11
Government and Education	4	3	7	14
Manufacturing	3	6	11	20
Trade, Transport and Logistics	3	3	3	9
Total	13	13	28	**54**

its competitors with regard to procurement. Both the time and competitor related questions specified performance in four dimensions: delivery times, price of goods, quality of goods, and transaction costs. As is known from earlier research, these four variables are reliable indicators for buyer-supplier performance in general (O'Toole, 2002; Humphreys, 2004)

As with the e-procurement maturity measurement, reliability analysis was performed to validate the aggregation of the eight questions, resulting into one latent indicator of buyer-supplier interaction performance. The Chronbach's Alpha score of 0.66 over the eight variables is acceptable. The resulting factor appears to be normally distributed.

DATA ANALYSIS AND RESULTS

To preliminary test the first hypothesis, a Spearman correlation between e-procurement maturity and procurement performance (as described above) was conducted. The result shows a significant correlation at the 0.05 level (2-tailed; r=0.361; p = 0.008). Next, to actually test our hypothesis, the correlation was re-estimated by controlling for company size and yearly procurement budget. This partial correlation analysis demonstrates that IT maturity, when taking firm size and procurement budget in account, is still significantly and positively related to buyer-supplier performance. This clearly supports our hypothesis, the claim that IS/IT and e-procurement investments do matter for the performance level organizations achieve in successfully managing their buyer-supplier relationships.

To investigate how strong the e-procurement maturity—compared to company size and the annual procurement budget—is related the buyer-supplier per-

formance a multiple regression analysis was executed. The regression model (with buyer-supplier performance as the dependent variable, and e-procurement maturity, company size and procurement spending as the independent variables) explains 24 percent of the variance in buyer-supplier performance (R^2 =0.24) and shows that the use of IS/IT and e-procurement determines performance in a significant, positive way ($\beta = 0.410$; p = 0.01). The effects of both company size ($\beta = 0.193$; p = 0.193) and procurement budget ($\beta = 0.135$; p = 0.374) are not significant. Figure 1 summarizes the outcomes of our estimates regression model and analysis.

The confirmation of our hypothesis provides a solid basis for formulating strategies for companies, no matter their company size or procurement budget, to start investments in IS/IT and e-procurement in order to improve the procurement business function.

FUTURE TRENDS AND GENERIC ADVICE

The results of the research provide strategies for companies willing to embark on e-procurement initiatives. Assessing a company's current situation using the E-PMF framework gives a good insight into the possible IT enhancements and investments a company could take to improve its procurement business function. It is a challenge not only to prove that IT investments can have tangible or intangible benefits (Brynjolfsson, 1993), but also to provide objective measurements and tools to plan, score and screen IT systems and strategies.

A solid IT base and a good understanding of the IT systems within the organization will better leverage the advances of e-procurement. Research in the fields of business/IT alignment (Scheper, 2002; Ward & Pep-

Figure 2. Results of the regression analysis

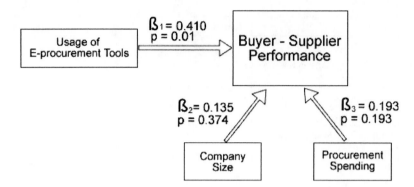

pard, 2002) and strategic management (Venkatraman 1989, 1993; Brynjolfsson & Hitt, 2000) also point in the direction of a balanced and well structured (Tallon, 2001) IT approach as a key success factor for IT.

There are a number of areas for further research. First of all more cases should be applied to further investigate the support of the hypothesis. The sample of procurement managers that participated in the survey was quite small to generalize the results to the Dutch, European, and world markets of e-procurement. In a larger research, companies from outside the Netherlands should also participate. Next, the measurement of cumulative scales like the CMM scale should improve. Taking an average of the procurement scores does not fully reflect the cumulative nature of maturity. It would be interesting to investigate other (mathematical) methods of measuring scores on a complementary scale.

CONCLUSION

The e-procurement maturity framework is a useful, empirically tested framework that provides objective tools and measurements for the development of the procurement business function. The E-PMF helped to answer the research question:

Does the implementation of e-procurement systems positively affect the performance of buyer-supplier interaction?

Using the E-PMF a questionnaire was built to measure the e-procurement maturity and the buyer-supplier interaction performance of 55 companies that differ in size and procurement budget. The questionnaire data

has been used to empirically test the hypothesis that e-procurement is positively related to an organizations' procurement performance. The results prove that IT indeed is beneficial for the performance of the procurement business function. From bivariate analysis the first result is that e-procurement systems implementation is positively correlated with the performance of buyer-supplier interaction.

To seek for robustness, a number of extended analyses were conducted to check the stability of the results and gather more information about the exact way buyer-supplier performance and e-procurement maturity coincided. Controlling for company size and yearly procurement spending did not change the results. The results of regression analysis further confirmed that e-procurement maturity is a strong and independent determinant of buyer-supplier interaction performance. This indicates the usage of e-procurement systems nowadays is a critical success factor for buyer-supplier interaction and the performance of the procurement business function in general.

REFERENCES

Adamson, J. Why is eProcurement failing? *Supply management.* Retrieved January 8, 2005 from http://www. tranmit.com/newsroom/0401_sm.htm

Anderson, K.V., Juul, N. C., Korzen-Bohr, S., & Pederson, J. K. (2003). *Fractional institutional endeavors and e-procurement in local government.* 16th Bled eCommerce Conference: eTransformation, Bled, Slovenia, June 9-11.

Batenburg, R., & Rutten, R. (2003). Managing innovation in regional supply networks: A Dutch case of "knowledge industry clustering." *Supply Chain Management: An International Journal, 8*(3), 263-270.

Batenburg, R., & Versendaal, J. (2004). Business alignment in the CRM Domain: Predicting CRM performance. In T. Leino, T. Saarinen, & S. Klein (Eds.), *Twelfth European Conference on Information Sytems*, ECIS 2004. Turku Finland.

Braun, C., & Winter, R. (2005). Classification of outsourcing phenomena in financial services. In D. Bartmann, F. Rajola, J. Kallinikos, D. Avos, R. Winter, P. Eindor, et al. (Eds.), *Thirteenth European Conference on Information Systems*, ECIS 2005. Regensburg Germany.

Brynjolfsson, E. (1993). The productivity paradox of information technology. *Communications of the ACM, 35*(12), 66-77.

Brynjolfsson, E., & Hitt, L. M. (1995). Paradox lost? Firm-level evidence on the returns to information systems spending. *Management Science, 42*(2), 541-558.

Brynjolfsson, E., & Hitt, L. M. (2000). Beyond computation: Information technology, organizational transformation and business performance. *The Journal of Economic Perspectives, 14*(4), 23-48.

Cavalla, D. (2003). The extended pharmaceutical enterprise. *Drugs Discovery Today, 8*(6), 267-274.

Cavinato, J. L. (1999). Fitting purchasing to the five stages of strategic management. *European Journal of Purchasing and Supply Management, 5*(2), 75-83.

Chen, I. J., Paulraj, A., & Lado, A. A. (2004). Strategic purchasing, supply management, and firm performance. *Journal of Operations Management, 22*(5), 505-523.

Fjermestad, J., & Hiltz, S. R. (2001). Group support systems: A descriptive evaluation of case and field studies. *Journal of Management Information Systems, 17*(3), 112-157.

Hagel, J. (2002). *Out of the box: Strategies for achieving profits today and growth tomorrow through Web services.* John Hagel III, US.

Humphreys, P. K., Li, W. L., & Chan, L. Y. (2004). The impact of supplier development on buyer-supplier performance. *Omega: The International Journal of Management Science, 32,* 131-143.

Johnson, M. E, & Whang, S. (2002). E-business and supply chain management: An overview and framework. *Production and Operations Management, 11*(4), 413-423.

Kauffmann, R. J., & Mohtadi, H. (2004). Proprietary and open systems adoption in e-procurement: A risk-augmented transaction cost perspective. *Journal of Management Information Systems, 21*(1), 137-66.

Kraljic, P. (1983). Purchasing must become supply management. *Harvard Business Review, 61*(5), 109-117.

Lamming, R. (1995). *Strategic procurement management in the 1990s: Concepts and cases.* Stamford, CT: Earlsgate Press.

Moe, K. E. (2004). Public e-procurement: Determinants of attitudes toward adoption. In R. Traunmüller (Eds.), *Proceedings of Electronic Government: Third International Conference.*

Narasimhan, R., & Das, A. (2001). The impact of purchasing integration and practices on manufacturing performance. *Journal of Operations Management, 19*(5), 593-609.

NIGP. (1996). National Association of State Purchasing Officials/National Association of Information Resource Executives Joint Force on Information Technology and Procurement. Reform.

Nunamaker, J. D., Valacich, J., Vogel, D., & George, J. (1991). Electronic meeting systems to support group work. *Communications of the ACM, 34*(7), 40-61.

O'Toole, T., & Donaldson, B. (2002). Relationship performance dimensions of buyer-supplier exchanges. *European Journal of Purchasing & Supply Management, 8,* 197-207.

Pan, G. S. C., Pan, S. L., & Flynn, D. (2004). De-escalation of commitment to information systems projects: A process perspective. *Journal of Strategic Information Systems, 13*(3), 247-270.

Paulk, M. C., et al. (1993). Capability maturity model, Version 1.1. *IEEE Software, July,* 18-27.

Paulk, M. C., et al. (1995). *The capability maturity model: Guidelines for improving the software process.* Reading, MA: Addison-Wesley.

Peppard, J., & Ward, J. (1999). "Mind the Gap": Diagnosing the relationship between the IT organisation and the rest of the business. *Journal of Strategic Information Systems, 8*(1), 29-60.

Porter, M. (1985). *Competitive advantage*. New York: Free Press.

Prahinski, C., & Benton, W. C. (2004). Supplier evaluations: Communication strategies to improve supplier performance. *Journal of Operations Management, 22*(1), 39-62.

Scheper, W. J. (2002). *Business IT alignment: Solution for the productivity paradox*. Deloitte & Touche, The Netherlands.

Speckman, R. (1981). A strategic approach to procurement planning. *Journal of Purchasing and Materials Management, Winter,* 3-9.

Subramaniam, C., & Shaw, M. J. (2004). The effects of process characteristics on the value of B2B e-procurement. *Information Technology and Management, 5,* 161-180.

Tallon, P. P., Kraemer, K. et al (2001). Executives' perceptions of the business value of information technology: A process-oriented approach. *Journal of Management Information Systems, 16*(4) 145-173.

Venkatraman, N. (1989). Strategic orientation of business enterprises: The construct, dimensionality, and measurement. *Management Science, 35*(8), 942-962.

Venkatraman, N. (1993). Continuous strategic alignment: Exploiting information technology capabilities for competitive success. *European Management Journal, 11*(2), 139-149.

Versendaal, J., & Brinkkemper, B. (2003). Benefits and success factors of buyer-owned electronic trading exchanges: Procurement at Komatsu America Corporation. *Journal of Information Technology Cases and Applications, 5*(4), 39-52.

Ward, J., & Peppard J. (2002). *Strategic planning for information systems* (3rd ed.).

Weele, A. J. van, & Rietveld, G. (2000). Professional development of purchasing in organisations: Towards a purchasing development model. *Global Purchasing & Supply Chain Strategies*. Retrieved July 8, 2005, from http://www.bbriefings.com

Womack, J. P., Jones, D. T, & Roos, D. (1991). *The machine that changed the world: The story of lean production*. New York: HarperPerennial.

Zenz, G., & Thompson, G. H. (1994). *Purchasing and the management of materials* (7th ed.). New York: John Wiley and Sons.

KEY TERMS

Capability Maturity Model: A model to measure, monitor and evaluate the professional development and engineering of software and related domains such as IT-governance, project management, and people management (Peppard & Ward, 1999).

Direct Goods: Direct goods and services are components and raw materials, which are used in the manufacturing process of a finished product (Lamming, 1995).

E-Procurement Systems: the application of a span of digital technologies, like electronic data interchange (EDI) and Internet technologies to enable exchanging partners smoothing and expanding the front-end and back-office integration of contracting, service, transportation, and payment of the products and services through processes, decisions, and transactions (Anderson, Juul, Korzen-Bohr, & Pederson, 2003).

Indirect Goods: Indirect procurement relates to products and services for maintenance, repair, and operations and focuses on products and services that are neither part of the end product nor resold directly (Zenz & Thompson, 1994).

IS/IT Maturity: The state of software systems and strategy of being fully planned and developed.

Procurement: Procurement combines the functions of purchasing, inventory control, traffic and transportation, receiving and inspection, storekeeping, and salvage and disposal operations (NIGP, 1996) of both direct and indirect goods.

Ambassadorial Leadership and E-Collaborative Teams

Richard R. Reilly
Stevens Institute of Technology, USA

Michael R. Ryan
Stevens Institute of Technology, USA

INTRODUCTION

If differences between virtual and traditional teams are bounded by the use of technology (Arnison & Miller, 2002; Griffith, Sawyer, & Neale, 2003), then virtual teams must not be considered a new phenomenon. What have changed are the tools which affect the breadth and depth of virtual teams. A global change in core communication technologies has occurred in the last 2 decades. Primary communication in the 1980s included letters, memos, telephone (one to one), and face-to-face meetings placing constraints on infrastructures supporting virtual teams; today's communication is based on technologies that transcend the physical constraints of the past but impose new and significant challenges in the interpersonal relationships.

The emergence of personal computers, the Internet, and wireless technologies created a communications revolution. E-mail proved as instantaneous as the telephone, as permanent as the written record, and capable of communicating simultaneously one to many. The Internet allowed written, audio, and visual collaboration. Cellular phones removed the tether of brick and mortar offices. Actors became always available. These technologies have reduced the world into a global neighborhood by relieving earlier constraints. In fact, all e-collaborations are virtual to some extent. The term *Virtual Distance*™ is used to indicate the team's position on a relationship scale that spans the purely virtual and the purely nonvirtual. Virtually distant teams are characterized by extensive use of electronic media, cultural differences, a lack of preexisting ties, and low perceived interdependence among other factors (Sobel-Lojeski, Reilly, & Dominick, 2006).

Effective teams develop high levels of trust and cohesion around the team mission and vision. But how can leaders build trust with virtually distant, global teams? As Bell and Kozlowski (2002) note, virtual teams do not fit in any existing typology for team leadership. Virtual leaders must cope with organizational, cultural, functional, and geographic boundaries; issues that do not figure into most leadership models. We suggest that virtual team leaders (VTLs) can apply the best aspects of transformational and transactional leadership by exhibiting a separate category of behavior that we call "ambassadorial behaviors."[1] Effective ambassadors create conditions for cooperation and collaboration between states. They use diplomacy to bridge the differences in cultural values and norms and establish greater communication while remaining sensitive to their differences and needs. Effective VTLs also act as ambassadors, in this case, between organizations, functions and cultures. This requires openness, empathy, and a certain level of "social intelligence" necessary for spanning the inherent boundaries (Ascalon, Schleicher, & Born, 2005). As a first step, effective VTLs recognize the factors that create Virtual Distance™ and mistrust between team members. Differences in cultural values and communication styles, for example, can impede trust levels and effective collaboration. For example, Gluesing et al. (2005) describe Celestial's French-American team in which the American leader tried to lead with traditional methods supplemented through travel and virtual meetings. These attempts proved unsuccessful leading to resistance, tension and finally outright conflict and hostility. Ultimately, the problem was solved by a facilitator who employed "shuttle diplomacy" in a series of one-on-one meetings to obtain the views of all team members so that shared understanding could be reached.

Ambassadorial behaviors promote trust and allow cohesion around common goals. As in diplomatic circles, the ambassador presents the values and norms of his or her "home" organization/culture to his or her host; concurrently, the ambassador seeks to understand the values and norms of this host and communicate that understanding to his or her "home" population.

LEADERSHIP THEORIES

Transformational and Transactional

Transformational and transactional leadership summarize behaviors that can be used to characterize the styles of different types of leaders (Bass, 1985), although effective leaders often exercise components of both (Yukl, 2001). Transformational leadership includes four behaviors: idealized influence, inspirational motivation, individualized consideration, and intellectual stimulation (Yukl, 2001).

Idealized influence supports member development of a strong positive identification with the leader. Transformational leadership theorists include charismatic behavior as a component of idealized influence. In developing the multifactor leadership questionnaire (MLQ), Avolio, Bass, and Jung (1999) have split idealized influence into two components—attributed to the leader and behavior of the leader. The behavioral component is closely aligned with charismatic leadership proposed by Conger, Kanungo, Menon, and Mathur (1997).

Individualized consideration supports followers by fostering personal efficacy. Inspirational motivation presents a collective purpose resulting from a clear vision articulated by the leader. Intellectual stimulation encourages member participation and contribution in developing a solution. These behaviors, individually and collectively, provide the foundation for member commitment and a sense of ownership (Ryan & Reilly, 2005). Bass (1985) postulates transformational leaders raise the level of personal awareness and encourage followers to transcend their self-interest resulting in increased motivation and greater effort.

Transactional leadership is based on an exchange between followers and leaders. This is defined by the task assigned and subsequent consequences (positive, negative, or neutral) for success or failure. Transactional leadership includes four behaviors: contingent reward, active management by exception (AMBE), passive management by exception (PMBE), and laissez-faire. Contingent reward, unequivocally, presents the expected output and resultant reward. AMBE exists when the leader actively monitors the follower and enforces guidelines designed to avoid mistakes. PMBE addresses mistakes after the fact and imposes a contingent punishment (negative reward). Laissez-faire is an extreme form of passive leadership and is unlikely

in a leader that assumes or exercises the role, but may be evident in an assigned leader's behavior.

Ambassadorial Leadership™

While traditional leadership models may work for traditional teams, as organizations change from traditional hierarchies to networked structures, new leadership behaviors are needed. Virtual collaborations, especially global virtual teams, offer a more complex and varied set of possible leader-follower relationships and require new approaches for understanding how to successfully lead e-collaborative teams. We suggest that a new model of Ambassadorial Leadership™ is essential for the effective leadership of virtual teams, especially global teams.

Ambassadorial Leadership™ focuses on behaviors that engage the team in building and expanding their relationships internally and externally. These behaviors include: internal boundary spanning; external boundary spanning; shared leadership; and impression management. Some of these behaviors are evident in traditional leadership models (i.e., impression management and external boundary spanning), but the traditional behavior has been limited, as we will discuss below.

Internal boundary spanning is concerned with the relationship between team units that are separated by some environmental, functional, or socio-economic barrier. External boundary spanning is concerned with the relationship between the team and/or its subunits and external entities that provide resources, are clients of the team, or both. Shared leadership allows the leader to leverage team resources by using team members as leads for specific parts of the project and in so doing aids in developing relationships within the team. Impression management addresses the communication between internal and external parties and helps manage the expectations of the team, sponsors, clients, and contributing parties. Ambassadorial Leadership™ complements the transformational goals of increased levels of personal awareness by drawing attention to team awareness and its potential to enhance the efforts of the individual.

Internal boundary spanning promotes team cohesion, understanding, and acceptance. The leader may exercise these behaviors directly or indirectly. Directly, the leader nurtures the team vision, advocates openness, and facilitates the development of relationships between distant and close members. Indirectly, the leader encourages other members to share leadership, actively engage fellow teammates in collaborative processes, and freely exchange ideas. It does not preclude disagreements, but limits them to content or approach rather than personal attributes, by emphasizing each individual's contribution to the team.

The virtual team leader (VTL) also bridges relationships between the virtual team and other individuals, intra-organizational units, and external entities. In addition to expanding the available knowledge base, boundary spanning is fundamental to securing tacit if not explicit support from these units (Ancona & Caldwell, 1992). The VTL promotes the team and its members, negotiates for additional resources (time, materials, personnel, etc.), manages expectations and impressions, and buffers demands. Indirectly, the leader challenges the team to extend their efforts to those tangent entities where they have established ties. In these situations, the team member may serve as a mediator to these groups in the same way as the team leader. Member awareness of intra-organizational support serves to develop trust in the organization. Expanding this authority to team members transcends the traditional leadership models that positioned the boundary spanning and impression management as the sole province of the team leader.

VIRTUAL DISTANCE™ AND LEADERSHIP

Most global virtual team research considers geographic distance a fundamental characteristic. But distance can also represent the emotional or psychological gap between team members who work in the same building and regularly meet FTF. For a team working primarily in virtual space, the socio-emotional "distance" may be a function of several other factors, in addition to the obvious ones of geography and computer mediation.

The socio-emotional distance between team members is influenced by a variety of factors. In addition to geographic and temporal differences, Virtual Distance™ incorporates differences in cultural values, communication styles, status, organizations, goal interdependence, and relationship history (Sobel-Lojeski et al., 2006). These factors collectively shape the perceptions of individuals engaged in collaborative work. Virtual Distance™ has a direct relationship to team member trust and vision clarity. Evidence also suggests that, as Virtual Distance™ increases, the behavior of the leader becomes more critical, not less (Reilly, Sobel-Lojeski, & Ryan, 2006). As in the Celestial team described above,

an appropriate leadership style can mean the difference between success and failure. Leading e-collaborative teams requires an understanding of how leadership behaviors can be used to reduce virtual distance.

Virtual Distance™ may, in fact, produce effects similar to those implied by leader-member-exchange (LMX) theory. LMX posits that interpersonal exchanges between leaders and followers are very different for "in-group" and "out-group" members. In-group members are more likely to encounter transformational behaviors while out-group members more frequently encounter transactional behaviors (Graen & Uhl-Bien, 1995). There is evidence that in-group, high-quality exchanges, may develop as a direct result of contributed effort. Additional factors contributing to high quality exchanges may exist outside the work environment, in the commonalities, norms, and shared values between the leader and follower (Liden & Maslyn, 1998). This places a burden on the VTL to promote a common set of norms and values between the virtually distant member (VDM) and virtually close member (VCM). Through such an engagement, the VDM's organizational citizenship behavior (OCB) may increase, stimulating an increase in collaborative effort. This effort may bring the member closer to in-group status as suggested by Liden and Maslyn and subsequently reduce the Virtual Distance™.

THE AMBASSADORIAL LEADERSHIP™ MODEL

Leading virtually distant e-collaborative teams demands both traditional transformational and transactional behaviors as well as new ambassadorial behaviors. Virtually distant members have different values and communication styles than the leader and VCMs. The ambassadorial virtual team leader (AVTL) promotes relationship building between close and distant members by encouraging sharing of personal information. The leader creates conditions for shared leadership within these virtually distant locations. He uses shared leadership to introduce diverse cultural values and communications styles. The AVTL establishes key relationships with members at remote locations, who serve as mentors and coaches. He ensures that communication of the vision is clear by working through these members who are conversant with the local customs. He establishes and reinforces common norms using these key members. The AVTL reinforces the vision and ensures that roles are clear through shared leadership and various forms of electronic communication. Some virtually distant members may have access to knowledge or data that might contribute new ideas to the team objectives. The AVTL works through shared leadership to exploit these resources. The leader also serves as a buffer for the virtually distant members, especially when organizational boundaries exist.

The AVTL establishes shared values with the distant member by expressing the vision in terms that are consistent with their values. The leader reinforces this vision by engaging members and providing an environment where they can contribute to the team leadership. Finally, the AVTL bolsters the relationship by acknowledging the distant member's importance through key face-to-face meetings. These behaviors forge a focal point for the member and as Howell (1988) indicates, the follower identifies with the leader creating a sense of common purpose and value. The leader emphasizes the interdependence of the team members. He acknowledges the VDM's diversity and empowers the member to assume a leadership role by introducing the team to elements from his cultural and environmental background. The leader uses this shared leadership to acknowledge the value of the distant member to the team. This sharing provides the team with more tools and avenues of expression and provides a basis for the leader to demonstrate the distant member's value to his respective functional superior.

The ambassadorial leader rewards distant members predicated on their performance. The AVTL pursues an understanding of the distant member's values and ensures that the rewards reflect those values. Part of the reward includes commending the individual to his functional group. This reinforces a shared vision by recognizing the member's contributions to the entire team. Besides reducing the Virtual Distance™ between members, the commendation illustrates desirable behaviors in pursuit of team tasks and goals.

The Virtual Distance™ dictates that initial exceptions be actively managed. The AVTL actively specifies assignments and monitors the member's effort towards their completion. As distance is reduced by shared experience, exceptions may be less actively managed.

CONCLUSION

The Virtual Distance™ that exists within an e-collaborative team requires leadership behaviors that transcend the traditional transformational and transactional models. Ambassadorial Leadership™ addresses the need for a more complete relationship between the team members; for an extended relationship between the team and its external stakeholders; and for the diversity of resources (collaborative, intellectual and physical) that can only be achieved through these relations.

REFERENCES

Ancona, D. G., & Caldwell, D. F. (1992). Bridging the boundary: External activity and performance in organizational teams. *Adminis. Science Quar., 37*(4), 634.

Arnison, L., & Miller, P. (2002). Virtual teams: A virtue for the conventional team. *Journal of Workplace Learning, 14*(4), 166.

Ascalon, M. E., Schleicher, D. J., & Born, M. P. (2005). Cross-cultural social intelligence: Development of a theoretically-based measure. *20th Annual Conf. of the Society for Industrial and Organizational Psyc.*, Los Angeles, CA.

Avolio, B. J., Bass, B. M., & Jung, D. I. (1999). Re-examining the components of transformational and transactional leadership using the multifactor leadership questionnaire. *Journal of Occupational and Organizational Psychology, 72*, 441.

Bass, B. M. (1985). *Leadership and performance beyond expectations*: New York: Free Press

Bell, B. S., & Kozlowski, S. W. J. (2002). A typology of virtual teams: Implications for effective leadership. *Group & Organization Management, 27*(1), 14.

Conger, J. A., Kanungo, R. N., Menon, S. T., & Mathur, P. (1997). Measuring charisma: Dimensionality and validity of the Conger-Kanungo Scale of Charismatic Leadership. *Revue Canadienne des Sciences de l'Administration, 14*(3), 290.

Gluesing, J. C., Alcordo, T. C., Baba, M. L., Britt, D., Wagner, K. H., McKether, W., et al. (2005) The Development of global virtual teams. In Gibson, C. B., Cohen, S. G. (Eds.), *Virtual Teams that Work*. San Francisco: Jossey Bass.

Graen, G. B., & Uhl-Bien, M. (1995). Relationship-based approach to leadership: Development of leader-member exchange (lmx) theory of leadership over 25

Figure 1. Traditional team: Ties are determined in relationship to the team leader

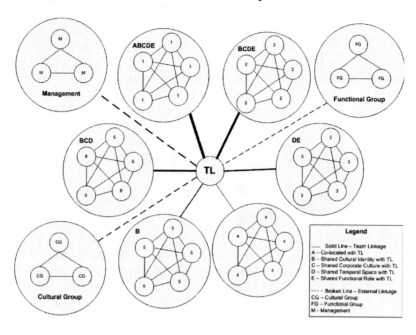

Figure 2. Ambassadorial Leadership™: The ambassadorial virtual team leader promotes establishment of direct ties within the team and between the team and external groups

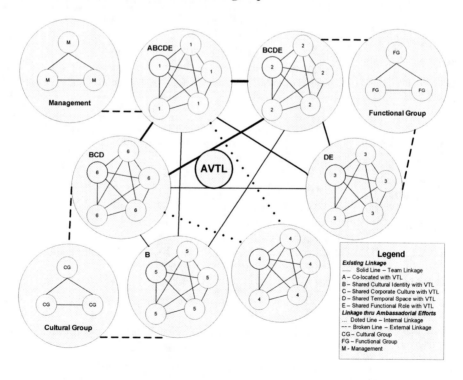

Ambassadorial Leadership™ behaviors for e-collaborative teams

Leadership Category	Behaviors
Ambassadorial Behaviors Internal Boundary Spanning	Promote relationship building between close and distant team members by encouraging sharing of personal information Educate local team members on differences in cultural values and communications styles of remote members Establish key relationships with members at remote locations who can serve as mentors and coaches Ensure that the vision is clear by working through key remote members Ensure that communications are consistent with local style Serve as a mediator in inter-organizational conflict Help establish and reinforce a common set of norms through members at remote locations Establish a member exchange between close and distant groups.
Ambassadorial Behaviors External Boundary Spanning	Formulate strategic plan with distant members to develop ambassadorial status with their close external groups. Promote strategy of team advocate with key distant members. Encourage team advocate to serve as a mediator in cases of conflict with external group. Encourage team advocate to buffer external demands placed on the team. Develop understanding of resources available from external groups. Communicate external group's suggestions, needs, and concerns to VTL. Establish communication channel to ensure knowledge base is kept current.
Ambassadorial Behaviors Shared Leadership	Create conditions for shared leadership at distant locations Establish key relationships with members at remote locations who can serve as mentors and coaches Establish a member exchange between close and distant groups.
Ambassadorial Behaviors Impression Management	Monitor conflicts and problems between close and distant members Serve as a mediator in cases of conflict. Recognize member's contribution to team efforts internally. Acknowledge team's contribution to organizational strategy externally. Monitor expectations (internal and external) and ensure alignment with reality.

Leadership Category	Behaviors
Transformational Inspirational Motivation	Clear explanation of goal interdependence Use of shared leadership to promote self-efficacy beliefs
Transformational Idealized Influence: Attributed Charisma	Clear expression of the vision in terms that are congruent with the values of the distant member Using shared leadership to reinforce and nurture the vision through personal interaction
Transformational Idealized Influence: Behavioral Charisma	Face-to-Face meetings at key points to present and reinforce the vision. Use of other media (e.g., video) to reinforce vision
Transformational Intellectual Stimulation	Encourage and recognize ideas through shared leaders Empower the member to engage in creative efforts consistent with vision.
Transformational Individualized Consideration	Promote the growth of the individual and team by reinforcing the synergies of a cultural and environmental exchange. Promote the member and emphasize their contribution to their functional superior. Recognize individual contributions through shared leader
Transactional Contingent Reward	Clarify tasks and goals and reward members for performance. Publicly commend the member within the team environment. Publicly commend the member within their functional department

continued on following page

Ambassadorial Leadership™ behaviors for e-collaborative teams, continued

Transactional		
	Active Management by Exception	Monitor the individual's efforts and actively intervene with any deviations from expected behavior.
Transactional		
	Passive Management by Exception	Not recommended
Transactional		
	Laissez-Faire	Not recommended

years: Applying a multi-level multi-domain perspective. *The Leadership Quarterly, 6*(2), 219.

Griffith, T. L., Sawyer, J. E., & Neale, M. A. (2003). Virtualness and knowledge in teams: Managing the love triangle of organizations, individuals, and information technology1. *MIS Quarterly, 27*(2), 265.

Howell, J. M. (1988). Two faces of charisma: Socialized and personalized leadership in organizations. In J. A. Conger & R. N. Kanungo (Eds.), *Charismatic leadership: The elusive factor in organizational effectiveness* (p. 352). San Francisco: Jossey-Bass.

Liden, R. C., & Maslyn, J. M. (1998). Multidimensionality of leader-member exchange: An empirical assessment through scale development. *Journal of Management, 24*(1), 43.

Reilly, R. R., Sobel-Lojeski, K., & Ryan, M. R. (2006). Leadership and organizational citizenship behavior in e-collaborative teams. *17th Annual IRMA International Conference.* Washington, DC.

Ryan, M. R., & Reilly, R. R. (2005). *Transformational and charismatic leadership in project management: A contingency model.* Paper presented at the Portland International Conference on Management of Engineering & Technology (PICMET), Portland, OR.

Sobel-Lojeski, K., Reilly, R. R., & Dominick, P. (2006). The role of virtual distance in innovation and success. In *Proceedings of the HICSS 39th Annual Conference.* Kauai, HI: HICSS.

Yukl, G. (2001). *Leadership in organizations* (5th ed.). Upper Saddle River, NJ: Prentice Hall.

KEY TERMS

Ambassadorial Leadership™: A set of behaviors that span the boundaries between cultures and organizations to create conditions for effective leadership

Charismatic Leadership: A leadership theory that concentrates on the leader's behavior and focuses on a shared vision.

Contextual Setting: In communication the relationship of context to content. In a high contextual setting the context is crucial and there is an underlying assumption of a shared background. In a low contextual setting, the content is essential and the information is conveyed explicitly in the communication; no shared background is assumed.

In-Group/Out-Group: A reference as to whether the leader-member exchange is high quality (in-group) or low quality (out-group). High quality refers to support with valued resources. Low quality implies a contractual like relationship. See leader-member exchange.

Leader-Member Exchange: A focus on the basic unit of leadership, the dyad. This exchange is a one-to-one measure of the leader's relationship with each follower.

Transactional Leadership: A leadership theory that is based on the transaction. It is often associated with reward or punishment for specific behaviors.

Transformational Leadership: A leadership theory that concentrates on the follower's behavior and growth.

Virtually Close Member (VCM): A team member that shares the same norms, environment, and cultural values as the virtual team leader.

Virtual Distance™ (Virtual Distance™ Index): A composite of physical, temporal, and personal elements that contribute to a perceived distance between team members.

Virtually Distant Member (VDM): A team member that has a different set of norms, environmental conditions, and cultural values as the virtual team leader.

ENDNOTE

[1] We use the term "ambassadorial" more broadly than Ancona and Caldwell (1992) who focused on boundary spanning behaviors between the team and the parent organization.

An Ambient Intelligent Prototype for Collaboration

Violeta Damjanovic
Salzburg Research, Austria

INTRODUCTION

The development of Web 2.0 has lead to a dramatic shift in the whole Web community. The decentralized and asynchronous applications are becoming prevalent both in Web-based and ubiquitous environments. In Web-based environments, each service (agent, application, Web service, learning object, mobile device, sensor) can be accessed through a single point, usually a Web portal, whereas in a ubiquitous environment, a service is dynamically composed by several agents that need to coordinate and negotiate with the aim of providing the most suitable adaptation for the user. In ubiquitous environments, distributed sensors follow the user's movements and based on either the user's typical tasks or the user's characteristics and preferences (learned from history and features of the context) appropriate adaptations of the interface, ambient features or functionality are made. Instead of having isolated user models for each application, a ubiquitous environment presumes the existence of a community of adaptive applications sharing user information.

In decentralized settings, each player maintains a small, locally created user model/profile, as needed for its own adaptation needs. Each person acts differently and needs to develop his/her own skills in own way. What makes the difference now is the availability of an advanced technology that enables us to rebuild learning and collaboration and make them more interactive, individualized, and adaptive. Nevertheless many difficult research challenges still remain, and much work is still needed if the existing relevant technologies are to be applied for the adaptation purposes in ubiquitous applications and the Semantic Web collaborative environment.

In this article, we explore the impact of ambient intelligence (AmI) on collaborative learning and experimental environments aiming to point out some new and upcoming trends in the professional collaboration on the Web. The article starts with some introductory explanations of both Web-based and ubiquitous en-

vironments. In addition, an overview of the relevant research issues is given. These issues represent the key paradigms on which the conceptual design of the AmIART prototype is based, and embrace the following facets: Ambient Intelligence, online experimenting, and personalized adaptation. The main idea of the AmIART prototype is to give users the feeling of being in training laboratories and working with real objects (paintings, artifacts, experimental components). Then, the AmIART prototype for fine art online experimenting is discussed in the sense of e-collaboration. When online experiments are executed in the Semantic Web environment, remote control of experimental instruments is based on knowledge that comes from domain ontologies and process ontologies (semantic-based knowledge systems). For these purposes, we present the ontology ACCADEMI@VINCIANA, as an example of a domain ontology (professional training domain), as well as the ontology GUMO (general user model and context ontology) that consists of a number of classes, predicates and instances aimed at covering all situational states and models of users, systems/devices and environments. In the following section, a collaborative scenario of using the AmiART prototype is given. The last section contains some conclusion remarks.

BACKGROUND

E-collaboration represents the convergence of technologies to allow people to work together. With the technology growing so fast, any technology should be available to meet the ongoing requirements from the field of e-collaboration.

In this section we explore the using of AmI as a developing technology that will increasingly make our physical environment sensitive and responsive to our presence. This section explores the key paradigms being used in the conceptual design of the AmIART prototype organized into the following three research issues: Ambient Intelligence, online experiments, and personalized adaptation.

Ambient Intelligence

The vision of AmI was first proposed by Philips Research in 1999, and was subsequently adopted in 2001 as the leading theme for the Sixth Framework Programme (FP6) on Research in Europe. This is the vision of a world in which technology is integrated into almost everything around us (ISTAG, 2005). AmI technology is intended to be in the service of humans; designed to adapt to the people's needs rather than making people adapt to the technology (Riva, 2005). In addition, this vision may be roughly described as being opposite to virtual reality (VR) (Riva, 2005): *until VR puts people inside a computer generated world, AmI puts the computer inside the human world to help us.*

In much of the general literature on emerging technologies, AmI is not clearly distinguished from earlier concepts such as *pervasive computing* or *ubiquitous computing*. The AmI paradigm builds on three recent key technologies (Alcañiz & Rey, 2005):

- **Ubiquitous computing:** It can be defined as the use of computers everywhere, but in the way that people are not aware of the presence of computers. Ubiquitous computing also means integration of microprocessors into everyday objects like furniture, clothing, white goods, toys, roads, smart materials;
- **Ubiquitous communications:** They enable these everyday objects to communicate with each other as well as with the user by means of *ad-hoc* and *wireless* networking (as recognized ongoing network technologies). Ad-hoc networking is a kind of a "mobile extension" of the Internet in environments with no network infrastructure. It is also a fundamental step towards achieving the goal of uninterrupted ubiquitous communications;
- **Intelligent user interfaces (user adaptive interfaces):** They enable the inhabitants of the AmI environment to control the environment and interact with it in a natural (voice, gestures) and personalized way (preferences, context). Key interface trends include the growth of agent communication languages (ACLs), the introduction of affect into the interface, and the growing focus on awareness and knowledge management. Affective

interfaces involve several different aspects like:

- Recognition of the user's affective state
- Generation of the computer's affective state
- Expression of the generated state by the computer

One of the main challenges recognized in ongoing intelligent user interfaces research efforts is building practically "invisible" interfaces (i.e., interfaces built in a human centered manner) through the ontology-based knowledge sharing. Intelligent products with intuitive interfaces will surround people, and both people and products will be always online, and connected by wireless communication technologies (Alcañiz & Rey, 2005).

Online Experimenting

When users work as a geographically distributed group, they communicate through computer-mediated channels like text chat, audio conferencing, video conferencing, application sharing (Faltin, 2004). These channels do not convey non-verbal messages (e.g., facial expressions) as faithfully as direct face-to-face communication. Also it is more difficult to refer to an object on the screen by application sharing than by pointing with a finger at it. This is why we introduce AmI concepts in online experimenting.

Online experiments are based on remotely controlling experiment equipment or software simulations of real experiments built for learning purposes. They enable users to get an experience without leaving their workplace and going to a traditional laboratory.

We distinguish local, remote and virtual experiments (Faltin, 2004):

- In a local experiment, users operate real devices and manipulate and measure real objects while being directly colocated with the devices and objects in the same room.
- In a remote experiment, users and devices are at different locations and the users manipulate the experiment devices through a computer connected (online) to the devices.
- Virtual experiments consist of software simulations of experiments and prerecorded measurements, pictures and videos but do include ma-

nipulations of real objects. They can be realized as local or distributed software applications.

Remote and virtual experiments are beneficial for a number of reasons, such as flexible and collaborative learning at any time and in any place, access to a large number of experiments and cost savings through experiment sharing (Faltin, 2004). When local experiments are too costly and/or take very long time to perform, online experiments may be the only alternative to provide hands-on experience.

Personalized Adaptation

As the AmI technology is designed to enhance people's experiences, the AmI requirements are not just technological. Accordingly, we propose personalized adaptation as an effective mean for improving the user's activities. Personalized adaptation is a key aspect of advanced, technology enhanced learning environments that supports ubiquitous, experimental and contextualized learning communities. In other words, the main aim of personalized adaptation is to support ubiquitous, decentralized, agent-based systems and devices for learning, training, and generally performing well in different environments.

At the same time, there is an opportunity for using new technologies and standards, such as metadata, Semantic Web, Semantic Web services, and Semantic Grid technologies. New technologies and standards have led to new achievements in order to provide the requisite global behaviour, without manual intervention (De Roure & Hendler, 2004).

THE AmIART PROTOTYPE

AmIART is an AmI prototype platform that can be used by different scientists to access collective skills and experiences, and to collaborate in a secure, reliable and scalable manner. It represents an AmI platform for fine art online experimenting, concerned with collaborating and learning about fine art materials, techniques and technologies.

The use of laboratories is essential for high quality education and collaboration not only in engineering and science, but also in the fine art related fields. Learning, socialize, working together, share ideas and engage in conversation are the main reasons why people join online collaborative environments and communities. Laboratories allow for the application and testing of theoretical knowledge in practical learning situations (Faltin, 2004). Active involvement in experiments and problem solving help learners collaborate and acquire working knowledge that can be used in real world situations (Faltin, 2004).

AmIART prototype environment is going to be:

- Able to perform collaboration scenario
- Aware of the specific characteristics of human presence and user personalities
- Able to adapt to the users' real needs

The human presence "as process" represents the continuous activity of the brain in separating "internal" from "external" within different kinds of signals (Riva, 2005). In addition, the presence "as process" can be divided into three different subprocesses (Riva, 2005):

- **Proto presence (self vs. nonself):** Proto presence is defined as a process of internal/external separation related to the level of perception/action coupling
- **Core presence (self vs. present external world):** Core presence is described as the activity of selective attention made by oneself on his/her perceptions
- **Extended presence (self relative to present external world):** Role of extended presence is to verify the significance of the events experienced in the external world to oneself

As shown in Figure 1, the *Context Manager* consists all of these three layers: proto presence, core presence, and extended presence. *Context Manager* receives context information from the user's internal world, as well external world, by using *endo* and *eso* sensors. Sensors (receptors) are used by the system to keep track of the specified data and variables of the working environment (external world), as well as of its own internal state (internal world) (Piva, Bonamico, Regazzoni, & Lavagetto, 2005). Our proposal for associating endo-sensors and eso-sensors, respectively, to the ACCADEMI@VINCIANA ontology and GUMO ontology is discussed in the upcoming subsections.

Figure 1. The AmIART prototype platform

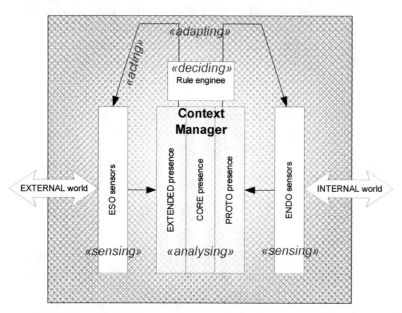

ACCADEMI@VINCIANA Ontology

Nowadays, experiments are more complicated and demand usage of specialized and expensive equipment. Such equipment, only some research centers, universities or institutes can afford. The online laboratories and experiments could be, therefore, the solution.

We present ACCADEMI@VINCIANA ontology that formalizes the knowledge for online fine art experiments. The knowledge aggregated in ACCADEMI@ VINCIANA can be used to support the following (Damjanovic, Kravcik, & Devedzic, 2005):

- Learning about fine art painting methods and materials (e.g., preventive conservation strategies, restoration, reproduction)
- Training in fine art painting methods and materials (e.g., painting damage diagnosis)
- Art fraud e-detection (e.g., original expertise)

The ontology includes the following categories of trainings (Kraigher-Hozo, 1991) (shown in Figure 2):

- Trainings based on using physical methods (dermatoscopes, micro abrasion equipments, microscopes, exploring the nanostructures of painting materials with X-rays, UV exploring, fluorescent exploring, analyzing small particles)

- Trainings based on using chemical methods (microchemistry approach with pigments identification, emission spectral analysis, the iodine probe, burning samples, exposing samples to the rays of the sun, high temperature)

All of these tools can be used to perform measurements and experiments, and to improve learning abilities about fine art painting methods and materials, as well as to explore originality, enable author identification, revile forgery, and much more.

GUMO (General User Model and Context) Ontology

The main idea behind GUMO ontology is to simplify the exchange of user model data between different user-adaptive systems (Heckmann, 2005). The current problem of syntactical and structural differences between existing user modeling systems could be overcome with a commonly accepted ontology specialized for user modeling tasks.

GUMO ontology integrates a user model ontology and a context ontology. A *Basic User Dimension* of this ontology is represented in Figure 3.

A decentralized user model allows each player (device, agent, networked application, Web-service, sensor) to share his/her own, locally created, user

Figure 2. A fragment of knowledge of the ACCADEMI@ VINCIANA ontology

Figure 3. GUMO' Basic User Dimension shown in the Ubis ontology browser

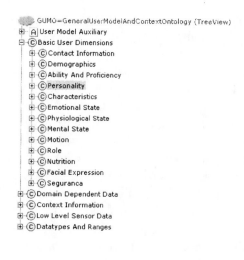

Figure 4. GUMO' Context Information: Location; physical environment; social environment

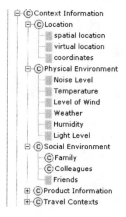

of the user's characteristics and to enable proper collaboration between different users bearing in mind their needs and personal preferences. The *Context Manager* also must have a location awareness module, as well as a component that provides data about enterprise policy.

A *Context Information* of GUMO ontology is represented in Figure 4.

THE AmIART COLLABORATIVE SCENARIO

The AmIART platform is aimed at identifying and supporting a number of innovative training, collaborative and knowledge management scenarios. For example, in a collaborative scenario, a number of knowledge workers might need to solve a particular problem from the domain of fine art investigations.

In a ubiquitous environment, users are followed by different devices (detection and sensing hardware) and the user information collected by the devices typically makes sense only in the given context (e.g., for a given period of time). An ideal scenario for ubiquitous computing system includes a real-time tracking mechanism that calculates the location of different components of the system. In addition, the system tracks the current state of each component and, using the acquired data, delivers messages to the user and interacts with him/her in an intelligent way (Alcañiz & Rey, 2005).

model and user data as well as to interpret that data in the context of the specific adaptation requirements of any ubiquitous application. GUMO ontology enables reusability and sharing of user models by various adaptive applications and user devices. Different detection and sensing hardware can be distributed in the environment or placed on the person to detect the different states of the user.

Ontology based sharing of knowledge must be supported by the AmIART *Context Manager* (shown in Figure 2) whose task is to detect any modification

Users will be able to select between different devices in order to have access to resources stored in any location on the network. Devices should be intuitive and adapt their behaviour to suit the current user and context. For example, fine art investigators (students, fine art conservators, restorers, technologists, fraud investigators) prepare for the collaborative, laboratory work in their local laboratories. They gather in an experimental room with a multimedia projector, sensor devices, lasers, and other mobile devices and equipment connected to the Web. If they are newbies, they fill a questionnaire for assessment of their knowledge, their personal characteristics and preferences. In the next step, the investigators connect to the remote laboratory, which consists of a dermatoscope, digital camera, and additional lighting that enables identification of fine changes in the structure of the artifact (e.g., painting) that is explored. The dermatoscope is moved across the painting canvas and the picture is shown on the investigators' Web browsers. The dermatoscope has a precision focusing mechanism, and investigators explore the painting's surface and take pictures of interesting fragments of the painting. In this way, the investigators from different laboratories can collaborate and detect together the painting's damages as well as make classical painting technology analysis, in a ubiquitous and a collaborative manner.

CONCLUSION

In the ambient environment, intelligence is provided through interaction and/or participation, and can be appreciated more as an assistive feature of the system, which addresses the real needs and desires of the user (ISTAG, 2005). Ambient intelligence presents *a set of properties of an environment that is in the process of creation*, so it is important to appreciate AmI as an *imaginary concept* and not as a set of specified requirements (ISTAG, 2005).

The knowledge-oriented and semantics-based approach to the ubiquitous environments, as well as the availability of the context data and the use of context in interactive applications (*context management*) can lead to the creation of new scientific results, new business, new collaborative ways of learning and experimenting, and even new research disciplines. In this article, an example of fine art professional training domain is represented.

The potential benefits of using the proposed AmIART prototype platform could be considered as follows:

- Enabling the use of all online available resources (learning materials, training devices, equipment) wherever the user is physically located
- Exploring and comparing ancient, as well as contemporary fine art technologies and techniques
- Analyzing results and deciding about the use of preventive painting strategies
- Collaboration aimed at achieving the original expertise and art fraud investigation.

REFERENCES

Alcañiz, M., & Rey, B. (2005). New technologies for ambient intelligence. In G. Riva, F. Vatalaro, F. Davide, & M., Alcañiz (Eds.), *Ambient intelligence* (pp. 3-15). IOS Press.

Damjanovic, V., Kravcik, M., & Devedzic, V. (2005). An approach to the realization of personalized adaptation by using eQ agent system. In *Proceedings of UM'2005 Workshop on Personalized Adaptation on the Semantic Web (PerSWeb'05)* (pp. 116-125).

De Roure, D., & Hendler, J. A. (2004). E-science: The Grid and the Semantic Web. *IEEE Intelligent Systems, 19*(1), 65-71.

Faltin, N. (2004). *Lab test with user and data collection on computer mediated-communication.* ProLearn Deliverable 3.1.1.

Heckmann, D. (2005). *Ubiquitous user modeling.* Unpublished doctoral dissertation, Universitat des Saarlandes, Germany.

ISTAG. (2005). Ambient intelligence: From vision to reality. In G. Riva, F. Vatalaro, F. Davide, & M. Alcañiz (Eds.), *Ambient intelligence* (pp. 45-68). IOS Press.

Kraigher-Hozo, M. (1991). *Painting/painting methods/materials.* Svjetlost.

Piva, S., Bonamico, C., Regazzoni, C., & Lavagetto, F. (2005). A flexible architecture for ambient intelligence systems supporting adaptive multimodal interaction with users. In G. Riva, F. Vatalaro, F. Davide, & M. Alcañiz (Eds.), *Ambient intelligence* (pp. 97-120). IOS Press.

Riva, G. (2005). The psychology of ambient intelligence: Activity, situation and presence. In G. Riva, F. Vatalaro, F. Davide, & M. Alcañiz (Eds.), Ambient intelligence (pp. 17-33). IOS Press.

KEY TERMS

Ambient Intelligence: Ambient intelligence represents the vision of a world in which technology is integrated into almost everything around us. It is provided through interaction and/or participation and can be appreciated more as an assistive feature of the system, which addresses the real needs and desires of the user.

Context Management: The goal of context management part is to design and implement a mechanism by which context information can be updated and distributed.

E-Collaboration: E-collaboration represents the convergence of technologies to allow people to work together.

Online Experiments: Online experiments present such kind of experiments that enable users to get an experience without leaving their workplace and going to a traditional laboratory. They are based on remotely controlling experiment equipment or software simulations of real experiments built for learning purposes.

Ontology: An ontology is a formal description of the meaning of the information stored in a system. Ontology provides explicit domain theories that can be used to make semantics of information explicit and machine processable.

Personalized Adaptation: Personalized adaptation is a key aspect of advanced, technology enhanced learning environments that supports ubiquitous, decentralized, agent-based systems and devices for learning, training, and generally performing well in different environments.

Semantic Web: The Semantic Web is a vision of an extension of the current web in which data are given meaning through the use of a series of technologies.

Ubiquitous Computing: The idea of ubiquitous computing as invisible computation was first articulated by Mark Weiser in 1988 at the Computer Science Lab at Xerox PARC. It can be defined as integration of microprocessors into everyday objects like furniture, clothing, toys, roads, smart materials.

Awareness Approaches of E–Collaboration Technology

Adriana S. Vivacqua
Federal University of Rio de Janeiro, Brazil

Jano M. de Souza
Federal University of Rio de Janeiro, Brazil

Jean-Paul Barthès
UTC – Université de Technologie de Compiègne, France

INTRODUCTION

Early field studies in collaborative work have shown that actors are capable of aligning and integrating their activities with those of others in an apparently seamless way (e.g., Heath & Luff, 1991). This is accomplished through the use of information gathered by overhearing others' conversations or surreptitiously monitoring their ongoing activities. To represent these practices of paying attention to what is going on in the environment, the term *awareness* was subsequently adopted (Schmidt, 2002). Researchers have dedicated much time to the study of how e-collaboration technologies might create some level of awareness between workers. Systems have been designed to enhance collaboration through the provision of information to create or maintain awareness of the working group. Even though different approaches have been introduced to address awareness, its creation and maintenance, researchers agree that most collaboration demands knowledge of others' activities (Dourish & Bellotti, 1992), and many have argued extensively that awareness is crucial for groups when performing their joint activities (Gutwin & Greenberg, 2004).

DEFINITIONS AND IMPACT ON COLLABORATIVE ACTIVITIES

Situation awareness research focuses on each individual's capacity to perceive elements and the cognitive processes involved in maintaining awareness of the environment. Endsley (2000) defines situation awareness (SA) as the process of perceiving environmental cues, interpreting their meaning and projecting their status in the near future. This information is used as a basis for individual decision making while working. Dourish and Bellotti (1992) define awareness as an understanding of activities of others, which provides a context for one's own activity. This information ensures that individual contributions are relevant to the group's activities, and enables individuals to assess others' actions with respect to group goals and work progress, which in turn allows individuals to adjust their behavior or take action according to the situation. Elaborating further on those concepts, Gutwin and Greenberg (1996) define workspace awareness as up-to-the-moment understanding of another person's interaction with the shared workspace. It is knowledge about the group's working environment, which creates an understanding of people within a workspace. These two definitions specialize the SA definition by defining the environment as a collaborative workspace, where information about other's activities and status is an important asset.

Rodden (1996) describes awareness as the overlap between nimbus and focus. Nimbus is the information given out by each element in space that can be perceived by others and focus describes the elements at which a user directs his or her attention. Thus, the awareness of individual A towards individual B is the intersection between the information being given out by B (B's nimbus) and the information A is interested in (A's focus). This model details how information to maintain awareness is obtained and shows that attention is a key aspect in the process, as it is affected by each individual's focus of attention and each element's provision of information.

In a collaborative environment, awareness involves knowledge about the people one is collaborating with (presence, identity, and authorship), the activities they are working on (actions, intentions and artifacts manipulated) and where (location of work, gaze direction, view and individual reach). Historical awareness information also includes action, artifact, and event history and should be provided in asynchronous work situations (Gutwin & Greenberg, 2002). This framework provides a starting point for designers to think about awareness elements and what information to provide in given situations.

In collocated environments, awareness information (i.e., information to maintain awareness), is gathered mainly through (1) intentional communication (i.e., communication intended by the sender, such as conversation and gestures); (2) consequential communication, or information transfer that happens as a consequence of the individual's activity within the environment, obtained by observing others' actions or body positions; and (3) feedthrough, which is the mechanism of determining a person's actions through cues given by the artifacts they interact with in the environment, such as position, orientation or movement (Gutwin & Greenberg, 2004).

Ethnographic studies have determined that awareness allows group members to manage the process of working together and is necessary for coordination of group activities (Dourish & Bellotti, 1992). Being aware of others' activities in a workspace allows participants to better understand the boundaries of their actions, which in turn helps them fit their own actions into the collaborative activity stream. This also enables groups to better manage coupling levels between their activities, helping people decide who they need to work with and when to make the transitions from looser to tighter coupling (Heath & Luff, 1991).

Furthermore, awareness simplifies communication by allowing individuals to reference the shared environment and elements within it: When discussing shared artifacts, the workspace can be used as a communication prop (Brinck & Gomez, 1992). This makes awareness an important building block for the construction of team cognition (Gutwin & Greenberg, 2004) and an enabler of shared understanding that allows individuals to get a better sense of the work that is being performed by others (Gutwin, Greenberg, Blum, & Dyck, 2005).

AWARENESS IN E-COLLABORATION APPLICATIONS

Maintenance of awareness is facilitated by physical proximity: it is possible to perceive a large amount of information simply by walking around the office, overhearing others or engaging in brief coffee break conversations (Kraut, Fussell, Brennan, & Siegel, 2002). Providing awareness information in e-collaboration environments, however, is far from simple, especially when participants are distant from one another. Input and output devices generate less information that face-to-face situations and users' interactions with computational workspaces also generate less information that physical environments (Gutwin & Greenberg, 2004). Given these limitations, a number of applications have been designed to provide awareness-enabling information in e-collaboration applications.

Early awareness work was heavily geared towards the use of video to support personal awareness and informal interactions. Experiments with the CRUISER (Root, 1988) and Portholes (Dourish & Bly, 1992) systems revealed that the possibility of easily engaging peers generated a number of new, spontaneous interactions. Through these systems, users could explore the virtual workplace, gaining awareness of their peers, which strengthened the sense of community between them. Negative reports involved the fact that everyday activities were not very exciting (users had to wait for a long time until something interesting happened), which led to loss of motivation to use the system. These early systems showed that awareness promotes group integration and helps people identify the right moment for starting a conversation by checking on their counterparts' availability. These systems continuously provide current identity and presence information, plus location and information on a member's actions (as it was possible to view whether one was answering the phone or chatting to a colleague).

The MAUI toolkit (Hill & Gutwin, 2004) is a Java-based toolkit that provides awareness-enhanced interface components for the construction of synchronous multiuser applications. These interface components (buttons, menus, windows) enable users to see when other users access them in their clients, providing current identity and activity information. User feedback on the interface widgets was positive, although they were concerned about the level of distraction caused by the added notifications.

CommunityBar (McEwan & Greenberg, 2005) is a configurable interface that enables users to keep abreast of group members' activities through media items (e.g., video, sticky notes, chat windows) providing identity, presence and some activity information. Initial observations revealed that users created separate spaces to support groups working on well defined, intense collaborative projects, which helped the group focus on the joint activities. The system also helped users move into closer interaction, providing awareness of others and context for their interactions. Users liked the ability to configure their focus and nimbus, but it was considered an overhead by authors, who state that automatic determination of focus and nimbus should be looked into.

A later study (Romero, McEwan, & Greenberg, 2006), showed that CommunityBar worked best for small cohesive groups, but was not a useful tool to support ad hoc groups, which formed outside the explicit structure provided by the system. It was considered useful for informal interactions, as it enabled easy transition from awareness into interaction. Users complained about the lack of historical information, the added effort necessary to explicitly define focus, and the coarse granularity of nimbus control. Feedback on the appropriateness of information was mixed, with many users indicating that Community Bar information was sometimes a distraction they did not want.

Most awareness technologies are implemented as notification servers, where users subscribe to certain events and are informed when those happen (Ramduny, Dix, & Rodden, 1998). This usually means that some effort must go into configuring the system, telling it what events to observe and send notifications. For instance, awareness in the PIÑAS (Morán, Favela, Martínez-Enríquez, & Decouchant, 2002) system is controlled via subscriptions to users and artifacts, which generate notifications whenever an element a user has subscribed to is activated. Thus, users were notified when they were working on the same document. Given that needs change according to the situation, configurations need to be constantly revised to keep in step with users' needs.

SELECTION OF AWARENESS INFORMATION

The large amount of awareness information potentially available could easily distract the individual rather than focus his/her attention on the tasks at hand, as noted by the feedback gathered in the aforementioned projects. Most people can only dedicate so much attention to surveying others' activities and assessing their impact on ongoing projects. This brings to the forefront the necessity for appropriate selection and delivery of awareness information. These considerations are similar to those existing in the field of artificial intelligence (AI), where researchers have been developing personalization and recommendation technology for over 10 years (e.g., see Maes, 1994). The difference is in that while AI researchers have been concerned mostly with individual information needs, awareness research looks to group needs.

Early on, Root (1988) suggests that the CRUISER system should be integrated with other applications, in order to provide information according to a user's current tasks, so that a user's activities would be inserted in their appropriate social contexts. Later research focused on providing information about others: Based on field research on information flow, Piazza (Isaacs, Tang, & Morris, 1996) provides awareness information about others who are working on similar tasks when using their computers. Similarity is determined by the data being accessed, the time in which this is (or was) done and the application used: The combination of these elements creates a conceptual "proximity" between users. By performing task based matches, Piazza provides a context for users to move into closer interaction. The authors note, however, that the system would need a better defined scope to function properly, or too many strangers might show up at any given moment, making the system useless.

A similar system, CUMBIA (Vivacqua, Moreno, & Souza, 2006) provides awareness information with the intent of finding new collaborators. This information is filtered according to the contents of documents being manipulated by each party. The provision of information about current work and level of expertise facilitates the location of experts. By finding people who are currently working on the same subject, the system maintains users in their current contexts, and draws on information that's fresh on their memory. Early user interviews showed that this provided useful matches, especially when a beginner was trying to get help. Experts, however, were concerned that they might be overrun with requests from beginners using the system.

Based on observations by Perer, Shneiderman, and Oard (2005), Vivacqua and Souza (2006) proposed an analysis of interaction records to determine identity, presence and activity information (who and what an individual should know about). An analysis of communication patterns elicits which of a user's contacts are currently receiving attention and should be tracked. Interactions with these collaborators are then processed for content and cross-referenced with ongoing activity information to determine which activities are related to which interactions. Interviews determined that an intensification of interaction corresponded to ongoing collaboration, and users recognized the need for awareness information in these cases. This investigation indicated that email rhythms can be used to determine current collaborators, whose activity information would be beneficial to the user.

To verify the utility of message content to obtain activity information, messages exchanged in two instances of collaborative activities were manually analyzed. When participants were physically close, and could easily meet in person, message bodies were not indicative of the collaboration themes, as they were short and related mostly to organizational issues. Actual discussion about project themes took place in face to face meetings. However, files sent as attachments, messages forwarded and links recommended were highly indicative of the topics of the collaborative activity, and can be seen as shared artifacts that can be tracked to follow collaboration. In the second situation (Vivacqua, Barthès, & Souza, 2006), participants were in different countries, which made physical contact harder, although occasional meetings did take place. In this case messages exchanged contained discussions on the subject itself that related to artifacts manipulated by users. These studies lead to the conclusion that it is possible to perform a textual match with ongoing activities, and determine which activities relate to the joint project based on information extracted from interaction records.

CONCLUSION AND FUTURE TRENDS

Appropriate selection and display of awareness information is an important issue in most e-collaboration environments. Our recent investigations have focused on the automatic determination of each individual's focus of interest (as it relates to ongoing collaboration), in order to provide awareness information to support joint efforts. In a large universe of acquaintances, it becomes important to distinguish the most relevant ones at any given moment, those that need to be watched in order to ensure the well-being of the collaborative project. An e-collaboration system should provide information that relates to the ongoing collaboration.

Future work should focus on developing more complex heuristics and inferences, for instance to identify the nature of the messages exchanged, the significance of ongoing tasks, the urgency of some information, to determine hierarchies, roles or levels of trust, or to infer the appropriate level of granularity of information. We anticipate that future work on user modeling and awareness will continue along these lines. Current technology requires too much effort on the users' part. More intelligent, automated and adaptable approaches should be designed for e-collaboration technologies to fully support awareness.

REFERENCES

Brinck, T., & Gomez, L. M. (1992). A collaborative medium for the support of conversational props. In M. Mantel & R. Baecker (Eds.), *Proceedings of the 1992 ACM Conference on Computer Supported Cooperative Work (CSCW'92)* (pp.171-178). New York: ACM Press.

Dourish, P., & Bellotti, V. (1992). Awareness and coordination in shared workspaces (1992). In M. Mantel & R. Baecker (Eds.), *Proceedings of the 1992 ACM Conference on Computer Supported Cooperative Work (CSCW'92)* (pp. 107-114). New York: ACM Press.

Dourish, P., & Bly, S. (1992) Portholes: Supporting awareness in distributed work group. In P. Bauersfeld, J. Bennett, & G. Lynch (Eds.), *Proceedings of the 1992 ACM Conference on Human Factors in Computing Systems (CHI'92)* (pp. 541-547). New York: ACM Press.

Endsley, M. R. (2000). Theoretical underpinnings of situation awareness: A critical review. In M. R. Endsley & D. J. Garland (Eds.), *Situation awareness analysis and measurement* (pp. 3-32). Mahwah, NJ: Erlbaum.

Gutwin, C., & Greenberg, S. (2002) A descriptive framework of workspace awareness for real-time groupware. *Computer Supported Cooperative Work, 11,* 411-446.

Gutwin, C., & Greenberg, S. (2004) The importance of awareness for team cognition in distributed collaboration. In E. Salas & S. M. Fiore (Eds.), *Team cognition: Understanding the factors that drive process and performance* (pp. 177-201). Washington, DC: APA Press.

Gutwin, C., Greenberg, S., Blum, R., & Dyck, J. (2005) *Supporting informal collaboration in shared-workspace groupware* (Tech. Rep. HCI-TR-2005-01). Saskatchewan, Canada: University of Saskatchewan, Interaction Lab.

Gutwin, C., Greenberg, S., & Roseman, M. (1996) Workspace awareness in real-time distributed groupware: framework, widgets and evaluation. In M. A. Sasse, R. J. Cunningham, & R. L. Winder (Eds.), *People and Computers XI: Proceedings of HCI'96* (pp. 281-298). British Computer Society.

Heath, C., & Luff, P. (1991). Collaborative activity and technological design: Task coordination in London Underground control rooms. In L. Bannon, M. Robinson, & K. Schmidt (Eds.), *Proceedings of the Second European Conference on Computer Supported Cooperative Work* (pp. 65-80). Amsterdam, The Netherlands.

Hill, J., & Gutwin, C. (2004) The MAUI toolkit: Groupware Widgets for Group Awareness. *Computer Supported Cooperative Work, 13,* 539-571.

Isaacs, E. A., Tang, J. C. & Morris, T. (1996). Piazza: A desktop environment supporting impromtu and planned interactions. In M. Ackerman (Ed.), *Proceedings of the 1996 ACM Conference on Computer Supported Cooperative Work (CSCW'96)* (pp. 315-324). New York: ACM Press.

Kraut, R. E., Fussell, S. R., Brennan, S. E., & Siegel, J. (2002). Understanding the effects of proximity on collaboration: Implications for technologies to support remote collaborative work. In P. Hinds & S. Kiesler (Eds.), *Distributed work* (pp. 137-162). Cambridge, MA: MIT Press.

Maes, P. (1994) Agents that reduce work and information overload. *Communications of the ACM, 37*(7) 30-40.

McCarley, J. S., Wickens, C. D., Goh, J., & Horrey, W. J. (2002). A computational model of attention/situation awareness. In *Proceedings of the 46th Annual Meeting of the Human Factors and Ergonomics Society* (pp. 1669-1673). Santa Monica, CA: Human Factors Society.

McEwan, G., & Greenberg, S. (2005). Supporting social worlds with the community bar. In *Proceedings of the 2005 International ACM SIGGROUP Conference on Supporting group Work (GROUP'05)* (pp. 21-30). New York: ACM Press.

Morán, A. L., Favela, J., Martínez-Enríquez, A. M. & Decouchant, D. (2002). Before getting there: Potential and actual collaboration. In J. M. Haake & J. A. Pino (Eds.), *Groupware: Design, implementation and use: Proceedings of the Eighth International Workshop in Groupware (CRIWG'02)* (LNCS 2440, pp 147-167). Berlin, Heidelberg, Germany: Springer-Verlag.

Perer, A., Shneiderman, B., & Oard, D. (2005) *Using rhythms of relationships to understand email archives* (Tech. Rep. HCIL-2005-08). College Park: University of Maryland, Department of Computer Science.

Ramduny, D., Dix, A., & Rodden, T. (1998) Exploring the design space for notification servers. In S. Poltrock & J. Grudin (Eds.), *Proceedings of the 2002 ACM Conference on Computer Supported Cooperative Work (CSCW'98)* (pp. 227-235). New York: ACM Press.

Rodden, T. (1996). Populating the application: A model of awareness for cooperative applications. In M. Ackerman (Ed.), *Proceedings of the 1996 ACM Conference on Computer Supported Cooperative Work (CSCW'96)* (pp. 87-96). New York: ACM Press.

Romero, N., McEwan, G., & Greenberg, S. (2006). *A field study of community bar: (Mis)-matches between theory and practice* (Report 2006-826-19). Calgary, Alberta, Canada: University of Calgary, Department of Computer Science.

Root, R. (1988) Design of a multi-media vehicle for social browsing. In I. Greif (Ed.), *Proceedings of the 1988 ACM Conference on Computer Supported Cooperative Work (CSCW'88)* (pp. 25-38). New York: ACM Press.

Schmidt, K. (2002). The problem with awareness: Introductory remarks on "Awareness in CSCW." *Computer Supported Collaborative Work, 11,* 285-298.

Vivacqua, A., Barthès, J. P., & Souza, J. M. (2006). A framework to support self governing design groups. In W. Shen, Z. Lin, J. P. Barthès, J. Luo, J. Deng, X. Li, J. Yong, et al. (Eds.), *Proceedings of the 10th International Conference on Computer Supported Cooperative Work in Design (CSCWD '06)* (pp. 375-380). Beijing, China: IEEE Press.

Vivacqua, A., Moreno, M., & Souza, J. M. (2006). Using agents to detect opportunities for collaboration. In W. Shen, K. Chao, Z. Lin, J. A. Barthès, & A. James (Eds.), *Computer supported cooperative work in design II* (**LNCS 3865,** pp. 244-253). Berlin, Heidelberg, Germany: Springer-Verlag.

Vivacqua, A., & Souza, J. M. (2006) Using email based network analysis to determine awareness foci. In Y. A. Dimitriadis, I. Zigurs, & E. Gómez-Sánchez (Eds.), *Groupware: Design, implementation and use: 12th International Workshop in Groupware (CRIWG '06)* (LNCS 4154, pp. 78-93). Berlin, Heidelberg, Germany: Springer-Verlag.

KEY TERMS

A

Awareness: An understanding of others' activities, which contextualizes one's own actions.

Awareness Information: Information to create or maintain a state of individual awareness.

Consequential Communication: Information transfer that happens as a consequence of an individual's activity within the environment. It is obtained by observing others' actions or body positions.

Feedthrough: Mechanism of determining a person's actions through cues given by the artifacts they interact with in the environment, such as position, orientation or movement.

Focus: elements at which an individual directs his or her attention.

Intentional Communication: Communication intended by the sender, such as conversation and gestures.

Interaction Analysis: A method for the investigation of the interactions of human beings with each other and with the objects in their environments.

Nimbus: Information given out by elements in space, which can be perceived by other people.

Situation Awareness: The process of perceiving and interpreting environmental cues, and projecting their future status in order to make decisions.

Workspace Awareness: Understanding of other participants' interactions with the shared workspace.

An Awareness Framework for Divergent Knowledge Communities

Farhad Daneshgar
University of New South Wales, Australia

Gerome Canal
LORIA – Nancy-Université, France

Alicia Diaz
UNLP, Argentina

INTRODUCTION

The Internet-based knowledge communities are considered today's main method for knowledge sharing in virtual communities. Zigurus and Qureshi (2001) suggest that collaborative systems and Web technologies have opened up myriad of possibilities for creating new and different types of relationships as well as increasing the reach of these relationships. On the other hand, knowledge logistics in Internet-enabled collaborative environments (i.e., who does what, how, using which resources, etc.) require novel conceptual abstractions and revised metaphors for collaboration and coordination as well as novel technological solutions, which go well beyond current collaborative software systems (Dustdar, 2004). From a technological perspective, the Web-based knowledge communities are special kind of today's Discussion Forums that are considered as today's main method for knowledge sharing in many virtual environments. These systems have their roots in the traditional structured messaging systems of the late 1980s initiated by researchers in the field of Computer-Supported Cooperative Work (CSCW). These platforms were designed to facilitate communication among remote participants at same/different times (Borenstein & Thyberg, 1993; Malone, Grant, Lai, Rao, & Rosenblit, 1993). The groupware technologies that support discussion forums are mainly Internet Web sites that use the Internet technology embedded within HTML. Like the newsgroups, discussion thread systems provide support for discussions and organize them according to topic and subtopic where users can participate. Unlike the newsgroups however, their services are not normally catalogued as part of the public Usenet service on the Internet, and therefore the Usenet search engine does not search what is written in a forum, although more

recent versions of these systems that support articulation of knowledge may use specialised protocols in order to provide their members with certain level of storage and search facilities.

With the recent growing body of literature on knowledge management and e-collaboration, much attention is now being given to creation of many novel conceptual/technological frameworks for managing knowledge creation in various kinds of e-communities. This article represents one such attempt with particular attention given to the *divergence occurrences* in knowledge communities, and their management as will be discussed later. A recent study by Diaz and Canals (2004) demonstrated that as the degree of people's involvement in various communications acts increases, so will the opportunities for divergence. Contrary to the general tendency within the CSCW community that regards conflicts as a synchronization and versioning problem in need of some solutions, the knowledge management community tends to live peacefully with such divergences and regards them as opportunities for interaction and therefore, sources of creating new knowledge.

BACKGROUND

A study by Diaz and Canals (2004) states that a natural consequence of the act of sharing knowledge in virtual knowledge communities is *divergence occurrences*. They also introduce a technological method called *DIVA* for management of divergence awareness in knowledge communities. According to this framework, divergences occur until such time when the community reaches a unique perspective.

The authors of this article share similar beliefs to those of the knowledge management community in the sense that conflicts and divergencies must be treated as a natural part of the knowledge sharing process that promotes emergence of new knowledge; and as a result, any technological method that supports knowledge-sharing activities must pay attention to divergences and conflicts as well as methods for managing these divergences. This article extends both the use as well as conceptual boundaries of e-collaboration research (Kock & Nosek, 2005) by integrating conceptual boundaries of two existing frameworks in order to provide a business context for management of the current virtual knowledge communities. This is shown in more details later on in the section Analysis of the Results.

The divergence awareness (DIVA) technological framework provides a management technological platform for managing *divergence occurrences* in *knowledge communities*. DIVA allows divergence to coexist within the community as a source of creating new knowledge while enabling community members to contribute while moving between private and shared knowledge spaces and managing contribution threads seamlessly. The DIVA workspace system is aware of both its members' profiles (skills, interests, etc.) as well as their evolution. As a result, it can deliver custom-made contributions to the community members. Divergence occurrences can be classified into (1) generation of alternatives, (2) arguments, and (3) different point of views about a topic of interest (Ibid).

The DIVA workspace consists of a *private knowledge workspace* (PKW) and a *shared knowledge workspace* (SKW). It is believed that community members must already enjoy a high level of knowledge usage before they are able to create PKWs for organising their knowledge artifacts, possibly by using files in folder hierarchies (Dustdar, 2004).

The PKW is articulated with personal view of the shared knowledge space. In the current version of DIVA there is no formal line that separates PKW and SKW; and these two concepts remain at conceptual levels and cannot be easily operationalised. As a result of the added formalism of the proposed integrated framework however, identification of these boundaries is greatly facilitated and formalised on the basis of the semantic concepts used in the added formalism called the *awareness net*. The proposed integrated framework integrates the people and process perspectives of the awareness net with the technological perspective of the DIVA in order to provide a solution that guides groupware design of collaborative systems in supporting divergence in knowledge communities more effectively.

THE AWARENESS NET

The awareness net is a major component of a conceptual framework called process awareness framework or PAF (Daneshgar, 1997; Ray, Shahrestani, & Daneshgar, 2005). The PAF was initially created for identification of the actors' awareness and knowledge sharing requirements in collaborative business processes. Subsequent studies revealed some additional capabilities for this framework including identification of the data modeling requirements of collaborative knowledge-based systems (Daneshgar et al., 2004), and identification of the user-interface design requirements for supporting knowledge-sharing processes (Daneshgar & van der Kwast, 2005). A summary of theoretical foundation of the awareness net is explained in the following paragraphs (adpted from Ray, Shahrestani & Daneshgar, 2005).

According to the awareness net, each actor (or, community member) is assigned one or more *roles* within the collaborative process (or, the community). Each role is associated with a set of *personal* and *collaborative tasks*. An actor can play various roles. In this study we extend the scope of knowledge sharing process from being a set of pairwise intellectual activities, as is the case for DIVA, into a larger collaborative process context where people perform various tasks and use various artifacts, as is defined by the awareness net model. The following sections demonstrate that by integrating the DIVA and awareness net frameworks the actors' knowledge-sharing requirements as well as their high-level PKW and SKW requirements can be defined more precisely. This in turn will guide groupware design for systems that support knowledge communities.

From a graphical perspective, the awareness net is a connected graph that represents a collaborative business process, in this case, the knowledge community. It consists of a set of collaborative semantic concepts and their relationships as its vertices and links respectively. When combined and linked together, these semantic concepts make up a connected graph called the awareness net that represents a collaborative process. Figure 1 shows awareness net for a typical DIVA knowledge community where members are involved in various

Figure 1. An awareness net for a DIVA knowledge community

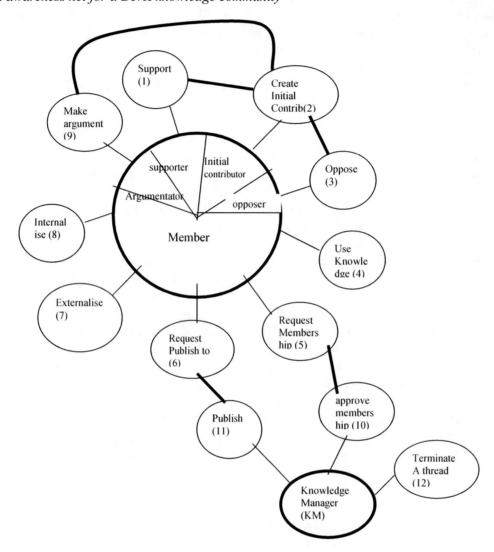

tasks using various artifacts (or knowledge resources) for their activities.

In Figure 1, thick-lined circles represent *roles*, thick lines represent *task artifacts*, narrow-lined circles represent *tasks*, and narrow lines represent *role artifacts*. A pair of *personal tasks* linked by a *role artifact* represents a *collaborative task*. The awareness net is a model of collaborative business process that facilitates identification and measurement of awareness and knowledge sharing requirements of the actors within the process. According to the underlying theoretical foundation of the awareness net, various levels of awareness can now be derrived for each of the roles in Figure 1. The overall definitions of these levels are:

Level-0 awareness for the role knowledge manager (KM) consists of the contextual knowledge about the tasks "approve membership," "publish," and "terminate a thread" as well as knowledge of how to approve membership, how to publish, and how to terminate a thread. These are shown by links between these tasks and the role knowledge manager.

Level-1 awareness for the role KM consists of the knowledge about the following additional objects: "request membership," "request to publish," "member" plus the two related tasks artifacts shared with the Member, plus the two related role artifacts used by the Member.

Various awareness paths/structures for some of the roles in Figure 1 are shown in Table 1.

Table 1. Various awareness paths for some of the roles in the awareness net

Role	Level (0) Awareness	Level (1) Awareness	Level (2) Awareness	Level (3) Awareness	Level (4) Awareness
Member	1+2+3+4+5+6+7+8+9 + Member + relevant role artifacts	Level (0) + 10+11+KM+ relevant artifacts	Same as level (1), as no roles left from the previous level	Same as level (2), no task artifacts left from previous level	Level (3) + 12
Knowledge Manager	10 + 11 + 12 + KM + all related role artifacts	Level (0) + 5 +6+Member + relevant role & task artifacts	Level (1)+1+ 2+3+9+ Argumentator +Supporter+ Initial Contrib + Opposer + relevant role artifacts	Level (2) + three task artifact connecting 1 to 2; 2 to 3; And 9 to 2	Level (3) + 4 + 7 + 8 + relevant role artifacts
Argumentator (Inherits awareness from its parent Member)	Member's level (0) + Argumentator	Member (1) + Level (0) + Initial Contributor + one related task artifacts	Member (2) + Level (1) + Supporter + Opposer + KM + relevant objects/ contexts on the path	Member (3) + Level (2) + Task artifacts connecting 1 to 2; 2 to 3; 5 to 10 and 6 to 11	Member's level (4)

AWARENESS AND KNOWLEDGE SHARING REQUIREMENTS OF ACTORS

According to the Figure 1, the *role* KM is involved in one *personal task* (task 12). The *role artifact* for this task consist of the technical knowledge of how to terminate a thread, as well as some privileged knowledge such as the administrator's password. This *role* is also involved in two *collaborative tasks* with the role Member. The *task artifact* involved in the first *collaborative task*, task 10, is knowledge infrastructure that enables Members to publish their contributions on the community board. The KM's second *collaborative task*, task 11, involves finalisation of a (new) Member's membership within the community, say, through filling up an online membership form. The *role* Member can play any of the *subroles* (also called as *specialised roles*) Argumentator, Supporter, Initial Contributor and Opposer; and as a result can interact with other specialised instances of the Member, as shown by the task artifacts connecting 9 to 2, 1 to 2, and 2 to 3. These *specialised roles* can also perform all the generic tasks that the parent role Member performs, both privately (tasks 3, 4, 7 and 8), and in collaboration with the KM (tasks 5 and 6).

Various subgraphs of the Figure 1 can be used to demonstrate, at conceptual level, various levels of awareness for the roles. For simplicity, Table 1 only shows these levels for three *roles* only.

ANALYSIS OF THE RESULTS

Figure 1 demonstrates an *awareness net* for the collaborative process of DIVA. On the other hand, DIVA framework provides a platform for management of divergence awareness in knowledge communities. Like any virtual community, DIVA-based knowledge community is also based on the assumption that various actors are assigned roles to play within the community, and to perform certain communication, interaction, and knowledge-sharing tasks within the community using available knowledge artifacts. However, in its current form, DIVA does not provide a formal method for modeling the tasks, activities, roles and knowledge artifacts used in the collaborative process; neither does it provide a method for identifying knowledge sharing requirements of the actors within the community.

By integrating awareness net and DIVA frameworks, we were able to draw an *awareness net* for DIVA as shown in Figure 1. From this awareness net, we were able to derive various awareness levels in the form of various pre-defined paths on the awareness net, as shown in Table 1. There are two measures for the awareness levels: *actual level of awareness* and *required level of awareness*. An excess of the *required level* over the *actual level of awareness* represents the awareness gap for the role. This awareness gap represents what the role needs to be aware of in order to be able to collaborate with others according to expectations. As a result of identifying the awareness gap for an actor, appropriate user interfaces can be provided to that role that facilitates the role's interaction and collaboration with others. This way, levels of awareness for each role can be maintained at required levels.

Another contribution of the proposed integrated framework is that by identifying various levels of awareness for each role, the community will be capable of providing custom-made *shared knowledge workspace* (SKW) to each member based on their required levels of awareness. In other words, the inflexibility of the traditional WYSIWIS interface design paradigm may now be relaxed and each community members will have his/her own SKW that may be a subset of what other members are allowed to access.

CONCLUSION

This article introduced an integrated conceptual framework for knowledge communities. The framework incorporates technological and process-related aspects of collaborative business process support systems with the aim of (1) creating new knowledge through management of divergence and arguments in these communities, and (2) providing a more flexible knowledge workspace than the traditional WYSIWIS method. The strengths of the proposed framework include its capability to model the collaborative process, as well as identify and externalise certain knowledge that is related to the storage and interface design requirements of the private and shared knowledge workspaces. In other words, it explicifies elements of knowledge logistics, and on that basis it creates design schemas for groupware systems that support members of knowledge communities. Under the proposed framework, the actors' various workspace requirements are directly related to the required levels of awareness within the knowledge community. These various levels of awareness correspond with the actors' ability in getting involved in various levels of knowledge sharing transactions with others. The belief foundation of the proposed formalism is that actors in knowledge communities are collaborating agents that need to have certain levels of awareness about various aspects of collaborative process in order to perform their knowledge-sharing tasks effectively. Such context is provided through various elements of knowledge logistic such as the tasks that others perform within the community, the roles involved in the process, and the artifacts and knowledge resources that these roles utilise in order to perform their tasks within the community.

REFERENCES

Borenstein, N. S., & Thyberg, C. A. (1993). Power, ease of use and cooperative work in a practical multimedia message system. In R. M. Baeker (Ed.), *Readings in groupware and computer-supported cooperative work* (pp. 485-500). Morgan Kaufmann.

Daneshgar, F. (1997). A CSCW approach to enhancing collaboration. In *Proceedings of the Third International Conference on Advanced Information Systems Engineering (CAISE'97)* (pp. 218-223). Barcelona, Spain.

Daneshgar, F. (2004). Awareness net: A modeling language for collaborative business processes. *Journal of Conceptual Modeling*, 32. Retrieved from www.inconcept.com/jcm

Daneshgar, F., Ray, P., Rahbi, F., Molli, H. S., Molli, P., & Godart. C. (2004). Knowledge sharing infrastructures for teams within virtual communities. In M. Fong (Ed.), *e-Collaborations and virtual organizations*. Hershey, PA: IGP.

Daneshgar, F., & van der Kwast, E. (2005). An awareness provisioning methodology for asynchronous virtual global forums. *Journal of Knowledge Management Practice, 6*. Retrieved from http://www.tlainc.com/articl78.htm

Diaz, A., & Canals, G. (2004). Supporting knowledge sharing in a community with divergence. In *Proceedings of I-KNOW '05 International Conference on Knowledge Management* (pp. 136-144). Graz, Austria.

Dustdar, S. (2004). Caramba: A process-aware collaboration system supporting ad hoc and collaborative processes in virtual teams. *Journal of Distributed and Parallel Databases, 15*(1), 45-66.

Kock, N., & Nosek, J. (2005). Expanding the boundaries of e-collaboration. *IEEE Transactions on Professional Communication, 48*(1), 1-9.

Malone, T. W., Grant, K. R., Lai, K. Y., Rao, R., & Rosenblit, D. A. (1993). The information lens: An intelligent system for information sharing and coordination. In R. M. Baecker (Ed.), *Readings in groupware and computer-supported cooperative work* (pp. 461-473). Morgan Kaufmann.

Nanoka, I. A. (1994). A dynamic theory of organizational knowledge creation. *Organization Science, 5*, 14-37.

Ray, P., Shahrestani, S., & Daneshgar, F. (2005). The role of awareness modelling in cooperative management. *Journal of Information Systems Frontiers, 7*(3), 215-225.

Zigurs, I., & Qureshi, S. (2001). Managing the extended enterprise, creating value from virtual spaces. In G. Dicksom & G. DeSactis (Eds), *Information technology and the future enterprise: New models for managers* (pp. 125-143). Prentice Hall.

KEY TERMS

Actors: Human agents that enact a set of *tasks* by assuming one or more *roles* within the process. In the awareness net there is no graphical representation for the "actors" and instead, actors are represented indirectly by relevant *role*(s) that they play within the process.

Actual Level of Awareness: Represents what an actor already knows of various aspects of collaboration (who is doing what, by whom, and how); the latter are defined within the awareness net as collaborative semantic concepts roles, tasks and task artifacts.

Collaborative Task: Is composed of two or more *personal tasks* that have a common goal and therefore (must) share a common *task artifact*.

Divergence Occurrence: Generation of alternatives, arguments, and different points of view about a topic of interest.

Levels of Awareness: Level-0 is a role's awareness about his/her *role artifacts* as well as the *tasks* that s/he plays within the community.

- Level-1 awareness is the role's *level 0 awareness*, PLUS all the objects on the awareness net that correspond to the related *tasks* and related *roles*.
- Level-2 awareness extends level 1 by including additional remaining *role* contexts within the

process. It is the role's *level 1 awareness* PLUS an awareness about all other remaining *roles'* contexts within the process.

- Level-3 awareness extends level 2 by including all the remaining *task artifact* contexts that exist within the process.
- Level-4 awareness extends level 3 by including all remaining semantic concepts (or, objects) on the *awareness net*; that is, everybody else's *personal tasks*, as well as their related *role artifacts*.

Personal Task: Consists of a sequence of "actions" or "execution steps" for achieving a specific process goal.

Private Knowledge Workspace (PKW): A non-public space that can be accessed by its owner only. It represents the private knowledge context and allows users to privately externalize any knowledge. It contains personal knowledge, point of view, and alternatives.

Required Level of Awareness: The level of awareness that is expected from a role based on both the nature of the task the role is playing as well as the organisational culture surrounding the community.

Role: A set of norms expressed in terms of obligations, privileges, and rights enabling actors to perform certain tasks within the process. A *subrole* or *specialised role* is a child of a ROLE concept that inherits many attributes from the concept ROLE.

Role Qartifact: Carries private knowledge/resources about how to perform *actions* associated with a *personal task.*

Shared Knowledge Workspace (SKW): Is a public space that can be accessed equally by any community member and represents shared knowledge context.

Task: Has two types: *personal task* and *collaborative task*.

Task Artefact: Carries shared knowledge resources about how various *actions* associated with a *collaborative task* are executed. It is shared by interacting *roles*.

A Basic Definition of E-Collaboration and its Underlying Concepts

Ned Kock
Texas A&M International University, USA

INTRODUCTION

Electronic collaboration (e-collaboration) is operationally defined here as collaboration using electronic technologies among different individuals to accomplish a common task (Kock & D'Arcy, 2002; Kock, Davidson, Ocker, & Wazlawick, 2001). This is a broad definition that encompasses not only computer-mediated collaborative work but also collaborative work that is supported by other types of technologies that do not fit most people's definition of a "computer." One example of such technologies is the telephone, which is not, strictly speaking, a computer—even though some of today's telephone devices probably have more processing power than some of the first computers back in the 1940s. Another example of technology that may enable e-collaboration is the teleconferencing suite, whose main components are cameras, televisions, and telecommunications devices.

The above operational definition, which I will use as a basis to discuss other related issues in this article, is arguably very broad, yet it is probably clearer than the general view of e-collaboration in industry, which some may also see as a bit unfocused. For example, some developers of e-collaboration tools, such as Microsoft Corporation and Groove Networks, emphasize their technologies' support for the conduct of electronic meetings over the Internet. There seems to be a concern by those developers with offering features that make electronic meetings as similar to face-to-face meetings as possible.

Industry information technology publications such as *CIO Magazine* and *Computerworld,* on the other hand, often tend to favor a view of e-collaboration technologies as tools to support business-to-business electronic commerce and virtual supply chain management over the Web. These are business activities that are arguably substantially different from electronic meetings, both in terms of scope and main goals. The primary audiences of industry information technology publications are information technology managers and professionals, who are the consumers of e-collaboration technologies. Given that, one can imagine the possible misunderstandings that may take place when those managers and professionals get together with developers' sales representatives to discuss possible e-collaboration technology purchases.

BACKGROUND

As far as buzzwords are concerned, *e-collaboration* is still in its infancy, even though the technologies necessary to make e-collaboration happen have been around for quite some time. Strictly speaking, e-collaboration could have happened as early as the mid-1800s, with the invention of the telegraph by Samuel F. B. Morse. The telegraph allowed individuals to accomplish collaborative tasks interacting primarily electronically. If one assumes that the telegraph was too cumbersome to support e-collaboration, it may be more reasonable to argue that the birth of e-collaboration could have been soon after that, in the 1870s, with the invention of the telephone by Alexander Graham Bell.

Yet, for a variety of reasons, true e-collaboration had to wait many years to emerge. Did the commercialization of the first mainframe computers in the 1950s, following the ENIAC project, help much in that respect? Not really, and that was not necessarily due to technological obstacles to developing e-collaboration systems for mainframes. The real reason seems to have been the cost of mainframes (Kock, 1999, 2005), which was then seen as too high for them to be used (a) by anyone other than very specialized workers, who often dressed like medical doctors; or (b) for anything other than heavy data-processing-intensive and/or calculation-intensive applications. Of course, e-collaboration was not seen as one of those applications. Moreover, worker collaboration was not even a very fashionable management idea by the time the mainframes hit the market big time in the 1960s (Kock, 2002).

Then the ARPANET, the precursor of today's Internet, happened in the late 1960s. The ARPANET Project's main goal was to build a geographically distributed network of mainframes within the United States that could withstand a massive, and possibly nuclear, military attack by what was then known as the Soviet Union. By that time, mainframes were used in ballistics calculations, without which intercontinental missiles would not be as effective in reaching their targets as they were expected to be. The Project was motivated by the Cold War between the United States and the Soviet Union, which reached a tense stage in the early 1960s. The main sponsor of the ARPANET Project was the U.S. Department of Defense.

One of the tools developed to allow ARPANET users to exchange data was called "electronic mail" (e-mail). E-mail was initially perceived as a "toy" system, which researchers involved in the ARPANET Project used to casually interact with each other. This perception gave way to one that characterizes e-mail as the father (or mother) of all e-collaboration technologies (Sproull & Kiesler, 1991). To the surprise of many, serious use of e-mail grew quickly, primarily as a technology to support collaboration among researchers, university professors, and students—the primary users of the ARPANET while it was in its infancy.

So, in spite of the fact that other technologies already existed that could have been used for e-collaboration, e-mail was arguably the first technology to be used to support e-collaborative work. Interestingly, e-mail's success as an e-collaboration technology has yet been unmatched—at least in organizational environments (college dorms do not qualify). This is somewhat surprising, given e-mail's granddaddy status as far as e-collaboration is concerned. Helping it hold that enviable position is e-mail's combination of simplicity, similarity to a widely used "low-tech" system (the paper-based mail system), and support for anytime-anyplace interaction.

E-COLLABORATION, CMC, AND CSCW

What I refer to in this article as e-collaboration research is in fact made up of several research streams, with different names and traditions. One such research stream is that of computer-mediated communication, also known as CMC, which has been traditionally concerned with the effects that computer mediation has on individuals who are part of work groups and social communities. One common theme of empirical CMC research is the investigation of the effects of computer mediation on group-related constructs by using as a control condition the lack of computer mediation—what some prefer to simply call "face-to-face interaction."

E-collaboration is not the same as computer-mediated communication. Earlier in this article, I defined e-collaboration as collaboration using electronic technologies among different individuals whose goal is to accomplish a common task. I would argue that, following from that definition, e-collaboration research should be seen as encompassing traditional CMC research as well as other lines of research that do not necessarily rely on computer-mediated communication to support collaborative tasks. One example would be the study of telephone-mediated communication. This argument also applies to another area of research normally referred to as computer-supported cooperative work (CSCW), for similar reasons. That is, e-collaboration research should also be seen as encompassing traditional CSCW research.

Another distinction that I would like to point out, and that may be seen as controversial by some, is that e-collaboration may take place in situations where there is no communication per se, much less computer-mediated communication. Let us consider for example a Web-based e-collaboration technology that allows different employees of an insurance company to accomplish the same collaborative task, namely the task of preparing a standard insurance policy for a customer. Since we are assuming that the collaborative work is on a standard insurance policy, it is not unreasonable to picture a case in which different employees would electronically input pieces of information through the e-collaboration technology that will become part of the final product (i.e., the policy) without those employees actually communicating any information to one another. In this case, the e-collaboration system would pull together different pieces of information from different individuals into what would in the end become an insurance policy, and in such a way that the individuals may not even have been aware of one another. Some, of course, will argue that this is not "really" e-collaboration. But it fits our definition of e-collaboration, presented earlier in this article: "...collaboration among different individuals to accomplish a common task using electronic technologies."

Today, many technologies exists that do not involve computer-mediated communication, and that nonetheless are becoming increasingly important as tools for e-collaborative work. Mobile e-collaboration devices, from cell phones to wireless personal digital assistants (PDAs), are a good example. Some may see those devices as computers, while others may not. Regardless of that, those devices are likely to be a key target of e-collaboration research in the near future.

CONCEPTS UNDERLYING E-COLLABORATION

What are the main "conceptual elements" that define an e-collaboration episode? This is a general question whose answer, I believe, can further shed light on what e-collaboration is (and what it is NOT). Moreover, identifying the key conceptual elements that make up e-collaboration will inevitable lead us to the identification of constructs that can be targeted in e-collaboration research.

Based on past research on e-collaboration, one could contend that the following conceptual elements define e-collaboration, in the sense that changes in those elements can significantly change the nature of an e-collaboration episode: (1) the collaborative task, (2) the e-collaboration technology, (3) the individuals involved in the collaborative task, (4) the mental schemas possessed by the individuals, (5) the physical environment surrounding the individuals, and (6) the social environment surrounding the individuals. Each of these elements is discussed next:

The collaborative task: An example of generic collaborative task that is often conducted with support of e-collaboration technologies today is that of writing a contract, particularly when the parties involved are geographically distributed. The nature of the collaborative task (e.g., whether it is simple or complex) can have a strong effect on its outcomes when certain e-collaboration technologies are used (Zigurs & Buckland, 1998; Zigurs, Buckland, Connolly, & Wilson, 1999).

The e-collaboration technology: This comprises not only the communication medium created by the technology but also the technology's features that have been designed to support e-collaboration. The implementation of a particular feature (e.g., video streaming) in a particularly type of e-collaboration technology (e.g., instant messaging) can have a strong effect on how the technology is actually used by a group of individuals to accomplish a given collaborative task (DeSanctis & Poole, 1994; Poole & DeSanctis, 1990).

The individuals involved in the collaborative task: This conceptual element refers primarily to certain characteristics of the individuals involved in the collaborative task, such as their gender and typing ability (which would be relevant in text-based e-collaboration contexts). This conceptual element also refers to the "number" of individuals involved in the e-collaboration episode, or the size of the e-collaborative group. An individual's gender, for example, may have a significant effect on how that individual perceives a particular e-collaboration technology (Gefen & Straub, 1997), which may affect that individual's behavior as part of a group of e-collaborators (Kock, 2001).

The mental schemas possessed by the individuals: This conceptual element refers to mental schemas (also referred to as "knowledge" or "background"; see, e.g., Kock, 2004; Kock & Davison, 2003) possessed by the individuals involved in the collaboration task, including socially constructed schemas that may induce the individuals to interpret information in a particular way (Lee, 1994). This conceptual element also refers to the degree of similarity of the mental schemas possessed by the individuals. The degree of similarity among the task-related mental schemas possessed by different individuals engaged in a collaborative task (e.g., whether task experts are interacting with other experts, or novices) may significantly affect the amount of cognitive effort required to successfully accomplish the task using certain types of e-collaboration technologies (Kock, 2004).

The physical environment surrounding the individuals: This comprises the actual tangible items that are part of the environment surrounding the individuals involved in the collaborative task as well as the geographical distribution of the individuals. Geographically dispersed individuals are more likely than colocated ones to use e-collaboration technologies that are perceived as "less rich" than face-to-face interaction and spend time and effort adapting the features of the technologies to their task-related needs (Kock, 2001; Trevino, Daft, & Lengel, 1990).

The social environment surrounding the individuals: This conceptual element refers primarily to aspects of the social environment surrounding the individuals involved in the collaborative task that can be characterized as being social influences on those individuals. Those aspects may involve expressed perceptions and/or behavior by peers, managers and other individuals (e.g., customers) toward e-collaboration technologies. For instance, an individual's behavior toward a particular e-collaboration tool, or certain features of that tool, may be significantly influenced by peer pressure (Markus, 1994), which may take the form of other individuals heavily using the e-collaboration tool and expressing positive opinions about the tool. That behavior may also be significantly influence by the position that the individual occupies in an organization's hierarchical management structure (Carlson & Davis, 1998).

The above discussion on key conceptual elements should be followed by a couple of caveats. First, the list of key conceptual elements presented is not comprehensive. There are certain elements that are relevant for e-collaboration research that are not covered by the above list. Second, the conceptual elements above may be (or have been) given different names by different researchers, or the same name but different meanings.

Nevertheless, I hope to have been able to accomplish one main goal by discussing the conceptual elements—to provide a glimpse at the complexity of e-collaboration and its many behavioral facets. Each of the conceptual elements above, if significantly manipulated in, say, a laboratory experiment or action research project (Kock, 2003), would potentially lead to variations in key variables. Among those key variables are two favorites of e-collaboration researchers: task outcome quality, and task efficiency. Task outcome quality is frequently assessed based on how "good" the task "product" is, often in terms of customer perceptions. Task efficiency is usually assessed based on how much time and/or cost is involved in accomplishing the task.

CONCLUSION

The field of e-collaboration has a promising future, in terms of both academic research and commercial software development. As an area of academic research,

e-collaboration has flourished since the 1980s and particularly the 1990s, which led to the need for new publications outlets—a need that, one recently launched journal, the *International Journal of e-Collaboration*, tries to address by its very existence. As an area of commercial software development, e-collaboration is likely to benefit from a critical assessment of how it can be applied to the benefit of individuals, organizations, and society.

In this article, I provided an operational definition of e-collaboration, and a historical glimpse at how and when e-collaboration emerged. I also argued that e-collaboration, as an area of research and industrial development, is broader than what is often referred to as computer-mediated communication— an argument that also applies to another field called computer-supported cooperative work (CSCW). Finally, I discussed key conceptual elements in connection with e-collaboration, which I hope will provide a relatively easy to understand conceptual basis for future research design and implementation. While the conceptual elements discussed have consistently been targeted individually in past research, rarely interaction effects among those conceptual elements have been investigated. There are tremendous research opportunities and challenges (mostly methodological) for researchers that decide to conduct research projects addressing those interaction effects.

The view that I propose here of e-collaboration is hopefully focused enough to allow for a clear understanding of what types of research would constitute e-collaboration research, and hopefully help shape this new and promising field of inquiry. At the same time, I hope that such view of e-collaboration is comprehensive enough to leave room for likely technological developments that are not seen today as enabling e-collaboration, but that may be seen as doing so in a not so distant future. One such likely development is that of virtual reality applications (Briggs, 2002) and their increasing use to support e-collaborative work. Other related technological developments are likely to arrive in other areas such as wearable computing and speech recognition, with significant impacts on how e-collaboration takes place in the context of certain collaborative tasks (Parente, Kock, & Sonsini, 2004).

REFERENCES

Briggs, J. C. (2002). Virtual reality is getting real: Prepare to meet your clone. *The Futurist, 36*(3), 34-42.

Carlson, P. J. & Davis, G. B. (1998). An investigation of media selection among directors and managers: From "self" to "other" orientation. *MIS Quarterly, 22*(3), 335-362.

DeSanctis, G., & Poole, M. S. (1994). Capturing the complexity in advanced technology use: Adaptive structuration theory. *Organization Science, 5*(2), 121-147.

Gefen, D., & Straub, D. W. (1997). Gender differences in the perception and use of e-mail: An extension to the technology acceptance model. *MIS Quarterly, 21*(4), 389-400.

Kock, N. (1999). *Process improvement and organizational learning: The role of collaboration technologies.* Hershey, PA: Idea Group.

Kock, N. (2001). Compensatory adaptation to a lean medium: An action research investigation of electronic communication in process improvement groups. *IEEE Transactions on Professional Communication, 44*(4), 267-285.

Kock, N. (2002). Managing with Web-based IT in mind. *Communications of the ACM, 45*(5), 102-106.

Kock, N. (2003). Action research: Lessons learned from a multi-iteration study of computer-mediated communication in groups. *IEEE Transactions on Professional Communication, 46*(2), 105-128.

Kock, N. (2004). The psychobiological model: Toward a new theory of computer-mediated communication based on Darwinian evolution. *Organization Science, 15*(3), 327-348.

Kock, N. (2005). *Business process improvement through e-collaboration: Knowledge sharing through the use of virtual groups.* Hershey, PA: Idea Group.

Kock, N., & D'Arcy, J. (2002). Resolving the e-collaboration paradox: The competing influences of media naturalness and compensatory adaptation [Special issue on electronic collaboration]. *Information Management and Consulting, 17*(4), 72-78.

Kock, N., & Davison, R. (2003). Can lean media support knowledge sharing? Investigating a hidden advantage of process improvement. *IEEE Transactions on Engineering Management, 50*(2), 151-163.

Kock, N., Davison, R., Ocker, R., & Wazlawick, R. (2001). E-collaboration: A look at past research and future challenges [Special Issue on E-Collaboration]. *Journal of Systems and Information Technology, 5*(1), 1-9.

Lee, A. S. (1994). Electronic mail as a medium for rich communication: An empirical investigation using hermeneutic interpretation. *MIS Quarterly, 18*(2), 143-157.

Markus, M. L. (1994). Electronic mail as the medium of managerial choice. *Organization Science, 5*(4), 502-527.

Parente, R., Kock, N., & Sonsini, J. (2004). An analysis of the implementation and impact of speech recognition technology in the heath care sector. *Perspectives in Health Information Management, 1*(5), 1-23.

Poole, M. S., & DeSanctis, G. (1990). Understanding the use of group decision support systems: The theory of adaptive structuration. In J. Fulk & C. Steinfield (Eds.), *Organizations and communication technology* (pp. 173-193). Newbury Park, CA: Sage.

Sproull, L., & Kiesler, S. (1991). Computers, networks and work. *Scientific American, 265*(3), 84-91.

Trevino, L. K., Daft, R. L., & Lengel, R. H. (1990). Understanding manager's media choices: A symbolic interactionist perspective. In J. Fulk & C. Steinfield (Eds.), *Organizations and communication technology* (pp. 71-94). Newbury Park, CA: Sage.

Zigurs, I. & Buckland, B.K. (1998). A Theory of Task-technology Fit and Group Support Systems Effectiveness. *MIS Quarterly.* 22(3), 313-334.

Zigurs, I., Buckland, B. K., Connolly, J. R., & Wilson, E. V. (1999). A test of task-technology fit theory for group support systems. *Database for Advances in Information Systems, 30*(3), 34-50.

KEY TERMS

ARPANET: The precursor of today's Internet; developed in the late 1960s through a projected sponsored by the U.S. Department of Defense.

CMC: Computer-mediated communication.

Collaborative Task: Task that is often conducted by a group of people with support of e-collaboration technologies.

CSCW: Computer-supported cooperative work.

E-Collaboration: Collaboration using electronic technologies among different individuals to accomplish a common task.

E-Collaboration Technology: Comprises not only the communication medium created by an e-collaboration technology but also the technology's features that have been designed to support collaborative work.

Mental Schemas: Mental structures possessed by the individuals involved in the collaboration task, including socially constructed mental structures that may induce the individuals to interpret information in a particular way.

Telegraph: Invention by Samuel F. B. Morse that allowed individuals to accomplish collaborative tasks interacting primarily electronically.

Blogging Technology and its Support for E-Collaboration

Vanessa Paz Dennen
Florida State University, USA

Tatyana G. Pashnyak
Florida State University, USA

INTRODUCTION

Weblogs, commonly known as "blogs," are Web sites that feature a series of dated posts appearing in reverse chronological order. They may be authored by individuals, groups, or organizations, and may be used to share writing and Web-based media of any kind. Although there is nothing inherent in the basic technology behind blogs specifically to facilitate community, as with many other technological media, blogs have been used to support e-collaboration and have become known as a form of Web-based discourse (Fleishman, 2002). This article discusses how blogs and their various enabling tools and technologies can be used to foster and maintain e-collaboration.

BACKGROUND

Blogs have evolved over time, from Web pages that were updated manually in the late-1990s to full sites with archives and labels that are generated using specific blogging tools. The original blogs were not collaborative in nature. Instead, they represented the work of individuals who posted dated entries to Web pages. Although these earliest blogs have been considered not all that different from personal home pages (Weiss, 2004), blogs are distinct from regular Web sites in a number of ways. The manner in which they are updated is different, with new items being added but not replacing older ones. Typically they have a clear sense of voice or authorship (Gill, 2004).

Significant growth of the blogging trend did not begin until around 1999 when several build-your-own-weblog tools, most notably Blogger (www.blogger.com), were released. These tools enabled the general public, specifically people with minimal technical knowledge, to create weblogs quickly and easily (Blood, 2000).

All that was needed was the ability to fill out simple forms on a Web page, much like sending e-mail. The practice of blogging has seen exponential growth since that time. Technorati (www.technorati.com/, 2006), a Web-based tool that tracks blogs, suggests that there are 75,000 new blogs created each day and, as of October 5, 2006, the tool is tracking 56.1 million blogs across an array of disciplines and fields, connecting people from all over the world.

The content of blogs varies widely. Some resemble online journals, whereas others may be more topic-focused. Herring, Scheidt, Bonus, and Wright (2005) found that most (70.4%) blogs could be categorized as journals; with filters, which provide links and metacommentary on other sites, constituting 12.6%; and k-logs, recording items relevant to a given project, also being found (3.0%). Additionally, they found blogs that served mixed functions (9.5%) as well as some that did not fit into any of these categories. While some of these blogs exist in relative isolation, many others have linked together. These blogrolls or lists of links to other blogs bring together bloggers who share common interests or enjoy each other's writing.

HOW BLOGS FOSTER E-COLLABORATION

Blogs can be used to support e-collaboration among people who know each other prior to blogging and are committed to engaging in a shared task, but they can also provide a forum through which individuals might find each other and form new collaborative partnerships. In the former case, a work group might elect to start a blog to document progress on a particular task or enable brainstorming and resource sharing. In the latter case, individuals with like interests may find each other through blogs and begin working together to achieve a particular goal.

Blogs enable people to interact and collaborate in four different ways: (1) publishing; (2) coauthoring/coediting; (3) social bookmarking; and (4) online discourse. In the most basic of senses, blogs provide a mechanism for an individual author to publish his thoughts or links to resources he has found in a forum where they can be accessed by others. Publishing may be considered one component of e-collaboration in that the blog author is submitting his work to a public audience for comment and use. In other words, it is an e-collaboration enabler.

Collaborative blog authoring and editing may take one of two forms. In the most common one, all authors can have access to the blog, compose their own posts and, depending on their permissions, may be able to edit all posts. Through the edit functions, collaborative writing can be done and feedback can be left in the comments regardless of one's permission status. Alternately, authors can have their own blogs, and these blogs can be combined via a feed into one master blog. In this second version, authors cannot readily access each other's posts to edit or delete them.

Social bookmarking is not about writing a blog but instead about finding, organizing, and sharing blog-based resources with others. Individuals can either compile their own lists of blogs and Web sites, and then share them with others, or they may collectively contribute to the same lists of bookmarks. Regardless of how it is done, the practice of social bookmarking greatly facilitates information searching and sharing.

Finally, blogs are an online forum in which discourse may be generated amongst many participants. Discourse may be initiated or prompted by remarks made in a blog post and further continued via the comments or other posts. It may look very much like dialogue, with iterations of statements and responses between particular interactants, or it may take the form of commentary without any direct form of responsive discussion.

TYPES OF BLOG-BASED E-COLLABORATION

Blog-based e-collaboration can be found in a variety of settings, some intentionally designed and others forming more informally. For example, blogs often are used to support learning in classrooms. Weblogs are useful tools for project-based learning because of the instant publishing capability and the ability to receive feedback via comments. In addition, blogging enhances students' technology and literacy skills (Ducate & Lomicka, 2005; Huffaker, 2005; Oravec, 2003; Ray, 2006). Blogs used for education are often called *edublogs*. It appears that blog-based communities of practice are especially important for distance-learning students who otherwise may suffer from isolation and alienation from their professional peers (Dickey, 2004).

Still in education, but outside the class context, blogs are being used to create communities of students, parents, teachers, and other local community members (Poling, 2005). Communities of university faculty and K-12 educators have developed, offering informal mentoring and support through the exchange of ideas and experiences (Dennen & Pashnyak, 2006; Pashnyak & Dennen, 2007). The online community can also help with cutting the isolation many first-year instructors feel (Poling, 2005). For academic bloggers, being a part of a blog-based community of practice is an essential part of both their professional and personal life because it provides a place to discuss both academic and nonacademic interests and experiences, seek and give advice, share their work in draft format and receive valuable critiques, and connect with colleagues from other universities and disciplines. It also provides a chance to be a public intellectual. Blogging had become a part of the professional identity and work of these bloggers (Farrell, 2005). However, blogging is not yet a highly regarded activity, and some academic bloggers worry that their colleagues and administrators may disapprove of their blogs and their jobs may be in jeopardy (Farrell, 2005; Hevern, 2004).

Blogging also has had a major impact on the media and political communities. Many newspapers, magazines, and radio and TV stations are maintaining blogs, and the format allows regular citizens to collaborate in the gathering and reporting of news. In 2003, "Baghdad blogger" was featured by news media around the world for his personal stories about the conditions in Baghdad during the bombing campaign and soliciting help in finding his missing friend. In 2004, presidential candidate Howard Dean pioneered blogging as a means of communicating with his supporters, thus demonstrating how blog-based communities can be created and used for political campaigns (Martindale & Wiley, 2005).

Blogging has had an influence on communication in a host of other professional disciplines. Medical professionals are using blogs to support e-collaboration, particularly where research is concerned (Boulos, Maramba, & Wheeler, 2006; Maag, 2005; Sauer et al., 2005). Blog-based communities are attracting the attention of U.S. intelligence officers who like the rich information swirling around the blogosphere and frequently mine it for the current on-the-ground situation. In addition, there are several attempts to establish e-collaboration projects among the officers (Andrus, 2005; Burton, 2005). Additionally, managers are discovering that blogging is a good way to share existing knowledge and practices as well as generate new ideas (McAfee, 2006).

BLOG CREATION TOOLS

Weblogging tools are flexible and easily can be used for a number of purposes. "Like a pen could be used to write a diary, a novel, a letter to a friend, or just a shopping list pinned to a fridge door, Weblogging tools can be used to publish a personal diary, to collect and share links, to facilitate a course, as an unfolding novel, a record of an experiment, a recipe book" (Efimova & Hendrick, 2005). Authors determine the best use of blogging for their purposes, thus creating people-driven rather than technology-driven blogs and blogging communities.

The basic function of blog creation tools is to enable the user to create and publish blog posts, although most blogging tools have additional features that add sophistication in terms of what can be posted, blog personalization, and blog functionality. In some instances, the blog creation tool serves a double purpose and also acts as a blog host, providing Web space and a URL for the blog. Blogger remains one of the most popular blogging tools that doubles as a host, with other popular sites including TypePad (www.typepad.com) and Live-Journal (www.livejournal.com). Other blogging tools such as Drupal (www.drupal.org; an opensource tool), WordPress (www.wordpress.org), and Moveable Type (www.sixapart.com/moveabletype), can be installed directly on an individual's or organization's own Web server for self-hosting and maintenance. Blogging has not been limited to standalone tools. Social networking sites such as MySpace (www.myspace.com) and Friendster (friendster.com) have integrated blogs as one

of their features. This list of blog creation tools is far from exhaustive but rather is representative of some of the most commonly used tools.

Although initially blog creation tools could be readily distinguished by their different features and focuses—for example, LiveJournal was considered a space for journal-style bloggers and TypePad was known for allowing posts to be categorized—today the differences between tools for the lay user are more about interface than features. Continuous development of the blogging tools themselves as well as third-party plug ins has greatly lessened the functional differences between different blogging tools.

The standard features of blogs include posts with a time and date stamp, comments, and archives. Comments often can be limited to registered members or may be open to anyone. Archives make it possible to review the history of blog posts organized by date of posting. Some blog tools also allow topical categorization of posts. All blog pages are built on templates, which typically contain a header area with a title and description of the blog and a sidebar area which the blog author may customize with profiles, links, photos, and other information.

TOOLS COMMONLY USED WITH BLOGS

In addition to the blogging tools themselves, there are a variety of complementary tools that can be used with blogs. These tools all can help support e-collaboration in different ways, increasing the efficiency with which individuals can engage in tasks such as tracking and reading blogs, searching blogs, and tracking comment threads. These tools tend to be created by third party vendors and many take the form of Web browser plug-ins. This section will describe various types of blog enhancement tools and provide examples. However, it should be noted that, as with any relatively new technology, the specific tools are ever-changing and sometimes sold or discontinued, and new ones are becoming available as well.

One category of tools makes blog reading more efficient. Blogs can be read, obviously, by directly accessing the blogs themselves, or via other tools that aggregate blog content and help centralize the blog monitoring process. Web-based tools such as Bloglines (www.bloglines.com) and Google Reader (www.google.com/reader) allow users to subscribe to syndication feeds of their favorite blogs. This technology is called Really Simple Syndication (RSS), and the aggregators read these feeds, pulling new blog content into their interfaces as it is published. Blog readers, then, need only check in one place to see if any of the blogs they read have been updated. Another related tool is Blogarithm (www.blogarithm.com), which sends an e-mail to users letting them know when blog content has been updated.

Although commenting features are built into most blogging tools, third party tools have been developed to make the commenting process more robust. Depending on the tool and particular settings, comments may be limited to particular users or open to anyone; certain users may be banned from commenting; and comments may be followed via RSS feed. Alternate commenting tools like Haloscan (www.haloscan.com) can be added to a blog to provide users with additional control over and layout options for comments compared to the generic comments supported by a blogging tool. Haloscan provides the added functionality of trackbacks, allowing people to make connections between related posts so that readers can easily find them. For example, if John reads Julie's post and writes his own post inspired by it, John can then provide a trackback link. The link will show up in the comment area of Julie's blog, showing that John has suggested his post would be of interest.

Other tools exist to help blog readers more efficiently track conversations that take place in comment threads. Without these tools, which include coComment (www.cocomment.com), Co.mments (www.co.mments.com), and Commentful (www.commentful.com), a reader would have to continuously revisit a blog to see if comments had been left following their own comments. With these tools, a reader can track subsequent comments to conversations they have entered via one centralized tool.

Another subset of tools helps users organize and search blog-based resources more effectively than a regular search engine. Technorati (www.technorati.com) works as a search engine specifically for blogs. Users can search Technorati for blogs that mention particular terms or link to specific URLs. Additionally, blog authors can add tags to their posts, which help classify them in the Technorati database.

Other tools help individual users manage and share their own lists of favorite blogs. On the most basic level, tools like Blogrolling (www.blogrolling.com) help bloggers more easily create and manage blogrolls, collections of links that can be posted to the sidebar of their blogs. Much more sophisticated than that is Del.icio.us (www.del.icio.us.com), a social bookmark manager that helps individuals manage their links to blogs of interest and share their lists with others.

Sometimes what users want to share is their thoughts about items found on a blog. Although one option for doing so is to provide a link via a social bookmarking tool and another is to create a blog of links to these sites, annotation tools allow users to leave commentary about blogs and Web pages in a manner similar to marking them up with a pen, highlighter, and sticky notes. These annotations exist as a layer on top of the blog, visible to other users of the annotation tool but transparent to everyone else. Examples of these tools are Diigo (www.diigo.com) and Fleck (www.fleck.com), which is an extension for the Firefox browser.

Organizational tools such as Airset (http://www.airset.com) can help create shared online calendars and include blogging functions and RSS feeds. Progress meters such as those found at Writertopia (www.writertopia.com/toolbox/meters) can help e-collaboration groups monitor their progress toward goals. These tools and others like them are enabling greater forms of e-collaboration through blogging.

FUTURE TRENDS

Blogging is only in its infancy and will likely continue to expand as more people become familiar with the technology and turn to blogs as tools for sharing. The biggest changes affecting e-collaboration via blogs include: (a) The tools, which are becoming increasingly sophisticated in terms of the interactive features they support and the indexing, searching, and social networking they allow; and (b) Increased acceptance of blogging as a legitimate form of communication, publishing, and shared work.

CONCLUSION

There are numerous opportunities for collaboration via blogging, which have the advantage of not being constrained by conventional boundaries such as time and location. Blog-based e-collaboration holds great promise for creating professional relationships, enabling and documenting discussions of new concepts and ideas, and offering informal support to online community members. As the technology continues to advance and becomes not only easier to use but more commonplace for regular computer users, it is likely we will see an increase in the everyday occurrence of e-collaboration facilitated by blogs.

REFERENCES

Andrus, C. (2005). The wiki and the blog: Toward a complex adaptive intelligence community. *Studies in Intelligence, 49*(3), 63-70.

Blood, R. (2000). *Weblogs: A history and perspective.* Retrieved October 1, 2006, from http://www.rebeccablood.net/essays/weblog_history.html

Boulos, M., Maramba, I., & Wheeler, S. (2006). Wikis, blogs and podcasts: A new generation of Web-based tools for virtual collaborative clinical practice and education. Retrieved October 1, 2006, from http://www.biomedcentral.com/1472-6920/6/41

Burton, M. (2005). How the Web can relieve our information glut and get us talking to each other. *Studies in Intelligence, 49*(3), 55-62.

Dennen, V., & Pashnyak, T. (2006). Informal academic mentoring via a Weblog community: A content analysis. In *Proceedings of the 10th Education and Virtuality Conference* (pp. 1-10).

Dickey, M. (2004). The impact of Web-logs (blogs) on student perceptions of isolation and alienation in a Web-based distance-learning environment. *Open Learning, 19*(3), 279-291.

Ducate, L., & Lomicka, L. (2005). Exploring the blogosphere: Use of Web logs in the foreign language classroom. *Foreign Language Annals, 38*(3), 410-421.

Efimova, L., & Hendrick, S. (2005). *In search for a virtual settlement: An exploration of Weblog community boundaries.* Retrieved October 1, 2006, from https://doc.telin.nl/dscgi/ds.py/Get/File-46041

Farrell, H. (2005). *Blogosphere as carnival of ideas.* Retrieved October 1, 2006, from http://chronicle.com/weekly/v52/i07/07b01401.htm

Fleishman, G. (2002) Been blogging? Web discourse hits higher levels. In J. Rodzvilla (Ed.), *We've got blog: How Weblogs are changing our culture.* Cambridge, MA: Perseus.

Gill, K. E. (2004, May 18). *How can we measure the influence of the blogosphere?* Paper presented at the WWW 2004 Workshop on the Weblogging Ecosystem: Aggregation, Analysis, and Dynamics, New York.

Herring, S., Scheidt, L. A., Wright, E., & Bonus, S. (2005). Weblogs as a bridging genre. *Information Technology & People, 18*(2), 142-171.

Hevern, V. (2004). Threaded identity in cyberspace: Weblogs & positioning in the dialogical self. *International Journal of Theory and Research, 4*(4), 321-335.

Huffaker, D. (2005). The educated blogger: Using Weblogs to promote literacy in the classroom. *AACE Journal, 13*(2), 91-98.

Maag, M. (2005). The potential use of "blogs" in nursing education. *Computers, Informatics, Nursing, 23*(1), 16-24.

Martindale, T., & Wiley, D. (2005). Using Weblogs in scholarship and teaching. *Tech Trends, 49*(2), 55-61.

McAfee, A. (2006). Enterprise 2.0: The dawn of emergent collaboration. *MIT Sloan Management Review, 47*(3), 21-28.

Oravec, J. A. (2003). Blending by blogging: Weblogs in blended learning initiatives. *Journal of Educational Media, 28*(2/3), 225-233.

Pashnyak, T. G., & Dennen, V. P. (2007). What and why do classroom teachers blog? In *Proceedings of the IADIS Web-Based Communities 2007 Conference* (pp. 172-178).

Poling, C. (2005). Blog on: Building communication and collaboration among staff and students. *Learning & Leading with Technology, 32*(6), 12-15.

Ray, J. (2006). Welcome to the blogosphere. *Kappa Delta Pi Record, 42*(4), 175-177.

Sauer, I., Bialek, D., Efimova, E., Schwartlander, R., Pless, G., & Neuhaus, P. (2005). "Blogs" and "wikis" are valuable software tools for communication within research groups. *Artificial Organs, 29*(1), 82-89.

Technorati. (2006). *About Technorati.* Retrieved October 5, 2006 from http://technorati.com/about

Weiss, A. (2004). Your blog? Who gives a @*#%! *netWorker, 8*(1), 40-ff.

KEY TERMS

Blog: Short for *Web log,* a frequently updated Web site containing date-stamped entries posted in reverse chronological order, often consisting of ideas, brief essays, photos, and hyperlinks to other Web sources, and allowing users to post comments.

Blogger: A person who maintains a blog.

Blogging: An act of posting entries to one's blog.

Blogroll: A list of favorite sites posted on the blog.

Comment: A message appended to a blog post which may be left by a reader.

E-Collaboration Technologies: Electronic technologies that enable collaboration among individuals engaged in a common task.

Edublogs: Blogs used in educational settings.

Feed: A feed is a source of syndicated information. Blog posts and comments may be available to users via feeds that are updated each time something new is published.

RSS: Really Simple Syndication, an easy way to track new entries in a Web log; once subscribed to RSS feed, the user simply clicks on the link conveniently located on the toolbar and receives a list of the site's most recent posts.

Social Bookmarking: The act of organizing one's personal Web bookmarks and sharing them with others.

Tag: A word or term used as a bookmark. Tags are used to help flag and retrieve Web-based content on particular topics

Trackback: A feature that helps bloggers link their posts to related posts written by other bloggers.

Web Annotations: Annotations left on Web pages by readers. The annotations exist in a layer separate from the Web page and do not change the page itself in any way.

Collaboration Engineering for Designing Self-Directed Group Efforts

Gert-Jan de Vreede
University of Nebraska at Omaha, USA
Delft University of Technology, The Netherlands

Robert O. Briggs
University of Nebraska at Omaha, USA
University of Alaska Fairbanks, USA

Gwendolyn L. Kolfschoten
Delft University of Technology, The Netherlands

INTRODUCTION

Collaboration is important to create organizational value. By collaborating, people can accomplish more than they could as separate individuals. Collaboration is the making of joint efforts towards a goal. Yet, achieving effective e-collaboration is easier said than done. Groups are not likely to overcome the challenges of collaboration by themselves (Nunamaker, Briggs, Mittleman, Vogel, & Balthazard, 1997; Schwarz, 2002).

To support organizations and groups in their e-collaboration efforts, a myriad of technical approaches for e-collaboration support have been developed, such as group (decision) support systems, video conferencing, and computer supported collaborative work. This technology support is often combined with process support called facilitation (Nunamaker et al., 1997). While research has demonstrated the added value of such groupware systems in the field (Fjermestad & Hiltz, 2001), and facilitation is regarded as a critical success factor for successful use of such systems (Nunamaker et al., 1997; Vreede, Boonstra & Niederman, 2002), their adoption and diffusion is difficult (Briggs et al., 1999).

Consequently, collaboration support in the form of technology and process support is not always available to groups that could benefit from such support. While collaboration support is often offered by experts its sustained implementation would be simplified if we could transfer the skills required for collaboration support to practitioners in the organization. When they can use such skills on a recurring basis, collaboration support could be availed to groups for recurring processes, without mounting-up costs from hiring group process professionals. This approach is named collaboration engineering (CE).

Collaboration engineering is defined as an approach to designing collaborative work practices for high-value recurring tasks, and deploying those designs for practitioners to execute for themselves without ongoing support from professional facilitators (Briggs, Kolfschoten, Vreede, & Dean, 2006; Briggs, Vreede, & Nunamaker, 2003; Vreede & Briggs, 2005). To enable the transition of collaboration support skills and their application by practitioners we need to be able to design easy to use, robust collaboration support, both in terms of process support and technology support. Collaboration Engineering research therefore addresses both a design and deployment challenge, that when overcome enable more sustained implementation of collaboration support. In this article, we will further explain the collaboration engineering approach; the challenge it addresses, the details of the approach and the research challenges it poses.

BACKGROUND

Collaborative efforts can be far more effective and efficient if they are explicitly designed, structured and professionally managed so as to minimize cognitive load and maximize the focus of purposeful effort (Nunamaker et al., 1997). This is often referred to as facilitation – the structuring and management of collaborative efforts. Research has discussed the tasks of a facilitator, which includes among other things; instructing and motivating the group, managing discussion and conflict, and employing tools and techniques to support groups in achieving their goals (Clawson, Bostrom, & Anson, 1993; Hayne, 1999; Schwarz, 2002). Facilitation not only provides structures with which people can work together, it also supports people in using available technologies, and points people to relevant information resources (Kolfschoten, Hengst, & Vreede, in press). Field research at IBM, Boeing, BP, EADS, and ING shows that facilitators can achieve reductions of over 50% in terms of labor costs and project time by applying groupware technologies to carefully designed collaboration processes (Vreede, Vogel, Kolfschoten, & Wien, 2003).

Unfortunately, without professional facilitation reaping the benefits of collaboration and collaboration technologies is difficult. Organizations often resort to implementing technologies, yet experiences show that technology alone seldom is the answer. What is needed is the conscious design of effective collaboration processes followed by the design and/or deployment of new collaboration technologies to support these processes. Such design effort requires extensive expertise on the appropriate use of technology (Dennis, Wixom, & Vandenberg, 2001), and experience with different challenges in group work (Schwarz, Davidson, Carlson, & McKinney, 2005)

The adoption and sustained use of collaboration support poses a further challenge. In order to implement successful collaboration support expertise in the design and facilitation of collaboration processes should be available to groups and teams in the organization. There are several approaches to the implementation of collaboration support in organizations, which despite their benefits are not always successful. One approach is, for instance to create internal support facilities with trained facilitators. Another approach is to hire (expensive) external facilitator/consultants. While the later is often expensive, internal facilitators are difficult to sustain as well. Facilitators are frequently promoted away, leaving nobody who knows how to use the technology. Likewise a budget crunch may mean facilitators are laid off, limiting access of teams to support capabilities (Agres, Vreede, & Briggs, 2005; Briggs et al., 1999; Briggs et al., 2003).

The aim of collaboration engineering is to offer an approach to implement collaboration support in organizations that is more likely to lead to sustained use and recurring benefits. Instead of training facilitators, who have extensive expertise, experience and skills, to support ad hoc "all round" processes, collaboration engineering focuses on training practitioners, i.e. domain experts in the organization, to support a frequently recurring high value collaborative task. It thus focuses on processes for mission-critical tasks that must be executed by teams rather than individuals, that must be executed frequently, and that have a high payoff if successful. The value of improving a recurring task is likely to be higher than that of improving an ad-hoc task, since the benefits will be derived in each instance of the recurring effort. Examples of such recurring collaboration processes can be found in various sectors, for example financial services, defense, and software development:

- Financial services
  - Risk assessment & mitigation (see example in Text Box 1)
  - Service product development
  - Sarbanes-Oxley assessments
  - Marketing focus groups
- Defense
  - Crisis response
  - Situational Awareness
  - Course of Action Analysis

Text Box 1. Example of a collaboration engineering effort in the field

Following industry guidelines, ING Group, an international financial services organization, was faced with the challenge to perform hundreds of operational risk management (ORM) workshops. They needed a repeatable collaborative ORM process that operational risk managers could execute themselves. Researchers applied Collaboration Engineering techniques to develop such a process, the Risk & Control Self Assessment (R&CSA) process. Collaboration Engineering was used to model and field test the R&CSA process in a pilot. After this pilot, the R&CSA process was fine-tuned and validated by a group of 12 ING ORM experts. Since then, over 350 ORM practitioners were trained to execute this process. To date, these ORM practitioners have successfully moderated hundreds of workshops where business participants identify, assess, and mitigate operational risks. While many consultancy organizations offer risk assessment workshops as part of their professional portfolio, ING now has the skills to run such workshops for themselves, without the support of external professionals.

- Software development
  - Requirements negotiation & specification
  - Usability testing
  - Requirements inspections
  - Code inspections

THE COLLABORATION ENGINEERING APPROACH

Collaboration engineering is located on a crossroad of disciplines, such as facilitation (process support for collaboration), information systems (technology support for collaboration), education (approaches for training and knowledge transition), systems engineering (design approach), organization science (deployment and implementation strategies), and group research in general. The collaboration engineering approach consists of three key phases: the investment decision phase, the design phase and the deployment phase (see Figure 1).

First, an investment decision should be made to determine the added value of:

- **The recurring task for the organization:** If the recurring task is not mission critical, then the (recurring) expenses to support the collaborative execution of the task may not outweigh the benefits.
- **A collaborative approach to the task:** A collaborative approach implies that several participants will be involved in each instance of the process. This makes the process costly (man hours) by definition, which is only justified if the task requires combined effort, sharing information and mutual learning, decision making and/or consensus building.
- **Training practitioners to offer this support:** Depending on the frequency of the task and the amount of groups that require support, one or more practitioners can be trained. Practitioners are domain experts in the organization, who perform the collaboration process as part of their task.

After a positive investment decision, the design phase starts. A collaboration engineer (an expert facilitator) uses reference knowledge on process support,

Figure 1. The collaboration engineering approach

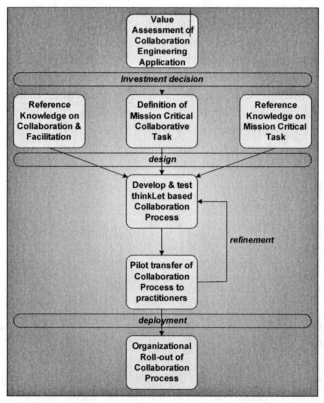

facilitation, and collaboration systems on to design a collaboration process for the recurring collaborative task. To create a collaboration process design that enables the practitioner to facilitate an organizational group to achieve the task's goal, the collaboration engineer faces a complex set of design challenges. He or she has to consider different stakes in the process and an optimal use of resources. Also, a high-quality collaboration process design is efficacious to the group goal, acceptable to all stakeholders, and transferable to the practitioner. Furthermore it should be reusable in each instance of the collaboration process and thus should consider the availability of resources (time, technology, effort and knowledge) in each instance of the task. A last and critical design challenge is that the practitioner will not know how to improvise when unpredictable effects of the process interventions occur. Therefore, the design should be predictable. To enable such designs, Collaboration Engineering uses thinkLets. ThinkLets are reusable and predictable design patterns that offer a prescription for the practitioner to execute each activity in the collaboration process. Furthermore, experiences from the field show that thinkLets can be easily and successfully transferred to practitioners.

Since the collaboration process design needs to be transferred to practitioners, it requires rigorous testing and refinement before it is deployed in the organization. First, the design is tested by a professional facilitator; next, a pilot is organized in which the design's transferability to practitioners is tested.

Finally, when the design is finished and tested satisfactorily, the Collaboration Engineer can deploy it in the organization. He or she will train the practitioners in the organization to guide the execution of the process, based on approaches in education and cognitive psychology. The success of this training is critical; if the process is unsuccessful, bad testimony and harmed self-efficacy of the practitioner might lead to abandonment of the process (Agres et al., 2005). Different supporting elements such as cue cards, an extensive manual and a process overview are provided to the practitioner not only for training, but also to support them during the process (Kolfschoten & Hulst, 2006; Kolfschoten, Pietron, & Vreede, 2006).

After the training, the practitioners should be able to perform the process by themselves without ongoing support from the collaboration engineer. However, they do need organizational and management support. During the further organizational roll-out, results of the

process' execution should be evaluated, and in larger projects, communities of practice can be formed to exchange experiences and improve or adapt the process to changes in the organization.

Initial experiences with collaboration engineering are very promising. The illustration on ING Group's R&CSA process in textbox 1 testifies to a successful deployment of a collaboration engineering process. Following the success of the R&CSA project, ING Group also commissioned the development of a Sarbanes-Oxley (SOX) assessment process, which was eventually recommended as company standard. The key perceived benefits of these efforts are to create and sustain control over the organization's collaboration processes. Says one ING Executive reflecting on the collaborative SOX approach: "Finally an approach that puts *us* in the driver's seat—and *not* the consultants." Other examples of collaboration engineering projects include, but are not limited to:

- The Rotterdam Port Authority used Collaboration Engineering techniques to support crisis response training and operational execution (Appelman & Driel, 2005)
- A process for collaborative usability testing was successfully employed for the development of a governmental health emergency management system (Fruhling & Vreede, 2005)
- Dozens of groups engaged in effective software requirements negotiations using the EasyWinWin process (Boehm, Gruenbacher, & Briggs, 2001; Briggs & Grunbacher, 2001)
- A repeatable process for collaborative standards writing and review was employed by a large number of biocontainment experts to derive a draft of a national standard for the planning, operation, and management of biocontainment units (Smith et al., 2006)
- A collaborative software code inspection process based on Fagan's inspection standards was successfully employed at Union Pacific (Vreede, Koneri, Dean, Fruhling, & Wolcott, 2006)
- A process for continuous end-user reflection on information systems development efforts was used in a large educational institution (Bragge, Merisalo-Rantanen, & Hallikainen, 2005)

In each of these projects, processes were employed that could be run by nonprofessional facilitators after a short training. This enables the organization to avoid the costs of professional facilitators while reaping the benefits of the training each time the process is executed as shown in the ING case.

FUTURE TRENDS

Despite encouraging results so far, many academic and practical challenges lie ahead to develop the area of collaboration engineering (see Figure 2). Research should focus on the theoretical foundations of collaborative work practices, group performance, facilitation, socio-technical engineering, creativity, technology acceptance, adoption & diffusion, and satisfaction. Insights on project strategies to inform practical advances such as toolkit prototypes with collaboration process building blocks (thinkLets) that can be used to design complex recurring collaboration processes are required. Furthermore, a rich variety of insights can be collected by developing and testing collaboration engineering

concepts in a number of diverse application domains, crossing boundaries between different social-cultural contexts. Last, collaboration engineering concepts, technologies, and methods are and will be developed and evaluated in the field through case studies and action research for several application areas.

Examples of research challenges that currently lie ahead for the collaboration engineering research field include, but are not limited to:

- Theory development in the areas of collaborative value creation, satisfaction with collaboration processes, collaborative work practice transition, and patterns of collaboration in group work (generation, clarification, reduction, organization, evaluation, consensus building).
- Developing modeling languages for collaboration processes, consisting of a metamodel and various aspect models. To support modeling activities, computer assisted collaboration engineering (CACE) tools will be designed and developed.

Figure 2. Overview of the collaboration engineering research area

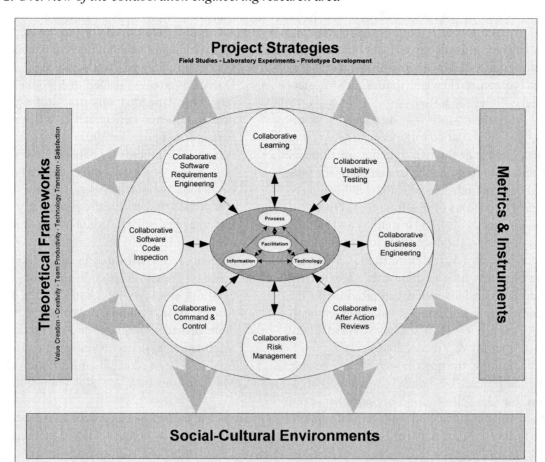

- Designing and deploying new collaboration technologies and processes based on the insights derived from theoretical and empirical findings of the research group.
- Designing and developing tools for facilitation support in both colocated and distributed settings.
- Developing an evaluation framework for the quality of collaboration processes designed using the collaboration engineering approach.

CONCLUSION

Collaboration engineering offers organizations an alternative to traditional collaboration support methods. It can be applied to both face-to-face collaboration, e-collaboration, synchronous, and asynchronous. The added value of the collaboration engineering approach lies in its rigorous design and deployment approach, using the thinkLets design pattern language, based on both expertise and an extensive body of theory and literature. This allows collaboration engineers to create more predictable and better transferable collaboration process. The approach to train practitioners to execute a specific recurring process, rather than training all-round facilitation experts has proved to provide a recurring added value which makes it easier to adopt and sustain collaboration support in the organization. A spin-off of the collaboration engineering approach is the collaboration engineering research program. The requirement to create predictable collaboration processes offered a new perspective on collaboration research. Instead of testing the effectiveness or efficiency of *technology in isolation*, collaboration engineering researchers focus on understanding the creation of successful collaboration *processes* through collaboration design patterns and supporting these with e-collaboration technology. This approach causes more relevant questions to be asked, for example, "Which brainstorming method causes more creativity?" rather than "Does technology X lead to more productivity?" This new perspective has advanced a deeper understanding of several research challenges and (seemingly) conflicting results (Santanen, 2005).

REFERENCES

Agres, A., Vreede, G. J. de, & Briggs, R. O. (2005). A tale of two cities: Case studies of GSS transition in two organizations. *Group Decision and Negotiation, 14*(4), 265-266.

Appelman, J. H., & Driel, J. van. (2005). Crisis-response in the port of Rotterdam: Can we do without a facilitator in distributed settings? In R. H. Sprague (Ed.), *Proceedings of the Hawaii International Conference on System Science* (pp.1-10). Waikoloa, HI: IEEE Computer Society Press.

Boehm, B., Gruenbacher, P., & Briggs, R. O. (2001). Developing groupware for requirements negotiation: Lessons learned. *IEEE Software, 18*(3), 45-66.

Bragge, J., Merisalo-Rantanen, H., & Hallikainen, P. (2005). Gathering innovative end-user feedback for continuous development of information systems: A repeatable and transferable e-collaboration process. *IEEE Transactions on Professional Communication, 48*(1), 55-67.

Briggs, R. O., Adkins, M., Mittleman, D. D., Kruse, J., Miller, S., & Nunamaker, J. F., Jr. (1999). A technology transition model derived from qualitative field investigation of GSS use aboard the *USS Coronado*. *Journal of Management Information Systems, 15*(3), 151-196.

Briggs, R. O., & Grunbacher, P. (2001). Surfacing tacit knowledge in requirements negotiation: Experiences using easywinwin. In R. H. Sprague (Ed.), *Proceedings of the Hawaii International Conference on System Sciences* [CD]. Wailea, HI: IEEE Computer Society Press.

Briggs, R. O., Kolfschoten, G. L., Vreede, G. J. de. & Dean, D. L. (2006). Defining key concepts for collaboration engineering. In N. C. Romano (Ed.), *Proceedings of the Americas Conference on Information Systems* [CD]. Acapulco, Mexico: AIS Press

Briggs, R.O., Vreede, G. J. de, & Nunamaker, J. F., Jr. (2003). Collaboration engineering with thinkLets to pursue sustained success with group support systems. *Journal of Management Information Systems, 19*(4), 31-63.

Clawson, V. K., Bostrom, R., & Anson, R. (1993). The role of the facilitator in computer-supported meetings. *Small Group Research, 24*(4), 547-565.

Dennis, A. R., Wixom, B. H., & Vandenberg, R. J. (2001). Understanding fit and appropriation effects in group support systems via meta-analysis. *Management Information Systems Quarterly, 25*(2), 167-183.

Fjermestad, J., & Hiltz, S. R. (2001). Group support systems: A descriptive evaluation of group support systems case and field studies. *Journal of Management Information Systems, 17*(3), 115-159.

Fruhling, A., & Vreede, G. J. de. (2005). Collaborative usability testing to facilitate stakeholder involvement. In S. Biffl, A. Aurum, B. Boehm, H. Erdogmus, & P. Grünbacher (Eds.), *Value based software engineering* (pp. 201-223). Berlin, Germany: Springer-Verlag.

Hayne, S. C. (1999). The facilitator's perspective on meetings and implications for group support systems design. *DataBase, 30*(3/4), 72-91.

Kolfschoten, G. L., den Hengst, M., & Vreede, G. J. de. (in press). Issues in the design of facilitated collaboration processes. *Group Decision and Negotiation.*

Kolfschoten, G. L., Pietron, L., & Vreede, G. J. de. (2006). A training approach for the transition of repeatable collaboration processes to practitioners. In S. Seifert & C. Weinhardt (Eds.), *Proceedings of the International Conference on Group Decision and Negotiation* [CD]. Karlsruhe, Germany: Universitätsverlag Karlsruhe.

Kolfschoten, G. L., & Hulst, S. van der. (2006). Collaboration process design transition to practitioners: Requirements from a cognitive load perspective. In S. Seifert & C. Weinhardt (Eds.), *Proceedings of the International Conference on Group Decision and Negotiation* [CD]. Karlsruhe, Germany: Universitätsverlag Karlsruhe.

Locke, E. A., & Latham, G. P. (1990). *A theory of goal setting and task performance.* Englewood Cliffs, NJ: Prentice Hall.

Nunamaker, J. F., Jr., Briggs, R. O., Mittleman, D. D., Vogel, D., & Balthazard, P. A. (1997). Lessons from a dozen years of group support systems research: A discussion of lab and field findings. *Journal of Management Information Systems, 13*(3), 163-207.

Santanen, E. L. (2005). Resolving ideation paradoxes: Seeing apples as oranges through the clarity of thinkLets. In R. H. Sprague (Ed.), *Proceedings of the Hawaii International Conference on System Sciences* [CD]. Waikoloa, HI: IEEE Computer Society Press.

Schwarz, R., Davidson, A., Carlson, P., & McKinney, S. (2005). *The skilled facilitator fieldbook: Tips, tools, and tested methods for consultants, facilitators, managers, trainers, and coaches.* San Francisco: Jossey-Bass.

Schwarz, R. M. (2002). *The skilled facilitator.* San Francisco: Jossey-Bass.

Smith, P. W., Anderson, A. O., Christopher, G. W., Cieslak, T. J., Vreede, G. J. de, Fosdick, G. A., et al. (2006). Designing a biocontainment unit to care for patients with serious communicable diseases: A consensus statement, biosecurity and bioterrorism. *Biodefense Strategy, Practice and Science, 4*(4), 351-365.

Vreede, G. J. de, Boonstra, J., & Niederman, F. A. (2002). What is effective GSS facilitation? A qualitative inquiry into participants' perceptions. In R. H. Sprague (Ed.), *Proceedings of the Hawaii International Conference on System Sciences* [CD]. Waikoloa, HI: IEEE Computer Society Press.

Vreede, G. J. de, & Briggs, R. O. (2005). Collaboration engineering: Designing repeatable processes for high-value collaborative tasks. In R. H. Sprague (Ed.), *Proceedings of the Hawaii International Conference on System Sciences* [CD]. Waikoloa, HI: IEEE Computer Society Press.

Vreede, G. J. de, Koneri, P. G., Dean, D. L., Fruhling, A. L., & Wolcott, P. (2006). Collaborative software code inspection: The design and evaluation of a repeatable collaborative process in the field. *International Journal of Cooperative Information Systems, 15*(2), 205-228.

Vreede, G. J. de, Vogel, D. R., Kolfschoten, G. L., & Wien, J. S. (2003). Fifteen years of in-situ GSS use: A comparison across time and national boundaries. In R. H. Sprague (Ed.), *Proceedings of the Hawaii International Conference on System Sciences* [CD]. Waikoloa, HI: IEEE Computer Society Press.

KEY TERMS

Collaboration: Joint effort towards a group goal (Briggs et al., 2006).

Collaboration Engineering: An approach to designing collaborative work practices for high-value recurring tasks, and deploying those designs for practitioners to execute for themselves without ongoing support from professional facilitators (Briggs et al., 2006).

Collaboration Engineering Design (noun): An artifact defining the sequence and logic of a set of steps for attaining some set of objectives, and the conditions under which these steps will be executed (Briggs et al., 2006).

Collaboration Engineering Designing (verb): To create, document and validate a collaboration process design (Briggs et al., 2006).

Deploying (verb): Implement the collaboration process and support in a way that it becomes a self-sustaining practice within an organization (Briggs et al., 2006).

Goal: A desired state or outcome (Locke & Latham, 1990).

High-Value Task: A task from which an organization derives substantial benefit or forestalls substantial loss by successful completion (Briggs et al., 2006).

Recurring Task: A task that must be conducted repeatedly, and that can be completed using a similar process design each time it is executed (Briggs et al., 2006).

Collaboration Methods and Tools for Operational Risk Management

Jürgen H. M. van Grinsven
AdvancedCollaboration.nl, The Netherlands
Delft University of Technology, The Netherlands

Marijn Janssen
Delft University of Technology, The Netherlands

Henk de Vries
AdvancedCollaboration.nl, The Netherlands

INTRODUCTION

Over the past years, various methods and tools have been used by financial institutions (FIs) to support operational risk management (ORM) (Brink, 2002). ORM supports decision-makers to make informed decisions based on a systematic assessment of operational risks (Brink, 2003; Cumming & Hirtle, 2001). By the end of the 1990s many FIs increasingly focused their effort on ORM. This was mainly motivated by the volatility of today's marketplace, costly catastrophes (e.g., Metallgesellshaft, Barings, Daiwa, Sumitomo, Enron, Worldcom) and regulatory-driven reforms such as the new Basel accord (Grinsven, Ale, & Leipoldt, 2006).

The main benefits of ORM are improving performance, reducing operational losses, enabling a more efficient use and allocation of resources, preventing major business disasters, optimizing the allocation of capital, increasing the chance of success, improving decision-making and creating a shared understanding for Operational Risk (OR) (Brink, 2002).

FIs often use loss data and expert judgment to estimate their exposure to operational risk (Brink, 2002; Cruz, 2002). Utilizing expert judgment is usually completed with more than one expert individually (often referred to as individual or self-assessments) or group-wise with more than one expert (often referred to as group-facilitated self-assessments). While individual self-assessments are currently the leading practice, the trend is more toward group-facilitated self-assessments. As a result, there is a need to support these group-facilitated assessments using e-collaboration methods and tools. However, there are a number of difficulties that financial institutions face which result in research issues that need to be bridged. The major difficulties are closely related to an effective, efficient and satisfying identification and estimation of the level of exposure to OR (Young et al., 1999).

In this article, collaboration methods and tools are discussed for supporting multiple expert judgement utilized in ORM. We also present the main research issues. First, an overview of ORM will be provided by discussing the main phases of ORM. Second collaboration methods and tools are presented that can be used to support the multiple experts' judgment and elicitation process and the role of experts is discussed. Finally research issues are discussed that need to be solved to leverage the full advantages of collaboration methods and tools as well as future developments.

BACKGROUND

An operational risk can be defined as the risk of direct or indirect loss resulting from inadequate or failed internal processes, people and systems or from external events (RMA, 2000). The management of OR involves a multitude of techniques that serve two main purposes: loss reduction and avoidance of catastrophic losses. Expert judgment is extremely important when internal and external loss data does not provide a sufficient, robust, satisfactory identification and estimation of the financial institutions exposure to OR (BCBS, 2003; Brink, 2002; Cruz, 2002).

Expert judgment is defined as the degree of belief, based on knowledge and experience that an expert makes in responding to certain questions about a subject (Clemen & Winkler, 1999; Cooke & Goossens, 2004). Expert judgment is increasingly advocated in various sectors for identifying and estimating the level of uncertainty about risk (Bigün, 1995; Kaplan, 1990; Muermann & Oktem, 2002). Expert judgment

Figure 1. Overview of collaboration methods and tools for ORM

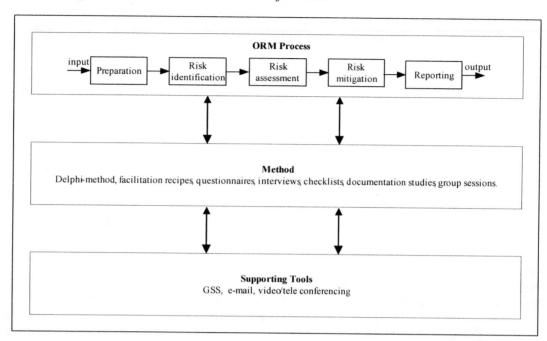

is often utilized by FIs to perform operational risk self-assessments. However, despite the fact that they use self-assessments, very few deploy them to provide them with a sufficient and satisfactory estimate of their exposure to OR (Finlay & Kaye, 2002).

Collaboration methods and tools can be used for the combined purposes of process improvement and knowledge sharing (Kock & Davison, 2003). Collaboration methods and tools exist in various forms. Figure 1 shows an overview of methods and tools, focusing on expert judgment utilized in ORM. The figure consists of three layers: a process layer, a method layer and a supporting tools layer. The *process* layer shows how the five main phases preparation, risk identification, risk assessment, risk mitigation and reporting phase, as proposed by Cooke and Goossens (2000), coincide with each other. Every phase is the input for the next phase resulting in an accurate estimate of exposure to OR as final output. The *method* layer shows the methods that can be used in the different phases of the expert judgment process. The *supporting tools* layer shows the supporting tools which can be used to support the methods used in the ORM process. The method and tool layer are discussed briefly in the following paragaphs.

E-collaboration is defined as collaboration using electronic technologies among different individuals to accomplish a common task (Kock, 2005). E-collabora-

tion technologies are increasingly used to support expert judgment activities (Brink, 2002; Cruz, 2002; Finlay & Kaye, 2002). E-collaboration technologies can support each of the five phases of the ORM process.

A group support system (GSS) is an electronic technology that supports a common collection of group tasks such as idea generation, organization and communication. GSS aim to improve (collaborative) group work (Vogel et al., 1990). Improvements are achieved by using information and communication technology to further structure multiple experts' exchange of ideas, opinions and preferences (Fjermestad & Hiltz, 2001; Turban et al., 2001). Turban and Aronson (2001) present an overview of GSS tools that can be used for supporting idea generation, idea organization, prioritizing, policy development and knowledge accumulation and representation. Table 1 presents a number of these GSS tools can be used in the ORM process to support the multiple experts (Weatherall & Hailstones, 2002).

Huber (1980) describes facilitation recipes as a detailed re-usable script that can be used for a group task such as brainstorming. In essence, these recipes help a facilitator to organize group communication and coordinate problem solving activities (Hacket, 1993; Hunter, Bailey, & Taylor, 1992; Ocker, Hiltz, Turoff, & Fjermestad, 1997). They are aimed at further increasing the effectiveness, efficiency and satisfaction of the collaborative group meeting (Grinsven, 2006). In the

Table 1. Examples of GSS Tools Used for ORM (Based on Weatherall & Hailstones, 2002)

GSS Tool	General description	Example
Electronic Brainstorming	Gathers ideas and comments in a unstructured manner	Identification of risks and control measures
Group Outliner	Allows groups to create and comment on a multilevel list of topics in a tree or outline structure	Reporting
Topic Commenter	Allows participants to comment on a list of topics	Precise formulation of risks and (alternative) control measures
Categorizer	Allows groups to generate a list of ideas and supporting comments	Mapping of control measures to risks
Vote	Supports consensus development through group evaluation of issues	Prioritizing of risks and control measures
Alternative Analysis	Allows groups to weight or rate a list of alternatives against a list of criteria	Assessing the frequency of occurrence and impact of risks and control measures

ORM context, recipes can be used to identify events, accurately describing an OR, or assessing an OR.

E-mail or electronic discussion forums are methods with which messages can be composed, send, and received using electronic communication systems using the Internet as an enabling platform (Massey et al. 2003). E-mail and discussion forums can be used to communicate, brainstorm and converge with other experts in the ORM process.

Video or (tele) conferencing systems can be used to support a number of tasks in distributed group sessions (Malhotra, Majchrzak, Carman, & Lott, 2001). Often multiple experts in the ORM process can not meet face-to-face due to large distances or scheduling problems. Then, such systems can be used to enable collaboration between the experts.

COLLABORATION METHODS AND TOOLS TO SUPPORT ORM

We discuss the collaboration methods and tools that can be used for each phase of the ORM process by presenting practical examples from a action research study in a large Dutch Financial Institution (DFI), and discuss future research.

The *preparation* phase provides the framework for the experts, taking into account the most important activities prior to the identification, assessment, mitigation, and reporting of operational risks. These activities can be divided in the subactivities: describing the context and objectives, choosing the method, identifying and selecting experts, defining the experts' roles, feedback and reporting and tryout the ORM exercise (Goossens

& Cooke, 2001). To avoid procedural inconsistency and biases it was suggested to make a list of criteria for identifying and selecting the experts and methods to be used. At DFI we used checklists, several structured interviews, documentation study and group facilitated sessions to make this list of criteria (Grinsven, 2006). Further, we learned that the Electronic Brainstorming Group Outliner and Categorizer tools can be used by the facilitator to support group facilitated sessions in order to clearly scope the context and objectives of the ORM exercise. Future research should aim to develop E-collaboration methods and tools to support the scoping activity in the preparation phase of the ORM process.

The *risk identification* phase aims to provide a reliable information base to enable an accurate estimation of the frequency and impact of OR in the risk assessment phase (Cooke & Goossens, 2000). The activities in the identification phase can be further divided in the identifying the events, formulating and categorizing the OR subactivities. One of the objectives of this phase is to arrive at a comprehensive and reliable identification of the OR to reduce the likelihood that an unidentified OR becomes a potential threat. Member status, internal politics, fear of reprisal and groupthink can make the outcome of the risk identification less reliable (Grinsven, 2006). Most risk management and GSS literature recommends to identify the risks anonymously. Useful methods to gain anonymous contributions are the Delphi method or structured group-facilitated self-assessment. At DFI, we facilitated a GSS workshop using Electronic Brainstorming method supported by a Categorizing and Vote tool to help the experts to build a shared view about the identified OR. In our case study, the experts appreciated the possibility to identify OR events anonymously. Future research should investigate the effects of anonymity and groupthink on the quality of the outcome from the risk identification phase.

The *risk assessment* phase aims for an accurate quantification of the frequency of occurrence and the impact associated with the potential loss of the identified OR and existing control measures (Brink, 2002). The activities in this phase can be divided in the following subactivities: assessing the OR (estimating frequency and impact) and aggregating the results. Most literature suggests to ensure the experts assess the OR individually, to minimize inconsistency and bias (see, e.g., Clemen & Winkler, 1999). Moreover, for the aggregation of the results, first use a simple mathematical aggregation method to calculate the results and then use a behavioral aggregation method to provide the rationales behind the individual assessments (Cooke & Goossens, 2004). In the DFI case study, we used the Alternative Analysis tool to anonymously assess the OR and promptly calculate the results. Then, we used the standard deviation of the results to further elicit the experts' opinion. We used several structured rounds from the Delphi method and facilitation recipes to enable the experts to provide the rationales behind the results. The facilitation recipes and GSS helped us to structure the interactions between the experts. Moreover, we learned that the GSS helped to prevent the results being influenced by groupthink and the fear of reprisal. Future research should focus on investigating the effect of using multiple aggregation methods on the quality of the results of the assessment of OR.

The *risk mitigation* phase aims to mitigate those OR that, after assessment, still have an unacceptable level of frequency and/or impact. The activities in this phase can be divided in three subactivities: identifying alternative control measures, reassessing the operational risks, and aggregating the results. The methods and tools that can be used in this phase are almost similar to the risk assessment phase. However, a more structured method and tool are suggested to identify alternative control measures that are more efficient and effective. At DFI we used the Topic Commenter tool to support this activity. Experts were enabled to elaborate / improve the existing controls and provide examples for them. Then, we used the Alternative Analysis tool to anonymously assess the OR again and calculated the results immediately. Again, we used the standard deviation of the results to further elicit the experts' opinion

regarding the rationales behind their assessment. We used facilitation recipes to structure the interaction between experts. Future research should focus on investigating the effect on the use of the existing methods and supporting tools on the effectiveness and efficiency of alternative control measures.

The *reporting* phase aims to provide the stakeholders such as the manager, initiator and experts with the relevant information regarding the ORM exercise. The activities in this phase can be divided in the subactivities: Documenting the results and providing feedback to the experts. Documenting is usually performed by two persons, each working on an individual basis on a part of the document. Cooke and Goossens (2000) suggest using a structured process to present the results to the experts. At DFI we used interviews and checklists to make sure the report complied to the relevant reporting standards. We facilitated a manual workshop to provide feedback to the experts which enabled them to leverage the experiences. Future research should investigate applying GSS to provide structured feedbacks.

FUTURE TRENDS

In the preceding section we discussed future research questions per process phase. ORM has gained major significance within financial institutions. Research indicates that there exists interdependence between the quality of data and good risk management (BCBS, 2003; Brink, 2003). Developments and trends concerning ORM indicate that FIs strive for (1) automation of their data aggregation and reporting from different business lines using single source of information, (2) single processes/tools leveraged for multiple purposes, and (3) regulatory requirements embedded in these processes and tools (Na, Coutou Miranda, van den Berg, & Leipoldt, 2005; RMA, 2005). The driving force behind these efforts are regulatory requirements such as Basel II and the Sarbanes-Oxley act which have already been put or will be put into operation in more than 100 countries (Arnold, 2005).

Web-based GSS is expected to become increasingly popular as a means to support the collaboration of distributed experts (Romano, Chen, & Nunamaker, 2002). This will allow multiple, distributed experts to collaborate in ORM activities from almost any location at any time. However, these tasks always have some form of inter-dependency and have to be coordinated (Sol, 1982).

Methods used for ORM, such as Delphi and facilitation methods are expected to be combined, resulting in multi-method and multi-tool approaches.

CONCLUSION

Expert judgment is extremely important when loss data does not provide a sufficient, robust, satisfactory identification and estimation of the FIs exposure to OR. The ORM process can be described using the preparation, risk identification, risk assessment, risk mitigation and reporting phase. We discussed methods and tools for each phase to support experts in order to achieve more effective, efficient and satisfying results. To achieve these results, it is important to create synergy between the activities in the ORM process, method and supporting tools.

E-collaboration methods and tools can be used to help gathering and processing information about operational risk. Moreover they can help enhance the effectiveness, efficiency and satisfaction of ORM sessions. Moreover, E-collaboration techniques have the potential to improve the ORM process by minimizing inconsistency and biases, reducing groupthink, gather and processing information.

REFERENCES

Arnold, K. (2005). AIM global data and risk management survey 2005.Retrieved from www.dmstudy.info/2005

BCBS. (2003). *Supervisory guidance on operational risk: Advanced measurement approaches for regulatory capital.* Office of the Comptroller of the Currency (OCC).

Bigün, E. S. (1995). Risk analysis of catastrophes using experts' judgements: An empiricalstudyonrisk analysis of major civil aircraft accidents in Europe. *European Journal of Operational Research 87,* 599-612.

Brink, G. J. v. d. (2002). *Operational risk: The new challenge for banks.* New York: Palgrave.

Brink, G. J. v. d. (2003). *The implementation of an advanced measurement approach within Dresdner Bank Group: IIR Conference Basel II: Best practices in risk management and-measurement.* Amsterdam, The Netherlands: Dresdner Bank Group.

Clemen, R. T., & Winkler, R. L. (1999). Combining probability distributions from experts in risk analysis. *Risk Analysis, 19*(2), 187-203.

Cooke, R. M., & Goossens, L. H. J. (2000). *Procedures guide for structured expert judgment.* Brussels/Luxembourg: European Commission.

Cooke, R. M., & Goossens, L. H. J. (2004). Expert judgment elicitation for risk assessments of critical infrastructures. *Journal of Risk Research, 7*(6), 643-156.

Cumming, C., & Hirtle, B. (2001). The challenges of risk management in diversified financial companies, Federal Reserve Bank of New York *Economic Policy Review.*

Cruz, M. (2002). Modeling, measuring and hedging operational risk. *Wiley Finance.*

Finlay, M., & Kaye, J. (2002). *Emerging trends in operational risk within the financial services industry.* London: Raft International.

Fjermestad, J., & Hiltz, S. R. (2001). Group support systems: A descriptive evaluation of case and field studies. *Journal of Management Information Systems, 17*(3), 115-160.

Goossens, L. H. J., & Cooke, R. M. (2001). Expert judgment elicitation in risk assessment. *Assessment and Management of Environmental Risks.* Kluwer.

Grinsven, J. H. M. v. (2006). *Improving operational risk management.* Unpublished doctoral dissertation, Delft University of Technology, The Netherlands.

Grinsven, J. H. M. v., Ale, B., & Leipoldt, M. (2006), Ons overkomt dat niet: Risicomanagement bij financiële instellingen. *Finance Incorporated, 6,* 19-21.

Hacket, D., Martin, & Charles, L. (1993). *Facilitation skills for team leaders.* Crisp.

Huber, G. P. (1980). *Managerial decision making.* Glenview, IL: Scott Foresman.

Hunter, D., Bailey, A., & Taylor, B. (1992). *The art of facilitation.* Tuscon, AZ: Fisher Books.

Kaplan, S. (1990). Expert information versus expert opinions: Another approach to the problem of eliciting/combining/using expert knowledge in probabilistic risk analysis. *Journal of Reliability Engineering and System Safety, 39,* 61-72.

Kock, N. (2005). What is e-collaboration? *International Journal of e-Collaboration, 1*(1), i-vii.

Kock, N., & Davison, R. (2003). Can lean media support knowledge sharing? Investigating a hidden advantage of process improvement. *IEEE Transactions on Engineering Management, 50*(2), 151-163.

Malhotra, A., Majchrzak, A., Carman, R., & Lott, V. (2001). Radical innovation without collocation: A case study at Boeing-Rocketdyne. *MIS Quarterly, 25*(2), 229-249.

Massey, A. P., Mitzi, M., Montoya-Weiss, & Hung, Y. T. (2003). Because time matters: Temporal coordination in global virtual project teams. *Journal of Management Information Systems, 19*(4), 129-155.

Muermann, A., & Oktem, U. (2002). The near-miss management of operational risk. *The Journal of Risk Finance, 4*(1), 25-36.

Na, H. S., Coutou Miranda, L., van den Berg, J., & Leipoldt, M. (2005). *Data scaling for operational risk modelling* (ERIM Tech. Rep. No. ERS-2005-092-LIS). Retrieved from http://hdl.handle.net/1765/7234

Ocker, R., Hiltz, S., Turoff, M., & Fjermestad, J. (1997). The effects of distributed group support and process structuring on software requirements development teams: Results on creativity and quality. *Journal of Management Information Systems, 12*(3), 127-153.

RMA. (2000, March). Operational risk: The next frontier. *The Journal of Lending & Credit Risk Management,* 38-44.

RMA. (2005). *Operational risk management: How emerging best practices can improve performance. A study from the Risk Management Association.* Retrieved from http://www.rmahq.org/RMA/OperationalRisk/

Romano, N. C., Chen, F. & Nunamaker, J. (2002). Collaborative project management software. In *Proceedings of the 35th Hawaii International Conference on System Sciences,* Hawaii.

Sol, H. G. (1982). *Simulation in information systems development.* Unpublished doctoral dissertation, Z. pl., Rijks Universiteit Groningen.

Turban, E., Aronson, J. E., & Liang, T. (2001). *Decision support systems and intelligent systems.* Upper Saddle River, NJ: Prentice Hall.

Vogel, D., Nunamaker, J. F., Martz, W. B., Grohowkisk, R., & McGoff, C. (1990). Electronic meeting systems experience at IBM. *Journal of Management Information Systems, 6*(3), 25-43.

Weatherall, A., & Hailstones, F. (2002). Risk identification and analysis using a group support system (GSS). In *Proceedings of the 35th Hawaii International Conference on System Sciences.* Hawaii.

Young, B., Blacker, K., Cruz, M., King, J., Lau, D., Quick, J., et al. (1999). Understanding operational risk: A consideration of main issues and underlying assumptions. *Operational Risk Research Forum.*

KEY TERMS

E-Collaboration Technique: Electronic technologies that enable collaboration among individuals engaged in a common task.

Expert Judgment: The degree of belief, based on knowledge and experience that an expert makes in responding to certain questions about a subject.

Facilitation Recipe: Specification of choices and actions that help a facilitator to organize group communication and coordinate problem solving activities.

Group Support Systems: A socio-technical system consisting of software, hardware, meeting procedures, facilitation support, and a group of meeting participants engaged in intellectual collaborative work.

Operational Risk: The risk of direct or indirect loss resulting from inadequate or failed internal processes, people and systems or external events.

Operational Risk Management: The identification, assessment and mitigation of operational risks.

Self-Assessment: Judgments by experts in financial institutions about their own work and processes to achieve a comprehensive and systematic review of the operational risks and control measures.

Supporting Tool: A tool that can be used to support methods and help achieve some goal.

A Collaborative Editor for Medical Learning Environments

Elisa Boff
Caxias do Sul University (UCS), Brazil
Federal University of Rio Grande do Sul (UFRGS), Brazil

Rosa Maria Vicari
Federal University of Rio Grande do Sul (UFRGS), Brazil

INTRODUCTION

Collaborative learning is changing the learning environments design. Intelligent tutoring systems (ITS), multiagent systems, affective computing, and virtual characters, are techniques and resources to improve individual and personalized learning.

ITS and ILE (intelligent learning environments) try to adapt to the characteristics of each student through the construction and the analysis of models that reflect both behavioral and cognitive aspects of the students. These systems represent more advanced pedagogical tools and provide more individualized learning experiences.

However, even with all the sophistication of these systems, cases can occur where the course material given by them is not sufficient to supply pedagogical necessities that a student comes to acquire during one learning activity. In these cases, is important to have tools that allow human-human interactions, where students can communicate themselves with tutors, or other students, and can jointly supply necessities and construct knowledge.

The importance of social interactions in the learning process is already known by the educational theoreticians. Some studies in this field are the Socio-Cultural approach of Vygostky (1999); some works of Piaget (1995); theories of Collaborative Learning (Dillenbourg, Baker, Blaye, & O'Malley, 1995); and others.

Recent advances in the ITS and ILE fields have proposed the use of agent's society-based architectures (Norman & Jennings, 2000). The principles of multiagent systems have showed a very adequate potential in the development of teaching systems due to the fact that the nature of teaching-learning problems is more easily solved in a collaborative way.

The collaborative learning systems development takes into account social factors, like those presented in Vassileva's and Cao's work (Cao, Sharifi, Upadrashta, & Vassileva, 2003). They concluded that is very important to consider sociological aspects of cooperation and to discover and describe existing relationships among people, existing organizational structures, and incentives for collaborative action. Hence, the learning environment can detect and solves some conflicts, help to perform tasks, and motivate learning and collaboration.

Based on presented ideas, our research group has been developing some intelligent learning environments to promote collaborative learning. The AMPLIA environment (Vicari et al., 2003) is a multiagent system that provides a collaborative Bayesian Net Editor to allow students build their own networks and compare them with the expert network. This collaborative construction happens between medical students.

This article presents the social agent that acts in the AMPLIA's collaborative editor in order to improve collaboration. AMPLIA is an intelligent probabilistic multiagent environment to support the diagnostic reasoning development and the diagnostic hypotheses modeling of domains with complex and uncertain knowledge, like the medical area. The social agent supports group formation and it makes a search among students of an ITS/ILE looking for suitable students to join in a workgroup. Hence, students can help others during a common learning task. For such, these agents takes into account some affective and social aspects of the students.

RELATED WORK

The group dynamic has been addressed by many researches and in different areas. The multiagent approach is adjusted to the problem of group formation and coordination.

Vassileva's (2001) research is about strategies and techniques of groups. Cheng and Vassileva (2005) proposed a motivation strategy for user participation based on persuasion theories of social psychology. In Cao et al. (2003), the goal is finding out how people develop attitudes of liking or disliking other people when interacting in a CSCW (computer supported cooperative work) environment in a collaborative-competitive situation, how they change their attitudes towards others when they realize their attitudes towards themselves, and how the design of the environment influences the emergent social fabric of the group.

A Bayesian network-based appraisal model was used by Conati (2002) to deduce a student's emotional state based on his or her actions.

Individuals' affective states have significant importance in the interaction process. For Scherer (2000), the affective states are divided in five categories: *Emotion* (episode related to synchronized responses for all or most organic systems to the evaluation of an external or internal event: anger, sadness, joy, fear, shame, pride, elation, and desperation); *Mood* (diffuse affective state that consists of the subjective feeling changing, with low intensity, but long duration, without apparent cause); *Interpersonal Stances* (affective position in relation to the other person in a specific interaction, such as distant, cold, warm, supportive, and contemptuous); *Attitudes* (attitudes are relatively tolerant, affectively colored beliefs, preferences, and predisposition in relation to objects or people, such as liking, loving, hating, desiring, and valuing); and *Personality Traits* (emotionally laden, stable personality dispositions and behavior tendencies, typical of a person, such as nervous, anxious, reckless, morose, hostile, envious, and jealous).

The social agent, described in this article, is based on social psychology ideas (to support social aspects) and affective states.

Most of medical software related to our application use knowledge-based models, while AMPLIA is a medical software that can be used for education purposes that considers cognitive and social states to build the student model, following an epistemological theory. Other medical software used for education purposes are Promedas (http://www.promedas.nl/), BioWorld (http://citeseer.ist.psu.edu/lajoie95establishing.html), Medikus (Möbus, 1995), and COMET (http://www.cs.ait.ac.th/~haddawy/pubs/iui04.pdf). Besides the knowledge-based model, BioWorld considers the self-confidence level. However, the strategies used in these

systems do not consider interactions between user and system based on cognitive models neither consider group interactions or group models.

INTELLIGENT PROBABILISTIC MULTI-AGENT ENVIRONMENT

AMPLIA is an intelligent multiagent learning environment designed to support training of diagnostic reasoning and modeling of domains with complex and uncertain knowledge (Vicari et al., 2003). AMPLIA focuses on the medical area. It is a system that deals with uncertainty under the Bayesian network approach, where learner-modeling tasks will consist of creating a Bayesian Net for a problem that the system will present. The construction of a network involves qualitative and quantitative aspects. The qualitative part concerns the network topology, that is, causal relations among the domain variables. After it is ready, the quantitative part is specified. It is composed of the distribution of conditional probability of the variables represented.

A negotiation process (managed by an intelligent *MediatorAgent*) will treat the differences of topology and probability distribution between the learner model built and the one built-in in the system. That negotiation process occurs between the agents that represent the expert knowledge domain (*DomainAgent*) and the agent that represents the learner knowledge (*LearnerAgent*). The *Social Agent* interacts with *Learner Agent* to suggest which classmate is recommended to work together with him or her. It also interacts with *Mediator Agent* that knows the domain and supports the negotiation process. All *Social Agent's* suggestions are sent to online students by chat tool.

The AMPLIA's pedagogical design was based on Piaget's (1995) and Vygostky's (1999) theories in order to support constructivist knowledge construction. In addition, the AMPLIA's Net Editor (see Figure 1) was designed as a collaborative editor to allow collaborative learning. Thus, the diagnostics hypotheses can be built by workgroups and, after all, can be compared with the expert network.

The collaborative editor is watched by the social agent, which main goal is improve student's learning stimulating his interaction with other students, tutors and professors. At AMPLIA, each user builds his/her own Bayesian Net for a specific pathology. The Bayesian Net corresponds to the student model for a particular problem solution in the health context. During this

Figure 1. Collaborative Bayesian Net Editor

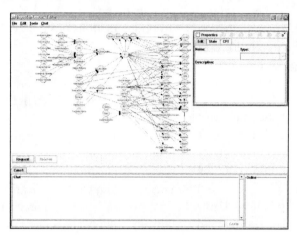

task, the social agent will recommend students to help other students. The social agent creates workgroups to solve tasks collaboratively. Besides, the agent uses the strategy of initiate a network construction to motivate students to interact with the environment.

The social agent reasoning is based on individual model and group model.

Social Agent's Individual Model

The individual model has the student features. The information collected that is important to define the right student to recommend are *Social Profile, Acceptance Degree, Sociability Degree, Mood State, Interest, Commitment Degree, Leadership,* and *Performance*.

The social profile (SP) is built during the students' interaction through a synchronous mechanism. The following information is collected during the students' interaction:

- **Initiatives of communication:** Number of times that a student had the initiative to talk with another.
- **Answers to initial communications:** Number of times that a student answered a communication request.
- **Interaction history:** Individuals with whom the student interacts or has interacted and number of interactions.
- **Friends group:** Individuals with whom the student interacts regularly, and number of interactions.

Based on Maturana and Varela (1998), we defined the acceptance degree (AD), which measures the acceptance between students. Such data are collected through a graphical interface that enables each student to indicate his/her acceptance degree for other students. This measurement may also be considered from a point of view of social networks. As the AD is indicated by the students themselves based on their affective structures, the measurement can indicate different emotions, such as love, envy, hatred, etc. The average of all ADs received by a student influences his/her sociability degree (SD).

The mood state (MS) represents our belief in the capability of a student to play the role of a tutor if he/she is not in a positive mood state (although the student may have all the technical and social requirements to be a tutor). We consider three values for the MS: "bad mood," "regular mood," and "good mood." These states are indicated by the students in a graphical interface through corresponding clip arts.

The socio-affective agent selects the action that maximizes this value when deciding how to act. The influence between nodes is shown in Figure 2.

The interest feature is given by the initiatives taken for the user (which material it had access without being recommended, with which student he/she initiated an interaction).

The commitment degree is based on the consistency theory (theory of persuasion of social psychology) places that after the people to assume a public commitment, they probably will act in more consistent way with its commitment. The idea is to use this theory as strategy to motivate the students to collaborate with

Figure 2. Decision network of student model

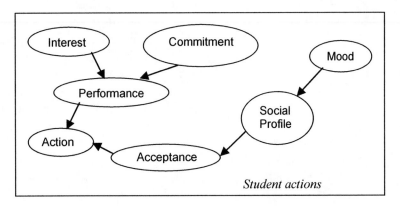

Figure 3. Decision network of social agent

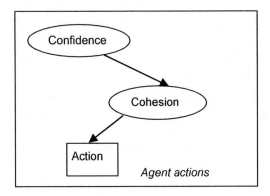

others. Thus, students who collaborate with other more actively (a quantitative measure of interactions can be made here, is not the ideal, but most viable at the moment) they will be inserted in groups with more frequency and will receive aid from better quality.

Social Agent's Group Model

The group model takes into account the cohesion and the confidence (or trust) in a workgroup. In the Figure 3, we represent the agent knowledge, which is supported by ontologies. The confidence node influences the cohesion node, like group cohesion is based on individual confidence in a group (Boff, Santos, & Vicari, 2006).

The group formation agent can decide who is joining in a group that brings more benefit to the group. This reliable type is interesting to help in the judgment of the agent regarding its confidence in the group that another agent belongs. The confidence is defined as the belief of an agent in the attributes such as the trustworthiness, honesty and ability of the agent "trusted" question.

The reputation of an agent defines an expectation on its behavior that is based on the comments of the agent or information on the last behavior of the agent in a specific context in data moment. He or she assumes that he or she has two agents, the agent A and the agent B. When the agent A does not have direct interaction with agent B or it is not certain on the trustworthiness of B, the agent A can take decisions based on the reputation of the agent B (gotten from other agents, or asking other agents about the reputation). Once agent A has interacted with agent B, it can verify or establish its confidence in the agent B in accordance with its degree of satisfaction in the interaction and use this confidence to take decisions for future interactions (Vassileva, 2001). Mechanisms on confidence and reputation can be used by agents to differentiate the good ones of bad (adequate or inadequate) the collaborating ones.

Joined groups have a great productivity. The cohesion is the attractiveness that the group exerts for its members, that of it they desire to continue to participate, resisting the idea to abandon it (Krüger, 1986). The cohesion is analyzed using the sociometric test of

Jacob Moreno (Moreno, 1993). We can define as group cohesion the solidarity and establishment of loyalty in a group and measure it for the amount of times that the same people choose to interact.

Collaboration Through Workgroups

As we noted before, AMPLIA has the purpose to be a constructivist learning environment. The first version of the AMPLIA editor was not collaborative. Hence, according to new learning trends in medicine, based on problem-based learning (Peterson, 1997), collaborative features were added to the editor. Thus, besides the individual online editing support, we added collaborative editing features and incorporated the social agent in order to improve the group activity.

Participation in a group depends on student approval. Students are invited to join the group, but they participate only if they want. When they refuse, they can be asked about the reason of the refusal. Information about actions performed by students, stored in the student model, is requested by the social agent to suggest new interactions. Groups are dynamically formed, based on the task that is being carried through. Students can participate simultaneously of several groups, following their interest. Each group must contain at least one student with leadership role. Social agent records the interactions among students during the learning session. This agent tries to identify which student collaborated more actively to the construction of the solution (BN model), and which students had their work modified less times.

It is better to create groups with democratic profile or with sharing roles, where all team members are able to lead the team. This can happens when responsibility for the operation of the team is shared. This technique is called *role-sharing*. Shared leadership leads to shared accountability and competencies. The leader of a team should focus on the process and keeps the team working well within a problem-solving process. Therefore, the democratic group presents better cohesion, independence and low frequency of aggressive behaviors. An important goal of Social Agent is to motivate shared leadership abilities among students.

CONCLUSION

The design of social and affective agents is been highlighted as an effective way to motivate and to improve group formation among students and also to promote collaborative learning.

Following this trend, we had built a socio-affective virtual character that interacts with students in order to motivate workgroup collaboration and the collective learning construction. The social agent identifies suitable students that can play the role of a tutor, and to recommend them to other students. Thus, the professor performs a secondary role, while students can be more autonomous.

According to Kock and McQueen (1996), asynchronous groupware shows promise as a medium of support for limited duration process improvement activities, it may not be suited to other types of activities, such as business process re-engineering and decision making, where high levels of interactivity are seen to be beneficial, and longer duration activities where group social development and communication protocols extend over a longer period of time. Therefore, AMPLIA's collaborative editor was developed as a tool to support synchronous work.

The tutor recommendation mechanism explores the social-affective dimension through the analysis of emotional states and students' social behavior. In this direction, we aim to contribute to the design of CLEs (collaborative learning environments) centered in students' features. In addition, the AMPLIA editor can be considered an intelligent collaborative tool that allows the creation of virtual workgroups to solve tasks in a collaborative way. Working in small groups, students can make your own decisions and prove their diagnostic hypotheses without the professor assistance. AMPLIA also allows individual work and access to medical tutorials.

In the near future we will model a mechanism to the automatic inference of students' affective states and personality, based on Zhou and Conati's model (2005).

REFERENCES

Boff, E., Santos, E. R., & Vicari, R. M. (2006). Social agents to improve collaboration on an educational portal. In Proceedings of the *Sixth IEEE International Conference on Advanced Learning Technologies*. Kerkrade, The Netherlands: IEEE Computer Society Press.

Cao, Y., Sharifi, G., Upadrashta, Y., & Vassileva, J. (2003, June 22). Interpersonal relationships in group interaction in CSCW environments. In *Proceedings of the User Modelling UM03 Workshop on Assessing and Adapting to User Attitudes and Affect*. Johnstown.

Cheng, R., & Vassileva, J. (2005, Jan 3-6). User motivation and persuasion strategy for peer-to-peer communities. In *Proceedings of HICSS'2005 (Mini-track on Online Communities in the Digital Economy/Emerging Technologies)*. Hawaii.

Conati, C. (2002). Probabilistic assessment of user's emotions in educational games. *Journal of Applied Artificial Intelligence, 16*(7-8), 555-575.

Dillenbourg, P., Baker, M., Blaye, A., & O'Malley, C. (1995). The evolution of research on collaborative learning. In P. Reimann & H. Spada (Eds), *Learning in humans and machines. Towards an interdisciplinary learning science* (pp. 189-211). London: Pergamon.

Kock, N., & McQueen, R. J. (1996). Asynchronous groupware support effects on process improvement groups: An action research study. In J. I. DeGross, S. Jarvenpaa, & A. Srinivasan (Eds.), *Proceedings of the 17ᵗʰ International Conference on Information Systems* (pp. 339-355). New York: The Association for Computing Machinery.

Krüger, H. (1986). *Introdução à psicologia social*. São Paulo, Brazil: EPU.

Maturana, H., & Varela, F. (1998). *Tree of knowledge: The biological roots of human understanding*. Boston: Shambhala.

Möbus, C. (1995). Towards an epistemology of intelligent problem solving environments: The hypothesis testing approach. In J. Greer (Ed.), *Proceedings of the World Conference on Artificial Intelligence and Education (AI-ED 95)* (pp. 138-145).

Moreno, J. L. (1993). *Psicoterapia de grupo e psicodrama: introdução à teoria e à prática*. São Paulo, Brazil: Editorial Psy.

Norman, T. J., & Jennings, N. R. (2000). Constructing a virtual training laboratory using intelligent agents. *International Journal of Continuous Engineering and Life-long Learning*.

Peterson, M. (1997). Skills to enhance problem-based learning. *Med Educ Online* [Serial online]. Retrieved from http://www.med-ed-online/

Piaget, J. (1950/1995). Explanation in sociology. In J. Piaget (Ed.), *Sociological studies*. New York: Routledge.

Scherer, K. R. (2000). Psychological models of emotion. In J. Borod (Ed.), *The neuropsychology of emotion* (pp. 137-162). Oxford/New York: Oxford University Press.

Vassileva, J. (2001). Multi-agent architectures for distributed learning environments. In *Proceedings of the AIED* (Vol. 12, pp. 1060-1069).

Vicari, R. M., Flores, C. D., Seixas, L., Silvestre, A., Ladeira, M. & Coelho, H. (2003). A multi-agent intelligent environment for medical knowledge. *Journal of Artificial Intelligence in Medicine, 27,* 335-366.

Vygotsky, L. S. (1999). *The collected works of L. S. Vygotsky* (c1987-c1999, Vol. 1-6). New York: Plenum Press.

Zhou, X., & Conati, C. (2003). Inferring user goals from personality and behavior in a causal model of user affect. *Proceedings of the International Conference IUI '03*, Miami, FL.

KEY TERMS

CLE: *Collaborative learning environments* are informatics systems designed to enable collaborative problem solving and to integrate e-collaboration tools or technologies.

Collaborative Bayesian Net Editor: A graphical Bayesian network editor that allows synchronous network construction. It can be also considered as a groupware tool that enables synchronous and asynchronous construction among individuals engaged in a common edition task.

Collaborative Learning: Method of teaching and learning in which students' explore a significant question or create a meaningful project. A group of students discussing a lecture or students from different schools working together over the Internet on a shared assignment are both examples of collaborative learning.

CSCW (Computer Supported Cooperative Work): Many authors consider that CSCW and *groupware* are synonyms. On the other hand, different authors claim that while groupware refers to real computer-based systems, CSCW focuses on the study of tools and techniques of groupware.

ILE: *Intelligent learning environments* are systems which apply artificial intelligence techniques to learning environments. A learning environment, in turn, is a category of educational software where the learner's task is not to answer a predefined series of questions but to explore a complex world.

Social Agent: A cognitive agent embodied with human social skills.

Student Model: Knowledge base that keeps the user (student) information. It also represents the computer system's belief about the learner's knowledge.

Collaborative Writing in E-Learning Environments

Neide Santos
IME/DICC – Universidade do Estado do Rio de Janeiro, Brazil

Flávia Maria Santoro
DIA – Universidade Federal do Estado do Rio de Janeiro, Brazil

Marcos R. S. Borges
IM/DCC&NCE – Universidade Federal do Estado do Rio de Janeiro, Brazil

INTRODUCTION

The writing of a collaborative document can be a constructive experience if the members of a group agree in sharing points view and knowledge in benefit of the final goal. In a true collaborative environment, each contributor has an equal ability to add, edit, and remove text. The writing process becomes a recursive task, where each change prompts others to make more changes. Therefore, collaborative writing can be more than representation and organization of ideas; it can help learning process by making participants build knowledge through group interaction. For that reason, during the writing process, participants must be encouraged to interact with others, sharing knowledge and discussing the theme in such way both individual and group learning can be assured. A supplementary benefit is the production of documents "enriched by the collective knowledge," reflecting the participants contributions on the subject matter.

The collaborative writing is a four-step process:

a.   **Brainstorming:** Survey of suggestions and ideas of each participant about the theme
b.   **Planning and Organization of Ideas:** Suggestions are stored, classified and organized, serving as database to later phases
c.   **Composition:** Each participant edits a complete text or is responsible for one of its parts, using the database
d.   **Review:** Texts edited by each participant are reviewed in order to conclude document

Some approaches can be adopted for setting up the Composition and Review phases of collaborative writing tasks: Each participants writes a complete text and the group argue the ideas contained, until they find a consensus; or each participants is responsible for one part of the document, using the ideas generated by all and later a single text is composed joining the parts; or the text can all be broken up in parts assigned for the participants, being constructed at the same time for the whole group. Any form of setting up this task must adopt a politics of annotations, comments and suggestions, in such way to stimulate interaction among participants.

We observed in literature that collaborative text editors are not focused on educational activities, thus they do not incorporate functionality to make participants discuss and interact while they build texts. The priority is document visualization and change notification. Alliance (Decouchant, Enríquez, & González, 1999) assigns roles and access rights to the various fragments of documents, while in IRIS (Koch & Koch, 2000) and PENCACOLAS (González et al., 1997), all group members write on the full text.

In this article, we review the main concepts, theories and supporting tools for collaborative writing and present and discuss our tool called EdiTex. Editex supports the following activities involved in coauthoring Edition, Perception, Coordination, Interaction, and Storage. A text has an author and an identification number, besides it is composed of fragments, which can be paragraphs, phrases, sections, chapters, and these comments can be associated to each fragment.

We assessed EdiTex in two case studies and the results point out that it helped the students in the task of writing an essay collaboratively. Even so, the general results showed that group composition, task nature, context and infrastructure for communication are key points for successful collaboration in groupware. In all case studies, the tasks performed were similar; all of them involving the collective production of short texts, despite the work processes have been different. The groups' members had similar academic formation, and anyway, we observed great differences in work dynamics. Individual characteristics have strong influence; therefore the environments must stimulate, as well as exploit individualities for the success of the work.

EDITEX: A COLLABORATIVE WRITING TOOL

We developed a tool for collaborative writing called EdiText (Santoro, Borges, & Santos, 2000, 2002, 2003). It was designed and implemented regarding the problems related to co-authorship in Computer-Supported Collaborative Learning environments (Bourguin & Derycke, 2001; George & Leroux, 2001; Tiessen & Ward, 1999; Wan & Johnson, 1994).

Many projects developed in collaborative learning environments involve document co-authorship. The definition of roles can contribute to stimulate interaction, ideas sharing and knowledge construction. It is also necessary to define the rules of interdependence in the accomplishment of the task to guarantee that this will be carried through in a collaborative way.

The editing activity must be described according to following criteria:

a. **Objective:** The activity aims to the coauthoring of a document that can contain texts, graphics and figures, for a group of students, in way that, at the ending of the activity, the students have acquired knowledge on the document subject. In addition, they must put into practice the interchange of information to reach the objective.

b. **Roles:** The writing task in group can involve: coordinator, writer and publisher.

c. **Products:** The expected product is a document that can contain texts, figures, and graphics.

d. **Interdependence elements:** The final product is presented by the group and not in separate parts of each one of its members.

e. **Interdependence rules:** Each member of the group must be responsible for the document as a whole, being able to edit one or more of its parts individually. All members must contribute with

suggestions and comments on the work. Individuals can use personal experience or specific knowledge to enrich the work.

The supporting tool must consider those aspects. The solution is a list of requirements that can be implemented through the addition of functionality. The requirements deal with the following issues: Edition, Perception, Coordination, Interaction, and Storage.

a. **Edition:** The participant edits the document asynchronously, or either, each participant can work independently of the presence of other group members, at the same moment. The document must be structuralized in fragments defined by the group. Every fragment must be associated to a member of the group that is responsible for its edition. A mechanism for annotations and comments on the fragments must be available.

b. **Awareness:** All group members must have the possibility of to visualize the full document. Every group members must have the possibility to get information on his responsibilities on the fragments. The group members must receive notification about changes made in any part of the document.

c. **Coordination of the activity:** Roles must be assigned, with different responsibilities on each fragment. Roles interchanges must be possible during the task execution. All members must have the same chances to contribute in the activity.

d. **Interaction:** Messages exchange among the members of the group must be possible, aiming at discussion of ideas on the document. Discussions must be registered.

e. **Storage of the document:** Storing versions of the document must be possible. Participants must save the document when the activity is finished. Participant can access previous versions of the document with the respective annotations.

The requirements specified had been extracted from the main works in collaborative writing found in literature (Decouchant, Enriquez, & González, 1999; Koch & Koch, 1998; Wan & Johnson, 1994). EdiTex implemented the following:

a. **Edition:** Synchronous and asynchronous edition

b. **Division of the document into fragments:** Addition of annotations and comments
c. **Awareness:** Total visualization of the document and list of logged participants
d. **Communication:** Messages and chats
e. **Storage:** Menu File (Save, Open, Close)

Implementation of EdiTex

In EdiTex, a text is composed of fragments that can be of any type—paragraphs, phrases, section, and chapter. The group can define the level of granularity of the fragments. Each fragment has an author (who created it), and an identification number, and to each one, comments can be associated. The main screen of the EdiTex has three areas of text (Figure 1):

a. **Edition area:** It corresponds to the place where each participant will go to edit its fragments, before to send it for the group. It work as a private space, therefore the content that is being edited is not visible for the other participants until the author sends it, through the Send command in the Fragment Menu, for all the group.
b. **All fragments:** It corresponds to the visualization of the complete text. In this area, the fragments are visualized as soon as they are sent, thus, in determined moments it is necessary to move the fragments in order to get a coherent text.
c. **Notes:** It represents the note pad, where each participant writes comments and annotations on fragment.

Following, the tool functionalities and forms of work are detailed.

a. **Edition:** The participant edits a part of the document fragment, using the work area designed for this function. Each participant is owner of parts of the document, which are of his entire responsibility, but he is aware that each fragment will contribute with the final document. Each member can work independently in asynchronous way or in synchronous way with the other members logged at the same moment. The window for visualization of the full document is not editable. The participant must write comments on fragments edited by other participants, making suggestions.
b. **Awareness:** The tool presents a "List of Participants" that shows who is logged at that moment. This mechanism is useful because it allows participants to know who is with them to interact synchronously in the collaborative writing task.
c. **Interaction:** Communication among the participants can be made through the Chat or a mechanism for message exchange. In the Chat, participants can talk to a specific member or all the participants. Talks are stored and made available.

EdiTex was implemented under an infrastructure framework [5]. It follows client-server architecture and the data model manipulated represents the data documents, fragments and annotations.

Figure 1. EdiTex interface

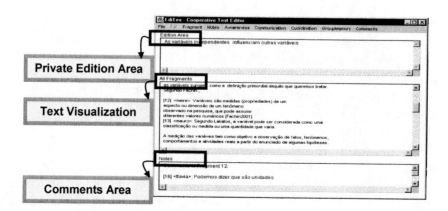

Figure 2. Server application schema

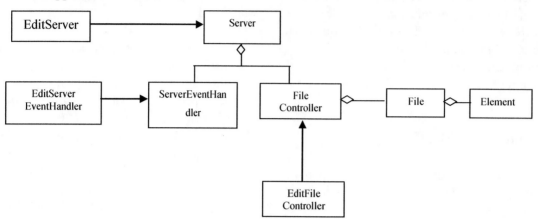

- **EdiTex: Server Application** (Figure 2): From the server framework, the classes of the ManagerServerEditex server and of the manipulation of specific events *ManipulatorEventsEditex* are specialized for the manipulation of the document process (package FILE.TEXT) and *ManagerManipulatorEventsEditex* for the manipulation of the environment events. The *ManipuladorEventsEditex* role is to manipulate the events that arrive in the server of the client application, where the session manager is responsible to notify all the participants which event was generated. The events that are handling in the server are Send Fragment and Send Commentary.

- **EdiTex: Client Application** (Figure 3): Each instance of the client application represents a participant logged in EdiTex. From the client framework, the ClientEditex class is specialized. The client application specializes the client framework in the two layers: management and interface. In the client layer, the extensions are similar to the server extensions. In the interface layer, it is necessary to extend the manager interface (ManagerGUIEditex), in order to create and to manage the adequate tool controls. The specific tool manager also needs to be specialized (ManagerGUIToolEditex). This class controls the presentation of the complete document in the area All Fragments, through a FrameInterface class. To update the interface in such way that the participant realizes the actions of the other participants, it is necessary a mechanism that allows the client layer notifies the interface on the occurred events. The events that are handling in the client are the same events handle in the server.

The tool has three controllers: ControlEdition, responsible for the text edition, as Copy, Cut, and Paste; ControlFragments, responsible for the sending of the edited fragments; and ControlComments, responsible for the comments on the fragments. The tool manager, using the interface manager, sends the event to be distributed for all the other client applications. The interface manager of EdiTex, ManagerGUIEditex, extends the Manager GUI class and is responsible for the creation of the common interface to all the integrated tools at the environment and for the activation of the ManagerGUIToolEditex that is the specific tool manager. The ManagerGUIEditex also has a helper that manages collaboration mechanisms tool (ManagerGUICooperation). This manager activates the interface controllers (GUIControl). The ManagerGUIToolEditex is responsible for the communication among the actions generated by the client and the participant interface. When the interface is created, it is added to the tool specific menus and it is generated the Editor interface, using the FrameInterface. The FrameInterface is responsible for the creation of the two text areas and of Note Pad, specializing the EdiTex interface. After the connection establishment, the collaborative activity to be developed by means of the Editor can be initiated.

EXPERIMENTAL FINDINGS WITH EdiTex

We made an experimental study with EdiTex in two case studies to evaluate the evolution of the collaborative process within the collective building of a text (Santoro, Borges, & Santos, 2003). We used the following

Figure 3. Client application schema

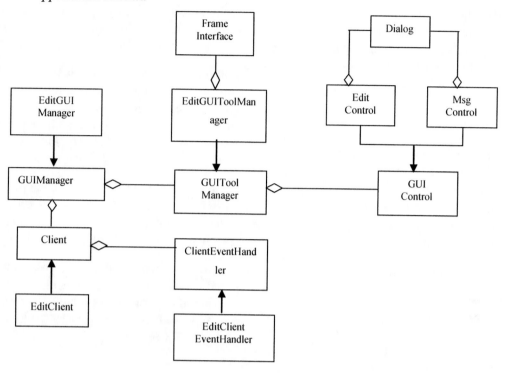

criteria: Communication, Contribution, Coordination, and Awareness, which represent the dimensions or increasing levels for the analysis of collaborative process. Four groups of students in post-graduate classes had to study a theme and write a brief report about it. All the groups used the speech as a communication resource. The writing communication induces the elaboration processes for the learning, so they are stimulated within the CSCL environments. Even so, it did not constitute a problem for the study, because all the interactions took place inside of the laboratory and the teacher, which performed also the observer, could register the most important events.

One of the interesting aspects observed is that only the fact of using a groupware tool, different from everything they had tried before, was already enough to establish collaboration state in all the groups. People are accustomed with the group work in traditional manner with division of tasks and they had to change the way they work. Another observation is the lack of experience in the definition of the work processes and more than that, the inability of evaluating the process was a constant in the groups. Thus, the use of a collaborative e-learning environment should necessarily

go by an adaptation process and first experiences will be full of uncertainties and lessons to be learned.

CONCLUSION

The general results show that the group composition, the task nature, the collaboration context and the infrastructure for communication are key points for successful collaboration in groupware. In both case studies, the tasks performed were similar; all of them involving the collective production of texts, despite the work processes have been different. The groups were composed of people with similar academic formation, and still thus, it observed great differences in work dynamics. Individual characteristics have strong influence; therefore the environments must offer conditions of stimulus, as well as to exploit the individualities for the success of the work. Now it is important to carry on our research in order to achieve as far as possible a generalization of the results achieved. A characteristic of the formative evaluation, such as the one used in our work, is that less *authentic* studies, from the point of view of a theory validation, can bring valuable preliminary results. In spite of the limitation in size of the accomplished studies, many problems were discovered and they can be applied in future experiences.

The results were very promising, clearly pointing the relationship between the functionally proposed in Editex and the level of collaboration stimulated. The tool induces a process of collaboration in which participants must perform the writing of a document looking at the others contributions and trying to mix ideas and understand group members thinking. The mechanisms implemented proved to address our main goals and configured real situations of developing a project and stimulating collaboration. Future works intend to carry on our research in order to achieve a generalization of the results achieved.

REFERENCES

Bourguin, G., & Derycke, A. (2001). *Integrating the CSCL activities into virtual campuses: Foundations of a new infrastructure for distributed collective activities.* Retrieved April 2, 2006, from http://www.ll.unimaas.nl/euro-cscl/Papers/18.pdf

Decouchant, D., Enríquez, A., & González, A. (1999). Alliance Web: Cooperative authoring on the WWW. In R. Baeza-Yates & P. Antunes (Eds.), *Proceedings of CRIWG'99 Conference* (pp. 286-295). Los Alamitos, CA: IEEE Computer Society.

George, S., & Leroux, P. (2001). *Project-based learning as a basis for a CSCL environment: An example in educational robotics.* Retrieved April 2, 2006, from http://www.ll.unimaas.nl/euro-cscl/Papers/57.pdf

González, O. M., Verdú, M. J., Dimitriadis, Y. A., Osuna, C. A., Iglesias, C. A., & López, J. (1997). PENCACOLAS: Groupware for learning. In *Proceedings of CRIWG'97 Conference* (pp. 71-80). Madrid, Spain: IEEE Computer Society.

Koch, M., & Koch, J. (2000). Application of frameworks in groupware: The Iris Group Editor environment. *ACM Computing Surveys, 32,* 28.

Santoro, F., Borges, M., & Santos, N. (2000). An infrastructure to support the development of collaborative project-based learning environments. In A. C. Salgado (Ed.), *Proceedings of the CRIWG'00 Conference* (pp. 78-85). , Madeira, Portugal: IEEE Computer Society Press.

Santoro, F. Borges, M., & Santos, N. (2002). Learning through collaborative projects: The architecture of an environment. *International Journal of Computer Applications in Technology, 16*(2/3), 127-141.

Santoro, F., Borges, M., & Santos, N. (2003). Experimental Findings with cooperative writing within a project-based environment. In M. Llamas-Nistal, M. J. Fernández-Iglesias, & L. Anido-Rifón (Eds.), *Computers and education: Towards a lifelong learning society* (pp. 179-190). London: Kluwer.

Tiessen, E., & Ward, D. (1999). Developing a *Technology of Use for Collaborative Project-Based Lear*ning. In *Proceedings of the CSCL'99 Conference* (pp. 631-639). Palo Alto, CA: Stanford University.

Wan, D., & Johnson, P. M. (1994). Computer supported collaborative learning using CLARE: The approach and experimental findings. In *Proceedings of the CSCW'94 Conference* (pp. 187-198). Chapel Hill, NC: ACM Press.

KEY TERMS

Collaborative Writing: Project where written works are created by multiple people together (collaboratively) rather than individually.

Collaborative Learning: mutual engagement of participants in a coordinated effort to solve the problem together.

Computer-Supported Collaborative Learning (CSCL): Technological support to supply collaborative learning.

Contribution: Particular view or piece of knowledge provided by a member of the group in the context of a work group.

Coordination: Set of actions that guarantee the consistency of shared objects in a collaborative environment.

Interdependence Rules: constrains that guide a group to work group in order to achive collaboration.

Perception: Act of becoming aware of other participants' work within a collaborative environment.

Collective Meaning in E-Collaborating Groups

Paul C. van Fenema
Netherlands Defense Academy, The Netherlands

Peter J. van Baalen
RSM Erasmus University, The Netherlands

INTRODUCTION

Organizations and groups have been typified as networks of shared meanings and systems of distributed knowledge (Tsoukas, 1996; Walsh & Ungson, 1991). The development of collective meaning is central to e-collaborating groups as well (Henderson, 1998). In this novel organizational form, professionals interact mostly through mediating technologies, and they may work from different sites and at different times (Kumar, van Fenema, & Von Glinow, 2005). Research has found that e-collaborating groups experience a wide variety of problems constructing and maintaining collective meaning (Cramton, 2001). Many aviation and ship navigation accidents are caused by collective meaning issues: misunderstandings, interpersonal collaboration problems, and technology-related failures (BFU, 2004; Hutchins, 1991a; NASA, 1999; Vaughan, 1996; Weick, 1993a). Current research commonly focuses on the complexity of specific cases for good reasons: large scale e-collaboration disasters take years of debate about causes of misunderstanding which have amongst others legal consequences. What is missing, however, is an inductive theory development process that sources from multiple cases and is aimed at explaining why e-collaborating groups failed to develop collective meaning.

The objective of this article is to categorize problems of developing collective meaning in e-collaborating groups, and to develop a theoretical analysis of these cases. We draw on a variety of qualitative studies from the areas of human factors, information systems, and organization studies that all focus on e-collaborating groups having difficulty to develop collective meaning. The article distinguishes problems of collective meaning in terms of expression and reflexivity. Next, an evolutionary perspective is developed that is used for analyzing these two categories. The article concludes with future trends relevant for academics and practitioners working in this area.

BACKGROUND: COLLECTIVE MEANING AS A PROBLEM OF EXPRESSION AND REFLEXIVITY

Collective meaning is defined as the socially constructed and meaningfully interrelated understandings of professionals working on the same group practice (Berger & Luckman, 1991; Schutz, 1967). When individuals develop collective meaning, they achieve a situated temporary state of mutual understanding. Collective meaning represents how group members' thinking interrelates (Weick & Roberts, 1993), usually without someone overseeing and controlling these processes (Hutchins, 1990; Van Baalen, Bloemhof-Ruwaard, & Heck, 2005). Collective meaning is situated because meaning construction is tied to particular roles, relationships, artifacts, times, and physical or information spaces (Kirsh, 1999). It is a temporary state in the sense that it can be disrupted by new experiences (Jones & Hinds, 2002; Weick, 1993a), new knowledge or by implementing new technologies (Edmondson, Bohmer, & Pisano, 2001). Collective meaning must be maintained to remain intact.

Our focus is on collective meaning in e-collaborating groups, defined as teams of professionals relying for their work communications mostly on information and communication technologies (ICT), and working usually from different locations and at different times. Examples include global teams of knowledge professionals working on products or services, emergency rescue workers, teams in the military, or aeronautics and space operations teams. These groups often struggle with the creation and maintenance of collective meaning, leading to inefficient processes, problematic outcomes, and sometimes dangerous consequences. Researchers have identified several instances of this phenomenon, relying on a case-based approach and focusing on topics such as communications, trust, and identity. Yet they have not categorized and explained these problems across multiple cases from a collective meaning point of view. We inductively identified two categories of situations where e-collaborating groups have difficulty developing collective meaning. These are characterized as problems of expression and reflexivity.

Expression. Members of e-collaborating groups openly struggle with collective meaning and they are aware of that struggle. This category concerns a lack of ability. First, examples of this category include cases where professionals involved in complex knowledge work cannot express and exchange thoughts easily using e-mail or groupware applications (Cramton, 1997; Kraut & Galegher, 1990; Malhotra et al., 2001). Professionals may lack skills and patience for crafting e-mail messages or expressing themselves effectively through videoconferencing (Egido, 1990; Nemiro, 2000). Their perspectives on tasks and resources differ, and understanding someone else's point of view is difficult (Jones & Hinds, 2002). Conflicts may arise due to incomplete understandings of a counterpart's situation, and frustration over technologies and e-collaboration processes (Armstrong & Cole, 1995; Hinds & Bailey, 2000). Second, in terms of opportunity, technologies may be unavailable, malfunctioning, or ill-suited for (multimodal) communications across distances (Weick, 1993a, 1993b). In short, professionals cannot express their thoughts well due to ability and/or opportunity constraints. This constrains the back and forth moving of ideas, the evolution of group thinking, and the creation and maintenance of joint situation awareness.

Reflexivity. In some groups, members make assumptions and interpretations that, unknown to them and others at the time, appear with hindsight to be incorrect. This gap is not resolved due to a lack of reflexivity. Compared to the first category, there is an additional layer of complexity, namely people assuming there is not a problematic dimension to their situation. We call this a problem of reflexivity, the human capability to reconsider their own and others' agency (Giddens, 1986). Individuals assume that their interpretation of a situation, of their own agency, and of their agency's relationship to others' agency is meaningful. For instance, professionals sometimes interpret rules applying to their role incorrectly or too rigidly, thereby failing to consider others (Chute, Wiener, Dunbar, & Hoang, 1995). Sometimes, professionals are sticking rigidly to their understanding of a task situation, even though they are presented with alternative perspectives

(Weick, 1993b). They may misunderstand communications but assume their interpretation is correct and that therefore further checking is considered unnecessary (Weick, 1993b). In other cases, someone makes a mistake (executing a task incorrectly, deviating from procedures, or not informing someone) but that person and the team members fail to notice (Bennett, 2000; Chute et al., 1995; Hutchins, 1991a; Jones & Hinds, 2002; Perrow, 1984). When professionals come from different units, sites, organizations, cultures, countries, and/or professional communities, they may interpret and remember the same words and symbols differently—without realizing their counterparts' point of view (Cramton, 1997; Dougherty, 1992; Meadows, 1996; Sole & Edmondson, 2002). Sometimes they use different standards for the same task (BFU, 2004; NASA, 1999; van Fenema & Simon, 2003). Under such conditions, professionals may incorrectly assume that their counterparts will understand and correct them if necessary (Snook, 2000; Weick, 2001). In many distributed groups, a coordinator liaises between subteams; for instance, air traffic controllers (Weick & Roberts, 1993), SWAT teams (Jones & Hinds, 2002), or a liaison between users in the United States and a vendor team in India (Meadows, 1996). That person may fail to receive and relay information, and bridge perspectives of each subteam (Meadows, 1996). Sometimes, the coordinator's initial understandings and plans may prove outdated, without people at a group level realizing this. Collective meaning is thus not updated (Jones & Hinds, 2002). A lack of reflexivity implies a gap between individuals' assumptions about collective meaning and the quality of collective meaning as it appears with hindsight. This wobbly basis of collective meaning may permeate a group (Bennett, 2000), remain uncorrected, and disrupts group processes. Sometimes they never get resolved, or too late for group survival. Few researchers have distinguished or offered conceptual explanations of the two categories. This has resulted in imprecise theorizing and recommendations. We address this issue after introducing an evolutionary perspective on the development of collective meaning.

COLLECTIVE MEANING: AN EVOLUTIONARY PERSPECTIVE

Collective meaning results from communication and negotiation processes (Donnellon, Gray, & Bougon,

1986; Goffman, 1983; Quinn & Dutton, 2005; Strauss, 1978). It can be conceived as the outcome of a sequence of interlocked behavior between individuals (Weick, 1979). Individuals construct collective meaning on the basis of minimal social situations (see Table 1) (Weick, 1995). Meaning construction processes hardly start from scratch; they are situated and resemble previous experiences as agency proceeds over time (Emirbayer & Mische, 1998). Hence, before elaborating on the collective meaning construction process, we must understand what a minimal social situation is. Such a situation requires, first, awareness of individuals involved. Awareness concerns answers to questions such as What is the type of professional situation we are engaged in, who is present, who pays attention (Weick & Roberts, 1993), who knows what (transactive memory, [Faraj & Sproull, 2000; Moreland, 1999]), and who has what responsibilities (work division)? Second, minimal social situations require structures for communications (Cooren, 2000; Donnellon et al., 1986; Quinn & Dutton, 2005) and redundancy of knowledge (Grant, 1996; Hutchins, 1996). And third, individuals activate and develop "expectation structures" (Luhmann, 1995), and mutual equivalence or expectation structure (Iannacci, 2003). These structures refer to an implicit contract between people that can be built and sustained without knowing necessarily the motives of another, and without people having to share goals (Donnellon et al., 1986; Weick, 1979). These three conditions result in minimal social situations that provide a certain level of predictability and an opportunity for engaging in social process and constructing collective meaning.

To understand the nature of disruptions of collective meaning and its implications for e-collaboration we distinguish three different phases of collective meaning development: emergence, habituation, and heedful rebuilding (see Table 1) (Weick et al., 2005). During the emergence phase, individuals enact a variety of perspectives as they seek with counterparts for a common approach to a collective task (Barinaga, 2002; Van Baalen et al., 2005). Habituation refers to a pattern of interaction which constitutes organizational routines. These routines can be disrupted by the implementation of new technologies (Edmondson et al., 2001). Collective meaning then needs to be rebuilt by heedful interactions between individuals (Weick & Roberts, 1993). Over the course of the evolutionary process of emergence, habituation, and heedful rebuilding, relational and cognitive distances may diminish as

people share and co-construct working relationships (Gabarro, 1990) and collective meaning (Donnellon et al., 1986). We now apply our theoretical perspective to the problems of expression and reflexivity.

EXPLAINING PROBLEMS OF COLLECTIVE MEANING: APPLYING THE EVOLUTIONARY PERSPECTIVE

The phases of collective meaning construction can be applied to the problem of expression and reflexivity (see Table 1). The first category suggests problems with the minimal social situation and the initial phase of constructing collective meaning. Individuals have no clear insight in basic questions such as who is around, who does what (Jarvenpaa et al., 1998). This characterizes oftentimes global student teams (Cramton, 2001). Moreover, during the emergence phase, the use of technology remains unadjusted and site specific, and individuals fail to develop a language for collaboration. This prohibits progress. Limited communications leave individuals functioning within their own context, rather than negotiating new structures, understandings, and processes that make their contribution useful to the group at large. The uncertainty people seem to experience in these e-collaborating groups (Cramton, 2001; Nemiro, 2000) can be explained by the unpredictability of others. Individuals do not know what to expect from others and they cannot make themselves more predictable to others. The process for creating collective meaning hardly takes off due to unresolved novelty of work practices, and low frequency and richness of (mediated) encounters. Relational and cognitive distances remain unresolved.

With the second category individuals assume a mature level of emergence, possibly even habituation (Snook, 2000; Weick, 1993b). Yet in reality they stay in the emergence phase, resulting in a gap between individual assumptions and collective reality. In other words, assumptions move faster across the phases than the underlying progress of constructing collective meaning as it becomes known with hindsight. This could be called a discrepancy of collective meaning progress. The resulting mental "bubble" of quasi collective meaning (i.e., the gap between assumed and real collective meaning) eventually bursts through reflexive feedback of others' understanding and the status of the work context providing new information. The

Table 1. An evolutionary perspective on collective meaning development in e-collaborating groups

	Precondition for collective meaning construction	Phases of collective meaning construction		
	Minimal social situation (Donnellon et al., 1986; Weick, 1995)	Emergence	Habituation	Heedful rebuilding
Problems of expression	- Ambiguity about basic organizational questions (who does what, who knows what, how does a group communicate)	- Groups get stuck in this phase - Isolated functioning of individuals; relational and cognitive distance - Unpredictability of others - Diverse use of technologies	- Lack of routine building - No minimal social situation	- N.A.
Problems of reflexivity	- Reasonable clarity of basic organizational questions	- Assumption of emergence - Discrepancy with real situation due to assumed stability, similarity	- Assumption of emergence and habituation - Discrepancy with real situation	- N.A.

mental bubble comes into existence when individuals assume that knowledge and structures for collaboration overlap. Relational and cognitive distances are downplayed and not addressed. In other words, people build on unchecked expectations with respect to others' thinking and work context. They may assume others do not make mistakes, thereby idealizing others' performances without solid grounds (Snook, 2000). Or they may start off with correct expectations, but forget to update these, assuming that situations remain the way they were (Jones & Hinds, 2002). Changes in the situation of counterparts, remaining unnoticed due to lack of communications and information processing, undermine the relevance of individual contributions (Kurland & Egan, 1999). Individuals interpret others through their own lens rather than eliciting information that could challenge that view and strengthen the embeddedness ("nesting") of their thinking in the larger group (Bigley & Roberts, 2001).

FUTURE TRENDS

The two categories of meaning problems in e-collaborating groups enable more precise theorizing and development of recommendations. Collective meaning increasingly evolves around artifacts, prototypes, and representations (Henderson, 1998; Hutchins, 1991b). These "boundary objects" show progress of thinking, and they absorb creative inputs (Bechky, 2003; Hender-son, 1991). Convergence of technologies has important implications for e-collaboration technologies: these tend to become integrated, multimodal, multisensory, and mobile. Globalization has made e-collaboration common, particularly for generations of workers currently playing around with digital games, mobile phones, e-communities, and the Internet (Friedman, 2005). The same style of interacting is permeating professional environments. "Disappearing" ambient technologies will make human-technology and human-technology-human-exchanges dominant working modes for creating value. Future research, must therefore pay attention to meaning negotiation in technology-intense environments with increasingly autonomous roles for ICT (Zuboff, 1988). Problems of collective meaning may take the form of an open struggle for mutual understanding (expression problem), or unknown discrepancy of understandings (reflexivity problem). Increasing experience with technology will reduce the occurrence of the first category of problems. Organizations will find the second type more difficult to resolve because it is caused by a distributed and diverse workforce experiencing life in different locations and cultures.

Research so far on e-collaborating groups suggests that "human factors" strongly influence success of workers. This article offers more specific insights in these factors. Innovative work processes cannot be understood in terms of information processing but rather as knowledge-intense social patterns that involve working relationships of some depth (Gabarro, 1990). As companies experience pressure to innovate and reduce cycle times, e-collaborating groups are under pressure to enhance creativity and accelerate the development of collective meaning. How groups proceed through the three phases of collective meaning must be further investigated. More research is needed to develop a fine-grained notion of these stages, and factors influencing these. A point of attention is the 'bubble' of quasi-collective meaning that plague e-collaborating groups. Avoiding or recovering from these is a top priority for high-performance groups, particularly those working under pressure or high risk.

differently educated. The complexity of knowledge work enhances the difficulty of developing collective meaning. For this situation, research should be initiated that reflects these challenges. Topics that seem worth to further explore include coordination, group development, communication, human-computer interaction, IT development and integration, and organizational structures. Research on the development of collective meaning should reflect the constituents of current work environments: decreasing cycle times, technology-intense environments, increased knowledgeability of workers, and complexity of collective tasks.

CONCLUSION

The proliferation of technologies has transformed the modern workplace from collocated hierarchically organized departments into a technology-intense, distributed environment of e-collaborating groups. Collective meaning, one of the *raisons d'être* of groups, is problematic in this environment. We propose two categories of problems plaguing these groups. First, individuals struggle openly as they try to express their thinking and interpret others' work (expression problem). Second, they experience situations of quasi-collective meaning as part of an unknown struggle for collective meaning (reflexivity problem). We reverted to organization science literature to develop inductively—based on empirical research in the area of e-collaborating groups—a more general theory of collective meaning construction. It is interpreted as a process of negotiation, building on minimal structures and common expectations, and proceeding through the following phases: emergence, habituation, and heedful rebuilding. When applied to e-collaborating groups experiencing problems of the first category, this process seems to stall early on. With respect to the second category, we suggest that individuals impose their thinking on situations. They assume stability in their counterparts' situation, and they suppose that others think like them and don't make mistakes. This results in a quasi-construction of collective meaning that usually collapses.

This categorization combined with the process model of collective meaning construction offers a powerful lens for studying e-collaborating groups. Professionals, facing high-pressure-to-perform environments must cope with the pitfalls of collective meaning as they work with group members who may be far away and

REFERENCES

Armstrong, D. J., & Cole, P. (1995). Managing distances and differences in geographically distributed work groups. In S. E. Jackson & M. N. Ruderman (Eds.), *Diversity in work teams: Research paradigms for a changing workplace* (pp. 167-186). Washington, DC: American Psychological Association.

Barinaga, E. (2002). *Leveling vagueness: A study of cultural diversity in an international project group*, Unpublished doctoral dissertation, Stockholm School of Economics, Stockholm, Sweden.

Bechky, B. A. (2003). Sharing meaning across occupational communities: The transformation of understanding on a production floor. *Organization Science, 14*(3), 312-330.

Bennett, S. (2000). *Tools of deconstruction? Understanding disaster Aetiology through cognitive theory: A case study of the Vincennes incident* (Working paper 17). University of Leicester, Scarman Centre.

Berger, P. L., & Luckman, T. (1991). *The social construction of reality: A treatise in the sociology of knowledge*. Harmondsworth, UK: Penguin Books.

BFU. (2004). *Investigation report* (Document #AX001-1/-2/02 [English version]). Braunschweig: German Federal Bureau of Aircraft Accidents Investigation. Retrieved from http://www.bfu-web.de

Bigley, G. A., & Roberts, K. H. (2001). The incident command system: High reliability organizing for complex and volatile task environments. *Academy of Management Journal, 44*(6), 1281-1299.

Chute, R. D., Wiener, E. L., Dunbar, M. G., & Hoang, V. R. (1995). *Cockpit/cabin crew performance: Recent research.* Paper presented at the 48th International Air Safety Seminar, Seattle, WA.

Cooren, F. (2000). *The organizing property of communication.* Philadelphia: John Benjamins.

Cramton, C. D. (1997). *Information problems in dispersed teams.* Paper presented at the Annual Meeting of the Academy of Management (Best Papers Proceedings), Boston.

Cramton, C. D. (2001). The mutual knowledge problem and its consequences for dispersed collaboration. *Organization Science, 12*(3), 346-371.

Crowston, K. (1996). *A taxonomy of organizational dependencies and coordination mechanisms.* Unpublished manuscript, Syracuse University School of Information Studies.

Donnellon, A., Gray, B., & Bougon, M. G. (1986). Communication, meaning, and organized action. *Administrative Science Quarterly, 31*(1), 43-55.

Dougherty, D. (1992). Interpretive barriers to successful product innovation in large firms. *Organization Science, 3*(2), 179-202.

Edmondson, A. C., Bohmer, R. M., & Pisano, G. P. (2001). Disrupted routines: Team learning and new technology implementation in hospitals. *Administrative Science Quarterly, 46*(4), 685-716.

Egido, C. (1990). Teleconferencing as a technology to support cooperative work: Its possibilities and limitations. In J. Galegher, R. E. Kraut, & C. Egido (Eds.), *Intellectual teamwork: Social and technological foundations of cooperative work.* Hillsdale, NJ: Erlbaum.

Emirbayer, M., & Mische, A. (1998). What is agency? *American Journal of Sociology, 103*(4), 962-1023.

Faraj, S., & Sproull, L. (2000). Coordinating expertise in software development teams. *Management Science, 46*(12), 1154-1568.

Friedman, T. L. (2005). *The world is flat: A brief history of the twenty-first century.* New York: Farrar, Straus & Giroux.

Gabarro, J. J. (1990). The development of working relationships. In J. Galegher, R. E. Kraut, & C. Egido (Eds.), *Intellectual teamwork: Social and technological foundations of cooperative work* (pp. 79-110). Hillsdale, NJ: Erlbaum.

Giddens, A. (1986). *The constitution of society: Outline of the theory of structuration.* Berkeley: University of California Press.

Goffman, E. (1983). The interaction order. *American Sociological Review, 48*, 1-17.

Grant, R. M. (1996, Winter). Toward a knowledge-based theory of the firm. *Strategic Management Journal, 17*, 109-122.

Henderson, K. (1991). Flexible sketches and inflexible data bases: Visual communication, conscription devices, and boundary objects in design engineering. *Science, Technology and Human Values, 16*, 448-473.

Henderson, K. (1998). *On line and on paper: Visual representations, visual culture and computer graphics in design engineering.* Cambridge, MA: MIT Press.

Hinds, P. J., & Bailey, D. E. (2000). *Virtual teams: Anticipating the impact of virtuality on team process and performance.* Paper presented at the Annual Meeting of the Academy of Management (Best Papers Proceedings), Toronto, Canada.

Hutchins, E. (1990). The technology of team navigation. In J. Galegher, R. E. Kraut, & C. Egido (Eds.), *Intellectual teamwork: Social and technological foundations of cooperative work.* Hillsdale, NJ: Erlbaum.

Hutchins, E. (1991a). Organizing work by adaptation. *Organization Science, 2*(1), 14-39.

Hutchins, E. (1991b). The social organization of distributed cognition. In L. B. Resnick, J. M. Levine & S. D. Teasley (Eds.), *Perspectives on socially shared cognition* (pp. 283-307). Washington, DC: American Psychological Association.

Hutchins, E. (1996). *Cognition in the wild* (2nd ed.). Cambridge, MA: MIT Press.

Iannacci, F. (2003). The Linux managing model. *First Monday Internet Journal, 8*(12). Available online from

http://www.firstmonday.org/issues/issue8_12/iannacci/#i3

Jarvenpaa, S. L., Knoll, K., & Leidner, D. E. (1998). Is anybody out there? Antecedents of trust in global virtual teams. *Journal of MIS, 14*(4), 29-64.

Jones, H. L., & Hinds, P. (2002). *Extreme work teams: Using swat teams as a model for coordinating distributed robots.* Paper presented at the CSCW 2002, New Orleans, LA.

Kirsh, D. (1999). *Distributed cognition, coordination and environment design.* Paper presented at the European Cognitive Science Society. Available online from http://cogsci.ucsd.edu/~kirsh

Kraut, R. E., & Galegher, J. (1990). Patterns of contact and communication in scientific research collaboration. In J. Galegher, R. E. Kraut, & C. Egido (Eds.), *Intellectual teamwork: Social and technological foundations of cooperative work* (pp. 149-173). Hillsdale, NJ: Erlbaum. Kumar, K., van Fenema, P. C., & Von Glinow, M. A. (2005). Intense collaboration in globally distributed work teams: Evolving patterns of dependencies and coordination. In D. L. Shapiro, M. A. Von Glinow, & J. L. C. Cheng (Eds.), *Managing multinational teams: Global perspectives.* Oxford: Elsevier/ JAI.

Kurland, N. B., & Egan, T. D. (1999). Telecommuting: Justice and control in the virtual organization. *Organization Science, 10*(4), 500-513.

Luhmann, N. (1995). *Social systems.* Stanford, CA: Stanford University Press.

Malhotra, A., Majchrzak, A., Carman, R., & Lott, V. (2001). Radical innovation without collocation: A case study at Boeing-Rocketdyne. *MIS Quarterly, 25*(2), 229-249.

Meadows, C. J. (1996). *Globework: Creating technology with international teams.* Unpublished master's thesis, Harvard University, Boston.

Moreland, R. L. (1999). Transactive memory: Learning who knows what in work groups and organizations. In L. Thompson, D. Messick, & J. Levine (Eds.), *Shared cognition in organizations: The management of knowledge* (pp. 3-31). Mahwah, NJ: Erlbaum.

NASA. (1999). *Mishap investigation board, mars climate orbiter, phase I report.* Retrieved November 1999, from ftp://ftp.hq.nasa.gov/pub/pao/reports/1999/MCO_report.pdf

Nemiro, J. E. (2000). The glue that binds creative virtual teams. In Y. Malhotra (Ed.), *Knowledge management and virtual organizations* (pp. 101-123). Hershey, PA: Idea Group.

Perrow, C. (1984). *Normal accidents.* New York: Basic Books.

Quinn, R. W., & Dutton, J. E. (2005). Coordination as energy-in-conversation. *Academy of Management Review, 30*(1), 36-57.

Schutz, A. (1967). *Phenomenology of the social world.* Evanston, IL: Northwestern University Press.

Snook, S. A. (2000). *Friendly fire: The accidental shootdown of U.S. Black hawks over northern Iraq.* Princeton, NJ: Princeton University Press.

Sole, D. L., & Edmondson, A. C. (2002). Bridging knowledge gaps: Learning in geographically dispersed crossfunctional teams. In N. Bontis & C. W. Choo (Eds.), *Strategic management of intellectual capital and organizational knowledge.* Oxford, UK: Oxford University Press.

Strauss, A. L. (1978). *Negotiations: Contexts, processes, and social order.* San Francisco: Jossey-Bass.

Tsoukas, H. (1996, Winter). The firm as a distributed knowledge system: A constructionist approach. *Strategic Management Journal, 17*, 77-91.

Van Baalen, P. J., Bloemhof-Ruwaard, J., & Heck, E. (2005). Knowledge sharing in an emerging network of practice: The role of a knowledge portal. *European Management Journal, 23*(3), 300-314.

van Fenema, P. C., & Simon, S. J. (2003). *Failure of distributed man-machine collaboration: An initial analysis of the midair collision over Germany on July 1, 2002.* Paper presented at the Ninth European Conference on Cognitive Science Approaches to Process Control (2003), Amsterdam (available through ACM Digital Library).

Vaughan, D. (1996). *The challenger launch decision: Risky technology, culture, and deviance at NASA.* Chicago: University of Chicago Press.

Walsh, J. P., & Ungson, G. R. (1991). Organizational memory. *Academy of Management Review, 16*, 57-91.

Weick, K. E. (1979). *The social psychology of organizing*. Reading, MA: Addison-Wesley.

Weick, K. E. (1993a). The collapse of sensemaking in organizations: The Mann Gulch disaster. *Administrative Science Quarterly, 38*, 628-652.

Weick, K. E. (1993b). The vulnerable system: An analysis of the Tenerife air disaster. In K. H. Roberts (Ed.), *New challenges to understanding organizations*. New York: Macmillan.

Weick, K. E. (1995). *Sensemaking in organizations*. Thousand Oaks, CA: Sage.

Weick, K. E. (2001, March). Friendly fire: The accidental shootdown of U.S. Black hawks over northern Iraq [Review of the book]. *Administrative Science Quarterly*, 158.

Weick, K. E., & Roberts, K. (1993). Collective mind in organizations: Heedful interrelating on flight decks. *Administrative Science Quarterly, 38*, 357-381.

Weick, K. E., Sutcliffe, K. M., & Obstfeld, D. (2005). Organizing and the process of sensemaking. *Organization Science, 16*(4), 409-421.

Zuboff, S. (1988). *In the age of the smart machine: The future of work and power*. New York: Basic Books.

KEY TERMS

Collective Meaning: The situated, temporary state of mutual understanding between people who share an experience, history or practice.

Collective Task: A task that requires contributions of more than one individual and must be accomplished in a limited period of time.

Coordination: The management of task dependencies (Crowston, 1996), implying the meaningful interrelating of activities.

E-Collaborating Group: Individuals charged with a collective task for which they rely on mostly on information technology tools.

Group: A number of individuals responsible for the joint completion of a task that exceeds their individual capabilities.

Mediating Technologies: Information processing artifacts that support individual and collective work processes.

Negotiated Order: Coherence of understandings and actions resulting from a process of mutual adjustment and interpreted as a process or outcome.

Concurrency Control in Real–Time E–Collaboration Systems

Wenbing Zhao
Cleveland State University, USA

INTRODUCTION

E-collaboration refers to the collaboration of a group of people sharing the same task, using electronic technologies (Kock & Nosek, 2005). In the Internet age, the interactions and communications among the collaborators, the management of related information, the recording of the progress, and the outcome of the task are primarily facilitated by modern software systems, often called e-collaboration systems or groupware. Such systems range from those enabling loosely coupled, asynchronous collaborations, such as e-mail and software source version control systems, to those supporting tightly coupled, synchronous (also termed as *real-time*) collaborations, such as group editors, e-classroom, and group-decision systems.

For all e-collaboration systems, some degree of concurrency control is needed so that two people do not step on each other's foot. The demand for good concurrency control is especially high for the tightly coupled, real-time e-collaboration systems. Such systems require quick responses to user's actions, and typically require a WYSIWIS (what you see is what I see) graphical user interface (Ellis, Gibbs, & Rein, 1991). This requirement, together with the fact that users are often separated geographically across wide-area networks, favors a decentralized system design where the system state is replicated at each user's site. This places further challenges on the design of concurrency control for these systems.

Furthermore, the users of e-collaboration systems often follow a social protocol during their work on the common task (Ellis et al., 1991). For example, if a number of users are collaborating on a shared document, one would not edit a paragraph if it is clear that another user is editing it. This is very different from other type of concurrent systems, such as database systems, where users' actions are completely independent and isolated. Therefore, real-time e-collaboration systems often favor an optimistic concurrency control approach.

There has been intense research on this issue in the past two decades or so. There are primarily two approaches: (1) locking-based, which uses locks to synchronize different users on access to a shared document (Greeberg & Marwood, 1994); (2) serialization based, which ensures a consistent order of operations on a shared document for all users. Both are initially derived from the concurrency control practices in database systems and have been significantly extended over the years. The most dominant approach appears to be the optimistic serialization based on operational transformation (Ellis & Gibbs, 1989).

BACKGROUND

Event Ordering

Assuming that several people are working together on a shared document using an e-collaboration system, a user might decide to insert or delete a character in a particular position in the shared document. Such an insertion or deletion will need to be propagated from the user who performed the operation to all other users. Consequently, we distinguish the *operation generation* (from a particular site by a user) from *operation execution* (locally or remotely). Depending on the concurrency control used, the local execution might not be identical to the remote execution.

To perform concurrency control, we often need to establish the order of the operations in the system. To determine such an order, a logical lock (Lamport, 1978) can be used to assign a logical timestamp for each operation. Following the notion of the *happened-before* relationship (see the terminology section for definition), a partial ordering, termed as *precedence property*, can be established for all operations in e-collaboration systems. Given two operations o1 and o2, o1 is said to precede o2 if and only if the execution of o1 at site S happened before the generation of o2 at S (Ellis & Gibbs, 1989).

Figure 1. Relationship between different operations

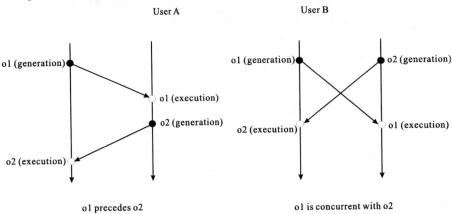

As shown in Figure 1, some operations have the precedence relationship, while some do not. The events that cannot be associated with a precedence relationship are called concurrent operations. Two concurrent operations are said to *conflict* with each other, if both operate on the same object. To further distinguish the order of concurrent operations, we might need to determine a *total order* of all operations.

CONSISTENCY AND CONCURRENCY CONTROL

Concurrency control is the activity of coordinating concurrent operations that might be conflicting with each other. Concurrency control determines whether to allow, postpone, or deny the execution of a user's operation, depending on the relative ordering of the operation under consideration with respect to other operations, and the object on which the operation applies. If an optimistic concurrency control is used, it may also include the tasks such as undo (roll back a user's change) and operation transformation if a conflict of two or more operations occur.

The purpose of concurrency control is to maintain the *consistency* of the system state replicated at different sites, at the end of each operation execution, or after each conflict resolution. It is easy to see why some form of control is needed to preserve the consistency of the replicas in the presence of concurrent activities. If two conflicting operations are applied at user A and user B in different order, the replica at user A might be different from the replica at user B. If this happens, the state is said to have *diverged*.

In addition to the consistency requirement, concurrency control should also preserve users' intent. This requirement is unique in e-collaboration systems. To understand this subject better, consider the following example. Two users A and B are working on a shared document using a group editor. Assume that the initial document (i.e., the system state) contains a string "collaboration system." User A wants to insert a character "e" and a hyphen "-" right before the word "collaboration". Concurrently, user B wants to insert a character "s" at the end of the word "system." If both users' intent is preserved, the concurrency control should ensure the final state at both user A's and B's sites to be "e-collaboration systems".

Locking-Based Concurrency Control

Locking-based concurrency control (Greeberg & Marwood, 1994), as the name suggests, uses a lock to control the order in which a shared object is updated. A lock is a software construct that guards the access of a shared object. To access the object, one must *acquire* the lock first. After the update is completed, the lock must be *released*. Subsequently, the lock can be granted to the next user waiting in line. Since a lock can be granted to only one user at a time, it ensures sequential access of the shared object.

Pessimistic Locking

In early e-collaboration systems, pessimistic locking is often used as a way of concurrency control. In pessimistic locking, the request-for-lock operation is blocking, *i.e.*, one must wait until the lock is granted to access the object.

Recall that e-collaboration systems are often implemented as distributed systems with full replication. Before a lock is granted, the system must ensure that no one else holds the lock for the associated object, and everyone is notified that the lock is going to be granted to a particular user. This can incur substantial overhead. Even if the lock is eventually granted, the response time is not negligible.

This obvious drawback can be remedied by a number of ways. For example, the user who could not be granted a lock immediately should not be blocked, but rather, a "permission denied" notice should be conveyed to the user. One can also envision that the objects whose locks have been granted to other users are marked with appropriate indicators so that the user does not ask for those locks in the first place.

Pessimistic locking is not without any merit. If the conflict resolution is too computationally expensive, pessimistic locking might be the desirable method to be used. Furthermore, some e-collaboration systems, such as video conferencing, require one active speaker at a time, and therefore fit the pessimistic locking well.

Optimistic Locking

In e-collaboration systems, users are normally conscious of what other users are working on and tend to avoid conflicting operations voluntarily. This observation suggests that a more optimistic locking strategy can be used to minimize the locking response time. The idea is that the system should give a *tentative approval* immediately to the user who requested a lock after the request is received. The user can therefore apply changes to the local copy of the object without substantial waiting time. In the mean time, the system executes the locking mechanism until the lock can be officially granted or denied. If eventually the lock-request should be denied, the user's change on the object is discarded and the original object state is restored.

There are two flavors of optimistic locking depending on if more locks can be requested while the user is waiting for the official outcome of the first lock request. In fully-optimistic locking, the user is allowed to request more than one lock, while in semi-optimistic locking the user is not allowed to do so.

To support fully-optimistic locking, the system must remember all the tentative operations while waiting for

the lock-request decisions. If one of the lock requests is denied, all objects touched by the user since the request for that lock must be restored to their original state. If semi-optimistic locking is used, only a one-step undo is needed.

Serialization-Based Concurrency Control

Concurrency control can also be done through enforcing a total order on all operations in an e-collaboration system. Similar to the locking-based concurrency control, serialization-based concurrency control can be further classified into pessimistic and optimistic approaches depending on if an out-of-order operation is ever allowed to happen (Greenberg & Marwood, 1994).

Pessimistic Serialization

In pessimistic serialization, an operation is not allowed to take place before the order of this operation with respect to other operations is determined by the system to avoid conflict. In an e-collaboration system with full replication, the order establishment must be done through message passing, and it must be done for every operation carried out by every user. This is rather similar to pessimistic locking. However, pessimistic serialization can be more restrictive than pessimistic locking because pessimistic locking allows concurrent activities if they are not conflicting (e.g., updates on different objects).

To implement pessimistic serialization, a reliable, totally ordered multicast protocol such as Totem (Moser et al., 1996) can be used. However, this approach works well only if a small number of users collaborate over a local-area network.

Optimistic Serialization

Optimistic serialization allows potentially conflicting operations to happen locally at each user's site and resorts to a conflicting resolution algorithm to address any conflicts. If the conflicts can be resolved properly, effectively a serialized execution according to some total order is achieved. There are two well-known conflicting resolution algorithms in optimistic serialization.

- **Serialization by undo/redo:** When a conflict is detected, the object is brought back to the last consistent state by undoing recently applied

Figure 2. Concurrent execution with and without operational transformation

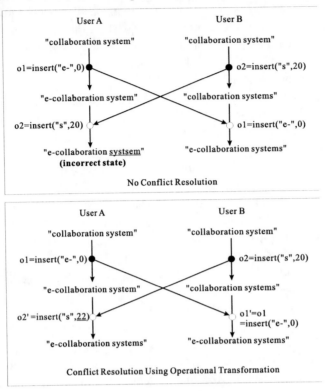

operations, and those operations are reapplied to the consistent state with redo in the correct order (Ignat & Norrie, 2004).

- **Serialization by operational transformation (OT):** A conflicting operation is transformed before it is applied to an object, as if the operation is executed in the correct order (Ellis & Gibbs, 1989).

The former approach has the advantage of allowing an application-independent implementation. However, the total order it imposes might not always match the user's intent. For example, if users A and B concurrently modified the same paragraph, this approach would either undo A's or undo B's changes. In effect, one of the users' changes is discarded. This does not follow the user's intent, because a merge of the changes from both users is normally expected.

The latter approach requires a carefully specified set of rules for the transformation. The rules typically are application-dependent and can be very complicated. To illustrate how the transformation can be done, consider the group editor example introduced in the background section again. At that time, we specified what should

be the correct outcome, without providing a concrete solution.

As shown in Figure 2, user A's operation is represented as o1=insert("e-", 0). The first parameter "e-" indicates the characters inserted. The second parameter indicates the position of the insertion, which is the beginning of the string. User B's insertion operation is represented as o2=insert("s", 20). Obviously, o1 conflicts with o2. Recall that the state is replicated, both users apply their changes immediately to their local copy. Without transformation, the state at user A's site would become "e-collaboration systsem" after executing o2. Apparently, user B's intent is violated. The problem is that the state referenced by o2 is no longer there when o2 is about to be executed at user B's site. We need to transform o2 with respect to all the concurrent operations that have executed locally since the last consistent state. In this case, user A has inserted two characters (in o1) before user B's insertion position. Consequently, we need to transform o2 from insert("s", 20) to insert("s", 22) before we apply it at user A's site.

The OT-based approach provides very fast response time and high concurrency because an operation is always executed immediately. Due to the use of application-dependent rules for transformation, the user's intent can often be preserved as well.

FUTURE TRENDS

The OT-based approach is perhaps one of the most viable concurrency control strategies in e-collaboration systems. It was originally used for a text-based group editor (Ellis & Gibbs, 1989). It has since been refined and extended (Li & Li, 2004; Ressel, Nitsche-Ruhland, & Gunzenhauser, 1996; Sun & Ellis, 1998), and incorporated in many other real-time e-collaboration systems, such as table-based (Palmer & Cormack, 1998), graphical-based (Chen & Sun, 1999), XML-based (Davis, Sun, & Lu, 2002) group editors, and collaborative multimedia authoring (Xiao, 2002).

As we mentioned before, the OT-based approach requires a set of rules for transforming conflicting operations. However, the design and verification of such rule set is often difficult. The correctness of the transformation rule set directly determines the validity of the OT-based concurrency control. A landmark-based transformation approach (Li & Li, 2005) is recently introduced to partially address this issue.

Because no one method will fit all e-collaboration systems, it is desirable to offer a suite of concurrency control algorithms to an e-collaboration system so that the best algorithm can be used for each task. To further increase the flexibility of concurrency control, a plug-in interface should be provided by the system for applications to provide their own if they desire. Yang and Li (2004) have recently proposed such an adaptive concurrency control framework.

Another open issue is related to the security and dependability of e-collaboration systems. Recall that replication is often used in e-collaboration systems for fault tolerance and fast response. In the current model, it is assumed that networks and the processes may be subject to benign fault only. However, this is no longer the case for e-collaboration systems that operate over the Internet and potentially across different administrative domains. We believe that there is an urgent need to extend the fault model to including arbitrary, potentially malicious faults. The change of the basic fault model will demand significant changes of existing concurrency control methods.

CONCLUSION

Concurrency control is an essential building block of e-collaboration systems. We provided a simple taxonomy of existing concurrency control strategies. They are locking-based and serialization-based concurrency control. Each can be further classified as pessimistic and optimistic approaches. For many real-time e-collaboration systems, the demand for quick response, the use of social protocols, and the tolerance of errors, favor an optimistic approach to concurrency control.

REFERENCES

Chen, D., & Sun, C. (1999). A distributed algorithm for graphic objects replication in real-time group editors. In *Proceedings of the ACM Conference on Supporting Group Work* (pp. 121-130). Phoenix, AZ: The Association for Computing Machinery.

Davis, A., Sun, C., & Lu, J. (2002). Generalizing operational transformation to the standard general markup language. In *Proceedings of the ACM conference on Computer Supported Cooperative Work,* (pp. 58/67), New Orleans, LA: The Association for Computing Machinery.

Ellis, C., & Gibbs, S. (1989). Concurrency control in groupware systems. In *Proceedings of the ACM SIGMOD Conference on Management of Data* (pp. 399-407). Portland, OR: The Association for Computing Machinery.

Ellis, C., Gibbs, S., & Rein, G. (1991). Groupware: Some issues and experiences. *Communications of the ACM, 34*(1), 39-58.

Greenberg, S., & Marwood, D. (1994). Real-time groupware as a distributed system: Concurrency control and its effect on the interface. In *Proceedings of the ACM Conference on Computer Supported Cooperative Work* (pp.; 207-217). Chapel Hill, NC: The Association for Computing Machinery.

Ignat, C., & Norrie, M. (2004). Grouping in collaborative graphical editors. In *Proceedings of the ACM Conference on Computer Supported Cooperative Work* (pp. 447-456). Chicago: The Association for Computing Machinery.

Kock, N., & Nosek, J. (2005). Expanding the boundaries of e-collaboration. *IEEE Transactions on Professional Communication, 48*(1), 1-9.

Lamport, L. (1978). Time, clocks, and the ordering of events in a distributed system. *Communications of the ACM, 21*(7), 558-565.

Li, D., & Li, R. (2004). Preserving operation effects relation in group editors. In *Proceedings of ACM conference on Computer Supported Cooperative Work* (pp. 457-466). Chicago: The Association for Computing Machinery.

Li, R. & Li, D. (2005). A landmark-based transformation approach to concurrency control in group editors. In *Proceedings of the International ACM SIGGROUP Conference on Supporting Group Work* (pp. 284-293). Sanibel Island, FL: The Association for Computing Machinery.

Moser, L., Melliar-Smith, P., Agarwal, D., Budhia, R., & Lingley-Papadopoulos, C. (1996). Totem: A fault-tolerant multicast group communication system. *Communications of ACM, 39*(4), 54-63.

Palmer, C., & Cormack, G. (1998). Operation transforms for a distributed shared spreadsheet. In *Proceedings of the ACM Conference on Computer Supported Cooperative Work* (pp. 69-78). Seattle, WA: The Association for Computing Machinery.

Ressel, M., Nitsche-Ruhland, D., Gunzenhauser, R. (1996). An integrating, transformation-oriented approach to concurrency control and undo in group editors. In *Proceedings of ACM Conference on Computer Supported Cooperative Work* (pp. 288-297). Cambridge, MA: The Association for Computing Machinery.

Sun, C., & Ellis, C. (1998). Operational transformation in real-time group editors: Issues, algorithms, and achievements. In *Proceedings of the ACM Conference on Computer Supported Cooperative Work* (pp. 59-68). Seattle, WA: The Association for Computing Machinery.

Xiao, B. (2002). Collaborative multimedia authoring: Scenarios and consistency maintenance. In *Proceedings of International Workshop on Collaborative Editing*, New Orleans, LA. Retrieved from http://dsonline.computer.org/portal/cms_docs_dsonline/dsonline/topics/collaborative/events/iwces-4/Xiao.paper.pdf

Yang, Y., & Li, D. (2004). Dynamic architectures: Separating data and control: Support for adaptable consistency protocols in collaborative systems. In *Proceedings of ACM conference on Computer Supported Cooperative Work* (pp. 11-20). Chicago: The Association for Computing Machinery.

KEY TERMS

Concurrency Control: The activity of coordinating concurrent operations that might be conflicting with each other. Concurrency control determines whether to allow, alter, postpone, or deny the execution of a user's operation, depending on the relative ordering of the operations under consideration, and the objects on which the operations apply.

Conflicting Operations: Two concurrent operations are said to conflict with each other if both operate on the same object. The granularity of the object is determined by the concurrency control strategy.

Consistency: All copies of replicated system state are identical to each other at each synchronization point. In e-collaboration systems, consistency also implies the preservation of user's intent. Consistency is achieved through concurrency control.

Group Editor: A type of e-collaboration system allowing multiple geographically distributed members of a team to work on the same document collaboratively in an interactive session.

Happened Before: Given two events, e1 and e2, we say e1 happens before e2, or e1 → e2, if one of the following conditions holds: (1) e1 and e2 happen in the same process, and e1 comes before e2; (2) if e1 is the sending of a message, and e2 is the receipt of the message. The happened-before relationship is transitive; for example, if e1 → e2 and e2 → e3, then e1 → e3.

Locking (Pessimistic, Optimistic): A software lock is associated with each shared object. One must acquire a lock before updating a shared object, and must release the lock once the update is completed. In pessimistic locking, a user must wait until the lock is granted explicitly before updating an object. In optimistic locking, the user can access the shared object as soon as a tentative approval is granted. If the lock request is eventually denied in optimistic locking, the object state must be restored.

C

Operational Transformation: When an out-of-order operation is detected, it is transformed before being applied to the system state to preserve consistency. The transformation requires a set of application-dependent rules.

Precedence Property: Given two operations o1 and o2, o1 is said to precede o2 if and only if the execution of o1 at site *S* happened before the generation of o2 at site *S*. The precedence property establishes a partial ordering on different operations.

Serialization (Pessimistic, Optimistic): Serialization relies on establishing a total order of all operations in the system to preserve consistency. In pessimistic serialization, all operations must be executed according to the total order. In optimistic serialization, an operation is executed immediately at the local site and then propagated to remote sites. If an out-of-order operation is detected, either the system rolls back to a consistent state and redoes with the correct total order, or the out-of-order operations are transformed so that an equivalent total order can be realized.

Consequences of IM on Presence Awareness and Interruptions

Jesus Carmona
Texas A&M International University, USA

INTRODUCTION

Technology has changed the way we communicate in the workplace; new and improved computer-mediated communication tools are available for our use, and media choice has become an issue (Cameron & Webster, 2005). Nowadays it is hard to decide what communication tool to use or how we convey messages when using certain media (Trevino, Daft, & Lengel, 1990).

Instant messaging (IM) is a computer-mediated tool that is used to send and receive text messages in a synchronous manner using the Internet. IM has become a common channel of communication between family members and friends (Goldsborough, 2001); almost 53 million adult Americans trade instant messages, and 24% of them swap IM more frequently than e-mail (Shiu & Lenhart, 2004). After seeing this tool's usefulness, managers are beginning to introduce it in the workplace as an informal way to communicate; at the same time, IM seems to bring unintended (though not necessarily negative) consequences like presence awareness (Cameron & Webster, 2005) and interruptions (Rennecker & Godwin, 2005).

Various theoretical frameworks have been used to study IM, mostly in the fields of communications and electronic monitoring (Cameron & Webster, 2005), many of which utilize qualitative methods. Very few empirical studies are published in this area, and those available are written by IM vendors or IM developers using colleagues as their main subjects of study (Cameron & Webster, 2005).

This article studies IM's effects on interruptions and presence awareness, as well as the effect presence awareness has on interruptions. For the statistical analysis a subset of 111 elements of the February 2004 PEW Internet and American Life surveys dataset was used as a sample. PLS Graph software was used to create the structural model and test the relationships between the constructs Interruptions (INT), Presence Awareness (AWA), and use of IM in the workplace (IMW).

BACKGROUND AND HYPOTHESES

The main tool used by managers to do their work is communication. Theories in the communications field suggest that media is as important as the conveyance of the message (Trevino et al., 1990). In other words, the content of the message is as important as the medium used to deliver it. The Symbolic Interactionist perspective has been utilized to explain symbolic cues conveyed by different media; for example, an official e-mail may imply formality while the use of IM may convey urgency but informality (Trevino et al., 1990).

Presence awareness is the ability to see who is online at specific times. Most IM systems display a list of users connected to the network (or Internet, depending on the IM software). This list helps conversation initiators judge if recipients are available for conversation (Nardi, Whittaker, & Bradner, 2000); most IM systems are able to post an "away" or "busy" message to let others know the IM user's status, reducing at the same time the interruption level by allowing recipients to negotiate availability. To compensate for privacy concerns, IM systems are also capable of blocking users from the list in order to "hide" from them; this gives the user complete control over who sees him/her as "online" (Cameron & Webster, 2005). IM users sometimes send short messages (e.g., "hello?") to initiate a conversation, and it is up to the recipient to either answer or wait for a more appropriate time. Privacy concerns are important for people, and even though presence awareness can be considered invasive, most users found the IM monitoring system less invasive than video cameras (Zweig & Webster, 2002).

Another important factor in presence awareness is the sense of social connection; experiencing connection with other people makes users feel socially engaged and gives them the confidence to know that somebody is available (Nardi et al., 2000). In general, presence awareness is an IM consequence that both the initiator and the recipient can see as beneficial because it gives both the ability to consent communication without the hassle of face-to-face negotiation. From the previous discussion, the following hypothesis can be inferred:

- *H1:* The use of IM in the workplace will have a positive effect on presence awareness.

O'Conaill and Frohlich (1995) define interruption as "a synchronous interaction which is not initiated by the recipient, is unscheduled, and results in the recipient discontinuing their current activity" (p. 262). Interruption does not necessarily mean disruption, but even the notification of an incoming message can cause interruption, which may or may not negatively affect performance (Cutrell, Czerwinski, & Horvitz, 2001). It has been hypothesized that interruptions derail the flow of activities directed toward accomplishing a task and delays can contribute to work disorganization when a worker is unable to move forward with a task due to insufficient information (Rennecker & Godwin, 2005). Consequently one can expect users, especially recipients, to perceive IM as interruptive. This leads to a second hypothesis:

- *H2:* The use of IM in the workplace will have a positive effect on interruptions.

Managers introduce the use of IM in the workplace as means to communicate; even though IM is considered informal, the sense of urgency has made IM the medium of choice when information is needed to complete a task (Nardi et al., 2000). One way in which a task can be delayed is by not having the information needed; a common way to obtain this information from a co-

worker or supervisor is to contact the person using either an asynchronous medium (e.g., e-mail) or a synchronous medium (face-to-face, telephone, IM) and each communication method would have its own advantages and disadvantages. With the former method, more delay can be experienced because of the nature of the medium. If the latter method is used and if the recipient is available, a faster response can be guaranteed (Rennecker & Godwin, 2005). IM is a method of choice because thanks to the presence awareness, a negotiation is possible between the initiator and the respondent. First, by taking a quick look at the "online" list a person can tell if a user is available; second, from the status on the list, one can determine if the user is "busy" or "away"; and third, a quick "are you there?" message can ensure that the user is available and ready to communicate (Nardi et al., 2000). This discussion leads to a third hypothesis:

- *H3:* Presence awareness will have no effect on interruptions

Since the hypotheses refer to a set of causal links involving three constructs, a representation can be provided in the form of a structural model (see Figure 1).

RESEARCH METHOD

The method employed for data analysis in this study was partial least squares (PLS), an alternative structural equation modeling technique (Chin, 1998). PLS was implemented through the PLS-Graph software V.03.00 (Chin, Marcolin, & Newsted, 1996; Chin, 1998). A sub-

Figure 1. Hypothetical model and hypotheses

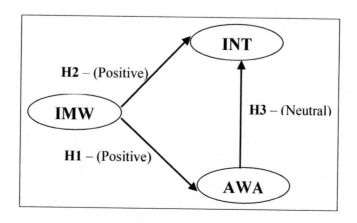

Table 1. Item loadings per construct

Construct	Item	Item loading
Interruption	INT1	0.7617
	INT2	0.8930
	INT3	0.8041
Presence Awareness	AWA1	0.7974
	AWA2	0.7989

Table 2. Reliability, convergent, and discriminant validity coefficients

	CR	AVE	INT	AWA
INT	0.861	0.675	**0.822**	
AWA	0.778	0.637	0.167	**0.798**
CR- Composite Reliability; AVE – Average Variance Extracted				
Diagonal elements are the square root of AVE. Off diagonal element is the correlation between constructs.				

set of 111 elements of the February 2004 PEW Internet and American Life surveys (http://www.pewinternet. org) dataset was used as a sample for the statistical analysis; this subset contained only respondents who acknowledge using IM in the workplace. This PEW Internet and American Life Project dataset was gathered through telephone interviews conducted by Princeton Survey Research Associates between February 3 and March 1 of 2004, from a sample of 2,204 adults, aged 18 and older.

The measurement model in PLS is assessed in terms of item loadings and reliability coefficients as well as the convergent and discriminant validity. Individual item loadings on each construct, greater than 0.7, are considered as adequate values (Fornell & Larcker, 1981). Reliability is measured through the composite reliability estimate, and a value of 0.7 or greater for each construct is considered acceptable. To justify using a construct (convergent validity), the average variance extracted (AVE) can also be used as a measure of reliability, and for a construct to be used, AVE should be greater that 0.5 (Fornell & Larcker, 1981). A satisfying level of discriminant validity is achieved when the square root of the AVE for a particular construct is larger than the correlations of the other constructs (Fornell & Larcker, 1981).

In Table 1, we can observe the loadings of each item for each construct. All items have a value greater than 0.7, making them suitable for use on each construct. Table 2 shows composite reliability, AVE, and correlations among constructs. Satisfying levels of convergent and discriminant validity can be observed for the presented reflective measurement model.

RESULTS

In PLS, the structural model is assessed by examining the path coefficients (standardized betas [β]), T statistics, and the R square (R^2) to indicate the overall predictive strength of the model. Figure 2 illustrates the structural model with the results of the PLS analysis, showing partial correlations (betas [β]) and R squares (R^2) or explained variance. Betas (β) followed by two asterisks were significant at $P < 0.01$ in a one-tailed T-test; Betas (β) not followed by an asterisk were non-significant. The $P < 0.05$ can be seen as the upper threshold of acceptability (Rosenthal & Rosnow, 1991). T values were calculated using the bootstrapping method.

From Figure 2, we can observe that the use of IM in the workplace (IMW) had a positive significant effect on interruptions (INT), which provides support

Figure 2. Results of the PLS analysis

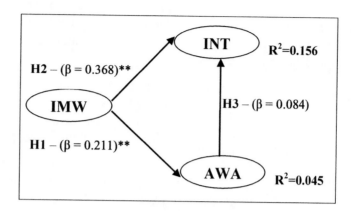

for hypothesis H2. In the same figure, we notice how the use of IM in the workplace (IMW) had positive significant effect on presence awareness (AWA), which also offers support for hypothesis H1. Additionally, Figure 2 suggests that presence awareness (AWA) had no significant effect on interruptions (INT), which in turn supports the last hypothesis H3.

Finally, Figure 2 also implies that the relationship described in the structural model accounts for approximately 16% of the variance in the interruptions (INT) construct and 5% of the variance in the presence awareness (AWA) construct.

DISCUSSION AND CONCLUSION

Recently, the use of IM has increased in the workplace; managers are introducing it as an informal synchronous communication tool, but at the same time unintended consequences, either positive or negative, can emerge from its use including interruptions and presence awareness. This study attempts to examine them according to the literature reviewed.

The findings in this paper are consistent with the presented hypotheses. Of special interest is that interruptions are directly and significantly affected by IM but not by presence awareness; in other words presence awareness is acting as a filter for interruptions. Presumably, the many features IM has to show users, like the status of the recipient (if he is "away" or "busy"), help in the negotiation of the interaction between parties. Another interesting result is how IM users find presence awareness useful and not intrusive, but instead

use it to their advantage. In fact, awareness can be seen from two different perspectives: the initiator and the recipient. When seen from the initiator's point of view, the results of this study become more obvious because the initiator is looking for information, and presence awareness can give that person a panoramic view of everybody who is online and available to answer questions or engage in conversation. Seen from the recipient's perspective, maybe the inquiry could be interruptive at first, but the IM user has the ability to negotiate and not answer the first request or use the "busy" or "away" feature of the IM software. This in turn helps initiators and recipients deal with privacy concerns; most IM software has the "block" function to completely hide the user from unwanted or unsolicited interaction.

One of the biggest limitations of this study was the dataset used for the PLS analysis. Since secondary data was used, the analysis was confined to the number and type of variables available in the set. A replication of this study with primary data could make a greater contribution to the very thin collection of scholarly empirical papers in this field. Additionally, the inclusion of some other variables of interest like achieved efficiency, effectiveness, multitasking, delays, and so forth will allow a more robust study.

Finally, while only some of the variables were measured in connection with IM and its uses in the workplace, interesting findings were revealed. A theoretical framework should be constructed through a more comprehensive study to explicitly and empirically measure IM, its components, and its effects.

REFERENCES

Cameron, A. F., & Webster, J. (2005). Unintended consequences of emerging communications technologies. *Computers in Human Behavior, 21*(1), 85-103.

Chin, W. W. (1998) Issues & opinion on structural equation modeling. *MIS Quarterly, 22*(1), 7-16.

Chin, W. W., Marcolin, B. L., & Newsted, P. R. (1996), A partial least squares latent variable modeling approach for measuring interaction effects: Results from a Monte Carlo simulation study and voice mail emotion/adoption study. In J. I . DeGross, S. Jarvenpaa, & A. Srinivasan (Eds.), *Proceedings of the 17th International Conference on Information Systems* (pp. 21-41). New York: The Association of Computing Machinery.

Cutrell, E., Czerwinski, M., & Horvitz, E. (2001) Notification, disruption, and memory: Effects of messaging interruptions on memory performance. In M. Hirose (Ed.), *Proceedings of Human-Computer Interaction* (pp. 263-269). Tokyo, Japan: IOS Press.

Fornell, C., & Larcker, D. (1981). Evaluating structural equation modeling with unobservable variables and measurement error. *Journal of Marketing Research, 18*(3), 39-50.

Goldsborough, R. (2001). Instant messaging for instant communications. *Link-Up, 18*(5), 7.

Nardi, B. A., Whittaker, S., & Bradner, E. (2000). Interaction and outeraction: Instant messaging in action. In W. A. Kellogg & S. Whittaker (Eds.), *Proceedings of the 2000 ACM Conference on Computer Supported Cooperative Work* (pp. 79-88). Philadelphia: ACM Press.

O'Conaill, B., & Frohlich, D. (1995). Timespace in the workplace: Dealing with interruptions. In *Proceedings of CHI Human Factors in Computing Systems* (pp. 262-263). Denver, CO: ACM Press.

Rennecker, J., & Godwin, L. (2005). Delays and interruptions: A self-perpetuating paradox of communication technology use. *Information and Organization, 15*(3), 247-266

Rosenthal, R., & Rosnow, R. L. (1991). *Essentials of behavioral research: Methods and data analysis.* Boston: McGraw Hill.

Shiu, E., & Lenhart, A. (2004). How Americans use instant messaging. In *Pew Internet & American Life Project*. Retrieved September 15, 2005 from http://www.pewinternet.org/

Trevino, L. K., Daft, R. L., & Lengel, R. H. (1990). Understanding managers' media choices: A symbolic interactionist perspective. In J. Fulk & C. Steinfield (Eds.), *Organizations and communication technology.* London: Sage Publications.

Zweig, D., & Webster, J. (2002). Where is the line between benign and invasive? An examination of psychological barriers to the acceptance of awareness monitoring systems. *Journal of Organizational Behavior, 23*(5), 605-633.

KEY TERMS

Asynchronous Medium: A communications medium that does not requires that both parties are present at the same time in the same space (for example: e-mail).

Initiator: The IM user that begins the communication request.

Instant Messaging (IM): A computer-mediated tool that is used to send and receive text messages in a synchronous manner using the Internet.

Interruption: A synchronous interaction that is not initiated by the recipient, is unscheduled, and results in the recipient discontinuing their current activity.

Presence Awareness: The ability to see who is online at specific times. Most IM systems display a list of users connected to the network using the IM system.

Recipient: The IM user that receives the communication request.

Symbolic Interactionist Perspective: A theory used to explain symbolic cues conveyed by different communication media.

Synchronous Medium: A communications medium that requires that both parties are present at the same time in the same space (for example: face-to-face or telephone).

Content-Based Searching in Group Communication Systems

Gábor Richly
Budapest University of Technology and Economics, Hungary

Gábor Hosszú
Budapest University of Technology and Economics, Hungary

Ferenc Kovács
Budapest University of Technology and Economics, Hungary

INTRODUCTION

The importance of real-time pattern recognition in streaming media is rapidly growing (Liu, Wang, & Chen, 1998). Extracting information from audio and video streams in an e-collaboration scenario is getting increasing relevance as the networking infrastructure develops. This development enables the use of rich media content. Shared archives of this kind of knowledge need tools for exploration, navigation and searching. As an example, to filter out redundant copies of an audio record, added by different members of an e-community, helps to keep the knowledge-base clean and compact.

The work presented here focuses on algorithms used for content-sensitive searching in audio. The main problem of such algorithms is the optimal selection of the reference patterns applied in the recognition procedure. The proposed method is based on distance maximization. It is able to quickly choose the reference pattern to be used by the pattern recognition algorithm (Richly, Kozma, Kovács, & Hosszú, 2001).

The article presents a novel approach of searching patterns in shared audio file storages such as peer-to-peer (P2P) based systems. The proposed method is based on the recognition of specific patterns in the audio contents extending the searching possibility from the description based model to the content based model.

The presented method identified as EMESE (experimental media-stream recognizer) is an important component of a light-weight content-searching system which is suitable for the investigation of network-wide shared file storages. The efficiency of the proposed procedure is demonstrated in the article.

BACKGROUND

From the introduction of Napster (Parker, 2004), Internet based communication has been developing toward the application level networks (ALN). Hosts are getting more and more powerful enabling various collaborative applications to run and create mutual logical connections (Hosszú, 2005). They establish a virtual overlay and, as an alternative to the older client/server model, they use the P2P communication. The majority of such system deals with file sharing, requiring the important task of searching in large, distributed shared file storages (Cohen, 2003; Qiu & Srikant, 2004).

Up to this time, searches have been based on the various attributes of the media contents (Yang & Garcia-Molina, 2002). The attributes called metadata can be the name of the media file, the name of authors, date of recording, type of media (genre) or some other keywords of descriptive attributes. However, if incorrect metadata were accidentally recorded, the media file may become invisible due to misleading descriptions.

Currently, powerful computers provide the possibility to implement and widely use pattern recognition methods. Due to the large amount of media files and their very rich content, limited pattern identification should be reached as a realistic goal. This article introduces the problem of media identification based on the recognition of well-defined patterns.

Another problem is introduced, if the pattern-based identification method must be extended from media files to real-time media streams. The hardness of this problem is the requirement that the pattern identification system should work real-time even in less powerful computing environment as well. For this purpose, full-featured media monitoring methods are not applicable since they require large processing power in order to run their full-featured pattern recognition algorithms.

The novel system called EMESE is dedicated for solving the special problem where a small but significant pattern must be found in a large voice stream or bulk voice data file in order to identify known sections of audio. Since this work is limited to sound files, the pattern is referred to as the soundprint of the specific media. It serves for uniquely identifying the media. This kind of patterns have also became known as the *audio fingerprint* (Haitsma, Jaap, & Kalker, 2002). The developed method is light-weight, meaning that its design goals were fast operation and the relatively small computing power. In order to reach these goals, the length of the pattern to be recognized should be very limited and the total score is not required.

This article deals mainly with the heart of the EMESE system, the pattern recognition algorithm, especially with the selection of the soundprints, the process called *reference selection*.

THE PROBLEM OF THE PATTERN RECOGNITION

In the field of sound recognition, there are many different methods and applications for specific tasks (Kondoz, 1994; Coen, 1995).

The demand for working efficiently with streaming media on the Internet increases rapidly. These audio streams may contain artificial sound effects besides the mix of music and human speech. The effects furthermore may contain signal fragments that are not audible by the ear. As a consequence, processing of this kind of *audio signal* is rather different from the already developed methods. For example, the short-term predictability of the signal is not applicable.

The representation of digital audio signal as individual sample values lacks any semantic structure to help automatic identification. For this reason, the audio signal is transformed into several different orthogonal or quasiorthogonal bases to enable detecting certain properties.

Already, there are solutions for classifying the type of broadcast on radio or television using the audio signal. An application (Akihito et al., 1998) makes basically a speech/music decision by examining the spectrum for harmonic content, and the temporal behavior of the spectral-peek distribution. Although it was applied

successfully to that decision problem, it cannot be used directly for generic recognition purposes. Liu et al. (1998) also describe a scheme classifying method where the extracted features are based on short-time spectral distribution represented by a bandwidth and a central frequency value. Several other features, for example the volume distribution and the pitch contour along the sound clip are also calculated. The main difficulty of these methods is their high computation-time demand. Therefore their application for real-time or fast monitoring is hardly possible, when taking the great number of references to be monitored into account.

A similar monitoring problem has already been introduced (Lourens, 1990) where the energy envelope is used as the signal feature and its section on the record signal as the *reference*. That was correlated with the input (*test*) signal. The demand on real-time execution drove the development of a novel recognition scheme (Richly et al., 2000) that is capable of recognizing a pattern of transformed audio signal in an input stream, even in the presence of level-limited noise. This algorithm selects a short segment of the signals in the case of each record in the set of records to be monitored (Richly, Kozma, Kovács, & Hosszú, 2001).

Tests carried out on live audio broadcasts showed that the success of identification process depends on the proper selection of the representative short segment. The position where this soundprint can be extracted is determined by the recognition algorithm of the proposed system EMESE. The selected references must be non-correlated to avoid false alarms.

The method applied in EMESE is analyzed in the following section in order to explain the synchronization of the monitoring system to the test stream under various conditions. The measured results are also presented.

THE SOUND IDENTIFICATION IN EMESE

The audio signal, sampled at $f_s = 16kHz$ is transformed into a spectral description. It is a block of data, where the columns are feature vectors of the sound corresponding to a *frame* of time-domain data ($N_f = 256$ samples, $T_f = 16ms$ long). First, the amplitude of the Fourier spectrum is computed from the frame. Then averaging is adapted to the neighboring frequency lines to project the spectrum onto the Bark-scale. The reason

Figure 1. The "distance-pit" around the reference position

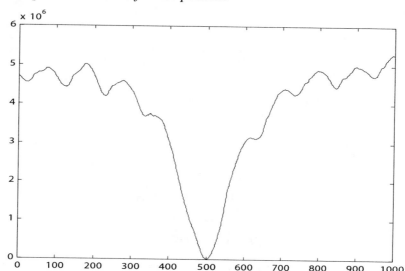

for this is to speed up the later comparison stage and to include a well established emphasizing tool used in audio processing, the perceptual modeling of the human auditory system. The resulting $N_B=20$ values build up a vector, which is normalized and quantized. Two levels are determined for each vector, namely 10% and 70 % of the peak value of the Bark-spectrum. The transformed frame is named a *slice*, since they build up the reference pattern. In every reference, there are $N_S=$ 50 slices of non-overlapping consecutive frames. The audio section, from which the reference was made, is called the *soundprint* of that specific record.

The scheme of the recognition algorithm is to grow the already identified parts of the reference patterns continuously according to the input. This means that the algorithm takes a frame from the input signal, executes the previously described transformation series and compares the resulting slice to the actual one of every reference. The actual slice is the first one in every reference initially. If it is found to be similar to the slice computed from the input stream, a slice-hit occurs and the actual slice is the next non overlapping input slice which will be compared to the next slice of that reference. If an input slice is found to be non-similar, the actual slice of that reference is reset to the first one. The similarity is evaluated by calculating the weighted Manhattan-distance of the two slices that is the sum of the absolute element-vise differences in the slice vectors.

To achieve more accurate alignment between the test and reference signals, the initial slice-hit in a reference is evaluated using a distance buffer. In this circular-memory, the distances of the first reference slice to overlapping test slices are stored and the middle of the buffer is examined whether it contains the lowest value in the buffer. In case it does, and it also satisfies the threshold criteria, the identification of the reference proceeds to the next reference slice. This method intends to align the reference's beginning to the best matching position in the identification process (Richly, Kozma, Hosszú, & Kovács, 2001).

After successfully identifying the last slice of a reference, the system successfully identified that record in the monitored input.

THE METHOD OF SELECTING THE REFERENCE PATTERNS

The selection algorithm also uses the previously described weighted Manhattan-distance for measuring the similarity of audio segments. In the vicinity of the reference's beginning, there have to be frames that vary a lot in the sense of the applied distance metric. This has to be fulfilled because the pattern recognition algorithm cannot synchronize to the given reference otherwise, since the record may appear anywhere in the monitored signal. This way a robust synchronization can be realized that is also successful in the presence of noise.

Figure 2. The width of the pits around the sound-print candidate

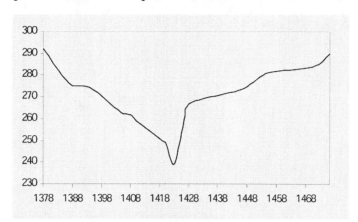

Table 1. Reference selection algorithm of the EMESE

1. In the first turn, the reference candidate is selected from the $\frac{w}{2}$ th sample of the first region of the given record (the *region* is $w=5000$ samples). The first region begins on the first sample of the record, and it will define the first sample of the frame.
2. The frame is compared to all possible frames in the region using the distance metric mentioned above. As a result $d(i)$ is obtained, where $i=0...w-N$, as shown on Figure 1.
3. The next region is selected $k*N$ samples forward in the record, and step 2 is repeated. Further regions are selected the same way and the corresponding $d(i)$ is calculated until we reach the end of the record.
4. The steepest pit and the corresponding i_{opt} frame position in the record is selected examining all the $d(i)$ functions for the narrowest pit.
5. In the $k*N$ vicinity of position i_{opt} the frame with the narrowest distance-pit is determined using a gradient search algorithm. This is the lowest point of the function on Figure 2.
6. The reference consisting of N_R slices (transformed frames) is extracted from the record beginning with the frame selected in the previous step.
7. This reference is tested for uniqueness using the recognition algorithm. If the reference appears in the record more than once, not only at the correct position, then the next best reference must be selected in the previously described way.
8. The reference is then tried against all the other records, to filter out in-set false alarms. If the reference is found in any other record, step 7 is used for reference reselection.
9. The above steps are applied to all other records.

If a long soundprint (as reference candidate) is taken from a record to be monitored and the distance of this section is calculated all along the record, then it can be observed that the distance function has a local minimum, a *pit* around the candidate's position (Richly, Kozma, Hosszú, & Kovács, 2001). This is demonstrated in Figure 1, where the x-axis shows which frame of the record is compared with the selected candidate, while the y-axis shows the Manhattan-distance values.

To achieve robust synchronization during the recognition, one must guarantee large Manhattan-distance between the candidate and its vicinity. This is assured, if the slope of the pit, as showed in Figure 1, is as big as possible. To select the best distance-pit and its

corresponding candidate section, the steepness of the pit-side should be determined. However, because it is generally not constant, as an alternative, the width at a given value can be calculated. Figure 2 shows pit-width of 100 candidate sections, where the sections are extracted from the same record so that their first samples are consecutive in the record. In Figure 2, the horizontal axis is the sample position of the candidate in the record, while the vertical axis is the width of the pits at a record-adaptive level.

The reference selection algorithm in EMESE is based on the same principle. Since the pattern recognition method uses the first frame as kernel and grows from the first record, the pit-width for single frame long candidates are observed. The minimum value has to be found in the above function without calculating every point.

It must also be assured that the selected reference does not occur any more in the record again or in any other records. Using database terminology, the reference must be a key. To avoid unambiguous identification, the selected reference must be tried against all the other records for identification. If it is not a unique key, a new reference must be selected.

The exact solution would require us to compare every one of the reference-candidates to every other one. This would mean a lot of comparisons even in the case of a few records that could not be done in a conceivable time period.

In the presented algorithm, the number of comparisons is tried to be kept as low as possible. To do so only the vicinity of the reference candidate is examined in a region having the width w, where w is expressed in number of samples. Also not all possible reference candidates are examined, but only every 100th. The algorithm is listed in Table 1.

RESULTS

Using this algorithm, references from 69 advertisements were selected that previously had been recorded from a live Internet audio stream. During the tests these advertisements were broadcast in test internet stream-media for monitoring. To test the robustness of the system, white noise was added to the input test signal. The duration of the test was 48 hours. The recognition results are shown in Figure 3.

Real-time performance of the system was also observed by the number of references handled. The computer used was equipped with a Pentium-II 350 MHz processor and 256 MB of RAM, and the maximum possible number of references was 258. If the record set to be monitored was added to, the reference selection for the added record had to be performed, and the new references have to be checked for false alarms. If a possible false alarm was detected due to representative signal similarity, the selection had to be repeated for the whole set. This took 52.5 minutes in case of 69 records. This is a worst-case scenario and it should

Figure 3. Percentage of patterns successfully identified by the recognition algorithm

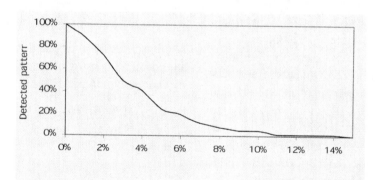

Figure 4. Result of the synchronization test

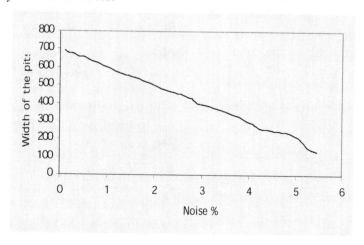

be very rare. The average selection time for every new record was 10 minutes.

The second test was synchronization test. Fifty frames from the stream were selected and the dependency of the pit width on the level of noise was observed. The noise level, where the monitoring algorithm cannot synchronize to the correct frame is shown in Figure 4.

FUTURE TRENDS

Based on the continuous development of host computers, the P2P-based file search systems will have increasing need for the content-based media identification and also the enhancements in the hardware will provide possibility to run light-weight but sophisticated pattern-based identification methods. Using Internet-oriented pattern identification tools as the EMESE, the content-based search methods will be inherent parts of the e-collaboration systems.

CONCLUSION

A reference selection method described in the article has been successfully applied in an existing real-time recognition algorithm that was used on live audio streams to identify specific sound signals. The selection algorithm takes the properties of the recognition algorithm into account. The algorithm was tested on Internet media streams with a prerecorded signal set and good results were obtained. Further tests should be carried out to determine the exact effects of the input noise level on the width of the distance-pit.

The experimental results and algorithms presented in the article proved that the pattern-fitting media identification methods can be implemented even in Internet-related environment, where the computing power and media data quality are limited.

REFERENCES

Adar, E., & Huberman, B. A. (2000). Free riding on gnutella. *First Monday, 5*(10). Retrieved June 6, 2005, from http://firstmonday.org/issues/issue5_10/adar

Akihito, M. A., Hamada, H., & Tonomura, Y. (1998, July-September). Video handling with music and speech detection. *IEEE Multimedia*, 16-25.

Coen, L. (1995). *Time-frequency analysis.* Upper Saddle River, NJ: Prentice Hall.

Cohen, B. (2003, May). *Incentives build robustness in bittorrent.* Retrieved June 6, 2005, from http://bitconjurer.org/BitTorrent/bittorrentecon.pdf

Haitsma, J., & Kalker, T. (2002). A highly robust audio fingerprinting system. In *Proceedings from the International Symposium on Musical Information Retrieval (ISMIR2002)* (pp. 144-148).

Hosszú, G. (2005). Mediacommunication based on application-layer multicast. In S. Dasgupta (Ed.), *Encyclopedia of virtual communities and technologies* (pp. 302-307). Hershey, PA: Idea Group.

Kondoz, A. M. (1994). *Digital speech: Coding for low bit rate communications systems.* Chichester, England: John Wiley & Sons, Inc.

Liu, Z., Wang, Y, & Chen, T. (1998, October). Audio feature extraction and analysis for scene segmentation and classification. *Journal of VLSI Signal Processing Systems for Signal, Image and Technology, 20,*61-79.

Lourens, J. G. (1990, September). Detection and logging advertisements using its sound. *IEEE Transactions on Broadcasting, 36*(3), 231-233.

Parker, A. (2004). *The true picture of peer-to-peer file sharing,* Retrieved June 8, 2005, from http://www.cachelogic.com

Qiu, D., & Srikant, R. (2004, August 30-September 3). *Modeling and performance analysis of BitTorrent-like peer-to-peer networks.* Paper presented at the ACM SIGCOMM'04, Portland, OR.

Richly, G., Kozma, R., Hosszú, G., & Kovács, F. (2001). A proposed method for improved sound-print selection for identification purposes. In N. Mastorakis, V. Mladenov, B. Suter, & L. J. Wang (Eds.), *Advances in scientific computing, computational intelligence and applications* (pp. 455-458). Danvers: WSES Press.

Richly, G., Kozma, R., Kovács, F., & Hosszú, G. (2001). Optimised soundprint selection for identification in audio streams. *IEE Proceedings-Communications, 148*(5), 287-289.

Richly, G., Varga, L., Hosszú, G., & Kovács, F. (2000). Short-term sound stream characterization for reliable, real-time occurrence monitoring of given sound-prints, *MELECON2000, 2,* 526-529.

Yang, B., & Garcia-Molina, H. (2002, July). Efficient search in peer-to-peer networks. In *Proceedings of the IEEE International Conference on Distributed Computing Systems (ICDCS'02),* 5-14. Vienna, Austria. Washington, DC:IEEE Computer Society.

KEY TERMS

Application Level Network (ALN): Applications running in the hosts can create a virtual network from their logical connections. The virtual network is called *overlay* (see below). Such software entities are not able to communicate with each other without knowing their logical relations. The most cases this ALN software entities use the *P2P model* (see below), not the *client/server* (see below) one for the communication.

Audio Signal Processing: Means coding, decoding, playing and content handling of audio data files and streams.

Bark-Scale: Nonlinear frequency scale modeling the resolution of the human hearing system. 1 Bark distance on the Bark-scale equals to the so called critical bandwidth that is linearly proportional to the frequency under 500Hz and logarithmically above that. The critical bandwidth can be measured by the simultaneous frequency masking effect of the ear.

Client/Server Model: Communicating method, where one host has more functionality than the other. It differs from the *P2P model* (see below).

Content-Based Recognition: The media data are identified based on its content and not on the attributes of its file. It is also called *content-sensitive searching*.

Manhattan-Distance: The L_1 metric for the points of the Euclidean space defined by summing the absolute coordinate differences of two points ($|x2-x1|+|y2-y1|+ \ldots$). Also known as "city block" or "taxi-cab" distance; a car drives this far in a lattice-like street pattern.

Overlay: Applications, that create an *ALN (see above)*, work together, and usually follow the *P2P communication model* (see below).

Pattern Recognition: Procedure of finding a certain series of signals in a longer data file or signal stream.

Peer-to-Peer (P2P) Model: Communication method where each node has the same authority and communication capability. They create a virtual network, overlaid on the Internet. Its members organize themselves into a topology for data transmission.

Synchronization: Procedure carried out to find the appropriate points of two or more streams to form correct paralleling.

Context-Based Explanations for E-Collaboration

Patrick Brezillon
University Paris 6, France

INTRODUCTION

E-collaboration is generally defined in reference to ICT used by people in a common task (Kock, 2005; Kock, Davison, Ocker, & Wazlawick, 2001). However, when speaking of e-collaboration, people seems to put more the emphasis on "e-" than on "collaboration"; that is, on the ICT dimension of the concept that on the human dimension. Along the human dimension, e-collaboration requires to revisit previous concept of cooperation, conflict, negotiation, justification, explanation, etc. to account for the sharing of knowledge and information in the ICT dimension. We discuss in this article of explanation generation in this framework.

Any collaboration supposes that each participant understands how others make a decision and follows the series of steps of their reasoning to reach the decision. In a face-to-face collaboration, participants use a large part of contextual information to translate, interpret and understand others' utterances use contextual cues like mimics, voice modulation, movement of a hand, etc. In e-collaboration, it is necessary to retrieve this contextual information in other ways. Explanation generation relies heavily on contextual cues (Karsenty & Brézillon, 1995) and thus would play a role in e-collaboration more important than in face-to-face collaboration.

Fifteen years ago, Artificial Intelligence was considered as the science of explanation (Kodratoff, 1987). However, there are few concrete results to reuse now from that time. There are several reasons for that. The first point concerns expert systems themselves and their past failures (Brézillon & Pomerol, 1997):

- There was an exclusion of the human expert providing the knowledge for feeding the expert systems. When an expert generally provided something like "Well, in the context A, I will use this solution," the knowledge engineer retained the pair {problem, solution} and forgot the initial triple {problem, context, solution} provides by the expert. The reason was to generalize in order to cover a large class of similar problems when the expert was giving a local solution. Now we know that a system needs to acquire knowledge within its context of use.

- On the opposite side, the user was excluded from the noble part of the problem solving because all the expert knowledge was supposed to be in the machine: the machine was considered as the oracle and the user as a novice (Karsenty & Brézillon, 1995). Thus, explanations aimed to convince the user of the rationale used by the machine without respect to what the user knew or wanted to know. Now, we know that we need to develop a user-centered approach (Brézillon, 2003).

- Capturing the knowledge from the expert, it was supposed to put all the needed knowledge in the machine, prior the use of the system. However, one knows that the exception is rather the norm in expert diagnosis. Thus, the system was able to solve 80% of the most common problems, on which users did not need explanations. Now, we know that systems must be able to acquire incrementally knowledge with its context of use.

- Systems were unable to generate relevant explanations because they did not pay attention to what the user's question was really, in which context the question was asked. The request for an explanation was analyzed on the basis of the available information to the system.

Thus, the three key lessons learned are (1) KM stands for management of the knowledge in its context; (2) any collaboration (including e-collaboration) needs a user-centered approach; and (3) an intelligent system must incrementally acquires new knowledge and learns corresponding new practices.

Focusing on explanation generation, it appears that a context-based formalism for representing knowledge and reasoning allows to introduce the end-user in the loop of the system development and to generate new types of explanations.

With new findings about context available now, a new insight is possible on past problems abandoned previously by lack of a relevant solution at that time, like incremental knowledge acquisition, practice learning and explanation generation. Previously, they were considered as distinct problems. Now their integration in the task at hand of the user offers new options, especially for e-collaboration.

Hereafter, the article is organized in the following way. First, we comment briefly previous works on explanations in order to point out what is reusable. Second, we discuss explanation generation potentialities in a context-based formalism called contextual graphs. Finally, we show what explanations can bring in e-collaboration, maybe more than in face-to-face collaboration.

BACKGROUND

Explanations in Knowledge-Based Systems

The first research on explanations started with rule-based expert systems. Imitating a human reasoning, the presentation of the trace of the expert system reasoning (i.e., the sequence of fired rules) was supposed to be an explanation of the way in which the expert system reaches a conclusion. Rapidly, it was clear that it was not possible to explain heuristics provided by human experts without additional knowledge. It was then proposed to introduce a domain model. It was the second generation of expert systems, called the knowledge-based systems. This approach reached also its limits because it was difficult to know in advance all the needed knowledge and also because it was not always possible to have models of the domain. However, the main weakness was the lack of consideration for the user and what the user wanted as explanation. The user's role was limited to be a data gatherer for the system. A second observation was that the goal of explanations is not to make identical user's reasoning and the system reasoning, but only to make them compatible: the user must understand the system reasoning in terms of his own mental representation. For example, a driver and a garage mechanic can reason differently and reach the same diagnosis on the state of the car. The situ-

ation is similar in e-collaboration where specialists of different domains and different geographical areas must interact in order to design a complex object. A third observation is that the relevance of explanation generation depends essentially on the context use of the topic to explain (Abu-Hakima & Brezillon, 1994; Karsenty & Brezillon, 1995). We discuss this point later in the article.

Beyond the need to make context explicit, first in the reasoning to explain, and, second, in the explanation generation, the most challenging finding is that lines of reasoning and explanation must be distinguished, but considered jointly, the line of explanation being able to modify the line of reasoning (Abu-Hakima & Brézillon, 1994). Thus the key problem for providing relevant explanations is to find a uniform representation of elements of reasoning and of context.

Explanations and Contexts

A frequent confusion between representation and modeling of the knowledge and reasoning implies that explanations are provided in a given representation formalism, and their relevance depends on explanation expressiveness through this formalism. For example, ordinary linear differential equation formalism will never allow to express—and thus explaining—the self-oscillating behavior of a nonlinear system. Thus, the choice of representation formalism is a key factor for generating relevant explanations for the user and is of paramount importance in e-collaboration with different users and several tasks.

A second condition is to account for, make explicit, and model the context in which knowledge can be used and reasoning held. For example, a temperature of 24°C in winter in Paris (when temperature is normally around 0°C) is considered to be hot in Paris and cold in Rio de Janeiro (when temperature is rather around 35°C). Thus, the knowledge must considered within its context of use for providing relevant explanations, like to explain to a person living in Paris why a temperature of 24°C could be considered as cold in some other countries. (We will not discuss in this article the problem of affordance such as the use of an umbrella to walk or to protect from the sun and not the rain.)

There is now a consensus around the following definition "context is what constrains reasoning without intervening in it explicitly" (Brezillon & Pomerol, 1999), which applies also in e-collaboration (although

Figure 1. Activity "exploitation of an information" (Brézillon, 2005)

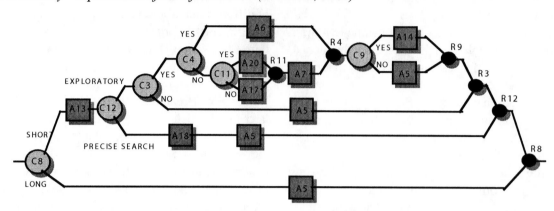

with more complex constraints) where reasoning is developed collectively. In e-collaboration, explanation generation is a means to develop a shared context among the actors in order to have a better understanding of the others (and their own reasoning), to reduce needs for communication and to speed up interaction.

From our previous works on context (e.g., see Brezillon, 2005), several conclusions have been reached. First, a context is always relative to something that we call the (current) focus of attention of the actors. Second, with respect to this focus, context is composed of external knowledge and contextual knowledge. The former has nothing to see with the current focus (but could be mobilized later, once the focus moves), when the former can be more or less related directly to the focus (at least by some actors). Third, actors address the current focus by extracting a subset of contextual elements, assembling and structuring them all together in a proceduralized context, which is a kind of « chunk of knowledge » (Schank, 1982). Fourth, the focus evolving, the status of the knowledge (external, contextual, into the proceduralized context) evolves too. Thus, there is a dynamics of context that plays an important role in the quality of explanations.

As the context exists with the knowledge, a context-based generation of explanations do not requires an additional effort if such an explanatory knowledge is integrated in the knowledge representation at the time of their acquisition and the representation of the reasoning (see Brézillon, 2005, on this aspect). However, this supposes to have a context-based formalism allowing a uniform way to represent elements of reasoning and of contexts.

CONTEXTUAL GRAPHS AND EXPLANATION GENERATION

A contextual graph represents the different ways to solve a problem. It is a directed graph, acyclic with one input and one output and a general structure of spindle (Brezillon, 2005). Figure 1 gives an example of contextual graph. A path in a contextual graph corresponds to a specific way (i.e., a practice) for the problem solving represented by the contextual graph. It is composed of elements of reasoning and of contexts, the latter being instantiated on the path followed (i.e., the values of the contextual elements are required for selecting an element of reasoning among several ones). Elements in a contextual graph are actions (square boxes in Figure 1), activities (complex actions like subgraphs), contextual elements (couples C-R in Figure 1) and parallel action groupings (a kind of complex contextual elements). A contextual element is a pair composed of a contextual node (e.g., C4 in Figure 1) and a recombination node (e.g., R4).

Some mechanisms of aggregation and expansion (like in conceptual graphs of Sowa, 2000) provide different local views on a contextual graph at different levels of detail by aggregating a subgraph in an item (a temporary activity) or expanding it. This representation is used for the recording of the practices developed by users, which thus are responsible for some paths in the contextual graph, or at least some parts of them.

In this context-based formalism of representation, we have established a typology of explanations, based on previous works and exploiting the capabilities of contextual graphs. By adding a new practice, several contextual information pieces are recorded automatically (date of creation, creator, the practice-parent) and others are provided by the user himself like a definition and comments on the item that is introduced. Such contextual information is exploited during the explanation generation. Thus, the richness of contextual-graph formalism leads in the expressiveness, first, of the knowledge and reasoning represented, and second of the explanations addressing different users' requirements. The main categories of explanations developed in contextual graphs are:

- **Visual explanations:** They correspond to a graphical presentation of a set of complex information generally associated to the evolution of an item, e.g. the contextual graph itself, the decomposition of a given practice, the series of changes introduced by a given user, regularities in contextual graphs, and so forth.

- **Dynamic explanations:** They correspond to the progress of the problem solving during a simulation addressing questions as the "what if" question. With the mechanisms of aggregation and expansion, a user can ask an explanation in two different contexts and thus received two explanations with different presentations (e.g., with the details of what an activity is doing in one of the two explanations). The dynamic nature of the explanation is also related to the fact that items are not introduced chronologically in a contextual graph. For example, the contextual element C12 in Figure 1 has been introduced after C3 when it is situated before it in the practice development (i.e., C3 is in the "future" of C12 but one of the reason of the introduction of C12). Finally, the proceduralized context along a practice is an ordered series of instantiated contextual elements, and changing the instantiation of one of them is changing of practice and thus changing of explanation.

- **User-based explanations:** The user being responsible of some practice changes in the contextual graph, the system uses this information to tailor its explanation by detailing parts unknown of the user and sum up parts developed by the user.

- **Micro- and macroexplanations:** Again, with the mechanisms of aggregation and expansion, it is possible to generate an explanation at different levels of detail. For complex item like an activity or a subgraph, it is possible to provide on them a microexplanation from an internal viewpoint on the basis of activity components. A macroexplanation from an external viewpoint is built with respect to the location of the activity in the contextual graph like any item.

- **Real-time explanations.** There are three types. First, the explanation is asked during a problem solving when the system fails to match the user's practice with its recorded practices. Then, the system needs to acquire incrementally new knowledge and learning the corresponding practice developed by the user (generally due to specific values of contextual elements not taken into account before). This is an explanation from the user to the machine Second, the user wished to follow the reasoning of a colleague having solved the problem with a new practice (and then we are back to simulation). Three, the system tries to anticipate the user's reasoning from its contextual graph and provides the user with suggestions and explanations when the user is operating.

Moreover, these different types of explanation (and others that we are discovering progressively) can be combined in different ways like visual and dynamic explanations.

FUTURE TRENDS

When the machine fails to address correctly a problem, the machine may benefit of its interaction with the human actors to acquire incrementally the missing knowledge and learn new practices. As a consequence, the machine will be able to explain later its choices and decisions. Contextual graphs are able to manage incremental acquisition and learning, and begin to provide some elementary explanations. After a while, a contextual graph is a kind of corporate memory for this specific problem solving.

As a general learned lesson, expressiveness of the knowledge and reasoning models depends essentially of the representation formalism chosen for expressing such models. This appears a key element of e-collaboration with multiple sources of knowledge and different lines of reasoning intertwined in a group work. This is a partial answer to our initial observation that e-collaboration would be better understood if we consider jointly its two dimensions, the human dimension and the technology dimension. Then, explanation generation would be revised in order to develop "collective explanations" for all the (human) participants in the e-collaboration, that is in each mental representation. Going one step further, it would be possible to compare with another view where ICT is controlled by an "intelligent agent" interaction in the e-collaboration with human agents.

CONCLUSION

Relevant explanations are a crucial factor in any collaboration between human actors, especially when they interact by computer-mediated means like in e-collaboration. First because an e-collaboration looses some advantages of a face-to-face collaboration in which a number of contextual elements are exchanged in parallel with the direct communication. Second because an e-collaboration can benefit of new ways to replace this "hidden exchanges" of contextual cues between actors by the use of the computer-means themselves. For example, it is possible to consider new types of explanation in an e-collaboration.

Explanation generation is very promising for e-collaboration because explanations use and help to maintain a shared context among actors of an e-collaboration. We are now in a situation in which computer-mediated interaction concerns human and software actors. Software must be able to react in the best way for human actors. For example, for presenting a complex set of data, a software piece could choose a visual explanation taking into account the type of information that human actors are looking for. We show that making context explicit allows the generation of relevant explanations. Conversely, explanations are a way to make contextual knowledge explicit and points out the relationships between context and the task at hand, and thus develop a real shared context.

A key factor for the success of relevant explanations is to use a context-based formalism that represent all the richness of the knowledge and reasoning that is in the focus. A good option is to consider context of use simultaneously with the knowledge. This supposes to have a context-based formalism like the contextual graphs introduced in this article. In such formalism, elements of reasoning and of contexts are represented in a uniform way. As a consequence, this allows developing new types of explanation like visual explanations, dynamic explanations, and real-time explanations.

Indeed, we have developed a new typology of explanations that include past works on explanations but goes largely beyond. Moreover, these different types of explanations are not independent and can be combined together to provide richer explanations.

REFERENCES

Abu-Hakima S., & Brézillon P. (1994). *Knowledge acquisition and explanation for diagnosis in context* (Research report 94/11). Paris: University Paris VI, LAFORIA.

Brézillon, P. (2003). Focusing on context in human-centered computing. *IEEE Intelligent Systems, 18*(3), 62-66.

Brézillon, P. (2005). Task-realization models in contextual graphs. *Lectures Notes in Artificial Intelligence, 3554*, 55-68.

Brézillon, P., & Pomerol, J.-Ch. (1999). Contextual knowledge sharing and cooperation in intelligent assistant systems. *Le Travail Humain, 62*(3), 223-246.

Brézillon, P., & Pomerol, J.-Ch. (1997). Lessons learned on successes and failures of KBSs (Special issue). *Failures and Lessons Learned in Information Technology Management, 1*(2), 89-98.

Karsenty, L., & Brézillon, P. (1995). Cooperative problem solving and explanation. *International Journal of Expert Systems with Applications, 8*(4), 445-462.

Kock, N., Davison, R., Ocker, R., & Wazlawick, R. (2001). E-collaboration: A look at past research and future challenges. *Journal of Systems and Information Technology, 5*(1), 1-9.

Kock, N. (2005) What is e-collaboration? (Editorial essay). *International Journal of E-Collaboration, 1*(1), i-vii.

Kodratoff, Y. (1987). Is artificial intelligence a subfield of computer science or is artificial intelligence the science of explanation? In I. Bratko & N. Lavrac (Eds.), *Progress in machine learning* (pp. 91-106). Cheshire, UK: Sigma Press.

Potier, D. (2005). *Génération de nouveaux types d'explications dans le formalisme des graphes contextuels.* Paris : University Paris 6, Rapport de DEA, LIP6.

PRC-GDR. (1990). *Actes des 3ᵉ journées nationales PRC-GDR IA organisées par le CNRS* (Textes réunis par Bernadette Bouchon-Meunier). Editions Hermes.

Schank, R. C. (1982). *Dynamic memory : A theory of learning in computers and people.* Cambridge University Press.

Sowa, J. F (2000). *Knowledge representation: Logical, philosophical, and computational foundations.* Pacific Grove, CA : Brooks Cole.

KEY TERMS

Context: Elements that constrain a problem solving without intervening in it explicitly. Two parts are distinguished in the context with respect to a focus, namely the contextual and external knowledge.

Contextual Graphs: A context-based formalism for representing elements of reasoning and of contexts in a uniform way.

Context-Based Reasoning: A reasoning often cannot be separated from the context in which it takes place. In a rule, the conclusion is intertwined with conditions. Such reasoning is decison making, interpretation, diagnosis, pattern recognition.

Explanation: A presentation by an explainer to an explainee in order to allow the explainee to link a striking information in his/her mental representation of the world based on contextual cues. The line of explanation is not the line of reasoning to explain.

Practice: The result of the transformation made by an actor of a procedure for taking into account the specificity of a given context. This is a contextualization of a procedure.

Proceduralized Context: The subset of contextual knowledge pieces that are selected, collected, assembled, organized structured in a chunk of knowledge to be used in the current focus (e.g., the current step of problem solving).

Shared Context: Part of the contextual knowledge that is elaborated progressively by actors and thus shared, even if not identical.

Cultural Influences on Virtual Teamwork Collaboration

Amyris Fernandez
Ibmec Educacional, Brazil

INTRODUCTION

This article discusses culture influence on virtual teamwork collaboration efficiency. We can understand culture as a framework of meanings, which gives a certain group of people the same linguistic framework, collective interpretation of the environment, its ways of self understanding, its customs, traditions, and values. Culture also involves the human spirit, personal expression, principles, and moral commitments. Culture shapes the way people interpret and send messages and the way they think about issues like contractual obligations, work procedures or professional ethics (Lavoire, 2001).

Dispersed ways of working started to exist to respond to expertise constraints, created by downsizing, mergers, and acquisitions, globalization, and employee preferences. Since the early 1980s, organizations started to offer their employees a combination of nontraditional work practices, settings, and locations to supplement traditional offices. Different terms try to cover different practices and policies of geographically dispersed work such as *teleworking, telecommuting, working-at-home, working-at-a-distance, remote work*, and *virtual work*. Although this subject has been discussed for many years, a universal definition still is not in place (Johnson, 2001). For practical and academic reasons we will use Nancy Johnson's (2001) definition, which states that **v**irtual work is a mode of work in which employees perform all or significant part of their roles from a base physically separated of their employers, and where Information technology plays an important role in virtual teamwork by supporting all business practices to create, share and communication among team members.

There are two main reasons why 46 million people that telecommuted in the United States in 2005: cost savings and the increase in productivity (Langhoff, 2005). But geographically dispersed, cross-functional teams also claim to enhance learning and innovation. They are expected to be more creative, since their members bring different intellectual and occupational backgrounds (Boutellier, Gassman, Macho, & Roux, 1998; Brown & Eisenhart 1995; De Meyer, 1993a, 1993b; Gordon & Motwani, 1996; Leonard, 1995; Madhavan & Grover, 1998). Empirical evidence suggests that diversity constrains effective sharing, because there are occupational and contextual constraints. The interpretive barriers rise from differences in perspectives, priorities, and typical approaches to problem solving, and even terminology, and they may come from the specific social or physical contexts. They hinder understanding and team cohesion among different functional units, occupational workgroups, and across localities (Sole & Applegate, 2001).

In other words, different cultural backgrounds bring different perspectives on the same subject, creating a great opportunity for knowledge sharing and learning, but it also creates barriers, because group participants may not share the same language code (even when they agree that English should be used as a common language among participants), norms, and procedures. Thus, work progress may face some challenges that will influence e-collaboration. Here we will discuss those challenges and opportunities on the face of cultural issues.

CULTURAL INFLUENCES ON WORK PATTERNS

Culture consists of specific learned norms based on attitudes, values, and beliefs, all of which exist in every society. There is evidence of culture pervasive importance. Culture cannot easily be isolated from such factors as economic and political conditions and institutions. Considerable evidence indicates that some aspects of culture differ significantly across national borders and have a substantial impact on how business is normally conducted in different countries. Lavoie (2001) believes that to determine the wealth generating potential of a society, we need to ask about mineral resources, but also about more subjective factors, such as the value of a society's stock capital, or society's entrepreneurial growth potential, both related with culture, here understood as a framework of meaning, that affect nations, business or group results.

Culture is transmitted by various patterns, such as from parent to child, from teacher to pupil, from social leader to follower, and from peer to peer. However, because of multiple influences, individual and societal values and customs may evolve over time. Change may come through choice or imposition. Change by choice may take place as a by-product of social and economic change or because of contacts with other cultures and their business practices that present reasonable alternatives.

Cultural tied behavioral practices affects business, such as the role of competence. In countries under the influence of the United States, a person's acceptability for jobs and promotions is based on competence. Thus, the workplace is characterized more in competition than in cooperation. In many cultures, competence is secondary importance, and the belief that it is right to place some other criterion ahead of competence is just as strong in those cultures as the belief in competence is in the United States. In some countries, an autocratic style of management is preferred; in others, a consultative style prevails. Interestingly, those preferring an autocratic style are also willing to accept decision making by a majority of subordinates (Daniels & Radebaugh, 1998).

We have also to consider that there is low-context cultures—that is most people consider relevant only information that receive firsthand and that bears very directly on the decision they need to make. They also spend little time on "small talk" in business situations. High context cultures—that is, most people consider that peripheral and hearsay information are necessary for decision making because they bear on the context of the situation. As we can imply, these differences between low and high context cultures may create different business process approaches, which may lead to a better conclusion or a great misunderstanding.

We can understand organizational culture as the set of fundamental assumptions about what products the organization should produce, how it should produce them, where and for which customers. Organizational culture is a powerful unifying force that restrains political conflict and promotes common understanding, agreements and procedures. The organizational culture is also able to restrain change, especially technological

changes, because it threatens commonly held cultural assumptions, and it creates a great deal of resistance (Laudon & Laudon, 2001). There may be some organizations that value change and technology, and others that do not pay attention to that. It may influence virtual work and technology adoption.

ALTERNATIVE WORKPLACES AND MANAGEMENT OF VIRTUAL TEAMS

Companies are investing in alternative workplaces since the 80's, and the most obvious reason is cost reduction. By eliminating offices, for example, business can save millions of dollars a year. Alternative workplaces, combined with communication technologies, and the use of personal computers, handhelds and other communication devices create the right environment to have teleworkers, people that may operate at home, or mobile. Virtual employees tend to devote less time to office routines, and more time to customers, improving productivity and effectiveness on the job. Alternative workplaces give companies an edge to retain talented, highly motivated employees who find flexibility to work from home especially attractive (Carr, 1999).

Potential benefits are clear, but at the same time, alternative workplaces are not for everyone. Organizations readiness to adopt the program is a must, in the cultural and technological point of view. Most of the times, managing cultural changes and systems improvements required by these programs are substantial, and the same happen to create virtual teams.

The management philosophy and style needs to be more informational, rather than industrial, in order to give room to alternative workplaces arrangements. Informational organizations operate mainly through voice and data communications. They are not necessarily high tech, but employees style are flexible, informal, change when necessary, have a sense of respect for personal time and priorities, and are committed to use technology to improve performance. Industrial, in this context, means that organization's structure, systems, and management processes are designed for intensive face-to-face interaction and that employees remain rooted to specific workplaces. Thus, a dynamic, non-hierarchical, technology driven organization is more likely to adopt alternative workplaces programs than a highly command-driven one, because its culture is more open and fosters proactive ways of doing a job.

It is also necessary to be committed to the new ways of operating, which means that it is necessary to review company's incentives and rewards policies in light of different ways in which work may be completed. In the traditional office environment it is very difficult not to reward people partly for they effort to do a job. In the virtual environment it is necessary to focus a lot more on results, which means that the cultural change are about focusing on processes and results, rather than interpersonal relationships. The alternative workplace programs also implies in a transition from conventional to new places for most employees. For those accustomed to a structured office environment, it may be hard to adjust to a self-directed schedule, and they may also feel lonely in a remote setting.

To make the alternative workplace initiatives work, it is necessary to consider technology issues, training, and an appropriate, flexible administration support. Virtual teams enable companies to create groups where diversity is possible, because there are no geographical barriers. But, if location and time are not an issue, differences in work procedures may be (Rennecker, 1999). High contextual cultures may be a challenge to low context cultures, because they do not share the same pace to observe the environment and deal with information. High context cultures also are less pragmatic, and may not know other cultures work procedures. It may create some misunderstandings, even when people have a project plan, and deadlines to deal with. Even in face of these problems, knowledge will be created. The program also needs to consider how the company will accumulate and share knowledge. One approach to facilitating knowledge combination and exchange is through geographically dispersed teams that uses computer mediated communications, and collaborative technologies. Those teams create ways to combine knowledge in new ways, creating competitive advantage (Apgar, 1998). Knowledge, as a collective creation develops through actions, resides on ways of interacting, something extremely related with national and business culture. Implicit sharing meanings rise from the community members, and gradually they take this knowledge for granted, which makes it difficult to interact with other communities (Cramton, 1999).

CREATING AN ENVIRONMENT FOR VIRTUAL COLLABORATION

Collaborative activities use cooperative task structures based on active participation, and peer interaction, to help the team in achieving its common goal of completing projects. Collaboration typically defines the "mutual engagement of participants solving a problem together," but scholars adopt a slightly different definition, which is the co-ordination of effort to build common knowledge.

In traditional work teams, physical proximity defines the workspace, but in virtual environments, communication defines the space. Every person has a unique communication style. The first virtual team task is to recognize and align styles, in order to make all members share clear expectations.

Virtual teams are forced to communicate through technology channels. In this case, the content of a message has two components: technical and emotional. The technical side includes all sort of measurable items such as date, time, path through the site, and bandwidth needs. The emotional content is closely tied with the cultural environment, though, because it refers to feelings, attitudes, and energy level of the individual who created the message and the emotional state of the recipient as well. Emotional messages tend to create a more sophisticated channel of communication, but it also requires a high sensibility from the team leader. The person in this role must coach members with positive and developmental feedback. The appropriate channel choice to interact, the length of the message, and the clear focus on task and performance minimizes the emotional impact in a positive way, avoiding wrong interpretations (Wardell, 1998).

Virtual work environments also requires explicit, transparent processes, comprised by approaches, methodologies, work plans, procedures, and outputs, deliverables, records, milestones, outcomes, documents. This enables consistency of procedures, results and performance measurement, which builds trust on the team. Mintzberg (1979) described five different types of organizational structures: entrepreneurial (small start-up business), machine bureaucracy (mid-size manufacturing firms), divisionalized bureaucracy (Fortune 500 firms), professional bureaucracy (law firms, school systems), adhocracy (consulting firms). Bureaucracies are more procedure driven, and outputs help to measure employee's performance, but their environment tend to restrain change, and technology acceptance. On the other hand, start-ups and adhocracies are more information intense, and their processes relies more in communication, which creates an appropriate environment to virtual work, but some level of procedures must be in place, otherwise virtual workers will not understand their roles, milestones and outputs.

Virtual projects success depends on the company commitment in providing technology, support and training to remote employees (Apgar, 1998). But choosing the right technology is a matter of identifying needs, analyzing the various options, and then making an educated decision. To make effective technology decisions, the team needs to consider: Who? What? Where? How?

It is especially important to know how people and their culture view technology. It means that are certain countries where people are more open to accept new technologies, specially the ones that enable information sharing, improving the interpersonal relationship, like Brazil. But we also have to consider the business culture. In hierarchical structures, collaboration is a more formal process, so synchronous environments like chat rooms are less likely to be available for participants, because they tend to be more informal.

On the other hand, not only the business culture participates on which sort of technology will be available, but the nature of work. It is important to know what they need to do, which kind of information they will work with, how often and fast they will need to communicate.

We must take in consideration that certain companies and cultures do not accept very well the telecommuting system. Collaborative tools must take the business issues into consideration, because it is possible that all collaboration will occur only inside the company, which makes security questions less sensitive. In this case, the data collection, selection and distribution issues are more relevant. It arise questions regarding technologies compatibility, and plans for growth (Duarte & Snyder, 1999).

If an organization decides to work virtually, it needs to know the time zone, and differences among locations, which means learning the available technology

at the traditional office, at home, in shared offices, and if there is any mobile connections available at planes or cars in other countries.

However, not all cultures are ready for the virtual communication. Technology mediated channels force people to be more focused on the task, more pragmatic, and less contextual. It may represent a cultural change, which means that we have to expect resistance, not only because change is difficult to people, but also because there is technology issues involved.

Virtual team management deals with people from different backgrounds, geographically dispersed, that may not report directly to the team manager, or maybe don't work in the same company. These virtual team members may not have the same technology knowledge, and may not share the same values. Munkvold and Anson (2001), found that, in order to improve project success rate during the change process, it will need:

- A high level champion from the organizational behavior department
- A close collaboration between the organizational behavior department and the IT department
- A real user design/implementation team
- A focus on learning and training throughout the organization

The two researchers believe that project must look for more coordination in the cultural change aspect, in ways to reduce internal competition for resources, and to improve internal diffusion and adoption. For them, the intangible nature of the virtual collaboration tools and implementation benefits can result in less organizational commitment, becoming the main reason why an internal champion is needed. The IT department can also provide support to find ways to measure success.

CONCLUSION

Virtual teams and collaboration in technology based environments requires a transition from industrial to a knowledge mindset, where physical presence is not really necessary, but only employees' knowledge and skills matter. It means that there will be an organizational cultural change that may impact on personal cultural beliefs. We have to understand that collaboration only happens over time because it requires participants to develop trusting relationships. Trust is built of many constructs including: good faith, honesty, commitment, competence, expertness, morality, which depends heavily on the person's cultural background, therefore, influencing on group member communication patterns (Gignac, 2004).

To accomplish the virtual team collaboration, employees must be involved from the start of the project. Organizations must promote cooperation and teamwork, and provide training. It is required to develop and publish policies and procedures. They have to offer incentive programs for participants. Because all these activities are related with change, a plan is required, and key performance indicators must be in place before the project begins. Tools like Balanced Scorecard can help integrate different departments involved in the process, and keep track of cultural change plan success.

So, it takes a lot of management effort to create the right work environment, in order to accommodate all diversity aspects, technology challenges, cultural differences, and get all the best of virtual team workers.

REFERENCES

Apgar, M., IV. (1998, May-June) The alternative workplace: Changing where and how people work. *Harvard Business Review,* 121-136.

Boutellier, R., Gassman, O., Macho, H., & Roux, M. (1998) Management of dispersed product development teams: The role of information technologies. *R&D Management, 28*(2), 13-26.

Brown, J. S., & Eisenhardt, K. (1995). Product development: Past research, present findings, and future directions. *Academy of Management Review, 20*(2), 343-378.

Carr, N. G. (1999, May-June). Being virtual: Character and new economy. *Harvard Business Review,* 3-7.

Cramton, C. D. (1999). *The mutual knowledge problem and its consequences in geographically dispersed teams* (Working chapter). George Manson University.

Daniels, J. D., and Radebaugh, L. H. (1998). *International business: Environments and operations.* New York: Addison-Wesley.

DeMeyer, A. (1993a). Internationalizing R&D improves a firm's technical learning. *Research of Technology Management, 36*(4), 42-49.

DeMeyer, A. (1993b). Management of an international network of industrial R&D lab. *R&D Management, 23*(2), 109-121.

Duarte, D. L., & Snyder, N. T. (1999). *Mastering virtual teams.* San Francisco: Jossey-Bass.

Gignac, F. (2004). *Building successful virtual teams.* Norwood, MA: Artech House Incorporated.

Gordon, I., & Motwani, S. (1996). Issues in cooperative software engineering using globally distributed teams. *Information and Software Technology, 38*(1), 646-656.

Laudon, K. C., & Laudon, J. P. (2004). *Management information systems: Managing the digital firm.* Upper Saddle River, NJ: Prentice Hall.

Lavoie, D. (2001). Culture & enterprise: The development, representation & morality of business. Oxford, UK: Routledge.

Leonard, D. A., Brands, P. A., Edmondson, A., & Fenwick, J. (1998). Virtual teams: Using communications technology to manage geographically dispersed development groups. In S. P. Bradley & R. L. Nolan (Eds.), *Sense and respond: Capturing value in the network era* (pp. 258-298). Boston: Harvard Business School Press.

Johnson, N. (2001). *Telecommuting and virtual offices: Issues & opportunities.* Hershey, PA: Idea Group.

Mitzenberg, H. (1979). *The structuring of organizations.* Englewood Cliffs, NJ: Prentice Hall.

Munkvold, B. E., & Anson, R. (2001, September-October). Organizational adoption and diffusion of electronic meeting systems: A case study. *GROUP '01,* 279-287.

Nelson, R., & Winter, S. *An evolutionary theory of economic change.* Cambridge: The Belknap Press.

Rennecker, J. (1999). *Overlap and interplay: Cultural structuring of work and communication in one virtual group* (Report). MIT, Sloan School of Management.

Sole, D., & Applegate, L. M. (2001). Knowledge sharing practices and technology use norms in dispersed development teams. In *Proceedings of the International Conference on Information Systems.*

Szulanski, G. (1995). Unpacking stickiness: An empirical investigation of barriers to transfer practice inside the firm. *Academy of Management Journal (Best Chapters Proceedings, 1995),* 437-441.

Wardell, C. (1998, November). The art of managing virtual teams: Eight key lessons. *Harvard Management Update*, 3-4.

KEY TERMS

Alternative Workplaces (AW): The combination of nontraditional work practices, settings, and locations that is beginning to supplement traditional offices.

Business Process: A set of interrelated activities performed in an organization with a goal of generating value in a connection with a product or service.

Collaborative Technologies: Technologies enabling individuals and groups to communicate, collaborate, and interact to share knowledge and information, focusing on those to facilitate dispersed interaction across time and space. Includes telephones, audio- and video-conferencing facilities, electronic discussions, online chat environments, application sharing, desktop conferencing, and shared document repositories.

Knowledge Management: The set of processes developed in an organization to create, gather, store, maintain, and apply the firm's knowledge.

Virtual Work: A mode of work in which employees perform all or significant part of their roles from a base physically separated of their employers, and where Information technology plays an important role in virtual teamwork by supporting all business practices to create, share and communication among team members.

Virtual Team: Geographically dispersed teams.

Definition, Antecedents, and Outcomes of Successful Virtual Communities

Anita L. Blanchard
University of North Carolina - Charlotte, USA

INTRODUCTION

Howard Rheingold's (1993) book *The Virtual Community: Homesteading on the Electronic Frontier* was the first to bring virtual communities to the attention of researchers and practitioners. Although virtual groups have been examined previously, Rheingold's descriptions of participating in the WELL, an Internet-based bulletin board, vividly portrayed the potential of online social groupings. Rheingold told stories of people who had never met face-to-face providing socio-emotional and even financial support to each other through times of crisis and celebration.

Since then, the popularity of virtual communities (also known as online communities) has increased. Interacting with others online became more common as organizations and society began to perceive it as a normal behavior and not one engaged in primarily by the socially inept. Indeed, virtual communities became a typical mode of interaction for both work and social purposes. At work, employees have organizationally sanctioned virtual communities such as the company listserv as well as virtual communities for professionals to interact with each other outside their organizations (e.g., Charity-HR, a listserv for HR professionals in non-profit organizations). Some organizations have even developed virtual communities for their customers. Some of these virtual communities are for users of particular products, like the wristwatch enthusiasts (Rothaermel & Sugiyama, 2001). Others, however, are designed to allow customers to provide input for the company's new products and services (Catterall & Maclaran, 2002).

Virtual communities have also become quite common in social interactions. Many neighborhoods have developed listservs as well as electronic bulletin boards to allow neighbors to interact and share information. Social groups who interact face-to-face (FtF) may also use virtual communities to keep members informed and connected between their meetings. The most common social virtual community, however, may consist of people who are physically dispersed and never interact FtF. These virtual communities are formed around a shared interest in a particular topic. These topics range from movies, to food and wine, to pets, to political topics, and even to aspects of parenthood as evidenced by the hundreds of interactive sites on Babycenter.com.

BACKGROUND

But what are virtual communities and what distinguishes them from mere virtual groups? Ironically, the definition of community has always been a bit difficult. Even among traditional, FtF communities, there are over 71 definitions (see Jones, 1997). Among the issues in defining FtF communities is been whether communities need to be colocated, like a neighborhood, or whether they can be dispersed like a community of interest (e.g., stamp lovers).

Currently, community researchers agree that both co-located and dispersed groups can be communities. However, members of these groups must have a *sense of community* to be considered a community (McMillan & Chavis, 1986). Sense of community is defined as group members' feelings of belonging, identity, attachment, and influence among each other. By using this criterion, virtual communities can be defined as groups of people who interact primarily through e-collaboration technologies and who have developed feelings of belonging, identity, attachment, and influence (i.e., a sense of virtual community) with each other.

Virtual communities have degrees of virtuality. At one extreme are dispersed virtual communities, which exist entirely online. Members of dispersed virtual communities live in many different locations and do not interact with each other FtF. At the other extreme are colocated virtual communities in which members primarily meet FtF, and the e-collaboration technology supplements their interactions. Virtual communities for employees co-located within a single organization as well as neighborhoods, and social/volunteer groups fall primarily into this type. In the middle are virtual communities that exist primarily online. Members may be dispersed or colocated; however, these members additionally interact FtF.

Virtual communities also exist over a variety of e-collaboration technologies (see Figure 1). These technologies can be asynchronous, in which communication is delayed like e-mail or bulletin boards, or synchronous, in which communication is instantaneous like instant messaging and chatrooms. Another key feature is whether the e-collaboration technologies allow one-to-one communication like instant messaging, one-to-many communication like blogs, other Web pages and some information distributing listservs or whether they allow many-to-many communications like bulletin boards and most interactive listservs. Other more advanced e-collaboration technologies allow avatars (pictorial representations of the communicators) as well as two-dimensional representations (e.g., rooms and parks) in which people can interact.

In general, virtual communities are valued because they are considered to have positive effects on both the organizations that sponsor them and within the general community in which they are used. In particular, they are believed to increase the amount of social and intellectual capital available in the organization or larger society. Social capital is defined as the networks, norms, and trust of a group (Putnam, 1996) while intellectual capital is defined as the knowledge that is created and shared within a group (see Bieber et al., 2002).

CURRENT ISSUES IN VIRTUAL COMMUNITIES

One of the most pressing current issues in virtual community research is to understand virtual community success. Virtual community success is defined as the ability for the virtual community to sustain itself while meeting its members' needs and maintaining member satisfaction within the community.

Jones (1997) was one of the first researchers to seek to identify the characteristics of a successful virtual community. He takes an anthropological perspective, arguing that one can identify a successful virtual community when one can identify objective components

of the community's existence. He calls these objective features a *virtual settlement* and argues that they are composed of: (a) a minimal level of interactivity, (b) by a variety of communicators, (c) with a minimum level of sustained membership, and (d) interacting in a common public space. When these four features exceed a minimal threshold, then Jones argues that the online group can be called a virtual settlement. A virtual settlement is distinct from a virtual community like buildings are distinct from a village. However, he argues that once one has identified a virtual settlement, one is likely to have identified a virtual community.

Within successful virtual communities, researchers have additionally identified three types of members: leaders, participants, and lurkers. Leaders have assumed some sort of prominence in the group. Often, they are informal leaders without any sort of formal authority. Instead, leaders are generally prototypical members who are more likely to provide help and assistance to other members.

Participants are members who contribute to the public communications, but are not considered leaders. Lurkers simply read messages but do not publicly contribute to them. Lurkers are sometimes considered negatively (Kollock & Smith, 1996) because they free-load off the other members' contributions. However, this may only be true if the number of active participants is very small and they have to engage in a disproportionate amount of activity for the community to survive. If the total number of participants is high and the number of active participants is adequate enough to spread out the communication effort, then lurkers are not freeloaders. Blanchard and Markus (2004) found in the study of their virtual community that there were approximately 250 active participants and 16,775 lurkers. If each one of these lurkers posted just once, the sheer volume of messages would overwhelm the cognitive capacities of the virtual community members.

Researchers have additionally focused on the social processes of the virtual community participants. They have noted that successful virtual communities have developed particular social processes that help the community function. These include the exchange of socio-emotional and informational support between members, the development of trust between members, and the development and enforcement of norms of behavior.

Figure 1. Types of e-collaboration technologies by communication timing and number of communication partners

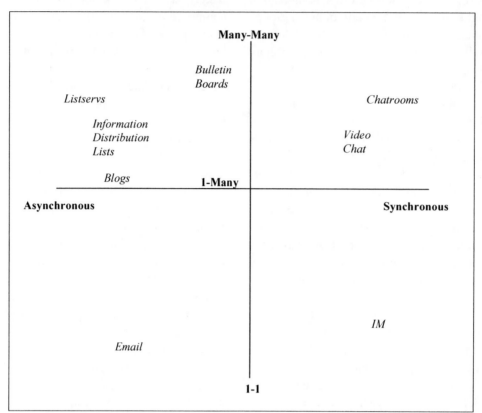

Exchange of Information and Support

The exchange of information and socio-emotional support is one of the most important aspects virtual communities. Indeed, the primary reason that virtual communities attract so much attention is that the exchange of information and socio-emotional support has been found in such a wide variety of social groups as well as organizational groups and in open source communities.

One of the primary issues in this area revolves around why people are compelled to help others, particularly when they don't know the other members. Constant and his colleagues (1997) found that members who provide online help and support to others in a work virtual community are more likely to have a higher regard for and to be good citizens of the sponsoring work organization. That is, their assistance in the online group could be another form of their organizational citizenship behavior (i.e., pro-social extra-role behavior that helps organizational functioning). Other researchers

have found that members provide information when it enhances their status and demonstrates their expertise (Wasko & Faraj, 2005). Certainly, more helpful members of the group are considered more positively by the group, even if they are extreme in the amount of help they provide.

Trust

The development and maintenance of trust is also considered very important in virtual communities (e.g., Boyd, 2002). One reason trust is an essential part of a successful virtual community is that deception is so easy online (Joinson & Dietz Uhler, 2002). Because communicators can remain anonymous, they can easily change relatively minor facets of themselves or their personalities such as their fudging their age, attractiveness or income level. With effort, they can change more major facets such as their gender or their persona. Although the minor forms of deception are not considered altogether inappropriate, the major forms of deception are (Utz, 2005).

Many virtual communities have developed ways of addressing this issue. Ironically, one of the more successful modes for developing trust is when virtual community members to meet face-to-face (Joinson, 2001; McKenna & Green, 2002). Even members who do not actually meet face to face, but hear of others who do believe that the group is more trustworthy (Blanchard & Markus, 2004). Other social processes include having members interact using their own name and their "real" e-mail addresses, as well as following the history of member's posts.

Norms

Finally, the development and enforcement of norms within virtual communities has generated a great deal of interest. Norms of behavior include the appropriate topics of conversations in the group as well as particular styles of communicating within the group. Successful virtual communities must satisfy their members by not only what is discussed but also how it is discussed. For example, health care communities should focus on health care while technological support communities should provide technological support. New members (i.e., newbies) should read the group's FAQs before asking the "same question" that has been answered again and again. Additionally, some virtual communities do not allow members to post advertisements or conduct financial transactions in the group whereas other groups allow it in particular instances.

The actual patterns of communication may also differ between virtual communities with groups (Postmes, Spears, & Lea, 2000). Some groups may be high users of emoticons such as ;-) and acronyms that require insider knowledge by group members. Thus, particular groups are able to distinguish themselves from each other by their unique communication styles, even if the different groups share some of the same members.

The reason members do (or do not) adhere to norms has been of special interest. One approach to understanding whether members follow the group's norms has to do with the salience of the group member's social or individual identity. Spears and Lea (1992) developed a model which argues that virtual community members either have their social identity salient (i.e., they identify as a member of the group) or their individual identity salient (i.e., they identify as a unique individual within the group). Members are going to be more susceptible to group processes when their group identity is salient.

Sanctions

In addition to understanding why members develop and follow group norms, it is also interesting how members enforce their group's norms, that is, how members sanction each other. This is particularly relevant because sanctioning is related to trust.

Norms of behavior are particularly difficult to enforce over e-collaboration technologies because people are removed from each other. If someone behaves inappropriately, it is difficult for other members to sanction them. Whereas in FtF groups, one could use nonverbal language (e.g., frowns, turning the body away) to inform others of their inappropriate behavior, this is not possible online. Instead, virtual community members can ignore the offending member, directly tell the member to stop their behavior, or try to have the member kicked out, which is difficult in non-moderated groups. Trolls, participants who enter groups specifically to challenge group norms, offer a particular challenge to both virtual community members to handle them and researchers in order to understand them.

Other sanctioning options depend on the e-collaboration technology. For example, members can be excluded in some technologies (particularly listservs) but not others (e.g., blogs). Content can be filtered by moderators in listservs but not bulletin boards.

The sanctioning options additionally highlight that although the social processes in virtual communities are quite important, the technological features of the different e-collaboration software are clearly important, too. Some researchers have approached this issue from a technological determinism approach. That is, if the technology is configured in a particular way, then a virtual community will likely develop upon it. The key then is to determine the correct configuration of e-collaboration technology that will support community. However, this approach to understanding virtual communities has fallen out of favor because it does not incorporate the social processes necessary to create and maintain a virtual community (e.g., Walther, 1996, 1997).

Others have thus argued that it is the interaction between the social processes and the technology that affect virtual community success. Indeed, Markus (2005) argues that the next stage of e-collaboration research must acknowledge that particular technological features do play an important role in how well the interactions are supported online. Her approach is not that technology causes behavior online, but that certain technological features can support (or hinder) particular types of interaction, and, by extension, support (or hinder) the success of the virtual community.

Technological features that are may affect virtual community success include the ability to save the community's messages, the ability to reply to messages and retain some part of the previous message, the ability to link threads of messages, the ability communicate both publicly and privately, the ability to limit membership (or not), the ability to track other members' participation including time on the group, number of postings and type of postings, and the ability to detect if others who are currently "in" the community. Future research will need to theoretically and empirically link these technological features to the social processes.

Additionally, the topics discussed here have mainly addressed the success of virtual communities that have already been established. A great deal of research needs to be done in order to determine how virtual communities develop. For example, how easily can virtual communities be designed? Or are they more likely to evolve from an initial group of like-minded people? Certainly, our understanding of what makes an established virtual community successful will provide virtual community developers a goal to which to strive. Nonetheless, developers will have their own unique issues in creating and then maintaining a successful virtual community.

FUTURE TRENDS

There are two future trends in virtual communities. First, the acceptance of e-collaboration technologies continues to increase. As a result, virtual communities are likely to continue develop, become more common, and become more accepted as a productive mode of developing and maintaining work and social relationships. Virtual communities of employees are likely to develop both within and between organizations. Businesses are also likely to pursue their development of virtual communities with customers as a way of promoting their products and developing brand loyalty.

Virtual communities will also continue to develop within society, particularly co-located virtual communities in which people use e-collaboration technologies as an additional mode of communication with their FtF partners. Thus, virtual communities will become

incorporated as another commonplace communication medium for neighbors, schools, and volunteer social and civic organizations.

The second future trend deals with the research on virtual communities. Researchers are just beginning to address how virtual communities influence organizations and society. Initial studies demonstrate that virtual communities generally have positive impacts on FtF social capital (Wellman, Haase, Witte, & Hampton, 2001) although there is little research of their effects on organizations. However, virtual communities of practice may provide a fruitful avenue of research in this area.

In order to better establish this line of research, it will continue to cross disciplinary and methodological boundaries. Because e-collaboration technologies cross the boundaries from work to home to community and to society, research on virtual communities that focuses on one particular discipline is likely to miss important components of virtual communities as well as great bodies of literature. Researchers from social psychology, community, and environmental psychology, organizational psychology and organizational behavior, information science and information technology, business and management, sociology, and communication all have a role to play in understanding virtual communities. This will require not only interdisciplinary thinkers but also training and academies that recognize the importance of cross-disciplinary approaches to understand this phenomenon.

Additionally, methodologies will continue to be diverse. Preece and Maloney-Krichner (2005) note in their special edition on virtual communities for the *Journal of Computer Mediated Communication* that virtual community methodologies range from ethnographies to linguistic analyses to case studies to survey research. Although experimentation is not as popular as it once was (Walther, 1996), clearly even laboratory experiments have a role in understanding virtual communities.

CONCLUSION

Virtual communities have become established in organizations and society as a way for groups of people to interact, help, and connect with each other. Although some researchers still argue whether or not virtual communities exist, others argue that this question is only relevant to those people who have not experienced it (Haythornwaite, Wellman, & Garton, 1998). As e-collaboration technologies become more integrated into people's everyday working and social lives, virtual communities will be become common as a way for the average person to connect with others.

The next phase of virtual community research will be to better examine the antecedents and outcomes of successful virtual communities. The research has moved beyond merely descriptive accounts of virtual communities and is establishing theoretical insights into successful virtual communities. However, a challenge for researches will be to identify the technical features of successful virtual communities without making claims about technological determinism. This will likely require a depth and breadth of knowledge of social processes and technology that is likely to cross traditional academic disciplines.

Finally, because of how quickly e-collaboration technologies develop and change, we can anticipate only that new forms of virtual communities are likely to develop and become popular. For example, blogs are a relatively new form of e-collaboration technology and already, they have become quite popular and have developed, in some cases, into virtual communities. The speed with which these technologies develop will challenge researchers to keep up with them and to identify the new technical features that contribute to or inhibit virtual groupings. Nonetheless, researchers must keep from becoming obsolete. Virtual communities now and in the future can provide important ways to contribute to the functioning and well-being of organizations and society.

REFERENCES

Bieber, M., Engelbart, D., Furuta, R., Hiltz, S. R., Noll, J., Preece, J., et al. (2002). Toward virtual community knowledge evolution. *Journal of Management Information Systems, 18*(4), 11-35.

Blanchard, A. L., & Markus, M. L. (2004). The experienced "sense" of a virtual community: Characteristics and processes. *The DATA BASE for advances in information systems, 35*(1), 65-79.

Boyd, J. (2002). In community we trust: Online security communication at eBay [Electronic version]. *Journal of Computer Mediated Communication, 7*(3).

Catterall, M., & Maclaran, P. (2002). Researching consumer in virtual worlds: A cyberspace odyssey. *Journal of Consumer Behaviour, 1*(3), 228-237.

Constant, D., Sproull, L., & Kiesler, S. (1997). The kindness of strangers: On the usefulness of electronic weak ties for technical advice. In S. Kiesler (Ed.), *Culture of the Internet* (pp. 303-322). Lawrence Erlbaum Associates.

Haythornwaite, C., Wellman, B., & Garton, L. (1998). *Work and community via computer-mediated communication*. New York: Academic Press.

Joinson, A. N. (2001). Self-disclosure in computer-mediated communication: The role of self-awareness and visual anonymity. *European Journal of Social Psychology, 31*(2), 177-192.

Joinson, A. N., & Dietz-Uhler, B. (2002). Explanations for the perpetration of and reactions to deception in a virtual community. *Social Science Computer Review, 20*(3), 275-289.

Jones, Q. (1997). Virtual-communities, virtual settlements & cyber-archaelogy: A theoretical outline. *Journal of Computer-Mediated Communication, 3*(3), 24.

Kollock, P., & Smith, M. (1996). Managing the virtual commons: Cooperation and conflict in computer communities. In S. Herring (Ed.), *Computer-mediated communication: Linguistic, social, and cross-cultural perspectives*. Amsterdam: John Benjamins.

Markus, M. L. (2005). Technology-shaping effects of e-collaboration technologies: Bugs and features. *International Journal of e-Collaboration, 1*(1), 1-23.

McKenna, K. Y. A., & Green, A. S. (2002). Virtual group dynamics. *Group Dynamics, 6*(1), 116-127.

McMillan, D. W., & Chavis, D. M. (1986). Sense of community: A definition and theory. *Journal of Community Psychology, 14*, (6-23).

Postmes, T., Spears, R., & Lea, M. (2000). The formation of group norms in computer-mediated communication. *Human Communication Research, 26*(3), 341-371.

Preece, J., & Maloney-Krichner, D. (2005). Online communities: Design, theory, and practice [Electronic version]. *Journal of Computer Mediated Communication, 10*(4).

Putnam, R. D. (1996). Bowling alone: America's declining social capital. *Journal of Democracy, 6,* 65-78.

Rheingold, H. (1993). *The virtual community: Homesteading on the electronic frontier.* Reading, MA: Addison-Wesley.

Rothaermel, F. T., & Sugiyama, S. (2001). Virtual Internet communities and commercial success: Individual and community-level theory grounded in the atypical case of TimeZone.com. *Journal of Management, 27*(3), 297-312.

Spears, R., & Lea, M. (1992). Social influence and the influence of the 'social' in computer mediated communication. In M. Lea (Ed.), *Contexts of computer-mediated communication.* New York: Harvester Wheatsheaf.

Utz, S. (2005). Types of deception and underlying motivation. *Social Science Computer Review, 23*(1), 49-56.

Walther, J. B. (1996). Computer-mediated communication: Impersonal, interpersonal, and hyperpersonal interaction. *Communication Research, 22,* 33-43.

Walther, J. B. (1997). Group and interpersonal effects in interactional computer mediated collaboration. *Human Communication Research, 23,* 342-369.

Wasko, M., & Faraj, S. (2005). Why should I share? Examining social capital and knowledge contribution in electonic networks of practice. *MIS Quarterly, 29*(1), 247-253.

Wellman, B., Haase, A. Q., Witte, J., & Hampton, K. (2001). Does the Internet increase, decrease, or supplement social capital? Social networks, participation and community commitment. *American Behavioral Scientist, 45,* 437-456.

KEY TERMS

Common Bond Communities: Virtual communities that form because of the members' relationships with each other.

Common Identity Ccommunities: Virtual communities that form because of the members' interest in a particular topic.

Co-Located Virtual Communities: Virtual communities which are associated with a physical location. Members interact both online and face-to-face.

Dispersed Virtual Ccommunities: Virtual communities which are not associated with a physical location. Members are dispersed around the globe and may not ever meet each other face-to-face.

Emoticons: Combinations of text that are believed to portray communicator emotions, for example, :-) and :-(.

Sense of Virtual Ccommunity: Members' feelings of belonging, attachment, identity and influence with each other in a group supported by e-collaboration technology.

Trolls: Members who enter virtual communities with the primary objective of stirring up trouble among the established members

Virtual Communities: Also known as online communities, these are groups of people who interact primarily through e-collaboration technologies and who have developed a sense of community with each other.

Virtual Settlements: An e-collaboration technology supported group identified by a minimum number of interactive, public interactions by a variety of sustained contributors.

Virtual Community Success: When virtual communities are self-sustaining and meet the needs of their members and maintain member satisfaction

A Design Framework for Mobile Collaboration

Pedro Antunes
University of Lisboa, Portugal

INTRODUCTION

Mobile collaboration involves people working together and moving in space. Research in mobile collaboration has primarily focused on technical issues like connectivity support or remote information access. We argue there is a lack of research on many nontechnical issues vital to design mobile collaboration systems, disentangling the relationships between collaboration, work context, and mobility.

Our fundamental concern is to go beyond the technical issues towards the assimilation of the mobility dimension in all processes shaping collaborative work, including information sharing, context awareness, decision making, conflict management, learning, etc. This article aims to codify into a design framework:

- Some fundamental human factors involved in mobile collaboration.
- Several guidelines for developing mobile collaboration systems.

The design framework provides general constructs identifying phenomena of interest necessary to inquire about the work context, human activities, and system functionality. The framework identifies *what* information may interest designers, bounding their relationships with the other stakeholders. The framework also guides the design process, identifying *how* user requirements may be applied during the implementation phase.

The framework has been validated in several real-world design cases. Two cases will be briefly described. This research contributes to the design of mobile collaborative systems. The most significant contributions are related to artifacts and emphasize that designers shall explore the potential of artifacts to support concerted work and sensemaking activities.

BACKGROUND

Several conceptual frameworks have been proposed in the group support systems (GSS) field (DeSanctis &

Gallupe, 1987; Nunamaker, Dennis, Valacich, Vogel, & George, 1991; Pinsonneault & Caya, 2005). However, these frameworks capture the notion of place in a very restrictive way, more tied to group proximity than mobility, where geographical references play a central role in tying information together (Mackay, 1999).

The above limitation is being tackled in two closely related research areas: collaborative spatial decision-making (CSDM) and spatial decision support systems (SDSS) (Nyerges, Montejano, Oshiro, & Dadswell, 1997). SDSS address the combination of DSS with geographical information systems (GIS), while CSDM studies the integrated support to collaboration, decision, mobility, and geographical information.

We find several studies on the infrastructural basis of SDSS. Zhao, Nusser, and Miller (2002) identify the infrastructural requirements for SDSS. Gardels (1997) and Touriño et al. (2001) contribute with the integration of multimedia with geo-referenced data. Hope, Chrisp, and Linge (2000) tackle the access to remote databases by fieldworkers, while Pundt (2002) addresses data visualization in the same context. All of these research projects do not directly address mobile collaboration but explore basic features necessary to support this functionality.

Regarding the human factors of SDSS, we account for studies of user interaction with multimodal and tangible GIS interfaces (Coors, Jung, & Jasnoch, 1999; Rauschert, Agrawal, Sharma, Fuhrmann, Brewer, & MacEachren, 2002). In the same line, we also cite developments in synthetic collaborative environments for geo-visualization (Grønbæk, Vestergaard, & Ørbæk, 2002; Manoharan, Taylor, & Gardiner, 2002). However, these research studies address fixed work settings.

More in line with collaboration studies, we find several research emphasizing the need to support group modeling in CSDM (Armstrong, 1994, 1997). Some propose very specific solutions, such as the integration of workflow management with SDSS (Coleman & Li, 1999).

Finally, addressing the broad-spectrum CSDM design, we find the work from Tamminen, Oulasvirta, Toiskallio, and Kankainen (2004), who propose an integrated framework with guidelines for eliciting innovative ideas for mobile technology based on context-awareness (although not collaboration). Nyerges et al. (1997) also propose an integrated framework for CSDM, but the framework is specific for the transportation context.

As demonstrated by the research previously cited, there is a whole new perspective over GSS brought by the mobility dimension, making CSDM quite distinct from GSS. However, the most important distinctions are not captured by current GSS and CSDM frameworks: (1) the central role of geo-references in the information architecture; (2) the interaction support to obtain, manage and share geo-referenced data while in the field; (3) the role of geo-references in modeling group work; and (4) the added impact of context awareness in the system design, regarding in particular work place mobility. Our perspective is that we need to integrate these various phenomena into a meaningful and purposeful framework.

THE FRAMEWORK

The framework is bounded by two major requirements: It has to be open for exploring and interpreting mobile collaboration in various settings, thus requiring relatively abstract elements and constructs, and it has to link them in a purposeful way. Our major goal is to set the initial boundaries for inquiring about mobile collaboration, setting at the same time a design roadmap.

The framework, shown in Figure 1, is structured around five basic elements and four design phases. The basic elements are teams, tasks, artifacts, and places, while the design phases consider data collection, work analysis, prototyping, and value determination. As described below in more detail, the basic elements have an important role throughout the design phases, structuring the various design activities taking place in each phase.

The relationships between the five basic elements are defined as follows. Teams manipulate artifacts to accomplish tasks in certain places. This combination of elements affords the most common spatial arrangements that we find in collaborative settings. The same argument applies to artifacts and tasks, were we may consider having artifacts/tasks fixed in a single place, distributed, or moving through several places. We assume these elements are consensual in the CSDM field, so that no further considerations are necessary.

In contrast, the relationship between artifacts and tasks, noted as *collaborative capability*, deserves further consideration. The notion of collaborative capability (Nunamaker, Romano, & Briggs, 2002) identifies several categories of increasing ability for successful creation of meaning, ranging from the individual, collective, and coordinated to the concerted creation of meaning. The theory is that organizations will increase their potential to create value by increasing their collaborative capability. Further details and validity tests of this theory can be found in Bach, Belardo, and Faerman (2004) and Qureshi and Briggs (2003). We realize this theory has an immediate impact in CSDM design, because work processes are affected by geographical constraints and thus there may be an opportunity for increasing the organizational effectiveness. From this

Figure 1. Design framework for mobile collaboration

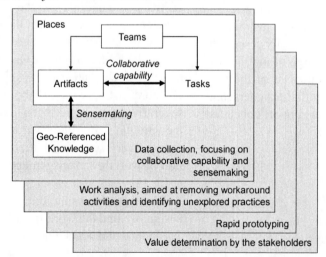

theory we draw an implication for design: The development of shared artifacts, supporting concerted tasks, should be preferred to the development of individualized artifacts so that work processes become independent of geographical constraints.

The final framework basic element is geo-referenced knowledge. We regard the manipulation of artifacts, in mobile collaboration, not an end in itself but a mean to construct and augment shared knowledge about the work space and the objects found on it. This shared knowledge is necessarily tied to geographical references and mediated through artifacts. We may characterize the relationship between artifacts and geo-referenced knowledge as *sensemaking*: an ongoing process aiming to create order and make retrospective sense of what occurs (Weick, 1993). We argue sensemaking precisely captures the fundamental nature of mobile collaboration: people handling together information in fluid contexts. As the sensemaking theory posits, the outcomes from mobile collaboration result from "thinking by doing" (Weick, 1993), since problems and solutions are highly context dependent. The presence of this element in the framework introduces one more implication for design: artifacts must enrich sensemaking by integrating mechanisms for searching, browsing, visualizing or summarizing geo-referenced information.

We now turn our attention to the design phases. The first phase concerns data collection aimed at understanding the work context. In this phase we adopt the contextual inquiry method (Beyer & Holtzblatt, 1998), which utilizes a mix of ethnography and interviews to understand the work. While contextual inquiry is context independent, this phase is structured around the framework basic elements, and specifically collects data about collaborative capability and sensemaking (how users organize themselves and make sense of geo-referenced data).

The second phase is dedicated to analyze work from the field data. Again, the framework plays an important role centering the analysis around places, artifacts and geo-referenced knowledge, focusing the modeling activity on the phenomena of most interest to mobile collaboration. We also suggest that attention to collaborative capability and sensemaking will raise new opportunities for removing workaround activities and identifying unexplored work practices, which are characteristic of innovative design solutions (Vicente, 1999).

The third phase is rapid prototyping. Here, low- or mid-fidelity prototypes serve to communicate with the

stakeholders and evaluate the feasibility of the design ideas. The prototypes are fundamentally built around artifacts, task support and geo-referenced knowledge management.

Finally, the last step concerns the value determination by the stakeholders. We have been using context interviews (Beyer & Holtzblatt, 1998) to gather feedback from the stakeholders about the design solutions. Next, we describe two cases where this framework has been applied.

CASE STUDY ONE

This case addressed work redesign at a national agency responsible for inventorying geological resources. One major problem with this organization was that an inventory process took a long time to complete, mostly because experts had to go repeatedly to the field to retrieve information and resolve conflicts.

The framework helped organizing the field observations and interviews with experts involved in the process. This way we came to understand how work moved between the office and the field, what artifacts were used, and how geological information was gathered, analyzed, organized and consolidated. The inventory process required a combination of individual and collaborative activities, since expertise from different fields had to be combined.

Then, we began to analyze the work process, focusing on the five basic framework elements: teams, tasks, artifacts, places, and geo-referenced knowledge. At this stage we realized that a typical geological inventory took about 2 years to complete, as a consequence of several visits to the field, multiple activities in the office and many gap periods. Several critical incidents concurred to this situation: (1) bad initial data; (2) the occurrence of doubts when in the field or in the office; (3) the occurrence of conflicts between experts, which could only be resolved by sending someone to the field for confirmation; and (4) the concurrent execution of multiple inventory processes, causing management and planning difficulties. The framework had also a crucial role in the identification of the major design requirements:

- Fieldwork evolved around two artifacts: the field book and the combination of a map with a transparent overlay. The map/overlay allowed drawing inventory data, while the field book was used to annotate supplementary information, including doubts and concerns arising in fieldwork. All relevant knowledge was geo-referenced, both in the map/overlay and field book.

- The field book was personal, signifying a reduced collaborative capability. This indicated that sharing the field book could increase the collaborative capability.
- Sensemaking was problematic because of the many unresolved doubts arising during fieldwork and difficulties reconstructing the field context in the office. Also, geo-referenced knowledge was distributed between the field book and map/overlay, which were difficult to co-relate. These observations indicated there was ample opportunity to develop information management mechanisms aiming to increase sensemaking.
- The inventory process was delayed by the need to swap work between the office and the field, a situation which could be resolved by increasing the team's collaborative capability: Bringing all relevant stakeholders together to resolve problems as they were appearing in the field or in the office.

These requirements lead us to prototype a digital artifact integrating the field book and map/overlay, and supporting cross-referencing and searching. We also allowed the fieldworker to contact the office workers using GPRS and an instant messaging mechanism. The redesigned work process allowed the fieldworker to get in contact with the office workers and immediately exchange comments on any occurring problem. The elements in the field book were synchronized to keep the conversation in context and facilitate sensemaking. Also, the fieldworker had an easier task when moving back to the office. Because doubts were resolved in the field, there was less time spent in the office. Addressing our observation that all knowledge were geo-referenced, the instant messages exchanged between the field and office workers were preserved in the field book with automatic associations to the geographical position of the fieldworker, thus keeping the doubts, comments or opinions in their context.

The prototype was evaluated with a field test and contextual interviews with several experts from the national agency. The obtained results indicate that the system increased sensemaking and collaborative capability. Related to sensemaking, the participants regarded very positively the expeditious way to locate

points and associate them in the field book. Related with collaborative capability, the participants were extremely favorable to the communication between field and office workers, effectively resolving problems occurring in the field and thus simplifying the whole inventory process. More details about this case study can be found in (Antunes & André, 2006).

CASE STUDY TWO

This case involved work optimization in a small accountancy company, where meetings were the primary coordination mechanism. The company was not satisfied with the meetings productivity and regarded technology as a silver bullet. Different alternatives were experimented, which included the use of GSS and workflow tools, but cultural factors contributed to an unenthusiastic view of these technologies since they imposed too much structure to meetings. We proposed an alternative approach, which would not conflict with their informal work organization. The proposal considered the use of personal digital assistants (PDA) in meetings.

The framework allowed organizing the several data collected from interviews and meeting observations. We observed that the company had three types of meetings: (1) briefings, aimed to discuss ongoing projects; (2) planning meetings, where tasks and personnel were allocated to new projects; and (3) process definition meetings, where the whole collection of projects was taken in perspective to ensure an adequate allocation of resources. Different teams participated in these meetings, accomplishing different tasks and using different artifacts and knowledge.

One issue raised by the framework during data collection was to identify people and information mobility related with meetings. During work analysis we characterized the specific nature of the artifacts moved by the accountants, such as meeting agendas, "to do" lists, and calendaring information. We came to understand two fundamental problems related with collaborative capability and sensemaking:

- It was difficult to move artifacts out of meetings. Sensemaking was affected by the lack of context (e.g., when a meeting outcome was delivered to someone that did not participate in the meeting).
- Meetings were affected by reduced collaborative capability, in particular the absence of a shared

whiteboard capable to integrate the data brought by the participants.

These problems lead to the development of a prototype with the following characteristics: use PDA to bring information into and out of meetings; integrate the meeting information in a shared whiteboard; and supply a sensemaking mechanism capable to display the information flows across several meetings in an integrated way.

This case study was evaluated in two dimensions: framework and prototype. Selected accountants participated in evaluation tasks carried out at each design stage, evaluating the quality of data collected, work analysis, design ideas and prototype. The obtained feedback indicated that the framework was useful to elicit the organizational context of the problem. The evaluators also considered the data collection phase very useful and efficient. The work analysis phase was also considered very useful to help them understand the possibilities and limitations of the proposed solution.

Concerning the prototype, the evaluators considered the sensemaking functionality very useful and adjusted to their needs and thought that the simplicity of the PDA role bringing information in and out of meetings was adequate to their expectations, provided that not much text editing was required. More details about this case study can be found in Antunes and Costa (2002) and Costa, Antunes, and Dias (2002).

CONCLUSION

One important advantage of design frameworks is codifying current knowledge and best practices into design guidelines directly pointing towards where innovation may emerge. Our framework leads designers to identify meaningful ways to articulate places, users, tasks, artifacts and geo-referenced knowledge. The framework also guides the design process, keeping the designer focused on the issues most relevant to mobile collaboration.

The presented case studies highlight two different contexts where the framework pointed directly towards these concerns and definitely was useful informing the adopted designs. The evaluations conducted within the case studies confirmed the relevance of the framework as well as the relevance of the adopted design solutions. Artifacts emerged as the most important area of

concern in mobile collaboration, mostly because they have potential to increase the collaborative capability and sensemaking.

REFERENCES

Antunes, P., & André, P. (2006). A conceptual framework for the design of geo-collaborative systems. *Group Decision and Negotiation, 15*, 273-295.

Antunes, P., & Costa, C. (2002). Handheld CSCW in the meeting environment. *LNCS, 2440,* 47-60.

Armstrong, M. (1994). Requirements for the development of GIS-based group decision support systems. *Journal of the American Society for Information Science, 45*(9), 669-677.

Armstrong, M. (1997). *Emerging technologies and the changing nature of work in GIS*. Bethesda, MD: American Congress on Surveying and Mapping.

Bach, C., Belardo, S., & Faerman, S. (2004). Employing the intellectual bandwidth model to measure value creation in collaborative environments. In *Proceedings of 37th Hawaii Int. Conference on System Sciences.*

Beyer, H., & Holtzblatt, K. (1998). *Contextual design: Defining customer-centered systems*. San Francisco: Morgan Kaufmann.

Coleman, D., & Li, S. (1999). Developing a groupware-based prototype to support geomatics production management. *Computers, Environment and Urban Systems, 23,* 1-17.

Coors, V., Jung, V., & Jasnoch, U. (1999). Using the virtual table as an interaction platform for collaborative urban planning. *Computers & Graphics, 23*(4), 487-496.

Costa, C., Antunes, P., & Dias, J. (2002). Integrating two organisational systems through communication genres. *LNCS, 2315,* 125-132.

DeSanctis, G., & Gallupe, R. (1987). A foundation for the study of group decision support systems. *Management Science, 33*(5), 589-609.

Gardels, K. (1997). Open GIS and on-line environmental libraries. *SIGMOD Record, 26*(1), 32-28.

Grønbæk, K., Vestergaard, P., & Ørbæk, P. (2002). Towards geo-spatial hypermedia: Concepts and prototype implementation. In *Proceedings of the 30th ACM Conference on Hypertext and Hypermedia.*

Hope, M., Chrisp, T., & Linge, N. (2000). Improving co-operative working in the utility industry through mobile context aware geographic information systems. In *Proceedings of the 8th ACM International Symposium on Advances in Geographic Information Systems.*

Mackay, S. (1999). Semantic integration of environmental models for application to global information systems and decision-making. *ACM SIGMOD Record, 28*(1), 13-19.

Manoharan, T., Taylor, H., & Gardiner, P. (2002). A collaborative analysis tool for visualisation and interaction with spatial data. In *Proceedings of the 7th International Conference on 3D Web Technology.*

Nunamaker, J., Dennis, A., Valacich, J., Vogel, D., & George, J. (1991). Electronic meeting systems to support group work: Theory and practice at Arizona. *Communications of the ACM, 34*(7), 40-61.

Nunamaker, J., Romano, N., & Briggs, R. (2002). Increasing intellectual bandwidth: Generating value from intellectual capital with information technology. *Group Decision and Negotiation, 11*(2), 69-86.

Nyerges, T., Montejano, R., Oshiro, C., & Dadswell, M. (1997). Group-based geographic information systems for transportation site selection. *Transportation Research, 5*(6), 349-369.

Pinsonneault, A., & Caya, O. (2005). Virtual teams: What we know, what we don't know. *International Journal of e-Collaboration, 1*(3), 1-16.

Pundt, H. (2002). Field data collection with mobile GIS: Dependencies between semantics and data quality. *GeoInformatica, 6*(4), 363-380.

Qureshi, S., & Briggs, R. (2003). Revision the intellectual bandwidth model and exploring its use by a corporate management team. In *Proceedings of 36th Hawaii International Conference on System Sciences.*

Rauschert, I., Agrawal, P., Sharma, R., Fuhrmann, S., Brewer, I., & MacEachren, A. (2002). Designing a human-centered, multimodal GIS interface to support emergency management. In *Proceedings of 10th ACM International Symposium on Advances In Geographic Information Systems.*

Tamminen, S., Oulasvirta, A., Toiskallio, K., & Kankainen, A. (2004). Understanding mobile contexts. *Personal and Ubiquitous Computing, 8*(2), 135-143.

Touriño, J., Rivera, F., Alvarez, C., Dans, C., Parapar, J., Doallo, R., et al. (2001). COPA: A GE-based tool for land consolidation projects. In *Proceedings of the 9th ACM International Symposium on Advances in Geographic Information Systems.*

Vicente, K. (1999). *Cognitive work analysis: Toward safe, productive, and healthy computer-based work.* Erlbaum.

Weick, K. (1993). The collapse of sensemaking in organizations: The Mann Gulch disaster. *Administrative Science Quarterly, 38*, 628-652.

Zhao, P., Nusser, S., & Miller, L. (2002). Design of field wrappers for mobile field data collection. In *Proceedings of 10th ACM international symposium on Advances in geographic information systems.*

KEY TERMS

Mobile Collaboration: People collaborating and moving through space.

Design Framework: A collection of general constructs identifying phenomena of interest and guiding the design process.

Collaborative Spatial Decision Making: The integrated study of collaboration, decision making, and mobility support.

Collaborative Capability: Defines four levels in increasing ability to create meaning: individual, collective, coordinated, and concerted.

Sensemaking: An ongoing process aiming to create order and make sense of what occurs.

Geo-Referenced Knowledge: Knowledge that is tied to a geographical reference.

Group Support System: A technological system supporting and mediating group work.

Design Patterns for Facilitation in E-Collaboration

Gwendolyn L. Kolfschoten
Delft University of Technology, The Netherlands

Robert O. Briggs
University of Nebraska at Omaha, USA
University of Alaska Fairbanks, USA

Gert-Jan de Vreede
University of Nebraska at Omaha, USA
Delft University of Technology, The Netherlands

INTRODUCTION

Collaboration is essential for the creation of organizational value (Hlupic & Qureshi, 2002, 2003). In our current global economy, there are many groups that have few possibilities to meet face to face, and therefore must hold their collaboration processes in a distributed electronic environment. Collaboration and e-collaboration can be challenging (Nunamaker, Briggs, Mittleman, Vogel, & Balthazard, 1997), especially in a distributed environment. When groups face complex tasks, they often find it difficult to follow a focused, effective, and efficient path to accomplish their goals. Groups therefore frequently resort to the use of facilitators and facilitation techniques. However, facilitation itself is a challenging undertaking (den Hengst & Adkins, 2005; Niederman, Beise, & Beranek, 1993; Romano, Nunamaker, Briggs, & Mittleman, 1999; Zhao, Nunamaker, & Briggs, 2002), particularly in a distributed setting.

Facilitators are group process professionals who design and conduct processes to help a group achieve its goals. Facilitation is a complex task. Facilitators must master a collection of techniques skills and interventions, and must attend to many simultaneous details in their work (Clawson, Bostrom, & Anson, 1993). Effective facilitation therefore requires extensive training and experience. Experienced facilitators typically know and use a larger set of techniques than novice facilitators (Kolfschoten, den Hengst, & de Vreede, 2005). Communities of facilitators often draw upon libraries with facilitation techniques (Briggs & de Vreede, 2001; FacilitatorU, 2005; Jenkins, 2005).

This article will discuss the added value of capturing and sharing facilitation techniques. Facilitation technique libraries can offer a learning source for novice facilitators, but can also function as a language among facilitators. In order to use facilitation techniques predictably, we need to capture techniques that are frequently used and that have predictable outcomes. In this research we will show research results in which collaboration patterns are identified on different levels. Patterns in collaboration can be recreated through the documentation of design patterns, scripts to capture reusable solutions to recurring problems. We will first explain what design patterns are, and how they are used in facilitation. Next we will present results from an analysis of the transcripts of 93 group support systems (GSS) sessions that took place between 2000 and 2002. In these sessions we identified patterns of facilitation interventions. We will explain these interventions and how they can be documented and used to recreate specific patterns in e-collaboration, and thus create predictable facilitation techniques.

BACKGROUND

Design patterns are reusable solutions to recurring problems. They were originally introduced by Alexander, Ishikawa, Silverstein, Jacobson, Fiksdahl-King, and Angel (1977), in the domain of architecture. However, design patterns can be created for many design disciplines. For example, design patterns were introduced in the object oriented software modeling in the beginning of the 1990s (Gamma, Helm, Johnson, & Vlissides, 1995), and have been applied to the development of communication software (Rising, 2001), productivity software (Harrison & Coplien, 1996), and e-learning (Niegemann & Domagk, 2005). When a number of patterns are collected in libraries, they constitute a pattern language.

Alexander (1979) suggests the following benefits for design patterns and pattern languages:

- As a common language. Design patterns are a language, a vehicle for communication. It enables users to name and share complex concepts without having to explain them over and over again.
- For design and as inspiration for new or improved design patterns. Design patterns describe solutions to problems that occur over and over again. These solutions can be used separately, or to inspire designers to create new solutions.
- To design solutions in a specific domain.
- For teaching, to capture and share expert knowledge.
- To enable anyone to design the specific solutions or objects. Alexander's idea was that with his books people could build houses by themselves.
- To enable the creation of objects that are lively and improve the quality of human life. Alexander's pattern language serves a higher purpose; the patterns he and his colleagues described should create morally sound objects.
- To enable the creation of a whole coherent system, instead of loose individual objects that are not in harmony with their environment.

DESIGN PATTERNS FOR FACILITATION

Recently, researchers have begun to document a design pattern language called *thinkLets* for collaborative work practices (Briggs, de Vreede, & Nunamaker, 2003). A thinkLet is a named, documented facilitation technique that produces a known pattern of collaboration among people working toward a goal. ThinkLets are meant to be the smallest unit of intellectual capital needed to reliably recreate a pattern of collaboration in a group (Briggs et al., 2003). A thinkLet provides a transferable, reusable, and predictable building block for the design of a collaboration process. The patterns of collaboration that thinkLets create can be classified in six general patterns of collaboration (Briggs, Kolfschoten, de Vreede, & Dean, in press). These patterns are:

- **Generate:** Move from having fewer to having more concepts in the pool of concepts shared by the group
- **Reduce:** Move from having many concepts to a focus on fewer concepts that the group deems worthy of further attention
- **Clarify:** Move from having less to having more shared understanding of concepts and of the words and phrases used to express them.
- **Organize:** Move from less to more understanding of the relationships among concepts the group is considering
- **Evaluate:** Move from less to more understanding of the relative value of the concepts under consideration
- **Build consensus:** Move from having fewer to having more group members who are willing to commit to a proposal

ThinkLets are facilitation techniques & interventions that are documented according to a specific format. While there is a current set of thinkLets created by documenting the best practices of expert facilitators, any facilitation technique for collaboration and e-collaboration can be documented as a thinkLet and added to the pattern language. Variations and modifications in the use of thinkLets can result in new thinkLets.

The resulting thinkLets may be useful for the following purposes:

- **Comparative research:** ThinkLets allow researchers to compare different patterns of collaboration. This will increase our understanding of collaboration efforts (Santanen, 2005).
- **Consistency and completeness of the thinkLet library:** The thinkLet documentation format enables the documentation of thinkLets in a way that enables others to use them, it forces the author of a thinkLet to be complete and consistent.
- **Classification:** Finding a good, taxonomic classification of thinkLets will simplify the choice of thinkLets at design time, and also enable collaboration engineers and researchers to find the gaps in the library of facilitation techniques.
- **Predictability:** ThinkLets create a predictable result in terms of the patterns of collaboration that the participants exhibit. This predictability is currently mostly based on expert facilitators' testimony of their best practices. A first empiri-

cal attempt to determine the predictability of facilitation interventions is described by Santanen (Santanen, de Vreede, & Briggs, 2004).

Facilitators and communities of facilitators in particular thus can benefit from thinkLets or facilitation design patterns in a number of ways. Documenting facilitation design patterns in libraries will enable a community of facilitators to share best practices and to select predictable appropriate facilitation techniques to support a group in their e-collaboration process.

METHOD: PATTERN HARVESTING AND RECOGNITION

In this article we corroborate several thinkLets and the patterns of collaboration they create through the analysis of patterns in the transcripts of collaboration processes. ThinkLets are facilitation techniques and interventions that are documented according to a specific format. While there is a current set of thinkLets, which is created through the rigorous documentation of best practices of experts, any facilitation technique for collaboration and e-collaboration can be documented as a thinkLet. A set of thinkLets can be used as a pattern language. When thinkLets are used they create predictable patterns of collaboration. Variations and modifications in the use of thinkLets result in new thinkLets. This process is described in Figure 1.

To date, over 50 different thinkLets have been described (Briggs & de Vreede, 2001). Many were captured from the experiences of expert facilitators around the globe. To determine if such patterns indeed emerge

in e-collaboration sessions, and whether there were also higher level patterns in the use of thinkLets, a set of 150 sessions conducted between 2000 and 2002 were initially selected to be analyzed for pattern recognition and harvesting. This set was reduced by only retaining 93 sessions that met the following criteria:

- *The session involved real problems from the workplace rather than academic exercises.* Academic exercises do not adequately represent the way real organizational use facilitation techniques. Demonstration sessions may serve to illustrate the value of an intervention, but may not reflect actual use patterns among people working together toward a real organizational goal.

- *The session consisted of at least one module.* A module consists of a series of activities that modify, extend or use the same data set in the GSS transcript. It is possible that recurring modules may exist that operate on more than one data set, but in *post hoc* analysis of transcripts, it is not possible to determine whether thinkLets that do not share a data set were deliberately juxtaposed as a thinkLet, or whether they were used to achieve two different goals.

- *The sequence of the activities in a module was clear.* We excluded sessions where it was not possible to determine by *post hoc* analysis the order in which the activities had been conducted. The sequence was not always clear from the agenda.

- *The thinkLets were clear.* The thinkLets could all be identified or reconstructed from the available information in the transcripts, and uncertainties

Figure 1. The relation between different types of patterns and facilitation techniques

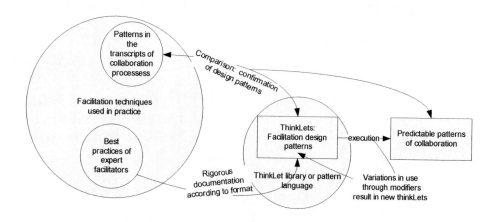

could be resolved by interviewing the session's facilitator.

PATTERN RECOGNITION

Based on the transcripts and discussion with the facilitators we identified 16 different thinkLets and 2 recurring variations on those thinkLets. Many of these thinkLets occurred several times in the data set. Most often used is an elaboration pattern. In this pattern a set of ideas, categories or propositions is use as a starting point for elaboration. Elaboration is a sub-pattern of the more-general generate pattern. In different variations, this patter occurred 285 times in the 93 sessions, meaning that in many sessions the elaboration pattern occurred several times, sometimes in further elaboration, sometimes on different topics.

Feedback from facilitators and investigation of their meeting agenda's revealed that facilitators often develop a preference for certain facilitation techniques. Based on the ones they know well, and are skilled in using, they tend to keep their set of frequently used patterns limited compared to the available libraries in their communities (Kolfschoten et al., 2005).

ThinkLets also tended to be used in a specific phase of the process (Kolfschoten, Appelman, Briggs, & de Vreede, 2004a). Based on the required input and the type of output, a thinkLet often has a typical place in the facilitation process. For instance generation thinkLets, thinkLets in which the group moves from having few ideas to having a large set of ideas, often occur at the beginning of a group process, where evaluation thinkLets, where the group determines the (relative) value of concepts, occur more at the end of the process. In the middle of a process, a frequently recurring pattern was an organizing thinkLet, in which ideas were classified in categories (Kolfschoten et al., 2004a).

Higher Level Patterns or Sequences

The data set described above also showed higher level pattern design, consisting of recurring sequences of 2 to 4 thinkLets. A sequence is a combination of thinkLets that describe an order in which they can be executed for a specific purpose (Lukosch & Schümmer, 2005). An example is the use of a sequence of generation thinkLets in the group elaborates on a topic and then elaborates further on the ideas that came from the first activity. Creating a detailed structure. Often this combination was followed by an evaluation thinkLet, to select the best ideas (Kolfschoten et al., 2004a). Sequences can also be used as a pattern for a recurring e-collaboration task. Combinations of thinkLets of then have a specific added value. In Figure 2 we visualized how thinkLets and sequences appear in a collaboration process.

Harvesting New Patterns

In the data set we also identified a few recurring variations on the thinkLets documented. We call these variations modifiers. A modifier is an intervention, which on its own does not create a repeatable pattern of collaboration, but which can be applied to different thinkLets to create a recurring variation (Kolfschoten et al., 2004a; Kolfschoten, Briggs, Appelman, & de Vreede, 2004b; Kolfschoten, Briggs, de Vreede, Jacobs, & Appelman, in press). Such interventions are powerful facilitation tricks that can be used to re-focus the process, to solve or prevent a conflict or confusion and to emphasize a specific type of outcome. Because the interventions are not creating a full pattern of collaboration, but rather adjust one, the same modifier can be applied to several thinkLets. An example is to stimulate participants to brainstorm ideas that are better than the ones that they already collected. This technique can be applied to different brainstorming thinkLets. In Figure

Figure 2. Visualization of the relation between thinkLets, sequences and modifiers in a collaboration process

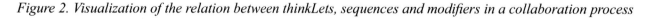
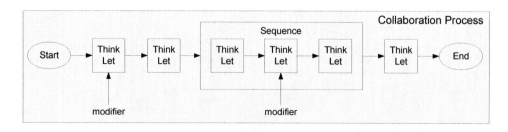

2 we visualized how thinkLets and modifiers relate in a collaboration process.

CONCLUSION

Patterns of facilitation techniques and interventions can be a valuable source of learning and design for the facilitation of e-collaboration processes. They can be used as a shared language, and capturing them allows facilitators to select among a larger library of facilitation techniques. To this end, thinkLets or facilitation design patterns should be documented with sufficient detail, and according to a specific template (see the article titled "Thinklets for E-Collaboration" for more detail). Different aspects of this template feature different functions of facilitation design patterns. For example, naming the thinkLet and providing a picture supports the memorization of the thinkLet. By giving the thinkLet a clear label, faster and more specific communication within the facilitation community can be accomplished. Other elements of the thinkLet documentation format support their transferability, the selection among thinkLets, their classification, their application in different environments and the comparison of thinkLets (Kolfschoten et al., 2004b; Kolfschoten et al., in press; Santanen, 2005).

One of the important features of Alexander's pattern language was that it was made to support users, not only in building a house, but more in building communities that are lively and coherent. When we translate this to facilitation design patterns, this would constitute that a good library of thinkLets should not only support the facilitator in choosing an appropriate facilitation technique for the specific e-collaboration process, but rather it should offer a library of techniques that foster good, pleasant and constructive collaboration. Currently there is no standard that describes good e-collaboration. Studying and designing collaboration processes through a pattern perspective is an exciting new frontier in e-collaboration research. Such research should not only aim to create a large and diverse library of facilitation design patterns, but also to test those patterns on a set of criteria, to increase the value of the pattern language, and to constitute a language that creates successful and pleasant e-collaboration.

REFERENCES

Alexander, C. (1979). *The timeless way of building.* New York: Oxford University Press.

Alexander, C., Ishikawa, S., Silverstein, M., Jacobson, M., Fiksdahl-King, I, & Angel, S. (1977). *A pattern language, towns, buildings, construction.* New York: Oxford University Press.

Briggs, R. O., Kolfschoten, G. L., de Vreede, G. J., & Dean, D. L. (2006). *Defining key concepts for collaboration engineering.* Paper will be presented at the Americas Conference on Information Systems, Acapulco, Mexico.

Briggs, R. O., & de Vreede, G. J. (2001). *ThinkLets, building blocks for concerted collaboration.* Delft, The Netherlands: Faculty of Technology Policy and Management, Delft University of Technology.

Briggs, R. O., de Vreede, G. J., & Nunamaker, J. F., Jr. (2003). Collaboration engineering with thinkLets to pursue sustained success with group support systems. *Journal of Management Information Systems, 19*(4), 31-63.

Clawson, V. K., Bostrom, R., & Anson, R. (1993). The role of the facilitator in computer-supported meetings. *Small Group Research, 24*(4), 547-565.

den Hengst, M., & Adkins, M. (2005). The demand rate of facilitation functions. In R. H. Sprague (Ed.), *Proceedings of the Hawaii International Conference on System Sciences* [Electronic version]. Washington, DC: IEEE Computer Society Press.

FacilitatorU. (2006). *Factivities.com.* Retrieved April 26, 2006, from http://www.factivities.com/exercises.html

Gamma, E., Helm, R., Johnson, R., & Vlissides, J. (1995). *Design patterns: Elements of reusable object-oriented software.* Reading, MA: Addison-Wesley Publishing Company.

Harrison, B. N., & Coplien, J. O. (1996). Patterns of productive software organizations. *Bell Labs Technical Journal, 1*(1), 138-145.

Hlupic, V., & Qureshi, S. (2002). What causes value to be created when it did not exist before? A research model for value creation. In R. H. Sprague (Ed.), *Proceedings of the Hawaii International Conference on*

System Sciences. Washington, DC: IEEE Computer Society Press.

Hlupic, V., & Qureshi, S. (2003). A research model for collaborative value creation from intellectual capital. In L. Budin, V. Luzar-Stiffler, Z. Bekic, & V. Hljuz-Dobric (Eds.), *Proceedings of the International Conference of Information Technology Interfaces* (pp. 431-438). Cavtat, Croatia: SRCE University Computing Centre

Jenkins, J. (2006). *IAF mehods database*. Retrieved April 26, 2006, from http://www.iaf-methods.org

Kolfschoten, G. L., Appelman, J. H., Briggs, R. O., & de Vreede, G.J. (2004a). Recurring patterns of facilitation interventions in GSS sessions. In R. H. Sprague (Ed.), *Proceedings of the Hawaii International Conference on System Sciences* [Electronic version]. Washington, DC: IEEE Computer Society Press.

Kolfschoten, G. L., Briggs, R. O., Appelman, J. H., & de Vreede, G. J. (2004b). ThinkLets as building blocks for collaboration processes: A further conceptualization. In G. J. de Vreede, L. A. Guerrero, & G. M. Raventos (Eds.), *Lecture Notes in Computer Science* (pp. 121-136). San Carlos, Costa Rica: Springer-Verlag.

Kolfschoten, G. L., Briggs, R. O., Vreede, G. J. de, Jacobs, P. H. M., & Appelman, J. H. (2006). Conceptual foundation of the thinkLet concept for collaboration engineering. *International Journal of Human Computer Science*.

Kolfschoten, G. L., den Hengst, M., & de Vreede, G. J. (2005). Issues in the design of facilitated collaboration processes. In R. Vetschera & S. Köszegi (Eds.), *Proceedings of the Group Decision and Negotiation Conference* [Electronic version]. Vienna, Austria: Organisation und Plannung, Universität Wien.

Lukosch, S., & Schümmer, T. (2006). Groupware development support with technology patterns. *International Journal of Human Computer Systems*.

Niederman, F., Beise, C. M., & Beranek, P.M. (1993). Facilitation issues in distributed group support systems. In M. R. Tanniru (Ed.), *Proceedings of the Conference on Computer personnel research* (pp. 299 - 312). St. Louis, MO: ACM Press.

Niegemann, H. M., & Domagk, S. (2006). *Elen project evaluation report*. Retrieved April 26, 2006, from http://www2.tisip.no/E-LEN

Nunamaker, J. F., Jr., Briggs, R. O., Mittleman, D. D., Vogel, D., & Balthazard, P. A. (1997). Lessons from a dozen years of group support systems research: A discussion of lab and field findings. *Journal of Management Information Systems, 13*(3), 163-207.

Rising, L. (2001). *Design patterns in communication software*. Cambridge, UK: Cambridge University Press.

Romano, N. C., Jr., Nunamaker, J. F., Jr., Briggs, R. O., & Mittleman, D. D. (1999). Distributed GSS facilitation and participation: Field action research. In R. H. Sprague (Ed.), *Proceedings of the Hawaii International Conference on System Sciences* [Electronic version]. Washington, DC: IEEE Computer Society Press.

Santanen, E. L. (2005). Resolving ideation paradoxes: Seeing apples as oranges through the clarity of thinkLets. In R. H. Sprague (Ed.), *Proceedings of the Hawaii International Conference on System Sciences* Electronic version]. Washington, DC: IEEE Computer Society Press.

Santanen, E. L., de Vreede, G. J., & Briggs, R. O. (2004). Causal relationships in creative problem solving: Comparing facilitation interventions for ideation. *Journal of Management Information Systems, 20*(4), 167 -197.

Zhao, J. L., Nunamaker, J. F., Jr., & Briggs, R. O. (2002). *Intelligent workflow techniques for distributed group facilitation*. In R. H. Sprague (Ed.), *Proceedings of the Hawaii International Conference on System Sciences* [Electronic version]. Washington, DC: IEEE Computer Society Press.

KEY TERMS

Pattern: "A pattern describes a problem which occurs over and over again and then describes the core of the solution to that problem, in such a way that you can use this solution a million times over, without ever doing it the same way twice." (Alexander et al., 1977, p. x)

ThinkLet: "A named, scripted collaborative activity that gives rise to a known pattern of collaboration among people working together toward a goal." (Briggs et al., 2006, p. 6)

Pattern of Collaboration: "The nature of a group's collaborative process when observed over a period of time as they move from a starting state to some end state." (Briggs et al., 2006)

Facilitation Technique: A method, used by facilitators to support a group process.

Facilitation Intervention: An action performed by a facilitator, aimed to change the group process in a specific way.

Sequence: "A specific combination of several thinkLets that is reused frequently in a variety of contexts" (Kolfschoten et al., 2006 p. 5).

Modifier: "Repeatable variation that can be applied to a set of thinkLets to create a predictable change in the patterns of collaboration that those thinkLets produce" (Kolfschoten et al., 2006).

Developing Synergies between E-Collaboration and Participant Budgeting Research

Kevin E. Dow
Kent State University, USA

Ralph H. Greenberg
Temple University, USA

Penelope Sue Greenberg
Widener University, USA

INTRODUCTION

E-collaboration, through group support systems (GSS) and other forms of computer-mediated communication (CMC), is increasingly used in organizations. GSS and CMC technologies offer organizations new ways to communicate information and knowledge, to interact synchronously or asynchronously, and to generate ideas, make decisions and solve problems. Although a sizable body of literature has developed that examines behavior in face-to-face (FtF) settings compared to CMC settings, one area that is relatively unexplored in the e-collaboration literature is participative budgeting, a widely used aspect of management control systems.

Budgeting is a topic of continuing interest to managers and scholars because of its important role in coordinating activities, allocating resources, motivating employees, and communicating goals and constraints in organizations. Participation by subordinates has long been considered an important aspect of the budgeting process. Participation has been assumed to lead to improved attitudes, communication, motivation, performance, and satisfaction (Shields & Shields, 1998). Participative budgeting has also been assumed to take place in face-to-face meetings between supervisors and subordinates.

Recently, however, budgeting software (a form of group support system) has been deployed in many companies and computer-mediated communication within virtual planning teams has begun to replace face-to-face meetings between superiors and subordinates (Smith, Goranson, & Astley, 2003). E-collaboration technologies have the potential to significantly impact many aspects of the budgeting process and the outcomes of that process. Because results of GSS and CMC research have shown that the effects of e-collaboration are dependent on multiple factors, the impacts on participative budgeting need to be investigated.

In addition, participative budgeting provides a new environment in which to study the role of e-collaboration technologies. Participative budgeting differs from many of the tasks examined in the GSS and CMC literature because, like participation in the systems development process (Hartwick & Barki, 1994), participation has a dual role (contributing to the development process and using the system after it is developed). It involves both the contribution to the budgeting process (e.g., information sharing) and the future effects of that participation (e.g., job satisfaction, motivation to achieve budget goals, and actual performance). In the GSS and CMC literature, the purpose of the collaboration is to accomplish the immediate task at hand.

Even though research in e-collaboration and participative budgeting have drawn from different theoretical backgrounds and have very different perspectives, the primary aim of both is to enhance group interactions, communication, and decision making. The purpose of this article is to identify potential synergies through which scholars in both areas can enhance future research.

BACKGROUND

E-Collaboration Research

The e-collaboration literature draws on theories of communication that are primarily concerned with social aspects (e.g., social presence, social influence),

Table 1. Characteristics e-collaboration and participative budgeting research

	E-collaboration	Participative Budgeting
Status of participants	• Varies	• Hierarchical (subordinate and superior)
Implications of outcomes	• Performance of the task itself • Participant's perceptions of and satisfaction with task and group interaction (immediate outcomes)	• Performance of the task itself • Motivation and commitment to future behavior
Collaborative/ participative tasks	• Idea generation • Decision making • Problem solving • Negotiation	• Communication of private information • Goal setting
Communication medium	• Extensive examination of face-to-face, CMC and GSS	• Not considered
Theories	• Task-technology fit • Psychobiological	• Psychology • Economics (principal-agent) • Sociology (organizational justice)
Independent variables	• Medium (CMC v. FtF) • Characteristics of the GSS • Characteristics of the task and group	• Level of participation • Incentives
Dependent variables and outcome factors	• Efficiency of the collaboration • Consensus • Effectiveness (number of comments, level of understanding) • Satisfaction • Usability (number of errors)	• Motivation • Future performance • Satisfaction • Perception of organizational justice • Willingness to communicate accurate information
Moderating, intervening, mediating and adaptation variables	• Numerous	• Numerous

on theories that are primarily concerned with technological aspects (e.g., media richness, task-technology fit), and on theories that integrate both aspects (e.g., psychobiological). Two common threads in most of these theories are concerned with the communication medium and the task.

Social presence theory (Short, Williams, & Christie, 1976), which predates the information super highway, has had a significant influence on GSS and CMC research. Under social presence theory, communication is more effective when the medium has the appropriate level of social presence for the level of interpersonal involvement necessary for the task. Social influence theory emphasizes the importance of social influence on attitudes toward communication media. However, under social influence theory, influences like peer pressure, cultural background, and mental schema may have a stronger effect on attitudes than characteristics of the medium itself.

Media richness theory (Daft & Lengel, 1986) classifies communication media according to its ability to convey nonverbal cues, immediate feedback, personality traits, and natural language. Under media richness theory, the criterion for matching the media to the collaborative task is based on the need to reduce uncertainty. Face-to-face communication is the richest

medium. The telephone is less rich. Most intranet- and Internet-based media are near the other end of the spectrum and are classified as lean. Task-technology fit theory (Zigurs & Buckland, 1998) proposes a set of ideal profiles composed of an internally consistent set of task contingencies and GSS elements that affect group performance. Like social presence theory, media richness and task-technology fit theories emphasize using the appropriate medium for the task at hand.

The psychobiological model (Kock, 2004) proposes that there is a negative causal link between the "naturalness" of a computer-mediated communication medium, which is the similarity of the medium to the face-to-face medium, and the cognitive effort required for an individual using the medium for knowledge transfer. This theory is a fresh perspective because it is integrative in that it encompasses previous theories, instead of attempting to negate them, and because it examines the *reasons* why face-to-face and CMC can lead to different outcomes. The task is an aspect of this theory, but the focus is on the cognitive effort required by the difference between 'natural' medium (face to face) and lean CMC mediums.

Empirical e-collaboration research, as classified by Fjermestad (2004), has typically modeled communication mode (face to face versus GSS or CMC) as the primary independent variable. Other independent variables include context, group, method, process structure, task, and task support. Moderating (adaptation) and intervening variables include communication dimensions (media richness, social presence), group member perceptions of task, adaptation process (rules, resources), process gains and losses, group role. Dependent variable categories include consensus, effectiveness, efficiency, satisfaction and usability.

Thus the e-collaboration literature, as shown in Table 1, has been concerned with the interaction of the task and the medium on the performance of the task itself. A wide variety of tasks (e.g., idea generation, problem solving and consensus) and myriad aspects of the medium (e.g., synchronous, asynchronous, structured interactions, support available) have been examined. While empirical results have been somewhat mixed on the relative benefits of CMC versus face-to-face (Fjermestad & Hiltz, 1999; Baltes, et al, 2002), it is widely accepted that communication medium and task characteristics do interact to impact outcome measures of the task and the collaboration.

Participative Budgeting

Participation by subordinates and the management style of the supervisor have been studied extensively since the pioneering works of Argyris (1952) in budgeting and Locke (1968) in goal setting. Participative budgeting, like e-collaboration, is a complex concept. It serves many roles within the management control system, including coordinating activities, allocating resources, motivating employees, and communicating goals and constraints in organizations. The roles of participation that are relevant here are the communication of information used to set goals and the motivation of employees to perform a task that will be evaluated using the goals as the benchmark.

Participative budgeting research has a rich tradition of drawing on aspects of economic, sociological and behavioral theories (Covaleski, Evans, Luft, & Shields, 2003; Shields & Shields, 1998). The economics-based research focuses on the value of budgeting practices to the organization. This research assumes the subordinate knows more about the task and the task environment than the superior. It relies heavily on the principle-agent (owner-employee, supervisor-subordinate) relationship and the optimal compensation practices that lead the agent's truthful communication of information and

the agent's performance (e.g., Fisher, Frederickson, & Peffer, 2000).

The sociology-based research has emphasized that individuals within the organization have conflicting interests and that organizations engage in diverse activities in an uncertain environment. Contingency theory is used to identify practices leading to improved (as opposed to optimal) performance depending on variables such as size, organizational justice, environmental uncertainty and technology (e.g., Fisher, 1995; Lindquist, 1995). The focus is usually on organizational performance and on why participation even exists.

The psychology-based research investigates the reaction of subordinates to specific budgeting practices, such as difficulty of budget targets, supervisor's performance-evaluation style (budget emphasis), and the how compensation is related to budget targets (e.g., Murray, 1990; Young, 1998). The focus is individual behavior.

Empirical budgeting research has typically modeled the level of participation as the independent variable with various moderating and intervening variables, such as characteristics of the employee (locus of control, management style), task characteristics (difficulty, uncertainty), performance evaluation criteria (budget emphasis) organizational characteristics (decentralization, procedural and distributive justice), and environmental characteristics (uncertainty). The dependent variables include communication (sharing information), goals, satisfaction, attitude, commitment, motivation, and subsequent performance (Shields & Shields, 1998).

Thus the participative budgeting literature, as shown in Table 1, has been concerned with the subordinate's communication of information, the impact of budget-related variables on subordinate's subsequent performance, and why participation even exists. Like the e-collaboration literature, results in the participative budgeting literature are somewhat mixed. But a widely accepted result is that many factors potentially influence the relationship of participation to motivation and performance.

PARTICIPATIVE BUDGETING AND E-COLLABORATION

Even though e-collaboration and participative budgeting are both concerned with group interactions, communication, and decision making, they have drawn on different theoretical backgrounds and have very different perspectives. These differences can lead to additional insights in each of the literatures if aspects of the other are examined. A broader, still more integrative theory could then be developed.

What Can E-Collaboration Learn From Participative Budgeting?

Lessons e-collaboration systems designers and researchers can learn from participative budgeting come from the three theoretical foundations of the literature, economics, sociology, and psychology.

In budgeting lab experiments based on economic principal/agent theory, the subordinate agent (subject) typically participates in setting goals and then performs a task, with the supervisor principal (experimenter or experimental rules) evaluating the performance and determining the subordinate's reward function. The subordinate is assumed to have private information (not available to the superior) about the assigned task, the task environment, their ability and willingness to perform the task. Participation is defined as communication of the subordinate's private information and the purpose of the participation to reduce subordinate-supervisor information asymmetry, which reduces uncertainty for the supervisor. The compensation contract (reward function) is intended to increase performance and to increase the truthfulness of the information communicated by the subordinate (Kanodia, 1993). Results indicate that compensation schemes can be devised that can both increase the accuracy of the information communicated and improve subsequent performance of the task (Waller, et al., 1988).

Nontruthful communication and the appropriate design of compensation schemes are both issues that extend beyond participative budgeting into other areas of collaboration. For example, an interesting hypothesis might be whether a subordinate's communication is more truthful and subsequent performance is improved when the participation is in a face-to-face setting, where multiple cues are available to the supervisor, than in a CMC setting.

In sociology-based budgeting research based on the contingency theory of organizations, Brownell (1982) identified four classes of variables: cultural, organizational, interpersonal and individual. The theory predicts that as these variables, along with the environment, become more uncertain, the organization uses integrating

mechanisms like participative budgeting to coordinate the actions within the organization (Shields & Shields, 1998). While the focus in most of these studies is on organizational performance, a primary thrust of this literature is on why organizations choose to allow subordinates to participate in the budgeting process. One reason is an attempt to control uncertainty.

Even though the e-collaboration literature has considered some contingent variables (i.e., aspects of context such as nationality, organizational size, and time pressure), the focus has not been on these contingent variables as an explanation for why firms invest in collaboration technology and why they allocate human resources to collaborative tasks, how the collaboration impacts organizational performance, and if uncertainty is, in fact, reduced.

The participative budgeting literature based on psychology theorizes an effect of participation on both satisfaction and subsequent performance. Like the economics-based and sociology-based research, this literature assumes that participative budgeting exists to reduce uncertainty and to reduce supervisor-subordinate information asymmetry (Shields and Shields 1998). The appropriateness of the technology to the task and the cognitive effort required to use that technology for the task may, in fact, affect subsequent behavior and attitudes.

Thus several aspects of the participative budgeting literature can be used to extend the e-collaboration literature. Consideration of incentives to increase truthful communication, the impact of collaboration/ participation on subsequent performance, and reasons why firms invest in e-collaboration are all interesting future research topics.

What Can Participative Budgeting Learn From E-Collaboration?

The primary lesson that participative budgeting can learn is that communication medium counts. As increasing numbers of organizations are deploying collaboration technologies, supervisors in the budgeting process and scholars designing research studies should recognize that communication medium and the characteristics of group support systems may change the behavior of both the supervisor and the subordinate in the budgeting process. These technologies also allow the formation of virtual planning teams not restricted to just the superior and the subordinate. They should become aware of the unintended consequences, as well as the intended consequences of moving from face-to-face to CMC. The psychobiological theory of computer-mediated communication (Kock, 2004) recognizes the importance of the difference in cognitive effort required by using leaner CMC mediums as opposed to the richer face-to-face. This cognitive effort could potentially have both immediate and subsequent impacts. This theory predicts that unnatural media will lead to perceived obstacles to communication, and that participants may adapt their behavior to compensate for those obstacles. However, the form of that adaptation and the impacts still need to be investigated.

Another lesson is that budgeting researchers can learn is that collaborating on the budget probably involves more than just communication. In the participative budgeting literature, the collaboration task is usually some form of communication relevant to setting goals and may include setting the goal. This is too narrow a view of participation. Generating ideas, problem solving, negotiating and decision making are all aspects of deciding on budget goals. Consideration of the task characteristics (e.g., structure, equivocality, complexity, analyzability, importance, etc.; Fjermestad, 2004) could also have an impact on the immediate and subsequent outcomes of participative budgeting.

FUTURE TRENDS

Collaborative technologies offer organizations options that have the potential to significantly impact many aspects of their management control systems. The widespread availability of collaborative technologies, such as Web-based chat tools, Web-based asynchronous conferencing tools, collaborative writing tools, workflow control systems, and document management applications, has the ability to significantly impact the budgeting process and the outcomes of that process. In addition, many of the applications are specially designed for budgeting tasks. This will make it even more important for systems designers and scholars in both management control systems and information systems to understand the immediate and subsequent impacts.

It is likely that there will be an increased use of cross-functional teams, as opposed to hierarchical dyads in budgeting, as organizations move to a process-centric focus. This will make some aspects of the budgeting

literature obsolete (hierarchical dyads) and some aspects of the e-collaboration literature (virtual teams) relevant to a broader range of situations. However, incentives may be needed to encourage communication and information sharing among departments that were previously more autonomous. Collaborative interorganization planning teams consisting of boundary-role persons have already begun to replace some intraorganizational budgeting activities.

Increased interorganizational information sharing and the associated risks make the communication process more delicate and sensitive. Management control and the measurement of performance are more difficult in these settings. The traditional social and cultural norms are not available for influencing behavior.

The recent top management focus on cost/benefit analysis of investments (assets and human resources) is not likely to disappear. This will force systems designers and scholars in GSS & CMC to focus on how group performance benefits firm and how technology benefits group performance.

CONCLUSION

Currently there are fundamental differences between the models of participative budgeting and e-collaboration. One difference is that participative budgeting assumes a dual role of participation, communication of information used to set goals and to motivate performance of a task that will be evaluated using the goals as the benchmark. Another difference is that models of participative budgeting have a very narrow view of the tasks involved in the budgeting process. Unlike budgeting, the e-collaboration literature examines a wider variety of tasks. Participative budgeting is also different from e-collaboration in that the act of participating is intended to have an impact on aspects other than the immediate task, such as to increase commitment to the budgeted goals, to motivate them to achieve those goals, and/or to promote positive attitudes.

Melding the theoretical approaches to and empirical findings from the each of these literatures and practices can be used not only to enhance theory development, but also to improve management control systems, group support systems and related research. Since the objectives of both are to enhance group interactions, communication, and decision making, there should be ample amounts of synergy arising from the integration of these two literatures.

REFERENCES

Argyris, C. (1952). *The impact of budgets on people.* New York: Controllership Foundation.

Baiman, S., & Lewis, B. L. (1989). An experiment testing the behavioral equivalence of strategically equivalent employment contracts. *Journal of Accounting Research, 27*(1), 1-20.

Baltes, B. B., Dickson, M. W., Sherman, M. P., Bauer, C. C., & Laganke, J. S. (2002). Computer-mediated communication and group decision making: A meta-analysis. *Organizational Behavior and Human Decision Processes, 87*(1), 156-179.

Covaleski, M. A., Evans, J. H., III., Luft, J. L., & Shields, M. D. (2003). Budgeting research: Three theoretical perspectives and criteria for selective integration. *Journal of Management Accounting Research, 15,* 3-49.

Daft, R. L., & Lengel, R. H. (1986). Organizational information requirements, media richness and structural design. *Management Science, 32*(5), 554-366.

Fisher, J. G. (1995). Contingency-based research on management control systems. *Journal of Accounting Literature, 14,* 24-53.

Fisher, J. G., Frederickson, J. R., & Peffer, S. A., (2000). Budgeting: An experimental investigation of the effects of negotiation. *The Accounting Review, 75*(1), 93-114.

Fjermestad, J. (2004). An analysis of communication mode in group support systems research. *Decision Support Systems, 37,* 239-263.

Fjermestad, J., & Hiltz, S. R. (1998-1999). An assessment of group support systems experimental research: Methodology and results. *Journal of Management Information Systems, 15*(3), 7-149.

Fulk, J., Schmitz, J., & Steinfield, C. W. (1990). A social influence model of technology use. In J. Fulk & J. Schmitz (Eds.), *Organizations and communications technology* (pp. 117-140). Newbury Park, CA: Sage.

Hartwick, J., & Barki, H. (1994). Explaining the role of user participation in information systems use. *Management Science, 40*(4), 440-465.

Kanodia, C. (1993). Participative budgets as coordination and motivation devices. *Journal of Accounting Research, 31,* 172-189.

Kock, N. (2004). The psychobiological model: Towards a new theory of computer-mediate communication based on Darwinian evolution. *Organization Science, 15*(3), 327-348.

Lindquist, T. (1995). Fairness as an antecedent to participative budgeting: Examining the effects of distributive justice and referent cognitions on satisfaction and performance. *Journal of Management Accounting Research, 7,* 122-147.

Locke, E. (1968, May). Toward a theory of task motivation and incentives. *Organizational Behavior and Human Performance, 3,* 157-189.

Mahenthiran, S., Greenberg, P. S., & Greenberg, R. H. (1993). The impact of computer-mediated communication on the processes and outcomes of negotiated transfer pricing. *Accounting, Management and Information Technologies, 3*(4), 229-248.

Milani, K. (1975). The relationship of participation in budget-setting to industrial supervision performance and attitudes: A field study. *The Accounting Review, 50*(2), 274-284.

Murray, D. (1990). The performance effects of participative budgeting: An integration of intervening and moderating variables. *Behavioral Research in Accounting, 2,* 104-123.

Shields, J. F., & Shields, M. D. (1998). Antecedents of participative budgeting. *Accounting, Organizations & Society, 23*(1), 49-76.

Short, J., Williams, E., & Christie, B. (1976). *The social psychology of telecommunications.* London: Wiley.

Smith, P. T., Goranson, C. A., & Astley, M. F., (2003). Intranet budgeting does the trick. *Strategic Finance, 84*(11), 30-33.

Waller, W., Slack in participative budgeting: The joint effect of truth-inducing pay scheme and risk preferences. *Accounting, Organizations and Society, 13,* 87-98.

Young, S. M., (1988). Individual behavior: Performance, motivation, and control. In K. Ferris (Ed.), *Behavioral accounting research: A critical analysis* (pp. 229-246). Columbus, OH: Century VII.

Zigurs, I., & Buckland, B. K.(1998). A theory of teak/ technology fit and group support systems effectiveness. *MIS Quarterly, 22*(3), 313-334.

KEY TERMS

E-Collaboration Technologies: Electronic technologies that enable collaboration among individuals engaged in a common task.

Management Control Systems: The process, procedures and policies by which managers influence other members of the organization to implement the organization's strategies.

Media Richness Theory: Theory that claims that lean media are not appropriate for knowledge and information communication (i.e., equivocality and un-

certainty reduction), and that the adoption of media and the outcomes of its use will usually reflect this fact.

Participative Budgeting: The process in which managers (subordinates) who are accountable to their supervisors for performance related to budget goals, participate in the determination of those goals.

Social Presence Theory: Theory that claims that communication is more effective when the communication medium has the appropriate level of social presence for the level of interpersonal involvement necessary for the task.

Digital Divide and E–Health Implications for E–Collaboration Research

Michele Masucci
Temple University, USA

INTRODUCTION

E-health has rapidly gained attention as a framework for understanding the relationship between using information and communication technologies (ICTs) to promote individual and community health, and using ICTs for improving the management of health care delivery systems. The use of e-collaborative tools is implicit to the delivery and access of e-health. Development of the capacity to transmit and receive digital diagnostic images, use video telecommunications for supporting the remote delivery of specialized care and surgical procedures, and the use of e-communication technologies to support logistical elements of medical care (such as scheduling appointments, filling prescriptions, and responding to patient questions) are just a few ways in which e-communications are transforming how medical care is embedded within institutional, organizational, family, and community settings.

The emerging field of e-collaboration focuses attention on the need for society to critically examine how electronic communication technologies facilitate, shape, and transform the ways in which organizations, groups, and communities interact. There are many works that explain how to (a) develop e-health systems, (b) assess the use of such systems, and (c) analyze the health outcomes that can be achieved with effective e-health applications (Brodie et al., 2000; Eder, 2000; Spil & Schuring, 2006). Less attention has been paid to how advances in e-collaboration research might inform e-health applications development and scholarly discourse. Because of this gap in the literature, few discussions pertain to understanding patient perspectives about the advantages and disadvantages that may result from rapidly emerging interconnections among access to health care, health information, health support systems, and ICTs (Berland et al., 2001; Hesse et al., 2005; Gibbons, 2005; Gilbert & Masucci, 2006).

E-HEALTH AND THE DIGITAL DIVIDE

Facilitating equitable e-health remains a difficult challenge because of persistent disparities in using and accessing ICTs among vulnerable and marginalized population groups (Atkinson & Gold, 2002; Brodie et al., 2000; Gibbons, 2005; Skinner, Biscope, & Poland, 2003; West & Miller, 2006). More research is needed to examine how differing experiences, self-efficacies, and adaptive styles among users of e-communication tools relate to the collaborative aspects of accessing and delivering health care (Atkinson & Gold, 2002; Hsu et al., 2005; Katz, Nissan, & Moyer, 2004). In particular, the collaborative aspects of implementing effective e-health policies could focus on such issues as (a) the role of educational training in using e-systems for accessing health information, health care provider consultations, and health management protocols, (b) the effects of alternative information delivery systems for enhancing patient care and community wellness, (c) the ways in which patient knowledge acquisition processes are related to the use of e-communication systems, (d) the privacy concerns related to e-collaboration strategies for accessing patient health care records, and (e) the tradeoffs associated with a movement to integrate e-communication approaches across the continuum of health care access by patients and health care providers.

In addition, e-collaboration research can lead to an understanding of the ethical implications of advances in e-health. Such methodological approaches as social action research applications in e-collaboration can result in creating tools for implementing e-health systems (such as using e-mail exchanges to foster system compliance) while also investigating the means by these approaches work to improve e-health outcomes (Kock, 2004, 2005). What society stands to gain from inquiry into these issues is a greater understanding of how e-collaborative approaches can enhance the rapid move toward using e-technologies in achieving patient health outcomes and managing the delivery of health care systems (Gibbons, 2005).

Gilbert and Masucci (2005, 2006) have examined the ICT use frameworks among such population groups as a basis for determining the most effective means of understanding and supporting empowerment goals for those groups. A focus on e-health suggests that a consideration of values and experiences with ICTs could connect an understanding of how individuals relate educational training, ICT access, health knowledge acquisition, and health care access to examine the ultimate value placed on the adoption of e-health approaches for one's personal as well as family health (i.e., Cline & Haynes, 2001; Cotten & Gupta, 2004; Houston & Allison, 2002; Kickbush, 2001; Kivits, 2006; Reddick, 2006). And, as e-collaboration tools are examined for their potential to support equitable access to e-health systems, it is important to understand that the context within which they are used relates directly to the potential outcomes that can be achieved.

For instance, an e-health system that is designed to use e-mail reminders for checking blood pressure at home among patients with diabetes may not be effective if the health care provider examines the e-mails once per week due to workplace constraints. E-mail messages sent from a privacy-secured e-mail system within a hospital may not be accessible from remote locations by health care providers, further delaying responses to patients. Patients may not have frequent access to e-mail systems as a basis for reporting blood pressure or other health characteristics. An understanding of the use of the tool for enhancing e-health delivery should examine context as well as how different ICT use patterns shapes perspectives about (a) the benefits of e-health systems, (b) the challenges associated with learning how to use such systems, and (c) the different ways in which patients and providers approach e-communications and other e-collaboration tools for implementing such systems.

IMPLICATIONS FOR USING INTERNET TELEMEDICINE TO MANAGE HEALTH CONDITIONS

Internet telemedicine refers to the use of health communication tools delivered through the use of Web interfaces for managing specific health conditions. Implicit in the use of Internet telemedicine is the goal of using such systems to foster collaborations among patients and health care providers to manage specific health conditions and procedures. Such collaborations by definition recreate the geographies of health care access and delivery (Cutchin, 2002). This reconstitution of patient-provider communications can involve such e-communication enabled tasks as (a) transmitting self-monitored information about specific health conditions to physician accessible data bases, (b) patient and provider tracking of variables related to specific health conditions, and (c) improved patient-provider communications about the implications of trends related to specific conditions.

The collaborative roles of patients and health care providers ultimately take shape around the use of specific e-communication tools for specific health conditions. Patient empowerment has the potential to increase as they are drawn into more proactive involvement in the gathering and examination of data pertaining to their health conditions (Prokosch, Ganslandt, Dumitru, & Übert, 2006).

A recent study of the use of an Internet telemedicine system to manage risk factors for cardiovascular disease illustrates the complexities involved in examining the relationships among geographic, social, and networked access to the Internet among low income patients (Masucci et al., 2006). The activities that were undertaken to improve the likelihood that individuals impacted by digital divide barriers would use the Internet telemedicine system included (a) developing and implementing an internet training protocol that addresses infrastructure and educational barriers to accessing ICTs among participants enrolled in the study, (b) assessing self-efficacy issues related to acquiring skills needed to use the internet communication tool developed for the study, and (c) analyzing social, demographic, and spatial patterns associated with health outcomes among patients who use the system (Masucci et al., 2006).

The participant group using the Internet telemedicine system was generally representative of people who have mitigated access to ICTs due to a combination of economic status, educational background and age (NTIA 1995, 1998, 1999a, 1999b, 2002, 2004). The 44 participants in the study were from inner-city Philadelphia and rural (non-suburban) northeastern Pennsylvania, with an average age of 60. Fifty-two percent of the study participants were African American; 73% were women; nearly 65% earned less than

$25,000. The study found that training participants to use the telemedicine system for self-monitoring such information as blood pressure, steps walked, weight, and number of cigarettes smoked, resulted in strong compliance using the system, with 84% sending data after being trained (Masucci et al., 2006).

Discussions held during the training sessions suggest that inexperience with using e-communications is closely associated with living at the economic and social margins of society for many individuals. Most participants (66%) had no prior Internet experience; those who did were often relying on settings with old computers and low bandwidths (Masucci et al., 2006). However, nearly all participants placed a high value on using telemedicine and e-communication tools for managing their health. This may account for their willingness to allocate time needed to negotiate the problem of accessing ICTs outside of the home and involving others to assist them with using the system. Among those who had prior experience, use of the telemedicine system relied on accessing ICTs through the support of family members, community groups, shelters, and local libraries in order to use the telemedicine system. This involved needing to learn about the locations and terms of use that would apply to each setting. It also required that their understanding of the Internet telemedicine system was transportable among the different settings they used.

Perhaps due to the patchwork of access described by participants, significant gaps in their conceptual understanding of how e-communications and the Internet function needed to be addressed in the training discussions. Nonetheless, each had specific interests in gaining information through using the Internet, indicated through the wide variety of Internet uses demonstrated through training sessions. Included were searches related to specific health conditions, applying for jobs, shopping, games, recipes, and e-mail. Few seemed to understand the implications of designing the system to be Health Insurance Portability and Accountability Act (HIPAA) compliant. However, the issue of information privacy was a central theme of training discussions. Despite the limited experiences using ICTs among many participants in the study, discussions were highly nuanced. Participants expressed concerns about protecting personal identifying information (such as study IDs, usernames and passwords, and social security numbers), maintaining privacy around Internet search procedures, and the potential for e-communications with health care providers to be read by unauthorized individuals.

CONCLUSION

The perspectives gained through this study about the use of ICTs for managing health are important ones to consider in the debates about equitable uses of e-collaboration among patients and health care providers. Particularly as e-health approaches are increasingly relied upon to deliver health care, the embedded nature of e-communications in Internet telemedicine systems has the potential to empower patients in their interactions with health care providers and improve their ability to effectively use information related to their health disseminated through the Internet. However, such factors as (a) the availability of computers and the Internet at home, (b) prior ICT use experiences of individuals, (c) awareness of health information resources available on the Internet, and (d) skill levels with using e-communication tools and overall technological literacy, can affect the overall ability of an individual to benefit from using such systems. Moreover, it is likely that individuals who come from the most disadvantaged groups will continue to manage a tightly knit set of circumstances including low education and literacy levels, low income, and marginal housing stability, to benefit from e-health approaches such as using Internet telemedicine systems. Individuals and families who are challenged in all of these ways may have specific ideas about how they can benefit from e-health yet struggle to achieve the technology self-efficacy required to gain optimal results.

These concerns form the basis for a significant societal ethics and policy debate. How can individuals from marginalized groups maximize opportunities to benefit equally from e-health advances? A dialogue about this issue must attend to the underlying health disparities that are associated with socio-economic vulnerability. Internet health systems need to be developed with the needs of underserved populations at the forefront of design considerations. Perhaps one approach is to increase efforts to develop systems and information resources that address the health concerns faced disproportionately by marginalized groups, such as increased risks for diabetes and cardiovascular disease, as compared with mainstream populations. Another consideration is to develop systems that can be accessed and used by those with limited experience and skill using ICTs. In the case of the telemedicine system developed for this study, the patient interfaces were purposefully designed to be easy to read, free of distracting hyperlinks, and intuitive for individuals with basic literacy levels.

These changes were implemented based on learning about the skill levels of likely users and their perspectives about the value of the system. Prior to the design changes being implemented, the system was information rich, to the extent that patients with little prior experience using ICTs did not know where or how to get started. It is critical for the development of equitable e-health systems that the system development strategies reflect a concern for the highly contextual ICT experiences of potential users. An understanding of the value of health information, e-health delivery systems, and telemedicine by marginalized groups can be accomplished through interacting around the use of ICTs. This requires breaking the social isolation that can exist among groups through finding opportunities for collaborating around the shared goals for developing e-health systems. The use of e-communications for such collaborative processes may seem implicit, but given the digital divide barriers that may constrain some, the form and meaning of e-collaboration may have to be adjusted to reflect existing disparities. This work shows that such adjustments can provide insights into the complex array of decision-making that forms the context for the use of e-communication tools for improving health.

ACKNOWLEDGMENT

This study was supported, in part, by grants from the Pennsylvania Department of Health and the National Science Foundation (ESI-0423242). Any opinions, findings, and conclusions or recommendations are those of the author and do not necessarily reflect the views of the granting agencies.

REFERENCES

Atkinson, N. L., & Gold, R. S. (2002). The promise and challenge of e-health interventions. *Proceedings of the 2nd Scientific Meeting of the American Academy of Health Behavior, 26*(6), 494-503.

Berland, G. C., Elliott, M. N., Morales, L., Algazy, J. I., Kravitz, R. L., Broder, M. S., et al. (2001). Health information on the Internet: Accessibility, quality, and readability in English and Spanish. *Journal of the American Medical Association, 285,* 2612-2621.

Brodie, M., Flournoy, R. E., Altman, D. E., Blendon, R. J., Benson, J. M., & Rosenbaum, M. D. (2000). Health information, the Internet, and the digital divide. *Health Affairs, 19*(6), 255-265.

Cline, R. J. W., & Haynes, K. M. (2001). Consumer health information seeking on the Internet: The state of the art. *Health Education Research, 16*(6), 671-692.

Cotten S. R., & Gupta, S. S. (2004). Characteristics of online and offline health information seekers and factors that discriminate between them. *Social Science and Medicine, 59*(9), 1795-1806.

Cutchin, M. P. (2002). Virtual medical geographies: Conceptualizing telemedicine and regionalization. *Progress in Human Geography, 26*(1), 19-39.

Eder, L. (2000). *Managing healthcare information systems with Web-enabled technologies*. Hershey, PA: Idea Group Publishing.

Gibbons, M. C. (2005). A historical overview of health disparities and the potential of e-health solutions. *Journal of Medical Internet Research, 7*(5), e50.

Gilbert, M., & Masucci, M. (2005). Research directions for information and communication technology and society in geography. *Geoforum, 36*(2), 277-279.

Gilbert, M., & Masucci, M. (2006). Geographic perspectives on e-collaboration research. *International Journal of E-Collaboration, 2*(1), i-v.

Hesse, B. W., Nelson, D. E., Kreps, G. L., Croyle, R. T., Arora, N. K., Rimer, B. K., et al. (2005). Trust and sources of health information: The impact of the Internet and its implications for health care providers: Findings from the first health information national trends survey. *Archives of Internal Medicine, 165,* 2618-2624.

Houston, T. K., & Allison, J. J. (2002). Users of Internet health information: Differences by health status. *Journal of Medical Internet Research, 2002-4*(2), e7.

Hsu, J., Huang, J., Kinsman, J., Fireman, B., Miller, R., Selby, J., et al. (2005). Use of e-health services between 1999 and 2002: A growing digital divide.

Journal of the American Medical Informatics Association, 12, 164-171.

Katz, S. J., Nissan, N., & Moyer, C. A. (2004). Crossing the digital divide: Evaluating online communication between patients and their providers. *American Journal of Managed Care, 10*(9), 593-598.

Kickbusch, I. S. (2001). Health literacy: Addressing the health and education divide. *Health Promotion International, 16*(3), 289-297.

Kivits, J. (2006). Informed patients and the Internet: A mediated context for consultations with health. *Professionals Journal of Health Psychology, 11*(2), 269-282.

Kock, N. (2004). The three threats of action research: A discussion of methodological antidotes in the context of an information systems study. *Decision Support Systems, 37*(2), 265-286.

Kock, N. (2005). Using action research to study e-collaboration. *International Journal of e-Collaboration, 1*(4), i-vii.

Masucci, M., Homko, C., Santamore, W., Berger, P., McConnell, T., Shirk, G., et al. (2006). Cardiovascular disease prevention for underserved patients using the Internet: Bridging the digital divide. *Telemedicine and E-health, 12*(1), 58-65.

National Telecommunications and Information Administration (NTIA). (1995). *Falling through the net: A survey of the "have nots" in rural and urban America.* Washington, DC: National Telecommunications and Information Administration.

National Telecommunications and Information Administration (NTIA). (1998). *Falling through the net ii: New data on the digital divide.* Washington, DC: National Telecommunications and Information Administration.

National Telecommunications and Information Administration (NTIA). (1999a). *Falling through the net ii: New data on the digital divide.* Washington, DC: National Telecommunications and Information Administration.

National Telecommunications and Information Administration (NTIA). (1999b). *Falling through the net: Defining the digital divide.* Washington, DC:
National Telecommunications and Information Administration.

National Telecommunications and Information Administration (NTIA). (2000). *Falling through the net: Toward digital inclusion.* Washington, DC: National Telecommunications and Information Administration.

National Telecommunications and Information Administration (NTIA). (2002). *A nation online: how americans are expanding their use of the Internet.* Washington, DC: National Telecommunications and Information Administration.

National Telecommunications and Information Administration (NTIA). (2004). *A nation online: entering the broadband age.* Washington, DC: National Telecommunications and Information Administration.

Prokosch, H., Ganslandt, T., Dumitru, R. C., & Ükkert, F. (2006). Telemedicine and collaborative health information systems. *Information Technology, 48*(1), 12-23.

Reddick, C. G. (2006). The Internet, health information, and managing health: An examination of boomers and seniors. *International Journal of Healthcare Information Systems and Informatics, 1*(2), 20-38.

Skinner, H., Biscope, S., & Poland, B. (2003). Quality of Internet access: Barrier behind Internet use statistics. *Social Science and Medicine, 57*(5), 875-80.

Spil, T., & Schuring, R. W. (2006). *E-health systems diffusion and use: The innovation, the user and the use IT model.* Hershey, PA: Idea Group Publishing.

West, D. M., & Miller, E. A. (2006). The digital divide in public e-health: Barriers to accessibility and privacy in state health department Web sites. *Journal of Health Care for the Poor and Underserved, 17*(3), 652-67.

KEY TERMS

Digital Divide: The National Telecommunications and Information Administration developed a report series (NTIA, 1995, 1998, 1999a and b, 2000, 2002, and 2004) that conducted an analysis of the ownership and use of computers by households in the U.S. The reports indicated that disparities between the households with the highest and lowest degrees of computer use and Internet access were related to income, race, and gender. The reports called the disparities a "digital divide."

E-Health: The use of information and communication technologies (ICTs) to support individual and community health through creating health management information systems, electronic scheduling systems, electronic prescription services, transmittal of health records and diagnostic imagery among providers, and the remote delivery of care and consultations.

E-Health Ethics: Also referred to as Health Internet Ethics (Hi-Ethics), considerations to ensure the equitable distribution of health benefits for individuals and families through using e-health to disseminate health services. Concerns include protecting information privacy, ensuring that current, reliable health information, and advice is shared using e-health approaches, and that access to services is available to all.

HIPAA: The Health Insurance Portability and Accountability Act created standards for regulating the transactional records related to the delivery of electronic health care. This involved creating national identifiers for health care providers, insurance plans, and places of employment. HIPAA requirements also exist to protect the privacy and security of health information.

Internet Telemedicine: The use of Web interfaces for accessing and delivering telemedicine services.

Technological Literacy: The knowledge, skills, and self-efficacy required to critically engage advancements in technologies and technological influences on daily life.

Telemedicine: The delivery of medical care between settings that are geographically separate through the use of telecommunication systems, including traditional telephone systems and electronic communication systems.

E-Collaboration and the Financial Auditor

Vincent J. Shea
Kent State University, USA

INTRODUCTION

Accounting firms are constantly seeking ways to improve the financial audit function within their practice. The audit industry, like many other service industries, is highly competitive and requires firms to complete the audit for their clients as efficiently as possible. In many instances, accounting firms cannot raise the price of their service and can therefore maximize profits only by reducing the cost of the audit. As such, firms are constantly searching for ways to improve the efficiency of the financial audit. Two customary means by which efficiency improvements for an accounting firm can manifest themselves are a decrease in the number of employees or a decrease in the time to complete the audit. While an accounting firm can only decrease its staff by so much, many of the efficiency improvements of the financial audit are produced by the adoption of new technology to facilitate better and faster collaboration by the employees within the firm.

The financial audit function has changed significantly over the past 100 years. While the theory and purpose of the audit has changed slightly, the methods and applications have modified with the use of technology over time. Thirty years ago, the most common tools used in audit were a red pencil and a blue pencil. With these tools, the auditor would cross-check and list all transactions to the financial statements. Today, the common tools for the auditor are his laptop and audit software. What used to take many hours for the audit team to complete can now be done in a fraction of the time and staff. This is an example of how the accounting firms have adopted technology to improve the efficiency and effectiveness of the audit. However, one form of technology audit firms have not been quick to adopt is e-collaborative technology.

E-collaboration is defined as "collaboration among individuals engaged in a common task using electronic technologies" (Kock & Nosek, 2005, p. 1). E-collaboration is commonly associated with and perceived as computer mediated communication (CMC); though strictly speaking, e-collaboration can be non-CMC, such as the telephone or other electronic communication device. However, the most common forms of e-collaboration used in accounting firms are net meeting, virtual workplaces, and Internet communication such as e-mail and instant messaging.

The purpose of this article is to address how accounting firms can adopt e-collaboration within their organizations. Examples of current e-collaboration adoption by accounting firms are examined. This article will also suggest how e-collaboration could benefit the audit function by making it more efficient and effective. A discussion of how e-collaboration research can be applied to audit research is explored as well as how audit research can help e-collaboration research.

BACKGROUND

E-Collaboration Research

Prior research has shown that the adoption of new technologies can have a positive effect on the financial performance of the firm. A common area of benefits within the firm is in supply chain management (SCM). For example, firms are relying more on the flow of information between the organization and the supplier to maximize efficiency and to maintain market competency. If the information does not flow smoothly between the channels, then the results can have a dramatic effect on the performance of the firm (Cassivi, Lefebvre, Lefebvre, & Leger, 2004). Studies have shown that firms that adopt e-collaborative technologies may sometimes have a negative effect on the performance of their business because the adoption was done incorrectly or was not used in the areas best suited (Marquez, Bianchi, & Gupta, 2004). Firms must understand that with any technology, there are not only benefits that arise from adoption of new technologies, but also barriers and limitations of adoption.

Many of the barriers and limitations of collaborative technology can be explained via media richness theory. This theory suggests that the use of collaborative technology will decrease the quality of the work as performed by the firm because individuals prefer face-to-face communication the most (Daft & Lengel, 1986). However, compensatory adaptation theory (Kock, 2005) proposes that in the long run, collaborative technology can be more beneficial than traditional forms of communication. This theory argues individuals prefer face-to-face because our biological development has been formed for this method of communication. Over time, through the use of collaborative technology, individuals overcompensate for barriers and obstacles to the use of collaborative technologies, therefore achieving a superior outcome compared to face-to-face.

Financial Audit Research

The traditional financial audit is the examination (usually by an individual certified public accountant [CPA] or an accounting firm comprised of CPAs) of the financial statements of a firm to provide an opinion as to whether the financial statements fairly portray an accurate picture of the firm's financial position. All publicly traded firms in the United States are required to have their financial statements audited by CPAs. Further, government agencies and financial institutions require audits for regulation and credit purposes.

Auditing research is a broad spectrum and can be broken into two areas: the individual audit and the accounting firm. While this subject is a wide open topic, there are many opportunities for research. For example, there is research examining how the number of staff in the accounting firm affects the quality of the audit (Glover, 1997; Margheim, Kelley, & Pattison, 2005). Another example of audit research is the impact of accounting firm tenure and audit quality (Carcello & Nagy, 2004). Firm tenure and audit quality have gained an increased level of attention with the Sarbanes-Oxley Act of 2002 (SOX). Some examples include increase focus of the company's internal controls, further documentation required by the auditor, and greater attention on the quality of the auditor and the accounting firm. Yet, with the increase of acceptance of technology by accounting firms, audit research has begun to focus on the success of technology implementation into the audit (Fischer, 1996).

Studies have examined how communication within the firm is affected by technology. For example, Murthy and Kerr (2004) compared the audit team effectiviness of those who use e-collaborative technologies (specifically,

bulletin boards and online chat) versus the traditional face-to-face collaboration. The results of their work demonstrate that auditors who use technologies for solving team task tend to be more efficent and effective than those performing the same tasks with traditional face-to-face meetings. Specifically, they find that the bulletin board method was the most effective, followed by the online chat and face-to-face communication, respectively.

Adoption of Collaborative Technologies by Accounting Firms

Two of the most important factors for an accounting firm to consider when adopting a new technology are efficiency and effectiveness. Efficiency can be thought of simply as outputs divided by inputs. Though the outputs and inputs are often different for each accounting firm and specific situation, often the output for the accounting firm is the audit work and the inputs are anything associated with producing the audit work. The effectiveness for an accounting firm can be thought of as the quality of the audit work performed. The dilemma accounting firms are having is trying to maximize efficiency while at the same time at least maintaining their current level of effectiveness. These are issues accounting firms find themselves discussing whenever they are considering adopting any new technology.

The most common forms of e-collaboration used in accounting firms are real-time communication (e-review), and collaborative content management (electronic workpapers). E-review allows auditors to communicate with colleagues the review notes concerning issues of the audit and to review several audit engagements concurrently while reducing costs. This method tends to decrease the time spent in face-to-face collaboration, while increasing the speed of decision making process. In addition, auditors who use e-review are more likely to meet their time budget for the audit (Brazel, Agoglia, & Hattfield, 2005).

A common method of documenting the audit, by auditors, is via workpapers. Traditionally, the workpapers have been done on paper, and stored in filling cabinets at the accounting firm's headquarters or offsite locations. Because of the shear volume of these workpapers, the costs associated with the maintaining and storing can be enormous (since the accounting firms are required to maintain their workpapers for a period of 5 years). To decrease this cost, many large accounting firms have adopted electronic workpapers (Bedard, Jackson, Ettredge, & Johnstone, 2003). Electronic workpapers allow accounting firms to provide their members instant access to the progress of the audit and easy access to prior years' work.

Barriers and Benefits of Collaborative Technology for the Accounting Firms

Prior research has shown that accounting firms, like any other service oriented business, have issues when adopting new technology. As Bedard et al. (2003) showed with electronic workpapers, members who have performed the task longer are less willing to adapt to new technology. They also show that those who are more involved with the process, given proper training, are more willing to adapt to the technology in the accounting firm than those who are not. If the users perceive the technology to be troublesome or not useful, then often the technology is unsuccessful within the firm. These are issues accounting firms must understand when adopting e-collaboration.

The first barrier that accounting firms must overcome is the acceptance of collaborative technology by experienced staff. Bedard et al. (2003) shows that experienced, more established auditors are less willing to adopt new technology than the inexperienced, younger auditors. For collaborative technology to succeed in the accounting firm, e-collaboration research needs to examine how to overcome this experience factors. The workpapers research suggests that extensive training may reduce the difficulties of using the technology, yet may not necessarily reduce the individual's perceived usefulness of the technology. Practitioners and researchers should examine how to inform experienced users of the usefulness of the collaborative technology.

Technology succeeds in accounting firms when it complements the traditional audit methods. Accounting firms may perceive the use of collaborative technology as "all or none" and have decided not to adopt the technology because they perceive the use of it as a decrease in the quality of the audit work. For example, Brazel et al. (2005) shows that preparers who anticipate a face-to-face meeting are more concerned with audit effectiveness, produce higher audit quality, and are less efficient than prior years. However, the compensatory adaptation theory suggests that this is a short term outcome and in the long run the audit work

will become more effective and efficient than the traditional face-to-face.

The final barrier accounting firms must overcome is the effect of collaborative technology on the effectiveness of the audit work. Theoretically, auditors could use collaborative technology to perform the audit from any location in the world. However, this could cause a dramatic effect on the perceived quality of the audit. Auditors are understood to maintain a certain level of audit quality. If the audit work is perceived to be ineffective, the accounting firm can be liable in court as a result of the incomplete job.

Accounting firms benefit from the adoption of collaborative technology through the increase of audit efficiency and the decrease in audit cost. To optimize these benefits, accounting firms must examine how to incorporate the technology into the audit function while maintaining the current level of audit effectiveness. For example, auditors generally perform all of their audit work on site at the client's office. Much of this work can easily be completed at the auditor's office or any other off site location through collaborative technology. Researchers and practitioners should examine how to maximize the benefits of collaborative technology for the accounting firm while maintaining the effectiveness of the audit.

What E-Collaboration Can Learn from Audit Research

The primary lesson that e-collaboration can learn from audit research is regarding boundaries. Audit research has shown that the adoption of a new audit method or an improvement of an audit process will only be accepted if the adopter understands the entirety of the existing boundaries. For example, electronic workpapers were only beneficial to the accounting firm once all of the staff levels of the firm accepted the changeover from paper to electronic (Bedard et al., 2003). While internal boundaries are a common matter for accounting firms, the accounting firm must be aware of the boundaries outside of the firm. Outside boundaries include the clients and legislative and regulations. For example, clients, stakeholders, or members of the court of law may perceive their auditor as performing an incomplete job if the auditor performs the entire audit function from the accounting firm's office. Thus, accounting firms are bound by this perception to complete the audit function in the traditional ways. Many of these barriers are not unique to the audit function as well as the accounting firm. Researchers of e-collaboration must examine these issues to maximize the potential benefits of e-collaboration for firms and industries.

An additional lesson e-collaboration can learn from audit research is the examination of different users. In many instances, audit research compares low-level to senior level staff on their reaction to an audit process or judgment to examine comparability and differences in their reactions. E-collaboration can use similar methods to examine the use of collaborative technologies by users and the differences in their acceptances.

What Audit Research Can Learn From E-Collaboration

E-collaboration research shows that the acceptance of e-collaborative technologies may produce an initial challenge for the users and firms. However, over time, these deficiencies and weaknesses dissipate and turn into strengths and benefits. Audit research can learn from the e-collaborative research stream called compensatory adaptation theory. Compensatory adaptation theory states that e-collaborative technologies are initially less effective compared to the traditional methods, such as face-to-face, because our biological makeup has been adopted and designed specifically for the traditional methods. However, overtime and through practice and patience, individuals will adjust to e-collaborative technologies and will produce better outcomes than the traditional methods (Kock, 2005). Like many industries, the accounting firms expect quick results from the adoption of a new technology or method. Thus, audit research can learn to develop patience in adoption, and widen their window of examination. This longer time may produce better quality audit research.

FUTURE TRENDS

Collaborative technologies offer many potential benefits that accounting firms can maximize to their advantage. Much of the future trends in research will examine how to overcome the perceived barriers of the technology. Accounting firms can use collaborative technologies to multi-task different audits, and perform the audits offsite. The use of the technology will decrease budgeted staff hours, travel cost and miscellaneous cost of the

audit. The Sarbanes-Oxley Act has increased the work and cost of the audit dramatically because of requirements to meet regulations. Collaborative technologies will allow firms to offer more services and possibly offer audit cost savings to the client.

However, all of these benefits will not be realized until accounting firms overcome their fear of adoptions. In addition to the accounting firms, audit clients must be willing to accept the technology as well. Researchers and practitioners will need to address how the audit can incorporate collaborative technologies, while maintaining an acceptable level of audit effectiveness. In addition, research will need to be addressed to express to clients how the audit function is evolving by the use of collaborative technologies and ensure these technologies are not only providing benefits to the auditor but to the client as well.

CONCLUSION

This paper addresses how collaborative technology is used and can be used in the accounting firm. Accounting firms are constantly searching for ways to improve audit efficiency because of the relationship between efficiency and cost. However, at the same time, firms are trying to maintain the effectiveness of the audit. Any decrease in audit effectiveness, will in turn increase the chance of being sued by the client, investors, or other stakeholders. Thus, with any changes in audit methods, there must be an increase in efficiency, and the method must at least maintain current audit effectiveness. Accounting firms should not adopt e-collaboration to replace the traditional audit methods, but should be used to compliment. E-collaboration provides many benefits to the auditor including: documentation of auditor-client discussions, timeless access to audit and client documents, and a decrease in potential cost of the audit including minimal auditor travel and auditor billing hours as well. Though the audit managers may be hesitant to adopt the e-collaboration because of their learning curve and required auditor training, with any technology adoption the firm will adapt and prosper.

REFERENCES

Bedard, J. C., Jackson, C., Ettredge, M. L., & Johnstone, K. M. (2003). The effect of training on auditors' acceptance of an electronic work system. *International Journal of Accounting Information Systems, 4*, 227-250.

Brazel, J. F., Agoglia, C. P., & Hattfield, R. C. (2005). Review methods matter. *CPA Journal, 75*(9), 36-38.

Carcello, J., & Nagy, A. (2004). Audit firm tenure and fraudulent financial reporting. *Auditing: A Journal of Practice & Theory, 23*(2), 5-5.

Cassivi, L., Lefebvre, E., Lefebvre, L. A., & Leger, P.-M. (2004). The impact of e-collaboration tools on firms' performance. *The International Journal of Logistics Management, 15*(1), 91-110.

Daft, R., & Lengel, R. (1986). Organizational information requirements, media richness and structural design. *Management Science, 32*(5), 554-571.

Fischer, M. (1996). "Real-izing" the benefits of new technologies as a source of audit evidence: An interpretive field study. *Accounting, Organizations and Society, 21*(2/3), 219-242.

Glover, S. M. (1997). The influence of time pressure and accountability on auditors' processing of nondiagnostic information. *Journal of Accounting Research, 35*(2), 213-226.

Kock, N. (2005). Media richness or media naturalness? The evolution of our biological communication apparatus and its influence on our behavior toward e-communication tools. *IEEE Transactions on Professional Communications, 48*(2), 117-130.

Kock, N., & Nosek, J. (2005). Expanding the boundaries of e-collaboration. *IEEE Transactions on Professional Communications, 48*(1), 1-9.

Margheim, L., Kelley, T., & Pattison, D. (2005). An empirical analysis of the effects of auditor time budget pressure and time deadline pressure. *Journal of Applied Business Research, 21*(1), 23-35.

Marquez, A. C., Bianchi, C., & Gupta, J. N. D. (2004). Operational and financial effectiveness of e-collaboration tools in supply chain integration. *European Journal of Operational Research, 159*, 348-363.

Murthy, U. S., & Kerr, D. S. (2004). Comparing audit team effectiveness via alternative modes of computer-mediated communication. *Auditing: A Journal of Practice & Theory, 23*(1), 141-152.

KEY TERMS

Audit Software: A tool used by audit firms to complete the audit. The software allows auditors to perform analytical tasks and to document the controls and procedures of the client.

Certified Public Accountant (CPA): Qualified accountants who are licensed to provide audited opinions on financial statements.

Electronic Workpapers: A paperless version of documenting the auditor's work.

E-Review: The process in which the auditor receives and responds to audit review notes via e-mail.

Financial Audit: The examination of the financial statements by an independent third party to provide an opinion on whether or not the financial statements are complete and accurate.

Sarbanes-Oxley Act of 2002 (SOX): A federal law that enhances auditor impendence and corporate responsibility. In addition, it requires firms to provide additional financial disclosures and provides additional powers and jurisdiction to the Securities and Exchange Commission.

Workpapers: The process used by auditors to document the auditors' work. It is the key documentation process of the audit.

E–Collaboration as a Tool in the Investigation of Occupational Fraud

Bobby E. Waldrup
University of North Florida, USA

INTRODUCTION

E-collaboration has become a staple of productivity in organizations of all sizes and types in the last decade. As the Fortune 500 companies have now moved to a 100% participation rate in a blended e-mail/instant messenger communication environment, it is estimated that 3% to 6% of the leaders, including such names as General Motors, Microsoft, and Boeing have even instituted official corporate blog sites (Bruner, 2005). As the techniques and uses of e-collaboration become more pronounced in firms, the need for related guidance in the occupational fraud examination field is growing. While both the theoretical and applied e-collaboration literatures are increasing, this is a relatively unexplored area in the field of fraud examination.

Fraud can be widely classified as either fraud *against* or *on behalf* of the organization. While fraud *on behalf* of the organization, such as financial statement fraud, has made the most sensational headlines recently (i.e., Enron, WorldCom, Arthur Anderson), it is occupational fraud *against* the organization that has the most widespread effects. In its most recent account on occupational fraud (ACFE, 2004), the Association of Certified Fraud Examiners reported that the average organization loses 6% of revenues to occupational fraud and abuse, and these losses cut across virtually every size and type of institution.

The default model typically used for reacting to occupational fraud against the organization follows four traditional steps (Albrecht, McDermott, & Williams, 1994). First, an *incident* occurs. Second, an *investigation* ensues centered around interviewing and document examination. Third, at the end of an investigation, a *resolution* is determined on what action, if any, is to be taken. Finally, the company chooses to *act* (or not act) publicly or legally related to the fraud occurrence.

The purpose of this article is to blend previous research in the fields of e-collaboration and fraud examination. Specifically, compensatory adaptation theory (Kock, 1998) is used as a lens through which

to model the optimal use of e-collaboration during the aforementioned second step of the default fraud model; the investigation process of occupational fraud. Finally, examples of investigatory pros and cons are presented to illustrate the model's approach towards local optima usage points.

BACKGROUND

E-Collaboration Research

The use of computer-mediated communication (CMC) has become a widespread tool in the application of work-related tasks (see Trevino, Daft, & Lengel, 1990; Trevino, Webster, & Stein, 2000 for a comprehensive and longitudinal discussion). A range of suppositions have been applied to explain the various dimensions of CMC, and how these facets interact to enhance or detract from task effectiveness. Three theories have developed over the past 40 years which have become a staple in the information delivery literature. These theories of social presence, media richness, and compensatory adaptation are the basis upon which the optimal use of e-collaboration techniques in fraud examination will be explored.

Social presence theory was first put forth in the 1970's by the Communications Studies Group. As described by Short, Williams, and Christie (1976), the theory describes "the degree of salience of the other person in the interaction and the consequent salience (and perceived intimacy and immediacy) of the interpersonal relationships" (p. 1).

Extending this generic interaction, the social presence theory essentially illustrates that differing communications media causes changes in the degree of social presence, and these changes cause a difference in interaction itself. In the context of CMC, this theory predicts that as a "nonverbal" form of communication, social presence is diminished when compared to face-to-face communication. CMC lacks facial expressions,

Figure 1. Compensatory adaptation model

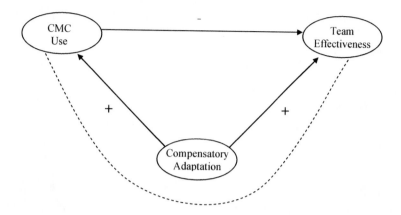

voice tone, inflection, gaze, posture, dress, and even environmental décor. Therefore, in the a priori, this theory predicts that the use of computer-mediated communication, in any form and to any extent, would lower the effectiveness of a fraud examination which employs it in lieu of a face-to-face investigatory technique.

During the following decade, the theory of social presence was refined and tested by researchers, bringing about a rival though complementary theory of media richness. Media richness theory (Daft & Lengel, 1986) classifies communication media along a "richness" continuum. Daft, Lengel, and Trevino (1987) categorizes "richness" as a media's ability to carry nonverbal cues, convey personality, provide rapid feedback, and support natural language. Like the social presence theory, media richness theory predicts that CMC conveys less "richness" than face-to-face communication. Therefore, this reinforces the notion that using CMC as a fraud investigation technique should produce suboptimal effectiveness at every level of incorporation.

Sallnas, Rassmus-Grohn, and Sjostrom (2000) illustrates that, while social presence theory and media richness theory were developed during the pre-Internet era of communication, they still widely influenced CMC research. However, a recently growing body of empirical studies has failed to support this linear relationship between richness/presence and task effectiveness (Lee, 1994; Markus, 1994; Ngwenyama & Lee, 1997). These studies illustrate that users may actually choose "lean" communication media for secondary reasons, and will modify their behavior independent of the richness/presence level in a compensating manner (Kock, 2004).

Compensatory adaptation theory (Kock, 1998, 2001) attempts to explain these recent empirical findings by a U-shaped effectiveness adaptation model.

This theory first recognizes the inherent negative effect that using CMC will have on task effectiveness (as predicted by the social presence and media richness theories). The compensatory adaptation theory then argues that a team (such as a fraud investigation team) will adapt to these negative effects by employing compensating behavior. According to Kock, Lynn, Dow, and Akgun (2006), this compensating behavior produces an offsetting indirect positive effect on task effectiveness to produce a neutral impact on the quality of team outcomes.

Paradoxically, the theory even suggests that in some situations, this compensatory adaptation behavior may even result in positive outcomes, causing team performance to actually be more effective using CMC than by using only face-to-face communication. It is this paradox first identified by Kock (1998) that explains how humans tend to overcompensate for computer-mediated obstacles and achieve even better outcomes than if those obstacles had never been present in a given task. The remainder of this paper will examine Kock's paradox of task effectiveness using compensatory adaptation behavior through the lens of an occupational fraud examination.

Fraud Examination Research

The root of the fraud examination process is the fraud triangle. While there are limitless ways to perpetrate a

Figure 2. The fraud triangle prerequisite elements

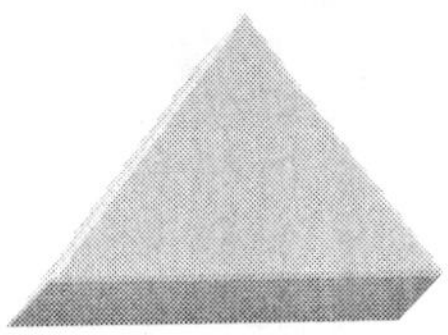

Figure 3. The fraud triangle necessary steps

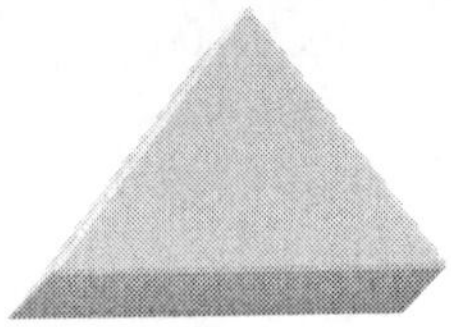

fraud, this triangle illustrates the three key prerequisite elements common to all of them. A fraud is perpetrated when the elements of opportunity, pressure, and rationalization come together to push a perpetrator over the proverbial edge.

In response to the Sarbanes Oxley Act of 2002, the American Institute of Certified Public Accountants released the Statement on Auditing Standards number 99 (AICPA, 2002). SAS 99, as it has come to be commonly known, is the official guidance for CPAs in the audit of financial fraud. The fraud triangle was vaulted to prominence when it was specifically codified as part of the required audit procedures in compliance with the Act.

An alternative version of the fraud triangle has emerged which better operationalizes the necessary steps that are common to initiate a complete fraud. To effect and complete an occupational fraud, a perpetrator must (1) enact a *theft*, (2) *conceal* that a theft has taken place, and (3) *convert* the stolen assets to personal assets. This version coined the *fraud triangle + inquiry* approach to the act of occupational fraud, is shown in Figure 3.

Albrecht (2003) recommends that a fraud investigation should follow a four-pronged tactic along these steps. His sample methodology of this reactive approach is presented in Table 1.

As can be seen from the table, there are a myriad of approaches that a fraud examiner can take to investigate the three steps of theft, concealment, and conversion. The final approach, inquiry, requires the examiner to move beyond investigating records, documents, and processes to actually talking to possible witnesses and suspects.

Figure 4 illustrates a sample approach (Albrecht, 2003) to a typical occupational fraud investigation.

If a team were investigating a purchasing employee suspected of taking kickbacks, they might utilize the methods shown. First, they might check the suspected employee's personnel records for evidence of financial difficulties, and simultaneously exam trends of price and purchasing records from various vendors. Next, former employees and unsuccessful vendors might be interviewed. Third, public records could be searched to gather evidence about the suspect's lifestyle, as well as surveillance over his work and home patterns.

Table 1. The fraud triangle + inquiry approach

(1) Theft Investigation • Surveillance and covert operations • Invigilation • Physical evidence	(3) Conversion Investigation • Public records searches • Net worth method
(2) Concealment Investigation • Document examination • Audits • Computer searches • Physical asset counts	(4) Inquiry Investigation • Interviews and interrogation • Honesty testing

Figure 4. Theft investigation methods

Forth, the team may perhaps interview other buyers and other coworkers about the possibility of kickbacks and related internal controls. Finally, in this ever circling scenario, team members will finally interview the suspect himself.

This typical theft investigation method illustrates two cogent points. First, several of these steps can be undertaken offsite without arousing suspicion of investigation. This typical methodology employs a tangential approach of investigating as much content as possible while simultaneously delaying the exposure of an investigation itself. Secondly, since occupational fraud is not a static instance, a team's investigative approach may in fact actually affect a fraudster's future actions. Once a fraudster realizes an investigation is under way he will take steps to conceal past actions, such as destroying documents or erasing computer files. It is this circular interplay between investigator and perpetrator in which the selective use of e-collaborative techniques can increase the effectiveness and efficiency of fraud examination.

FRAUD EXAMINATION AND E-COLLABORATION

The choice of e-collaborative communication techniques generically affects the *immediacy* of that communication (Kock, 2001). As stated by Short et al. (1976), "immediacy is one dimension of social presence—a measure of the psychological distance a communicator puts between himself and the object of his communication."

Specific to occupational fraud, an investigatory team consciously chooses the level of immediacy that they wish to employ based upon specific facts of each case. This choice of immediacy in turn affects the level of e-collaboration that the team will employ (see Malone & Crowston, 1994, for a comprehensive discussion of this coordination effect).

As with most other team-related tasks, the choice of fraud examination techniques involves an inherent tradeoff between task effectiveness and task efficiency. As illustrated in Figure 5, this tradeoff can be portrayed by an inverted u-shaped relationship.

This model depicts an example of the "paradox" that Kock (1998) and related papers have theorized in the MIS literature.

As an investigation team undertakes the steps necessary to track an occupational fraud, it will naturally adopt communication choices that are partially electronic in nature. A simple illustrative example of this is the use of e-mail amongst team members and between the team and vendors. The use of e-mail serves the dual purposes of allowing distributed team members to collaborate without the need to physically meet, and also allows for a continuously archived record of team and vendor communications related to the alleged fraud. These dual benefits simultaneously increase the team's investigative effectiveness (documentation) and efficiency (distributed deployment).

As Kock et al. (2006) illustrated, there are also instances when compensatory adaptation behavior adopted by team members may simultaneously increase task effectiveness and efficiency. According to the Investigative Professionals (2006), using CMC

Figure 5. Fraud examination technique e-collaboration tradeoff model

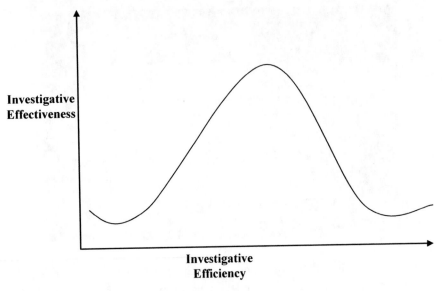

Figure 6. E-collaboration technique for fraud examination blended model

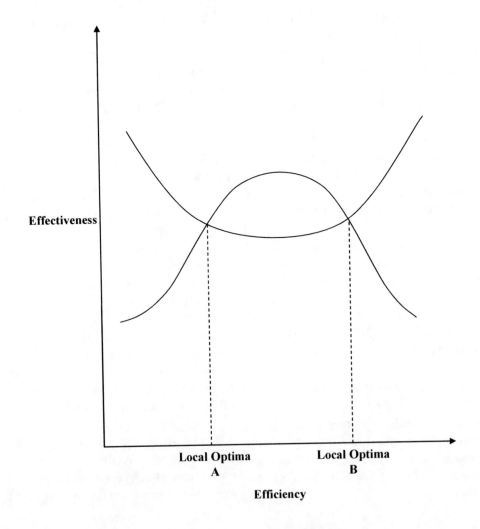

techniques in the investigative process may help to overcome perceived or real shortcomings of team members. These shortcomings are manifested in the psycho-social effects within the organization presented in group dynamics. By using CMC, an investigator can portray themselves as a person of a different sex, age, or experience level. These demographic attributes are difficult to mask in person, but relatively simple online. The use of CMC could increase investigative effectiveness by shielding a suspect from these verbal and visual distracters, while at the same time increasing efficiency by allowing for note-taking and referencing without concern for appearance in front of the suspect.

However, a peak comes in the investigative process in which these dual benefits fail to materialize by the further use of computer mediated communication techniques. As was shown in the example from Figure 4, there comes time in a fraud examination when face-to-face interviewing is necessary to increase investigative effectiveness, but this will, in turn, yield a lower efficiency of the task steps. At this point the natural tradeoff relationship between effectiveness (increased) and efficiency (decreased) reasserts itself.

Figure 6 presents a blended model of investigative effectiveness and efficiency which overlays the Compensatory Adaptation Model (Figure 1) and the Fraud Examination Tradeoff Model (Figure 5).

This blended model shows that, when taken together, these theories predict that two local optima appear when the tradeoff of task efficiency and effectiveness is examined for an occupational fraud investigation team that is employing compensating behavior throughout the use of computer mediated communication techniques.

These two local optima represent the theoretically most advantageous points at which a fraud team can use CMC. As seen in Figure 6, each optima provides an equally beneficial use of CMC techniques, just at two different points in the tradeoff model. Between these two levels of usage, there is either an increase or decrease of effectiveness and efficiency, but they are offsetting in their benefits. While these benefits are presented as *offsetting*, it can be assumed that distance from the local optima is exponentially related to a decrease in effectiveness and efficiency. Therefore, while outside the bounds of this model presentation, the model does not address this relationship when far from the local optima.

LIMITATIONS

This model presentation lacks field testing in a controlled environment. Therefore, this research suffers from a lack confirmation that would normally be desired. Secondly, the conditions under which the local optima are formed or dissipate in real simulations are yet to be documented.

FUTURE TRENDS

The literature has explored and debated the merits of CMC. The various techniques of CMC have been tested as to their merits through the lens of the competing (and evolving) theories of social presence, media richness, and compensatory adaptation. While these theories have become the springboard of empirical studies to test the use of CMC on task effectiveness and efficiency, there have been no studies which profile an occupational fraud examination team.

An occupational fraud examination team is different from most "teams" due to the problem of *immediacy* that is not so prevalent in other team-related tasks. Inherent in the investigation of fraud is the oscillating need to both examine the facts and circumstances of a fraud while at the same time not changing the behavior of the fraudster(s).

Future research in this area could result in empirical studies of fraud examination teams. A four-pronged approach is suggested for these future studies. First, additional model building should be performed to further refine the inflection points of the efficiency/effectiveness tradeoff model. Second, case studies should be examined which further disaggregates the categories of potential occupational frauds. Third, experiments which simulate occupational fraud could be conducted to further test the tenets of task effectiveness and efficiency tradeoffs specific to this type of fraud. Finally, overall empirical studies could then be undertaken to provide generalizable and useful rules for fraud examiners.

CONCLUSION

The proliferation of e-collaborative communication techniques has become virtually ubiquitous in the business world. Likewise, the complexity and effects of occupational fraud has paralleled this growth.

Research in these fields has rarely intersected in any meaningful way.

Occupational fraud investigation is generally unlike other more commonly explored team-related tasks in the management and accounting fields. This uniqueness poses distinctive challenges for meaningful research to address tradeoffs in these team tasks. This study has attempted to illustrate these challenges, and suggest future springboards for consequential research which addresses the unique needs of occupational fraud examination teams.

REFERENCES

AICPA. (2002). *Consideration of fraud in a financial statement audit: Statement on auditing standards # 99*. New York: American Institute of Public Accountants.

Albrecht, W. S. (2003). *Fraud examination*. Mason, OH: Thomson South-Western.

Albrecht, W. S., McDermott, E. A., & Williams, T. L. (1994, February). How companies can reduce the cost of fraud. *The Internal Auditor*, 28-35.

Association of Certified Fraud Examiners. (2004). *The report to the nation on occupational fraud and abuse*. Austin, TX: Author.

Bruner, R. E. (2005). *Percentage of Fortune 500 blogging*. Retrieved March, 7, 2005, from www.business-blogconsulting.com

Daft, R. L., & Lengel, R. H. (1986). Organizational information requirements, media richness and structural design. *Management Science, 32*(5), 554-571.

Daft, R. L., Lengel, R. H., & Trevino, L. K. (1987). Message equivocality, media selection, and manager performance: Implications for information systems. *MIS Quarterly, 11*(3), 355-366.

Investigative Professionals. (2006). *Interviewing techniques*. Retrieved July 12, 2007, from www.how-toinvestigate.com/investigation/interview.htm

Kock, N. (1998). Can communication medium limitations foster better group outcomes? An action research study. *Information & Management, 34*(5), 295-305.

Kock, N. (2001). Compensatory adaptation to a lean medium: An action research investigation of electronic communication in process improvement groups. *IEEE Transactions on Professional Communication, 44*(4), 267-285.

Kock, N. (2004). The psychobiological model: Towards a new theory of computer-mediated communication based on Darwinian evolution. *Organization Science, 15*(3), 327-348.

Kock, N., Lynn, G. S., Dow, K. E., & Akgun, A. E. (2006). Team adaptation to electronic communication media: Evidence of compensatory adaptation in new product development teams. *European Journal of Information Systems, 00*(1), 1-11.

Lee, A. S. (1994). Electronic mail as a medium for rich communication: An empirical investigation using hermeneutic interpretation. *MIS Quarterly, 18*(2), 143-157.

Malone, T., & Crowston, K. (1994). The interdisciplinary study of coordination. *ACM Computing Surveys, 26*(1), 87-119.

Markus, M. L. (1994). Electronic mail as the medium of managerial choice. *Organization Science, 5*(4), 502-527.

Ngwenyama, O. K., Lee, A. S. (1997). Communication richness in electronic mail: Critical social theory and the contextuality of meaning. *MIS Quarterly, 21*(2), 145-167.

Sallnas, E. L., Rassmus-Grohn, K., & Sjostrom, C. (2000). Supporting presence in collaborative environments by haptic force feedback. *ACM Trans. Computer-Human Interaction, 7*(4), 461-476.

Short, J., Williams, E., & Christie, B. (1976). *The social psychology of telecommunications*. London: Wiley.

Trevino, L. K., Daft, R. L., & Lengel, R. H. (1990). Understanding manager's media choices: A symbolic interactionist perspective. In J. Fulk & C. Steinfield (Eds.), *Organizations and communications technology* (pp. 71-94). Newbury Park, CA: Sage.

Trevino, L. K., Webster, J., & Stein, E. W. (2000). Making connections: Complementary influences on communication media choices, attitudes, and use. *Organization Science, 11*(2), 163-182.

KEY TERMS

Compensatory Adaptation Theory: Teams will adapt their behavior to positively counteract the inherent reduction in task effectiveness caused by computer medicated communications.

Computer Mediated Communication: Any form of communication between two or more individuals who interact and/or influence each other via computer-supported media.

Fraud Triangle: An illustrative model of the three prerequisites of pressure, rationalization, and opportunity that must be present to allow an occupational fraud to take place.

Immediacy of Communication: A measure of the psychological distance a communicator puts between himself and the object of his communication.

Media Richness: A media's ability to carry nonverbal cues, convey personality, provide rapid feedback, and support natural language.

Occupational Fraud: The use of one's occupation for personal enrichment through the deliberate misuse or misapplication of the employing organization's resources or assets.

SAS 99: The statement on auditing standard that guides external auditors in investigation of occupational fraud in compliance with the Sarbanes-Oxley Act of 2002.

E–Collaboration Enhanced Host Security

Zoltán Czirkos
Budapest University of Technology and Economics, Hungary

Gábor Hosszú
Budapest University of Technology and Economics, Hungary

Ferenc Kovács
Budapest University of Technology and Economics, Hungary

INTRODUCTION

The importance of the host security problems come into prominence by the growth of the Internet, since the network means a breaking point to the intruders (Wang, Jha, McDaniel, & Livny, 2004). The article presents the e-collaboration related security questions, the main concepts of the **intrusion detection** and the different classes of the system protection methods.

As an example of the application for non-conventional purposes, a security system is presented in the article that utilizes just the network for protecting the operating system of the computers. The software maintains a database about the experienced intruding attempts. Its entities working on each computer share their experiments among each other on the *peer-to-peer* (P2P) overlay network created by self organizing on the Internet. In such a way the security of the participants is increased, and then they can take the necessary steps.

BACKGROUND

Currently the *application-level networking* (ALN) has increasing importance. In this communication technology the applications running on host directly create connections among them and they use these connections in order to exchange information and packets. Their communication way is different from the more traditional networking model, where the communicating software entities create connections among them for solving certain task (e.g., downloading a file). In case of the ALN the applications produce more stable virtual network, called *overlay*, which can be used complex file-management and application level routing functionalities (e.g., making *application-layer multicast*;

ALM). The ALN overlays use typically the P2P communicating model oppositely to the more traditional *Client/Server* model (Hosszú, 2005).

A special kind of the P2P networking is the Grid-computing (shortly Grid), where the registered participants actively collaborate to produce new results (Uppuluri, Jabisetti, Joshi, & Lee, 2005). The notion of a Grid has gained popularity as a metaphor and guiding principle for system architectures designed to permit large-scale resource sharing across widespread heterogeneous collections of systems (Foster, Kesselman, & Tuecke, 2001). An important feature is the notion of a dynamic Virtual Organization (VO) in which a collection of individuals or organizations share resources in an ad hoc way for a period of time, with minimal effort required to set up or finalize the organization (Martin & Cook, 2004).

It has been clear that careful consideration of security issues is central to the successful deployment of Grids. Potential resource providers will be reluctant to participate if the possibility of misuse of their resource is too great; potential customers will not use Grid services if they cannot achieve an adequate guarantee of quality of service (QoS)—including integrity, confidentiality, and availability (Martin & Cook, 2004).

Although Internet connections are now almost ubiquitous, and of very low cost, different applications and organizations find good reasons to employ leased private networks. They may use the Internet protocol (IP) for their implementation, in such a way virtual private network (VPN) can be realized. VPNs can also be used to link separate sites, by use of private leased lines, so that network traffic may travel over a long distance as if it were within a single site. Multinational corporations implement internal networks in this way; the long-distance links may be well-protected, but the attached nodes are necessarily more accessible (Martin & Cook, 2004).

To the end user, VPNs are something of a marvel. They allow a roaming device—a laptop or a personal digital assistant (PDA)—using any Internet connection to behave as if it were part of their home corporate network, apparently on the inside of any firewall protection, and with potentially full access to sensitive network data and resources. Moreover, this solution is completely sanctioned and even supported by their system and network administrators. Those administrators are also able to use a VPN to connect remote sites using the Internet. Although the traffic travels over the public IP network, the encryption prevents clear-text eavesdropping or tampering (Martin & Cook, 2004).

In order to use a VPN over each connection between a user and a resource node, a potentially enormous number of VPNs will be needed, with associated key management challenges for each. This will almost certainly render the enterprise unmanageable. Even a model in which only the Grid nodes (compute, data, broker, and other resource) participate, VPNs will exhibit an exponentially rising set-up cost for adding new nodes (Martin & Cook, 2004).

The VPN itself cannot provide perfect security. Whilst VPNs support strong encryption of data over a shared medium, they do not provide complete undetectability of activity. It must be noted that VPNs do not attempt to shield the presence of users within the network, or hide identity of endpoints and hosts, the types of data, and the frequency of data exchanges. If the primary uptake of VPNs within Grid infrastructures is to provide security, the designers of such Grids must be aware of the above vulnerabilities.

To address the problem of traffic snooping within VPNs, an accepted form of defense is the use of a single IPSec channel between VOs, as described in Herscovitz (1999). This provides a degree of secrecy and a degree of immunity to traffic analysis, but as illustrated in Cohen (2003), single IPSec channels between endpoints are sometimes not possible.

Based on the reviewed properties of the networking technologies of the e-collaboration it can be stated that a single solution has not solved the problem. The host-based and the network-based intrusion detections are equally important. In the followings the various intrusion detection methods will be analyzed.

THE INTRUSION DETECTION

This section describes basic security concepts, dangers threatening user data and resources. We describe different means of attacks and their common features one by one, and show the common protection methods against them.

Information stored on a computer can be personal or business character, private or confidential. An unauthorized person can therefore steal it. Stored data can not only be stolen, but changed. Information modified on a host is extremely useful to cause economic damage to a company.

Resources are also to be protected. Resource is not only hardware. Typical type of attack is to gain access to a computer to initiate other attacks from it. This is to make the identification of the original attacker more difficult.

Intrusion attempts, based on their purpose, can be of different methods. But these methods share things in common, scanning networks ports or subnetworks for services, and making several attempts in a short time. This can be used to detect these attempts.

With attempts of downloading data, or disturbing the functionality of a host, the network address of the target is known by the attacker. He scans the host for open network ports, in order to find buggy service programs. This is the well-known port scan. The whole range of services is probed one by one. The object of this is to find some security hole, which can be used to gain access to the system (Teo, 2000). The most widely known software application for this purpose is Nmap (Nmap, 2006).

Unfortunately, not every attack is along with easily automatically detectable signs. For example the abusing of a system by an assigned user is hard to notice.

The oldest way of intrusion detection was the observation of user behavior (Kemmerer & Vigna, 2002). With this some unusual behavior could be detected. For example, somebody on holiday still logged in the computer. This type of intrusion detection has the disadvantage of being casual and non-scalable for complex systems.

Next generation intrusion detection systems utilized monitoring log files, mainly with Unix type operating systems. Of course this is not enough to protect a system, because many types of intrusions can only be

Table 1. Types of firewalls

Firewall type	Description
Packet level	Filtering rules are based on packet hearers, for example the address of the source or the destination.
Application level	content of network packets are also examined to identify unwanted input.
Personal	Designed for workstations and home computers. With these the user can define, for which applications he grants access to the network.

detected too late. Supervising a system is only worth this expense if the intrusion detection system also analyzes the collected information. This technology has two main types: anomaly detection and misuse detection.

Anomaly detection has a model of a properly functioning system and well behaving users. Any deviation it founds is considered a problem. The main benefit of anomaly detection is that it can detect attacks in advance. By defining what is normal, every break of the rules can be identified whether it is part of the threat model or not. The disadvantages of this method are frequent false alerts and difficult adaptability to fast-changing systems.

Misuse detection systems define what is wrong. They contain intrusion definitions, alias signatures, which are compared with the collected supervisory information, searching for the signs of the known threats. Advantage of these systems is that investigation of already known patterns rarely leads to false alerts. At the same times these can only detect known attack methods, which have a defined signature. If a new kind of attack is found, the developers have to model it and add to the database of signatures.

PROTECTION METHODS

Computers connected to networks are to be protected by different means (Kemmerer & Vigna, 2002), described in detail as follows.

The action taken after detecting an intrusion can be of many different types. The simplest of these is an alert which describes the observed intrusion. But the reaction can be more offensive, like informing an administrator, ringing a bell or initiating a counterstrike.

The counterstrike may reconfigure the gateway to block traffic from the attacker or even attack him. Of course an offending reaction can be dangerous; it may be against an innocent victim, as the attacker may

load the network with spoofed traffic. This appears to come from a given address, but in reality it is generated somewhere else. Reconfiguring the gateways to block traffic from this address will generate a denial of service (DoS) type attack against the innocent address.

No system can be completely secure. The term of a properly skilled attacker (Toxen, 2001) applies to a theoretical person, who by his infinite skills can explore any existent security hole. Every hidden bug of a system can be found, either systematically, or accidentally.

The more secure a system is, the more difficult the use it (Bauer, 2005), so a trade-off between security and usability has to be made. Before initiating medium and large sized systems it is worth making up a so-called security policy.

The simplest style of network protection is a firewall. This is a host which provides a strict gateway to the Internet for a subnetwork, checking traffic and maybe dropping some network packets. The three main types of firewalls are listed in the Table 1.

HOST AND NETWORK BASED DETECTION AND PROTECTION SYSTEMS

Network intrusion detection systems (NIDS) are capable of supervision and protection of company-scale networks. One commercially available product is RealSecure (2006), while Snort is an open source solution (Snort, 2006). Snort mainly realizes a probe. It is based on a description language, which supports investigation of signatures, application level protocols, anomalies, and even the combination of these.

Investigation of network traffic can sometimes use uncommon methods. One of these is a network card without an IP address (Bauer, 2002). The card, while it is connected to the network (through a hub or a switch), gets all the traffic, but generates no output, and therefore can not be detected. The attacker already broken into the subnetwork cannot see that he is monitored. Peters (2006) shows a method of special wiring of Ethernet connectors, which makes up a probe that can be put between two hosts. This device is unrecognizable with software methods.

Information collected by probes installed at different points of the network is particularly important for protection against network scale attacks. Data collected by one probe alone may not be enough, but an extensive analysis of all sensors' information can reveal the fact of an attack (RealSecure, 2006). For the aid of sensors communicating in the network has the Intrusion Detection Working Group (IDWG) of the Internet Engineering Task Force (IETF) developed the Intrusion Detection Message Exchange Format (IDMEF) (Debar et al., 2006).

A NOVEL NETWORK-BASED DETECTION SYSTEM

This section introduces a novel system, which uses just the network, to protect the hosts and increase their security. The hosts running this software create an application level network (ALN) over the Internet. The clients of the novel software running on individual hosts organize themselves in a network. Nodes connected to this ALN check their operating systems' log files to detect intrusion attempts. Information collected this way is then shared over the ALN, to increase the security of all peers, which can then make the necessary protection steps by oneself. This collaboration also helps them to discover network-size attacks. That could be impossible for single-standing hosts.

The developed software is named Komondor, which is a famous Hungarian guard dog.

The speed and reliability of sharing the information depends on the network model and the topology. Theory of P2P networks has gone through a great development since the last years. Such networks consist of peer nodes. The parts of an applications realizing

a peer-to-peer based network can be seen on Figure 1 (Hosszú, 2005). The lower layer is responsible for the creation and the maintenance for the overlay network, while the upper one for the communication.

As one host running the Komondor detects an intrusion attempt and shares the address of the attacker on the overlay network, the other ones can prepare and await the same attacker in safety. Komondor nodes protect each other this way. This is shown on Figure 2.

Different hosts run the uniform copies of Komondor, monitoring the occurring network intrusion attempts. If one of the peers detects an attempt on a system supervised, takes two actions:

1. Strengthens the protection locally, by configuring the firewall to block the offending network address
2. Informs the other peers about the attempt

The first working version of Komondor monitors system log files. These can contain various error messages, which may refer to an intrusion attempt, for example login attempt with an inexistent user name.

Figure 3 presents the algorithm of the software Komondor. It checks the log files every second, while the database should be purged only on hourly or daily basis (Czirkos, 2005).

The Komondor network aided security enhancement system is new in principle. It was under extensive testing for months. With its aid not only simulated, but real intrusion attempts were blocked. Effectiveness is determined by the diversity of peers. Intrusion attempts exploiting security holes are software and version specific. The more different peers participate the network, the more likely is an invulnerable system to protect other vulnerable ones.

It is important to emphasize, that the proposed protection system is intended to *mask* the security holes of

Figure 2. Attack against a Komondor node

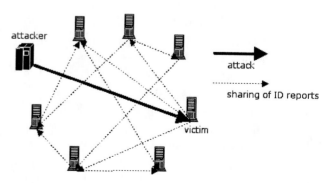

Figure 1. Block diagram of a P2P application

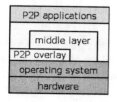

Figure 3. Algorithm of the Komondor entity

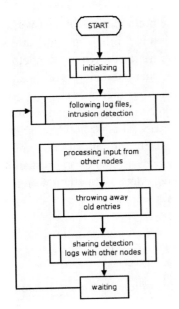

collaboration tool. This second property of the novel system provides enhanced security to the networked hosts, where the light-weight Komondor client has been implemented.

CONCLUSION

The article overviewed the different aspects of host security. Typical attacks were reviewed, along with methods of intrusion detection furthermore, the implemented and widely-used host protection methods have been presented.

The article also proposed a novel application, which utilizes the P2P networking model in order to improve the effectiveness of operating system security. The system is easy to use; its clients on different networked nodes organize a P2P overlay automatically, and do not need any user interaction.

services provided by the host, not to *repair* them. It can provide protection in advance, but only if somewhere on the network an intrusion was already detected. It does not fix the security hole, but keeps the particular attacker from further activity. If the given security hole is already known, it is worth rather fixing that itself.

FUTURE TRENDS

Attackers are more successful than ever. They share vulnerability information very rapidly, and practice building on others' work. The growth of the Internet works to their advantage. Contrary to this, defenders usually view security as local responsibility, and do not share information outside their organization for privacy and business reasons (Lincoln, 2004).

Research is focused on collaborative systems to handle this situation. Distributed intrusion detection systems like *DShield* (DShield, 2006) are built to enhance protection of hosts. However, as it is pointed out by Lincoln (2004) that system administrators do not have the right tools to share information. The lack of standards also slows down their spread of collaborative systems. That is why software Komondor was developed, which is not simply an easy-to-use intrusion detection software but also a network-based

REFERENCES

Bauer, M. (2002). *Stealthful sniffing, intrusion detection and logging.* Retrieved January 10, 2006, from http://www.linuxjournal.com/article/6222

Bauer, M. (2005). *Linux server security* (2nd ed.). O'Reilly.

Cohen, R. (2003). On the establishment of an access VPN in broadband access networks. *IEEE Communications Magazine, 41*(2), 156-163.

Czirkos, Z. (2005). Development of P2P based security software. In *Proceedings of the Conference of Scientific Circle of Students (Second Award).* Budapest, Hungary: Budapest University of Technology and Economics, Faculty of Electrical Engineering and Informatics (in Hungarian).

Debar, H., Curry, D., & Feinstein, B. (2006). *The intrusion detection message exchange format.* Retrieved July 31, 2006, from http://www.ietf.org/

DShield distributed intrusion detection system. (2006). Retrieved March 14, 2006, from http://www.dshield.org/

Foster, I., Kesselman, C., & Tuecke, S. (2001). The anatomy of the grid: Enabling scalable virtual organization. *International Journal of High Performance Computing Applications, 15*(3), 200-222.

Herscovitz, E. (1999). Secure virtual private networks: The future of data communications. *International Journal of Network Management, 9*(4), 213-220.

Hosszú, G. (2005). Mediacommunication based on application-layer multicast. In S. Dasgupta (Ed.), *Encyclopedia of virtual communities and technologies* (pp. 302-307). Hershey, PA: Idea Group.

Kemmerer, R. A., & Vigna, G. (2002). Intrusion detection: A brief history and overview. *Security & Privacy a Supplement to IEEE Computer Magazine, 35*(4), 27-30.

Lincoln, P. (2004). *Privacy-preserving sharing and analysis*. Retrieved March 22, 2006, from http://www.cis.upenn.edu/~timedc/oct04/

Martin, A., & Cook, C. (2004). Grids and private networks are antithetical. *UK Workshop on Grid Security Experiences, IV,* 9-16.

Nmap free security scanner, tools & hacking resources. (2006). Retrieved January 5, 2006, from http://www.insecure.org/

Peters, M. (2006). *Construction and use of a passive ethernet tap*. Retrieved January 5, 2006, from http://www.snort.org/docs/tap/

RealSecure. (2006). *Internet security systems.* Retrieved January 5, 2006, from http://www.iss.net/

Snort. (2006). *Snort: The de facto standard for intrusion detection/prevention*. Retrieved January 4, 2006, from http://www.snort.org

Teo, L. (2000). *Port scans and ping sweeps explained.* Retrieved January 10, 2006, from http://www.linux-journal.com/article/4234

Toxen, B. (2001). *Real world Linux security*. Indianapolis, IN: Prentice Hall PTR.

Uppuluri, P., Jabisetti, N., Joshi, U., & Lee, Y. (2005, July 11-15). P2P grid: Service oriented framework for distributed resource management. In *Proceedings of the IEEE International Conference on Web Services*, (pp. 347-350). Orlando, FL: IEEE Computer Society Press.

Wang, H., Jha, S., McDaniel, P. D., & Livny, M. (2004, June 7-9). Security policy reconciliation in distributed computing environments. In *Proceedings of the IEEE 5th International Workshop on Policies for Distributed Systems and Networks* (pp. 137). IBM Thomas J. atson Research Center. Yorktown Heights, NY: IEEE Computer Society Press.

KEY TERMS

Application Level Network (ALN): The applications, which are running in the hosts, can create a virtual network from their logical connections. This is also called *overlay network*. The operations of such software entities are not able to understand without knowing their logical relations. Most cases of these ALN software entities use the *P2P model*, not the *client/server* model for the communication.

Client/Server Model: A communicating way, where one hardware or software entity (server) has more functionalities than the other entity (the client), whereas the client is responsible to initiate and close the communication session towards the server. Usually, the server provides services that the client can request from the server. Its alternative is the *P2P model*.

Collaborative Security System: A network-based system in which the participants share information about intrusion attempts and other events noticed. This way the security of each peers is increased, and the system is also able to discover network-size attacks.

Data Integrity: The integrity of a computer system means that the host behaves and works as its administrator intended it to do so. Data integrity must therefore be always monitored.

Firewall: This is a host or router which provides a strict gateway to the Internet for a subnetwork, checking traffic and maybe dropping some network packets.

Overlay Network: The applications, which create an *ALN* work together and usually follow the *P2P communication model*.

Peer-to-Peer (P2P) Model: A communication way where each node has the same authority and communication capability. Each peer provides services the others can use, and each peer sends requests to other ones.

Security Policy: A set of rules to act, in which the expectations and provisions of accessibility of the computer for the users and the administrators are also included. It is worth it to be made up before initiating medium- or large-sized computer networking systems.

E-Collaboration for Internationalizing U.S. Higher Education Institutions

Jaime Ortiz
Texas A&M International University, USA

INTRODUCTION

The current globalization wake requires countries to cope with their deficiencies in international competencies to become credible political actors and sustained economic players (Stiglitz, 2003). They must develop expertise on foreign policies, emphasize functional knowledge, and make concerted efforts to narrow technological gaps to successfully unleash global competitiveness (McGrew, 2005). There is a high correlation between the level of global competence acquired by U.S. students during their undergraduate and graduate education and the success of the country as a key international player (American Council of Education, 1995). U.S. higher education institutions are compelled to increase their stature by constantly seeking ways to expand their network of partnerships with foreign counterparts. Any steps into that direction will broaden and strengthen the internationalization of their curriculum to facilitate and improve students' understanding of the world.

According to the Association of Governing Boards of Universities and Colleges (2001), U.S. institutions of higher learning need a major transformation of their governance structures around international issues. Among the most urgent changes are advances in information technology, shifting focus of education from teaching to learning, increased competition from corporate, for-profit institutions, and the creation of online educational enterprises (Green & Olson, 2003). Within that context and given that physical distance is a constant component in international relations, electronic technologies prove critical for U.S. higher education institutions to establish an efficient and effective communication flow with foreign academic entities. It becomes then pertinent to explore the role of electronic collaboration (e-collaboration) and its impact in the internationalization process, from initial contacts to subsequent agreement negotiations.

BACKGROUND

Developing international programs implies fostering globally oriented campuses with an emphasis on students and faculty (Olson, Green, & Hill, 2005). International programs units are the primary organizational cells responsible for internationalizing higher education institutions. Within those units, chief international officers are in charge of facilitating institutional linkages. They have university-wide responsibility for establishing agreements and coordinating activities with foreign counterparts. In particular, they (a) encourage the development of new programs and enhance existing ones, (b) assist students interested in expanding their education abroad and faculty eager to teach awarding credit courses at cross-border institutions, and (c) develop visiting scholar programs in which faculty from affiliated institutions spend much longer time on teaching or research assignments. Once agreements begin to materialize, university presidents committed to the international scene need to place a higher priority on enhancing strategic partnerships that allow their institutions to become internationally competent entities (Summers, 2002).

Understanding e-collaboration is critical to achieve a desirable level of global competitiveness. E-collaboration is generally understood as communication established by different individuals through the use of computer or non-computer electronic technologies to accomplish a common task (Kock & Nosek, 2004). Typical examples of e-collaboration technologies involve the use of computer-mediated communication over the internet or other computer network infrastructures and also non-computer electronic technologies such as telephones or teleconferencing suites (Kock & Hantulla, 2005). E-collaboration tasks with an international focus address issues ranging from securing tight alliances to setting up informal working relationships in order to foster academic competencies. These types of e-collaboration allow engagement with foreign institutions of higher education to better prepare students to function within an interdependent and integrated economy. Therefore, e-collaboration in its many forms plays a key role in determining the extent to which these relationships among educational organizations are nurtured.

EXPLORING E-COLLABORATION

A comprehensive analysis of the components of e-collaboration is provided by Kock (2005). He discusses six main conceptual elements to identify the nature of an e-collaboration episode. They are: the collaborative task, the e-collaboration technology, the individuals involved in the collaborative task, their mental schemes, and the physical as well as social environments that surround them. It becomes then relevant to explore the extent to which these accepted criteria of e-collaboration manifest themselves in the internationalization process being sought out by the U.S. higher education community.

The Collaborative Task

The international task in which e-collaboration is mostly used is the creation of documents that describe the terms of cooperation among institutions. These documents are called letters of intent or affiliation agreements with foreign academic institutions, professional associations, scientific communities, non-governmental organizations, private businesses, or government entities. In Klasek's (1992) words, they are the formal testimony to the history of the relationship, identify initial areas of common interest, and serve as a basis

for additional cooperative programs in various disciplines. Once completed, a related task comes along by which memoranda of understanding are conducted to outline specific implementation stages and provide transparency to the relationship. These documents bond institutions together and spell out the different clauses under which student and faculty exchanges, study abroad opportunities, curriculum enhancement, joint research and development, or technical assistance will take place.

Formal completion of the above knowledge-intensive collaborative task follows guidelines and contents which could make it increasingly complex depending on the nature and origin of the relationship being sought (Scott, 2000). For instance, legal conditions in developing countries tend to be more convoluted than those observed in the US. Relatively fragile institutional structures often discourage attempts to foster academic relationships due to potential financial mishaps and liability issues. In these cases, U.S. higher education institutions must ensure full compliance with regulations enacted overseas and reduce misinterpretation on how the legal system is being enforced. In such a context, available information obtained electronically reduces uncertainty and allows a prompt clarification of murky regulations.

The E-Collaboration Technology

A second conceptual element is purely technological and it is based on the communication medium and the technological features available to engage into the international collaborative task. Technology presents itself differently not only in each country but also in each city (Chapman & Radmondt, 1998). Institutions of higher education support a vast array of broadband levels. These range from primitive technologies and obsolete software packages that frequently prevent partners from maintaining a flowing communication to fiber-optic networks over which top-of-the-line communication services can be provided (Bradshaw, Gee, & Powell, 2002). Thus, chief international officers should realize that the process of outlining memoranda of understanding may be slower with some counterpart institutions not because of lack of interest but because of outdated communication channels for disseminating information in an electronic form.

Nowadays, e-collaboration is enhanced by the power of the internet which has evolved to the point where it can basically fulfill the most outlandish requests (Crockett, 2006). The pervasiveness of the internet era has gained relevance at an exponential rate over other types of contacts. Each next internet stage empowers users with instant messaging, Web-based discussion boards, or teleconferencing services in such a way that electronic materials are steadily replacing printed ones. It is crucial for U.S. institutions aspiring to attract international partners that their technological resources are constantly updated and understood by their personnel (Green & Shoenberg, 2006). However, despite the benefits in achieving a successful completion of the task, one must be cautious about an entire reliance on technology. A relationship fully electronically mediated might not foster the trust and commitment attained by a face-to-face encounter between interested individuals. For example, an excessive flow of e-mail messaging certainly encourages amicable relations between institutions but also potentially adds an element of triviality to the communication. Such a casual environment can be balanced by an at least once a year face-to-face contact. Therefore, chief international officers must make themselves aware of the various technologies available and the impact of those technologies on the collaborative task in order to rationally allocate money and time.

Individuals Involved in the Collaborative Task

This third conceptual element deals with the characteristics of the individuals involved in the international collaborative task. These characteristics could be demographic, based on age or gender, and/or intrinsic, based on skills or expertise. Outlining memoranda of understanding requires active communication among different individuals both within and outside of participating institutions. Faculty members must timely communicate their intentions to engage their departments and colleges in programmatic activities with an international institution to their chief international officers. Systematic verbal and electronic updates from the latter to senior administrators are necessary in order to ensure their support after an agreement is finally signed. Afterwards, a communicational flow is initiated by chief international officers to parties such as deans, faculty members, or study abroad program leaders to create a chain of contacts with their foreign counterparts at an intensity which depends on the speed and accessibility of e-collaboration.

Faculty members and students of the home institution are then informed of the opportunities available. In the case of students, e-mails addressed to the entire student body or video-conferences targeting specific student groups prove successful marketing tools (Crockett, 2006). Once faculty members decide to engage in an international enterprise, they communicate with the host institution for both academic and non-academic information. At this stage e-collaboration allows for a fluid dialog with foreign students, faculty and students under exchange, and visiting scholars (Green & Shoenberg, 2006). Lastly, it is important for U.S. institutions of higher learning to seek community involvement when an international task culminates with the visit of eminent foreign dignitaries or business executives. The most expedite way to invite local authorities and secure media coverage for debates and discussions with local audiences are through e-mails. These instances of e-collaboration do not require *per se* a team of tech-savvy individuals, rather a relatively capable team to assist with programming upgrades and system compatibilities (Eraut, 2000).

Mental Schemes Possessed by the Individuals

A fourth conceptual element entails the level of technological sophistication attained by the individuals involved in an international collaborative task as well as the homogeneity of their mental schemes for completing it. E-collaboration is designed to reach a much wider set of individuals representing prospective partner institutions than is possible on face-to-face encounters (Kock & Hantulla, 2005). A critical analysis of their possibly different mental schemes consists of assessing the extent to which each party involved in the collaborative task welcomes the available technology. Some may be eager to take advantage of the innovative culture of the internet while others may favor a more personal approach.

Mental schemes of individuals also depend on political ideology and cultural background. It is easy to misinterpret lack of interest as unwillingness to engage in an international collaborative task. Recently, a faculty member from an institution in Alaska proposed a study abroad program to Cuba that seemed to fit the needs of several academic departments. Initial e-mails asking for input were sent to various senior administrators. It was noticeable the lack of involvement of one department chair after repeated announcements. Given the fact that the individual in question was of an advanced age, the intended program leader thought that a lack of response was due to his reaction to text-intensive communication or poor technological training. It was not until a phone conversation that it became evident that his lack of involvement was ideologically based. As a Cuban exile, he simply opposed any institutional initiative to approach Cuba under the current political regime.

Institutional Web sites gain relevance as providers of a first impression about the institution as e-collaboration becomes the rule rather than the exception. They allow prospective partners to assess the institutions' willingness to engage in educational endeavors and test their own understanding against an ideal relational framework (Bradshaw et al., 2005). U.S. higher education institutions with a clear strategic commitment to expand their international dimension should clearly state it on their Web page and provide accessible links to their international programs offices. In addition, they differ greatly in terms of the nature of their research being created. Such a disparity severs the ties by which individuals relate to each other and ultimately alters their attitudes and performance. Chief international officers need to adapt their mindsets to deal with an assortment of technical expertise, political

ideologies, research interests, and personal opinions from which campus internationalization strategies can later be drawn. Eventually, an increased degree of intellectual sophistication for handling e-collaboration facilitates their decision making processes in a context of uncertainties.

The Physical Environment Surrounding the Individuals

This fifth conceptual element refers to those physical assets and geographical distribution influencing the individuals responsible for completing the international task. A widespread and intertwined network of people, places, and institutions signals the complexities of international education. E-collaboration among higher education institutions is more closely dependent on the infrastructural quality rather than the geographical proximity faced by institutional partners overseas. Both condition chief international officers to resort to e-collaboration technologies to ensure that task-related educational services are properly rendered. E-collaboration under adequate infrastructure and affordable connection compensates for remoteness or physical inaccessibility at the time of identifying suitable partner institutions. Geographic dispersion forces individuals at each end to rely on the support of computer-mediated communication or computer-supported cooperative work. In the past, technological scarcity and geographic dispersion delayed completion of the international task. Delays that lasted anywhere between one to two years were common depending on postal mail reliability and travel allowance for personal visits. In these cases, e-collaboration bridges geographical distances by allowing for a type of communication which increases the chances of reaching successful memoranda of understanding and implementing them effectively.

Nevertheless, a complete reliance on physical assets and geographical distribution at the time of deciding to approach foreign universities is risky. It would divide knowledge in two areas: one which embraces institutions that have invested in technical change and closed geographical gaps and another one which contains less capital intensive and geographically dispersed institutions. In doing so, U.S. higher education institutions may prevent their faculty and students from acquiring culturally rich experiences coming from developing countries. Such an approach would further isolate institutions and individuals located in certain geographical areas from a productive academic dialogue and research sharing. The dilemma becomes: if one of the main purposes of U.S. institutions of higher learning is to contribute to the advancement of knowledge, it would then be unethical to ensure that progresses are achieved by excluding less privileged institutions that might not be deemed capable of a productive e-collaboration experience. Otherwise, they may prefer physically endowed and geographically appealing options and privilege e-collaboration over longer and more expensive means such as personal meetings, letters, or phone calls.

The Social Environment Surrounding the Individuals

A last conceptual element covers the social aspects and how they modify behavioral patterns of those in charge of completing the international task. Chief international officers must be aware of individuals from different cultures to understand their perceptional cues and communicate their ideas (Woolf, 2002). E-collaboration changes the ways they interact with counterparts in countries in which joint programs already exist or in which such programs ought to be developed. However, it is hard to imprint a global intercultural experience when synchronous or a-synchronous e-collaboration is chosen as the communication media. Initially, communicational behavior was mainly correlated with professional experience and foreign language proficiency among counterparts. Nowadays, e-collaboration requires a set of norms that regulate a courteous online interaction. These social codes are commonly known as netiquette (Mann & Stewart, 2002). Many issues related to netiquette in the international arena still remain unexplored though. Especially, in terms of the communication channel being used, the recommended set of online behavior, or the home and host-country practices that might affect it.

Social codes are indeed relevant in e-collaboration. In fact they are even more carefully scrutinized given that its mostly written interaction does not allow for perceptional cues that trigger immediate corrections or further explanation. In some countries a prompt response to e-mails signals efficiency and willingness to cooperate while in others a few days delay or not answering at all is socially tolerable due to possible understaffing and/or accessibility problems. Unawareness of those parts of the world in which institutions are interested often leads to interactions which may offend counterparts, misunderstand their messages, or even take offense when it is not intended.

It is imperative for chief international officers to have face-to-face communication to create a sense of belonging which first translates into credibility and later maintains successful institutional relationships. Despite the unquestionable importance of e-collaboration it must remain clear that it does not intend neither to dismantle nor replace face-to face encounters. Rather, e-collaboration attempts to enhance in a meaningful, thought provoking, way its contribution to the social interaction obtained from them. Caution should be exercised though in cases when impersonality arises because of an excessive reliance on collaborative electronic endeavors. Foregone opportunities to observe body language or inflections in the voice tone can easily misinterpret meanings and discourage participation and discussion.

FUTURE TRENDS

The globalization process predicts a greater convergence towards e-collaboration practices in the higher education community. Social, legal, economic, political, and technological conditions signal U.S. institutions of higher education the way to interact with overseas institutions (Ortiz, 2005). In particular, economic and technological conditions are the ones that will provide either further restrictions or impulses on the ways e-collaboration is exercised at the campus level. Therefore, a great challenge for U.S. higher education institutions will be to widen appreciation and understanding of those conditions among chief international officers and other relevant stakeholders.

U.S. higher education institutions are growing at such a high pace that they must be cognizant of the pros and cons offered by e-collaboration. Failure to identify them will distort the ultimate purpose of establishing successful and long-lasting international academic relationships. The use of anytime-anyplace e-collaboration technologies such as e-mail or Web-based chat tools for supporting international relations will undoubtedly increase. Even though chief international officers will still rely on face-to-face contacts to strengthen linkages with counterpart institutions they will realize that e-collaboration complements rather than substitutes such a face-to-face interaction by building up on those encounters. In turn, e-collaboration will allow them to more effectively reach consensus and establish workable frameworks in ways that inquires and statements will expedite the core cognitive learning process about a partner institution (Rovai, 2002).

CONCLUSION

E-collaboration stands out as a useful tool for U.S. higher education institutions to unfold their vision for a comprehensive campus internationalization process. It becomes imperative for them to successfully exploit it in their quest to prepare an internationally knowledgeable citizenry. Unveiling a truly international education dimension requires a tight network of institutions that crosses academic disciplines and transcends institutional boundaries. E-collaboration reinforces the dialogue leading to long-term educational partnerships among agencies, consortia, and organizations by developing an interactive approach to engage academic units, school, or colleges into meaningful specific international learning and research opportunities.

ACKNOWLEDGMENT

Suggestions made by Dr. María P. Mosquera M. are gratefully acknowledged.

REFERENCES

American Council of Education. (1995). *Educating Americans for a world in flux*. Washington, DC: Author.

Association of Governing Boards of Universities and Colleges. (2001). *Institutional governance*. Washington, DC: Author.

Bradshaw P., A., Gee, A., & Powell, S. (2002). *Virtual communities and professional learning across a distributed, remote membership* (Research report). University of Navarra.

Chapman, C., & Ramondt, L. (1998). *Online learning communities*. Retrieved May 21, 2006 from http://www.ultralab.net/papers/online_learning_communities/

Crockett, R. (2006, March). Why the Web is hitting a wall. *Business Week, 20,* 90-92.

Eraut, M. (2000). Non-formal learning and tacit knowledge in professional work. *British Journal of Educational Psychology, 70,* 113-136.

Green, M., & Olson, C. (2003). *Internationalizing the campus: A user's guide*. Washington, DC: American Council on Education.

Green, M., & Shoenberg, R. (2006). *Where faculty live: Internationalizing the disciplines*. Washington, DC: American Council on Education.

Klasek, C. (1992). Inter-institutional cooperation guidelines and agreements. In Klasek, C., (Ed.) *Bridges to the future: Strategies for internationalizing higher education*. Carbondale, IL: Association of International Education Administrators.

Kock, N. (2005). What is e-collaboration? *International Journal of e-Collaboration, 1*(1), i-vii.

Kock, N., & Hantulla, D. (2005). Do we have e-collaboration genes? *International Journal of e-Collaboration, 1*(2), i-ix.

Kock, N., & Nosek, J. (2005). Expanding the boundaries of e-collaboration. *IEEE Transactions on Professional Communication, 48*(1), 1-9.

Mann, C., & Stewart, F. (2002). *Internet Communication and qualitative research: A handbook for researching online*. London: Saga. .

McGrew, A. (2005). The logics of globalization. In J. Ravenhill (Ed.), *Global political economy* (pp. 207-234). Oxford University Press.

Olson, C., Green, M., & Hill, B. (2005). *Building a strategic framework for comprehensive internationalization*. Washington, DC: American Council on Education.

Ortiz, J. (2005). Toward the internationalization of business education in Latin America. In J. R. McIntyre & I. Alon (Eds.), *Business and management education in transitioning and developing countries: A handbook* (pp. 227-241). M. E. Sharpe.

Rovai, A. (2002). Building sense of community at a distance. *International Review of Research in Open and Distance Learning, 3,* 1-15.

Scott, P. (2000). Globalization and higher education: Challenges for the 21st century. *Journal of Studies in International Education, 6*(1), 59-77.

Summers, L. (2002, February 18). An ambitious agenda to remake the nation's leading university. *Business Week,* 22-30.

Stiglitz, J. (2003). *Globalization and its discontents.* New York: Norton Press.

Woolf, M. (2002). Harmony and dissonance in international education: The limits of globalization. *Journal of Studies in International Education, 6*(1), 5-15.

KEY TERMS

Affiliated Institution: Institution with academic expectations, accreditation curricula, and quality standards similar to the ones offered in the home country. It also becomes the *in situ* facility to receive students, faculty, and scholars to use a broad range of delivery models to offer a foreign language of choice, discuss issues of certain countries, and engage into collaborative research.

Campus Internationalization: A strategically integrated process of adequately funded actions, projects, and programs related to international student services, international educational exchanges, international language institutes, and overseas technical cooperation projects implemented into the teaching, research, and service functions of academic departments and colleges.

Chief International Officers: Worldly experienced, functionally knowledgeable, and culturally sensitive individuals entrusted with primary responsibilities related to strategically serve the mission of administering the internationalization of higher education institutions.

Curriculum Internationalization: Institutional efforts made by faculty members and administrators, especially at general education courses, to develop, include, and/or enhance international elements into the curriculum. It requires a careful planning and implementation and needs to be crafted through strong leadership to effectively raise the quality of higher education

Globalization: An ongoing process understood as the increasing flow of expertise, information, productive factors, and technology leading to the interrelatedness and interconnectedness of countries around the world and explained by lower trade and investment barriers and technology advances.

International Education: A conscious and orchestrated initiative undertaken by higher education institutions to offer academic, extracurricular, and nontraditional activities related to the social, legal, economic, political, and technological aspects of countries around the world to greatly enhance the knowledge and productivity of college graduates.

Study Abroad Initiative: Summer, semester, or year-long academic and/or practical experience undertaken by undergraduate or graduate college students overseas to enhance and expand their formal education. Terrorism threats, health issues, political instability, and immigration restrains hinder such opportunities despite institutional efforts to raise awareness about their importance.

An E-Collaboration Overview of Behavior and its Relationship with Evolutionary Factors

Vanessa Garza
Texas A&M International University, USA

INTRODUCTION

The collaboration of individuals across large geographic distances began some time ago, perhaps as far back as the 19th century, with the invention of the telegraph, due to its significant impact on communication (Teresko, 2000). Today, with the increased use of computers, the Internet, and the World Wide Web, electronic communication (e-communication), as well as **electronic collaboration** (e-collaboration) offer individuals around the world the possibility of working together. The wide use of tools such as e-mail and instant messaging, among others, captured the attention of scholars, who began searching for theories that could explain the behavior surrounding the use of electronic media (Kock, 2005b; Simon, 2006).

Throughout the years, researchers have provided a number of explanations in order to offer a better understanding of the factors influencing the use of technology in communication. For instance, the media richness theory holds that face-to-face communication is the richest media available, therefore other forms of communication (such as e-mail) are leaner types of media (Daft & Lengel, 1986). On the other hand, in task-technology fit theory, the outcomes do not necessarily depend on the media being used. In this theory, outcomes do not depend on the technology itself, but vary according to how appropriate the technology is for the task being accomplished (Dennis, Wixom, & Vandenberg, 2001). Other explanations involve social context and its influence on the use of technology (Kock 2005b; Markus, 1994; Simon, 2006). A more recent explanation is the media naturalness model, which goes a step forward and presents evolution as a means to understanding human communication. This view holds that, throughout evolution, humans have become adapted to certain elements of communication, which today are considered "natural" and relates these elements to the use of electronic communication tools (Kock, 2004, 2005b). The objective of this article is to provide an overview of some of the views surrounding e-collaboration, focusing on possible evolutionary explanations of behavior toward it.

BACKGROUND

In order to understand what e-collaboration encompasses one must begin with **computer mediated communication** (CMC) as well as **computer-supported cooperative work** (CSCW). The former deals with the use of computers for any type of interaction, while the latter involves all instances in which humans use technology for any type of activity. **E-collaboration** is described by Kock (2005b) as including both CMC and CSCW, since it does not exclusively deal with computers but may involve other electronic devices (such as a telephone). Kock also states that e-collaboration may involve instances in which there is no actual communication; an example of this would be the collaboration of individuals in creating an online resource without ever directly communicating with each other. Taking these aspects into consideration, **e-collaboration** can be defined as the use of electronic technologies by individuals who are working together to reach a common goal (Kock, 2005a; Kock & D'Arcy, 2002).

There are a number of explanations, dealing with communication, as well as technology in general, which are related to e-collaboration. The well-known **media richness theory** suggests that the use of electronic media depends on how rich or lean the media is (Daft & Lengel 1986). The benchmark for such "richness" is face-to-face communication, which is considered the most effective method for communication because it involves important factors like immediate feedback, tone of voice, and facial expression (Lee, 1994). This means that individuals' behavior towards certain collaboration tools may be explained by their level of "richness" as defined by this theory. While the media richness theory has been supported by a number of studies it has also been challenged by others (Kock, 2005a; Lee, 1994; Simon, 2006). Another available theory is the **task-technology fit theory**, which suggests that outcomes will vary in any given situation depending on the type of technology used and on the fit between the technology and the task (Dennis et al., 2001, on Goodhue & Thomson, 1995). This can be applied to e-collaboration because some of the available technologies may be perceived as more appropriate than others for achieving specific tasks. The **social influence view** holds that behavior toward a particular technology may be affected by social influences and not by the technology itself. This would occur in the case of an employee who is required to provide prompt responses using instant messaging, therefore forcing the technology to increase in "richness" (Kock, 2005a, on Fulk, Schmitz, & Steinfield, 1990).

While these views attempt to explain human behavior by focusing either on the technology itself, or on social influences surrounding it, they do not seem to provide a scientific explanation for such behavior. This has led to the creation of a more recent explanation: the **media naturalness model**. The main idea in this model is that evolution has prepared the human body for certain types of communication that are perceived as being more "natural." This view holds that as more elements of face-to-face interaction are used in a medium, the "naturalness" of such medium will increase. The basis of this argument is found in Darwinian evolution, which may provide some insight into human behavior towards e-collaboration (Kock, 2001).

EVOLUTIONARY PSYCHOLOGY, DARWINIAN EVOLUTION, AND E-COLLABORATION

Darwin's theory of evolution, studied in evolutionary biology, was extended, not too long ago, into the field of psychology, creating what is now known as **evolutionary psychology**. In more recent years, this theory has also been expanded to the field of information systems and has been used to provide more in-depth explanations of human behavior towards areas such as communication, technology and e-collaboration. **Evolutionary psychology** searches for the "origin of behavior," which is presumed to have developed over millions of years (Dunn, 2004, p. 126). Cosmides and Tooby (2001) state that "a complete causal explanation of any behavior-rational or otherwise-necessarily invokes theories about the architecture of [humans']

computational devices" (p. 327). In other words, humans possess built-in "devices" which have been formed by millions of years of biological evolution and the creation of theories is required in order to find an explanation for these devices. All of these devices, or mechanisms, are believed to have evolved in order to solve particular problems faced by humans throughout the evolutionary process. However, these problems are not necessarily in existence today (Buss, 1995). Because the world we live in today has only existed for a relatively short time and evolution has taken place "after millions of years of gradual change" (Lindahl, 2000, p. 28), most of the built-in devices found in humans were developed for a completely different hunting and gathering world (Cosmides & Tooby, 2001; Jones, 1999; Kock, 2005a). An example of this theory would be women's greater spatial-location memory, an evolutionary adaptation useful for gathering, or men's superior upper-body strength, an adaptation useful for hunting. Both of these are examples of a built-in mechanism, which, in the past, was used for survival (Buss, 1995).

In his **evolutionary theory**, Darwin (1859, 1998) argued that human facial muscles are not pointless divine creations. On the other hand, Darwin saw that these muscles are mainly used for facial expression of emotion and that the expressions used are relatively similar across different populations. This similarity across cultures, religions and other groups, implies that, through evolution, humans have become adapted to certain forms of communication which include nonverbal cues (like facial expressions) and which have been embedded in humans' internal mechanism. Hence, the majority of these expressions appear to be unaffected by factors such as culture and distance. In addition, while most of these facial muscles are used for communication, only a few are used for other physical activities (like chewing). Many of these muscles are used primarily for communication through facial expression. Therefore, humans have been programmed by evolution to take part in face-to-face communication involving nonverbal cues, which provide additional meaning to the message being conveyed (Jones, 1999; Kock, 2005a). In other words, because of evolutionary reasons, face-to-face communication is the most effective and effortless communication media for human beings.

These evolutionary theories are some of the basic ideas espoused by the **media naturalness model**, in which face-to-face communication is presented as the most natural communication medium. Contrary to other theories, this theory attempts to provide a reason which can explain why people prefer certain types of communication media over others (Kock, 2005a). Kock (2005a) also states that human beings currently use a "biological communication apparatus [which] has been used for co-located and synchronous communication using facial expressions, body language, and sounds" (p. 120). Hence, since the type of communication used in e-collaboration does not necessarily imitate these natural conditions, human behavior may demonstrate some level of resistance to collaborate electronically. In addition, the use of **e-collaboration** tools can increase ambiguity and frustration for those who are not capable of overcoming such resistance (Bordia, 1997). In a sense, it can be said that our "computational devices," as defined by Cosmides and Tooby (2001), are "mal-adaptations" to today's world, since they were developed for certain needs encountered throughout evolution, most of which no longer exist (Kock, 2005a, p. vii).

However, not all is lost, there is a possibility that the underlying mechanisms of human behavior can become adapted to the use of modern collaborative tools and that users have the possibility of overcoming the problems encountered in the use of e-collaboration tools. There are a number of factors that are believed to influence the performance and adequacy of existing tools. Factors like typing speed, experience with using previous electronic technologies, and having enough time to adapt to the technology could increase the effectiveness of e-collaboration tools and allow them to become just as effective as face to face. In addition, when using these tools people tend to become more task oriented, minimizing problems faced when using e-collaboration tools (Bordia, 1997). For example, since typing takes more time and effort than face-to-face interaction one might compensate for this by typing only useful and task-oriented statements. Because of such compensation, the quality of the contributions to the collaborative effort can increase, helping individuals overcome the difficulties existent in the use of **e-collaboration**. In addition, adapting to electronic tools might require a higher degree of effort, both physical and mental. This increased effort may be significant enough that it allows a person to achieve a more successful collaboration while using less natural media (Kock, 2005a).

Throughout the course of millions of years of evolution, humans became adapted to communication through the use of sounds, expressions, and other nonverbal communication (Kock, 2004). This adaptation may be a cause of the problems and frustration individuals often encounter while attempting to use e-collaboration technologies such as e-mail, teleconferencing, messaging tools, and so forth. Through the adoption of ideas from evolutionary psychology, some researchers now believe that evolution may be the answer to many of the questions concerning human behavior toward electronic media. However, it is also believed that there are ways in which humans can become better adapted to present communication technologies overcoming many of the existing problems.

FUTURE RESEARCH

The study of electronic collaboration is still relatively new. Many of the e-collaboration technologies currently available to most people were not as widely used or as readily available a few years ago. Because of the availability of new technologies and the increased worldwide use of the Internet, educational institutions and businesses located in distant geographic regions can now communicate and collaborate to achieve common goals. Research related to e-collaboration still has a long way to go as it is continuously being conducted to analyze different types of collaborative tools and the behavior toward them. A better understanding of such behavior can maximize the benefits which e-collaboration can yield.

The study of evolution has already been related to a number of fields and may be an indication that the current separation among fields such as information systems, economy, psychology and sociology could become smaller. All of these fields could benefit from the study of the origin of human behavior because, after all, at the core of many fields lies human behavior. The study of **Darwinian evolution** may provide a common platform for all these fields to achieve a better understanding of behavior (Buss, 1994; Cosmides & Tooby, 2001).

A better understanding of evolution and of the built-in mechanisms (devices) humans are believed to possess may lead the way to better methods of analyzing human behavior towards **e-collaboration**. In order to formulate adequate experiments, researchers must first understand the underlying reasons for human behavior. Researchers must have some idea of what to look for, something that the study of evolution may be able to provide (Buss, 1994).

CONCLUSION

This article has provided a brief overview of some of the views surrounding e-collaboration. Although other theories were presented, the main theory discussed involves evolutionary explanations of behavior. Because of the increased use of computers, the Internet and other collaborative tools, research related to electronic collaboration has become more popular in recent years. For years, researchers have presented a number of theories, providing some insight into the factors that influence human behavior, a few of which were briefly explained here. The media richness theory sees face-to-face communication as the richest media available and holds lean media as yielding lower quality outcomes. The task-technology fit theory suggests that outcomes differ depending on the fit between the technology and the task being completed. The social influence view differs in that it does not focus on technology itself but on social influences that can affect the behavior toward technology. Finally, the media naturalness model was explained. This model holds that face-to-face interaction is the most natural communication media and bases some of its arguments in **Darwinian evolution**.

From evolutionary biology to evolutional psychology to e-collaboration, **Darwin's theory of the evolution of the species** has influenced the study of behavior. According to evolutionists, human's ability to communicate was developed throughout millions of years of biological evolution. Because of this, humans have a built-in adaptation to face-to-face communication that can influence behavior toward e-collaboration technologies. All of this leads to the conclusion that humans are adapted to a world that no longer exists and must become readapted to modern life. Although the adaptation to modern technology may be difficult, it is not impossible. Factors like typing speed, previous experience with the use of electronic technologies, and the availability of sufficient time to become adapted to new technologies may increase the effectiveness of e-collaboration tools. In addition, humans may be able to overcome the difficulties through greater effort which may result in successful electronic collaboration.

Finally, this article recognizes that research in e-collaboration is relatively new and has plenty of growth possibilities. The study of the origin of behavior might be the key to understanding modern human behavior. It may provide researchers with the ability to design more adequate experiments that will aid them in acquiring a deeper understanding of human attitudes toward electronic collaboration.

REFERENCES

Bordia, P. (1997). Face-to-face versus computer-mediated communication: A synthesis of the experimental literature. *The Journal of Business Communication, 34*(1), 99-120.

Buss, D. M. (1995). Evolutionary psychology: A new paradigm for psychological science. *Psychological Inquiry, 6*(1), 1-30.

Cosmides, L., & Tooby, J. (2001). Better than rational: Evolutionary psychology and the invisible hand. *AEA Papers and Proceedings, 84*(2), 327-332.

Daft, R. L., & Lengel, R. H. (1986). Organizational information requirements, media richness and structural design. *Management Science, 32*(5), 554-571.

Darwin, C. (1859). *On the origin of species by means of natural selection*. London: Murray.

Darwin, C. (1998). *The expression of the emotions in man and animals*. Oxford, UK: Oxford University Press.

Dennis, A. R., Wixom, B. H., & Vandenberg, R. J. (2001). Understanding fit and appropriation effects in group support systems via meta-analysis. *MIS Quarterly, 25*(2), 167-193.

Dunn, D. S. (2004). Teaching about the origins of behavior: A course on evolutionary and cultural psychology. *Teaching of Psychology, 31*(2), 126-127.

Fulk, J., Schmitz, J., & Steinfield, C. W. (1990). A social influence model of technology use. In J. Fulk & C. Steinfield (Eds.), *Organizations and communications technology* (pp. 117-140). Newbury Park, CA: Sage.

Goodhue, D., & Thompson, R. L. (1995. Task-technology fit and individual performance. *MIS Quarterly, 19*(2), 213-236

Jones, D. (1999). Evolutionary psychology. *Annual Review of Anthropology, 28,* 553-575.

Kock, N. (2001). The ape that used email: Understanding e-communication behavior through evolution theory. *Communications of the Association for Information Systems, 5*(3), 1-29.

Kock, N. (2004). The psychobiological model: Towards a new theory of computer-mediated communication based on Darwinian evolution. *Organization Science, 15*(3), 327-348.

Kock, N. (2005a). Media richness or media naturalness? The evolution of our biological communication apparatus and its influence on our behavior toward e-communication tools. *IEE Transactions on Professional Communication, 48*(2), 117-130.

Kock, N. (2005b). What is e-collaboration? *International Journal of E-Collaboration, 1*(1), i-vii.

Kock, N., & D'Arcy, J. (2002). Resolving the e-collaboration paradox: The competing influences of media naturalness and compensatory adaptation [Special issue]. *Information Management and Consulting, 17*(4), 72-78.

Kock, N., & Hantola, D. (2005). Do we have e-collaboration genes? *International Journal of E-Collaboration, 1*(2), i-ix

Lee, A. S. (1994). Electronic mail as a medium for rich communication: An empirical investigation using hermeneutic interpretation. *MIS Quarterly, 18*(2), 143-157.

Lindahl, B. I. B. (2000). Health and evolution. *Scandinavian Journal of Public Health, 28,* 309-311.

Simon, A. F. (2006). Computer-mediated communication: Task performance and satisfaction. *The Journal of Social Psychology, 146*(3), 349-379.

Teresko, J. (2000). E-collaboration. *Industry Week, 249*(11), 31-35.

KEY TERMS

Computer Mediated Communication (CMC): How humans use computers to form, support, and maintain relationships with others, regulate information flow, and make decisions.

Computer Supported Cooperative Work (CSCW): All contexts in which technology is used to mediate human activity such as communication, coordination, cooperation, competition, entertainment, games, art, and music.

Electronic Collaboration: The collaboration among different individuals who are using electronic technologies to accomplish a common task.

Evolutionary Psychology: Psychological approach which searches for the origin of behavior. In this approach, mental traits are seen as adaptations developed throughout the evolution of the species.

Media Naturalness Model: Model which is based on the Darwinian evolution of human communication. Face-to-face interaction is seen as the most natural communication medium.

Media Richness Theory: Theory which claims that communication media which does not possess nonverbal cues present in face-to-face communication are considered to be lean media. The use of lean media is seen as yielding lower quality outcomes.

Social Influence View: Holds that social influences can affect behavior toward technology.

E-Collaboration Technologies Impact on Learning

Saurabh Gupta
University of North Florida, USA

Robert Bostrom
University of Georgia, USA

INTRODUCTION

Universities and corporate training facilities have been investing in information technologies (IT) to improve education and training at an increasing rate during the past decade. Many new companies as well as educational units are emerging to provide tools, services and content to enable the effective design of IT-based learning solutions (ASTD, 2004). Although research on technology-mediated learning has increased in recent years, it still lags behind developments in practice. Many predict that the biggest growth in the Internet, and the area that will prove to be one of the biggest agents of change, will be online learning, or e-learning (Bostrom, 2003). The boom in the application of technology to education and training underscores a fundamental need to understand how these technologies can be used to improve the learning process.

E-learning research has only recently attracted the attention of information system (IS) scholars, although the topic has been consistently of interest to educational researchers. In spite of the interest, research in this area has been fragmented (Alavi & Liedner, 2001; Bostrom, 2003). One of the reasons for this fragmentation is the lack of agreement on definitions and terms, especially e-learning. In this article, we focus on the definition given by Alavi and Liedner (2001)—"Technology-mediated learning (or e-learning) is defined as an environment in which the learner's interactions with learning materials, peers, and/or instructor are mediated through advanced information technology"

Although the initial focus of e-learning in the Educational literature has been at the individual level, a review of Education literature points out that learning strategies are shifting towards a more active and group-oriented learning referred to as cooperative or collaborative learning (Alavi et al., 1995; Kelley, 1998). Collaborative learning (CL) evolved from the work of psychologists such as Johnson (1981) and Slavin et al. (1985). It refers to instructional methods that encourage students to work together to accomplish shared goals, beneficial to all. It involves social (interpersonal) processes where participants help each other to understand as well as encourage each other to work hard to promote learning (Johnson & Johnson, 1999).

CL is a versatile procedure and can be used for a variety of purposes ranging from teaching specific content to ensuring active cognitive processing of information during a lecture or demonstration (Johnson et al., 1992, 1994). CL procedures have also been found to be more effective than traditional instructional methods in promoting student learning and academic achievement (Johnson et al., 1981; Slavin et al., 1985). In a comparison of CL vis-à-vis traditional classroom learning, Education researchers found that a collaborative approach increases student involvement with the course as well as with each other, increases the level of critical & active thinking, promotes problem-solving skills and increases student satisfaction (Gupta & Bostrom, 2004).

E-collaboration technologies facilitate collaborative learning by offering a rich, shared, virtual workspace in which instructors and students can interact one-to-one, one-to-many, and many-to-many in order to learn together anytime and anyplace (Bostrom et al., 2003). These technologies can be broadly classified as asynchronous/online anywhere tools such as email, discussion databases, streaming audio/video; or synchronous/online live (real-time) tools such as instant messaging, chat, audio/video conferencing.

In spite of the growing importance of e-learning and CL, important research is lacking in collaborative e-learning (CEL). Most of the research in the Education literature has concentrated on face-to-face forms of collaboration or using minimal technology to support it. With advances in information systems, there have been rapid advances in distance learning and virtual team learning. Greater amount of learning is now done using synchronous or asynchronous technology than ever before and there is a need to understand this phenomenon in detail. Finally, the research is lacking good grounding in theory and has focused on input-output models rather than focusing on the process involved in attaining the learning outcomes.

In this article, we first review the IS and Education literature. Next, we identify the primary reasons for the inconsistency in findings in both literatures. Finally, we present a theoretical model for investigating collaborative e-learning. The conclusion section briefly provides directions for future research.

BACKGROUND

E-collaboration technologies are broadly defined as electronic technologies that enable co Research in collaborative e-learning (CEL) has two strong reference disciplines: IS and Education. As mentioned earlier, IS e-learning research has been very limited especially in the CEL area, with only a limited set of chapters focused on CEL. The empirical research in IS stems from the long tradition of Group Support System (GSS), an early e-collaboration technology, research with its focus on process gains/process losses in collaborative settings. Some studies have explored the use of GSS to foster case discussions in a traditional classroom (Hashaim, Rathnam, & Whinston, 1991; Leidner & Fuller, 1997). Others have examined the use of GSS to enable collaboration in small teams of students in traditional classes. As summarized in Table 1, some studies have reported a positive effect of e-collaboration technologies (Alavi, 1994; Drummond, Boldyreff, & Ramage, 2001), while others have not (Alavi et al., 2002; Hiltz, Coppola, Rotter, & Turoff, 2000).

Substantial research in the area of technology supported to learning groups has been done in the area of education. This research domain is known as computer-supported collaborative learning or CSCL. In a recent metareview, Lou, Abrami, and d'Apollonia (2001) examined 122 studies for comparison between small groups versus individual learning when students learn using computer technology. The meta-analysis indicates that, on average, small group learning has significantly more positive effects than individual learning on student individual achievement, group task performance and

several process and affective outcomes. However, the meta-analysis pointed out a wide variation in the results of the experiments (Lehtinen, Hakkarainen, Lipponen, Rahikainen, & Muukkonen, 2003). Post-hoc analysis suggests that the important structures accounting for the variance in the outcomes were technology, task, group and learner characteristics. For details refer to Lou et al. (2001) and Lehtinen et al. (2003).

Given the potential and pervasiveness of computing technology, it is important to understand the reasons for the variance in results in both IS and education. We highlight four important limitations:

1. Research in the area of CSCL uses both technology as well as collaboration to enhance learning. However, these studies do not differentiate between the effect of collaboration or technology. Most of the studies have compared CSCL to individual learning without technology. To establish the effectiveness of CSCL, studies need to analyze the incremental benefit of collaboration and/or technology.

2. Studies have been done in different contexts using different e-collaboration technologies making it impossible to compare experiments. The studies also do not distinguish between different pedagogical ideas on how computers have been implemented in the learning environment. In addition, most of instructional technology research in Education has focused on content-delivery, designed for individuals, whereas, most IS research has focused on technology to support collaboration, not content-delivery. In a typically education study, two-person team would sit around computer system going through content together. We are starting to see much richer blended technologies environments being used but there is little research on these new environments.

3. There has been a lack of well-controlled experiments hampering internal validity of results. Only a few longitudinal studies have been conducted. Studies are also limited in the number of participants and amount of content covered. Most of the studies described the systems and conditions as well as the participants' conversation processes but presented no data on learning outcomes. Education researchers also point out the variance in results that exists in these studies.

Table 1. IS research in technology mediated collaborative learning

Author	Learning context	Major findings
Alavi et al. (2002)	Distributed learning environment Comparing two kinds of electronic mediation—simple and sophisticated.	Simple systems users exchanged more messages about learning task whereas sophisticated system users spent more time on sense-making about the technology. Learning outcomes were higher for simple system users.
Drummond et al. (2001)	Qualitative study with software engineering students using groupware.	Groupware showed a potential for benefit when used correctly.
Hiltz et al. (2000)	Experiment to set up meeting date where participates worked together.	Participants who are actively involved in collaborative (group) learning on-line, the outcomes can be as good as or better than those for traditional classes, but when individuals are simply receiving posted material and sending back individual work, the results are poorer than in traditional classrooms.
Leidner & Fuller (1997)	Quasiexperimental design for case analysis.	Students involved in IT-based collaborative learning showed higher levels of interest in learning than individual learners, but lower levels of performance.
Lim et al. (1997)	Experiment of setting up a meeting date with participants.	Codiscovery forces learner to engage in deeper level thinking Codiscovery learners generated more occurrences and larger proportion of deeper level utterances than did self-discovery subjects
Alavi et al. (1995)	Comparing collaborative learning in one campus and teams spanning across campuses.	Different learning environment were found to be equally effective, however, higher critical thinking was found in distant technology mediated environment. Distant students using technology mediation were more committed and attracted to their groups compared to local students who worked face to face or were technology mediated.
Alavi (1994)	Comparing groups using GDSS versus groups not using and impacts on student learning and classroom experiences.	Learning outcomes of GSS supported student teams is superior to non-supported teams
Hashaim et al. (1991)	Classroom setting for case discussion.	Groupware system helped students and instructors guide the discussion towards its ultimate conclusion, keeping in mind the specifics of the situation as well as experience gained from past case discussions.

4. Large numbers of chapters in education show that CSCL has the potential to influence learning. However, conditions under which gains of CSCL can be obtained elude current literature.

Recently, Alavi and Liedner (2001) provided a framework for e-learning research that highlights the importance of focusing on the learning processes and the lack of theoretical base for understanding in this area.

Fjermestad, Hiltz, and Zhang (2005) and Sasidharan and Santhanam (2006) also provide a similar framework. As these frameworks suggest, focus needs to change to the learning process to investigate the following research question: How does technology enhance learning in a given context (students, instructor/mentor, instructional method, environmental factors)?

We extend these basic frameworks and integrate them with adaptive structuration theory (AST) to provide a holistic model for focusing on e-learning. AST takes a socio-technical perspective of the learning process. It provides the ability to provide a specific theoretical perspective for e-learning, while encompassing other theories. AST provides the ability to focus on the learning method and structures simultaneous as well as independently, while keeping the focus on the learner. Next we briefly outline this theoretical model.

THEORETICAL MODEL OF COLLABORATIVE E-LEARNING

Poole and DeSanctis (2003) identify seven requirements for applying AST effectively. These seven requirements guided the development of the proposed model, shown in Figure 1. Given the space limitations, we will briefly overview the model, for a detailed description of the model see Gupta and Bostrom (2004, 2005).

Epistemology establishes overarching beliefs about the nature of knowledge and about what it means to "know" something (Hannafin, Kim, & Kim, 2004). These provide a design strategy for designing the learning method needed accomplish the learning outcomes. The epistemological perspective can be described in terms of spirit (Poole & DeSanctis, 1990), or the general

intend with regard to values and goals underlying the choice of learning method. Spirit is the "official line" which the learning method presents to participants regarding how to act, interpret the capabilities and how to fill in gaps in procedure which are not explicitly specified (DeSanctis & Poole, 1994). Spirit of the epistemological perspective and the learning goals provide general guidelines regarding the design and use of the learning method.

The learning method is viewed as a collection of three forms of structures: collaborative structures which refer to the social setup of the group (Johnson & Johnson, 1999); e-collaboration technologies present an array of structures for possible use in interpersonal interaction and cognition (DeSanctis & Poole, 1994); and learning technique provide specific procedures to attain learning goals (Schunk, 2004). These three structural sets together provide a basis for group interaction and subsequent learning.

The model highlights the fact that learning achievements are governed by a fit between the structures within a learning method: collaborative, technology and learning techniques. Structures are formal and informal procedures, techniques, skills, rules and technologies, embedded in a learning method, which organize and direct individual or group behavior. The

Figure 1. Conceptual model of learning process

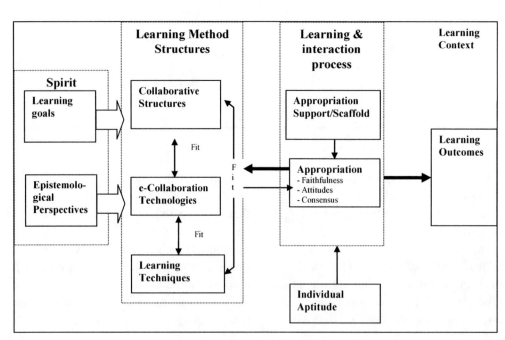

model also argues that these fits are a necessary but not a sufficient condition to improve performance. Without proper appropriation of these structures, learning outcomes are less likely to improve even if fit exists (Dennis et al., 2001).

Appropriation is the process in which the group uses, adapts and reproduces a structure (Poole & De-Sanctis, 2003). Within AST, structures are produced and reproduced in the interaction of participants. As structures are repeatedly reproduced, system tends towards stability (Giddens, 1984). When well-designed and relevant structures are successfully appropriated (i.e. used in the spirit they were designed to be used), they will contribute to higher learning outcomes. An unfaithful appropriation of well-designed and relevant structures, on the other hand, might result in lower learning outcomes. As shown in Figure 1, faithfulness is one of three key variables involved in appropriation, the other two being attitude and consensus (DeSanctis & Poole, 1994).

Appropriation may be supported by providing help, though facilitation or other forms of support or scaffolds. Scaffolds can be used to enhance understanding of the content or the process of appropriating the learning method structures (Hannafin, Hannafin, Hooper, Rieber, & Kini, 2001).

Learning outcomes are the result of the learning process. These can be broadly classified into four dimensions namely, skill, cognitive, affective and meta-cognitive outcomes (Bloom & Krathwohl, 1984). These learning outcomes are also affected by initial levels of individual aptitude or difference. Individual aptitude also affects the level of appropriation, moderating the relationship between learning method and learning outcomes.

The theoretical model presented can also be applied in contexts where one or two structural sets are not present. For example, when applied to individual learning with technology, collaboration structures would be removed from model. Since each individual goes through the process alone, consensus is no longer a relevant dimension of appropriation in this context.

CONCLUSION

This article draws from two different streams for research to provide a common frame for investigating collaborative e-learning. Most importantly, we present a theoretical model that can be used to investigate this phenomenon. Gupta and Bostrom (2005) outline key research areas and propositions based on this model.

The ultimate contribution of this line of research is to develop a specific model for collaborative e-learning. Such an understanding will not only contribute to science, but also provide much needed guidelines for instructional designers and educators.

REFERENCES

ASTD. (2004). *Blended learning survey 2004.* Woburn, MA: Training & Development Magazine, Training Magazine (UK).

Alavi, M. (1994). Computer-mediated collaborative learning: An empirical evaluation. *MIS Quarterly, 18*(2), 159-174.

Alavi, M., & Liedner, D. E. (2001). Research commentary: Technology-mediated learning: A call for greater depth and breadth of research. *Information Systems Research, 12*(1), 1-10.

Alavi, M., Marakas, G. M., & Yoo, Y. (2002). A comparative study of distributed learning environments on learning outcomes. *Information Systems Research, 13*(4), 404-415.

Alavi, M., Wheeler, B. C., & Valancich, J. S. (1995). Using it to reengineer business education: An exploratory investigation of collaborative telelearning. *MIS Quarterly, 19*(3), 293-211.

Bloom, & Krathwohl. (1984). *Taxonomy of educational objectives Handbook i: Cognitive domain.* New York: Addison-Wesley.

Bostrom, R. P. (2003). *Tutorial: E-learning: Facilitating learning through technology.* Chapter presented at the Americas Conference on Information Systems, Tampa, Florida.

Bostrom, R. P., Kadlec, C., & THomas, D. (2003). Implementing and use of collaboration technology in e-learning: The case of a joint university-corporate MBA. In B. E. Munkvold (Ed.), *Implementing collaboration technologies in industry: Case examples and lessons learned* (pp. 211-245). London: Springer.

Dennis, A. R., Haley, B. J., & Vandenberg, R. J. (2001). Understanding fit and appropriation effects in group support systems via meta-analysis. *MIS Quarterly, 25*(2), 167-193.

DeSanctis, G., & Poole, M. S. (1994). Capturing the complexity in advanced technology use: Adaptive structuration theory. *Organization Science, 5*(2), 121-147.

Drummond, S., Boldyreff, C., & Ramage, M. (2001). Evaluating groupware support for software engineering students. *Computer Science Education, 11*(1), 33-54.

Fjermestad, J., Hiltz, S. R., & Zhang, Y. (2005). Effectiveness for students: Comparisons of "in-seat" and ALN courses. In S. R. Hiltz & R. Goldman (Eds.), *Learning together online: Research on asynchronous learning networks* (pp. 39-80). Mahwah, NJ: Erlbaum.

Giddens, A. (1984). The constitution of society: Outline of the theory of structuration. Berkeley: University of California Press.

Gupta, S., & Bostrom, R. P. (2004). *Collaborative e-learning: Information systems research directions.* Chapter presented at the Americas Conference on Information Systems, New York.

Gupta, S., & Bostrom, R. P. (2005). *Working chapter: Theoretical model for investigating collaborative e-learning.* Athens, GA.

Hannafin, M. J., Hannafin, K. M., Hooper, S. R., Rieber, L. P., & Kini, A. S. (2001). Research on and research with emerging technologies. In D. H. Jonassen (Ed.), *Handbook of research for educational communications and technology* (pp. 378-402). Mahwah, NJ: Erlbaum.

Hannafin, M. J., Kim, M. C., & Kim, H. (2004). Reconciling research, theory, and practice in Web-based teaching and learning: The case for grounded design. *Journal of Computing in Higher Education, 15*(2), 3-20.

Hashaim, S., Rathnam, S., & Whinston, A. (1991). *Catt: An argumentation based groupware system for enhancing case discussions in business schools.* Chapter presented at the International Conference on Information Systems, New York.

Hiltz, S. R., Coppola, N., Rotter, N., & Turoff, M. (2000). Measuring the importance of collaborative learning for the effectiveness of ALN: A multi-measure, multi-method approach. *Journal of Asynchronous Learning Networks, 4*(2), 103-125.

Johnson, D. W. (1981). Student-student interaction: The neglected variable in education. *Educational Research, 10*(1), 5-10.

Johnson, D. W., & Johnson, R. T. (1999). Making cooperative learning work. *Theory into Practice, 38*(2), 67-74.

Johnson, D. W., Johnson, R. T., & Holubec, E. J. (1992). *Advanced cooperative learning* (Rev. ed.). Edina, MN: Interactive Book Co.

Johnson, D. W., Johnson, R. T., & Holubec, E. J. (1994). *The new circles of learning: Cooperation in the classroom and school.* Alexandria, VA: Association for Supervision and Curriculum Development.

Johnson, D. W., Maruyama, G., Johnson, R. T., Nelson, D., & Skon, N. L. (1981). Effects of cooperative, competitive, and individualistic goal structures on achievement: A meta-analysis. *Psychological Bulletin, 89*(1), 47-62.

Kelley, D. S. (1998). *Cooperative learning as a teaching methodology to develop computer-aided drafting problem-solving skills.* Unpublished doctoral dissertation, Mississippi State University.

Lehtinen, E., Hakkarainen, K., Lipponen, L., Rahikainen, M., & Muukkonen, H. (2003). *Computer supported collaborative learning: A review.* Unpublished manuscript.

Leidner, D., & Fuller, M. (1997). Improving student learning of conceptual information: GSS supported collaborative learning vs. individual constructive learning. *Decisions Support Systems Journal, 20*(2), 149-163.

Lim, K., Ward, L., & Benbasat, I. (1997). An empirical study of computer system learning: Comparison of co-discovery and self-discovery methods. *Information Systems Research, 8*(3), 254-272.

Lou, Y., Abrami, P. C., & d'Apollonia, S. (2001). Small group and individual learning with technology: A meta-analysis. *Review of Educational Research, 71*(3), 449-521.

Poole, M. S., & DeSanctis, G. (1990). Understanding the use of group decision support systems: The theory of adaptive structuration. In J. Fulk & C. W. Steinfield (Eds.), *Organizations and communication technology* (pp. 175-195). Newbury Park, CA: Sage.

Poole, M. S., & DeSanctis, G. (2003). Structuration theory in information systems research: Methods and controversies. In M. E. Whitman & A. B. Woszczynski (Eds.), *The handbook of information systems research.* Hershey, PA: Idea Group.

Sasidharan, S., & Santhanam, R. (2006). Technology-based training: Toward a learner centric research agenda. In P. Zang & D. F. Galletta (Eds.), *Human-computer interaction and management information systems.* M. E. Sharpe.

Schunk, D. H. (2004). Learning theories: An educational perspective (4th ed.). Upper Saddle River, NJ: Pearson/Merrill/Prentice Hall.

Slavin, R. E., Sharon, S., Kagan, S., Hertz Larzarawitz, R., Webb, C., & Schmuck, R. (1985). *Learning to cooperate, cooperating to learn.* New York: Plenum Press.

KEY TERMS

Collaborative Learning: Groups of students work together in searching for understanding, meaning or solutions or in creating a product.

Technology-Mediated Learning or E-Learning: An environment in which the learner's interactions with learning materials, peers, and/or instructor are mediated through advanced information technology.

E-Collaboration Technology: Technologies that offer a rich, shared, virtual workspace in which instructors and students can interact one-to-one, one-to-many, and many-to-many in order to learn together anytime and anyplace

Epistemological Perspective or Spirit: General intend with regard to values and goals underlying the choice of structure.

Learning Outcomes: Learning outcomes are the result of the learning process. These can be broadly classified into four dimensions, namely, skill, cognitive, affective and metacognitive outcomes.

Learning Process: It is the process of in which an individual or group uses, adapts and reproduces structures or appropriates the structures.

Learning Method Structures: Learning method structures are formal and informal procedures, techniques, skills, rules and technologies, embedded in a learning method, which organize and direct individual or group behavior.

E-Collaboration Through Blogging

Murli Nagasundaram
Boise State University, USA

INTRODUCTION

On September 8, 2004, about 2 months prior to the U.S. presidential elections, the national TV network, CBS, broadcast a segment on its *60 Minutes* program that cast doubts on George W. Bush's service in the Air National Guard. The program host shared four documents obtained by the network which appeared to suggest that Mr. Bush had not adequately satisfied his service requirements. Within hours of the broadcast, the Internet *blogosphere*—Webspeak for the community of *webloggers* who share information through online Internet journals—was abuzz with discussions regarding the veracity of the documents, and it soon began to emerge that the documents might have been forged. On September the 20th, Andrew Heyward, President of CBS News, apologized on behalf of the network for disseminating unverified information. Six months later, CBS's longtime anchor, Dan Rather, was compelled to resign (Thornburgh & Boccardi, 2005).

Numerous individuals—many unacquainted with each other, and geographically dispersed—*electronically collaborated* in dissecting and dismantling the CBS report through the medium of blogs. They precipitated a crisis in a powerful media organization that required the intervention of its chief for its resolution. All of this occurred in a matter of less than two weeks. Blogging had established itself as a media force to reckon with.

This article is a brief introduction to blogging with its application to *e-collaboration*. We learn what blogs are, their history and evolution, how they might be categorized and in what manner they could be used to support e-collaboration. We also discuss some issues and concerns regarding blogs.

BACKGROUND

The term *blog* is short for *Web Log* (also, *weblog*), a frequently updated, online, Web-based journal or diary (Blood, 2000), with postings presented, typically, in reverse chronological order. The term *Web Log* was coined by Jon Barger (Blood, 2000). Blogs are "owned" by one or more *bloggers* who access their blogs via a Web browser. A *blogger* can log in, type up, and publish a blog entry containing their ideas, comments or other information within a matter of minutes. Blogs are distinguished from other websites and other e-collaboration technologies in that:

- The blogger makes fairly frequent postings to the site, which are listed, typically, in reverse chronological order, allowing visitors to read the latest post first
- Blog entries have date and time stamps, allowing the reader to assess its historical value
- Readers may post comments, engaging in a *conversation* with the blogger as well as other readers
- Bloggers may post images, audio, and video documents

After a slow start less than a decade ago, blogging has been growing very rapidly in recent times (Sifry, 2006). Blogging has permeated into corporations, politics, news—practically every segment of society (e.g., Greenwood, 2006). In 1998, there were just a handful of blogs (Blood, 2000), 23 in 1999 (Garrett, 2002), and only 200 in early 2002. Subsequently, blogging rapidly grew to such prominence that in December 2004, the *Communications of the ACM* devoted a special issue to the subject (vol. 47, Issue 12) titled, *The Blogosphere*. Blogger (www.blogger.com), a free and easy-to-use blogging tool-cum-blog hosting service was the most responsible for the growth in blogging. In October 2006, the *blogosphere* tracking blog, Technorati, was tracking 57 million blogs (up from 27.2 million in February) and estimated that the number was doubling every 236 days (down from 5.5 months in February); the "blogosphere" grew to 60 times its size in 3 years. About 100,000 new blogs were coming online everyday (up from 75,000 in February)—more than one new blog every second. About 1.3 million postings are made daily to all the blogs—about 54,000 per hour (Sifry, 2006). Blood (2004) notes that the rapid evolution of blogging technology since 1997 has been driven by widespread practice: "When any sizable number of bloggers start doing something, someone, it seems, will construct a tool to automate it—further popularizing the activity" (p. 55).

E-COLLABORATION THROUGH BLOGGING

An important distinction between a blog and other types of Web sites is its *interactive* nature. Blogging technology allows every blog to be turned into an interactive discussion space and in fact, many blogs are used as such (e.g., as a community forum; Nardi, Schiano, Gumbrecht, & Swartz, 2004). Blogs are, in essence, a fusion of Web pages and Usenet-style discussions and combine the advantages of both. One of the first—and well-known—bloggers was the software developer Dave Winer (2006). Winer's widely read blog, *Scripting News* (www.scripting.com), included news and commentary on technology, politics, and society. Winer engaged intensively with his readership, and their discussions were publicly available on his blog. Some of those discussions led to business relationships or the development of new kinds of software or standards. Blogs are, therefore, well-suited for *e-collaboration*.

Blood (2000), a pioneering blogger, writes on the collaborative nature of blogging:

While weblogs had always included a mix of links, commentary, and personal notes, in the post-Blogger explosion increasing numbers of weblogs eschewed this focus on the Web-at-large in favor of a sort of short-form journal. These blogs, often updated several times a day, were instead a record of the blogger's thoughts: something noticed on the way to work, notes about the weekend, a quick reflection on some subject or another. Links took the reader to the site of another blogger with whom the first was having a public conversation or had met the previous evening, or to the site of a band he had seen the night before. Full-blown conversations were carried on between three or five blogs, each referencing the other in their agreement or rebuttal of the other's positions. Cults of personality sprung up as new blogs appeared, certain names appearing over and over in daily entries or listed in the obligatory sidebar of "other weblogs" (a holdover from Cam's original list). It was, and is, fascinating to see new bloggers position themselves in this community, referencing and reacting to those blogs they read most, their sidebar an affirmation of the tribe to which they wish to belong.

As with individuals in the real world, many bloggers, over time, become members of blogging communities that have common interests, for instance Japanese anime films, AJAX programming or national-level politics, and often coordinate real-world activities through their blogs. A convention in the blogging community is for each blog to include a *blogroll*, which is a list of the blogger's favorite blogs, some of whose authors may be friends or acquaintances, at least over the Internet. Blogger A might read Blogger B's post and respond with either a comment on Blog B or through a posting on her own blog. Blogger C, might read one or both postings and respond with comments on Blogs A and/or B or a post a blog entry in his own blog. This process evolves much like a face-to-face conversation, albeit in a far more complex fashion, with comments from other readers (cf. Kumar, Novak, Raghavan, & Tomkins, 2004). From such conversations, a blog network or community evolves which resembles the communication structure of an offline social group. This makes blogs well-suited for the purposes of *e-collaboration*. Herring et al. (2005) found that a few key blogs in a blog network served as lynchpins binding a network together.

Meyer (2006) noted the emergence of distinct patterns in the content of blogs. He found that bloggers in a collaborative network assume different roles as reflected in the blog entries. He classified blogs as Producers, Reviewers, and Pointers:

- **Producers:** These are blogs that create original content (for the most part). The posts you see tend to be a little longer although that doesn't have to be true. I would count my own blog as part of this category. Producers are the source of much of the original material.
- **Reviewers:** These blogs take topics that are originated elsewhere and put their own spin on the material. Often they will expand the topic or take it off in a different direction. The hallmark of these blogs is that they don't simply point to the source material; they riff on it somehow.

- **Pointers:** These blogs serve to connect readers to content found elsewhere. Sometimes these are aggregators on a single, well-defined topic area. Sometimes they can be quite broad. Usually, there is very little (or no) commentary on the linkages. (Meyer, 2006)

Note the close interdependence among these blog categories: The *producer* generates content; the *reviewer* serves as a gatekeeper, filtering and interpreting what the *producer* generates; and the *pointer* ensures that the content—filtered and unfiltered—reaches a wider audience. In addition, there are the blog visitors (who may or may not have their own blogs) that leave comments on these and other blog sites, thereby communicating with the bloggers and also with other blog visitors. This is a very rich and complex communication structure.

The power of blogging for e-collaboration is illustrated through a real-life example of a planned and structured *interblog conversation*. Four experienced user interface designers engaged in a formal, public conversation on the subject of *Design Vision*. They employed their own blogs to support and structure the conversation (Wroblewski, 2006). They agreed on a process in which they would take turns posting their perspectives as blog entries in each others' blogs. Eventually, all four blogs included entries by all four participants; once the conversation ended, a separate set of blog entries was created with links pointing to the entire conversation in a structured sequence. Also, the entire conversation was turned into a single PDF file publicly posted to all four blogs. This blog conversation proceeded in the form of a panel discussion with four key discussants and a general reading audience. This conversation was *closed* at one level in that only the four designers could contribute to it. It was also *open* and *public* so that any Internet user could follow the conversation. This conversation could have been further structured to accept comments from Internet blog visitors, akin to questions posed by an audience to panelists at a conference.

Academic institutions have recognized the usefulness of blogging to support *collaborative learning,* and are beginning to incorporate them into course curricula, where feasible (Lin et al., 2006; Snowden, 2006). Blogging has been suggested as a means of building community among globally dispersed members of a specific profession (Wiebrands, 2006) and also to serve

as a shared knowledge repository in the intelligence community (Andrus, 2005).

Here are some ways in which blogs may be used for *e-collaboration* in an organization:

1. **Individual blog:** Each individual or employee is assigned a blog and is tasked with posting information regarding any project on which she is working that would be useful to others. Others may, as necessary, post comments or queries. By making the organizational blogosphere searchable, the entire blog space becomes the organization's knowledge repository, as well as a record of discussions regarding various issues. The blog is used as an alternative to e-mail which can get "lost' in a user's mailbox. It is also possible to make blog postings that are in addition, e-mailed to select recipients.

2. **Team blog:** Each team is assigned a single blog to which all team members make postings, answering each others' questions, solving each others' problems, etc. This is especially useful in the case of teams whose members are geographically or temporally distributed.

3. **Project blog:** Each project (rather than team) is assigned a blog. The project blog becomes its knowledge repository. The project blog may be used to both record ideas, information, and conversations during the term of the project as well as lessons learned, when it is wrapped up. Once the project is completed or terminated, the blog is officially "closed" to any further modification and made available to future projects for reference.

4. **Departmental/divisional blog:** Each department or division maintains a blog to which postings are made by select members of the division including the divisional head. Others may read and comment on postings but not make any postings of their own.

5. **Corporate blog:** Corporate blogs are used in myriad ways. One approach, as exemplified by Paul Otellini, CEO of Intel Corp., is to use it as a means of interacting with Intel employees (Piquepaille, 2006). Another approach, as seen in Sun President Jonathan Schwartz's blog is to connect with the marketplace including customers, developers and the lay public (Schwartz, 2006). Robert Scoble of Microsoft used his blog as an informal way of linking with the world outside

Microsoft while still maintaining credibility by also offering personalized opinions on things that go wrong at Microsoft and accepting informal feedback from employees, developers, users, customers, and others.

6. **Blog networks:** Over time, networks of blogs might emerge in an organization and begin to resemble the communication structure of the organization. An organization's blogs could eventually include a complex blog ecosystem made up of Producers, Reviewers, and Pointers. A blogger's interests, competencies and responsibilities are likely to shape the nature of each blog.

Many corporations are turning to blogs as an alternative medium for interacting with customers, prospects, employees and partners (Gordon, 2006). Gordon points out improving market status, personalizing customer relationships, boosting public relations and improving recruitment as being among the key benefits. To many stakeholders, the homey, casual nature of blogs—ironically—gives them a credibility lacking in formal PR announcements. The *Wall Street Journal* (Warren & Jurgensen, 2007) reports that regular visitors to blogs such as Digg.com and Reddit.com (*Pointer* blogs, in Meyers' taxonomy) have been responsible for the dramatic increase in the popularity of certain Websites, corporations, products and ideas. They note:

The opinions of these key users have implications for advertisers shelling out money for Internet ads, trend watchers trying to understand what's cool among young people, and companies whose products or services get plucked for notice. It's even sparking a new form of payola, as marketers try to buy votes. (Warren & Jurgensen, 2007, p. P1)

Former Microsoft employee Robert Scoble's *Scobleizer* blog (Scoble, 2006) attained celebrity status in the *blogosphere*. *Scobleizer* candidly discussed both the strengths and shortcomings in of his employer and its products, thereby garnering respect and admiration. Scoble was mainly a *producer*, but also did some *reviewing* and *pointing*. Another *producer*, Jensen Harris (2006), releases information periodically on work being done on the next release of Microsoft Office and dialogues with and collects feedback from his readers, most of whom are developers for the Microsoft technology platform. Thousands of corporate blogs are for internal use only. Intel Corp. chairman Paul Otellini's very first blog posting generated 350 employee comments (Piquepaille, 2005).

BLOGGING: SOME PITFALLS

Organizational blogging may lead to some unintended consequences. Some corporations such as Microsoft and Sun Microsystems have chosen to host thousands of employee blogs, most of which are for internal consumption only, with a few open to the public. Once a blog posting is made, it is difficult to control the spread of information posted therein. This leaves organizations with the difficult task of both determining and monitoring what kinds of information may be safely disseminated through their employees' blogs. Might corporate secrets be revealed inadvertently? Is it at all feasible to monitor the blogs of thousands of employees? An anonymous Microsoft blogger who calls his blog Mini-Microsoft became a celebrity by being highly critical of the company (*A Rendezvous,* 2005). How might the organization guard against exposure to lawsuits stemming from a blog posting? These issues are in some ways the same as those that exist for e-mail, but the relative openness of blogs (which appear as Web pages, rather than e-mail, which is directed to specific recipients) makes the situation even more problematic.

Permitting reader comments results, on occasion, in abuse. *Comment spam* is one form of abuse where unscrupulous vendors promote their products via blog comments. A more serious problem is the posting of unacceptable comments such as expressions of hatred for certain social groups, the use of bad language, or postings that create a legal liability for the blogger. Casual comment abuse may be deterred by requiring registration and then logging in before postings. Emerging technologies might help fix more serious problems.

CONCLUSION

Starting at the fringe, blogging is rapidly maturing into a technology that has great potential as a tool for organizational communication, cooperation, and collaboration. With increasing globalization blogs can break down the barriers of time and space and help corporations stay close to the customer.

Blogs marry the ease and familiarity of browsing the Web with the efficient simplicity of e-mail to deliver a benefit superior to either e-mail or a normal Web site. They are a combination of information generation, aggregation, presentation and publication, as well as interpersonal and inter-group communication systems.

Blogging systems are constructed from standard, off-the-shelf Web development tools. Many systems are open sourced and may be therefore customized in a variety of ways. The blogging explosion means that the tools and facilities available within blogging systems will grow and change dramatically in the coming years in much the manner that simple, early word processing programs have morphed into today's complex document production systems. Future blogging systems will incorporate tools for creating graphics, tables, databases, audio and video.

REFERENCES

A rendezvous with Microsoft's Deep Throat. (2005, September 26). *BusinessWeek Online.* Retrieved April 20, 2006, from http://www.businessweek.com/magazine/content/05_39/b3952009.htm

Andrus, D. C. (2005). The wiki and the blog: Toward a complex adaptive intelligence community. *Studies in Intelligence, 49*(3).

Blood, R. (2000, September 7). Weblogs: History and perspective. *Rebecca's pocket.* Retrieved April 24, 2006, from http://www.rebeccablood.net/essays/weblog_history.html.

Blood, R. (2002). *The weblog handbook: Practical advice on creating and maintaining your blog.* New York: Perseus Books Group.

Blood, R. (2004). How blogging software reshapes the online community. *Communications of the ACM, 47*(12), 53-55.

Garrett, J. J. (2002). *The page of only weblogs.* Retrieved April 15, 2006, from http://www.jjg.net/retired/portal/tpoowl.html

Gordon, S. (2006). Rise of the blog. *IEE Review, 52*(3), 32-35.

Harris, J. (2006). *Jensen Harris: An office user interface blog.* Retrieved April 21, 2006, from http://blogs.msdn.com/jensenh

Herring, S. C., Kouper, I., Paolillo, J. C., Scheidt, L. A., Tyworth, M., Welsch, P., et al. (2005). Conversations in the blogosphere: An analysis "from the bottom up." In *Proceedings of the 38th Annual Hawaii International Conference on System Sciences* (p. 107b). Washington, DC: IEEE Computer Society Press.

Kumar, R., Novak, J., Raghavan, P., & Tomkins, A. (2004). Structure and evolution of blogspace. *Communications of the ACM, 47*(12), 35-39.

Lin, W.-J., Yueh, H.-P., Liu, Y.-L., Murakami, M., Kakusho, K., & Minoh, M. (2006). Blog as a tool develop e-learning experience in an international distance course. In *Proceedings of the Sixth International Conference on Advanced Learning Technologies* (pp. 290-292). Washington, DC: IEEE Computer Society Press.

Myers, E. (2006). The three types of blogs: Producers, reviewers and pointers. *ICE: Improving customer experience.* Retrieved March 12, 2006, from http://www.egmstrategy.com/ice/direct_link.cfm?bid=F1C806E8-A81C-4D15-58A7EB5FA7EFF8C6

Nardi, B. A., Schiano, D. J., Gumbrecht, M., & Swartz, L. (2004). Why we blog. *Communications of the ACM, 47*(12), 41-46.

Piquepaille, R. (2005, February 16). The blog of Intel president Paul Otellini. *Roland Piquepaille's Technology Trends.* Retrieved April 23, 2006, from http://www.primidi.com/2005/02/16.html

Schwartz, J. (2006). *Jonathan's blog.* Retrieved May 1, 2006, from http://blogs.sun.com/roller/page/jonathan

Scoble, R. (2006). *Scobleizer.* Retrieved April 10, 2006, from http://scobleizer.wordpress.com/

Sifry, D. (2006, November 6). *State of the blogosphere, October, 2006.* Retrieved February 13, 2006, from http://www.sifry.com/alerts/archives/000443.html

Snowden, C. (2006). MIT's class blog pilot: Adapting blogs for class use: technical, pedagogical, and practical issues. In *Proceedings of World Conference on Educational Multimedia, Hypermedia and Telecommunications 2006 (pp. 2873-2877).* Chesapeake, VA: AACE.

Thornburgh, D., & Boccardi, L. D. (2005, January 5). *Report of the independent review panel on the September 8, 2004, 60 Minutes Wednesday segment "For the Record" concerning President Bush's Texas Air National Guard Service.* Retrieved April 30, 2006, from http://www.rathergate.com/CBS_report.pdf

Warren, J., & Jurgensen, J. (2007, February 10). The wizards of buzz. *The Wall Street Journal*, p. P1.

Wiebrands, C. (2006, September 19-22). *Creating community: The blog as a networking device.* ALIA 2006 Biennial Conference, Perth, Western Australia, Australia.

Winer, D. (2006). *Scripting news.* Retrieved April 17, 2006, from http://www.scripting.com/

Wroblewski, L. (2006). Design vision: Introduction. *Functioning Form.* Retrieved April 27, 2006, from http://www.lukew.com/ff/entry.asp?266

KEY TERMS

Blog: A frequently updated, online Web journal (abbreviation of *weblog*).

Blogger: An individual who makes postings to one or more *blogs*. Also the name of the most popular blog hosting service, now owned and operated by Google, Inc.

Blogosphere: The entire universe of *blogs* as well as community of *bloggers* on the Internet.

Blogroll: A list of a *blogger's* favorite *blogs* and listed prominently in his or her *blog*.

Blogspace: Same as *blogosphere*.

E-Collaboration: Collaboration among individuals engaged in a common task using modern networking technologies.

Pointer: A *blog* that connects readers to content generated elsewhere.

Producer: A *blog* that creates original content.

Reviewer: A blog that comments or expands on original content generated elsewhere.

Technorati: A *blogosphere* tracking site.

Web Log or Weblog: Expansion of *blog*.

E–Collaboration Using Group Decision Support Systems in Virtual Meetings

Jamie S. Switzer
Colorado State University, USA

Jackie L. Hartman
Colorado State University, USA

INTRODUCTION

When e-collaborating, there is often a need to bring everyone involved together for a meeting. With potential meeting participants often widely dispersed geographically, the meeting could be conducted virtually by utilizing technology known as groupware. Procedures for conducting successful face-to-face meetings have been in place for many years. However, with the rise in the number of computer-mediated virtual meetings being held amongst e-collaborators, there are additional considerations to take into account when conducting virtual meetings using groupware. This article discusses the use of a particular type of groupware (GDSS) in virtual meetings conducted by participants collaborating in an electronic environment.

GROUPWARE

Groupware is defined as any technology that improves group productivity (Briggs & Nunamaker, 1994). It is a generic term for specialized computer aids designed for use by collaborative work groups (Johansen, 1988).

Types of Groupware

The term *groupware* can in actuality stand for many different things. Briggs and Nunamaker (1994, p. 61) have identified several names and concepts defined as equivalent to groupware: group decision support systems, electronic conferencing, team databases, computer supported cooperation, video teleconferencing, shared drawing, workflow automation, information filtering, coordination support, collaboration support, electronic meeting systems, and team scheduling and project management.

Regardless of what it is called, groupware supports e-collaboration (Kock & McQueen, 1997), communication, and coordination (Orlikowski & Hofman, 2003) and allows people to work together to perform the following types of functions in an electronic environment (Liff, 1998):

- Management support, including meeting facilitation
- Document sharing and management
- Group calendaring and scheduling
- Project management
- Information sharing and threaded discussion forums
- Real-time interactions, including audio and video conferencing and whiteboard collaboration
- Knowledge management, which allows organizations to create a corporate memory

Group Decision Support Systems

One category of groupware increasing in popularity is group decision support systems (GDSS). GDSS encourage such activities as group idea generation, voting, brainstorming, decision making, and consensus reaching (Holtham, 1994) by removing common communication barriers. Huber defines GDSS as "a set of software, hardware, and language components and procedures that support a group of people engaged in a decision-related meeting" (1984, p. 195). DeSanctis and Gallupe offer a similar definition, calling it "an interactive, computer-based system that facilitates the solution of unstructured problems by a set of decision makers working together as a group" (1985, p. 379). Pollock and Kanachowski (1993) define GDSS as a system where group members use computers interactively to support the group's decision-making capacity.

Studies have shown that technology is essential to the success of e-collaborations (see Cai, 2005). According to Poole and Holmes (1995), the strength of GDSS comes from its ability to enhance communication and information exchange, complex information processing tasks, and coordination and organization of group collaborations. GDSS facilitate e-collaboration by combining the use of computer technology (both hardware and software), video, audio, and telecommunication systems (Barnes & Greller, 1994).

There are different levels of GDSS involved in e-collaboration (DeSanctis & Gallupe, 1987). At its most basic, GDSS provide features that facilitate common communication behaviors such as voting and electronic message exchange. The next level of GDSS provide a means to model decisions and group decision techniques to reduce the uncertainty that can occur in the decision making process. At its highest level, GDSS are tools to manage group communication patterns in e-collaboration and can include expert advice in the selection and arrangement of procedures to be followed during a virtual meeting.

CONDUCTING VIRTUAL MEETINGS USING GDSS

The primary purposes of meetings are to exchange work-related information, to make decisions, or to accomplish tasks. Guidelines for conducting successful face-to-face meetings have been in place for many years. For each face-to-face meeting, four stages of meeting protocol should be adhered to:

1. Determine the need for a meeting
2. Prepare for the meeting
3. Conduct the meeting
4. Follow-up after the meeting

Even when proper procedures are followed, there are several problems that can arise in traditional face-to-face meetings. Issues not related to the relevant task can sidetrack the group. Dominant personalities can monopolize the group's time and attention. The free flow of creative thought may be discouraged by ideas being attacked or the fear of retribution. There can be premature closure of the meeting to avoid conflict. The record of the meeting can be subjective, incomplete, or lost.

Compounding traditional face-to-face meeting complexities is the rise in the number of virtual meetings, which necessitate additional considerations. As with face-to-face meetings, organizers of virtual meetings should also follow the four stages of meeting protocol mentioned previously. But because of the very nature of the virtual meeting, there are additional considerations that need to be taken into account to conduct effective virtual meetings using GDSS. By utilizing e-collaboration technologies such as GDSS, interference with collaborative activities can be reduced and problems inherent in traditional meetings eliminated (DeSanctis, 1993).

To determine if a meeting is warranted is particularly important in a virtual environment. With e-collaborators scattered around the world in several different time zones, conducting meetings virtually presents a challenge that goes beyond the issues associated with face-to-face meetings. With participants thousands of miles apart geographically, scheduling the virtual meeting can be a difficult task. The use of an e-collaboration technology such as GDSS to conduct electronically mediated meetings is very effective in reaching those geographically dispersed team members (Munter & Netzley, 2002).

Technological considerations are the most crucial part of preparing for the virtual meeting. Determining the most appropriate technology to conduct the virtual meeting depends on the meeting agenda as well as organizational resources. Consider if the information to be delivered or the task to be achieved could best be accomplished via GDSS. The technology should serve the meeting, not dominate it (Duarte & Snyder, 1998). Nunamaker, Briggs, Mittleman, Vogel, and Balthazard (1996) argue that technology cannot make up for poor planning or ill-conceived meetings, and could even make the situation worse.

The technology must be in good working order, and a back-up plan must be in place in the (very likely) event the technology will fail. All of the people at the virtual meeting need to be trained and experienced in the technology, otherwise they will not participate. Additionally, a trained GDSS facilitator must be present at the virtual meeting to ensure its success. According to Munkvold, "the use of a facilitator is an absolute necessity for running an effective, co-located electronic meeting" (2003, p. 18).

Distributing the agenda in a timely fashion is also an important part of preparing for the virtual meeting.

Besides the agenda, any additional materials must also be distributed to the e-collaborators before the scheduled meeting. If the virtual meeting is held to discuss a particular report, for instance, a person at a remote site would be at a definite disadvantage if the report was not sent in time for the virtual meeting.

When conducting a virtual meeting, facilitators need to be extremely aware of the fact that words or phrases can quite easily be misinterpreted. Some GDSS are text based only and therefore do not allow e-collaborators to view nonverbal cues. The lack of nonverbal cues such as body language and facial gestures can make communication more difficult in virtual meetings, thus accentuating language, culture, and style differences (Henry & Hartzler, 1998). To avoid creating tension among the e-collaborators, everyone must make certain the communication is clear and precise with no chance for a misunderstanding.

It is easy for people to "fade into the background" during virtual meetings, particularly those facilitated by GDSS. If the virtual meeting participants can't see each other, some people may never contribute to the discussion. In a computer-mediated meeting environment, it could simply be that the e-collaborator's typing skills are poor, and keeping up with the "discussion" is too difficult. Regardless of the reason, virtual meeting facilitators must make certain that everyone is participating and has equal opportunity to share thoughts and ideas. Virtual meeting participants should also identify themselves each time they contribute to the discussion.

When the virtual meeting ends, the communication and e-collaboration do not (Hoefling, 2001). GDSS meetings tend to generate quite a bit of information (Arkesteijn, DeRooij, & VanEekhout, 2004). It is critical to follow up after virtual meetings to make sure everyone understood the results of the meeting and knows what action to take. GDSS allows e-collaborations to be recorded electronically, thereby facilitating follow-up activities from the virtual meeting. The use of GDSS can also be very valuable in collecting a myriad of data for further consideration by the e-collaborators.

Using GDSS and other types of groupware "symbolizes new ways of integrating information technology with innovations in management and group process to produce more effective forms of collaboration" (Creighton & Adams, 1998, p. 13), including virtual meetings. But while groupware tools are very effective in many e-collaboration circumstances, there are potential downfalls (Burnett, 1994; Creighton & Adams, 1998; Johansen, 1988).

The technical problems are the most obvious, since technology is vulnerable to failure. The technology may fail during the virtual meeting, effectively ending the session without completing the agenda; computers do crash. The software may not function properly. If the technology is not working, the meeting cannot be held and valuable time and energy is lost.

The other concerns are social in nature. Using GDSS to facilitate virtual meetings may result in the over-control of the e-collaborators by focusing too much on the technology and process. People may not be computer literate. Creativity may be stifled because the meeting is too structured around the groupware requirements. Meeting participants may also lose a sense of community and personal touch.

MIXED RESULTS IN THE USE OF GDSS

Results of years of research on e-collaborating using GDSS are proving to be inconclusive, with studies supporting both positive effects on collaboration using GDSS and negative effects from the use of the tool (see Limayem, Banerjee, & Ma, in press). Jessup, Connolly, and Galegher (1990) also cite numerous studies that obtained mixed results in examining the effectiveness of GDSS.

Despite the fact that decades of research have yielded inconsistent findings, overall results "suggest that the use of a GDSS improves decision quality, depth of analysis, (and) equality of participation" (Fjermestdad, 2004). The use of GDSS tends to produce more positive effects compared to face-to-face methods, particularly with respect to the type of task the participants in the virtual meeting are engaged in (Fjermestdad & Hilz, 1998/99).

TRENDS IN CONDUCTING VIRTUAL MEETINGS USING GDSS

New technologies supporting e-collaboration are being developed and evaluated in virtual meeting contexts. "Smart meeting rooms" take GDSS technologies to a much higher level and allow for augmented reality meeting support and virtual reality generation of meetings, both in real time or off-line (Nijholt, Akker, & Heylen, 2006). Meeting participants in these "smart" or "virtual" meeting rooms will conduct all the virtual meeting activities currently facilitated by GDSS (brainstorming, discussing, voting, etc.), only with virtual reality representations of themselves, instead of e-collaborating in a strictly textual environment.

CONCLUSION

GDSS assist in focusing the efforts of e-collaborators toward the task or problem at hand. They enable productivity in virtual meetings through technology-facilitated collaboration and avoid information overload. Inputs can be anonymous, so many ideas can be quickly generated to solve problems or identify opportunities. GDSS facilitate distilling those ideas to the very best ones, clarify exactly what is intended, organizing the ideas, and evaluating and prioritizing them. It can help e-collaborators build consensus and produce deliverables that result in specific actions.

Using GDSS to facilitate a virtual meeting ensures the overall quality of effort put into the process is increased, as is the probability of reaching consensus. It guarantees no single person dominates the procedure. The e-collaborators can easily stay on task and not get distracted. An objective, precise documentation of the entire meeting is generated. Virtual meetings can also be much more efficient using GDSS.

The use of GDSS is intended, among other things, to promote e-collaboration by providing the opportunity to generate a wealth of ideas from virtual meeting participants since they can input their thoughts using the computer without interrupting another person (Olson & Olson, 2003). Using GDSS in a technology-mediated collaborative meeting environment can greatly increase efficiency. As Lococo and Yen observe, "Efficient sharing of ideas can be transformed into shared understanding and into shared priorities" (1998, p. 91). Smith (2001) argues that by improving how meetings are conducted, communication, morale and productivity will also improve.

GDSS can be very powerful tools for conducting effective virtual meetings. Used properly, as previously discussed, e-collaboration technologies such as GDSS have tremendous potential to successfully facilitate virtual meetings. While "the same technology will not provide the same results with each group and in each

setting" (Boiney, 1998, p. 343), e-collaborating using GDSS has many benefits.

Duarte and Snyder (1999) emphasize the importance of understanding the issues regarding e-collaboration by making five points regarding virtual meetings:

1. Facilitating a virtual meeting includes managing the agenda, the participants, and the technology
2. Select the technology that is appropriate for the outcome of the meeting. Match the use of technology to specific agenda items
3. Leverage the agenda and the use of technology to maximize participant recall, the opportunity for participants to contribute, motivation of the participants, and to reduce social pressure on the participants
4. Make use of social protocols and best practices for using the selected technology
5. Make certain that logistics cover issues such as compatibility of technology training in using new systems as well as backup plans

For virtual meetings to be effective when they are conducted via GDSS, traditional meeting procedures need to be adjusted and redefined to take into consideration the special circumstances involved in using technology that facilitates e-collaboration. As Grenier and Metes (1995) observe, "This meeting's not over, it's just gone to another medium."

REFERENCES

Arkesteijn, H., DeRooij, J., & VanEekhout, M. (2004). Virtual meetings with hundreds of managers. *Group Decision and Negotiation, 13,* 211-221.

Barnes, S., & Greller, L. M. (1994). Computer-mediated communication in the organization. *Communication Education, 43,* 129-142.

Boiney, L. G. (1998). Reaping the benefits of information technology in organizations: A framework guiding appropriation of group support systems. *The Journal of Applied Behavioral Science, 34*(3), 327-367.

Briggs, R., & Nunamaker, F. (1994). Getting a grip on groupware. In P. Lloyd (Ed.), *Groupware in the 21ˢᵗ century: Computer supported cooperative working toward the millennium* (pp. 61-72). Westport, CT: Praeger.

Burnett, A. (1994). Computer assisted creativity. In P. Lloyd (Ed.), *Groupware in the 21ˢᵗ century: Computer supported cooperative working toward the millennium* (pp. 197-205). Westport, CT: Praeger.

Cai, J. (2005). A social interaction analysis methodology for improving e-collaboration over the Internet. *Electronic Commerce Research and Applications, 4*(2), 85-99.

Creighton, J., & Adams, J. (1998). *CyberMeeting: How to link people and technology in your organization.* New York: American Management Association.

DeSanctis, G. L. (1993). Shifting foundations in group support systems research. In L. M. Jessup & J. S. Valacich (Eds.), *Group support systems: New perspectives* (pp. 97-111). New York: Macmillan.

DeSanctis, G. L., & Gallupe, R. B. (1985). Group decision support systems: A new frontier. *Database, 16*(1), 377-387.

DeSanctis, G. L., & Gallupe, R. B. (1987). A foundation for the study of group decision support systems. *Management Science, 33*(5), 587-609.

Duarte, D. L., & Snyder, N. T. (1999). *Mastering virtual teams: Strategies, tools, and techniques that succeed.* San Francisco: Jossey-Bass.

Fjermestdad, J. (2004). An analysis of communication mode in group support systems research. *Decision Support Systems, 37*(2), 239-263.

Fjermestdad, J., & Hiltz, S. R. (1998/99). An assessment of group support systems experimental research: Methodology and results. *Journal of Management Information Systems, 15*(3), 7-149.

Grenier, R., & Metes, G. (1995). *Going virtual: Moving your organization into the 21ˢᵗ century.* Upper Saddle River, NJ: Prentice Hall.

Henry, J., & Hartzler, M. (1998). *Tools for virtual teams: A team fitness companion.* Milwaukee, WI: American Society for Quality Press.

Hoefling, T. (2001). *Working virtually: Managing people for successful virtual teams and organizations.* Sterling, VA: Stylus.

Holtham, C. (1994). Groupware: Its past and future. In P. Lloyd (Ed.), *Groupware in the 21ˢᵗ century: Computer*

supported cooperative working toward the millennium (pp. 3-14). Westport, CT: Praeger.

Huber, G. P. (1984). Issues in the design of group decision support systems. *MIS Quarterly, 8*(3), 195-204.

Jessup, L. M., Connolly, T., & Galegher, J. (1990). The effects of anonymity on GDSS group process with an idea-generating task. *MIS Quarterly, 14*(3), 313-321.

Johansen, R. (1988). *Groupware: Computer support for business teams.* New York: The Free Press.

Kock, N. F., & McQueen, R. J. (1997). Using groupware in quality management programs. *Information Systems Management, 14*(2), 56-62.

Limayem, M., Banerjee, P., & Ma, M. (in press). Impact of GDSS: Opening the black box. *Decision Support Systems.*

Liff, A. (1998). Fostering online collaboration and community. *Association Management, 50*(9), 33-38, 92.

Lococo, A., & Yen, D. C. (1998). Groupware: Computer supported collaboration. *Telematics and Informatics, 15*, 85-101.

Munkvold, B. E. (2003). *Implementing collaboration technologies in industry: Case examples and lessons learned.* London: Springer.

Munter, M., & Netzley, M. (2002). *Guide to meetings.* Upper Saddle River, NJ: Prentice Hall.

Nijholt, A., Akker, R., & Heylen, D. (2006). Meeting and meeting modeling in smart environments. *AI & Society, 20*(2), 202-220.

Nunamaker, F., Briggs, B., Mittleman, D., Vogel, D., & Balthazard, P. (1996). Lessons from a dozen years of group support systems research: A discussion of lab and field findings. *Journal of Management Information Systems, 13*(3), 163-207.

Olson, G. M., & Olson, J. S. (2003). Groupware and computer-supported cooperative work. In J. A. Jacko & A. Sears (Eds.), *The human-computer interaction handbook: Fundamentals, evolving technologies and emerging applications* (pp. 583-595). Mahwah, NJ: Erlbaum.

Orlikowski, W. J., & Hofman, J. D. (2003). An improvisational model for change management: The case of groupware technologies. In T. W. Malone, R. Laubacher,

& M. S. Morton (Eds.), *Inventing the organizations of the 21ˢᵗ century* (pp. 265-281). Cambridge, MA: The MIT Press.

Pollock, C., & Kanachowski, A., (1993). Application of theories of decision making to group decision support systems (GDSS). *International Journal of Human-Computer Interaction, 5*(1), 71-94.

Poole, M. S., & Holmes, M. E. (1995). Decision development in computer-assisted group decision making. *Human Communication Research, 22*(1), 90-127.

Postmes, T., & Lea, M. (2000). Social processes and group decision making: Anonymity in group decision support systems. *Ergonomics, 43*(8), 1252-1274.

Smith, T. E. (2001). *Meeting management.* Upper Saddle River, NJ: Prentice Hall.

KEY TERMS

Computer-Mediated Collaboration: Working together by using technology-facilitated means.

Computer-Mediated Communication: Communicating by using technology-facilitated means.

E-Collaboration: Working together in an electronic environment.

Group Decision Support Systems: A category of groupware that encourages activities such as group idea generation, voting, brainstorming, decision making, and consensus reaching.

Groupware: Technology that supports collaboration, communication, and coordination in an electronic environment.

Meeting Protocol: The rules or conventions of correctly conducting a meeting.

Virtual Meetings: Meetings conducted electronically, not face-to-face.

Virtual Meeting Protocol: The rules or conventions of correctly conducting a virtual meeting.

E-Collaboration within Blogging Communities of Practice

Vanessa Paz Dennen
Florida State University, USA

Tatyana G. Pashnyak
Florida State University, USA

INTRODUCTION

The act of blogging is not an inherently social or community-oriented one, but increasingly blogs have been used both as a forum for sharing information with and interacting with others and as a huge repository of information and opinions to be collected, organized, and shared. New blogs are being created at a very quick rate; Technorati, a blog-indexing company, estimates that they are tracking over 70 million blogs as of February, 2007, a number that is continuously growing (Technorati, 2007). Combined with current trends toward online social networking and learning communities, it is not surprising that blogs have been used to support e-collaboration.

Blogging communities have developed in two ways. First, through this proliferation of blogs, individuals with like interests have found each other and built online connections. Second, people with real-life connections have realized the potential of blogging technology to facilitate collaboration and have purposefully created blogs to support their efforts. This article provides an overview of how blogging communities of practice are defined, have developed, and have come to use the tools for e-collaboration.

BACKGROUND

The Internet was developed in part to bring together groups of people who are separated by time and space, and the concept of people working together in virtual communities stretches back to the pre-Web days of this technology (Rheingold, 2000). Blogging is an act that can be done individually, but has come to be done collaboratively by many people over the years. It is essentially involves writing and publishing brief posts to a Web page in reverse chronological order.

The definition of blog contains no requirement of an external audience, interactants, or discourse; a blog could be a personal diary, a log of work, or a collection or personal links intended for the authors' eyes only. That blogging became a community-oriented activity in some contexts is perhaps more indicative of a desire for technology-mediated communication than of anything inherent in the technology itself, which did not extend beyond basic Web-page creation.

Communities, in the general sense, may be cultural, geographical, or based on interest. An online community might be any of these three, but defining community can prove a big challenging. The word community often is used rather casually with reference to people communicating via the Web. Any collection of people who might communicate online has come to be called a virtual community, and corporate entities have used the concept of online community as a marketing strategy. However, online community is based on more than just the ability to communicate with others, reading the same blog or having signed up for the same bulletin board or email list. Just because people read the same online newspaper, for example, does not make them a community; individual readers likely do not know of each other's identity or presence, nor do they gain anything from it.

A true community requires that additional criteria be met, and evidence of community can be found in many forms. Kling and Courtwright (2003) have criticized overly general use of the term community, suggesting that a true community extends beyond mere ability to communicate with others. They highlight interpersonal elements such as a developing sense of trust among participants and the resulting willingness to take risks as indicators of community. Josefsson (2005) provides a framework of social dynamics to be explored, with community being demonstrated through the interrelationship of forms of expression, individual identity, relationships, and behavioral norms. Baym (1998) suggests that community is dependent on having a fertile external context, temporal structure, system infrastructure, group purposes, and participant characteristics. When a sense of interdependence can be achieved based on these elements, we can say that community likely exists. Thus the determination of community extends beyond just noting that people are interacting via blog posts.

BLOGGING AND COMMUNITIES OF PRACTICE

Blogs have been viewed as a forum for sharing one's knowledge and points of view. A study by the Pew Internet Foundation (Lenhart & Fox, 2006) showed that 64% of bloggers surveyed engaged in blogging to share practical knowledge or skills, and 76% did so to document and share personal experiences. Both of these types of sharing are critical elements of a community of practice, in which people not only pass along information but also how they use, understand, and feel about this information. Additionally, 50% of the bloggers surveyed indicated that they blog to network or meet new people, showing how it can be a community-expanding experience.

Essentially, blogs or groups of blogs authored by people with commonalities or like interests have evolved into communities unto themselves, often becoming communities of practice (CoPs). Communities of practice are groups of people who engage in social learning about a given subject or profession (Lave & Wenger, 1991; Wenger, 1998). It is this shared practice that differentiates them from any other type of community (e.g., a cultural community or geographic community). The process through which they learn is called a cognitive apprenticeship (for more information, see Dennen & Burner, in press). Those engaged in a community of practice are not merely acquaintances; rather, the members are usually practitioners who share similar backgrounds, tools, experiences, goals, and ways of thinking. As a community, these individuals engage in discussions, exchange ideas, ask and answer questions, and share what they have learned individually, thus building relationships and group knowledge. Although these are all visible acts described as part of a community of practice, the community of practice model allows for those who interact largely by observ-

ing. Such engagement, termed legitimate peripheral participation (Lave & Wenger, 1991), often leads to more formal and tangible community involvement.

Wenger (1998) describes the various trajectories or paths that one might be on within a community of practice. There are five trajectories: (1) peripheral, or a newcomer who generally observes and may or may not become more involved; (2) inbound, or one who is joining the community and learning much about it in the process; (3) insider, or one who has been a member for a while and is engaged in supporting new members and community maintenance; (4) boundary, or one who participates as a member of another related community; and (5) outbound, or a member who is preparing to exit the community. In a blog-based community of practice these trajectory designations take on two meanings; a member may be on one trajectory with regards to the practice itself and another with regards to the blogging community. An experienced practitioner may have much to contribute, but be new to the norms that govern blog-based interactions and thus find the need to observe at first, and then gradually increase participation over time before becoming an insider in a blog-based e-collaboration.

It is important to note that just as communities are not created by putting a group of people in the same room, online communities of practice are not simply created by linking a certain number of related weblogs; rather, these communities are gradually built through the interaction of members who may never meet each other in person, yet enjoy lasting informal relationships. Relatively little is known about how CoPs naturally form and evolve, given that most are studied after they are fairly well formed. Additionally, the sub-set of intentionally designed communities is different from those that develop organically (Schwen & Hara, 2003).

Tapped In® (www.tappedin.org) is an example of an intentionally designed CoP, and its designers still struggle with terminology in that the technology may be supporting several communities engaged in a variety of practices (Schlager & Fusco, 2003). In other words, this site, which was designed to support e-collaboration and professional development for teachers, has grown through community involvement into something somewhat different from its original plan. The deviation or drift from the original design is not a bad thing; meeting the needs of community members is more important than following a prescribed path.

TYPES OF BLOG-BASED COMMUNITIES

Blog-based e-collaboration can support practice in any number of existing physical settings in addition to enabling new relationships among people who are otherwise not colocated. Blogging has transformed work practices in many disciplines, allowing groups of professionals from the same field to easily share their expertise, experiences, and knowledge to foster new approaches to problems, regardless of distances separating them (e.g., *Information World Review* at http://blog.iwr.co.uk). For example, blogging has been used to support e-collaboration in the medical research community (Boulos, Maramba, & Wheeler, 2006; Sauer et al., 2005; Maag, 2005), intelligence community (Andrus, 2005; Burton, 2005), business settings (McAfee, 2006), and education (Dennen & Pashnyak, 2006; Ducate & Lomicka, 2005; Oravec, 2003; Poling, 2005).

Not all blog-based communities need to be professionally-oriented. Personal communities of practice may be formed by people with a particular shared interest, background, or need. For example, there are online communities for mothers who exchange experiences and ideas on child-rearing, often called "mommybloggers" (see www.mommybloggers.com for an example); knitters who exchange patterns and techniques (see Wei, 2004 for a discussion of knitting blogs); and travelers who exchange photos and travelogues. Some communities of practice serve dual purposes—while addressing professional topics, they also allow users to share comments about their personal lives, interests, experiences, news, and social issues. In other instances, an individual blog may span two or more communities. The blogger need not segment her identity into topic-specific blogs, but instead may combine interests on one blog.

Blogging communities tend to fit one of three formats: blog/blogger-centric, connecting topic, or boundaried (White, 2006). The first model refers to a community of posters and commenters centered around a single blog. In such a community, the posting author(s) maintain complete control and may enforce certain rules or censor comments if deemed necessary. The second model is a community of blogs united by a common interest or topic. As these communities grow, minicommunities often are formed within as people find more specific interests. The third model is centered

around a platform (e.g., myspace.com), with access for nonregistered users greatly limited. However, many blog-based communities may not fit such categories so neatly (Efimova & Hendrick, 2005). Blog-based communities of practice that have been successful tend to more organically reflect the needs and interests of the members than categories set forth by researchers.

COMMUNITY DEVELOPMENT AND MAINTENANCE

Online blog-based communities of practice form around various themes, interests, disciplines, and subjects. Each community has a purpose and is guided by certain norms and policies. The development of these norms and policies is spearheaded by the original creators of the community, whether implicitly or explicitly, and they evolve as community grows and bloggers add individual contributions (Preece & Maloney-Krichmar, 2005; Wei, 2004). Within a blogging community, not all blogs are alike. Each blogger can create a site that reflects the author's purpose, interest, and personality. Still there are commonalities; together, through a collection of linked blogs, members may develop and follow communal rules and norms (Blood, 2004). Some bloggers may post under pseudonyms, often to allow a true expression of thoughts, ideas, and values without fear of repercussions to their real-life identities; in order for pseudonyms to work within online communities, there must be reasonable boundaries of how this free thought is expressed (Reed, 2005). Again, the elements of trust amongst participants and risk-taking are components of virtual communities.

To form community and establish collaborative relationships, individual bloggers must make a conscious effort to reach out to others with similar interests and backgrounds. Once exchanges start flowing, deep relationships may evolve, allowing for successful and meaningful e-collaboration. However, communicating within blog-based communities is not a simple process. While the author of a particular blog has an absolute power over her domain, a certain tone is needed to engage others. Being positive or constructive, interactive, and willing to respect the opinions of others all are crucial requirements of online community membership. Also important is for the established community members to acknowledge new members and make them feel welcome. It is the new members who enrich the community with new perspectives and voices. At the same time, newcomers must earn the acceptance of existing members by learning and respecting their norms and expectations.

The connections between different blogs can be seen through links, particularly reciprocal links (Herring, Scheidt, Wright, & Bonus, 2005), which serve as one indicator that a community might be present. Links have been used in social network analysis (Marlow, 2004) because they demonstrate awareness, but they are not necessarily markers of e-collaboration. E-collaboration often can be seen in the comments of blogs. Comments can provide evidence of the relationships between individuals and provide a space for sharing stories and advice, asking questions, and refining ideas (Dennen & Pashnyak, 2007; 2006). However, e-collaboration that is enabled or supported by blogs may not always be visible to the outside observer. Blogs may serve as an information sharing repository for a community of practice, with the significant interactions taking place via other forums.

Archives are another feature of blogs that help support the development of communities of practice. Blogging software automatically archives both posts and comments, making them readily available to participants at any times. The presence of archives means that newcomers – specifically, peripheral, inbound, and boundary participants—to the community may see what has transpired at earlier points in time, quickening the pace with which they may integrate into the community and engage in effective e-collaborations. Additionally, archives help provide a document of earlier interactions that may be useful for revisiting decisions and improving the collaborative process.

FUTURE TRENDS AND CONCLUSION

Many companies and organizations have demonstrated an interest in developing online communities of practice to support e-collaboration efforts such as workplace learning, knowledge management, and professional development. The next several years will likely see an increase in the use of blogs to support e-collaboration within communities of practice; the development of blog-based tools and plug ins that support increasingly sophisticated forms of e-collaboration; research methods that help better assess evidence of online community

and document and measure the social networks and e-collaborations developed across blogs.

In closing, this technology has proven its potential both for bringing together people who are not otherwise co-located and enabling new relationships to develop between them and for supporting communication within existing communities of practice. Whether blogs are the mechanism through which communities of practice exist and develop or extensions of existing CoPs, they are yet another form of Web-based communication that promotes interaction between people separated by time and space.

REFERENCES

Andrus, C. (2005). The wiki and the blog: Toward a complex adaptive intelligence community. *Studies in Intelligence, 49*(3), 63-70.

Baym, N. (1998). The emergence of online community. In S. G. Jones (Ed.), *Cybersociety 2.0*. Thousand Oaks, CA: Sage.

Blood, R. (2004). How blogging software reshapes the online community. *Communications of the ACM, 47*(12), 53-55.

Boulos, M., Maramba, I., & Wheeler, S. (2006). Wikis, blogs and podcasts: A new generation of Web-based tools for virtual collaborative clinical practice and education. *BMC Medical Education, 6*(41). Retrieved October 1, 2006, from http://www.biomedcentral. com/1472-6920/6/41

Burton, M. (2005). How the Web can relieve our information glut and get us talking to each other. *Studies in Intelligence, 49*(3), 55-62.

Dennen, V. P., & Burner, K. (in press). The cognitive apprenticeship model in educational practice: Research on uses and instructional strategies. In J. M. Spector, M. D. Merrill, M. P. Driscoll, & J. vanMerrionboer (Eds.), *Handbook of research in educational technology* (3rd ed.). Mahwah, NJ: Erlbaum.

Dennen, V., & Pashnyak, T. (2006). Informal academic mentoring via a weblog community: A content analysis. In *Proceedings of the 10th Education and Virtuality Conference* (pp. 1-10).

E

Dennen, V. P., & Pashnyak, T. (2007). Finding community in the comments: The role of reader and blogger responses in a weblog community of practice. In *Proceedings of the IADIS International Conference on Web-based Communities* (pp. 11-17).

Ducate, L., & Lomicka, L. (2005). Exploring the blogosphere: Use of web logs in the foreign language classroom. *Foreign Language Annals, 38*(3), 410-421.

Efimova, L., & Hendrick, S. (2005). In search for a virtual settlement: An exploration of weblog community boundaries. Retrieved October 1, 2006, from https://doc.telin.nl/dscgi/ds.py/Get/File-46041

Herring, S., Scheidt, L. A., Wright, E., & Bonus, S. (2005). Weblogs as a bridging genre. *Information Technology & People, 18*(2), 142-171.

Kling, R., & Courtright, C. (2003). Group behavior and learning in electronic forums: A sociotechnical approach. *The Information Society, 19,* 221-235.

Josefsson, U. (2005). Coping with illness online: The case of patients' online communities. *The Information Society, 21,* 141-153.

Lave, J., & Wenger, E. (1991). *Situated learning: Legitimate peripheral participation.* Cambridge: Cambridge University Press.

Lenhard, A., & Fox, S. (2006). *Bloggers: A portrait of the Internet's new storytellers.* Washington, DC: Pew Internet & American Life Project.

Maag, M. (2005). The potential use of "blogs" in nursing education. *Computers, Informatics, Nursing, 23*(1), 16-24.

Marlow, C. (2004). *Audience, structure and authority in the weblog community.* Paper presented at the International Communication Association, New Orleans, LA.

McAfee, A. (2006). Enterprise 2.0: The dawn of emergent collaboration. *MIT Sloan Management Review, 47*(3), 21-28.

Oravec, J. A. (2003). Blending by blogging: Weblogs in blended learning initiatives. *Journal of Educational Media, 28*(2-3), 225-233.

Poling, C. (2005). Blog on: Building communication and collaboration among staff and students. *Learning & Leading with Technology, 32*(6), 12-15.

Preece, J., & Maloney-Krichmar, D. (2005). Online communities: Design, theory, and practice. *Journal of Computer-Mediated Communication, 10*(4). Retrieved October 1, 2006, from http://jcmc.indiana.edu/vol10/issue4/preece.html

Ray, J. (2006). Welcome to the blogosphere. *Kappa Delta Pi Record, 42*(4), 175-177.

Reed, A. (2005). 'My blog is me': Texts and persons in UK. *Ethnos, 70*(2), 220-242.

Rheingold, H. R. (2000). The virtual community: Homesteading on the electronic frontier (Rev. ed.). Cambridge, MA: MIT Press.

Sauer, I., Bialek, D., Efimova, E., Schwartlander, R., Pless, G., & Neuhaus, P. (2005). "Blogs" and "wikis" are valuable software tools for communication within research groups. *Artificial Organs, 29*(1), 82-89.

Schlager, M. S., & Fusco, J. (2003). Teacher professional development, technology, and communities of practice: Are we putting the cart before the horse? *The Information Society, 19,* 203-220.

Schwen, T. M., & Hara, N. (2003). Community of practice: A metaphor for online design? *The Information Society, 19,* 257-270.

Technorati. (2007). *About technorati.* Retrieved February 10, 2007, from http://technorati.com/about

Wei, C. (2004). Formation of norms in a blog community. *Into the Blogosphere: Rhetoric, Community, and Culture of Weblogs.* Retrieved October 1, 2006, from http://blog.lib.umn.edu/blogosphere/formation_of_norms.html

Wenger, E. (1998). *Communities of practice: Learning, meaning, and identity.* Cambridge: Cambridge University Press.

White, N. (2006). Blogs and community: Launching a new paradigm for online community? *The Knowledge Tree, 11.* Retrieved October 3, 2006, from http://kt.flexiblelearning.net/au/edition-11-editorial/blogs-and-community

KEY TERMS

Blog: Short for *weblog,* a frequently updated Web site containing date-stamped entries posted in reverse chronological order, often consisting of ideas, brief essays, photos, and hyperlinks to other Web sources, and allowing users to post comments.

Blogger: A person who maintains a blog.

Blogging: An act of posting entries to one's blog.

Boundary Trajectory: A community of practice trajectory in which a member of one community of practice offers insight and expertise to another related community of practice.

Comment: A message left in response to a blog post.

Community of Practice: A group of people who create a community around a shared practice. Members of a community of practice engage in social learning, sharing knowledge, experiences, and reflections.

E-Collaboration Technologies: Electronic technologies that enable collaboration among individuals engaged in a common task.

Inbound Trajectory: A community of practice trajectory in which a person is becoming an increasingly involved member of a community of practice, learning about the norms and expectations.

Insider Trajectory: A community of practice trajectory in which one is a full, participating member of a community of practice, familiar with the norms and expectations, and engaged in community maintenance.

Legitimate Peripheral Participation: The act of observing a community without necessarily engaging in active participation for the purposes of learning about norms and expectations as well as determining whether one wishes to become a member.

Outbound Trajectory: A community of practice trajectory in which a member prepares to diminish or end participation. This trajectory is often characterized by passing on responsibilities and knowledge to those who will continue in the community.

Peripheral Trajectory: A community of practice trajectory in which a person tends to mostly observe, learning about community norms and expectations while determining whether to make a greater investment in the community.

Reciprocal Link: When two bloggers provide hyperlinks to each other's blogs on their own blogs as part of a mutual exchange.

E-Collaboration: A Dynamic Enterprise Model

Eric Torkia
Technology Partnerz, Ltd., Canada

Luc Cassivi
University of Quebec – Montreal, Canada

INTRODUCTION

Over the last 10 years or so, we have been witnessing a major paradigm shift from the information age to the relationship age (see Table 1). According to Galbreath (2002), the relationship age is truly about the value of the relationships a firm maintains and manages. Customers, employees, suppliers and partners all contribute synergistically to the economic output of the firm. Ashkenas, Ulrich, Todd, and Kerr (1995) put it more formally by saying that many organizations were faced with a rate of change that exceeded their capability to respond, and that they had to attempt to retool their organization in order to meet an entirely new set of criteria for success. Hence, a firm carrying out business in the Relationship Age is essentially focusing on improving one or more of the success factors such as speed of execution, process and product flexibility, knowledge integration, and ability to innovate new and profitable processes, products, or services (Ashkenas et al., 1995).

The primary focus of the information age firms was internal integration of resources and information. The companies evolving in the relationship age are different because they also focus on external integration with its partners, suppliers, employees, regulators, customers, investors, market analysts, trade associations, and other entities that influence the general business climate in which a given firm operates (Galbreath, 2002).

The concept of the relationship age is an important one because it underscores the many changes that have been prompted by the Internet and global competitiveness requirements. It also highlights the need for an organization to develop new competencies in the areas of internal and external collaboration. In order to be able to develop a collaborative organization, it is important to understand what it should look like. To this affect, Logan and Stokes (2004) state that a collaborative organization comprises the following characteristics:

1. The values and objectives of employees and management are aligned
2. A climate of mutual trust and respect exists
3. The knowledge of all the staff, customers, and suppliers is shared and pooled to optimize the organization's operations and opportunities
4. Decision making is more decentralized than it is in most current organizations and more stakeholders in the organization play a role in defining the direction in which the organization moves
5. Hierarchical structures are kept to a minimum; the company is managed democratically by consensus rather than by command and control

Browne and Zhang (1999) also define the collaborative organization and its context:

Today's organizational boundaries are blurring, partnerships with suppliers, clients and even competitors are commonplace, and quality and efficiency issues extend well beyond the traditional enterprise boundary. [...] Information flow between business partners can be seamlessly and effectively facilitated. Individual companies work together to form inter-enterprise networks across the product value chain, in order to survive and achieve business successes.

Furthermore, most current approaches to e-collaboration seek to promote some or all of the following:

- The leveraging of partners business capabilities to achieve flexibility or performance within operations

Table 1. Characteristics of the various economic ages (Source: Galbreath, 2002)

	Industrial Age 1880 - 1985	Information Age 1955 - 2000	Relationship Age 1995 + Beyond
Characteristic	**Industrial Age**	**Information Age**	**Relationship Age**
Basis for value creation	Products	Information	Knowledge
Strategic Planning Cycle	5 years	3 years	Continuous
Management Structure	Centralized	Decentralized	Virtual
Key investments	Land & Machines	IS, IT and network/telecom infrastructures	People and knowledge tools
Primary Strategic Resource	Raw materials	Information	Relationship Assets
Nature of production	Mass production	Specialization	Mass Customization & personalization
Economic output	Goods	Services	Experiences
Marketing, Sales and Service	Uniformity (push)	Segmentation	1-to-1 relationships
Pricing	Fixed	Flexible	Dynamic
Nature of competition	Distrust + Barriers to entry	Cooperation & Loose affiliations	Trust & Collaboration
Basis for market valuation	Book value	Revenue/earnings multiples	Market-to-book ratio and market capitalization

- Decisions are business driven and not technology driven
- The use of cross-organizational processes
- The use of the Internet and Web technologies as process and capability enablers
- Tighter relationships with customers and suppliers to improve overall market economics and business performance

These similarities are not a coincidence; they are a "natural reaction to today's business environment!" (Champy, 2002). In line with the seminal works on interorganizational collaboration Galbreath (2002), Venkatraman (1994), and Venkatraman and Henderson (1998) further define Relationship Age businesses (collaborative organizations) as those that:

- Leverage knowledge about their network of relationships, including customers, employees, partners, suppliers and even investors
- Realize that the ability to create memorable experiences is often dependent on external resources,

leveraging the specialization of other market participants via outsourcing, risk/reward-sharing arrangements, and new alliances in order to create more compelling value propositions in the market

Developing relationship age capabilities cannot occur in a vacuum. Some structure is required to create value, as was the case in the Industrial Age with Taylor's scientific analysis of work, and in the information age with systems thinking and processes. Now that we have moved into the relationship age, it seems appropriate to revisit some of the collaboration and e-collaboration models that are characteristic of the relationship age.

A brief overview of organizational theory relating to collaboration is presented in the next section. This overview touches upon resource and systems based theory as well as collaborative organization archetypes. Finally, the subsequent section introduces the dynamic enterprise alignment (DEA) model, which tries to catalyze alignment between the various resources of the organization.

BACKGROUND ON COLLABORATIVE ORGANIZATIONAL THEORY

Curtis, Hefley, and Miller (2001) define an organization as "a collection of units for which an executive management is responsible. An organization could constitute an entire company or agency, or it could constitute only a component of a larger organizational entity, such as a division or branch." Modern enterprise models are far richer in theory and have translated into two major schools of thought: resource based theory (RBT) to define what constitutes a firm, and systems theory to define how they work.

The resource-based view suggests that a firm's unique resources (including knowledge) and capabilities provide the basis for a strategy. In contrast with more traditional views, the resource-based view is grounded in the perspective that a firm's internal environment, in terms of its resources and capabilities, is more critical to the determination of strategic action than is the external environment. Moreover, a resource based approach allows a firm to optimally exploit its core competencies relative to the opportunities available in the external environment.

On the other hand, the school of systems theory treats the organization as a collection of parts unified to accomplish an overall goal. If one part of the system is removed, the nature of the system is changed as well (McNamara, 1999). If we draw a simple analogy to the human body—the organization and the human body are both a collection of resources, processes, infrastructures and systems that enable responsive adaptation to changes in the environment.

Essentially, systems theory revolves around two currents of thought—knowledge driven and process driven. A knowledge driven organization achieves success in the marketplace by learning and sharing knowledge across organizational boundaries in order to align people to a shared vision. Conversely, a process driven organization takes a much more operational perspective by focusing on the processes and mechanics that enable a firm to achieve success in the marketplace.

However, when defining a collaborative organization, a new layer of concepts are superimposed. According to Browne and Zhang (1999), independently of the approach used (resource or systems), electronic collaborative organizations can summarily be split into two categories: extended enterprises and virtual organizations.

In sum, they suggest that an extended enterprise, such as proposed by Womack and Jones (1994), seeks to improve performance by cultivating and optimizing long-term relationships within the supply chain. The benefit of this approach lies in ability of an organization to better leverage the knowledge and capabilities of it suppliers and partners by focusing on long-term benefits accrued through the codevelopment and sharing of continuous operational improvements throughout the value chain. Adopted by firms such as Toyota, Honda, and Bose, the extended enterprise is a model that seems more adapted to manufacturing, and more specifically, to firms where operational efficiencies are the key to success.

Conversely, the virtual organization, as proposed by Gomes-Casseres (1994), Baldwin and Clark (1997), or Chesborough and Teece (1996), seeks to readapt/reconfigure a firm's structures, processes, or products to meet punctual market opportunities. Firms who employ this model seek to reduce fixed costs, obtain the best expertise in short order, and promote flexibility and elasticity in their operations by obtaining resources on an as-needed basis. Primarily suited to service and technology firms, successful examples of firms who have adopted a virtual organizing model are many, but the most notable are Dell, Li and Fung, Solectron, Cisco, as well as a plethora of service firms.

Others such as Venkatraman and Henderson (1998) claim that no one formally defined structure will ensure the success of a virtual (collaborative) organization. In fact, they do not make a distinction between activities carried out in a virtual organization versus those in an extended one. According to their virtual organizing model, a firm may adopt highly dynamic collaborative practices in one area of the firm and implement highly stable ones in others. For this reason they reject a virtual organization as a distinct structure (like functional, divisional, or matrix) or in Browne and Zhang's (1999) case, virtual and extended business archetypes. Instead, they consider virtualness as a strategic characteristic that is applicable to any organization, including century-old companies that manufacture cement, chemicals, and automobiles as well as new entrants in the fast-changing high-technology marketplace. Therefore, based on Venkatraman and Henderson (1998) virtual organizing model, a third "hybrid" collaborative organization can be defined alongside the virtual organization and the extended enterprise.

As we have seen, many authors have different perspectives on what a collaborative firm should look like and how it should operate. The purpose of this section was to harmonize these visions and theories. It also outlined some e-collaboration features required to build an organizational model (presented in the next section) that supports the understanding and adoption of collaboration and its related practices.

THE DYNAMIC ENTERPRISE ALIGNMENT MODEL

Largely based on modern knowledge management and resource based theory, the dynamic enterprise alignment model serves to catalyze alignment between the various levels and resources of the organization. The dynamic enterprise alignment model (see Figure 1) incorporates the concept of organizational, strategic and operational alignment as well as systems thinking and RBT. The model seeks to expose areas of misalignment or development by making them easy to identify in either a knowledge or process driven rationale.

At the center of the model lie the organization's vision and business objectives. Within the model are people, operations, and technology representing the areas within a process or organization that can and should be leveraged to achieve targeted results. On the outside perimeter of the model is a continuous improvement/alignment cycle.

Alternatively, by adopting a resource-based approach, eliminating gaps between people, operations, and technology becomes a priority. Table 2 summarizes potential gaps that lead to misalignment and suboptimal results.

Alignment means a lot of thing to a lot of people and worse, it's a moving target for both organizations and individuals. Several authors cite the universal importance of alignment at strategic, organizational and operational levels (Logan & Stokes, 2004; Venkatraman & Henderson, 1993). With this in mind, the DEA is designed to be applicable in all three alignment contexts.

Organizational alignment implies that the firm's stakeholders (management team, employees, partners, clients, etc.) share a universal understanding of the objectives and issues facing the firm. Organizational alignment also means that the different suborganizations must harmonize their objectives for the overall success of the firm (see Figure 2). In most cases, these cultures, metrics and objectives are at odds with each other; especially difficult when they are all interpretations of the same global business objectives using different filters and mental models (Senge, 1990). It is equally important to note that without a shared understanding of the issues and opportunities, each suborganization will dictate its own priorities for its people, operations, and technology resources.

Strategic alignment, according to Venkatraman and Henderson (1993), is based on two fundamental assumptions:

- Economic performance is directly related to the ability of management to create a strategic fit between the position of an organization in the competitive product-market and the design of an appropriate administrative structure to support its execution.
- Strategic fit is dynamic. The choices made by one firm or enterprise (if fundamentally strategic), will over time evoke immediate actions, which necessitate subsequent responses. Thus, strategic alignment is not an event but a process of continuous adaptation and change.

Strategic alignment within a collaborative firm means the same thing as it does in a more conventional one. In essence, organizations seeking strategic alignment are looking to focus all their resources on specific objectives. According to Mankins and Steele (2005), companies loose an average 37% of their strategies' overall effectiveness due to strategic alignment issues such as unavailable resources, poorly communicated strategy, organizational silos, etc.

Operational alignment implies that the firm's resources, both intellectual and physical, are targeted towards achieving specific operating objectives such as quality, inventory control, customer service, and so forth. Operational alignment is often referred to as management by objective and includes all the planning, monitoring, leading, and control processes such

Figure 1. Dynamic enterprise alignment (DEA) model (Adapted from Barney, 1991; Brown & Eisenhardt, 1999; Garelle & Stark, 1988; Logan & Stokes, 2004; Nohria, Joyce, & Robertson, 2003; Venkatraman & Henderson, 1993)

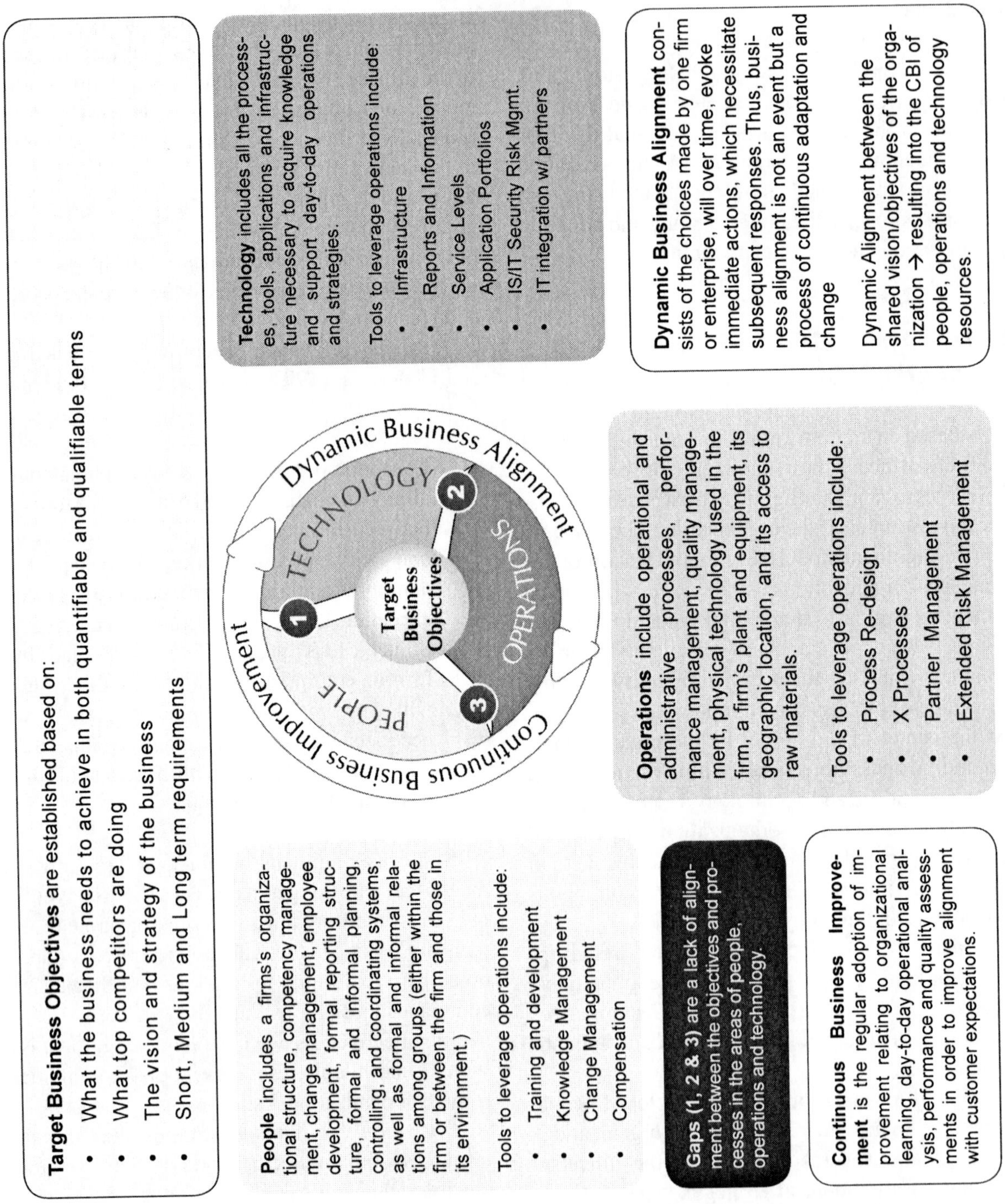

Target Business Objectives are established based on:

- What the business needs to achieve in both quantifiable and qualifiable terms
- What top competitors are doing
- The vision and strategy of the business
- Short, Medium and Long term requirements

Technology includes all the processes, tools, applications and infrastructure necessary to acquire knowledge and support day-to-day operations and strategies.

Tools to leverage operations include:

- Infrastructure
- Reports and Information
- Service Levels
- Application Portfolios
- IS/IT Security Risk Mgmt.
- IT integration w/ partners

Dynamic Business Alignment consists of the choices made by one firm or enterprise, will over time, evoke immediate actions, which necessitate subsequent responses. Thus, business alignment is not an event but a process of continuous adaptation and change

Dynamic Alignment between the shared vision/objectives of the organization → resulting into the CBI of people, operations and technology resources.

People includes a firms's organizational structure, competency management, change management, employee development, formal reporting structure, formal and informal planning, controlling, and coordinating systems, as well as formal and informal relations among groups (either within the firm or between the firm and those in its environment.)

Tools to leverage operations include:

- Training and development
- Knowledge Management
- Change Management
- Compensation

Operations include operational and administrative processes, performance management, quality management, physical technology used in the firm, a firm's plant and equipment, its geographic location, and its access to raw materials.

Tools to leverage operations include:

- Process Re-design
- X Processes
- Partner Management
- Extended Risk Management

Gaps (1, 2 & 3) are a lack of alignment between the objectives and processes in the areas of people, operations and technology.

Continuous Business Improvement is the regular adoption of improvement relating to organizational learning, day-to-day operational analysis, performance and quality assessments in order to improve alignment with customer expectations.

Dynamic Business Alignment

TECHNOLOGY

OPERATIONS

PEOPLE

Continuous Business Improvement

Target Business Objectives

1 2 3

Table 2. Examples of operational alignment gaps

GAP	Symptoms	Potential benefits of addressing GAP
1 **People / Technology** gaps exist when technology (systems and applications) are not enabling your people to do more.	• Mechanistic use of information technologies and systems. • Lack of skills/ competencies to properly use tools in place. • Complaints that technology and systems do not fulfill user requirements. • Users work-around the system to get their job done.	• Lower operating costs and higher employee productivity. • Better communication and integration with partners, suppliers, distributors and customers. • Better return on technology investments. • Individuals within the organization have a solid understanding of the technology in place and usefulness, and seek to use it to improve business performance.
2 **Technology / Operations** gaps exist when your systems and technology do not reflect the organization's processes, strategies or target business objectives.	• Reporting processes and tools do not provide insight into business operations. • Technology is perceived as a cost center and not as a process enabler. • People have to manually input the same information into several systems resulting in duplication of work, costly errors and unreliable business information.	• Increased profitability due to more effective and efficient operations. • Opportunities for operational innovation are identified and capitalized on. • Opportunities for collaboration and process outsourcing are identified and capitalized on. • Improved quality and customer satisfaction. • Reduction in errors and operating costs.
3 **Operations / People** gaps exist when your people's capabilities do not reflect strategy or process requirements.	• Workers have a task-oriented view of their work. • Top/Down information flows - i.e. Functional Silos. • Low morale. • Compensation systems do not reflect business objectives. • High Turn-over.	• People have the skills and competencies to understand and operate optimally in their business environment. • Operations are designed with people in mind. • Employees are happier and more productive.

as budgets, marketing, and operating plans be clearly linked to corporate strategy. Collaboration, supply chain management, as well as other e-business initiatives often serve as catalysts for internal business change and alignment, most notably business process reengineering. The need for coherent and optimized internal processes does not go away when a firm decides to integrate across organizational boundaries; it is amplified (Champy, 2002; Hammer, 2001; Venkatraman, 1994).

Beyond alignment, the model also has a strong impetus towards execution that is necessary to ignite and maintain the velocity of the continuous improvement/ alignment cycle. In Table 3, we break down the model into its respective parts and present the logic behind the model as well as how it can serve as a management framework that supports the consistent achievement of business benefits.

CONCLUSION

In the relationship age, it would appear that the benefits of adopting e-collaboration are pretty stark and should,

Figure 2. DEA model and Hagel's (2002) unbundled organization

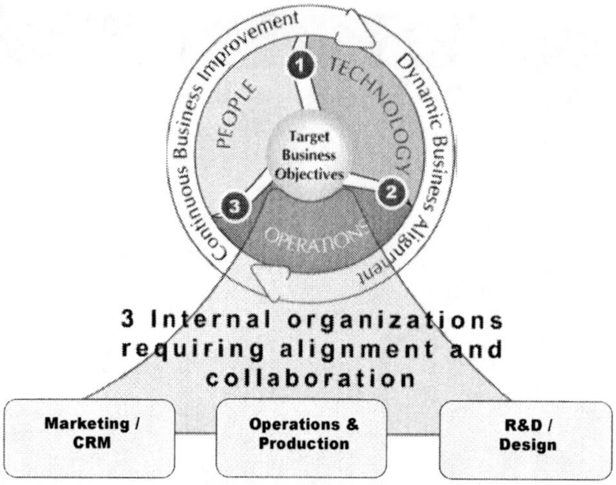

Focus	Customer Relationships	Infrastructure Management	Innovation and commercialization
Skills	Direct Marketing	Operations	Product Innovation
Economics	Economies of scope	Economies of scale	Efficiencies of speed
Culture	Service-oriented	Cost conscious	Creative Culture

by extension, create a strong impetus to move towards this paradigm.

E-collaboration has played a pivotal role in supporting organizations move towards the relationship age. Examples of those who have made the transition to the relationship age and e-collaboration include many of the world's top performing firms—who are also known as the most collaborative (e.g., Dell, Toyota, Nike, Wal-Mart, etc.). However, e-collaboration cannot happen with technology alone. Therefore, by integrating the people and operations with technology, a holistic and dynamic approach thus ensues; paving the way for the initial development of the DEA model.

The current trends and theories relating to the collaborative organizations have been presented and synthesized into a collaborative model known as the DEA model. The benefits of adopting the DEA model are several. Firstly, by structuring the model around simple concepts and vocabulary, it makes building a dialogue and alignment between different stakeholders much easier. Secondly, through the application of the

DEA model, an organization can easily map out how it is using its resources and optimize their allocation. Thirdly, the model highlights gaps and areas of misalignment within the firm that, if dealt with promptly, will translate into a superior capability to realize performance improvements available through targeted internal and external collaboration.

The DEA model can serve as a framework for research that is rooted in simplicity, as it can be adapted to almost any set of circumstances. Theoretical approaches to closing organizational, strategic and operational gaps (presented in Table 2) are necessary to achieve performance levels in a collaborative environment. Therefore, further research in the tools and approaches derived or adapted from the DEA model should be conducted to refine some of the components of the model. Other research paths or practical applications include assessing the how the DEA can be used to benchmark an organization's collaborative capability against best practice.

Table 3. Components of the DEA model

Model Component	General Description	Objective	Organizational Alignment	Strategic Alignment	Operational Alignment
Target Business Objectives	The definition of the objectives an organization is pursuing to achieve its vision.	Build alignment by clearly setting and communicating objectives and business agenda	Set a shared vision by involving all stakeholders	Set business objectives to carry out shared vision	Set operating and planning objectives to realize strategic objectives
People	People includes a firm's organizational structure, competency management, change management, employee development, formal reporting structure, its formal and informal planning, controlling, and coordinating systems, as well as informal relations among groups within a firm and between a firm and those in its environment. (Becker, 1964; Tomer, 1987; Kaplan and Norton, 2004)	Resources, processes and activities that seek to develop or leverage the competencies and relationships embedded in a firm's human capital.	Create a culture and clear expectations that foster performance	Obtain the competencies required to carry out objectives, strategies and processes.	Develop targeted OCM strategies and competency development to foster rapid adoption of change and innovation in systems and processes.
Operations	Operations include Operational and administrative processes, performance management, quality management, physical technology used in a firm, a firm's plant and equipment, its geographic location, and its access to raw materials. (Williamson, 1975)	Resources, processes and activities that seek to develop or leverage a firm's operational expertise in the attainment of business objectives.	Create understanding of processes, capabilities, operating culture, principals and assumptions.	Obtain competencies and develop tangible and intangible assets.	Develop performance, flexibility and efficiency of business processes.
Technology	Technology includes all the processes, tools, applications and infrastructure necessary to acquire knowledge and support day-to-day operations and processes. (Venkatraman, 1994, Kaplan and Norton 2004)	Resources, processes and activities that seek to develop or leverage information and knowledge as enablers for innovation, tighter relationships and operational excellence.	Full understanding of capabilities of the firm's business technology.	Alignment between business strategy, IS/IT strategy, and Stakeholder Needs	Rapid deployment of new systems and Innovation.
GAPS 1 2 3	Gaps are a lack of alignment between the objectives and processes in the areas of people, operations and technology.	1 **People / Technology** gaps exist when technology (systems and applications) are not enabling your people to do more. 2 **Technology / Operations** gaps exist when your systems and technology do not reflect the organization's processes, strategies or target business objectives. 3 **Operations / People** gaps exist when your people's capabilities do not reflect strategy or process requirements.	There is a lack of understanding of the relationships between people, operations and technology.	The objectives in one area do not support those in the others. (e.g Budget Allocation Process to strategic initiatives)	The processes and activities in one area impede the performance of processes and activities in the other 2. Examples are provided in Table 2.
DYNAMIC BUSINESS ALIGNEMENT	**Dynamic Business Alignment** are choices made by one firm or enterprise (if fundamentally strategic), will over time evoke immediate actions, which necessitate subsequent responses. Thus, strategic alignment is not an event but a process of continuous adaptation and change (Venkatraman and Henderson, 1993)	Dynamic Alignment between the shared vision/objectives of the organization and customer expectations → resulting into CBI of people, operations and technology resources.	Understand customer expectations and marketplace requirement	Generate knowledge that enables the firm to respond rapidly to changing customer requirements and marketplace conditions.	Put the right information in the hands of the right people at the right time.
CONTINUOUS BUSINESS IMPROVEMENT	**Continuous Business Improvement** is the regular adoption of improvement relating to organizational learning and day-to-day operational analysis, performance and quality assessments to improve alignment with customer expectations		Continuous integration of organizational learning / knowledge into strategy and operations	Identification of improvement opportunities to better meet customer expectations	Implementation and monitoring of business changes to assess benefits and impacts on organizational, strategic and operational alignment.

REFERENCES

Ashkenas, R., Ulrich, D., Todd, J., & Kerr, S. (1995). A new world order: Rising to the challenge of new success factors. In *The boundaryless organization: Breaking the chains of organizational structure* (pp. 1-30). San Francisco: Jossey-Bass.

Baldwin, C. Y., & Clark, K. B. (1997). Managing in an age of modularity. In *Harvard Business Review on managing the value chain* (pp. 1-28). Boston: Harvard Business School Press.

Barney, J. (1991). Firm resources and sustained competitive advantage. *Journal of Management, 17*(1), 99-120.

Becker, G. S. (1964). *Human capital*. New York: Columbia.

Brown, S. L., & Eisenhardt, K. M. (1998). *Competing on the edge*. Boston: Harvard Business School Press.

Browne, J., & Zhang, J. (1999). Extended and virtual enterprises: Similarities and differences. *International Journal of Agile Management Systems, 1*(1), 30.

Champy, J. (2002). *X-engineering the corporation*. New York: Warner Books.

Chesborough, H. W., & Teece, D. J. (1996). When is virtual virtuous? Organizing for innovation. In *Harvard Business Review on Strategic Alliances* (pp. 151-172). Boston: Harvard Business School Press.

Curtis, B., Hefley, W. E., & Miller, S. A. (2001, July). *People capability maturity model [P-CMM]*. Software Engineering Institute.

Galbreath, J. (2002). Success in the relationship age: Building quality relationship assets for market value creation. *The TQM Magazine, 14*(1), 8-24.

Garelle, E. G. R., & Stark, J. (1988). *Integrated manufacturing: Strategy, planning, and implementation*. New York: McGraw-Hill.

Georgakopoulos, D., & Tsalgatidou, A. (1998). *Technology and tools for comprehensive business process lifecycle management, workflow management systems and interoperability* (NATO ASI Series F). Springer Verlag.

Gomes-Casseres, B. (1994). Group versus group: How alliance networks compete. In *Harvard Business Review on Strategic Alliances* (pp. 75-96). Boston: Harvard Business School Press.

Hagel, J. (2002). *Out of the box: Strategies for achieving profits today and growth tomorrow through Web services*. Boston: Harvard Business School Publishing.

Hammer, M. (2001). The super-efficient company. *Harvard Business Review, 79*(8), 82-93.

Kaplan, R. S., & Norton, D. P. (2004). Measuring the strategic readiness of intangible assets. *Harvard Business Review, 82*(2), 52-63.

Logan, R. K., & Stokes, L. W. (2004). *Collaborate to compete: Driving profitability in the knowledge economy*. New York: Wiley.

Mankins, M. C., & Steele, R. (2005). Turning great strategy into great performance. *Harvard Business Review, 83*(7), 64-73.

Nohria, N., Joyce, W., & Robertson, B. (2003). What really works? *Harvard Business Review, 80*(7), 42-52.

Senge, P. M. (1990). *The fifth discipline: The art and practice of the learning organization* New York: Currency-Doubleday.

Tomer, J. F. (1987). *Organizational capital: The path to higher productivity and well-being*. Preager.

Venkatraman, N. (1994). IT-enabled business transformation: From automation to business scope redefinition. *Sloan Management Review, 35*(2), 73-87.

Venkatraman, N., & Henderson, J. C. (1993). Strategic alignment: Leveraging information technology for transforming organizations. *IBM Systems Journal, 32*(1), 4-17.

Venkatraman, N., & Henderson, J. C. (1998). Real strategies for virtual organizing. *Sloan Management Review, 40*(1), 33-49.

Williamson, O. (1975). *Markets and hierarchies*. New York: Free Press.

Womack, J. P., & Jones, D. T. (1994). From lean production to the lean enterprise. In *Harvard Business Review on Managing the Value Chain* (pp. 221-250). Boston: Harvard Business School Press.

KEY TERMS

Best Practice: A practice contained in a process area that describes an essential activity to, in part or in whole, accomplish a goal of the process area.

Business Objectives: Strategies devised by executive management to ensure an organization's continued existence and to enhance its profitability, market share, and other factors influencing the organization's success.

Business Process Reengineering (BPR): "Is the activity of capturing business processes starting from a blank sheet of paper, a blank computerized model, document, or repository. Once an organization captures its business in terms of business processes, it can measure each process to improve it or adapt it to changing requirements" (Georgakopoulos & Tsalgatidou, 1998).

Continuous (Business) Process Improvement (CPI): "Involves explicit measurements, reconsideration, and redesign of the business process…CPI may be performed after BPR and before information systems and computers are used for automating a process" (Georgakopoulos & Tsalgatidou, 1998).

Organizational Maturity: The extent to which an organization has explicitly and consistently deployed practices or processes that are documented, managed, measured, controlled, and continually improved. Organizational process maturity may be measured via a process appraisal. [P-CMM v2.00, July 2001]

Performance Alignment: The congruence of performance objectives and the consistency of performance results across the individuals, workgroups, units, and organization.

Performance Management: The process of establishing objective criteria against which unit and individual performance can be measured, providing performance feedback, managing performance problems, rewarding and recognizing outstanding performance, and enhancing performance continuously (P-CMM v2.00, 2001; Curtis et al., 2001).

Process Improvement: A program of activities designed to improve the performance and maturity of the organization's processes, and the results of such a program.

Process Lifecycle: The lifecycle of a business process involves everything from capturing the process in a computerized representation to automating the process. This typically includes specific steps for measuring, evaluating, and improving the process.

Process Performance Baseline: A documented characterization of the actual results achieved by following a process, which is used as a benchmark for comparing actual process performance against expected process performance.

Relationships (Network Of): Includes customers, employees, partners, suppliers, and even investors.

E-Collaboration-Based Knowledge Refinement as a Key Success Factor for Knowledge Repository Systems

T. Rachel Chung
University of Pittsburgh, USA

Kwangsu Cho
University of Missouri, Columbia, USA

INTRODUCTION

Electronic knowledge repository systems are fundamental tools for supporting knowledge management (KM) initiatives (Alavi, 2000; King, Marks, & McCoy, 2002). The KPMG Consulting Knowledge Management Research Report 2000 (KPMG, 2000) shows 61% of 423 firms surveyed in the United States and Europe have either implemented or expected to implement repository systems. A follow-up KPMG survey (KPMG, 2003) shows that more than 70% of the firms have either implemented knowledge repositories in the last 2 years or planned to implement them in the next 2 years. Compared to other IT systems for KM, repositories are one of the most widely implemented and used KM tools (KPMG, 2000).

While increasing availability of digitization has minimized the cost and effort needed to create and maintain knowledge repositories, it also results in an overflowing amount of knowledge codified with varying degrees of quality. Without an efficient and effective approach to manage knowledge quality and relevance, knowledge repositories can easily collect large numbers of documents that receive little use (Haas & Hansen, 2005; Hansen & Haas, 2001), especially when contribution leads to tangible rewards (Garud & Kumaraswamy, 2005), or when other competing sources of knowledge are more attractive (Gray & Durcikova, 2005).

Existing KM research suggests two dominant design options for knowledge refinement processes. A common practice advocated by KM researchers is *expert-centralized knowledge refinement*. This approach is characterized by the commission of a centralized review committee composed of domain experts to refine and approve knowledge before the knowledge enters a repository system (Goodman & Darr, 1998;

Markus, 2001; Tobin, 1998; Zack, 1999). The other option with emerging presence is decentralized knowledge refinement, where the decision-making process is decentralized across refiners and the quality of contributed knowledge is determined collaboratively among participating refiners. When such a "collaborative refinery" (Ackerman & McDonald, 1996) is supported by electronic media, including telephone, e-mail, or computer technologies such as groupware, e-collaboration (Kock, 2005) becomes the foundation of the refinement process.

Compared to the dominant expert-centralized knowledge refinement, a decentralized approach can be a viable alternative to design and implement knowledge refinement processes, primarily because e-collaboration makes it possible to incorporate diverse perspectives in the process of knowledge refinement from knowledge user perspectives. Here we examine how e-collaboration tools have been applied to support both models of knowledge refinement.

KNOWLEDGE REFINEMENT

Knowledge refinement is the process of evaluating, analyzing and optimizing the quality of knowledge to be stored in a repository (Alavi, 2000; Cho, Chung, King, & Schunn, in press; Zack, 1999). Refinement mechanisms based on e-collaboration serve as a critical factor that determines the success of knowledge repository systems.

Codifying knowledge that is otherwise tacit provides many benefits, but achieving optimal usage is not easy (Conner & Prahalad, 1996; Hansen, Nohira, & Terney, 1999; Nonaka, 1994). Only when the content of a knowledge repository is accurate (Tobin, 1998), relevant and of high quality (Sussman & Siegal, 2003) are users motivated to access and reuse the content. Taking raw contribution as input material, refinement processes create value added by optimizing raw contribution for maximal usage, rendering the output refined knowledge—a more potent resource for KM efforts. As such, knowledge refinement supports quality assurance of knowledge repositories, an issue that stands as one of the most critical issues for KM practitioners and corporate executives (King et al., 2002).

E-Collaboration for Knowledge Refinement

Knowledge refinement is inherently a collaborative task between refiners and knowledge authors. Adapting Zigurs et al.'s (1998) definition of a task, knowledge refinement can be viewed as a set of behavioral requirements for accomplishing the goal of evaluating, analyzing and optimizing knowledge contribution for repository storage, using some process and given information. The process can involve one or more individuals. When more than one individual are involved in the process, knowledge refinement becomes a collaborative task. Information given in the knowledge refinement task includes the knowledge contribution, the target audience, and the purpose of the contribution.

The quality evaluation component of the knowledge refinement task can be conceptualized as a collaborative judgment task (Campell, 1988; Zigurs et al., 1998). When refining a knowledge object for repository storage, the refiner must consider and integrate information presented in the knowledge object, and to make a judgment about its quality, or to predict the likelihood that it will be useful to repository users for their tasks and in new contexts. If the knowledge object demonstrates room for improvement, the refiner then devises methods to improve the knowledge object.

Many information technologies can serve as e-collaboration media for knowledge refinement. These technologies vary with respect to the amount of communication, collaboration and process structuring they support (Kock, 2005; Zigurs et al., 1998). Research suggests that system features supporting communication and information processing best fit judgment tasks such as knowledge refinement (Zigurs et al., 1998). However e-collaboration technologies for knowledge refinement support more than simply the judgment process. They make democratic knowledge refinement processes possible by engaging authors and refiners from diverse backgrounds in improving knowledge when quality has been determined to be suboptimal. With the expert-centralized approach, the author is required to follow the expert refiner's' decision. In contrast, collaborative knowledge refinement assumes power balance between refiners and authors. For example, e-collaborative technology can hide the identity and status of the refiner, which prevents authors from being biased by refiners' authority and allows them to focus on the content of refinement.

E-collaborative technologies for knowledge refinement can be classified into two categories. The first is *direct refinement technologies*, where multiple participants refine and edit a codified document directly. The second is *indirect refinement technologies,* where participants refine the document indirectly by providing feedback to the author. The author then integrates the feedback and makes improvement to the knowledge object accordingly.

Direct refinement e-collaboration tools are exemplified by the wiki technology (Leuf & Cunningham, 2001). The best-known and most successful e-collaboration project using the wiki technology is Wikipedia (Wikipedia, n.d.), an online encyclopedia that allows anyone to edit the content. Wiki allows people from different functional, expertise, and cultural backgrounds to directly refine codified knowledge. Although wiki supports communication among users through the discussion threads behind the scene, for most documents users are allowed to directly modify the content without communication. For those documents, all users have the same role in refining knowledge quality, regardless of whether they are domain experts or not.

Indirect refinement tools for e-collaboration are widespread. SWoRD (Cho & Schunn, 2007) is a distributed system that allows non-expert reviewers to anonymously evaluate and comment on codified documents. These evaluations then help the authors improve document quality. Because the reviewers only provide feedback on the document, and do not directly edit the content, this approach to refinement is indirect. Indirect refinement using SWoRD has proven to be effective in terms of increasing document quality and creativity, especially if the reviewers are peers of the target knowledge users (Cho, Chung, King, & Schunn, 2006). Tools that support indirect refinement must also scaffold refinement process.

Industry leaders in knowledge management initiatives have implemented e-collaboration tools for indirect refinement. Bain and Company, for example, have built mechanisms for reviewers and users to provide feedback on knowledge objects available in the knowledge management system Global Experience Center (Terra & Gordon, 2003). These mechanisms enable knowledge community members to collaborate with the author in improving knowledge quality without formal coordination mechanisms. Siemen's knowledge sharing system ShareNet provides similar functionalities. In contrast, other systems such as Texaco's best practices databases and Xerox's Eureka knowledge sharing system are designed to allow expert reviewers to validate and refine knowledge submissions in collaboration with authors.

Collaborative refinement, whether direct or indirect, can be a labor intensive and even cost prohibitive process (Markus, 2001). Therefore, tools have been developed to automate some of the refinement activities. For example, Answer Garden 2, a second generation architecture for organizational memory and collaborative help support, implements a collaborative refinery that automatically collects, culls, organizes and distills information from individual and collective spaces (Ackerman et al., 1996). Such automation tools, however, simplify refinement tasks and do not completely replace refinement by human authors and editors.

MENTAL SCHEMAS OF INDIVIDUAL PARTICIPANTS

E-collaboration technologies are particularly beneficial for knowledge refinement because they help alleviate perspective differences that challenge knowledge reuse. Authors that codify knowledge for repositories can be conceptualized as *knowledge sources*. Knowledge is often contributed with the intention that it will be applied to future tasks by other *knowledge users*. Knowledge sources and users can differ in expertise, functional background, culture, among other features. These differences lead to diverse perspectives among sources and users which can make knowledge reusability particularly challenging to achieve, because perspective differences make it difficult to reutilize a piece of knowledge codified by someone else. The most drastic differences in perspective occur between experts and novices, where expertise-seeking novices

experience significant challenges attempting to use knowledge codified by experts that is inaccessible to users without the relevant expertise (Markus, 2001).

Perspective differences between experts and novices make experts underestimate the level of task difficulty for novices (Hinds, 1999), understand complex systems differently from novices (Hmelo-Silver & Pfeffer, 2004), and prefer different types of explanations than novices do (Arnold, Clark, Collier, & Leech, 2006; Cho, Schunn, & Charney, 2006). *Expertise gaps*, or discrepancies between the source and the recipient in a knowledge transfer process in their levels of expertise, can make knowledge transfer from experts to novices a difficult task to accomplish (Chung, Bateman, & Cho, 2005).

In contrast, individuals with more similar levels of expertise share knowledge bases, and find it easier to relate to each other with respect to problems and solutions (Bernardin, 1986; DeNisi & Mitchell, 1978; Rogoff, 1998). Consequently, non-experts are able to improve written accounting instructions for their peers (McIsaac & Sepe, 1996). Research suggests that non-expert peer reviewers improve document quality of a knowledge repository to a greater extent than a single expert does (Cho, Chung, King, & Schunn, 2006a).

Perspective differences can arise from functional or structural diversity, as well as cultural differences. For example, various stakeholders on the same software development project can have different perspectives on the project's requirements (Basili, Green, Laitenberger, Lanubile, Shull, Sorumgard & Zelkowitz, 1996), because different stakeholders have different opinions, impressions, and viewpoints about what the requirements are. Moreover, people with different cultural backgrounds can develop polarizing views about tasks, work, and authority (Alavi, Kayworth, & Leidner, 2005; Cramton & Hinds, 2005; Shore & Venkatachalam, 1996). Cultural differences can be organizational, national, or both. Perspective discrepancies as a result of cultural differences can make documents authored with one cultural frame difficult to comprehend and apply in a different culture.

E-collaboration that engages appropriate "knowledge intermediaries" (Markus, 2001, p. 61) can help reduce perspective differences and produce codified knowledge that is more easily accessible to novices. For example, to minimize the impact of expertise gap, it is important that these knowledge intermediaries possess a moderate level of expertise, which allow them to take a perspective closer to that of novices'.

FUTURE TRENDS

One fundamental assumption for KM benefits is that organizations realize significant gains simply by mining, disseminating and applying the knowledge that has accumulated internally over time (O'Dell & Grayson, 1998). Efforts to collect contribution pay off only when the contributed knowledge is applied to new tasks in new contexts. As such, the repository-based approach to knowledge management is successful only to the extent that its content is actively utilized by organizational members, either for replication (Dixon, 2000; Markus, 2001) or for innovation (Majchrzak, Cooper, & Neece, 2004). As KM research sheds light on factors that facilitate knowledge contribution (Garud et al., 2005; Kankanhalli, Tan, & Wei, 2005), it is now important to allocate more research attention to motivating access, retrieval, and reuse of contributed knowledge (Dixon, 2000; Majchrzak et al., 2004; Markus, 2001).

As document management systems play an increasingly important role in supporting organizational memory and knowledge management (Sprague, 1995), refinement mechanisms based on e-collaboration should become an integral component of such systems. Document management systems that collect contributions (Sprague, 1995) provide a static approach to KM. Incorporating e-collaboration technologies for knowledge refinement would significantly improve these systems' dynamic KM capabilities.

While collaboration-based knowledge refinement serves as a necessary pathway to knowledge reusability, in order to achieve maximal reusability, additional supporting mechanisms must be in place. In software development, for example, formal and systematic software reuse is less than optimal despite its well understood benefits, often as a result of incentive conflict, difficulties searching for and locating relevant software components, lack of trust in components not developed locally, and coordination and ownership issues prevent developers from reusing existing software components (Fichman & Kemerer, 2001; Sherif, Zmud, & Browne, 2006). These findings suggest that knowledge reusability probably is contingent upon managerial, motivational and incentive factors. Future studies should investigate incentive structures and usability factors in knowledge reuse.

CONCLUSION

Refinement mechanisms based on e-collaboration serve as a critical success factor for knowledge repository systems by providing a quality assurance mechanism. To the extent that e-collaboration enables refinement of knowledge contributions by users with different perspectives, or stakeholders in different functions, e-collaboration can serve a critical role in sustaining knowledge repositories. Without e-collaboration tools and techniques, knowledge refinement from divergent perspectives would be difficult to implement.

KM efforts have been established in many corporations. The management, maintenance, and support for KM, however, remain costly because automation can be achieved only to a limited extent. Quality assurance is still a largely labor-intensive process (Voelpel, Dous, & Davenport, 2005). The e-collaboration model of knowledge refinement discussed here may provide one way to optimize knowledge quality cost-effectively.

REFERENCES

Ackerman, M. S., & Mcdonald, D. W. (1996). *Answer Garden 2: Merging organizational memory with collaborative help.* Paper Presented At The Computer Supported Collaborative Work 1996, Cambridge, MA.

Alavi, M. (2000). Managing organizational knowledge. In R. W. Zumd (Ed.), *Framing the domains of it management: Projecting the future......through the past.* Cincinnati, OH: Pinnaflex Educational Resources.

Alavi, M., Kayworth, T. R., & Leidner, D. E. (2005). An empirical examination of the influence of organizational culture on knowledge management practices. *Journal Of Management Information Systems, 22*(3), 191-224.

Arnold, V., Clark, N., Collier, P. A., & Leech, S. A. (2006). The differential use and effect of knowledge-based system explanations in novice and expert judgment decisions. *MIS Quarterly, 30*(1), 79-97.

Basili, V. R., Green, S., Laitenberger, O., Lanubile, F., Shull, F., Sorumgard, S., et al. (1996). The empirical investigation of perspective-based reading. *Empirical Software Engineering.*

Bernardin, H. J. (1986, Fall). Subordinate appraisal: A valuable source of information about managers. *Human Resource Management, 25,* 420-438.

Campell, D. J. (1988). Task complexity: A review and analysis. *Academy of Management Review, 13*(1), 40-52.

Cho, K., Chung, T. R., King, W. R., & Schunn, C. (in press). Peer-based computer-supported knowledge refinement: An empirical investigation. *Communications of the ACM.*

Cho, K., Chung, T. R., King, W. R., & Schunn, C. (2006). *Creating peer-based knowledge refineries.*

Cho, K., & Schunn, C. (2007). Scaffolded writing and rewriting in the discipline: A Web-based reciprocal peer review system. *Computers and Education, 48*(3), 409-426.

Cho, K., Schunn, C., & Charney, D. (2006). Commenting on writing: Typology and perceived helpfulness of comments from novice peer reviewers and subject matter experts. *Written Communication, 23*(3), 260-294.

Chung, T. R., Bateman, P., & Cho, K. (2005, December 11). *Expertise gaps and profiles: An integrated view of expertise on knowledge transfer.* Paper presented at the the Fifth Annual PRE-ICIS Information Systems Cognitive Research Workshop, Las Vegas, NV.

Conner, K. R., & Prahalad, C. K. (1996). A resource-based theory of the firm: Knowledge versus opportunism. *Organization Science, 7*(5), 477-502.

Cramton, C., & Hinds, P. (2005). Subgroup Dynamics in internationally distributed teams: Ethnocentrism or cross-national learning. In B. Staw & R. Kramer (Eds.), *Research in organizational behavior*: Elsevier.

Denisi, A. S., & Mitchell, J. H. (1978). An analysis of peer ratings as predictors and criterion measures and a proposed new application. *Academy Of Management Review, 3,* 369-374.

Dixon, N. (2000). *Common knowledge: How companies thrive by sharing what they know.* Cambridge, MA: Harvard Business School Press.

Fichman, R. G., & Kemerer, C. F. (2001). Incentive compatibility and systematic software reuse. *Journal of Systems and Software, 57,* 45-60.

Garud, R., & Kumaraswamy, A. (2005). Vicious and virtuous circles in the management of knoweldge: The case of Infosys Technologies. *MIS Quarterly, 29*(1), 9-33.

Goodman, P. S., & Darr, E. D. (1998). Computer-aided systems and communities: Mechanisms for organizational learning in distributed environments. *MIS Quarterly, 22*(4), 417-440.

Gray, P. H., & Durcikova, A. (2005). The role of knowledge repositories in technical support environments: Speed versus learning in user performance. *Journal of Management Information Systems, 22*(3), 159-190.

Haas, M. R., & Hansen, M. (2005). When using knowledge can hurt performance: The value of organizational capabilities in a management consulting company. *Strategic Management Journal, 26,* 1-24.

Hansen, M. T., & Haas, M. R. (2001). Competing for attention in knowledge markets: Electronic document dissemination in a management consulting company. *Administrative Science Quarterly, 46*(1), 1-28.

Hansen, M. T., Nohira, N., & Terney, T. (1999). What's your strategy for managing knowledge? *Harvard Business Review,* 106-116.

Hinds, P. J. (1999). The curse of expertise: The effects of expertise and debiasing methods on predictions of novice performance. *Journal of Experimental Psychology-Applied, 5*(2), 205-221.

Hmelo-Silver, C. E., & Pfeffer, M. G. (2004). Comparing expert and novice understanding of a complex system from the perspective of structures, behaviors, and functions. *Cognitive Science, 28*(1), 127-138.

Kankanhalli, A., Tan, B. C. Y., & Wei, K. K. (2005). Contributing knowledge to electronic knowledge repositories: An empirical investigation. *MIS Quarterly, 29*(1), 113-143.

King, W. R., Marks, P. V., & Mccoy, S. (2002). The most important issues in knowledge management. *Communications of the ACM, 45*(9), 93-97.

Kock, N. (2005). What Is e-collaboration? *International Journal of E-Collaboration, 1*(1), I-Vii.

KPMG. (2000). *Knowledge management research report.* Retrieved August 1, 2006, from http://www.

providersedge.com/docs/km_articles/kpmg_km_research_report_2000.pdf

KPMG. (2003). *Insights from KPMG's European knowledge management survey 2002/2003*. Retrieved August 1, 2006, from http://www.providersedge. com/docs/km_articles/insights_from_kpmg_european_km_survey_2002-03.pdf

Leuf, B., & Cunningham, W. (2001). *The wiki way: Collaboration and sharing on the Internet*: Addison-Wesley Professional.

Majchrzak, A., Cooper, L. P., & Neece, O. E. (2004). Knowledge reuse for innovation. *Management Science, 50*(2), 174-188.

Markus, M. L. (2001). Toward a theory of knowledge reuse: Types of knowledge reuse situations and factors in reuse success. *Journal of Management Information Systems, 18*(1), 57-93.

Mcisaac, C. M., & Sepe, J. F. (1996). Improving the writing of accounting students: A cooperative venture. *Journal of Accounting Education, 14*(4), 515-533.

Nonaka, I. (1994). A dynamic theory of organizational knowledge creation. *Organization Science, 5*(1), 14-37.

O'dell, C., & Grayson, C. J. J. (1998). *If only we knew what we know: The transfer of internal knowledge and best practice*. New York: The Free Press.

Rogoff, B. (1998). Cognition as a collaborative process. In D. Kuhn & R. S. Siegler (Eds.), *Cognition, perception & language, Vol. 2*, (pp 697-744). New York: Wiley.

Sherif, K., Zmud, R. W., & Browne, G. J. (2006). Managing peer-to-peer conflicts in disruptive information technology innovations: The case of software reuse. *MIS Quarterly, 30*(2), 339-356.

Shore, B., & Venkatachalam, A.R. (1996). role of national culture in the transfer of information technology. *Journal of Strategic Information Systems, 5*(1), 19-35

Sprague, R. H. J. (1995). Electronic document management: Challenges and opportunities for information systems managers. *MIS Quarterly, 19*(1), 29-49.

Sussman, S. W., & Siegal, W. S. (2003). Informational influence in organizations: An integrated approach to knowledge adoption. *Information Systems Research, 14*(1), 47-65.

Terra, J. C., & Gordon, C. (2003). *Realizing the promise of corporate portals*. Burlington, MA: Butterworth-Heinemann.

Tobin, D. (1998). Networking your knowledge. *Management Review, 87*(4), 46-48.

Voelpel, S. C., Dous, M., & Davenport, T. H. (2005). Five steps to creating a global knowledge-sharing system: Siemens' sharenet. *Academy of Management Executive, 19*(2), 9-23.

Wikipedia. (2006). Retrieved April 17, 2006, from http://en.wikipedia.org/wiki/wikipedia

Zack, M. H. (1999). Managing codified knowledge. *Sloan Management Review, 40*(4), 45-58.

Zigurs, I., & Buckland, B. K. (1998). A theory of task-technology fit and group support systems effectiveness. *MIS Quarterly, 22*(3), 313-334.

KEY TERMS

Decentralized Knowledge Refinement: A knowledge refinement approach that involves both experts and nonexperts in quality judgment and improvement processes.

Direct Refinement: A knowledge refinement process in which multiple participants refine and edit a codified document directly.

Expert-Centralized Knowledge Refinement: A knowledge refinement approach that involves experts only in quality judgment and improvement processes.

Expertise Gap: The discrepancy between the knowledge source and user in their levels of expertise.

Knowledge Refinement: The process of evaluating, analyzing and optimizing knowledge to be stored in a repository.

Indirect Refinement: A knowledge refinement process where reviewers refine the document indirectly by providing qualitative or quantitative feedback to the author.

Knowledge Source: An individual who provides content to a knowledge repository.

Knowledge User: An individual who accesses documents stored in a knowledge repository and applies the knowledge to his or her tasks.

E–Collaborative Knowledge Construction

Bernhard Ertl
Universität der Bundeswehr München, Germany

INTRODUCTION

Knowledge has become an important factor in the success of organizations. Several authors reflect this in their use of terms such as *knowledge society* (e.g., Nonaka, 1994) or *knowledge age* (e.g., Bereiter, 2002). The role of knowledge has changed fundamentally with the development of a knowledge society. Knowledge is still an indispensable resource for the individual as well as for an organization, but the emphasis lies on the creation of new knowledge (see Nonaka, 1994). This change also has consequences for the individual acquisition of knowledge and, in turn, for learning. In traditional learning scenarios, knowledge was seen as a commodity that could be transferred directly from one brain to another. This resulted in an interaction between teacher and learner, in which the teacher had an active role and presented parts of his knowledge to the learners, who passively received and memorized them (see Ertl, Winkler, & Mandl, 2007). However, studies have shown that whilst learning by such presentations of explicit knowledge enabled learners to reproduce it in tests, they failed to transfer it to new situations and often failed to apply it in the creation of new knowledge—the knowledge learners acquired remained inert (Renkl, Mandl, & Gruber, 1996).

BACKGROUND

Innovative approaches to teaching and learning no longer only focus on the transfer of explicit knowledge, but pay more attention to tacit knowledge (Nonaka, 1994). Tacit knowledge is often not conscious and therefore almost impossible to teach explicitly. It may comprise of situational, conceptual, procedural and strategic skills (see De Jong & Fergusson-Hessler, 1996; Nonaka, 1995). It is an important key for the application of existing knowledge and the creation of new knowledge. Constructivist approaches postulate that each learner has to construct new knowledge actively to appreciate the applicability of knowledge.

Approaches such as the *cognitive apprenticeship* (Collins, Brown, & Newman, 1989) or *situated learning* (Lave & Wenger, 1991) place learners in a collaborative scenario that enables them to construct knowledge actively in collaboration with learning partners. Four different processes can be seen as particularly beneficial for collaborative knowledge construction (see Fischer, Bruhn, Gräsel, & Mandl, 2002): Learners' *externalization* and *elicitation* of knowledge, their *conflict-oriented negotiation,* and their *consensus-oriented integration.* Learners' externalization requires them to elaborate knowledge comprehensibly to their learning partners. This challenges them to actively use their knowledge. Elicitation describes a request for new knowledge to the learning partners. Learning partners are required to externalize their knowledge and the learner himself has the chance to fill gaps in his knowledge based on these externalizations. Conflict-oriented negotiation describes learners' discussion of divergent perspectives on the content, whereby consensus-oriented integration comprises of learners' efforts to find a synthesis of their different viewpoints. Consequently, the processes of externalization and elicitation primarily facilitate the acquisition and application of knowledge while negotiation and integration focus more on the creation of new knowledge. To sum up, collaborative knowledge construction is attributed with many benefits for learners (see, e.g., Cohen & Lotan, 1995; Ertl, Fischer & Mandl, 2006; Lou, Abrami & d'Apollonia, 2001; Roschelle & Teasley, 1995).

E-collaborative knowledge construction shifts these processes to scenarios of computer mediated communication. However, the term "e-collaboration" is associated with several different meanings or styles of collaboration and it is necessary to distinguish between them for conceptual clarity (see Dillenbourg, 1999; Gräsel, Fußangel & Pröbstel, 2006). One facet of e-collaboration can be described as the *exchange* of information and working material (see Gräsel et al., 2006). This style of collaboration takes place in a more casual manner and has mutual benefit from the material of the respective collaboration partners as its main goal. Another aspect concerns a professional division of work.

Dillenbourg (1999) also calls this quality *cooperation*. Collaboration partners share a goal and have a joint plan for reaching it. In order to do this, they split the work into different steps and work individually within each step. Collaborating partners' interaction relates in this case to the planning and division of work rather than to collaboration on the content. However, e-collaborative knowledge construction requires *collaboration* in a style in which collaboration partners interact frequently with *content-specific* activities. This means that they work together at the same (virtual) place to construct one joint product or mental artifact (see Bereiter, 2002). Such collaboration does not necessarily have to happen synchronously—however, the collaboration partners' timing and their commitment has to be solid enough for the processes of collaborative knowledge construction to take place.

ENVIRONMENTS FOR E-COLLABORATIVE KNOWLEDGE CONSTRUCTION

Environments for e-collaborative knowledge construction rely on the computer, which features collaboration partners' communication; for example, by the provision of newsgroups, chats, or audio-visual communication. Furthermore, the computer screen has to provide the instructional design, e.g. instructional elements and learning material for the learners (see Kirschner, Sweller & Clark, 2006). Learners *share* this computer screen—even if located in different places. They may share the same interface structure and contents but not necessarily see the same picture simultaneously when accessing the learning environment (see Weinberger, 2003). However, in some situations they may also share one application and work simultaneously (*application sharing*). In such cases, they can see the moves of their collaboration partners during collaboration (see Dillenbourg & Traum, 2006; Ertl et al., 2006; Pata, Sarapuu, & Lehtinen, 2005). Environments for e-collaborative knowledge construction do not necessarily require fully synchronous communication, yet they require collaboration partners to be simultaneously on task. In the following, we will show two different environments for e-collaborative knowledge construction: One learning environment using discussion boards and a videoconferencing one.

An Environment Using Discussion Boards

Environments that use discussion boards, forums or newsgroups are quite common in the domain of virtual seminars in higher education (see Koschman, Suthers, & Chan, 2005; Schnurer, 2005; Weinberger, 2003). This communication is asynchronous, which means that there is no immediate reply to a contribution and collaboration partners have enough time for thoughtful replies to colearners' contributions (see Schnurer, 2005; Weinberger, 2003). Furthermore, many systems allow learners to edit and improve contributions (see Clark & Brennan, 1991; Dennis & Valacich, 1999). However, when applying discussion boards for e-collaborative knowledge construction, the instructional design of the learning environment has to ensure that they have *similar paces* (see Fischer & Waibel, 2002)—their activities have to be synchronized to a certain degree.

Weinberger (2003) describes an example of such an environment. He chose the asynchronous environment because the instructional design of his study focused on elaborate individual case analyses, which develop during the ongoing collaboration. In this environment, three learners deepened their understanding regarding an educational theory. They worked collaboratively on a problem-solving task based on three learning cases. For the collaborative case solutions, the environment provided three discussion boards, one for each case. In collaboration with their teammates and referring to individual resources, learners negotiated to find a suitable solution for each case. They wrote messages about case diagnoses and commented on each other's contributions. This negotiation requested them to externalize and apply their content-specific knowledge as well as case-solving strategies. At the end, one learner prepared synthesis of their perspectives as a final solution for each case. In this scenario, the asynchronous learning platform enabled learners to communicate and to reply to each other's comments with a temporal delay, yet because of the fixed timeframe provided for working in the learning environment, they could correspond timely enough to collaborate in knowledge construction and come to a joint case solution.

An Environment Using Videoconferencing

In videoconferencing, learners communicate in spoken words through an audio and a video channel (see Ertl et al., 2006). The audio channel transmits spoken discourse and the video channel usually provides an image of the head and the chest of the learning partners. To support e-collaborative knowledge construction in videoconferencing, learners find a shared application on their screen in such scenarios. This functions as a tool for making contents of the spoken communication permanent.

Ertl, Reiserer, and Mandl (2005) describe a study, in which two learners were negotiating on collaborative theory learning. The instructional design requested a synchronous communication to support learners' dialogue and an immersive interaction of elicitations and externalizations. The audio communication facilitated the learners' elaborations by the natural flow of language, while the video was not essential to the task. However, it increased the awareness of the communication partners and made learners feel more comfortable. In this learning environment, each learner had knowledge about one particular theory of educational psychology and the goal of the scenario was that both learners should understand both theories. This scenario required the learners to each teach their respective theory to their partner. Therefore, they had to externalize their theory knowledge. Furthermore, they had to understand their partner's theory and to elicit knowledge from their partners. Both learners used the shared application for taking notes, making visualizations and providing a collaborative summary of both theories.

Outcomes of Both Environments

Both studies found beneficial effects for the collaborative work on the task and individual learning outcomes with respect to each learning environment (Ertl et al., 2005; Weinberger, 2003). These results are in line with several other studies using similar learning environments (see, e.g., Bromme, Hesse, & Spada, 2005; Koschmann et al., 2005). Learners improved their knowledge about the particular learning material during their activity in the learning environment (see Ertl et al., 2005; Weinberger, 2003). Furthermore, they also acquired several skills which could be seen as tacit

knowledge in this scenario: Weinberger (2003) emphasized that learners acquired beneficial collaboration strategies, and Ertl et al. (2005) stress that learners get skilled in discriminating between conceptual aspects and evidence for theories.

MEDIA AFFORDANCES FOR E-COLLABORATIVE KNOWLEDGE CONSTRUCTION

Considering that the environments described are quite different with respect to the communication scenario, one might wonder if one particular communication scenario could be superior to others. There are a number of theories and taxonomies about media choice (see Daft & Lengel, 1984; Dennis & Valacich, 1999; McGrath & Hollingshead, 1994) and some empirical studies comparing different media with respect to learners' performance on different tasks, which may help in answering this question (see Anderson et al., 1997; Fischer, Bruhn, Gräsel & Mandl, 2000; Pächter, 2003; Piontkowski, Böing-Messing, Hartmann, Keil & Laus, 2003; Weinberger, Ertl, Fischer, & Mandl, 2005). In general, the theories and taxonomies are somewhat lacking in evidence and the studies report heterogeneous results. The explanation for these blurry answers lies in the fact that researchers used different tasks and conceptualizations and measures for the outcome of e-collaborative knowledge construction. To resolve this heterogeneity and to make clear predictions, researchers sometimes tried to investigate which medium is best suited to a particular task and with respect to defined goals (see, e.g., Anderson et al., 1997). However, this kind of issue raises the question as to the goal of such studies. Might it be valuable to know which communication is best for one particular context or might it be more sensible to think about how to realize elements required by the instructional design of an environment for collaborative knowledge construction using different tools. This was already Clark's (1994) argumentation, in which he states that the type of instruction influences the learning much more than the medium.

We will exemplarily illustrate this claim using the aspect of the synchronicity of communication, which is one of the categories in Dennis and Valacich (1999). In synchronous scenarios, the communication happens immersively. Learners talk or "chat" with each other during e-collaborative knowledge construction.

They can react to their partners' statements quickly. In contrast, communication partners have to wait until a statement has arrived in asynchronous scenarios and thereby the communication flow is not immersive (see Weinberger et al., 2005). This means that synchronous communication features frequent interaction and coordination while asynchronous communication evokes more thoughtful and comprehensive replies. Consequently, tasks requiring highly frequent interaction—for example collaborative teaching—may be solved better in synchronous scenarios and thoughtful case analyses may in turn require asynchronous communication (see also McGrath & Hollingshead, 1994; Pächter, 2003). Nevertheless, the instructional design of the scenario may reduce such effects: for example when introducing individual phases in synchronous communication scenarios. Designing a videoconferencing task using a sequence of collaborative and individual phases may give learners the chance for exchange as well as for individual reflection (see Ertl et al., 2005; Rummel & Spada, 2005). In contrast, instructional design could give learners in an asynchronous learning environment a strict timeframe for their activities (see Weinberger, 2003). This could synchronize learners' activities when they are working with discussion boards and improve the exchange of the learning partners (see Fischer & Waibel, 2002).

Conclusion

The focus of this article was on e-collaborative knowledge construction. In contrast to the broad concept of learning, e-collaborative knowledge construction relates to an interactive process of collaborative knowledge acquisition or the collaborative creation of new knowledge. This article has shown two examples of environments for collaborative knowledge construction in different communication scenarios. These environments were not restricted to the transfer of explicit knowledge—learners also had the opportunity to socialize tacit knowledge. To reach this goal both environments provided different elements of instructional design to overcome limitations of the media: The asynchronous environment of Weinberger (2003) provided learners with a strict timeframe to facilitate tight collaboration and timely contributions of the collaboration partners which are a prerequisite for successful collaborative knowledge construction. In

contrast, instructional design of the videoconferencing environment provided learners with the task of creating a collaborative summary of their respective theories in the shared application. This enabled them to work on their shared mental artifact. To sum up, the design of the environment for e-collaborative knowledge construction may compensate for the differences in the various communication scenarios.

E-collaborative knowledge construction is of major significance for e-collaboration. This significance applies mainly to situations that require e-collaboration in interdisciplinary teams, in which different experts are collaborating to find the best solution for a problem (see Rummel & Spada, 2005). To do this, each of them has to bring in his/her particular expertise, yet they also have to construct a shared problem space together (see Fischer et al., 2000). They then have to learn about the perspectives of their e-collaboration partners and to construct a team knowledge about the problem, e-collaboratively.

E-collaboration and e-collaborative knowledge construction is no trivial task. Collaboration partners must learn to work together (see Rummel & Spada, 2005). Therefore, e-collaboration partners require facilitation to improve the results of their collaboration (see Mandl, Ertl, & Kopp, 2006) and to avoid undesired group effects (see, e.g., Salomon & Globerson, 1989). It is important that tools and workspaces comprise of several scaffolds for e-collaboration in future.

REFERENCES

Anderson, A. H., O' Malley, C., Doherty Sneddon, G., Langton, S., Newlands, A., Mullin, J., et al. (1997). The impact of VMC on collaborative problem solving: An analysis of task performance, communicative process, and user satisfaction. In K. E. Finn, A. J. Sellen, & S. Wilbur (Eds.), *Video mediated communication* (pp. 133-155). Mahwah, NJ: Erlbaum.

Bereiter, C. (2002). *Education and mind in the knowledge age.* Mahwah, NJ: Erlbaum.

Bromme, R., Hesse, F.-W. & Spada, H. (2005). *Barriers and biases in computer-mediated knowledge communication: And how they may be overcome.* Dordrecht, The Netherlands: Kluwer.

Clark, H. H., & Brennan, S. E. (1991). Grounding in communication. In L. B. Resnick (Ed.), *Perspectives on socially shared cognition* (pp. 127-149). Washington, DC: American Psychological Association.

Clark, R. E. (1994). Media will never influence learning. *Educational Technology Research and Development, 42,* 21-29.

Cohen, E. G., & Lotan, R. A. (1995). Producing equal-status interaction in the heterogeneous classroom. *American Educational Research Journal, 32*(1), 99-120.

Collins, A., Brown, J. S., & Newman, S. (1989). Cognitive apprenticeship: Teaching the crafts of reading, writing, and mathematics. In L. B. Resnick (Ed.), *Knowing, learning, and instruction: Essays in honor of Robert Glaser.* Hillsdale: Erlbaum

Daft, R. L. & Lengel, R. H. (1984). Information richness: A new approach to managerial behavior and organizational design. *Research in Organizational Behaviour, 6,* 191-233.

De Jong, T., & Ferguson-Hessler, M. G. M. (1996). Types and qualities of knowledge. *Educational Psychologist, 31,* 105-113.

Dennis, A. R., & Valacich, J. S. (1999). *Rethinking media richness: Towards a theory of media synchronicity.* Paper presented at the 32nd Annual Hawaii International Conference on Systems Sciences.

Dillenbourg, P. (1999). What do you mean by 'collaborative learning'? In P. Dillenbourg (Ed.), *Collaborative-learning: Cognitive and computational approaches* (pp. 1-19). Oxford, England: Elsevier.

Dillenbourg, P., & Traum, D. (2006). Sharing solutions: Persistence and grounding in multimodal collaborative problem solving. *Journal of the Learning Sciences, 15*(1), 121-151.

Ertl, B., Fischer, F., & Mandl, H. (2006) Conceptual and socio-cognitive support for collaborative learning in videoconferencing environments. *Computers & Education, 47*(3), 298-315

Ertl, B., Reiserer, M., & Mandl, H. (2005). Fostering collaborative learning in videoconferencing: The influence of content schemes and cooperation scripts on shared external representations and individual learning outcomes. *Education, Communication & Information, 5*(2), 147-165.

Ertl, B., Winkler, K., & Mandl, H. (2007). E-learning: Trends and future development. In F. M. M. Neto & F. V. Brasileiro (Eds.), *Advances in computer-supported learning* (pp. 122-144). Hershey, PA: Information Science.

Fischer, F., Bruhn, J., Gräsel, C., & Mandl, H. (2000). Kooperatives Lernen mit Videokonferenzen: Gemeinsame Wissenskonstruktion und individueller Lernerfolg [Cooperative learning in videoconferencing. Collaborative knowledge construction and individual learning outcomes]. *Kognitionswissenschaft, 9,* 5-16.

Fischer, F., Bruhn, J., Gräsel, C., & Mandl, H. (2002). Fostering collaborative knowledge construction with visualization tools. *Learning and Instruction, 12,* 213-232.

Fischer, F., & Waibel, M. C. (2002). Wenn virtuelle Lerngruppen nicht so funktionieren, wie sie eigentlich sollen [If virtual learning groups don't work as they should]. In U. Rinn & J. Wedekind (Eds.), *Referenzmodelle netzbasierten Lehrens und Lernens–Virtuelle Komponenten der Präsenzlehre* (pp. 35-50). Münster, Germany: Waxmann.

Gräsel, C., Fußangel, K., & Pröbstel, C. (2006). Lehrkräfte zur Kooperation anregen–Eine Aufgabe für Sisyphos? [Encouraging teachers to collaborate—a Sisyphus task?]. *Zeitschrift für Pädagogik, 52*(2), 205-219.

Kirschner, P. A., Sweller, J., & Clark, R. E. (2006). Why minimal guidance during instruction does not work: An analysis of the failure of constructivist, discovery, problem-based, experiential, and inquiry-based teaching. *Educational Psychologist, 41*(2), 75-86.

Koschmann, T., Suthers, D., & Chan, C. (Eds.). (2005). *Computer supported collaborative learning 2005: The next 10 years!* Mahwah, NJ: Earlbaum.

Lave, J., & Wenger, E. (1991). *Situated learning: Legitimate peripheral participation.* New York: Cambridge University Press.

Lou, Y., Abrami, P. C., & d'Apollonia, S. (2001). Small group and individual learning with technology: A meta-analysis. *Review of Educational Research, 71*(3), 449-521.

Mandl, H., Ertl, B., & Kopp, B. (2006). Computer support for collaborative learning environments. In L.

Verschaffel, F. Dochy, M. Boekaerts, & S. Vosniadou (Eds.), *Instructional psychology: Past, present and future trends. Fifteen essays in honor of Erik De Corte* (pp. 223-237). Amsterdam: Elsevier.

McGrath, J. E., & Hollingshead, A. B. (1994). *Groups interacting with technology: Ideas, evidence, issues, and an agenda.* Thousand Oaks, CA: Sage.

Nonaka, I. (1994). A Dynamic theory of organizational knowledge creation. *Organization Science, 5*(1), 14-37.

Nonaka, I. (1995). *The knowledge-creating company.* Oxford, UK: Oxford University Press.

Pächter, M. (2003). *Wissenskommunikation, kooperation und lernen in virtuellen gruppen* [Knowledge communication, cooperation and learning in virtual groups]. Lengerich, Germany: Pabst.

Pata, K., Sarapuu, T., & Lehtinen, E. (2005). Tutor scaffolding styles of dilemma solving in network-based role-play. *Learning and Instruction, 15*, 571-587.

Piontkowski, U., Böing-Messing, E., Hartmann, J., Keil, W., & Laus, F. (2003). *Transaktives gedächtnis, informationsintegration und entscheidungsfindung im medienvergleich* [Transactive memory, integration of information and decision making with respect to different media]. *Zeitschrift für Medienpsychologie, 15*, 60-68.

Renkl, A., Mandl, H., & Gruber, H. (1996). Inert knowledge: Analyses and remedies. *Educational Psychologist, 31*(2), 115-121.

Roschelle, J., & Teasley, S. D. (1995). The construction of shared knowledge in collaborative problem solving. In C. O'Malley (Ed.), *Computer supported collaborative learning.* (pp. 69-97). Berlin, Germany: Springer.

Rummel, N., & Spada, H. (2005). Learning to collaborate: An instructional approach to promoting collaborative problem-solving in computer-mediated settings. *Journal of the Learning Sciences, 14*, 201-241.

Salomon, G., & Globerson, T. (1989). When teams do not function the way they ought to. *International Journal of Educational Research, 13*(1), 89 - 99.

Schnurer, K. (2005). *Kooperatives lernen in virtuell-asynchronen hochschulseminaren. Eine Prozess-Produkt-Analyse des virtuellen Seminars „einführung in das wissensmanagement" auf der basis von felddaten* [Cooperative learning in virtual-asynchronous university courses. A process-product analysis of the virtual course "introduction to knowledge management" based on empirical data]. Unpublished doctoral dissertation, Ludwig-Maximilian-University, Munich.

Weinberger, A. (2003). *Scripts for computer-supported collaborative learning.* Retrieved July 13, 2007, from http://edoc.ub.uni-muenchen.de/archive/00001120/01/Weinberger_Armin.pdf

Weinberger, A., Ertl, B., Fischer, F., & Mandl, H. (2005). Epistemic and social scripts in computer-supported collaborative learning. *Instructional Science, 33*(1), 1-30.

KEY TERMS

Application Sharing: Mechanism that allows e-collaboration partners to work with the same application on the same document simultaneously.

Collaborative Teaching: Method of education in which a group of learners acquire knowledge by alternately assuming the role of teachers.

E-Collaborative Knowledge Construction: Synchronized e-collaboration with the goal to acquire or create new knowledge.

Environment for E-Collaboration: Working place of an e-collaborator. The environment provides all the tools and resources applied during e-collaboration.

Explicit Knowledge: Knowledge that can be intentionally expressed and quantified. Examples of this kind of knowledge include facts and descriptions.

Learning Case: Description of a real-world scenario, which helps learners to apply their knowledge.

Mental Artifact: Immaterial product, which e-collaboration partners construct during the process of e-collaboration.

Mutual Dependency: Requirement for successful e-collaboration. Mutual dependency ensures that both partners can benefit from e-collaboration and it may reduce undesired group effects.

Shared Problem Space: The shared knowledge of e-collaboration partners which is necessary to solve a problem collaboratively.

Tacit Knowledge: Knowledge, which is acquired rather unconsciously by socialization or practice. It may be seen as complementary to explicit knowledge.

Enhancing E-Collaboration Through Culturally Appropriate User Interfaces

Dianne Cyr
Simon Fraser University, Canada

INTRODUCTION

Prior to the Internet, forms of social expression, communication, and collaborative behavior are known to be sensitive to cultural nuances. According to researcher Geert Hofstede (1991), a widely used definition of culture is proposed where "Every person carries within him or herself patterns of thinking, feeling, and potential acting which were learned through their lifetime" (p. 4). Hofstede referred to such patterns as mental programs or "software of the mind." It is expected that such mental programming related to cultural differences will affect perceptions of the electronic medium as well (Raman & Watson, 1994). Related to the topic of this volume, culture has a place in the consideration of e-collaboration when individuals come together to work toward a common goal using electronic technologies. This may include various domains including e-business, e-learning, distributed project management, working in virtual teams of various forms, to name a few.

While there is little work to date on the explicit topic of culture and e-collaboration, there is evidence that creating culturally appropriate user interfaces (Cyr & Trevor-Smith, 2004) contributes to a better perception of the interface (Kondratova & Goldfarb, 2005), and indeed to enhanced levels of Web site trust and satisfaction (Cyr, 2006). In e-commerce settings, Web sites that are perceived as appropriate to the user have also resulted in greater commitment (Oliver, 1999). In this article, and building on previous work in related areas, it is argued that the development of culturally appropriate electronic interfaces can enhance user involvement, ultimately resulting in enhanced e-collaboration.

In the following sections, culture as a context for e-collaboration is outlined followed by considerations of the Web used as a communication tool, and how trust and satisfaction are related to the online collaborative process. The article ends with concluding remarks.

CULTURE AS CONTEXT

Over the decades sociologists have proposed that socially shared meanings are culture specific. These shared meanings are grounded in language, geographical proximity, and history as shared by members of nations or those who have lived within the same social environment (Hofstede, 1980). In both commercial and noncommercial settings, culture has been found to have implications for information systems research. More specifically, culture is proposed to affect online trust (Jarvenpaa, Knoll, & Leidner 1999), Web site development (Sun, 2001), use of group support systems (Reining & Mejias, 2003), predisposition for type of electronic communication media (Straub, 1994), among other topics.

Related to how individuals operate together in groups, mental schemas for knowledge construction (Kock, 2004) influence the impact of e-collaboration technologies on the individuals involved. Schemas can be socially constructed, causing groups to interpret information in specified ways (Lee, 1994)—thus influencing perception and interaction. As Kock (2004) elaborates, the degree to which members of a task group share similar schemas, then less cognitive effort is required to successfully accomplish the task. Members of the same cultural group are more likely to share similar mental maps or schemas than with members of external groups.

In decision making, Tseng and Stern (1996) found significant differences in the information gathering behavior between Asians and North Americans. Different online communication strategies were uncovered in a study that included Japan, Spain, and the United States (Okayazaki & Rivas, 2002). Further, cultural differences exist in instant messaging between Asia and North America (Kayan, Fussell, & Setlock, 2006). For example, North Americans reported significantly less multiparty chat and rated emoticons lower in importance than Asians. Ethnicity has been established as a factor in electronic brainstorming (Tan, Wei, Watson, Clapper, & McLean1998). Based on the foregoing, it is a natural extension that culture influences e-collaboration behavior.

THE WEB AS A COMMUNICATION MEDIUM

Bordia claims that "computers and electronic networks have revolutionized communication" (1997, p. 99). Although it is not always clear that the electronic medium has enhanced the communication process since nonverbal cues that form a significant portion of the transmitted message are mostly missing. In fact, research has demonstrated that an absence of nonverbal cues that serve to "embellish meaning or social context regarding gender, age or status" can potentially hamper communication efficiency (Bordia, 1997, p. 9). Alternately, new capabilities for communicating content and collaborating using the Web are created (Tsao & Lin, 2001).

Simon (2001) used media richness theory (from Daft & Lengel, 1986) to examine how information richness might enhance user perceptions of the interface. Various design characteristics of Web sites were considered such as shapes, colors, language, site layout, and quality of information. It was expected that information rich Web sites would reduce user ambiguity, increase trust, reduce perception of risk, and encourage users to utilize the site. Significant differences between cultural groups in the study were uncovered, with Asians registering higher levels of trust with information across all Web sites in the study than European and North American groups. This finding suggests that not only is creating information rich interfaces useful generally, but that across cultures different preferences for the user interface prevail.

In studies on graphical user interfaces (GUI) in group support systems (GSS), benefits of using icons over text have been established, although this research does not explore aesthetics or usability of the system (Sia, Tan, & Wei, 1997). One would expect that computer supported, interactive, visual representations contribute to a user's assessment of the medium as more effective, and may vary across cultures. As already noted, in one study on instant messaging differences in perceptions of emoticons across cultures were discovered. In other work, German users were found to value hierarchy and verbal components of a Web page, while Asia users prefer visuals (Sun, 2001). Color connotes different cultural meaning (Barber & Badre, 2001; Cyr & Trevor-Smith, 2004). Red means happiness in China but danger in the United States. In a study that compared Canadian, American, German and Japanese users, Japanese favored a more visual and "emotional" approach to user interface design (Cyr et al., 2005).

Related to the preceding, and in the context of e-collaboration, it is expected that if user interfaces are culturally appropriate then users are more likely to become engaged in the communication process, and with one another. This would be the case whether cross-cultural collaboration occurs in e-business, e-learning, or distributed work settings.

TRUST AND E-COLLABORATION

The use of computer-mediated and Web-based communication technologies has created new virtual environments in which trust plays a significant role. Online trust between collaborating parties is more difficult to elicit than trust in traditional settings since users have fewer cues than in a face-to-face or brick-and-mortar context. The primary communication interface for the user is an information technology artifact rather than a person.

Despite the complexity involved in studying online trust, in recent years a body of knowledge has been developed around this topic. David Gefen (2000, 2003) has contributed significantly in this area and identifies trust as a combination of: (1) beliefs about integrity, benevolence of another party; (2) general belief that another can be trusted; (3) feelings of confidence and security in the response of another; or (4) a combination of these various characteristics. Further, Corritore et al. (2003, p. 740) provide a definition of online trust that includes cognitive and emotional elements. They suggest that trust encompasses "an attitude of confident expectation in an online situation or risk that one's vulnerabilities will not be exploited". In e-collaboration, the user ideally seeks confidence in the computer mediated experience and that over time a trusting relationships with other users can be established.

There is a growing reservoir of research on the role of trust in computer-mediated communication (CMC) including virtual collaborations and long-distance communications in instant messaging, blogs, e-mail, teleconferencing, video conferencing, or virtual teams (Bos et al., 2002; Jarvenpaa et al., 1998; Walther, 1992). Trust research in these contexts has focused on interpersonal relationships as they pertain to collaborative work (Hossain & Wigand, 2004). More specifically, how initial trust develops is important to understanding how long distance relationships are developed and nurtured (McKnight et al., 1998). Taken collectively, it is expected that if Web based technologies are able to elicit trust in users, then communication as in GSS and other forms of e-collaboration will be enhanced.

Alternately, it has been proposed that CMC can potentially delay trust formation related to a slower rate at which clues about a partner can be gathered in a virtual setting (Walther, 1992). One might expect trust formation to be especially difficult in a CMC context that transgresses cultural boundaries. Not only is there difficulty in obtaining cues about a partner due to spatial distance, but due to cultural distance or unfamiliarity as well. Little work has been undertaken on cross-cultural effects and online trust, and studies completed in this area have generally yielded inconclusive results. For instance, trust in e-commerce was examined between American and Taiwanese participants and systematic differences were found concerning privacy, but no differences were evident related to culture (Lui et al., 2004). In another study on culture and trust, differences existed between American, Canadians, and Germans with Japanese (Cyr et al., 2005).

In sum, trust formation is important in virtual relationship building just as it is in face-to-face encounters. In settings where visual cues are absent, and when cultural differences prevail, e-collaboration is likely to be more challenging. Once again, it is expected that cultural sensitivity to user values and expectations is a precursor to the development of successful working relationships using technology. In some instances face-to-face encounters prior to virtual collaboration may be useful both within and between cultures.

SATISFACTION AND E-COLLABORATION

For decades satisfaction has been important to the information systems research agenda, and has frequently been used as a facsimile for success. If users are satisfied, then they have positive attributions to

the system, service, or user encounter. Numerous determinants of satisfaction have been proposed and confirmed including system quality and information quality (Doll & Torkzadeh, 1988). With the advent of the Internet many other studies have considered the user interface as a potential contributor to online satisfaction. This has included the "ambience associated with the site itself and how it functions" (Szymanski & Hise, 2000, p. 313).

Disconfirmation theory has applicability to e-collaboration. This theory stipulates that satisfaction is determined by a comparison between the perception of performance and a cognitive standard (Oliver & DeSarbo, 1988). In other words, does the object meet a user's expectations? Marketing studies have a history of using disconfirmation theory to explain or predict satisfaction in the context of products or services. The theory has also been used to measure manager satisfaction with group decision support systems (DeSanctis & Gallupe, 1987). In this instance, user satisfaction is broadly defined as the "multidimensional attitude towards various aspects of MIS such as output quality, man-machine interface, EDP staff and services, and various user constructs such as feelings of participation and understanding" (Khalifa & Lui, 2004, p. 39).

In other work, Mahmood et al. (2000) compiled studies on user satisfaction over a ten year period and determined that satisfaction is mainly affected by user background, perceived benefits, and organizational support. User background is determined by user experience, user skills, and user involvement in the system development process. Perceived benefits are measured by user expectations, ease of use, and perceived usefulness of the system or object. Organizational support is driven by user attitudes toward information systems, organizational encouragement, and perceived attitude of senior management. Expectations about the system are thus related to disconfirmation theory and are shaped by personal experience and environmental factors (Khalifa & Lui, 2004). It is anticipated that the above determinants of satisfaction are tempered by culture, and the user experience of in-country system characteristics and support.

Taken collectively, it is feasible that culture factors into the satisfaction equation and is based on different values that prevail across cultural groups (Hofstede, 1980). As one example, Hofstede outlines different levels of uncertainty avoidance in diverse cultures, when members of a group seek to reduce personal risk. Such risk aversion occurs in traditional settings, but could be extrapolated to online encounters as well. Also part of Hofstede's framework, certain cultures such as those in North America prefer to operate more independently, while Asian cultures tend to operate in groups and collectively. While these are cultural generalizations, in the context of e-collaboration there are indications that different expectations regarding online interactions are likely. If a discrepancy exists between cultural values and the collaborative format, then user satisfaction is less likely to occur.

Pulling together the threads of research as it affects various forms of e-collaboration, it appears that user values and expectations including those that are culturally based will influence perceptions of user satisfaction. User values, norms, and experience are likely to have an impact on reactions to technologies used for computer mediated communication.

CONCLUSION

The introduction and use of e-collaboration technologies has met with mixed levels of success (DeSanctis et al., 1993; Kock, 2004). In this article, it has been argued that culture is important to better understanding how e-collaboration operates. It is an area that has been largely overlooked to date. However culturally sensitive user interface design is described to potentially increase user trust and satisfaction, therefore enhancing the user e-collaborative experience.

In the future it is expected that designers of e-collaboration tools or interfaces will factor in cultural characteristics to the technology system. As one example there is growing evidence that culturally appropriate user interfaces are important for educational courseware (Pfremmer, 2004). Cultural acknowledgement can impact user impressions of usability, accessibility and acceptability of software and will influence factors such as ease of learning, efficiency of use, memory ability, error frequency and severity, and subjective satisfaction (Kondratova & Glodfarb, 2005). Also related to e-learning, collaboration and learning can be hindered by teaching style differences, problems related to different educational values, or diverse language and semantics (McLoughlin, 1999).

Similarly, cultural characteristics should be considered as they relate to effective functioning of project management teams, group support systems, and other forms of e-collaboration. A challenge for researchers is how to best adapt various technologies for diverse cultures. In an area ripe for future enquiry, investigations might focus on how trust and satisfaction are enhanced in e-collaboration, and factors that influence this process.

REFERENCES

Barber, W., & Badre, A. N. (2001). Culturability: The merging of culture and usability, In *Proceedings of the Fourth Conference on Human Factors and the Web.* Basking Ridge, NJ.

Bordia, P. (1997). Face-to-face versus computer-mediated communication: A synthesis of the experimental literature. *The Journal of Business Communication, 34*(1), 99-120.

Bos, N., Olson, J., Gergle, D., Olson, G., & Wright, Z. (2002). Effects of four computer-mediated communications channels on trust development. In *Proceedings of CHI* (pp. 135-140).

Corritore, C. L., Kracher, B., & Wiedenbeck, S. (2003). On-line trust: Concepts, evolving themes, a model. *International Journal of Human-Computer Studies, 58,* 737-758.

Cyr, D. (2006). *Modeling website design across cultures: Relationships to trust, satisfaction, and e-loyalty* (Working paper). Simon Fraser University.

Cyr, D., Bonanni, C., Bowes, J., & Ilsever, J. (2005). Beyond trust: Website design preferences across cultures. *Journal of Global Information Management, 13*(4), 24-52.

Cyr, D., & Trevor-Smith, H. (2004). Localization of Web design: An empirical comparison of German, Japanese, and U.S. website characteristics. *Journal of the American Society for Information Science and Technology, 55*(13), 1-10.

Daft, R. L., & Lengel, R. H. (1986). Organizational information requirements, media richness and structural design. *Management Science, 32*(5), 554-571.

DeSanctis, G., & Gallupe, R. B. (1987). A foundation for the study of group decision support systems. *Management Science, 33*(5), 589-609.

DeSanctis, G., Poole, M. S., Dickson, G. W., & Jackson, B. M. (1993). Interpretive analysis of team use of group technologies. *Journal of Organizational Computing, 3*(1), 1-29.

Doll, W. J., & Torkzadeh, G. (1988). The measurement of end-user computing satisfaction. *MIS Quarterly, 12*, 258-273.

Gefen, D., Karahanna, E., & Straub, D. W. (2003). Trust and TAM in online shopping: An integrated model. *MIS Quarterly, 27*(1), 51-90.

Gefen, D. (2000). E-commerce: The role of familiarity and trust. *The International Journal of Management Science, 28*, 725-737.

Hofstede, G. H. (1991). *Cultures and organizations: Software of the mind.* London: McGraw Hill.

Hofstede, G. H. (1980). *Culture's consequences.* Beverly Hills, CA: Sage.

Hossain, L., & Wigand, R. T. (2004). ICT enabled virtual collaboration through trust. *Journal of Computer Medicated Communication, 10*(1).

Jarvenpaa, S. L., Knoll, K., & Leidner, D. E. (1998). Is anybody out there? Antecedents of trust in global virtual teams. *Journal of Management Information Systems, 14*(4), 29-64.

Jarvenpaa, S. L., Tractinsky, N., Saarinen, L., & Vitale, M. (1999). Consumer trust in an Internet store: A cross-cultural validation. *Journal of Computer Mediated Communication, 5*(2).

Kayan, S., Fussell, S. R., & Setlock, L. D. (2006, November 4-8). Cultural differences in the use of instant messaging in Asia and North America. In *Proceedings of the ACM Computer Supported Collaborative Work.* Banff.

Khalifa, M., & Liu, V. (2004). The state of research on information satisfaction. *Journal of Information Technology Theory and Application, 5*(4), 37-49.

Kock, N. (2004). The psychobiological model: Toward a new theory of computer-mediated communication based on Darwinian evolution. *Organization Science, 15*(3), 327-348.

Kondratova, I., & Goldfarb, I. (2005). Cultural visual interface design. In *Proceedings of EdMedia* (pp. 1255-1262). Montreal: National Research Council Canada.

Lee, A. S. (1994). Electronic mail as a medium for rich communication: An empirical investigation using hermeneutic interpretation. *MIS Quarterly, 18*(2), 143-157.

Lui, C., Marchewka, J., & Ku, C. (2004). American and Taiwanese perceptions concerning privacy, trust, and behavioral intentions in electronic commerce. *Journal of Global Information Management, 12*(1), 18-40.

Mahmood, M. A., Burn, J. M., Geomoets, L. A., & Jacquez, C. (2000). Variables affecting information technology end-user satisfaction: A meta-analysis of the empirical literature. *International Journal of Human-Computer Studies, 52*(4), 751-771.

McKnight, D. H., Cummings, L. L., & Chervany, N. L. (1998). Initial trust formation in new organizational relationships. *The Academy of Management Review, 23*(3), 473-490.

McLoughlin, C. (1999). *Culturally inclusive learning on the Web.* Retrieved July 13, 2007, from http://lsn.curtin.edu.au/tlf/tlf1999/mcloughlin.html

Okayazaki, S., & Rivas, J. A. (2002). A content analysis of multinationals' Web communication strategies: Cross-cultural research framework and pre-testing. *Internet Research: Electronic Networking Applications and Policy, 12*(5), 380-390.

Oliver, R. L. (1999). Whence customer loyalty? *Journal of Marketing, 63*, 33-44.

Oliver, R. L., & DeSarbo, W. S. (1988). Response determinants in satisfaction judgments. *Journal of Consumer Research, 14*, 495-507.

Pfremmer, R. (2004). Content *design considerations for localizing e-learning projects.* Retrieved July 13, 2007, from http://www.multilingual.com

Raman, K. S., & Watson, R. T. (1994). National culture, information systems, and organizational implications. In P. C. Deans & K. R. Karwan (Eds.), *Global information systems and technology: Focus* on the Organization and its Function Areas (pp. 493-513). Hershey, PA: Idea Group.

Reining, B. A., & Mejias, R. (2003). A cross-cultural study of the effects of group support systems on meeting outcomes. In *Proceedings of the First International Management Systems Conference*. San Diego: CA.

Sia, C., Tan, B., & Wei, K. (1997). Effects of GSS interface and task type on group interaction: An empirical study. *Decision Support Systems, 19*, 289-299.

Simon, S. J. (2001). The impact of culture and gender on Web sites: An empirical study. *The Data Base for Advances in Information Systems, 32*(1), 18-37.

Straub, D. W. (1994). The effect of culture on IT diffusion: E-mail and fax in Japan and the U.S. *Information Systems Research, 5*(1), 23-47.

Sun, H. (2001, October). Building a culturally-competent corporate Web site: An explanatory study of cultural markers in multilingual Web design. *SIGDOC*, 95-102.

Szymanski, D., & Hise, R. T. (2000). E-satisfaction: An initial examination. *Journal of Retailing, 76*(3), 309-322.

Tan, B. C. Y., Wei, K. K., Watson, R. T., Clapper, D. L., & McLean, E. R. (1998). Computer-mediated communication and majority influence: Assessing the impact in an individualistic and a collectivistic culture. *Management Science, 44*(9), 1263-1278.

Tsao, H., & Lin, K. (2001). A two-level approach to establishing a marketing strategy in the electronic marketplace. In *Proceedings of the 34ᵗʰ Hawaii International Conference on System Sciences*.

Tseng, L. P., & Stern, B. L. (1996). Cultural difference in information obtainment for financial decisions–East versus West. *Journal of Euro-Marketing, 5*(1), 37-48.

Walther, J. B. (1992). Interpersonal effects in computer-mediated interaction: A relational perspective. *Communication Research, 19*(1), 52-90.

KEY TERMS

Culture: Patterns of thinking, feeling, and potential acting learned throughout a lifetime, and applicable to groups of people often from the same nation state.

E-Collaboration: Collaboration among individuals involved in a common task using electronic technologies.

E-Learning: Learning in groups or individually as facilitated by online interfaces.

Online Satisfaction: Positive attributions by a user toward an online system, service, or encounter.

Online Trust: A belief that an online user has confidence in a computer mediated experience.

User Interface: The use of technology to connect users in an online environment.

Enhancing Electronic Learning for Generation Y Games Geeks

Sophie Nichol
Deakin University, Australia

Kathy Blashki
Deakin University, Australia

INTRODUCTION

This article explores a purpose built learning community, or that which Bruffee (1999) refers to as "conversational community," of University students. The community functions primarily as a collaborative learning environment, specifically for students studying games design and development at Deakin University. Specifically this article focuses on the electronic, or online learning, "Web community" of the games students. The students typically use the online environment as a supplement to face-to-face lectures and tutorials in games design and development. The games students at the centre of this article are affectionately referred to as "Games Geeks" (with their approval!), and are demographically considered, by virtue of their age, to be Generation Y (those born between 1979 and 2000). Generation Y, and the games students in particular, are collaborative learners with an increased disposition for peer learning and social relationships. Communication amongst Generation Y is continually shifting between face-to-face to online modes, and culturally specific languages such as Leet Speak (Blashki & Nichol, 2005) have evolved as part of these slippery social negotiations and hierarchies. Within the game students' social and educative milieu, learning via traditional "transmission" forms (the hallmark of university education), is eschewed for a more collaborative and participatory method supplemented by mentor relationships and constructive conversation amongst peers. Active participation and a sense of belonging to a community of knowledgeable peers, allows students to grant authority to their peers for "constructive, reacculturative conversation" (Bruffee, 1999, p. 12) of their work and ideas. Acceptance into the community is dependant upon students being willing to submit to this authority.

Participation in a community, such as that of the games students, assists in building the skill of interdependence in students, which helps them in turn to build creative skills (Blumenthal, Inouye, & Mitchell, 2003). Game students are not normally associated with creativity, and are perceived to be more comfortable dealing with numerical bits and bytes. However, creativity is a primary skill required by the game students for success in their chosen studies and career. Creativity comprises of four key components in interaction: person, process, product and environment. Whilst all of these factors are of equal importance in the exploration of creativity, the focus of this article is the underrated "environment," also known as in this article, a purpose built learning community. Often the environment is determined as merely a creativity support tool deemed an integral part of the technology (IT) component. The authors prefer the term creativity support system, as the goal of our environment is to support creativity, not emulate it (as is the goal of artificial intelligence). The creativity support system (purpose built learning community) in this article embraces three crucial components: creativity, reflection and environment. Web communities are constantly evolving and adapting to the use made of them in much the same way as any community of people, and reflection is a skill that aids the process of change in Web communities. As Martins states:

Building a Web based community does not consist of merely placing software on the Internet. The way people interact in a community contributes strongly to its long-term evolution. (Martins, 2006. p. 284)

Furthermore, technological design of the Web community is also facilitated via reflection within each games student (Nichol & Blashki, 2006). Not only does reflection help to continue engaging the students within the community, but it also draws in new members.

This article will explore the establishment of a purpose built collaborative electronic environment for university students, specifically games design and development students, via a participatory design process, in which the students are responsible for the design, development and some implementation of the environment (in both social and technical dimensions). From assimilation within the purpose built environment the games students have become creative e-collaborators. The over arching goal of these purpose built environments is to facilitate creative and reflective activities and behaviours within the games students, thus enhancing their learning within the games discipline. The type of activities the purpose built learning community aids the game students with, includes: communication with peers, mentors and teachers, access to learning material and assignment submission, face-to-face collaboration, brainstorming, and "play."

BACKGROUND

The "purpose built" learning community of this article comprises an online Web community in conjunction with a face-to-face "games room." The online Web community is implemented and facilitated within Deakin University's Deakin Studies Online (DSO) online teaching and learning Web site. The online community is nominally supervised by the staff involved in teaching the game students, however as discussed later and in more detail, the presence of the staff does not effect the natural "flow" of conversation in the community. The game students of this article use DSO for units such as Fundamentals of Games and Audio and Visual Elements of Games. Within each unit, various areas of

discussion are available for the students to use, from general "student talk" to "current affairs." The Web community encompasses multicampus communication between the games students. The games room is a physical environment that comprises a number of desktop computers as well as a game console playing area (see Figure 1).

Note boards, desks, and whiteboards are also available to offer students a place for work as well as play. It is located in a quiet room with sufficient space to accommodate more than 20 people. This environment has its functional and aesthetic roots in the studio environment first described by Blashki (2000) and used to facilitate student-oriented learning in information technology students during the 1990s. Within such an educational setting, the studio environment "provides students with the 'practical' subjects that establish closer connection/links between experience, knowledge and practice" (Blashki, 2000). Schőn (1987) similarly explores this with his Reflective Practicums, suggesting that such environments assist students in overcoming intermediate zones of practice. In this article, the "intermediate zones of practice" (games design and development) combines traditional programming with elements of narrative in the design of games. The combination of these two environments in which students collaborate face to face in addition to the online CSS for learning and creativity is known as a "hybrid space" (McDonald, 2005). The games students participate in the hybrid space because of their automatic membership into the community based on their choice of the games major, and also to improve their skills and competencies in the study of games (McDonald, 2005). In addition the high level of motivation experienced by the games students was an overwhelming factor in their continued

Figure 1. Games room

interest in the community. The essential e-collaboration components that define our community of games students include:

- Mental proximity of some sort is required between members, as opposed to the physical proximity (Weakley & Edmonds, 2004) that characterises face-to-face communication.
- Nonspecialist members form the basis of community members. Members are respected in the community for their "street credibility"; for example, the games students' use of Leet Speak within their communications in the games community (Blashki & Nichol, 2005, p. 72). Weakley and Edmonds (2004) define this as "establishment of common ground."
- Predominantly comprises asynchronous and text based communications.
- A degree of trust is integral to the environment. For example, ideas are shared among students in the knowledge that they will be read and valued by other games students.
- Sharing knowledge that is both visible and tacit (Coenen, 2005). Weakly and Edmonds (2004) define this as "a way of recording and reviewing past decisions."
- A stable identity for group members (Weakley & Edmonds, 2004, p. 40). This is possible within the games students' environment, as access into the system is subject to their enrolment at University, which cannot be changed either quickly or easily.

Further to these factors that characterise e-collaboration within the community of games students, learning is also facilitated via four key learning elements: immersion, engagement, risk/creativity and agency. A significant corollary to the establishment of the games community is the concomitant creation of a unique learning hybrid space characterised by "play." Such a learning environment is distinguished from traditional classroom and other digital learning, e-learning, environments by the four key elements of successful interactive learning; *immersion* (the active involvement of physical, emotional and cognitive processes and concentration), *engagement* (the ability to attract and sustain the user's prolonged interest), *risk/creativity* (the ability to move beyond the expected and experiential boundaries of stasis and safety required in order to overcome habits) and *agency*

(the user's active control over the learning and playing process). Creativity and reflection are described next, in addition to the design and implementation of the purpose built learning community (creativity support system) via a participatory design process with the students.

WHAT IS CREATIVITY?

Creativity is often perceived as an individual characteristic. However, as argued by Paulus (1999) and O'Neill and Warr (2005) creativity is essentially a social phenomenon and almost all creativity involves collaborative and/or group processes. Fischer and Nakakoji (1997) also suggest that "creative activity grows out of the relationships between an individual and the world of his or her work, and out of the ties between an individual and other human beings" (p. 22). In addition, as Florida (2004) highlights in his book *The Rise of the Creative Class*, creativity has social, financial, and economic benefits in communities. However, the issue with developing creativity in a community is that it is difficult, if not almost impossible, to define. In games design and development, creativity is perceived as an individual skill; however, influences such as the environment and a collective understanding are the foundation to individual growth of creative skills. The authors are particularly interested in the development of creativity in those perceived by self and others as "not creative" and we are keen to explore Fischer's (2005) suggestion that *the power of the unaided mind is highly overrated.* That is, we intend to refute the assertion that creativity is a "gift" bestowed upon a chosen few.

The purpose built learning community of the games students is referred to as a creativity support system (CSS) in this article. The CSS aims to facilitate creativity and reflection within the games students. The CSS assists the development of creativity and reflection by the provision of an environment rich in the four elements of; resources, personal motivation, exploration and social, discussed further below. The components of the CSS are shown in Figure 2, and are in a constant state of interaction.

The concept of environment encompasses both social and technical aspects interacting with reflective and creative abilities within the game students (users of the creativity support system). On the technical side, information technology (IT) such as computers,

Figure 2. Components of the games students creativity support system

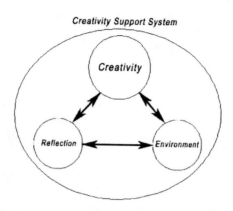

games consoles, and the Internet are used as a means to support the students in their creativity support system, and as discussed, occurs via online discussions and the games room. The difference in the creativity support system shown in Figure 2 is the way in which the games students (social actors within the community) interact and influence the design of the creativity support system, in addition to having their creative and reflective skills supported. This is facilitated, in particular, through the learning elements of immersion, engagement, risk/creativity and agency. The students' engagement in the community and its content often results in a level of engagement where students feel comfortable providing feedback and reflection on both the social and technical design of the environment. For example, at the beginning of semester 2 in 2005, one games student initiated a discussion thread in the online environment called "What's going on?" The student was disgruntled by his perception of a lack of teaching input into the class and thus initiated this discussion. Not only did this student's input and reflection effect changes in the design of the social and technical specifications within the CSS, he also effected modifications to the teaching of the unit.

The elements that facilitate such an environment in enhancing creativity, in addition to the learning elements of immersion, engagement, risk/ creativity and agency, can be defined according to the following four components:

1. *Resources* (idea time, idea support, challenge and involvement, sufficient resources (tools and information)
2. *Personal motivation* (independence, trust and

openness, tolerance for uncertainty and ambiguity, playfulness and humour, absence of interpersonal conflicts)
3. *Exploration* (risk taking, debate about the issues, freedom, reflection)
4. *Social* (supervisory arrangements, work group supports, team work (collaboration). (Isaksen, Lauer, Ekvall, & Britz, 2001, p. 175)

The constraints imposed by the confines of a article do not allow for the inclusion of many specific examples of creativity and reflection in action, as harnessed by the CSS. For such examples refer to the authors' research (Nichol & Blashki, 2005, 2006).

ENHANCING E-COLLABORATION FOR GENERATION Y GAMES GEEKS

Within the traditional computer science education, students are routinely embedded in the technical/scientific rationalist paradigm and actively instructed to perform as instrumental problem solvers who select technical means best suited to particular purposes (Schőn, 1987, p. 8). Technical/scientific rationalism (existing knowledge) has a significant function in the training of scholars and is considerably important in the teaching of Science and IT; however, any paradigmatic response cannot hope to be either definitive or comprehensive. Research into university education is not proceeding along the pragmatic social philosophy of knowledge generation through action and experimentation (with an emphasis on participative democracy), but by separating reflection from praxis and segregating method from application (Levin & Greenwood 2006). In particular, such a one-size-fits-all solution is inappropriate for the purpose-built learning environments of the games students of this article. As students compelled to converge the technical knowledge of computer systems and the sociological and psychological knowledge of people and play with the design knowledge of visual communication, these young people are the first in a new generation of creative digital practitioners. They are exemplars of Schon's "intermediate zones of practice" (1987, p. 13), within which they are "reflective practitioners." Further to their role as reflective practitioners, the games students' online collaboration within the community is particularly aligned with their cultural and social experiences as Generation Y. The CSS factors discussed above are known to assist in nurturing creativity and reflection; however, such factors are also known to encourage and nurture the education and training of Generation Y. As an example, McCrindle (2006) characterises Generation Y as particularly susceptible to influence by their peers, and less so by "authority." The authority for the games students is the teaching staff; however, the staff deliberately attempted to promote themselves as nonauthoritarian, and avoided imposing restrictions on topics of conversation that occurred in the online community. Indeed, the staff often initiated conversations regarding topics such as Leet Speak (Blashki & Nichol, 2005) and current affairs in games, in an attempt to promote and encourage debate. Initially students found it difficult to accept the nonauthoritarian role of the staff; however, with persistence staff successfully assimilated into the culture of the games students. The communities are owned by the games students. In a CSS conducive to creativity, *social* factors need to be present and this includes interaction with peers. In addition McCrindle discusses the prevalent cultural values of Generation Y as characterised by tolerance and expectations of freedom (McCrindle, 2006). These values relate to the CSS factors of *freedom* and also to the learning element of *agency*. McCrindle further suggests that Generation Y require a training environment that is unstructured, spontaneous and interactive. Such conditions are integral to the success of the CSS. In addition, key characteristics of the learning environment such as immersion and agency further support the specific learning needs of the Generation Y.

FUTURE TRENDS

The convergence of e-collaboration and creativity support systems is an emerging and thus currently limited area. Studies such as this are the precursor to generalised acceptance of the convergence of e-collaboration, learning and creativity. The future trend of e-collaboration is evident in the growing literature documenting the use of Web-based communities, not only for business but also for learning (Blashki & Nichol, 2005; McDonald, 2005; Weakley & Edmonds, 2004). The emergence of that which McDonald (2005) refers to as e-learning communities is the next step in education. Current versions of e-collaboration that exist in many universities remain confined within the paradigmatic framework of e-learning. Any e-collaboration remains confined to individual courses or units. Learning communities will incorporate "learners, educators and business professionals in a contextual, creative space with the goal of developing not only the knowledge and skills of learners, but also of creating ideas, synergies and opportunities" (McDonald, 2005, p. 109). From a design perspective, Web-based communities require a clear and defined set of interaction principles. Existing face-to-face interaction between humans is very well established; however, human interaction within Web communities is still evolving and continues to do so with increasing alacrity. Furthermore, as Weakley and Edmonds (2004) suggest, "rather than trying to make system that fakes the effect of 'being there' we should create systems that offer new facilities, ones that we might continue to use even when working face to face" (p. 241).

CONCLUSION

This article explores the establishment of a purpose built collaborative electronic environment for University students, specifically games design and development students, via a participatory design process, in which the students are responsible for the design, development and some implementation of the environment (in both social and technical dimensions). The over arching goal of these purpose built environments is to facilitate creative and reflective activities and behaviours within the games students, thus enhancing their learning within games. Other research by Nichol and Blashki (Blashki & Nichol, 2005; Nichol, 2005; Nichol & Blashki, 2005,

2006) outlines the results and discussion of our study in more detail. Further work in this area includes many areas such as: facilitation of a creativity support system designed around results gathered from the previous work into the games geek community, with the goal to facilitate that which McDonald (2005) refers to as a "learning community." Further, work also includes the testing of students' creative potential before and after their immersion within the game geek community.

REFERENCES

Blashki, K. (2000, May, 21-24). *Bending the paradigm: The use of the studio as an essential component of the structure of the bachelor of multimedia.* Paper presented at the I.R.M.A. Information Resource Management Association, Anchorage, AK.

Blashki, K., & Nichol, S. (2005). Games geek goss: Linguistic creativity in young males. *Australian Journal of Emerging Technology and Society, 3*(1), 71-80.

Blumenthal, M. S., Inouye, A. S., & Mitchell, W. J. (2003). *Beyond productivity: information, technology, innovation, and creativity.* Washington, DC: The National Academics Press.

Bruffee, K. A. (1999). *Collaborative learning.* London: The Johns Hopkins University Press.

Coenen, T. (2005). *How social software and rich computer mediated communication may influence creativity.* Paper presented at the IADIS International Conference Web Based Communities, Lisbon, Portugal.

Fischer, G., & Nakakoji, K. (1997). Computational environments supporting creativity in the context of lifelong learning and design. *Knowledge-Based Systems, 10*(1), 21-28.

Fisher, G. (2005). Distances and diversity: Sources of social creativity. *Creativity and Cognition,* 128-136.

Florida, R. (2004). *The rise of the creative class.* North Melbourne, Australia: Pluto Press.

Isaksen, S. G., Lauer, K. J., Ekvall, G., & Britz, A. (2001). Perceptions of the best and worst climates for creativity: Preliminary validation evidence for the situational outlook questionnaire. *Creativity Research Journal, 13*(2), 171-184.

Martins, H. F. (2006). *Social issues as success factors for Web-based communities.* Paper presented at the IADIS International Conference on Web Based Communities, San Sebastian, Spain.

McCrindle, M. (2006). *Bridging the gap: Generational diversity at work* Baulkham Hills: McCrindle Research Pty Ltd.

McDonald, D. (2005). *Complex learning communities.* Paper presented at the IADIS International Conference on e-Society, Qawra, Malta.

Nichol, S. (2005). *Creative geeks? Facilitating the creative growth of games students using engaging environments.* Paper presented at the OZCHI Australian Computer-Human Interaction, Canberra, Australia.

Nichol, S., & Blashki, K. (2005). *The realm of the game geek: Supporting creativity in an online community.* Paper presented at the IADIS, San Sebastian.

Nichol, S., & Blashki, K. (2006). *Games geeks in context: Developing the environment to engage creative expression.* Paper presented at the ED-MEDIA: World Conference on Educational Multimedia, Hypermedia and Telecommunications, Orlando, Florida.

Paulus, P. B. (1999). Group creativity. In S. Pritzker & M. A. Runco (Eds.), *Encyclopedia of creativity* (Vol. 1, pp. 779-784). San Diego, CA: Academic Press.

Schön, D. A. (1987). *Educating the reflective practitioner.* San Francisco: Jossey-Bass.

Warr, A., & O'Neill, E. (2005). Understanding design as a social creative process. *Creativity and Cognition,* 118-127.

Weakley, A., & Edmonds, E. (2004). *Web-based support for creative collaboration.* Paper presented at the IADIS International Conference Web Based Communities, Lisbon, Portugal.

KEY TERMS

Community: A social group of any size whose members have a *community of interest* and *cultural background* within the games design and development area.

Creativity: Consists of four components interacting: person, process, product, and environment. Many skills make up what is known as *creativity,* such as motivation and reflection.

E-Collaboration: The collaboration of students and staff as well as the facilitation of learning within a Web-based environment.

Environment: The physical and social elements that surround students in their learning at university as well as their day to day life. In this study known as *purpose built learning environment.* Environment is a component of the student's *creativity support system.*

Generation Y: The generation of people born between 1979 and 2000. They have beliefs and values such as having fun, an enjoyable lifestyle, and having few absolutes for their future.

Purpose Built: An environment (encompassing physical, technical, and social components) designed with a specific educational goal in mind.

Reflection: A skill within each individual. Thoughts are considered and reflected upon before further investigation.

E-Scheduling

Gerhard F. Knolmayer
University of Bern, Switzerland

INTRODUCTION

Collaboration between business partners can take many forms, ranging from simple exchange of elementary data to collaborative work on product development and division of labor in production and distribution processes. This article describes concepts, systems, and experiences with computer-aided collaborative scheduling.

Scheduling is the allocation of resources over time to perform a collection of tasks (Baker, 1974). A schedule maps activities to resources, together with their planned start and end times. It determines what activities will be realized with what resources at what time.

Scheduling is traditionally seen primarily as an activity geared to a specific workshop or factory. Increased division of labor and globalization of manufacturing activities demand the coordination of distributed production activities. As scheduling decisions are often short term and taken close to execution, real-time information exchange, seamless task collaboration, and contingency management among geographically dispersed factories may be beneficial (Jia, Fuh, Nee, & Zhang, 2002).

E-Scheduling can be defined as the application of computer and network technology as devices for coordinating tasks that are somehow related. With the evolution of the Web, eScheduling systems also became available for use in business-to-business (B2B) and business-to-consumer (B2C) relationships, for coordinating appointments, meetings, and reservations. In the remainder of the article we discuss e-scheduling first in production systems and then in office and service environments.

BACKGROUND

At the core of scheduling problems there is a coordination issue: Resources must be available for working together at the same time, and mostly also at the same place. Information and tools may be needed to execute the tasks and have to be considered in the scheduling process. Thus, several resources must be available at the right time at the right place in the right quality, resulting in a typical logistical problem.

Scheduling procedures have traditionally been regarded primarily in a manufacturing context. Several objectives may be relevant in scheduling. Much effort has been spent on developing algorithms for solving the associated assignment and sequencing problems. Mathematical complexity theory has shown that most sequencing problems are NP-hard (Lenstra, Rinnooy Kan, & Brucker, 1977). Therefore research has been focused on developing heuristics and interactive systems for finding good, but not necessarily optimal solutions.

Some operations management concepts recommend segmenting shops into autonomous work groups (Baines, 1993; Schuring, 1992). These groups are scheduled to execute certain tasks within predetermined time frames (e.g., 1 week). The assignment of tasks to certain members of the group and machines and the sequences in which the tasks are executed on them are not decided centrally but by some member(s) of the group. Experience has shown that the actors in the groups are heavily interdependent in their activities and, as a consequence, spend a great deal of time on coordinating and negotiating their activities. This is also true for relationships with other working groups, central planners, purchasing agents, process support technicians, and others. In many situations, the actors had no efficient means of supporting this interaction. They kept a number of "private" logbooks, but little was based on well-defined procedures and supported by tools (Carstensen & Schmidt, 2005). Results like this show that there is a need for a well-defined collaboration on the shop floor and for coordinating schedules in distributed systems.

From an information systems viewpoint, scheduling is part of or closely related to manufacturing resource planning (MRP II) and enterprise resource planning (ERP) systems. Earlier MRP systems either excluded detailed scheduling tasks or proposed detailed but inflexible schedules in the form of lists, which might have been updated only once a week. When PCs with graphical user interfaces became available, systems were developed that allowed an interactive definition and improvement of schedules, for example by presenting machine- and task-oriented Gantt charts and allowing schedule modifications by drag-and-drop. These systems originated in Germany, and the German term *leitstand* (Adelsberger & Kanet, 1991; Kurbel, 1993) is sometimes also used in English for such an electronic control unit. Leitstands work as part of a computer hierarchy, receiving short-term data from the ERP system, supporting the scheduler interactively in his tasks, acquiring data from the shop floor, and transmitting basic data or data aggregates to the ERP system. In a medium-sized or large enterprise, typically several control units are implemented for scheduling different workshops. These decentralized leitstands are usually not directly connected and do not offer other schedulers real-time information that may be relevant for their decisions.

Manufacturing execution systems (MES) are designed for shop floor control, statistical process control, and management of work in progress (Chang, 2005). Scheduling is sometimes seen as functionality offered by MES. Some sources mention potential advantages of communication between MES (Kratzer & Erhard, 2004).

Some vendors do not provide functionality for detailed scheduling in their ERP system but offer a supply chain management (SCM) system that includes an advanced planning and scheduling (APS) system. For instance, SAP's APO (advanced planner & optimizer) includes a "production planning/detailed scheduling" (PP/DS) module, which allows scheduling of production orders with very fine granularity. Although such modules are offered as part of an SCM system, they are not designed for collaborative use by several units, but provide their advanced functionalities primarily for one particular unit.

COLLABORATIVE PRODUCTION SCHEDULING

In this section, we assume that scheduling decisions of one unit influence the schedules of other units. These units may be different shops at a certain plant or different plants within a group or different companies within a supply chain. The units depend on their suppliers to provide the right materials at the right place at the right time and in the right quality. Owing to this dependence the schedulers of the receiving unit may wish to know details of how the preceding production and distribution operations have progressed. We distinguish between a simple exchange of information between these units and more sophisticated types of collaboration.

Information Exchange

The SCM literature emphasizes the importance of information exchange, among other reasons to avoid the bullwhip effect (Lee, Padmanabhan, & Whang, 1997). Information exchange is discussed primarily for point of sales data, inventory data, and machine utilization data.

Information needed to support scheduling decisions may be exchanged via reports about (potentially real-time) data that are published on a Web site or sent to a PDA. Such reports may be designed to present production data by time-frame, stock-keeping unit, production area, and operation.

In contract manufacturing the schedule of transports may determine production schedules and a need for exchanging information between distribution and production schedulers' results (Chang & Lee, 2004; Chen & Vairaktarakis, 2005). With respect to distribution schedules and their fulfillment, Track&Trace systems (Hannon, 2004) have become quite popular. They show the progress made in bridging the spatial distance between supplier and recipient, allow the recipient to prepare for arrivals, but also to adjust production schedules if the item required should arrive too late.

The customer may receive information about successfully finished operations and when the remaining operations are scheduled for execution. As usual, this could be done via alerting mechanisms (e.g., sending e-mails), by providing information on the Web, or even by allowing access to (parts of) the scheduling system. Visibility of real-time data for business partners is regarded as one of the main properties of the "real-time enterprise" (Rabin, 2003).

In the chemical industry, changes in the schedule of one plant can affect several other plants, and ripple effects may increase the magnitude of changes in plants downstream. For instance, in the Bayer company the plant schedules are highly interdependent. The results of the nightly centralized scheduling run are broken-down into plant-specific models where decentralized planners use these models for local changes. The local scheduler should:

- Be able to work on a smaller model of the facilities he is allowed to schedule but at the same time be able to share data with and view information from other plants
- Be able to see the schedule changes of relevant production steps in other plants
- Make other plants aware of schedule changes
- Reduce conflicts and find a mutually agreeable solution for product chains running through multiple plants with the help of a chain planner

Complex communication mechanisms are set up to achieve these goals. Central coordination mechanisms are combined with complementary information exchange amongst decentralized decision makers between the scheduling runs (Berning, Brandenburg, Gürsoy, Mehta, & Tölle, 2002).

In coordinating highly complex, costly, and closely interrelated projects, such as space missions, problems between different organizations are compounded as the schedule approaches real-time operation, and the impacts of last-minute changes create problems for other missions because their ability to update the large quantity of data support products is limited. The concept of e-scheduling involves seeking to provide timely and accessible information about a variety of schedule products via an end-to-end, version-controlled schedule database. The architecture allows schedulers to focus on scheduling rather than on data management. Main results are:

- Faster turnaround of schedule changes through Web-based publication of updated schedules
- Fewer meetings of conflict resolution personnel
- Conflict resolution activities shifted from central meetings to missions in conflict
- Reduced management of redundant files and reformatting through common schedule data
- Shared knowledge through open publication of schedules
- Automation of accountability for schedule changes through versioning of the schedule database
- Expanded capabilities and tool developments via common application programming interfaces to schedule data

- Reduced cost of scheduling activities through automation of large components of the scheduling process (Smith & Wang, 2001)

Advanced Forms of Collaboration

Information exchange is only a first step in collaboration, because no formal coordination procedures are established. More sophisticated types of collaboration are coordinated activities in forecasting and planning, such as those proposed in the CPFR approach (Seifert, 2003).

Establishing an isolated central scheduling mechanism for several members of a supply chain will typically fail because of differing goals of the companies, secrecy issues, and the complexity of determining optimal schedules. If central planning is inappropriate, the total planning model may be decomposed into a distributed planning system in which a central agency coordinates proposals that result from local planning efforts and modifies guidelines and constraints iteratively. This planning approach, based on Dantzig/Wolfe's decomposition principle (1960) for solving very large linear programs and the hierarchical planning concept suggested by Hax and Meal (1975), has been transferred to coordinating organizational units in a Supply Chain. Dudek and Stadtler (2005) describe the iterative adjustment of production plans and schedules between two partners of a supply chain. The sum of the objective function values of the two partners is remarkably improved by exchanging and adjusting preliminary schedules so as to consider each partner's situation. Similar results are obtained for a cooperation between a manufacturer and a distributor in a "supply chain scheduling" context (Dawande, Geismar, Hall, & Sriskandarajah, 2006).

E-SCHEDULING OF OFFICE AND SERVICE TASKS

Today's work processes are highly collaborative, meaning that several employees of an organization, and often also persons outside the organization, must be available at the same time to exchange information, elaborate plans, and make decisions. Thus, schedules are relevant not only in manufacturing but also in many office and service tasks.

If a client wants to meet a service provider, he may have to make an appointment. Such appointments are typically agreed upon in phone calls with the secretary of the service provider. Today, asynchronous communication procedures are available for initiating or making appointments. E-scheduling systems can be applied to book time slots for human or physical resources:

- Meeting rooms and other shared facilities or devices can easily be booked via an e-scheduling system. In a broad sense, every electronic reservation system (Mitev, 1999) can be regarded as an e-scheduling application, because a resource (hotel room, car, seat in an aircraft, etc.) is scheduled for a certain client and a given time interval.
- Human resources can also be booked via the Web. A person interested in using services of this resource may seek contact via a Web browser. Authentication procedures may apply to avoid misuse.
  - Open schedules of service providers allow clients to select a certain day and time interval. A client can be authorized to retrieve additional information for selecting a certain person and also be asked to provide additional information (e.g., to define the purpose of the consultation).
  - Car-service activities can be self-scheduled by customers.
  - Companies can post time intervals when they will accept visits from sales representatives or deliveries and allow booking of time slots.
- Co-workers and external persons may coordinate physical meetings or audio/video conferences via a Web-based calendar. Several tools are available for generating such calendars, many of them developed in the open source community. Social, individual, and technological issues of groupware calendar systems have been discussed (Palen, 1999); the use of Web calendar systems may lead to coordination benefits in terms of communality and connectivity and result in more efficient contact between coworkers (van den Hooff, 2004).
- Assigning teachers to classes is a major coordination effort in educational institutions. Automatically generated schedules typically result in conflicts because of different, unexpected, or not clearly expressed goals of the teachers (Nepal

& Ally, 2002). An e-scheduling system could provide interactivity by distributing proposals or making them accessible via the Web and by allowing change requests or comments.

Scheduling systems typically show only dichotomous information for each time interval: The resource is available or not. An associated problem is concurrency control: person A may retrieve information from the system and need some time to come to a decision. In the meantime, person B books the resource so that it is no longer available in the relevant time interval. Some airline reservation systems give advice if only few seats are free, mentally preparing the client that the option may soon become unavailable.

A problem may result if a scheduled person finds the automatically generated schedule inappropriate from the point of her local needs, constraints, or goals. Therefore, bookings of human resources are often not automatically confirmed. The client may have to make several suggestions for convenient appointment times and a scheduler of the service provider will make the selection. The reason for this procedure is that constraints and preferences of scheduled persons may not be explicitly defined, in which case they only become visible when the schedule is compiled or presented for validation. In some cases the request may result in additional communication needs to determine the detailed type of service requested, to allow preparations, and provide estimates for the duration of the activity. These issues may be solved in phone calls; in other cases lean media, such as an e-mail conversation, may be sufficient. Agent systems have been proposed in which the computerized agents negotiate with each other to solve complex coordination issues (Kaplansky & Meisels, 2002).

FUTURE TRENDS

In manufacturing applications we assume that future scheduling systems will support automated and event-driven information exchange. A more sophisticated coordination of decentralized scheduling activities will emerge. There are strong indicators that the use of e-scheduling beyond manufacturing applications, in office and service applications, will gain much in relevance. Co-workers and consumers will become conversant with the use of Web-based calendars and scheduling tools. However, special needs and individual goals of humans should be better supported by future tools and interfaces to foster acceptance. Agent systems may contribute to solve these issues.

CONCLUSION

Problems encountered in determining schedules centrally are missing or biased information from other units, the complexity of solving an overall problem and, in the case of scheduling human resources, clandestine goals and constraints of the persons involved. Thus there exists a need for decentralized decision making. However, decentralized scheduling decisions involve many interdependencies. Information exchange about schedules and more sophisticated efforts in coordinating the schedules are worthwhile in many situations. The Web offers many options for supporting collaborative scheduling, inside a company, but also in B2B and B2C relationships.

REFERENCES

Adelsberger, H. H., & Kanet, J. J. (1991). *The Leitstand: A new tool in computer-integrated manufacturing. Production and Inventory Management Journal, 32*(1), 43-48.

Baines, A. (1993). Autonomous work groups. *Work Study, 42*(1), 6-7.

Baker, K. R. (1974). *Introduction to sequencing and scheduling.* New York: Wiley.

Berning, G., Brandenburg, M., Gürsoy, K., Mehta, V., & Tölle, F.-J. (2002). An integrated system solution for supply chain optimization in the chemical process industry. *OR Spectrum, 24*(4), 371-401.

Carstensen, P., & Schmidt, K. (2005). *Autonomous working groups in manufacturing: Core activities and requirements for IT support.* Retrieved April 21, 2006, from http://www.itu.dk/~schmidt/papers/idak_summary.pdf

Chang, J. (2005). *IET MES (manufacturing execution system) for IC packing & testing.* Retrieved April 21, 2006, from http://hosteddocs.ittoolbox.com/JC051305.pdf

Chang, Y.-C., & Lee, C.-Y. (2004). Machine scheduling with job delivery coordination. *European Journal of Operational Research, 158*(2), 470-487.

Chen, Z.-L., & Vairaktarakis, G. (2005). Integrated scheduling of production and distribution operations. *Management Science, 51*(4), 614-628.

Dantzig, G. B., & Wolfe, P. (1960). Decomposition principle for linear programs. *Operations Research, 8*(1), 101-111.

Dawande, M., Geismar, N. H., Hall, N. G., & Sriskandarajah, C. (2006). Supply chain scheduling: Distribution systems. *Production and Operations Management, 15*(2), 243-261.

Dudek, G., & Stadtler, H. (2005). Negotiation-based collaborative planning between supply chains partners. *European Journal of Operational Research, 163*(3), 668-687.

Hannon, D. (2004). Track and trace technology. Improved shipment visibility produces inventory savings. *Purchasing, 133*(4), 43-46.

Hax, A. C., & Meal, H. C. (1975). Hierarchical integration of production planning and scheduling. In M. A. Geisler (Ed.), *Logistics* (pp. 53-69). Amsterdam, The Netherlands: Elsevier.

Jia, H. Z., Fuh, J. Y. H., Nee, A. Y. C., & Zhang, Y. F. (2002). Web-based multi-functional scheduling system for a distributed manufacturing environment. *Concurrent Engineering, 10*(1), 27-39.

Kaplansky, E., & Meisels, A. (2002). Negotiation among scheduling agents for distributed timetabling. In E. Burke & P. De Causmaecker (Eds.), *Proceedings of the 4th International Conference on the Practice and Theory of Automated Timetabling (PATAT)* (pp. 517-520). Gent, Belgium: KaHO St.-Lieven.

Kratzer, G., & Erhard, P. (2004). Manufacturing to manufacturing (M2M) integration: Schnittstellen beherrschen—Produktionszeiten verkürzen. Retrieved April 21, 2006, from http://www.sps-magazin.de/artikel/itp-artikel.asp?key=DNRWl1I38Ev6z4nN3aqPWfPsx

Kurbel, K. E. (1993). Production scheduling in a leitstand system using a neural-net approach. In E. Balagurusamy & B. Sushila (Eds.), *Artificial intelligence technology: Applications and management:*

Proceedings of the Fourth International Computing Congress (pp. 297-305). New Delhi, India: Tata Mc-Graw-Hill.

Lee, H. L., Padmanabhan, V., & Whang, S. (1997). Information distortion in a supply chain: The bullwhip effect. *Management Science, 43*(4), 546-558.

Lenstra, K., Rinnooy Kan, A. H. G., & Brucker, P. (1977). Complexity of machine scheduling problems. In P. Hammer et al. (Eds.), *Annals of discrete mathematics, 1* (pp. 343-362). Amsterdam, The Netherlands: North-Holland.

Mitev, N. N. (1999). Electronic markets in transport: Comparing the globalization of air and rail computerized reservation systems. *Electronic Markets, 9*(4), 215-225.

Nepal, T., & Ally, M. I. (2002). Possible models for timetabling at tertiary institutions. In E. Burke & P. De Causmaecker (Eds.), *Proceedings of the Fourth International Conference on the Practice and Theory of Automated Timetabling (PATAT)* (pp. 228-237). Gent, Belgium: KaHO St.-Lieven.

Palen, L. (1999). Social, individual & technological issues for groupware calendar systems. In *Proceedings of the SIGCHI Conference on Human Factors in Computing Systems* (pp. 17-24). New York: ACM.

Rabin, S. (2003). The real-time enterprise, the real-time supply chain. *Information Systems Management, 20*(2), 58-62.

Schuring, R. (1992). Reasons for the renewed popularity of autonomous work groups. *International Journal of Operations & Production Management, 12*(4), 61-68.

Seifert, D. (2003): *Collaborative planning, forecasting, and replenishment: How to create a supply chain advantage.* New York: Amacom.

Smith, J. H., & Wang, Y.-F. (2001). E-scheduling the deep space network. *IEEE Aerospace Conference Proceedings, 7*, 3385-3391. Piscataway, NJ: IEEE

van den Hooff, B. (2004). Electronic coordination and collective action: Use and effects of electronic calendaring and scheduling. *Information & Management, 42*(1), 103-114.

KEY TERMS

Agent: Software that works toward goals in a dynamic environment on behalf of another entity (human or computational) without continuous direct supervision or control and exhibits a significant degree of flexibility and even creativity in how it seeks to transform goals into tasks.

Enterprise Resource Planning (ERP) System: An accounting-oriented, multimodule, integrated information system for identifying and planning the enterprise-wide resources needed to take, make, ship, and account for customer orders.

Groupware Calendar: A Web-based system that allows an authorized group of persons or (software) agents to access a common calendar and to modify its entries.

Leitstand: Decentralized control unit for detailed planning and scheduling of production orders and workflows.

Manufacturing Execution System (MES): A shop floor control system that includes either manual or automatic labor and production reporting as well as online inquiries and links to tasks that take place on the production floor.

Scheduling: Allocation of resources over time to perform a collection of tasks. A schedule maps activities to resources, together with their planned start and end times.

Supply Chain Management (SCM): SCM tries to improve the flow of materials, information, and financial resources within the company and among collaborating companies by sharing information, by concerted planning and scheduling, by coordinated execution, and by concerted monitoring and controlling to raise the competitiveness of the entire supply chain.

Track & Trace: Remote monitoring system providing information about the position of vehicles, containers, palettes, or other resources.

Evolving Gender Communication Issues in E-Collaboration

Cathy L. Z. DuBois
Kent State University, USA

INTRODUCTION

Much has been written about gender differences in communication. Gender stereotypes propose that men communicate in a direct manner and focus on information; women communicate in an indirect manner and focus on relationships. Tannen (1995) suggests that gender differences in communication contribute to the "glass ceiling." Further, Eubanks (2000) noted that the Internet and the World Wide Web are actively and aggressively hostile to women. Such discourse fosters gender stereotypes of the past and paints a gloomy picture for women with regard to participation and success in the realm of workplace e-collaboration.

However, I propose that e-collaboration provides a medium with the potential to allow women to work outside the realm of traditional gender stereotypes. Gender stereotypes are not an accurate reflection of reality, for both women and men have a wide array of communication styles. Activation of gender stereotypes constrains rather than facilitates communication. Face-to-face communication is rich with cues that activate gender stereotypes. Women can benefit from opportunities in the workplace to operate free from the restrictions placed upon them by gender stereotypes, and to further develop their communication competence.

My argument rests on two premises. First, e-collaboration offers a means through which women can communicate with men without being encumbered by many of the usual social trappings that activate gender stereotypes. It also offers opportunities for mindful articulation of ideas, expanded access to organizational members, clear tracking of idea generation, all of which increase one's opportunities to be heard.

Second, the growing presence of women in traditionally male workplace roles, as well as in cyberspace, is changing workplace culture and the norms of the past. Women and men continue to learn from one another and expand their competence and comfort in a variety of roles. This article will outline relevant theory and statistics that support both arguments.

BACKGROUND

Social Presence and Media Richness Theories

Social presence theory (Short, Williams, & Christie, 1976) proposes a one-dimensional continuum of social presence that reflects the degree of awareness of the other person in a communication interaction. Text-based communication anchors the low end of this continuum, and face-to-face communication anchors the high end. This theory posits communication will be most effective when the level of social presence is appropriate for the interpersonal involvement requirements of the task.

Daft and Lengel's (1986) media richness theory places media upon a continuum with regard to capacity to provide rapid feedback, convey non-verbal cues and personality, and support the use of natural language. They suggest that media should be selected as appropriate for the need to reduce ambiguity in collaborative tasks, and that face-to-face communication is the most effective for reducing discussion ambiguity.

Both theories suggest the degree of social presence or media richness should be matched to the situation at hand. Face-to-face communication is not always ideal, and has some negative aspects. It carries both financial and emotional costs, as people travel to be together and wait for one another, and succumb to the need for small talk to establish rapport. Face-to-face communications require time-consuming meetings and can create interruptions to work accomplishment. Because face-to-face communication can decrease effective use of time in the workplace, people actually seek to avoid it at times (Nardi & Whittaker, 2002).

They close office doors, choose to work offsite, and opt for lean electronic communication.

Gender Stereotypes in Communication

Gender stereotypes are associated with societal roles, which have spilled over into work roles. The workplace and its rules were forged by men and for men (Connell, 1995). Men assumed positions of power; they were bosses, problem solvers, good with numbers and technology; they looked to one another for ideas and innovation; efficiency and productivity became the basis of success. Accordingly, workplace communication norms highlight the importance of technical and informational exchange, as well as the importance of direct communication. These norms continue to play an important role in effective task accomplishment, and they continue to be associated with men.

Much of what is written on communication gender stereotypes is closely linked to social role stereotypes. For example, Simon and Pederson (2005) provide an overview of communication styles that limits women to cooperative, intuitive, relational and feeling communications, and attributes to men the workplace necessary communications of asserting information with a focus on intellect/facts/reason/logic/order/structure. Similarly, Herring (1994) notes that women and men use language differently; women use apologies, questions, and supportive language; men use strong assertions, self-promotion, and challenges.

However, if women were limited to relational communications at work they could never accomplish their tasks successfully. Information-focused communication is not simply the domain of men; women are both comfortable and competent in this domain. In fact, recent evidence suggests that when men and women share the same role or the same task, and have the same status, both genders adopt similar communication strategies (Basow, 2004).

When dealing with stereotypes, it's important to remember that there is greater variance in behavior within any given group than there is between groups. Thus, some men communicate in a manner more consistent with female stereotypes (Tannen, 1996), and some women communicate in a manner that's more consistent with male stereotypes. Tannen (1994) prefaces her book *Talking from 9 to 5* by recognizing that although she presents gender-stereotypical patterns of communication, each individual's style is unique and shaped by their personal history. "Patterns I describe are always a matter of degree, of a range on a continuum, not of absolute difference" (Tannen, 1996, p. 13). Thus, stereotypes oversimplify and potentially misrepresent reality; they can bring to bear in any situation inappropriate assumptions and conclusions.

Stereotype Activation

Gender communication stereotypes are deeply held and widely accepted. Stereotypes simplify cognitive processing by simplifying our world. As noted, stereotypes consist of generalizations that may or may not be situationally accurate. Yet once evoked, these generalizations constrain how communication is perceived and interpreted. When the receiver of communication has active generalizations that are not accurate, their cognitive processing of the communication will likely distort the content and tenor of the communicator's message.

Rich media offer high social presence and richness, and contain visual and auditory cues that are likely to cognitively engage stereotypes. Tannen (1994) discusses the variety of ways in which women are "marked," which range from clothing and makeup choices to surname choices. These markers become social cues, from which attributions are made. The mere physical image of a woman thereby communicates a variety of messages and distractions to anyone in her presence. Attributions are made about her before she has a chance to speak, and these attributions impact how her communications are received. To complicate matters, women are expected to exhibit normative behavior, and those who violate stereotypical norms often face negative consequences (Heilman, 2001)

Communication that occurs without evoking gender stereotypes is less bound by the constraints and potential distortion of gender stereotypes and attributions. Text-based media offer low social presence and richness, and thereby contain fewer social cues that activate gender stereotypes. When stereotypes are not activated, the communication sent by women is more likely to be accepted at face value, rather than filtered through a gender screen. Thus, women are likely to find that the use of lean media in e-collaboration can free them from the stereotypical gender roles. For example, when stereotypes are active, men who expect women to be submissive might be offended by assertive communication from a woman. But when assertive text is not accompanied by the visual and auditory cues that bring gender stereotypes to mind, men are likely to accept assertive communication from a female and work moves ahead without disruption.

WOMEN AND WORKPLACE E-COLLABORATION

The Changing Workplace

Norms regarding workplace communication changed dramatically in the last century. The arrogance of scientific management yielded to worker empowerment; managers transitioned from rudely directing workers to asking for and acting upon their input. Executives with poor interpersonal skills are less well tolerated than they were in the past; they are often sent to interpersonal skills coaching to keep their careers from derailing. Clearly, the use of communication for the purpose of relationship building underlies these workplace changes.

This communication evolution has paralleled the entry of women into the workforce. Women's social role as relationship builders carried into the workplace. Women entered in subservient, support roles, but have increasingly risen to roles of influence over time. Their attention to relationships has gradually permeated the workplace, and is becoming as much the domain of men as it is of women. The U.S. Department of Labor (2005) reports that a larger percentage of employed women than employed men held jobs in the category of management, professional, and related occupations in 2004. Clearly, women now share in the professional realm that once belonged to men.

Similarly, a decade ago computers and the Internet were largely the domain of males (Ono & Zavodny, 2002). However, BLS statistics indicate that more women than men used a computer and the Internet at work in 2003, and that women and men are now equal in their use of the Internet from work (Fallows, 2005). Computer and Internet use among managers and professionals is about 80% (U.S. Department of Labor, 2005); these are the workers most likely to utilize e-collaboration.

E-Collaboration, Compensatory Adaptation, and Mindful Communication

E-collaboration includes an array of technologies that provide varying degrees of communication richness. At the low/lean end, they include asynchronous text; at the high/rich end they include electronic meeting support with audio and images that rival face-to-face meetings. Although most e-collaboration utilizes computers, other technologies (such as telephone communications) are electronically supported (Kock & Nosek, 2005). The richer e-collaborative technology is currently costly, and limited to those organizations with the expertise and funds to support it. Thus, most e-collaboration is currently at the less rich end of the spectrum.

The use of lean media isn't necessarily limiting, and can be perceived as advantageous. Fulk, Schmitz, and Steinfield (1990) and Markus (1994) explicated how social influence and geographic distribution can motivate users to choose lean communication media. The compensatory adaptation model (Kock, 1998, 2001) demonstrated how individuals compensate for cognitive obstacles inherent in the use of lean media for collaborative tasks. In fact, compensatory adaptation can lead to higher quality task outcomes than would have been achieved without the barriers posed by lean media. Subjects reported that their individual contributions were of higher quality because their writing was supported by higher quantity and quality of thought than would be possible in face-to-face communication.

Lean media increase the need for mindful communication. Mindful communication is "planful, effortfully processed, creative, strategic, flexible and/or reason based (as opposed to emotion based)" (Burgoon, Berger, & Waldron, 2000, p. 112). Mindful communication through lean media can offer benefits that face-to-face communication lacks. The writer has an opportunity to organize and reorganize thoughts before sending them, to edit phrases for clarity, and to remove or use emotion as is situationally appropriate. Thereby the writer has more control over the impression that is conveyed by the written words.

However, lean media communications that lack mindfulness, such as overly emotional or poorly conceived messages, can lead to misunderstandings or even spark 'e-mail wars' that damage relationships and interfere with task accomplishment. Having a written record of ineffective or abusive communications can bring consequences to the sender (e.g., lawsuit evidence) that are more harsh than those that might result from similarly spoken comments. This can have a chilling effect on inappropriate written communications in the workplace, which benefits both males and females.

Workplace Electronic Communication: Restrictions and Opportunities

The gender gap in computer use has closed significantly over the past decade; I contend that the gender gap in workplace communication has begun to close in a similar manner. Gender-stereotypical communication styles are simply less relevant in workplace e-communication.

Much of what has been researched and written about how females are disadvantaged on the Internet and with computers has been outside the realm of professional and managerial work (e.g., Eubanks, 2000; Herring, 1994). Writers who emphasize gender communication differences have focused on male-female participation and communication styles in chat-rooms or listservs, and cite primarily social rather than work-related exchanges. Although it might be true that males are controlling, rude, engage in flaming, and make sexual comments in such contexts, these types of communication are less likely to occur in workplace e-collaboration. Not only are they inappropriate and distract from task accomplishment, they carry the potential of sexual harassment complaints and legal action.

Thus, e-collaboration for work purposes poses inherent restrictions on the extremes of both male and female stereotypical communication styles. The focus of workplace e-collaboration should be primarily on the work, on the exchange of ideas, on organizational outcomes. This informational focus places workplace e-collaboration communication in the stereotypical male domain, but stops short of tolerating inappropriate communication.

However, workplace e-collaboration cannot be limited to information. Whenever people interact on a regular basis the need for relationship building is present. Collaborators must let one another know when they have processed others' ideas and extend exchanges of appreciation and recognition. Courtesy must be maintained; rude e-mails may not be fully read, and it's entirely obvious if the contributions of any group member are ignored. This relational focus places workplace e-collaboration communication in the stereotypical female domain, but stops short of distracting touchy-feely communication.

E-communication offers additional advantages to women in the workplace. The use of electronic media allows women to expand their audience beyond those to whom their access is usually restricted. Women who do not have easy access to powerful others can include them in an e-communication distribution list and have their ideas heard. Women can also use e-communication to stake a claim to their ideas. The ideas offered by a woman in a face-to-face meeting can either intentionally or unintentionally be attributed to the men in the room (Tannen, 1994). Perhaps this happens because the women talks softly or in a gentle tone of voice, because she speaks in an indirect manner, or simply because gender stereotypes dictate that men are the ones who come up with ideas. However, when a woman offers a new idea in writing, with her name and a date on the communication, it is clear that this is her idea. This written record of proof makes it less likely that women will have their ideas 'stolen' by or inappropriately attributed to men when using e-collaboration.

Finally, women who lack well developed information-focused, assertive workplace communication skills, but understand their importance and functionality, can find opportunity to practice communicating in this manner by using lean media in e-collaboration. Asynchronous text does not require an immediate response, and this lag time can be used productively. Women can take time to mindfully organize their e-communications to shape the impression they want to make, which in some ways will mirror the stereotypically male tendencies to be direct and focus on logic and information. After all, these tendencies are highly functional for both genders in workplace communication. Women can turn rambling paragraphs into numbered lists, make sure their main points are stated up front and summarized at the end, and include a list of action items. This editing capacity allows them to create a message that will have optimum impact to the readers. Finally, women can remove the markers of traditionally female language, such as replace *we* with *I*, remove apologies and tone down emotion.

FUTURE TRENDS

The use of lean media in e-collaboration to increase the freedom with which women communicate in the workplace has practical implications that are twofold.

First, women are encouraged to continue to stretch in their use of technology for e-collaboration. Women still lag behind men in their awareness of tech-related issues and in comfort with computer technology. As the gender gap in Internet use closed, so will the technology gap.

Second, organizational leaders and managers will continue to embrace culture change. The most functional organizational norms are those that support broad collaboration and data-based decision-making within an organization. In our increasingly knowledge driven economy, the communication component across jobs will continue to grow. Restricting workers to gender-driven stereotypes is simply not functional. It impedes, rather than fosters success. E-collaborators will forge communication styles that facilitate effective work accomplishment. These norms will be neither stereotypically male nor female in nature, but a combination of the best of both styles.

CONCLUSION

Gender is a socially constructed phenomenon (Fausto-Sterling, 1985). And what is socially constructed can be reconstructed in a socially determined environment. Irrespective of one's gender, the keys to successful communication are the capacity to perceive what is situationally appropriate and the skill to flexibility to adjust one's style to be consistent with what is situationally appropriate. "Effective interpersonal communication calls for directness on some occasions, and subtlety on others, by both women and men" (LaFrance & Harris, 2004, pp. 149-150). As organizational culture and normative workplace roles continue to evolve, socially constructed norms will change and it will become easier for both males and females to effectively utilize a wider range of communication styles. In the meantime, both genders can enjoy the freedom of role-restriction available through e-collaboration.

REFERENCES

Basow, S. (2004). The hidden curriculum: Gender in the classroom. In M. A. Paludi (Ed.), *Preager guide to the psychology of gender* (pp. 117-131). Westport, CT: Praeger.

Burgoon, J. K., Berger, C. R., & Waldron, V. R. (2000). Mindfulness and interpersonal communication. *Journal of Social Issues, 56*(1), 105-127.

Connell, R. W. (1995). *Masculinities.* Berkeley: University of California Press.

Daft, R. L., & Lengel, R. H. (1986). Organizational information requirements, media richness and structural design. *Management Science, 32*(5), 554-571.

Eubanks, V. (2000). Paradigms and perversions: A woman's place in cyberspace. Retrieved January 6, 2006, from http://www.cpsr.org/prevsite/publications/newsletters/issues/2000/winter 2000/eubanks.html

Fausto-Sterling, A. (1985). *Myths of gender. Biological theories about women and men.* New York: Basic Books.

Fletcher, J. K. (1999). *Disappearing acts: Gender, power, and relational practice at work.* Cambridge, MA: MIT Press

Fulk, J., Schmitz, J., & Steinfield, C. W. (1990). *A social influence model of technology use. Organizations and Communications Technology.* Newbury Park, CA: Sage.

Heilman, M. E. (2001). Description and prescription: How gender stereotypes prevent women's ascent up he organizational ladder. *Journal of Social Issues, 57*(4), 657-675.

Herring, S. (1994). Bringing familiar baggage to the new frontier: Gender differences in computer-mediated communication. In V. Vitanza (Ed.), *CyberReader* (pp. 190-201). Needham Heights, MA: Allyn & Bacon.

Kock, N. (1998). Can communication medium limitations foster better group outcomes? An action research study. *Information & Management, 34*(5), 295-305.

Kock, N. (2001). Compensatory adaptation to a lean medium: An action research investigation of electronic communication in process improvement groups. *IEEE Transactions on Professional Communication, 44*(4), 267-285.

Kock, N., & Nosek, J. (2005). Expanding the boundaries of e-collaboration. *IEEE transactions on professional communication, 48*(1), 1-9.

LaFrance, M., & Harris, J. L. (2004). Gender and verbal and nonverbal communication. In M. A. Paludi

(Ed.), *Preager guide to the psychology of gender* (pp. 133-154). Westport, CT: Praeger.

Markus, M. L. (1994). Electronic mail as the medium of managerial choice. *Organization Science, 5*(4), 502-527.

Nardi, B. A., & Whitaker, S. (2002). The place of face-to-face communication in distributed work. In P. Hinds & S. Kiesler (Eds.), *Distributed work* (pp. 83-110). Cambridge, MA: MIT Press.

Ono, H., & Zavodny, M. (2002). *Gender and the Internet* (Working Paper #495). Stockholm, Sweden: SSE/EFI Working Paper Series in Economics and Finance.

Short, J., Williams, E., & Christie, B. (1976). *The social psychology of telecommunications*. London: Wiley.

Simon, V., & Pedersen, H. (2005). Communicating with men at work: Bridging the gap with male co-workers and employees. Retrieved December 5, 2005, from http://www.itstime.com/print/mar2005p.htm

Tannen, D. (1994). *Talking from 9 to 5: How women's and men's conversational styles affect who gets heard, who gets credit, and what gets done at work*. New York: William Morrow.

Tannen, D. (1996). Gender gap in cyberspace. In V. Vitanza (Ed.), *CyberReader* (pp. 141-143). Needham Heights, MS: Allyn & Bacon.

U.S. Department of Labor. (2005). *Women in the labor force: A databook* (Report 985). Washington, DC: Author.

Fallows, D. (2005). *How women and men use the Internet*. Washington, DC: Pew/Internet & American Life Project.

KEY TERMS

E-Collaboration: People working together using electronic technologies such as computers, the Internet, video conferencing, and wireless devices.

Gender Stereotypes: Generalizations made about how and why men and women differ in what they choose to do and how they choose to do it.

Lean E-Communication: Primarily text-based communication, void of voice and image capability.

Marking: The presence of cues from which attributions are made.

Media Richness Theory: A communication theory that places media upon continuum with regard to their capacity to provide rapid feedback, convey non-verbal cues and personality, and support the use of natural language.

Mindful Communication: Communication that is planful, processed with effort, creative, strategic, and reason (rather than emotion) based.

Social Presence Theory: A communication theory that proposes a one-dimensional continuum of social presence, which reflects the degree of awareness of the other person in a communication interaction.

Extending TAM to Measure the Adoption of E–Collaboration in Healthcare Arenas

Fernao H.C. Beenkens
Delft University of Technology, The Netherlands

Robert M. Verburg
Delft University of Technology, The Netherlands

INTRODUCTION

A number of developed countries are experiencing significant demographic changes as a result of an ageing population (Grumbach et al., 2002). In order to cope with this structural social change, radical improvements in the healthcare process are vital in order to gain more efficiency. The use of e-collaboration may be an excellent way to improve the service levels and efficiency in the healthcare domain. With regard to the adoption of information technology (IT) and information systems (IS), the healthcare sector has traditionally lagged behind other sectors but this is gradually changing (Wu, Wang, & Lin, 2005). The use of technology acceptance models (TAM) in order to explain the adoption of technology-based applications in the healthcare arena, however, is expected to lead to much better insights (Raitoharju, 2005). Both the acceptance of new technologies by healthcare professionals and the usefulness of explanatory models are therefore vital in order to explain the current lack of progress in the adoption of e-services. In this article we will review the currently available models with regard to the adoption, acceptance, and adaptation of e-collaboration services in the healthcare domain.

The goal of this article is to propose a new model based upon the conclusions of this review. We argue that there are no models available with sufficient explanatory power as network aspects, the role of recipients, and cross-border issues are not sufficiently taken into account. We propose therefore a number of guidelines for a new process-like model that should incorporate insights from previous models. Rather than viewing technology adoption as a single static snapshot, the new model should encompass a more continuous evaluation of the adoption process and should evaluate the service continuously throughout several phases in time. In order to address these issues, we developed a new holistic model that we hope, solves the points of attention named above.

BACKGROUND

The acceptance of e-collaboration services in healthcare is not directly comparable to the acceptance of these services in other sectors. Healthcare supply often takes place in broader networks of care-related actors, which increases the interdependency of organizations, and which makes the diffusion and adoption of new technology even more complicated (Andriessen, 2003). This is supported by Hu, Chau, and Sheng (2002), who state it is typical for telemedicine services to span organizational boundaries. As a result of this inter-organizational nature, Hu et al. (2002) remark that technology adoption investigations must include multiple organizations simultaneously. The essential characteristics of users and technologies in the context of professional healthcare differ greatly from those in customary commercial context (Wu, Wang, & Lin, 2005).

Succi and Walter (1999) provide some examples of typical differences between physicians and managers. Medical professionals constitute a separate community as a result of their specific knowledge and their achieved professional status associated with special power and prestige. A second difference is the large professional autonomy of healthcare professionals in comparison to business professionals. Since outsiders do not have the knowledge to evaluate the practices, professionals have to protect themselves against incompetence, carelessness and exploitation. Healthcare professionals, therefore, proclaim that they themselves are in the best position to operate, control, and also to regulate their own practices (Succi & Walter, 1999).

Figure 1. Interrelations of various healthcare related actors within a healthcare environment

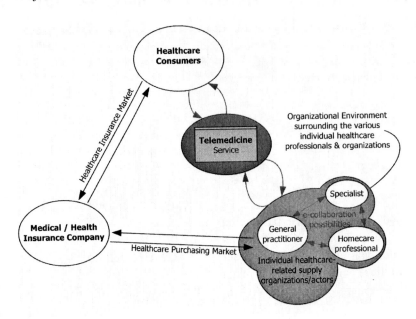

For the healthcare professionals in the hospital and homecare, Raitoharju (2005) concludes that more time spent with IT means less time spent with the patients which is not very popular among care personnel. This again shows the importance of paying adequate attention to the process of technology acceptance when introducing e-collaboration services in healthcare settings. In this paper we focus in particular on the group of causality oriented models based upon Davis' technology acceptance model (TAM) (Davis, 1993), because of the availability of literature with regard to healthcare-related studies of TAM.

TECHNOLOGY ACCEPTANCE MODEL AND EXTENDED TAM

The original technology acceptance model (Davis, 1989) as well as its extensions explain and predict the eventual use pattern of new (software) technology by end-users. TAM was extended in order to take relevant determinants of perceived usefulness into account (Venkatesh & Davis, 2000). Figure 2 shows TAM as well as the integrated extended TAM.

Various studies have shown that when using the extended TAM in a healthcare sector, only the perceived usefulness plays an important role and the perceived ease of use is negligible (Chau & Hu, 2002; Chismar & Wiley-Patton, 2002; Hu et al., 1999). Jayasuriya

(1998) also concluded that the most important factors with respect to the acceptance of PC-usage among healthcare professionals (mainly nurses in this study) are the perceived usefulness in combination with their knowledge and expertise of a PC.

Hu, Chau, Sheng, and Tam (1999) examined TAM using physicians' acceptance of technology. In their conclusion they state that in order to foster individual intentions to use a technology, it is important to try and achieve a positive attitude towards using the technology. This means that a positive perception of the technology's usefulness is crucial, while the ease of use of the technology itself may be less important for physicians.

According to Raitoharju, "the TAM-model has many useful points but it is not ideal when evaluating the healthcare sector (and) the use of IT is not necessarily a convenient goal when measuring user acceptance" (2005, p. 4). Another important characteristic which is not taken into account in TAM is that the supply of healthcare and homecare takes place especially in broader networks of care-related actors. This implies that the interdependency of organizations is increased by interorganizational diffusion and adoption processes (Andriessen, 2003).

Venkatesh and Davis (2000) conclude that because of the increasing trend of organizations changing from hierarchical command-and-control structures towards more network-structures of autonomous teams, the

Figure 2. Integrated overview of the extended TAM where the original TAM is shown in the box

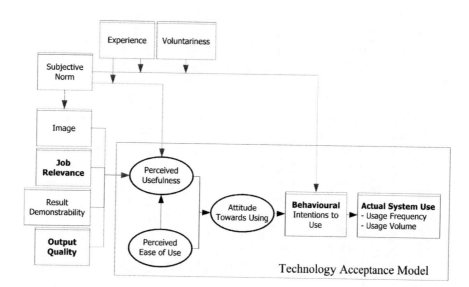

organizational mandate as a "lever for increasing usage" is experiencing its limits. According to them, the nature and role of social influence processes will need to be elaborated beyond the extended TAM, as the decision to adopt becomes more a team than an individual organizational-level decision.

For example, homecare is normally provided within a network of healthcare professionals and healthcare consumers. Therefore the decision of a partner whether or not to adopt a certain technology or service has implications for other partners and possible future users in the network. These network-aspects are comparable with the two-stage adoption process described by Andriessen (2003). Hu et al. (2002) also stress the importance of multi-organizational technology acceptance considering that e-health and homecare services are being delivered in a network environment across individual boundaries. TAM and the extended TAM focus on the behavioral intentions of a single person and not specifically on a network of healthcare organizations, consumers, or teams within a healthcare organization.

TECHNOLOGY TRANSITION MODEL

The reason for studying the usability of the technology transition model (TTM) (Briggs et al., 1998) in this article is that TTM is developed and based upon TAM to try and explain technology usage for a longer period

of time than TAM. Like (extended) TAM, TTM also focuses on causal relationships in order to explain and predict the adoption of technology.

The TTM on the other hand was developed in order to explain the stagnating use and sometimes sudden ending of the usage of group support systems (GSS), based upon following premises:

- TAM has a focus on the individual use of technology while GSS has a clear focus on the usage of new technology within the different teams of an organization
- TAM has been used to explain the acceptance of technology from the moment of introduction until several weeks after the introduction of the product
- TAM is not able to explain why successful usage of GSS technology or another application within an organization stops all of a sudden after several years

As can be seen in Figures 2 and 3, both TAM and TTM focus on the system use. The system use is modeled by Briggs et al. (2003) as a positive function of the behavioral intentions. Although TTM is extended with a number of factors and also has a focus on long-term use of technology, the TTM does not explain adoption of e-collaboration. TTM already takes more factors into account, but not enough to explain the use of new e-collaboration technology in an organization or net-

Figure 3. Technology transition model (TTM) (After Briggs et al., 1998)

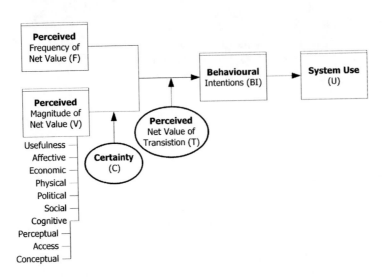

work of actors. Bangert and Doktor (2000) conclude that there are still important factors missing in TTM such as the type of organization or specific organizational characteristics. In order to construct a new model that does incorporate network aspects, the next section clarifies the importance of multiorganizational characteristics.

IMPORTANCE OF ORGANIZATIONAL CHARACTERISTICS

According to Hu et al. (2002) the ultimate success of telemedicine in an adopting organization requires adequate attention to both technological and managerial issues. Tanriverdi and Iacono (1998) investigated the role of knowledge barriers (Attewell, 1992) in order to explain why diffusion of e-health and e-collaboration in healthcare remains low. While Attewell (1992) suggests that a knowledge barrier should be seen as a one-dimensional construct focusing on lack of technical knowledge, the research of Tanriverdi and Iacono (1999) extends the knowledge barrier theory by showing that a knowledge barrier is comprised of four different dimensions: technical, economic, organizational, and behavioral knowledge barriers.

The theory of knowledge barriers can be seen as a sort of process-model; however, the model also has ele-

ments of overlaps, feedback, and feed-forward cycles. Although the study gives no fixed order in which the barriers should be overcome, the most successful case studies first tried to fix technical problems. When the organizational barrier was overcome, the physicians were more willing to use the new technology, so that the behavioral barriers were also lowered.

Typical telemedicine services span organizational boundaries (Hu et al., 2002) and this requires technology adoption investigations to consider multiple organizations simultaneously. According to Tornatzky and Fleischer (1990), technology adoption that takes place in an organization is jointly influenced by the technological context, the organizational context, and the external environment. Organizational context again refers to the internal conditions necessary for an organization to adopt a technology (Iacovou, Benbasat, & Dexter, 1995). The organizational readiness, which is described by Hu et al. (2002) as the collective attitude of medical staff in the case of a healthcare application, is related to organizational culture as has been described by Bangert and Doktor (2000), and consists of seven factors. One of these factors is, "the interdependence among organizational units: the organization focuses strongly on the interdependence among units to optimize organizational goals at all levels" (Bangert & Doktor, 2000, p. 357).

GUIDELINES AND PROCESS-ORIENTED ADOPTION MODEL

Based upon the literature discussed in the previous sections, the following guidelines for the development of a new process-oriented model for healthcare technology acceptance can be introduced:

- The adoption of e-collaboration services in healthcare takes place between different intra- or interorganizational teams instead of a single end-user.
- The causality aspects of TAM-based models and time as a variable should be changed into a process-oriented technology adoption. In other words, there is a need for a more active and engaged approach to adoption.
- Cross-cultural issues become more important as e-collaboration services in the healthcare arena are often multiorganizational.
- The current models do not pay sufficient attention to the specific level of the technology adoption by healthcare service recipients. The difference between organizational and individual adoption decisions should therefore be recognized as a difference of crucial importance.

The guidelines offer a viable starting point for the revision of currently available models for the assessment of the acceptance of e-collaboration services in healthcare contexts. These guidelines are summarized in a new process-oriented model for technology acceptance. Figure 4 provides an overview of the variety in time and actors. The factor time consists of three phases. The second factor deals with the differences between healthcare consumers as well as healthcare professionals and institutions. Figure 4 should be seen as a holistic model and presents a possible integration of time and multiactor characteristics in technology acceptance.

FUTURE TRENDS

More and more people want to stay in their familiar environments longer when they grow older and become in need of more comprehensive care. Therefore their surroundings should be made more "intelligent" in order to secure their own safety. This means that people can be monitored at a distance by care professionals with the use of technology based on ambient intelligence (AmI). Ambient intelligence (AmI) refers to technology being used to make an environment aware of the presence of people and to interact with them. Ambient intelligent technology is embedded context aware, personalized, adaptive, and anticipatory (Aarts, 2004). However, in order to maximize the usability of AmI, adequate attention should be paid to how the elderly interact with these advanced technologies. As they often have difficulty with learning to cope with advanced innovative technologies. Therefore when revising current technology acceptance models, more attention should be paid to people's individual acceptance capacities based upon specific individual impairments and capabilities. The second trend has a more organizational focus. As the provision of healthcare becomes more integrated, different teams or individual care professionals have

Figure 4. Suggested process-oriented model with multiactor integration throughout time

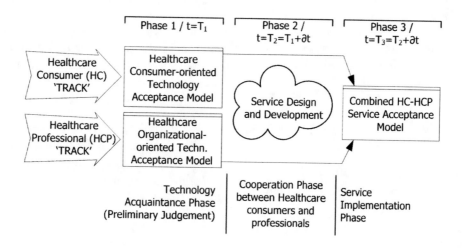

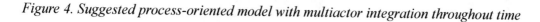

to cooperate even closer with each other in order to ensure a seamless provision of healthcare. Finally, we foresee an overall increase in e-collaboration in the healthcare domain as more and more applications become available in the near future.

CONCLUSION

Current models for technology acceptance based upon TAM are not sufficient enough to assess the acceptance and adoption of e-collaboration in healthcare areas. The main reason for this is that the current models omit time as a factor, and do not focus on the network-characteristics of e-collaboration and organization of the working environment. Therefore this article suggested a new model for technology acceptance in healthcare on the basis of four guidelines as starting point. This model differentiates between two dimensions: time on the one hand and the possibility of multi-actor presence on the other hand. The model suggests three phases during the development of a new service. This process of development starts with a technology acquaintance phase, followed by a cooperation phase, and ended by the implementation phase of the service. When looking at the related actors, the suggested model distinguishes between healthcare consumers and healthcare professionals. While they are identified as separate actors during the first phase, they are considered as closely interrelated during the latter two phases. Further research still has to fill in the different phases with the relevant constructs, but this holistic model should function as a new start in explaining the acceptance of technology in the homecare area more accurately.

REFERENCES

Aarts, E. (2004). Ambient intelligence: A multimedia perspective. *IEEE Computer Society/Multimedia, 11(1)*, 12-19.

Andriessen, J. H. E. (2003). *Working with groupware: Understanding and evaluating collaboration.* London: Springer.

Attewell, P. (1992). Technology diffusion and organizational learning: The case of business computing. *Organizational Science, 3*(1), 1-19.

Bangert, D., & Doktor, R. (2000). Implementing store-and-forward telemedicine: Organizational issues. *Telemedicine Journal and e-Health, 6*(3), 355-360.

Briggs, R., Adkins, M., Mittleman, D., Kruse, J., Miller, S., & Nunamker, J.F., Jr. (1998). A technology transition model derived from field investigation of GSS use aboard the USS Coronado. *Journal of Management Information Systems, 15*(3), 151-195.

Briggs, R. O., Vreede, G. J. D., & Nunamaker, J. F. (2003). Collaboration engineering with ThinkLets to pursue sustained success with group support systems. *Journal of Management Information Systems, 19*(4), 31-64.

Chau, P. Y. K., & Hu, P. J.-H. (2002). Investigating healthcare professionals' decisions to accept telemedicine technology: An empirical test of competing theories. *Information & Management, 39*(4), 297-311.

Chismar, W. G., & Wiley-Patton, S. (2002). *Does the extended technology acceptance model apply to physicians.* Paper presented at the 36th Hawaii International Conference on System Sciences, Hawaii.

Davis, F. D. (1989). Perceived usefulness, perceived ease of use and user acceptance of information technology. *MIS Quarterly, 13*, 319-340.

Davis, F. D. (1993). User acceptance of information technology: System characteristics, user perceptions and behavioral impacts. *International Journal of Man-Machine Studies, 38*(3), 475-487.

Doktor, R., Bangert, D., & Valdez, M. (2005). *Organizational learning and culture in the managerial implementation of clinical e-health systems: An international perspective.* Paper presented at the 38th Annual Hawaii International Conference on System Sciences.

Grumbach, C., Heikelä, M., Skilton, T., Okkonen, A., Kaukipuro, K., & Huisman, R. (2002). *Independent living market in Germany, UK, Italy, Belgium and Netherlands* (No. 133/2002). Helsinki: TEKES-National Technology Agency of Finland.

Hu, P. J.-H., Chau, P. Y. K., & Sheng, O. R. L. (2002). Adoption of telemedicine technology by health care organizations: An exploratory study. *Journal of Organizational Computing and Electronic Commerce, 12*(3), 197-221.

Hu, P. J.-H., Chau, P. Y. K., Sheng, O. R. L., & Tam, K. Y. (1999). Examining the technology acceptance model using physician. *Journal of Management Information Systems, 16*(2), 91-112.

Iacovou, C. L., Benbasat, I., & Dexter, A. S. (1995). Electronic data interchange and small organizations: Adoption and impact of technology. *MIS Quarterly, 19*(4), 465-485.

Jayasuriya, R. (1998). Determinants of microcomputer technology use: Implications for education and training of health staff. *International Journal of Medical Informatics, 50*(1/3), 187-194.

Meyer, A., & Goes, J. B. (1988). Organizational assimilation of innovations: A multi level contextual analysis. *Academy of Management Journal, 31*(4), 899-923.

Raitoharju, R. (2005). *When acceptance is not enough: Taking TAM-model into healthcare.* Paper presented at the 38th Annual Hawaii International Conference on System Sciences, Hawaii.

Rogers, E. M. (2003). *Diffusion of innovations* (5th ed.). New York: Free Press.

Succi, M. J., & Walter, Z. D. (1999). *Theory of user acceptance of information technologies: An examination of health care professionals.* Paper presented at the 32nd Annual Hawaii International Conference on System Sciences.

Tanriverdi, H., & Iacono, C. S. (1998). Knowledge barriers to diffusion of telemedicine. In *Proceedings of the International Conference on Information Systems* (pp. 39-50). Helsinki, Finland: Association for Information Systems.

Tanriverdi, H., & Iacono, C. S. P. (1999). Diffusion of telemedicine: A knowledge barrier perspective. *Telemedicine Journal, 5*(3), 223-244.

Tornatzky, L. G., & Fleischer, M. (1990). *The process of technological innovation.* Lexington, MA: Lexington Books.

Venkatesh, V., & Davis, F. D. (2000). A theoretical extension of the technology acceptance model: Four longitudinal field studies. *Management Science, 46*(2), 186-204.

Wu, J.-H., Wang, S.-C., & Lin, L.-M. (2005). *What drives mobile health care? An empirical evaluation of technology acceptance.* Paper presented at the 38th Annual Hawaii International Conference on System Sciences, Hawaii.

KEY TERMS

E-Health: The (remote) monitoring and management of patients as well as the use of systems that provide access to expert advice and patient information by care professionals. On the one hand, e-health focuses on the provision of care between care consumers and professionals at a distance (telemedicine). On the other hand, e-health can comprise ICT-based services between healthcare professionals in order to maximize efficiency (health information).

Health Information: The use of information technology in order to maximize the efficiency of healthcare provision through increasing the accessibility of healthcare related information. Health information is not especially oriented on the provision of healthcare between care consumers and professionals.

Technology Acceptance: The decision of possible users whether to use a specific technology or service.

Technology Adaptation: The fact that the originally intended use of a technology or service is sometimes changed or adapted by people or organizations in practice.

Technology Appropriation: After being introduced to the end-user, the intended use and meaning of a technology is sometimes altered by the user. Even if carefully introduced, the original situation is sometimes adapted. With regard to adaptation, appropriation has an accent on the fact that the user adjusts the technology for his own best practice, while adaptation relates to the fact that the use sometimes changes in general.

Technology Diffusion: The process that describes the takeover of a certain technology by other actors or organizations other than the organization that originally started using the technology or service. Such actors or organizations are usually related to the original actor.

Telemedicine: The provision of healthcare between healthcare recipients and providers in an innovative way supported by ICT. The focus is hereby on the healthcare provision itself instead of on the efficiency-oriented IT systems and services (health information).

Facilitation of Technology–Supported Communities of Practice

Halbana Tarmizi
University of Nebraska at Omaha, USA

Gert-Jan de Vreede
University of Nebraska at Omaha, USA
Delft University of Technology, The Netherlands

INTRODUCTION

Communities of practice (CoP) has gained on prominence since it emerged as a concept in early 1990s, introduced by Lave and Wenger (1991) as situated learning. They argue that knowledge is acquired through active participation in a community as a new member moves from peripheral to full participation in the community. Since then, the CoP concept has evolved (Kimble & Hildreth, 2004), as Wenger (2004) defined CoP as "groups of people who share a passion for something that they know how to do, and who interact regularly in order to learn how to do it better" (p. 2).

While the original concept of CoP is based on face-to-face interaction (e.g., in traditional apprenticeship) (Lave & Wenger, 1991), the advancement in information and communication technology (ICT) has widened this concept as space and time barriers collapsed. Establishing and sustaining CoPs then becomes the key challenge for those individuals involved in CoPs development and operation (e.g., members and sponsors). The introduction of a facilitator role is believed to help CoPs in overcoming some of the challenges and in achieving their goals. However, what roles and tasks a facilitator would face in a CoP is an important question to be addressed. In this article, we identify those potential roles and tasks that a CoP facilitator can face in a community.

THE NATURE OF COMMUNITIES OF PRACTICE

CoPs can serve several purposes within organizations, including providing a forum for sharing ideas, solving problems, disseminating best practices, and organizing knowledge (Wenger, McDermott, & Snyder, 2002).

A defining feature of a CoP is that they, more or less, emerge spontaneously ("bottom-up") from the informal networking among groups of individuals who share similar interests or passions (Lave & Wenger, 1991). In recent years, however, CoPs are increasingly initiated by a sponsor in senior management level ("top-down") (Fontaine, 2001) and get full organizational support in terms of human resources, policy, and technology. The price drop in information and communication technology, combined with the advancement of e-collaboration technologies, and the Internet has resulted in an increasing number of CoPs going virtual with minimal or no face-to-face interaction. This in turn creates new challenges in building and sustaining CoPs. Virtual environments increase interaction barriers. These barriers hinder effective coordination and communication (Jarvenpaa, Knoll, & Leidner, 1998), make people less attentive and less receptive to contextual cues (Sproull & Kiesler, 1986), and hinder the creation of trust (van House, Butler, & Schiff, 1998).

As a community, a CoP has a different nature than a team. Ferran-Urdaneta (1999) highlighted some of those characteristic differences (see Table 1).

These characteristics, especially the large number of members, less clear goals, and low levels of coordination; increase the challenges in sustaining a CoP. Large numbers of members could become a threat when members start forming splinter cells, which could lead to the break down of the CoP. A large number of members could also endanger the sense of identity and members' commitment toward the community (Wenger et al., 2002). Low coordination and unclear goals within a community could effect members' "sense of community" feeling, which in turn has direct relation to members' participation (Yoo, Suh, & Lee, 2002). Low member interdependency, i.e. no member is crucial for community survival, could be seen as an

Table 1. Characteristic differences between community and team

CHARACTERISTICS	COMMUNITY	TEAM
Member	Large	Small and identifiable
Coordination	Low	High
Member interdependency	Low	High
Lifespan	Infinite	Finite / short
Goal	less clear	Clear and measurable

advantage. However, this low member interdependency could increase the effect of social loafing (Steiner, 1972); that is, reduction in individual effort that often occurs in groups. This effort reduction could stem from member's belief that community could survive without his or her contribution (Comer 1995). If this belief is adopted by most of the members, then the survival of the community would be questionable.

A CoP may have synchronous and asynchronous interactions among its members, utilizing various available e-collaboration technologies, such as bulletin board, email, online chat, phone, or video conference. Those technologies could facilitate communication and overcome time and/or distance barriers inherent in distributed CoP. Johnson (2001) argues that establishing discussion should be one of the main functions of a community because it is a means to expand knowledge (Bielaczyc & Collins, 1999). Expanding the knowledge of employees, which in turn expands the knowledge of organization, is the ultimate goal of the organization in establishing a CoP.

However, a community's characteristics, as described earlier, create a challenge for both sustaining a CoP over time and establishing a successful fully participatory discussion. The key issue for sustaining a CoP is maintaining the participation of its members. This is also one of the indicators as to whether the approach used in developing a community is effective (Gongla & Rizzuto, 2001). The greater the number of participating members, i.e., members that are enthusiastic participants in activities and topics created by others, the higher the chance that the community will thrive. Hildreth et al. argue: "participation is central to the evolution of the community and to the creation of relationships that help develop the sense of trust and identity that defines the community" (2000, p. 30). One

of the supporting CoP roles, that has the potential to play crucial role in reaching this goal, is the facilitator role. A CoP facilitator could help the community and its members navigate through existing obstacles (Fontaine, 2001) and keep the community flourishing. The importance of facilitator in helping groups or meetings to move forward and meet their goals more effectively has been researched and proven, especially in group support systems (GSS) field. Extensive research has been done and published (e.g., Anson et al., 1995; Dickson et al. 1996; Hayne, 1999).

CoP AND GSS FACILITATION

Despite the potential of facilitation as a way to overcome some of the challenges in a CoP, little is known about facilitator tasks within CoP. Hardly any research addressing facilitation issues in CoP has been published (Tarmizi & de Vreede, 2005). Therefore, learning from other fields, i.e., the GSS field, could enhance our understanding regarding the facilitator role in CoPs. In order to apply those findings into a CoP, we have to consider existing processes within CoPs and understand differences between processes used in CoPs and GSS meetings. By understanding those differences, we are in position to develop a base for research into CoPs facilitation.

Facilitation used in GSS meetings help structure the process of a meeting, so that participants are able to reach optimal outcomes in an efficient manner. A traditional GSS session (i.e., a face-to-face interaction), usually takes relatively shorter time than CoP interaction, since in CoP communication is mostly asynchronous and the content is fragmented, so that members have to devote more concentration, time and energy to organize things in a meaningful way (McFadzean & McKenzie, 2001). This longer period of interaction in a CoP is accompanied by arrival of new members and departure of existing members, in contrast to a GSS meeting where traditionally the participants stay through the entire meeting. A CoP facilitator would be involved in not only in structuring the process within the community, but s/he also needs to deal with departure of key members and arrival of new members. Additionally, since most of the time a CoP is embedded in an organization, we believe a CoP facilitator would also have to deal with other entities within its organization, such as management and other communities. The facilitator needs to be aware of an organization's policies, management, and strategic goals. To bring maximal benefit to an organization, a CoP should match to organizational processes (Gongla & Rizzuto, 2001). Furthermore, a community should engage in cooperation with other communities for the sake of organization's goals. For instance, a facilitator could help in facilitating these border crossing activities and could also offer advice to struggling CoPs. Border crossing activities are defined as all activities pertaining to community and involving community members and entities outside the community. Therefore, we believe that facilitation tasks in CoP would be broader and have a longer duration than those used in GSS meetings.

In the recent decade, we have also seen more research in GSS field extended to distributed setting (e.g., Anson & Munkvold, 2004; Mittleman et al., 2000; Niederman et al., 1993). The qualitative study by Niederman et al. indicated several concerns raised by facilitators regarding GSS in distributed setting. Those concerns include: restricted facilitator capability, controlling and coordinating multiple sites, retaining task focus and technology reliability. Those concerns reflect the more complex nature of distributed setting. Mittleman et al. (2000) listed several lessons learned and best practices to address them. Those lessons learned include: (1) more challenge for participants to follow the meeting; (2) participants don't get feedback; (3) participants aren't aware who is on the meeting; (4) technology is unreliable; and (5) it's difficult to converge. Anson and Munkvold (2004) studied four different types of meeting based on same/different time and place. They found that practices in face-to-face meeting can be extended to distributed environment. However, distributed meeting modes impose new challenges regarding facilitation and participants' participation and engagement. All of

those studies are supporting our earlier argument that facilitation in distributed setting, such as in CoP, is more complex and challenging than face-to-face facilitation, such as in traditional GSS session. Still, none of those studies really go into details in describing facilitator's tasks in distributed environment. Therefore, our CoP facilitation study could also help in advancing research in GSS field, since in practice there are some similarities in facilitating CoP and distributed groups.

FACILITATION TASKS AND COP

A good starting point for developing CoP facilitation tasks is a model proposed by Clawson and Bostrom (1996). It can be considered the most specific and comprehensive, covering the facilitation tasks proposed by other researchers (Vreede et al., 2002). This model listed 16 facilitator tasks based on a study among experienced facilitators. However, we argue that two of those tasks; that is, (1) demonstrate flexibility and (2) demonstrate self-awareness and self-expression, are more about the facilitator's characteristics than tasks. In understanding the processes within CoPs, we consider the case-study based work about CoPs within IBM Global Services by Gongla and Rizzuto (2001) as the current most comprehensive body of knowledge. Through mapping Clawson and Bostrom's (1996) model to the identified CoPs processes, we end up with 19 additional tasks that are not listed in GSS model. Combined the with 16 previously identified tasks, there are 33 tasks that we believe are crucial for facilitation in CoP. The more detailed explanation of this extrapolation process can be found in Tarmizi and de Vreede (2005). The following table shows those tasks, classified into several categories (see Figure 1).

The CoP facilitation task taxonomy presented above shows several roles that a facilitator potentially has to fulfill as part of his or her duties. Some of those roles are external to the community, while others are internal to the community. Those external roles are required since CoP is embedded within an organization. The role as *external information source* is important in order to connect his or her CoP with other organizational CoPs and to keep in touch with management and key decision makers. Through this role, a facilitator of a successful CoP would also have the opportunity to help in building new CoPs. The nature of a CoP, compared to a GSS meeting, requires a facilitator to interact more

Figure 1. The CoP facilitation task taxonomy[1]

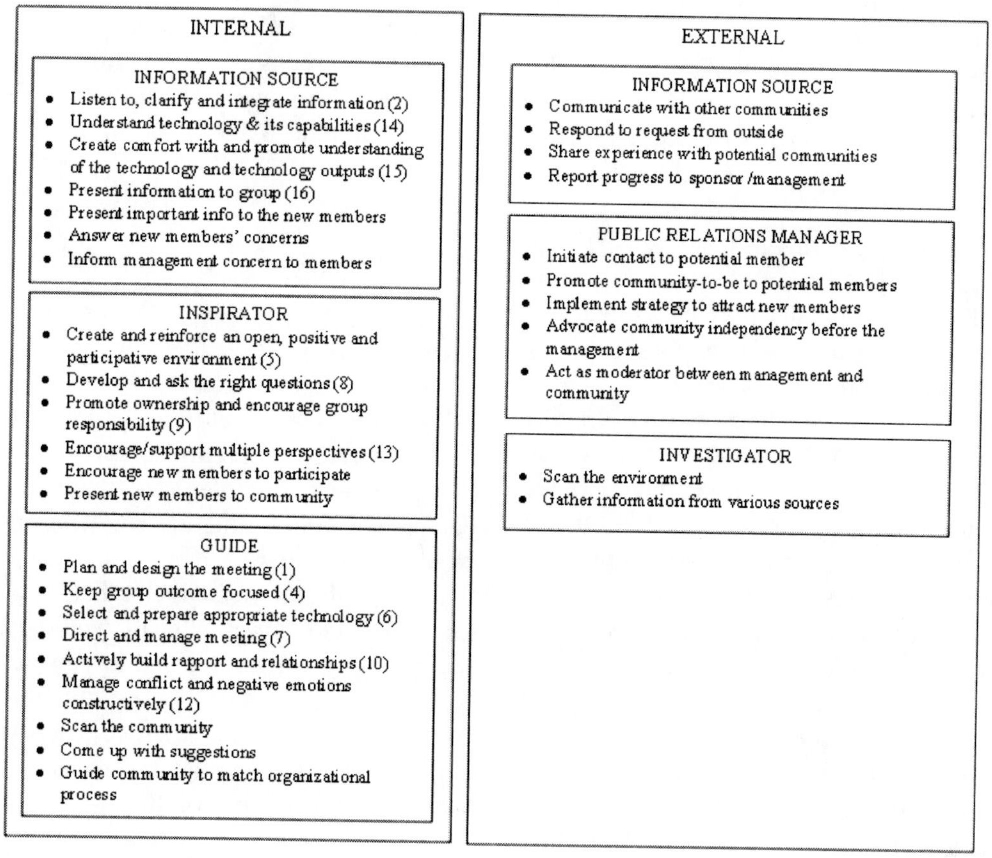

```
┌─────────────────────────────────────┐   ┌─────────────────────────────────────┐
│              INTERNAL               │   │              EXTERNAL               │
│                                     │   │                                     │
│  ┌───────────────────────────────┐  │   │  ┌───────────────────────────────┐  │
│  │      INFORMATION SOURCE       │  │   │  │      INFORMATION SOURCE       │  │
│  │ • Listen to, clarify and      │  │   │  │ • Communicate with other      │  │
│  │   integrate information (2)    │  │   │  │   communities                 │  │
│  │ • Understand technology & its │  │   │  │ • Respond to request from     │  │
│  │   capabilities (14)            │  │   │  │   outside                     │  │
│  │ • Create comfort with and      │  │   │  │ • Share experience with       │  │
│  │   promote understanding        │  │   │  │   potential communities       │  │
│  │   of the technology and        │  │   │  │ • Report progress to sponsor/ │  │
│  │   technology outputs (15)      │  │   │  │   management                  │  │
│  │ • Present information to        │  │   │  └───────────────────────────────┘  │
│  │   group (16)                   │  │   │                                     │
│  │ • Present important info to     │  │   │  ┌───────────────────────────────┐  │
│  │   the new members              │  │   │  │    PUBLIC RELATIONS MANAGER   │  │
│  │ • Answer new members' concerns │  │   │  │ • Initiate contact to         │  │
│  │ • Inform management concern     │  │   │  │   potential member            │  │
│  │   to members                   │  │   │  │ • Promote community-to-be to  │  │
│  └───────────────────────────────┘  │   │  │   potential members           │  │
│                                     │   │  │ • Implement strategy to       │  │
│  ┌───────────────────────────────┐  │   │  │   attract new members         │  │
│  │           INSPIRATOR          │  │   │  │ • Advocate community          │  │
│  │ • Create and reinforce an      │  │   │  │   independency before the     │  │
│  │   open, positive and           │  │   │  │   management                  │  │
│  │   participative environment (5)│  │   │  │ • Act as moderator between    │  │
│  │ • Develop and ask the right     │  │   │  │   management and              │  │
│  │   questions (8)                │  │   │  │   community                   │  │
│  │ • Promote ownership and         │  │   │  └───────────────────────────────┘  │
│  │   encourage group             │  │   │                                     │
│  │   responsibility (9)            │  │   │  ┌───────────────────────────────┐  │
│  │ • Encourage/support multiple    │  │   │  │         INVESTIGATOR          │  │
│  │   perspectives (13)            │  │   │  │ • Scan the environment        │  │
│  │ • Encourage new members to      │  │   │  │ • Gather information from     │  │
│  │   participate                  │  │   │  │   various sources             │  │
│  │ • Present new members to        │  │   │  └───────────────────────────────┘  │
│  │   community                    │  │   │                                     │
│  └───────────────────────────────┘  │   └─────────────────────────────────────┘
│                                     │
│  ┌───────────────────────────────┐  │
│  │             GUIDE             │  │
│  │ • Plan and design the          │  │
│  │   meeting (1)                  │  │
│  │ • Keep group outcome focused(4)│  │
│  │ • Select and prepare            │  │
│  │   appropriate technology (6)   │  │
│  │ • Direct and manage            │  │
│  │   meeting (7)                  │  │
│  │ • Actively build rapport and    │  │
│  │   relationships (10)           │  │
│  │ • Manage conflict and negative  │  │
│  │   emotions constructively (12) │  │
│  │ • Scan the community           │  │
│  │ • Come up with suggestions     │  │
│  │ • Guide community to match      │  │
│  │   organizational process       │  │
│  └───────────────────────────────┘  │
└─────────────────────────────────────┘
```

with external entities. This interaction would shape a facilitator into the role of *public relations manager*. In this role, a facilitator could engage in promoting his or her CoP to potential new members in order to compensate the attrition among the existing CoP's members. A departure of a key member (e.g., the one with extensive knowledge in certain domain) would leave a hole in a CoP. A facilitator should assume his or her role as public relations manager to find, contact, and attract potential new members that could fill the gap caused by the departure of the key member. The public relations manager role also includes tasks of promoting the CoP to management and being a moderator between management and the CoP. Since the facilitator also needs to be aware of what is going on outside his or her CoP, s/he should assume the *external investigator* role. This role will make sure that his or her CoP is always informed and on cutting edge. In this role, s/he will actively scan the environment outside

of his or her CoP and gather information from various sources that are important for his or her CoP.

Those outside oriented roles are only a small part of a CoP facilitator's duties. Most of the time, a facilitator has to assume roles internal to the CoP. First and foremost, a facilitator should provide guidance for the community. This role as a *guide* is important due to the characteristics of a CoP (e.g., less clear goal, large number of members and low coordination). Through this role, a facilitator could provide some structures to members' interactions in term of planning, designing, directing, and managing. Matching a CoP to organizational process in order to bring maximal benefit to organization could also fall into this role. A facilitator should also present information to the CoP. This role of *internal information source* would include answering members' questions or concerns, listening and clarifying members view points or information, presenting important information to new members, and helping members with the supporting technology. Last but not least, there is a role as *inspirator*. A facilitator should inspire members to actively participate in the community, since participation level is key issue for sustaining a CoP and is also one of the indicators of whether the approach used in developing a community is the right one or not (Gongla & Rizzuto, 2001). In this role, a facilitator should create an environment conducive to members' participation. This can be achieved, for example, by establishing norms geared toward encouraging contribution from any member in the community. Then the facilitator can help in enforcing those norms effectively, for example by cautioning those who try to disparage contributions from other members. Asking the right questions (Wang, 2005) and encouraging responsibility and multiple perspectives (McFadzean & McKenzie, 2001) are ways for triggering more active participation among members. To achieve this, a facilitator can use open questions, comparison questions or probe and synthesis questions. A good facilitator will use the right question at the right moment. As Wang (2005) in her study of online discussion found that (1) open questions promote participation; (2) comparison questions provoke the intellectual moves by the audience; and (3) probe and synthesis questions facilitate the process of constructing knowledge. By using comparison questions, a facilitator can also encourage members to view an issue from different perspective, triggering multiple perspectives discussion. For example, a facilitator can ask a comparison question that engages members from different backgrounds cognitively: 'I'm wondering if our engineers have the same perspective as we do?' This question will encourage members with engineering background to voice their views and help the community in benefiting from multiple perspectives. As an inspirator, a facilitator should also encourage new members to participate after introducing them to the community.

COPs involving geographically dispersed participants cannot exist without technology, and there are different forms of such technology. Technology should help facilitators in performing their duties in a COP, including improving members' participation. Technology can play an important role in helping facilitators to fulfill some of their tasks. While there are a number of potential e-collaboration technologies that can be used in COPs, several features are thought to be useful for helping facilitators in fulfilling their tasks. Tarmizi et al. (2007) listed some of those features. Two examples include notification alert (Millen & Patterson, 2002) and a contribution-based reputation system (Kelly et al., 2002). Those two features can help in encouraging new members to participate. A new community member can build his reputation through active participation and he becomes aware about the ongoing activities in the community. All of these in turn will help community to thrive.

CONCLUSION

In this article, we have identified facilitation tasks in communities of practice. Those tasks lead to several roles that a CoP facilitator can assume within the community. All these different roles for a facilitator in a CoP reflect the more complex nature of CoPs facilitation compared to GSS facilitation. Knowing this complexity would help in training and preparing facilitators for their duty in CoP. Having a facilitator that is well trained and knows what to expect in a CoP would definitely help the community to evolve and to prosper and in turn, would help the organization benefit from the CoP. A facilitator with in-depth expertise in the subject domain would bring additional advantage to the community, since it will help him to fulfill some of the identified tasks (e.g., developing and asking the right questions or in promoting community to potential members). For organizations, knowing CoP facilitation tasks would be helpful for evaluating facilitator's performance. This in turn will help organizations in finding the right persons to facilitate their CoPs.

REFERENCES

Anson, R., Bostrom, R., & Wynne, B. (1995). An experiment assessing GSS and facilitator effects on meeting outcomes, *Management Science, 41*(2), 189-208.

Anson, R., & Munkvold, B. E. (2004). Beyon face-to-face: a field study of electronic meetings in different time and place modes. *Journal of Organizational Computing and electronic Commerce, 14*(2), 127-152.

Bielaczyc, K., & Collins, A. (1999). Learning communities in classrooms: A reconceptualization of educational practice. In C. Reigeluth (Ed.), *Instructional-design theories and models: A new paradigm of instructional theory* (Vol. 2, pp. 269-292). Mahwah, NJ: Erlbaum .

Clawson, V. K., & Bostrom, R. P. (1996). Research driven facilitation training for computer supported environments. *Group Decision and Negotiation, 1,* 7-29.

Commer, D. R. (1995). A model of social loafing in real work groups. *Human Relations, 48*(6), 647-667.

Dickson, G., Limayem, M., Lee Partridge J., & De-Sanctis, G. (1996). Facilitating computer supported meetings: A cumulative analysis in a multiple criteria task environment. *Group Decision and Negotiation, 5*(1), 51-72.

Ferrán-Urdaneta, C. (1999). Teams or communities? Organizational structures for knowledge management. In *Proceeding of SIGCPR'99* (128-134). New Orleans, LA.

Fontaine, M. (2001). Keeping communities of practice afloat: Understanding and fostering roles in communities. *Knowledge Management Review, 4*(4), 16-21.

Gongla, P., & Rizzuto, C. R. (2001). Evolving communities of practice: IBM Global Services experience. *IBM Systems Journal, 40*(4), 842-862.

Hayne, S. C. (1999). The facilitator's perspective on meetings and implications for group support systems design. *The DATA BASE for Advances in Information Systems, 30*(3/4), 72-91.

Hildreth, P., Kimble, C., & Wright, P. (2000). Communities of practice in the distributed international environment. *Journal of Knowledge Management, 4*(1), 27-38.

Jarvenpaa, F., Knoll, K., & Leidner, D. (1998). Is anybody out there? Antecedents of trust in global virtual teams. *Journal of Management Information Systems, 14,* 29-64.

Johnson, C. M. (2001). A survey of current research on online communities of practice. *Internet and Higher Education, 4,* 45-60.

Kelly, S. U., Sung, C., & Farnham, S. (2002). Designing for improved social responsibility, user participation and content in on-line communities. In D. Wixon (Ed), *Proceedings of the SIGCHI Conference on Human Factors in Computing Systems* (pp. 391-398). New York: ACM Press.

Kimble, C., & Hildreth, P. (2004, May 26-28). *Communities of practice: Going one step too far?* Paper presented at Ninth AIM Conference, Evry, France.

Lave, J., & Wenger, E. (1991). *Situated learning: Legitimate peripheral participation.* Cambridge, UK: Cambridge University Press.

McFadzean, E., & McKenzie, J. (2001). Facilitating virtual learning groups: A practical approach. *Journal of Management Development, 20*(6), 470-494.

Millen, D. R., & Patterson, J. F. (2002). Stimulating social engagement in a community network. In *Proceedings of ACM 2002 Conference on Computer Supported Cooperative Work* (pp. 306-313). New York: ACM Press.

Mittleman, D. D., Briggs, R. O., & Nunamaker, J. F., Jr. (2000). Best practices in facilitating virtual meeting: Some notes from initial experience. *Group Facilitation, 2*(2), 5-14.

Niederman, F., Beise, C. M., & Beranek, P. M. (1993). Facilitation issues in distributed group support systems. In *Proceedings of the 1993 ACM SIGCPR Conference* (pp. 299-312).

Sproull, L., & Kiesler, S. (1986). Reducing social cues: Electronic mail in organizational communication. *Management Science, 32*(11), 1492-1512.

Steiner, I. D. (1972). *Group process and productivity.* New York: Academic Press.

Tarmizi, H., & Vreede, G.-J. (2005, August 11-14). A facilitation task taxonomy for communities of practice. In *Proceedings of the 11ᵗʰ Americas Conference of Information Systems,* (pp. 3545-3554), Omaha, Nebraska.

Tarmizi, H., Vreede, G.-J., & Zigurs, I. (2007). Leadership challenges in communities of practice: Supporting facilitators via design and technology. *International Journal of e-Collaboration, 3*(1), 18-39.

Van House, N. A., Butler, M. H., & Schiff, L. R. (1998). Cooperative knowledge work and practices of trust: Sharing environmental planning data sets. In S. Poltrock & J. Grudin (Eds.), *CSCW'98: Proceedings of the ACM Conference on Computer Supported Cooperative Work* (pp. 335-343). Seattle, WA: ACM Press.

Vreede, G. J. De, Boonstra, J. A., & Niederman, F. (2002). What is effective GSS facilitation? A qualitative inquiry into participants' perceptions. In *Proceedings of the 35th Hawaiian International Conference on System Sciences,* (pp.616-627). Los Alamitos, CA: IEEE Computer Society Press.

Wang, C. H. (2005). Questioning skills facilitate online synchronous discussions. *Journal of Computer Assisted Learning, 21*(4), 303-313.

Wenger, E. (2004, January-February). Knowledge management as a doughnut: Shaping your knowledge strategy through communities of practice. *Ivey Business Journal*, 1-8.

Wenger E., McDermott, R., & Snyder, W. M. (2002). *Cultivating communities of practice*. Boston: Harvard Business School Press.

Yoo, W.-S., Suh, K.-S., & Lee, M.-B. (2002). Exploring the factors enhancing member participation in virtual communities. *Journal of Global Information Management, 10*(3), 55-71.

KEY TERMS

Border Crossing: Activities pertaining to the community involving individuals outside the community.

Communities of Practice: Groups of people who share a passion for something that they know how to do, and who interact regularly in order to learn how to do it better.

E-Collaboration: Collaboration among individuals engaged in a common task using electronic technologies.

Facilitation: Activities carried out to help groups achieve its own outcomes

Facilitator Role: Different roles that a facilitator can take in performing his or her tasks in the community.

Group Support Systems: Integrated computer based systems that facilitate the solution of semi-structured or unstructured group problems.

Knowledge Management: A process through which organizations create, store, and utilize their collective knowledge.

Factors for Effective E-Collaboration in the Supply Chain

Sharon A. Cox
Birmingham City University, UK

John S. Perkins
Newman College of Higher Education, UK

INTRODUCTION

The exchange of information is recognised as a major enabler of effective supply chain management. The potential business benefits of using technology to improve communication between trading partners are dependent upon two main factors. Firstly, the information sent must be timely and accurate as this determines the speed and efficiency with which customer orders are fulfilled (Gattorna & Walters, 1996). Secondly, business processes must be in place to respond efficiently to the information received.

This article examines the transition from cooperation to collaboration in the supply chain. The manufacturer-retail supply chain is analysed from three perspectives: strategic policy, process integration and community practice. Critical success factors for implementing effective e-collaboration in the supply chain are defined from the authors experience of action research (Perkins & Dingley, 2001; Cox, Krasniewicz, Perkins, & Cox, 2006). The article concludes by considering the future trends and challenges of e-collaboration in the supply chain.

BACKGROUND

A supply chain is defined as two or more organisations working together to create a competitive advantage through the sharing of information, making joint decisions, and sharing benefits (Simatupang & Sridharan, 2005). The supply chain incorporates functions of purchasing, order fulfilment and inventory but supply chain management extends beyond these functions (Croom, 2005). Supply chain management focuses on the external links, feedback linkages and collective learning between organisations (Saad & Patel, 2006). It is characterized by control across functional,

organisational and geographical boundaries (van Hoek, 1998).

The supply chain focuses on the co-ordination of supplying goods and services in a reliable and timely manner. Organisations in the supply chain cooperate for the mutual benefit of the chain (Sahay, 2003). Benefits of supply chain cooperation may include improvements in customer service (Tan, 2001); understanding of future product demand (Sahay, 2003); reduced transaction costs at the customer-supplier interface (Burnes & New, 1998); reduced time to market (Graham & Hardaker, 2000); revenue enhancements, operational flexibility (Simatupang & Sridharan, 2005); and efficiency improvements (Dingley & Perkins, 1999).

Interorganisational cooperation required for supply chain management can take many forms with different degrees of cooperation and commitment. As supply chains evolve, product and process innovation is needed in order to remain competitive (Cassivi, 2006). This requires a move from supply chain cooperation to supply chain collaboration.

Supply chain cooperation requires the co-ordination and alignment of activities of two or more separate organisations in the supply chain. Supply chain collaboration goes beyond cooperation; it involves the collective contribution of organisations to a deeper committed relationship through the integration and embedding of activities, resources and processes to form a jointly-owned emerging system. Collaboration enables organisations to provide greater value to customers than if the organisations acted alone (Simatupang & Sridharan, 2002) and enables competitive advantage to be gained over rival value chains (Archer & Yuan, 2000).

Internet technologies are the major enabler of improvements in supply chain management (Kirchmer, 2004). Dingley and Perkins (1999) propose a distinction between *forging* links and *tempering* links in the supply chain. Forging links refers to the establishment of co-operative partnerships; tempering links secures and strengthens the relationship making it more difficult for the collaboration to breakdown as a result of brittleness in the join between partners. E-collaborative systems temper supply chain relationships.

E-collaborative systems can improve relationships between manufacturers and retailers in the supply chain by distributing product information, automating order entry, tracking order status and attaining customer feedback (Bhatt & Emdad, 2001). Integrating business processes through collaboration improves the quality of customer service by improving corporate decision making (Agarwal & Shankar, 2003) which increases the speed of service delivery, improving responsiveness to the market demand and reducing time to market (Favilla & Fearne, 2005). Internet technology enables data to be shared between partners in the supply chain, creating a virtual supply chain, which is information based rather than inventory based. However, shared information can only be leveraged through collaboration, integrating processes and using common systems (Barratt, 2004). E-collaboration is not just about developing mechanisms with which to exchange information at the operational level, but also needs to implemented at the tactical and strategic levels in organisations across the supply chain (Barratt, 2004).

IMPLEMENTING E-COLLABORATION IN THE SUPPLY CHAIN

E-collaboration changes trading relationships and requires business models to change to enable and facilitate the collaboration. Pateli and Giaglis (2005) suggest that the transformation of new business models initiated by technological change should be modelled within seven categories. The authors (Cox et al., 2006) conducted action research in a manufacturer-retailer supply chain. Previously, major UK supermarket chains used electronic data interchange (EDI) to exchange information enabling supply chain cooperation with manufacturers. E-collaborative systems were developed to initiate supply chain collaboration. The systems provided process integration at the manufacturer-retailer interface (order processing and promotion management) that became embedded within internal value chains. From this experience, the categories proposed by Pateli and Giaglis (2005) can be examined from three perspectives.

The first perspective is the *strategic* view of collaboration which defines the rationale for engaging in and committing to the collaborative partnership (incorporating Pateli & Giaglis, 2005, categories of business objectives, core competencies, market scope, critical success factors). Secondly, the *integration* view of collaboration considers what form the collaboration takes and the value that emerges from the collaboration (relationship model and value exchange of Pateli & Giaglis, 2005). Finally, the *community* perspective of collaboration examines communities of practice (involving actors, roles and responsibilities of Pateli & Giaglis, 2005).

The three perspectives emphasise that e-collaboration in the supply chain, extends beyond the operational integration of IT systems and encourages organisations to consider the wider issues of business transformation, whilst retaining a focus on business process and practice. These perspectives are discussed in the following sections.

Strategic Policy Perspective

At the strategic level, the fundamental decisions for an organisation relate to issues of with whom to collaborate, why and how (Barratt, 2004). The high investment needed to develop and sustain e-collaboration means that organisations cannot collaborate with all their partners in the supply chain and one supply chain strategy cannot meet the different needs of each chain within which an organisation participates (Barratt, 2004). Opportunities and challenges arise from both participating and not participating in a collaborative relationship. For example, fiercer price negotiations can be fuelled by the imbalance of power within a collaborative relationship (Perkins & Dingley, 2001).

Collaborating organisations need to share common goals in terms of, for example, service, quality, technological innovation, time to market and cost drivers (Hughes, Ralf, & Michels, 1998). Organisations also need to understand their own role in the collaboration (Hibbert & Huxham, 2005) and supply chain transformation should be aligned with the strategic direction of the organisation (Favilla & Fearne, 2005). Individual collaborating organisations may have different expectations of the distribution of benefits to which the relationship may lead. These expectations need to be understood so partners can work together to achieve the benefits that they mutually identify (Hughes et al.,

1998). Agreement of the resources to be committed by each party is needed to enable the realisation of expectations, which will then contribute to creating a sense of reliability, promoting trust between collaborating organisations (Dawson, 2000). In the authors' experience, the pressure to collaborate can be such that insufficient time is given to agreeing the expectations of the collaboration, resulting in imbalance of benefits and lack of trust.

At the strategic policy level, critical success factors include:

1. Decide who to collaborate with
2. Develop a different strategy for each supply chain collaboration
3. Define clear expectations of the collaboration
4. Identify strategic needs of both parties
5. Agree shared benefits from the collaboration
6. Assess pressures to engage in the collaboration
7. Recognise the importance of the collaborative relationship and secure leadership and commitment from senior management
8. Maintain the balance of power in the collaboration
9. Communicate collaborative strategy and publicise early success
10. Annually evaluate and negotiate processes
11. Ensure policies are in place to seize opportunities arising from the collaboration
12. Seek to institutionalise the collaboration

Process Integration Perspective

Supply chain collaboration can differ in terms of the depth of system coupling, the degree of information shared and the range of processes supported (Caglliano et al., 2003). Internet technology provides the communication interface between partners but supply chain collaboration cannot be achieved through IT investment alone; it requires processes to be re-engineered to use and act upon the inter-company information communicated.

The introduction of an e-collaboration system required the manufacturer to change finance, sales, purchasing, production planning, distribution forecasting, procurement and production systems to be effectively integrated with the system. Changing these systems involved changing the IT, IS, business processes and working practices of the organisation to support e-collaboration in the supply chain.

In a collaborative relationship between a manufacturer and independent retailer (Perkins & Dingley, 2001), the impact on the retailer's processes was such that the retailer withdrew from the collaboration. The retailer demonstrated their greater power in the relationship by not using the e-collaboration system installed by the manufacturer as it was based on espoused requirements of business processes rather than collaborative practice.

At the process integration level, it is recommended that companies:

1. Agree clear expectations of business processes
2. Define clear roles and responsibilities
3. Address issues of compatible technology
4. Assess the potential impact on internal business processes
5. Agree scheduling of collaborative activities and establish realistic time-scales
6. Establish procedures to respond to and resolve problems
7. Develop communication practices to improve understanding of the different trading pressures experienced by trading partners
8. Determine agreed content of shared information
9. Establish definitions of shared data and agree quality measures
10. Minimise the impact of a proliferation of collaborative systems
11. Provide time to develop the collaborative relationship
12. Document and communicate new procedures

Community Practice Perspective

Internet technology provides the means to integrate collaborative processes but culture and behavioural change remain the challenge of the supply chain (Boddy, Cahill, Charles, Fraser-Kraus, & Macbeth, 1998). E-collaboration has ostensibly removed the need for personal contact in the sharing, communication and processing of information, but personal relationships are critical to the effectiveness of e-collaboration (Hughes et al., 1998). Personal interaction provides the arena for the socialisation and internalisation stages that are necessary to provide the knowledge transfers required for trust to be gained (Dawson, 2000). Attempts to enforce collaboration processes by meetings and reporting mechanisms neglect the attitudinal issues of collaborative practice, such as joint goals, shared resources and commitment to a shared vision (Khan & Mentzer, 1996).

The introduction of the e-collaboration system in the manufacturer-retail supply chain (Dingley & Perkins, 1999), emphasised written (electronic) communication, processes rather than oral communication skills which changed the nature of the communities of practice. This resulted in problems of data semantics and issues of trust and reliability.

Memory distinguishes collaborative relationships from co-operative business transactions (Zwass, 2003). Effective collaboration requires a collaborative culture to be developed which comprises of information exchange, trust, mutuality, openness and communication (Barratt, 2004). Generating trust is essential to overcome the suspicions of each partner (Todeva & Knoke, 2005), demonstrating commitment to the collaboration through continued accountability (Nahapiet, Gratton, & Rocha, 2005). Trust can be considered in terms of: contracts: trust that parties will do what they say they will do; self disclosure: trust in terms of reciprocal sharing of information and physical trust: relating to physical safety of information (Agarwal & Shankar, 2003).

Fairness of exchange in collaborative relationships can be examined in terms of distributive justice, procedural justice and interactional justice (Fearne, Duffy, & Hornibrook, 2005) which can sometimes conflict. Distributive justice considers the degree to which the outcomes of the collaborative partnerships are shared equally and can be measured by factors such as the perceived fairness of price or payment terms, the distribution of costs and charges and the degree to which contributions and discounts are demanded (Fearne et al., 2005). Procedural justice refers to the processes of managing the collaboration and in achieving its outcomes. It includes the degrees of: bilateral communication, impartiality, refutability (the opportunity to appeal about decisions and policies taken by one of the partners in the supply chain), rationale provided for decisions, familiarity demonstrated by each party about the market conditions affecting the other partners and the degree of courtesy demonstrated in conducting the collaboration (Fearne et al., 2005). Interactional justice refers to the fairness of each partner's behaviour during the process of collaboration.

At the level of community practice collaborating partners need to:

1. Seek to establish a culture of openness
2. Define clear expectations
3. Work together to solve problems
4. Identify changes in skill sets required
5. Acknowledge cultural differences
6. Encourage adherence to new ways of working
7. Respond to and resolve problems quickly
8. Ensure reliable and consistent communication
9. Ensure a consistent approach is adopted by individuals
10. Seek to avoid the development of a blame culture
11. Establish definitions of shared data and agree quality measures
12. Maintain professional integrity

FUTURE TRENDS

E-collaboration in supply chain transformation is a business project not an IT project (Favilla & Fearne, 2005). The continuing challenges relate to how to change business processes and practices to accommodate different e-collaboration systems. Organisations need to address the structural, process and cultural barriers that restrict effective collaboration in multiple supply chains.

A further challenge relates to the development of metrics for e-collaboration to help managers identify areas in the supply chain that need attention (van Hoek, 1998). A problem with measuring the effectiveness of e-collaboration is that the activities an organisation needs to measure are not under their control (van Hoek, 1998). Supply chain performance measures focus on uni-dimensional financial measures (such as return on investment) that neglect the holistic nature of supply chain management (Saad & Patel, 2006). Systematic approaches to collaboration only focus on tangible aspects of process, structure and technology, neglecting the rich tacit knowledge that is embedded in communities of practice. For organisations to move from cooperation to collaboration, future work needs to acknowledge the important role of communities of practice in e-collaboration and facilitate ways to enrich the practice to enable the potential benefits of e-collaboration to emerge.

CONCLUSION

The transition from supply chain cooperation to supply chain collaboration offers benefits to the sustainability of organisations. Advances in internet technologies have removed the technical barriers to collaboration but sometimes the rationale for, and the impact of, the collaboration can be neglected. Internet technology is a necessary component for e-collaboration in the supply chain but is insufficient as a facilitator of collaboration. Cases of a manufacturer-retail supply chain highlighted problems with the introduction of e-collaboration systems such as strategic positioning and the balance of power in the customer-supplier relationship.

Critical success factors have been identified to address some of the problems with e-collaboration in the supply chain encountered in the manufacturer-retail supply chain. It is proposed that future work on e-collaboration systems needs to consider three levels of: strategic policy, process integration and community practice in order to maximise the benefits that can emerge from effective supply chain collaboration moving beyond cooperation to the formation of integrated e-communities.

REFERENCES

Agarwal, A., & Shankar, R. (2003). On-line trust building in e-enabled supply chain. *Supply Chain Management: An International Journal, 8*(4), 324-334.

Archer, N., & Yuan, Y. (2000). Managing business-to-business relationships throughout the e-commerce procurement lifecycle. *Internet Research, 10*(5), 385-395.

Barratt, M. (2004). Understanding the meaning of collaboration in the supply chain. *Supply Chain Management: An International Journal, 9*(1), 30-42.

Bhatt, G. D., & Emdad, A. F. (2001). An analysis of the virtual supply chain in electronic commerce. *Logistics Information Management, 14*(1/2), 78-84.

Boddy, D., Cahill, C., Charles, M., Fraser-Kraus, H., & Macbeth, D. K. (1998). Success and failure in implementing supply chain partnering: An empirical study. *European Journal of Purchasing and Supply Management, 4*(2/3), 143-151.

Burnes, B., & New, B. (1998). Developing the partnership concept for the future. In B. Burnes & B. Dale (Eds.), *Working in partnership: Best practice in customer-supplier relations* (pp. 101-109). Aldershot: Gower.

Caglliano, R., Caniato, F., & Spina, G. (2003). E-business strategy: How companies are shaping their supply chain through the internet. *International Journal of Operations & Production Management, 23*(10), 1141-1162.

Cassivi, L. (2006), Collaboration planning in a supply chain. *Supply Chain Management: An International Journal, 11*(3), 249-258.

Cox, S. A., Krasniewicz, J. A., Perkins, J. S., & Cox, J. A. (2006, September 12-14). Modelling the organisational transformation associated with implementing e-business collaborative systems in the supply chain. In *Proceedings of the British Academy of Management Conference (BAM2006) in Association with the University of Ulster and Queen's University*, [CD ROM]. Belfast, Ireland.

Croom, S. R. (2005). The impact of e-business on supply chain management: An empirical study of key developments. *International Journal of Operations and Production Management, 25*(1), 55-73.

Dawson, R. (2000). *Developing knowledge-based client relationshipsThe future of professional services*. Woburn: Butterworth-Heinemann.

Dingley, S., & Perkins, J. (1999). Tempering links in the supply chain with collaborative systems. In *Proceedings of the Ninth Annual Conference of BIT: Generative Futures*, [CD ROM]. Manchester: Machester University.

Favilla, J., & Fearne, A. (2005). Supply chain software implementations: Getting it right. *Supply Chain Management: An International Journal, 10*(4), 241-243.

Fearne, A., Duffy, R., & Hornibrook, S. (2005). Justice in UK supermarket buyer-supplier relationships: An empirical analysis. *International Journal of Retail & Distribution Management, 33*(8), 570-582.

Gattorna, J. L., & Walters, D. W. (1996). *Managing the supply chain: A strategic perspective*. London: Macmillan.

Graham, G., & Hardaker, G. (2000). Supply-chain management across the Internet. *International Journal of Physical Distribution and Logistics, 30*(34), 286-296.

Hibbert, P., & Huxham, C. (2005). A little about the mystery: Process learning as collaboration evolves. *European Management Review, 2*(1), 59-69.

Hughes, J., Ralf, M., & Michels, B. (1998). *Transform your supply chain.* London: International Thomson Business Press.

Khan, K. B., & Mentzer, J. T. (1996). Logistics and inter-departmental integration. *International Journal of Physical Distribution & Logistics Management, 26*(8), 6-19.

Kirchmer, M. (2004). E-business process networks: Successful value chains through standards. *Journal of Enterprise Information Management, 17*(1), 20-30.

Nahapiet, J., Gratton, L., & Rocha, H. O. (2005). Knowledge and relationships: When cooperation is the norm. *European Management Review, 2*(1), 3-14.

Pateli, A. G., & Giaglis, G. M. (2005). Technology innovation-induced business model change: A contingency approach. *Journal of Organizational Change Management, 18*(2), 167-183.

Perkins, J., & Dingley, S. (2001). Collaborative systems architecture to reduce transaction costs in e-business: changing the balance of power. In *Proceedings of the Third International Conference on Enterprise Information Systems* (pp. 1155-1161). Setúbal, Portugal.

Saad, M., & Patel, B. (2006). An investigation of supply chain performance measurement in the Indian automotive sector. *Benchmarking: An International Journal, 13*(1/2), 36-53.

Sahay, B. S. (2003). Supply chain collaboration: The key to value creation. *Work Study, 52*(2), 76-83.

Simatupang, T. M., & Sridharan, R. (2002). The collaborative supply chain. *International Journal of Logistics Management, 13*(1), 15-30.

Simatupang, T. M., & Sridharan, R. (2005). The collaboration index: A measure for supply chain collaboration. *International Journal of Physical Distribution & Logistics Management, 35*(1), 44-62.

Tan, K. C. (2001). A framework for supply chain management literature. *European Journal of Purchasing and Supply Management, 7*(1), 39-48.

Todeva, E., & Knoke, D. (2005). Strategic alliances and models of collaboration. *Management Decision, 43*(1), 123-148.

van Hoek, R. I. (1998). Measuring the unmeasurable: Measuring and improving performance in the supply chain. *Supply Chain Management, 3*(4), 187-192.

Zwass, V. (2003). Electronic commerce and organizational innovation: Aspects and opportunities. *International Journal of Electronic Commerce, 7*(3), 7-37.

KEY TERMS

Collaborative Practice: The ways of working that emerge between communities of practice as they work towards the achievement of common goals.

Community of Practice: The emergent process of social learning as a group of people with shared values, beliefs, and goals work together towards a common aim.

E-Collaboration: The use of information technology to establish, facilitate and sustain cooperation between two geographically dispersed parties, who have common goals, to enable them to work together for mutual benefit.

E-Collaboration System: A computer-based system that is accessed and used by more than one organisation to support business transactions in the supply chain. The system allows data to be automatically updated in a partner organisation's systems during the processing of a transaction.

Internet Technologies: Communication based information technology, including network protocols and communication mechanisms that enable data transmission within and between geographically dispersed organisations to support formal business processes across interfirm networks.

Supply Chain: Two or more directly dependent organisations vertically aligned such that the outputs of one provides goods or services that are essential inputs to the value chain of the other. Physical supply chains are inventory-based; virtual supply chains are information-based.

Supply Chain Collaboration: Integrating the business processes of two or more organisations in the supply chain, embedding activities to form a jointly-owned emerging system with shared goals.

Supply Chain Cooperation: Aligning the activities of two or more organisations in the supply chain to coordinate the supply of goods or services, creating a competitive advantage through improved service or efficiency improvements.

Supply Chain Management: Aligning internal business processes to respond to information received from the customer-supplier interface, and to provide accurate information to the interface, in order to co-ordinate the reliable and timely supply of goods and services to members of the supply chain.

Faculty Perceptions of Traditional and Electronic Communications Channels

Rolando Pena-Sanchez
Texas A&M International University, USA

Ibrahim Mescioglu
Texas A&M International University, USA

Richard C. Hicks
Texas A&M International University, USA

INTRODUCTION

Teaching involves the transfer of knowledge at two levels—group communication and personal communication. Personal communication between the faculty and the student is a valuable component of the learning experience which supplements in-class lectures. This paper examines the perceptions of the faculty of a small state university about their usage of personal communication in a face to face, e-mail, and telephone context.

The traditional approach to personal communication between faculty and students has been through face-to-face communication, usually in the form of office hours. Virtually all faculty at the college level hold scheduled Office Hours, which may be supplemented by appointments. However, the communications revolution which has occurred because of the wide spread availability of computers and e-mail has had some impact upon the personal communication between faculty and students. This survey examines the usage patterns of faculty.

BACKGROUND

Previous research into communication channels falls into two basic streams (Marcus, 1994). The first perspective focuses on the communication channel itself, as in Daft (1986). The second perspective focuses on the social context of the communication (Falk, 1987).

Beside these two streams other issues related to the communication channels have been considered as well. In a study of managers and executives, Carlson determined that executives selected communications media either by the ease of use or by the richness or social presence of the media (Carlson, 1998). In other research, Gefen and Straub found that women perceived e-mail as richer (Trevino, 1987) than their male counterparts (Gefen, 1997).

One shortcoming of these perspectives is that they focus on the selection process used by the sender of the communication, instead of the receivers of the communication (Sitkin, 1992). In faculty to student personal communications, the selection of a communication channel is usually made by the students as the senders of communication. However, the faculty has a significant input to the selection process because of the difference in status. Because of this difference, it is hoped that this research will widen the current body of communication research. The findings can be generalized to the relationship between customers and service agents of a company as well as the relationship between employees and managers.

In this study we are interested in the differences of faculty members' perception of alternative communication channels. Each communication channel can present different advantages that are unique to that channel. Similarly, each channel might have shortcomings that other channels do not have. This article follows from (Pena-Sanchez, 2005). Accuracy, convenience, timeliness and confidentiality are some of the important factors that differentiate the possible set of communication channels when compared against each other. Students might prefer one channel (i.e., e-mail) when convenience is more important, while preferring another channel (i.e., office hours) when accuracy of the communication's content is more important. The perceived importance of these factors when comparing the possible communication channels is the focus of this study.

DATA AND METHODOLOGY

A random sample of size $n = 49$ was used to test several hypotheses. The sample represents 32% of the entire faculty population ($N = 153$) of Texas A&M International University. The survey was pilot tested by a small group of Management of Information Systems and Decision Science faculty before its administration.

Given the limited number of observations, the discrete scale of the measurements, and the fact that data do not meet parametric F-test assumptions like normality and homocedasticity of the variances, the statistical techniques used consist of some nonparametric methods based on ranks such as the Spearman rank correlation coefficient test, and the Mann-Whitney test (Conover, 1980).

RESULTS OF THE SURVEY

The Cronbach's alpha coefficient estimates for internal consistency as an evaluation of the survey's reliability (Pena-Sanchez, 2005) was 0.7053 for the set of variables convenience, retention, and efficiency under e-mail communication, 0.7718 for the set confidentiality, confrontation, and emotional support using office hours as a communication channel, and 0.7540 for accuracy, reaching a consensus, and overall effectiveness when the communication channel was office hours and/or e-mail. These values are shown in Table 1.

$E(x)$ is the expected value or average for the variable x.

The next plot, Figure 1 explores the behavior between age and overall effectiveness using e-mail.

The survey results support the theory that communications media are selected by the richness required by the task (Fann, 1989; Reinsch, 1990; Rice, 1993).

Office hours are the preferred communications channel for tasks requiring rich communication media, especially confrontation and emotional support. On the other hand, the faculty considered e-mail to be superior for convenience, retention, and efficiency (see Table 3).

According to Table 2, there is no significant statistical relationship between age and communication media preference, except for confrontation criterion.

MANAGERIAL IMPLICATIONS TO INNOVATION AND LEARNING

Hilton (1999) foresees that in the future email can automate the human face and become part of Internet enhanced education. Only recently has e-mail has reached the critical mass where innovation can occur in both education and business. The nearly universal availability of email to both students and employees is changing communication preferences.

Managers and educators alike depend on effective communication. To provide additional insight into the preferences for communications channels, we will cluster the hypothesis by the preferred channel and see if broader conclusions for managerial implications to innovation and learning may be reached.

- **E-mail' strengths:** E-mail is preferred for convenience, retention, and efficiency, which indicates that e-mail is the preferred medium for communication that is unambiguous and asynchronous, especially if wide dissemination is desired. Unlike face to face communication, e-mail is asynchronous, allowing each party to communicate independently of the other party. It also creates documentation, and may be sent to many people

Table 1. Cronbach's alpha coefficient estimates for the indicated set of variables (cluster) and preferred media

Cluster	Preferred media	Cronbach's alpha correlation coefficient estimate
Convenience, retention, and Efficiency	e-mail	0.7053
Confrontation, confidentiality, and emotional support	office hours	0.7718
Accuracy, receiving work, reaching a consensus, and overall effectiveness	e-mail and office hours	0.7540

Table 2. The Spearman rank correlation coefficient ρ between age and the indicated variable (criterion) associated to communication by e-mail, to test the null hypothesis $H_{12,0}$: $\rho = 0$, vs. the alternative hypothesis $H_{12,a}$: $\rho < 0$; which is equivalent to say that younger faculty members will show a higher preference for e-mail; $n=44$ in all cases, as 5 faculty declined to submit their ages.

Variable	Statistic ρ	*p*-value	Decision
Timeliness	-0.183	0.234	Do not reject $H_{11,0}$
Accuracy	-0.088	0.572	Do not reject $H_{11,0}$
Convenience	-0.189	0.219	Do not reject $H_{11,0}$
Retention	-0.123	0.428	Do not reject $H_{11,0}$
Confidentiality	-0.057	0.713	Do not reject $H_{11,0}$
Efficiency	-0.206	0.180	Do not reject $H_{11,0}$
Receiving work	-0.169	0.271	Do not reject $H_{11,0}$
Reaching a consensus	-0.265	0.082	Do not reject $H_{11,0}$
Confrontation	-0.331	**0.028**	Reject $H_{11,0}$ at $\alpha=0.05$
Emotional support	-0.013	0.936	Do not reject $H_{11,0}$
Overall effectiveness	-0.244	0.110	Do not reject $H_{11,0}$

Table 3. Mann-Whitney test for the null hypothesis $H_{13,0}$: E(freshmen_overall eff. e-mail) = E(graduate_overall eff. e-mail), versus the alternative hypothesis $H_{13,a}$: E(freshmen_overall eff. e-mail) > E(graduate_overall eff. e-mail); which is equivalent to say that faculty teaching to freshmen students will show a higher preference for e-mail than those that teach to graduate students.

Courses	n	E(x)	Median	Sum of ranks	Statistic U	p-value	Decision ($\alpha=0.05$)
freshmen	$n_f = 13$	4.85	5	126.50	35.5	0.06	Do not reject $H_{12,0}$
graduate	$n_g = 10$	5.80	6	149.50			

at the same time. The asynchronous nature of e-mail may help reduce ambiguity because the sender may research and edit the e-mail until it conveys the desired message. In contrast, the synchronous nature of face to face communication implies the need for an immediate response, which limits the amount of research and editing that can be performed.

- **E-mail's weaknesses:** Office hours are preferred for confidentiality, confrontation, and emotional support, which indicate that face to face communications are preferred for situations that involve emotions. Face to face communication may be considered less confidential because there is no inherent documentation that could be read by others. In situations involving confrontation and

emotional support, some the richness of communication clues, such as tone of voice and body language, allow additional levels of communication beyond those of the printed word. The synchronicity of face to face communication means that the sender expects to receive an immediate reply, which may be especially undesirable in confrontational situations.

There is no clear preference between face to face communication and e-mail for overall preference, accuracy, receiving work, and reaching a consensus. This is perhaps the most interesting cluster because both channels have strengths and weaknesses.

In the ratings for overall preference, the equal support for both face to face communication and e-mail indicates

Figure 1. Plot of age and perceived effectiveness of e-mail

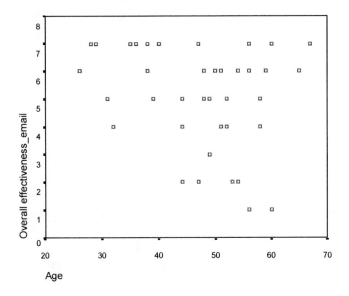

to both academics and managers that it is important to keep all lines of communication open. This finding also supports the theory that managers choose a communication channel based on the characteristics of the task; which will be relevant to obtain the achievements during the "learning process" (McEntee, 1997).

The ratings for accuracy and receiving work also showed similar preferences. The accuracy of e-mail is enhanced because it is self-documenting, while the accuracy of face to face communication is enhanced by the increased availability of clues. Receiving work by e-mail is more convenient, while receiving work face to face offers the immediate opportunity for clarifications.

The inclusion of reaching a consensus in the cluster with same preference between face to face communication and e-mail (see Table 1) is interesting. As reaching a consensus is concerned with removing ambiguity, it would appear to belong in the face to face (office hours) cluster. One interpretation of this survey would be that professors are becoming more comfortable with making decisions and removing ambiguity by e-mail because of the ability to research and edit their responses. It is also of note that reaching a consensus does not have the emotional implications of a task involving confidentiality, conflict, or emotional support.

This survey indicates that preferences for e-mail and face to face communication are influenced by certain types of tasks. Face-to-face communication is the preferred channel for tasks involving emotions and when an immediate response is required. It is the richest of the communication channels, so face to face communication has the potential to convey more information than e-mail. E-mail is the preferred channel if the communication is unambiguous and can be asynchronous. The lack of overall preference indicates that both types of communication are employed for a wide variety of other tasks, with the choice of channel being influenced by the need for synchronicity and the amount of ambiguity inherent in the task.

Providing more alternative communication channels to the customers will certainly improve customer satisfaction levels. A company that provides customer service only through their in-store customer service departments has important opportunities to improve its customer satisfaction levels. If this company starts providing customer support through 1-800 phone lines, e-mail, online chat with customer service agents and instant messaging, customers will have a choice of communication channels. As presented in this article and in (Pena-Sanchez, 2005), customers can choose different communication channels for different purposes; just like students choose different communication channels in different situations. From this perspective, selection of which communication channels to provide and how to maintain them are crucial decisions for managers. The needs and expectations of more tech savvy customers might be different from those of less technically oriented customers. These differences will also reflect in selection of their communication channels. If the first group prefers e-mail and online chat and the second group prefers in-store support, managers will have to make decisions regarding the skills of agents that they will assign to those departments.

CONCLUSION

Our conclusions supported by the nonparametric statistical analysis are that faculty expressed a preference for e-mail for the variables convenience, retention, and efficiency. Office hours were preferred for confrontation, confidentiality, and emotional support. Equal preference for e-mail and office hours were expressed for accuracy, receiving work, reaching a consensus, and overall effectiveness. The same preference for the three communication channels (office hours, e-mail, and phone) was expressed for timeliness criterion (see Table 3).

The Cronbach's alpha coefficient estimates for internal consistency as an evaluation of the survey's reliability were acceptable values, which are shown in Table 1. From which, we can conclude that the degree to which survey items are homogeneous and reflect the same underlying constructs is acceptable.

From Figure 1, our conclusion is that the younger faculty perceives e-mail as more effective than older faculty. The range of effectiveness perception of faculty members under the age of 40 ($n = 11$) was from 4 to 7; and from 27 faculty over 40 years old, 11 of these indicate preference of 4 or lower. Except for confrontation criterion, there was not significant statistical relationship between age and communication media preference, but we discovered conclusive empirical evidence about the pattern between age and all of the criteria (timeliness, accuracy, etc.) for using e-mail, which is indicated by the negative signs of rank correlation coefficients for all cases in Table 2. Thus, young faculties tend to prefer using e-mail.

At a significance level of $\alpha = 5\%$, the survey results (p-value $= 0.06$) do not provide evidence to conclude that faculty teaching to freshmen students will show a higher preference for e-mail than those that teach to graduate students. At $\alpha = 10\%$, the conclusion will be different, but there was not significant difference among faculty preferences to use e-mail with freshmen or graduate students. The preference for e-mail of faculty teaching to graduate students tends to be higher than those that teach to freshmen students (see Table 3).

As a result of the statements presented in the managerial implications section, and supported by empirical evidence (see Table 1), and because "better retention" means "better learning," therefore we can conclude that the e-mail can be conceived as an innovative e-learning tool.

DIRECTIONS FOR FURTHER RESEARCH

These findings should impact both managers and educators and their choice of communications media. As educators, we all should seek the most effective and efficient media for the various communication tasks. The importance of communication between employees as well as faculty and students outside of the classroom environment should provide ample motivation for additional research into this topic.

This survey is limited in scope, and should be expanded to cover a larger sample of faculty. The demographics of the studied population may not be typical of the entire educational environment. The culture studied by this survey may be atypical, or further research may find that there a number of sub-cultures within an organization or university, each with its own communications practices.

This survey indicates that e-mail is becoming a more important media for communications. Further research is needed into the effectiveness of various communications media and tasks so that both business professionals and faculty may communicate more effectively.

It is hoped that this paper will foster more research into the relationships between communications media and tasks, so that more effective and efficient communications will occur both in organizations and universities.

The survey may be obtained by email from the authors.

REFERENCES

Carlson, P. J., & Davis, G. B. (1998). An investigation of media selection among directors and managers: From "self" to "other" orientation. *MIS Quarterly, 22*(3), 335-352.

Conover, W. J. (1980). *Practical nonparametric statistics*. New York: Wiley.

Daft, R. L., & Lengel, R. H. (1986). Organizational information requirements: Media richness and structural design. *Management Science*, 32(5), 554-571.

Gefen, D., & Straub, D. W. (1997). Gender differences in the perception and use of e-mail: An extension to the technology acceptance model. *MIS Quarterly*, 21(4), 389-400.

Hilton, T. S. E. (1999). A model for internet-enhanced education systems derived from history and experiment. *Journal of Computer Information System, 39*(3), 6-17.

Johnson, J. D., Chang, H., Pobocok, S., Etherington, C., Reusch, D., & Wooldridge, J. (2000). Functional work groups and evaluations of communication channels: Comparisons of six competing theoretical perspectives. *Journal of Computer-Mediated Communication, 6*(1).

Marcus, M. L. (1994). Electronic mail as the medium of managerial choice. *Organizational Science, 5*(4), 502-527.

McEntee, E., & Pena-Sanchez, R. (1997). The learning process of cooperation by reciprocity. In *Proceedings of the 15th Studies experiences exchange meeting upon education (Third Congress of Academic Quality, ITESM, 1)* (pp. 184-188).

Pena-Sanchez, R., & Hicks, R. C. (2005). Students' communications media usage for competitive academic tasks outside the classroom environment. *Competition Forum. American Society for Competitiveness, 3*(2), 357-373.

Reinsch, N. L., & Beswick, R. W. (1990). Voice mail versus conventional channels: A cost minimization analysis of individual's preferences. *Academy of Management Journal, 33*(4), 601-616.

Rice, R. E. (1993). Media appropriateness: Using social presence theory to compare traditional and new organizational media. *Human Communications Research, 19*(4), 451-484.

Sitkin, S. B., Sutcliffe, K. M., & Barsios-Choplin, J. R. (1992). A dual-capacity model of communications media choice in organizations. *Human Communication Research, 18*(4), 563-598.

Trevino, L. K., Lengel, R., & Daft, R. L. (1987). Media symbolism, media richness, and media choice in organizations: A symbolic interactionist perspective. *Communications Research, 14*(5), 553-574.

KEY TERMS

Collaboration: Interactions that take place between two individuals, such as negotiation, information sharing, and communication.

Communications Media: The channels through which communication travels, such as e-mail, face to face, or telephone.

E-Collaboration: Online communication and cooperation with the aid of appropriate tools, generally through electronic communications.

E-Mail: The use of computers and networks to pass messages from sender to receiver.

Face to Face: Communication that take place verbally between individuals.

Office Hours: An example of face to face communication in the academic arena.

Telephone: Communication that takes place over the telephone.

Faculty Preferences for Communications Channels

Rolando Pena-Sanchez
Texas A&M International University, USA

Richard C. Hicks
Texas A&M International University, USA

INTRODUCTION

In higher education, there are two distinctly different means of communication. The first is group communication, which normally takes place in the classroom. Most of the communication in the classroom uses the face-to-face media. Outside of the classroom, however, various communications media may be used. In this article, we examine the preferences for face to face, e-mail, and telephonic communication for a variety of tasks.

Traditionally, communication outside of the classroom has been accomplished through face-to-face communication, usually in the form of office hours. Virtually all faculty at the college level hold scheduled office hours, which may be supplemented by appointments. In addition, the widespread availability of computers and e-mail has had a significant impact on the communication between faculty and students.

As Marcus (1994) explains, there are two basic streams of research into communication channels. The first stream, as exemplified by the research of Daft and Lengel (1986) and others, focuses on the communication channel. The second perspective focuses on the social context of the communication (Fulk, Stienfield, Schmitz, & Power, 1987).

Outside of these two streams, other factors are considered. In a study of managers and executives, Carlson determined that executives selected communications media either by the ease of use or by the richness or social presence of the media (Carlson & Davis, 1998). In other research, Gefen and Straub found that women perceived e-mail as richer (Trevino, Lengel, & Daft, 1987) than their male counterparts (Gefen & Straub, 1997).

Most studies of communications channels focus on the preference of the sender of the communication instead of the receiver (Sifkin, 1992). In faculty-to-student personal communications, as in the selection of channels of employees to management, the selection of a communication channel is usually made by the senders of communication. However, the faculty (and management) have a significant input to the selection process because of the difference in status. It is hoped that this research will widen the current body of communication research and can be generalized to the relationship between managers and employees found in business.

When considering the choice of a communications channel, three factors that must be evaluated are the richness of the communication channel, the immediacy of the channel, and the social context of the task to be performed by the communication. We will next consider these factors.

Richness of Communications Channels

Face-to-face communication is considered to be the richest of these communication channels. As face-to-face communication uses all of the senses, gives immediate feedback, and is more spontaneous, it is the richest of these communication channels (Durlak, 1987). In addition to words, communication is performed by facial expression, body language, and clothes. The expression of humor and sarcasm are far easier to convey in face-to-face communication.

Telephone communication is the next richest of the communication channels studied. Besides words, communication is enhanced by the inflection of the speaker's voice. Humor and sarcasm are less apparent but are still perceivable.

E-mail filters out all but verbal clues to meaning (Karahanna & Straub, 1999). E-mail communication is limited to words, so it is the least rich of the studied communication channels. Words are the predominate means of communication. Emoticons may be used to indicate emotional components, such as humor, but with less richness than the spoken word.

Immediacy of Communications Channels

Both face-to-face and telephone channels receive immediate responses after they have been initiated because of their synchronous nature. However, this assumes that the communication has been successfully initiated. A student has to wait until the scheduled opportunity (usually office hours) to initiate the communication. Often, this requires a wait of several days.

E-mail is asynchronous because of its unscheduled nature. The student first sends the e-mail, and then waits until the faculty member receives the communication and responds. The waiting period may be from seconds to days, depending on the circumstances. On the other hand, e-mail is not bound by geographical constraints, so a student and faculty member may be in different countries and have rapid communication.

Privacy of e-mail communications may be problematic (Clyde, 1999), especially when traveling. The perception that the university may read a faculty members e-mail was reported by as many of 50% of the faculty members in one survey (Beheruz, Barnes, Burst, & Kaye, 1999).

Social Context of Communications Tasks

Selection of a communication channel has many components. As many types of communication take place between faculty and students, different channels may be selected for different types of communication. Social information processing takes the position that the individual's social environment impacts on the selection of communications channels (Karahanna & Straub, 1999). Some of the characteristics of this task are imparting the feeling of group membership, representing diversity of viewpoints, and providing information that can be passed to others. Social presence indicates the degree to which a channel simulates face-to-face communication (Durlak, 1987). Cost minimization is determined by three factors: access, errors, and delays (Reinsch & Beswick, 1990). Effort costs can be associated with the distance between the two parties (Trevino et al., 1987), familiarity with the channel (Steinfeld, 1987) and length and complexity of the message (Daft & Lengel, 1986).

In a 2000 survey, Johnson et al. classified the choice of communications media by the following tasks: social presence, uncertainty reduction, appraisal, social information processing, decision making, and cost reduction (Johnson 2000). They measured the perceived value of written, interpersonal, and e-mail for these tasks. We will contrast our results with Johnson's in the discussion portion of the article.

In the next section, we will examine faculty—student communication and develop hypotheses about the impact of communications channel choice on the various components of these communications.

FACULTY-STUDENT COMMUNICATION

Timeliness is an important component of any communication. In this environment, it must be recognized that students do not have unfettered access to faculty. Many faculty members are available to students only during scheduled office hours. However, many faculty members will answer e-mail outside of office hours.

- H_1: E-mail will be considered as the most timely communication channel.

The accuracy of the communication is of paramount concern. In face-to-face communication, the richness of the channel offers more clues as to the meaning conveyed. However, no documentation of the conversation is created except for when the student takes notes. E-mail is inherently self-documenting.

- H_2: E-mail will be considered as the most accurate communications channel.

The convenience of the communications channel is important to both parties. It may be very difficult for the student to be present during office hours because of work or other classes. It is not always possible to make alternative arrangements for face-to-face or telephone communication. E-mail may be received or sent in an asynchronous manner without a prearranged meeting time and place.

- H_3: E-mail will be considered the most convenient communications channel.

Retaining the contents of the communication is also very important. The student may not ask all of the relevant questions or remember all of the responses. E-mail, as mentioned previously, is inherently self-documenting.

- H_4: E-mail will be considered as the channel offering the best retention.

Another aspect of communication is confidentiality. While most office hours are held in private, other students may overhear the conversations between student and faculty. E-mail may be read by other students, especially if it is received in a public place such as a computer lab, but the student is in control of the receiving environment.

- H_5: E-mail will be considered as the most confidential communications channel.

Another reason for student-faculty communication is for the turning in of previously assigned work by the student. This may be accomplished during office hours or by e-mail. However, as no actual interaction is required by this task, it is probably more convenient for both student and faculty to perform this task by e-mail.

- H_6: E-mail will be the preferred channel for turning in previously assigned work.

Students and faculty often interact about the assignment of work. In some cases, this simply involves the student receiving the assignment, which may be performed at a class meeting. It is often the case that there is considerable interaction and discussion about the assignment. As e-mail is asynchronous, it may take many e-mail communications for the assignment to evolve.

- H_7: Face-to-face communication will be the preference for assigning work.

In many circumstances, it may be necessary for the student and faculty to achieve a consensus about the assignment. Face-to-face communication allows for rapid evolution of the task, where e-mail may involve considerable delays in reaching a consensus.

- H8: Face to face will be the preferred channel for reaching a consensus.

Some meetings between student and faculty involve a confrontation. In face-to-face communication, it is possible that emotions will be involved. E-mail, by its asynchronous nature, allows each party to restate their positions before communicating them.

- H_9: E-mail will be the preferred channel for confrontational meetings.

It may also be necessary to offer emotional support to students, especially when the student's performance is below their expectations. The richness of face-to-face communication enable faculty to respond more appropriately than e-mail.

- H_{10}: Face-to-face communications will be preferred for offering emotional support to students.

- H_{11}: E-mail will be the preferred media for overall communication between faculty and students.

The choice of communications channel is affected by the familiarity of the sender with the channel (Rice, 1993). As younger faculty members are more likely to be technologically adept, it is likely that they will show a higher preference to e-mail communication.

- H_{12}: Younger faculty members will show a higher preference for e-mail for all criteria (timeliness, accuracy, etc.).

E-mail allows the communication to be performed at a distance. Freshmen students are less likely to wish to perform direct communication with faculty as they may be more intimidated than more experienced students. From the faculty perspective, e-mail would seem more appropriate for communications with freshmen than graduate students because of the volume of communication and the depth of the topics covered by the communication.

- H_{13}: Faculty teaching to freshmen students will show a higher preference for e-mail than those that teach to graduate students.

DATA AND METHODOLOGY

A random sample of size $n = 49$ was used to test several hypotheses. The sample represents 32% of the entire faculty population ($N = 153$) of Texas A&M International University. The survey (shown at the end of this article as Appendix A) was pilot tested by a small group of Management of Information Systems and Decision Science faculty before its administration.

Given the limited number of observations, the discrete scale of the measurements, and the fact that data do not meet parametric F-test assumptions like normality and homocedasticity of the variances, the statistical techniques used consist of some nonparametric methods based on ranks such as the Friedman test (Conover, 1980).

Results of the Survey

Table 1 shows the distribution of academic ranks participating in the survey. This distribution is fairly consistent with the distribution of academic ranks at the studied university.

Table 2 shows the distribution of communications between students and faculty that occur out of the classroom environment. This table shows that office hours (face to face) and e-mail are the dominant means of communication.

Table 3 shows the preferences for each of the tasks. The same preference was shown for timeliness. E-mail was the preferred media for convenience, retention, and efficiency. Office hours were preferred for confidentiality, confrontation, and emotional support. E-mail and office hours were equally preferred for accuracy, receiving work, reaching a consensus, and overall preference. The last column in this table shows comparisons for the three media channels used in the hypotheses stated earlier in the article. H1 (e-mail preferred for timeliness), H5 (e-mail preferred for confidentiality), and H9 (e-mail preferred for confrontation) were not supported. The other hypotheses are supported, although the same preferences were noted in four hypotheses.

Limitations of the Survey

The survey is limited by the small sample size. One e-mail request and written request for participation was sent to each faculty member. However, the sample size is sufficient for the statistical techniques used by the authors.

The survey is also limited by the characteristics of the sample population. All of the survey participants are faculty at one university. This e-mail system used by this university has only recently been accessible from off campus. In addition, the university services a population which is characterized by low income, which implies that some students will not have external e-mail access for communicating with the faculty. These two characteristics may cause the survey to understate the importance of e-mail.

Table 1. Frequency description for the faculty rank

Faculty rank	Frequency	Percentage
Lecturer	7	14.3
Adjunct	6	12.2
Assistant	22	44.9
Associate	8	16.3
Full	6	12.2
Total	**49**	100

Table 2. Percentage of student-faculty communication time about course work per media channel

Media:	Office hours	e-mail	Phone	Fax
Percentage:	44.53	41.29	14.02	0.16

Table 3. The Friedman statistic (T_f) test for each null hypothesis: $H_{i,0}$: All three channel-communications (office hours, e-mail, and phone) population distribution are identical versus the alternative hypothesis $H_{i,a}$

Null $H_{i,0}$ & Varaiable	Statistic T_f	p_value	Decision	Prefered media (Equation {1})
$H_{1,0}$ Timeliness	2.958	0.228	Do not reject $H_{1,0}$	Same preference
$H_{2,0}$ Accuracy	15.167	**0.001**	Reject $H_{2,0}$ at α=0.01	office hs. & e-mail
$H_{3,0}$ Convenience	15.642	**0.001**	Reject $H_{3,0}$ at α=0.01	e-mail
$H_{4,0}$ Retention	26.195	**0.001**	Reject $H_{4,0}$ at α=0.01	e-mail
$H_{5,0}$ Confidentiality	12.014	**0.002**	Reject $H_{5,0}$ at α=0.01	office hs.
$H_{6,0}$ Efficiency	26.270	**0.001**	Reject $H_{6,0}$ at α=0.01	e-mail
$H_{7,0}$ Receiving work	49.205	**0.001**	Reject $H_{7,0}$ at α=0.01	e-mail & office hs.
$H_{8,0}$ Reaching a cons.	17.554	**0.001**	Reject $H_{8,0}$ at α=0.01	office hs. & e-mail
$H_{9,0}$ Confrontation	25.396	**0.001**	Reject $H_{9,0}$ at α=0.01	office hs.
$H_{10,0}$ Emotional sup.	68.675	**0.001**	Reject $H_{10,0}$ at α=0.01	office hs.
$H_{11,0}$ Overall effect.	43.111	**0.001**	Reject $H_{11,0}$ at α=0.01	office hs. & e-mail

CONCLUSION

Our conclusions supported by the nonparametric statistical analysis are that faculty expressed a preference for e-mail for the variables convenience, retention, and efficiency. Office hours were preferred for confrontation, confidentiality, and emotional support. Equal preference for e-mail and office hours were expressed for accuracy, receiving work, reaching a consensus, and overall effectiveness. The same preference for the three communication channels (office hours, e-mail, and phone) was expressed for timeliness criterion (see Table 3).

DIRECTIONS FOR FURTHER RESEARCH

These findings should impact both managers and educators and their choice of communications media. As educators, we all should seek the most effective and efficient media for the various communication tasks. The importance of communication between employees as well as faculty and students outside of the classroom environment should provide ample motivation for additional research into this topic.

This survey is limited in scope, and should be expanded to cover a larger sample of faculty. The demographics of the studied population may not be typical of the entire educational environment. The culture studied by this survey may be atypical, or further research may find that there a number of subcultures within an organization or university, each with its own communications practices.

This survey indicates that e-mail is becoming a more important media for communications. Further research is needed into the effectiveness of various communications media and tasks so that both business professionals and faculty may communicate more effectively.

It is hoped that this article will foster more research into the relationships between communications media and tasks, so that more effective and efficient communications will occur both in organizations and universities.

A previous version of this article was presented at the College Teaching and Learning Conference, which was held in Orlando, Florida, from January 5-9, 2004, and will appear in the *International Journal of Innovation and Learning*.

The survey may be obtained by e-mail from the authors.

REFERENCES

Beheruz, S., Barnes, C. , Burst, M., & Kaye, L. (1999, July-August). E-mail communications in colleges: Are they private? *Journal of Education for Business, 74*(6), 347-350.

Carlson, P. J., & Davis, G. B. (1998, September). An investigation of media selection among directors and managers: From "self" to "other" orientation. *MIS Quarterly, 22*(3), 335-352.

Clyde, A. (1999, October). The traveler's guide to e-mail access. *Teacher Librarian, 27*(1), 64-66.

Conover, W. J. (1980). *Practical nonparametric statistics* (2nd ed.). New York: John Wiley & Sons.

Daft, R. L., & Lengel, R. H. (1986). Organizational information requirements: Media richness and structural design. *Management Science, 32*, 554-571.

Durlak, J. T. (1987). A typology of interactive media. In M. L. McLauglin (Ed.), *Communication yearbook 10* (pp. 743-757). Beverly Hills, CA: Sage.

Fenn, G. L., & Smeltzer, L. R. (1989) Communication attributes used by small business owner/managers for operational decision making. *Journal of Business Communication, 26*, 305-321.

Fulk, J., Stienfield, C. W., Schmitz, J., & Power, J. G. (1987). A social information processing model of media use in organizations. *Communications Research, 14*, 529-552.

Gefen, D., & Straub, D. W. (1997, December). Gender differences in the perception and use of e-mail: An extension to the technology acceptance model. *MIS Quarterly, 21*(4), 389-400.

Johnson, J. D., Chang, H., Pobocok, S., Etherington, C., Reusch, D., & Wooldridge, J. (2000). Functional work groups and evaluations of communication channels: Comparisons of six competing theoretical perspectives. *Journal of Computer-Mediated Communication, 6*(1).

Karahanna, E., & Straub, D. (1999).The psychological origins of perceived usefulness and ease-of-use. *Information Management, 35*(4), 237-250.

Marcus, M. L. (1994), Electronic mail as the medium of managerial choice. *Organizational Science, 5*, 502-527.

Reinsch, N. L., & Beswick, R. W. (1990). Voice mail versus conventional channels: A cost minimization analysis of individual's preferences. *Academy of Management Journal, 33*, 601-616.

Rice, R. E. (1993). Media appropriateness: Using social presence theory to compare traditional and new organizational media. *Human Communications Research, 19*, 451-484.

Sitkin, S.B., Sutcliffe, K. M., & Barsios-Choplin, J. R. (1992) A dual-capacity model of communications media choice in organizations. *Human Communication Research, 18*, 563-598.

Steinfeld, C. W., Jin, B., & Ku, L.L. (1987) A preliminary test of a social information processing odel of media use in organizations. *Proceedings of the International Communications Association,* Montreal, Canada.

Trevino, L. K., Lengel, R., & Daft, R. L. (1987). Media symbolism, media richness, and media choice in organizations: A symbolic interactionist perspective. *Communications Research, 14*(5), 553-574.

KEY TERMS

Communications Media: The channels through which communication travels, such as e-mail, face to face, or telephone.

E-Mail: The use of computers and networks to pass messages from sender to receiver.

Face to Face: Communication that take place verbally between individuals.

Office Hours: An example of face-to-face communication in the academic arena.

Telephone: Communication that takes place over the telephone.

Collaboration: Interactions that take place between two individuals, such as negotiation, information sharing, and communication.

APPENDIX A

Survey

Faculty Perceptions of Communication Channels

1. What is your academic rank?   Lecturer ❑   Adjunct ❑   Assistant ❑   Associate ❑   Full ❑

2. What is your gender?   F ❑   M ❑

3. What is your age? _____ years

4. What college / department do you teach in?

❑   College of Nursing
❑   College of Arts and Sciences
❑   Language and literature
❑   Psychology and Sociology
❑   Social Sciences
❑   Fine and Performing Arts
❑   Mathematical and Physical Sciences
❑   Biology and Chemistry
❑   College of Business Administration
❑   MIS and Decision Sciences
❑   Accounting, Finance, and Economics
❑   Management, Marketing, and International Business
❑   College of Education
❑   Curriculum and Instruction
❑   Professional Programs
❑   Special Populations

5. Which of the following are included in your syllabus?

❑   E-mail
❑   Office Phone
❑   Fax Phone
❑   Home Phone

6. How many hours of scheduled office hours do you have per week? _____

7. What percentage of the total time allocated for office hours have you spent with students in course related discussions? _____%

continued on following page

8. What percentage of your student—faculty communication about course work out of the classroom is by? (Please break down the percentages by courses taught; row sum=100%):

Course	Office hours %	E-mail %	Phone %	Fax %

Please rate each of the communication media on a scale of 1 to 7:

1	2	3	4	5	6	7

UNIMPORTANT   IMPORTANT

with 1 being lowest (unimportant) and 7 being highest (important), for the following questions

9. How do you rate each media for timeliness?
   Office Hours _____ *E-mail* _____ *Telephone*

10. How do you rate each media for accuracy of the communication?
    Office Hours _____ *E-mail* _____ *Telephone*

11. How do you rate each media for your convenience?
    Office Hours _____ *E-mail* _____ *Telephone*

12. How do you rate each media for retention of the communication (after 2 or 3 weeks)?
    Office Hours _____ *E-mail* _____ *Telephone*

13. How do you rate each media for confidentiality?
    Office Hours _____ *E-mail* _____ *Telephone*

14. How do you rate each media for efficiency in assigning new work?
    Office Hours _____ *E-mail* _____ *Telephone*

15. How do you rate each media for receiving work from students?
    Office Hours _____ *E-mail* _____ *Telephone*

16. How do you rate each media for reaching a consensus between student and faculty?
    Office Hours _____ *E-mail* _____ *Telephone*

17. How do you rate each media for conducting meetings where a confrontation is likely?
    Office Hours _____ *E-mail* _____ *Telephone*

18. How do you rate each media for meetings in which you will be providing emotional support or motivation to students?
    Office Hours _____ *E-mail* _____ *Telephone*

19. How do you rate each media for overall effectiveness in communication?
    Office Hours _____ *E-mail* _____ *Telephone*

Gender Differences and Cultural Orientation in E-Collaboration

G

Yingqin Zhong
National University of Singapore, Singapore

Zhen Wang
National University of Singapore, Singapore

John Lim
National University of Singapore, Singapore

INTRODUCTION

Electronic collaboration (e-collaboration) is defined as collaboration among individuals engaged in a common task using electronic technologies (Kock & Nosek, 2005). Examples of technologies used in e-collaboration are email, groupware and chat tools, which support communication, information sharing and coordination among team members synchronously or asynchronously. Collaborative learning is regarded as an important information processing activity in e-collaboration. E-collaboration technologies enable affective learning related to interactive communication and teamwork to be achieved. Members learn from one another by actively engaging in exchanging knowledge and information based on their understanding as well as individual experiences (Harasim, Hiltz, Teles, & Turoff, 1995). New information is in this way integrated with existing cognitive structures (Leidner & Jarvenpaa, 1995). Growing interest in supporting the needs of active learning, along with concurrent improvements in e-collaboration technologies, has prompted research on computer-supported collaborative learning (CSCL).

CSCL research focuses on the interaction of computer-supported learning systems and collaborative systems by integrating collaborative learning and e-collaboration (O'Malley, 1995). Mediating through e-collaboration technologies, members' perceptions on status, roles and power repartition perceived by group members affect their participation in collaborative learning activities (Rutkowski, Vogel, van Genuchten, Bemelmans, & Favier, 2002). Gender has been considered as one of the fundamental personal characteristics having profound influences on individual perceptions, attitudes, and performance, particularly in the collaborative learning setting (Morris, Venkatesh, & Ackerman, 2005). Morris et al. also promoted the importance of studying such influences. Gender differences are more profound in individualistic culture than in collectivistic culture. Hence, users' cultural orientation is another pertinent factor impacting jointly with gender on distributed e-collaboration teams (Rutkowski et al., 2002).

CSCL technologies make it possible for globally distributed teams to work on projects across cultures. Previous studies have highlighted the importance of individualism versus collectivism, and considered this dimension as the most distinguishing characteristic of culture[1] influencing group process (e.g., Oetzel, 2001; Triandis, 1995). People with different cultural backgrounds differ in perceptions on learning and working (Choong & Salvendy, 1999). Differences stemming from the contrasting cultural backgrounds would lead to different communication processes (Oetzel, 2001); this underscores the importance in building a set of group norms, which, if established, could potentially overcome the challenges (Levi, 2001). A set of collaborative group norms, through promoting equal participation and group wellbeing, leads to the learning achievement of every group member.

We adapt the expectation status theory in this study to investigate how and to what extent gender, cultural orientation and group norms affect users' self-perceived influential status and their participation in the context of CSCL.

BACKGROUND

The expectation states theory (Berger, Rosenholtz, & Zelditch, 1980) suggests that group member tend to evaluate other members on the basis of stereotypical performance expectation, which is influenced by status characteristics, particularly in the initial collaboration phase in CSCL activities. Status characteristic is a characteristic of a member associated with distinct performance expectation (Berger et al., 1980). For instance, gender is a diffuse status characteristic defined in the society, and it entails expectations for normatively appropriate behaviors in social contexts (Herschel, 1994). According to Augustinova, Oberlé, and Stasser (2005), it is generally recognized that people may approach impending group work quite differently depending on how they view themselves relative to other members. In this paper, we term this perceptual status as *self-perceived influential status.*

Members with collectivistic culture background value more the group needs and goals, social norms and duty, and group cooperation (Cox, Lobel, & McLeod, 1991). In contrast, members with individualistic culture background emphasize on self-interest and belief (Bontempo & Rivero, 1992). They tend to value more personal time and freedom (Massey, Montoya-Weiss, Hung, & Ramesh, 2001). Gender differences appear to be more pronounced in individualistic culture, as people in such culture are expected to act according to their own interest (Hofstede, 1991). In Mortenson's (2002) experiment, gender based behavior was only supported in subjects associated with individualistic culture. Watkins' study (1998) also showed that the notion of women valuing social relationship is more significant in countries with individualistic culture.

Group norms are expected behaviors that each group member has implicitly or explicitly accepted to follow to the best of their ability (Hare, 1976). Group norms can be defined and created explicitly in the group; moreover, they can be influenced and changed by members' past experiences (e.g., cultural orientation), initial collaboration behaviors among members, and critical events in the group history (Feldman, 1984). Group norms are considered as an effective measure in coordinating distributed teams in the use of new media (Levi, 2001). By establishing a set of group norms that promotes equal participation and group wellbeing, it can encourage all members to participate actively in CSCL activities.

MAIN THRUST OF THE ARTICLE

A Conceptual Framework

The expectation states theory highlights gender differences on members' self-perceived influential status, and in turn their participation in collaborative learning; this study expands the theory to the e-collaboration context by incorporating the analysis on technology capability. The effects of self-perceived influential status on participation are moderated by group norms; however, group norms tend to be highly influenced by members' cultural orientation, especially during initial contacts among members. Deducing from the preceding discussion, we put forth a conceptual framework that identifies the inter-relationships among technology, gender, cultural orientation, group norms and self-perceived influential status and consequently participation in e-collaboration (see Figure 1). The following deliberates on the individual components of the proposed framework.

Gender Differences on Self-Perceived Influential Status

Owing to the lower social presence in computer-mediated communication, members have relative less information and understanding about other members (Bargh & McKenna, 2004). The availability of social context cues may determine a learner's perception of other group members. Gender is one of the salient status characteristics, which influences members' own perception about their status in terms of competence in attaining group goals comparatively to other group members (Pelled, 1996).

According to the literature, men are generally task-oriented, independent, and value self-sufficiency; they prefer conflict as its outcome largely determines one's status in the interaction (Hofstede, 1991). In contrast, women are relationship-oriented, affiliable, and nurturing (Kray, Galinsky, & Thompson, 2002). At the presence of conflicts, women are likely to be accommodating with the purpose of ending the conflicts quickly to preserve the relationship. It has been found that women tend to use stricter standards on themselves and other women but employ looser standards on men in group activities, whereas males reserved more lenient standards for themselves but applied the more stringent criteria to their female members (Foschi, 1996). Con-

Figure 1. Framework of gender differences and cultural orientation on participation in e-collaboration

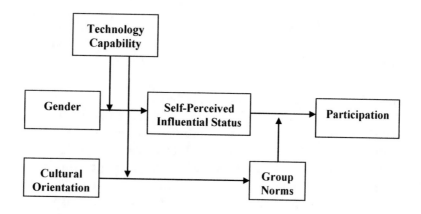

sequently, women tend to report lower self-perceived influential status as compared to the men, regardless of their actual abilities.

Effects of Technology on Gender Differences

The nature of the collaboration content embedded in the CSCL environment influences members' self-perceived influential status in mixed gender groups. In particular, the nature of the e-collaboration topic, task and related activities tend to affect female members' interest in participation substantially (Brunner, Bennett, & Honey, 2000). For example, female members are very interested in social dilemmas solving, intensive writing, and flexible pursuit of problem solving when involving in CSCL. Therefore, in order to foster equal participation of all members, it is important to support gender traits profoundly in designing interfaces for CSCL (Choong & Salvendy, 1999).

Attention is needed to introduce the new technologies and computer software in gender neutral or even "girl-friendly" settings so as to more actively encourage female participation, as well as accord them with more prominent positions (Hakkarainen & Palonen, 2003). To overcome the low status triggered by gender differences perceived by female members in e-collaboration, anonymity is one possible way to overcome their inner restraints and evaluation apprehension, and in turn encourage more efforts and participation in the collaborative learning (Hakkarainen & Palonen, 2003). Moreover, the use of anonymity functions, without indicating the author of a message, encourages more

equal participation between women and men due to the minimization of status- and gender-marked cues, which are often biased, in the collaboration process. This is because all members, especially female members, perceive themselves to be judged based on the quality of their contributions rather than on gender and other stereotypical dues (Rutkowski et al., 2002).

Self-Perceived Influential Status and Participation

People may approach impending group work quite differently depending on how they view themselves relative to other group members (Augustinova et al., 2005). In particular, members' participation is related to their self-perceived influential status which describes members' own perception about their status in the group in terms of competence in attaining group goals as compared to other group members. According to Campbell (1990), self-initiated efforts and responses to others' requests are two forms of individual behaviors caused by perception in group participation. Self-initiated effort refers to contribution which a group member expends on the group at his/her own inutility and continue working even under adverse conditions. Response to others' requests is defined as an attempt in helping others to solve problems towards the group goal. Self-enhancement is a basic need for people to see themselves in a positive light in relation to relevant others (Pratt, 1998). As a result, if a member perceives himself/herself to be of a low influential status, (i.e., he/she tends to perceive one's contribution in the collaborative learning would have relatively low influential power in the group), this member demonstrates less self-initiated efforts and less responses to others' requests.

Effects of Group Norms on Participation

It is important to establish a set of *collaborative* group norms that promote equal participation and group well-being to lead group members to active participation in sharing knowledge with other members (Kock & Davison, 2003). Before establishing of the group norms, individual members' cultural backgrounds influence how members collaborate and communicate (Feldman, 1984). Oetzel (2001) has identified the importance of a member's cultural orientation as an important factor influencing his/her participation in groups under adverse conditions. The underlying reason is that people from a collectivistic culture are presumed to care for the development of other members, whereas individualistics only care for their self-development (Hofstede, 1991). For example, Gabrenya, Wang, and Latane (1985) found that individualistics are found to withhold efforts in group activities, whereas collectivistics demonstrate more efforts to contribute in collaboration. Consistently, collectivistics are motivated to find a way to fit into the group, to fulfill and create obligation, and in general to become part of various interpersonal relationships (Markus & Kitayama, 1991). This connotes the potential and importance of cultivating a set of collaborative group norms that encourage equal participation and emphasizes the learning opportunities of all members in CSCL, especially when members are from individualistic culture.

To promote equal participation from both female and male members in CSCL, the importance of a set of collaborative group norms should be highlighted (Hakkarainen & Palonen, 2003; Mayer-Smith, Pedretti, & Woodrow, 2000). The collaborative group norms promote cooperation and inclusion for both genders in collaboration, information sharing to achieve complex goals, and further self-development among all members regardless the differences in members' perceiving their influential status (Goncalo & Staw, 2006; Tong & Klecun, 2004). In particular, the collaborative group norms will engage female members to active participation substantively, as they have higher affiliation needs and are thus more likely to conform to group norms (Morris et al., 2005).

Effects of Technology in Developing Group Norms

In distributed teams, CSCL technologies play an essential role in cultivating coordination among members by providing communication support. Rich communication media enable faster information exchange, more personal cues, and greater variety of communication channels and languages (Daft & Lengel, 1984). The use of computer-mediated communication channels can help to increase the level of perceived commitment and trust in the other party. This communication support also helps members to acquaint with one another in terms of interests and habits, thus smoothing the building of normative links among members. Moreover, by establishing the collaborative group norms with technical tools to structure the group process, collaborative learning would be enhanced in terms of more evenly distributed information, better understanding of the task, and clearer task division and responsibilities allocation as compared to traditional face-to-face settings (Cramton, 2001). Besides, the use of anonymity feature in certain discussion sessions potentially facilitates building group norms which encourage group members to contribute (Dennis, Wixom, & Vandenberg, 2001).

FUTURE TRENDS

Future work should enhance the framework proposed by examining other factors including learning task characteristic and time. Task characteristic affects members' motivation and self-evaluation in participation; a match between interests and task would strengthen one's motivation to participate in collaborative learning. Besides, research could also study task and technology fit in terms of communication and information processing support (Zigurs & Buckland, 1998). Furthermore, in designing group task, researchers may explore in what ways team members are likely to establish commitment and consensus on decisions. More studies are needed to look into how IT can facilitate in this process. A richer medium can be chosen, as it enables faster feedbacks, richer language, and more personal cues for members to reach a higher level of trust and internalized clear mutual goals. Lastly, time is another factor which receives growing attention from researchers, such as McGrath's (1991) TIP (time, interaction, performance) model, and Walther's (1992) SIP (social information processing) theory. In distributed teams, whether and how the length of the collaboration period affects member's initial disposition and expectation would be an interesting issue for future studies.

CONCLUSION

CSCL enables promising mechanisms to improve learning in distributed groups. In view of globalization, distributed learning teams have evoked the need to understand issues and effects related to gender differences and cultural orientation (in terms of individualism and collectivism) among users, as well as consequences including user participation—all in the realm of e-collaboration. Based on the expectation state theory, this study looks into how and to what extent gender, cultural orientation and group norms affect users' self-perceived influential status and their participation in the context of CSCL. This paper synthesizes related works in Information System research and group literature, and would help uncover and highlight pertinent elements in relation to supporting distributed groups in collaborative learning; a conceptual framework is formulated to provide guidelines in eliminating undesirable effects on members' perceptions triggered by gender differences.

REFERENCES

Augustinova, M., Oberlé, D., & Stasser, G. L. (2005). Differential access to information and anticipated group interaction: Impact on individual reasoning. *Journal of Personality and Social Psychology, 88*(4), 619-631.

Bargh, J., & McKenna, K. (2004). The Internet and social life. *Annual Review of Psychology, 55*(1), 573-590.

Berger, J., Rosenholtz, S. J., & Zelditch, M., Jr. (1980). Status organizing processes. *Annual Review of Sociology, 6,* 479-508.

Bontempo, R., & Rivero, J. C. (1992). *Cultural variation in cognition: The role of self-concept in the attitude behavior link.* Paper presented at the meeting of the American Academy of Management, Las Vegas, NV.

Brunner, C., Bennett, D., & Honey, M. (1998). Girl games and technological desire. In J. Cassell & H.

Jenkins (Eds.), *From Barbie to Mortal Kombat: Gender and computer games* (pp. 72-88). Cambridge, MA: MIT Press.

Choong, Y. Y., & Salvendy, G. (1999). Implications for design of computer interfaces for Chinese users in mainland China. *International Journal of Human Computer Interaction, 11*(1), 29-46.

Compbell, J. P. (1990). Modeling the performance prediction problem in industrial and organizational psychology. In M. D. Dunnette & L. M. Hough (Eds.), *Handbook of industrial organizational psychology* (Vol. 1, 687-732). Palo Alto, CA: Consulting Psychologists Press.

Cox, T., Lobel, S., & McLeod, P. (1991). Effects of ethnic group cultural differences on cooperative and competitive behavior on a group task. *Academy of Management Journal, 34*(4), 827-847.

Cramton, C. D. (2001). The mutual knowledge problem and its consequences for dispersed collaboration. *Organization Science, 12*(3), 346-371.

Daft, R. L., & Lengel, R. H. (1984). Information richness: a new approach to managerial behavior and organizational design. In L. L. Cummings, & B.M. Staw (Eds.), *Research in organizational behavior* (Vol. 6, pp. 191-233). Homewood, IL: JAI Press.

Davison, R., & Vreede, G. J. de (2001). The global application of collaborative technologies, *Communications of the ACM, 44*(12), 69-70.

Dennis, A. R., Wixom, B. H., & Vandenberg, R. J. (2001). Understanding fit and appropriation effects in group support systems via meta-analysis. *MIS Quarterly, 25*(2), 167-193.

Feldman, D. C. (1984). The development and enforcement of group norms. *Academy of Management Review, 9*(1), 47-53.

Foschi, M. (1996). Double standards in the evaluation of men and women. *Social Psychology Quarterly, 59*(3), 237-254.

Gabrenya, W. K., Wang, Y., & Latane, B. (1985). Social loafing on an optimizing task: Cross-cultural differences among Chinese and Americans. *Journal of Cross-Cultural Psychology, 16,* 223-242.

Goncalo, J. A., & Staw, B. M. (2006). Individual-collectivism and group creativity. *Organizational Behavior and Human Decision Processes, 100,* 96-109.

Hakkarainen, K., & Palonen, T. (2003). Patterns of female and male students' participation in peer interaction in computer-supported learning. *Computers & Education, 40,* 327-342.

Harasim, L., Hiltz, S. R., Teles, L., & Turoff, M. (1995). *Learning networks: A field guide to teaching and learning online.* MA: The MIT Press.

Hare, A. P. (1976). *Handbook of small group research.* New York: The Free Press.

Herschel, R. T. (1994). The impact of varying gender composition on group brainstorming performance in a GSS environment. *Computers in Human Behavior, 10*(2), 209-224.

Hofstede, G. (1991). *Cultures and organizations.* Berkshire: McGraw-Hill.

Kock, N., & Davison, R. (2003). Can lean media support knowledge sharing? Investigating a hidden advantage of process improvement. *IEEE Transactions on Engineering Management, 50*(2), 151-163.

Kock, N., & Nosek, J. (2005). Expanding the boundaries of e-collaboration. *IEEE Transactions on Professional Communication, 48*(1), 1-9

Kray, L. J., Galinsky, A., & Thompson, L. (2002). Reversing the gender gap in negotiations: An exploration of stereotype regeneration. *Organizational Behavior and Human Decision Processes, 87*(2), 386-409.

Leidner, D., & Jarvenpaa, S. L. (1995). The use of information technology to enhance management school education: A theoretical view. *MIS Quarterly, 19*(3), 265-291.

Levi, D. (2001). *Group dynamics for teams.* Thousand Oaks, CA: Sage.

Markus, H., & Kitayama, S. (1991). Culture and the self: Implications for cognition, emotion, and motivation. *Psychological Review, 98*(2), 224-253.

Massey, A., Montoya-Weiss, M., Hung, C., & Ramesh, V. (2001). Cultural perceptions of task-technology fit. *Communications of the ACM, 44*(12), 83-84.

Mayer-Smith, J., Pedretti, E., & Woodrow, J. (2000). Closing of the gender gap in technology enriched science education: A case study. *Computers & Education, 35*(1), 51-63.

McGrath, J. E. (1991). Time, interaction, and performance (TIP): A theory of groups. *Small Group Research, 22*(2), 147-174.

Morris, M. G., Venkatesh, V., & Ackerman, P. L. (2005). Gender and age differences in employee decisions about new technology: an extension to the theory of planned behavior. *IEEE Transactions on Engineering Management, 52*(1), 69-84.

Mortenson, S. T. (2002). Sex, communication values, and cultural values: Individualism-collectivism as a mediator of sex differences in communication values in two cultures. *Communication Reports, 15,* 57-70.

Oetzel, J. G. (2001). Self-construals, communication processes, and group outcomes in homogeneous and heterogeneous groups, *Small Group Research, 32*(1), 19-54.

O'Malley, C. (1995). Designing computer support for collaborative learning. In C. O'Malley (Ed.), *Computer supported collaborative learning,* (pp. 283-297). Berlin, Germany: Springer Verlag.

Pelled, L. H. (1996). Demographic diversity, conflict, and work group outcomes: An intervening process theory. *Organization Science, 7*(6), 615-631.

Pratt, M. C. (1998). To be or not to be? Central questions in organizational identification. In D. A. Whetten & P. C. Godfrey (Eds.), *Identity in organizations: Building theory through conversations* (pp. 171-207). Thousand Oaks, CA: Sage.

Rutkowski, A. F., Vogel, D. R., van Genuchten, M., Bemelmans, T. M. A., & Favier, M. (2002). E-collaboration: The reality of virtuality. *IEEE Transactions on Professional Communication, 45*(4), 219-231.

Tong, A., & Klecun, E. (2004). Towards accommodating gender differences in multimedia communication. *IEEE Transactions on Professional Communication, 47*(2), 118-129.

Triandis, H. C. (1995). *Individualism & collectivism.* Boulder, CO: Westview Press.

Walther, J. B. (1992). Interpersonal effects in computer-mediated interaction: A relational perspective. *Communication Research, 19*(1), 52-90.

Zigurs, I., & Buckland, B. K. (1998). A theory of task/technology fit and group support systems effectiveness. *MIS Quarterly, 22*(3), 313-334.

KEY TERMS

Anonymity: Function to enable group members to contribute comments without being identified.

Collaborative Learning: Activity in which a group of people learn from one another by actively engaging in exchanging knowledge and information based on their understanding and experience.

Computer Supported Collaborative Learning (CSCL): Collaborative learning activities that are supported by electronic technologies.

Distributed Team: Team supported by electronic technologies to enable the collaboration of members across different locations and time zones.

Electronic Collaboration (E-Collaboration): Collaboration among individuals engaged in a common task using electronic technologies.

Group Norms: Expected behaviors that group members have agreed to follow to the best of their ability.

Status Characteristic: Characteristic of a group member that is associated with distinct performance expectation.

A Generic Definition of Collaborative Working Environments

Karl A. Hribernik
Breman Institute of Industrial Technology and Applied Work Science (BIBA), Germany

Klaus-Dieter Thoben
Breman Institute of Industrial Technology and Applied Work Science (BIBA), Germany

Michael Nilsson
Luleå University of Technology, Sweden

INTRODUCTION

Collaborative working environments (CWE) are widely recognised as representing the next step in the development of collaborative technologies over the coming decade (European Commission, 2004, 2005; Schaffers, Brodt, Pallot, & Prinz, 2006). According Isidro Laso Ballesteros of the New Working Environments unit of the European Commission, CWE is a RTD domain with the objective "to develop technologies that will allow synchronous and asynchronous real time seamless interactions between individuals who define common objectives and work actively and effectively to achieve these common goals, participating in agreed business processes." However, the term has yet to be defined in a manner upon which a common perspective towards research and development in the field can be suitably adopted. Against that background, this article attempts to develop such a perspective on the topic by elaborating a definition rooted in widely acknowledged works on the subjects of collaboration and e-collaboration.

BACKGROUND

In European Commission (2005), Ulf Dahlsten states that:

... collaborative tools combined with remote management techniques offer a new field of opportunities for borderless co-operation between corporations and skilled knowledge workers across Europe and around the world. Those geographical outsourcing initiatives which have generated fears for job losses have also proved, in well managed situations, to be a source for new investment and business opportunities. Cases of what could be called 'reverse outsourcing' are emerging and represent an unexpected return for those countries which have explored new ways of collaborative working. (European Commission, 2005)

In today's emerging knowledge society, the ability to interact and collaborate with others is becoming an increasingly important factor in capitalizing on the arising opportunities. Tools, technologies and work processes must be further developed to facilitate the greater flow of information, ideas and media between actors involved in collaborative work. The recent developments in the field of ICT technology allow individuals involved in such collaborations to become increasingly liberated in their role as knowledge workers, being enabled to interact seamlessly with colleagues, resources and information facilitating improved work processes and value generation. ICT tools, services, applications and platforms supporting collaborative work have been adopted into commercial, private, and scientific areas with varying success over the past decades. CWE represent the next step in the development of these technologies, moving from disparate, insular tools, applications, systems and services to seamlessly integrated ICT environments for collaborative work from a multi-domain perspective.

In order to provide a theoretical framework for the scope and functional requirements of CWE, the following sections analyse the concepts of collaboration and collaborative work in relation to ICT environments, to arrive at a definition for generic CWE.

DEFINING COLLABORATION

The first step in acquiring a perspective on CWE is to arrive at a common understanding of the core terms *collaboration* and *collaborative work*, respectively. According to Stoller-Schai (2003), *collaboration* is typically employed in diverse areas of applied work, and is a term which is only seldomly defined precisely. Nevertheless, many organisations use it liberally when referring either to technology or strategic orientation. Furthermore, Stoller-Schai (2003) contends that it is used in disparate contexts, often in combination or synonymously with other, similar, "co-" terms such as *coordination, communication,* or *cooperation.* The Latin prefix *co-*, common to all of these terms, may be translated with "together" or "with." In *collaboration,* the prefix is concatenated with the Latin term *laborare,* which means "to work, toil or suffer." Thus, the original meaning of the word *collaboration* is "to work together." Collaboration occurs wherever it is necessary for individuals to pool resources to complete a given task. It can apply to both the act of collaboration as well as the result. Sinclair, Fox and Bullon (1995) accordingly define *collaboration* in the former sense as "the act of working together to produce a piece of work, especially a book or some research" and in the latter as "a collaboration is a piece of work that has been produced as the result of people or groups working together." From the Latin root of the term and the presented definition, it is obvious that *collaboration* is synonymous with *collaborative work* and will be treated accordingly in the following. For the purposes of this article, the authors will not differentiate between different types of work, such as commercial, not-for-profit, scientific, private, etc.

Stoller-Schai (2003) proposes a more specific definition of collaboration as "a mutually influencing activity by one or more persons oriented toward common goals and the solution or completion of a problem or task. This takes place within a mutually agreed and created context (common syntactical space, cooperative setting) in physical co-presence using common resources."

The characteristics of what activities "collaboration" encompasses in practice can be differentiated according to individual domains. In business domains especially, the goal of collaboration is, as stated above, for a group of individuals to achieve a specific goal. From a manufacturing industry perspective, collaboration according to Thoben, Hribernik, Kirisci, and

Eschenbaecher (2003) can represent a concept which exceeds the ambition of approaches such as Supply Chain Management or Customer Relationship Management. However, regarding collaborations in the science and research domains, the emphasis is often less on the completion tasks with common goals but on knowledge sharing activities in which the goals of the individual participants may well be divergent.

FROM COLLABORATION TO E-COLLABORATION

Further clarification is necessary when looking at collaboration and its related terms from a perspective of information and communications technologies (ICT) supported collaboration. For example, Scheer, Grieble, Hans, and Zang (2005) introduce the term *C-Business* (collaborative business) in their process excellence approach, whereas the consulting company G5 Technologies Inc. uses the term *cCommerce* (collaborative commerce). The term *cCommerce* has since been developed by the Gartner Group. Both definitions can be put in relationship to terms of e-commerce and e-business, which are both widely accepted. Furthermore, Stoller-Schai (2003) extends his definition of collaboration to encompass ICT support by introducing the term e-collaboration. The inclusion of "computer mediated" differentiates e-collaboration from collaboration as follows:

This takes place within a mutually agreed and created computer-mediated context (common syntactical space, cooperative setting) in physical co-presence using common resources." However, does not restrict collaboration only to activities between individual persons: „Collaboration must be viewed as interpersonal process between one or more people or groups. As a prerequisite collaboration needs a continuous exchange on information and data. (Stoller-Schai, 2003)

The ESA study THE VOICE (Grenham, Le Duc, & Fusco, 2005) brings a further definition of the term *e-collaboration* into the discussion. In this context, it defines two layers of collaboration. The first layer represents "human-to-human interaction, which has to be supported by a 'lower level' kind of collaboration among 'machines,'" whereas the second is "pure computer-to-computer collaboration, possibly without any user awareness of this which has also the role of supporting the first level above described" (Grenham, Le Duc, & Fusco, 2005). In this definition, e-collaboration is viewed as any activity performed by means of or supported by electronic tools or infrastructure, which is aimed at achieving a goal/meta-goal and possibly involving human interaction. This is analogous to the above definitions; however, we see the concept is extended by the introduction of "pure computer-to-computer collaboration." This implies that the actual interaction can take place solely between electronic tools. However, for *e-collaboration* not to contradict the above definitions of collaboration, "pure computer-to-computer collaboration" needs to be initiated by human beings in the course of an activity oriented toward common goals and the solution or completion of a problem or task.

A further dimension of collaboration with "artificial entities" is brought into the discussion by the analysis of both traditional and innovative collaborative work scenarios as carried out in Hribernik, Nilsson, Fusco, and Niitamo (2005). Specific collaboration scenarios such as those that take place for example in domains of the construction industry (Thoben et al., 2005), emergency management (Hribernik, Ganzer, & Schmidt, 2003), maintenance or Customer Relationship Management demonstrate that collaboration often takes place around an individual product at various stages of its product lifecycle, regardless of whether virtual or physical products are concerned. In combination with emerging concepts such as intelligent products, product-instance specific information management (Hribernik, Rabe, Schumacher, & Thoben, 2005), "the Internet of things" (Fleisch & Mattern, 2005) and intelligent sensor webs, the possibility of including "intelligent" artificial entities in collaborative work is feasible within practical boundaries. Thus, the inclusion of artificial entities in the spectrum of actors of collaborative work is intended to prefigure these developments.

The above definitions stress three key aspects of ICT-supported collaboration. Firstly, collaboration is viewed as communication between actors (enterprises, institutions, groups, individuals, machines, etc.). Secondly, e-collaboration describes cooperation on a technical level which enables machines and computers to exchange data. The third aspect is coordination which is described by the exact coordination of communication, cooperation and coordination processes. These three aspects differentiate the concept of ICT-

supported collaboration, or e-collaboration, from other approaches

STAGES, ASSETS, AND FORMS OF COLLABORATION

Thoben et al. (2003) provide a multidimensional approach to collaboration, encompassing four dimensions of collaborative business with respect to extended products. The approach adopts a multidisciplinary and multisector perspective, in which human skills and competencies, heterogeneous knowledge assets, independent business models and proprietary ICT systems are transformed into cooperative and concurrent working teams, shared and accessible resources, synchronized and interconnected processes, and interoperable and open enterprise. Excluding the dimension phases of extended products' lifecycles, the remaining three dimensions provide a suitable framework for the identification of generic requirements for collaborative working environment functionality:

1. Stages of collaboration (initiation, management, operation and dissolution)
2. Forms of collaboration (ad hoc, mediated, planned and hybrid forms)
3. Assets of collaboration (people, ICT-systems, knowledge and processes)

Previous models of collaboration generally focused on a single stage and indeed mainly on the operation of collaboration through concepts such as end-to-end supply chain synchronisation, business process integration, collaborative planning or co-operative design. The remaining phases of collaboration (initiation, management and end-of-life), however, remain underrepresented.

The term *collaboration assets* refers to the resources required to carry out collaboration. These include individual people, knowledge resources, the processes involved, and finally the ICT systems employed. Within this scope, software systems are required to provide an integrated, holistic approach to support all dimensions—human cooperation, ICT interoperability, processes cross-operation and knowledge asset joint-operation (Thoben et al., 2003).

Three different forms of collaboration can be identified according to Thoben et al. (2003). The forms constitute a continuum running from long-term, planned collaboration to short-term, spontaneous forms of collaboration. In addition, a fourth form can be constructed constituting a hybrid of any of the former:

1. Planned collaboration: Long-term—collaborative enterprise processes are planned, defined and implemented.
2. Mediated collaboration: Medium-term—a mediator is employed to facilitate collaboration
3. Ad hoc collaboration: Short-term—on demand, spontaneous and task-specific collaboration
4. Hybrid **collaboration:** A collaboration including mixed planned, mediated, and/or ad hoc collaborations

In planned collaboration, institutional collaborative business processes are defined and implemented in order to drive towards well-defined, interenterprise goals.

In a mediated collaboration, a facilitator, mediator or brokerage service is in charge of mapping the demands and offers within a specific collaboration. The mediator is furthermore responsible for mitigation and the solution of conflicts, driving the collaboration process towards a win-win situation that is not defined a priori, but is subject to constant fine-tuning along time.

In the ad hoc form, collaboration is sought and realised by opportunity, on-demand and on-the-fly without specific preparatory phase or strategic objective.

However, in many cases, a given collaboration will often not be discretely identifiable as one of the above three. For example, a large construction project may superficially appear to be a planned collaboration. On a closer look, certain short-term subcontracts may prove to be mediated collaborations, and certain tasks may even be carried out as ad hoc collaborations as shown in Thoben et al. (2005). This leads to the conclusion that a fourth form of is needed which may contain a mixture of the forms identified above. This form is termed hybrid collaboration. In fact, the study of collaborative scenarios carried out in Hribernik et al. (2005) shows that 80% of the scenarios exhibit the characteristics of hybrid collaboration.

COLLABORATIVE WORKING ENVIRONMENTS

The Application Management White Paper of ITIL (Information Technology Infrastructure Library), a widely recognised customizable framework of best practices that promote quality computing services in the IT sector, defines the term *environment* in its Glossary of Term as follows:

A collection of hardware, software, network communications and procedures that work together to provide a discrete type of computer service. There may be one or more environments on a physical platform, e.g. test, production. An environment has unique features and characteristics that dictate how they are administered in similar, yet diverse manners. (ITIL, 2004)

The foundation of the environment consists of network communications functionality, upon which layers of hardware and software reside to provide a "discrete type of computer service." In this case, *support for collaborative work* is the discrete type of computer service in question. Accordingly, CWE must exhibit unique features and characteristics inherent to that specific type of environment.

Summarizing the aspects of definitions of "collaborative work" and "ICT environments" presented in the previous sections, as well as the identification of the discrete stages, forms and assets of collaboration, an ICT environment can be said, in a broad sense, to specifically support collaborative work if it fulfils the following conditions:

I.   The ICT environment provides suitable support one or more of the four stages of collaboration, with clear focus on the operation stage:

1. Initiation
2. Management
3. Operation
4. Dissolution

II. The ICT environment adequately supports one or more forms of collaboration:

1. Planned collaboration
2. Mediated collaboration
3. Ad hoc collaboration
4. Hybrid collaboration

III. The ICT environment integrates the four collaboration assets:

1. People
2. Processes
3. Knowledge
4. ICT systems

A generic collaborative working environment could thus be seen as "a collection of hardware, software, network communications and procedures" (ITIL 2004) that provides full support for optimal interactions in the space enumerated above. An actual CWE implementation can realistically be expected to provide support for a subset of conditions of the matrix.

Furthermore, the actors involved in collaborative work need to be specified within a definition of CWE. On the one hand, collaboration can justifiably be viewed as a solely human-to-human interaction process, as proposed in the definitions of Sinclair et al. (1995) and Stoller-Schai (2003). However, the emergence of technologies such as sensor webs, intelligent embedded systems, intelligent products, Smart Tags as well as their expected wide-spread adoption by industry indicate collaboration scenarios in which artificial entities may well be autonomous or passive collaborative actors. Thus, a restriction of CWE to human actors is too limiting and does not encompass the full spectrum of current development and emerging possibilities. The definition by Grenham et al. (2005) successfully integrates these possibilities in the two layers, which also emphasizes the supporting function of machine interaction.

CONCLUSION

CWE can thus be said to represent integrated collections of hardware, software, network communications and procedures which enable collaborative work across the four stages of collaboration, for all forms of collaboration and integrating all collaboration assets. CWE support distributed actors, whether individuals, teams, organisations or machines.

Assuming the details of *collaborative work* can safely be omitted from the definition for the sake of clarity, a definition of CWE is proposed as follows: "CWEs are defined as integrated and connected resources providing shared access to contents and allowing distributed actors to seamlessly work together towards common goals."

REFERENCES

European Commission. (2004). *Next generation collaborative working environments 2005-2010: First Report of the Expert Group on Collaboration@Work* (Research report). Belgium, Brussels: European Commission Information Society Directorate-General.

European Commission. (2005). *Collaboration@Work: The 2005 report on new working environments and practices* (Research report). Luxembourg, Luxembourg: Office for Official Publications of the European Communities.

Fleisch, E., & Mattern, F. (2005). *Das Internet der Dinge*. Berlin, Germany: Springer.

Grenham, A., le Duc, I., Beco, S., & Fusco, L. (2005). *ECollaboration survey technical note* (Research report). Frascati, Italy: European Space Agency.

Hribernik, K. A., Ganzer, M., & Schmidt, T. (2003). Mobile communications for emergency management. In A. Gameiro (Ed.), *Proceedings of the 12th IST Summit on Mobile and Wireless Communications—Enabling a Pervasive Wireless World* (pp. 382-386). Aveiro, Portugal: Instituto Telecomunicacoes.

Hribernik, K. A., Nilsson, M., Fusco, L., & Niitamo, V.-P. (2005). *BrainBridges IST-015982 D2.2 A set of high-level objectives, definitions and concepts* (Research report). Luleå, Sweden: Luleå University of Technology, Centre for Distance-Spanning Technology.

Hribernik, K., Rabe, L., Schumacher, J., & Thoben, K.-D. (2005). The product avatar as a product-instance-centric information management concept. In A. Bouras, B. Gurumoorthy, & R. Sudarsan (Eds.), *Proceedings of PLM'05, International Conference on Product Lifecycle Management* (pp. 10-21). Lyon, France: Inderscience Enterprises Ltd.

ITIL Office of Government Commerce IT Infrastructure Library. (2004). *OGC ITIL application management glossary of terms*. London: Office of Government Commerce.

Laso Ballesteros, I. (2003) Collaboration @ work. Grid-enabled collaborative mobile working environments. In Luczak, H. & Zink, K.J. (Eds.), *Human factors in organizational design and management -VII. Re-designing work and macroergonomics - fu-*

ture perspectives and challenges (pp. 365-371). Santa Monica, CA: IEA Press

Schaffers, H., Brodt, T., Pallot, M., & Prinz, W. (2006). *The future workspace: Perspectives on mobile and collaborative working.* Enschede, The Netherlands: Telematica Instituut.

Scheer, A.-W., Grieble, O., Hans, S., & Zang, S. (2002). Geschäftsprozessmanagement—The 2nd wave. *Information Management & Consulting, 17,* 9-15.

Sinclair, J., Fox, G., & Bullon, S. (Eds.). (1995). *Collins cobuild English dictionary.* London: HarperCollins.

Stoller-Schai, D. (2003). *E-collaboration: Die Gestaltung internetgestützter kollaborativer Handlungsfelder* (Research report). Bamberg, Germany: Difo-Druck GmbH.

Thoben, K.-D., Hribernik, K. A., Kirisci, P., & Eschenbaecher, J. (2003). Web services to support collaborative business in manufacturing networks. In F. Weber, K. S. Pawar, & K-D. Thoben (Eds.), *Enterprise engineering in the networked economy; Proceedings of the 9th International Conference on Concurrent Enterprising* (pp. 453-462). Espoo, Finland.

Thoben, K.-D., Kirisci, P., Hribernik, K., Steinmann, R., Kalbitzer, T., & Eggers, T. (2005). Die Bauorganisation im Wandel: Möglichkeiten der Prozessoptimierung durch den Einsatz mobiler Informations- und Kommunikationstechnologien auf Baustellenumgebungen. *ZWF, 100*(6), 359-364.

KEY TERMS

Ad Hoc Collaboration: Short-term, on demand, spontaneous, and task-specific collaboration.

Collaboration: A mutually influencing activity by one or more persons oriented toward common goals and the solution or completion of a problem or task. This takes place within a mutually agreed and created context (common syntactical space, cooperative setting) in physical copresence using common resources.

Collaboration Assets: The resources required to carry out collaborations. Specifically, these are defined as people, process, ICT systems and knowledge.

Collaboration Stages: The individual stages of the life cycle of any given collaboration. Specifically, these are the initiation, management, operation and dissolution stages.

Collaborative Working Environments: Collaborative working environments (CWE) are defined as integrated and connected resources providing shared access to contents and allowing distributed actors to seamlessly work together towards common goals.

Environment: A collection of hardware, software, network communications and procedures that work together to provide a discrete type of computer service. There may be one or more environments on a physical platform (e.g., test, production). An environment has unique features and characteristics that dictate how they are administered in similar, yet diverse manners.

Mediated Collaboration: Medium-term collaborations in which a mediator is employed to facilitate collaboration.

Planned Collaboration: Long-term collaborations in which collaborative enterprise processes are planned, defined, and implemented.

Global Funding of E-Collaboration Research

Ned Kock
Texas A&M International University, USA

INTRODUCTION

The term *e-collaboration* refers to collaboration among individuals involved in the execution of common tasks using electronic technologies (Kock, 2005, 2007). Therefore, e-collaboration can be seen as a broad term that refers to a range of technology-supported activities, such as those using computer-mediated communication technologies, telephone and telephone-like devices, and group support systems (Kock, 2005a, 2005b). Those technologies are generally referred to as e-collaboration technologies.

E-collaboration technologies that build on the infrastructure provided by the Internet have undoubtedly revolutionized business (Kock, 2005; Kock & Nosek, 2005; Sproull & Kiesler, 1991). They support a vast number of business transactions, whether they are business-to-business or business-to-consumer transactions (Gefen & Straub, 2003). E-collaboration technologies also support the creation of communities of consumers (Van Alstyne & Brynjolfsson, 2005), a trend that is becoming increasingly common among certain interest groups; for example, personal health product buyers, music aficionados, avid book readers.

The current trend toward increased global trade owes much of its existence to e-collaboration technologies (Standing & Benson, 2000). Many of the information and knowledge exchanges that precede the flow of goods and services within a country take place in large part—and in some cases in their entirety—through e-collaboration technologies. This is also true in connection with the flow of goods and services across national boundaries. Large U.S. automakers, for example, jointly design engine parts with offshore contractors using sophisticated e-collaboration suites. Those parts are then manufactured by the contractors, shipped to the automakers, and incorporated into car engines.

The rising price of oil has added another advantage to the use of e-collaboration technologies in business. As oil prices go up, so does the cost of face-to-face interaction between individuals located in different cities, states, or countries. The farther those individuals are, geographically speaking, the more expensive it is to have them interact face-to-face. Since almost no trade can effectively take place without the exchange of information and knowledge, the potential return on investment in e-collaboration technologies is likely to increase as time goes by.

Of course, the above scenario may not become a reality if oil prices were to go down, or cheap oil alternatives hit the market, in the next few years. Even if that were the case, there would also be other related drivers toward an increasing use of e-collaboration technologies as alternatives to commuting and travel for face-to-face interaction. One such driver is the growing body of evidence that burning fossil fuels leads to a rise in global temperatures, with potentially disastrous consequences looming on the horizon.

Alternatives to fossil fuels have their problems as well. One of them is that they regularly end up consuming a great deal of the very same fossil fuels that they are meant to replace. Electricity, for example, which is used to power hybrid cars, is often produced by burning coal or natural gas. Ethanol may be an exception, but recent studies suggest that its production on a scale large enough to replace fossil fuels may have a dramatic negative impact on the availability of grains used for human and animal food consumption. It seems that instead of trying to reduce fossil fuel emissions related to a higher demand for transportation, the use of e-collaboration technologies should be promoted as a replacement for at least some of the face-to-face interaction among geographically distributed workers.

DIVERSE NATIONAL GOVERNMENT FUNDING AGENDAS

Technologies with great e-collaboration potential usually attract government interest and soon become the target of organized government research funding. This is particularly true in countries like the United States and New Zealand, and country groups like the European Union (EU). A significant amount of government funding is channeled to research on e-collaboration every year. This is often done indirectly through the creation of funded research programs in much broader areas such as information and communication technologies.

Because different countries and country groups vary in their industry composition and natural culture, it is no surprise that they end up having markedly different agendas. For example, a number of companies that develop and commercialize e-collaboration software are primarily based in the United States—much more than in Europe. Thus, it would be natural to see a stronger emphasis on e-collaboration research using open source software in Europe than in the United States—which seems to be what is currently happening.

A good example of clearly divergent agendas in government research funding in the area of e-collaboration is the comparison between the EU and the U.S. models. In the EU, emphasis is placed on applied results, such as interconnection of rural businesses, in terms of government funding of e-collaboration research. In contrast, the U.S. tends to favor projects that will lead to original findings, which are expected to be published in selective academic publication outlets such as conferences and journals.

FUNDING IN THE EU AND THE U.S.

Table 1 provides an example of divergent approaches for funding founding in the different countries and country groups. It summarizes key characteristics of the funding model adopted by the EU, and contrasts that with the funding model employed by the main equivalent funding agency in the United States, namely the National Science Foundation. The term *principal investigator* is used to refer to the researcher who is the main coordinator of a research project.

As it can be seen from Table 1, there are key differences in the funding models employed by the EU and the United States. It is beyond the scope of this article to provide a detailed discussion of the merits of each funding model, or a detailed analysis of the likely consequences of each model in terms of research impacts on ICT development in the EU and the United States. While such discussion would undoubtedly add value to the article, the complexities associated with such a broad comparison would probably be better addressed through a book-length publication. Also, much more consultation is needed with researchers in the EU and

United States to produce such a detailed discussion. Hopefully this article will provide the motivation for this and other related initiatives.

Interestingly, one could argue that the EU model fosters research that is better aligned with the "action research" tradition (see, e.g., Kock, 2003, 2006), in which inquiry is seen as aimed at having a positive impact on the participating organizations and society at the same time as the investigation is being conducted. The U.S. model arguably fosters research that is better aligned with the "experimental research" model, whereby inquiry is guided by the goal of testing theory and related hypotheses either in laboratories or the field.

It is important to note that comparing the European Commission with the National Science Foundation presents several challenges, which means that the discussion presented in the article should be examined with some caution. One of the problems is that there are other research funding bodies in the EU other than the European Commission. The situation is the same in the United States, with several other research funding organization other than the National Science Foundation; for example, DARPA, Office of Naval Research, Army Research Institute, Air Force Research Laboratory. Nevertheless, it seems that the National Science Foundation, due to its breadth of research coverage, is the organization the fits best the notion of a U.S. counterpart of the European Commission in the EU.

It is also important to note that comparing the EU with the United States leads to some unavoidable limitations in the conclusions drawn from that comparison. While the EU and the United States present some macro-level similarities, such as economic size and overall level of development, they also are different in many aspects. While the EU is a multination body with diverse constituents, the United States is one single country. (Although some would argue that there is a lot of diversity among the States that make up the United States—e.g., a visit to the southern part of Texas may conceivably look like an overseas trip to a New York State resident.) Also, many different languages are spoken in the EU, whereas in the U.S. English is spoken by the vast majority of the population. (Spanish is also spoken but is far behind English, and is often spoken among bilinguals.)

A more detailed comparison of the EU and U.S. models for government funding of e-collaboration research is provided in an article by Kock and Antunes (2007).

Table 1. Priorities and criteria for funding in the EU and the U.S.

EU (European Commission)	U.S. (National Science Foundation)
Funding is provided to a consortium involving several organizations from different EU countries. Often more than 10 organizations and countries are represented.	Funding is provided to a principal investigator and coinvestigator, which usually are based in universities and/or research centers. Often less than three organizations from the United States are represented.
Emphasis is placed on integration with other research projects and broader EU initiatives.	Self-contained projects of high scientific impact are quite welcome.
Explicit emphasis is placed on desirable peripheral impacts of the project, such as impact on SMEs, gender diversity, and rural development.	Explicit emphasis is placed on the research component of the project, with some interest in minority inclusion and diversity, and the project's relationship with education activities.
Publication of results is not seen as very important. Development of tools-methods and their practical use, is. The expectation is that practical use will be part of the project.	Publication of results is seen as fairly important. Less emphasis is placed on practical use of tools-methods as part of the project. The idea here is that the tools-methods will be disseminated through publications, and later used by nonproject participants.
Less emphasis is placed on a controlled or semicontrolled empirical evaluation of impacts of the research project. Controlled laboratory experiments are not very welcome.	More emphasis is placed on a controlled or semi-controlled empirical evaluation of impacts of the research project. Controlled laboratory experiments are welcome.
No strict limit on number of pages in proposal.	Usually limited to 15 pages, with additional material provided in appendices.
More guidance is provided to expert reviewers, and stricter interpretation of rules is expected.	Less guidance and more leeway on what to look for as "good" elements are left at the discretion of the expert reviewers.
Evaluation is by consensus meetings, where all expert reviewers have to unanimously agree on a score in connection with an aspect of a proposal (e.g., quality of consortium, impact, etc.). Unanimous consensus is required for a funding decision to be made.	Evaluation is conducted by expert reviewers independently at first, based on an online version of the proposal. A group discussion is conducted at the end. Unanimous consensus is not needed for a funding decision to be made.
Evaluation process is very expensive and time-consuming. Expert reviewers are brought in from several countries (including countries outside the EU), and work together for several days or more (often several weeks).	Evaluation process is relatively inexpensive and not very time consuming for reviewers. Expert reviewers work together for a day, after they produce their independent reviews.

CHALLENGES AND OPPORTUNITIES

The challenges stemming from divergent government funding agendas can be placed in three main categories—divergent standards, unfair subsidies, and poor security. The challenge of divergent standards can already be seen in many governmental efforts, such as the support for different digital mobile communication methods, which in turn may make it impossible (or very costly) for different e-collaboration technologies to interact with each other.

When a country imposes a relatively high tariff on a good in order to protect and stimulate a local industry, more often than not it is accused of unfair competition. This is particularly true when a trading partner country has no similar tariff. Another more subtle way of boosting a local industry is to subsidize it through research funding, which may happen with e-collaboration technologies as well as related products and services. This is what the unfair subsidies challenge refers to. This may lead to inferior e-collaboration technologies being offered in the marketplace, because of their reduced cost enabled by subsidies, and adopted by various organizations. The final outcome will likely be a poor return on investment on those e-collaboration technologies.

Finally, the challenge of poor security refers to the difficulty of e-collaboration technology developers to secure their technologies (e.g., make them less prone to attacks by viruses, worms, spyware) if other developers use different standards. Even if the same standards are used, but different enforcement approaches exist (e.g., more relaxed laws in some countries than others), secure multicountry e-collaboration will be difficult to support in an effective manner. This will pose obstacles to multicountry use of e-collaboration technologies, which is unfortunate because of those technologies' ability to bridge large physical and temporal distance.

Divergent funding of e-collaboration research can also create a number of opportunities, particularly if the challenges above are properly addressed. A key opportunity that comes to mind here is that of innovation. That is, divergent approaches to funding may lead to different ideas to be developed, which may in turn lead to technological breakthroughs and findings in the area of e-collaboration.

CONCLUSION

Global funding of e-collaboration research seems to be on the rise. There are many possible reasons for this phenomenon, one of which is that e-collaboration technologies support key contemporary economic trends. A key trend is that of international outsourcing, which itself is part of another more generic trend—the one toward increasing global trade. This article looks at what seem to be some of the current trends in global funding of e-collaboration research, and explores challenges and opportunities associated with these trends.

Different approaches to funding adopted by various countries may have their pros and cons. The fact that those approaches differ may bring about challenges and opportunities. It seems unlikely that funding models used by different countries will converge in the foresee-

able future. Therefore, governments may want to start funding a new form of e-collaboration research, namely the study of global integration of different national e-collaboration tools development and investigation initiatives. This meta-research topic has been largely neglected in the past.

Acknowledgment

That author would like to thank Jaime Ortiz for comments and suggestions on an earlier version of this article. The opinions expressed in this article, as well as any errors and omissions, are the sole responsibility of the author.

REFERENCES

Gefen, D., & Straub, D. W. (2003). Managing user trust in B2C e-services. *e-Service Journal, 2*(2), 7-24.

Kock, N. (2003). Action research: Lessons learned from a multi-iteration study of computer-mediated communication in groups. *IEEE Transactions on Professional Communication, 46*(2), 105-128.

Kock, N. (2005a). *Business process improvement through e-collaboration: Knowledge sharing through the use of virtual groups.* Hershey, PA: Idea Group.

Kock, N. (2005b). What is e-collaboration? *International Journal of e-Collaboration, 1*(1), i-vii.

Kock, N. (Ed.) (2006). *Information systems action research: An applied view of emerging concepts and methods.* New York: Springer.

Kock, N. (Ed.). (2007). *Emerging e-collaboration concepts and applications.* Hershey, PA: CyberTech.

Kock, N., & Antunes, P. (2007). Government funding of e-collaboration research in the European Union: A comparison with the United States model. *International Journal of e-Collaboration, 3*(2), 36-47.

Kock, N., & Nosek, J. (2005). Expanding the boundaries of e-collaboration. *IEEE Transactions on Professional Communication, 48*(1), 1-9.

Standing, C., & Benson, S. (2000). An effective framework for evaluating policy and infrastructure issues for e-commerce. *Information Infrastructure & Policy, 6*(4), 227-237.

Sproull, L., & Kiesler, S. (1991). Computers, networks and work. *Scientific American, 265*(3), 84-91.

Van Alstyne, M., & Brynjolfsson, E. (2005). Global village or cyberbalkans: Modeling and measuring the integration of electronic communities. *Management Science, 51*(6), 851-868.

KEY TERMS

Action Research: Research approach in which inquiry is seen as aimed at having a positive impact on the participating organizations and society at the same time the investigation is being conducted.

Collaborative Task: Task that is often conducted by a group of people with support of e-collaboration technologies.

E-Collaboration: Collaboration using electronic technologies among different individuals to accomplish a common task.

E-Collaboration Technology: Comprises not only the communication medium created by an e-collaboration technology but also the technology's features that have been designed to support collaborative work.

European Union: Country group comprising 25 European countries—Austria, Belgium, Cyprus, Czech Republic, Denmark, Estonia, Finland, France, Germany, Greece, Hungary, Ireland, Italy, Latvia, Lithuania, Luxembourg, Malta, The Netherlands, Poland, Portugal, Slovakia, Slovenia, Spain, Sweden, and the United Kingdom.

Global Warming: Effect suggesting that burning fossil fuels leads to a rise in global temperatures, with potentially disastrous consequences looming on the horizon.

National Science Foundation: Main federal government agency of the United States focusing on science and technology research funding.

Governance Mechanisms for E-Collaboration

Anupam Ghosh
ICFAI Institute for Management Teachers, India

Jane Fedorowicz
Bentley College, USA

INTRODUCTION

E-collaboration, defined as "collaboration among individuals engaged in a common task using electronic technologies" (Kock, Davison, Ocker, & Wazlawick, 2001), is increasingly gaining relevance at the interorganizational level because of the growing practice of working with dispersed project teams across the globe. E-collaboration links together partners on projects and business processes that cross legal boundaries, as is the case, for example, in supply chains and in product lifecycle management (PLM) teams. General purpose computer-based collaboration tools like the Internet, e-mails, instant messaging, discussion boards, groupware, portals, blogs, and wikis are commonly used for e-collaboration (Fichter, 2005), while task-specific tools exist for many interorganizational activities such as PLM or collaborative planning, forecasting, and replenishment (CPFR).

A primary purpose of interorganizational e-collaboration is sharing of information among business partners to attain predetermined objectives. However, sharing information can be risky as other partners in the relationship may behave opportunistically, having gained access to sensitive information or intellectual property. To facilitate information sharing and succeed in e-collaboration, firms engaged in partnerships need to agree on a common governance mechanism—a set of responses to conditions of uncertainty, dependence, and opportunism that exists in a business relationship (Alvarez, Barney, & Douglas, 2003; Heide, 1994). Trust, bargaining power, and contracts are three important governance mechanisms that shape interorganizational relationships and operational performance (Alvarez et al., 2003).

This article discusses the role of these three governance mechanisms (trust, bargaining power, and contracts) in support of information sharing in an e-collaboration environment. The operational performance of a collaborative team will be dependent on how effectively the members in the team share information and coordinate their activities. To allay the fears associated with sharing sensitive information, firms participating in the collaborative effort can manage their business relationships by introducing appropriate governance mechanisms. The following sections will describe the three governance mechanisms and discuss the interdependencies among them.

BACKGROUND

Information Sharing

Sharing information enables partners to integrate shared activities and improve their collective performance (Lee, Padmanabhan, & Whang, 1997; Lee & Whang, 2000; Simatupang & Sridharan, 2002; Yu, Yan, & Chang, 2001). Information sharing leads to many real benefits within a relationship. For example, in a supply chain setting, it can help to reduce the bullwhip effect, cut stock levels, reduce the cash conversion cycle, help to locate weak partners in the chain, provide cost savings, utilize unused capacity of other chain partners, enable risk taking and postponement, and so forth (Lee et al., 1997; Yang, Burns, & Backhouse, 2005). There are many attributes of information that must be considered when partners determine the nature and quality of a relationship. These attributes tend to have a dramatic impact on the ability of the collaboration to succeed in their cooperative effort. These attributes include: accuracy, understandability, relevance, timeliness, accessibility, completeness, appropriate amount, reliability, ease of use, degree of electronic integration, mode of data transfer, frequency of information sharing, and the cost of sharing information (Davis, 1989; Epstein & King, 1982; Fedorowicz & Lee, 1998-99; Gendron, Shanks, & Alampi, 2004; Wang & Strong, 1996; Zahedi, Pelt, & Song, 2001).

In addition to ensuring that the information being shared itself meets the needs of the individuals and organizations on a project team, those setting up a relationship must also manage the inherent risks associated with sharing information (Handfield & Bechtel, 2002). These risks include the potential for losing control over strategic information, identifying others' weaknesses, sharing competitive data, and using information to interfere with others' business processes.

To reduce such risks and succeed in information sharing, firms in the partnership need to agree on a common governance mechanism that will direct their relationship. Interfirm governance mechanisms, considered collectively, serve as a strategic response to conditions of uncertainty and dependence that exist in any business relationship and work towards reducing the threat of opportunism in an exchange (Alvarez et al., 2003; Heide, 1994). The presence of governance mechanisms in interorganizational partnerships also positively affects their collective performance (Dyer, 1996; Saxton, 1997; Zaheer, McEvily, & Perrone, 1998; Wathney & Heide, 2004). Such mechanisms are used for initiating, maintaining, and terminating business relationships (Heide, 1994). We will discuss three types of governance mechanisms in the remainder of this article. These are trust, bargaining power, and contracts (Alvarez et al., 2003; Dyer, 1996; Dyer & Chu, 2003).

Trust

Trust provides a foundation for collaboration (Kramer, 1999; Komiak & Benbasat, 2004; Rousseau, Sitkin, Burt, & Camerer, 1998; Whitener, Brodt, Korsgaard, & Werner, 1998) and is an important factor in determining the success of many business relationships (Jones & George, 1998; Paul & McDaniel, 2004; Scheer & Stern, 1992). Trust is defined as a psychological state that rests upon the expectations and beliefs of one party that another party will act in a certain manner, given that the trusting party is in some way vulnerable under conditions of risk and interdependency to actions by the other party (Sako, 1991; Sako & Helper, 1998). It is a complex construct that applies both to individuals (e.g., customers, Internet users), groups of individuals (e.g., communities of practice), companies, industry groups, political entities, and multi-organizational partnerships (Kramer, 1999; Sako, 1991; Sako & Helper, 1998; Svensson, 2001, 2004). There are many types of trust that can be applicable in an interorganizational

relationship setting. Four of the most relevant include calculative, competence, integrity, and predictability trust (Komiak & Benbasat, 2004; Newell & Swan, 2000; Paul & McDaniel, 2004). Calculative trust is an ongoing, market-oriented, economic calculation where each party assesses the benefits and costs to be derived from creating and sustaining a relationship. Competence trust is the ability of a party to perform a task that it claims it can perform and that covers technical, operational, human, and financial abilities. Trust in integrity is the belief that a trustee makes good faith agreements, tells the truth, and fulfills promises. Trust in predictability is the truster's belief that a trustee's actions (good or otherwise) are consistent enough that the truster can forecast them in a given situation. Each type of trust, as discussed, applies to different aspects or lifecycle phases of an interorganizational relationship.

Bargaining Power

The bargaining power of a firm gives it the "ability to bring about cost-free, intended changes in a (partner's) behavior" (Ramsay, 1996, p. 129). It is difficult to quantify, as it is a subjective concept. There are various sources of bargaining power (Cho & Chu, 1994; Cool & Henderson, 1998; Handfield & Bechtel, 2002; Lusch & Brown, 1982; Malony & Benton, 2000; Scheer & Stern, 1992). Bargaining power develops due to a partner (member) exerting control over critical resources and processes; when a member constitutes a large proportion of business for its partner; when it controls the largest share of the total value added to the final product or project; when a partner owns critical expertise or information; or when one partner has the ability to mediate punishments and rewards. Handfield and Bechtel (2002) contend that partners often possess power without exerting it. In fact, a stronger relationship emerges between the collaborators over time if member parties limit their exercise of power.

Contracts

As a governance mechanism, contracts help parties to delineate each others' authority and responsibility, and include actions to be performed and redress mechanisms. This governance mechanism encourages information sharing by formally minimizing risks. Risk is minimized by imposing penalties for opportunistic behavior (Barney & Hansen, 1994). According to Roxenhall and Ghauri (2004), though companies may not ultimately enforce a contract in the court of law, they are generally drawn up for three main reasons: to communicate information regarding commitments made to or by other parties, to support situations where a high degree of uncertainty exists about a party's ability to perform according to the contract, and because it is customary, to signify the presence of a business deal.

Certain key issues in a contract that affect information sharing include the actual existence of a written, legal document (Sen & Mitra, 2000), risk-sharing arrangement (Giannoccaro & Pontrandolfo, 2004; Greve, 2000; Holland & Fremault, 1997), length of coverage period (period for which the contract is in force) (Greve, 2000; Holland & Fremault, 1997), and responsibilities alignment (Greve, 2000; Pattison & Heron, 2003). Another key issues addressed in a contract is incentive alignment, considered by Simatupang et al. (2002) as a key coordination node in a multi-organizational relationship (Greve, 2000; Ibbs, 1991).

Trust, Bargaining Power, and Contract: Interdependencies

Prior research has revealed the interdependencies among these three governance mechanisms. Handfield and Bechtel (2002) conclude that when one party's dependence on another is high, giving rise to increased bargaining power of the latter, the use of formal contracts by the dependent party will increase. Their study does not support the idea that formal contracts increase the dependent party's level of trust in its partner. When one member's bargaining power is high, partner's trust is negatively affected. Ring and Van de Ven (1994) observe that "even when a high level of trust is present in a relationship, a reliance on trust at the interpersonal level may be conditioned by the legal systems or organizational role responsibilities, mitigating the ability of the parties to rely on trust as a matter of first preference." But this does not mean that a contract is always the precursor to trust. Foorman (1997) opines that even in an ongoing relationship, when there is solid foundation for trust the parties will still refer to the contract to test the fundamental principles on which the relationship is based. However, contracts have become such a commonality in business that Alvarez et al. (2003) contend that a contract is used independent of trust.

CONCLUSION

The combination of pervasive technological collaborative capabilities, the widespread move of outsourcing of business processes, and the creation of virtual teams that span organizational boundaries lead to the need for more formal mechanisms to govern e-collaborative relationships. Prior research has shown that uneven power distribution among collaborative partners diminishes the ability of these teams to rely on trust to govern the exchange of information and the reduction of risk in highly sensitive situations. In these cases, contracts are created to weaken power discrepancies and either substitute for or build trust among the parties.

In the future, as we learn more about how collaborative teams function in a virtual world, we will need to reexamine the role of trust, power, and contracts as protective tools for sensitive information and intellectual property. There are many research possibilities in this space. Initially, we need to document best practices where interorganizational information sharing has been deemed successful among virtual team members. Once we have a good understanding of the conditions under which virtual teams succeed (and indeed, have identified good measures of their success), we should study the relative importance of power, trust, and contracts in the success of a team's information sharing behavior, and subsequently, how these three mechanisms relate to the end performance of the team itself.

ACKNOWLEDGMENT

This article is drawn in part from Ghosh, A., & Fedorowicz, J. (2005, August). *Governance mechanisms for coordination and information sharing in supply chains: the role of trust,* Proceedings of the 11th Americas Conference on Information Systems, Omaha, NE.

REFERENCES

Alvarez, S. A., Barney, J. B., & Douglas, A. B. (2003). Trust and its alternatives. *Human Resource Management, 42*(4), 393-404.

Barney, J. B., & Hansen, M. H. (1994). Trustworthiness as a source of competitive advantage. *Strategic Management Journal, 15*(Winter), 175-190.

Cho, D.-S., & Chu, W. (1994). Determinants of bargaining power in OEM negotiations. *Industrial Marketing Management, 23*(4), 343-355.

Cool, K., & Henderson, J. (1998). Power and firm profitability in supply chains: French manufacturing industry in 1993. *Strategic Management Journal, 19*(10), 909-926.

Davis, F. D. (1989). Perceived usefulness, perceived ease of use, and user acceptance of information technology. *MIS Quarterly, 13*(3), 318-340.

Dyer, J. H. (1996). Does governance matter? *Keiretsu* alliances and asset specificity as sources of Japanese competitive advantage. *Organization Science, 7*(6), 649-666.

Dyer, J. H., & Chu, W. (2003). The role of trustworthiness in reducing transaction costs and improving performance: empirical evidence from the United States, Japan and Korea. *Organization Science, 14*(1), 57-68.

Epstein, B. J., & King, W. R. (1982). An experimental study of the value of information. *Omega, 10*(3), 249-258.

Fedorowicz, J., & Lee, Y. W. (1998-1999). Accounting information quality. *Review of Accounting Information Systems, 3*(1), 1-7.

Fichter, D. (2005). The many forms of e-collaboration: Blogs, wikis, portals, groupware, discussion boards, and instant messaging. *Online, 29*(4), 48-50.

Gendron, M., Shanks, G., & Alampi, J. (2004). Next steps in understanding information quality and its effect on decision-making and organizational effectiveness. *Proceedings of the IFIP International Conference on Network and Computing*, Wuhan, China (pp. 283-294).

Giannoccaro, I., & Pontrandolfo, P. (2004). Supply chain coordination by revenue sharing contracts. *International Journal of Production Economics, 89*(2), 131-139.

Greve, C. (2000). Exploring contracts as reinvented institutions in the Danish public sector. *Public Administration, 78*(1), 153-164.

Handfield, R. B., & Bechtel, C. (2002). The role of trust and relationship structure in improving supply chain responsiveness. *Industrial Marketing Management, 31*(4), 367-382.

Heide, J. B. (1994). Interorganizational governance in marketing channels. *Journal of Marketing, 58*(1), 71-85.

Holland, A., & Fremault, A. (1997). Features of a successful contract: Financial futures on LIFFE. *Bank of England Quarterly Bulletin, 37*(2), 181-186.

Jones, G. R., & George, J. M. (1998). The experience and evolution of trust: Implications for cooperation and teamwork. *The Academy of Management Review, 23*(3), 531-546.

Kramer, R. M. (1999). Trust and distrust in organizations: Emerging perspectives, enduring questions. *Annual Review of Psychology, 50*, 569-598.

Kock, N., Davison, R., Ocker, R., & Wazlawick, R. (2001). E-collaboration: A look at past research and future challenges. *Journal of Systems and Information Technology, 5*(1), 1-9.

Komiak, X. S., & Benbasat, I. (2004). Understanding customer trust in agent-mediated electronic commerce, web-mediated electronic commerce, and traditional commerce. *Information Technology and Management, 5*(1-2), 181-207.

Lee, H. L., Padmanabhan, V., & Whang, S. (1997). The bullwhip effect in supply chains. *Sloan management Review, 38*(3), 93-102.

Lusch, R. F., & Brown, J. R. (1982). A modified model of power in the marketing channel. *Journal of Marketing Research, 19*(3), 312-323.

Malony, M., & Benton, W. C. (2000). Power influences in the supply chain. *Journal of Business Logistics, 21*(1), 49-73.

Newell, S., & Swan, J. (2000). Trust and interorganizational networking. *Human Relations, 53*(10), 1287-1328.

Pattison, P., & Herron, D. (2003). The mountains are high and the emperor is far away: Sanctity of contract in China. *American Business Law Journal, 40*(3), 459-510.

Paul, D. L., & McDaniel, R. R. J. (2004). A field study of the effect of interpersonal trust on virtual collaborative relationship performance. *MIS Quarterly, 28*(2), 183-227.

Ramsay, J. (1996). Power measurement. *European Journal of Purchasing & Supply Management, 2*(2/3), 129-143.

Rousseau, D. M., Sitkin, S. B., Burt, R. S., & Camerer, C. (1998). Not so different after all: A cross-discipline view of trust. *The Academy of Management Review, 23*(3), 393-404.

Sako, M. (1991). The role of trust in Japanese buyer-supplier relationships. *Ricerche Economiche, 45*(2-3), 449-474.

Sako, M., & Helper, S. (1998). Determinants of trust in supplier relations: Evidence from the automotive industry in Japan and the United States. *Journal of Economic Behavior & Organization, 34*(3), 387-417.

Saxton, T. (1997). The effects of partner and relationship characteristics on alliance outcomes. *Academy of Management Journal, 40*(2), 443-456.

Scheer, L. K., & Stern, L. W. (1992). The effect of influence type and performance outcomes on attitude toward the influencer. *Journal of Marketing Research, 29*(1), 128-142.

Simatupang, T. M., & Sridharan, R. (2005). The collaboration index: A measurement for supply chain collaboration. *International Journal of Physical Distribution & Logistics Management, 35*(1), 44-62.

Svensson, G. (2001). Extending trust and mutual trust in business relationships towards a synchronized trust chain in marketing channels. *Management Decision, 39*(6), 431-440.

Svensson, G. (2004). Vulnerability in business relationships: The gap between dependence and trust. *Journal of Business and Industrial Marketing, 19*(7), 469-483.

Wang, R. Y., & Strong, D. M. (1996). Beyond accuracy: What data quality means to data consumers. *Journal of Management Informations Systems, 12*(4), 5-34.

Wathney, K. H., & Heide, J. B. (2004). Relationship governance in a supply chain network. *Journal of Marketing, 68*(1), 73-89.

Whitener, E. M., Brodt, S. E., Korsgaard, M. E., & Werner, J. M. (1998). Managers as initiators of trust: An exchange relationship framework for understanding managerial trustworthy behavior. *The Academy of Management Review, 23*(3), 513-530.

Yang, B., Burns, N. D., & Backhouse, C. J. (2005). An empirical investigation into the barriers to postponement. *International Journal of Production Research, 43*(5), 991-1005.

Yu, Z., Yan, H., & Cheng, T. C. E. (2001). Benefits of information sharing with supply chain partnerships. *Industrial Management and Data Systems, 101*(3/4), 114-119.

Zahedi, F., Pelt, W. V. V., & Song, J. (2001). A conceptual framework for international web design. *IEEE Transactions on Professional Communication, 44*(2), 83-103.

Zaheer, A., McEvily, B., & Perrone, V. (1998). The strategic value of buyer-supplier relationships. *International Journal of Purchasing and Materials Management, 34*(3), 20-26.

KEY TERMS

Bargaining Power: Refers to the ability of a firm to bring about cost-free, intended changes in its partner business firm's behavior.

Contract: A written or oral instrument of relationship governance that delineates the authority and responsibility of parties engaged in an exchange relationship, and includes actions to be performed by parties and redress mechanisms.

E-Collaboration: Collaboration among individuals or groups engaged in a common task using electronic technologies.

Governance Mechanisms: A set of responses to conditions of uncertainty, dependence, and opportunism that exists in a business relationship.

Information Sharing: Sharing of relevant data that fit the organization or project team's task in hand.

Interorganizational Collaboration: The use of information and communication technologies to support sharing of information among business partners to attain predetermined objectives.

Trust: A psychological state that rests upon expectations and beliefs of one party that another party will act in a certain manner, given that the trusting party is in some way vulnerable under conditions of risk and interdependency to actions by the other party.

Governing E-Collaboration in E-Lance Networks

Robert Hooker
Florida State University, USA

Carmen Lewis
Florida State University, USA

Hugh Smith
Florida State University, USA

Molly Wasko
Florida State University, USA

James Worrell
Florida State University, USA

Tom Yoon
Florida State University, USA

INTRODUCTION

The close of the twentieth century witnessed unprecedented advances in information and communication technology (ICT), which brought about tremendous changes to almost every facet of society. Although these advances dramatically changed the way we keep in touch, perhaps the biggest change could be in the way that we organize and conduct business transactions. Some would argue that, for the first time in human history, technology has progressed to the point where individuals can now achieve the same benefits as large organizations, without giving up the benefits of freedom, flexibility, and control (Malone, 2004). This revolution has been dubbed the "dawn of the e-lance economy," and the purpose of this article is to define the e-lance phenomenon and elaborate on how ICT enables individuals and organizations to engage in e-collaboration for the purposes of economic exchange without the strong reliance on formal contracts and control mechanisms normally associated with market exchanges, or hierarchical structures associated with formal organizations.

While there are many forms of "freelance" or networked organizations, this research focuses on e-lance networks that are aggregations of autonomous e-lancers (freelance employees integrating their efforts through networked ICTs) communicating and collaborating primarily through information and communication technologies to achieve common goals. Based on this definition, e-lancers are autonomous in that they do not share a common organizational affiliation, are goal-directed as they come together to accomplish a specific task, and are virtual due to reliance upon computer-mediated communications to coordinate efforts. While networked organizational forms are not new (i.e., the film industry), what is new about e-lance networks is the ability to coordinate work without same-time and same-place interactions through e-collaboration tools. In the e-lance economy, projects are posted by customers, requests for proposals or online bidding is transmitted electronically from suppliers, and individuals or small teams accomplish work based on their unique personal skills. Once the project is completed, the network disbands and participants pursue other opportunities.

In this article, we focus on the role of brokers as the essential facilitators of e-collaboration. E-lance brokers are Web-based and serve as online clearinghouses for information about customers and their projects, as well as suppliers of services seeking work, allowing knowledge work to be traded like a commodity. Brokers bring together those seeking services and those who can provide those services to meet the particular needs of the customer. The study of the different e-collaboration tools used by e-lance brokers provides important insights into how loosely coupled, autonomous agents exchange services through e-lance forms of organization. Examining the different e-collaboration mechanisms and how these mechanisms translate into successful transactions, is essential for understanding the future of knowledge work. Since knowledge-based work can be codified and shared electronically, such as software development, consulting, translation, and accounting, e-collaboration tools enabled through ICTs present viable alternatives to traditional models of organizing.

One organization that has been able to capitalize on the concept of e-lance to support innovation is pharmaceutical giant Eli Lilly. Recognizing that it was impossible to "own" more than a small fraction of all of the greatest scientists/scientific discoveries in the world, Eli Lilly and Company launched InnoCentive LLC, to create an open network of scientists and researchers and accelerate innovation. Through its Web site, innocentive.com, innovation-driven companies can post scientific problems to be solved by a global community of scientists and researchers in the areas of biology and chemistry. To date, innocentive.com has over 90,000 registered scientists worldwide, has awarded over $1.5 million to solvers, and notes that the success rate has been far higher than in-house performance, at around one-sixth of the cost (http://www.innocentive.com/about/newsandpress.html).

The key contribution of this article is to examine how e-collaboration between customers and suppliers is facilitated by the technical features offered by the brokers. This article will unfold as follows. First, we will define and describe network forms of governance, explaining how e-lance differs from more traditional mechanisms for exchange. Next, we will explore how e-lance brokers use ICTs to augment market controls (formal contracts and payment systems) with the social controls associated with network forms of governance to safeguard against opportunistic behavior and failure to perform. We follow with examples from one of the more popular e-lance Web sites, www.elance.com.

BACKGROUND

Transaction cost economics (TCE) proposes that there are costs associated with conducting economic exchanges, and the purpose of selecting a form of governance is to minimize those transaction costs (Williamson, 1994). The main focus of TCE is how to best minimize three sources of transaction costs: (1) the costs associated with adapting exchanges, (2) the costs associated with coordinating exchanges, and (3) the costs incurred to safeguard exchanges. For a governance form to emerge and thrive, it must address these exchange problems more efficiently than other governance forms (Williamson, 1994). The basic economics of organizations suggest that when it is cheaper to conduct transactions within the boundaries of an organization, the organization will grow. Conversely, when it is cheaper to transact externally with independent entities in the open market, organizations remain small or shrink. Therefore, current governance theories focus primarily on examining internal organizational dynamics (e.g., the resource-based view of the firm), the creation of alliances between firms (e.g., research on when to merge, acquire or create alliances), and industry structure (e.g., Porter's five forces) to improve a firm's competitive advantage.

For any governance form to emerge and thrive it must address the problems of adapting, coordinating, and safeguarding exchanges more efficiently than other forms of governance under certain exchange conditions (Williamson, 1994). For instance, demand uncertainty makes vertical integration risky, and drives the need for adaptation. Customized, or asset-specific exchanges create dependencies between entities, requiring high levels of coordination and rigid safeguards to protect against miscommunication and/or opportunism. Task complexity refers to the number of specialized inputs needed to create a product or service, and heightens the need for coordination. These exchange conditions make it untenable to use either markets or hierarchies as governance forms. The need for adaptation is best handled through markets, and inhibits the use of hierarchies. On the other hand, market mechanisms are not efficient for coordinating and safeguarding complex tasks that are customized.

The combination of these specific exchange conditions requires a more flexible form of governance that more closely resembles networks than hierarchies, where clusters of firms or specialist units coordinate interactions through relationships rather than chains of commands or contracts (Miles & Snow, 1986). Interorganizational networks provide an alternative between either relying on the open market or vertical integration for conducting business activities (Thorelli, 1986). This network form of governance has been increasingly adopted since the 1980s, when organizations were faced with increasingly competitive global demands (Miles & Snow, 1992). Coupled with these changing competitive demands, ICTs were driving down the costs associated with acquiring, processing, and sharing information. Effective ICTs reduce external coordination costs, leading to the ability to coordinate exchanges more effectively through market transactions with value-added partnerships (Gurbaxani & Whang, 1991). The ability of ICTs to significantly reduce transaction costs, especially the costs of coordination and the costs associated with risks, facilitates the development of stable, tightly coupled relationships among firms (Clemons & Row, 1992). In networks, "governance" is based on *social controls*, rather than bureaucratic structures and/or formal contractual relationships. Therefore, network governance theory is built upon the premise that the social and relational structures underlying the network create effective social controls that facilitate adapting, coordinating, and safeguarding exchanges among independent entities. Therefore, network governance has emerged as a viable alternative to markets and hierarchies for coordinating exchanges.

However, the question that needs to be addressed in the e-lance economy is how transactions can be governed when the unit of the economy is the individual, not the firm. Because information can be shared instantly and inexpensively among many people regardless of location, the value of centralized decision-making and expensive bureaucracies decreases—individuals can manage themselves, coordinating their efforts through electronic links with other independent parties (Friedman, 2005). Yet, because knowledge-intensive work often requires extensive coordination, markets are not the most efficient means to conduct these transactions. Network governance balances the competing demands of these exchange conditions, suggesting that e-lance forms of organization, supported by e-collaboration technologies, will become an increasingly important alternative for the coordination of complex knowledge work.

E-lance networks help create and support an efficient link between customers and suppliers, without incurring many of the costs associated with negotiating formal contracts, and aid in the coordination of customer needs with the services offered by suppliers. Given that the Internet has created a situation whereby information is often disorderly and inconsistent, search costs as they apply to overall transaction costs are higher for firms in the open market. When taking into account the needs for specific goods and services that an organization may have, it is no surprise that many firms try to mitigate the risk of opportunistic behavior by attempting to fulfill service and component needs in-house, thereby managing interactions through the organizational hierarchy. E-lance technology is now helping individuals, as well as large firms, learn how to further reduce transaction costs through a more coordinated flow of information.

In the e-lance economy, knowledge work is largely accomplished through fluid, temporary networks where individual e-lancers band together to exploit specific business opportunities (Malone & Laubacher, 1998). In this fluid environment, tasks are not coordinated through hierarchies, but rather by the individuals themselves. In contrast to hierarchical structures based on formal controls, e-lance networks are transient associations that are limited in duration and largely self-regulated based on social controls, such as reputation and collective sanctions (Jones, Hesterly, & Borgatti, 1997). E-lance is essentially a hybrid between traditional market forms and network governance, enabling the combination of market and network forces and creating a new form of hybrid network-market governance. This new form relies on the social controls in network governance to augment decisions made while engaging in market exchanges. These social controls include concern for reputation, collective sanctions, and restricting access to certain types of exchanges. Electronic exchanges through e-lance brokers help make these social controls more effective, as well as reduce the overall costs of transactions.

Concern for reputation helps to safeguard exchanges by increasing the amount of information network members have regarding the behavior of other network members. Because of this increased information, uncertainty regarding other members is reduced, helping to deter deceptive behavior. Although reputations can have negative economic consequences, they can also serve to enhance knowledge sharing by increasing the status of those members who make valuable contributions. E-lance environments are inherently uncertain partially because they lack face-to-face interaction that enables e-lancers to form opinions of each other. Because of this uncertainty, the e-lancer's reputation within this network plays a crucial role in future economic exchanges. At the end of a project, reputations are affected when the customer and supplier rate one another. These ratings are important because they assist other customers in deciding which supplier to select, and vice versa. Hence, in an e-lance environment, customer feedback helps to form e-lancers' reputations, which improves the coordination of future exchanges. Along with enhanced coordination, this process also facilitates safeguarding because e-lancers are less likely to act inappropriately for fear of loss in future exchanges.

Collective sanctions help to safeguard exchanges as well, by increasing the costs of opportunistic behavior and encouraging the monitoring of network members. Collective sanctions can take many forms and are employed to punish those network members who fail to conform to established norms and values. These efforts aimed at self-regulating the network can range from "flaming" (leaving poor feedback) to exclusion from the network. Collective sanctions are most effective when the norm is to punish those members who act opportunistically or fail to perform (Ostrom, 1990). One benefit of electronic environments is that the amount of monitoring is decreased for each member because interactions are transparent and available to everyone in the network. By establishing the consequences of stepping outside the bounds of what is allowed, e-lancers are deterred from misconduct. In extreme instances, e-lancers can be banned from the network. Thus, collective sanctions help to further deter opportunistic behavior in an e-lance environment by increasing the likelihood of exclusion from future transactions.

The final social control, restricting access, facilitates network governance by reducing the number of individuals requiring coordination. This decrease in the number of active network members enhances coordination by reducing the amount of variance brought to an exchange. Restricting access also encourages safeguarding exchanges by reducing transaction costs and minimizing the need for monitoring. Thus, stronger ties develop between networked individuals as a result of restricting access and safeguarding exchanges. In an e-lance environment, restricting access reduces the number of e-lancers invited to engage in an exchange. There are many benefits associated with this type of social control in e-lance networks. The first benefit is that it increases the number of interactions among the network members, leading to higher levels of cooperation. An e-lance site also benefits from restricting access by minimizing efforts associated with monitor-

ing, which enhances rational behavior between the customers and suppliers. Finally, restricting access reduces the amount of coordination required by establishing routines through repeated exchanges, allowing more efficient transactions when posting and bidding on projects.

E-LANCE IN ACTION

One of the most successful e-lance brokers is www.elance.com. Inspired by Malone and Laubacher's (1998) *Harvard Business Review* article "The Dawn of the E-lance Economy," elance.com was founded in 1998 to build the infrastructure to support Malone and Laubacher's vision. In the short time since its inception, elance.com has reached an excess of $7 billion in services under management by 2006, all within the services sector. Elance.com boasts over 2,000 e-lancers in more than 50 categories. Service sectors represented by elance.com postings include graphic design, Web design, software development, writing, translation, and market research. The success of elance.com is such that over 100,000 projects are posted annually, with this success being reported in industry and business periodicals such as *Business Week, Information Week,* and *Forbes.*

As a broker that connects customers and e-lancers, elance.com provides the information necessary to allow members of the various networks to assess each other and evaluate their trading opportunities. For e-collaboration to succeed and facilitate social controls, customers and e-lancers alike must have both access to information necessary to evaluate potential trading partners, as well as the means necessary to sanction those who do not conform to accepted norms. This is provided in several ways. First, both customers and e-lancers complete profiles that detail their location and other relevant demographics. As a broker, elance.com captures and provides statistics regarding past transactions and trading partner satisfaction. By displaying this information for all members of the network, elance.com minimizes the need for monitoring, as trading partners are presented with sufficient information necessary to evaluate whether or not to consider someone for potential services and employment.

Many of the features of elance.com lend themselves to supporting the network controls previously discussed. Because reputation is a critical social control within the e-lance economy, elance.com provides for both customers and e-lancers to evaluate their respective trading partners at the conclusion of the transaction. By capturing free-form feedback and ratings, and displaying these for both 6-month and life-to-date intervals, the network itself facilitates self-governance and either rewards or punishes participants based on service and quality. Because these transactions are computer-mediated rather than face-to-face, the threat of damage to one's reputation is a strong incentive to minimize opportunistic behavior and ensure performance.

Another key mechanism to safeguard transactions is the threat of collective sanctions. In order for collective sanctions to be effective, norms of behavior must be established. Although many of these norms are likely negotiated by the members of the network, elance.com does explicitly state some of the expected norms for "Select Status" customers and e-lancers. Those norms consistent across both customers and e-lancers include posting a complete profile, engaging in professional communication, and working with partners to define requirements. Moreover, customers are expected to post projects in good faith and pay in a timely manner, while e-lancers are to provide detailed bids and submit and negotiate written contracts as requested. For members who fail to conform to these norms, network members may impose collective sanctions. These collective sanctions may be as minor as flaming or as severe as being suspended from the network. For example, elance.com lists suspension of membership as a penalty for repeated poor feedback. In this manner, the broker facilitates the network members to be self-policing and exclude members who routinely provide suboptimal services.

Lastly, by restricting access on certain transactions, both customers and e-lancers reduce the costs of coordination and monitoring. By effectively reducing the number of trading partners available for an exchange, there is less variance in terms of services provided, which serves to increase relationships among a smaller set of partners. Elance.com enables restricted access by allowing customers to limit bidding to "Select" status e-lancers. Since these "Select" e-lancers have also committed to abide by a code of conduct, restricting access in this manner reduces uncertainty surrounding the exchange, thereby minimizing transaction costs.

CONCLUSION

In contrast to hierarchies with formal power structures and markets with contractual obligations, network governance relies on social controls as the primary source of influence on the behavior of actors in the network. In some instances, social controls are even more effective for controlling individual behavior than authority, bureaucratic rules, or standardization (Ouchi, 1979). In e-lance networks, network governance is more likely to emerge and thrive as an alternative governance form when social controls are present. Primary sources of social control include concerns about reputation, imposing collective sanctions, and restricting access to certain types of exchanges. Through the use of different e-collaboration tools, and based on the study of one of the largest e-lance brokers, it appears that interactions supported by ICTs create certain advantages that leverage and make more salient the different social controls. Therefore, it is expected that the future of knowledge work will change significantly, where work is managed through decentralized, fluid, self-managing networked forms of organization supported by e-collaboration in an e-lance economy.

NOTE

Each author contributed equally to this article.

REFERENCES

Clemons, E. K., & Row, M. C. (1992). Information technology and industrial cooperation: The changing economics of coordination and ownership. *Journal of Management Information Systems, 9*(2), 9-29.

Friedman, T. (2005). *The world is flat: A brief history of the twenty-first century.* New York: Farrar, Straus and Giroux.

Gurbaxani, V., & Whang, S. (1991). The impact of information systems on organizations and markets. *Communications of the ACM, 34*(1), 59-73.

Jones, C., Hesterly, W. S., & Borgatti, S. P. (1997). A general theory of network governance: Exchange conditions and social mechanisms. *Academy of Management Review, 22*(4), 911-945.

Malone, T. W. (2004). *The future of work*. Boston: Harvard Business School Press.

Malone, T. W., & Laubacher, R. J. (1998). The dawn of the e-lance economy. *Harvard Business Review, 76*(5), 144-153.

Miles, R. E., & Snow, C. C. (1986). Organizations: New concepts for new forms. *California Management Review, 28*(3), 62-73.

Miles, R. E., & Snow, C. C. (1992). Causes of failure in network organizations. *California Management Review, 34*(4), 53-72.

Ostrom, E. (1990). *Governing the commons: The evolution of institutions for collective action*. Cambridge: Cambridge University Press.

Ouchi, W. G. (1979). A conceptual framework for the design of organizational control mechanisms. *Management Science, 25*(9), 833-848.

Thorelli, H. B. (1986). Networks: Between markets and hierarchies. *Strategic Management Journal, 7*(1), 37-51.

Williamson, O. E. (1994). Transaction cost economics and organization theory. In N. J. Smelser & R. Swedberg (Eds.), *The handbook of economic sociology* (pp. 77-107). Princeton, NJ: Princeton University Press.

KEY TERMS

Demand Uncertainty: The lack of a predictable level of demand for services.

E-Lance Economy: New economy where knowledge work is largely accomplished through fluid, temporary networks where individual e-lancers band together to exploit specific business opportunities.

E-Lance Broker: Brings together those seeking services and those who can provide those services to meet the particular needs of the customer through e-collaboration tools.

E-Lancers: Freelance employees integrating their efforts through networked information and communication technologies.

Human Asset Specificity: The knowledge assets individuals possess in the form of specific skills or expertise regarding a particular area of interest for the firm.

Network Governance: A form of governance based on *social controls*, rather than bureaucratic structures and/or formal contractual relationships.

Social Controls: Informal mechanisms to encourage self-governance and self-control, such as concerns about reputation, imposing collective sanctions, and restricting access to certain types of exchanges.

Task Complexity: The need for specialized inputs and processes, resulting either from an increased scope of activities, number of unique functions that need

Group Size Effects in Electronic Brainstorming

Alan R. Dennis
Indiana University, USA

Michael L. Williams
Pepperdine University, USA

INTRODUCTION

Collaboration is a fundamental element of group brainstorming. Researchers have long considered how to improve collaboration to improve group brainstorming, but the general conclusion of this research is that due to problems in the communication process, people generate fewer ideas when they work together in groups than when they work separately and later pool their ideas (i.e., in "nominal groups") (Mullen, Johnson, & Salas, 1991; Paulus, Larey, & Ortega, 1995).

The goal of this brief article is to analyze the effect of group size on idea generation in both verbal and electronic brainstorming (EBS) groups. Group size effects were analyzed by a meta-analysis of 21 previously published articles. Section one reviews how group size impacts the communication process in group idea generation. Section two and three present the methods and results of our meta-analysis. Section four is a discussion of the results and implications for future research and practice.

BACKGROUND

Group brainstorming may be conducted several ways. Verbal groups allow participants to call out ideas simultaneously. Nominal groups encourage participants to first conduct individual brainstorming and compile separate lists of ideas. These ideas are then aggregated to compile a list of ideas. Finally, EBS groups use a variety of technologies to collaborate and generate ideas. EBS groups may be co-located or across time and space. In this article we compare the effect of group size on the number of ideas generated by each type of group.

Much prior EBS research follows the processes gains and losses framework (Hill, 1982; Steiner, 1972). Simply put, communication among group members introduces performance improvements (process gains) and restrictions (process losses) (see Table 1).

Potential Process Gains

Synergy is the ability of an idea from one participant to trigger a new idea in another participant, an idea that would otherwise not have been produced (Dennis & Valacich, 1993; Lamm & Trommsdorff, 1973).

Social facilitation is the ability of the presence of others to affect one's performance (Allport, 1920; Levine, Resnick, & Higgins, 1993; Zajonc, 1965). Social facilitation can have both positive and negative effects on performance (Robinson-Staveley & Cooper, 1990; Sanna, 1992).

Potential Process Losses

Production blocking refers to the need to take turns speaking in verbal brainstorming (Diehl & Stroebe, 1987). Production blocking is the single most important source of process losses in verbal brainstorming groups (Diehl & Stroebe, 1987; Gallupe, Cooper, Grise, & Bastianutti, 1994).

Evaluation apprehension may cause participants in verbal brainstorming to withhold ideas because they fear a negative reaction from other participants (Diehl & Stroebe, 1987; Lamm & Trommsdorff, 1973).

Anonymity has been shown to affect behavior in several studies (Diener, 1979; Saks & Ostrom, 1973; Siegel, Dubrovsky, Kiesler, & McGuire, 1986). Anonymity has influenced group participants to share ideas that might otherwise be withheld due to evaluative apprehension. Prior research on anonymity, however, is equivocal.

Social loafing is the tendency for individuals to expend less effort when working in a group than when working individually (Karau & Williams, 1993). Social loafing may arise because participants believe their

Table 1. Potential process gains and losses

	Nominal Group Brainstorming	Verbal Brainstorming	Electronic Brainstorming
Process Gains			
• Synergy	None	Increases as the size of the group increases	Increases as the size of the group increases
• Social Facilitation	Depends upon group structure	Some effect	Some effect
Process Losses			
• Production Blocking	None	Increases with group size	None
• Evaluation Apprehension	None	Increases with group size	None
• Social Loafing	Depends upon group structure	Increases with group size	Increases with group size
• Cognitive Interference	None	Increases with group size	Some effect
• Communication Speed	Some Effect	None	Some Effect

contributions to be dispensable and not needed for group success because responsibility for completing the task is diffused among many participants (Latane, Williams, & Harkins, 1979).

Cognitive interference is in many ways the inverse of synergy. Cognitive interference occurs when the ideas generated by other participants interfere with an individual's own idea generation activities (Pinsonneault & Barki, 1999; Straus, 1996).

Communication speed is influenced by the need to type or write rather than speak. It is found in both nominal and EBS groups. For most people, speaking is faster than typing or writing so the need to type may inhibit idea generation by slowing down communication (Nunamaker, Dennis, Valacich, Vogel, & George, 1991; Williams & Karau, 1991).

THE ROLE OF GROUP SIZE

Group size is an important moderator of idea generation because the balance of process gains and losses changes depending upon the size of the group.

Verbal brainstorming groups experience process gains of synergy and social facilitation as the group size increases. But they also suffer from process losses that increase with the size of the group due to production blocking, social loafing, evaluation apprehension, and cognitive interference. Nominal group brainstorming experiences process gains from social facilitation but no gains from synergy. Likewise, if nominal groups anonymously sum up the product of their work, they may experience some social loafing and communication speed problems, but no production blocking, evaluation apprehension, or cognitive interference.

EBS groups should experience synergy that increases with the size of the group as well as some social facilitation effects. EBS groups are also likely to suffer from cognitive interference, lower communications speed, and some social loafing that increases with group size.

Figure 1 offers a shorthand summary of these patterns. The figure does not attempt to display the detailed effects of individual process gains and losses on brainstorming methods, but merely indicates the overall trend effects for each method and the effects of group size. For example, overall process gains for both verbal and EBS groups should increase with group size to some threshold level where the value of adding another participant will be only minimally positive. Process losses in verbal brainstorming groups should increase fairly quickly as the size of the group increases; previous research suggests that losses increase more quickly than gains, because nominal groups have outperformed verbal brainstorming groups. It should be noted, however, that some of the process losses incurred by verbal brainstorming groups do not follow a linear trend. For instance, although the effect of evaluative apprehension should increase with group size, social impact theory (Latane, 1981) suggests that this effect will level off when the group reaches a threshold size.

Figure 1. Expected process gains and losses

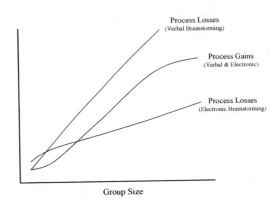

Additionally, the effect of production blocking by a single participant in a small group (e.g., 3-5 participants) will have a larger proportional effect than in a large group (e.g., 25 or more).

In contrast, process losses for EBS groups start higher (because of communications speed problems and the inherent loss of typing vs. speaking) but increase much more slowly with group size.

The primary message of Figure 1 is the indication that at some size process gains should exceed process losses, and EBS groups should outperform nominal groups.

Figure 1 suggests two conclusions. First, for verbal vs. EBS groups, we would expect large EBS groups to generate more ideas than similarly sized verbal brainstorming groups. Our expectations for smaller groups are less clear. EBS introduces a fixed communication speed process loss due to the need to type, regardless of the size of group. Conversely, small verbal groups suffer few production blocking losses, and—depending upon on the group—would also likely suffer few losses due to evaluation apprehension. We believe that for most small groups, EBS process losses should exceed the process losses in a verbal brainstorming group. Therefore, small verbal brainstorming groups should generate more ideas than similarly sized EBS groups.

Our second set of conclusions pertains to nominal vs. EBS groups. Nominal groups experience gains only from social facilitation, the same gains we could logically expect for EBS groups. EBS groups, however, experience synergy that increases with group size. Both nominal and EBS groups can be expected to experience similar losses due to social loafing (modest), production blocking (none), and evaluation apprehension

(none). EBS groups may suffer more from cognitive interference and communications speed effects. Thus we expect that small nominal brainstorming groups should generate more ideas than small EBS groups. Conversely, large EBS groups should generate more ideas than large nominal brainstorming groups.

The pattern of gains and losses in EBS versus verbal and nominal group brainstorming (see Figure 1) suggest that there should be a curvilinear relationship between size and the relative performance differences between EBS and both alternate groups.

METHOD

Selection of Studies

To analyze the effect of group size on brainstorming, we conducted a meta-analysis of 21 published articles. To locate studies, we performed computer searches on 11 databases, did manual searches through likely MIS, psychology, and management journals and conference proceedings, and read previous literature reviews[1]. We focused only on published refereed papers, including journal articles and conference papers but omitting dissertations and working papers. We also omitted studies of e-mail and "chat" software. All selected papers compared a treatment group (EBS use) to a control group (verbal brainstorming or nominal groups brainstorming) and provided means and standard deviations (or other statistics) for the number of ideas produced.

Studies containing several experiments or multiple conditions with different group sizes were disaggregated and treated as separate data points (Hunter & Schmidt, 1990). For example, Gallupe, Cooper, Grise, and Bastianutti (1992) reported the comparison between EBS and verbal brainstorming groups for five levels of group size. Since individuals were randomly assigned to conditions, we treated the five levels of group size as independent studies for the analysis. This procedure resulted in a data set with 22 usable data points for the EBS versus verbal brainstorming comparison and 14 usable data points for the EBS versus nominal group brainstorming comparison.

Analysis Technique

It is beyond the scope of this article to describe in detail the statistical algorithm underlying a meta-analysis (see Hunter & Schmidt, 1990). In short, the mean number of ideas for the verbal brainstorming groups (or nominal brainstorming groups) is subtracted from the mean number of ideas for the EBS groups, and this difference is divided by the pooled standard deviation from both conditions to produce a weighted average effect size for each study. A positive effect size means that the EBS groups generated more ideas, while a negative effect size means that the verbal or nominal groups generated more ideas.

We then analyzed the data using two separate regressions, one for the EBS vs. verbal brainstorming comparison, and one for the EBS vs. nominal brainstorming comparison. We used group size as the independent variable and the effect size as the dependent variable. Because we theorized a curvilinear relationship, the regression equation had four terms:

$$EffectSize = constant + \beta_1 * GroupSize + \beta_2 * GroupSize^2 + \beta_3 * GroupSize^3$$

RESULTS

EBS vs. Verbal Brainstorming

Table 2 shows the results from the EBS versus verbal brainstorming regression. All three beta terms were statistically significant. The model had an adjusted R^2 of 43.7%, suggesting a good fit.

EBS vs. Nominal Group Brainstorming

Table 3 shows the results from the EBS versus nominal group brainstorming regression. All three beta terms were statistically significant. The model had an adjusted R^2 of 77.3%, suggesting a good fit.

DISCUSSION

The results show a clear pattern: Group size is a significant factor in explaining the performance differences between EBS and verbal groups. Simply put, size matters; it explains almost half of the differences relative to verbal groups and about three quarters of the differences relative to nominal groups.

The data support our arguments that there is a curvilinear relationship between size and relative benefits, but the patterns are different between the verbal brainstorming and nominal group brainstorming. Figure 2 shows a plot of the predicted performance differences between EBS and verbal and nominal groups, using the coefficients from the two regression models to produce the curves. We plotted the curves in this figure over the range of group sizes for which we had the most data (2-12 member groups).

From Figure 2, we can see that for very small groups (2-3 members), verbal brainstorming is a superior

Table 2. EBS vs. verbal brainstorming regression results

Parameter	Estimate	t	p
Intercept	-6.220	-2.45	0.026
Group size	3.440	2.59	0.019
Group size^2	-0.502	-2.36	0.030
Group size^3	0.024	2.32	0.033

Table 3. EBS vs. nominal group brainstorming regression results

Parameter	Estimate	t	p
Intercept	2.031	1.36	0.191
Group size	-1.289	-2.49	0.023
Group size^2	0.174	3.21	0.005
Group size^3	-0.006	-3.30	0.004

strategy. In verbal groups of these sizes, there are few process losses, while the need to type introduces noticeable losses in the EBS groups. However, once group size reaches four members, EBS groups produce more ideas than verbal groups. The effect size in this size range is about 1.5, which is a very large effect size (Cohen, 1988 labels an effects size of 0.8 as "large"). The relative performance differences remain at about this level until group size reaches 10 members, whereupon the differences begin to dramatically rise again.

The pattern in Figure 2 for EBS vs. nominal group brainstorming is slightly different. Two-member EBS and nominal groups produce about the same number of ideas, but as size increases (from 3-8 members), EBS groups produce fewer ideas (a medium effect size of about -0.4 to -0.8, average about -0.6). As group size increases, the performance differences begin to reverse and EBS gradually starts to outperform nominal groups, so that with 10-12 member groups, EBS produces increasingly more ideas (an effect size of about 1.0).

Although the two curves are shaped differently, they suggest the same basic conclusion: Relative to idea generation, EBS groups under perform verbal and nominal brainstorming groups initially, with a gradual increase in performance as group size increases. For verbal brainstorming, the break-even point comes quickly, so that EBS groups produce more ideas in groups with just four members. For nominal group brainstorming, the break-even point comes later; it is not until group size hits about 10 members that EBS groups begin to produce more ideas than nominal groups.

These results generally support our arguments that EBS can be a useful form of e-collaboration.

FUTURE TRENDS

We believe that e-collaboration through the use of EBS can enhance idea generation, particularly for larger groups. We recommend that groups larger than 8 members e-collaborate to maximize idea generation. For smaller groups, verbal and nominal brainstorming still seem appropriate.

CONCLUSION

While this article has focused on the effect of e-collaboration on idea generation, the implications of EBS go beyond simple idea generation (Dennis & Reinicke, 2004). The essence of creativity is the development and exchange of ideas that ultimately find their way into later stages of organization processes such as decision-making or planning. Continued research into the use and usefulness of EBS will provide a richer understanding of how e-collaboration can enable more efficient and effective group processes and idea generation.

REFERENCES

Allport, F. H. (1920). The influence of the group upon association and thought. *Journal of Experimental Psychology. 3*(3), 159-182.

Collins, B. E., & Guetzkow, H. (1964). *A social psychology of group processes for decision-making.* New York: Wiley.

Cohen, J. (1988). *Statistical power analysis for the behavioral sciences* (2nd ed.). Hillsdale, NJ: Lawrence Earlbaum Associates.

Dennis, A. R., & Reinicke, B. A. (2004). Beta versus VHS and the acceptance of electronic brainstorming technology. *MIS Quarterly, 28* (1), 1-20.

Dennis, A. R., & Valacich, J. S. (1993). Computer brainstorms: More heads are better than one. *Journal of Applied Psychology, 78*(4), 531-537.

Diehl, M., & Stroebe, W. (1987). Productivity loss in brainstorming groups: Toward the solution of a riddle. *Journal of Personality and Social Psychology, 53,* 497-509.

Diehl, M., & Stroebe, W. (1991). Productivity loss in idea-generating groups: Tracking down the blocking effect. *Journal of Personality and Social Psychology, 61,* 392-403.

Diener, S. C. (1979). Deindividuation, self-awareness, end disinhibition. *Journal of Personality and Social Psychology, 37,* 1160-1171.

Gallupe, R. B., Cooper, W. H., Grise, M. L., & Bastianutti, L. (1994). Blocking electronic brainstorms. *Journal of Applied Psychology, 79*(1), 77-86.

Hill, G. W. (1982). Group versus individual performance: Are N + 1 heads better than one? *Psychology Bulletin, 91*(3), 517-539.

Hunter, J. E. & Schmidt, F. L. (1990). *Methods of meta-analysis: Correcting error and bias in research findings.* Newbury Park, CA: Sage Publications.

Karau, S. J., & Williams, K. (1993). Social loafing: A meta-analytic review and theoretical integration. *Journal of Personality and Social Psychology, 65,* 681-706.

Lamm, H., & Trommsdorff, G. (1973). Group versus individual performance on tasks requiring identical proficiency (brainstorming): A review. *European Journal of Social Psychology, 3,* 361-388.

Latane, B. (1981). The psychology of social impact. *American Psychologist, 36,* 343-356.

Latane, B., Williams, K., & Harkins, S. (1979). Many hands make light the work: The causes and consequences of social loafing. *Journal of Personality and Social Psychology, 37,* 823-832.

Levine, D., Resnick, L. B., & Higgins, E. T. (1993). Social foundations of cognition. *Annual Review of Psychology, 44,* 585-612.

Mullen, B., Johnson, C., & Salas, E. (1991). Productivity loss in brainstorming groups: A meta-analytic integration. *Basic and Applied Social Psychology, 12,* 3-23.

Nijstad, B. A., Diehle, M., & Strobe, W. (2003). Cognitive stimulation and interference in idea-generating groups. In P. B. Paulus & B. A. Nijstad (Eds.), *Group creativity: Innovation through collaboration* (pp. 137-159). Cambridge: Oxford University Press.

Nunamaker, J. F., Dennis, A. R., Valacich, J. S., Vogel, D. R., & George, J. F. (1991). Electronic meeting systems to support group work. *Communications of the ACM, 34*(7), 41-61.

Paulus, P. B., Larey, T. S., & Ortega, A. H. (1995). Performance and perceptions of brainstormers in an organizational setting. *Basic and Applied Social Psychology, 17,* 249-265.

Pinsonneault, A., & Barki, H. (1999). The illusion of electronic brainstorming productivity: Theoretical and empirical issues. *Information Systems Research, 10*(4), 378-382.

Pinsonneault, A., Barki, H., Gallupe, R. B., & Hoppen, N. (1999). Electronic brainstorming: The illusion of productivity. *Information Systems Research, 10*(4), 110-133.

Robinson-Staveley, K., & Cooper, J. (1990). Mere presence, gender, and reactions to computers: Studying human computer interaction in the social-context. *Journal of Experimental Social Psychology, 26*(2), 168-183.

Saks, M. J., & Ostrom, T. M. (1973). Anonymity in letters to the editor. *Public Opinion Quarterly, 37,* 417-422.

Sanna, L. J. (1992). Self-efficacy theory: Implications for social facilitation and social loafing. *Journal of Personality and Social Psychology, 62*(5), 774-786.

Siegel, J., Dubrovsky, V., Kiesler, S., & McGuire, T. (1986). Group processes in computer-mediated communication. *Organization Behavior and Human Decision Processes, 37*(2), 157-187.

Steiner, I. D. (1972). *Group process and productivity.* San Diego, CA: Academic Press.

Straus, S. G. (1996). Getting a clue: The effects of communication media and information dispersion on participation and performance in computer-mediated and face-to-face groups. *Small Group Research, 57*(3), 448-467.

Valacich, J. S., Dennis, A. R., & Connolly, T. (1994). Idea generation in computer-based groups: A new ending to an old story. *Organization Behavior and Human Decision Processes, 57*(3), 448-467.

Williams, K. D., & Karau, S. J. (1991). Social loafing and social compensation: The effects of expectations of coworker performance. *Journal of Personality and Social Psychology, 61,* 570-581.

Zajonc, R. B. (1965). Social facilitation. *Science, 149,* 269-274.

Ziegler, R., Diehl, M., & Zijlstra, G. (2000). Idea production in nominal and virtual groups: Does computer-mediated communication improve group brainstorming? *Group Processes & Intergroup Relations, 3*(2), 141-159.

KEY TERMS

Anonymity: A condition where individual contributions to a brainstorming session are not attributed to individual participants.

Evaluation Apprehension: A process loss that causes participants in verbal brainstorming to withhold ideas because they fear a negative reaction from other participants.

Cognitive Interference: A process loss that occurs when the ideas generated by other participants interfere with an individual's own idea generation activities.

Communication Speed: A potential process loss caused by the need to type or write rather than speak.

Production Blocking: A process loss caused by the need to take turns speaking in verbal brainstorming.

Social Facilitation: A process gain caused by the ability of the presence of others to affect one's performance.

Social Loafing (or Free Riding): A process loss caused by the tendency for individuals to expend less effort when working in a group than when working individually.

Synergy: A process gain that occurs when an idea from one participant triggers a new idea in another participant, an idea that would otherwise not have been produced.

GSS Research for E-Collaboration

Sathasivam Mathiyalakan
Winston-Salem State University, USA

INTRODUCTION

Group Support Systems (GSS) are a subset of Decision Support Systems and address the organizational task related needs of groups in areas such as brainstorming, alternatives generation, alternative organization, prioritization, & evaluation, group memory, and communication through the use of information and communication technologies. Group Decision Support Systems (GDSS), a subset of GSS address the decision making needs of groups. The roots of GSS go back to early and middle 1980s. During the 1990s, the GSS research stream (i.e., using decision, information, and communication technologies to support the needs of individuals, groups, and organizations) was one of the popular topics amongst researchers and as a consequence, more than 200 GSS related publications exist (Prasad & Tata, 2005).

Initial GSS research focused on comparing face-to-face groups with that of computer-supported groups. Subsequent research dealt with distance and time dispersion issues. It is at this time, that the term *Distributed Group Support System* (DGSS) was defined to identify distributed groups that use GSS. The late 1990s and early 21st century saw a tremendous increase in the use of computer networks (Internet, Intranet, Extranet, etc.). The widespread availability of Internet at many parts of the world has led many GSS developers and researchers to focus on collaboration through the Internet. Although the terms e-collaboration and virtual teams precede the widespread availability and use of the Internet, the Internet has given new collaboration opportunities and challenges.

Kock and Nosek (2005) identify two trends within the field of e-collaboration research. First is the development of subcommunities devoted to the examination of a particular topic. The second trend is on integrating prior research and identifying topics of relevancy for e-collaboration. Our study falls within the second trend noted above. It is our argument that findings of GSS/GDSS research cannot be blindly transferred to e-Collaboration research and there is a need to determine the extent of applicability, relevancy, and boundaries. The purpose of this study is to examine the previous GSS findings and assess their implications and applicability within in the e-collaboration environment. We begin by providing a brief background to e-collaboration and then proceed to assess the GSS research findings. After identifying important future trends, we provide concluding remarks.

BACKGROUND

E-collaboration is "collaboration among individuals engaged in a common task using electronic technologies" (Kock, Davison, Ocker, & Wazlawick, 2001). The growth and development of GSS and its specialized applications such as GDSS and GDSS provided one of the foundations for the growth, use, and acceptance of e-collaboration tools. Benefits of e-collaboration benefits include

a. Any place/any time meetings and training/learning
b. Process loss reduction associated with face to face communication (see Marsden & Mathiyalakan, 1999; Steiner, 1972)
c. Advanced communicational, information, and decisional support
d. Cost reduction
e. Productivity enhancement
f. Increase in profits and customer/business partner support through better planning, lead and delivery time reduction
g. Process integration
h. Strategic effects

To achieve these stated benefits, e-collaboration tools provide modules for e-mail/automated e-mail/unified messaging, instant messaging, whiteboards, audio/video conferencing, threaded discussion, blogs, calendaring and scheduling, directory services, document sharing, knowledge and data repositories, databases, content management, group/individual decision making support, alternative generative and voting, hardware/ software sharing, project management, and others.

DOMINANT ISSUES OF GSS/GDSS

Over the last 20 years, researchers have focused their attention on a number of topics within the GSS research stream. Initial focus was on comparing outcome measures for computer supported and non-computer-supported groups. Subsequent research has focused on alternative GSS system designs, trust, leadership, conflict management, group history and membership, media richness, and so forth. While initial results suggest mixed findings on GSS usefulness, later results do show that GSS can be valuable tools for collaboration.

Given that GSS/GDSS technologies of the late 1980s and 1990 is vastly different to today's e-collaboration technologies. Technology adoption, implementation, and use has been one of the dominant theme of IS research over the past 40 years has been. Even though technology changes over time, it is important to examine past practices and extrapolate guidelines and suggestions for use with the newer technology. The issue then becomes how should the system be designed and how to structure the process and inputs to maximize organizational benefits. In this respect, with limited availability of space, we identify the following as important issues that need to be reexamined in the e-collaboration environment: (a) system design, appropriate or relevant technology toolbox, and task-technology fit; (b) group structure; and (c) process issues.

THE APPLICABILITY OF GSS/GDSS RESEARCH TO E-COLLABORATION

System Design, Appropriate or Relevant Technology Toolbox, and Task-Technology FIt

The technology landmark has changed dramatically, both in terms of number of players and the level of technological sophistication within the last decade. Two main findings and their applicability to the e-collaboration environment emerge from an analysis of GSS literature. First, the GSS technology and system design play an important role in meeting outcomes (Benbasat & Lim, 1993; Sambamurthy & Poole, 1992; Zigurs,

DeSanctis, & Billingsley, 1991). Previous research also suggests the need for task and the technology fit (Easton & Nunamaker, 1990; Goodhue & Thompson, 1995). Information richness is the ability of a medium to facilitate shared meaning or convey information (Daft & Lengel, 1984). This suggests that some media have the ability to transmit more cues than others. Daft and Lengel (1984) propose that different media could be ranked in terms of feedback, multiple cues, language variety, and personal focus. Prior GSS studies rarely used multimedia capabilities. The interaction with other users was primarily through keyboard typing and where permissible nonverbal cues. Some of the later GSS studies did use rudimentary audio conferencing capabilities. Nowadays with bandwidth becoming a nonissue and the availability of Web-based software, e-collaboration have multimedia capabilities and provide rich media for collaboration. The different e-collaboration tools available in the marketplace vary in terms of capabilities, cost, features, and so forth, and it is commonplace for organizations to employ diverse technologies. A group engaged in e-collaboration may use different tools at different collaboration points. Participants may also be engaged in multi tasking during the collaboration process. In this environment, there is a need to examine the applicability and relevance of previous work on task-technology fit.

Second, while IS research is rich with literature of technology adoption and implementation, such studies with emphasis on strategic perspectives are rare. Previous GSS research has generally focused on individual and group level. There is a need to examine technology adoption and implementation from a strategic perspective. Issues of inhibitors, cost-benefit analysis, organizational transformation, formation and use of virtual teams, technology and business process re-configuration, design and integration of interorganization systems need to be examined at greater depth if e-collaboration is to be a success.

Group Structure

Group structure refers to issues related to group size, status/power structure, group norms, history, and so forth (Pinsonneault & Kreamer, 1990). Initial interest in groups arose due to the belief that pooling the knowledge and abilities of several individuals is beneficial to organizations. The primary rationale for using GSS is to minimize process losses. Processes losses occur due to the pressures due to conformity, fear of negative appraisal by superiors, domination by more senior/ranking members of the group (status differences), communication monopolization, ability of only one member to speak at a time (production blocking), lack of information support, fear of ridicule/reprisals for speaking out (evaluation apprehension), and information overload (Steiner, 1972).

Dennis and Valacich (1994) emphasized that while many factors affect group performance, group size is a key ingredient since it places a limit on the knowledge available to the meeting group. Early research suggest that the optimal group size is five (Hackman & Vidmar, 1970; Hare, 1952; Slater, 1958). The determination of the optimal group size was based on factors such as participant satisfaction and participation opportunities. Much of this work was based on the concept that as the size of a group increases, meeting outcome measures (net value) increase until a maximum point is reached. Any further increases in group size yield negative net benefits. Prior GSS study (Chidambaram & Bostrom 1993; Chidambaram, Bostrom, & Wynne 1990; Tan, Raman, & Wei, 1994) use groups of five based on this determination. Subsequent work suggest otherwise (Dennis, Valacich, & Nunamaker 1990; Marsden & Mathiyalakan, 1999).

Anonymity was seen as a desirable issue and many GSS incorporate the means to ensure participant anonymity. Many e-collaboration tools promote features for video conferencing or other means for reducing participant anonymity. Therefore these systems could be seen as an extension of face to face meetings with electronic support and where participants may be dispersed both in time and location. We argue that there is a need to examine whether process losses do occur within e-collaboration environment as the theoretical foundation on which GSS studies were based (i.e., process loss may not be valid).

Process Issues

Prior research findings on GSS are based largely on laboratory studies. Organizational groups may be different from laboratory groups in several dimensions including task type and complexity, nature of groups and duration, nature of incentives. McGrath (1991) notes that while researchers have studied groups for over a half a century, there are limitations on inferences that can be made as much of the work focused on the use of

groups that met only once. Researchers tend to avoid longitudinal studies due to cost, subject attrition, and career impacts. Longitudinal studies provides permits the investigation of learning, attitudes and changes in attitudes over time, effect of technology on work, use of technology, development of group cohesiveness, conflict resolution, group composition and changes therein.

McGrath (1991) further notes that group membership continuity or change are defining aspects of group outcomes. But such investigations in practice are rare. The few studies that have studied groups over time suggest that not all groups are alike and that group developmental patterns differ. Chidambaram et al. (1990) find that there were significant differences in developmental patterns between manual (non-GDSS) groups and GDSS groups. Initially, the GDSS groups exhibited lower conflict management ability and cohesiveness than manual groups. But, by the end of the four meetings, this pattern had reversed, suggesting a potential benefit (enhanced conflict resolution skills) for GDSS groups. Dennis et al. (1990) find that established groups exhibited different personalities than ad hoc groups. The established groups were more likely to express critical comments and uninhibited comments and had a less even distribution of participation. Mathiyalakan (2002) investigates group composition change and continuity in groups of three and five. The initial experiments yielded results formally verifying that membership continuity produces better meeting outcomes (increased performance and productivity, and decreased solution time) while membership change produced the exact opposite (decreased performance and productivity, and increased solution time).

Two major findings emerge from above discussion. First, these studies indicate the importance of group dynamics and how it changes over time as over time, GDSS groups may exhibit a different pattern of behavior and task outcomes than manual groups, and group developmental patterns over time may not be the same. Thus, the results of a study taken during the formative stages of a group may not be indicative of its future, suggesting the potential value of longitudinal investigations. Second, if group dynamics change over time and groups become more cohesive with time, then trust building measures especially among business partners are needed to ensure that e-collaboration leads to fruitful and beneficial outcomes for both organizations and its constituents.

FUTURE TRENDS

The trend toward greater use and reliance on e-collaboration tools is expected to accelerate in the next decade. Several factors contribute toward this acceleration. These include movement toward market orientation in several parts of the world; global production and consumption; increasing technology savynesses of employees, firms, and consumers; decreasing technology usage costs; and the increasing use of the Internet for business and personal use. We also expect that other institutions most notably universities and colleges where e-learning is gaining prominence to contribute to this trend.

Prior GSS studies with few exceptions have usually focused on examining GSS within a single organization. Globalization is likely to bring the issue of cultural issues to the forefront. As individuals and suppliers/business partner firms begin using e-collaboration as a way to reduce costs and increase productivity and profitability, system design, multi language communication capabilities, and participant/team characteristics are likely to play an important role.

Currently, more than 100 vendors serve the e-collaboration market. Market consolidation is likely to occur impacting the choice of technology availability. The growth of Web services is also likely to affect B2B collaboration. Factors such as use of web services within private exchanges, P2P technologies, Web services network, the movement toward mainstream products due to integration complexity, decline of ERP vendors may contribute to changes in collaboration among business partners (Chan, Kellen, & Nair, 2004).

Another trend that is likely to be observed is the changing nature of collaboration. The last years has seen the shift from same time/same place to any time/any place meetings with instant communication capabilities. This will have a profound impact on technology use and outcomes as not all participants are likely to use the same system for collaboration. Thus, some participants at a location may use a set of technologies while others who are mobile may be using PDAs, cell phones, and so forth to communicate and interact with other participants. Thus, while technologies may be integrated, the use of different applications entails, thus, participants may experience differing outcomes due to the use of differing communication media.

CONCLUSION

The use of e-collaboration tools for both business and personal use is a defining issue in today's global, interconnected, networked world. The roots of e-collaboration goes back to GSS and Computer Supported Cooperative Work (CSCW), or perhaps even further. The purpose of this study was to examine the previous GSS findings and assess their implications in the e-collaboration environment. This article contributes to both the academic research community and practicing managers but bringing to attention the issues and challenges that need to be understood fully. Given the space limitations, we identified three areas: System design, appropriate or relevant technology toolbox, and task-technology fit, Group structure, and Process issues as vital to developing a greater understanding of e-collaboration.

As discussed previously, not all GSS findings are relevant and directly transferable to e-collaboration. Future research should examine the basic premises and theory behind GSS and assess their relevancy and implication for the e-collaboration environment. In this study, we identify system design, group structure, and process issues are three of the major issues that need to be reexamined. Today's technology is vastly different and superior to the technology of the late 1980s and 1990 in terms of multimedia capabilities, bandwidth, networked environment, multitasking, and so forth. Earlier theories on task-technology fit may not be relevant and thus the need to examine this issue in an environment rich with technology capability, availability, complexity, and level of sophistication. Issues of inhibitors, cost-benefit analysis, organizational transformation, formation and use of virtual teams, technology and business process reconfiguration, design and integration of interorganization systems need to be examined at greater depth if e-collaboration is to be a success.

Key group structure assumption such as group size and provision of anonymity have been altered and may not be valid within the e-collaboration environment. Due to outsourcing, off-shoring, and globalization, there has been a tremendous growth in the use of virtual groups. Group dynamics do change over time, and groups composition does play an important part in group cohesion. This indicates that trust building measures are needed to overcome lack of familiarity. More longitudinal experimental and case studies are needed to fully understand what works, what does not, and how within the e-collaboration environment.

REFERENCES

Benbasat, I., & Lim, L. (1993). The effects of group, task, context, and technology variables on the usefulness of group support systems: A meta-analysis of experimental studies. *Small Group Research, 24*(4), 430-462.

Chan, S. S., Kellen, V., & Nair, C. (2004). *Adopting Web services for B2B e-collaboration: Research directions.* Retrieved April 8, 2006, from http://www.kellen.net/WebServices.htm

Chidambaram, L., & Bostrom, R. P. (1993). Evolution of group performance over time: A repeated measures study of GDSS effects. *Journal of Organizational Computing, 3,* 443-469.

Chidambaram, L., Bostrom, R. P., & Wynne, B. E. (1990). A longitudinal study of the impact of group decision support systems on group development. *Journal of Management Information Systems, 7,* 7-25.

Daft, R. L., & Lengel, R. (1984). Information richness: A new approach to managerial information processing and organization design. In B. Staw & L. Cummings (Eds.), *Research in organization behavior* (pp. 6, 191-233). Greenwich, CT: JAI Press.

Dennis, A. R., & Valacich, J. S. (1994). Group, subgroup, and nominal group idea generation: New rules for a new media? *Journal of Management Information Systems, 20*(4), 723-736.

Dennis, A. R., Valacich, J. S., & Nunamaker, J. F., Jr. (1990). An experimental investigation of small, medium, and large groups in an electronic meeting system environment. *IEEE Transactions on Systems, Man, Cybernetics, 20*(5), 1049-1057.

Easton, G. K., & Nunamaker, J. F. Jr. (1990). Using two different electronic meeting system tools for the same task: An experimental comparison. *Journal of Management Information Systems, 7*(1), 85-100.

Goodhue, D. L., & Thompson, R. L. (1995). Task-technology fit and individual performance. *MIS Quarterly, 19*(2), 213-236.

Hackman, J. R., & Vidmar, N. (1970). Effects of size and task type on group performance and member reactions. *Sociometrika, 33,* 37-54.

Hare, A. P. (1952). A study of interaction and consensus in different-sized groups. *American. Sociology Review, 17,* 261-267.

Holsapple, C.W.H. & Whinston, A.B. (1996) *Decision support systems: A knowledge-based approach,* St. Paul, MN: West Publishing

Kock, N., & Nosek, J. (2005). Expanding the boundaries of e-collaboration. *IEEE Transactions on Professional Communication, 48*(1), 1-9.

Kock, N., Davison, R., Ocker, R., & Wazlawick, R. (2001). E-collaboration: A look at past research and future challenges. *Journal of Systems and Information Technology, 5*(1), 1-9.

Marsden, J. R., & Mathiyalakan, S. (1999). A comparative study of individuals and groups over time in an electronic meeting system environment. *IEEE Transactions on Systems, Man, and Cybernetics, 29*(2 Part C), 169-185.

Mathiyalakan, S. (2002). A methodology for controlled empirical investigation of membership continuity and change in GDSS groups. *Decision Support Systems, 32,* 279-295.

McGrath, J. E. (1991). Time, interaction, and performance (TIP) a theory of groups. *Small Group Research, 22,* 147-174.

Pinsonneault, A., & Kraemer, K. L. (1990). The effects of electronic meetings on group processes and outcomes: An assessment of the empirical research. *European Journal of Operations Research, 46,* 143-161.

Prasad, S., & Tata, J. (2005). Publication patterns concerning the role of teams/groups in the information systems literature from 1990 to 1999. *Information and Management, 42,* 1137-1148.

Sambamurthy, V., & Poole, M. S. (1992). The effects of variations in capabilities of GDSS designs on management of cognitive conflict in groups. *Information Systems Research, 3*(3), 224-251.

Slater, P. E. (1958). Contrasting correlates of group size. *Sociometrika, 21,* 129-139.

Steiner, I. D. (1972). *Group process and productivity.* New York: Academic Press.

Tan, B. C. Y., Raman, K. S., & Wei, K. (1994). An empirical study of the task dimension of group support system. *IEEE Transactions on Systems Man, & Cybernetics, 24,* 1054-1060.

Townsend, A., DeMarie, S., & Hendrickson, A. (1998) Virtual teams: Technology and the workplace of the future. *The Academy of Management Executive 12*(3), 17-29

Zigurs, I., DeSanctis, G., & Billingsley, J. (1991). Adoption patterns and attitudinal development in computer-supported meetings: an Exploratory Study with SAMM. *Journal of Management Information Systems, 7*(4), 51-70.

KEY TERMS

Computer Supported Collaborative Work: The use of computers to support cooperative work among multi participant (e.g., collaborative authoring), as distinct from work that may not be cooperative (Holsapple & Whinston, 1996).

E-Collaboration: Collaboration among individuals engaged in a common task using electronic technologies (Kock et al., 2001).

Electronic Meeting Systems (EMS): Designed to make meetings more productive through the use of IT. The technology is designed to directly impact and change the behavior of groups to improve group effectiveness, efficiency, and satisfaction.

Group: A simple kind of organization characterized by participants having comparable authority about the group's task (e.g., decision making), little in the way of formal division of labor, and few restrictions on who can communicate with whom. (Holsapple & Whinston, 1996).

Group Decision Support System: A tool used in supporting decision making needs of groups.

Groupware bComputer/communication technology used to facilitate the work of the group (e.g., a GDSS) (Holsapple & Whinston, 1996).

Team: A hierarchically designed organization in which there is one deciding participant and one or more supporting participants. In contrast to a group, there is clear differentiation of decision making authority, a division of labor into distinctly specialized duties, and a restricted pattern of communication. (Holsapple & Whinston, 1996).

Virtual Teams: Groups of geographically and/or organizationally dispersed coworkers that are assembled using a combination of telecommunication and information technologies to accomplish an organizational task. (Townsend et al., 1998)

Human and Technology Leadership Roles in Virtual Teams

Ilze Zigurs
University of Nebraska at Omaha, USA

Terrance Schoonover
University of Nebraska at Omaha, USA

INTRODUCTION

Effective leadership can make a difference in the success of a team, and virtual teams are no exception. However, virtual teams present special challenges, particularly in the expression of the context-rich and personal influence that is such an important part of leadership (Bell & Kozlowski, 2002). Communication through computer-mediated channels requires a different awareness of leadership roles and how they can be expressed. In fact, the communication system itself has roles to play, some of which might be perceived by team members as leadership roles. We begin with a brief summary of the historical context of leadership and existing research on leadership in computer-supported teams. We then discuss opportunities for research on enhancing leadership in virtual teams, particularly from the perspective of integrating human with technology roles.

BACKGROUND

Leadership has a long and interesting history of theoretical perspectives and practical application. Leadership has been examined in terms of personality traits, as specific behaviors or behavior patterns, in terms of different styles, as typologies, or with respect to the contingencies of the situation (Bass, 1990). The different perspectives present increasingly complex views that take into account not just the individual leader but the impact and importance of the context, both social and organizational. In virtual teams, the context also includes technology, since virtual teams rely on computer-mediated environments.

Leadership can be defined as the exercise of influence for the purpose of achieving goals (Bass, 1990). In virtual teams, "e-leadership" has been defined as a social influence process that is mediated by advanced information technologies "to produce a change in attitudes, feelings, thinking, behavior, and/or performance with individuals, groups, and/or organizations" (Avolio, Kahai, & Dodge, 2001, p. 617). The process of influence is central to exercising leadership. Influence is traditionally exercised through face-to-face communication, but in computer-mediated environments, the influence process is quite different. One of the most interesting aspects of leadership in virtual teams arises from the situation itself, namely the integration of the role of technology in what is traditionally a personal influence situation. Thus, it is particularly relevant to ask how human and machine roles affect leadership in virtual contexts.

Virtual teams can be defined as a group of individuals who are geographically and/or organizationally and otherwise dispersed and who rely on collaboration technologies to carry out team activities (Dubé & Paré, 2004; Zigurs, 2003). Collaboration technology can be defined as an integrated and flexible set of tools that support team communication, process, and information sharing.

Table 1 summarizes key perspectives on leadership, showing the development of ideas from a fairly simple trait perspective to more complex interaction and systemic views. The table shows the implications of each characterization for virtual teams, that is, the types of challenges that arise in this new environment, and the resultant implications for technology design.

LEADERSHIP ROLES IN VIRTUAL TEAMS

An early study of human and machine roles in face-to-face computer-supported teams provides a foundation for an analysis of leadership roles (Zigurs & Kozar, 1994). The study examined three categories of group

Table 1. Characterizations of leadership and implications for virtual teams

Leadership Perspective	Overview	Selected Observations	Implications in Virtual Teams	Implications for Technology Design
Trait (Stogdill, 1948)	Effectiveness is defined in terms of required personal traits, such as physical characteristics, personality, ability.	Managerial motivation and managerial skills appear to be most promising predictors.	Traits might be difficult to assess in virtual environments, since cues are reduced.	Media-rich technologies such as video conferencing provide more verbal and non-verbal cues, though task type must be taken into account.
Power-influence (French & Raven, 1959)	Effectiveness is a function of the source and amount of power available to the leader. Types include reward power, position power, personal power, coercive power, referent power.	Leaders are more effective when they have at least a moderate amount of position power and develop personal power as a supplement.	Power and influence are more difficult to express in virtual environments, where cues are reduced.	Media-rich technologies can help to communicate position power, e.g., via video cues and voice inflection; personal profiles of participants can remind the team of positions.
Behavior (Blake & Mouton, 1964)	Emphasizes what leaders do in terms of effective and ineffective behaviors, e.g., consideration and initiating structure.	Context is a moderator for conditions in which consideration or structure work best.	Behavior is expressed through communication and needs to be explicit.	Communication tools should be designed to provide diverse formats for structuring messages that will support different types of behaviors.
Situational-Contingency (Hersey & Blanchard, 1977)	Success depends on situational factors. Factors are many, e.g., nature of task, external environment, role expectations, authority relationships.	Task-oriented style tends to be more effective in extremely favorable or unfavorable conditions, while relational style works in moderate conditions.	Technology environment introduces another contingency factor.	Technology should include tools to support the group in determining the characteristics of the situation, e.g., mood meter, social agents, awareness alerts.
Transformational vs. transactional (Burns, 1978)	Transformational leadership focuses on developing followers into leaders. Transactional leaders focus on exchange of rewards for performance and punishment for non-compliance.	Transformational leadership generates higher levels of effort, satisfaction, commitment, performance.	Skills for developing followers are likely to be different in a virtual environment, e.g., leader behaviors would be modeled differently in distributed environments than in person.	Technology that identifies both individual and group accomplishment and growth is needed, e.g., personal and group expertise profiles, role simulations.
Self-leadership (Manz, 1986)	Emphasizes developing positive thinking patterns and work practices in discipline, competence, self-confidence.	Focusing on giving workers more control over their tasks increases motivation and satisfaction.	Virtual environments might create more opportunities for controlling communication.	Collaborative technologies should be designed to give users a feeling of autonomy and control, e.g., ability for each team member to control agenda.
Full-range leadership (Avolio, 1999)	Leadership is a total system in which leaders and followers interact, influence, and develop each other and the context.	Transformational leadership has a consistently positive effect on developing rewarding contexts.	Virtual environments are consistent with this concept because technology is part of the system and a resource for leadership.	Design tools that build user confidence and autonomy, e.g., built-in process expertise.

roles from a classic typology by Benne and Sheats (1948): (1) group task, (2) socio-emotional group-building and maintenance, and (3) individual roles. That typology reflects the well-accepted recognition that teams need to focus on team development as well as task accomplishment, while managing individually-oriented behaviors that detract from team functioning.

Several roles from this study can be identified as leadership types of roles, namely, Information/Opinion Giver, Information/Opinion Seeker, Motivator, Mediator, and Gate-Keeper. Teams met in a group decision support room and, at the beginning and end of each session, were asked about roles they played and roles that the collaboration technology played. The technology was viewed as playing several leadership roles, including Gate-Keeper, Information/Opinion Seeker, and Motivator. The study shows that a collaboration technology tool can be perceived as playing a leadership role within a team. The question in virtual teams is how such technology can be leveraged to overcome the inherent difficulties of communicating leadership over distance. A look at the existing research in this area provides insight on this issue.

CONTINUING ISSUES

The research on leadership of virtual teams has examined such phenomena as leadership style, emergent versus assigned leadership, trust, motivation, and leader practices. We discuss the main findings briefly, focusing on studies that have relevance to leader roles and technology.

Leadership style has been examined in computer-mediated teams, with the finding that both participative and directive leadership styles were positively related to performance (Kahai, Sosik, & Avolio, 2004). The phenomenon of emergent leaders in virtual teams has also been examined (Sarker, Grewal, & Sarker, 2002; Yoo & Alavi, 2004). One study was conducted over ten weeks with virtual teams of senior executives who were participating in an executive development program (Yoo & Alavi, 2004). E-mail messages of all team members were coded for content related to leadership roles, to determine how the behaviors and roles of emergent leaders contrasted with non-leaders. Emergent leaders had more task-oriented messages and enacted the roles of initiator, scheduler, and integrator. Interestingly, there were no significant differences

between emergent leaders and other team members in the number of relationship-oriented messages. In that case at least, emergent leaders were not supporting the essential socio-emotional side of team development. We can speculate that socio-emotional development is particularly difficult to do in virtual environments, but it should be no less important. Another interesting aspect of this study was the introduction of the phrase "distributed leadership system" (Yoo & Alavi, 2004, p. 50), a term that relates well to the ideas of self-leadership (Manz, 1986), team member roles, and the role of technology (Zigurs & Kozar, 1994).

The role that trust plays in leadership is also important. DeRosa, Hantula, Kock, and D'Arcy (2004) used media naturalness theory (Kock, 2001) to explain discrepancies among several studies of trust. The media naturalness perspective consists of three principles. First, a decrease of naturalness in a communication medium will lead to an increase in cognitive effort, an increase in ambiguity levels, and a decrease in physiological arousal. Second, the innate similarity principle states that there should be innate influences that are common to all regardless of cultural backgrounds. Third, the learned schema principle emphasizes that individuals acquire communication schemas through the environment, and as a result will have individual learning differences. The perspective can be used to make predictions on how trust will be developed and modified in virtual teams.

Regarding motivation, Hertel, Konradt, and Orlikowski (2004) looked at four components that they titled VIST. These were valence, which is the member's motivation or buy-in to contributing to the team's goals, instrumentality, which is the perceived importance of the member's contribution to the team, self-efficacy or the team member's perceived ability to perform the job, and trust that they will be respected and not exploited by the team. The way that management approached these components had a direct correlation to team motivation and their consequent effectiveness.

A type of leadership in which there has been relatively little virtual team research is management by objective (MBO). This approach emphasizes goal setting, participation, and feedback with the manager's role being one of intermediary and coach. Hertel et al. (2004) used a field study of virtual teams to look at the effect of MBO. They reported a strong correlation between the team members' perceptions of goal setting quality and the manager's perception of the effective of the team. Motivational indicators of the team members were a modifying variable. Team members in the more successful virtual teams reported being better informed than those in the less successful teams.

Drawing on five years of research and project analysis, Burtha and Connaughton (2004) pointed out several practices of effective virtual leadership. These included the need for face-to-face events to build an environment of trust, communicating often and through multiple media outlets, establishing and reinforcing roles and responsibilities, providing structure and process, understanding and respecting cultural differences, and expecting and embracing change.

Leadership practices have also been examined with respect to different phases of virtual teams. Beranek, Broder, Reinig, Romano, and Sump (2005) identified four project phases and the leadership processes that should take place in each phase for project success. In the pre-project phase, leaders need to select team members and communicate the project mission, priority, and success criteria. In the project initiation and mid-term phases, leaders need to establish and manage team boundaries, develop shared mental models, and manage communication processes. In the wrap-up phase, leaders need to annotate successes and address lessons learned.

Two conclusions can be drawn from this brief review. First, many studies take the approach of examining existing leadership perspectives and applying them in the new context of virtual teams. This approach helps to understand the extent to which our current understanding translates to new environments, but there are inherent limitations. Second, collaboration technology often remains in the background in these studies, even though it is the medium through which roles are expressed.

RESEARCH ISSUES AND FUTURE TRENDS

Theoretical perspectives in existing research on leadership were summarized in Table 1. A new perspective that has emerged in the study of collaboration technology is that of adaptive structuration theory (DeSanctis & Poole, 1994), which describes team interaction as an ongoing process of adaptation of existing rules and resources. The use of structuration theory as a theoretical framework for studying leadership in teams is

interesting because leadership can be viewed as one source of structure for the team. Leaders with different styles would be expected to create different rules and resources for the team. The interaction of the styles (and their associated rules and resources) with the process structures that might be available to or used within the team is a fruitful area for future research, especially in the interaction of leadership with technology.

Collaboration technologies have evolved to provide more sophisticated tools and features, including file-sharing, chat, shared document editing, meeting management, voting and other forms of idea evaluation, project management tools, and awareness alerts. The latter tool is particularly interesting, in that the awareness that current tools provide is generally related to whether a member is in a workspace or whether a document has been changed. But teams need awareness on many dimensions and especially related to leadership roles. Research needs to address how tools can support better awareness not only of the roles of other members in the team, but of the ongoing process and leadership of the team. Commitment and a sense of belonging could be enhanced by collaboration tools that provide enhanced awareness.

The distribution of leadership roles has been mentioned from several perspectives, including self-leadership, full-range leadership, and a distributed leadership system. These different terms all express the idea that everyone plays a role in moving the team forward and in influencing each other and the process. Collaboration technology has the potential to play a much larger role in such a leadership system than current tools are capable of doing. For example, there is a body of research on facilitation and facilitator roles in computer-mediated collaboration (e.g., Clawson & Bostrom, 1996; Dickson, Limayem, Partridge, & DeSanctis, 1996). This research has identified specific task interventions and facilitator roles for effective team functioning in a computer-mediated environment. Examples include structuring team activities, guiding the agenda, directing the meeting, and integrating ideas, to name a few. Recent attention has focused on standardizing these types of facilitator tasks within a body of knowledge and perspective known as collaboration engineering (Briggs, de Vreede, & Nunamaker, 2003). Clusters of process knowledge, known as thinkLets, are used as objects to be combined to create effective team processes. Although the perspective does not imply technology support per se, thinkLets can be instantiated in collaboration technology tools. Another area of research has been to provide expert system help with facilitation, via automated meeting procedures and meeting counselors (Aiken, Liu Sheng, & Vogel, 1991).

Further evolution of technology might include developing integrated human-technology leadership systems that could redefine traditional practices of leaders. For example, rather than a leader monitoring the progress of team members, all members could be given the ability to see the current state of each team member's contribution. The system could also provide views to individuals of their comparative contribution and thus play a motivational role. Another role of the technology could be to help reduce social distance and build trust by, for example, greater personalization of communication and increased visibility and awareness among team members. All these ideas imply greater focus on the socio-emotional aspects of team functioning, as opposed to the task focus that has been the largest part of computer-mediated tool development to date.

CONCLUSION

Leadership remains an important phenomenon, especially so in virtual teams, where the computer-mediated environment changes traditional forms of expression of influence. Leadership roles include such important activities as providing direction, building trust, influencing opinion, developing norms, encouraging participation, and structuring process. If virtuality means dispersion, then virtual team members are by definition harder to "bring together." The challenge for future research and practice is to develop collaboration technology tools that support and even carry out leadership roles, thus bridging the gap between traditional and virtual leadership effectiveness.

REFERENCES

Aiken, M., Liu Sheng, O., & Vogel, D. (1991). Integrating expert systems with group decision support systems. *ACM Transactions on Information Systems, 9*(1), 75-95.

Avolio, B. (1999). *Full leadership development: Building the vital forces in organizations.* Thousand Oaks, CA: Sage Publications.

Avolio, B. J., Kahai, S., & Dodge, G. E. (2001). E-leadership: Implications for theory, research, and practice. *Leadership Quarterly, 11*(4), 615-668.

Bass, B. M. (1990). *Bass & Stogdill's handbook of leadership: Theory, research, and managerial applications.* New York: Free Press.

Bell, B. S., & Kozlowski, S. W. J. (2002). A typology of virtual teams: Implications for effective leadership. *Group & Organization Management, 27*(1), 14-49.

Benne, K. D., & Sheats, P. (1948). Functional roles of group members. *Journal of Social Issues, 4*(2), 41-49.

Beranek, P. M., Broder, J., Reinig, B., Romano, N.C., Jr., & Sump, S. (2005). Management of virtual project teams: Guidelines for team leaders. *Communications of the Association for Information Systems, 16*, 247-259.

Blake, R. R., & Mouton, J. S. (1964). *The Managerial Grid.* Houston, TX: Gulf.

Briggs, R. O., de Vreede, G. J., & Nunamaker, J.F., Jr., (2003). Collaboration engineering with thinkLets to pursue sustained success with group support systems. *Journal of Management Information Systems, 19*(4), 31-63.

Burns, J. M. (1978). *Leadership.* New York: Harper & Row.

Burtha, M., & Connaughton, S. (2004). Learning the secrets of long-distance leadership. *KM Review, 7*(1), 24-27.

Clawson, V. K., & Bostrom, R. P. (1996). Research driven facilitation training for computer supported environments. *Group Decision and Negotiation, 5*(1), 7-29.

DeRosa, D. M., Hantula, D. A., Kock, N., & D'Arcy, J. (2004). Trust and leadership in virtual teamwork: A media naturalness perspective. *Human Resources Management, 43*(2 & 3), 210-232.

DeSanctis, G., & Poole, M. S. (1994). Capturing the complexity in advanced technology use: Adaptive structuration theory. *Organization Science, 5*(2), 121-147.

Dickson, G., Limayem, M., Lee Partridge J., & DeSanctis, G. (1996). Facilitating computer supported meetings: A cumulative analysis in a multiple criteria task environment. *Group Decision and Negotiation, 5*(1), 51-72.

Dubé, L., & Paré, G. (2004). The multi-faceted nature of virtual teams. In D. J. Pauleen (Ed.), *Virtual teams: Projects, protocols, and processes* (pp. 1-39). Hershey, PA: Idea Group Publishing.

French, J., & Raven, B. H. (1959). The bases of social power. In D. Cartwright (Ed.), *Studies of social power* (pp. 150-167). Ann Arbor, MI: Institute for Social Research.

Hersey, P., & Blanchard, K. H. (1977). *The management of organizational behaviour* (3rd ed.). Upper Saddle River, NJ: Prentice Hall.

Hertel, G., Konradt, U., & Orlikowski, B. (2004). Managing distance interdependence: Goal setting, task interdependence, and team-based rewards in virtual teams. *European Journal of Work and Organizational Psychology, 13,* 1-28.

Kahai, S. S., Sosik, J. J., & Avolio, B. J. (2004). Effects of participative and directive leadership in electronic groups. *Group & Organization Management, 29*(1), 67-105.

Kock, N. (2001). The ape that used e-mail: Understanding e-communication behavior through evolution theory. *Communications of the Association for Information Systems, 5*(3), 1-29.

Manz, C. C. (1986). Self-leadership: Toward an expanded theory of self-influence processes in organizations. *Academy of Management Review, 11,* 585-600.

Sarker, S., Grewal, R., & Sarker, S. (2002). Emergence of leaders in virtual teams: What matters? *Proceedings of the 35th Hawaii International Conference on System Sciences* (p. 273). Washington, DC: IEEE Computer Society Press.

Stogdill, R. M. (1948). Personal factors associated with leadership: A survey of the literature. *Journal of Psychology, 25,* 35-71.

Yoo, Y., & Alavi, M. (2004). Emergent leadership in virtual teams: What do emergent leaders do? *Information and Organization, 14*(1), 27-58.

Zigurs, I. (2003). Leadership in virtual teams: Oxymoron or opportunity? *Organizational Dynamics, 31*(4), 339-351.

Zigurs, I., & Kozar, K. (1994). An exploratory study of roles in computer-supported groups. *MIS Quarterly, 4*(1), 277-297.

KEY TERMS

Collaboration Technology: An integrated and flexible set of tools that support team communication, process, and information sharing.

Distributed Leadership System: Virtual team leadership that is distributed among several members and potentially collaboration technology (see Yoo & Alavi, 2004).

Emergent Leadership: Leadership that occurs in a team that does not have a formally appointed leader, in which one or more members of the team take on various leadership duties with the implicit consent of the other team members.

Full-Range Leadership: A total system in which leaders and followers interact, influence, and develop each other and the context (see Avolio, 1999).

Leadership: The exercise of influence for the purpose of achieving goals.

Self-Leadership: Leadership of one's self by development of individual strengths and motivation to perform; a self-influence process (see Manz, 1986).

Virtual Team: A group of individuals who are geographically and/or organizationally and otherwise dispersed and who rely on collaboration technologies to carry out team activities.

IM Support for Informal Synchronous E-Collaboration

Stefan Hrastinski
Jönköping International Business School, Sweden

INTRODUCTION

Instant messaging (IM) is a synchronous communication medium that can be used to maintain a list of "friends." These friends can be contacted whilst online and running the software, by text messages or initiating a chat, audio or video conferencing session.

Figure 1 shows a screenshot of Microsoft's MSN Messenger. Three persons are online and can be contacted by right-clicking their name and choosing what kind of interaction (e.g., send message or file) that is to be initiated. It has been argued that IM enables more informal interaction compared with common synchronous media, such as chat and videoconferencing since meetings do not need to be scheduled (e.g., Cameron & Webster, 2005; Contreras-Castillo et al., 2004). Instead, users can spontaneously e-collaborate synchronously with others when they are online.

There has been a "mass adoption" of IM around the world. The popularity of IM may be explained by the fact that it has been adopted in various settings:

- Managers have begun to recognize IM systems' potential to support informal interaction. This has led to corporations installing IM software on employees' workstations (Cameron & Webster, 2005). In 2004, 53 million American adults exchanged instants messages and 24% of them used IM more often than e-mail (Shiu & Lenhart, 2004).

- Students at European and American universities commonly use IM (Beuschel et al., 2003). For example, 63% of the students at Jönköping University use an IM system (Andersson & Azadi, 2004). Out of these, 90% use it at least weekly and 52% have been using it for more than three years.

- Adolescents commonly use IM to interact socially with friends (Boneva et al., 2006). As they become older, it is probable that they will bring IM to other settings, such as work and education.

Even though many use IM, few have actually examined it in research, which implies that our knowledge

Figure 1. Screenshot of MSN Messenger

of how IM is used in practice is limited (Cameron & Webster, 2005; Quan-Haase, Cothrel, & Wellman, 2005). However, there are some notable exceptions that have been conducted in work and in educational settings, which are reviewed in this article.

The aim of this article is to review how IM may be used to support e-collaboration. This is addressed by reviewing studies on IM in both work and in higher education settings. These two settings "share the problem of creating and sustaining a positive work and learning environment" (Haythornthwaite, 2000, p. 201) and by including research from both areas a deeper understanding may be obtained.

The article is structured as follows. First, definitions of informal interaction and e-collaboration are discussed. Then, studies on IM at work and in higher education are reviewed. Finally, conclusions derived from the reviewed studies are put forth.

INFORMAL E-COLLABORATION AND INSTANT MESSAGING

Informal interactions "take place at the time, with the participants, and about the topics at hand" (Fish, Kraut, & Chalfonte, 1990, p. 2). Nardi, Whittaker, and Bradner (2000) argue that informal interaction generally is "impromptu, brief, context-rich, and dyadic" and "support joint problem solving, coordination, social bonding, and social learning—all of which are essential for complex collaboration" (p. 79). It has been recognized for many years that informal interaction is vitally important in organizations (Kraut, Fish, Root, & Chalfonte, 1990; Mintzberg, 1973). In this article, *informal e-collaboration* is defined as spontaneous, brief collaboration mediated by technology among individuals engaged in a common task.

A characteristic that distinguishes IM from other commonly used communication media, is its potential to enable informal e-collaboration. Fish et al. (1990) developed criteria for appropriate characteristics of technology for *informal interaction*. These were (a) access to a suitable population of others, (b) an environmental mechanism that brings people together, (c) the effort needed to initiate and conduct a conversation should be low, and (d) a visual channel. Interestingly, most available IM systems, both commercial (e.g., MSN

Messenger, ICQ, Yahoo Messenger, AOL Messenger) and those designed for collaborative knowledge work (e.g., Lotus Workplace Messaging, Jabber Extensible Instant Messaging), have these characteristics. Even though some technologies seem more beneficial than others in supporting informal interaction and e-collaboration, it needs to be noted that it is the users, and not the technology per se, that decide if it is to be used for informal interaction.

INSTANT MESSAGING AT WORK

Recently, IM has received increased attention in research on e-collaboration. Since 2000, exploratory studies have mainly been presented at computer-supported cooperative work and human-computer interaction conferences. Journal publications more focused on organizational consequences of IM use began to appear in 2005 (Cameron & Webster, 2005; Cho, Trier, & Kim, 2005; Quan-Haase et al., 2005). Below, research findings are reviewed and categorized.

IM enables informal work-related e-collaboration. IM is generally used for spontaneously asking specific work-related questions (Cho et al., 2005; Handel & Herbsleb, 2000; Quan-Haase et al., 2005) but also to "coordinate and arrange meetings, and inquire about colleagues' availability for discussion" (Quan-Haase et al., 2005). The messages are characterized by an informal tone (Cameron & Webster, 2005) and usually only two employees are involved in a conversation (Quan-Haase et al., 2005).

IM has also been reported to be used to support in-depth problem solving, where various strategies to find a solution to a problem are discussed (Isaacs, Walendowski, Whittaker, Schiano, & Kamm, 2002; Quan-Haase et al., 2005). Isaacs et al. (2002) compared heavy and light IM users. They argued that heavy users use IM to support collaboration, such as to "discuss a broad range of topics via many fast-paced interactions per day" (p. 11), while light users use IM to support coordination, like scheduling.

IM not only replaces but also complements other media and face-to-face meetings. Cameron and Webster (2005) found that many employees felt that IM was used as a replacement of telephone, e-mail and face-to-face meetings. However, in contrast, others perceived IM as an additional method to reach others, which has been labeled outeraction (Nardi et al., 2000). Notably, IM has been found to be related with two types of polychronic use, as illustrated by the following examples (Cameron & Webster, 2005):

- *Communicating with one individual by several media.* "An individual might communicate with another using two media. For example, John may use the telephone to communicate verbally with Jane and use IM to quickly send her URLs to websites" (p. 12).
- *Communicating with more than one individual by several media.* For example, "an individual might be talking on the telephone with Jane while engaging in an IM conversation with Mary" (p. 12).

IM enables both internal and external e-collaboration. As noted above, IM is primarily used to support work-related e-collaboration. An assumption in the literature has been that IM is primarily used for e-collaboration in distributed teams. However, Quan-Haase and colleagues (2005) showed that IM was often used by co-located teams and as much for internal interaction as for external interaction. They found that co-located workers formed dense networks supported by IM and coined the term local virtuality to describe such new forms of work. This finding is supported by Cho et al. (2005) who argued that employees perceived IM as more useful for communication inside than outside a company.

IM enhances employees' sense of community but may also function as a barrier. As argued above, IM is commonly used to support both internal and external e-collaboration related with work-tasks. This may explain why IM has been suggested to enable higher connectivity and sense of community (Quan-Haase et al., 2005). However, it has also been reported that IM may be used to create distance to other employees. Quan-Haase and colleagues found that the medium was used to create distance to their superiors, especially when "difficult decisions have to be made or sensitive topics discussed."

INSTANT MESSAGING IN HIGHER EDUCATION

Most studies on IM in educational settings have focused on students engaged in higher education. These studies have been conducted in three settings: (1) online education (Hrastinski, 2006a, 2006b; Nicholson, 2002), (2) blended education (Contreras-Castillo et al., 2004), defined as the combination of traditional and online educational approaches (Hrastinski & Keller, 2007), and (3) outside formal educational settings (Chen, Yen, & Huang, 2004). Below, research findings are reviewed and categorized.

IM enables informal e-collaboration within groups. Hrastinski (2006a, 2006b) found that IM was primarily used to collaborate on work and exchange information. Similarly, Contreras-Castillo et al. (2004) reported that IM was used for collaboration among course participants. However, previous research on IM has concluded the opposite: "[IM was used] for social interaction and discussion about the school, rather than course material and group work" (Nicholson, 2002, p. 371). Chen, Yen, and Huang (2004) argued that students use IM for both social reasons and coursework. Together, these studies underline that IM will not by default be used for particular types of e-collaboration—this is probably dependent on many so far unknown factors. However, one consistent finding across several of these studies is that many students find the medium useful for informal or spontaneous interaction (Contreras-Castillo et al., 2004; Hrastinski, 2006a, 2006b; Nicholson, 2002).

In Nicholson's (2002) study of an online course the students suggested that IM might be particularly useful to support group work. Hrastinski (2006b) investigated this further by mapping perceived interaction patterns during an online course and concluded that IM was almost solely used to support interaction within groups. Interestingly, when comparing groups that adopted IM with those that did not, it was found that the adopters communicated more frequently and were related with a higher level of perceived participation.

IM enhances students' sense of community. Several studies on IM in educational settings suggest that IM may boost collaboration and community: (a) Contreras-Castillo et al. (2004) argued that IM helped reduce students' feeling isolated and increased collaboration; (b) Hrastinski (2006a, 2006b) suggested that complementing asynchronous media with IM enabled a higher level of participation among students and groups who voluntarily decided to use the medium; and (c) Nicholson (2002) argued that a master class that used IM felt a stronger sense of community since students found it easier to communicate and had more venues for informal and social interaction.

IM is not a silver bullet. Most studies on IM have explored possible beneficial uses and therefore few disadvantages of using the medium have been identified. However, IM should not be considered as a silver bullet—simply introducing IM will, of course, not enable interaction between all students of a class. Some students simply prefer to work individually, especially those who choose blended or online courses. Thus, some may prefer to be "witness learners" and stay in the periphery throughout a course (Bento & Schuster, 2003; Lave & Wenger, 1991) despite e-collaboration technologies being introduced (Hrastinski, 2006a).

IM adoption may be more likely if students know each other beforehand. As stated above, Hrastinski (2006a, 2006b) found that IM was rarely used to support social exchanges, while Nicholson (2002) reported the opposite. One difference between these two studies is that students in the latter group met face-to-face in the beginning of the course. A majority of social bonding in online courses has been reported to occur during on-campus sessions and then maintained via e-mail and synchronous media (Haythornthwaite & Kazmer, 2002). This is also supported by a study on IM use among adolescents, where it was argued that IM simulates spending time with friends (Boneva et

al., 2006). Consequently, a higher level of social support might have occurred if students had previously met face-to-face.

FUTURE TRENDS AND CONCLUSION

Employees, students, and adolescents commonly use IM. All studies that have been reviewed focus on the exchange of text messages. This is probably the most common way of communicating by IM today. However, most IM systems also enable spontaneous audio and video conferencing with one or several others. Researchers have a unique possibility to begin researching these richer applications before they become more widely used. Moreover, since rather few have studied IM, despite having been adopted in various settings, there is also a need for research that test and build on the findings reviewed in this article.

The aim of the article has been to review how IM may be used to support e-collaboration. In addressing this aim, I have discussed the use of IM in both work and educational settings. By including these two settings, it was aimed to achieve a more thorough understanding of the benefits and limitations of using IM to support e-collaboration. What can then be learnt from the two brief reviews presented in this article?

More or less all findings need to be interpreted carefully since only a handful of studies have been conducted on IM use. Even though most of the early studies suggest that IM is beneficial in enabling e-collaboration, some prefer to work individually and may choose not to use IM or may even use it to distance themselves from others. In Table 1, three conclusions that are supported by several indicators derived from the reviewed stud-

Table 1. Conclusions from studies on IM in work and education settings

Conclusion	Work setting	Educational setting
IM enables informal e-collaboration	Ask questions spontaneously	Interact spontaneously
	Informal tone	One-to-one or group-based communication
	One-to-one communication	
IM is primarily used for work-related (as compared with social) e-collaboration	Ask questions about work	Collaborate on work
	Coordinate and arrange meetings	Exchange information
	In-depth problem solving	Discuss group projects
	Internal and external e-collaboration	
IM enables a sense of community	Additional method of reaching others	Sometimes used to interact socially
	Internal e-collaboration	Enables collaboration and community

ies are listed. The key finding of this review is that IM may be recommended for those who want to support informal work-related e-collaboration.

REFERENCES

Andersson, J., & Azadi, O. (2004). *Instant messaging: The future communication tool among students?* Unpublished bachelor's thesis, Jönköping International Business School.

Bento, R., & Schuster, C. (2003). Participation: The online challenge. In A. Aggarwal (Ed.), *Web-based education: Learning from experience* (pp. 156-164). Hershey, PA: Idea Group Publishing.

Boneva, B., Quinn, A., Kraut, R., Kiesler, S., Cummings, J., & Shklovski, I. (2006). Teenage communication in the instant messaging era. In R. Kraut, M. Brynin, & S. Kiesler (Eds.), *Domesticating information technology*. Oxford: Oxford University Press.

Cameron, A. F., & Webster, J. (2005). Unintended consequences of emerging communication technologies: Instant messaging in the workplace. *Computers in Human Behavior, 21*(1), 85-103.

Chen, K., Yen, D. C., & Huang, A. H. (2004). Media selection to meet communication contexts: Comparing e-mail and instant messaging in an undergraduate population. *Communications of the Association for Information Systems, 14*, 387-405.

Cho, H.-K., Trier, M., & Kim, E. (2005). The use of instant messaging in working relationship development: A case study. *Journal of Computer-Mediated Communication, 10*(4).

Contreras-Castillo, J., Favela, J., Perez-Fragoso, C., & Santamaria-del-Angel, E. (2004). Informal interactions and their implications for online courses. *Computers & Education, 42*(2), 149-168.

Fish, R. S., Kraut, R. E., & Chalfonte, B. L. (1990). *The VideoWindow system in informal communications.* Paper presented at the Conference on Computer-Supported Cooperative Work, Los Angeles.

Handel, M., & Herbsleb, J. D. (2000). *What is chat doing in the workplace?* Paper presented at the Computer Supported Cooperative Work Conference, New Orleans, LA.

Haythornthwaite, C., & Kazmer, M. M. (2002). Bringing the Internet home: Adult distance learners and their Internet, home, and work worlds. In B. Wellman & C. Haythornthwaite (Eds.), *The Internet in everyday life* (pp. 431-463). Malden, MA: Blackwell Publishing.

Hrastinski, S. (2006a). Introducing an informal synchronous medium in a distance learning course: How is participation affected? *Internet and Higher Education, 9*(2), 117-131.

Hrastinski, S. (2006b). The relationship between adopting a synchronous medium and participation in online group work: An explorative study. *Interactive Learning Environments, 14*(2), 137-152.

Hrastinski, S., & Keller, C. (2007). Computer-mediated communication in education: A review of recent research. *Educational Media International, 44*(1).

Isaacs, H., Walendowski, A., Whittaker, S., Schiano, D. J., & Kamm, C. (2002). *The character, functions, and styles of instant messaging in the workplace.* Paper presented at the Computer-Supported Cooperative Work, New Orleans, Louisiana.

Kraut, R. E., Fish, R. S., Root, R. W., & Chalfonte, B. L. (1990). Informal communication in organizations: Form, function, and technology. In S. Oskamp & S. Spacapan (Eds.), *Human reactions to technology: The Claremont Symposium on applied social psychology.* Beverly Hills, CA: Sage.

Lave, J., & Wenger, E. (1991). *Situated learning: Legitimate peripheral participation.* Cambridge: Cambridge University Press.

Mintzberg, H. (1973). *The nature of managerial work.* New York: Longman.

Nardi, B. A., Whittaker, S., & Bradner, E. (2000). *Interaction and outeraction: Instant messaging in action.* Paper presented at the Computer Supported Cooperative Work Conference, Philadelphia.

Nicholson, S. (2002). Socialization in the "virtual hallway": Instant messaging in the asynchronous web-based distance education classroom. *Internet and Higher Education, 5*(4), 363-372.

Quan-Haase, A., Cothrel, J., & Wellman, B. (2005). Instant messaging for collaboration: A case study of a high-tech firm. *Journal of Computer-Mediated Communication, 10*(4).

Shiu, E., & Lenhart, A. (2004). *How Americans use instant messaging*. Retrieved April 24, 2006, from: http://www.pewinternet.org/pdfs/PIP_Instantmessage_Report.pdf

KEY TERMS

Blended Education: The combination of traditional and online educational approaches.

Chat: Text-based synchronous communication medium.

Informal E-Collaboration: Spontaneous brief collaboration mediated by technology among individuals engaged in a common task.

Informal Interaction: "Take place at the time, with the participants, and about the topics at hand" (Fish et al., 1990, p. 2).

Instant Messaging: Communication medium that can be used to spontaneously interact with others that are online through various types of synchronous media.

Local Virtuality: Co-located people that use technology to communicate (Quan-Haase et al., 2005).

Online Education: Distance education primarily delivered online.

Outeraction: "Communicative processes people use to connect with each other and to manage communication, rather than to information exchange" (Nardi et al., 2000, p. 79).

Polychronic Media Use: Using several media simultaneously.

Synchronous Interaction: Real-time interaction that may be place-independent.

Impact of Collaborative Delivery of Enterprise ICT Services

Jiri Vorisek
University of Economics Prague, Czech Republic

George Feuerlicht
University of Technology, Sydney, Australia

INTRODUCTION

Most organizations today are looking for more cost effective approaches to delivering enterprise applications to their user base. Among the alternatives that are becoming increasingly popular are various forms of e-collaboration that involve the sharing of information between organizations, integration of interenterprise business processes among partner organizations, and the delivery of software services by external application service providers (ASPs). Such recent trends are likely to produce a situation where most enterprise applications will be implemented collaboratively or supplied as services, making the Software-as-a-Service model the dominant method of enterprise application delivery. The extensive use of externally supplied software and information services will change the shape of the ICT (Information and Communication Technologies) market and impact on management decisions about the deployment of enterprise ICT (Harber, 2004). These changes will affect both user organizations and organizations supplying ICT products and services. In this paper we analyze the above trends and discuss the impact of the Software-as-a-Service model on ICT user organizations and ICT suppliers. We first discuss the key enterprise computing trends and the strategic importance of ICT.

BACKGROUND: KEY ENTERPRISE COMPUTING TRENDS

This section is a discussion of key trends that we consider to have significant impact on future use of ICT in organizations, and, consequently, on the composition of the ICT market.

Strategic Importance of ICT

Nicolas Carr (2003) argues that ICT is today accessible to most organizations and therefore is losing its strategic significance. We dispute this claim, and argue that ICT cannot be considered in isolation from entrepreneurial activities, business processes, and company culture. It is the close alignment between ICT and business processes that can provide competitive advantage to organizations, and produce high quality products and services at lower cost, resulting in a strategic advantage (McCabe, 2003; Nevens, 2002)

In general, there are two types of enterprise applications: applications that support business processes (e.g., logistics, CRM, etc.) and applications that directly implement business processes (e.g., electronic banking, mobile telephony, airline e-tickets, etc.). For the first type of enterprise applications, it is possible to gain competitive advantage by combining ICT with unique company culture and knowledge. This unique combination enables the company to function effectively and utilize key assets such as organizational knowledge and culture. The second type of enterprise application provides a service or product to customers and its timely deployment and unique features can result in competitive advantage (Vo í ek, 2005). There are many recent examples of ICT providing competitive advantage to organizations. For example, the courier service eKurýr (www.ekuryr.cz) that operates in the Czech and Slovak Republics has been highly successful principally because of its unique electronic system, eKurýr. Similarly, while not every new technology is important, there are situations where missing out on a new technological development can be fatal. For example, today most suppliers of accommodation services must provide Web-based applications to allow worldwide access to booking and other services in order to avoid losing a significant market share.

Another factor that supports the argument for the strategic importance of ICT is the unremitting growth in the demand for timeliness and quality of information for decision making from all levels of management. According to Gartner (2004), the required response time to important events has decreased from 2 months in 2002 to 1 month in 2004, and will further decrease to 1 day in 2010. While this forecast may not be entirely accurate, the requirement to react faster to important external events is clearly evident.

Increasing Process Orientation of the Enterprise

Toward the end of the last century, it was becoming clear that managing enterprises based on business functions could lead to conflicts between the goals and interests of individual departments and organizational goals as a whole, resulting in numerous problems including unpredictable responses to important events. Many organizations have adopted the process management approach in order to address such problems. The importance of process-management is still growing. The principal aim of a process-managed enterprise is to achieve real-time response to important events. This requires that the organization has active sensors (usually using ICT) that indicate new events (e.g., arrival of an order, time to send Value Added Tax [VAT] returns, production line failure, etc.). As soon as the event occurs, the correct process is activated as a response.

This trend also affects enterprise ICT management, with many enterprises adopting process management of their ICT. ITIL and COBIT methods are de facto standards in this area.

The ICT marketplace has responded to the transition to process-management with a relatively wide choice of tools for business process modeling and for optimization, monitoring, and management of business processes in real time. The Organization for the Advancement of Structured Information Standards (OASIS) has defined a number of standards in this area (OASIS, 2005).

The success of process-management depends on a number of critical success factors. The most important of these are (Vo í ek & Dunn, 2001):

- Appropriately chosen detail of business process definition and its alignment with the knowledge of the employees undertaking the process. A detailed definition of a business process enables the use of less-qualified but well-trained employees. On the other hand, it prevents utilization of employees' creativity and reduces the flexibility of the process.
- Appropriately chosen process maturity. CMM (Compton et al., 2002) defines six levels or process maturity. The lowest level is for a nonexistent process, the highest describes an optimized process. However, it is not sensible to plan for the highest level for each process. This would be too expensive for processes that are not vital to the enterprise and occur infrequently.
- Appropriate utilization of process methods and standards. When implementing process-management it is essential to use appropriate methods and standards (e.g., ITIL or COBIT). Recent experience indicates that applying these methods and standards mechanically can lead to problems and that methodologies must be tailored to the specific conditions of the enterprise.

Management of the Relationship Between Business and ICT Using ICT Services

For over 50 years, computer professionals and end-users have been searching for an optimal way to communicate with one another, and for an optimum division of responsibility for the costs and benefits of ICT projects. A new approach for managing the relationship between business and ICT is emerging based on the concept of ICT service described using an SLA (service level agreement). Service is a basic element that defines the boundary between business and ICT activities. Methodologies such as SPSPR (Vo í ek & Dunn, 2001), which define the responsibilities of different types of managers and the content of the communication between business and ICT managers without excessive use of technological concepts, are required to define services and their interfaces to business.

The management of ICT services has a number of critical success factors. The most important of these are:

- The ability of the owners of business processes to define SLA for ICT requirements
- The focus of ICT services. ICT services should be derived from the requirements of business

processes, not from the interests of individual departments

- The ability of ICT managers to specify and manage ICT infrastructure that facilitates provision of agreed-upon scalable ICT services

Emphasis on Management of the Return on ICT Investment

The ICT crisis lasting since the beginning of this decade resulted in increased emphasis on the management of the return on ICT investment. Well-managed enterprises no longer invest in ICT without a thorough analysis of the return on investment and refuse to finance risky long-term projects.

Increasingly, CEOs require that their CIOs ensure that increases in the investment in ICT closely correlate with increased revenues, and that no project is started unless an improvement in the performance of the enterprise can be expected. The question is how a CIO can fulfill such expectations.

The first requirement is a scalable ICT infrastructure and scalable ICT processes (Feuerlicht & Vo í ek, 2004). When an enterprise operates its information systems using its own ICT infrastructure, scaling up or down may not be a realistic option. An enterprise has to plan its ICT infrastructure for the maximum anticipated load and incremental increases and reductions in capacity (e.g., disposing of surplus hardware, software licenses and ICT specialists) are often impossible in practice. The solution to this problem is to purchase external software services delivered by providers with scalable, multi-tenant architecture.

Another requirement is improvement in the enterprise performance, and that cannot be achieved by the deployment of ICT solutions alone. This requirement can be only met by an appropriate distribution of responsibilities of ICT and business managers, for example using the SPSPR model. Using this model, the benefits of the commissioned ICT services will be the responsibility of the business process owner. The owner of the business process has to add the cost of each ICT service to the cost of other (non-ICT) processes and then evaluate the effectiveness of the overall process. If the cost of an ICT service is too high, the requirements should be reconsidered (e.g., reducing functionality, the number of users or availability). By contrast, the CIO is responsible for ensuring that the cost of an ICT service is competitive with similar ICT services on the market.

Purchasing External ICT Services Instead of Purchasing ICT Products

The desire to concentrate enterprise activities on core business leads companies to outsourcing of supporting business processes and to considerations of ICT outsourcing in order to achieve scalability in ICT services based on the pay-as-you-use principle.

Making decisions about what to own and what to purchase as an external service presents a number of challenges. An enterprise must have a sourcing strategy and use it to make such decisions. The development of a sourcing strategy and its use for decision-making is a complex process since a large number of alternatives with different critical success factors need to be considered (Feuerlicht & Vo í ek, 2003). A recent analysis (Vo í ek & Feuerlicht, 2004) and predictions (Cohen, 2004) show that the dominant forms of outsourcing will be business process outsourcing, complex outsourcing of ICT, and ASP.

Effective utilization of outsourcing depends on a number of critical success factors:

- Choosing an appropriate variant of outsourcing.
- Choosing an appropriate granularity of ICT services. At one extreme an ICT service can include all of the functionality of an ERP system, while at the other extreme a service could be a single transaction (e.g., ordering an airline ticket using a Web service).
- Monitoring of ICT services to be able to carry out a detailed analysis of the cost of the services, processes and resources. Without good monitoring it is not possible to find out what provides a better value - internal or external provider.
- Quality of decision-making when deciding whether to outsource or not depends on the quality of information about the ICT market (services on offer) and on the quality of the sourcing strategy.

IMPACT ON END-USER ORGANIZATIONS

If the above-mentioned trends continue, we can anticipate the following impact on end-user organizations:

- Increasingly decision about utilizing ICT will be made by the owners of business processes and as a part of strategic management. This will require changes in their qualification. Managers, who understand how to use ICT to develop new products or services, or how to gain new customers, will become indispensable members of the top management team in most enterprises (Santosus, 2005).
- Employees of an ICT department will need to demonstrate the value of ICT for the business and offer new ways of utilizing ICT by the business.
- Because of outsourcing, the number of technologically oriented specialists (e.g., programmers, ICT administrators) in companies will decrease. However, the number of employees involved with the relationship between the business and ICT services will increase. Not all of these employees will be working in the ICT department. In 2004, Gartner predicted that in North American and European companies the fraction of employees working in ICT will reach 8% in 2006 and in the "ICT driven industries" it will be as high as 15-20%.
- The integrative and innovative role of ICT departments will grow. This is particularly the case for the "ICT driven industries." This is because ICT processes are not like standard support processes such as accounting or purchasing, they have an immediate impact on the effectiveness of most core business processes.
- The volume of ICT services will be scalable, and ICT costs will correlate with the level enterprise activities and the turnover.

Even though many ICT services will be sourced from external providers, the number of employees concerned with the utilization of ICT is not expected to decrease significantly. However, the structure of their qualifications will need to change in order for the enterprise to maintain the following skills and knowledge:

- How to gain a competitive advantage using ICT
- How to design the overall architecture for ICT services
- Which services, processes and resources should be owned and which should be outsourced

- Selection of the best supplier of an ICT service
- Monitoring of ICT services and measurement of the benefits of ICT for business processes

IMPACT ON SUPPLIER ORGANIZATIONS

If the above-mentioned trends are realized, particularly outsourcing of ICT services and close monitoring of the relationship between costs and benefits, we can anticipate the following impact on ICT supplier organization:

- The sale of new software licenses to end-user organizations will decrease. Tables 1-3, compiled using annual reports of several large software companies, confirm this trend, and indicate that while the income from new licenses decreased, income from maintenance and related services grew. Furthermore, according to Haber (2004), 80% of software cost can be attributed to maintenance of applications and related activities. The increasing cost of maintenance could be another factor that will contribute to the greater focus of end-user organizations on outsourcing of their applications.
- Because of the decrease in the sales of new licences and the growth in outsourcing, the software industry is likely to undergo further rationalization, with some software vendors who do not adapt their businesses to this new environment failing (Dubie, 2004).
- Instead of software and hardware products, the vendors will need to focus on providing scalable ICT services. In effect, hardware and software products will be managed and deployed by their suppliers in order to provide scalable services to clients. The gradual transition of large ICT suppliers from products to services is illustrated in the tables Table 4, 5, and 6:
- There will be changes in the structure and company culture of ICT vendor companies, reflecting the fact that supplying ICT services requires a different type of enterprise than the supply products and licences.

Outsourcing of services is often offered in the context of the Utility Computing model, where the cost of

Table 1. Software license vs. support (Oracle)

Oracle	1999	2000	2001	2002	2003
License					
annual growth		20%	6%	- 25%	- 6%
as a fraction of the total income	41%	43%	42%	36%	34%
Support					
annual growth		27%	20%	8%	8%
as a fraction of the total income	27%	29%	33%	40%	44%

Table 2. Software licenses vs. maintenance (SAP)

SAP	1999	2000	2001	2002	2003
License					
annual growth		27%	5%	-11%	-6%
as a fraction of the total income	38%	39%	35%	31%	31%
Maintenance					
annual growth		44%	27%	15%	6%
as a fraction of the total income	23%	27%	29%	33%	37%

Table 3. Software licenses vs. maintenance (Siebel)

Siebel	1999	2000	2001	2002	2003
License					
annual growth		118%	- 4%	- 34%	- 31%
as a fraction of the total income	62%	61%	51%	43%	36%
Services, Maintenance, etc.					
annual growth		126%	44%	- 8%	- 7%
as a fraction of the total income	38%	39%	49%	57%	64%

Table 4. Revenue composition (IBM)

IBM % revenue	2000	2001	2002	2003	2004
servers	22,68	22,32	20,04	18,72	18,89
PCs	17,83	14,51	13,78	12,97	13,47
HW total	40,51	36,83	33,82	31,68	32,35
SW	14,81	15,58	16,10	16,06	15,68
IT services	38,96	42,08	44,79	47,83	47,99
Financial services	4,07	4,12	3,98	3,17	2,71
others	1,65	1,39	1,31	1,26	1,27

Table 5. Revenue composition (HP)

HP % revenue	2000	2001	2002	2003	2004
servers	26,00	25,26	22,33	21,03	20,12
PCs	35,85	33,04	30,23	29,03	30,81
Printers, scanners	22,47	24,01	28,20	30,89	30,28
HW total	84,32	82,31	80,75	80,95	81,22
IT services	14,05	15,84	17,10	16,91	17,24
Financial services	2,00	2,62	2,89	2,63	2,37

Table 6. Revenue composition (SUN)

SUN Microsystems % revenue	2000	2001	2002	2003	2004
Servers		67,88	59,19	54,60	52,34
Storage		14,39	13,58	13,56	13,42
Products total	85,37	82,27	72,77	68,16	65,76
Support services		11,99	20,31	24,87	26,81
Professional and Knowledge services		5,74	6,92	6,97	7,43

services is derived based on unit costs (per user in a particular category, per server, etc.). The customer can change the volume of the service as required and pay only for the actual number of users or supported servers in a given time period (e.g., one month). The migration to ASP-provided software services will not happen over the short-term, as there is an inherent distrust of external providers of software services, and unwillingness to relinquish control of data and key ICT infrastructure.

The increase in the required number of specialists who integrate business processes with ICT services will not affect only end-user organizations. It will provide an opportunity for new consulting firms that specialize in this area. There are specific considerations that apply to different countries and geographical regions (e.g., India, China, Czech Republic, etc.).

There is a general agreement that the impending ICT market changes will affect the organizational structure and company culture of large ICT suppliers. The key factors in the future development of ICT will include:

- Labour costs, where large global companies such as HP or IBM will need to compete for market share with smaller, locally active companies that typically have lower labor costs and more extensive knowledge of the customer base. This will be most apparent in the SME market.
- Ability of start-up companies to enter new untried market segments and to offer more innovative as well as more risky types of services.

CONCLUSION

Strategic advantage cannot be achieved simply by deploying new technology, but requires effective integration of ICT with entrepreneurial activities and enterprise business processes. Such integration enables the company to speed up its responses to important events, reduces cost and increases the quality of information used for decision-making. It is therefore essential that ICT projects focus on integration and customization of enterprise applications to closely reflect important business processes. Enterprise process-management and process-managed ICT are essential today. Implementing process-management is a long-term activity that requires specific knowledge and skills, with success depending on specific critical factors.

Management of the relationship between business and ICT using ICT services has proved the best solution to the long-standing problem of communication between business and ICT professionals. However, this requires new knowledge and skills on both sides.

Well-managed enterprises must be able to control their ICT costs and align the investment in ICT to reflect the growth in the revenues of the enterprise. Preconditions for success in this area are scalable ICT services and a clear allocation of responsibilities for the benefits and costs of ICT between the business and ICT managers.

Notwithstanding these challenges, it is highly probable that by the end of this decade outsourcing, particularly ASP outsourcing, will be the dominant form of acquiring ICT services.

The emerging trends described in this paper will have a significant impact on both the suppliers and users of ICT. Collaborative relationships will form the basis for the interaction between ICT user and supplier organizations, allowing business to focus on their core activities with a corresponding impact on the composition and qualifications of their workforce.

REFERENCES

Carr, N. G. (2003). IT doesn't matter. *Harvard Business Review, 81*(5).

Cohen, P. (2004). *Twelve technical and business trends shaping the year ahead.* http://www.babsoninsight.com/contentmgr/showdetails.php/id/687

Compton, N. L., et al. (2002). *Capability Maturity Model® Integration (CMMISM), Version 1.1, Staged Representation* (Technical Report CMU/SEI-2002-TR-012). The Software Engineering Institute, 2002, http://www.sei.cmu.edu/cmmi/

Dubie, D. (2004). Vendors make the utility computing grade. *Network World Fusion.* 22/03/2004, http://www.nwfusion.com/news/2004/0322dellsummit.html

Feuerlicht, G., Vo í ek, J. (2003). Key success factors for delivering application services. In V. E. Praha (Ed.), *Systems Integration 2003 Conference* (pp. 274-282).

Feuerlicht, G., Vo í ek, J. (2004). Utility computing: ASP by another name, or a new trend? In V. E. Praha (Ed.), *Systems Integration 2003 Conference* (pp. 269-280).

Haber, L. (2004). ASPs still alive and kicking. Retrieved January 30, 2004, from http://www.aspnews.com /trends/article.php/3306221

McCabe, L. (2003). *A winning combination: Software-as-services plus business consulting and process services* (Summit Strategies Market Strategy Report, ID#: 3SS-07). Retreived http://www.summitstrat.com/store/3ss07detail

Nevens, M. (2002). The real source of the productivity boom. *Harvard Business Review*, 23-24.

OASIS Standards. (2005). 20/04/2005, http://www.oasis-open.org

Oracle Financial Reports. (2005). http://www.oracle.com/corporate/investor_relations/analysts/

Santosus, M. (2005). Inferiority complex, 14/04/2005, http://www.cio.com/archive/031504/ reality.html

SAP Financial Reports. (2005). http://www.sap.com/company/investor/reports/

Ross, W., & Weill, P. (2002) Six IT decisions your IT people shouldn't make. *Harvard Business Review*, *80*(11).

Vo í ek J. (2005). Verbal communications with Martin Bednár, SAP ČR, Radim Hradílek, IBM ČR, Jan Kameníček, HP ČR, and Jiří Polák, Deloitte&Touche, Praha.

Vo í ek, J., & Dunn, D. (2001). Management of business informatics: opportunities, threats, solutions. In V. E. Praha (Ed.), *Systems Integration 2001 Conference*.

Vo í ek, J., & Feuerlicht, G. (2004). Is it the right time for the enterprise to adopt software-as-a-service model? *Information Management*, *17*(3/4), 18-21.

KEY TERMS

ASP (Application Service Provider): A company that offers individuals or enterprises access over the Internet to application programs and related services.

Business Process: A collection of related, structured activities that produce a specific service or product for a particular customer.

Enterprise Applications: Application software that implements a set of business functions, for example ERP (enterprise resource planning) or CRM (customer relationship management).

ICT Service: A basic element that defines the boundary between business and ICT activities. ICT service is described using an SLA (service level agreement).

SLA (Service Level Agreement): A contract between the provider and customer of the ICT service. The main parts of the SLA are content of the service, volume of the service, quality of the service, and price of the service.

Software-as-a-Service: Model for application delivery. Application service provider controls all necessary ICT infrastructure (HW+SW) and delivers application functionality as a service via Internet to many customers.

Multi-Tenant Architecture: Software architecture that is designed to support a large number of users from different user organizations on a scalable technological infrastructure.

The Impact of Personality on Virtual Team Creativity and Quality

Rosalie J. Ocker
The Pennsylvania State University, USA

INTRODUCTION

A series of experiments investigated creativity and quality of work-product solutions in virtual teams (Ocker, 2007, 2005; Ocker & Fjermestad, 1998; Ocker, Hiltz, & Johnson, 1998; Ocker, Hiltz, Turoff, & Fjermestad, 1996). Across experiments, small teams with about five graduate students interacted for approximately two weeks to determine the high-level requirements and design for a computerized post office (Goel, 1989; Olson, Olson, Storrosten, & Carter, 1993). The means of interaction was manipulated in these experiments such that teams interacted via one of the following treatments: (1) asynchronous computer-medicated communication (CMC), (2) synchronous CMC, (3) asynchronous CMC interspersed with face-to-face (FtF) meetings, or (4) a series of traditional FtF meetings without any electronic communication.

A repeated finding across experiments was that teams interacting *only* using asynchronous CMC—that is, teams without any FtF or synchronous communication—produced significantly more creative results than teams in the other treatments. Additionally, asynchronous virtual teams rated high in creativity were generally not the same teams that were judged high in terms of the quality of their deliverable.

To further examine these findings, this article presents results of an exploratory study designed to investigate the impact of individual personality facets on team outcomes. The objective of this study is to determine whether differences in team outcomes—in terms of the level of creativity versus the quality of the team deliverable—can be predicted by individual member personality. Specifically, two research questions are investigated:

Do individual member personalities predict virtual team creativity?

Do individual member personalities predict virtual team quality?

BACKGROUND

Personality traits, which are persistent across situations and time, distinguish an individual from others. In the domain of psychology, it is readily accepted that there are five broad factors or dimensions of personality traits (Costa & McCrae, 1992; Goldberg, 1993). An individual falls somewhere along the continuum of a given dimension.

Extraversion, openness, agreeableness, conscientiousness, and negative emotionalism (also known as neuroticism) comprise the five dimensions. Extraversion encompasses an individual's tendency for sociability and interactivity as opposed to solitude and seclusion. Openness encompasses an individual's tendency for abstract or original ideas versus tangible facts. Agreeableness encompasses an individual's tendency for cooperative versus competitive interaction with others. Conscientiousness encompasses an individual's tendency for convergent, task-oriented versus divergent, process-oriented work styles. Finally, negative emotionalism encompasses how an individual responds to stress, from a wide-range of emotions to a narrow range of emotions.

Dimensions are broadly defined. Personality facets were developed to more precisely measure the particular attributes subsumed within the broad domains. Thus, each factor is comprised of multiple facets. Each facet includes a common "portion" attributable to the associated factor, as well as a portion attributable to that particular facet. McCrae and Costa (1992) developed six 8-item facet scales for each dimension. As a means of assessing the discriminant validity of the facet scales, they related each scale to various items from the Adjective Check List (Gough & Heilbrun, 1983). Twenty-six of these "ACL-defined" facets achieved discriminant validity. These are depicted in Table 1.

Table 1. ACL personality facet scales

Negative Emotionalism	
Anxiety	anxious, fearful, worrying, tense, nervous, -confident, -optimistic
Depression	Worrying, -contented, -confident, -self-confident, pessimistic, moody, anxious
Self-Consciousness	shy, -self-confident, timid, -confident, defensive, inhibited, anxious
Vulnerability	-clear-thinking, -self-confident, -confident, anxious, -efficient, -alert, careless
Extraversion	
Warmth	friendly, warm, sociable, cheerful, -aloof, affectionate, outgoing
Gregariousness	sociable, outgoing, pleasure-seeking, -aloof, talkative, spontaneous, -withdrawn
Assertiveness	aggressive, -shy, assertive, self-confident, forceful, enthusiastic, confident
Activity	energetic, hurried, quick, determined, enthusiastic, aggressive, active
Excitement Seeking	pleasure-seeking, daring, adventurous, charming, handsome, spunky, clever
Positive Emotions	enthusiastic, humorous, praising, spontaneous, pleasure-seeking, optimistic, jolly
Openness	
Fantasy	dreamy, imaginative, humorous, mischievous, idealistic, artistic, complicated
Aesthetics	imaginative, artistic, original, enthusiastic, inventive, idealistic, versatile
Feelings	excitable, spontaneous, insightful, imaginative, affectionate, talkative, outgoing
Actions	interests wide, imaginative, adventurous, optimistic, -mild, talkative, versatile
Ideas	idealistic, interests wide, inventive, curious, original, imaginative, insightful
Values	-conservative, unconventional, -cautious, flirtatious
Agreeableness	
Trust	forgiving, trusting, -suspicious, -wary,-pessimistic, peaceable, -hard-hearted
Straightforwardness	-complicated, -demanding, -clever, -flirtatious, -charming, -shrewd, -autocratic
Compliance	-stubborn, -demanding, -headstrong, -impatient, -intolerant, -outspoken, -hard-hearted
Modesty	-show-off, -clever, - assertive, -argumentative, -self-confident, -aggressive, -idealistic
Conscientiousness	
Competence	efficient, self-confident, thorough, resourceful, confident, -confused, intelligent
Order	organized, thorough, efficient, precise, methodical, -absent-minded, -careless
Dutifulness	-defensive, -distractible, -careless, -lazy, thorough, -absent-minded, -fault-finding
Achievement Striving	thorough, ambitious, industrious, enterprising, determined, confident, persistent
Self-discipline	organized, -lazy, efficient, -absent-minded, energetic, thorough, industrious
Deliberation	- hasty, -impulsive, -careless, -impatient, -immature, thorough, -moody

METHOD

Data Set

The data set consisted of 47 participants from 11 teams comprising the asynchronous CMC treatment in Ocker (2005). Females comprised 37% of the participants. The average work experience was approximately 8 years. Fifty-five percent of participants were between the ages of 23 and 30, while 28% were between 31 and 35. Participants indigenous to the United States accounted for 93% of the data set.

Task

The Computerized Post Office (CPO) was the task used in this experiment. This task was adapted from Goel (1989) and Olson et al. (1993). Olson et al. characterize this task as incorporating planning, creativity, decision-making, and cognitive conflict (McGrath, 1984). Teams were required to reach consensus on the requirements and high-level design for the CPO. Specifically, teams were to address four areas: (1) the functionality of the CPO (i.e., the services to be offered customers), (2) the user interface, (3) advantages and disadvantages of the CPO, and (4) major implementation considerations that would feed into the next phase of developing an implementation plan. Each team produced a single written at the end of the experiment.

Subjects and Team Composition

Subjects consisted of graduate students in the MBA and MSIS program at a branch campus of a large university. For their participation, all subjects received course credit. Subjects were assigned to teams randomly. All teams had a zero history of working together.

Technology and Facilitation

Virtual teams communicated electronically using the FirstClass computer conferencing system (www.softarc.com). Each team communicated in its own conference on FirstClass. The conferences were minimally facilitated. The conference facilitator's role was that of a technical assistant, helping teams with equipment problems and answering questions of a technical nature.

Training

Subjects met within their respective classes for training on the essential aspects of FirstClass. Within classes, subjects were randomly assigned to training conferences. Training was completed within one hour. All subjects were trained using the same practice problem, a modified version of "Entertainment for Dutch Visitors" (Olson et al., 1993).

Procedure

Teams did not follow a predetermined process or structure. They were instructed to develop and submit a report by the end of the experimental period. Leaders were designated for each team.

Each team exchanged only *asynchronous* electronic comments within its own computer conference. All members were explicitly instructed to communicate solely within their respective conferences.

All subjects completed a consent form and a background survey prior to the start of the experiment. All teams had access to the FirstClass computer conferencing system throughout the entire experimental period.

At the end of the experiment, subjects completed the postexperiment survey and the ACL personality measure.

Personality Measure

The Adjective Check List (ACL) (Gough & Heilbrun, 1983) was used to assess participant personality. The ACL is comprised of 300 descriptor words. An individual checks the words that are self-descriptive (there is no limit to the number of words that one can check).

Measure of Team Creativity

To ascertain the degree of creativity of teams' CPO design, a process that resulted in an objective measure of creativity was developed (Ocker, 2005). Using each team's report deliverable, a list of the unique ideas was compiled with regard to services contained in the team's CPO design. The unique list of services from each team was merged into one combined list of unique services and duplicate services were eliminated.

Based on this list, the number of *original* ideas for each team was counted. Statistically, original ideas are unique ideas generated by less than 5% of a given sample (Thompson, 2003). As there were 47 subjects participating in the experiment, an idea was considered original if it occurred one or two times (i.e., was contained in no more than two groups' reports). Because teams generated different length reports and team size varied between four and five participants, *percentages* (rather than the actual count of ideas) were deemed to be a more accurate way to objectively measure creativity. Thus, using the total number of unique ideas within each team's report as the denominator, the percentage of original ideas was calculated.

Measure of Team Quality

Two expert judges measured the quality of each team's solution as contained in the report deliverable. Judges rated various aspects of the report content, including the feasibility of the solution (i.e., realistic or unrealistic), as well as the clarity of the written presentation and the completeness of report (as per the task description). Judges also rated the overall quality of each team's solution.

ANALYSIS AND RESULTS

Team Creativity and Quality

The results of the assessment of team creativity (based on original ideas) and quality (overall) are shown in Table 2. Three teams—A7, A8, and A10—topped the creativity rankings with over 20% of their ideas falling into the original category. However, A10's solution was judged to be unrealistic. In terms of overall quality, A3 and A5 were judged highest with scores of 98 and 95, respectively.

Personality Dimensions and Facets

Based on responses to the ACL, participant scores were calculated for each of the five personality dimensions as well as 26 facets. The facets were calculated based on the facet adjectives contained in Table 1. That is, for each adjective that a participant selected, a point was added or subtracted to the facet score, according to the sign of the adjective. The five factors were calculated by summing the facet scores associated with each factor.

Regression Analysis

Testing for a group effect. The data for this study have a multilevel structure since participants are nested within teams and there are variables describing participants (personality traits) and variables describing teams (creativity and quality). Because members interacted with one another in` teams, there is a lack of independence and the potential for a team or *group effect* (Gallivan & Bebunan-Fich, 2005). To test for a group effect, a dummy variable was created for each team. A series of regression analyses were conducted where the dependent variable was the personality factor or facet, with the teams constituting the independent variables. No *between team* significant differences were found for any personality factor or facet, indicating that a group level effect was not evident. Thus, an analysis at the individual member level was permissible.

Exploring the effect of member personality. Two sets of regression analyses were conducted to explore the impact of member personality on the creativity and quality of team reports. First, the five personality dimensions were regressed on each dependent variable (percent original ideas were used for creativity). The results, as contained in Table 3, indicate that none of the five broad factors were predictive of either creativity or quality.

Second, the facet scores were regressed on team creativity, again, using percentage of original ideas and overall quality. Summary statistics for the creativity regression model, as contained in Table 4, show that the model is marginally significant at the .10 level.

As shown in Table 5, four personality facets were found to significantly predict team creativity. (The related personality dimension is indicated in brackets.) The idea, assertiveness, and anxiety facets are positively related to creativity while the achievement striving facet is negatively related. Based on the standardized beta coefficients, the assertiveness and idea facets account for more variance than do the achievement and anxiety facets.

Summary statistics for the quality regression model, as contained in Table 4, show that the model is mar-

Table 2. Team creativity and quality scores

Team	A1	A2	A3	A4	A5	A6	A7	A8	A9	A10
Total Unique Ideas	35	37	41	40	44	15	49	36	34	37
Original	6	6	4	7	6	0	18	9	4	8
% Original	0.17	0.16	0.10	0.18	0.14	0.00	0.37	0.25	0.12	0.22
Overall Quality (Out of 100%)	93	85	98	75	95	75	85	85	90	93[a]

[a] unrealistic solution

Table 3. Factor regression model for creativity and quality

Creativity Factors						
Model	Sum of Squares	df	Mean Square	F	Sig.	Adj. R Square
Regression	507.63	5	101.53	1.25	0.31	0.03
Residual	3002.78	37	81.15			
Total	3510.41	42				
Quality Factors						
Model	Sum of Squares	df	Mean Square	F	Sig.	Adj. R Square
Regression	43573	5	87.15	1.5	0.21	0.06
Residual	2144.96	37	57.97			
Total	2580.7	42				

Table 4. Facet regression model for creativity and quality

Creativity Facets						
Model	Sum of Squares	df	Mean Square	F	Sig.	Adj. R Square
Regression	603.90	26	23.23	1.87	0.10	0.36
Residual	186.22	15	12.42			
Total	790.11	41				
Quality Facets						
Model	Sum of Squares	df	Mean Square	F	Sig.	Adj. R Square
Regression	21722.06	26	835.46	1.88	0.08	0.33
Residual	8899.60	20	444.98			
Total	30621.66	46				

Table 5. Facet regression coefficients for creativity and quality

Creativity			
Facet (dimension)	Std. Beta	t	Sig.
Idea (openness)	1.42	3.26	0.005
Assertive (extraversion)	1.92	2.48	0.03
Achievement striving (conscientiousness)	-0.74	-2.22	0.04
Anxiety (neuroticism)	0.79	2.21	0.04
Quality			
Facet (dimension)	Std. Beta	t	Sig.
Trust (agreeableness)	-0.46	-2.43	0.03
Deliberation (conscientiousness)	0.79	2.53	0.02

ginally significant at the .08 level. As shown in Table 5, two personality facets were found to significantly predict team quality. (The related personality dimension is indicated in brackets.) The deliberation facet is positively related to quality while the trust facet is negatively related. Based on the standardized beta coefficients, deliberation accounts for more variance than does trust.

SUMMARY OF FINDINGS

At the factor level, personality did not predict either team creativity or quality. This lack of significance at the factor level is not an unusual finding, given the broad nature of the measures (e.g., Paunonen & Ashton, 2001). Hence, researchers have increasingly turned to personality facets as a more revealing measure of personality differences. In line with other research, the results of the analysis of personality facets were more revealing. Two quite different member profiles emerge from this facet analysis, both of which have face validity.

Different personality facets were predictive of team level creativity and quality. In terms of creativity, we see an individual who is an imaginative and original thinker, who enthusiastically expresses his/her ideas (without being overbearing), and who is more concerned with ideas than the project grade. Interestingly, each significant facet comes from a different personality dimension (the only dimension not represented it Agreeableness). With regard to quality, the analysis points to an individual that is deliberate, thorough and careful, and not terribly trusting of his or her teammates. That is, this individual is more apt to rely on him- or herself to complete the project work rather than rely on other teammates.

FUTURE RESEARCH

There is a lack of research on the impact of team member personality on the performance of virtual teams. Thus, there is a need for larger scale studies of the role of personality in virtual teams. Future studies should incorporate different measures of personality, and in particular, measures of personality facets.

CONCLUSION

Study results indicated a different set of personality facets to predict team creativity versus quality. Deliberate, thorough individuals (deliberate facet within the conscientious factor) who did not trust that their teammates would do the work (trust facet within the agreeableness factor) were more apt to produce high-quality, polished deliverables, seemingly at the expense of creativity. Individuals that were original thinkers (idea facet within the openness factor) and who were not particularly thorough or determined (achievement striving facet within the conscientiousness factor) were predictive of team creativity. Given a time delimited context, these results help to explain a general finding across experiments that teams high in creativity were not the same teams that were high in quality.

REFERENCES

Costa, P. T., Jr., & McCrae, R. R. (1992). Normal personality assessment in clinical practice: The NEO personality inventory. *Psychological Assessment, 4,* 5-13.

Gallivan, M. J., & Benbunan-Fich, R. (2005). A framework for analyzing levels of analysis issues in studies of e-collaboration. *IEEE Transactions on professional Communication, 48*(1), 87-104.

Goel, V. P. (1989). Motivating the notion of generic design within information-processing theory: The design problem space. *AI Magazine, 10*(1), 18-35.

Goldberg, L. R. (1993). The structure of personality. *American Psychologist, 48,* 26-34.

Gough, H. G., & Heilbrun, A. B. (1983). The adjective check list manual. Palo Alto, CA: Consulting Psychologists Press.

McCrae, R. R., & Costa, P. T., Jr. (1992). Discriminant validity of NEO-PIR facet scales. *Educational and Psychological Measurement, 52,* 229-237.

McGrath, J. E. (1984). *Groups, interaction and performance.* Englewood Cliffs, NJ: Prentice Hall.

Ocker, R. J. (2005). Influences on creativity in asynchronous virtual teams: A qualitative analysis of experimental teams. *IEEE Transactions on Professional Communication, 48*(1), 22-39.

Ocker, R. J. (2007). Creativity in asynchronous virtual teams: Putting the pieces together. In S. P. MacGregor & T. Torres-Coronas (Eds.), *Higher creativity for virtual teams: Developing platforms for co-creation.* Hershey, PA: Idea Group.

Ocker, R. J., & Fjermestad, J. (1998, January). Web-based computer-mediated communication: An experimental investigation comparing three communication modes for determining software requirements. In *Proceedings of the 31st Hawaii International Conference on System Sciences (HICSS-31)* (CD ROM). Hawaii: IEEE Computer Society.

Ocker, R. J., Fjermestad, J., Hiltz, S. R., & Johnson, K. (1998). Effects of four modes of group communication on the outcomes of software requirements determination, *Journal of Management Information Systems, 15*(1), 99-118.

Ocker, R. J., Hiltz, S. R., Turoff M., & Fjermestad, J. (1996). The effects of distributed group support and process structuring on software requirements development teams, *Journal of Management Information Systems, 12*(3), 127-154.

Olson, J. S., Olson, G. M., Storrosten, M., & Carter, M. (1993). Groupwork close up: A comparison of the group design process with and without a simple group editor. *ACM Transactions on Office Information Systems, 11,* 321-348.

Paunonen, S. V., & Ashton, M. C. (2001). Big five factors and facets and the prediction of behavior. *Journal of Personality and Social Psychology, 81*(3), 524-539

Thompson, L. (2003). Improving the creativity of organizational work groups. *Academy of Management Executive, 17*(1), 96-109.

KEY TERMS

Asynchronous CMC: Computer-mediated communication (see below) that occurs at different times (e.g., e-mail); that is, not instantly.

Computer-Mediated Communication (CMC): Communication between individuals that occurs using information and communication technology.

Five Factors or Dimensions of Personality: A theory of personality; five factors are extraversion, openness, agreeableness, conscientiousness, and negative emotionalism (also known as neuroticism); factors are broadly defined.

Personality Facets: Developed to more precisely measure the particular attributes subsumed within the broad personality factors; each factor is composed of multiple facets, where a facet includes a common 'portion' attributable to the associated factor as well as a portion attributable to that particular facet.

Personality Traits: Distinguish an individual from others; traits are persistent across situations and time.

Synchronous CMC: Computer-mediated communication (see above) that occurs at the same time (e.g., instant messaging); that is, in real time.

Virtual Team: Team of individuals who collaborate across space, time, and/or organizational boundaries supported by information and communication technology.

Implementing Varied Discussion Forums in E-Collaborative Learning Environments

Jianxia Du
Mississippi State University, USA

George Pate
Mississippi State University, USA

INTRODUCTION

E-collaboration designs are more successful for online learning environments than pedagogical approaches that emphasize students working alone with materials posted online. Software can be constructed in such a way as to support online group collaboration. The design can only facilitate the desired behavior, not produce it. For the students to adapt a structure of interaction that is collaborative in make-up, the instructor must shape, reproduce, and encourage desired behavior, and the students must be able and willing to participate on a regular basis (Hiltz & Benbunan-Fich, 1997).

Despite earlier uncertainties, online students and instructors can provide emotional support and sociability, as well as information and instrumental assistance to one another. For such an educational environment, it takes the correct software to support group communication, with an emphasis upon collaborative learning approaches rather than on individual learning (Hiltz & Wellman, 1997).

Energetic approaches present learning as a social process that constructs knowledge by formulating ideas into words. These ideas are built upon the reactions and responses of others allowing learning to not only to be energetic, but also interactive (Mead, 1934).

Collaborative refers to instructional methods that support students working together on academic tasks. Collaborative learning is basically different from the traditional classroom situations in which the instructor is the primary source of knowledge or skills (Harasim, 1990).

Studies have shown that collaborative group learning strategies result in more student participation with the course (Hiltz, 1994) and more engagement in the learning process (Harasim, 1990). Collaborative group learning methods are more effective than traditional methods in promoting students' learning and achievement and enhancing student satisfaction with the learning and classroom experience (Johnson, 1981).

According to a study conducted by Hiltz and Benbunan-Fich (1977), working in groups drastically increases motivation, perception of skill development, and solution satisfaction. With reference to self-reported learning, there is an interaction between medium of communication and group vs. individual learning. The results of their study also discovered that conditions with or without both factors, for example individuals-manual and groups online, perceived higher learning than in situations where only one of the factors are present. According to Hiltz, (1986), online discussions create new kinds of possibilities for collaboration and for learning.

Creating quality online instruction is a challenging task for most online instructors, with promoting engaging online discussions being the most difficult part of the instruction. Instructors frequently struggle with creating online discussions that will promote "critical thinking skills" (Toledo, 2006, p. 150) in an asynchronous environment instead of simply presenting dead-end questions that go nowhere. This article will review several suggested variances in online discussions that allow engaged critical thinking, promote subject matter understanding along with group member and individual online discussion participation, and assist instructors in choosing appropriate methods for their particular instructional goals.

BACKGROUND

Interactions between the students and instructor and among the students themselves are significant to the process of e-learning (Pallof & Pratt, 1999), because interaction is associated with students' learning and their perceptions of online courses (Berge, 1999; Flottemesch, 2000). A caution should be added that using the technology incorrectly can result in students becoming bored, inattentive, or even frustrated with the online discussion experience (Berge, 1999), and many instructors have indicated a lack of student participation in online discussions (Jin, 2005). It is important to structure the asynchronous discussions in order to provide a foundation for critical discussions and critical thinking (Jeong, 2000). Jiang (1998) found that students displayed higher levels of achievement when online interactions were an important component of the course. The use of technology as an online discussion tool allows the online instructor to use the tool in facilitating insight and understanding rather than as a one-way dispenser of knowledge. When used to facilitate learning, the possibilities for technology implementation and integration are broadened.

Importance of Group Work

Faculty use group projects and discussions to engage students in a cooperative and/or collaborative learning environment. In examining group dynamics in an online environment, Fisher, Thompson, and Silverberg (2005) indicate that one of the strengths of group work is that it helps a student explore his or her thinking, providing opportunities for knowledge construction with their peers. Distance learners have indicated experiencing a sense of social isolation (Lally & Barrett, 1999; Du, Zhang, Olinzock, & Adams, in press). This sense of isolation can be addressed by having group members work together in unique ways, providing opportunities for students to attend to the academic and social components of the online class (Du, Zhang, Olinzock, & Adams, in press; Gabelnick, MacGregor, Matthews, & Smith, 1990). Students have indicated that group work provides them opportunities to have deeper analysis of topics, to reflect on their learning, discover different approaches to tasks, and to discover points they missed in their preparation for the discussion.

Researchers are beginning to examine online groups from a systems perspective. A systems perspective recognizes and studies every component in terms of how that component affects the system and how the system affects each component (Carabajal, LaPointe, & Gunawardena, 2002). Online groups are complex systems that are dynamic and adaptive (McGrath, Arrow, & Berdahl, 2000). With online groups there is the additional component of the technology tools, which can't be ignored when examining online groups (Fisher, Thompson, & Silverberg, 2005; McGrath, Arrow, & Berdahl, 2000).

Group Size

One component of online groups relates the group size. The size of the group has a significant impact on group success (Fisher, Thompson, & Silverberg, 2005). Fisher, Thompson, and Silverberg indicate that large groups are better for discussions where the aim is exploring and collecting information. To facilitate coordination, small groups of three to five are better for these types of projects. Mennecke and Valacich (1998) found that a critical group size is approximately seven members. The use of a smaller group size is designed to allow for greater idea flow and development (Mennecke & Valacich, 1998; Fisher et al., 2005).

As group size increases, group members feel the group has a harder time obtaining or reaching its desired effect or goals (Carabajal, LaPointe, & Gunawardena, 2000). Bonito and Hollinghead (1997) found that as group size increases active members maintain their level of contribution, but less active members postings decrease in proportion. The key is to have a group size large enough to provide different perspectives, but still small enough so that each member of the group has a voice (Fisher et al., 2005).

Prior Preparation

Another important component to groups and online discussions deals with the prior preparation of the group members. Prior preparation by group members is an important component for successful group participation (Petress, 2004; Havard, Du, & Xu, in press). Jonnasen (1996) refers to computer conferencing as a "mindtool" that prompts a larger amount of reflection and analytical thinking while still connecting learners. Students have found group projects more rewarding when they were actively involved in the pre-planning, reading, and implementation (Fisher, Thompson, & Silverberg, 2005).

Johanning (2000) found that using writing as a way to prepare for small group discussions provides opportunities for rich learning experiences. Tai-Seale and Thompson (2000) used "assigned conversation," a focused study of reading assignments, and found that this method increased students' level of preparation, active participation, and the amount learned. Cohen (1994) adds a word of caution that preparation that is suitable for interaction in more routine learning tasks may have an opposite effect and actually constrain the discussion when the task is less structured and the learning objective is more conceptual.

Characteristics of Group Members

The characteristics of group members are another important component of online group work. Teachers use various methods in forming online groups. Some will mix students into groups attempting to balance technology skills, leadership ability, content knowledge, and diversity based on their personal knowledge of the students. Other teachers randomly assign students to groups. Carabajal et al. (2002) indicate the importance of balance in online discussions. Online discussions foster equal participation among the participants, but it doesn't lend itself to patterns of leadership where one person dominates what is designed to be a shared space. This poses a conflict, because Havard, Du, and Xu (in press) state that online groups need an effective moderator or the group loses coherence and becomes a group of individuals formulating their thoughts online. In addition, if one member is particularly adept at the skills required by the group task, that individual's skills overshadow the group's ability to succeed. The online discussions need to allow each group member to bring his or her knowledge, abilities, backgrounds, and experiences to the group process as they construct new knowledge.

Group Purpose

When developing group tasks the quality of the interaction needs to be a specific design goal in order to promote deeper learning (Garrison & Cleveland-Innes, 2005). Online groups have a greater proportion of task-related messages and are conducive to brainstorming tasks. (Hillman, 1999; Hollingshead, McGrath, & O'Connor, 1993). Jin (2005) found that when students believed the discussion was personally relevant and applicable to the class, they were more engaged in the discussions.

IMPLEMENTING VARIED ONLINE DISCUSSION FORUMS

Course Design and Context

A completely Web-based class, Design and Evaluation of Instructional Software, was delivered through WebCT with 28 students. Upon completion of the course, students are expected to be able to understand the relationship between human learning and multimedia instructional design and development through intensive readings, collaborative and interactive discussions, reflective writings, and hands-on projects.

All the advisement was conducted online, using discussion boards, e-mails, or telephone. The content of the course related to the design and development of multimedia applications, which was very challenging to teach and to learn in an online environment. The course was designed to be student-centered, interactive, and collaborative. Most assignments required collaborative efforts using the available WebCT asynchronous or synchronous communication tools. The instructor designed instructional materials, learning activities, group projects, and assessment instruments.

As a final group project, the student groups are required to collaboratively develop a complete instructional design portfolio project which involves the selection of a real instructional problem and presenting an entire evaluative design and solution for the instructional problem selected. Because of the complexity, inclusive requirements, and interactivity and collaboration involved in completing this project, students have to attend the multiple discussion activities: (1) experience discussion, (2) debate discussion, (3) panel discussion, and (4) symposium discussion. Finally, students will receive 50% of their final score from their performance on this project.

Experience Discussion

1. **Definition:** Groups (small or large) who review and discuss a subject such as a book, movie, writings, or experience

2. **Use:** To bring out differing view points or new issues for thought provoking discussions

3. **Set up:** Involve group in planning approach to the review and discussion then have open discussion of the subject

4. **Limitations:** Some members may have difficulty relating to the subject for purpose of discussion

Implementation of Experience Discussion

After individual group members conduct intellectual research regarding the topic for discussion, the group coordinator leads the individual group members in reaching a consensus on ideas to be presented by the group to the other groups in asynchronous online discussions. The groups then begin presenting their review of the subject to the rest of the class through WebCT's discussion section, either through textual format or multimedia presentations. Student engagement in these thought-provoking asynchronous discussions frequently bring out differing viewpoints from the other groups. The groups then engage in additional stimulating dialogue in order to better understand the subject under investigation by practicing critical thinking and investigation of the other group's ideas—sometimes reaching a consensus and sometimes not—at the conclusion of the discussion sessions. Experience discussions involving critical thought and research are not intended to reach a concrete conclusion regarding the subject under discussion, but rather to engage the students in thinking beyond their normally rigid patterns of thought and reasoning.

Panel Discussion

1. **Definition:** A leader and a limited selection of participants have a discussion in a conversational format, after which the larger group later joins.

2. **Use:** Provokes and stimulates better discussion.

3. **Set up:** A leader picks no more than 8 members for the initial group, then invites an information gathering approach from the members on the subject. The leader then asks for the entire group to join in summarizing or adding to the results of the discussion.

4. **Limitations:** A charismatic person can dominate the discussion or monopolize the presentations, as well as easily get off subject.

Implementation of Panel Discussion

After each class member is assigned to a group, the group leader then arranges for each member to be responsible for an area of the discussion subject for reading and research. The coordinator then arranges online chats or group discussions with his or her individual group members to bring each individual's research findings and ideas into a formal document or multimedia presentation to be shared online with the other groups. The groups are then scheduled by the instructor for stimulating exchanges among the groups regarding their subject matter research and information. These discussion sessions can be scheduled for a single session or multiple discussion sessions according to the discretion and intended goals of the instructor. The panels or groups are organized as information-gathering units that prepare themselves to share their ideas and findings as a group with other groups in online discussion sessions that follow their information gathering stage. They can choose as groups to reach a common conclusion regarding the subject matter under review, choose to continue discussions after further research and review as groups, or choose to agree that the subject may not have a definitive answer and is an area for continued debate, discussion, research, or review. These are certainly desirable goals in online panel discussion forums that promote critical questioning and thought.

Symposium Discussion

1. **Definition:** Progression through phases of a subject by expert speakers who provide concise, short speeches on their part of subject.

2. **Use:** Preferred for gaining information on a specific issue.

3. **Set up:** Leader seeks out experts among the larger group and plans an outline for presenting subject information. The experts give reports on the subject. The leader invites questions to the specific experts. At the end of the symposium the leader summarizes the discussions.

4. **Limitations:** A charismatic speaker can dominate the gathering and monopolize the discussions.

Implementation of Symposium Discussion

The discussions can be implemented by involving the whole class at once, or the class can be divided into groups with coordinators handling the discussions for their group. The possibility also exists that the chat room available in WebCT can be used for synchronous discussions with all involved in real-time discussion, either with the subject matter expert present or just about his or her subject matter presentation. But generally accepted practices dictate asynchronous discussions, because normally course members have different work schedules since the majority of the class members are nontraditional students with families and full-time careers.

Debate Discussion

1. **Definition:** Pro and con discussion presented by teams on opposite sides of a subject with the objective of swaying the audience.
2. **Use:** A friendly manner for bringing up perspectives of opposing groups.
3. **Set Up:** The opposing groups should have time limits on their presentations and direct their efforts to influencing the audience rather than attacking the opponents.
4. **Limitations:** Difficult to find objective participants.

Implementation of Debate Discussion

After groups have been chosen, rules for debate must be presented by the instructor and adhered to by the participants in order for the debate to be orderly and successful. First of all, a time limit for each group to present its views at one time should be set. Normally around 5 minutes per presentation should be sufficient, but each instructor will have to evaluate time for specific schedules and purposes. The instructor cannot stress enough that each participant is to engage in the debate in a friendly, courteous manner, and that the intent of presenting their views on the subject up for debate is not to attack the opposing teams, but rather to attempt to influence the thinking of the other teams through their presentations, fact finding research, and persuasive skills. If engagement becomes out of order for any reason, the instructor should step in immediately to remind the participants of the rules and to suggest alternative ways of debating the topic. As an outside assignment, the instructor can require that the course members research rules for proper debate practices either on the Internet or through library resources and

post a compiled set of rules from each group, then allow the groups to reach a consensus on rules for the whole class in conducting the debate in a mannerly and respectful fashion.

CONCLUSION

By creating flexible learning environments that provide online students with a variety of Web forums, the online discussion sessions can avoid becoming boring, burdensome, mundane, and ineffective means of critical thinking, learning, and group collaboration. By creating a variety of online discussion methods or forums, we can as instructors assist students in developing critical thinking skills that will assist them in deeply understanding the course material and experiencing a positive feeling about the online learning environment through collaborative group efforts, as opposed to rote memorization and lone presentations that unfortunately do not facilitate successful learning in online instructional settings. The primary point to take from this article is to recognize and deal with the conflict inherent in the different students' need in a collaborative online learning environment. We believe it is possible to create more flexible online discussion forums, though we recognize that this innovation may be more time intensive for instructors who count on e-collaboration for certain scalability in online environments. A possible outcome might be to consider differentiating spaces for various sorts of needs to help the students in their quest to focus interactions on learning rather than support.

REFERENCES

Benbast, I., & Lim, L. (1993). The effects of group, task, context, and technology variables on the usefulness of group support systems: A meta-analysis of experimental studies. *Small Group Research, 24*(4), 430-462.

Havard. B., Du, J. X., & Xu, J. Z. (in press). Collaborative learning and communication media. *Journal of Interactive Learning Research.*

Berge, Z. L. (1999). Interaction in post-secondary Web-based learning. *Educational Technology, 39*(1), 5-11.

Bonito, J. A., & Hollingshead, A. B. (1997). Participation in small groups. In B. R. Burleson (Ed.), *Communication yearbook 20* (pp. 227-261). Beverly Hills, CA: Sage Publications.

Carabajal, K., LaPointe, D., & Gunawardena, C. N. (2002). Group development in online learning communities. In M. G. Moore, & W. G. Anderson (Eds.), *Handbook of distance education* (pp. 217-234). Mahwah, NJ: Lawrence Erlbaum Associates.

Cheung, W. S., & Hew, K. F. (2004). Evaluating the extent of ill-structured problem solving process among pre-service teachers in an asynchronous online discussion and reflection log learning environment. *Journal of Educational Computing Research, 30*(3), 197-227.

Cohen, E. G. (1994). Restructuring the classroom: Conditions for productive small groups. *Review of Educational Research, 64*(1), 1-35.

Du, J., Zhang, K., Olinzock, A., & Adams, J. (in press). An investigation of students'

perspectives on the meaningful nature of online discussion. *Journal Interactive Learning Research.*

Fisher, M., Thompson, G. S., & Silverberg, D. A. (2005). Effective group dynamics in e-learning: Case study. *Journal of Educational Technology Systems, 33*(3), 205-222.

Flottemesch, K. (2000). Building effective interaction in distance education: A review of the literature. *Educational Technology, 40*(3), 46-51.

Gabelnick, F., MacGregor, J., Matthews, R., & Smith, B. (1990). Learning community models. *New Directions for Teaching and Learning, 41*, 19-37.

Gallini, J. K., & Zhang, Y. (1997). Socio-cognitive constructs and characteristics of classroom communities: An exploration of relationships. *Journal of Educational Computing Research, 17*(4), 321-339.

Garrison, D. R., & Cleveland-Innes, M. (2005). Facilitating cognitive presence in online learning: Interaction is not enough. *The American Journal of Distance Education, 19*(3), 133-148.

Hillman, D. C. A. (1999). A new method for analyzing patterns of interactions. *The American Journal of Distance Education, 13*(2), 37-47.

Hollingshead, A. B., McGrath, J. E., & O'Connor, K. M. (1993). Group task performance and communication technology: A longitudinal study of computer-mediated versus face-to-face work groups. *Small Group Research, 24*(3), 307-333.

Jiang, M. (1998). Distance learning is a Web-based environment: An analysis of factors influencing students perceptions of online learning. *Dissertation Abstracts International, 59*(11), 4044A. (UMI No. 9913679)

Jeong, A. (2000). *Bulletin boards support critical thinking in small group discussions.* University of Wisconsin-Madison, Learning Technology and Distance Education. Retrieved January 24, 2006, from http://ltde.tripod.com/groupware.bbcriticalthinkingsurvey.htm

Jin, S. H. (2005). Analyzing student-student and student-instructor interaction through multiple communication tools in Web-based learning. *International Journal of Instructional Media, 32*(1), 59-67.

Johanning, D. I. (2000). An analysis of writing and postwriting group collaboration in middle-school pre-algebra. *School Science and Mathematics, 100*(3), 151-160.

Johnson, D. W., & Johnson, R. T. (1994). Cooperative learning in the culturally diverse classroom. In R. A. DeVillar, C. J. Faltis, & J. P. Cummins (Eds.), *Cultural diversity in schools* (pp. 57-73). Albany: State University of New York Press.

Jonassen, D. H. (1996). *Computers in the classroom: Mindtools for critical thinking.* Englewood Cliffs, NJ: Prentice Hall.

Lally, V., & Barrett, E. (1999). Building a learning community online: Toward social-academic interaction. *Research Papers in Education, 14*(2), 147-163.

McGrath, J. E., Arrow, H., & Berdahl, J. L. (2000). The study of groups: Past, present, and future. *Personality and Social Psychology Review, 4*(1), 95-105.

Mennecke, B. E., & Valacich, J. S. (1998). Information is what you make of it: The influence of group history and computer support on information sharing decision quality and member perceptions. *Journal of Management Information Systems, 15*(2), 173-197.

Palloff, R. M., & Pratt, K. (1999). *Building learning communities in cyberspace: Effective strategies for the online classroom.* San Francisco: Jossey-Bass.

Pavitt, C., & Johnson, K. (1999). An examination of the coherence of group discussions. *Communication Research, 26*(3), 303-321.

Petress, K. C. (2004). The benefits of group study. *Education, 124*(4), 587-589.

Postmes, T., Spears, R., & lea. M. (1998). Breaking or building social boundaries? SIDE-effects of computer-mediated communication. *Communication Research, 25*, 689-715.

Stake, R. E. (1995) *Art of case study research.* Thousand Oaks, CA: Sage.

Tai-Seale, T., & Thompson, S. B. (2000). Assigned conversations. *College Teaching, 48*(1), 15-18.

Yin, R. K. (2002) *Case study research: Design and methods* (3rd ed.). Thousand Oaks, CA: Sage.

KEY TERMS

Experience Discussion: Groups (small or large) who review and discuss a subject such as a book, movie, writings, or experience.

Panel Discussion: A leader and a limited selection of participants have a discussion in a conversational format after which the larger group later joins.

Symposium Discussion: Progression through phases of a subject by expert speakers who provide concise, short speeches on their part of subject.

Debate Discussion: Pro and con discussion presented by teams on opposite sides of a subject with the objective of swaying the audience.

Web Forum: A facility on the Internet for holding discussions, or the Web application software used to provide the facility. A sense of virtual community often develops around forums that have regular users. Technology, computer games, and politics are popular areas for forum themes, but there are forums for a huge number of different topics.

Online Learning: Can be quite varied in their overall approach to the teaching and learning process, but they often have certain characteristics in common. In most online courses, students use a computer to connect to a course site on the World Wide Web.

Scalability: Frequently used as a magic incantation to indicate that something is badly designed or broken. Often you hear in a discussion "but that doesn't scale" as the magical word to end an argument. This is often an indication that developers are running into situations

Induced Cooperation in E-Collaboration

Reza Barkhi
Virginia Tech, USA
American University of Sharjah, UAE

INTRODUCTION

Groups are increasingly using collaborative technology such as a group decision support system (GDSS) to communicate electronically. Electronic communication channels may influence the behavior and strategy that individuals in a group employ to share information and collaborate (Barua & Whinston, 1995). When members in a group are not cooperative, they can become competitive; they may play games to maximize their rewards at the expense of others in the group; some individuals may even exchange untruthful information and utilize the information asymmetry to maximize their own rewards at the expense of others.

A task that is often faced by organizational decision makers is a mixed-motive negotiation task (McGrath, 1984) where the parties have mixed motives: cooperate and compete. For a mixed-motive task, a GDSS should anticipate the games that members may play and provide decision-modeling tools and incentive structures that discourage dysfunctional gaming behavior to encourage truthful information exchange.

This paper presents a game theoretic view of collaborative work and suggests that the design of effective GDSS tools should be guided by the way the tool discourages dysfunctional gaming behavior. We present an illustrative experimental study that investigates the influence of communication channel, incentive structure, and problem modeling tools on decision performance, diversity of solutions, and information exchange truthfulness in collaborating groups.

BACKGROUND

A GDSS may help disseminate information to group members, but it cannot force the group to "think" (Dennis, 1996). Thinking is a form of information processing and providing a problem-modeling tool that will process the input data parameters to generate information can aid the group members to "think." The effectiveness of a problem-modeling tool depends on the quality of the underlying parameters.

Each individual in a group has his/her own set of objectives, private information, and interpretation of the problem. Nevertheless, each must develop a consistent shared interpretation of the problem. A GDSS may help the group members develop a shared interpretation, but it cannot help identify an optimal solution if group members do not share their private information or if they exchange untruthful information.

Problem-modeling tools of a GDSS can incorporate incentive structures that affect the decision-making process and outcomes. Incentive structures can influence the strategy that individuals employ to protect their stakes in the organization, the decision of whether or not to share information, and the decision of what type of information to share. Electronic communication may affect the truthfulness of the information that members in a group exchange in an effort to develop a shared cooperative context (Zack, 1993). In the absence of a shared cooperative context, members may misrepresent their information and engage in deceptive behavior to engage in a game to mislead others in an effort to maximize their individual payoffs, sometimes at the expense of the group's payoffs.

A GAME THEORETIC VIEW

We view a GDSS group interaction as an N-person game where $v(N)$ is the worth of the N individuals working together for a *grand* coalition. This coalition involves all N members and provides the highest total payoff. It is possible for some members to form their own coalition instead of working within the grand coalition scheme. In this case, $v(S)$ for $S \subseteq N$ is the worth of coalition S.

A group is often created because the members can achieve a higher outcome if the N individuals work together than if they work alone. That is, there is synergy between group members and the effort of members has a super-additive property. The payoff that results from the grand coalition, $v(N)$, is divided among the N members in such a way that the allocations to all members referred to as imputations $(x_1, ..., x_i, ..., x_N)$ do not exceed the total payoff generated from the grand coalition. Each member has an incentive to cooperate and engage in the grand coalition if his share of the grand coalition exceeds what she can achieve by working alone. A grand coalition can be broken because often in many types of cooperative games the unanimous consent of all players is needed to achieve the joint payoff *v(N)*.

Incentives can induce members to adopt a *cooperative* orientation or an *individualistic* orientation (Deutsch, 1973). With the cooperative orientation, one has an incentive to do well while being concerned about the payoff that others receive. With an individualistic orientation, one has an incentive to do as well as he can without concern for the payoff for others in conflict situations. Deutsch (1973) suggests that mutual awareness of a shared cooperative orientation is likely to help establish mutual trust. Mutual trust can positively influence information sharing; whereas, mutual awareness of a shared individualistic orientation is likely to result in a relationship of mutual suspicion that can negatively influence information sharing. Members in a group generally negotiate on the basis of their perceptions of the other person's trustworthiness and fairness and communication channel can influence these perceptions.

COMMUNICATION IN GDSS

Communication channel may affect group interaction and information exchange. A GDSS designed for distributed members interacting at the same time is a distributed GDSS or DGDSS. Prior research suggests that DGDSS groups communicate differently than do face-to-face GDSS groups (FGDSS), and that design requirements for a GDSS that supports each of these groups are substantially different (Hightower & Sayeed, 1996).

FGDSS uses a rich communication channel while DGDSS uses a lean electronic communication channel (Short, Williams, & Christie, 1976). Lean channels of communication and low social presence characterize insensitive, cold, and impersonal environments. Low social presence, however, makes it more difficult to establish a shared cooperative context (Zack, 1993). With a shared cooperative context, members perceive

higher levels of cooperation that can lead to more truthful exchange of information. Much of the nonverbal and verbal communication cues that form a normal part of human interaction are filtered in a DGDSS group, resulting in lower social presence and potentially less truthful exchange of information.

Social influence theory (SIT) (Fulk, Schmitz, & Steinfield, 1990) questions the basic assumptions of media richness theory (MRT) (Daft & Lengel, 1986) and postulates that media perceptions are, in part, socially constructed, vary by user and the social context, are subjective, and that making choices about media is retrospectively and subjectively rational.

Incentive structure may induce a social context (i.e., cooperative or individualistic). Social context can, in turn, mitigate media perceptions. Incentive structures can also influence the degree of equivocality in a task, and communication channels may intensify the level of conflict that an incentive structure promotes. Hence, a rich channel may be a better fit for group-based incentive (i.e., higher equivocality), and a lean channel a better fit for individual-based incentive (McGrath & Hollingshead, 1993). In addition, high levels of social presence may promote truthful information exchange, making it more appropriate for group-based incentive. FGDSS groups using a face-to-face communication channel may find it easier to define issues together to develop common ground (Nyhart & Dauer, 1986). This implies that for group-based incentive, where building common ground is more crucial than it is for individual-based incentive, a FGDSS may be more appropriate than a DGDSS.

A DGDSS may have a negative effect on mixed-motive tasks in that it can foster the view of negotiation as a win-lose situation (Rhee, Pirkul, Jacob, & Barkhi, 1995). Negotiators with a win-lose orientation become more individualistic and may not exchange truthful information. By depersonalizing communication, "lean" media induce group members to exchange minimal information because the parties may not think the information is important to communicate. We study the impact of communication mode and incentive structures on group interaction, performance, and information exchange in groups using either a Level One GDSS (i.e., providing communication support) or a Level Two GDSS (i.e., providing communication and problem modeling support) (DeSanctis & Gallupe, 1987).

PROBLEM MODELING

Incentive structures can be embedded into the GDSS problem-modeling tools to induce a social context in virtual groups. Problem-modeling tools may improve process gains by reducing the problems arising from incomplete task analysis, incomplete use of information, and information overload. Appropriate models identify the important decision variables, the objective function, and the constraints, and may reduce information overload. Although problem modeling and quantitative tools have been studied in the individual decision-making literature, they have been overlooked in the GDSS literature.

A problem-modeling tool may impose a two-phase approach to problem solving: a convergence and a divergence approach. Convergent approaches promote high consensus while divergent approaches can lead to conflict that hinders the ability of the group to achieve consensus and commitment to a final solution (Janis & Mann, 1997). A divergent approach promotes improved quality of outcomes, as it encourages a more comprehensive understanding of the issues. However, because it hinders the ability of the group to achieve consensus, the problem-solving process may not converge. The dilemma resulting from the tradeoff between decision quality and convergence may be overcome by following a two-phase strategy.

In the divergent phase (phase I), each individual generates solutions from his or her perspective, and in the convergent phase (phase II) each individual shares his or her solutions with others and evaluates other individuals' proposed solutions (Niederman & DeSanctis, 1995). In phase I, each member models the problem with the best available information, and finds the individual optimal solution using a problem-modeling tool. In phase II, each member proposes solutions to others, checks the neighborhood of solutions, evaluates the solutions that others propose, checks the neighborhood of the solutions proposed, and updates private information with the new information.

Each time the group members optimize with new parameters, they land in a new region of the solution space. They perform a local search around that optimal solution to see if there are any solutions that are in the vicinity of the optimal solutions and are also acceptable to others. This process continues until the group decides that the best solution identified thus far should be accepted, so that the problem-solving process may converge. Groups using a GDSS with a problem-modeling tool can more effectively search the solution space if they exchange truthful information.

METHOD

We conducted a 2x2x2 (communication channel x incentive structure x problem-modeling tool) experimental design. Communication channel had two levels: face-to-face (FGDSS) and distributed computer-mediated-communication (DGDSS). Incentive structures had two values: group-based and individual-based, and problem-modeling tools had two values: Level Two GDSS with problem-modeling tool and Level One GDSS without problem-modeling tool. Individuals were randomly assigned into the eight experimental cells.

One hundered and fifty advanced college students (juniors and seniors) and executive MBA students were recruited from business decision-making courses at a large American university to participate in this study. The task was adopted from previous research (Barkhi, 2005). The subject pool consisted of approximately 60% male and 40% female participants. Fifteen percent of the course grade was allocated for the experiment and this provided strong motivation to do well in the experiment.

RESULTS

We found that performance of members in DGDSS groups was lower than that of members in FGDSS groups when the incentive was group-based. Members in DGDSS groups with limited communication cues developed the impression that others would take free rides. In an attempt to protect themselves from free riders, they became free riders themselves, and hence the group productivity suffered as a result. However, when incentive was individual-based, the performance of members in DGDSS was statistically equal to members in FGDSS.

Members in DGDSS groups had lower performance on average when incentive was group-based than when it was individual-based. There is less free riding behavior among members in FGDSS groups compared to those in DGDSS groups, but still the incentive scheme overrides the effects of the communication channel. We found that communication modes do not equally influence different incentive structures. When incentive is individual-based, members deviate from the grand coalition (optimal effort) as frequently in FGDSS groups as in DGDSS groups. The results indicate that when incentive is individual-based, communication channel does not affect the strategy that members employ.

We also found that when incentive is group-based the ratio of selecting the cooperative strategy is higher for FGDSS groups than for DGDSS groups. This means that the "rich" channel in FGDSS groups help group members to develop a cooperative context to select the cooperative strategy more often than will DGDSS groups. The FGDSS members with group-based incentive, on average, identified the cooperative strategy as the best strategy more frequently than did their DGDSS counterparts. The "rich" channel in FGDSS helped decision makers to correctly identify the cooperative strategy more frequently than did the "lean" channel in DGDSS. This may imply that group-based incentive structures may be dysfunctional in a computer supported GDSS setting that provides a "lean" electronic channel of communication. On the other hand, the individual-based incentive structure is likely to provide a better fit" for DGDSS settings.

The results of the analysis of truthfulness suggest interesting results. The members in FGDSS groups with group-based incentive structure reported untruthful information substantially less than the members of DGDSS groups with group-based incentive. Under the individual-based incentive structure, members in FGDSS groups engaged in untruthful information exchange substantially less frequently than did the members in DGDSS groups. Hence, regardless of the incentive structure, members in FGDSS groups are more truthful than their DGDSS counterparts.

We also found that the number of diverse solutions generated is larger for groups using a Level Two GDSS with a problem-modeling tool that promotes a two-phase solution strategy than those using a Level One GDSS without a problem-modeling tool. Member task performance, subsequently, is higher for groups using a Level Two GDSS than those using a Level One GDSS. This means that the GDSS tool that supports problem modeling influences the individuals' task performance and possibly his/her strategy (i.e., more diverse solutions). Also, the members are more truthful in their information exchange when they use a Level Two GDSS that helps members to provide their information incrementally as model parameters.

CONCLUSION

We found that incentive structure can mitigate the effect of communication channels. The individual-based incentive structure, in the absence of task interdependence across members, resulted in improved group performance, and more truthful information exchange when members use a DGDSS. On the other hand, group-based incentive structure promotes free riding more in DGDSS groups than in FGDSS groups. Hence, there may be a better fit between individual-based incentives and computer supported distributed work environments than there is between these environments and group-based incentives, when there is little task interdependence across group members.

Under the "lean" channel used in DGDSS, on the average, the decision makers found it more difficult to develop mutual awareness of a shared cooperative orientation to develop a shared cooperative context. Instead, they seemed to develop mutual awareness of a shared individualistic orientation that resulted in a relationship of mutual suspicion, free riding, and untruthful information exchange. Hence, if group-based incentive is used, it is even more critical for management to promote an atmosphere that is conducive to a cooperative orientation so that members do not engage in free riding behavior. It is important for managers to pay particular attention to free riding behavior in a distributed DGDSS environment.

Given that GDSS users are likely to adopt their strategies to the available features (Todd & Benbasat, 1991), it should be possible to design GDSS features to promote user behavior to become aligned with organizational objectives. It should make it cognitively easier for decision-makers to make decisions that are aligned with organizational objectives and cognitively harder for each individual to maximize his or her own interests at the expense of the organizational interest. This is an issue regarding the design of incentive structures but it also directly influences decisions related to the design of specific DGDSS features, and hence the effectiveness of the DGDSS when communication channel is "lean."

Management may consider using incentives that rely on individual performance rather than group performance when members meet at different places and communicate using lean channels. This result is very managerially relevant as the design of incentive system is a variable, controllable by organizational policy makers that should be set properly to achieve organizational objectives. This means that the effectiveness of the DGDSS is dependent on the underlying organizational structural variables such as incentive structures as incentives can mitigate the influence of communication channels on decision strategies.

The proliferation of information-based and decision support technologies into virtual organizations that use DGDSS may not result in successful implementation and usage unless proper incentives are designed that prevent free riding, promote a cooperative context, and induce truthfulness in a faceless setting when decision makers are distributed.

The problem-modeling tool enhanced the truthfulness of the information that was exchanged among the group members. The information exchanged among group members was incorporated, as model parameters, in the modeling of the problem. That is, a problem-modeling tool did provide a vehicle for group members to communicate their private information and aid them in processing the information that was received from each member. This implies that the lack of a problem-modeling tool may be one explanation of why group members do not "process" the information they receive from others in their group—they may not understand how to use this information or may assume that it is not important to exchange it with others.

The results suggest that communication mode mitigates the nature of information that is exchanged. This implies that communication mode and incentive structures may be designed to compound or reduce the negative influences of each other. Also, incentives can be incorporated into GDSS problem modeling tools to encourage truthful information exchange and improved decision quality.

REFERENCES

Barkhi, R. (2005). Information exchange and induced cooperation in group decision support systems. *Communication Research, 32*(5), 646-678.

Barua, A., Lee, C.-H. S., & Whinston, A. B. (1995). Incentives and computing systems for team-based organizations. *Organization Science, 4*(2), 487-504.

Daft, R. L., & Lengel, R. H. (1986). Organizational information requirements, media richness and structural design. *Management Science, 32*(5), 554-571.

Dennis, A.R. (1996). Information exchange and use in group decision making: You can lead a group to information, but you can't make it think. *MIS Quarterly, 20*(4), 433-457.

DeSanctis, G., & Gallupe, R. B. (1987). A foundation for the study of group decision support systems. *Management Science, 33*(5), 589-609.

Deutsch, M. (1973). *The resolution of conflict.* New Haven, CT: Yale University Press.

Fulk, J., Schmitz, J., & Steinfield, C. W. (1990). Social influence model of technology use. In J. Fulk & C. W. Steinfeld (Eds.), *Organizations and communication technology* (pp. 117-140). Beverly Hills, CA: Sage Publications.

Janis, I. L., & Mann, L. (1977). *Decision making: A psychological analysis of conflict, choice, and commitment.* New York: The Free Press.

Hightower, R. T., & Sayeed, L. (1996). Effects of communication mode and prediscussion information distribution characteristics on information exchange in groups. *Information Systems Research, 7*(4), 451-465.

McGrath, J. E. (1984). *Groups: Interaction and performance.* Englewood Cliffs, NJ: Prentice-Hall.

McGrath, J. E., & Hollingshead, A. B. (1993). Putting the "group" back in group support systems: Some theoretical issues about dynamic processes in groups with technological enhancements. In L.M. Jessup & J. S. Valacich (Eds.), *Group support systems* (pp. 78-96). New York: McMillan.

Niederman, F., & DeSanctis, G. (1995). The impact of a structured-argument approach on group problem formulation. *Decision Sciences, 26*(4), 451-474.

Nyhart, J. D., & Dauer, E. A. (1986). A preliminary analysis of the uses of scientific models in dispute prevention, management, and resolution. *Missouri Journal of Dispute Resolution*, 29-54.

Rhee, H. -S., Pirkul, H., Jacob, V. S., & Barkhi, R. (1995). Effects of computer-mediated communication on group negotiation: An empirical study. *Proceedings of the 28th Annual Hawaii International Conference on Systems Sciences* (pp. 270-279).

Short, J., Williams, E., & Christie, B. (1976). *The social psychology of telecommunications*. New York: John Wiley & Sons.

Todd, P., & Benbasat, I. (1991). An experimental investigation of the impact of computer based decision aids on decision making strategies. *Information Systems Research, 2*(2), 87-115.

Zack, M. H. (1993). Interactivity and communication mode choice in ongoing management Groups. *Information Systems Research, 4*(3), 207-239.

KEY TERMS

Communication Richness: The number and capacity of channels used for communication and the higher the number of channels, the richer is the communication.

Free Riding: When members in a group benefit from the effort of others without contributing their fair share.

Game Theory View of GDSS: Views GDSS users and their interaction as a cooperative or non-cooperative *N-person* game where coalitions of *S* members where *S<N* are possible.

GDSS: A computer-based tool that aids decision makers to collaborate in group interaction.

Incentive Structure: Describes the method that the members receive rewards for the quality of their decisions.

Individual/Group Performance: For group performance the unit of analysis is group and for individual performance the unit of analysis is individual.

Level One GDSS: A GDSS that supports the communication aspects of group interaction.

Level Two GDSS: A GDSS that supports the communication as well as decision aiding aspects of group interaction.

Instant Messaging as an E-Collaboration Tool

Qinyu Liao
University of Texas at Brownsville and Texas Southmost College, USA

Xin Luo
Virginia State University, USA

INTRODUCTION

Technology-enabled e-collaboration has been a common practice in modern organizations. Advances in interenterprise software and communication technologies, along with globalization, networking, and digitization have led organizations to look for collaboration tools that will be effective for people to exchange information in business environment. Corporate instant messaging (IM) use has been exploding in recent years. According to Gartner forecasts, worldwide spending on enterprise IM will almost triple from $231 million in 2004 to nearly $640 million in 2009. By the end of the decade, Gartner anticipates 90% of corporate e-mail users will also have IT-controlled IM accounts. IM can provide the kind of presence elements for real-time interaction that can be integrated with other corporate collaboration tools.

This article introduces the background of IM as an e-collaboration tool, discusses the utilization of IM in organizations for different e-collaboration tasks, includes solutions for pitfalls and concerns in IM enabled e-collaboration, and provides the future trend of IM for e-collaboration.

BACKGROUND

There are three major types of e-collaboration systems (McLaren, 2002): (1) Message-based systems that transmit information to partner applications using technologies such as fax, e-mail, EDI, or XLM messages, (2) electronic procurement hubs, portals, or marketplaces that facilitate purchasing of goods or services from electronic catalogues, tenders, or auctions, and (3) shared collaborative systems that include collaborative planning, forecasting, and replenishment capabilities in addition to electronic procurement functionality.

Although a wide range of tools is available, many of them still rely heavily on e-mails for communication for e-collaboration, complemented by peer-to-peer communication, and calendaring.

Although the literature on e-collaboration is filled with mixed findings (Orlikowski, 1992), three major theories have been proposed to understand the e-collaboration behavior and outcomes: media richness theory (Daft & Lengel, 1986), task-technology theory (Zigurs & Buckland, 1998; Zigurs, Buckland, Connolly, & Wilson, 1999), and the mental schemas impact framework (Kock, 2004; Kock & Davison, 2003; Lee, 1994). Developed in the 1980s, the media richness theory claims that different communication media can be classified as lean or rich, according to their ability to convey knowledge and information. Lean media are not appropriate for information communication, which can be reflected in the adoption of media and the outcome of its use (Daft, Lengel, & Trevino, 1987; Lengel & Daft, 1988). Task-technology fit theory focuses on the nature of the collaboration task and its strong impact on its outcomes when certain e-collaboration technologies are used (Zigurs & Buckland, 1998; Zigurs et al., 1999). The mental schemas impact framework suggests that the mental schema possessed by individuals and the individuals' interpretation of information can significantly affect the amount of cognitive effort required to successfully accomplish the task using certain types of e-collaboration technologies (Lee, 1994).

As a strategic e-collaboration tool, instant messaging has been used by millions of individuals for business negotiation, real-time reminders, medical emergencies, or any time e-mail (Richardson, 2002). Some of the unique features of instant messaging include presence awareness, immediate closed loop communication, multi-party collaboration, anytime, anywhere access, opportunistic interaction, broadcasting of information or questions, negotiation of availability for interaction, within-medium polychromic communication, pop-up recipient notification, silent interactivity, and ephemeral transcripts (Rennecker & Godwin, 2003; Avrahami & Hudson, 2004; Marshak, 2004).

IMMEDIATE/CONTROLLED RESPONSES FOR EMPLOYEE-EMPLOYEE COLLABORATION

IM eliminates the time typically lost to "telephone tag" or wasted trips to the office of a coworker who is absent or otherwise occupied. Employees can use IM to exchange information, pose quick questions and clarification, arrange and coordinate meetings, conduct simultaneous conversation over multiple media, solicit immediate responses, and clear up isolated issues that come up unpredictably during the day. For example, if a secretary in a law office had a question concerning a final report, she can instant-message two lawyers in the office. Both contribute to the solution. The whole process may take two minutes and everyone stays at his/her desk. It would have taken at least 10 minutes with someone walking down the hall (Beckman & Hirsch, 2001). Clients of law firms are pressured to use technologies like IM to be more responsive. Studies have found that IM can make the firm even more responsive to clients without having to spend more time to meet them in person (Krause, 2004). IM users can even reach a person on the run when the instant messaging is transferred to a mobile device for immediate responses.

Although not a rich media, IM provides cues about the status of interactants and their behaviors over time. IM displays the online presence of the employee to all other members of the collaboration. The applications include (1) a "pop-up" mechanism to display messages the moment they are received, (2) a visible list of other users, compiled by the user, and (3) a method for indicating when "buddies" are online and available to receive a message. IM applications also allow users to change parameters in the system in order to provide a more detailed view of their availability. Other users are made aware of this status via automated replies from the user or by indications visible on the buddy list. This gives other employees detailed information about each other's availability and they can then

decide whether to contact the person later or send an e-mail, voicemail, or other message that the recipient can respond to later.

NEAR-VISIBLE EMPLOYEE-CUSTOMER INTERACTION

When a communication interface is rich in expressiveness and closely emulates real face-to-face interactions, a Web merchant will be able to enhance its perceived trustworthiness in the eyes of online consumers. As mentioned in a report in *The Wall Street Journal* (Wingfield 2002), although consumers generally enjoy the convenience of online shopping, long-distance purchases made without human contact still make them cautious before putting things into their shopping carts. To attract and maintain online customers, pioneer online retailers are using instant messaging to give "live help" to assist online shoppers when they have questions about their products or services through real-time communication with their sales associates. Currently, most live help services are implemented through instant text chatting between shoppers and customer service representatives. These conversations can be initiated either by the shoppers or the customer service representatives who engage customers proactively by greeting them or inviting them to chat online. Lands' End's average value of an order increases by 6% when a potential customer uses the live help function, and an online visitor who uses Lands' End instant messaging is 20% more likely to complete a purchase than the one who does not (Ducevich, 2002).

Although vendor-supported "1-800" telephone numbers and Internet calls through VoIP are available through some Web sites, most small- and medium-sized vendors cannot afford dedicated call centers and service staff. Real-time video conferencing is even rarer. Besides the technical constraints, such as transmission bandwidth and video compression, the other concern of live help providers is that the provision of audio and video output makes a customer service representative (CSR) unable to multitask. With instant text messaging, a CSR can interact with several shoppers simultaneously, which significantly reduces the running cost while maintaining satisfactory response latency.

SENSE OF COMMUNITY BETWEEN COLLABORATORS

IM technology lets users communicate across networks, in remote areas, and in a highly pervasive and ubiquitous manner. According to a research carried out for Microsoft by Vanson Bourne Ltd., it is useful for helping to make quick decisions at work (Weekes, 2006). IM is used most commonly for one-on-one exchange in which an individual contacts another individual to discuss needed information. This enables the user to create a private chat room with one or more users. It also supports one-to-many and many-to-many communication (where a user can initiate a session in which all invitees can interact with one another) for collaboration between group/team members. When a problem is difficult or demands expertise in another area, employees reach out to other colleagues for help. This occurs on an ongoing basis, creating a culture where most problems are solved by a network of people drawn into the problem solving process whenever necessary. In this case, only those who can help respond and save time on the part of other colleagues, who can work on their own projects and are only asked to participate when needed. The presence indication allows users to become more conscious of and potentially more involved with the dynamics of their groups and social networks.

Social networking sites using instant text messaging, like Friendster and Meetup.com, have become popular for things like networking and organizing political events. On a personal level, studies found that IM are not as useful for developing and sustaining social relationships as phone calls or face-to-face meetings (Cummings, 2002), but increased IM hours are found to be associated with decreased depressive symptoms (Morgan & Cotten, 2003). Social influence, such as pressure from managers on their subordinates, can make a medium that is seen as lean based on the media richness hypothesis become richer than face-to-face (Markus, 1994). In this case, the subordinates who disagree with the managers about certain issues would choose to instant message the managers rather than talk to them directly.

CONCERNS AND PITFALLS

The biggest concerns about instant messaging are security, documentation, and control. Unlike other electronic communications, instant messages are generally unprotected. Since all data is stored on the server (a third party) during an instant messaging session, the information could be intercepted and less secure than if the users keep data on their own servers where they at least know what sort of security is in place. It is suggested that online meetings using instant messaging system should reinforce the security by requesting passwords and log-ons. In addition, according to a Microsoft research, 71% of those using IM are doing so via software they have downloaded and installed themselves rather than as part of company-wide policy. This can be a potential security risk.

Documentation is another managerial concern, as the interaction transcript is either ephemeral or archived in IM systems. Earlier versions of instant messaging software allow different parties to send instantaneous electronic messages in a virtual conversation, but these messages are rarely archived anywhere. Therefore, there is absolutely no record of any conversation that took place. The ephemerality allows for a level of privacy not available by either telephone or e-mail, potentially impacting communication and media use patterns (Rennecker & Godwin, 2003). For example, if lawyers and firms use IM to give advice to their clients without archive, there will be no record of any conversation or advice that took place. All this makes it nearly impossible to defend oneself if a client accuses a lawyer of giving bad advice via IM. It is also problematic for those who work in regulated industries like securities, since federal laws require that a record be kept of all communications with clients and fellow employees. Recently developed IM systems are capable of archiving transcripts, which can help organizations comply with such newly established policies as HIPAA (Health Insurance Portability and Accountability Act) and Sarbanes-Oxley Act, thereby alleviating managers' concerns about information security.

Sometimes people working under time pressure do not want to show their availability to avoid interruptions, or they only want to be available for certain group. Some worry that IM will contribute to the information overload that staff already suffer from in the form of e-mail. If they choose to "hide" their presence, the communication will be blocked. The availability of the names on the IM list could be used by companies as a means for monitoring, and in some cases accountability. This could lead to misunderstanding or privacy violation. Companies need to decide what IM software and tools will be permitted and how sophisticated these should be when it comes to areas such as security and compliance. In the case of customer service representatives, if instant messaging with multiple customers at the same time, the quality of customer service can be affected by repetitive interruptions.

We understand that to make IM a valuable and powerful business tool without the trouble of security breach and information overload, companies should establish IM usage policies and some basic etiquette should be followed:

1. If initiating the IM, state the topic and ask the other person if they have time to IM with you.
2. A terse response, such as "in a meeting" or "talk later," is merely a concise way of letting you know they are currently unavailable.
3. Emoticons, graphical representations of facial expressions, can help to provide context and support the more nature form of communication.
4. If a contact has set their status as "busy," respect this and do not send them a message unless it is urgent.
5. Make sure the sound accompanying an IM does not disturb others.
6. As with e-mail, using capital letters is the equivalent of shouting, so avoid them.
7. Don't invite someone to join a multi-party IM session without asking the others first.
8. Don't make lengthy private exchanges at work. Time quotes may be an appropriate corporate control.
9. Use only big consumer brands IM services like AOL Instant Messenger, MSN Messenger, and Yahoo Messenger.
10. Never send attachments by IM, but use the corporate e-mail system for this.

FUTURE TRENDS

Face-to-face is considered richer communication than e-mails, with the ability to carry non-verbal cues, provide rapid feedback, convey personality traits, and support the use of natural language (Lee, 1994). According to Horvath (2001), e-collaboration network should have the following attributes:

- Open, low cost connectivity
- Very large, flexible, multimedia data storage capabilities
- Systems and channel integration
- Higher-level self service capabilities
- Intelligence gathering and analysis
- Supply chain collaboration exchanges
- Sophisticated security capabilities
- New electronic commerce capabilities

The first major step in the advances of IM was the inclusion of audio communication: Second generation IM clients allow talking with users anywhere in the world using the computer microphone and speakers. Other typical features of second-generation IM clients is real-time file sharing and e-mail support. A further step in this trend is the creation of video instant messaging, which also allows both video chat live, and the video messages recording/sending to the users who are in the chat rooms. Professional versions of instant messaging software are offering some higher security and message archiving features. To use the video features of a video IM the user needs a video capture card and a normal camera or a webcam. The third generation IM clients have even more features. Below are listed the more interesting features for e-collaboration:

- **Remote assistance:** A user can see and, if permitted, take control of the computer of another user. That is like sitting next to each other and looking at the same screen.
- **Application sharing:** Use the same application in real-time. If the user opens a program, for example a word processor, the connected users can work on the document together.
- **Shared sketching:** Draw diagrams with other users at the same time. Communicators can sketch their ideas simultaneously as if they were both drawing on the same whiteboard (Riva, 2002).

CONCLUSION

Because of its technological innovativeness, such as immediacy, synchronous interaction, and sense of community, IM has become an ideal tool for e-collaborations that require quick response and certain level of control in the workplace. The IM e-collaboration can be used for employee-to-employee and employee-to-customer communications. With added features like remote assistance and video capability, richer, more visible and effective communication can be facilitated, especially for geographically dispersed organizational constituents. If proper etiquette is followed to address the social concerns and avoid the common pitfalls, IM will be embraced by more organizations as a strategic weapon to improve traditional business relationships and the quality of business processes.

REFERENCES

Avrahami, D., & Hudson, S. E. (2004). Balancing performance and responsiveness using an augmented instant messaging client. *Proceedings of the ACM's Conference on Computer Supported Cooperative Work*, Chicago.

Beckman, D., & Hirsch, D. (2001). Same time, different place. *ABA Journal, 87*(9), 70-71.

Cummings, J. N. (2002). The quality of online social relationships. *Communications of the ACM, 45*(7), 103-108.

Daft, R.L., & Lengel, R. H. (1986). Organizational information requirements, media richness and structural design. *Management Science, 32*(5), 554-571.

Daft, R. L., Lengel, R. H., & Trevino, L. K. (1987). Message equivocality, media selection, and manager performance: Implications for information systems. *MIS Quarterly, 11*(3), 355-366.

Ducevich, D. (2002). Lands' End's instant business [Electronic Version]. *Forbes*. Retrieved July from http://www.forbes.com/2002/07/22/0722landsend.htm

Horvath, L. (2001). Collaboration: The key to value creation in supply chain management. *Supply Chain Management: An International Journal, 6*(5), 205-207.

Kock, N. (2004). The psychobiological model: Toward a new theory of computer-mediated communication based on Darwinian evolution. *Organization Science, 15*(3), 327-348.

Kock, N., & Davison, R. (2003). Can lean media support knowledge sharing? Investigating a hidden advantage of process improvement. *IEEE Transactions on Engineering Management, 50*(2), 151-163.

Krause, J. (2004). The top ten in tech. *ABA Journal, 90*(12), 34-40.

Lee, A. S. (1994). Electronic mail as a medium of rich communication: An empirical investigation using hermeneutic interpretation. *MIS Quarterly, 18*(2), 143-157.

Lengel, R. H., & Daft, R. L. (1988). The selection of communication media as an executive skill. *Academy of Management Executive, 2*(3), 225-232.

Markus, M. L. (1994). Electronic mails is the medium of managerial choice. *Org. Sci., 5*(4), 502-527.

Marshak, D. (2004). *Instant messaging at work*. Retrieved October 14, 2006, from http://interruptions.net/literature/Marshak-PSGP1-22-04CC.pdf

McLaren, T. (2002). Supply chain collaboration alternatives: Understanding the expected costs and benefits. *Internet Research, 12*(4), 348-364.

Morgan, C., & Cotten, S. R. (2003). The relationship between Internet activities and depressive symptoms in a sample of college freshmen. *CyberPsychology & Behavior, 6*(2), 133-142.

Orlikowski, W. J. (1992). Learning from notes: Organizational issues in groupware implementation. In J. Turner & R. Kraut (Eds.), *Proceedings of CSCW '92 Conference* (pp. 362-369). New York: The Association for Computing Machinery.

Rennecker, J., & Godwin, L. (2003). Theorizing the unintended consequences of instant messaging for worker productivity. *Sprouts, 3*(3), 137-168.

Richardson, R. (2002). *Instant messaging goes to work*. Retrieved October 14, 2006, from http://www.callcentermagazine.com/GLOBAL/stg/commweb_shared/shared/article/showArticle.jhtml?articleId=8701235&pgno=2

Riva, G. (2002). The sociocognitive psychology of computer-mediated communication: The present and future of technology-based interactions. *CyberPsychology & Behavior, 5*(6), 581-598.

Weekes, S. (2006). Instant gratification. *Personnel Today, June 27*, 27-27.

Wingfield, N. (2002). A question of trust. *Wall Street Journal-Eastern Edition, 240*(54), 6.

Zigurs, I., & Buckland, B. K. (1998). A theory of task-technology fit and group support systems effectiveness. *MIS Quarterly, 22*(3), 313-334.

Zigurs, I., Buckland, B. K., Connolly, J. R., & Wilson, E. V. (1999). A test of task-technology fit theory for group support systems. *Database for Advances in Information Systems. 30*(3), 34-50.

KEY TERMS

Computer-Mediated Communication: A cluster of interpersonal communication systems used for conveying written text, generally over the Internet.

E-Collaboration: Electronic technologies that enable collaboration among individuals engaged in a common task.

Globalization: The growing interdependence of countries worldwide through increasing volume and variety of cross-border interactions and transactions in goods and services, free international capital flows, and more rapid and widespread diffusion of technology.

Instant Messaging: An Internet-based application that provides semi-synchronous communication between participants.

Media Richness: A classification of media based on its ability to carry nonverbal cues, provide rapid feedback, convey personality traits, and support the use of natural language.

Virtual Workspace Technologies: An integrated set of tools that offer a variety of communication support capabilities including a well-organized and searchable common team repository and group discussion forums.

Interaction and Context in Service-Oriented E-Collaboration Environments

Christoph Dorn
Vienna University of Technology, Austria

Schahram Dustdar
University of Leicester, UK

Giovanni Giuliani
HP Italiana SRL, Italy

Robert Gombotz
Vienna University of Technology, Austria

Ke Ning
National University of Ireland, Ireland

Sébastien Peray
European Microsoft Innovation Center, Germany

Stephan Reiff-Marganiec
University of Leicester, UK

Daniel Schall
Vienna University of Technology, Austria

Marcel Tilly
European Microsoft Innovation Center, Germany

INTRODUCTION

As it has been observed in the recent decade, collaborating teams become ever more unstable, less tightly coupled and more distributed and mobile. Workers participate in multiple teams that pursue different goals that need not be related in any way. This radical way in which the workplace is changing for the individual and the team requires highly adaptable groupware and intelligent support for the individual in order to minimize the time lost for management and coordination when switching between different teams, different workspaces, and different contexts. Thus, a service-oriented approach seems promising to provide individual, context-aware building blocks for adaptable groupware.

To achieve this goal, we start by analyzing patterns of human interaction. Together with a context meta-model, such patterns enable effective selection, adaptation, and invocation of services.

BACKGROUND

Many context frameworks target specific groups such as mobile users (Tang et al., 2001; Bardram & Hansen, 2004) or small mobile groups (Pokraev et al., 2005) acting independently of others. More generic frameworks try to cover a wider area but hence lack explicit support for things like group interaction. Exemplary tools focusing mostly on context and hardly on the interaction between people on a broader scope are CASS (Fahy & Clarke, 2004), CoBra (Chen, Finin, & Joshi, 2003), CORTEX (Biegel & Cahill, 2004), Gaia (Roman et al., 2002), Hydrogen (Hofer, Schwinger, Pichler, Leonhartsberger, & Altmann, 2002), and

SOCAM (Gu, Pung, & Zhang, 2004). An overview of these and additional frameworks can be found in Baldauf et al. (2006).

On the other hand, groupware systems such as the file-orientated BSCW (Bentley, Horstmann, Sikkel, & Trevor, 1995), virtual office-like Groove (www.groove. net), or the process-aware ad hoc collaboration tool Caramba (Dustdar, 2004) focus on collaborative functions while taking context only to some extent into account. Moreover, highly synchronous collaboration tools such as coediting neglect long-term team interactions.

Most teamwork tools are tightly integrated applications, whereas a service-oriented collaboration architecture reflects the notion of scoped functions wrapped as services that are provided, described, published, found, invoked and aggregated. Hence, a SOA oriented approach to context awareness (Gu et al., 2004) and collaboration (Jørstad, Dustdar, & van Do, 2005) seems to be very promising, but the notion of context as proposed by Dey and Abowd (1999), needs to be extended beyond involved services (Dorn & Dustdar, 2006) to explicitly include teams as a first order entity.

The exhaustive review of current literature by Powell, Picolli, and Ives (2004) reveals that research efforts have merely focused on distributed teams as a whole without analyzing the internal interaction. Hence, three interaction patterns provide the basis to improve our understanding of e-collaboration requirements.

INTERACTION PATTERNS

The term *interaction pattern* refers to a common, reoccurring interaction scenario between actors. The term *relation* refers to a tie or link between two actors within a pattern. We take three initial interaction patterns that are well known in the domain of software engineering (SE) and apply them to the domain of human collaboration.

PROXY PATTERN

Originally, the Proxy pattern was introduced by Gamma, Helm, Johnson, and Vlissides (1994, p. 207) as a structural pattern in software design. The intention for using a proxy is "to provide a surrogate or placeholder for another object to control access to it." Besides forwarding the clients' requests and sending back the response, a proxy can do pre- or post-processing depending on its type. A real-life example of a proxy in human collaboration is a secretary. He or she receives e-mails, phone calls, messages, etc., which are actually intended for a different entity, the boss. The secretary pre-processes these client requests by, for example, filtering out unwanted requests—protection proxy—or even answering simple requests without having to involve the boss—cache proxy (Dustdar & Hoffmann, 2006).

A proxy pattern usually describes a 1:1 relationship between proxy and original as depicted in Figure 1a. However, there are two exceptions, remote proxies and firewall proxies, where a proxy is responsible for multiple originals (Figure 1b).

BROKER PATTERN

The Broker architectural pattern can be used to structure distributed software systems with decoupled components that interact by remote invocations. "A broker component is responsible for coordinating communication, such as forwarding requests, as well as

Figure 1. Proxy pattern

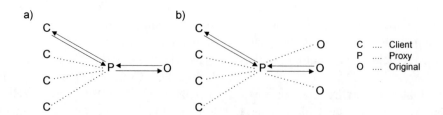

Figure 2. Broker pattern

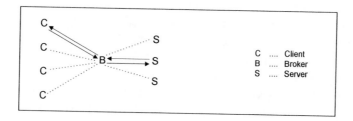

Figure 3. Master/slave pattern

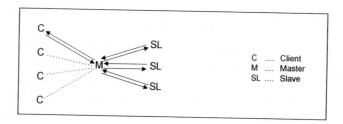

for transmitting results and exceptions" (Buschmann, Meunier, Rohnert, Sommerlad, & Stal, 1996, p. 101). According to Dustdar and Hoffmann (2006, p. 190), "a broker's foremost goal is to achieve location transparency of servers/services…The broker is responsible to locate a server/service that can handle a given request. Then the broker forwards the request to the appropriate component, receives its response and delivers the response to the client." In contrast to a proxy, a broker as depicted in Figure 2 does not perform any pre- or post-processing. Broker B sends messages received from a set of clients C to a pool of servers S.

MASTER/SLAVE PATTERN

The SE domain defines a Master/Slave (M/S) pattern as follows: "The Master-Slave design pattern supports fault tolerance, parallel computation and computational accuracy. A master component distributes work to identical slave components and computes a final result from the results these slaves return" (Buschmann et al., 1996). The M/S pattern is illustrated in Figure 3.

CONTEXT MODELING

Context aware systems have been discussed for some time as a way to enhance computer systems—this happens mostly on two levels: either at the system management level or to enhance the user experience. In both cases relevant context information is being gathered and the system adapts to the respective context. One example of context use to enhance the user's experience stems from the telecommunications domain. The aim was to develop a framework to allow end-users to express how they want their communications to be conducted, and it was identified that this depends very much on the context of the users (Reiff-Marganiec & Turner, 2002).

Up to now, context models discussed in the literature have been either targeting concrete situations or have been very abstract. However, we believe that a context model should provide the scope for encapsulating all kinds of context information such as individual settings and team environments, short-lived coordination activities and long-running complex projects. At the same time the model needs to provide ways of managing context: The model should have notions of how contexts can change and which changes influence other areas of context. An example here is, if users change their location by going home from work they might also change their roles from say, a team leader to wife and mother.

Hence, our suggested context model consists of several layers, ranging from concrete instantiations (the system level; M0) via a domain layer (M1) to a meta-model layer (M2). An overview is shown in Table 1. The meta-model layer represents concepts that are relevant for all collaborative working environments, while the domain layer instantiation layer contains concepts that are required for collaborative working environments (CWE) in a specific domain. Finally, the instance layer provides concrete instances for a particular situation. We also have a layer (M3) which provides notations for expressing the concepts at level M2—however, this is very generic and for example UML or OWL provide the relevant mechanisms.

Let us consider just two examples: roles and locations. At the meta-model layer we are aware that users play certain roles, however roles differ very much from domain to domain. In a medical context we might have doctors and nurses; in academia we find teachers and students. For locations we might have grid coordinates when considering outdoor locations, or we might have room numbers if we are inside a building. At the instance level we find that Jim might be a manager and that he is currently in a meeting in a specific building/room.

As a W3C standard, OWL (Web ontology language) is a promising technology for context modeling. Its syntax is based on RDF and XML, which guarantees human and machine readability, compatibility, and extensibil-ity, thus providing a generic framework for storage, management, and exchange of context information. We are developing an OWL-based context model, which can be used to describe the M0, M1, and M2 layers, illustrated in Table 1, within a coherent ontol-ogy. In that sense OWL can be used as generic model-ing notation at model level M3. Figure 4 represents a simplified user context model, which can be mapped to RDF/XML syntax.

At the very top layer, we model the generic collabo-ration environment concepts and their relationships. There are three concepts or classes in our example, *User, Role,* and *Location*, and two object properties (relationships), has_role and has_location. In the do-main specific layer, we define specific concepts and relationships. Since we are talking about collaboration between two universities, *Student* and *Staff* are two suit-able subclasses of *Role*, while two exemplary subclass of *Location* are *UK* and *Ireland*. Staff in turn features two subclasses: *Researcher* and *Technician*. In the same way, we create subclasses for the location branch. In our example, there are two *User* instances: Tom and David. "Tom_Role" is instance of Student. "David_Role" is instance of Researcher. "Tom_Location" is instance of ULEICES. "David_Location" is instance of NUIG. The relationships between the instances depict that Tom is a Researcher and located in ULEICES, and David is a Student and located in NUIG.

W3C has provided a standard way to query such models, which is SPARQL protocol. In our example, if we want to query if David is located in NUIG, we can just issue the query sentence:

Table 1. Layered context model

M3	M2 Model of Meta Classes	M1 Model of System Level Objects	M0 System Level
	CWE	Domain Specific	Concrete
	Role	**Managers, Technical Staff**	**Manager::Jim**
Language for system models	All possible system models	Concrete system model	Concrete System

Figure 4. OWL-based context model

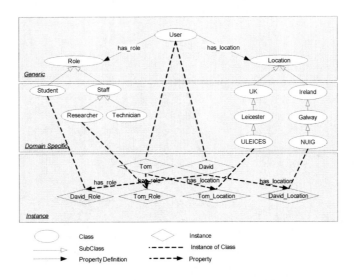

ASK { David   has_location   ?location.
     ?location   rdf:type NUIG.}
The result is: YES.

Organizing context in form of hierarchies has proven to be a suitable method for representing and modeling context information (Dorn & Dustdar, 2006). Thus modeling hierarchies as OWL subclass relationships enables to automatically derive high level information by using an OWL parser.

SOA ARCHITECTURE FOR COLLABORATIVE WORKSPACES

SOA (service-oriented architecture) is becoming a more and more established paradigm. The fundamental concept of SOA is the notion of a *service,* which is defined by the following basic properties:

- The service provides well-defined functionality
- The interface contract for the service is platform-independent
- The service can be dynamically discovered and invoked
- The service is self-contained and loosely coupled

SOA and Web services-based solutions have been used in many domains, and have proven the advantages promised in the definitions mentioned above—flexibility, interoperability, loose coupling—and have become mature technologies. Still, due to the dynamic nature of teams and the large variety of group types, additional research efforts need to be put in the following areas:

- Enabling flexibility in terms of dynamic discovery of available services, which can frequently change based on user and team settings
- Extended description of services—syntax is not sufficient—for context-based autonomous selection of alternative services
- Support mechanisms to facilitate composition and orchestration of higher level services, to support the typical operations of collaboration
- Dynamic adaptation of services compositions based on the context of users and teams

Before we can approach dynamic, context-aware service composition, we need to enable context-aware service selection. Existing approaches either do not take context into account at all or focus only on individuals (Cuddy, Katchabaw, & Lutfiyya, 2005) rather than teams.

Through the dynamic selection of services it is possible to cater to different requirements of collaboration depending on the team structure, interaction pattern, or context. This approach allows composition at different levels, like concrete and abstract workflows and non-coupled tasks that have to be combined during

Figure 5. Context-based service selection for a task

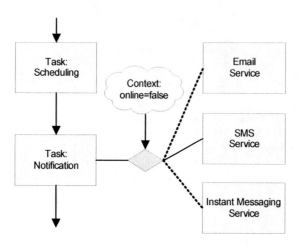

Figure 6. Relevance-based service selection

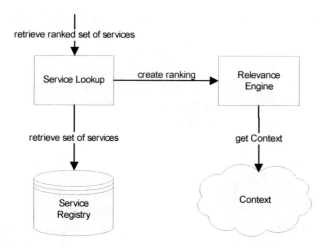

runtime. Thus, an architecture supporting collaborative workspaces must be able to react on changing context settings; for example, to contact a user depends on his location, his current online status, and also on the device he is currently using. In situations when he is sitting in front of his PC it might be possible to contact him directly via e-mail, but if he is currently on the move—his context would say that he is offline—a possible notification for contacting would be sending an SMS to his mobile phone. Hence, as SOA helps to abstract from the underlying technology and infrastructure, a platform will offer different contact services, such as e-mail, SMS, or instant messaging (see Figure 5) depending on context information.

The decision is not made by the context itself, but rather the collaboration platform selects the most relevant service based on context information. For this purpose all services are registered in a service registry such as UDDI. Figure 8 displays the process of a lookup service extracting a set of services featuring context-based metadata and applying the relevance engine to create a ranking of these services. To do so, the relevance engine retrieves relevant context information.

FUTURE TRENDS

We believe that more research is necessary to study the nature of human collaboration itself and to use this knowledge to design software that supports the user and yet remains transparent as possible. We also see a growing need for collaborative devices to automatically sense the user's context and to provide information, notification, and support based on that context.

A trend in the not so distant future in service-oriented application development is not only the construction of large-scale software systems, but also the execution of small, everyday tasks. As users are expected to become ever more flexible, they will need to dedicate a significant amount of their time to the discovery of services that are the most fitting for their individual needs. This in turn calls for the greatest possible amount of system support during service discovery. We believe that our human-centerd approach of service-interaction mining will be a significant contribution to useful service provisioning and, during service lookup, for ranking of discovered services.

CONCLUSION

Computer-supported or -mediated human-to-human collaboration offers the great flexibility of working on joint activities, together with other teammates across space and time. People work on numerous activities in different projects and teams at the same time. Context helps people to focus their attention on relevant pieces of information. Interruptions can be minimized by promoting important requests. Various contextual information such as location, presence, and so forth helps to establish team awareness in disparate teams. By using human interaction patterns we can reveal hidden interactions and make collaboration more effective.

A SOA-based approach to challenges in CWE allows us to support changing needs of distributed teams. These requirements become increasingly dynamic as teams are dispatched in a multitude of collaborative environments in respect to team coupling, time of existence, location dynamics, and so on. A Web services-based architecture allows collaboration services to be discovered, assembled, aggregated, and adapted according to users' context. Hence, a flexible and reconfigurable collaboration environment that supports team specific requirements can be realized.

REFERENCES

Baldauf, M., Dustdar, S., & Rosenberg, F. (2006). A survey on context aware systems. *International Journal of Ad Hoc and Ubiquitous Computing, 2*(4), 263-277.

Bardram, J. E., & Hansen, T. R. (2004). The AWARE architecture: Supporting context-mediated social awareness in mobile cooperation. In *Proceedings of the 2004 ACM Conference on Computer Supported Cooperative Work* (pp. 192-201). New York: ACM Press.

Bentley R., Horstmann T., Sikkel K., & Trevor J. (1995). Supporting collaborative information sharing with the World-Wide Web: The BSCW shared workspace system. In *Proceedings of the 4th International WWW Conference* (pp. 63-74). Boston: ACM Press.

Biegel, G., & Cahill, V. (2004). A framework for developing mobile, context-aware applications. In *Proceedings of the Second IEEE international Conference on Pervasive Computing and Communications* (pp. 361-364). Washington, DC: IEEE Computer Society.

Buschmann, F., Meunier, R., Rohnert, H., Sommerlad, P., & Stal, M. (1996). *Pattern-oriented software architecture: A system of patterns*. West Sussex, England: John Wiley & Sons.

Chen, H., Finin, T., & Joshi, A. (2003). An ontology for context-aware pervasive computing environments. *Knowledge Engineering Review, 18*(3), 197-207.

Cuddy, S., Katchabaw, M., & Lutfiyya, H. (2005). Context-aware service selection based on dynamic and static service attributes. In *Proceedings of the IEEE International Conference on Wireless and Mobile Computing, Networking and Communications* (pp. 13-20). Washington, DC: IEEE Computer Society.

Dey, A. K., & Abowd, G. D. (1999). *Towards a better understanding of context and context-awareness* (GVU Technical Report GITGVU-99-22) Georgia Institute of Technology, College of Computing.

Dorn, C., & Dustdar, S. (2006). Sharing hierarchical context for mobile Web services. *Distributed and Parallel Databases, 21,* 85-111.

Dustdar, S. (2004). Caramba: A process-aware collaboration system supporting ad hoc and collaborative processes in virtual teams. *Distributed and Parallel Databases, 15*(1), 45-66.

Dustdar, S., & Hoffmann, T. (2006). Interaction pattern detection in process oriented information systems. *Data and Knowledge Engineering, 62,* 138-155.

Fahy, P., & Clarke, S. (2004). CASS: A middleware for mobile context-aware applications. In *Proceedings of the Workshop on Context Awareness, MobiSys 2004.* Retrieved January 1, 2007, from http://sigmobile.org/mobisys/2004/context_awareness/papers/cass12f.pdf

Gamma, E., Helm, R., Johnson, R., & Vlissides, J. (1994). *Design patterns: Elements of reusable object-oriented software.* Boston: Addison-Wesley.

Gu, T., Pung, H. K., & Zhang, D. Q. (2004). A middleware for building context-aware mobile services. In *Proceedings of IEEE Vehicular Technology Conference (VTC 2004)* (pp. 2656-2660). Washington, DC: IEEE Computer Society.

Hofer, T., Schwinger, W., Pichler, M., Leonhartsberger, G., & Altmann, J. (2002). Context-awareness on mobile devices: The hydrogen approach. In *Proceedings of the 36th Annual Hawaii International Conference on System Sciences* (pp. 292-302). Washington, DC: IEEE Computer Society.

Jørstad, I., Dustdar, S., & van Do, T. (2005). A service-oriented architecture framework for collaborative services. In *Proceedings of the 3rd International Workshop on Distributed and Mobile collaboration (DMC) IEEE WETICE'05.* Washington, DC: IEEE Computer Society Press.

Pokraev, S., Koolwaaij, J., van Setten, M., Broens, T., Costa, P.D., Wibbels, M., et al. (2005). Service platform for rapid development and deployment of context-aware, mobile applications. In *Proceedings of the International Conference on Web Services* (pp. 639-646). Washington, DC: IEEE Computer Society Press.

Powell, A., Picolli, G., & Ives, B. (2004). Virtual teams: A review of current literature and directions for future research. *ACM SIGMIS Database, 35*(1), 6-36.

Reiff-Marganiec, S., & Turner, K. J. (2002). Use of logic to describe enhanced communication services. In

D. A. Peled & M. Y. Vardi (Eds.), *LNCS2529: Formal Techniques for Networked and Distributed Systems - FORTE2002.* Berlin: Springer.

Roman, M., Hess, C., Cerqueira, R., Ranganathan, A., Campbell, R. H., & Nahrstedt, K. (2002). A middleware infrastructure for active spaces. *IEEE Pervasive Computing, 1*(4), 74-83.

Rosenberg, F., Platzer, C., & Dustdar, S. (2006). Bootstrapping performance and dependability attributes of Web services. In *Proceedings of the IEEE International Conference on Web Services (ICWS'06)* (forthcoming). Washington, DC: IEEE Computer Society Press.

Tang, J. C., Yankelovich, N., Begole, J., Van Kleek, M., Li, F., & Bhalodia, J. (2001). ConNexus to awarenex: Extending awareness to mobile users. In *Proceedings of the SIGCHI Conference on Human Factors in Computing Systems* (pp. 221-228). New York: ACM Press.

KEY TERMS

Ad-Hoc Collaboration: The type of collaboration that emerges spontaneously and does not follow a predefined process.

Context Meta Model: Meta model describing properties of a context model, thereby enabling context transformation and aggregation across application domains and different forms of representations.

Human Interaction Pattern: The term human interaction pattern refers to a common, reccurring interaction scenario between collaborating actors situated in CSCW environments.

Service-Composition: The process of aggregating a number of heterogeneous services to create more sophisticated functionality. In the domain of e-collaboration, service composition enables peers to combine simple services to generate highly complex functions.

Service-Oriented Architecture (SOA): A paradigm in distributed systems that introduces the concept of describing, publishing, finding, binding, composing, and invoking (Web) services.

Team Context: Context that describes the situation of a group of individuals, including information on interdependencies and relations between team members such as coupling, vision, spatial clustering, and dynamics.

Web Service: A Web service is a self-contained software component offering a well-defined interface, accessible through standardized protocols by remote clients. Loose coupling of Web services enables flexibility and reusability.

Interaction Model in Groupware Use for Knowledge Management

Jessada Panyasorn
University of Bath, UK

Niki Panteli
University of Bath, UK

Philip Powell
University of Bath, UK

INTRODUCTION

In the modern and highly changing marketplace, organizations need to find strategies to achieve sustainable competitive advantage. Knowledge management is employed in many different functions and business processes in organizations. As the organizations realize that by identifying, extracting, and capturing the "knowledge assets" of the firm, they can be both fully exploited and fully protected as a source of sustainable competitive advantage (Newell, Robertson, Scarbrough, & Swan, 2002). Suggestions that specialized knowledge has become an essential ingredient for business success are becoming commonplace since productivity is dependent on the application and development of new knowledge (Blacker, 1993; Drucker, 1993; Zack, 1999).

This article focuses on the use of Lotus Notes, a well-known groupware application studied in the literature for cooperation, because its features cover the broad definition of groupware as technologies providing "electronic networks that support communication, coordination and collaboration through facilities such as information exchange, shared repositories, discussion forums and messaging" (Orlikowski & Hofman, 1997, p. 12). It advances the debate on the potentials of groupware in knowledge management. It posits that although Lotus Notes has been the focus of existing research, a paucity of studies has examined its functionality in relation to knowledge management. Therefore, the authors aim to illustrate the use of Notes for cooperation, which leads to organizational knowledge management.

BACKGROUND

Whereas information technologies (IT) are considered as basic requirements for individuals and organizations to necessitate flexibility and responsiveness in today's competitive and dynamic business environments, organizations are increasingly compelled to invest in IT that enable them to manage knowledge-based resources. This is because knowledge-based resources enable sustainable competitive advantage (Alavi & Leidner, 1999), and a variety of IT have been designed and implemented to facilitate knowledge management. The advancement of IT leads to boundless cooperation. IT enables access to broader information and knowledge, which increase creativity and innovation (Gurteen, 1998). The new technologies are making it possible for organizations to operate relatively independently from geographical location, thereby blurring the boundaries between one organization and another, and freeing internal communications within organizations (Blackler, 1995). Groupware is an exemplar of these technologies (Ciborra, 1996; Hayes, 2001). As mentioned by Vandenbosch and Ginzberg (1996), groupware systems such as Lotus Notes enable organizations to create intra-organizational memory in the form of both structured and unstructured information and to share this memory across time and space.

LOTUS NOTES AS A KNOWLEDGE MANAGEMENT TOOL

The use of Lotus Notes for knowledge management activities has been studied in different types of organizations. Lotus Notes, a combination of document

Table 1. Summary of the previous studies on the use of Lotus Notes

Study	Site and size	Lotus Notes functions	Method	Length
Robertson, Sorensen, and Swan (2001)	Universal consulting: Medium	E-mail Discussion databases	Interviews Non-participant observation Documentation	Over 2 years (1996-Spring 1998)
Vandenbosch and Ginzberg (1996)	American insurance firm: Large	Lotus Notes databases	Interviews Surveys	Ten months (began after decision to expand use of Lotus Notes from 200 users to whole firm)
Orlikowski (1993)	Alpha: Large consulting firm: competitive culture	Electronic mail, discussion database, some databases for browsing	Unstructured interviews Documentation Participant observation	Five months (began prior to Lotus Notes installation)
Hayes (2001)	Compound UK: Large multinational pharmaceutical	E-mail Strategic selling databases Discussion databases Contact recording databases	Semi-structured interviews Informal discussions and interactions	Two-and-a-half year period (18 months after first Lotus Notes implementation)
Brown (2000)	Narajo: Large oil firm	Public forum databases: firm notice board and "challenge" database Workflow database Tracking database	Participant observation Interviews	3 months (began after Lotus Notes implementation)
Ciborra and Suetens (1996)	EDF: Large, an international distribution part of a French energy provider	E-mail Discussion forum and databases such as world culture, news forum, expert databases.	Interviews	Over a year and a half (began after Lotus Notes was implemented)
Ciborra (1996)	Roche: Large Diagnostic division of multinational pharmaceutical	Cosis applications: multidisciplinary knowledge base	Interviews	Over 2 years (began after Lotus Notes was implemented)
Orlikowski (1996)	Zeta: Large software firm	Incident Tracking Support System Training database six firm-wide bulletin boards with electronic mail	Unstructured and semi-structured interviews. Non-participant observation Documentation	6 months (began two years after the ITSS developed on Lotus Notes.)
Karsten and Jones (1998)	CCC: Small computer consulting firm	Discussion and news databases, project databases	Participant observation Interviews Documentation	3 years (began prior to Lotus Notes implementation)

creator and indexer, database generator and manager, and messaging platform (Vandenbosch & Ginzberg, 1996), allows information to be distributed between different users in a structured or semi-structured way (Brown, 2000). Drawing upon the existing literature, this section presents the main functions of Lotus Notes in knowledge management. Nine major case studies discussing the use of Lotus Notes and its potential to knowledge management are reviewed. The selection of these studies is based on the rich descriptions of how Lotus Notes was implemented and used for facilitating information and knowledge management in the cases. Table 1 provides a synopsis of these studies by taking into account the research site, the functions of Lotus Notes that were used, methods for collecting data and length of the study.

Seven cases were carried out in large organizations, whereas the other two cases (Robertson, Sorenson, & Swan, 2001; Karsten & Jones, 1998) were conducted in SMEs. The sectors are diversified, however, the major sector being consulting (Robertson et al., 2001; Orlikowski, 1993; Karsten & Jones, 1998). In terms of research approach, these studies employ an in-depth case study approach of which interview is the main method of data collection. The main benefit identified

of using Lotus Notes is that shared information and knowledge on Lotus Notes can be accessed and retrieved by users regardless of time and location. Most studies found that factors embedded in the organizational context have a major influence on the successful use of Lotus Notes: collaborative culture (Orlikowski, 1993), incentive structure (Robertson et al., 2001; Orlikowski, 1993), homogeneous group (Hayes, 2001), management style (Karsten & Jones, 1998), and dispersion of organization (Ciborra & Suetens, 1996).

Based on this review, Lotus Notes has been developed and used in different ways for supporting knowledge management activities. Table 2 synthesizes these different use modes of Lotus Notes in knowledge management as identified in the literature; the five use modes are explained next.

First, Lotus Notes is used to publish information. Published information is disseminated in different forms and for different purposes. For example, an interactive newsletter was published to disseminate news within an international department of the French energy provider, EDF (Ciborra & Suetens, 1996). Technical documents were published and disseminated outside the customer support department of Zeta (Orlikowski, 1996). Meeting minutes were published to allow those not present

Table 2. Summary of use modes of Lotus Notes

Use Mode	Use Description	Source
Publishing	Publishing information (e.g., newsletter, technical documents, product catalogues, employee directories).	Orlikowski (1996), Ciborra and Suetens (1996), Karsten and Jones (1998)
Searching	Searching for or acquiring organization information (e.g., full text search capabilities, document indexer).	Robertson et al. (2001), Orlikowski (1993, 1996), Ciborra and Suetens (1996)
Maintaining	Recording and maintaining a computer-based "organizational memory" (e.g., best practices, business process, frequently asked questions).	Robertson et al. (2001), Vandenbosch and Ginzberg (1996), Orlikowski (1993, 1996), Hayes (2001), Brown (2000), Ciborra and Suetens (1996), Ciborra (1996), Karsten and Jones (1998)
Sharing	Discussing and sharing ideas, experience, information and knowledge with other individuals and groups in the organization (e.g., via e-mail, discussion databases, public fora).	Robertson et al. (2001), Vandenbosch and Ginzberg (1996), Orlikowski (1993), Hayes (2001), Brown (2000), Ciborra and Suetens (1996), Ciborra (1996), Karsten and Jones (1998)
Creating	Understanding and creating individuals' and groups' tacit knowledge (e.g., shared databases within homogeneous group).	Hayes (2001)

in the meeting of CCC to keep informed (Karsten & Jones, 1998).

The second use mode of Lotus Notes is searching. Lotus Notes comprises capabilities such as full text search and document indexer in searching or acquiring information. In universal consulting (Robertson et al., 2001), project leaders used the indexing and search facilities of Lotus Notes to acquire specific information found in e-mail and discussion databases. In Zeta (Orlikowski, 1996), the provision of a powerful search capability within the incident tracking support system (ITSS), an application on Lotus Notes, allowed the specialists to search quickly and easily their database of well-documented incident histories. Searching ITSS provided potentially reusable problem resolutions as well as knowledge about problem-solving processes. Similarly, expertise in Alpha (Orlikowski, 1993) used Lotus Notes for organization database browsing. Using Lotus Notes for searching information was also found in EDF (Ciborra & Suetens, 1996). Managers at EDF could search for others' experience about foreign cultures before they left for missions abroad.

The third use mode of Lotus Notes is maintaining. This mode focuses on using Lotus Notes to record and maintain a computer-based "organizational memory" such as best practices, business process, and frequently asked questions. For example, in Compound UK (Hayes, 2001), contact recording database enabled employees to record the views, interests, and requirements of particular doctors that could be retrieved for future use. Training database in Zeta (Orlikowski, 1996) maintained sample problems extracted from the ITSS database which new hires worked with to try and resolve problems. Quality project documentation, which is a valuable by-product of using Lotus Notes for discussion on project work across countries, was maintained on Lotus Notes databases of Universal consulting (Robertson et al., 2001).

The fourth use mode is sharing. With this mode, individuals, and groups in the organization use Lotus Notes to discuss and share ideas, experience, information, and knowledge with each other. This use mode exists in all the cases as Lotus Notes provides several mechanisms including e-mail, discussion databases, and public fora. In American insurance company (Vandenbosch & Ginzberg, 1996), people who were geographically dispersed participated in discussions about process change aimed at standardizing the company's key activities across its several regional divisions. Similarly, a strategic selling database was created in Compound UK (Hayes, 2001) to enable employees in different functions to input their views and information in a structured way with the aim of bringing together the employees' shared knowledge so that they might contribute to a successful sale.

The final use mode of Lotus Notes is creating. This use mode addresses the potential of Lotus Notes for creating individuals' and groups' tacit knowledge. This mode is different from the other modes in that individuals' tacit knowledge is made explicit on the shared database of Lotus Notes, whereas others can use the shared knowledge database, as the tacit knowledge within groups allows them to understand the subtleties that underlie the meaning expressed on the shared database (Hayes, 2001). Hence, tacit knowledge is created in individuals' mind. However, the creating mode of Lotus Notes is likely to depend on the organizational context. For instance, within the same functions in Compound UK, employees could draw on their shared tacit knowledge to interpret skillfully and make judgements concerning the views recorded on the shared databases by members of their own functions (Hayes, 2001). On the other hand, the attempt to create a common knowledge pool for the global organization of EDF was not satisfied due to the misalignment to virtual organization context with a highly scattered structure and based on the strong competence and autonomy of the agents and experts. As a result, it is difficult to reconcile the style of working and knowing with prescriptions to share information (Ciborra & Suetens, 1996).

Having explained the use modes of Lotus Notes existing in the literature, the five use modes can be regarded as the role of knowledge management systems in that they support four sets of socially enacted "knowledge processes" introduced by Alavi and Leidner (2001). This includes the processes of knowledge creation, storage and retrieval, transfer, and application. Therefore, Lotus Notes is considered as a potential tool supporting knowledge management.

INTERACTION RICHNESS MODEL OF GROUPWARE

This section presents the "interaction richness" model based on the case study findings. The model taxonomizes the use modes of Lotus Notes on two dimensions. The first dimension entails the types of interaction that may take place in a Lotus Notes environment, which may either be human-Notes interactions or human-human, Notes-mediated interactions. The second dimension involves the types of knowledge management processes experienced in a Lotus Notes environment, namely coordination, communication, and collaboration. These processes were found in the study of Lotus Notes by Karsten (1995) and Orlikowski (1996).

Coordination is regarded as the direction of individuals' efforts toward achieving common and explicitly recognized goals (Blau & Scott, 1963). The use modes that fall into the coordination category are "searching" and "publishing." Publishing is concerned with human-human mediated interactive coordination, since humans use Lotus Notes as an information-sending channel to receivers, contributing to the exchange of knowledge. Publishing use mode is mainly utilized by the e-mail feature that organizations use to distribute news and information among staff. Searching leads to a human-Notes interactive coordination process. Searching could be utilized in relation to publishing. This is because Notes enables users to search for the information published in the databases in order to complete their individual tasks such as answering customers' enquiries.

The communication process emphasizes the exchange of information between dispersed individuals and it mainly includes increasing connectivity, bandwidth, and protocols for the exchange of many types of information such as text, graphics, and voice (Ellis, Gibbs, & Rein, 1991). "Maintaining" and "sharing" use modes can be put into this category. As the "maintaining" use mode focuses on records of information and knowledge retrieved by users, databases act as an agent that communicates the information maintained to individual receivers. Therefore, maintaining is regarded as the use mode stimulating human-Notes interactive communication process. For example, users can retrieve presentation files, documents, or the history that they can learn from Notes databases. On the other hand, the sharing mode emphasises the exchange of knowledge between individuals who are both senders and receivers, thus it is an exemplar of human-human mediated interactive communication process. For example, Notes databases such as customer contact details and solutions provided to the customers are used to share information among staff, which helps them to fulfil a project.

Finally, collaboration is a process of shared creation: two or more individuals with complementary skills interacting to create a shared understanding (Schrage, 1990). Thus, the fifth use mode of Notes, "creating,"

belongs to the human-human mediated interactive collaboration process as it refers to understanding the shared knowledge database by drawing on individuals' shared tacit knowledge. From the existing literature, it is not possible to identify a use mode that belongs to the human-Notes interactive collaboration process. However, due to the interpretive flexibility embedded in IT (Orlikowski, 1992), it is argued here that users will, over time, learn to explore the potentials of Notes contributing to a collaboration between humans and technology where each benefits from the other. Based on this, the authors add another use mode, "exploring," which humans can consult and collaborate with Notes to achieve a common goal. Such use mode may emerge when Notes is adjusted or integrated with other tools for the KM benefits. However, more evidences are still needed to prove this.

Accordingly, organizations may employ the interaction richness model in order to recognize the use modes they utilize and to assess the extent to which Notes brings them benefits. This will urge organizations to use Notes in a way that supports the 3Cs processes that lead to knowledge created in individuals' actions rather than to use it to maintain "information," which provides less value to the organizations. Finally, the model will also be worthwhile for those organizations that have already implemented Notes in order to find an approach to develop their KM practices.

Figure 1 illustrates the interaction richness model of Lotus Notes use modes for knowledge management.

CONCLUSION

The aim of this article was to explore the different use modes of Lotus Notes as a groupware system for knowledge management. Nine previous case studies of Notes were reviewed. The authors found the five use modes existing in the literature, which are publishing, sharing, maintaining, searching, and creating. A sixth use mode, exploring, was identified in this article in order to provide a further potential of Notes, which could be developed in relation to knowledge management. The authors, therefore, created the interaction richness model which divided the six use modes into two main categories, namely human-human mediated interaction and human-Notes interaction. The model is expected to provide a better understanding of how Notes is used to support knowledge management in organizations and lead to a more effective use of Notes.

Figure 1. Interaction richness model of Lotus Notes use modes

	Coordination	Communication	**Collaboration**
Human-Human mediated interaction	Publishing	Sharing	Creating
Human-Notes interaction	Searching	Maintaining	Exploring

Having reviewed the literature, mostly in large organizations, not all the use modes have been found in a single organization. It may be because only limited functions of Lotus Notes were utilized in those large organizations, or Lotus Notes was implemented only in some departments within these large organizations. Since the technology itself may not fulfill knowledge management projects, the support of other factors such as strategies, organizational structure, and processes (Boland & Tenkasi, 1995; Robertson et al., 2001; Zack, 1999) need to be taken into account. In addition, the use of Notes is more strongly influenced by aspects of the organizational context, internal social structure, culture and the users' capabilities than by any intrinsic logic of the technology (Karsten & Jones, 1998; Orlikowski, 1993) and the importance of these should not be underestimated in understanding the implementation of KM systems such as groupware in organizations. Future research is required from different organizational contexts in order to identify the relationship between each use mode in supporting knowledge management, to contribute to a better understanding of the use of groupware for knowledge management, and to increase generalizability. In particular, it may be worthwhile to conduct a study of Notes use in SME and developing country contexts as it was neglected in the previous studies. As Lotus Notes is adaptable and customised over time, the temporal dimension needs to be taken into account in order to better examine its impact, and the interactions between technology and organizations in terms of improvement and deterioration and thus contributing to the "exploring" potential of Lotus Notes. This therefore should be in the agenda for future research.

REFERENCES

Alavi, M., & Leidner, D. E. (1999). Knowledge management systems: Issues, challenges, and benefits. *Communications of the Association for Information Systems, 1*(2), 1-37.

Alavi, M., & Leidner, D. E. (2001). Review: Knowledge management and knowledge management systems: Conceptual foundations and research issues. *MIS Quarterly, 25*(1), 107-136.

Blacker, F. (1993). Knowledge and the theory of organizations: Organizations as activity systems and the reframing of management. *Journal of Management Studies, 30*(6), 863-884.

Blacker, F. (1995). Knowledge, knowledge work and organizations: An overview and interpretation. *Organization Studies, 16*(6), 1021-1046.

Blau, P., & Scott, W. R. (1963). *Formal organizations: A comparative approach.* London: Routledge & Kegan Paul Ltd.

Boland, R. J., & Tenkasi, R. V. (1995). Perspective making and perspective taking in communities of knowing. *Organization Science, 6*(4), 350-372.

Brown, B. (2000). The artful of groupware: An ethnographic study of how Lotus Notes is used in practice. *Behaviour & Information Technology, 19*(4), 263-273.

Ciborra, C. U. (1996). Mission critical: Challenges for groupware in a pharmaceutical company. In C. U. Ciborra (Ed.), *Groupware and teamwork: Invisible aid or technical hindrance?* Chichester, UK: John Wiley & Sons.

Ciborra, C. U., & Suetens, N. T. (1996). Groupware for an emerging virtual organization. In C. U. Ciborra (Ed.), *Groupware and teamwork: Invisible aid or technical hindrance?* Chichester, UK: John Wiley & Sons.

Drucker, P. (1993). *Post-capitalist society*. Oxford: Butterworth-Heinemann.

Ellis, C. A., Gibbs, S. J., & Rein, G. L. (1991). Groupware: Some issues and experiences. *Communications of the ACM, 34*(1), 38-58.

Gurteen, D. (1998). Knowledge, creativity and innovation. *Journal of Knowledge Management, 2*(1), 5-13.

Hayes, N. (2001). Boundless and bounded interactions in the knowledge work process: The role of groupware technologies. *Information and Organization, 11,* 79-101.

Karsten, H. (1995). "It's like everyone working around the same desk": Organisational Readings of Lotus Notes. *Scandinavian Journal of Information Systems, 7*(1), 3-32.

Karsten, H., & Jones, M. (1998). The long and winding road: Collaborative IT and organisational change. *Proceedings of the CSCW '98*, Seattle, USA.

Newell, S., Robertson, M., Scarbrough, H., & Swan, J. (2002). *Managing knowledge work*. Palgrave Macmillan.

Orlikowski, W. J. (1992). The duality of technology: Rethinking the concept of technology in organizations. *Organization Science, 5*(3), 398-427.

Orlikowski, W. J. (1993). Learning from Notes: Organizational issues in groupware implementation. *The Information Society, 9*(3), 237-250.

Orlikowski, W. J. (1996). Evolving with Notes: Organizational change around groupware technology. In C. U. Ciborra (Ed.), *Groupware and teamwork: Invisible aid or technical hindrance?* Chichester, UK : John Wiley & Sons.

Orlikowski, W. J., & Hofman, J. D. (1997). An improvisational model for change management: The case of groupware technologies. *Sloan Management Review, Winter,* 11-21.

Robertson, M., Sorensen, C., & Swan, J. (2001). Survival of the leanest: Intensive knowledge work and groupware adaptation. *Information Technology & People, 14*(4), 334-353.

Schrage, M. (1990). *Shared minds*. New York: Random House.

Vandenbosch, B., & Ginzberg, M. J. (1996). Lotus Notes and collaboration: Plus ca change. *Journal of Management Information Systems, 13*(3), 65-81.

Zack, M. H. (1999). Managing codified knowledge. *Sloan Management Review, 40*(4), 45-58.

KEY TERMS

Groupware: Technology that provides electronic networks that support communication, coordination, and collaboration through facilities such as information exchange, shared repositories, discussion forums, and messaging.

Information: A flow of messages, results from placing data within some meaningful content.

Interaction: Defined as a process by which two or more things have an effect on each other and work together.

Interaction Richness Model: Defined in this study as the extent to which depth of interaction leads to the exchange of information or knowledge.

Knowledge: Knowledge is what is believed and valued on the basis of the meaningfully organized accumulation of information (messages) through experience, communication, or inference. Individuals use their knowledge to perform actions such as creating information for other individuals, while knowledge is created in practice, in the activities of and interactions between individuals.

Knowledge Management: A strategy for managing knowledge that includes people, processes, and technology for creating, capturing, categorizing, disseminating, and using knowledge to generate value to the organization.

Lotus Notes (Notes): One of the most well-known groupware products in the market by IBM, it allows information to be distributed between different users, using the shared databases integrated with e-mail.

Interrelationships between Web-GIS and E-Collaboration Research

Michele Masucci
Temple University, USA

INTRODUCTION

Geographic information systems (GIS) refers to the computer hardware and software that supports the management and analysis of spatial information. There has been a recent increase in the development of Internet accessible GIS applications, called Web-GIS (Al-Kodmany, 2001; Carver, Evans, Kingston, & Turton, 2000). Web-GIS facilitates participation among stakeholders through disseminating user interfaces for storing, accessing, and analyzing spatial information using the Internet (Al-Kodmany, 2001; Carver et al., 2000; Dragicevic & Balram, 2004). Participatory and community GIS approaches focus on system design that supports collaboration among organizations serving and representing interests of many constituent groups, including nontechnical users (Carver, 2003; Craig, Harris, & Weiner, 2002; Drew, 2003; Elwood & Ghose, 2004; Elwood & Leitner, 2003; Ghose, 2005; Ghose & Elwood, 2003; Kyem, 2004; Seiber, 2003).

The emphasis on participatory aspects of GIS is important for many decision processes, including community planning, environmental management, and citizen advocacy uses (Carver, 2003; Drew, 2003; Ghose, 2005). Improving the management of collaborative aspects GIS design, referred to as geocollaboration, has also been gaining attention among researchers (Al-Kodmany, 2001; Brewer, MacEachren, A. M., Abdo, H., Gundrum, J., & Otto, 2000; Carver et al., 2000; Churcher & Churcher, 1999; MacAcherin, 2001; Schafer, Ganoe, Xiao, Coch, & Carroll, 2005). The creation of virtual GIS environments encompasses both trends, fusing geovisualization and e-communication techniques to improve multiple user exchanges and experiences in GIS computing.

E-collaboration research has the potential to advance Web-GIS because of the focus on evidence-based strategies for using e-technologies to distribute tasks and access shared information resources. For instance, consideration of the use of groupware for managing the consolidation of field surveys that could comprise a spatial data set might involve (a) assessing e-technologies for tracking documents, (b) improving document portability, and (c) integrating spatial data with other information resources. The specific use of decision support-ware, e-communication applications, and database integration tools is often directly related to the overall computing environment within an organizational setting (Kock & Nosek, 2005; Markus, 2005). However, many organizational approaches for administrating computer support for all kinds of management tasks fail to account for specific collaborative frameworks used in conducting GIS enabled research (MacEachran, Gahegan, & Pike, 2004; Schafer et al., 2005).

THE NATURE OF SPATIAL INFORMATION

Information is geographic if it pertains to (a) a specific location on the Earth's surface, (b) to knowledge of *where* something is, or (c) knowledge about *what* is at a given location (Goodchild, 2000). Italics are used to emphasize the terms *where* and *what*. Both terms underscore the empirical roots of the field of geographic information science (GIScience). Geographic information technologies (GITs) consist of (a) global positioning systems (GPS), (b) remote sensing, and (c) geographic information systems (Goodchild, 2000). They are used for gathering, interpreting, managing, and analyzing geographic information. An important factor in examining how e-collaboration strategies are interrelated with GIS use is that the datasets involved are extremely large, often measured in gigabytes and terabytes (Goodchild, 2000). In addition, there are extensive hardware requirements for manipulating such datasets with embedded spatial analytical software due to the complexity of systems that may be modeled. Improvements in e-communications and computing infrastructure enabled GIS developers and users to collaborate across work environments, setting the conditions for the emergence of Web-GIS.

The current prevalence of Web-GIS applications reflects the extent of hybridization among spatial information resources, e-communication technologies, and analytical applications (Zook, Dodge, Aoyama, & Townsend, 2004). GPS capabilities are now integrated with handheld information technologies, such as cell phones and personal digital assistants (PDAs). PDAs and cell phones can display one's exact geographic position in real time. Such information can be integrated, along with descriptive characteristics of specified locations, directly into a GIS. This makes it possible to send, receive, and adapt spatial information from organizational settings or field locations, despite the fact that the datasets can be so large. These combined technologies are commonly used for tasks like (a) monitoring local environmental systems using remote sensing techniques, (b) assessing parcel characteristics in cities, and (c) managing emergency services (e.g., Haag & Haglund, 2002). Moreover, improvements in the ability to share and distribute spatial information resources among collaborators addresses the challenges associated with fostering participation in complex decision-making processes noted by Poore (2003).

A CASE STUDY OF WEB-GIS DEVELOPMENT AND USE

Technical experts guide many participatory models for information system design. This can result in what are sometimes referred to as *technocratic* participatory models for GIS development and use. Many GIS scholars and professionals have observed that such systems can quickly become the domain of GIS experts who may advocate on behalf of communities (Gilbert & Masucci, 2006; Harvey, 2001; Kyem, 2004; Masucci, 2000; McLafferty, 2002; Pipek, Märker, Rinner, & Schmidt-Belz, 2000; Schroeder, 1999). But this can result in the loss of an authentic community perspective because of the technological experience and expertise gaps that exist among GIS developers and nontechnical community users.

In 2004, the Information Technology and Society Research Group of Temple University (formerly called the E-Collaboration Research Center) initiated a program to develop a GIS that could be used by community partners and high school students. The aim was to draw on hybrid technologies to support participation in developing a GIS that would be of value to community participants. The approach taken for developing the system, called bITS-GIS, was to improve communication among all participants with the purpose of better aligning GIS resources with community management issues and spatial analysis at the local scale.

Partners in the project included local community serving organizations, residents, and students situated in North Philadelphia communities. Specifically, these included Asociación de Puertorriqueños en Marcha, United Way of Southeastern Pennsylvania, the School District of Philadelphia, the Temple University Telemedicine Research Center, and Delawarevalley.org. The partnership aimed to (a) facilitate collaboration among university professors and students, high school teachers and students, community organizations, and family members; (b) assess the viability of this collaborative approach for developing a GIS; and (c) use the geographic information technology resources developed through bITS for the analysis of environmental quality and community development.

The participatory design of the GIS was initiated through a series of meetings to discuss (a) the collaborative approach that would be undertaken for identifying spatial information needs of constituent partners, (b) how spatial information would be shared and how maps generated by the GIS would be used and disseminated, and (c) the relationship between developing a community GIS and advancing community educational goals for high school youth.

The criteria that guided the GIS development approach included (a) representing and reflecting the community in terms of its information needs and management goals in the design of the GIS, (b) representing and reflecting participation that advances community capacities in the realm of GIScience, and (c) providing the institutional capacity for conducting geospatial analysis of environmental quality, community development, and other local scale problems.

One of the most effective ways to accomplish these goals was to implement a Web-GIS approach. An important technical element needed to accomplish this was to facilitate the analysis of spatial patterns and associations at the most disaggregate community scale. This involved developing software that would integrate information resources provided by individuals using Web-enabled map applications such as Google Earth with advanced GIS applications such as ArcGIS.

This approach permitted the use of georeferenced and nongeoreferenced spatial information. One example of georeferenced information included in bITS-GIS is the identification of specific locations pertaining to field observations taken by project participants. Nongeoreferenced information incorporated in the system included digital archives of qualitative observations, such as digital photographs, oral histories, and field notes. In alignment with the community goals articulated in the community meetings, the bITS-GIS ultimately supported the archiving, cataloging, and analysis of (a) local environmental quality indicators and environmental quality and health characteristics at the local scale; (b) local community development characteristics, including access to jobs, the effects of economic development initiatives, and the role of community organizations in stimulating development; and (c) information pertaining to community identity, including important historical settings and landmarks.

E-COLLABORATION ASPECTS OF bITS-GIS

One of the most important tasks undertaken for developing the bITS-GIS was to complete an inventory of existing digital spatial data records based on input from community partners. Spatial data resources relevant to bITS-GIS were integrated into a project file according to the design criteria established in collaboration with project community partners. The bITS-GIS project file was created using ArcGIS. ArcGIS was also used to conduct spatial analysis and to produce maps resulting from the analysis. Technical support staff members converted data and map files developed through the collaboration into mini-project files that are used by community partners and high school project participants for education and advocacy purposes. E-communication tools and information delivery methods, including web pages and web logs, file sharing software, and transportable data storage devices were used to facilitate the inclusion of student projects in the community GIS and to share outcomes with the community.

For instance, a software application called Map Wizard was used to create an Internet presentation of maps created in ArcGIS on the project website. Map Wizard enables the creation of web-enabled map layers and attribute data files. These can be accessed solely through the use of a mouse, making it unnecessary to learn specific GIS software commands. High school students involved in developing bITS-GIS added data

to the system that was gathered during educational field experiences. Web-based forms designed specifically for this purpose, as well as Google Earth map point, path, and polygon tools, were used by the students to locate features such as sites and itineraries associated with their field work, field observations, and spatial data gathered.

Map Wizard was used to distribute information consolidated through the use of the other applications using the Internet. The entire approach was designed to meet the collaboratively defined criteria of ease of use of system resources. In addition, new software was developed that enabled project partners to use open access Internet Applications, such as Google-Earth, to input location information and associated narrative descriptions and other attributes into the GIS project file. The software, called bITSY, is a data conversion tool that creates a text file from kml format so it can be integrated with the ARC-GIS project file.

The information resources developed by community partners and high school students were consolidated with baseline digital spatial data available from secondary sources to fulfill the goals of the community GIS. The system integrated the following types of spatial information pertaining to the area of interest of the study, including: (a) area geographies, (b) line geographies, (c) point geographies, (d) socio-demographic variables, and (e) environmental quality and (f) public health information. These information resources were obtained using web-based spatial data libraries and through data sharing among community partners.

In addition, the bITS-GIS incorporated qualitative information collected by high school students, including the following: (a) narrative descriptions related to educational themes of public health, daily life activity patterns, industrial history and architecture, disabilities, environmental quality, and community social history; (b) digital documentation of locations visited, including digital photographs, narrative descriptions, and video streams, and (c) attribute data pertaining to locations visited.

Instructional materials and sample exercises disseminated through the project website were accessed by program participants in both formal and informal computing settings as part of their educational experience. These materials were developed to foster basic competencies related to interpreting and visualizing geographic information. The procedure for providing students and community members with acquiring knowledge and experiences that supported their engagement in developing bITS-GIS involved the: (a) interpretation of maps, digital aerial photographs, and satellite imagery, (b) creation of mental maps, (c) creation of maps of field locations and field work visited by students, (d) creation of map documentation to represent and classify field observations, (e) documentation and encoding of field observations in digital format, (f) linking digital attribute data to field sites using Google Earth, bITS-Blog, and bITS-GIS tools; and (g) creation of meta-data descriptions in digital format and their incorporation into the bITS-GIS.

Maps and information resources that have resulted from partnership activities and high school student fieldwork are accessible on the Internet. The software applications used also addressed the problem of interoperability among the spatial information resources. The objective in addressing the challenge of creating a GIS that is interoperable across the Internet and project file platforms was to increase the ease of use of spatial information among community participants and high school students. This Web-GIS approach permitted program participants to (a) work in teams in different remote locations at the same time, (b) update information resources included in bITS-GIS in asynchronous time, and (c) complete these tasks in a distributed work environment. In addition, it permitted technical experts involved in the project to support data integration collaboratively with program participants, no matter which work or field setting was utilized at any given time.

In order to accomplish the main goal of creating an easy to use approach for student learning and engagement in the development of the bITS-GIS, technical issues needed to be resolved. The use of Internet applications supported the document sharing strategy and group e-communication approach related to the further development of system resources. And the Web-based applications chosen for engaging high school students and community members in the task of developing system resources related to their baseline skill sets and ages.

A protocol for entering data and locating field observations that would eliminate the challenge high school students and nontechnical community users would face in the mastery of command structures within the ArcGIS software suite was also developed. The Web interface consists of a Web form that has various fields, and it can be customized to conform to the data entry fields needed to input observations gathered during student fieldwork. Information entered by the students and community members using the Web forms is archived within the bITS-GIS ArcGIS project file and is used to generate Web-accessible map layer systems and attributes databases. Students retrieved the collective information archived by the system using the Web-based maps generated from extracting layers archived within the ArcGIS project file and served to the Internet using Map Wizard.

SUMMARY

E-collaboration researchers focus on how the use of electronic communication technologies can support multidimensional group tasks (Dasgupta, 2003). The field encompasses inquiry related to informal group, small organizational and large institutional settings—and often considers the approaches taken to perform tasks that strengthen ties among institutions of different capacity and scale. This emphasis relates directly to the communication and electronic collaboration context for many GIS users and researchers. It has been suggested here that GIScientists concerned with developing systems that involve community collaborations may wish to draw from the field of e-collaboration to assess strategies for finding innovative solutions for interoperability, data sharing, and dissemination practices. The emphasis placed on the relationship between organizational settings and group tasks can inform GIS users of evidence-based criteria for selecting tools and strategies for tasks as diverse as compiling large spatial datasets to disseminating project outcomes using the Internet. In particular, with community Web-GIS initiatives, thoughtful implementation of e-collaboration strategies can support that the system developed reflects the partnership arrangements for system development and accessing system resources.

ACKNOWLEDGMENT

This manuscript was made possible by a grant from the National Science Foundation (ESI-0423242). Any opinions, findings, and conclusions or recommendations are those of the author and do not necessarily reflect the views of the National Science Foundation.

REFERENCES

Al-Kodmany, K. (2001). Visualization tools and methods for participatory planning and design. *Journal of Urban Technology, 8*(2), 1-37.

Brewer, I., MacEachren, A. M., Abdo, H., Gundrum, J., & Otto, G. (2000). Collaborative geographic visualization: Enabling shared understanding of environmental processes. *IEEE Symposium on Information Visualization,* 137-141.

Carver, S. (2003). The future of participatory approaches using geographic information: Developing a research agenda for the 21st century. *URISA Journal, 15*(APAI), 61-71.

Carver, S., Evans, A., Kingston, R., & Turton, I. (2000). Accessing Geographical Information Systems over the World Wide Web: Improving public participation in environmental decision-making. *Information Infrastructure and Policy, 6*(3), 157-170.

Churcher, C., & Churcher, N. (1999). Realtime conferencing in GIS. *Transactions in GIS, 3*(1), 23-30.

Craig, W., Harris, T., & Weiner, D. (Eds.). (2002). *Community participation and Geographic Information Systems.* London: Taylor and Francis.

Dasgupta, S. (2003). The role of controlled and dynamic process environments in group decision making: An exploratory study. *Simulation & Gaming, 34*(1), 54-68.

Dragicevic, S., & Balram, S. (2004). A Web GIS collaborative framework to structure and manage distributed planning processes. *Journal of Geographical Systems, 6*(2), 133-153.

Drew, C. (2003.) Transparency: Considerations for PPGIS research and development. *URISA Journal, 15*(APA 1), 73-78.

Elwood, S., & Ghose, R. (2004). PPGIS in community development planning: Framing the organizational context. *Cartographica, 38*(3/4), 19-33.

Elwood, S., & Leitner, H. (2003). GIS and spatial knowledge production for neighborhood revitalization: Negotiating state priorities and neighborhood visions. *Journal of Urban Affairs, 25*(2), 139-157.

Fusco, L., van Bemmelen, J., & Guidetti, V. (2005). Emerging technologies in support of long-term data and knowledge preservation for the earth science community: Experiences with digital libraries and grid at ESA. In *Proceedings of Ensuring Long-term Preservation and Adding Value to Scientific and Technical data, (PV2005).*

Haag, F., & Haglund, S. (2002). The application of remote sensing techniques to landscape level environmental research: A hybrid approach combining visual and digital interpretation. *Norsk Geografisk Tidsskrift, 56*(4), 265-270.

Ghose, R. (2005). The complexities of citizen participation through collaborative governance. *Space and Polity, 9*(1), 61-75.

Ghose, R., & Elwood, S. (2003). Public participation GIS and local political context: Propositions and research directions. *URISA Journal, 15*(APA II), 17-24.

Gilbert, M., & Masucci, M. (2006). The implications of including women's daily lives in a feminist GIScience. *Transactions in GIS, 10*(5), 751-761.

Goodchild, M. (2000). NCGIA core curriculum-Unit 2. Retrieved from http://www.ncgia.ucsb.edu/giscc/units/u002/

Harvey, F. (2001). Constructing GIS: Actor Networks of Collaboration. *URISA Journal, 13*(1), 29-37.

Kock, N., & Nosek, J. (2005). Expanding the boundaries of e-collaboration (Special Issue on Expanding the Boundaries of E-Collaboration). *IEEE Transactions on Professional Communication, 48*(1), 1-9.

Kyem, P. A. K. (2004). Of intractable conflicts and participatory GIS applications: The search for consensus amidst competing claims and institutional demands. *Annals of the Association of American Geographers, 94,* 37-57.

MacEachren, A. M. (2001). Cartography and GIS: Extending collaborative tools to support virtual teams. *Progress in Human Geography, 25*(3), 431-444.

MacEachren, A. M., Gahegan, M., & Pike, W. (2004). Visualization for constructing and sharing geo-scientific concepts. *Proceedings of the National Academy of Sciences, 101*(Suppl 1), 5279-5286.

Markus, M. L. (2005). Technology-shaping effects of e-collaboration technologies: Bugs and features. *Inter-*

national Journal of e-Collaboration, 1(1), 1-23.

Masucci, M. (1999). Virtuality and reality: Navigating the power relationships within geographic/information technology partnerships to discover the terrain of empowerment. In *Papers of GISOC'99. The First International Conference on GIS and Society*. Minneapolis: The University of Minnesota.

Masucci, M. (2000). Institutional partnerships in using and developing information technology for community environmental monitoring. In I. Viadana & M. Lombardo (Eds.), *Universidade e Comunidade na Gestão do Meio Ambiente* (pp. 65-75). São Paulo, Brazil: UNESP (State University of São Paulo) Press.

McLafferty, S. (2002). Mapping women's worlds: Knowledge, power and the bounds of GIS. *Gender, Place and Culture, 9*(3), 263-269.

Pipek, V., Märker, O., Rinner, C., & Schmidt-Belz, B. (2000). Discussions and decisions: Enabling participation in design in geographical communities. In M. Gurstein (Ed.), *Community informatics: Enabling communities with information and communications technologies* (pp. 358-375). IDEA.

Poore, B. (2003). The open black box: The role of the end user in GIS integration. *The Canadian Geographer/Le Géographe canadien, 47*(1), 62-74.

Schafer, W. A., Ganoe, C. H., Xiao, L., Coch, G., & Carroll, M. J. (2005). Designing the next generation of distributed, geocollaborative tools. *Cartography and Geographic Information Science, 32*(2), 81-100.

Schroeder, P. (1999). Changing expectations of inclusion, toward community self-discovery. *URISA Journal, 11*(2), 43-51.

Seiber, R. E. (2003). Public participation Geographic Information Systems across borders. *The Canadian Geographer, 47*(1), 50-61.

Zook, M., Dodge, M., Aoyama, Y., & Townsend, A. (2004). New digital geographies: Information, communication and place. In S. D. Brunn, S. L. Cutter, & J. W. Harrington (Eds.), *Geography and technology* (pp. 155-176). Kluwer Academic Publishers.

KEY TERMS

Geocollaboration: The collaborative aspects of designing and using spatial information resources.

Geographic Information Systems (GIS): The computer hardware and software that supports the management and analysis of spatial information.

Geovisualization: The visual communication aspects of displaying spatial information.

GIScience: The science associated with developing and advancing GIS and technologies.

Participatory GIS: A GIS that has been designed to involve multiple stakeholders and seeks to increase engagement in advocacy, planning and management processes among participating groups.

Spatial Data: Information that identifies the location of features in alignment with georeferencing systems, along with attributes qualifying that information.

Web-GIS: A GIS that is developed using Web resources, disseminated through the Web, or involves Web-based communication.

Levels of Adoption in Organizational Implementation of E-Collaboration Technologies

Bjørn Erik Munkvold
University of Agder, Norway

INTRODUCTION

While the e-collaboration term only dates back a few years, its roots can be traced back at least two decades to the research and development in areas such as groupware, computer supported cooperative work (CSCW), group support systems (GSS), and computer-mediated communication (CMC). As defined by Kock (2005), the e-collaboration term can be seen to encompass a wide range of technologies supporting collaboration among indivduals engaged in a common task. In this article, the e-collaboration term thus incorporates previous research and practice within the areas mentioned above. The term organizational implementation is used to denote the process of introducing the technology in an organizational setting (Walsham, 1993).

Ever since the first organizational applications of e-collaboration technologies, such as videoconferencing and group decision support systems, there has been a focus on the process related to how organizations and user communities adopt these technologies. Early research pointed to how adoption of e-collaboration technologies may be more challenging than other types of IT, as the effects and benefits from its use are dependent on the common adoption and use among all members of a group or user community (Grudin, 1989). Over the years, a rich base of empirical studies has developed, illustrating the complexity often involved in the process of organizational implementation of e-collaboration (see Munkvold, 2003, for a review of this research). Examples of issues influencing this process include the potential disparity in work and benefit among different adopters (Grudin, 1989), the users' mental models of the technology (Orlikowski, 1992), the need for a supportive technological and behavioral infrastructure (Palen & Grudin, 2003), and user training that also emphasizes the collaborative nature of the technology (Orlikowski, 1992).

A problem with accumulating and comparing the findings from the research on organizational implementation of e-collaboration technologies is that these studies may include adoption at various levels: individual, group, organizational, and even interorganizational. As such, e-collaboration practices may cover the whole span from two persons collaborating on a joint document, collaboration in teams and projects, enterprise-wide collaboration, and interorganizational collaboration as in virtual supply chains. Illustrating this problem, an analysis of 36 studies of e-collaboration published in seven information systems (IS) journals during the period 1999-2003 found that over two-thirds of the studies contained one or more problems of levels incongruence related to the level of the theory, the level of the data analysis, and the unit of analysis (Gallivan & Benbunan-Fich, 2005). Adding to this complexity is also the wide range of e-collaboration technologies and applications possibly incorporated within the e-collaboration term (Munkvold, 2003), and the potentially inherent flexibility in use of these. Finally, the multi-disciplinary nature of the e-collaboration area also implies challenges in developing a common terminology for describing phenoma related to e-collaboration adoption and use.

This article defines and discusses key concepts related to implementation of e-collaboration technologies in organizations, with main focus on the different levels of adoption that can be identified in this process. The aim is thus to contribute to a shared vocabulary and understanding of different adoption levels in organizational implementation of e-collaboration.

BACKGROUND

The term implementation is used differently in different research communities. In areas such as computer science, human-computer interaction (HCI), and software engineering, this term basically refers to the actual coding of the system, while in IS research and practice the term denotes the process of introducing the technology in an organizational setting. Grudin (1993) discusses how differences in terminology may constitute a barrier to effective communication between these communities.

In the research on information technology (IT) implementation, one of the most influential perspectives has been the Diffusion of Innovations (DOI) theory (Fichman, 2000; Rogers, 1995). Diffusion is here defined as "the process by which an innovation is communicated through certain channels over time among the members of a social system" (Rogers, 1995, p. 5). In the context of e-collaboration technologies, this refers to the process by which the adoption and use of the technology spreads throughout an organization, both as a result of planned distribution as well as emerging social mechanisms such as peer pressure. According to DOI theory, different attributes of an innovation may affect the rate of adoption. Examples of such attributes are relative advantage (the degree to which an innovation is perceived as better than the idea it supersedes), and compatibility (the degree to which an innovation is perceived as being consistent with the existing values, past experiences, and needs of potential adopters) (Rogers, 1995, p. 15). Another important term related to adoption and diffusion of e-collaboration technologies is critical mass, denoting the number of users that have to adopt a technology before the adoption of the technology becomes self-sustaining (Markus, 1987). Before a critical mass of users is reached the benefit from the system for the individual user will be limited, thus implying a risk that the early adopters may discontinue its use.

Based on the DOI perspective, Cooper and Zmud (1990) introduced a model for the IT implementation process covering all stages from project initiation and acquisition of a new technology (through purchase or in-house development) to the final stage where the technology is "internalized" in the daily work practices and full benefits from the technology may be realized. In this model, the term adoption is used at the organizational level, referring to "the decision to make full use of an innovation" (Rogers, 1995, p. 21). The adoption by individual users, referred to as acceptance in the model, takes place after adaptation of the technology in the organization, with adaptation including the following

activities: acquiring and/or developing the technology, installing it, training the users, and developing routines for use. According to this perspective, organizational adoption does not necessarily imply adoption by the individual users. In other words, an organization may decide to invest in a technology and make it available for use throughout the organization, while the individual users for various reasons may still decide not to adopt the technology. Research on implementation of e-collaboration technologies in organizations includes several examples of adoption failing at the individual level (Munkvold, 2003).

While IS research refers to adoption as taking place both at the organizational and individual level, the more "user-centric" communities such as HCI only use adoption in the meaning of an individual's decision to use a technology (Palen & Grudin, 2003). The organizational decision to implement a technology is instead referred to as deployment of the technology, in the sense of making it available for use (op. cit. Palen & Grudin, 2003). While this can be seen as another example of the potential for misunderstanding among these different communities (Grudin, 1993), it also illustrates how both the IS and the HCI communities acknowledge that IT implementation involves adoption decisions at both the organizational and the individual level.

In addition to issues related to adoption at different organizational levels, the scope of application also affects the implementation process. Earlier classifications of the DOI-based IT implementation literature have used concepts such as locus of adoption (individual/organizational) (Fichman, 1992) or locus of technology impact (internal to the IS unit, intraorganizational, and interorganizational) (Prescott & Conger, 1995). In characterizing different contexts of adoption and diffusion, Leonard-Barton (1990) introduced the concepts of technology span (few users vs. many users) and technology scope (diverse applications vs. homogeneous applications).

ADOPTION LEVELS AND SCOPE IN E-COLLABORATION

The former sections introduced key concepts related to adoption levels and scope in organizational IT implementation. This section briefly summarizes research on e-collaboration implementation related to these concepts, including examples of empirical studies.

Much of the research on individual adoption of e-collaboration technologies has applied a HCI-perspective, focusing on user interface issues for collaboration tools (Ellis, Gibbs, & Rein, 1991). In addition, some studies have applied the Technology Acceptance Model (TAM) (Davis, 1989) as the basis for studying the relationship between perceived ease of use, perceived usefulness, and adoption of different types of e-collaboration technologies (e.g., Dasgupta, Granger, & McGarry, 2002). However, for e-collaboration technologies to become useful requires adoption by a group or community of users. This poses additional research challenges for evaluating group dynamics and interaction related to technology adoption and use (Grudin, 1989). Thus, most of the research to date on adoption and use of e-collaboration technologies in groups has been conducted as student experiments in controlled settings. However, there is also an increasing number of field studies reporting on e-collaboration adoption in organizational groups (Fjermestad & Hiltz, 2001). Several of these studies apply an emergent perspective, explaining adoption and use of technologies in groups as a social process framed by structural characteristics of the group, task, and technology. One of the most influential of these perspectives is adaptive structuration theory (DeSanctis & Poole, 1994).

The research on adoption of e-collaboration technologies at the organizational level can be characterized as rather heterogeneous, with few unifying theoretical frameworks. Theoretical perspectives applied include DOI, socio-technical systems theory, socio-cognitive perspectives, and structuration theory (Zigurs & Munkvold, 2006). For example, building on DOI and based on fieldwork in ten companies, Applegate (1991) presented a framework viewing implementation of e-collaboration technology as alignment of group, task, and technology within a given organizational and environmental context.

According to the process perspective of IT implementation introduced in the previous section, e-collaboration implementation involves transition through adoption at different levels. A multiple case study of implementation of various e-collaboration technologies for supporting collaboration among distributed organizational units illustrated how the implementation process involved adoption at several organizational levels (Munkvold, 1999, 2002). While the initial decision to acquire the technology was made at the corporate level, the decision to adopt the technology was decentralized to each organizational unit. Adoption was here dependent on the ability to identify perceived relative advantage of the new technology compared to the existing work practices. The further deployment of the technology was then contingent on a positive decision to adopt the technology in each organizational unit. After establishing the e-collaboration infrastructure, the next step in the process involved gaining individual adoption of the technology among the target users in each organizational unit. This again required a robust technological infrastructure, and being able to relate the functionality offered in the technology to perceived needs in the employees' daily work practices. As demonstrated in this study and several others (e.g., Ciborra, 1996), this stage in the process is critical in that lack of perceived benefits among the target users may often result in non-use. Further, even when routine use of the technologies is established, developing effective collaborative work practices can be a lengthy process. In general, several studies show how the implementation of e-collaboration in organizations tends to be a long process, spanning several years from the initial decision to acquire/develop a new system to reaching a stage of advanced, full-scale use (Karsten & Jones, 1998; Orlikowski, 1996). These findings in particular relate to the implementation of enterprise-wide systems offering a flexible set of e-collaboration tools (Ciborra, 1996).

Research on implementation of e-collaboration has also shown that for flexible technologies, including a set of tools that can be applied at the users' discretion, the adoption of these tools may differ. For example, field studies of the implementation of Lotus Notes, one of the most widely researched e-collaboration technologies, show how the level of adoption in terms of sophistication and scope of use of the different tools in this system varies greatly between different organizations, with most of the organizations only making limited use of the technology (Karsten, 1999). In a similar vein, King (1996) distinguished between the following levels of groupware deployment: limited focused application, enterprise-wide application, and enterprise-redefining application.

The use of pilots is a common strategy for managing a stage-wise implementation through expanding adoption levels and scope of application. Opper and Fersko-Weiss (1992) distinguished between two types of pilots for e-collaboration technologies. The first type is the experimental pilot, where a small group of experienced IT users are experimenting with the technology to identify how the technology can support improved work practices. Typically, the importance of this pilot project is not critical to the business, to allow for innovative and flexible use. The experiences from this pilot then form the basis for an expanded pilot, where a larger number of representative organizational members use the technology for a more complex business task of higher importance and visibility in the organization. The results from the expanded pilot then provide the basis for the adoption decision regarding further deployment and use of the technology in the various parts of the organization.

An example of a planned pilot approach involving both experimental and expanded pilots is the implementation of data conferencing technologies in the Boeing company (Poltrock & Mark, 2003). Despite successful pilots demonstrating potential benefits for the company from full-scale adoption and use of the technology, the implementation team encountered unexpected barriers in their efforts of transferring this system to become part of the IS architecture in Boeing. These barriers were both technical (e.g., network load and security issues) and organizational (e.g., developing a business case, and establishing ownership for a completely new technology in the central IS organization), showing the complexity involved in large scale e-collaboration implementations.

There are also several examples of bottom-up or "grassroots" adoption of e-collaboration technologies. For example, studies in Sun and Microsoft showed how widespread user adoption of electronic calendars was driven mainly by peer pressure, and without managerial mandate (Palen & Grudin, 2003). This again can be seen as examples of the social influence model of technology use (Fulk, Schmitz, & Steinfield,1990). The explanatory power of this perspective is illustrated by a study of the adoption and use of two equivalent video telephone systems in one organization, where social influence mechanisms combined with critical mass theory were able to explain why only one of the systems survived (Kraut, Rice, Cool, & Fish, 1994).

FUTURE TRENDS

With the trend being increased integration of e-collaboration technologies, further research should focus on organizational practice related to implementing a portfolio of e-collaboration technologies (Munkvold, 2003; Munkvold & Zigurs, 2005). This includes sequencing and coordinating a staged implementation of interrelated e-collaboration services in a manner that establishes ownership and effective work practices. Further, despite the increasing focus on interorganizational applications of e-collaboration technologies, there is still little empirical research on implementation in this type of context. This may also include a focus on challenges related to adoption of e-collaboration technology in different organizational cultures, and possibly also national cultures in the case of international projects (Munkvold, 2005). In general, there is a need for more research on the relationship between adoption of e-collaboration technologies at the different organizational levels discussed in this article.

CONCLUSION

This article has provided a brief discussion on how organizational implementation of e-collaboration technologies involves adoption at multiple levels (individual, group, organizational), and an overview of key terms used for analyzing this process. The article intends to support the accumulation of findings from research on e-collaboration implementation, as well as to provide a conceptual basis for further empirical research.

REFERENCES

Applegate, L. M. (1991). Technology support for cooperative work: A framework for studying introduction and assimilation in organizations. *Journal of Organizational Computing, 1,* 11-39.

Ciborra, C. (1996). *Groupware & teamwork: Invisible aid or technical hindrance?* Chichester, UK: John Wiley & Sons.

Cooper, R. B., & Zmud, R. B. (1990). Information technology implementation research: A technological diffusion approach. *Management Science, 36*(2), 123-139.

Dasgupta, S., Granger, M. & McGarry, N. (2002). User acceptance of e-collaboration technology: An extension of the technology acceptance model. *Group Decision and Negotiation, 11,* 87-100.

Davis, F. D. (1989). Perceived usefulness, perceived ease of use and user acceptance of information technology. *MIS Quarterly, 13*(3), 319-339.

DeSanctis, G. & Poole, M. S. (1994). Capturing the complexity in advanced technology use: Adaptive structuration theory. *Organization Science, 5*(2), 121-147.

Ellis, C. A., Gibbs, S. J., & Rein, G. L. (1991). Groupware: Some issues and experiences. *Communications of the ACM, 34*(1), 39-58.

Fichman, R. G. (1992). Information technology diffusion: A review of empirical research. In J. DeGross et al. (Eds.), *Proceedings of the Thirteenth International Conference on Information Systems* (pp. 195-206). Dallas, Texas.

Fichman, R. G. (2000). The diffusion and assimilation of information technology innovations. In R. W. Zmud (Ed.), *Framing the domains of IT management: Projecting the future...through the past* (pp. 105-127). Cincinnati, OH: Pinnaflex Educational Resources.

Fjermestad, J., & Hiltz, S. R. (2001). Group support systems: A descriptive evaluation of case and field studies. *Journal of Management Information Systems, 17*(3), 115-160.

Fulk, J., Schmitz, J., & Steinfield, C. W. (1990). A social influence model of technology use. In J. Fulk & C.W. Steinfield (Eds.), *Organizations and communication technology* (pp. 117-140). Newbury Park, CA: Sage Publications.

Gallivan, M. J., & Benbunan-Fich, R. (2005). A framework for analyzing levels of analysis issues in studies of e-collaboration. *IEEE Transcations on Professional Communication, Special Issue on Expanding the Boundaries of E-Collaboration, 48*(1), 87-104.

Grudin, J. (1989). Why groupware applications fail: Problems in design and evaluation. *Office: Technology and People, 4*(3), 245-264.

Grudin, J. (1993). Two communities, two languages. *Communications of the ACM, 36*(4), 113.

Karsten, H. & Jones, M. (1998). The long and winding road: Collaborative IT and organisational change. *Proceedings of CSCW '98* (pp. 29-39). New York, NY: ACM Press.

Karsten, H. (1999). Collaboration and collaborative information technologies: A review of the evidence. *The DATA BASE for Advances in Information Systems, 30*(2), 44-65.

King, W. R. (1996). Strategic issues in groupware. *Information Systems Management, 13*(2), 73-75.

Kock, N. (2005). What is e-collaboration? *International Journal of e-Collaboration, 1*(1), 1-6.

Kraut, R. E., Rice, R. E., Cool, C., & Fish, R. S. (1994). Life and death of new technology: Task, utility, and social influences on the use of a communication medium. In *Proceedings of the ACM 1994 Conference on Computer Supported Cooperative Work (CSCW '94)* (pp. 13-21). New York, NY: ACM Press.

Leonard-Barton, D. (1990). The intraorganizational environment: Point-to-point versus diffusion. In F. Williams & D. V. Gibson (Eds.), *Technology transfer. A communication perspective* (pp. 43-61). Newbury Park, CA: Sage Publications.

Markus, M. L. (1987). Toward a "critical mass" theory of interactive media: Universal access, interdependence and diffusion. *Communication Research, 14*(5), 491-511.

Munkvold, B.E. (1999). Challenges of IT implementation for supporting collaboration in distributed organizations. *European Journal of Information Systems, 8*, 260-272.

Munkvold, B. E. (2002). Alignment of collaboration technology adoption and organizational change: Findings from five case studies. In M. Khosrowpour (Ed.), *Collaborative information technologies* (pp. 141-153). Hershey, PA: IRM Press.

Munkvold, B. E. (2003). *Implementing collaboration technologies in industry: case examples and lessons learned*. London: Springer-Verlag.

Munkvold, B. E. (2005). Experiences from global e-collaboration: Contextual influences on technology adoption and use. *IEEE Transactions on Professional Communication, Special Issue on Expanding the Boundaries of E-collaboration, 48*(1), 78-86.

Munkvold, B. E., & Zigurs, I. (2005). Integration of e-collaboration technologies: Research opportunities and challenges. *International Journal of e-Collaboration, 1*(2), 1-24.

Opper, S., & Fersko-Weiss, H. (1992). *Technology for teams: Enhancing producivity in networked organizations*. New York: Van Nostrand Reinhold.

Orlikowski, W. J. (1992). Learning from Notes: Organizational issues in groupware implementation. In *Proceedings of CSCW'92* (pp. 362-369). New York, NY: ACM Press.

Orlikowski, W. J. (1996). Improvising organizational transformation over time: A situated change perspective. *Information Systems Research, 7*(1), 63-92.

Palen, L., & Grudin, J. (2003). Discretionary adoption of group support software: Lessons from calendar applications. In B. E. Munkvold (Ed.), *Implementing collaboration technologies in industry: Case examples and lessons learned* (pp. 159-180). London: Springer-Verlag.

Poltrock, S. E., & Mark, G. (2003). Implementation of data conferencing in the Boeing Company. In B. E. Munkvold (Ed.), *Implementing collaboration technologies in industry: Case examples and lessons learned* (pp. 129-158). London: Springer-Verlag.

Prescott, M. B., & Conger, S. (1995). Information technology innovations: A classification by IT locus of impact and research approach. *The Data Base for Advances in Information Systems, 26*(2-3), 20-41.

Rogers, E. M. (1995). *Diffusion of innovations* (4th ed.). New York: The Free Press.

Walsham, G. (1993). *Interpreting information systems in organizations*. Chichester, UK: John Wiley & Sons.

Zigurs, I., & Munkvold. B. E. (2006). Collaboration technologies, tasks, and contexts: Evolution and opportunity. In D. Galletta & P. Zhang (Eds.), *Human-computer interaction and management information systems: applications. Advances in management information systems, Vol. 6,* (pp. 143-169). Armonk, NY: M.E. Sharpe.

KEY TERMS

Adoption: the decision to make full use of an innovation (Rogers, 1995).

Critical Mass: the number of users having to adopt a technology to make the further adoption process self-sustaining.

Diffusion: the process by which an innovation is communicated through certain channels over time among the members of a social system (Rogers, 1995).

Organizational IT Implementation: an organizational effort directed towards diffusing appropriate information technology within a user community (Cooper & Zmud, 1990).

Relative Advantage: the degree to which an innovation is perceived as better than the idea it supersedes (Rogers, 1995).

Technology Acceptance Model (TAM): a model of individual acceptance of IT, stating that an individual's adoption of IT is dependent on the perceived ease of use and perceived usefulness of the technology.

Technology Deployment: acquiring and installing the technology to make it available for use.

A Macro-Level Approach to Understanding Use of E-Collaboration Technologies

Sanjiv D. Vaidya
Indian Institute of Management Calcutta, India

Priya Seetharaman
Indian Institute of Management Calcutta, India

INTRODUCTION

The term *e-collaboration technologies* (ECT) in an organization refers to the collective system of interactive computer-based tools that facilitate a variety of group tasks. It thus includes among others, electronic mailing systems, bulletin boards, intranets and extranets, messaging systems, group support systems, decision rooms, computer conferencing tools, and computer based video-conferencing systems, etc. ECTs have often been referred to in the literature, using various terms to highlight specific uses such technologies have been put to. These include *group decision support systems, group support systems, computer supported collaborative work, groupware,* and *collaborative technologies.*

ECTs are among the many IT applications that have seen a rapid deployment in organizations due to greater use of task-teams and groups. There is thus an increased use of inter-departmental and cross-functional teams (Sarker, Valacich, & Sarker, 2005). Groups are hence viewed as a "basic unit of the formal organization structure" (Applegate, 1991). Second, the coming of the PCs, the advent of easy-to-use software and the developments in network technologies constitute additional impetus for the use of computers to support collaborative work in organizations.

Use of technology support for collaborative work is believed to increase productivity in organizations. It is hence important to examine the use of such technologies in greater depth (Markus, 1994). Increasing access to communication technologies without adequately understanding the task requirements and the potential change in the work environment and processes may lead to information overload and may not benefit the user. This study therefore, aims at enhancing our understanding of the broad factors influencing the use of ECT in organizations.

The article is organized in the following manner. The first section reviews concerned literature relevant to the use of ECT in organizations. The second section describes a framework depicting the drivers of ECT use in organizations. The framework represents a macro level perspective of the phenomenon. The subsequent section highlights the implications of such a framework and the potential for further research.

BACKGROUND

Researchers have argued that adoption and use of communication technologies arise from changes in the organization itself. Three perspectives have often been used to highlight this. The first is the technological perspective which views technology as an enabler of organizational forms; the second is the organizational perspective which views technology as being designed to fit organizational structures and forms while the third perspective is an "emergent perspective" which views technologies as "occasions" for structuring (Dutton, 1999). In all these three perspectives, adoption and use of technology is subsumed.

The organizational perspective involves development of integrative frameworks which encompass various factors which help design the "fit" between the organization and the technology. Such integrating frameworks in research entail knitting together varied sets of variables which represent or manifest the same underlying or superjacent construct (Gladstein, 1984). This is mainly due to a diversity of perspectives and standpoints assumed by different researchers.

Current theories which have primarily focused on groups and teams in laboratory settings may not adequately explain behavior seen in complex, interdependent task and technology settings in organizations. Some authors have pointed out this inadequacy

Figure 1. Background

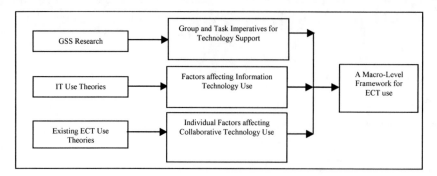

of current theory in the area (Gallivan, 2001; Van den Hooff, Groot, de Jonge, 2005). Among the research issues suggested by many GDSS researchers and authors (e.g., Dennis, Nunamaker, & Vogel, 1990; DeSanctis & Gallupe, 1985; Gray & Mandviwalla, 1999), "an integrated framework for...understanding field use of GSS" has been highlighted as an essential direction of research.

IT Use Theories

In order to be able to understand use of e-collaboration technologies, one can essentially draw upon theories of information technology adoption and use in organizations. Early IT use research, till around the 1980s has dominantly focused on motivation and perceptions regarding the technology and its potential as a dominant factor in influencing why and how people use IT (see, e.g., Robey, 1979; Trice & Treacy 1986). With the coming of a simple yet presumably powerful technology acceptance model (TAM) in the mid 1980s (Davis, 1989), the shift to technology characteristics and attitudinal effects of perceptions gained predominance (see, e.g., Adams, Nelson, & Todd, 1992; Igbaria, Guimaraes, & Davis, 1995; Viswanath & Davis, 2000). The onset of computer networks in early 1990s and their active use in organizations forced IS researchers to give more importance to the overall organizational environment, the task related factors and the IT management issues. The 1990s and beyond also saw the use of multilevel constructs-based theories attempting to explain technology use. For example, Venkatesh, Morris, Davis, and Davis (2003) presented a unified view of user acceptance of IT by developing a theory using performance expectancy, effort expectancy,

social influence and facilitating conditions as predictors of behavioral intention and actual use. They also included gender, age, experience and voluntariness to use as moderators. A similar progress line can also be drawn for research in groupware and ECT use, though the same level of maturity is yet to be attained.

Determinants of ECT Use

Drawing upon the IT use theories, IS researchers have examined various factors which influence use of e-collaboration technologies including task characteristics (Pinsonneault & Kraemer, 1990; Maznevski & Chudoba, 2000), accessibility (Rice & Shook, 1988; Lou & Scamell, 1996), top management support, awareness of the potential and other technology characteristics such as complexity, group supportability (Sarker et al., 2005), to name a few.

Van den Hooff et al. (2005) found, after a meta-analysis and a test of various theories on use of communication technologies, that use of electronic mail is determined by a combination of medium characteristics, individual user characteristics, users' social environment, and task characteristics.

The dominant measures of medium characteristics have often been the richness of the medium and its appropriateness to the task to be performed. Some researchers have argued that even low rich media such as e-mail may be used quite frequently for managerial tasks (Markus, 1994). Some others have argued that richer media may not necessarily lead to better performance in more equivocal tasks (Dennis & Kinney, 1998; Suh, 1999).

Another factor which has been given considerable importance in the literature on e-collaboration tech-

Figure 2. Factors influencing ECT use: Summary of existing literature

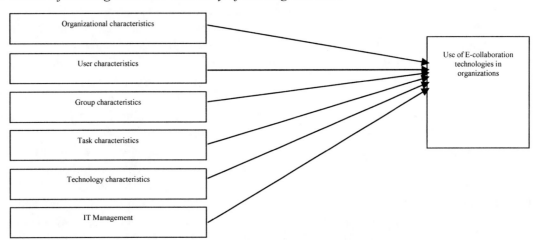

nologies concerns the characteristics of individuals. Viewing adoption of computer supported cooperative work, from an end-user perspective, Turner and Turner (2002), for instance, discussed the importance of individuals' characteristics in influencing the uptake of collaborative technology.

Authors have also examined learning in the context of collaborative technology use. DeSanctis and Poole (1994) attempted to understand the adaptive structuration in organizations due to use of advanced information technologies. They argued that differences in people's adaptation to computer use and the consequent effects on decision making and other outcomes is a result of the interplay between the two types of structures, viz. social structures embedded in the technology and those set in the action. Carlson and Zmud (1999) highlighted the need for managers to understand that use of communication technologies is strongly influenced by channel experiences and that users' must go through a learning process.

In summary, it can be said that use of ECT is a result of the combined effect of various characteristics of task, group, organization, technology and the management environment. The following section aims to present a macro-level perspective encompassing these factors. While the framework presented below does not delineate and define many lower level constructs and variables in a specific manner, it satisfies the aim of presenting a macro-level framework to render possible a structured analysis of e-collaboration technology environments.

THE MACRO LEVEL PERSPECTIVE

An action performed or to be performed by an individual or a group of people (organized or otherwise) essentially requires an impetus often referred to as need or pressure and a tendency to behave in a particular manner often referred to as inclination or proclivity. For this demand and desire to take effect, the environment in which the action is to be performed must provide the scope or the potential. This set of triadic constructs of potential, pressure and proclivity form the basis of the macro level framework of use of e-collaboration technologies in organizations that we have developed and is presented and explained in the following sections.

Current research in this direction lacks a macro-level framework but mainly presents individual lower level factors which influence various aspects of ECT use. In an attempt to fill this gap, we present here a framework which explains the phenomenon from a wider perspective. While this framework is based mainly on existing literature and our understanding of the phenomenon, results from some practical field studies have also reiterated the constructs of the framework (see, e.g., Vaidya & Seetharaman, 2005).

In the context of use of ECT, the organizational environment and its requirements for the use of groups and the need for communication therefore, offers the scope for use of collaborative technologies. The pressures to use such e-collaboration technologies arise from characteristics of the task, group and the organization which render use of technology support for group task execution a necessity. In the same vein, the inherent inclination and drive towards use of IT in general and collaborative technology in particular, defines the proclivity towards use of e-collaboration technology. We will now examine the individual factors which constitute each of these constructs.

POTENTIAL FOR ECT USE

Certain organizational situations may naturally require members of the organization to collaborate and work jointly on particular tasks. These situations generate certain drivers which allow individuals in organizations to collaborate. Among the many reasons for use of groups in organizations Lee, Newman and Price (1999) suggested, were increasing complexity in decision making situations, need for coordination across divisions, diverse knowledge or information requirements to increase task performance, increasing necessity for communication and coordination, and growing belief in involving concerned people in the decision-making process. These specific reasons for organizations to use groups result in the need to collaborate.

Factors contributing to the felt need for collaboration in organizations can thus be determined from these. It is possible to analyze a given organizational situation and identify the degree to which each of these factors contributes to the scope for collaboration. This felt need for collaboration in organizational tasks, translates into potential for use of ECT. In other words, where there is a scope for collaboration to execute the task, there is a potential for using ECT support.

Now, we examine some of the sources of pressures to use e-collaboration technology.

PRESSURES TO USE ECT

Groups are likely to collaborate using technology under two conditions—when there is sufficient compulsion to utilize it for their task and/or when there is a tremendous drive or inclination to use technology support. This is especially true when alternative media are available to complete the task. Under such circumstances, task, group and organization related pressures to use collaborative technology play an important role in determining the extent to which groups use collaborative technology. These include aspects such as geographical and temporal dispersion of the group and the response time requirements imposed by the task.

A task team may have members who are geographically spread across multiple countries (see, e.g., Malhotra, Majchrzak, Carman, & Lott, 2001). Internationalization of organizations mainly due to economies of scale; availability of resources such as urban infrastructure, natural or intellectual resources and low costs of resources such as labor have led to increasing geographically dispersed organizational task groups.

Temporal separation refers to differences in the time of work of different members of the group. Two kinds of temporal dispersion may exist: (1) temporal dispersion due to locational difference and (2) temporal separation due to differences in working hours. While the former is mainly due to geographic dispersion, the latter is fairly common in organizations attempting to maximize capacity utilization or those working 24x7 such as hospitals.

Certain task environments necessitate faster responses to external stimuli than others. For instance, in stable environments, organizations are well aware of market situations and can predict factors of influence with a certain level of ease. But in dynamic organizational environments, the resultant uncertainty is likely to lead to a need for faster responses and therefore greater coordination requirements. Use of e-collaboration technologies may be seen as a means of meeting the need for such faster responses and increased coordination. Thus such situations create pressures for the group to collaborate using the ECT.

Finally, pressures experienced by the group for adopting and using ECT translate into actual use beyond the basic essential level of use, only if individuals in the group are positively inclined towards use of technology support.

PROCLIVITY TOWARDS ECT USE

IT drive is the orientation of users—individually and collectively—towards use of IT in the organization. Drawing upon this definition of *IT Drive,* the group inclination to use e-collaboration technologies can be described as "the tendency of the members of the group individually and collectively, to favor the use of e-collaboration technologies over alternative means, to accomplish the specified group task."

The drive to use e-collaboration technologies would encompass two broad factors, viz., group orientation

Figure 3. A macro-level framework for ECT use

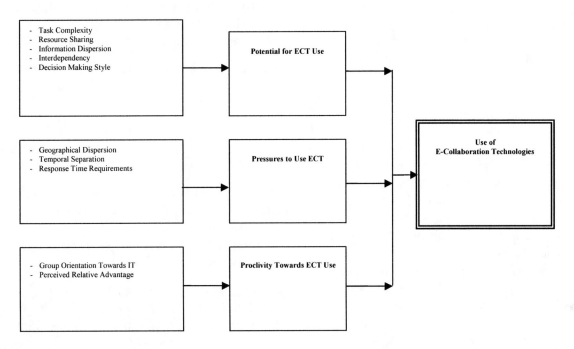

towards information technology and perceived relative advantage of using e-collaboration technologies. The group orientation reflects the natural tendency of the group towards application and use of information technology for various organizational activities. Positive orientation of groups towards IT can be a result of specific factors such as top management orientation towards IT or long term factors such as positive orientation of the organization in general towards IT especially due to effective IT management. Negative orientation of the group towards IT, similarly, can be a result of importance given by the top management to IT, weak orientation of the group leadership towards IT or improper management of IT in the organization. Demographic factors of the members such as age, educational qualification, and so forth also impact the orientation of the group towards IT.

Perceived relative advantage refers to the degree to which an innovation is superior to ideas it supersedes or replaces (Rogers, 1962). Alternative means of communication amongst the members of the group include face-to-face communication, written memos, telephone, voice-based media, etc. Groups often exhibit a tendency to choose a communication medium depending on the different interpersonal, informational or decision tasks they perform. It is also possible that the social and cultural context in which the group operates defines the choice and use of a particular medium. In essence, the relative advantage of using collaborative technology over alternative means of task execution, as perceived by the group members, influences the use of such technology in organizations.

A diagrammatic representation of the framework thus far presented appears in Figure 3.

CONCLUSION

Burgeoning network infrastructure and escalating environmental pressures are compelling organizations to depend more on technology support to achieve organizational tasks. Yet, organizations are far from wringing the maximum value from their IT investments. One such investment is network infrastructure which renders possible e-collaboration to support group tasks in organizations. This framework aims at enhancing our understanding of the reasons for the inability of individuals and groups to appropriately use such e-collaboration technologies.

IMPLICATIONS

The framework proposed in this article has implications for researchers and managers alike. Managers can use the framework to decipher the reasons for ineffective use of ECT in their organizations, especially in specific task-groups. It is indeed possible to analyze a group's use of ECT in a structured manner using the framework broadly described above and uncover the impediments to and inhibitors of adequate and appropriate use of ECT. One feasible approach could be to identify the task requirements in terms of the technology support necessary, evaluate the available technology infrastructure and then analyze the gap between the actual use of e-collaboration technologies available and the desirable level of use.

On the other hand, IS researchers have the arduous task of examining more closely the micro-level factors which constitute the broad macro-level constructs described here. Such studies would enable us to tackle individual factors in a more focused manner leading to an effective use of ECT in organizations.

REFERENCES

Adams, D. A., Nelson, R. R., & Todd, P. A. (1992). Perceived usefulness, ease of use, and usage of information technology: A replication. *MIS Quarterly, 16*(2), 227-248.

Applegate, L. M. (1991). Technology Support for cooperative work: A framework for studying introduction and assimilation in organizations. *Journal of Organizational Computing, 1*(1), 11-39.

Carlson, J. R., & Zmud, R. W. (1999). Channel expansion theory and the experiential nature of media richness perceptions. *Academy of Management Journal, 42*(2), 153-170.

Davis, F. D. (1989). Perceived usefulness, perceived ease of use and user acceptance of information technology. *MIS Quarterly, 13*(3), 319-340.

Dennis, A. R., & Kinney, S. T. (1998). Testing media richness theory in the new media: The effects of cues, feedback, and task equivocality. *Information Systems Research, 9*(3), 256-274.

Dennis, A. R., Nunamaker, J. F., Jr., & Vogel, D. R. (1990). A comparison of laboratory and field research in the study of electronic meeting systems. *Journal of Management Information Systems, 7*(3), 107-135.

DeSanctis, G., & Gallupe, R. B. (1985). Group decision support systems: A new frontier. *Database for Advances in Information Systems, 16*(2), 2-10.

DeSanctis, G., & Poole, S. (1994). Capturing the complexity in advanced technology use: Adaptive structuration theory. *Organizational Science, 5*(2), 121-147.

Dutton, W. H. (1999). The virtual organization: Tele-access in business and industry. In G. DeSanctis & J. Fulk (Eds.), *Shaping organizational form: Communication, connection, and community* (pp. 473-495). Thousand Oaks, CA: Sage.

Gallivan, M. J. (2001). Organizational adoption and assimilation of complex technological innovations: Development and application of a new framework. *Database for Advances in Information Systems, 32*(3), 51-85.

Gladstein, D. (1984). Groups in context: A model of task group effectiveness. *Administrative Science Quarterly, 29*, 499-517.

Gray, P., & Mandviwalla, M. (1999). New directions for GDSS. *Group Decision and Negotiation, 8*, 77-83.

Igbaria, M., Guimaraes, T., & Davis, G. (1995). Testing the determinant of microcomputer usage via a structural equation model. *Journal of Management Information Systems, 11*(4), 87-114.

Lee, D., Newman, P., & Price, R. (1999). *Decision making in organizations*. London: Financial Times.

Lou, H., & Scamell, R. W. (1996). Acceptance of groupware: The relationships among use, satisfaction, and outcomes. *Journal of Organizational Computing and Electronic Commerce, 6*(2), 173–190.

Malhotra, A., Majchrzak, A., Carman, R., & Lott, V. (2001). Radical innovation without collocation: A case study at Boeing-Rocketdyne. *MIS Quarterly, 25*(2), 229-249.

Markus, L. M. (1994). Electronic mail as the medium of managerial choice. *Organization Science, 5*(4), 502-527.

Maznevski, M. L., & Chudoba, K. M. (2000). Bridging space over time: Global virtual team dynamics and effectiveness. *Organization Science, 11*(5), 473-492.

Pinsonneault, A., & Kraemer, K. L. (1990). The effects of electronic meetings on group processes and outcomes: An assessment of the empirical research. *European Journal of Operations Research, 46*(0), 143-161.

Rice, R. E., & Shook, D. E. (1988). Access to, usage of, and of from an electronic messaging system. *ACM Transactions on Office Information Systems, 6*(3), 255-276.

Robey, D. (1979). User attitudes and MIS use. *Academy of Management Journal, 22*(3), 527-538.

Rogers, E. M. (1962). *Diffusion of innovations.* New York: The Free Press.

Sarker, S., Valacich, J. S., & Sarker, S. (2005). Technology adoption by groups: A valence perspective. *Journal of the Association for Information Systems, 6*(2).

Suh, K. S. (1999). Impact of communication medium on task performance and satisfaction: An examination of media-richness theory. *Information & Management, 35*(5), 295-312.

Trice, A. W., & Treacy, M. E. (1986). *Utilization as a dependent variable in MIS research.* Paper presented at the International Conference on Information Systems.

Turner, P., & Turner, S. (2002). End-user perspectives on the uptake of computer supported cooperative working. *Journal of End User Computing, 14*(2), 3-16.

Vaidya, S. D., & Seetharaman, P. (2005). *Collaborative technology use in organizations: A typology.* Paper presented at the 11th Americas Conference on Information Systems.

Van den Hooff, B., Groot, J., & de Jonge, S. (2005). Situational influences on the use of communication technologies: A meta-analysis and exploratory study. *Journal of Business Communication, 42*(1), 4-27.

Venkatesh, V., Morris, M. G., Davis, G. B., & Davis, F. D. (2003). User acceptance of information technology: Toward a unified view. *MIS Quarterly, 27*(3), 425-478.

Viswanath, V., & Davis, F. D. (2000). A theoretical extension of the technology acceptance model: Four longitudinal field studies. *Management Science, 46*(2), 186-204.

M

KEY TERMS

E-Collaboration Technologies (ECT): The collective system of interactive computer-based tools that facilitate a variety of group tasks.

Geographic Dispersion: Physical dispersion of group members across geographically distant locations, thus necessitating use of technology support for group task.

Perceived Relative Advantage: Refers to the degree to which an innovation is perceived as being superior to the ideas it supersedes or replaces.

Potential for ECT Use: Refers to the opportunity for the use of ECT arising out of task characteristics and organizational environment in which the group is embedded.

Pressures to Use ECT: Captures the felt need for technology support for group task experienced due to group and task environment factors.

Proclivity Towards ECT: Captures the natural tendency of the group members to use ECT arising mainly from inclination towards technology and the relative advantage arising out of its use.

Temporal Separation: Differences in the time of work of different members of the group arising due to geographic dispersion or due to differences in working hours.

Managing E-Collaboration Risks in Business Process Outsourcing

Anne C. Rouse
Deakin University, Australia

INTRODUCTION

A marked development in the last decade has been the growth of "virtual organizations" (or "extended enterprises"), where a network of service supplier and vendor firms cooperates to create customer value. One form of cooperation is described as business process outsourcing (BPO). A business process involves several interrelated activities performed with the goal of generating customer value. Because of the growth in e-collaboration tools, it is now possible for firms to outsource even core business processes to external vendors. Examples of processes typically outsourced include logistics, customer support, human resources, and back-office accounting functions. BPO and the value networks created by vendors and purchasers hold the promise of substantial business benefits associated with specialization and scale. These include reduced costs, greater business flexibility, and higher service quality. According to the Gartner Group, the world market for BPO services is likely to increase from $100 billion in 2002 to $173 billion by 2007 (Gartner, 2004).

E-collaboration is a core aspect of BPO, as vendor and purchaser are physically separated, and without this collaboration, the level of integration needed between vendor and client would be impossible. Maturing IT capabilities, and in particular e-collaboration tools, were important drivers of the large growth in outsourcing witnessed since 1989. Yet the e-collaboration that enables BPO also introduces new corporate risks, particularly those associated with sharing of data, and with the change from face-to-face interactions based on propinquity to computer-mediated interactions. Drawing on a series of focus groups, this paper summarizes the promises e-collaboration holds for BPO, but also highlights risks that need to be managed. These risks have increased with recent legislative demands like the US Sarbanes-Oxley and EU privacy legislation.

The findings reported here are based on ten focus groups and individual interviews with practitioners involved in outsourcing IT/IS or BPO services. These were conducted between 1999 and 2004. In all, 46 informants were interviewed in the focus groups, and a further five informants were interviewed individually. While most informants were from purchaser organizations, one focus group involved informants from outsourcing vendors. Services supplied within these outsourcing arrangements included back-end bank processing, scientific data collection, call centre operations, delivery of ongoing mainframe services, software development, help desk operations, and desktop support. Details are reported in Rouse (2002) and Rouse and Corbitt (2004).

BACKGROUND

There are three major classes of outsourcing: (1) outsourcing of IT/IS services, or "ITO" (where the services to be supplied involve the development and delivery of technology and information systems), (2) BPO, where relatively complex, IT-supported businesses services are involved, and (3) simple outsourcing (such as cleaning) where no IT support is involved. This paper is concerned with the first two forms, BPO and IT/IS outsourcing, which can be considered a particular form of BPO. Both involve complex business processes supported by IT, and the handing over of sensitive data resources to a third party. It is the complexity, business impact, and the integral role of IT that distinguishes these from other, simpler forms of outsourcing.

In practice, BPO involves the delivery of a service, rather than a physical product (manufacturing). Consequently, the service delivery (or production) process has several characteristics: these include intangibility, variability, and the fact that the output is perishable—if not delivered at the right time it has no value (Langford & Cosenza, 1998). Another important characteristic of services is labeled "inseparability"—in other words, the service is created by the coordinated (and so inseparable) activities of the deliverer and receiver.

Because outsourcing involves industrial services, a large number of vendor and purchaser employees can perform part of the delivery process, and for the process to work well their actions have to be articulated, communicated, and coordinated (Bitner, Faranda, Hubbert, & Zeithaml, 1997).

Complex outsourcing (like ITO and BPO) cannot exist without some form of e-collaboration to effect coordination and communication. Such outsourcing also requires fast data communications capabilities and mechanisms for easily moving data between client and vendor databases. These technologies overcome geographical distance, so BPO now often involves supply of services across national boundaries, allowing western firms to use lower cost labor from India, China, or other developing countries—described as offshore outsourcing, or "offshoring."

E-collaboration technologies are electronic technologies that enable collaboration among individuals engaged in a common task (Kock, Davison, Ocker, & Wazlawick, 2001). A range of these are used to coordinate the actions of participants in the outsourcing-based service production process. Examples include e-mail, tele-, video-, and data-conferencing, groupware, electronic meeting systems, Web-based chat and asynchronous conferencing tools, collaborative document preparation, document management technologies, and shared databases.

The nature of BPO is illustrated in Figure 1. Outsourced business processes involve transforming purchaser data, often using specialized software packages, and automated routines. With outsourcing, whenever e-collaboration tools are used this data is transmitted in digital form. "…Final products supplied to the client are available digitally via network connections, e.g. a processed payroll list, or a new inventory list…"(Gewald & Dibbern, 2005, p 2). The flows of e-collaboration data, particularly when distributed/accessed over the Internet, represent a point of vulnerability, as do the databases controlled by the vendor/subcontractor. Once data leaves the purchaser organization, strategies for protecting it (such as passwords, encryption, virtual private networks, etc.) become the responsibility of the vendor, and the value of these protections is only as good as the integrity of vendor staff and processes. The mechanisms for the purchaser to ensure protection of key data become problematic—the purchaser must rely on a contract with the vendor to guarantee security—a very different management approach.

BPO RISKS

BPO potentially results in a number of benefits to purchasers, including cost savings, flexibility, improved quality, and allowing the purchaser to concentrate on core business (Lacity & Hirschheim, 1995). However, outsourcing has also been described as "risky business" (Aubert, Patry, & Rivard, 2002) because empirical research to date has revealed that purchasers who outsource frequently fail to obtain the theoretical benefits (Rouse & Corbitt, 2003) and often encounter unexpected downsides. In practice, in deciding on whether or not to outsource decision makers must weigh up the potential benefits against the risks of negative outcomes.

Figure 1. E-collaboration between vendor and purchaser creates additional points of vulnerability

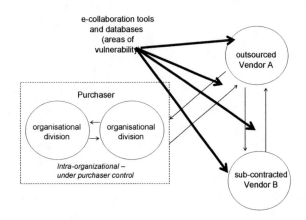

Not all BPO risks are related to e-collaboration tools. For example a major and commonly encountered risk is that the costs projected for the outsourced arrangement are poorly thought through, leading to a flawed business case (Lacity & Hirschheim, 1995). However, several risks are associated with the replacement of social, face to face coordination and control mechanisms with computer-mediated e-collaboration. Other risks are associated with sharing data across organizational boundaries.

The e-collaboration risks commonly raised by informants fall into two categories: those (1 and 2 below) associated with the handing over data to the vendor, and (3 and 4) those associated with replacing in-house, largely cultural control processes, with formal, contractual controls supported by computer-mediated coordination and communication.

Interception Risks

Because data needs to be moved between vendor and purchaser, it is subject to the typical risks of data interception and theft. These may occur through accessing data in transit, or through contamination of workstations with "malware" code that, for example, captures data being keyed. This is probably the least difficult risk to manage, as there are a range of technical solutions. These include encryption, virus protection, and software that tests for and removes malicious software. These risks are present for many in-house business processes too, when data has to be moved between sites, and are not inherently due to outsourcing.

The main issue for purchasers to be aware of is that outsourcing does not devolve responsibility for managing this risk. The purchaser remains responsible, in the eyes of customers, and increasingly, legislators, for ensuring that operations are properly performed and controlled, regardless of whether this is done in-house or by an external vendor. This responsibility needs to be actioned by specifying security requirements in the outsourcing contract, and by establishing appropriate consequences for breaches. Several focus group informants were caught out in their initial outsourcing experiences, because they had not specified detailed requirements related to security in their contract, expecting that (as in their own organization) security measures were automatically applied by the vendor. However, just having specific terms in the contract will not ensure that the vendor acts in the specified way—purchasers

need to verify the vendor's security measures, and to ensure that consequences are sufficiently negative to ensure the vendor is deterred by their possibility.

Privacy, Confidentiality, "Identify Theft," and "Loss of Intellectual Property" Risks

In a recent survey of German banks' outsourcing, Gewald & Dibbern (2005) found that respondents were less concerned with privacy risks than with financial or operational risks. Yet recent European, US, and Australasian legislation now heavily penalize organizations for failing to safeguard customer and corporate records. Issues of privacy and particularly "data theft" are important outsourcing risks because digital data exchanged with the vendor is relatively easily accessed (and copied) by the vendor's front-line operators. BPO has significantly higher risks than does the delegation of factory-based manufacturing, where the risks of losing key data and intellectual property (IP) are mitigated by the fact that shop-floor staff rarely has easy access to them. The recent security breach at CardSystems Solutions—when 40 million MasterCard, Visa, and American Express cardholders woke up to find that their account details had been stolen—is an example of how important this outsourcing risk is in terms of customer trust (Rouse & Watson, 2005).

Because outsourced business processes usually require that vendor staff access non-encrypted data in order to process it (Gewald & Gellrich, 2005), encryption alone will not address this risk. The vendor needs to institute a range of behavioral controls to avoid relatively easy, but potentially large-scale, data theft through use of technologies like flash drives and external hard disks. Many of these controls are social, and are dependent on a range of cultural values (backed up by the likelihood of institutional repercussions, such as threats of jail). One explanation for the low concern of the German bankers about privacy risks might be that decision makers knew that vendor staff would work within the same national culture as the purchaser companies, and that purchaser staff would know that, in Germany, security breaches would likely result in prosecution. However, where BPO involves cross-national outsourcing, differences in cultural values and legal infrastructure may not be appreciated. This issue is discussed further, below.

Managing the risks of data theft and loss of IP relies largely on a process of recognition, planning, and verification. Purchasers should consider the marketing and reputation risks associated with such losses as well as the increasing risks of penalties associated with privacy and governance-related legislation. Purchasers should work with the vendor to determine management strategies to avoid data theft. As for interception risks, the contract needs to explicitly outline requirements, and should include appropriate penalties. An important third step is for the purchaser to continue to verify the procedures and safeguards used by the vendor through spot checks.

Failing to Understand and Codify the Behaviors Required of the Various Collaborators

As articulated above, the quality of the service supplied to the purchaser depends on the coordinated actions of a range of staff working for both vendor and client. Yet outsourcing involves replacing the face-to-face supervision of staff, selected and vetted by a client firm, with arms-length supervision of an external vendor through an outcomes-based contract. Day-to-day coordination is achieved largely through computer-mediated interactions between staff working for vendor and purchaser. For the service-delivery process to work, all players need to know what behaviors are required, and when to act. This is an area where many misunderstandings and problems occur with outsourcing (Rouse, 2002). Replacing controls based largely on tacit understandings and social mores with explicitly codified requirements, converted into documents and transmitted via e-collaboration tools, involves a process of attenuation. Many focus group informants were surprised to find how many of the controls exercised prior to outsourcing were tacit, relying largely on cultural understandings and expectations. Thus "cultural compatibility" between vendor and purchaser has come to be viewed as an important, but elusive success factor for outsourcing (Hancox & Hackney, 2000).

Managing this risk involves recognizing the richness and complexity of the control processes that occur within organizations, and the extent to which these rely on cultural expectations. Vendor and client must also devote the required attention, skill, and resources to surfacing and articulating what needs to be done by each player in the service delivery process. Another management strategy is to recognize and alert staff that communication and coordination will be more attenuated, and so open to misunderstanding, when

relying on computer-mediated tools. Choosing culturally compatible partners is desirable, as is frequent face-to-face dealings (Kern & Willcocks, 2001), though the latter is not always possible, particularly where the vendor and purchaser are operating in different countries. It is not yet clear whether e-collaboration tools like videoconferencing will adequately substitute for such encounters.

An important aspect of managing this risk is to recognize the difficulties imposed by this replacement, even with recent developments in e-collaboration tools (particularly shared documents, teleconferencing, and e-mail, which have become critical to supporting this process). Many of our informants told us how unrealistic it is to expect that staff will consult, or even be aware of, complex procedure manuals even if they are available online. It is here that evolving tools, such as listservs, FAQs, and hypertext are proving helpful. However, the process of abstracting, codifying, and converting usually un-articulated behavioral requirements into carefully-simplified, computer-mediated forms is a highly skilled task, as is training participants in the required behaviors and sensitizing them to potential differences in worldviews that can lead to misunderstandings. Informant reports indicate that the costs associated with this consequence of outsourcing are substantially underestimated (Rouse, 2002).

Cultural Misunderstandings Associated with Different Social and Legal Settings

Cultural differences (and consequent misunderstandings) between vendor and purchaser are magnified where offshore outsourcing is involved. An additional risk dimension is added because contracts entered into may not be practically enforceable in foreign jurisdictions. The focus groups revealed that many outsourcing clients did not think seriously about cross-national issues because they were so embedded in their own national culture, where legal protections are taken for granted. They assumed that contracts would be enforced, and that staff would be deterred from large-scale theft because of legal sanctions. However, the behavior of vendor staff depends on cultural values and a robustly enforced legal system. Managers from countries with well established legal systems and corruption-free societies (like Australasia, Singapore, the UK, and US) do not necessarily recognize that a contract entered into in countries where corruption is prevalent (Transparency

International, 2004) may have little practical effect on the behavior of staff.

Another area where cross-national differences can be magnified, particularly when computer-mediated communication occurs, is the use of English as a "common" language. Vendors in India, in particular, are cognizant of this and are instituting training programs for staff to "Americanize" conversational styles, but this is an issue that can still cause problems and misunderstandings. Unwitting use of writing conventions that have different meanings in the other culture (e.g., related to formality, or forms of politeness) can hamper relationships between players and contribute to misunderstandings. The extent to which e-collaborators are unpredictable (because they do not use expected conventions) decreases trust and cooperation (Kasper-Fuehrer & Ashkenasy, 2001).

As with risks described above, risk management requires recognizing and acknowledging these risks, and preparing for them. Issues associated with legal systems and cultural values related to corruption need to be considered as part of the outsourcing strategic decision. Issues associated with the potential for miscommunications can in part be addressed by alerting staff to the potential problems, and by training. However, as with previous risks, the costs associating with managing the risk can be substantial and need to be identified and factored into the "business case" for whether to outsource or not.

FUTURE TRENDS

Research into outsourcing to date has largely been exploratory (despite being a topic for almost 15 years) and at this stage much is still theoretical or anecdotal. The study of e-collaboration between vendor and purchaser, particularly given the shortcomings of existing outsourcing research, is an important potential research topic, and one that is ripe for investigation. New communication and organizational forms can lead to new risks that need to be understood and managed. In future, more studies into the practical mechanisms for increasing relationship quality and trust should be conducted.

CONCLUSION

Outsourcing requires people from different organizations to collaborate synchronously or asynchronously to achieve common tasks, and to ensure the effectiveness of the service delivery process. Most of this collaboration is done through a range of electronic tools. The digitized nature of the data exchanged introduces substantial risks that must be managed. The change from largely tacit, cultural controls to codified, explicit controls demands substantial effort from the purchaser in recognizing and addressing potentially false assumptions and misunderstandings. Issues of trust, cooperation, and relationship quality are of paramount importance to outsourcing researchers yet there is a paucity of empirical investigations, and the important role of e-collaboration technologies in supporting vendor-purchaser relationships has been considered only peripherally. There are significant opportunities for future e-collaboration research in the BPO domain.

REFERENCES

Aubert, B. A., Patry, M., & Rivard, S. (2002). Managing IT outsourcing risk: Lessons learned. In R. Hirschheim, A. Heinzel, & J. Dibbern (Eds.), *Information systems outsourcing: Enduring themes, emergent patterns and future directions* (pp. 155-176). Berlin: Springer.

Bitner, M. J., Faranda, W., Hubbert, A. R., & Zeithaml, V. A. (1997). Customer contributions and roles in service delivery. *International Journal of Service Industry Management. 8*(3), 193-205.

Gartner Group. (2004). *Outsourcing market view: What the future holds.* Press Release 09.06.2004. Retrieved November 12, 2004, from www.gartner.com

Gewald, H., & Dibbern, J. (2005). *The influential role of perceived risks versus perceived benefits in the acceptance of business process outsourcing.* E-Finance Labs Working Paper 2005-9, University of Frankfurt. Retrieved February 22, 2006 from www.efinancelab.de/results/pubs/pubscluster1.php.

Gewald, H., & Gellrich, T. (2005, July 7-10). Impact of perceived risk on the capital market's reaction on outsourcing announcements. In *Proceedings of the 9th Pacific Asia Conference on Information Systems (PACIS 2005)*, Bangkok, Thailand.

Hancox, M., & Hackney, R. (2000). IT Outsourcing: Frameworks for conceptualizing practice and perception. *Information Systems Journal, 10,* 217-237.

Kasper-Fuehrer, E., & Ashkenasy, N. (2001). Communicating trustworthiness and building trust in interorganizational virtual organizations. *Journal of Management, 27,* 235-254.

Kern, T., & Willcocks, L. P. (2001). *The relationship advantage: Information technologies, sourcing and management.* Oxford: Oxford University Press.

Kock, N., & McQueen, R. J. (1996). Asynchronous groupware support effects on process improvement groups: An action research study. In J. I. DeGross, S. Jarvenpaa, & A. Srinivasan (Eds.), *17th International Conference on Information Systems* (pp. 339-355). New York: Association for Computing Machinery.

Kock, N., Davison, R., Ocker, R., & Wazlawick, R. (2001). E-collaboration: A look at past research and future challenges. *Journal of Systems and Information Technology, 5*(1), 1-9.

Lacity, M. C., & Hirschheim, R. (1995). *Beyond the information systems outsourcing bandwagon: The insourcing response.* New York: Wiley.

Langford, B. E., & Cosenza, R. (1998). What is service-good analysis? *Journal of Marketing Theory and Practice, 6*(1), 16-26.

Rouse, A. C. (2002). *Information technology outsourcing revisited: Success factors and risks.* Unpublished doctoral thesis, University of Melbourne, Australia.

Rouse, A. C., & Corbitt, B.J. (2004, July 8-11). IT supported business processes(BPO): The good, the bad, and the ugly. In *Proceedings of the 8th Pacific Asia Conference on Information Systems (PACIS 2004),* Shanghai, China.

Rouse, A. C., & Corbitt, B. (2003) *Revisiting IT outsourcing risks: Analysis of a survey of Australia's Top 1000 organizations.* 14th Australasian Conference on Information Systems, Perth, Australia.

Rouse, A. C., & Watson, D. J. (2005). Cyberfraud and identity theft: The hidden costs of business process outsourcing. *Monash Business Review, 1*(2), 30-33.

Transparency International (2004). *Corruption perception index.* Retrieved June 13, 2005, from www.transparency.org/cpi/2004

M

KEY TERMS

BPO: Business process outsourcing. Outsourcing of relatively complex business processes or activities that are supported by information technologies.

Business Process: A set of interrelated organizational activities performed by a number of individuals with the goal of generating customer value. This process can be intra-organizational or interorganizational.

Offshoring: Offshore outsourcing or cross-national outsourcing, where the vendor and client operate in different countries.

Outsourcing: Assigning, to required results, responsibility for all or a portion of the activity and tasks involved in a business process to an external provider. The provider is responsible for management of organizational assets, resources, and/or activities.

Risk: The likelihood or probability of some consequence seen as undesirable, although the term is often used synonymously with "downside." Risk exposure is calculated as the probability of an undesirable consequence multiplied by its (negative) magnitude.

Risk Management: The process of evaluating and selecting alternative responses to risk.

Service Delivery Process: the set of activities that take place to perform a service. Performance involves the coordinated actions of both the provider and user (customer) of the service. In industrial settings, there may be many types of users, and the number of performers increases.

Managing Intercultural Communication Differences in E-Collaboration

Norhayati Zakaria
Universiti Utara Malaysia, Malaysia

INTRODUCTION

With the heightened trends of globalization and increased sophistication of computer-mediated communication (CMC) technologies, people can collaborate anywhere, at anytime, and with anyone. Thus, it can be argued that distance no longer matters. Yet at the same time, people will continue to be confronted with different cultural backgrounds that present conflicts in terms of value systems, attitudes, beliefs, and basic assumptions. In this respect culture does matter, even at a distance. As such multinational corporations (MNCs) need to ascertain the compatibility between the types of technology to be selected and used, and their employees' cultural values when they assemble global virtual teams from all parts of the world.

Global virtual teams can be defined as people who work in a geographically and organizationally dispersed locations, composed of heterogeneous team members, and they use computer-mediated communication technologies during e-collaboration (Zakaria, Amelinckx, & Wilemon, 2004). Due to the increasing use of global virtual teams as a new working structure, MNCs need to manage intercultural communication, defined as interaction between people of diverse cultural backgrounds with distinct communication patterns, preferences, and styles (Novinger, 2001; Gudykunst, 1997). Edward Hall (1976), an intercultural communication theorist, has established that different cultures communicate using different styles that impact face-to-face communication and collaboration. In addition, manifestations of culture are often shown in a person's intercultural communicative behaviors.

Several studies have established that communicative behaviors vary across and within cultures, and that these variations can be explained by Hall's concept of cultural diversity. In his theory called high vs. low context, he explained that communicative behavior is strongly rooted in one's cultural background. For example, in high context cultures (e.g., Malaysia, Korea, Japan, France, etc.), people put more emphasis on non-verbal cues, and in low context cultures (e.g., USA, UK, Italy, Australia, etc.), people rely more on words spoken or written.

In order to avoid misunderstanding and misinterpretations, it is important to comprehend the meaning in what a person says and also how things are said—that is, the communication style one uses for generating ideas, exchanging opinions, sharing knowledge, and expressing ideas. Therefore, this article presents two key research questions:

1. What are the impacts of culture on the global virtual teams' performance during e-collaboration?
2. How do MNCs build intercultural communication competencies to manage cultural differences among global virtual teams?

This article will be organized as follows: in the first section, I will introduce the phenomenon of globally distributed collaboration, or what I term e-collaboration, to point out the significance of a new working structure—global virtual teams. Next, I will present the research gaps that are identified between cross-cultural management, intercultural communication, and CMC to provide concrete background to the phenomenon. Third, I will highlight the potential cultural impacts on e-collaboration. Fourth, I will provide a conceptual framework of building intercultural communication competencies, with suggestions on how to manage the cultural differences in global virtual teams. Finally, I will conclude the article by providing some managerial and theoretical implications of e-collaboration.

BACKGROUND

Based on the past studies, substantial empirical research in cross-cultural management and intercultural communication literature has established that numerous challenges arise when people of different cultures collaborate and communicate at an interpersonal level (Adler, 2002; Gudykunst, 2003; Hooker, 2003). The findings suggest that the challenges that exist in one's communicative behaviors can lead to potential managerial problems such as communication misunderstandings and misinterpretations, intensified conflicts, failure to coordinate, ineffective decision-making, anxieties and uncertainties, and many more (Adler, 2002; Gudykunst & Kim, 2005; Ting-Toomey, 2005).

In a similar vein, CMC literature has observed that technology may facilitate or hinder effective communication (Daft & Lengel, 1986; Kiesler & Sproull, 1992; Sproull & Kiesler, 1986; Walther, 1996) depending on the compatibility of values such as task fit vs. culture fit vs. technology fit. Daft and Lengel's (1984) theory of media richness explains that whether a technology is appropriate for a given managerial task depends on the technology's richness or leanness. E-mail is considered a lean medium, since it relies solely on written text, and videoconferencing is considered a rich medium, since it has verbal, audio, and visual components. Daft and Lengel argue that e-mail fails to evoke sufficient and necessary social and contextual cues and that such technology may therefore not be desirable or effective in a culture that is highly dependent on non-verbal cues when communicating, as in the high context cultures. In contrast, for a culture that is dependent on words or the content of a message such as low context culture, e-mail would be an appropriate tool that facilitates distributed collaboration and communication.

It is well established that computer-mediated communication (CMC) allows people to communicate and collaborate unrestricted by barriers of time and space. Additionally, given the distributed and non-hierarchical nature of global virtual teams, CMC is an ideal method of communication among the members. CMC is defined as a process whereby messages are electronically transmitted from senders to receivers in both synchronous and asynchronous settings (Elton, 1982; Olaniran, 1994). CMC allows for both dyadic and conference (multiuser) interactions. CMC does not include the methods by which two computers communicate, but rather how people communicate using computers.

CMC technologies are just tools that facilitate the communication that takes places between the members. Yet, what is more important is to understand the human dynamics when they use such collaborative technologies to work more effectively across geographical distance.

Therefore, cultural barriers stemming from different managerial aspects and communication styles may adversely affect various elements of collaboration such as negotiations, deliberation of ideas, self-disclosure, conflict resolution, coordination, decision making, and so on (Thorne, 2003). Potential culture-related management problem areas include overcoming high anxiety and uncertainty of feelings (Gudykunst, 1997), managing conflicting and frustrating situations (Adler, 2002), saving face in confrontational situations (Ting-Toomey, 1999), making effective group decisions (Oetzel, 2005), using language and non-verbal communication (Lim, 2003; Tayeb, 2003), and adjusting to and acculturating in a new environment (Kim, 1990).

Without doubt, the need to recognize cultural problems stemming from this new collaborative working structure is crucial for MNCs. However, little empirical research has attempted to bridge the areas of intercultural communication and cross-cultural management, and CMC (Amant, 2002; Olaniran, 2001). Therefore, there is a crucial need to fill the gap in the literature by exploring the cultural influence on the use of CMC in order to provide a concrete understanding of the phenomenon of working together apart as in the context of global virtual teams.

CULTURAL IMPACTS ON E-COLLABORATION

As previously mentioned, global virtual teams are using CMC at an intensified rate. Although team members no longer need to meet face-to-face and confront varied cultural shocks from their first-hand experience, nonetheless they will still encounter anxieties and uncertainties through CMC because of the differences that exist in their communication contextual values (Hall & Hall, 1990). Edward Hall (1976) developed an intercultural communication theory called high vs. low context. He introduces this dimension as a continuous spectrum which illustrates the degree to which a person pays attention to non-verbal cues in a communicative situation. In this article, however, the concept of context will only be elaborated in term of the two extreme values—high vs. low context. In essence, high context culture emphasizes settings or the environment (i.e., context) while low context culture emphasizes words (i.e., content—verbal or written).

As noted by Bresnahan, Shearman, Lee, Ohashi, and Mosher (2002), high context communication styles use a non-assertive approach; they place less value on talk and more on the non-verbal aspect. On the contrary, low-context communication styles are assertive and value straightforward talk. Following from that, Bresnahan et al. observed that high context communication styles have two different underlying assumptions. When a communication takes place with the in-group members, there is a shared understanding among the members. The form of a message that high context individuals send to the in-group members is thus terse, containing restricted codes (Bernstein, 1964; Gudykunst et al., 1996). Restricted code means that a message does not contain verbose words; rather, the message is composed using shortened words, phrases, and sentences (like a secret code). The messages rely more on non-verbal elements such as tone of the message, nature of relationship, social context, and use of silence. Only the receiver, or possible another in-group member would be successful in decoding or interpreting the meaning of the message.

Similarly, when a high context communication takes place with an out-group member with whom the message sender has no strong or prior relationship (for example, an acquaintance or a stranger), an individual does not usually provide a lot of personal information; his message is written at a superficial level, again producing a short message where no details are included.

Despite this tendency to terseness, in some other situations high context people do produce messages that are lengthy, inexact, and ambiguous. This is strongly evident in situations where an individual wants to avoid telling the truth about some situation for fear of hurting someone's feelings, embarrassing a person or himself, or confronting a conflicting situation (LeBaron, 2003; Ferraro, 2003; Hall, 1976; Ting-Toomey, 1999). The ambiguous style is also used to protect the feelings of people that an individual is in contact with, especially his in-group members or the group that he belongs to (Triandis, 1994).

On the other hand, low context communication style—verbal or written—relies on the heavy use of words. Hence, low context communication is content-dependent where words are the primary strategy to effective communication (Gudykunst & Kim, 2002; Hall, 1976). Basically, low context people send short, succinct, and terse message when they express their opinions, or state their feelings. Low context people

value openness. Speaking their minds and telling the truth are some of their key communication characteristics. The terse messages are often used in situations when they want to assert a point, without the fear of hurting the feelings of others because they believe in truth (Grice, 1975; Gudykunst & Nishida, 1986).

When communicating, the distinction between the in-group vs. out-group is less applicable in the low context culture as compared to high context culture. Moreover, low context people value individualism and base their behavior on true feelings (Bainbridge-Frymier, Klopf, & Ishii, 1990). They are also more inclined to express bluntly and talk freely than high context people. Although the truth might hurt, this strategy is useful in being precise and accurate. Sometimes high context audience views this strategy as lack of tact or diplomacy, hence the message is interpreted as harsh, rude, or blunt (Zakaria, 2006; Zakaria et al., 2003).

As mentioned, context alludes to what, why, how, when, and to whom a message is sent. In essence, high context communication is "context-dependent." The decision on how much information is disclosed (amount of information, short or long messages) in high context communication depends largely on the receiver of the message (who) and the topic of the message (what) is to be disclosed and what is to be kept private or confidential). Quite the opposite, low context communication is 'content-dependent' and people would normally place stronger emphasis on words spoken and written. Additionally, the communication that people are engaged in is based on contractual agreements rather than personal relationships.

BUILDING INTERCULTURAL COMMUNICATION COMPETENCIES

As much as MNCs need to create effective cross-cultural training for their global managers in order to develop intercultural communication competencies, global virtual teams need a new set of online communication competencies. Chen and Starosta (1998) suggested that intercultural communication can be developed based on three main competencies: cognitive, affective, and adroitness. As illustrated in Figure 1, these skills requires three different phases of development: (1) creating awareness and educating people on cultural differences, (2) developing sensitivity towards cultural differences, and consequently (3) modeling and producing appropriate behaviors to facilitate acculturation in the new and foreign environment. These three phases takes place in a cyclical process and not linear. In other words, once a person has developed a competency in one of the phases, such competency will be built into the next phases until the team members form a solid foundation of the three required intercultural communication competencies. In brief, the following discussions describe the three different phases:

- **Cultural awareness:** Understanding cultural differences usually take place at the cognitive or thinking level. This stage triggers the team members' mindsets in order to establish common ground among them. At this initial stage, knowledge and information regarding the culture in question needs to be fully learned and understood so that people can logically and rationally become aware of and understand the basic differences such as food, climate, language, geography, time, and many others. This cognitive process also includes understanding self-awareness and cultural awareness—learning about oneself and others. With such understanding, people can create higher self-awareness and can learn to appreciate

others' differences better as well as predict the effects of their behaviors on others.

- **Cultural sensitivity:** Accomplishing the level of cultural sensitivity means having effective skills where a person is required to look into deeper aspect than the cognitive or logical level. This aspect involves the emotive behaviors such as inculcating feelings of empathy, tolerance, caring, patience, respect—a depiction of the benevolent behaviors when dealing with cultural differences. Additionally, people need to learn to make inferences, and deeper interpretations of the meaning of communication. For example, people need to be sensitive to the verbal and non-verbal communication patterns that make distinction between people that depend heavily on words or people that rely on cues such as facial expressions, body movements, gestures, space, and many others. Once they become more sensitive to the body languages shown by other people, they will increase their level of understanding and tolerance.

- **Cultural adroitness:** Behaving appropriately can only be carried out once people have both the general and specific knowledge about a culture as well as when people have learned to be tolerant,

Figure 1. A conceptual framework of online intercultural communication competence (adapted from Zakaria, 2000, and Chen & Starosta, 1998)

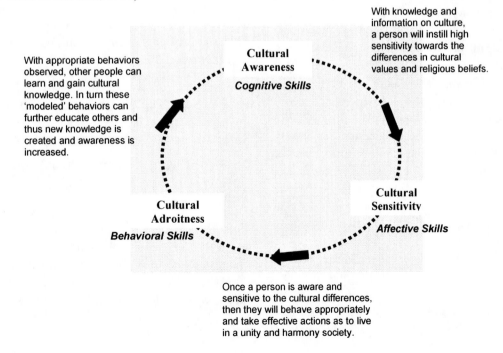

sensitive, responsive, and open to differences shown in one's behavioral change. Only when a person is equipped with adequate knowledge and sensitivity, can he or she demonstrates and engages in effective cultural behaviors as well as acquired the needed behavioral skills. Additionally, with high knowledge and sensitivity, people will perform more effectively and efficiently at work place when dealing with diversity that exists among members of the global virtual teams.

CONCLUSION

In a nutshell, substantial empirical research in cross-cultural management and intercultural communication literature has established that numerous challenges arise when people of different cultures collaborate and communicate at an interpersonal level. Furthermore, new challenges also arise in e-collaboration, such as when people use CMC to collaborate, because there are inherent limitations that arise from technologies. Thus, there are theoretical and managerial implications for understanding the effectiveness of global virtual teams when MNCs use this new working structure to develop new projects and to foster effective intercultural collaboration across geographical boundaries.

As for the theoretical implication, this article offers insights on managing effective e-collaborative behaviors and the significance for developing intercultural communication competencies for global virtual teams. By applying high context vs. low context theoretical lens, this article introduces and applies an intercultural communication theory in a new working context which is global virtual teams. The theoretical implications signify the need for building a culturally-attuned theory for globally distributed collaboration (Zakaria, 2006). This theory will also provide the understanding on how to effectively manage the members who are dependent on CMC yet still encounter cultural differences at a distance. As a result, this article attempts to bridge the interdisciplinary research fields that can provide a more meaningful and explanatory power to the new working phenomenon.

As for the managerial perspective, MNCs need to develop innovative cross-cultural training programs for building online intercultural communication competencies as to correspond with the distributed form of working structure together with the cultural flavor of its employees. The training program needs to consider the new patterns and styles of communicative and collaborative behaviors of the global virtual teams. In addition, information technology experts also need to build and design culturally-sensitive IT applications to fulfill the needs and demands of people from varied cultural background. For instance, MNCs need to consider and understand that certain medium of CMC (e.g., synchronous vs. asynchronous) are more appropriate than the other depending on one's cultural values and preferences. Thus, the issue of culture fit vs. technology fit vs. task fit need to be carefully measured by MNCs.

REFERENCES

Adler, N. J. (2002). *International dimensions of organizational behavior* (3rd ed.). Cincinnati, OH: South-Western.

Amant, K. S. (2002). When cultures and computer collide: Rethinking computer-mediated communication according to international and intercultural communication expectations. *Journal of Business and Technical Communication, 16*(2), 196-214.

Bainbridge-Frymier, A., Klopf, D. W., & Ishii, S. (1990). Japanese and Americans compared on the affect orientation construct. *Psychological Reports, 66,* 985-986.

Bernstein, B. (1964). Elaborated and restricted codes: Their social origins and some consequences. *American Anthropologist, 66,* 55-69.

Bresnahan, M. J., Shearman, S. M., Lee, S. Y., Ohashi R., & Mosher, D. (2002). Personal and cultural differences in responding to criticism in three countries. *Asian Journal of Social Psychology, 5*(2), 93-105.

Chen, G. M., & Starosta, W. J. (1998). *Foundations of intercultural communication.* Needham Heights, MA: Allyn & Bacon.

Daft, R. L., & Lengel, R. H. (1984). Information richness: A new approach to managerial behavior and organizational design. In L. L. Cummings & B. M. Staw (Eds.), *Research in Organizational Behavior, 6* (pp. 191-233). Homewood, IL: JAI Press.

Daft, R. L., & Lengel, R. H. (1986). Organizational information requirements, media richness and structural design. *Management Science, 32*(5), 554-571.

Elton, M. C. J. (1982). *Teleconferencing: New media for business meetings*. New York: American Management Association.

Ferraro, G. P. (2003). *The cultural dimension of international business* (3rd ed.). Upper Saddle River, NJ: Prentice Hall.

Grice, H. P. (1975). Logic and conversation. In P. Cole & J. L. Morgan (Eds.), *Speech acts* (pp. 41-58). New York: Academic Press.

Gudykunst, W. B. (1997). Cultural variability in communication. *Communication Research, 24*(4), 327-348.

Gudykunst, W. B. (Ed.). (2003) *Cross-cultural and intercultural communication*. Thousand Oaks, CA: Sage Publications.

Gudykunst, W. B., & Kim, Y. Y. (2002). *Communicating with strangers: An approach to intercultural communication* (4th ed.). London: McGraw Hill.

Gudykunst, W. B., Matsumoto, Y., Ting-Toomey, S., Nishida, T., Kim, K., & Heyman, S. (1996). The influence of cultural individualism-collectivism, self-construal, and individual values on communication styles across cultures. *Human Communication Research, 22*(4), 510-543.

Gudykunst, W. B., & Nishida, T. (1986). Attributional confidence in low-and high-context cultures. *Human Communication Research, 12*, 525-549.

Hall, E. T. (1976). *Beyond culture*. Garden City, NJ: Anchor Books/Doubleday.

Hall, E. T., & Hall, M. (1990). *Understanding cultural differences: Germans, French, and Americans*. Boston: Intercultural Press.

Hooker, J. (2003). *Working across cultures*. Stanford, CA: Stanford University Press.

Kiesler, S., & Sproull, L. (1992). Group decision-making and communication technology. *Organizational Behavior and Human Decision Processes, 52*, 96-123.

Kim, Y. (1990). Intercultural communication competence: A systems-theoretic view. In S. Ting-Toomey & F. Korzenny (Eds.), *Cross-cultural interpersonal communication*. Newbury Park, CA: Sage.

Kim, Y. Y. (2005). Adapting to a new culture: An integrative communication theory. In W. B. Gudykunst (Ed.), *Theorizing about intercultural communication* (pp. 375-400). Thousand Oaks, CA: Sage Publications.

Lim, T. S. (2003). Language and verbal communication across cultures. In W. B. Gudykunst (Ed.), *Cross-cultural and intercultural communication* (pp. 53-72). Thousand Oaks, CA: Sage.

Novinger, T. (2001). *Intercultural communication: A practical guide*. Austin: University of Texas Press.

Oetzel, J. G. (2005). Effective intercultural workgroup communication theory. In W. B. Gudykunst (Ed.), *Theorizing about intercultural communication* (pp. 351-372). Thousand Oaks, CA: Sage.

Olaniran, B. A. (1994). Group performance in computer-mediated and face-to-face communication media. *Management Communication Quarterly, 7*, 256-281.

Olaniran, B. A. (2001). The effects of computer-mediated communication on transculturalism. In V. H. Milhouse, M .K. Asante, & P. O. Nwosu (Eds.), *Transcultural realities: Interdisciplinary perspectives on cross-cultural relations* (pp. 55-70). Thousand Oaks, CA: Sage.

Sproull, L. S., & Kiesler, S. (1986). Reducing social context cues: Electronic mail in organizational communication. *Management Science, 32*, 1492-1512.

Tayeb, M. (2003). *International management: Theories and practices*. Essex: Pearson Education Limited.

Thorne, S. L. (2003). Artifacts and cultures-of-use in intercultural communication. *Language, Learning and Technology, 7*(2), 38-67.

Ting-Toomey, S. (1999). *Communicating across cultures*. New York: Guilford.

Ting-Toomey, S. (2005). Identity negotiation theory: Crossing cultural boundary. In W. B. Gudykunst (Ed.), *Theorizing intercultural communication* (pp. 211-233). Thousand Oaks, CA: Sage Publication.

Triandis, H. C. (1994). *Culture and social behavior*. New York: McGraw-Hill.

Tylor, E. (1871). *Origins of culture*. New York: Harper & Row.

Walther, J. B. (1996). Computer-mediated communication: Impersonal, interpersonal, and hyperpersonal interaction. *Communication Research, 23*(1), 3-43.

Zakaria, N. (2000). The effects of cross-cultural training on the acculturation process of the global workforce. *International Journal of Manpower, 21*(6), 492-510.

Zakaria, N. (2006). *Culture matters? The impact of context on the globally distributed decision making processes during World Summit on Information Society.* Unpublished doctoral dissertation, Syracuse University.

Zakaria, N., Amelinckx, A., & Wilemon, D. (2004). Working together apart? Building a knowledge sharing culture for global virtual teams. *Creativity and Innovation Management, 13*(1), 15-29.

Zakaria, N., Stanton, J. M., & Sarkar-Barney, S. T. M. (2003). Designing and implementing culturally-sensitive IT applications: The interaction of culture values and privacy issues in the Middle East. *Information Technology & People, 16,* 49-75.

KEY TERMS

Collaboration: Collaboration occurs through a purposive relationship when there is a desire to solve a problem, or to create or discover something within a set of constraints" (Schrage, 1990, p 36).

Computer Mediated Communication (CMC): The process whereby messages are electronically transmitted from senders to receivers in both asynchronous (e.g.. e-mail, discussion forums, etc.) and synchronous (Internet relay chat, videoconferencing, etc.) settings (Elton, 1982; Olaniran, 1994).

Culture: One of the earliest and most widely cited definitions of culture is "that complex whole which includes knowledge, belief, art, morals, law, custom, and any other capabilities and habits acquired by a man as a member of society" (Tylor, 1871, p.1).

Globally Distributed Collaboration: Distributed collaboration implies that participants are geographically dispersed and using computer-mediated communication technologies and the concept "global" highlights the presence of individuals from different cultural backgrounds.

Global Virtual Teams: Groups with members who are (1) culturally diverse, (2) organizationally and geographically dispersed, and (3) use electronic communication when collaborating.

High Context: Most of the information is in the physical context or internalized in the person. (Hall, 1976).

Intercultural Communication Competence: The skills to negotiate mutual meanings, rules, and positive outcomes as well as be equipped with three basic competencies—cognitive, affective, and behavioral.

Low Context: Mass of information is bested in the explicit code (Hall, 1976).

Managing Online Discussion Forums for Collaborative Learning

Wing Lam
Universitas 21 Global, Singapore

Eu-Jin Kong
Universitas 21 Global, Singapore

Alton Chua
Nanyang Technological University, Singapore

INTRODUCTION

In recent years, there has been significant growth in online education (Schrum & Hong, 2002; Evans & Haase, 2001). The number of academic journals devoted to online education also suggests that researchers are paying much attention to advancing online educational methods. One promising area of investigation is collaborative learning, which involves students learning as a group (Zhang & Nunamaker, 2003), much of which takes place electronically without face-to-face interaction (Townsend, DeMarie, & Hendrickson, 1998). One popular tool used to support collaborative learning is the online discussion forum (ODF), which allows asynchronous interaction between participants. This paper describes the experiences of using ODFs for collaborative learning at Universitas 21 Global (U21G), a newly established e-university.

LITERATURE REVIEW

A key advantage of ODFs is that they take away the apprehension and embarrassment students often have about contributing to physical face-to-face discussions which can, and often are, dominated by a small number of individuals (Lieblein 2000). ODFs allow individuals to express their ideas in a free manner, providing them with more time for reflection before interacting with others (Hew & Cheung, 2003). The exchange of ideas in an ODF can also support a constructivist approach to learning where students are able to create new knowledge (Gilbert & Dabbagh, 2005). Furthermore, encouraging students to lead and moderate discussion groups has also been shown to have a positive effect on student participation (Poole, 2000; Corner & Corner, 2003). ODFs therefore potentially serve as an excellent environment for collaborative learning.

There are, of course, some drawbacks to the use of ODFs, one of which is reliance on technology (Pérez Cereijo, Young, & Wilhelm, 2001). Poor Internet access can hamper access to ODFs. The usability of ODFs can also be comprised by poor navigation, excessive page loading and refresh times, and the technical design of the ODF. Song, Singleton, Hill, and Koh (2004) identified comfort with online technologies as one determining factor to a satisfactory online learning experience.

However, making effective use of ODFs for collaborative learning is not simply a technological challenge, but one that involves careful management and facilitation on the part of educators. One challenge is stimulating discussion (Clark, 2001). Providing students with ODF does not necessarily mean students will make use of them. Educators must therefore think about student motivations for using ODFs. Hammond (2000) suggests that the lack of social inclusion in an online environment can also be the cause of non-participation in ODFs. Another issue is the quality of the discussion in an ODF. The number of postings to an ODF does not necessarily correlate well with student learning or effort (Salter 2000). Desanctis, Fayard, Roach, and Jiang (2003) emphasises the importance of deep discussion which involves challenge, reflection and debate, as opposed to surface discussion, which is shallow or repetitive in nature. Students can also struggle to make meaningful contributions to an ODF because of a lack of structure and guidance (Black, 2005).

COLLABORATIVE LEARNING

U21G Background

U21G is an e-university established as a joint venture between Universitas 21 (U21), a network of 18 international universities, and Thomson Learning. U21G began offering its first program, the MBA, in July 2003. The typical profile of a U21G MBA student is an individual in their mid-30s, holding a middle management position.

Pedagogy Overview

Programs offered by U21G are delivered entirely online, there is no face-to-face classroom study. Students interact amongst themselves and with professors using a range of Web-based communication tools such as ODFs, e-mail, online chat, and audio conferencing. Students are provided with online courseware specifically designed by U21G containing all the necessary content and resources. Class sections are 12 weeks in duration and class sizes vary between 20 to 35 students. Each class section is led by an adjunct professor.

Use of ODFs for Collaborative Learning

Students are assessed on their participation and performance in collaborative learning activities on the ODFs, which can account for up to 40% of a student's final mark. Although there is some debate around whether ODFs should not be assessed at all, U21G believes that the assessment of ODFs provides a strong motivation for students not only to participate, but to participate to the best of their abilities.

In these collaborative learning activities, there is a strong emphasis on peer learning (Bailey & Luetkehans, 1998). An example of how ODFs are used for collaborative learning is shown in Figure 1.

Here, students are required to download a Harvard business case, ponder over a couple of case questions, and post their answers to the ODF. Such questions are typical of the way ODFs are used for collaborative learning at U21G.

Observed Problems with ODFs

Since the launch of the MBA program, U21G has conducted over 100 class sections. Early on, several problems were observed with the use of ODF for collaborative learning, namely:

Figure 1.

Instructions
Read and analyze the following case about the Eastman Chemical company and answer the questions that follow.
 Title:
Constructing an e-Supply Chain at Eastman Chemical Co.

Case No.:
#HKU222

Date:
25 September 2002

Authors:
Yen, B., Farhoomand, A.F., and Ng, P..

Publisher:
Harvard Business School case (Field) study

Questions
  1. What were the key success factors facing Craig Knight in his quest to sell Eastman's approach to e-Business?
  2. Describe the risks facing Knight and Eastman in their e-business initiative and suggest what you would have done to reduce or mitigate that risk.
Post your answer to the questions in a discussion board posting of about 100 words. Alternatively, provide a critique of the postings by at least two of your fellow students. Your critique should further develop points already made, provide alternative perspectives, or introduce new supporting evidence.

- Students had a tendency to regurgitate points already made in previous postings
- Students made postings that were irrelevant to the main thread of discussion or diverged away from the important issues
- Students made postings that were incomprehensible or incoherent
- Some students did not participate at all in the ODF, or kept their participation to a bear minimum
- Discussion threads were sometimes too long that they became unwieldy and disorganized
- Students posted too late when the heat of the discussion had already dissapated

U21G also came to quickly realize that assessing students on the volume or the frequency of postings tended to result in a large number of low-quality postings, as characterized by those that either:

- Provide little new insight
- Are of a trivial nature
- Show little evidence of critical analysis
- Repeat what has already been stated
- Have little relevance to the main line of discussion
- Are factually incorrect

A successful collaborative learning activity in the ODF is therefore one that meets three criteria, namely

- Has a high proportion of high-quality postings
- Involves the active and timely participation of all students in the class
- Is well-organized

U21G's experience suggests that the use of ODFs for collaborative learning activities is more likely to succeed when professors take an active role in their management and facilitation, and the next section describes the management and facilitation practices U21G has adopted as a consequence.

MANAGEMENT AND FACILITATION

A Phased Approach

At U21G, the collaborative learning activities in the ODFs are seen as having a lifecycle comprising of four distinct phases, as given in Table 1.

Preparation Phase

The preparation phase is concerned primarily with the initial design of the collaborative learning activity.

- **Question design:** U21G favors collaborative learning activities whereby students are given an article or case to read, followed by case questions which require posting to the ODF. By presenting a specific set of case questions, the discussion starts off in a precise and directed manner. Without such direction, there is a risk that very little discussion will emerge at all, or conversely, that there will be too much unfocused discussion. U21G also encourages students to comment on the postings of fellow students, so that students are not all coerced into answering the same set of questions.

- **Assessment guidelines:** Students should be provided with guidelines on how participation in collaborative learning activities will be assessed. This in turn has a positive effect in helping students think carefully about how they formulate their contributions and responses to the ODF. Students at U21G, for example, know high marks will be awarded to postings that:

  - Introduce new ideas or perspectives not discussed in the course material
  - Make an original contribution that no one else in the class has made
  - Is backed up by prior research or elucidated from the student's own personal work experience

Importantly, the assessment is more weighted towards the quality, rather than quantity, of a student's postings on the ODF.

- **ODF tool usability:** Students can potentially spend many hours a week on an ODF, so it is important that the ODF tool has a high level of usability.

Kick-Off and Ramp-Up Phase

The aim in this phase is to generate sufficient momentum on the ODF that it becomes self-sustaining.

Table 1. A phased-approach to managing ODFs

Phase	Salient issues
Preparation	Discussion topics need to be anchored around a specific set of questions. This provides a platform for the confluence of thoughts.
	ODF tools must possess a high level of usability and an easy learning curve.
Kick-off and ramp-up	Self-introductions can be used as socialization mechanisms. Gaining familiarity among students help create a conducive environment to peer learning.
	Ground rules need to be articulated. Students' expectations can be better managed when social norms and decorum online have been defined.
	Response from reticent students needs to be elicited deliberately. Early interventions from the professor could engender students' participation.
Cruising	The professor needs to maintain an ongoing presence.
	Appropriate encouragement and feedback to students' postings needs to be given. This has a positive effect of spurring other students on.
	The professor ought to play the role of a guide on the side. This develops independent learning in students.
	Discussion threads need to be systematically organized as they grow. This facilitates searching on the topics discussed in the ODF.
	Trigger questions may be injected into flagging discussion questions. New but related topics for discussion are thus generated.
	Informal discussion threads can be created as part of community building efforts. This helps foster a sense of togetherness.
	Reporting tools can be used to monitor students' participation. This helps identify reticent students.
Closure	Discussions on ODF ought to be wind down gradually. This signals a planned rather than an abrupt closure of the ODF.
	Key points and outstanding issues in the discussions ought to be summarised. This helps students in future collaborative learning activities
	Discussions on the ODF could be archived. They could serve as sources of useful reference or teaching tools.

- **Self-introductions:** If the collaborative learning activity is one where students are meeting for the very first time, students should post an introductory message on the ODF. This is important in creating a sense of community and to avoid the feeling of isolation that students often experience with distance education (Song et al., 2004). The professor may begin by posting his or her own (fairly detailed) introductory message. To avoid overly-brief introductions, the professor should guide students as to what information students should be included in their introductions. For example:

  ○ Preferred name
  ○ Current job role and the organization they work for
  ○ A brief history of their work experience

  ○ Interesting experiences or observations they have had in the subject area
  ○ What they hope to gain from the subject

- **Ground rules:** The professor should make clear to students basic ground rules and expectations. For example:

  ○ How often such students check the ODFs? Do they need to check every day?
  ○ Are postings voluntary or mandatory?
  ○ Are postings required to be of a certain length?

Students should also be reminded of the code of conduct—that is, language and tone of postings, respect for the opinions of others, and academic honesty and plagiarism. When students have questions that would be of interest to the rest of a class, they should be asked to post those questions on the ODF rather than via private e-mail.

- **Discussion kick-start:** There may be occasions in the kick-off and ramp-up phase when the professor will need to kick-start some initial discussion on the ODF. One approach a professor might use is to simply prompt those students who have not already contributed. Another approach is for professors to assign a team of students to lead the discussion on a particular topic. The approach is particularly useful for including students who have a tendency to sit at the periphery of a discussion.

Cruising Phase

The cruising phase is reached when a high level of participation on the ODF has been reached and where the ODF is largely self-sustaining. In the cruising phase, the professor no longer needs to make a proactive attempt to generate discussion.

- **Professor presence:** It is important for the professor to maintain a presence (Anderson, Rourke, Garrison, & Archer, 2001) in the ODF during the cruise phase. Although it is not necessary to respond to every posting, the professor should continue to make occasional postings to the ODF to assure students of the guidance available.
- **Giving encouragement and feedback:** The professor should provide encouragement and feedback. One approach is to congratulate students whose postings make a significant contribution.
- **Become the "guide on the side" rather than the "sage on the stage":** In traditional on-campus lectures, the professor often assumes the role of a "sage on the stage" (Mazzolini & Maddison, 2003). This can have a counter-productive effect in ODFs, as students become dependent upon the professor to provide answers rather than discovering answers through peer learning. Instead, the professor needs to become a "guide on the side," where he or she:

  ○ Refrains from posting answers to the ODF
  ○ Refrains from stating whether answers are right or wrong

  ○ Avoids posting too much, too often
  ○ Facilitates the sharing of ideas and opinions

- **Organizing the discussion:** Even with threaded discussions, long discussions can become hard to navigate. It is therefore important for the professor to keep the discussion well organized. The professor should create separate discussion threads when new topics sprout from the discussion of an existing topic, and break large discussions into separate threads.
- **Trigger questions:** Trigger questions can be used by the professor to maintain the momentum of a discussion or inject life into a flagging discussion. An example of such a trigger question is given below:

So far we have had a good deal of discussion about the process of risk management in software projects. Much of this discussion has been based around conceptual models of risk management. In your experience, how well do you think risk management is actually practiced on real software projects?

- **Community-building:** Socialization takes places in an online environment as it does in a face-to-face environment. Special informal discussion areas can be created within the ODF to support socialization.

- **Monitoring student participation:** As the number of students in a class grows, the monitoring of student participation in the ODFs will be increasingly reliant upon reporting tools. At U21G, the reporting tools enable a professor to track student logins, search for postings made by a particular student, and identify students who have not posted at all or posted infrequently.

Closure Phase

The closure phase represents the concluding of the collaborative learning activity.

- **Winding down:** Although online discussions may stretch over weeks and months, there comes a point at which a collaborative learning activity needs to be concluded. A discussion should not be abruptly ended. Instead, the professor should begin to wind-down the discussion gradually, which can be done by announcing a date when the discussion will end.

- **Summary and feedback:** A discussion should be closed by summarizing the key points and providing feedback on the overall quality of the discussion that took place.
- **Archiving:** In some cases, if may be helpful to archive all or part of the discussion in the ODF for review or use at a later stage.

FUTURE TRENDS

The growth in online education means that the management and facilitation of ODFs will become an increasingly important skill for online educators, just like the face-to-face teaching skills in a traditional classroom setting. Online educators should avoid paying too much attention to the underlying technology of ODF tools, and instead focus on the collaborative learning activities made possible through the use of such tools.

CONCLUSION

U21G's experience strongly suggests that collaborative learning is more likely to be achieved when ODFs are deliberately managed and facilitated. A phased approach is advocated comprising of four phases, namely the preparation phase, the kick-off and ramp-up phase, the cruising phase, and the closure phase. Such a phased-approach to managing ODFs enables professors involved in online education to identify and develop the necessary management and facilitation skills.

REFERENCES

Anderson, T., Rourke, L., Garrison, D. R., & Archer, W. (2001). Assessing teaching presence in a computer conferencing context, *Journal of Asynchronous Learning Networks, 5*(2), 1-17.

Bailey, M. L., & Luetkehans, L. (1998). Ten great tips for facilitating virtual learning team. In *Proceedings of the Annual Conference on Distance Teaching and*

Learning, [Electronic Version], Retrieved from http://www.psdcorp.com/dislearn.htm

Black, A. (2005). The use of asynchronous discussion: Creating a text of talk. *Contemporary Issues in Technology and Teacher Education, 5*(1). Retrieved from http://www.citejournal.org/vol5/iss1/languagearts/article1.cfm

Clark, J. (2001). Stimulating collaboration and discussion in online learning environments. *Internet and Higher Education, 4,* 119-124.

Corner, J., & Corner, P. D. (2003). Teaching OR/MS using discussion leadership. *Interfaces, 33*(3), 60-69.

Desanctis, G., Fayard, A., Roach, M., & Jiang, L. (2003). Learning in online forums. *European Management Journal, 21*(5), 565-577.

Evans, J. R., & Haase, I. M. (2001). Online business in the twenty-first century: An analysis of potential target markets. *Internet Research: Electronic Networking Applications Policy, 11*(3), 246-260.

Gilbert, P. K., & Dabbagh, N. (2005). How to structure online discussions for meaningful discourse: A case study. *British Journal of Educational Technology, 36*(1), 5-8.

Hammond, M. (2000). Communication within online forums: The opportunities, the constraints, and the value of a communicative approach. *Computers & Education, 35,* 251-262.

Hew, K. F., & Cheung, W. S. (2003). An exploratory study on the use of asynchronous online discussion in hypermedia design. *Journal of Instructional Science & Technology, 6*(1). Retrieved February 23, 2006, from http://www.usq.edu.au/electpub/e-jist/docs/Vol6_No1/an_exploratory_study_on_the_use_.htm

Lieblein, E. (2000). Critical factors for successful delivery of online programs. *Internet and Higher Education, 3,* 161-174.

Mazzolini, M., & Maddison, S. (2003). Sage, guide or ghost? The effect of professor intervention on student participation in online discussion forums. *Computers and Education, 40,* 237-253.

Pérez Cereijo, M. V., Young, J., & Wilhelm, R. W. (2001). Factors facilitating student participation in Web courses. *Journal of Computing in Teacher Education, 18*(1), 32-39.

Pool, D. M. (2000). Student participation in a discussion-oriented online course: A case-study. *Journal of Research on Computing in Education, 33*(2), 162-177.

Salter, G. (2000). Making use of online discussion groups. *Journal of Australian Council for Computers in Education, 15*(2), 5-10.

Schrum, L., & Hong, S. (2002). From the field: Characteristics of successful tertiary online students and strategies of experienced online educators. *Education and Information Technologies, 7,* 5-16.

Song, L., Singleton, E. S., Hill, J. R., & Koh, M. H. (2004). Improving online learning: Student perceptions of useful and challenging characteristics. *Internet and Higher Education, 7,* 59-70.

Townsend, A. M., DeMarie, S. M., & Hendrickson, A. R. (1998). Virtual teams: technology and the workplace of the future. *Academy of Management Executive, 12*(3), 17-29.

Zhang, D., & Nunamaker, J. F. (2003). Powering e-learning in the new millennium: An overview of e-learning and enabling technology. *Information Systems Frontiers, 5*(2), 207-218.

KEY TERMS

Deep Discussion: Discussion that involves challenge, reflection and debate.

Collaborative Learning: Learning activities that are performed as part of a group.

Online Discussion Forum (ODF): An online asynchronous collaboration tool that enables students to communicate through threaded discussion.

Online Education: The use of the Internet and other online technologies to deliver educational programs.

Online Presence: Being visible within an online team or community.

Peer Learning: Learning conducted via the exchange of ideas with fellow students.

Universitas 21 Global (U21G): An e-university owned by the 19 member universities in the Universitas 21 network that delivers its academic programs entirely online.

M

A Matrix for E-Collaboration in Rural Canadian Schools

Ken Stevens
Memorial University of Newfoundland, Canada

INTRODUCTION

This is a case study of interinstitutional e-collaboration in a rural part of Canada, based on e-teaching and e-learning for senior high school students. In the process of developing e-collaboration between institutions, new structures and processes were created that complemented traditional schools. Through this initiative, e-collaboration provided extended educational and, indirectly, vocational opportunities for senior students in small schools in Atlantic Canada.

In a pilot study in 1999, selected advanced placement (AP) subjects were made available online for what was believed to be the first time. Furthermore, the AP subjects were made available online to students in senior classes in small rural Newfoundland and Labrador secondary schools ($N = 8$). There were no online AP courses to guide the initiative at the time and the curriculum at this level had not previously been provided to students in small rural schools. Through interinstitutional e-collaboration senior students in rural schools in a small part of Atlantic Canada were provided with extended learning opportunities. AP instruction was developed both on site and online, synchronously and asynchronously within new collaborative structures organized around e-teaching and e-learning.

BACKGROUND

It has always been difficult to provide senior students in rural schools with curriculum opportunities equal to their peers who are educated in larger, usually urban institutions. Governments find it difficult to justify the appointment of specialist teachers to rural schools when there are small numbers of students on site requiring instruction in their areas of specialization. Accordingly, many senior rural students leave home to be educated in urban schools. In some cases they are taken by bus to larger institutions on a daily basis while others are educated in boarding schools. The issue of equality of educational and, indirectly, vocational, opportunities is at the heart of most initiatives to provide extended learning opportunities for rural students (Hawkes & Halverson, 2002; Information Highway Advisory Council, 1997).

Newfoundland and Labrador is one of the smallest Canadian provinces in terms of population. The province is characterized by its predominantly rural social structure, its distinctive history, and its unique culture. Approximately two thirds of schools in this province are in rural communities, almost all of which are located on the coast. A few of the smallest and most isolated communities have no road access. The reorganization of primary, elementary, and secondary education in Newfoundland and Labrador into 10 school districts provided an opportunity to develop the first digital intranet in the province. School District 8 contained 18 schools ranging in student enrolment from 40 to 650. School District 8 was approximately 2 hours by road from the capital city, St. Johns, which is the location of Memorial University of Newfoundland. Eight schools within the school district, together with the TeleLearning and Rural Education Centre of Memorial University of Newfoundland, formed a digital intranet within which senior science courses were taught in open classes. By including this research and development centre (located within the Faculty of Education of the only university in the province) in the new rural school structure, collaborative research between academics and teachers was encouraged. The synergy of professional development between schools and a professional faculty addressed a concern recently raised by Thompson, Bakken, and Clark (2001) that "seldom are classroom teachers required to become involved in research/scholarly inquiry." As schools in the school district and academics at the University were involved in the organization of teaching and learning in the new electronic structure, collaboration quickly followed. Collaboration between participating schools and the Centre for TeleLearning and Rural Education at Memorial University of Newfoundland led to the development of Web-based Advanced Placement courses.

A MATRIX FOR E-COLLABORATION BETWEEN SMALL RURAL SCHOOLS

The creation of the first digital intranet in Newfoundland and Labrador was with a group of eight small schools in the same rural education district and was a pilot study. Schools in a single district were academically and administratively integrated so that teacher expertise could be shared by designated senior students for whom on-site instruction was not otherwise available. This involved collaborative teaching and learning in an environment within which schools were competing for a declining number of students.

The development of the first school district digital intranet in Newfoundland and Labrador involved a matrix of technological, pedagogical, organizational, and conceptual change. In rural Newfoundland and Labrador, this matrix supported the creation of a rural school district digital intranet of four interconnected dimensions:

Technology	Pedagogy
Organization	Conceptual change

Technologically, the development of the school district digital intranet was difficult. In many parts of the province telecommunications infrastructure was barely adequate to link schools within such a structure. Minimum specifications were adopted for computer hardware and network connectivity. All schools involved in the project had DirecPC satellite dishes installed to provide a high-speed down-link. In most rural communities in this part of Canada, digital telecommunications infrastructures do not enable schools to have a high-speed up-link to the Internet. For real-time instruction, Meeting Point and Microsoft NetMeeting were selected. This combination of software enabled a teacher to present real-time interactive instruction

to multiple sites. An orientation session was provided for students prior to the implementation of this project. Students had to learn how to communicate with each other and with their instructor using these new technologies before classes could begin. Although there was some shared technical support in the district, principals expressed a desire to have a technician, if possible, on each site. In some schools, teachers with expertise in computer technology were undertaking technician roles in addition to their classroom activities to keep computers operating. This was accepted by teachers and principals as necessary in emergencies, but it was not considered to be an ideal situation. At the time of the pilot study there was a fragile technical infrastructure in some parts of the intranet, depending in some instances on the good-will of particular teachers to ensure that it was maintained. This situation was not conducive to the expansion of the intranet and was not likely to encourage new teachers to begin using IT in their classrooms. The need for students to be comfortable with the technologies to which they were introduced became apparent early in the school year. One teacher subsequently recommended training in the use of technology prior to a student undertaking an online course in future, including the use of WebCT (My Records, chat rooms, lessons, quizzes, Bulletin Board, private mail and using attachments to transfer assignments); Netmeeting and Meetingpoint (logging in, whiteboard, chat, efficient collaborative use of the microphone). Students were almost evenly divided between those who had computers at home with Internet access and those who did not. During the third research visit, 25 students were interviewed: 13 indicated that they had this technology in their homes while 12 did not.

Pedagogically, the integration of schools in a single district meant teaching in ways that were different from traditional classroom practices. Instead of providing instruction exclusively within their own classrooms, teachers had to consider teaching collaboratively from one site to another in what became shared teaching and learning space. The challenge of teaching between rather than exclusively in schools focused attention on what Van Manen (2002) terms "the pedagogical task of teaching." For some teachers this was difficult to accept when a colleague on another site had the role of teaching AP students on line in his or her school, from another school in the district intranet. For those teachers who taught the initial AP subjects of chemistry, mathematics, physics and biology within the new collaborative structure (the school district digital intranet) there was little pedagogy to guide them other than, for one of the teachers, previous experience as a distance education instructor. Issues arose in the delivery of classes between schools (or sites) involving the balance of synchronous and asynchronous instruction, motivation, control, and student lack of confidence based on inexperience in learning other than by formal classroom instruction and assessment. One of the first challenges for students was how to interact online with peers they had not met. In some cases this led to awkwardness and embarrassment that threatened to impede learning. Teachers had to adjust to talking less during lessons and to prepare questions very carefully so that all students could participate. The development of judicious questioning by teachers helped students learn from one another as they all considered how to respond.

Students acknowledged that their online AP courses made new demands on their study skills. Many were confused by the requirements of what was for almost everyone, considerably more difficult content in each of the four disciplines, combined with a new delivery system with teachers they did not meet in a traditional classroom situation. By midyear, students had learned a good deal about learning in an intranet with computer technologies and interactive software. Four field trips to visit all participating sites in the intranet were made by the author during the academic year and during the final visit, principals were interviewed about the provision of AP science to senior students in their schools. AP teachers provided written reports at intervals throughout the academic year. At the conclusion of the school year, senior managers in the Board Office were interviewed about the administration of the Internet within the school district. During student interviews, information about teachers' teaching styles and their use of hardware and software inevitably became central issues. For example, students noted in their final interview at the end of the school year that a good e-teacher was able to engage all students who were online together and had a logical layout with the white board, upon which numerous examples were expected to be provided to facilitate learning. A good e-teacher was thought to be someone who "encourages thinking" by students, develops collaborative learning and "occasionally visits us at our schools."

The four e-teachers in the intranet each provided the author with four written reports at intervals throughout the school year. Earlier reports indicated a preoccupa-

tion with coming to terms with a new way of teaching, using new technologies. In the latter part of the year teacher reports were increasingly reflective and less concerned with day to day technological issues. The most experienced e-teacher in the intranet used three ways of teaching. His first method was the lecture, making use of the whiteboard and video. This was useful for giving information to students but frequent questioning was needed by the teacher to maintain student attention. His second method was the question and answer model without a formal lesson plan. Students were asked in advance to bring to the lesson pertinent questions for resolution. Few of these classes were successful. The only times they were considered by the teacher to be successful was when they were initiated by the students. Occasionally a third approach was used. The application sharing and collaboration features of NetMeeting were found to work well with the windows version of MPLI (multi purpose laboratory interface) software. It was therefore possible to do some labs across the web, in this case, in physics. Laboratories in chemistry made considerable use of CDs and of videoed experiments that students could watch. In both biology and chemistry, laboratory classes were undertaken at a central site by the instructor with students attending from several intranet schools.

Organizationally, the integration of schools in a district digital intranet, involved institutional collaboration beginning with the coordination of senior class timetables so that students located on multiple sites could be taught together online. The first rural district digital network was organized by the board office, including the selection of online teachers together with technical and organizational support. The administration of the intranet was evaluated by principals ($N = 8$) of participating sites and senior managers ($N = 2$) in the board office. Principals noted a variety of changes that the digital intranet had brought to their schools in the last year including an expanded range of subjects that the intranet provided in their schools for senior students. Principals usually decided on the scheduling of subjects in their schools. However, in bringing schools together on-line for collaborative teaching in an intranet, there was general agreement of the need for a common timetable. The variety of timetables that existed in different schools in the school district was considered to be a major problem in the operation of the intranet in its first year. This became the top priority for the future development of e-collaboration. At the end of the school year, principal's comments included two areas of unease with the emerging interinstitutional collaborative environment. One principal pointed out that he did not see himself as the director of a virtual school and that it was necessary to appoint someone to handle this role in the Board Office. Another principal expressed unease about online teachers entering and leaving his school. He noted that there was a need for much better communication between online teachers and principals so that parents could be kept informed of developments in this pioneering venture in e-collaboration.

Conceptually, the development of a school district digital intranet involved a different way of thinking about teaching, learning and the organization of schools. One principal reported at the end of the school year that parents in his school district were "excited, confused, and afraid" of the technological developments in their local school. They were excited by the possibilities, confused because they did not understand how the technology worked, and afraid that this might be a glimpse into the future of education. The first conceptual change for teachers was that rather than being appointed to teach their own classes, in their own classrooms, in the schools to which they were appointed, a selected few were asked to teach other teacher's students, located in other classrooms beyond their own school and community, on-line. A second conceptual change that the first school district digital intranet initiated was the introduction of learning at a distance in traditional classrooms. In the first digital intranet in Newfoundland and Labrador some teachers were introduced to the notion of teaching both in a traditional classroom and online, through the Internet. This was an important step in the integration of virtual and actual teaching. Students were also introduced to the notion of learning both on site and on line during a school day. These were significant steps in the integration of on-site and online education or the merging of actual and virtual classes (Stevens & Stewart, 2005). A third conceptual change was that schools that were small in terms of the number of students attending in person, on site, and the number of teachers who were appointed to them, could become relatively large schools in terms of the range of subjects they could offer with the addition of online instruction. Finally, there was a conceptual change in the realization by students, parents and teachers that school location was not necessarily a barrier to accessing areas of the curriculum that had

not traditionally been provided on site in traditional classrooms. Advanced Placement subjects, taught in large urban schools throughout North America, could be made available to senior students in small and geographically isolated schools.

Each of the four parts of the matrix was shaped though collaboration between each of the other parts. The organization of the intranet depended on the connectivity provided by Internet-based technologies, assisted by the installation of satellite dishes in Newfoundland and Labrador schools. The technological dimension depended on the organizational skills of administrators in the school district office collaborating with on-site administrators and teachers. Several times during the first year of operation there were technological problems that were solved collaboratively by administrators and teachers working alongside technicians. Technological changes introduced by the Internet and its application to school district organization to facilitate the administrative and academic linking of classrooms in small and dispersed schools, encouraged teachers to consider new, collaborative ways of teaching in open learning environments. The introduction of the first school district digital intranet in Newfoundland and Labrador challenged the exclusivity of traditional classrooms in which a defined number of students were taught, in person, by a single teacher.

FUTURE TRENDS

The future of the initiative outlined above has been secured by the creation of the Centre for Distance Learning and Innovation (CDLI) of the Newfoundland and Labrador Department of Education, following extensive research (Brown, Sheppard, & Stevens, 2000; Stevens, 2002) and a ministerial inquiry into the implications of the initial school district digital intranet (Government of Newfoundland and Labrador, 2000). Since its inception, CDLI (http://www.cdli.ca/) has considerably extended the range and the level of involvement in e-learning in schools throughout the province of Newfoundland and Labrador. Much of the work undertaken by CDLI involves program development to extend the curriculum for the province's high schools, but there is also an innovation component to its work that explores new technological and pedagogical possibilities and assesses their suitability for the province's schools.

An overriding consideration is the enhancement of learning opportunities that interinstitutional collaboration has produced in rural Newfoundland and Labrador. Schools that are small in size when measured in terms of the number of students attending, in person, on a daily basis, have become large educational institutions in terms of the range of curriculum offerings they can provide on site, both synchronously and asynchronously. Instruction can be provided from collaborating institutions as well as directly from CDLI e-teachers. The introduction of CDLI as part of the Department of Education of Newfoundland and Labrador supports the integration of on site and on line teaching and learning within collaborative structures.

New teaching professionals are emerging in Newfoundland and Labrador: A proliferation of e-teachers, who teach online and m-teams who support them on site by mediating with e-learners while providing traditional face-to-face instruction in their participating schools (Barbour & Mulcahy, 2005). There is considerable demand for instructional designers and for Web site developers. These are likely to become increasingly important areas of educational expertise in future. The growing use of handheld and laptop technology by students and teachers is likely to provide increased flexibility in both teaching and learning (Griffin & Sherrod, 2005; Mathiasen, 2004; Norris & Soloway, 2004).

CONCLUSION

The changes inaugurated by a single school district digital intranet have challenged the dominance of traditional classrooms that perpetuate on-site, synchronous teaching. The implementation of the initial four AP courses and the establishment of CDLI to advance and support e-learning has changed high school education in Newfoundland and Labrador. In the collaborative structures of digital intranets linking schools throughout the province, together with collaborative teaching that has been made possible within them, there is now the possibility of further integration of onsite and online instruction as e-learning becomes part of traditional classrooms. Arguably, the most significant educational contribution of the pilot school district digital intranet developed in one of the most rural Canadian provinces has been to demonstrate that computers in classrooms are, essentially, collaborative tools.

REFERENCES

Barbour, M., & Mulcahy, D. (2005, September). The role of mediating teachers in Newfoundland's new model of distance education. *The Morning Watch.*

Brown, J., Sheppard, D., & Stevens, K. (2000). *Effective schooling in a tele-learning environment.* St. John's, NL, Centre for Tele-Learning and Rural Education, Faculty of Education, Memorial University of Newfoundland.

Government of Newfoundland and Labrador. (2000). *Supporting learning: Report on the ministerial panel on educational delivery in the classroom,* St John's, NL, Department of Education.

Griffin, D., & Sherrod, B. (2005) *Technology use in rural high schools improves opportunity for student achievement.* Atlanta, GA: Southern Regional Education Board. Available online from http://www.sreb.org/programs/EdTech/pubs/PDF/05T01-TechnologyUseinRuralHS/pdf

Hawkes, M., & Halverson, P. (2002). Technology facilitation in the rural school: An analysis of options. *Journal of Research in Rural Education, 17*(3), 162-170.

Information Highway Advisory Council. (1997). *Preparing Canada for a digital world.* Ottawa, Canada: Industry Canada.

Mathiasen, H. (2004) Expectations of technology: When the intensive application of IT in teaching becomes a possibility. *Journal of Research on Technology in Education, 36*(3), 2273-294.

Norris, C., & Solloway, E. (2004) Envisioning the handheld-centric classroom. *Journal of Educational Computing Research, 30*(4), 281-294.

Stevens, K. J. (2002). The expansion of educational opportunities in rural communities using Web-based resources. In G. A. Santana Torrellas & V. Uskov (Eds.), *Computers and advanced technology in education* (pp. 221-225). Anaheim, CA: ACTA Press.

Stevens, K. J., & Stewart, D. (2005) *Cybercells: Learning in actual and virtual groups.* Melbourne: Thomson-Dunmore Press.

Thompson, J., Bakken, L., & Clark, F. L. (2001) Creating synergy: Collaborative research within a professional development school partnership. *The Teacher Educator, 37*(1), 49-57.

Van Manen, M. (2002) The pedagogical task of teaching. *Teaching and Teacher Education, 18*(2), 135-138.

KEY TERMS

AP (Advanced Placement): High school courses administered from Baltimore, Maryland, that are of post–high school curriculum standard. Many North American universities provide credit towards first year courses, depending on the standard of pass obtained.

Asynchronous: In delayed time (e.g., learning from a Web site at a time that is personally convenient).

CDLI (The Centre for Distance Learning and Innovation of the Department of Education of Newfoundland and Labrador): CDLI promotes e-learning teams and strives to provide access to educational opportunities for students, teachers and other adult learners in both rural and urban communities within Newfoundland and Labrador.

Digital Intranet: Schools, usually located in rural communities, that are linked through the Internet for collaborative teaching and learning.

E-Teachers: Teachers who teach online, through the Internet.

Matrix: A place in which a thing is developed (*Concise Oxford Dictionary*).

M-Teams: Mediating teachers who support e-teachers on sites where students receive instruction from e-teachers. Sometimes the entire staff of a small school will act as an m-team.

Open Classes: Classes in schools that are academically and administratively integrated so that teachers and learners can collaborate.

Synchronous: In real time (e.g., face-to-face instruction).

M

Multilevel Modeling Methods for E–Collaboration Data

Sema A. Kalaian
Eastern Michigan University, USA

INTRODUCTION

By design, most e-collaboration research has multilevel structures where e-collaboration groups of individuals may be created for different purposes (e.g., productivity improvements, or organizational decision-making). E-collaboration data usually has multilevel structures such as individuals (e-collaboration group members) are nested within groups and we have variables describing the individuals as well as the groups are an example of multilevel design.

Over the course of the past few decades, the multilevel modeling (hierarchical linear modeling, mixed-effects, random coefficients, variance components) methods have been developed (Hedeker & Gibbons, 1996; Longford, 1993; Rasbash et al., 2000; Raudenbush & Bryk; 2002). As a result of computer advances, the use of the multilevel methods to examine the effects of groups or contexts on individual outcomes has simply exploded across all disciplines (e.g., business, medicine, psychology, social and behavioral science).

E-collaboration data usually consists of multiple groups and levels, for example, individual-level data (micro-level), which consists of the characteristics of the individuals within e-collaboration groups and group-level data (macro-level), which consists of the characteristics of the e-collaboration groups. However, many E-collaboration theories center on the presumption that individual measurements at one level are usually influenced by the group dynamics. Yet, despite the multilevel structure of the e-collaboration data, the multilevel statistical methods have not been used in e-collaboration research to address questions of critical significance for decision-making purposes. Gallivan and Benbunan-Fich (2005) reviewed 36 e-collaboration empirical studies published from 1999 to 2004 in six IS journals and found that over two-thirds of these studies contained one or more problems of levels of analysis that cast doubts about the validity of the results of these studies. They stated in their article that one methodological issue of particular concern in e-collabo-

ration research seems to be the researchers' decision to analyze data at either the individual level or the group level, even when the theory that provides the basis for the research is formulated at both the individual and group levels and the research setting featured individuals working in e-collaboration groups. In such settings, the observations for individuals within the same group are correlated to some extent because these individuals share the same experiences and environments.

One of the common mistreatments of e-collaboration multilevel data in e-collaboration research is to disaggregate the data to the individual-level (micro-level) and ignore the existence of group-level (macro-level). Ignoring the multilevel structure and the grouping structure of the e-collaboration data has serious methodological consequences. Inaccurate and biased parameter estimates and biased standard errors of these estimates are examples of such methodological problems.

The other mistreatment of the e-collaboration multilevel data is to aggregate the individual-level (micro-level) to the group-level (macro-level) by using aggregated outcome and explanatory measurements such as the mean or the total values of these measurements. Similar to the disaggregation treatment of the multilevel data, the aggregation practice will lead to serious methodological problems. One of the significant problems is biased results and inaccurate conclusions because the analysis results of the aggregated measurements are different from the analysis results of the original individual-level measurements. Also, these biased results from analyzing the aggregated data leads to "ecological fallacy" (Robinson, 1950) where correlations between aggregated variables at the group level are used to make conclusions about individual level relationships.

Thus, one of the primary advantages of multilevel models is that they allow one to simultaneously investigate relationships within a particular hierarchical level as well as relationships between variables across hierarchical levels. This leads to valid and unbiased results and conclusions.

Multilevel modeling methods can be applied to different kinds of hierarchically structured e-collaboration data. E-collaboration data with continuous, binary, ordinal, or count outcomes are a few examples of such different applications. However, the two-level multilevel e-collaboration data with continuous outcomes (dependent variables) and individuals are clustered (nested) within e-collaboration groups is one of the most basic and common applications in e-collaboration research. Thus, the field of e-collaboration research that deals with using e-collaboration and virtual teams needs special and rigorous research methods to meet the challenges and the complexities of the multi-group e-collaboration data.

The present article aims to (1) conceptualize and present the two-level multilevel model for e-collaboration research, (2) conceptualize the Intra-Class Correlation Coefficient (ICC), (3) conceptualize R^2 in e-collaboration multilevel modeling, (4) present centering methods that can be used in e-collaboration multilevel modeling, (5) present parameter estimation and hypothesis testing methods for e-collaboration multilevel modeling, and (6) list some of the existing commercial software packages that can be used for analyzing the e-collaboration multilevel data.

TWO-LEVEL MULTILEVEL MODEL

The two-level multilevel model is characterized as having two levels where individuals (e-collaboration group members) are nested within e-collaboration groups and there are predictors for each of the two levels. Hence, in multilevel modeling with two levels, each level is represented by its own regression equation. In this multilevel modeling application, e-collaboration researchers are primarily interested in assessing the effects of the individual characteristics (e.g., experience, age, education) within e-collaboration groups as well as e-collaboration group characteristics (e.g., location, size) on the continuous outcome variable (e.g., performance, accomplishment) and the interactions between the individual and group characteristics. Thus, these multilevel models express relationships among variables within each of the levels and specify how variables at one level influence relations occurring at another level (cross-level interaction).

It is important to note that understanding the technical conceptualization of the two-level model as

presented below is needed for multilevel data analysis purposes using multilevel software packages. For example, this technical presentation clarifies the need for two data files to be inputted to the HLM software package. One is the Individual-Level (Level-1, Micro-Level) data file and the other is the Group-Level (Level-2, Macro-level) data file.

Individual-Level (Level-1) Model

The Individual-Level, Level-1, or Micro-level model specifies the relationships among various individual characteristics as independent explanatory variables (predictors), X_{ij}, for each of the j e-collaboration groups and the dependent variable, Y_{ij}. This Level-1 model takes the form of:

$$Y_{ij} = \beta_{oj} + \beta_{pj} X_{pij} + r_{ij}, \tag{1}$$

where, $i = 1, 2, 3, \ldots, n_j$ individuals within E-collaboration group j. $j = 1, 2, 3, \ldots J$ E-collaboration groups. β_{oj} represents the intercept for the individual and β_{pj} represents p regression coefficients (slopes) capturing the effect of the p predictors X_{ij} on the outcome, Y_{ij}. In multilevel modeling, these Individual-Level (Level-1, Micro-Level) regression coefficients are assumed to be random and vary from one e-collaboration group to another. r_{ij} represents the Level-1 random error and assumed to be normally distributed with mean zero and a common variance, σ^2. These errors are assumed to be uncorrelated with the Level-1 predictor variables. Also, the variances of the random errors (σ^2) are assumed to be equal (homogeneous) across the e-collaboration groups. Thus, Individual-Level (Level-1) model yields j separate set of regression estimates for the intercept and each of the p slopes.

Group-Level (Level-2) Model

In multilevel modeling, the intercept and the slopes (regression coefficients) estimates from the Individual-Level (Level-1) model are conceived as outcome (dependent) variables in Group-Level (Level-2, Macro-Level) model. These Level-2 dependent variables (intercept and slopes) from Level-1 are modeled by the q Group-Level (Level-2) characteristics (predictors). The Group-Level (Level-2) intercept and slope models (Equations 2 and 3) take the form of:

$$\beta_{0j} = \gamma_{oo} + \gamma_{oq} G_{qj} + U_{oj} \qquad (2)$$

$$\beta_{pj} = \gamma_{po} + \gamma_{pq} G_{qj} + U_{pj} \qquad (3)$$

G_{qj} represents Group-Level (Level-2) predictors (e.g., group size, group type- academicians vs. non-academicians). γ_{00} represents the average intercept for the j groups. γ_{0q}, γ_{p0}, and γ_{pq} represent Group-Level (Level-2, Macro-Level) fixed effects regression coefficients that capture the influence of group characteristics (G_{qj}) on the Individual-Level (Level-1) relationships (in this case, β_{0j} and β_{pj}). U_{oj} and U_{pj} are the Group-Level (Level-2) random residuals, which have a multivariate normal distribution with a zero mean vector and variance-covariance matrix T with $\tau_{00}, \tau_{pp}, \tau_{0p}$ and $\tau_{p'p}$ components where $p' \neq p$. τ_{00} represents the population variance among the intercepts of the j groups. τ_{pp} represents the population variance among the slopes of the j groups for the pth predictor variable. τ_{0p} is the population covariance between the intercepts and the pth slope. These Level-2 residuals and predictors are assumed to be uncorrelated.

No-Predictors Individual-Level (Level-1) Model

This model is a much simpler model than the Level-1 model presented above. In this model, there are no explanatory variables (predictors) at both levels. The basic model is extremely useful in multilevel modeling to assess the need for multilevel modeling and subsequently the need for more complex models, where predictor variables are included in the Individual-Level and Group-Level models (as presented above). A no-predictors individual-level model is represented as:

$$Y_{ij} = \beta_{oj} + r_{ij}, \qquad (4)$$

where, Y_{ij}, represents the outcome for individual i, $i=1,2, \ldots, n_j$ in group j, $j = 1,2, \ldots, J$. B_{oj} represents the mean of the outcome in group j. r_{ij} is Level-1 random error with a mean of zero and a variance, σ^2.

No-Predictors Group-Level (Level-2) Model

At the Group-Level (Level-2), B_{oj} allowed to vary randomly across groups, with a mean γ_{00} and error term U_{oj} as follow:

$$\beta_{oj} = \gamma_{oo} + U_{oj}, \qquad (5)$$

where, γ_{00} represents the overall grand mean of the outcome across all schools. U_{oj} is the error term representing the deviation of each group's mean from the grand mean, γ_{00}. The variance of U_{oj} is τ_{00}.

INTRA-CLASS CORRELATION COEFFICIENT (ICC)

The no-predictor model can be used to assess the Intra-Class Correlation Coefficient (ICC). As is mentioned earlier that the multilevel models are needed in e-collaboration research data because of the groupings of individuals, which violate the assumption of independence. Thus, Intra-Class Correlation Coefficient measures the degree of dependency among individuals within groups due to common knowledge and experiences of individuals within e-collaboration groups (Kalaian & Kasim, 2006). Thus, ICC measures the extent of homogeneity of the groups. Also, it provides an index of the proportion of the variance explained by the grouping structure (Hox, 2002). ICC is represented as the ratio of the Group-Level (Level-2) variance from fitting the two-level no-predictors multilevel model, τ_{00} to the total variance of Individual-Level and Group-Level models as follow:

$$ICC = \tau_{00} / (\tau_{00} + \sigma^2) \qquad (6)$$

An ICC close to zero indicates that e-collaboration groups are similar to each other and the grouping structure of the data can be ignored. Therefore, multilevel modeling is not needed to analyze the e-collaboration data.

R² IN MULTILEVEL MODELING

As in traditional regression analysis, the amount of variance explained in the two-level multilevel modeling is represented by R^2. The variance estimates from the no-predictors and the two-level models with predictors provide estimates to be used for calculating R^2 (Hox, 2002; Kalaian & Kasim, 2006; Raudenbush & Bryk, 2002). There are at least three types of R^2 depending on the level of the multilevel analysis and the formulation of the multilevel model. Three basic kinds of R^2 are presented.

1. **Individual-Level R²:** R^2 for the Level-1 model represents the percentage of variance in the Level-1 outcome accounted for by the Level-1 predictors and represented as:

    $$R^2_{\text{Individual-Level Model}} =$$
    $$\frac{\sigma^2_{\text{No-predictors Individual-Level Model}} - \sigma^2_{\text{Individual-Level Model}}}{\sigma^2_{\text{No-predictors Individual-Level Model}}}$$

    (7)

2. **Group-Level R² for the intercept:** R^2 for the Level-1 random intercept represents the percentage of the random intercept variance accounted for by the Level-2 predictors and is estimated as:

    $$R^2_{\text{Group-Level Intercept Model}} =$$
    $$\frac{\tau_{00 \text{ - No Predictors Group-Level Model}} - \tau_{00 \text{ - Group-Level Model}}}{\tau_{00 \text{ - No Predictors Group-Level Model}}}$$

    (8)

3. **Group-Level R² for the slopes:** There are p estimates of R^2 for the Level-2 model, one for each of the p predictors used to predict the outcome in Level-1. Each of the p R^2s for the Level-1 random slopes represents the percentage of a particular p slope variance accounted for by the q Level-2 predictors and represented as:

    $$R^2_{\text{Group-Level Slope Model}} =$$
    $$\frac{\tau_{00 \text{ - No Predictors Group-Level Model}} - \tau_{pp \text{ - Group-Level Model}}}{\tau_{00 \text{ - No Predictors Group-Level model}}}$$

    (9)

CENTERING INDIVIDUAL AND GROUP PREDICTORS

Centering of the Individual-Level and Group-Level predictors has significant implications for having more meaningful and interpretable multilevel results, especially when the predictor variable does not have meaningful zero value. For example, employees' yearly income do not have meaningful zero value because there is no employee in the group/organization with no income. Therefore, centering has a significant influence on Individual-Level and Group-Level coefficients and their variances. Three types of Individual-Level centering are presented below and these methods can be easily conceptualized and extended for Group-Level (Level-2) models

1. **Grand-mean centering:** In this centering method, the grand mean, $\bar{X}$, of a particular Level-1 predictor variable is subtracted from that particular predictor value for each individual, X_{ij}, across all j E-collaboration groups. This is the most frequently used centering method where all values of a particular predictor are centered on the grand mean and it is represented as:

    $$X_{ij} - \bar{X} \tag{10}$$

2. **Group-mean centering:** This method of centering is recommended when there is a substantial differences between the group means. In this method, all values of a particular predictor, X_{ij}, is centered on its group mean, $\bar{X}_j$, by subtracting the Group-Level (Level-2) mean of the predictor from each of the predictor values for each individual as follow:

    $$X_{ij} - \bar{X}_j \tag{11}$$

3. **Specific-value centering:** All values of a particular predictor, X_{ij}, is centered on a theoretically meaningful value by subtracting a specific and theoretically meaningful value (e.g., specific age) from each predictor value for each individual as follow:

    $$X_{ij} - SpecificValue \tag{12}$$

Centering of Group-Level (Level-2) predictor variables is less critical than centering the Individual-Level predictors, except when interaction terms between the Group-Level predictors are included at Level-2 (Raudenbush & Bryk, 2002).

MULTILEVEL ESTIMATION AND HYPOTHESIS TESTING

The two-level multilevel model analyses provide three kinds of point and interval parameter estimates with different hypothesis testing methods to make inferences from these multilevel parameter estimates.

1. **Group-Level fixed effects:** Multilevel modeling provides optimal weighted estimates for the Group-Level fixed effect parameters, γ's. In this estimation method, the fixed effect estimates are weighted by their precision, which is the reciprocal of the total Individual-Level and Group-Level variances (e.g., σ^2 and τ_{00} for the two-level no-predictor model). Multilevel analysis provides a t-test for testing the hypothesis that each of the Group-Level fixed effects parameters is significantly different from zero (e.g., $H_0: \gamma_{pq} = 0$).
2. **Individual-Level random effects:** Multilevel modeling provides empirical Bayes ("shrinkage" or "optimal weighted") estimates for the Individual-Level random regression coefficients, β's. In this estimation method, the fixed effect estimates are weighted by the reliability of each group and the reliabilities of all J groups. These reliabilities are functions of the Individual-Level (Level-1) and Group-Level (Level-2) variances. A Z-test can be used for testing the hypothesis that each of the Individual-Level random coefficients is significantly different from zero (e.g., $H_0: \beta_{pj} = 0$).
3. **Individual-Level and Group-Level variances:** Multilevel modeling also provides empirical Bayes estimates of the Individual-Level residuals variance (σ^2) and the variances-covariances of the Group-Level residuals (τ_{00} representing the variance of U_{0j}, τ_{pp} representing the variance of U_{pj}, τ_{0p} representing the covariance between U_{0j} and U_{pj}, and $\tau_{p'p}$ representing a covariance between $U_{p'j}$ and U_{pj} where $p' \neq p$. A Chi-square test statistic tests the null hypothesis that a particular Group-Level residual variance is significantly different from zero (e.g., $H_0: \tau_{00} = 0$ or $H_0: \tau_{pp} = 0$).

MULTILEVEL SOFTWARE PACKAGES

Many well-known commercial specialized software packages for analyzing multilevel structured data have been developed. Hierarchical linear modeling (HLM) and its latest version is HLM 6 (Raudenbush, et al., 2004) is one of the widely used packages. MLn, MLwIN (Rasbash et al., 2000), VARCL (Longford, 1993), SAS Proc Mixed (Littell, Miliken, Stroup, & Wolfinger, 1996), SPSS command VARCOMP, MIXOR (Hedeker & Gibbons, 1996) are examples of other existing software packages for analyzing multilevel data. These commercial software packages are used to analyze multilevel data in many disciplines (e.g., business, social science, medicine, education) and it can be used as well for analyzing e-collaboration multilevel data that is presented in this article.

It is important to note the existence of the software "PinT," which is a specialized program for Power calculations of two-level designs and can be downloaded from http://stat.gamma.rug.nl/snijders/multilevel.htm. This software can be used in the design phase of a multilevel E-collaboration study (Snijders & Bosker, 1999).

CONCLUSION

Multilevel models are increasingly recognized in most disciplines (e.g., business, medicine, health, psychology) as a potentially important analytical method for research designs with multilevel structures where individuals are nested within groups/organizations and we have variables describing the individuals as well as the groups/organizations. Generally, multilevel models aim to draw simultaneously Individual-Level (Micro-Level) and Group-Level (Macro-Level) inferences and the interactions between them. By design, most e-collaboration research has multilevel structures where e-collaboration groups of individuals may be created for different purposes (e.g., productivity improvements, or organizational decision-making). Gallivan and Benbunan-Fich (2005) called for the consistency of the levels of the theory and the levels of e-collaboration data to avoid problems of levels incongruence.

Therefore, it is necessary to use multilevel modeling to meet the challenges and complexities of the e-collaboration multilevel theories, research designs, and data structures.

It is important to note that the basic two-level multilevel model for analyzing e-collaboration and virtual team data with continuous outcomes that is presented in this article, can be easily extended to three or more levels. Individuals nested within e-collaboration groups, which in turn are nested within organizations is an example of a three-level model. Also, other e-collaboration multilevel models can be formulated for other types of outcome metrics, such as binary, ordinal, count, and multinomial outcomes and can be analyzed using commercially available multilevel software packages such as HLM6 software, which is designed for analyzing different kinds of multilevel models, including the one that is presented in this article.

In conclusion, due to the multilevel nature and structure of e-collaboration research and because of the severe methodological consequences of ignoring the multilevel structure of the e-collaboration data (e.g., non-valid results), it is important that the e-collaboration researchers pay special attention to multilevel issues as they design e-collaboration studies, formulate e-collaboration theories, and collect the multilevel structured data. This can be achieved by using multilevel modeling methods and software to analyze the multilevel structured e-collaboration data. Thus, the harmony of the e-collaboration research, the multilevel structured e-collaboration data, and multilevel modeling methodology and analysis lead to unbiased estimates and valid e-collaboration research findings and conclusions.

REFERENCES

Hedeker, D., & Gibbons, R. D. (1996). MIXOR: A computer program for mixed-effect ordinal regression analysis. *Computer Methods and Programs in Biomedicine, 49,* 157-176.

Gallivan, M. J., & Bebunan-Fich, R. (2005). A framework for analyzing levels of analysis issues in studies of e-collaboration. *IEEE Transactions on Professional Communication, 48*(1), 87-104.

Hox, J. (2002). *Multilevel analysis: Techniques and applications.* Mahwah, NJ: Lawrence Erlbaum Associates.

Littell, R. C., Miliken, G. A., Stroup, W. W., & Wolfinger, R. D. (1996). *SAS system for mixed models.* Cary, NC: SAS Institute, Inc.

Longford, N. T. (1993). *VARCL: Software for variance component analysis of data with nested random effects (maximum likelihood). Manual.* Groningen: ProGAMMA.

Kalaian, S. A., & Kasim, R. M. (2006). Hierarchical linear modeling. In Neil Salkind (Ed.), *Encyclopedia of measurement and statistics.* Sage Publications.

Rasbash, J., Browne, W., Goldstein, H., Yang, M., Plewis, I., Healy, M., et al. (2000). *A user's guide to MlwiN.* London: Multilevel Models Project, University of London.

Raudenbush, S. W., & Bryk, A. S. (2002). *Hierarchical linear models: Applications and data analysis methods.* Sage Publications.

Raudenbush, S. W., Bryk, A. S., Cheong, Y., & Congdon, R. T. (2004). *HLM 6: Hierarchical linear and nonlinear modeling.* Chicago: Scientific Software International.

Robinson, W. (1950). Ecological correlations and the behavior of individuals. *American Sociological Review, 15,* 351-357.

Snijder, T., & Bosker, R. (1999). *Multilevel analysis.* London: Sage Publications.

KEY TERMS

τ_{00}: Is the Level-2 (group-level) variance among the random intercepts, β_{0j}, from the Individual-Level (Level-1) model.

τ_{11}: Is the Group-Level (Level-2) variance among the random slopes, β_{pj}, from the Individual-Level (Level-1) model.

Amount of Variance Explained at Level-1 (R^2): Is an index of the amount of variance explained at Level-1 by the p Level-1 predictors. Here, R^2 is estimated by comparing the Level-1 variance, σ^2, estimate from the basic two-level model with the Level-1 variance, σ^2, estimate from the no-predictor model. (See Equation 7)

Amount of Variance Explained at Level-2 in each Regression Coefficient (R^2): Is an index of the amount of variance explained in each of the Level-1 random regression coefficients (intercept and slopes) at Level-2 by the q Level-2 predictors. Here, R^2 is estimated by comparing the Level-2 variance (τ_{00} or τ_{pp}) estimate from the basic two-level model with the Level-2 variance, τ_{00}, estimate from the no-predictor model. (See Equation 8 and 9).

Centering: Is changing the location of the Level-1 and Level-2 predictors in the multilevel model by subtracting the grand mean, group mean, or specific value from each predictor variable.

Empirical Bayes Estimate of a Random Level-1 Coefficient: Is an estimate of the Individual-Level random intercept or slope(s) for a particular e-collaboration group. This empirical Bayes estimate utilizes the data from that specific e-collaboration group and the data from all the e-collaboration groups.

Intra-Class Correlation (ICC): It is the proportion of the total variance explained by the group differences.

Multilevel Modeling: Refers to a set of different statistical methods for analyzing data with two or more levels of hierarchical or nested structures.

Two-Level Multilevel Model: Refers to analytic methods for data with two levels where individuals are nested within e-collaboration groups and there are predictors characterizing individuals as well as groups.

Multilingual Collaboration in Electronic Meetings

Milam Aiken
University of Mississippi, USA

INTRODUCTION

Groups in which participants do not speak the same language frequently find communication difficult. Yet, multilingual meetings are common as a form of collaboration. To overcome this language barrier, banks, government agencies, hospitals, the courts, and many other institutions have relied upon human translators to enable meeting participants to exchange ideas and opinions. For example, the United Nations General Assembly's discussions, conferences within the European Union, multinational corporations' business negotiations, and many other meetings are conducted almost daily, requiring large amounts of interpreters' scarce expertise and time.

In addition to the problems of translation efficiency and effectiveness, these meetings have the same limitations as those involving a single language: (1) only one participant can speak at a time, (2) comments must be transcribed manually for a permanent record, and (3) many group members do not participate because of shyness or because other speakers monopolize the conversation.

Group support systems (GSS) have automated the meeting process and improved the productivity of groups needing to share ideas (Dennis, George, Jessup, Nunamaker, & Vogel, 1988). By integrating machine translation (MT) with a GSS, multilingual groups can enjoy the same benefits as monolingual groups. This paper summarizes research conducted using automatic and semi-automatic natural language translation in electronic meetings and shows how a multilingual GSS (MGSS) can improve communication and collaboration.

MULTILINGUAL MEETINGS

Meetings involving more than one language (e.g., English and Spanish) typically incorporate human translators who work either synchronously (simulta-

neous interpretation) or asynchronously (consecutive interpretation). Using the first approach, an interpreter sitting in a soundproof booth listens using a headset to the speech in one language from the meeting room and translates the speech into a different language, speaking into a microphone connected to the headsets of selected participants in the meeting. With multiple interpreters, a speech or discussion can be translated into several languages almost simultaneously. Using the asynchronous approach, an interpreter takes notes or transcribes the comments from one language as they are spoken. When the speaker pauses or finishes, the interpreter then delivers the speech again in another language.

Besides the cost (an interpreter might charge 0.16 to 0.40 cents per word, depending on the target language) and problem of scheduling (interpreters with skills in the required language translation-pairs might not be available at the desired meeting time), even skilled interpreters can make errors because of poor perception of the source language (the speaker might slur words, speak softly, have an unusual accent, etc.), lack of familiarity with slang or technical terms peculiar to the meeting topic, or just human fallibility. Further, multilingual meetings suffer from the same problems as monolingual meetings in terms of productivity and participant satisfaction.

GROUP SUPPORT SYSTEMS

Group support systems (also called electronic meeting systems or groupware) automate traditional, oral meetings by allowing group members to type and view comments on computers connected via a network (LAN or WAN). These systems improve the efficiency and effectiveness of meetings primarily by: (1) allowing participants to submit and view all comments simultaneously, (2) automatically recording all submitted comments, and (3) providing anonymity (no participant can determine who wrote a particular comment). As a consequence of these provisions, studies have shown that for groups of more than seven participants who share ideas with a GSS, meeting times are shorter, more comments are generated, better quality comments are contributed, group members participate more and more equally, and participants are more satisfied with the meetings (Nunamaker, Briggs, Mittleman, Vogel, & Balthazard, 1997).

The use of the technology has become widespread, and one leading commercial GSS product, *Group-Systems* (www.groupsystems.com), currently is used on every continent except Antarctica. However, with some exceptions (e.g., Aiken, Hwang, Paolillo, & Lu, 1994a; Lagumdzija, 1996; Lim, Raman, & Wei, 1990; Mejias, Lazeneo, Rico, Torres, & Vogel, 1996; Morales, Moreira, & Vogel, 1995), most electronic meetings and nearly all research in the area have been conducted using English-speaking groups (Pervan, 1998). Even though English is used in many locations throughout the world, most people prefer to use their native language when communicating, and some means should be provided to accommodate multilingual groups using a GSS.

MULTILINGUAL E-COLLABORATION

The idea of translating typed comments within an electronic meeting was first proposed by Gray and Olfman (1989). The first completely automatic, multilingual GSS was developed in 1991 (Aiken, Martin, Reithel, Shirani, & Singleton, 1992), followed by another version (Aiken, Martin, Paolillo, & Shirani, 1994b) developed with the goal of improving translation accuracy. In these fully automatic MGSS meetings, participants were able to type in one language and submit the comment while translations automatically appeared on other terminals. Based upon software configuration, group members could be allowed to view comments only in their language (e.g., Spanish) or comments in a mixture (e.g., Spanish and English). In the latter approach, if a translation was inaccurate and a participant knew a little of the other language, he or she could possibly make a more accurate guess as to the correct meaning of the comment.

A third, semi-automatic version (Aiken, Rebman, Vanjani, & Robbins, 2002) was developed with the goal of supporting languages other than only English and Spanish (33 different languages) and allowing group members to participate any where in the world via the Web. In this semi-automatic MGSS, a staff member

played an active role in the translation process (cutting and pasting results from the PC-based MT program to the Web-based GSS software).

The final MGSS (Aiken, Wang, & Vanjani, 2003) was developed with the goal of making the system more portable, and was consequently, a completely Web-based system. In the newest version, both the GSS and the MT software are Web-based, allowing the staff members who participated in the translation process to be anywhere.

Automatic and semi-automatic GSS meetings are likely to have less accurate translations than electronic meetings served by human interpreters. As the group size increases and more participants submit comments nearly simultaneously, several interpreters are required to read a comment in one language, translate, and type again in another language. One study (Rebman, Aiken, & Cegielski, 2003) showed that undergraduate business students are able to type 36 "easy" words per minute (commonly occurring words with few syllables) and type 24 "difficult" words per minute. The final version of the MGSS, however, is able to translate and submit to the group 600 words per minute. In addition, a virtually unlimited number of the completely automatic MT programs could run during a meeting (even one per participant), making it possible to support even very large groups with a negligible lag in translation time. Thus, in terms of efficiency alone, the MGSS is superior.

Translation accuracy has been the major barrier to greater MT acceptance, but high accuracy might not be needed in a GSS meeting. For example, if one comment is not understood, there are likely to be other similar, if not redundant, adjoining comments that might be clearer or could aid the understanding of the earlier comment. In addition, a participant can submit a new comment asking for clarification from the group.

MACHINE TRANSLATION

Machine translation is the basis of an MGSS meeting, and its accuracy is vital to the success of the discussion. However, natural language translation is very difficult, and even expert humans are inconsistent. In addition, there are no universally accepted and reliable measures of translation accuracy (Balkan, Netter, Arnold, & Meijer, 1994). Some studies focus on the percentage of sentences with minor or major errors, some focus on the percentage of text that is understood by subjective evaluators, and others compute a measure automatically using N-grams (Papineni, Roukos, Ward, & Zhu, 2002).

Further, comparisons of MT evaluation studies are difficult due to differences in the number of subjects evaluating translations, the language skill of the subjects, the difficulty and nature of the source text, and the specific software used. In one representative study (Bezhanova, Byezhanova, & Landry, 2005), 17 English sentences were translated into Spanish using *LogoMedia, SYSTRAN,* and *PROMT.* All three of the MT systems produced usable translations, but none had an obvious advantage.

MGSS TRANSLATION

The previous section reviewed studies of the translation accuracy of existing documents, but none of these MT systems were used "on the fly" by a group. Only eight studies have investigated the use of language translation within an electronic meeting (Table 1). Most of the translations within the meetings had some errors (with accuracy ranging from 0% to 76%), but group members were able to understand the majority of the translations (with accuracy ranging from 40% to 100%). There are several explanations why this high variation occurred:

1. **Subject differences:** Humans have different levels of language fluency and mental acuity, and thus a phrase or sentence that is understood by some might not be apparent to others. For example, "What you eat?" might be understood by some as "What do you eat?" while others might understand it to mean, "What, you eat?" Even sentences that seem obvious but are grammatically incorrect such as "What be your name?" might not be understood by some English speakers.

2. **Software differences:** Four different machine translation systems were used in the MGSS studies. Even when using a single program, translations between some languages (e.g., German to French) were not as good as between others (e.g., English to German) due to peculiarities in the languages (e.g., the frequent use of contractions in French, adjectives follow nouns in Spanish but precede them in English, etc.).

3. **Input errors:** Comments written in electronic meetings often contain grammatical errors (e.g., missing punctuation, lack of capitalization for proper nouns, etc.) and typographical errors (e.g., incorrect spelling). As the well-known saying goes, "garbage in, garbage out." It is unreasonable to expect a language translation program to achieve high accuracy if the source comment contains errors. Studies in which the input accuracy was high showed high translation accuracy, and vice-versa.

4. **Use of slang, jargon, and idioms:** The MT system dictionaries do not contain many uncommon words. In addition, idioms such as "Are you pulling my leg?" are notoriously difficult to translate to another language. Literal translations are not always equivalent.

5. **Source text difficulty:** As the reading difficulty of the source text increases (as measured by average word and sentence length), it is reasonable to expect that translation also will become more difficult. For example, many beginning lessons in introductory language textbooks typically have short sentences and words.

6. **Chance:** There is a certain amount of unpredictability in how good a translation will be. In fact, the same source comment given to three professional, expert, human interpreters often results in three different translations.

Generalizations from these studies are difficult due to the samples of convenience and small sample sizes within ad hoc meetings. Future research should attempt to replicate the studies with larger groups. In addition, the use of an MGSS within actual multilingual organizational meetings should be investigated.

MANAGERIAL IMPLICATIONS

The question remains, "How accurate must a translation be?" (Hutchins, 2001). Many comments in a typical monolingual meeting are ungrammatical, and some might not even be understood by all group members. Therefore, 100% accuracy is probably an unreasonable expectation for lengthy samples of text.

The amount of accuracy required depends upon the formality or seriousness of the meeting and the degree to which participants are multilingual (Hacken,

2001). For example, Mexicans and Americans who are moderately bilingual but prefer using their native languages could probably understand mistranslations, especially if the translation is shown along with the source comments. If the meeting is only to exchange initial opinions in a focus group, perhaps high accuracy is not needed. However, if the meeting's purpose is to develop a legal contract, almost perfect translation accuracy is required.

A range of MGSS options are available to accommodate these different meeting scenarios:

1. **Completely automated MGSS:** In this form of meeting, one or many computer programs run on the same network as the electronic meeting and no human interpreters are involved. Translations are provided very quickly, but are often poor, mostly because of source comment errors in translations. Few MT systems support hands-off, batch-mode translation, however.

2. **Semi-automated MGSS:** In this form of meeting, staff members cut-and-paste MT-generated text into the GSS software with no editing. Thus, there is an extra few seconds of delay.

3. **Computer-assisted MGSS:** In this form of meeting, trained human interpreters edit the MT-generated text and then submit the translation to the group. Translation speed is limited by the reading, thinking, and editing speeds of the staff members, but this problem can be ameliorated by adding several interpreters, all sharing the work. Accuracy is higher with the addition of interpreter intervention.

4. **Non-automated MGSS:** In this form of meeting, human interpreters read a comment, prepare a translation without the use of an MT system or even a dictionary, and type the new text in a different language for submission to the group. The translations are provided slowly (unless there are many interpreters), but the translations are generally very good. However, the sequence of comments within the meeting might not be important. That is, it might not matter if the translations are several minutes late and appear near the end of the transcript.

5. **Asynchronous MGSS:** In this form of meeting, several sessions are conducted at different times using a Delphi or Nominal Group Technique (Dowling & St. Louis, 2000). For example, Eng-

lish-speaking group members could exchange typed ideas in an electronic meeting. During a short break, human interpreters (perhaps with the assistance of MT) translate and record all of the comments to Spanish. Spanish-speaking group members could then replace the earlier English-speaking participants and submit comments only in Spanish. After several turns using alternate languages, group members will have viewed all comments.

CONCLUSION

Eight studies of automatic and semi-automatic translation during electronic meetings involving seven different languages show a wide range of absolute and understanding accuracies. Despite the fact that meeting comments generated with Multilingual Group Support Systems lack 100% accuracy, in most cases, these systems can assist polyglot groups by providing nearly simultaneous translation of multiple comments in addition to parallel communication, anonymity, and automatic recording of the discussion.

REFERENCES

Aiken, M., Ablanedo, J., & Vanjani, M. (2004a). An analysis of electronic meeting comment translation. *Communications of the International Information Management Association, 4*(4), 13-24.

Aiken, M., Hwang, C., Paolillo, J., & Lu, L. (1994a). A group decision support system for the Asian Pacific rim. *Journal of International Information Management, 3*(2), 1-13.

Aiken, M., Kim, S., & Vanjani, M. (2004b). Chinese, Korean, and Japanese translation of English electronic meeting comments. *Asian Journal of Information Technology, 3*(12), 1291-1303.

Aiken, M., Martin, J., Paolillo, J., & Shirani, A. (1994b). A group decision support system for multilingual groups. *Information and Management, 26,* 155-161.

Aiken, M., Martin, J., Reithel, B., Shirani, A., & Singleton, T. (1992). Using a group decision support system for multicultural and multilingual communication. In

Proceedings of the 23rd Annual Meeting of the Decision Sciences Institute, 2 (pp. 792-794).

Aiken, M., Rebman, C., Vanjani, M., & Robbins, T. (2002). Meetings without borders: A multilingual Web-based group support system. In *Proceedings of the America's Conference on Information Systems* (pp. 146-149), Dallas, TX.

Aiken, M., Sloan, H., & Martin, J. (1998). Using a bilingual group support system. *Behaviour & Information Technology, 17*(3), 141-144.

Aiken, M., Ulrich, M., & Singleton, T. (1997). A bilingual group decision support system. *International Journal of Intelligent Systems in Accounting, Finance, and Management, 6,* 279-285.

Aiken, M., Vanjani, M., Martin, J., Young, C., & Govindarajulu, C. (1994c). Experiences with a bilingual group decision support system. *International Business Schools Computing Quarterly, 6*(1), 4-9.

Aiken, M., Wang, W., & Vanjani, M. (2003). Automatic comment translation in a Chinese-English electronic meeting. In *Proceedings of the 34th Annual Southwest Decision Sciences Institute Conference (SWDSI),* (pp. 30-34), Houston, TX.

Aiken, M. & Wong, Z. (2001). The influence of textual complexity on automatic speech recognition accuracy. In *Proceedings of the 32nd Annual Meeting of the Decision Sciences Institute,* (pp. 28-30). San Francisco, CA.

Aiken, M., Wong, Z., & Vanjani, M. (2002). Speech complexity and automatic recognition accuracy. *Academy of Information and Management Sciences (AIMS), 5*(1-2), 1-12.

Balkan, L., Netter, K., Arnold, D., & Meijer, S. (1994). Test suites for natural language processing. *Translating and the Computer,*(pp. 51-58). London: Aslib.

Bezhanova, O., Byezhanova, M., & Landry, O. (2005). *Comparative analysis of the translation quality produced by three MT systems.* Research report, McGill University, Montreal.

Dennis, A., George, J., Jessup, L., Nunamaker, J., & Vogel, D. (1988). Information technology to support electronic meetings. *MIS Quarterly, 12*(4), 591-624.

Dowling, K., & St. Louis, R. (2000). Asynchronous implementation of the nominal group technique: Is it effective? *Decision Support Systems, 29,* 229-248.

Gray, P., & Olfman, L. (1989). The user interface in group decision support systems. *Decision Support Systems, 5,* 119-137.

Hacken, P. (2001). Has there been a revolution in machine translation? *Machine Translation, 14,* 1-19.

Hutchins, W. (2001). Machine translation and human translation: In competition or in complementation? *International Journal of Translation, 13*(1-2), 5-20.

Lagumdzija, Z. (1996). Mission impossible: GroupSystems in Sarajevo 1995. In *Proceedings of the Seventh Annual GroupSystems Users' Conference*, Tucson, Arizona.

Lim, L., Raman, K., & Wei, K. (1990). Does GDSS promote more democratic decision-making? The Singapore experiment. In *Proceedings of the 23rd Annual Hawaii International Conference on System Sciences,* (pp. 59-65). Kailua-Kona, HI.

Mejias, R., Lazeneo, L., Rico, A. Torres, A., & Vogel, D. (1996). A cross-cultural comparison of GSS and non-GSS consensus and satisfaction levels within and between the U.S. & Mexico. In *Proceedings of the 29th Hawaii International Conference on System Sciences,* (p. 408). Kailua-Kona, HI.

Morales, B., Moreira, H., & Vogel, D. (1995). Group support for regional development in Mexico. In *Proceedings of the 28th Annual Hawaii International Conference on System Sciences,* (pp. 232-239). Kailua-Kona, HI.

Nunamaker, J., Briggs, R., Mittleman, D., Vogel, D., & Balthazard, P. (1997). Lessons from a dozen years of group support systems research: A discussion of lab and field findings. *Journal of Management Information Systems, 13*(3), 163-207.

Papineni, K., Roukos, S., Ward, T., & Zhu, W. (2002). BLEU: A method for automatic evaluation of machine translation. In *Proceedings of the 40th Annual Meeting of the Association for Computational Linguistics,* (pp. 311-318). Philadelphia, PA.

Pervan, G. (1998). A review of research in group support systems: Leaders' approaches and directions. *Decision Support Systems, 23,* 149-159.

M

Rebman, C., Aiken, M., & Cegielski, C. (2003). Speech recognition in the human-computer interface. *Information & Management, 40*(6), 509-519.

KEY TERMS

Delphi: A written, group, brainstorming technique characterized by a number of iterative stages through which a consensual view is reached.

Group Support System: A computer-based system that automates a meeting.

Idiom: An expression with a meaning that does not follow from the meaning of the individual words of which it is composed.

Machine Translation: Automated translation of a natural language pair (e.g., English to Spanish)

Multilingual Group Support System: A group support system with a machine translation interface allowing comments generated in a meeting to be translated.

Multilingual Meeting: A meeting involving several people who do not speak a common language such as English.

Nominal Group Technique: A written, group, brainstorming technique characterized by silent idea generation, round robin idea collection, grouping, and ranking.

Table 1. Accuracy of electronic meeting comment translations

Reference	Translation Software	Group Members	Raw Accuracy	Understood Accuracy
Aiken et al., 1992	Locally developed*	24 English in 4 groups. Facilitator added Spanish comments	NA	Spanish→English: 100%
Aiken et al., 1994b	MicroTac Spanish Assistant*	3 Spanish, 6 English	English→Spanish: 74% Spanish→English: 45%	English→Spanish: 96% Spanish→English: 85%
Aiken et al., 1994c	MicroTac Spanish Assistant*	3 Spanish, 5 English	English→Spanish: 70% Spanish→English: 46%	NA
Aiken et al., 1998	MicroTac Spanish Assistant*	2 Spanish, 2 English	English→Spanish: 71% Spanish→English: 76%	English→Spanish: 91% Spanish→English: 81%
Aiken et al., 2002	LanguageForce Universal Translator**	1 German, 1 French, 2 English	NA	German→English: 100% German→French: 40% French→German: 40% French→English: 60% English→German: 80% English→French: 80%
Aiken et al., 2003	SYSTRAN**	2 English, 2 Chinese	NA	English→Chinese: 93% Chinese→English: 100%
Aiken et al., 2004a	SYSTRAN**	Translation of 2 English-only-meeting comments. 3 Spanish evaluators	English→Spanish: 29%-48%	English→Spanish: 72%-74%
Aiken et al., 2004b	SYSTRAN**	Translation of 2 English-only-meeting comments. 3 Chinese, 3 Japanese, 3 Korean evaluators	English→Chinese: 0% English→Japanese: 0% English→Korean: 0%	English→Chinese: 43%-74% English→Japanese: 75%-78% English→Korean: 55%-78%

* *automatic: no human intervention in translation*
** *semi-automatic: facilitator cuts-and-pastes comment to translation software, runs the program, and cuts-and-pastes the translation back into the on-going meeting*

A New Model and Theory of Asynchronous Creativity

Dorrie DeLuca
University of Delaware, USA

INTRODUCTION

Global organizations necessarily operate in different countries with differing time zones, making face-to-face creativity costly in terms of travel and time expended. Yet for many other types of tasks, organizations would not hesitate to collaborate via e-mail, bulletin boards, file sharing, and other low-cost, readily available asynchronous media. Asynchronous e-collaboration is defined as collaboration among individuals engaged in a common task using electronic technologies that allow input at different times.

Many organizations automatically assume that synchronous, face-to-face (FTF) communication is best for creative tasks, a point of view propagated by outdated theory that suggests that FTF is the "richest" medium (Daft & Lengel, 1986). Recent research on process innovation (DeLuca & Valacich, 2006; Kock, 2006) suggests that certain characteristics of asynchronous e-collaboration are highly desirable. Organizations have postponed realization of benefits obtainable by virtual teams, a situation the author aims to improve by using the new model to develop new theory. With readily available, low-cost, simple Internet-based media, virtual teams have convened either purposefully or out of necessity, realizing cost savings and "surprisingly" good results.

There is a need for and call for theory which can explain and predict effects on creativity from use of asynchronous e-collaboration by dispersed teams (Hamilton, 2004; Kock, 2005b; Paulus, 2000; Weber, 2002). Asynchronous creativity theory (ACT), explained in this article, contributes to filling that need.

This article is divided into sections designed to introduce the theory to the readers. In the Background section, previous theory is summarized and the need for a new model and theory addressed. The model and theory are presented in four sections: a general model and theory overview and then one section on each set of influences on creativity—cognitive, social, and media influences. Those three sections contain the propositions of the theory. In the last section, Conclusion, consistent with the central mission of a good theory (Van de Ven, 1989), ACT contributes to knowledge on creativity of teams which use asynchronous e-collaboration and applies the knowledge to developing and managing teams for organizational innovation.

BACKGROUND

Most research on creativity is based on face-to-face (synchronous) interaction from a psychological trait perspective (Guilford, 1950), a social influences perspective (Amabile, Hennesey, & Grossman, 1986), or synchronous brainstorming (Dennis & Valacich, 1993; Dennis & Williams, 2005; Grise & Gallupe, 2000; Pinsonneault, Barki, Gallupe, & Hopper, 1999). Since existing creativity models have not generally considered the possibility of using asynchronous e-collaboration for creative tasks, a more inclusive model was developed by DeLuca (2006). It is an integrative, three-dimensional model which adapts and synthesizes Paulus' (2000) constructs of social and cognitive influences on creativity and integrates them with characteristics of various media from FTF to asynchronous e-collaboration as discovered in the author's applied research on virtual team innovation (DeLuca, Gasson, & Kock, 2006; Kock & DeLuca, 2006), and other research on group innovation (Kock, 2006) and media synchronicity (DeLuca & Valacich, 2006; Dennis & Valacich, 1999).

Research on e-collaboration and creativity together is primarily focused on brainstorming benefits gained through *synchronous* electronic media (Dennis & Valacich, 1993; Dennis & Williams, 2005; Grise & Gallupe, 2000; Pinsonneault et al., 1999).

Several existing theories explain part of the influences observed by the author. Media richness theory (Daft & Lengel, 1986) suggests that FTF communications would be necessary and richest for creative tasks but does not explain higher creativity using asynchronous e-collaboration. Media Synchronicity Theory (Dennis & Valacich, 1999) delves into media characteristics but with only partial examination of the psychological and sociological literature. Compensatory adaptation theory (Kock, 2005a) partly explains virtual teams' creative successes by demonstrating that users of convenient asynchronous e-collaboration adapt their communication behavior and effectuate some of the richness found in FTF (DeLuca et al., 2006), but the theory does not account for weaknesses of FTF media and strengths and desirability of asynchronous e-collaboration or adaptations made when communicating FTF.

The media-cognitive-social (MCS) model of creativity and related asynchronous creativity theory (ACT) were proposed by the author (DeLuca, 2006) and are summarized in this article. The propositions provide a more complete framework for researchers in information systems, psychology, and sociology to perform formal testing of a theory for creative tasks performed in a relatively novel way—asynchronously. The results may apply to a variety of organizational situations. ACT has the potential to change practice, by providing a basis for breaking down the walls of reluctance

of managers to use asynchronous e-collaboration for creative and/or innovation tasks by showing that there are built-in advantages.

The next section of this article presents the MCS model of creativity and the propositions for ACT, using the constructs in the model.

MEDIA-COGNITIVE-SOCIAL MODEL OF CREATIVITY AND ASYNCHRONOUS CREATIVITY THEORY

The essential constructs of ACT are shaped by the dimensions of the MCS model of creativity: media influences, cognitive influences, and social influences, all of which contribute to the generation of creative associations as shown in Figure 1. The model and theory are explained in greater detail in DeLuca (2006). Cognitive and social influences are adapted from Paulus' (2000) summary of existing creativity research, which is largely based on FTF, synchronous experiments. These relationships are shown in normal text. Media characteristics are adapted from media synchronicity theory (DeLuca & Valacich, 2006; Dennis & Valacich, 1999). Media influences and their relationship to cog-

Figure 1. Media-cognitive-social model of creativity (Adapted from DeLuca, 2006)

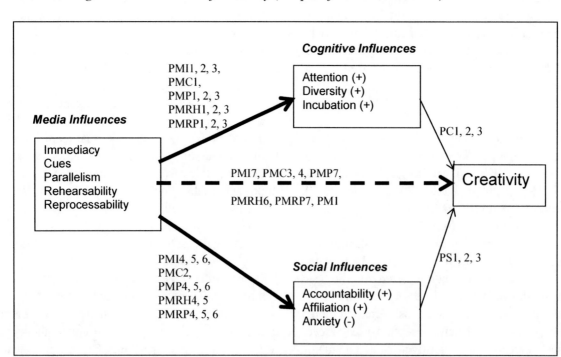

nitive and social influences and creativity make the model applicable to teams using any media for which the media characteristics can be defined as herein. These relationships are shown in bold. A dash is used to indicate derivation from other propositions. Cognitive and social influences are based on research that is primarily FTF, but found consistent across all media. The effect of media influences and thus the direct application of asynchronous media is found in the Media Influences section. ACT applies the model to the scope of virtual teams using asynchronous media for creative tasks. The abbreviations for the propositions of ACT are also shown on the figure.

COGNITIVE INFLUENCES ON CREATIVITY

Cognitive influences are summarized into three constructs which have a positive effect on creativity: Attention (amount of focus on the task at hand), Diversity (variety of frameworks/ideas), and Incubation (time away from the task at hand). Detractors from *Attention* (production blocking, cognitive interference/load, distraction, and negative conflict) negatively affect creativity.

PC1: Creativity is a positive function of Attention.

Assuming that creative ideas come from association to something else and more associations will mean more potential for ideas, then *Diversity* of environment and people would generally promote creativity.

PC2: Creativity is a positive function of Diversity.

Assuming that individuals must manage a variety of tasks and that they desire to accomplish their work tasks, having more time to manage those tasks, set them aside, and return to them would promote time available for creativity. Wallas (1926) indicates that *incubation* (rest or period without conscious attention to the problem or doing other activities) allows *illumination* (a "click" or "flash" of a new idea or solution often with only remote association to the problem. Letting the mind relax and not work on the task allows greater creativity when attention is returned to the task than if focus was attempted for a protracted period. This author asserts that the positive impacts of incubation are undervalued

in their current representation in the literature, largely due to the fact that much creativity research takes place FTF where incubation is less possible. It is certainly possible that extended incubation could lead to forgetting all about it, which could be mitigated with proper structure and timelines (DeLuca, 2003).

PC3: Creativity is a positive function of Incubation.

SOCIAL INFLUENCES ON CREATIVITY

Social influences on creativity include Accountability (a team member is associated with the quality of his/her input), Affiliation (the level of connection with the team) generally thought to have a positive effect on creativity until the point at which socializing distracts from the task, and Anxiety (a sense of unease) which has a negative effect on creativity. Assuming that individuals desire to accomplish their work tasks, *Accountability* (as seen in responsibility, competing with teammates to provide quality input, meeting goals) and lack of accountability (as seen in social loafing, free riding, illusion of productivity) affect creativity.

PS1: Creativity is a positive function of Accountability.

Affiliation (as seen in satisfaction, well-being) is necessary for the health of a team and may be as important for creativity as production goals. However, over-socializing during a meeting may weigh in more as Distraction from Attention mentioned above.

PS2: Creativity is a positive function of Affiliation.

Anxiety (as seen in evaluation apprehension, social inhibition, status, fear, avoidance) generally inhibits creativity. Some individuals may be inhibited in social settings, while others may experience anxiety due to a public deadline.

PS3: Creativity is a negative function of Anxiety.

MEDIA INFLUENCES ON CREATIVITY

The benefits of asynchronous e-collaboration are revealed by virtual team members (DeLuca et al., 2006;

Kock & DeLuca, 2006; DeLuca & Valacich, 2006), consistent with media capabilities of media of varying synchronicity (Dennis & Valacich, 1999) as adapted and summarized in Table 1.

Media influences on creativity are operationalized as: (1) *Immediacy*—rapid transmission and rapid feedback from their communications; (2) *Cues*—format by which information is conveyed, verbal and nonverbal language (e.g., tone, eye contact) as well as social cues (e.g., status, intimidation); (3) *Parallelism*—effectively working in parallel (simultaneous conversations in process, multiple people can communicate all at once, channel capacity); (4) *Rehearsability*—fine tuning a message before sending; and (5) *Reprocessability*—readdressing a message within the context of the communication event (e.g., rereading, printing).

Asynchronous, electronic, written media are high in parallelism, rehearsability, and reprocessability, characteristics that may be more desirable than immediate feedback or cues. Note how the FTF medium and e-mail are high in opposite characteristics. This would explain the fallacy of calling one medium the "richest," a shortcoming of earlier theories. Note also how synchronous electronic conferencing, the primary media in decision support system (DSS) brainstorming studies, does not allow full advantage of parallelism, rehearsability, and reprocessability, thereby limiting the potentiality of creativity.

Immediacy

Immediacy promotes creativity when one idea stimulates another and those ideas can be expressed (attention). High Immediacy inhibits creativity when two people have an idea at the same time and one interferes with the expression of the other (low communication channel capacity). We may be most creative in peace and quiet. Information overload and/or productivity drops may not occur in asynchronous environments, at least not at the same level, because the ideas can be managed over time and the broadened contexts can come from a variety (diversity) of environments over time. Many synchronous studies measure only ideation fluency, without attention to the usefulness of the ideas generated. Low immediacy allows time (incubation) for more integrative thinking which would improve the probability that an idea would be useful.

PMI1: Diversity is a negative function of Immediacy.

PMI2: Attention is a negative function of Immediacy.

PMI3: Incubation is a negative function of Immediacy.

Rapid feedback (as in a FTF environment) may be socially stimulating (serving affiliation), but interfere

Table 1. Media capabilities

Media	Immediacy	Social and Language Cues	Parallelism	Rehearsability	Reprocessability
Synchronous Face-to-Face	High	High	Low	Low	Low
Synchronous Video Conference	Med-high	Medium	Low	Low	Low
Synchronous Telephone Conference	High	Medium	Low	Low	Low
Synchronous Instant Messaging	Med-high	Low-med	Low-Med	Medium	Med-High
Synchronous Electronic Conferencing	Med-High	Low-med	Low-Medium	Low-med	Medium
Asynchronous Bulletin Board	Low-med	Low-med	High	High	High
Asynchronous e- mail	Low-med	Low-med	High	High	High
Asynchronous Written mail	Low	Low-med	High	High	High

with thinking or be emotionally charged (contributing to anxiety) which creates distractions from the ideation task. In a FTF situation it may be easier to ride on the efforts of the more vocal (negatively affect accountability). Also, the deadline for an asynchronous entry may not cause as much anxiety because it can be managed over a longer time period.

PMI4: Accountability is a negative function of Immediacy

PMI5: Affiliation is a positive function of Immediacy.

PMI6: Anxiety is a positive function of Immediacy.

PMI7: Creativity is a negative function of Immediacy.

Cues

Social cues promote creativity when they serve to recognize fruitful ideas or inhibit damaging behaviors. Social cues inhibit creativity when they intimidate or otherwise increase Anxiety. Someone who is shy or of a lower "rank" may be afraid to contribute in front of those of higher rank or just be shy or anxious in a group. This may be exacerbated by intimidating behaviors (e.g. "death stare," standing over someone, disapproving gestures) and interfering behaviors (e.g., sidebars). The net effect is that as social cues increase, so does anxiety, making anxiety a positive function of social cues.

PMC1: Attention is a negative function of Social Cues.

PMC2: Anxiety is a positive function of Social Cues.

PMC3: Creativity is a negative function of Social Cues.

Language cues promote creativity when the cues enhance the clarity of an idea and when that clarity serves to generate new ideas. Since humans are physiologically more suited to FTF communications than asynchronous, written, e-collaboration, most language and social cues are designed for FTF delivery (Kock, 2004). DeLuca et al. (2006) report that teams who use simple asynchronous, written, e-collaboration adapt

their communications to capture the cues necessary to communicate and filter out the cues interfering with effectiveness.

PMC4: Creativity is a positive function of Language Cues.

Parallelism

Parallelism reduces conflict for the communication channel and thus mitigates production blocking (Murthy & Kerr, 2003) which improves Attention.

PMP1: Attention is a positive function of Parallelism.

Howard-Jones and Murray (2003) conclude that Incubation "may be considered as a strategy to improve productivity" (p. 165). Parallelism contributes to Diversity by allowing more people to multitask, thereby increasing group size. For groups of nine or more, electronic brainstorming has been shown to be superior to FTF (Dennis & Williams, 2005; Pinsonneault et al., 1999).

PMP2: Incubation is a positive function of Parallelism.

PMP3: Diversity is a positive function of Parallelism.

Parallelism promotes creativity when participants can manage conflicting priorities by working on them without conflicting activity. Anytime, anyplace convenience is mentioned as a success factor by virtual teams (DeLuca & Valacich, 2006), enabling all team members, especially managers, to participate more fully (DeLuca, 2003). Freedom around processes is sited by Amabile (1998) as a "lever to pull" for managers to intrinsically motivate creativity and likely reduce Anxiety. Parallelism may inhibit creativity when working in parallel reduces the ability to feed off other's ideas. However, parallelism may also help, when group members cannot be available at a specific moment. Studies show (DeLuca, 2003; DeLuca & Valacich, 2006) that most virtual teams are able to perform convergence tasks while using solely asynchronous e-collaboration.

PMP4: Accountability is a positive function of Parallelism.

PMP5: Affiliation is a positive function of Parallelism.

PMP6: Anxiety is a negative function of Parallelism.

PMP7: Creativity is a positive function of Parallelism.

Rehearsability

Rehearsability promotes creativity when the ability go offline to refine a contribution forms a more productive (or Accountable) contribution or when the ability to refine reduces the Anxiety of contributing "on the fly," perhaps by Diverse group members with less language ability. Rehearsability inhibits creativity when repeating or refining takes too much energy away from other associations. The ability to rehearse may reduce Anxiety about contributing (DeLuca, 2003; Kock and DeLuca, 2006; DeLuca et al., 2006).

PMRH1: Attention is a positive function of Rehearsability.

PMRH2: Diversity is a positive function of Rehearsability.

PMRH3: Incubation is a positive function of Rehearsability.

PMRH4: Accountability is a positive function of Rehearsability.

PMRH5: Anxiety is a negative function of Rehearsability.

PMRH6: Creativity is a positive function of Rehearsability.

Reprocessability

Reprocessability promotes creativity when rereading messages helps generate new ideas. Reprocessability inhibits creativity when the amount of material becomes overwhelming or the work is postponed past a period of productivity. A distinguishing feature of asynchronous, written, e-collaboration is "group memory" which aids in mitigating evaluation apprehension (Murthy & Kerr, 2003), which reduces Anxiety. This written (and Accountable) log of contributions serves to allow new members to "catch up" and an existing member to "listen" to all colleagues. FTF teams sometimes enhance performance by adding this feature as a written or aural recording.

PMRP1: Attention is a positive function of Reprocessability.

PMRP2: Diversity is a positive function of Reprocessability.

PMRP3: Incubation is a positive function of Reprocessability.

PMRP4: Accountability is a positive function of Reprocessability.

PMRP5: Affiliation is a positive function of Reprocessability.

PMRP6: Anxiety is a negative function of Reprocessability.

PMRP7: Creativity is a positive function of Reprocessability.

The majority of the propositions above indicate that the media influences in which asynchronous e-collaboration is rated "high" are those having a positive influence on creativity. This is supported by studies where groups using computer-mediated communication were at least as good, if not better, at the creative task of idea-generation than FTF groups (Kock, 2006; Murthy & Kerr, 2003; Ocker, Fjermestad, Hiltz, & Johnson, 1998). DeLuca and Kock (2006) and DeLuca and Valacich (2006) found higher innovation success rates than FTF rates (Malhotra, 1998).

PM1: Creativity will be greater in groups using asynchronous e-collaboration than in groups using synchronous collaboration.

CONCLUSION

A new theory, asynchronous creativity theory (ACT), is proposed. The main proposition of the theory is that teams which use asynchronous e-collaboration have greater creativity than teams which use synchronous collaboration. This theory helps explain the results from studies on successful creativity tasks performed by virtual teams, a success that is not anticipated by traditional theories.

ACT is based on a new model of creativity, which is also proposed, the media-cognitive-social model of creativity. The model takes a cross-disciplinary look at the effects of five characteristics of communication media on three cognitive and three social factors traditionally viewed as affecting creativity. The focus here is applying the model to create a theory of creativity for teams which use primarily asynchronous electronic communication media. The three-dimensional model provides great potential as a basis for theory development relevant to business, psychology, and sociology.

This article meets criteria for new theory with the parsimonious organization provided by the MCS model of creativity and by clearly communicating the constructs and propositions (Bacharach, 1989). ACT is a viable theory based on the two criteria for evaluating a theory (Bacharach, 1989, p. 500): (1) its propositions can be *falsified* through operationalization and test; and (2) it has *utility* by *explaining* the relationships among the constructs in the MCS model of creativity, and predictions made favorably compare to the empirical evidence.

Asynchronous e-collaboration is used routinely for transmittal of information, but is less often considered for tasks requiring creativity. This is probably because, compared to traditional FTF communications, feedback is less immediate and interactions carry less language and social cues. Yet the ultimate effects of these asynchronous e-collaboration "disadvantages" may serve to induce creativity by (1) reducing distraction from Attention on the task; and (2) reducing Anxiety caused by status, fear, intimidation, extraneous socializing, or avoidance.

ACT contends that compensations must be made to use any media, but for creative tasks, we may actually need to compensate *more* when we *are* FTF than when we are *not*. Parallelism, Rehearsability, and Reprocessability characteristics of media appear to be more important to the creative process than Immediacy and Cues, and appear to enhance creativity by (1) increasing Attention to the task because other work can be set aside; (2) increasing Diversity of potential team members; (3) allowing for rest and Incubation of thoughts; (4) providing written Accountability; (5) increasing Affiliation with more team members; (6) rereading (Reprocessing) input; (7) reducing Anxiety from delivering an idea because it can first be refined and Rehearsed; and (8) providing the motivation of the additional freedom of anytime, anyplace communications (Parallelism).

ACT is a significant contribution to multidisciplinary creativity research. The final proposition is a succinct summary of the contribution of the theory, that asynchronous media may indeed be better for creativity tasks than the traditional synchronous media. The details of the other propositions may be applied to any media by analyzing its characteristics and their effects on creativity, thereby explaining more variance than any known theory.

ACT has potential to impact organizations by encouraging use of asynchronous e-collaboration for creative teams, simultaneously reducing travel and time expenditures—especially for global and/or 24/7 organizations—and increasing creativity effectiveness. Understanding the interplay of the three dimensions of the model also provides cautionary insight into the influences affecting creativity in virtual teams for considerations unique to each environment. Empirical testing of the explicit propositions of ACT is invited.

ACKNOWLEDGMENT

The author would like to gratefully acknowledge the support of a University of Delaware General University Research (GUR) grant for development of the ACT. The author wishes to thank Bob Briggs, Mike Ginzberg, Fred Niederman, and Ilze Zigurs for their valuable input. Any opinions expressed or mistakes in the article are the sole responsibility of the author.

REFERENCES

Amabile, T. M. (1998). How to kill creativity. *Harvard Business Review, 76*(5), 76-87, 186.

Amabile, T. M., Hennessey, B. A., & Grossman, B. S. (1986). Social influences on creativity: The effects of contracted-for reward. *Journal of Personality and Social Psychology, 50,* 14-23.

Bacharach, S. B. (1989). Organizational theories: Some criteria for evaluation. *Academy of Management Review, 14*(4), 496-515.

Daft, R. L., & Lengel, R. H. (1986). Organizational information requirements, media richness and structural design. *Management Science, 32*(5), 554-571.

DeLuca, D. C. (2003). Business process improvement using asynchronous e-collaboration: Testing the compensatory adaptation model. *Dissertation Abstracts International.* (UMI No.)

DeLuca, D. C. (2006). Virtual teams rethink creativity: A new theory using asynchronous communications. *Proceedings of the 12ᵗʰ Americas Conference on Information Systems* (pp. 129-138).

DeLuca, D. C., Gasson, S., & Kock, N. (2006). Adaptations that virtual teams make so that complex tasks can be performed using simple e-collaboration technologies. *International Journal of e-Collaboration, 2*(3), 64-85.

DeLuca, D. C., & Valacich, J. S. (2006). Virtual teams in and out of synchronicity. *Information Technology & People, 19*(4), 323-344.

Dennis, A. R., & Valacich, J. S. (1993). Computer brainstorms: More heads are better than one. *Journal of Applied Psychology, 78*, 531-537.

Dennis, A. R., & Valacich, J. S. (1999). Rethinking media richness: Towards a theory of media synchronicity. *Proceedings of the 32ⁿᵈ Annual Hawaii International Conference on System Sciences* (pp. 1-10).

Dennis, A. R., & Williams, M. L. (2005). A meta-analysis of group side effects in electronic brainstorming: More heads are better then one. *International Journal of e-Collaboration, 1*(1), 24-42.

Grise, M. L., & Gallupe, R. B. (2000). Information overload: Addressing the productivity paradox in face-to-face meetings. *Journal of Management Information Systems, 16*(3), 157-186.

Guilford, J. P. (1950). Creativity. *American Psychologist, 5*, 444-454.

Hamilton, D. (2004). The social and academic standing of the information systems discipline: General theory considered as cultural capital. *JITTA, 6*(2), 1-12.

Howard-Jones, P. A., & Murray, S. (2003). Ideational productivity, focus of attention, and context. *Creativity Research Journal, 15*(2/3), 153-166.

Kock, N. (2004). The psychobiological model: Towards a new theory of computer-mediated communication based on Darwinian evolution. *Organization Science: A Journal of the Institute of Management Sciences (INFORMS: Institute for Operations Research)*, 327-348.

Kock, N. (2005a). Compensatory adaptation to media obstacles: An experimental study of process redesign dyads. *Information Resources Management Journal, 18*(2), 41-67.

Kock, N. (2005b). What is e-collaboration? *International Journal of e-Collaboration, 1*(1), i-vii.

Kock, N. (2006). *Business process improvement through e-collaboration: Knowledge sharing through the use of virtual groups.* Hershey, PA: IGI Global.

Kock, N., & DeLuca, D. C. (2006). Improving business processes electronically: A positivist action research study of groups in New Zealand and the U.S. *Proceedings of the 11ᵗʰ Annual Conference of the Center for the Study of Western Hemispheric Trade* (Session VII, pp. 1-35).

Malhotra, Y. (1998). Business process redesign: An overview. *IEEE Engineering Management Review, 26*(3), 27-31.

Murthy, U. S., & Kerr, D. S. (2003). Decision making performance of interacting groups: An experimental investigation of the effects of task type and communication mode. *Information & Management, 35*1.

Ocker, R., Fjermestad, J., Hiltz, S. R., & Johnson, K. A. (1998). Effects of four modes of group communication on the outcomes of software requirements determination. *Journal of Management Information Systems, 15*(1), 99-118.

Paulus, P. B. (2000). Groups, teams, and creativity: The creative potential of idea-generating groups. *Applied Psychology: An International Review, 49*(2), 237-262.

Pinsonneault, A., Barki, H., Gallupe, R. B., & Hopper, N. (1999). Electronic brainstorming: The illusion of productivity. *Information Systems Research, 10*(2), 110-133.

Wallas, G. (1926). *The art of though.*, New York: Harcourt, Brace and World.

Weber, R. (2002). Editor's comments. *MIS Quarterly, 26*(1), iii-viii.

Van de Ven, A. H. (1989). Nothing is quite so practical as a good theory. *Academy of Management Review, 14*(4), 486-489.

KEY TERMS

Accountability: A team member is responsible for the quality of his/her input.

Affiliation: The level of relational development of the team.

Anxiety: A sense of unease.

Asynchronous E-Collaboration: Collaboration among individuals engaged in a common task using electronic technologies that allow input at different times.

Asynchronous Creativity Theory (ACT): Theory that considers the effects of communication media capabilities on the traditional social and cognitive factors affecting creativity, the net effect of which is that teams using asynchronous e-collaboration may have greater potential for creativity than do synchronous face-to-face teams.

Attention: Amount of focus on the task at hand.

Creativity: The development of novel ideas which are useful.

Cues: Format by which information is conveyed, verbal and nonverbal language (e.g., tone, eye contact) as well as social cues (e.g., status, intimidation).

Diversity: Variety of frameworks/ideas.

Immediacy: Rapid transmission and rapid feedback from their communications.

Incubation: Rest or period without conscious attention to the problem or doing other activities.

Media-Cognitive-Social (MCS) Model of Creativity: The model of media, cognitive, and social factors affecting creativity on which asynchronous creativity theory is based.

Parallelism: Effectively working in parallel (simultaneous conversations in process). Channel capacity so that multiple people can work all at once.

Rehearsability: Fine tuning a message before sending.

Reprocessability: Readdressing a message within the context of the communication event (e.g., rereading, printing).

The Practice and Promise of Virtual Project Management

Ilze Zigurs
University of Nebraska at Omaha, USA

Deepak Khazanchi
University of Nebraska at Omaha, USA

Azamat Mametjanov
University of Nebraska at Omaha, USA

INTRODUCTION

Virtual projects are essential components of modern organizations that seek to be flexible and take advantage of distributed resources. A virtual project is a project in which team members are dispersed geographically and potentially on other dimensions, and are working together to accomplish a specific task under time and resource constraints. Because of their dispersion, team members have to rely on computer-mediated communication tools to do their work. Virtual projects are prevalent in software development and increasingly common in research and development, marketing, and customer relationship management, hence they are an important phenomenon for study. The challenge is to discover which practices and perspectives help to enhance the effectiveness of virtual projects, so that team members can leverage the advantages of virtuality while avoiding its pitfalls.

The relevant literature on this topic comes from many areas, including studies of virtual teams as well as the body of knowledge in project management. In this article, we bring together disparate fields and provide an integrated view of virtual project management. We begin by defining key terms and concepts in the context of an overall framework and briefly describe relevant knowledge from current research. We then discuss key issues and future trends for research, and conclude with overall observations and implications.

BACKGROUND AND FRAMEWORK

Figure 1 shows the overall framework for the discussion of key concepts that are relevant to the study of virtual projects. The classic input-process-output approach identifies factors that are relevant to effective project management in a virtual environment. The following sections define and briefly discuss each of the factors (see Khazanchi & Zigurs, 2005 for a more detailed discussion). The purpose is not to be comprehensive in all the factors that might affect project management, but instead to focus specifically on management issues that are particularly salient or problematic in virtual contexts.

Organizational and Social Context

The management of virtual projects does not occur in a vacuum. The characteristics of the organization itself are important to how such projects are managed, as is the larger social context, which can include governmental and environmental issues. Organizational norms affect how technology is adopted and used (Orlikowski, 1993; Orlikowski & Robey, 1991) and therefore are relevant to the setting for virtual projects. A detailed discussion of these contextual factors is outside the scope of this article.

Input Factors

Virtuality is a term that is defined in a variety of ways, but typically with respect to dispersion. Virtual teams can be dispersed on many dimensions, most often geographically, and also in time, organizational affiliation, culture, and technology. The greater the dispersion, the greater is the virtuality of the team (Katzy, Evaristo, & Zigurs, 2000; Watson-Manheim, Chudoba, & Crowston, 2002). Other views of virtuality include dynamic switching among defined requirements and

Figure 1. Framework for the study of virtual projects

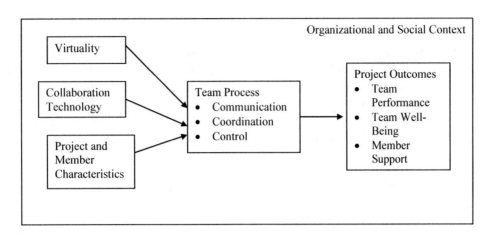

services (Mowshowitz, 1997) and extent of reliance on communication and information technologies (Dubé & Paré, 2004). Dispersion is an intuitively appealing characteristic by which to define virtuality, and technology is also an essential component of the ability to be virtual. Thus, we define *virtuality* as the extent to which project members are dispersed geographically and on other dimensions *and* rely on information and communication technologies for carrying out team processes and achieving project goals.

The second major concept is collaboration technology, which has also been characterized in different ways. The major challenge in defining collaboration technology is to avoid a monolithic view and be able to capture variability in technology features. Key perspectives on collaboration technology have defined it in terms of characteristics of media (Carlson & Zmud, 1999; Dubé & Paré, 2004), levels of support for information exchange or communication or information sharing (DeSanctis & Gallupe, 1987; McGrath & Hollingshead, 1994), and time-space configuration (Johansen, 1988). Consistent themes across different views of technology are that it must provide support for communication, for information exchange, and for structuring a team's process. Thus, we define *collaboration technology* as an integrated and flexible set of tools for structuring process, supporting task requirements, and communicating among project members. These characteristics are not fixed, but instead can be adapted by team members as they develop knowledge of the task, each other, and the technology itself (Carlson & Zmud, 1999).

The third major concept in the framework is *project and member characteristics*. Existing typologies of projects are based on such factors as the domain of the project (e.g., software engineering or construction), the extent of globalization, and project complexity, risk or scope (Palmer & Speier, 1997; Project Management Institute, 2004). The latter three factors are the most consistent ones, i.e., complexity, risk, and scope, and it is reasonable to classify most project characteristics under one of these three factors. Project complexity can be affected by the extent to which teams have variety in their size, culture, language, member characteristics, resources, and knowledge. Project scope can be affected by the extent of duration, innovation, and breadth. Project risk is defined typically in terms of different categories of risk in different phases of the project. Thus, project characteristics vary widely but can be examined across these common factors.

Several findings related to these input factors and their relationships with team process factors are worth highlighting. It is well established that the on-going process of team member communication can re-define input factors by a process of adaptation of both tools and team characteristics (DeSanctis & Poole, 1994). Virtuality clearly impacts the complexity of a project, in that the greater the temporal and geographic dispersion of team members, the greater the degree of communication, coordination, and control required. Research has shown that the impact of virtuality on project managers can be diminished by reducing temporal distance through collaboration with organizations in closer time zones and by reducing the intensity of collaboration by giving up some part of control to localized teams (Dubé & Paré, 2004).

Previous research has shown that choice of collaboration technology and the nature of communication media (video, audio, text) affects communication characteristics such as concurrency and feedback (Baker, 2002). In fact, early use of rich media can alleviate some of the problems with team cohesion (Burke, Aytes, & Chidambaram, 2001).

Project and member characteristics can affect media choice; for example, different cultures have varying preferences for different forms of communication (Massey, Hung, Montoya-Weiss, & Ramesh, 2001). The diversity of project members in terms of language, culture, and knowledge of the project domain needs to be effectively managed because the diffusion of these differences can remove barriers to communication processes (Dubé & Paré, 2004). The complexity, scope and risk of a project dictate the degree of required coordination (Godart et al., 2001) and control mechanisms (implicit vs. explicit) that would be more effective (Godart et al., 2001; Kirsch, 1997).

Team Process Factors

Three key process factors are essential to the effective management of virtual projects: communication, coordination, and control. These three factors appear repeatedly in the literature on project management, and they are particularly vital for virtual projects. *Communication* is the process through which people convey meaning to one another through the exchange of messages and information to carry out project activities. *Coordination* can be defined as the mechanisms through which people, technology, and other resources are combined to carry out the activities required to attain project goals (Crowston, 1991). Finally, *control* is the process of monitoring and measuring project activities to anticipate and manage variances from plans and goals (Project Management Institute, 2004).

The research on these process factors suggests the following. The start-up process for virtual teams is especially important; it should include training (Chinowsky & Rojas, 2003) and appropriate "sandwiching" of same time/same place meetings with distributed work, to help teams get started and to develop effective communication (Rutkowski, Vogel, Bemelmans, & Van Genuchten, 2002; Chinowsky & Rojas, 2003). Management can mitigate the effects of the distributed nature of virtual

teams by reducing temporal and cultural distance to the extent possible (Carmel & Agarwal, 2001). One way to deal with temporal distance is by the use of temporal coordination mechanisms, that is, specific process structures (Massey, Montoya-Weiss, & Hung, 2003). In terms of control, the idea of a portfolio of control modes may be the most appropriate, namely, control mechanisms should be adjusted based on such factors as task characteristics, role expectations, project-related knowledge and skills, and availability of pre-existing mechanisms (Kirsch, 1997).

Team Outcome Factors

The output of any team activity is multi-dimensional, that is, it includes effectiveness as well as team member satisfaction and commitment. Time-interaction-performance (TIP) theory (McGrath, 1991) provides a systemic and multi-dimensional picture of team output. TIP theory posits three simultaneous aspects of team functioning: (1) *team performance*, which means getting the task done; (2) *team well-being*, or the relations among team members, and (3) *member support*, which is the relationship of the individual to the team. Each dimension has implications for team output, as Figure 1 shows. Thus, an effective team not only gets the task done, but it also has members who perceive they have contributed to the team, and there is a team-level sense of commitment and accomplishment.

The research on outcome factors can be highlighted as follows. Effectiveness requires a focus on human aspects and not just technology (Rutkowski et al., 2002), which reinforces the effectiveness as multi-dimensional. Team cohesion and process satisfaction are dynamic factors and teams are able to develop them by adapting to the use of computer-mediated tools over time (Burke et al., 2001). Member support also needs time to grow, and cultural compatibilities have to be carefully managed (Rutkowski et al., 2002). One consistent observation from this research is that managers of virtual projects must allow time for adaptation, both within the team and to the technology. Team development occurs over time through building team trust, establishing processes for conflict resolution, communicating expectations clearly, and providing structure for computer-mediated communication (Chinowsky & Rojas, 2003).

RESEARCH ISSUES AND FUTURE TRENDS

Patterns of Effectiveness

The management of virtual projects is fundamentally different from that of traditional projects. Intuitively one would expect consistent differences in how managers deal with coordination, communication, and control in virtual projects. Using Alexander's notion of patterns (Alexander, 1965; Alexander, Ishikawa, Silverstein, Jacobson, Fiksdahl-King, & Angel, 1977), our research has focused on describing patterns of effective virtual project management that reflect these differences (Khazanchi & Zigurs, 2005). Given Alexander's notion of a pattern as a *three-part rule that expresses a relationship among a specific context, a problem, and a solution*, we identified an initial set of patterns with regard to coordination, control and communication that are common across virtual projects. This stream of research needs further development. If such effective patterns exist, we need to know how they can be translated into processes that impact project management in a virtual or non-virtual setting. Further, there is a need for more research into the possibility that patterns can be used as critical checks and/or design principles for guiding virtual projects.

Role of Communication

Another issue that is reinforced throughout this research is the critical importance of communication for effective management of virtual projects. Communication is important in and of itself, as well as through its relationship to coordination and control. Communication is essential for building trust, providing feedback, and communicating context (Avolio, Kahai, & Dodge, 2001; Khazanchi & Zigurs, 2005; Maznevski & Chudoba, 2000; Sarker & Sahay, 2002).

Collaboration Technology

The nature of collaboration technology for virtual projects is a continuing challenge. Fundamental leaps in technology developments can be impossible to foresee, often because of an inherent focus on incremental change from existing practices and tools. E-mail is still widely used as the "lowest common denominator," even with all its shortcomings in the face of complex needs for communication, coordination, and control. Many issues remain unresolved.

Ease of use of technology has already been mentioned, including the need to trade off comprehensiveness against accessibility and rapid adoption. Integration of tools is also essential. As e-mail spam challenges the attention and patience of users, the trend of using more lean and focused technologies such as wikis (editable Web sites), blogs, instant messaging, and/or Web-based file sharing may provide a potential technological answer for virtual project teams (Conlin, 2005; Hof, 2005). Our study of virtual project management in five global firms confirmed many previous observations that e-mail is still the dominant tool of choice for communication and coordination among virtual team members (Khazanchi & Zigurs, 2005).

In addition, the issue of team member "presence" and the need to communicate context in virtual project team interactions continues to be critical to understanding how project teams can build trust and shared understanding. It seems that lean tools such as wikis (versus feature-rich groupware tools) may provide the flexibility and ownership that create an ideal circumstance for engendering improved communication, enhancing credibility of shared information, building trust, and developing a shared understanding of the project.

Along these same lines, peer-to-peer applications such as Groove® have also emerged as potential tools that can provide anywhere/anytime access and an integrated environment for communication, coordination, and project management. But such tools are not a panacea. Team members must still develop norms for seemingly simple issues like posting and responding, as well as more complex aspects of coordination. Awareness of the presence of other members in a collaborative workspace is possible with a variety of tools, but awareness of team member attributes must still be managed and explicitly shared. Attributes may include culture, language, expectations, and communication style.

The issue of whether information should be "push versus pull" remains an open debate and highly individual in terms of preference. A "push" tool would automatically send information to all team members, while a "pull" tool would wait for members to retrieve that information. E-mail is the classic example of a push tool, while intranets and other team spaces are pull tools. Push tools create higher awareness, but can be intrusive and time consuming to manage.

Software Project Management

Finally, there is a growing need to better understand how communication, control and coordination vary across stages or phases of virtual software development projects. This is particularly important in the current context where software off-shoring, dual shoring and near-sourcing is an important component of a firm's information technology project portfolio. Whether one views the software development process in terms of the popular PMI PMBOK® model (Project Management Body of Knowledge) (Project Management Institute, 2004) or IBM® Rational® Unified Process® (RUP®), it seems evident that extensive communication and co-ordination and, to a lesser extent, control are needed in the early stages of software development, and substantively greater coordination and work synchronization are needed during the construction phase of software development. This assertion needs further investigation since it can lead to best practices that impact virtual project management.

CONCLUSION

The management of virtual projects remains a challenging endeavor. Existing research provides many guidelines on different pieces of the puzzle. We have presented an integrated framework for understanding those key pieces. Researchers have a full agenda of interesting questions, and managers have both problems and opportunities. The promise of virtual teams lies in the manager's ability to identify what is different and what remains the same, and to be able to integrate human with technological resources in a way that makes use of unique capabilities of all available resources.

REFERENCES

Alexander, C. (1965). *Notes on the synthesis of form.* Cambridge, MA: Harvard University Press.

Alexander, C., Ishikawa, S., Silverstein, M., Jacobson, M., Fiksdahl-King, I., & Angel, S. (1977). *A pattern language: Towns, buildings, construction.* New York: Oxford University Press.

Avolio, B. J., Kahai, S., & Dodge, G. E. (2001). E-leadership: Implications for theory, research, and practice. *Leadership Quarterly, 11*(4), 615-668.

Baker, G. (2002). The effects of synchronous collaborative technologies on decision making: A study of virtual teams. *Information Resources Management Journal, 15*(4), 79-93.

Burke, K., Aytes, K., & Chidambaram, L. (2001). Media effects on the development of cohesion and process satisfaction in computer-supported workgroups: An analysis of results from two longitudinal studies. *Information Technology & People, 14*(2), 122-141.

Carmel, E., & Agarwal, R. (2001). Tactical approaches for alleviating distance in global software development. *IEEE Software,* March/April, 22-29.

Carlson, J. R., & Zmud, R. W. (1999). Channel expansion theory and the experiential nature of media richness perceptions. *Academy of Management Journal, 42*(2), 153-170.

Chinowsky, P. S., & Rojas, E. M. (2003). Virtual teams: Guide to successful implementation. *Journal of Management in Engineering, 19*(3), 98-106.

Conlin, M. (2005). E-mail is so five minutes ago. *BusinessWeek, November 28*, 111-112.

Crowston, K. (1991). A coordination theory approach to organizational process design. *Organization Science, 8*(2), 157-175.

DeSanctis, G., & Gallupe, R. B. (1987). A foundation for the study of group decision support systems. *Management Science, 33*(5), 589-609.

DeSanctis, G., & Poole, M. S. (1994). Capturing the complexity in advanced technology use: Adaptive structuration theory. *Organization Science, 5*(2), 121-147.

Dubé, L., & Paré, G. (2004). The multi-faceted nature of virtual teams. In D. J. Pauleen (Ed.), *Virtual teams: Projects, protocols, and processes.* Hershey, PA: Idea Group Publishing.

Godart, C., Bouthier, C., Canalda, P., Charoy, F., Molli, P., Perrin, O., et al. (2001). Asynchronous coordination of virtual teams in creative applications. *Australian Computer Science Communications, 23*(6), 135-142.

Hof, R. D. (2005). Teamwork supercharged. *BusinessWeek, November 21*, 90-94.

Johansen, R. (1988). *Groupware: Computer support for business teams.* New York: Macmillan.

Katzy, B., Evaristo, R., & Zigurs, I. (2000). Knowledge management in virtual projects: A research agenda. In *Proceedings of the 33rd Hawaii International Conference on System Sciences*[Electronic version]. Washington, DC: IEEE Press.

Khazanchi, D., & Zigurs, I. (2005). *Patterns of effective management of virtual projects: An exploratory study.* Newtown Square, PA: Project Management Institute.

Kirsch, L. J. (1997). Portfolios of control modes and is project management. *Information Systems Research, 8*(3), 215-239.

Massey, A. P., Hung, Y. C., Montoya-Weiss, M., & Ramesh, V. (2001). When culture and style aren't about clothes: Perceptions of task-technology "fit" in global virtual teams. In *Proceedings of the 2001 International ACM SIGGROUP Conference on Supporting Group Work* (pp. 207-213). New York: ACM Press.

Massey, A. P., Montoya-Weiss, M. M., & Hung, Y. (2003). Because time matters: Temporal coordination in global virtual project teams. *Journal of Management Information Systems, 19*(4), 129-155.

Maznevski, M. L., & Chudoba, K. M. (2000). Bridging space over time: Global virtual team dynamics and effectiveness. *Organization Science, 11*(5), 473-492.

McGrath, J. E. (1991). Time, interaction, and performance (TIP): A theory of groups. *Small Group Research, 22*(2), 147-174.

McGrath, J. E., & Hollingshead, A. B. (1994). *Groups interacting with technology: Ideas, evidence, issues, and an agenda.* Thousand Oaks, CA: Sage Publications.

Mowshowitz, A. (1997). Virtual organization. *Communications of the ACM, 40*(9), 30-37.

Orlikowski, W. J. (1993). Learning from notes: Organizational issues in groupware implementation. *The Information Society, 9*(3), 237-250.

Orlikowski, W. J., & Robey, D. (1991). Information technology and the structuring of organizations. *Information Systems Research, 2*(2), 143-169.

Palmer, J. W., & Speier, C. (1997, August 15-17). A typology of virtual organizations: An empirical study. In *Proceedings of ACIS*. Indianapolis, IN.

Project Management Institute. (2004). *A guide to the project management body of knowledge (PMBOK)* (3rd ed.). Newtown Square, PA: Project Management Institute.

Rutkowski, A. -F., Vogel, D., Bemelmans, T. M. A., & Van Genuchten, M. (2002). Group support systems and virtual collaboration: The HKNet project. *Group Decision and Negotiation, 11*(2), 101-125.

Sarker, S., & Sahay, S. (2002). Information systems development by US-Norwegian virtual teams: Implications of time and space. In *Proceedings of the 35th Annual Hawaii International Conference on System Sciences* (Vol. 1, p. 18). Washington, DC: IEEE Press.

Watson-Manheim, M. B., Chudoba, K. M., & Crowston, K. (2002). Discontinuities and continuities: A new way to understand virtual work. *Information Technology & People, 15*(3), 191-209.

KEY TERMS

Collaboration Technology: An integrated and flexible set of tools for structuring process, supporting task requirements, and communicating among project members.

Communication: The process through which people convey meaning to one another through the exchange of messages and information to carry out project activities.

Control: The process of monitoring and measuring project activities to anticipate and manage variances from plans and goals.

Coordination: The mechanisms through which people, technology, and other resources are combined to carry out the activities required to attain project goals.

Task Effectiveness: A measure of successful attainment of a team's goals with respect to the deliverable; one of three key dimensions of time-interaction-performance theory.

Team Well-Being: The relationship of the individual to the team; one of three key dimensions of time-interaction-performance theory.

Virtual Project: A project in which team members are dispersed geographically and potentially on other dimensions, and are working together to accomplish a specific task under time and resource constraints.

Virtuality: The extent to which project members are dispersed geographically and on other dimensions *and* rely on information and communication technologies for carrying out team processes and achieving project goals.

Prerequisites for the Implementation of E-Collaboration

Thorsten Blecker
Hamburg University of Technology (TUHH), Germany

Ursula Liebhart
Alpen-Adria-University of Klagenfurt, Austria

INTRODUCTION

E-collaboration refers to a task-oriented cooperation between individuals by using electronic technologies. In the company's practice, e-collaboration mainly describes each form of collaboration between two or more locations and/or organizations via electronic channels in order to support electronic commerce and supply chain transactions.

However, a successful e-collaboration can be achieved only if many requirements are satisfied. We distinguish between three main fields that should be addressed in order to benefit of this concept. The advances in information and communication technology (ICT) are the core of e-collaboration since they make up the 'e' in e-collaboration. However, technology alone is not sufficient to make e-collaboration work. The second requirement is the appropriate organizational integration that supports the implementation of e-collaboration in practice. As a third requirement, we discuss human resource management and leadership. Employees should be capable of using e-technologies and cooperating via the electronic channel to carry out common tasks.

BACKGROUND

Understanding e-collaboration requires clarifying two components: the term "collaboration" and the prefix "e-." The term "collaboration" here is understood in a very traditional manner. By analogy to the *Heritage Dictionary of the English Language* (2004) we define collaboration as working together, especially in a joint intellectual effort. In today's high competitive processes this means that two or more people are sharing complex information on an ongoing basis for a specific goal or purpose. As customary for many current concepts companies often use ICT for an efficient and effective implementation of the necessary information sharing and information processing. On the one hand the application of ICT enables personal communication such as e-mail, video-conferences and chats. On the other hand ICT is often a necessary precondition to get access to common information or rather to allow people to interactively share information at a distance which is used in and arise during the collaboration process. Therefore we call this type of collaboration e-collaboration as it is facilitated by ICT.

In the technical literature, there are many definitions of e-collaboration. A widely accepted definition was provided in one of the first special issues on e-collaboration published by the *Journal of Systems and Information Technology*: E-collaboration is collaborating "among individuals engaged in a common task using electronic technologies" (Kock et al., 2001). Furthermore, beside others Monplaisier/Haji (2002) delineate an enabling function of ICT for collaboration and concentrate on its efficient use to support interaction too. Yet e-collaboration does not mean that all processes are based on the application of ICT. E-Collaboration today entails conventional ways of working such as meetings, phone and fax to facilitate information sharing, discussion and agreement as well as new services and technologies such as project collaboration, workflow tools, and e-conferencing.

Main purposes of applying ICT in collaboration are increasing productivity and efficiency by reducing unproductive travel time, creating shorter and more structured meetings, and providing faster exchange of information for an increased number of participants. In detail, companies strive for:

- Accessing and combining distributed assets and resources

- Reducing travel, project management, and administrative costs
- Speeding up product-development cycle time
- Reducing supply lead times
- Gaining real-time pricing and availability information, real-time order management, continuous demand management
- Improving agility, flexibility and speed of strategic relevant action across locations and company boundaries
- Synchronizing activities across teams leading to efficient coordination within an entire supply chain
- Reducing inventory-carrying costs and logistics costs
- Intensifying collaboration with suppliers, customers and partners
- Improving the supplier visibility

In spite of the multitude of benefits and the promising concepts developed over the last years, there are quite a number of problems to be solved in order to successfully implement e-collaboration in practice. For example Ackermann (2001) argues that in the last years researchers have "identified a base set of findings [to which] human activity is highly flexible, nuanced, and contextualized and that computational entities such as information transfer, roles, and policies need to be similarly flexible, nuanced, and contextualized ... (but) we do not know how to build systems that fully support the social world uncovered by these findings ... (this is) the social-technical gap ... the divide between what we know we must support socially and what we can support technically." Therefore beside the well known and often discussed technological challenges of e-collaboration companies have to concentrate the more "hidden challenges" such as social and cultural structures.

PREREQUISITES

Technology

The first and widely discussed prerequisite of e-collaboration is the technological infrastructure because information sharing is the core element of all forms of collaboration. This fact is often cited as evidence for the importance of ICT and as a reason that only the advances in ICT are breaking down geographical barriers in relation to sharing and accessing information.

In order to analyze the technological prerequisites we initially have to differentiate two general classes of ICT systems for collaboration purposes: messaging-based or repository-based. A *messaging-based infrastructure* is used to send process assignments through an e-mail/messaging system (e.g., Microsoft Exchange/SharePoint or Novell GroupWise). However these systems are limited by the e-mail capabilities and policies. In particular, storing documents in e-mail folders instead of dedicated collaboration servers prohibits a close interaction of the participants and may lead to slower, less secure, and less repeatable business processes. The second class of collaboration tools is the *repository-based infrastructure* (e.g., Lotus Notes, FileNet, Documentum eRoom, and OpenText.) These systems store project relevant documents in central repository and expose them to a team by a digital workspace. Further services of a repository-based infrastructure are for example notifications in the case of document changes, synchronous features (awareness, chat, application sharing, whiteboard, etc.) as well as asynchronous functionalities (workspace, task lists, mailing forum, document publication, versioning, member management, etc.).

Although it is (theoretically) possible to use heterogeneous network infrastructure to generate e-collaboration infrastructures, today most e-collaboration systems apply Internet Technologies. Nevertheless, in the case of e-collaboration we have to introduce another differentiation: *browser-based* and *nonbrowser-based systems*. While the first one use the widely accessible internet browser as front end (e.g., WebEX and eRoom), the latter one use proprietary software as client (e.g., Groove and ICQ).

E-collaboration makes high demands on the precise attributes of suitable technologies. Therefore there have been a large number of research articles addressing the importance of ICT for e-collaboration and analyzing the needed features. However beside the obvious qualities such as all-time and all-place availability, stable infrastructures, high-speed interconnections, device-unifying services and security there are much more prerequisites which are often neglected:

- **Interconnectivity:** The number of companies joining a project may cause numerous different IT-Infrastructures and company specific premises. This often complicates the data exchange between the partners. Therefore a main problem to solve in e-collaboration is to guarantee interconnectivity. For example, different CAD-systems and communication tools decrease the efficiency of collaboration in distributed product development processes by complicating data exchange and interaction between the partners (Gronau & Kern, 2004).

- **Flexibility:** A high organizational and technological flexibility is necessary in order to achieve fast and efficient changes in the configuration of partners participating in the different phases of collaboration projects with a low effort of hierarchical interventions.

- **Simplicity:** Modern ICT in e-collaboration is often not simple enough. The employees demand fewer choices or specific templates for e-collaboration purposes and do not use provided surplus features (Kock & Nosek, 2005). In many cases provided surplus features may reduce acceptance of ICT and trigger uncertainty for the whole process of interacting and coordinating. Thus, aligning supplied ICT with the really needed e-collaboration features is a main problem in order to reduce the complexity for the users. This context is well known as a standard problem of information management from the information supply and need model (e.g. Wigand, et al., 1997; Blecker et al., 2005).

- **Usability:** One the one hand it is very important for the acceptance of e-collaboration that it is possible to work with provided functions without the necessity to install proprietary client software. All asynchronous functions should be available as broswer-based system using TCP/IP. On the other hand it is often very important that in special collaboration types even non skilled worker without computer knowledge can use the systems. This leads to the next prerequisite.

- **Interface:** Here one must answer the questions what are the requirements on ICT depending on different collaboration types and how the system should be presented to the users in order to generate a suitable human-computer interaction (HCI). For example collaborative ICT is implemented more and more in machinery on distributed shop floors (Blecker & Graf, 2006), (e.g., voice over IP interactions on the shop floor). Thus, an enrichment

of the interface in the sense of a human-centered modification of HCI is necessary.

- **Autonomy:** The ICT architecture defined as a planned cooperation of technical, organizational, cost-related and social aspects, needs a high ability of self-organizing and partial decision autonomy to be able to integrate new cooperative partners (Kersten & Kern, 2003).

Of course, these prerequisites characterize a vision of collaboration and the future of doing business in a society where knowledge and intangibles are the main raw material. However, even today the design of the ICT infrastructure must be aligned more and more with these prerequisites instead of focusing only technical performance characteristics. Furthermore the importance and requirements for organizational patterns as well as are human resources and leadership slowly but profoundly changing with a more and more generalized use of ICT in collaboration processes.

ORGANIZATIONAL INTEGRATION

The next important component to discuss is on the one hand the integration of the aspects of e-collaboration inside the existing organizational and operational structure as well as the overall patterns of structural components and arrangements used to manage the complete organization. On the other hand it is the organizational culture through which the e-collaboration will get alive. Both components organizational structure and organizational culture have to be harmonized with the overall strategy.

First of all, the organization of distributed processes and their integration in the existing structure must be defined. However collaboration not only means sharing information, resources and skills but also knowledge. Therefore implementing e-collaboration is a question of traditional organization structures and procedures as well as of knowledge and resource management. Additionally the broad use of different ICT is not common standard in business. There are still lots of wait-and-see and negative attitudes from employees and also management members towards modern technologies. Engineers often overemphasize the technical solutions itself and oversee the user with their emotional concerns and other "hidden challenges" that arise from working in distributed teams (West Pole, 2005).

The valence of collaboration, the working together with business partners across organizational boundaries is a basic prerequisite for e-collaboration and is anchored in the organizational culture. The organizational culture is the entirety of all associations, f.e. norms, values, beliefs, interpretations, paradigms aso. based implicitly or explicitly on the patterns of behavior of a social system. It is mainly developed and stamped by management and offers some valuable functions for the whole members of the organization, like reduction of complexity, coordinative acting, employees identification a.s.o. Organizational culture is able to support e-collaboration through accepted norms and rules about the expected behavior in dislocated teams. Therefore organizations have to develop and share cooperative norms within the own business areas, the own organization and transfer them to collaborative interorganizational systems. Cooperative values like win-win-situations, consideration and respectfulness, trust, etc. have to be anchored in the organizational culture.

As shown above organizational structures and cultural anchorages have to be adopted and/or developed for the components "e-terms" and "collaboration." But "e-collaboration" takes a larger step further by combining these two perspectives to a new interdependent one:

- **Top management commitment:** Managerial support and establishing trust between managers and works are one of the key ingredients to the success of e-work. Top Management has to commit itself to e-collaboration by defining the mission, strategy and the concrete business case. Furthermore, it has to provide resources and adequate infrastructure for implementation and use of ICT for training and development of people as well as the necessary conditions to work efficiently. But one of the most important factors of success is the positive role model of top and lower management. They have to be present in world of e-collaboration, they have to understand their employees' problems by e-collaborating, and they have to e-collaborate themselves and show concrete advantages of e-collaboration.
- **Organizational structures and procedures:** E-collaboration needs turning away from strong hierarchies to flat, collaborative, team-orientated and project-centered structures of leadership. E-collaboration changes roles and expands responsi-

bilities, creates new jobs and changes the ways of communication between people. Employees have to be given more responsibility, a greater scope of decision making and more tools and procedures for a self-organization of teams. E-collaboration does not stop at organizational boundaries. Therefore organizational differences e.g. in decision making processes and decision structures become possible disturbances in e-collaboration.

- **Support of e-collaboration-teams:** E-collaboration-teams are virtual teams, which have to cross physical, temporal and organizational boundaries. Their common goal is to work better which means they have to be more productive, quicker and more innovative than physically present teams. Management has to support these teams by guaranteeing working conditions, e.g. technical support, flexible working hours because of different time zones, a greater scope of development for decisions and activities, as well as other resources, etc.
- **Change management and organizational culture:** Although, management defines the strategy, e-collaboration forces changes in the existing organizational structures and procedures. Thus, e-collaboration is also an important issue for management of change which forces a mindshift in some important parts of the daily way of working. The missing tradition of cooperation, different cultures and missing shared goals and shared knowledge are great barriers to greater cooperation (Riege, 2005).
- **Interorganizational trust:** Cultural interfaces due to differences in language, national attitudes, values, working mentalities, ways of thinking, problem solving, levels of education or measurement systems influence the efficiency of collaboration. On the one hand the technical quality of the product may suffer; on the other hand collaboration becomes more difficult due to misunderstanding and complications on a personal level. Trust in e-relationships gradually evolves from 'technology trust' to 'relationship trust' which enables different perspectives to grow from the technological, to economic, to behavioral and at last to an organizational perspective (Ratnasingam, 2005). Trust is especially elusive and can be fleeting in e-collaboration especially when communications are sparse and cultural differences are large (Qureshi, Liu, & Vogel, 2005).

Therefore a number of strategies to reinforce trust should be built up (e.g., proactive behavior, empathetic task communication) (Jarvenpaa, Knoll, & Leidner, 1998).

Social organizations are highly complex formations with a strategic orientation, derived structures and concerted procedures based on a high degree of communication and interaction among its members. Each intervention modifies the social construct and potentially produces opposition. A change management perspective is necessary for a reflected technological implementation in the social organization as well as the depth and broad of the change itself.

Human Resources and Leadership

The great challenge for human resource management concerning e-collaboration is to enable people to successfully work together and create competitive advantages by using ICT. With regard to e-collaboration three major working fields within the human resource management can be identified:

- **Collaborative competence:** To make the potential of e-collaboration useful and gain the multifaceted advantages from e-collaboration, employees need collaborative competencies, because the value added processes are increasingly based on communication, dialogs and team work. Collaborative competence manifests itself through high levels of collaboration among the employees and team members. It is characterized by mutual respect, shared work values and cooperation and is related to aspects such as open communication, trust, support, learning on the job, getting results and job satisfaction.
- **Loss of physical presence of team members:** The significant difference between collaboration and e-collaboration is the loss of physical presence. Many forms of the verbal, gesture and visual coordination of activities cease to exist as well as the possibility to physically move toward another or vice versa. The physical space offers more information which serves the physical and social orientation. In the virtual context the available and trained factors of the virtual coordination of social interaction disappear so that the prerequisites for a successful computer based communication have to be given.

- **Communication through ICT:** People with less or no experiences with ICT have a high degree of uncertainty in handling them. This may lead to hesitant activities, fear of failures, exaggerated expectations, a low tolerance of frustration, rapid loss of orientation, and so on. Additionally ambiguity concerning the various communication media or the own behavior (e.g., to strike the right tone) complicate the common being capable of acting in the virtual space.

The working fields of collaborative competencies, loss of physical presence in teams and communication through ICT display a high challenge for the development of human resources. These are not the small percentage of early missionaries who exist in every company and are ready for every kind of change but the larger amount of people in the organization, who cannot see the advantage und possibilities of e-collaboration. They have to be convinced, trained and supported by positive working conditions until they accept the new way of working. Thus, prerequisites and constraints of human resources and leadership for a successful e-collaboration are:

- **Technological readiness:** First of all employees have to develop technological readiness which means the competencies for a successful adaptation of the appropriate technologies for distance work. Therefore people have to learn the ideal use of various communication media and mix which optimally covers the communication requirements as well as different functions of awareness in virtual space. But more importantly for organizations is that their employees require the habits to use the technology for preparation, regular access, attention to others need for information, etc. (Olson & Olson, 2000).
- **E-collaboration-competence:** E-collaboration can be trained most effectively by the learning-by-doing-concepts, which forces people to use the technology and in work and offers reflection of experiences. Mentors and coaches on location and via internet or intranet can offer valuable mental and physical support. Virtual communication behavior, different forms of awareness (Caroll et al., 2006) and all kind of ambiguities and problems can be discussed, reflected and methods of solutions can be generated to create a better inter-

action competence. A further dimension for the development of the e-collaboration-competence is to enable the users to shape the virtual space by contributions, to assume responsibility for some e-collaborative activities as well as some role models.

- **E-project-management-competence:** The difficulties of effective and efficient virtual teamwork are underestimated in many ways. We know from "normal" groups, that they pass through typical stages of group development until they are well done for work. Virtual groups also have a lifecycle with typical phases like preparations, launch, performance management, team development and disbanding and within them some key activities (Hertel et al., 2005). Especially the first phases have to be supported by several useful technical tools which help to shape the common ground and built for trust through a trained and experienced e-leader.
- **Human development:** E-learning is a well-known form of learning by using modern forms of technologies, which offer a number of important advantages and enable competencies (e.g., just-in-time and self-paced online learning courses). Traditional trainings are more and more inadequately to offer the didactic dramaturgy. A step by step program to develop the potential of benefits from e-collaboration can be designed by different communication media, different time zones, by a mix between collocated and dislocated learning settings, etc. Furthermore there are some implications and recommendations for successful leadership in virtual teams, which should be reconsidered in training programs for e-collaboration (Cascio & Shurygailo, 2003; Zigurs, 2003).
- **Incentives to e-collaboration:** Incentive systems make co-workers more or less willing to share and seek or avoid collaboration technologies (Olson & Olson, 2000). Therefore an adopted team-based incentive system for a gradual use of different user levels and new features and functions of e-collaboration techniques according to users' competencies is a strong association to team effectiveness and may support e-collaboration.
- **Personal job career on e-collaboration:** It's a personal attitude if people want to develop themselves, to broaden one's mind and to climb up the career ladder. To combine e-collabora-

tion with new career paths within and between organizations, e.g., by creating new jobs would motivate employees to foster cooperation and to use e-collaboration.

A study among European human resource managers shows, that the next generation of technological human resource issues—especially e-learning, knowledge management and B2E/ESS/MSS-employee platforms—will become important (Capgemini, 2004, 2006). With this result the above discussed prerequisites will get more valences to support company's competitiveness. Therewith an important step to successful e-collaboration can be done.

FURTHER TRENDS

It is often said that the implementation of e-collaboration in an organization is not only a technical or information system project. This means e-collaboration projects are very complex because they deal not only with technology but also with humans and the joint outcomes of groups and organizations. As challenges for a successful implementation of e-collaboration still remain for example:

- An effective management support is needed to provide a vision for collaborating employees and to support implementation of the technical and organizational infrastructure.
- Incentives to use the ICT in every day business must be aligned with existing organizational, cultural and professional incentives.
- Social practices need to be cultivated to compensate for the loss of natural social cues because the new electronic communication channels are limited and less rich.
- Barriers to adoption need to be removed.
- Any lack of trust between group members as well as lacks of common vocabulary must be reduced.
- Manager must be trained to be able to allocate and to coordinate distributed work loads as well as to support the needs of their distributed project teams.

From this it follows that functionally driven ICT in e-collaboration may waste resources and time as long as it is not integrated within the human resources and organization. However, according to the so-called e-collaboration paradox it is possible that less medium naturalness (i.e., more dissimilarity with the face-to-face medium) induces compensatory adaptation and, in some cases, e-collaboration outcomes of better quality than in more natural media (Kock & D'Arcy, 2002). This shows that at present no sufficient theory on e-collaboration exists. Subsequently, research on e-collaboration has to focus organizational studies, human resource management, education, psychological questions, etc., too.

CONCLUSION

Recapitulating e-collaboration electronic media are not only another "channel" for exchanging material, value and information but new and promising ways of doing business in a turbulent environment. Therefore e-collaboration will develop to a core competence of future work. This includes a collaborative behavior as well as the knowledge of rules and attributes of computer based communication. But both are new concepts for organizations and it will take time to implement these. Especially many perquisites in the fields of technology, human resources and organizations must be fulfilled simultaneously. Therefore the discussed boundaries do not only demark current limits but also guide possible directions for future research.

REFERENCES

Ackerman, M. S. (2001). The Intellectual Challenge of CSCW: The gap between social requirements and technical feasibility. In J. Carroll, (Ed.), *HCI in the new millennium*, Addison-Wesley.

Blecker, T., Friedrich, G., Kaluza, B., Abdelkafi, N., & Kreutler, G. (2005). *Information and management systems for product customization*. Springer's Integrated Series in Information Systems, vol. 7. Springer Publishing: New York.

Blecker, T., & Graf, G. (2006). Web-based human-machine interaction in manufacturing. In C. Ghaoui (Ed.), *Encyclopedia of human computer interaction*. Idea Group Publishing, Hershey, PA.

Capgemini Consulting (2004, 2006). HR-Barometer. Bedeutung, Strategien und Trends in der Personalarbeit.

Caroll, J. M., Rosson, M. B., Convertino, G., & Ganoe, C. H. (2006). Awareness and teamwork in computer-supported collaborations. In *Interacting with Computers*, *18*(1), 21-26.

Cascio, W.F., & Shurygailo, S. (2003). E-leadership and virtual teams. In *Organizational Dynamics*, *31*(4), 362-376.

Gronau, N., & Kern, E. -M. (2004, August 22-27). Collaborative engineering communities in shipbuilding. In L. M. Camarinha-Matos, (Ed.), *Virtual enterprises and collaborative networks—IFIP 18th World Computer Congress TC5 / WG5.5 - 5th Working Conference on Virtual Enterprises,* pp. 329-338. Toulouse, France: Kluwer Academic Publishers.

Hertel, G., Geister, S., & Konradt, U. (2005). Managing virtual teams: A review of current empirical research. *Human Resource Management Review, 15,* 69-95.

Jarvenpaa, S. L., Knoll, K., & Leidner, D. E (1998). Is anybody out there? Antecedents and trust in global virtual teams. *Journal of Management Information Systems, 14*(4), 29-64.

Kersten, W., & Kern, E. -M. (2003). Distributed Product Development as a Core Element of Supply Chain Management. In *Proceedings of the 2003 IEEE Conference on Emerging Technologies and Factory Automation,* vol. 2 (pp. 35-39). Lisbon.

Kock, N., & D'Arcy, J. (2002). Resolving the e-collaboration paradox: The competing influences of media naturalness and compensatory adaptation. *Information Management Consulting, 17*(4), 72-78.

Kock, N., Davison, R., Ocker, R., & Wazlawick, R. (2001). E-collaboration: A look at past research and future challenges. *Journal of System Information Technology, 5*(1), 1-9.

Kock, N., & Nosek, J. (2005). Expanding the boundaries of e-collaboration. *IEEE Transactions on Professional Communication, 48*(1), 1-9.

Monplaisir, L., & Haji, T. (2002). Fundamental concepts in collaborative product design and development. In L. Monplaisir & N. Singh, (Ed.), *Collaborative engineer-*

ing for product design and development (pp. 2-10). GBV, Stevenson Ranch.

Olson, G. M., & Olson, J. S. (2000). Distance matters. *Human Computer Interaction, 15*, 139-179.

Qureshi, S., Liu, M., Vogel, D. (2005). A grounded theory analysis of e-collaboration effects for distributed project management. In *Proceedings of the 38ᵗʰ Hawaii International Conference on System Sciences.*

Ratnasingam, P. (2005). Trust in inter-organizational exchanges: A case study in business to business electronic commerce. *Decision Support Systems, 39*, 525-544.

Riege, A. (2005). Three-dozen knowledge-sharing barriers managers must consider. *Journal of Knowledge Management, 9*, 18-35.

The American Heritage Dictionary of the English Language (2004). Houghton Mifflin Company.

West Pole (2005). *Virtual workgroups: Hidden challenges to supporting distributed teams.* A whitepaper 1996-2005.

Wigand, R., Picot, A., & Reichwald, R. (1997). *Information, organization and management.* Chichester: John Wiley & Sons.

Zigurs, I. (2003). Leadership in virtual teams: Oxymoron or opportunity? *Organizational Dynamics, 31*(4), 339-351.

Zolin, R., Hinds, P. J., Fruchter, R., & Levitt, R. E. (2004). Interpersonal trust in cross-functional, geographically distributed work: A longitudinal study. *Information and Organization, 14*, 1-26.

KEY TERMS

Collaboration: Two or more people or organizations are working together, especially in a joint intellectual effort, and therefore sharing complex information on an ongoing basis for a specific goal or purpose.

Common Ground: Knowledge which the participants (of groups) have in common, and that they are aware that they have it in common.

E-Collaboration: Collaboration among individuals or organizations using electronic technologies.

Human Resource Management: The entire cycle of people in organizations, from personal planning and marketing, recruiting and selection, leadership and motivation, performance appraisal, personal development, compensation and benefits until the discharge of labor.

Internet Technologies: A family of ICT suitable for exchanging structured data about package-oriented transmissions on heterogeneous platforms, in particular protocols, programming languages, hardware, and software.

Leadership: Process of interaction of attitudes and behavior from single persons as well as interactions in and between groups for the purpose of the achievement of a common goal. Leadership focuses on what the leader does to encourage personnel and organizational performance by inspiring a shared vision, enabling others to act, modeling the way and encouraging the heart.

Organizational Culture: The organizational culture is the entirety of all associations, f.e. norms, values, beliefs, interpretations, paradigms aso. based implicitly or explicitly on the patterns of behavior of a social system. It is the set of values of an organization that helps to understand its members what the organization stands for, how things should be done and what it considers as important.

Presence–Based Real–Time Communication

Frank Frößler
University College Dublin, Ireland

Kai Riemer
Muenster University, Germany

INTRODUCTION

Presence-based real-time communication (RTC) presents itself as a new and emerging technology in the E-collaboration arena with a wide range of new products currently entering the market. Originally created through the integration of instant messaging, with its text chat functionality and presence awareness information, with voice-over IP (VoIP) communication RTC has been maturing over the past three years. Further information and communication channels have been added and RTC technology shows significant potential for integration with other collaborative application as well as general purpose systems like office software. By introducing RTC, its features, potential usage scenarios, and the main players and future trends, this article names several aspects which might inspire future research in this area.

BACKGROUND

Today's work practices have been undergoing significant changes, leading to new forms of organizing and collaborating. The virtualization of organizations and work contexts on the one hand, and the emergence of new information and communication devices on the other hand, are potential causes for this development. These two drivers lead to an all but perfect communication situation from the point of view of the user, as well as those organizations that rely heavily on dispersed collaboration across organizational units.

Firstly, new virtual forms of organizing present new challenges for the management and people working in these increasingly dispersed setups. These changes in the workplace are fueled by trends toward inter-firm partnering and organizational flexibility. As a consequence workplaces are increasingly fragmented and dispersed with teams being spread over several locations. This development clearly is enabled by capabilities of modern information systems and infrastructures like the Internet. Hence, people rely more and more on media and groupware-supported collaboration with participants who are geographically dispersed.

Secondly, over the last two decades, the number of electronic communication devices and channels at our disposal has increased, creating a heterogeneous accumulation of technologies that are available for the average user. With new communication technologies entering the arena the communication options have mushroomed. Nowadays, people may communicate via several telephones or Internet channels. In addition, many people do not just possess one e-mail address, messenger account, or phone number, but rather they use several similar channels. Consequently, the communicative complexity increases drastically for both the initiator and the recipient of a communication request. For initiators situations are characterized by a high uncertainty as they have to think about the recipient's location and context, the appropriate medium and the relevant contact details. Generally, all required information is not at the disposal of the initiator, resulting in failed communication attempts that are time consuming and costly. The recipient on the other hand is confronted with a myriad of communication devices as well as several addresses and numbers, creating a fragmented communication landscape whose coordination is time consuming and tedious.

These two trends bring about significant structural changes to today's working environment and the workplace situation of mobile professionals (cf. Kakihara, 2003). People are potentially confronted with a level of interaction that might exceed their personal preferences (Sørensen, Mathiassen, & Kakihara, 2002). Furthermore, today's work conditions are marked by increased fluidity of interactions with others. While fluidity offers benefits, such as interacting remotely and flexibly with others, it also creates interruptions and disturbances as asymmetries of interaction become more likely (Kakihara, Sørensen, & Wiberg, 2002).

Asymmetries of interaction occur if "the time and topic are convenient for the initiator, but not necessarily the recipient. This asymmetry arises because while initiators benefit from rapid feedback about their pressing issue, recipients are forced to respond to the initiator's agenda, suffering interruption" (Nardi, Whittaker, & Bradner, 2000, p,83). Current technologies such as the mobile phone offer only limited support for people in managing their increased communicative volume. Specifically, the effect of decreasing communication delays on the part of the initiator of a communication request often translates into a work interruption on the part of the recipient (Rennecker & Godwin, 2005). Information and communication requests reach each person unfiltered and people do not have gatekeepers that might help to manage and control the communicative volume. People have to fall back on tactics for minimizing interruptions, such as not answering their telephone, working away from their desk, or by signalling "away" or "busy" deliberately in their instant messaging tool. While this situation is unsatisfying on the individual level it also translates into organizational frictions in that information processes do not run as smoothly as they should or that the lack in availability of key personnel causes coordination problems in projects.

REAL-TIME COMMUNICATION TECHNOLOGY

Real-time communication technology can be seen as a technological attempt to mitigate the problems portrayed above. RTC presents itself as the result of the convergence of the telecommunications market and the market for groupware systems. Hence, of particular interest is the integration of communication channels with computer systems. RTC is based on the idea of unified communication (UC), which describes the combination and management of communication channels according to user preferences. Besides the integration of communication technology with computer systems, the merit of RTC lies in the provision of status information in regards to user availability and communication devices. Hence, two main components of RTC can be distinguished (see Table 1).

The idea behind UC is to relieve the user of the burden of juggling with a large number of devices and channels in different contexts. Thus, UC systems aim at integrating different information and communication

channels, such as e-mail, telephone, instant messaging, or SMS in order to reduce the fragmentation and complexity of today's information and communication landscape. UC is an extension of unified messaging (UM). The aim of UM systems is to manage and coordinate a user's asynchronous communication through a single portal in which all incoming messages of various channels such as e-mail, audio messages, fax, or SMS are collected and which allows for a conversion of messages between these media types: fax and short messages can be forwarded via e-mail, all text messages (SMS, e-mail, fax) or calendar entries can be read to the user by a machine voice, and the user can decide which device to use to access her/his messages of various types. Users are notified about missed calls and their subjects via e-mails.

UC extends the UM idea to synchronous communication. Users are aided by a communication middleware in the management of channels and devices through a rule-based coordination and filtering system. The user can define preferred channels (text, audio, video) and devices (landline, mobile, or IP phones). Incoming calls can thus be diverted and transferred between channels and devices according to a set of filters and rules that can be related to contexts/situations ("in the office," "at home"), time, caller, or caller group ("colleagues, "customers"). An incoming phone call can thus be transferred dynamically to the preferred audio device in a particular situation, as in when the user is not logged in to the office computer, all incoming calls from colleagues are transferred to the mobile phone, while after hours any caller will be diverted to the voice box. For doing so, it is required that each device is registered with the UC system. UC systems enable users (i.e., recipients) to manage the communication volume corresponding to their preferences and contextual demands. The locus of control is shifted from the initiator to the recipient who can decide what is important for organizing his/her work or which requests need instantaneous consideration. UC products from different suppliers are currently entering the market and it is predicted that, by 2007, 80% of enterprise communication purchase decisions will require support for this type of communication solution (Elliot, Blood, & Kraus, 2005).

The second defining feature of RTC is the presence-awareness information, which in RTC is not just limited to people in a buddy list, as is the case in instant messaging systems, but can also refer to

Table 1. Characteristics of RTC technology

Real-time communication technology	
Unified communication	• Various communication channels: E-mail, Instant Messaging, Telephones, Voice Messaging, Web conferencing, SMS, etc.
	• Channel combination and integration: management of synchronous communication
	• Unified Messaging
	• Priority and preference specific filtering of incoming communication requests
	• Definition of preferred devices
Presence awareness information	• Covering different devices
	• Separate awareness status for devices
	• Automatic adaptation of presence awareness information

other devices a user possesses. Thus, the presence awareness information of a person can be differentiated according to channels or devices in that for each device or for a particular channel (text, audio, video) a presence status is provided. For example, the status for audio communication might be "available," if one of the user's audio devices is registered being "active" by the RTC system. To the contrary, both audio and video communication status might show "temporarily unavailable" whenever the user is talking on one of the registered audio/video devices. In the latter situation, synchronous text communication via instant messenger might still be possible, as this does not have the same disruptive impact on the recipient.

In summary, potential benefits of RTC are a better management of personal communication complexity, a better availability, a better control over incoming requests, and less unpredictable disruptions of the work situation by incoming communication requests.

USAGE SCENARIO

While being aware of others' activities comes almost naturally in collocated settings, it is much more difficult to maintain awareness in dispersed work settings. By expanding the notion of presence-availability across space, RTC tries to address these issues. Although implications of the proliferation of presence-awareness information on work practices are not clearly predictable, expectations are that presence-awareness information might affect people's perception of being part of the organizational core (Sahay, 1997). Presence-

awareness information gives people a higher awareness of others availability. However, to avoid interaction overload, recipients can filter incoming information and communication requests as they assign priorities and preferences to particular tasks or persons. The following example aims at illustrating the use of RTC in an organizational setting.

A consultant can organize the communication request on different devices with criteria such as priority, presence-awareness-status, time-of-day, day-of-week, or device. If the consultant, for example, decides to work at home, all incoming calls from team members on his office phone will automatically be forwarded to his private phone number and if that fails, to his mobile phone. All other calls will be forwarded to a self-service-portal. The self-service-portal allows, depending on the initiator's access properties, to access the consultant's calendar, schedule appointments, and read and retrieve documents stored on an exchange folder. The consultant can check his e-mails, voice-mails, the calendar, and appointments over a voice-portal while travelling, being at home or at a client.

The example also illustrates that RTC does not shift control to the recipients without taking the initiator's needs into account. Rather, initiators are assured that they can close the bracket of a task, as RTC either allows direct communication or enables initiators to leave a message, schedule an appointment or access requested documents. Therefore, RTC can contribute, on the one hand, to minimizing delays on the side of the initiator as s/he can continue doing his/her work, and, on the other hand, gives recipients the control over organizing their work.

REAL-TIME COMMUNICATION PRODUCTS

RTC products address mainly two market segments, namely the mass market for private customers and the market for business customers. Skype is a good example for the first one, as it is both a popular and creative provider, regularly releasing new tools and features. At the moment, more than 61 million people are currently registered and more than 3 million people are using Skype simultaneously at any one time. Skype's first free available beta version with VoIP and instant messaging features was released in 2003. Since then, Skype tried to include other communication channels and now offers SkypeOut (calls to landline or mobile phones), SkypeIn (regular phone numbers but incoming calls are received in Skype), the forwarding from Skype calls to landline or mobile phones, video calls, conference calls, and voicemail services. Although p2p calls are for free, Skype charges for extra services such as SkypeOut or SkypeIn. Furthermore, Skype makes efforts to integrate its product with other software applications such as Microsoft Outlook or Mozilla FireFox. As the result, users can use integrated toolbars to see whether the sender of an e-mails or a person, whose SkypeID or telephone number is mentioned on a web site, is currently available.

The business segment is served by players such as Siemens, Alcatel, Nortel, Microsoft, Oracle, or IBM (Elliot et al., 2005). Siemens Hipath OpenScape is targeted at businesses that want to set up a comprehensive RTC environment with the option to integrate the traditional telephone infrastructure. Siemens OpenScape contains a personal unified messaging portal, a voice portal for accessing messages over the phone and further conferencing and workgoup functionality. RTC products partly reflect the history of their providers. IBM for example bundles and extends existing solutions in its IBM Workplace Collaboration Services, offering team collaboration, document, messaging, Web content management, and learning services as its main features.

FUTURE TRENDS

Over the next five years, a maturation and refinement of existing RTC products can be expected. Strong players such as IBM are expected to strengthen hitherto underdeveloped areas such as voice services that count as "must" features for RTC products. The standard set of features of RTC products different providers are offering is therefore expected to converge in the future. It is also expected that RTC will be integrated with enterprise resource planning (ERP) systems in order to facilitate on-demand communication among participants.

In addition, the definition of identity may be reconsidered. Currently, identity stands for an individual, and presence-awareness information is always related to one particular person. Reconsidering the role of identity in RTC might comprise skills, teams, locations, duty rosters, or responsibilities for persons. Identities are then integrated in documents, ERP, or other workflow applications and allow people to access, on an on-demand basis, responsible individuals without knowing in advance who they are. In doing so, presence awareness information can be attached to objects (e.g., a file) and indicate if one of the people, who can provide further

Table 2. Overview of some main players in each market segment

Market segment	Companies
Private Customers	• Skype • Google Talk • AOL Messenger • Yahoo! Messenger • MSN Messenger
Business solutions	• Siemens Hipath OpenScape • Alcatel OmniTouch Unified Communication • Nortel Multimedia Communication Server 5100 • IBM Workplace Collaboration Services • Microsoft Office Communicator 2005 • Oracle Collaboration Suite

information in regards to the object, is available for direct communication via the RTC system. Possible scenarios are hospital settings, service recovery settings, journalism, logistics, and field services, where information is critical and the availability of relevant people paramount.

So far, no empirical results exist in regards to the implementation and use of RTC. However, it can be argued that the study of RTC can be informed by work done on earlier groupware, such as Lotus Notes or Quick Place, as similar implications of RTC and groupware are likely to occur. Generally, groupware is best described as general-purpose-technology that needs to be adapted to the organizational context to match with users' work practices, communication norms, and local conditions (cf. Bansler & Havn, 2004). Its properties are dependent upon the context and are enacted by individual or collective, intended or unintended activities. The implementation process is never completed but should rather be understood as a continuous process with anticipated, emergent and opportunity-based changes (cf. Orlikowski, 1996). Taking such an understanding of groupware as a starting point, IT-researchers investigate, among other things, the use of groupware for knowledge management (cf. Malhorta, Majchrzak, Carman, & Lott, 2001; Qureshi & Keen, 2004), the role of mediators during the implementation process (cf. Bansler & Havn, 2004), or socio-political issues (cf. Hayes & Walsham, 2001). We argue that practitioners and academics should learn form the experiences gained from the study of earlier groupware and take them into account as this is the closest understanding of the organizational implications of RTC one can get at the moment. However, it will be required to return to and re-evaluate some of the aspects mentioned above in order to fully appreciate the implications of RTC.

CONCLUSION

We hope that the article contributes to elucidating the potential of RTC to tackle some of the issues of today's complex working environment. RTC might help people to organize their work by integrating information and communication channels, balancing delays and interruptions of work, and supporting people to cope with the informative and communicative volume.

Over the next few years, we expect RTC to become closely integrated with existing legacy and ERP systems. Currently, no empirical studies exist on the implications RTC has in organizations and researchers therefore are "'dreaming' and 'creating problems' as much as they are solving problems and recording and theorizing about effects." We dared to risk an outlook on the consequences of RTC and argued that people may contact others with the needed skills, resources or job roles, depending on their presence availability rather than previously established contacts.

However, we are cautious of any technological deterministic claims. Benefits that are often mentioned in line with mobile technology and RTC, such as minimization of idle time, faster response time, or more freedom and higher quality of life, are not an automatic outcome of technologies. People should be aware that the implications and properties of RTC will depend on the enactment by its users. The most collaborative software is futile if people are not willing to share their ideas (Dyson, 2004). Currently, research in the groupware domain offers a pool of findings one can use as starting point and sensitising devices. Nonetheless, more rigor and in-depth analysis are needed in the future to make sense of what the implications of RTC are on organizing dispersed work. RTC offers scholars a rich field for future research as aspects analysed for earlier groupware need to be revisited.

REFERENCES

Bansler, J. P., & Havn, E. (2004). Technology-use mediation: Making sense of electronic communication in an organisational context. *Scandinavian Journal of Information Systems, 16,* 57-84.

Dyson, E. (2004). Cultural change in real-time enterprise. In M. Bednarek (Ed.), *Real time: A tribute to Hasso Plattner*. Indianapolis: Wiley Publishing.

Elliot, B., Blood, S., & Kraus, D. (2005, February). Magic quadrant for unified communications. *Gartner Research Note;* 14.02.2005. Retrieved March 10, 2007 from www.skc.com/share/Documents/polyvideo/gartner.pdf

Hayes, N., & Walsham, G. (2001). Participation in groupware-mediated communities of practice: A socio-political analysis of knowledge working. *Information and Organization, 11,* 263-288.

Kakihara, M. (2003). *Emerging work practice of ict-enabled mobile professionals.* London: London School of Economics.

Kakihara, M., Sørensen, C., & Wiberg, M. (2002, May). *Fluid interaction in mobile work practice.* Paper presented at the First Global Roundtable, Tokyo, Japan.

Malhorta, A., Majchrzak, A., Carman, R., & Lott, V. (2001). Radical innovation without collocation: A case study at Boeing-Rocketdyne. *MIS Quarterly, 25*(2), 229-249.

Nardi, B. A., Whittaker, S., & Bradner, E. (2000). *Interaction and outeraction: Instant messaging in action.* Paper presented at the Conference on Computer Supported Cooperative Work, Philadelphia.

Orlikowski, W. J. (1996). Improvising organizational transformation over time: A situated change perspective. *Information Systems Research, 7*(1), 63-92.

Qureshi, S., & Keen, P. (2004). *Activating knowledge through electronic collaboration: Vanquishing the knowledge paradox.* Retreived March 10, 2007, from http://hdl.handle.net/1765/1473

Rennecker, J., & Godwin, L. (2005). Delays and interruptions: A self-perpetuating paradox of communication technology use. *Information and Organization, 15*(3), 247-266.

Sahay, S. (1997). Implementation of information technology: A time-space perspective. *Organization Studies, 18*(2), 229-260.

Sørensen, C., Mathiassen, L., & Kakihara, M. (2002, May). *Mobile services: Functional diversity and overload.* Paper presented at the New Perspectives On 21st-Century Communications, Budapest, Hungary.

KEY TERMS

Dispersed Collaboration: Collaboration of geographically separated individuals, generally supported by information and communication technology.

Interaction Overload: Level of interaction requests exceeding individual's preferences or capacity.

Presence Awareness Information: Being an integrated feature of different devices, presence awareness information signals people's situational availability for communication

Real-Time Communication (RTC): Based on unified communication and presence-awareness information, RTC is the result of the convergence of telecommunication and groupware systems. RTC enables ad hoc communication based on an integrated set of communication channels and devices and signals presence-awareness information.

Unified Communication (UC): Extends unified messaging by integrating asynchronous and synchronous communication. UC is based on a middleware platform that manages communication channels and devices and allows for a rules-based filtering and diversion of communication requests between channels and devices.

Unified Messaging (UM): Management and coordination of asynchronous communication through a single portal in which messages from different channels are bundled. Messages can be converted between media and accessed from various devices via a single portal.

Voice-Over IP (VoIP): Synchronous audio communication via the Internet. Instead of talking over a dedicated phone line, the communication stream is digitalized and transferred as data packets via TCP/IP or SIP protocols.

Prospects for E-Collaboration with Artificial Partners

Kathleen Keogh
The University of Ballarat, Australia

Liz Sonenberg
The University of Melbourne, Australia

INTRODUCTION

Recent work shows that there is interest in how individual artificial agents can work in successful competitive and collaborative teams including people and other agents. Applications involving competing agents include online auctions. Applications for collaborative teams include remote space missions, disaster recovery (e.g., to coordinate a rescue mission) and helping organize appointments for a team of people (Pynadath & Tambe, 2003); as an aid to independent living developing teams of health carers, including artificial carers (Wagner, Guralnik, & Phelps, 2002); in command and control as coordination and communication assistants (Fan et al., 2005); and pedagogical agents in teaching systems (e.g., Shaw, Ganeshan, Johnson & Millar, 1999; Feng, Shaw, Kim & Hovy, 2006).

There is considerable interest in developing multi-agent systems and teams involving people and/or artificial agents collaborating to achieve a common goal. This article will outline some current issues in software agent design with respect to team communication, coordination, and sharing of team situation awareness regarding the current state of a dynamic world. Topics covered include:

- How might artificial agents be defined?
- What mechanisms are helpful to enable agents to form and work in teams in a dynamic world?

Artificial agents are computer programs that have some inbuilt attributed human-like "intelligence" and that operate autonomously with some ability to choose whether or not to perform a task. Wooldridge defines an agent as "a computer system that is capable of independent action on behalf of its user or owner" (Wooldridge, 2002, p, 3). Such agents are able to reason dynamically and make decisions regarding tasks to be completed, so the solution is not preprogrammed in a deterministic way. Agents are usually programmed in terms of defining goals, tasks the agent is capable of performing, and how the agent should react or interpret the data about the world that is available to it. Multi-agent systems (MAS) comprise a set of agents interacting: involving cooperation, coordination and negotiation (Wooldridge, 2002)

BACKGROUND

How are Artificial Agents Defined?

Typical agent attributes to include and inform design include: beliefs, desires, and intentions (BDI). The BDI architecture (Rao & Georgeff, 1991) is not the only model for agency used, however due to space limitations, it is the only model discussed in this article. It is not uncommon to also find discussion and definition of agent roles, responsibilities, obligations, trust, commitment, and protocols for communication and negotiation in agent systems research literature.

When defining an agent, the following agent attributes are defined:

- **Beliefs** about the world encoded as a database of defining attributes and values that are accessible in some way to the agent
- **Desires** or goals that the agent is trying to satisfy
- **Intentions** adopted plans of action that have been chosen in order to achieve a goal.

The designer/programmer empowers the agent with a set of predetermined plans of action and the agent reasons based on predefined factors such as expected utility or specific preferences, to choose a plan to execute. The plan is essentially a predefined script that outlines a series of actions or sub-goals to be performed in order to satisfy a goal. The programmer creates a library of potential plans based on the domain expertise and how human experts would behave. Roles can be predefined as a set of goals and objectives—responsibilities that must be met by the agent accepting that role. In some systems roles are more explicitly defined in terms of hierarchical positions and expected behavior (Zambonelli, Jennings, & Wooldridge, 2003).

Agent reasoning is often non-monotonic, that is, the world is dynamically changing, so beliefs currently held may not be true in the future. Agent programming languages are generally created to enable agents to dynamically change their plans—for example, if a goal is no longer relevant, it should be ignored. Allowing agents to collaborate requires a meta level of additional self-knowledge in the agent to enable agents to negotiate. Agents need to know and possibly negotiate around their adopted roles and what actions they are capable of performing. An agent role can be defined statically at design time—in terms of goals to be performed or the role might be more flexible and negotiated dynamically—to enable more flexible and adaptive team reorganization at run time. Providing the infrastructure to enable an agent to be more flexible and to enable the reorganization of teams requires a more sophisticated agent design than the BDI approach of itself provides and more resources. According to the domain and level of sophistication and reorganization needed, the decision to "keep it simple," or to include more complicated structures is a trade off between flexibility and extra resources and structure required.

Agent Models for Cooperation

An additional component in the agent model that is made explicit by Griffiths and colleagues is an agent's motivation (Griffiths, Luck, & d'Inverno, 2003). They describe motivation in terms of intensity, a threshold for when it applies and functions for goal generation. Agents can chose from a library of partial plans. The motivation component provides a utility measure that an agent can use to decide upon their adopted plans.

Agents can be self-interested or collaborative. Collaborative agents have *obligations* toward other agents in their collaboration team. Self-interested agents make decisions and take actions only in their own interest. When agents have goals that they cannot

achieve alone, or that can be achieved more efficiently by sharing the workload, agents can be motivated to collaborate with other trusted agents to work together on a joint plan (Grosz & Kraus, 1999). When an agent receives conflicting requests from other agents, it has been suggested that these can be resolved by using explicit knowledge regarding relationships existing between agents and agent roles (McCallum, Norman, & Vasconcelos, 2004). Another approach to resolving conflict is to permit agents to engage only in group activities that don't conflict with personal goals (Findler & Malyankar, 2000).

There are different approaches to how agents may communicate and form into groups to work together: central command and control, dynamically allocated teams created "on the fly," negotiated team membership—by invitation and commitment. In some multi-agent systems, the team membership is agreed upon prior to beginning a collaborative task, other models allow for the team to be formed at the time of need and commitment prior to that time is not required. Team formation and coordination issues are discussed further in the next section.

INFRASTRUCTURES FOR AGENT TEAMWORK

Formation of Agent Groups: Team Membership, Team Plans, Motivation to Join In

Institutional organizations are collections of (human) agents that have roles, rights, and obligations. Multi-agent systems are collections of agents that interact within a dynamic situation. Agents are *situated* in the world and have available information via sensors or other input mechanisms to inform them about the current state of the world. When agents begin to collaborate and interact, they need an awareness of the limitations over time and space of their current knowledge, and an ability to share with other team agents possibly different information about the world.

One view of situation awareness (SA) is described with three main elements: (Endsley, Bolte, & Jones, 2003)

1.   Perception of the environment in time and space

2. Comprehension of the meaning in the perceived information
3. Projection of the expected state of the environment in the near future

Agents collaborating on a task need to have a shared understanding of the current world state. This is referred to as shared situation awareness (Endsley et al., 2003) or common ground (Klein, Feltovich, Bradshaw, & Woods, 2005). Common ground is the knowledge believed to be shared between communicating parties (knowledge, beliefs, and suppositions). See also Zhang and Hill (2000) for one approach to representing shared SA. Fan and colleagues have developed a collaborative SA simulation with agents and people in command and control situations based on a collaborative-RPD model (Fan et al., 2005).

Three different types of agent groups have been defined: alliances, teams, and coalitions. Agents in an alliance share similar goals and fully cooperate, giving up on personal goals whilst in the alliance; team agents are recruited under a leader of some kind; coalition agents engage only in group activities that don't conflict with personal goals (Findler & Malyankar, 2000).

Multi-agent *interactions* have been defined in various terms such as cooperative, collaborative, coordinated, competitive, and coherent. Inter-agent relationships might be centralized or decentralized (peer based) and information flow can be direct message based or via an indirect mechanism such as shared access to a workspace. Parunak and colleagues have made a start on defining a taxonomy of multi-agent interactions toward a language to distinguish between different types of interaction and correlation between agents (Parunak, Brueckner, Fleischer, & Odell, 2004).

There are different approaches to group formation. Some work involves using a central controlling agent or a central accessible directory or manager to coordinate groups in the agent world, such as Flores and Wijngaards (1999). Others enable groups to form and collapse independently and membership is by invitation only (Griffiths et al., 2003). Issues to consider include:

- When an agent is deciding to form a cooperating agent group (ask other agents for help to achieve a goal), how does that agent decide who to ask? Do they ask more agents than they need (to allow for some not being able to help)?

- At what point does the agent ask for help? Who is in control of the (cooperative) plan? Is there a leader agent?
- Does a team goal have precedence over an agent's individual goal?
- What information is shared between trusted agents?
- How do agents share situation awareness?
- How does an agent know which agents can be trusted? An internal notion of reliability or trust of other agents is needed.

Griffiths and colleagues discuss agents forming a cooperative intention as a group toward achieving goals (Griffiths et al., 2003). Agent(s) respond to an initiator agent's request for help and adopt the *already chosen* plan and together perform group action on that plan. There is not a central directory or control point from which outside agents can inspect/look up to find existing groups.

This initiator approach contrasts with the approach of Wooldridge and Jennings: they form a cooperative group and then they together *generate* a group plan. One possible group model involves the organization itself taking on an identity with clear organizational goals, rules and structure (Zambonelli et al., 2003). Roles and Interactions are clearly defined at design time.

Another model for teams is the formation of clans (Griffiths, 2003). Agents that are self interested are motivated by a special kinship motivation factor to join or form a group of agents that satisfy a level of trust and can collaborate. Agents in the clan reveal their most recent goals and according to a utility measure based on current motivations, decide whether to group, that is, to join a clan.

Interacting Agents: Knowledge, Commitment, and Trust

When people converse, they share a public perspective on the situation being discussed. Each individual has their own private perspective (internal mental "agency"—beliefs, intentions, etc.) but conforms to public normative rules for engaging with others. The interaction is based within a social context (Singh, 1998).

Approaches to conversation have been divided into two basic categories (Jones & Parent, 2004): intention-based, mentalistic approaches—focusing on the effect communicative acts have on mental states; or commitment based approaches.

Commitment based approaches rely upon there being an explicit reference to commitment by an agent. This commitment is kept separate to the agent's internal structure (e.g., Jones & Parent, 2004). This makes it possible to represent deceit, and communication acts about something without necessarily assuming belief that it is true (just because it was said). This approach is becoming more popular as it enables context and social commitment to be incorporated into an agent communication language (ACL). Taking account of this human quality is increasingly important as agents and people join together in collaboration, for such interactions rely upon a set of conventions governing the social interaction that need to be made explicit in the case of human-agent collaboration.

If agents are to work together collaboratively to successfully complete a task, they need to have a level of **trust** between them. Griffiths includes trust as an explicit component of an agent model. When an agent is considering potential collaborators, it needs to consider the likelihood of those other agents contributing and resulting in success (Griffiths, 2003). An agent's own view of another agent's reliability is its own trust of that other agent, an agent's view of the reputation of another agent is based on some combination (average of) the trust measures provided by other agents.

Organizations have been described as complex, computational and adaptive systems (Carley & Hill, 2001). Based on a view of organizational structure and emerging change in organizations with time, Carley and Hill have suggested that relationships and connections between agents in a network impact on the behavior in organizations. Relationships and interactions are claimed to be important to facilitate access to knowledge. This work may suggest that for teams involving artificial agents involved in a dynamic and emerging organizational structure, it might well be worth investigating the significance of relationship awareness to enable appropriate interactions between agents. A formal model of MAS organization that makes explicit agent roles with associated obligations, relationships between roles and influence due to relationships has been proposed by McCallum et al. (2004).

Coordinated Communication

One approach for the support and control of an agent team is to use policy management to govern agent behavior. This enforces a set of external constraints on behavior—external to each agent. This enables simpler agents to be used. Policies define the "rules" that must be adhered to in terms of obligations and authorizations granting permissions to perform actions (Bradshaw et al., 2003).

When software agents interact, they need to know what to communicate, to whom, and how. They might achieve coordinated communication by either:

- Following an interaction protocol (Paurobally & Cunningham, 2003) to define rules" in terms of content and messaging between agents
- Interacting using an emergent conversation protocol—a dynamic protocol that is developed from within the agent(s) as they interact (Flores & Wijngaards, 1999)
- Using decision theory to choose from available alternative communications (Gmytrasiewicz & Durfee, 2001)
- Controlling interactions by using external artifacts, such as shared data space between agents (Findlar & Malyankar, 2000)
- Adhering to an external policy that governs behavior (Bradshaw et al., 2003)

Tambe and colleagues have made significant progress toward integrating teams of *heterogeneous* multi-agent systems to work together (Pynadath & Tambe, 2003). In order to facilitate the (re)use of heterogeneous agents, Tambe has introduced a special teamwork model Teamcore, in which there are specialized abstract team proxy agents that perform all the team coordination and communication tasks and these interface with other heterogeneous agents.

FUTURE TRENDS

Further research questions that arise in the development of multi-agent teams are: How do agents share information and situation awareness? Can agent teams self organize and reorganize in an emergent team situation? Can agents hand over responsibilities (e.g., at the end of a shift, or to a more capable agent)? How self-aware should agents be of their position in a team hierarchy? How is a team coordinated and controlled? Can agents manage joint/team decisions? How should tasks be (optimally) allocated to agents in a team? Agents with more sophisticated attributes might be used in coordinated teams to work in simulated and real situations to improve performance and success.

CONCLUSION

Further work is needed to explore structures to be incorporated into agent architectures to enable sophisticated human-like behavior. Current research includes exploring communication and language protocols to coordinate agent interactions. Cognitive simulation studies are researching human qualities to be included in artificial agents to enable useful adoption of artificial agents. Multi-agent systems and hybrid teams involving humans and artificial agents provide exciting opportunities to extend on current technology toward improving human performance, especially in complex and dynamic situations.

REFERENCES

Bradshaw, J., Beautement, P., Breedy, M., Bunch, L., Drakunov, S., Feltovich, P., et al. (2003). Making agents acceptable to people. In N. Zhong & J. Liu (Eds.), *Handbook of intelligent information technology.* Amsterdam: IOS Press.

Carley, K. & Hill R. (2001). Structural change and learning within organizations. In A. Lomi (Ed.), *Dynamics of organizational societies: Models, theories and methods.* Cambridge, MA: MIT/AAAI Press.

Endsley, M., Bolte, B., & Jones, D. (2003). *Designing for situation awareness: An approach to user-centered design.* CRC Press.

Fan, X., Sun, S., Sun, B., Airy, G., McNeese, M., & Yen, J. (2005). Collaborative RPD-enabled agents assisting the three-block challenge in command and control in complex and urban terrain. In *Proceedings of 2005 BRIMS Conference Behavior Representation in Modeling and Simulation* (pp.113-123). Universal City, CA.

Feng, D., Shaw, E., Kim, J., & Hovy, E. (2006). An intelligent discussion-bot for answering student queries in threaded discussions. In *Proceedings of the 2006 international conference on intelligent user interfaces. 2006* (pp. 171-177).

Findler, N., & Malyankar, M. (2000). Social structures and the problem of coordination in intelligent agent societies. *Invited talk at the special session on agent-based simulation, planning, and control in 16th IMACS World Congress 2000.* Lausanne, Switzerland.

Flores, R., & Wijngaards, N. (1999). Primitive interaction protocols for agents in a dynamic environment. In B.R. Gaines, R. C. Kremer, & M. Musen (Eds.) *Proceedings of the 12th workshop on Knowledge Acquisition Modeling and Management (KAW'99)* (pp.3.2.1.-3.2.20). Banff, Canada.

Griffiths, N. (2003). Supporting cooperation through clans, cybernetic intelligence: Challenges and advances. In *Proceedings IEEE Systems, Man and Cybernetics, 2nd UK&RI Chapter Conference* (pp. 87-96).

Griffiths, N. Luck, M. d'Inverno, M. (2003). Annotating cooperative plans with trusted agents. In R. Falcone, S. Barber, L. Korba, & M. Singh (Eds.) *Trust, Reputation, and Security: Theories and Practice* (pp. 87-107). Lecture notes in artificial intelligence, 2631. Springer-Verlag.

Grosz, B., & Kraus, S. (1999). The evolution of shared plans. In A. Rao & M. Wooldridge (Eds.), *Foundations and theories of rational agency* (pp. 227-262).

Gmytrasiewicz, P. J., & Durfee, E. J, (2001). Rational communication in multi-agent environments. *Autonomous Agents and Multi-Agent Systems, 4,* 233-272.

Jones, A. J. I., & Parent, X. (2004) Conventional signaling acts and conversation. In F. Dignum (Ed.), *Advances in Agent Communication, International Workshop on Agent Communication Langauages (ACL2003),* (pp. 1-17). Lecture notes in computer science 2922. Melbourne, Australia 2004.

Klein, G. (1993). A recognition-primed decision-making model of rapid decision making. In G. Klien, J. Orasanu, R. Calderwood, & C. Zsambok (Eds.), *Decision making in action: models and methods* (pp. 138-147).

Klein, G., Feltovich, P. J., Bradshaw, J. M., & Woods, D. D. (2005). Common ground and coordination in joint activity. In W. R. Rouse & K. B. Boff (Eds.), *Organizational simulation*. New York: Wiley.

McCallum, M., Norman, T., & Vasconcelos, W. (2004, August). A formal model of organizations for engineering multi-agent systems. In G.A. Vouros (Ed.) *Proceedings of the International workshop on coordination in emergent agents societies CEAS '04, ECAI 2004*, Valencia, Spain.

Parunak, H. H., Brueckner, S., Fleischer, M., & Odell, J. (2004). A preliminary taxonomy of multi-agent interactions. In P. Giorgini, J. Müller, & J. Odell (Eds.), *Lecture notes on computer science, vol. 2935*. Berlin: Springer.

Paurobally, S., & Cunningham, J. (2003). Achieving common interaction protocols in open agent environments. In *Proceedings of the Workshop on Agentcities: Challenges in Open Agent Environments held in conjunction with the Second International Joint Conference on Autonomous Agents and Multi-Agent Systems, (AAMAS '03)*.

Pynadath, D., & Tambe, M. (2003). An automated teamwork infrastructure for heterogeneous software agents and humans. *Autonomous Agents and Multi-Agent Systems, 7*, 71-100.

Rao, A. S., & Georgeff, M. P. (1991). Modeling rational agents within a BDI-architecture. In R. Fikes & E. Sandewall (Eds.), *Proceedings of Knowledge Representation and Reasoning (KR&R-91)* (pp. 473-484). San Mateo, CA: Morgan Kaufmann.

Shaw, E., Ganeshan, R., Johnson, W. L., & Millar, D. (1999). Building a case for agent-assisted learning as a catalyst for curriculum reform in medical education. In *Proceedings of the Int'l Conf. on Artificial Intelligence in Education* (pp. 70-79).

Singh, M. P. (1998). Agent communication languages: Rethinking the principles. *Computer, 1998 IEEE* 40-47.

Sun, R., & Naveh, I. (2004). Simulating organizational decision-making using a cognitively realistic agent model. *Journal of Artificial Societies and Social Simulation, 7*(3).

Wagner, T., Guralnik, V., & Phelps, J. (2002). Achieving global coherence in multi-agent caregiver systems: Centralized vs. distributed response coordination. In ILSA *AAAI02 Workshop Automation as caregiver: The role of intelligent technology in elder care*.

Wooldridge, M. (2002). *An introduction to multi-agent systems*. Sussex: John Wiley & Sons.

Zambonelli, F., Jennings, M., & Wooldridge, M. (2003). Developing multi-agent systems: The Gaia methodology. *ACM Transactions on Software Engineering and Methodology, 12*(3), 317-370

Zhang, W., & Hill, R. (2000). A template-based and pattern-driven approach to situation awareness and assessment in virtual humans. In *Proceedings of the Fourth International Conference on Autonomous Agents*, Barcelona, Spain.

KEY TERMS

ACL: Agent communication language.

Agent: a software entity situated in an environment and capable of acting to change the environment.

BDI: Beliefs, desires, and intentions, an architecture for agent design.

Collaboration: cooperative work between two or more individuals/agents.

KB: Knowledge base.

MAS: Multi-agent system, a set of agents interacting, potentially involving cooperation, coordination and negotiation.

RPD: Recognition primed decision making, a model of expert decision making (Klein, 1993).

SA: Situation awareness, awareness of current state of the world situated in time and space.

Psychological Contracts' Influence on E-Collaboration

Vicki R. McKinney
University of Arkansas, USA

Mike Allen
UW–Milwaukee, USA

INTRODUCTION

Organizations pursing global opportunities find advantage in requiring the use of e-collaboration. While organizations employ e-groups strategically to accomplish tasks, empirical reports indicate a large number of group failures (Levi, 2001). E-collaboration groups may experience an increased chance of failure since establishing trust (Jarvenpaa & Leidner, 1999) and instituting shared work practices (Chudoba, Wynn, Lu, & Watson-Manheim, 2005) is often difficult.

Studies investigating group failure have produced dismal findings using traditional information-based views (Timmerman & Scott, 2006). Information-based theories feature a dominant focus on information flows and density of communication channels (Daft & Lengel, 1986). These theories focus on reducing uncertainty and equivocation when transferring information between parties. The approaches focus on the content of the communication and the means of message transmission as explanations for channel preference and message effectiveness.

One reason for disappointing research findings about the effectiveness of virtual groups is that focus on information and communication channel characteristics fail to reflect the important elements of communication. Personal relationships are meaningful and valued by people but studies fail to examine how communication methods relate to the development of relationships among group members. Information-based approaches examine choices made among the means of communication. Online and other forms of mediated communication replace the face-to-face options in work settings requiring employees to rely on technological or mediated communication. Haythornthwaite (2002) finds no difference between f2f and mediated communication when good prior relationships exist among group members.

Research findings suggest a change in focus for the study of work groups. We propose a relation-based focus incorporating the concept of psychological contracts to understand virtual group behavior. Psychological contracts, used to explain relationships between employers and employees (Rousseau, 1995), provide rich insights to examining e-group behavior. The psychological contract established between an individual and the team in an f2f environment transfers to the online environment. Group members perceiving a breach to their psychological contract in f2f settings perceived the same violation with a team in an e-collaboration setting.

BACKGROUND

Face-to-face communication provides the preferred channel for communicative exchanges (Daft & Lengel, 1986). While virtual team members have successfully combined the use of information communication technologies (ICTs) to effectively achieve the richness associated with f2f communication (Zack & McKenny, 1995), f2f communication remains the richest and preferred medium for communication (McKinney & Whiteside, 2006).

Media richness theory (MRT) (Daft & Lengel, 1986) and social presence theory (SPT) (Short, 1976) represent two dominant theories for studying the influence of specific communication media on individual's interpretation and information exchange. Studies based on technology choice theories find that consequences of use depend on contextual factors such as experience with the media and familiarity of the people engaged in the communication activity (Carlson & Zmud, 1999), thus reinstating that communication is more than just an exchange between sender and receiver (Katz & Kahn, 1978). Interpretation of the exchange within the social environment plays an important role (Falk, Schmitz, & Steinfield, 1990). This system of personal and professional relationships in which communication occurs plays a role in the formation of team mental models. One mental model is the psychological contract an individual develops towards the team.

PSYCHOLOGICAL CONTRACTS IN GROUPS

E-groups incorporate individuals selected for reasons such as availability, skills, experience, or politics. Concerns over what group membership means may cause an individual to experience uncertainty when joining a group (McCollum, 1995). After joining a team the initial experience may involve uncertainty about the group and task since membership is often diverse and team members may be meeting each other for the first time. For an e-group to succeed, reciprocal relationships must exist between the group and the individual. Reciprocal relationships involve individuals using background and experiences to develop expectations about the group and task.

Psychological contracts provide beliefs that individuals use to reduce uncertainty associated with entering group relationships (Rousseau & Parks, 1992). These beliefs become instrumental to the unique psychological contract the individual develops. Group members' psychological contracts define the member and group obligations to each other.

E-collaborations usually have legal contracts associated with them (Sabherwal, 1999). On the other hand, psychological contracts in group relationships are perceptual (Robinson, Kraatz, & Rousseau, 1994). Group members bring unarticulated expectations to the first group meeting, then behave according to expectations of appropriateness (McCollom, 1995). The psychological contract does not include all of the individual's expectations about the group; instead the psychological contract is a subset of expectations based on the individual's conveyed promise to the group (Rousseau & Tijoriwala, 1998). In other words, beliefs are formed from incomplete information about what an individual's obligations are to the group and what the individual will receive from the group (Rousseau, 2001). The accuracy of these beliefs reflects the quality of information available to the individual (Rousseau, 2001). Promises are constructed from fragments of

information received when communicating (Rousseau & Parks, 1992) creating an agreement that "exists in the eye of the beholder" (Rousseau, 1995, p. 6).

E-group members' psychological contracts address what individuals need to do and what they receive in return, yet each group member's perception of what specifically needs to be done and what specifically will be received remains unique. Individuals may not share similar views of the psychological contract (Rousseau & Tijoriwala, 1999), yet contracts sharing some beliefs permit the achieving of interdependent goals and provides group members a basis for aligning behaviors with the actual commitments made to the e-group (Dabos & Rousseau, 2005). As members interact and new information is presented, individuals use the new information to reevaluate the existing contract. The new information increases alignment between the individual's contract and contracts held by other e-group members (Arrow, McGrath, & Berdahl, 2000).

External influences operating in the e-group (e.g., social cues) and the individual's internal predisposition (e.g., cognitive styles, self-schemas) influence the individual's psychological contract (Dabos & Rousseau, 2005). This paper focuses on an individual's psychological contract with an e-group, yet we acknowledge that e-group members may concurrently hold other contracts (Rousseau, 1995).

PSYCHOLOGICAL CONTRACTS IN E-COLLABORATION

Psychological contracts have been acknowledged as significant in situations involving exchange agreements (Rousseau, 1995), suggesting that individuals develop a type of psychological contract when entering group relationships (Kramer, Hanna, Su, & Wei, 2001). Since e-collaboration relationships are based on exchange agreements psychological contracts can be useful in providing new insight when studying behavior in groups relying on ICTs. The major difference between f2f groups and groups in e-collaboration relationships is environmental. The individual's perceived psychological contract should not change when transferred from an f2f environment to an online environment since decision makers successfully continue relationships using ICTs after establishing f2f relationships (McKinney & Whiteside, 2006).

As e-groups collaborate and new experiences occur, gradual modifications may occur to an individual's psychological contracts. New external or internal events can cause modification to be made to each individual's "contract" (Rousseau, 1995). The strength of the contract under evaluation can be influenced by the coexistence of multiple contracts, such as employer/employee and other concurrent group member contracts.

When organizational decisions cause changes to occur among e-collaboration relationships, individuals may experience unmet expectations. Often e-group members maintain prior beliefs by selectively filtering information (Alavi & Tiwana, 2002). However, when individuals have a high degree of dependency on the expectations inherent in the psychological contract a perceive change may be interpreted as acceptable or interpreted as a change that reneges on perceived obligations (Robinson & Morrison, 2000).

Individuals, based on the psychological contract, expect group reciprocation in response to voluntary actions to fulfill perceived obligations (Turnley, Bolino, Lester, & Bloodgood, 2003). Effectively, a social exchange relationship becomes established among individuals within the group (Shore & Barksdale, 1998). Balance is the objective and ideal in this type of "contract" situation (Blau, 1964) since individuals evaluate the contract by measuring how the group is performing compared to "contract" expectations. An individual might perceive a significant difference between what is expected and what is received and view the "contract" as unfair, providing a justification for individual dissatisfaction (Robinson, 1996). An e-group member becoming dissatisfied can exhibit negative behaviors and feeling toward the group lowering their performance and contribution (Wanous, Poland, Premack, & Davis, 1992).

When an individual perceives that the terms of their psychological contract have not been met a breach may occur. A psychological contract breach means an individual believes in group failure to meet one or more perceived contractual obligations. The perception of a breach is influenced by the level of trust in the group (Robinson & Morrison, 1996). Trust mediates an individual's psychological contract breach and the motivation to continue making contributions (Robinson, 1996).

E-collaboration projects risk failure when trust is not established between the individual and e-group since trust has been suggested as influencing the ef-

fects other factors have on behaviors and attitudes (e.g., Jarvenpaa, Shaw, & Staples, 2004). This influence can occur through how an individual interprets or evaluates the information received (Dirks & Ferrin, 2001) or may result from accumulated social capital (Adler & Kwon, 2002). Individuals rely on established trusted relationships to obtain richer information (Krachhardt & Hanson, 1993). Depending on ICTs information selectively confirms or disconfirms prior beliefs (Robinson 1996) and group members decide consciously or subconsciously the influence communication activity on the existing e-collaboration psychological contract.

IMPLICATIONS

By focusing on psychological contracts in e-collaboration researchers may gain insight to the negative behaviors that often impact e-group success. Group members perceiving unmet expectations may feel a tendency to withdraw, believing a psychological contract breech exists that negates the individual's obligation (Rousseau, 1995). An employee wishing to remain with the organization will not exit, but may exhibit other signs, such as voicing discontent, a pessimistic attitude, or neglect of group obligations (Rousseau, 1995). A research agenda focusing on group conflict resolution provides a basis to understand the response of groups to dissension and disagreement.

When an e-group member uses negative behavior after feeling the psychological contract is breached the problem can escalate when other members respond. Antisocial group behavior evokes the response of antisocial individual behavior and increases in proportion to the length of time of the membership of the individual in the group (Robinson & O'Leary-Kelly, 1998). Antisocial group behavior may increase dissatisfaction with the e-group in individuals choosing not to participate in the antisocial behavior (Robinson & O'Leary-Kelly, 1998).

If multiple e-group members perceive breaches to their contract and align with each other as a subgroup a fault line could be created (Lau & Murnighan, 2005). The strength of relationships among the unhappy e-group members can cause the subgroup members to identify with the subgroup instead of the e-group leading to changes in communication patterns (Lau & Murnighan, 2005). If subgroups start to communicate mainly with subgroup and not with the e-group as a whole the environment will not foster trust and connectedness, creating difficulties with information exchange and mutual understanding.

This pattern of communicating with only a subgroup also jeopardizes the psychological safety of the e-group (Edmundson, 1999). The environment is considered psychologically safe when members share beliefs creating a climate of confidence and mutual understanding positively associated with e-group learning behavior (Edmundson, 1999). To address these set of issues requires an evaluation of both the substantive content of the message about the task and the relational implications of the messages sent by e-group members. Research designs should treat messages as multifunctional and address multiple goals as interpreted by members of the e-group.

Negative behaviors could arise from not identifying with one's e-group. Identification refers to the extent to which an individual senses oneness with the characteristics and actions of the group (Timmerman & Scott, 2006). Once identity with an e-group develops communicative behavior is produced leading to the group coordinating actions and decision making (Scott, Corman, & Cheney, 1998). A perceived breach in a member's psychological contract may prevent developing an identity with the e-group. The geographically dispersed nature of e-collaboration may make it easy for the non-identifying individual to become invisible. At the same time, an e-group generates a written record reviewable by members to ascertain the level and effectiveness of group participation by any particular group member.

An example of a negative behavior that could result from an individual perceiving a breached or violated psychological e-group contract is social loafing. A function of group size (Hogg, 1992), social loafing occurs when individuals in small groups reduce the effort they put into collective task performance (Latane, Williams, & Harkins, 1979). As mentioned above, an individual perceiving a "contract" breach may experience a decline in individual identity resulting in reduced motivation (Hogg, 1992). The lack of effort by the individual may negatively effect the motivation of other e-group members by reducing the acceptable norm of participation/productivity (Geen, 1991). Karau and Williams (1993) demonstrated that social loafing diminishes when the identifiability of individual contributions increases.

A strong issue for consideration represents the perception of the e-members that the communication represents a form of interaction (Fulford & Zhang, 1993). When e-group members perceive that the collaboration among the members represents an interaction requiring contribution and participation, then the advantages of a workgroup emerge. The e-group members begin to find the group a source of inspiration (Allen, Halone, Mabry, & Gamble, 2003). The group finds support and recognition, not in individual accomplishment, but in the success of the e-group (team).

CALL FOR RESEARCH

The reliance on theories (MRT, SPT) focused on the characteristics of the communication channel fail to provide an adequate understanding of the success and failures for virtual groups. The role of psychological contracts in e-collaboration offers a new perspective for examining the results of e-group efforts and needs to be investigated. Studies identifying and examining behaviors of e-groups need to expand the theoretical focus from the technology used by group members to include the relationships developed among the group members to gain insight on group behaviors. Understanding the success and failure of e-groups requires future research that examines how messages sent among members accomplish the relational dimensions of group development as well as fulfilling the requirements set forth by the task.

REFERENCES

Alavi, M., & Tiwana, A. (2002). Knowledge integration in virtual teams: The potential role of KMS. *Journal of the American Society for Information Science and Technology, 53*(12), 1029-1037.

Alder, P. S., & Kwon, S. (2002). Social capital: Prospects for a new concept. *Academy of Management Review, 27*(1), 17-40.

Allen, M., Halone, K., Mabry, E., & Gamble, S. (2003, November). *Communication correlates of social loafing: Initial tests of a functional competence theory*. Paper presented at the National Communication Association Convention, Miami, FL.

Arrow, H., McGrath, J. E., & Berdahl, J. L. (2000). *Small groups as complex systems: Formation, coordination, development, & adaptation.* Thousand Oaks, CA: Sage Publications.

Blau, P. M. (1964). *Exchange and power in social life.* New York: J. Wiley.

Carlson, J. R., & Zmud, R. W. (1999). Channel expansion theory and the experiential nature of media richness perceptions. *Academy of Management Journal, 42*(3), 153-170.

Chudoba, K. M., Wynn, E., Lu, M., & Watson-Manheim, M. B. (2005). How virtual are we? Measuring virtuality and understanding its impact in a global organization. *Information Systems Journal, 15*(4), 279.

Dabos, G. E., & Rousseau, D. L. (2005, August). *Psychological contracts and the informal social structure of organizations: Systemic and local effects.* Paper presented at the Academy of Management Annual Meeting, Honolulu, HI.

Daft, R. L., & Lengel, R. H. (1986). Organizational information requirements: Media richness and structural design. *Management Science, 32*(5), 554-571.

Dirks, K. T., & Ferrin, D. L. (2001). The role of trust in organizational settings. *Organization Science, 12*(4), 450-467.

Edmondson, A. (1999). Psychological safety and learning in work teams. *Administrative Science Quarterly, 44*(2), 350-383.

Falk, J., Schmitz, J. A., & Steinfield, C. (1990). A social information model of technology use. In J. Falk & C. Steinfield (Eds.), *Organizations and communication technology* (pp. 117-140). Newbury Park, CA: Sage.

Fulford, C., & Zhang, S. (1993). Perceptions of interaction: The critical predictor in distance education. *American Journal of Distance Education, 7*(3), 8-21.

Geen, R. G. (1991). Social motivation. *Annual Review Psychology, 42*, 377-399.

Haythornthwaite, C. (2002). Strong, weak, and latent ties and the impact of new media. *Information Society, 18*(5), 385-401.

Hogg, M. A. (1992). *The social psychology of group cohesiveness: From attraction to social identity.* New York: New York University Press.

Jarvenpaa, S. L., & Leidner, D. E. (1999). Communication and trust in global virtual teams. *Organization Science, 10*(6), 791-815.

Jarvenpaa, S. L., Shaw, T. R., & Staples, D. S. (2004). Toward contextualized theories of trust: The role of trust in global virtual teams. *Information Systems Research, 15*(3), 250-267.

Karau, K. J., & Williams, K. D. (1993). Social loafing: A meta-analytic review and theoretical integration. *Journal of Personality and Social Psychology, 65*(4), 681-706.

Katz, D. (1966). *The social psychology of organizations.* New York: J. Wiley.

Katz, D., & Kahn, R. L. (1978). *The social psychology of organizations* (2nd ed.). New York: J.Wiley.

Krachhardt, D., & Hanson, J. R. (1993). Informal networks: The company behind the chart. *Harvard Business Review, 71*(4), 104-111.

Kramer, R. M., Hanna, B. A., Su, S., & Wei, J. (2001). Collective identity, collective trust, and social capital: Linking group identification and group cooperation. In M. F. Turner (Ed.), *Groups at work: Theory and research* (pp. 173-196). Mahwah, NJ: Lawrence Erlbaum Associates.

Latane, B., Williams, K., & Harkins, S. (1979). Many hands make light the work: The causes and consequences of social loafing. *Journal of Personality and Social Psychology, 37*, 822-832.

Lau, D. C., & Murnighan, J. K. (1998). Demographic diversity and faultlines: The compositional dynamics of organizational groups. *Academy of Management Review, 23*(2), 325-340.

Levi, D. (2001). Understanding teams. In D. Levi (Ed.), *Group dynamics for teams* (pp. 3-18). Thousand Oaks, CA: Sage.

Lipnack, J., & Stamps, J. (1999). Virtual teams: The new way to work. *Strategy & Leadership, 27*(1), 14-18.

McCollum, M. (1995). Group formation: Boundaries, leadership, and culture. In J. Gillete & M. McCollum (Eds.), *Groups in context: A new perspective on group dynamics* (pp. 34-48). Lanham, MD: University Press of America.

McKinney, V. R., & Whiteside, M. M. (2006). Maintaining distributed relationships. *Communications of the ACM, 49*(3), 82-86.

Robinson, S. L. (1996). Trust and breach of the psychological contract. *Administrative Science Quarterly, 41*(4), 574-599.

Robinson, S. L., Kraatz, M. S., & Rousseau, D. M. (1994). Changing obligations and the psychological contract: A longitudinal study. *Academy of Management Journal, 37*(1), 137-152.

Robinson, S. L., & Morrison, E. W. (2000). The development of psychological contract breach violation: A longitudinal study. *Journal of Organizational Behavior, 21*(5), 525-546.

Robinson, S. L., & O'Leary-Kelly, A. M. (1998). Monkey see, monkey do: The influence of work groups on the antisocial behavior of employees. *Academy of Management Journal, 41*(6), 658-672.

Rousseau, D. M. (1995). *Psychological contracts in organizations: Understanding written and unwritten agreements.* Thousand Oaks, CA: SAGE Publications.

Rousseau, D. M. (2001). Schema, promise and mutuality: The building blocks of the psychological contract. *Journal of Occupational and Organizational Psychology, 74*(4), 511-541.

Rousseau, D. M., & Parks, J. M. (1992). The contracts of individuals and organizations. In L. L. Cummings & B. M. Staw (Eds.), *Research in organizational behavior* (pp. 1-43). Greenwich, CN: JAI Press.

Rousseau, D. M., & Tijoriwala, S. A. (1999). What's a good reason to change? Motivated reasoning and social accounts in promoting organizational change. *Journal of Applied Psychology, 84*(4), 514-528.

Sabherwal. R. (1999) The role of trust in outsource IS development projects. *Communications of the ACM, 42*(2), 80-86.

Scott, C. R., Corman, S. R., & Cheney, G. (1998). Development of a structural model of identification in the organization. *Communication Theory, 8*(3), 298-336.

Shore, L. M., & Barksdale, K. (1998). Examining degree of balance and level of obligation in the employment relationship: A social exchange approach. *Journal of*

Organizational Behavior, 19(S1), 731-744.

Short, J. (1976). *The social psychology of telecommunications.* New York: Wiley.

Timmerman, C. E., & Scott, C. R. (2006). Virtually working: Communicative and structural predictors of media use and key outcomes in virtual work teams. *Communication Monographs, 73*(1), 108-136.

Turnley, W. H., Bolino, M. C., Lester, S. W., & Bloodgood, J. M. (2003). The impact of psychological contract fulfillment on the performance of in-role and organizational citizenship behaviors. *Journal of Management, 29*(2), 187-206.

Wanous, J. P., Poland, T. D., Premack, S. L., & Davis, K. S. (1992). The effects of met expectations on newcomer attitudes and behaviors: A review and meta-analysis. *Journal of Applied Psychology, 77*(3), 288-297.

Zack, M. H., & McKenny, J. L. (1995). Social context and interaction in ongoing computer-supported management groups. *Organization Science, 6*(4), 394-422.

KEY TERMS

E-collaboration: The use of electronic mediated communication to coordinate a group.

Group: A set of between 3 to 25 interdependent individuals that share a common task or goal.

Identity: The membership a person feels or expresses with some larger group or culture.

Media Richness: The level of information available in the channel of communication.

Psychological Contract: A mental model held by a group member of the perceived obligations they owe the group and the obligations they perceive the group owes them.

Relationships: Tthe development of emotional, psychological, and personal connections among members of a group.

Virtual Group: A group whose communication practices do not include copresent f2f methods.

Reconsidering IT Impact Assessment in E-Collaboration

Az-Eddine Bennani
Université de Technologie de Compiègne, Reims Management School, France

INTRODUCTION

The question of the impact of information and communication technology (ICT) on company performance and its evaluation is a much discussed topic and is often analyzed ex-post after its implementation.

The literature review shows two main research groups using various models making it possible to explore this impact. The first falls under the economic production theory and the information and decision theory, often referring to econometric models (Alpar & Kim, 1990; Due, 1994; Brynjolofson & Hitt, 1993). It raises the question of ICT contribution in terms of efficiency and tries to show the existence of a relation between the investments made in this technology and the operational and financial performance of companies. The second group can be divided into three subgroups. The first subgroup examines performance as a dependent variable centered on ICT success perception (DeLone & McLean, 1992, 2002, 2003; Seddon, 1997). The second considers ICT effects on operational and managerial processes (Crowston & Treacy, 1986; Bakos 1987; Mooney, Gurbaxani, & Kraemer, 1995). Finally, the third bases its research works on contingency models (Henderson & Venkatraman, 1993; Iivari 1992).

Moreover, the permanent evolution of this technology and consequently its diversity of uses are accompanied by new explanatory factors of its impact. This leads to the appearance of new dependent variables that express the potential of a company in terms of strategic flexibility or competitive advantage (Porter & Millar 1985; Bakos & Treacy, 1986). In addition, some research refers to the MIT90 approach that regards the organization as an open system to study the impact of ICT.

In spite of the variety of answers to be found in the literature on the subject, it remains one of the concerns of researchers and managers. Still, one may wonder if it should not be considered ex-ante when defining company strategy, if it should it not be considered in the complex and evolutionary environment of a company, and it it should not take the impact generated by new phenomena such as e-collaboration into account. The objective of this article is to highlight one of the principal limitations to the ICT success evaluation model recommended by DeLone and McLean, and proposes a new, more global model, whose finality aims at considering the question of the impact of ICT and the various phenomena like e-collaboration that have resulted from e-business.

BACKGROUND

DeLone and McLean (1992) stressed that in 1980, at the time of the first ICIS conference, Peter Keen identified five questions management information systems (MIS) should fit into a coherent field of research. These questions were: (1) what are the disciplines of reference for MIS? (2) Which is the dependent variable? (3) How do MIS establish the cumulative tradition? (4) What is the relation of MIS research with computer science and MIS practice? (5) Which types of journals should MIS researchers publish their results in? Among these five questions, DeLone and McLean were more particularly interested in those relating to the dependent variable. In the introduction of their article, these authors specify that to ensure the contribution of MIS research to the professional world, it was essential to define the measurements of the awaited results accurately. They stressed that it was not very useful to measure the various independent variables, such as the degree of user participation or the investment level in ICT, if the dependent variable could not be measured with the same degree of accuracy. Here, the authors were particularly interested in ICT success in term of effectiveness: ICT produced the awaited effects and results. They recommended a model that was one of the first models to have established a link between multiple levels: the system level, the individual level, and the organization level. In this model success is determined by six categories: the quality of the system, the quality of information, the use of ICT, the satisfaction of the user, the individual, and the organizational impact. The link between the various levels results in the impact (joint or independent) of the quality of the system as well as that of information on the satisfaction of the user and the use of ICT. The latter constitutes a determinant of the impact on the individual, consequently generating an impact on the organization.

Ten years later, DeLone and McLean (2002) carried out a retrospective analysis based on the most significant research publications that apply, validate, challenge, and recommend modifications to improve their model of origin. The authors consequently proposed a renovated version of this model that incorporated the new recommendations. In 2003, a second modification was needed to take the e-commerce context into account. Subsequent to the adjustments brought in 2002 and 2003 to their model of origin, DeLone and McLean (2004) tested its relevance and assessed its success in e-commerce using two case studies.

METHODOLOGY

In order to show the relevance of the proposed new global model, the methodology adopted is based on the review, the analysis, and the criticism of the literature in parallel with the author's former research work on the representation and the modeling of the e-business phenomenon. The literature concerned is mainly that which focuses primarily on DeLone and McLean's research work (1992, 2002, 2003, 2004). Furthermore, it also draws on the research by Seddon and Kiew (1994), Seddon (1997), Etezadi-Amoli and Farhoomand (1996), and Wilkin and Hewett (1999), which uses, applies, and/or criticizes the DeLone and McLean model of origin since its first recommendation. Firstly, a meta-analysis of the major theoretical, conceptual, and empirical research work on DeLone and McLean ICT success models is used; one key limitation that is common to the various versions resulting from the model of origin is explained. Thorough analysis of these authors' articles shows the difficulty of applying this model in an e-business environment. Secondly, the conceptual relevance of a new model suggested apprehending not only the assessment of information and communication technology and its impact but also that of e-collaboration is emphasized. Finally, before concluding, e-supply chain management (e-SCM) is described as an illustration of e-collaboration.

THE DELONE AND MCLEAN MODEL: THE LIMITATIONS

The DeLone and McLean model stresses that the success of ICT and e-commerce depends on the satisfaction of individuals in the organization. It allows the evaluation of this success by considering the human being at the center of this question. But, managers are individuals who not only use ICT, but also decide on the investments to be made in it. Notice that this significant point is not considered in this model; neither is assessment considered in an environment where new phenomena related to e-business are permanent.

Moreover, successive updates in 2002 and 2003 lead us to question the universality of this model. They can be interpreted as a sort of recognition by DeLone and McLean of the limitations of their model. Is it necessary to update this model each time company environment changes, or should we not look for yet other unexplored ways and propose a suitable alternative?

All this highlights the need for a new, more "universal" conceptual model, in which information and communication technology reaches beyond the technical aspect, and handles the impact and assessment issues of this technology at a time of company global strategy definition, considering the e-business context where new phenomena such as e-collaboration appear permanently. The following describes a model that makes it possible to include both these two questions at the same time. Thus, it shows the relevance of this new suggested model.

E-MODEL: WHAT MAKES THIS MODEL RELEVANT?

The e-model presented in the following figure 1.1 is a representation model of the e-business. Its preliminary testing was conclusive in the hotel industry (Bennani, Lhajji, & Aihie, 2005) and in the pharmaceutical industry (Bennani, 2004). It lies within the scope of a new manner of carrying out business in an economy where ICT is increasingly omnipresent. It is an open and dynamic model characterized by plural complexity: technical, organizational, and strategic (Bennani, 2003). It is recommended as a model for the e-business phenomenon and based on three essential components: ICT, organization structure, and strategy. The latter are considered jointly, in "fusion," to allow one to formulate a company strategic vision. Their continuous interactions produce e-business modules such as e-commerce, e-SCM, e-CRM, and so on, that form the fundamental matrix of this model, thanks to which value creation for the customers occurs and the competitive advantage for the company is reinforced.

In this model the perception of value does not solely boil down to (1) the acceptance of ICT and its best use (Davis, 1989) or its good fit with the task (Goodhue & Thompson, 1995), (2) user satisfaction (Karahanna & Straub, 1999), and (3) a strategic alignment of ICT with company business strategy (Henderson & Venkatraman, 1993); but rather it extends to integration of various built-in modules using some contingency factors such as organizational structure, strategy, the environment, information and communication technology, quality of management, quality of coordination between information systems managers, business managers, and the executive...

The e-model falls under a complex, interactive, and evolutionary context. It is singular compared to the DeLone and McLean model and its derivatives, mainly by (1) the definition of the e-business modules, and (2) the faculty to make analysis at the industry level feasible. Its specificity results from several aspects, the first of which being the multiplicity of the levels of analysis. It thus establishes a link among various levels such as

the system level, the individual level, the organization level, the strategy level, the interorganization level, and the industry level. The system level is strongly influenced by the dynamism of the ICT industry—innovations and technologies, techniques and methods of distribution, data-processing staff, offers in the marketplace, competition, manufacturers of computer, software publishers, suppliers of services, and so forth. The individual level concerns the users, their skills and know-how, and their work attitudes. The organization level clarifies the organizational structure, the skills and competencies, and the strategy level translates into business strategies, business processes, and so on. Last of all, the industrial level relates to the sector the company belongs to, the current practices in this sector, the standards retained, the usual ICT solutions, the expected form of ICT performance, and so on.

The e-model makes it possible to take into account the issue of ICT impact at a time of company global strategy definition, by merging ICT with the various current and potential determinants to confront them with the strategic stakes created by the new business environment. It respects the approach of the ICT impact based on the contingency models (Iivari, 1992; Venkatraman, 1989) in strategic management and organizations theory, and accounts for the continual evolution of company processes. It proves to be the first

Figure 1. The e-model is a representation of e-business

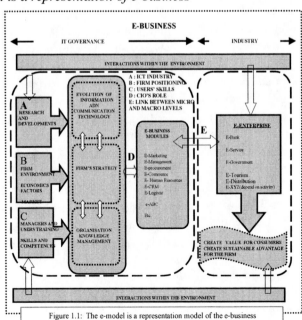

Figure 1.1: The e-model is a representation model of the e-business

model that integrates the dynamics and evolutionary aspects of its construction while making it possible to maintain a connection with the industry. It is based on the idea according to which the various modules, resulting from the process of interaction between its components, adapt to the changes in the environment in order to maintain and reinforce company competitive advantage.

The e-model is made up of three fundamental variables: ICT exceeding its role as a tool and becoming a parameter of influence on (1) the strategic formulation and the control of the business, (2) the organizational structure integrating various functions of the company (R&D, production, distribution, finance) and its know-how, and (3) the strategy implying the company's positioning in the marketplace. It proposes to break with the dissociation between these components established by a majority of experts and researchers. Indeed, these components are continuously in interaction, making possible the emergence of new modules that contribute to the transformation of a company thanks to various ICT solutions. The availability of a technology such as the Internet facilitates and reinforces the interactions between these modules. This supposes that companies indicate the latter, with diverse configurations and/or some arrangements to meet the requirements of their industry: e-banks, e-tourism, e-distribution, e-service, and so forth. Thus, exchanges of information in the systems value chains binding the partners, the customers, and the suppliers become possible and give shape to the e-business modules such as e-marketing, e-management, e-purchasing, e-commerce, e-CRM, and e-SCM. These modules, whose improvements are not necessarily known at the beginning, occur in time by taking account of the strategy and the organizational structure. In other words, they evolve jointly in technical, strategic, and organizational terms, respecting a fit defined in the sense of Venkatraman (1989). Their evolution is founded on a common system of shared beliefs and takes place within a process of training the stakeholders—customers, suppliers, partners, and so on. It is carried out within a dynamic framework generating continual interactions allowing the creation of multiple combinations, starting from the specific modules, which facilitates strategic deployment.

Focusing on the interaction issue between various components of the e-model and the e-business modules provides a field of materialization of the phenomenon describing the e-collaboration. This is what is illustrated next.

E-MODEL, E-COLLABORATION AND E-SUPPLY CHAIN MANAGEMENT

Kock and Nosek (2005) and Kock, Daison, Wazlawick, and Ocker (2001) define e-collaboration as collaboration based on information and communication technology, i.e. the use of this technology to create shared knowledge (Qureshi, Van Der Vaart, Kaulingfreeks, De Vreede, Briggs, & Nunamker, Jr ,2002). It results in the sharing of information between individuals and/or organizations for planning, coordination, and decision purposes. The e-model makes intralevel e-collaboration possible, at the microeconomic level: e-collaboration inter-components (ICT, strategy, and organization), module-to-module e-collaboration (e-commerce, e-CRM, e-SCM, etc.), and intramodule e-collaboration (for example, the collaboration between various stakeholders and processes of the e-SCM) using Internet/Intranet/Extranet. Here, intercomponent e-collaboration is explained by the continual interactions between the three main components of the e-model, namely, ICT, strategy, and organization.

E-SCM, as a particular e-business module makes the integration (Cassivi, Lefebvre, & Léger 2004) of company information and that of its partners and the various processes necessary to powerful management of the supply chain possible. It illustrates well an example of intramodule e-collaboration translating all the relations and activities facilitated by the use of ICT (Johnson & Whang, 2002; Williams, Esper, & Ozment, 2002). Here, e-collaboration exceeds e-sales and/or e-purchases and contributes to the performance of the company. It integrates several activities that require sharing information, decisions, processes and resources between the various partners. According to Folinas, Manthou, Sigala, and Vlachopoulou (2004), e-SCM contributes to the development of e-collaboration and makes the transition of the exchange of operational information towards strategic information possible. E-SCM is more than just a coordination and optimization function, it implies the integration of a network of stakeholders including producers, suppliers, customers, and so forth who are involved together in the strategic activities of the company thanks to the sharing of business processes.

CONCLUSION

The abundant literature on ICT success and its contribution to company performance reveals a great diversity in conceptualization and research methods as well as in the levels of analysis and dependent variables. Since the beginning of the eighties, the issue of ICT impact is ambiguous and difficult to define. It is not limited to productivity or to financial profitability; it extends to other factors such as structure, strategy, the environment, quality of management, and its implications. Adequate methodology and a suitable analysis framework considering all the above-mentioned factors and forthcoming ones to be identified would contribute to explain the legendary "productivity paradox." In addition to the five relevant questions raised by Peter Keen, we add another one: Which analysis framework and which model should be retained to register MIS in an e-business environment that is not limited only to the issue of effective and efficient use of information and communication technology? Note that the latter must be analyzed jointly with the business strategy and the organizational structure, and this ex-ante. The e-model recommended in this article contributes to this, and allows one to consider the ICT impact issue at a time of company global strategy definition. It also considers, as an example, that the question of e-collaboration must be examined as a parameter of ICT assessment owing to the fact that it is defined as collaboration based on information and communication technology.

REFERENCES

Alpar, R., & Kim, M. (1990). A microeconomic approach to the measurement of information technology value. *Journal of management Information Systems, 7*(2), 55-69.

Bakos, J. Y. (1987). Dependent variables for the study of firms and industry-level impacts of information technology. In *Proceedings of the 8th International Conference of Information Systems* (pp.10-23).

Bakos, J. Y., & Treacy, M. (1986). Information technology and corporate strategy: A research perspective. *MIS Quarterly, 10*(2), 107-119.

Bennani, A. (2003). Identifying and managing the complexity of e-business project leads to understand information technology complexity. In *Proceedings of the 12th International Conference on Management of Technology (IAMOT)*, Nancy, France, May 13-15.

Bennani, A. (2004). La réalité de l'alignement stratégique. Le cas des entreprises pharmaceutiques internationales : une étude exploratoire. In *Proceedings of European & Mediterannean Conference on Information Systems (EMCIS)*, Tunis, Tunisia, July 25-27.

Bennani, A., Lhajji, D., & Aihie, O. (2005). How does IT governance contribute to e-bBusiness paradigm? The case of tourism and hotel industry. *Actes du 10ème Colloque de l'Association Information Management*, Toulouse, France, 21-23 Septembre.

Brynjolofson, E., & Hitt, L. (1993). Is information systems spending productive? New evidence and new results. In *Proceedings of the 13th International Conference on Information Systems*, Orlando, Florida, (pp. 47-64).

Cassivi, L., Lefebvre, L., Léger, P. (2004). The impact of e-collaboration tools on firms' performance. *The International Journal of Logistics Management, 15*(1), 91-110.

Crowston, K., & Treacy, M. E. (1986). Assessing the impact of information technology on enterprise level performance. In *Proceeding of the 7th International Conference of Information Systems*, San Diego, (pp. 227-239).

Davis, F. D. (1989). Perceived usefulness, perceived ease of use, and user acceptance of information technology. *MIS Quarterly, September, 13*(3), 318-340.

DeLone, W. H., & McLean, E. R. (1992). Information systems success: The quest for the dependent variable. *Information Systems Research, 3*(1), 60-95.

DeLone W. H., & McLean E. R. (2002). Information systems success revisited. In *Proceedings of the 35th Hawaii International Conference on System Sciences*.

DeLone, W. H., & McLean, E. R. (2003). The DeLone and McLean model of information systems success: A ten-year update. *Journal of Management Information Systems, Spring*, 9-30.

DeLone, W. H., & McLean, E. R. (2004). Measuring e-commerce success: Applying the DeLone & McLean information systems success model. *International Journal of Electronic Commerce, 9*(1), 31-47.

Due, R. T. (1994). The productivity paradox revisited. *Information Systems Management, 4*(1), 74-76.

Etezadi-Amoli, J., & Farhoomand, A. (1996). A structural model of end user computing satisfaction and user performance. *Information Management, (30),* 65-73.

Folinas, D., Manthou, V., Sigala, M., Vlachopoulou, M. (2004). Evolution of a supply chain: Cases and best practices. *Internet Research, 14*(4), 274-283.

Goodhue, D. L., & Thompson, R. (1995). Task-technology fit and individual performance. *MIS Quarterly, June,* 213-236.

Henderson J. C., & Venkatraman N. (1993). Strategic alignment: Leveraging information technology for transforming organizations. *IBM Systems Journal, 32*(1), 4-16.

Iivari, J. (1992). The organizational fit of information systems. *Journal of Information Systems, 2*(1), 3-29.

Johnson, M. E., & Whang, S. (2002). E-business and supply chain management: An overview and framework. *Production and Operations management, 11*(4), 413-422.

Karahanna, E., & Straub, D. W. (1999). The psychological origins of perceived usefulness and ease of use. *Information & Management, 35*(4), 237-250.

Kock, N., Daison, R., Wazlawick, R., Ocker, R. (2001). E-collaboration: A look at past research and future challenges. *Journal of System Information Technology, 5*(1), 1-9.

Kock, N., & Nosek, J. (2005). Expanding the boundaries of e-collaboration. *IEEE Transactions On Professional Communication, 48*(1), 1-9.

Mooney, J. G., Gurbaxani, V., & Kraemer, K. (1995). A process oriented framework for assessing the business value of information technology. In *Proceedings of the International Conference on Information Systems,* Amsterdam, The Netherlands, (pp. 17-27).

Porter, M. E., & Millar, V. E. (1985). How information gives you a competitive advantage. *Harvard Business Review, 63*(4), 149-160.

Qureshi, S., Van Der Vaart, A., Kaulingfreeks, G., De Vreede, G., Briggs, R., Nunamker, J., Jr (2002). What does it mean for an organization to be intelligent?

Measuring intellectual bandwidth for value creation. In *Proceedings of the 35th Hawaii International Conference on System Sciences.*

Seddon, P. B. (1997). A respecification and extension of the DeLone and McLean model of IS success. *Information Systems Research, September,* 240-253.

Seddon, P. B., & Kiew, M. Y. (1994). A partial test and development of DeLone and McLean's model of IS success. In *Proceedings of international conference in information systems,* Vancouver, Canada, (pp. 99-110).

Venkatraman, N. (1989). The concept of fit in strategy research: Toward verbal and statistical correspondence. *Academy of Management Review, 14*(3), 513-525.

Wilkin, C., & Hewett, B. (1999). Quality in a respecification of DeLone and McLean's IS success model. *Proceedings of 1999 IRMA International Conference,* (pp. 663-672).

Williams, R. L., Esper, T. L., & Ozment, J. (2002). The electronic supply chain: its impact on the current and future structure of strategic alliances, partnerships and logistics leadership. *International Journal of Physical Distribution and Logistics Management, 32*(8), 703-719.

KEY TERMS

E-Business: is a new way of doing business in an economy where information and communication technology is increasingly ubiquitous and new phenomena like e-collaboration are emerging.

E-Model: defines the representation model of e-business. It is made up of three fundamental variables: information and communication technology, company strategy, and organizational structure.

E-Business Module: is an e-business unit resulting from a continuous process of interactions between information technology, strategy, and organizational structure. For example, e-management, e-purchasing, e-commerce, e-marketing, e-CRM, e-SCM, and so forth.

E-Collaboration: collaboration, based on information and communication technology, between individuals and/or organizations engaged in a common business strategy and/or business processes.

E-Collaboration Intercomponents: Is an e-collaboration due to interactions between the three principal components of the e-model, namely, information and communication technology, strategy, and organization.

E-Collaboration Module-to-Module: Is e-collaboration between modules such as e-commerce, e-CRM, e-SCM, and so on.

E-Collaboration Intramodule: Is collaboration between various stakeholders and between processes using information and communication technology, and this within the same e-business module. For example, the collaboration between various stakeholders and e-SCM processes using Internet/Intranet/Extranet.

E-Collaboration Intralevel: Is inter-component e-collaboration, module-to-module e-collaboration (e-commerce, e-CRM, e-SCM, etc.) or intra-module e-collaboration.

A Reflection on E-Collaboration Infrastructure for Research Communities

Lydia M. S. Lau
The University of Leeds, UK

Peter M. Dew
The University of Leeds, UK

INTRODUCTION

E-collaboration amongst researchers requires not only the people working together but all the layers of a collaborative system working together as well, starting at the point where people interface with the system. Although this article concentrates more on the technical infrastructure required for e-collaboration, the influence of the social and people issues on the conceptual design of the interface and the functionalities of the collaborative system will also be discussed. Often, the interface/interaction between the 'soft' and the 'hard' issues generates some interesting and dynamic effects between the layers of the infrastructure (Dourish, 1999).

As a starting point, this article presents a framework to illustrate the dependencies of the different layers of the infrastructures for a collaborative system. Against this background, a case study undertaken since the early 1990s on an e-collaborative environment will be discussed. The system, named the Virtual Knowledge Park (VKP), developed from a number of research projects and grew into a commercial platform. The developmental journey for the infrastructure used in the case study will be reviewed alongside the lessons learned.

The article concludes with an extrapolation of the infrastructure that is required for future research collaborations.

BACKGROUND

The common reference point to the beginning of collaborative systems can be dated back to 1984 when the term *computer-supported cooperative work* (CSCW) was coined by Irene Grief and Paul Cashman (Grudin, 1994). With the coining of terms such as *e-commerce*, *e-government* and *e-communities*, e-collaboration has become the latest label for CSCW.

The starting point of e-collaboration, however, can be traced back even before 1984 when e-mail, teleconferencing systems and office automation were the emerging technologies (Hiltz, 1984; Wainwright & Francis, 1984). The label groupware (a term first used by Peter and Trudy Johnson-Lenz) took over around 1990 to refer to the technologies for CSCW and included a wide range of applications (Ellis, Gibbs, & Rein, 1991). Advances in networking provided a driving force for the development of these collaborative systems. In particular, the system architecture could then evolve from being monolithic and centralised to becoming client-server and decentralised. Collaborative systems are becoming more pervasive (Chung & Dewan, 2004) in both the work and the social spaces of users.

Another driving force came from the World Wide Web (WWW), which became available to the commercial world in the 1990s. This in turn introduced various collaborative tools to a wider user base. These tools originally had been restricted to academic or large commercial organisations. This opening up of access fuelled the pace of development and the deployment of tools such as e-mail software on clients, discussion forum, instant messaging, and file sharing. These tools have become defacto standard functionalities in any collaborative systems today.

The availability of bandwidth has also made multimodal interaction a common experience in today's collaborative environments. Users do not just interact in text, but also in audio, graphical, and video modes. This has offered researchers new ways of working together, and richer media for the exchange of information, knowledge and ideas.

The next section charts the evolution of collaborative systems against the evolution of the other different layers of the underlying infrastructure. This can provide some insights into the shape of things to come.

EVOLUTION OF THE INFRASTRUCTURE FOR E-COLLABORATION

The technical infrastructure for a collaborative system consists of a number of layers interworking with each other. A framework is devised to analyse the evolution of each of these layers (namely access devices, collaboration metaphors, tools, architectures, and networks) and how each layer might impact on the development of the others over time (see Table 1). The timeline in the table only charts the period when the selected collaborative tools had a commercial presence.

In the mid-1980s, the human-computer interface changed from being text-based to becoming graphics-based, with icons representing items that were familiar on or near a desk at work. A "collaborative" environment typically consisted of e-mail and office tools. More complex systems such as Lotus Notes also started to emerge (IBM, 2005) for improving document management and internal communications. Workflow management systems (Kobine, 1986) were used to route forms from one person to another and to maximise the degree of automation. A mainframe mentality was still dominant hence architectures were typically centralised. Local area networks (LAN) began to make an impact as a common network infrastructure. Only the large international organisations had wide area

data networks (WAN) in place. These kinds of systems are still evolving—mostly being incorporated into an enterprise wide system and ported to the Internet with a browser front-end for access.

In the early 1990s, desktop computers gained more processing power. This encouraged the increasing use of graphics and video. There were experiments with various different metaphors on the computer screen, such as a graphical representation of buildings, rooms, workspaces, and so on. The workspace metaphor seems to have had the most long lasting appeal as it can still be found in today's products such as BSCW, eRoom, Groove, and VKP (see later sections). The World Wide Web started to make an impact but mainly for sharing hypertext documents. The range of applications increased when the client-server architecture was extended to the n-tier architecture (Lubich, 1995). After the mid-1990s, it was possible to run more complicated applications with database technologies and wrappers over the Internet. W3C standards became more mature. Virtual environments such as online communities mushroomed with a range of tools at their disposal (Benford, Brown, Reynard, & Greenhalgh, 1996; Pfister, Schuckmann, Beck-Wilson, & Wessner, 1998). There were attempts at mixing 3-D graphics into these online communities and the provision of desktop videoconferencing facilities, but the bandwidth was not high enough to meet user expectations regarding performance. There were

Table 1. Adoption of commonly known technologies in the infrastructures for e-collaboration

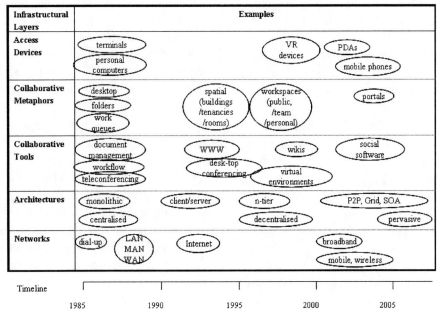

also experiments with augmented reality and different immersion devices (Streitz, Konomi, & Burkhardt, 1998).

In the 21st century, another leap was driven by the availability of high bandwidth connections and the increasing mobility of networked digital devices. Today, it is possible to receive up-to-date multimedia information on a PDA without the need to be near a socket. A portal (usually via a Web browser) has become a common metaphor for a gateway to a pool of resources. Collaborative software is also further permeating our social space (e.g., blogs and Yahoo groups). The use of peer-to-peer (P2P), Web services and grid computing in the architecture can in theory facilitate the delivery of resource-on-demand whenever and wherever desired. Music sharing activities using Kazaa is an example of using P2P. The realisation of the full potential of these emerging architectures is still an ongoing challenge. Integration and open standards are crucial.

A CASE STUDY: THE VIRTUAL KNOWLEDGE PARK (VKP)

The VKP, formerly known as the Virtual Science Park (VSP), grew out of a research and development programme at the interdisciplinary Centre for Virtual Working Systems at the University of Leeds (Dew, Leigh, Drew, Morris, & Curson, 1995; Leigh, Dew, Drew, Morris, & Curson, 1996). The vision for the development of a knowledge-based collaborative environment was to improve support for knowledge transfer activities and research collaborations between academics and industries. The system was intended to provide a professional service to its user communities, hence a spin-off company and an internal support unit emerged during this period to oversee the quality of deliveries and support.

The next few sections describes how the underlying infrastructure of the VKP evolved over a period of more than 10 years and the main lessons learned from a number of evaluation exercises.

ARCHITECTURE AND NETWORK

Since its inception, the VKP has been based on a client-server architecture over the Internet. There is a central repository for documents and a data feed

from another centralised database of expertise within the university. This centralised approach has largely remained unchanged.

COLLABORATION TOOLS: PHASE ONE

In the earliest release (i.e., the VSP in the mid-1990s), the role of VKP was mainly for providing support for information sharing and a means for electronic communications amongst group members. The following functionalities were provided:

- **Resource rooms:** For users to place online publications and resources with access permissions set by the administrator(s) of the tenancy
- **Communication tools:** Included discussion boards, notice boards, e-mail links plus synchronous conferencing tools
- **Personal offices:** Used as private working areas by individuals and also as a means for other users to locate and communicate synchronously and asynchronously with the individual owner of the office
- **Administrator's management toolkit:** For a tenancy administrator to set up access control and to create special subgroups as appropriate

Trials of the VSP(v1) with real users within and outside the university were conducted at the end of phase one. As a result, the seed of an "expertise matcher" was sown. This feature was to be built in the next version, VSP(v2).

A usability evaluation (Lau, Curson, Drew, Dew, & Leigh, 1999) also revealed that the rigid centralised/hierarchical access control was not liked by the users. (For example, a user needed the administrator's authorisation to allow a new member to join a resource room.) There was also a need for better document version control, awareness features, and integration between the shared area and a user's own work area. Even a "group" trash bin was suggested to act as a temporary bin just in case there was any disagreement regarding which files could be removed from the shared area. These issues could not be addressed in the VSP(v2) and had to wait for a redesign to the VKP as they required a major paradigm shift away from viewing groups as "tenancies" towards defining them as "users."

Figure 1. REPIS and VSP(v2)

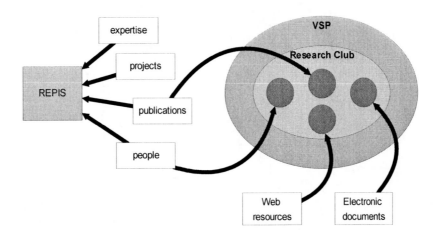

COLLABORATION TOOLS: PHASE TWO

In the VSP(v2), the expertise matcher feature should enable users to identify new potential collaborators. The notion of bringing people together also emerged so that the benefits of chance meetings in the real world could be replicated in the virtual version.

To enable users of the VSP(v2) to locate research expertise within the university, an expertise matcher was developed based on the University of Leeds Research Expertise and Publications Information System (REPIS). REPIS was a structured database which stored information on individual academics publications, current and completed research projects, and research expertise. Data were acquired from either university central sources, existing legacy publications databases, or by direct input by departmental administrators and individual academic members of staff. REPIS later evolved into the University of Leeds Publication Database (ULPD). More details could be found in Liu (2004).

Initially there were issues about the validation and quality of the data. However, the currency and accuracy of the information were assured when the data were integrated with the university's research support processes regarding preparation for the research assessment exercise (RAE). The RAE was a periodic peer review exercise for ranking the research standing of UK's higher education institutions. Figure 1 shows the conceptual view of the integration of REPIS in VSP(v2).

Another round of evaluations took place at the end of phase two. One example was the Packaging Executives Tutorial Information Service (PETIS) Tenancy on the VSP(v2). This project studied the requirements based on the print and packaging industry and developed a community of interest. VSP(v2), with REPIS input, provided a common information space to support the participants before, during and after face-to-face workshop events. The participants were researchers and senior executives. Experience showed the invaluable contributions of face-to-face interactions (via the workshops) and that the effective use of an on-line environment for further collaboration would require follow-on facilitation and cultural changes (Curson et al., 2001).

COLLABORATION TOOLS: PHASE THREE

A major overhaul of the design took place in this phase which formed the basis for the latest version, the VKP. Analysis from the previous evaluation activities conducted with end users over a number of years showed that a collaborative environment must be able to provide an organisational structure which can support four distinct levels:

- **The Organisational Level:** A park-wide services (metadata) should be available to all users

Figure 2. An overview of the VKP

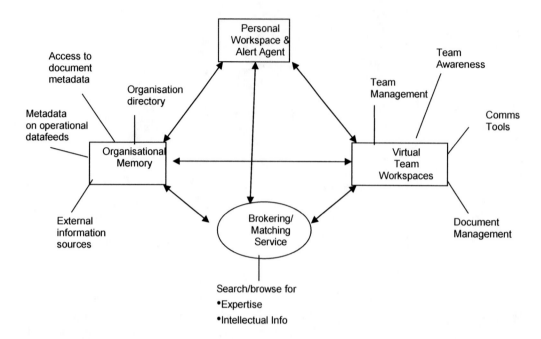

such as directories of information about people, organisations, services, and resources (i.e., the "organisational memory" and "brokering service" in Figure 2).

- **The Community Workspace:** A collaborative area where all facilities (the document management system, discussion boards, whiteboards, chat rooms, online messaging) would be available (i.e., some of the tools in the "virtual team workspaces" in Figure 2).

- **The Project Workspace:** A project area which could be created by any user based around themselves as well as named individuals selected from the Contact Management System. The creator of the project space would have the ability to assign access rights to documents and resources (i.e., the "virtual team workspaces" in Figure 2).

- **The Personal Workspace:** Personalised services including one's own customised "place," search spaces, links to project workspaces and communities. This could then be used as the home entrance for registered members (i.e., see the "personal workspace" in Figure 2).

To improve the awareness within the VKP, alert services could be set up by individuals so that messages would be sent to the individual's usual mailbox when a tracked event took place.

Before rolling out the deployment of the revamped VKP, an evaluation exercise took place on the prerelease version to identify any final refinement and the type of user training/support required (Lau, Adams, Dew, & Leigh, 2003). By this time, the VKP was a well-polished commercial system and the user expectation adjusted accordingly. A support unit was also set up to offer initial training and a help desk service.

To date, the VKP has been mainly used for its document management functions and project workspaces. The vision of facilitating the formation of communities/social networks has not been realised.

COLLABORATION METAPHORS AND ACCESS DEVICES

These two layers of the infrastructure are the ones closest to the end users. Getting them right had been challenging as user experiences and expectation have changed rather rapidly over the last decade. The only stable factor has been that the main access was via the web browser, with the option of adding a webcam and microphone for videoconferencing.

Underlying the VSP(v1) was the concept of the park metaphor and the notion of shared spaces. Clients of the VSP(v1) rented space on the Park in order to establish a tenancy which supported the development of their "business," drawing upon the functionalities provided in a suite of "rooms," including reception areas, workspaces, and knowledge resource rooms. An example of the metaphor used is shown in Figure 3.

This metaphor might be appropriate for corporate users but might be slightly alienating for grass root researchers. Moreover, mimicking the physical metaphors of a park/tenancies had in fact caused difficulties for users to navigate around the "virtual walls" in the environment. This was particularly the case when a user had access to more than one tenancy. In addition the desktop had too much nonfunctional space.

Hence in the VKP, a "grass root" view was provided without a 3-D graphical interface. Once logged on, a user would be able to go directly to his/her own workspace, but also able to see and access all the relevant team workspaces. A familiar navigation tree structure was provided on the navigation pane on the left (common in many desktop applications). The main display area remained on the right. By this stage, most users were familiar with the web browser interface and the window-based metaphors. The 3-D graphical interfaces did not seem to be missed. User-friendliness was not a problem anymore.

Figure 3. An office in the first version of VSP

WHAT DO RESEARCH COMMUNITIES WANT

Evaluating the VKP showed it had a mixed reception. On one hand, a working system had been delivered with useful functionalities. On the other, the activity level had been much lower than expected. Sustainability was identified as the main issue. An empirical study confirmed that there might be a conflict between the corporate (centralised) view of knowledge sharing and the individual researcher's (decentralised) perception on the cost and benefit of such activities (Tian, Lau, & Dew, 2004). Several myths were discovered on the way:

- **Researchers want to share:** An over simplistic view of sharing can limit the benefits of using an e-collaboration system. There is definitely a difference between sharing published/completed material and sensitive/work-in-progress material. The former will largely fall into the Library remit where sharing has been taking place for a long time. The latter, however, is more problematic and the circulation of this material needs to be tightly controlled.

- **Research resources are free:** There is a belief that undertaking research is a form of public service and research resources should be freely sharable. However, the reality is that research resources are expensive (e.g., licenses for software, equipment, time spent by a researcher) and increasingly a charge has to be levied for access to these resources. Ignoring this aspect might prohibit the willingness to share.

- **Researchers want to change the way they collaborate:** There is always some inertia or even resistance to changing working practices. Researchers are not exempt. If the current provision for collaboration seems adequate and is well understood, any new collaborative tools will be an unwelcome distraction for the individuals. Moreover, different types of material being shared in the e-collaboration system (e.g., documents, datasets, models) might dictate the means of sharing in a way researchers dislike. For example, certain datasets might be in a specific format only suitable for use by a specific piece of software.

- **Researchers all have the same level of IT proficiency and expectation:** While huge effort have been made to ensure the "user-friendliness" of the collaborative environment, it turns out that this could be a turn-off for some users. For example, some researchers in certain disciplines might be used to command line driven applications and would find the supposedly user-friendly interface cumbersome and limiting. A better understanding of the competencies and needs of specific groups of researchers might help us provide the correct level of customisation for them.

A couple of other lessons were also learned:

- 3-D metaphors offered little value-adding information for knowledge-based activities. Simply replicating "real-world" metaphors in a virtual world does not entice users.
- The value-statement of a system must be clearly articulated and understood by all in the development process as well as intelligibly communicated to the users. An inherently weak value-statement will inhibit the uptake of a system.

FOLLOW-ON EXPERIMENTAL SYSTEMS

The lessons learned from the case study has led to a belief that the knowledge sharing infrastructure should be more decentralised and should be directed towards empowering the individual researchers. The peer-to-peer (P2P) architecture appears to have the potential for a community-based knowledge market infrastructure (Tian, 2005). The service-oriented architecture (SOA) also supports the vision of market-based paradigm. A complete rethink on the underlying infrastructure might be timely.

In parallel, Grid technologies (Foster, Kesselman, & Tuecke, 2001) have gathered momentum in the scientific research communities (Atkins et al., 2003; Hey & Trefethen, 2002) and are spreading to the social and behavioural sciences (Berman & Brady, 2005; ESRC, 2004). Hence, an experimental prototype using a service-oriented architecture for integrating P2P and the Grid is being developed to better support scientific research communities (Pham, Lau, & Dew, 2004; Pham, Lau, Dew, & Pilling, 2005).

CONCLUSION

It has been an interesting two decades for e-collaboration. This article has provided an account of the developmental journey of a particular collaborative system for research communities—the Virtual Knowledge Park. Valuable lessons have been learned. Some could be addressed immediately at the application level, others seem to point to the need for a more fundamental change to the infrastructure. This article suggested a few developments which could trigger a very different approach to collaboration in the future.

ACKNOWLEDGMENT

We would like to thank all who have worked with us on the various projects which contributed to the experiences that were reflected in this article. In particular, Professor Christine Leigh, Dr. Jayne Curson, Dr. Richard Drew, and Dr. Craig Adams.

REFERENCES

Atkins, D. E., Droegemeier, K. K., Feldman, S. I., Garcia-Mollna, H., Klein, M. L., Messerschmitt, D.G., et al. (2003). *Revolutionizing science and engineering through cyberinfrastructure: Report of the National Science Foundation Blue-Ribbon Advisory Panel on Cyberinfrastructure* (Tech. Rep.). National Science Foundation.

Benford, S., Brown, C., Reynard, G., & Greenhalgh, C. (1996). Shared spaces: Transportation, artificiality, and spatiality. In *Proceedings of the 1996 ACM Conference on Computer Support Cooperative Work* (pp. 77-86). ACM Press.

Berman, F., & Brady, H. (2005). *Final report: NSF SBE-CISE Workshop on Cyberinfrastructure and the Social Sciences.* Retrieved from www.sdsc.edu/sbe/

Chung, G., & Dewan, P. (2004). Towards dynamic collaboration architectures. In *Proceedings of the 2004 ACM Conference on Computer Supported Cooperative Work* (pp. 1-10). New York: ACM Press.

Curson, J., Leigh, C., Dixon-Hardy, J., Lewis, W., Winnett, C., Lau, L., et al. (2001). Requirements of a lifelong learning environment for executives and academics—The PETIS experience. *International Journal of Continuing Engineering Education and Life-Long Learning (IJCEELL), 11*(1/2), 79-94.

Dew, P., Leigh, C. M., Drew, R., Morris, D., & Curson, J. (1995). Collaborative working systems to support user interaction within a Virtual Science Park. *Information Services and Use, 15*(3), 213-227.

Dourish, P. (1999). Software architectures. In M. Beaudouin-Lafon (Ed.), *Computer supported co-operative work: Vol 7. Trends in software* (pp. 195-219). Wiley.

Economic & Social Research Council, ESRC. (2004). E-social science background information. Retrieved from www.esrcsocietytoday.ac.uk/

Ellis, C. A., Gibbs, S. J., & Rein, G. L. (1991). Groupware: Some issues and experiences. *Communications of the ACM, 34*(1), 39-58.

Foster, I., Kesselman, C., & Tuecke, S. (2001). The anatomy of the grid: Enabling scalable virtual organisations. *International Journal of Supercomputer Applications, 15*(3), 200-222.

Grudin, J. (1994). Computer-supported cooperative work: History and focus. *IEEE Computer, 27*(5), 19-26.

Hey, A. J. G., & Trefethen, A. E. (2002). The U.K. e-science core programme and the grid. *Future Generation Computing Systems, 18*(8), 1017-1031.

Hiltz, S. R. (1984). *Online communities: A case study of the office of the future.* Norwood, NJ: Ablex. IBM. (2005). *The history of notes and domino.* Retrieved June 2006, from http://www-128.ibm.com/developerworks/lotus/library/ls-NDHistory/

Kobine, T. (1986, March 24-26). Staffware: A software system for the automation of routine administrative procedures. In *1986 office automation conference digest* (pp. 211-220). Houston, TX: AFIPS Press.

Lau, L. M. S., Adams, C. A., Dew, P. M., & Leigh, C. (2003). Use of scenario evaluation in preparation for deployment of a collaborative system for knowledge transfer. In *Proceedings of the 12ᵗʰ IEEE International Workshops on Enabling Technologies (WETICE 2003): Infrastructure for Collaborative Enterprises* (pp. 148-152). IEEE Computer Society Press.

Lau, L. M. S., Curson, J., Drew, R., Dew, P. M., & Leigh, C. (1999). Use of Virtual Science Park resource rooms to support group work in a learning environment. In *Proceedings of GROUP'99 International ACM SIG-GROUP Conference on Supporting Group Work* (pp. 209-218). ACM Press.

Leigh, C. M., Dew, P. M., Drew, R., Morris, D., & Curson, J. (1996). The Virtual Science Park. *British Telecommunications Engineering Journal, 1*(4), 322-329.

Liu, P. (2004). *An empirical investigation of expertise matching within an academia.* Unpublished dissertation, School of Computing, The University of Leeds, UK.

Lubich, H. P. (1995). *Towards a CSCW framework for scientific cooperation in Europe.* Berlin Heidelberg, Germany: Springer-Verlag.

Pfister, H., Schuckmann, C., Beck-Wilson, J., & Wessner, M. (1998). The metaphor of virtual rooms in the cooperative learning environment clear. In N. A. Streitz, S. Konomi, & H. Burkhardt (Eds.), *Cooperative building* (pp. 107-113). Berlin Heidelberg, Germany: Springer.

Pham, T. V., Lau, L. M. S., & Dew, P. M. (2004). The integration of grid and peer-to-peer to support scientific collaboration. In D. Michaelides & L. Moreau (Eds.), *Proceedings of GGF11 Semantic Grid Applications Workshop,* Honolulu (pp. 71-77).

Pham, T. V., Lau, L. M. S., Dew, P. M., & Pilling, M. J. (2005). Collaborative e-science architecture for reaction kinetics research community. In *Proceedings of the Challenges of Large Applications in Distributed Environments (CLADE) Workshop* (pp. 13-22). NC: IEEE Computer Society Press.

Streitz, N. A., Konomi S., & Burkhardt, H. (Eds.). (1998). Cooperative building: Integrating information, organization, and architecture. *Proceedings of the 1ˢᵗ International Workshop, CoBuild'98.* Berlin Heidelberg, Germany: Springer.

Wainwright, J., & Francis, A. (1984). *Office automation, organisation and the nature of work.* Hampshire, UK: Gower.

Tian, Y., Lau, L., & Dew, P. (2004). Importance of mutual benefits in online knowledge sharing communities. In D. Remenyi (Ed.), *Proceedings of the 4ᵗʰ European*

Conference on Knowledge Management (pp. 823-831). Academic Conferences International.

Tian, Y. (2005). *Towards a sustainable knowledge sharing environment for online research communities.* Unpublished dissertation, School of Computing, The University of Leeds, UK.

KEY TERMS

Architecture: A blueprint of the underlying structure for linking the components of a computerised system and associated digital resources. It is more common to present a logical view of the architecture in which the components are specified at the conceptual level.

Client-Server Architecture: A two-tier model which split the presentation for input/output (at the client) from the heavy-duty processing of information (at the server).

Collaboration Metaphor: A representation (on a computer screen) by which users can identify as a means for collaborative working, synchronously or asynchronously.

Community: A group of people who share common goals/interests and participate in joint activities.

CSCW (Computer-Supported Cooperative Working): The use of computer-based technologies to undertake joint activities by two or more people.

E-Collaboration: The latest term for CSCW (see above).

Groupware: The technology that support collaborative working.

Infrastructure: The basic framework for a system.

A Research Agenda for Identity Work and E-Collaboration

Niall Hayes
Lancaster University Management School, UK

Mike Chiasson
Lancaster University Management School, UK

INTRODUCTION

Many recent management programmes have sought to establish organisation-wide collaborations that connect people in different functional and occupation groups (Blackler, Crump, & McDonald, 2000). Typically, these programmes are made possible through the deployment and use of e-collaboration technologies such as groupware, workflow systems, intranets, extranets, and the internet (Ciborra, 1996; Hayes, 2001). Examples of these technologies include the use of shared folders for reports, coauthored documents, completed electronic forms, and discussion forums. Through the use of such technologies, work and views are made accessible to staff working within and between functional and occupational groups. Such management programmes are reported to have brought about significant changes in the nature of work within and between intra organisational boundaries, including the erosion of functional and community boundaries (Blackler et al., 2000; Easterby-Smith, Crossan, & Nicolini, 2000; Knights & Willmott, 1999).

A research topic that has been largely ignored in literature that has examined work-change is the formation of individual and group identities group identities through e-collaborations within and between occupational groups. The ways in which e-collaboration technologies are implicated in shaping the production and reproduction of identity is the focus of this article. We ask what identity work is, and why it is important. We then consider the effect of electronic collaboration on identity work both within and across occupational groups. Focusing on cross-occupational collaboration, we then consider how individuals rehearse their identity with others in their own group before collaborating with others, the role of hierarchy in identity work, and the role of technology and visibility in preserving dominant occupational groups. Finally we conclude the article with directions for future research.

WHAT IS IDENTITY AND WHY "IDENTITY WORK"?

We collectively refer to the things that distinguish one group from another—its culture, history, and language and interpersonal relationships—as its *identity*. As dictionary.reference.com (2006) defines it, an identity is a "set of behavioural or personal characteristics by which an individual is recognizable as a member of a group." These identities are viewed as capturing the essential features of an individual in a specific domain or group (Gioia, 1998), and are important in the production and reproduction of *social capital*, which is "the existence of a certain set of informal values or norms shared among members of a group that permit cooperation among them" (En.wikipedia.org, 2006). Given this, the construction of individual and group identities in general, but also during e-collaboration, is an important question.

How one addresses the study of identity depends on one's philosophical viewpoint. From a functional perspective, individual and group identities can be stable and measurable, produced through cognitive processes. In contrast, a relational perspective considers identities as unstable and immeasurable, produced through dynamic interaction amongst individuals within complex and specific contexts (Ashforth, 1998). Harqual (1998) exemplifies this perspective noting that "an identity must be negotiated between the self and others (the social public) in order for that person to be said to have an identity" (p. 230). Sveningsoon and Alvesson (2003) refer to this process of constructing identities as *identity work*, which they argue is an ongoing accomplishment that "is fluid, uncertain, in movement" (p. 1164). They claim that in only highly stable circumstances will individual and group identity work be limited and invisible. In contrast, identity work will be significant and visible in complex and fragmented contexts, where identities are unclear, and where individuals feel especially anxious, insecure or defensive in their interactions with others.

However, even in stable circumstances, identity work is constant and on-going. As Knights and Will-mott (1999, p. 163) explain, individuals both look inwards at themselves and at the identities that have been constructed by others. They argue that as a consequence of this, an individual's perception of their identity is shaped by the perception of others, and by the individual's presentation of his or her identity to others. Identities thus depend upon the judgements of colleagues in particular circumstances, which are uncontrollable and unstable (Alvesson & Willmott, 2004; Knights & Willmott, 1999; Sveningsoon & Alvesson, 2003). Given this, individuals are constantly involved in the construction and reestablishment of their identity with others, even in what appears to be relatively stable situations (Parker, 2000).

It is this relational and unstable view of identity construction which we adopt in this article and in our research, because there is significant and on-going work involved in identity construction and preservation, especially when individuals encounter people in different groups during e-collaboration, because identities are destabilized by these encounters. Given the expanded use of e-collaboration tools to cross boundaries in the formation and dissolution of cross-occupational teams, it is these unstable circumstances which are increasingly numerous and of interest to us in this article.

E-COLLABORATION AND IDENTITY WORK

The introduction of electronic collaboration technologies has led to the increased intersection of many different occupational groupings in organisations. It is at such intersections that identity work is most noticeable (Alvesson & Willmott, 2004). Increasingly, who collaborates with whom and for what purpose, is sparsely captured in the form of workflow or business process models, and only after the fact. Perhaps through such process models, the preexisting need to collaborate is established and as such there is a formal bond created between both the sender and receiver to collaborate in the first place. However, even in such formal circumstances, identity work was required before the formal model could have been constructed. Identity work between people in different groups may start

with an individual (the sender in most cases) needing to clarify his or her group membership. For example, a sender could say "I'm Bob from accounting and I need information from you (Mark in the Logistics department) about what happened to order x." When people are familiar with each other, this introductory material may be less detailed, and could simply say "we need some information about order x". In either case Mark will construct his identity in relation to Bob's by commenting on his identity through comments about his group and role.

However, identity work did not start with the message delivery. Bob's message has been constructed in a way to order to develop a caricature of his referent group, relying upon Mark's perceptions and stereotypes about this group in order to influence his response (hopefully, Mark responds with the information, and does not report him to the police). For example, Bob may rely upon the stereotype of accounting departments and logistics groups to stimulate a typical response. Through this process, identities are used, reinforced, changed and simplified through a collaborative process, in order to get work done. As a result, communication can produce unexpected surprises and shifts in interpretation, which can also transform the individual identities themselves. To understand this intergroup identity formation, we need to start with the important construction of identity between individuals in the same group.

Identity Construction within Occupational Groupings

Identifying with certain occupational groups requires considerable identity work. To belong to the accounting group requires individual staff to be intimately familiar with the remit of their own group, a detailed understanding of the fundamental nature of the group's roles within the organisation, and the sharing of knowledge with other staff in the group. It is through the detailed understanding of the roles and tasks that they perform in collaboration with others that identity work takes places within occupational groups. In this sense we believe that this core identity work is not to be examined through transcripts of face to face interactions only, but through what group members do—their practice (Brown & Duguid, 1998, 2000). Given the increasing role of electronic systems to carry out routine and nonroutine work, e-collaboration is as important part of identity work as face-to-face interaction. In the case of within-group identity construction, e-collaboration can be used to both reinforce and change group members' collaborations and their shared practice. As they come to reinforce the view of the accounting group for example, they strengthen their own identity and that of their group's identity and position with respect to other groups.

Identity Construction Between Occupational Groupings

When electronic collaborations take place across occupational groups, considerable identity work occurs that both preserves and changes group and individual identities. The context is now heterogeneous and uncertain, and so the possibilities for revised individual, group and intergroup identities are greater. This is in part due to the limited understandings of the detailed practices of staff in other occupational groups. When practice is not shared, then considerable identity work takes place in order to determine what each individual does and wishes to do, and how this newly created cross-occupational group will work. This may be both *productive* and *unproductive*. For productive examples, Bob's (in accounting) explanation to Mark (in Logistics) as to why he needs the information, in a specific format, by a particular date, is an attempt to get his work done. This request could generate a response from Logistics explaining that due to their own ways of operating this is not entirely possible, but through some detailed discussion, work-arounds or compromises could take place. In the process of such collaborations, detailed identity work takes place that are central to current and future collaborations.

In the process of producing this intergroup identity, both rely upon both *secondary* and *primary* influences (Ritzer, 1992). Secondary influences on identity include formal titles and roles specified by organizational labels, which may indicate similarity or differences between sender and receiver. Primary influences on identity include individuals known by the individual, such as interdepartmental workmates or contacts in other departments. Secondary and primary influences yield different uses of electronic and face-to-face forms of communication. For example, e-mail communication between individuals who know each other may be shorter and punchier because they understand the tacit meanings in their communication. In contrast, e-mail communication between individuals who are

unfamiliar with each other and in different departments, may require extensive clarification of the request and positions in order to be understood.

In many secondary situations, initial encounters between individuals, collaborations may be perceived to be more contentious and uncertain because the initial legitimacy of a request and the sender's intentions may be subject to question, time and effort. In many cases, this could result in disputes about the boundaries between the occupational groups. For example, in the interaction between accounting and logistics, there may be some discussion about how goods are accounted for and how they are dispatched. Both occupational groups may believe that it is within their remit of expertise and authority to bring about changes in this regard. Thus discussions would take place about whether this is something someone in their group can and should be doing. Through expressing their unease about the request with members of another occupational group, such identity work is occurring, however transitory, across the boundary between the different occupational groupings. Electronic collaboration thus increases the possibility for both intergroup collaboration and disputes to take place. However, such disputes are unlikely to be productively resolved through electronic collaboration, given the importance of face-to-face communication in developing initial trust and recognition. Such disputes highlight the political nature of the ways in which identities between different groups are produced and reproduced. They also highlight how domination and control are central to both intra and intergroup identity work (Alvesson & Willmott, 2004; Collinson, 2003).

Rehearsing One's Projected Identity

Importantly, such disputes about the boundaries between occupational groups are rehearsed within occupational groups prior and during their articulation in electronic or face-to-face format to members of other groups. For example, in the scenario outlined above, one would expect some discussion to occur with staff who share a common identity with each other, discussing the request from a member in another group within the organisation. This may produce a discussion and confirmation that "they" do not do that, which would then produce a negative response to the request. In this sense, primary influences are likely to take place back stage—where identity work about what we are and do, is worked out and rehearsed, before being presented to individuals in other groups. As such identity work is at least in part dramaturgical (Goffman, 1959).

Hierarchy and Identity Work

Importantly, identity work often involves groups and individuals with differing hierarchical positions in an organisation. This renders certain positions powerful and others powerless, depending on your philosophical viewpoint. For example, requests may be blocked or unanswered by someone occupying a particular organisational position, where he or she may feel the request is inappropriate or disrespectful of his or her individual and group's position. Especially for e-collaborations within and between occupational groups, this blockage is a way for the individual to reinforce their own view of their position in the hierarchy, as well as to project this identity to outsiders. For example, the accountant may believe he or she is an experienced accountant, and would react negatively to a request for something that they consider to be mundane. Alternatively, they may request something mundane of someone else in order to reassert their role in conducting higher level work, and in preserving a dominant-subordinate relationship. Due to the lack of familiarity with the practices of accounting, this misplaced request is more likely to come from someone outside of the accounting occupational grouping, unfamiliar with the nuances of their identity.

Technology and Visibility

The visibility provided by collaborative technology may also become a resource to ensure that the identity of one occupational grouping is preserved and dominant, and another one fluid and subordinate. For example, groupware technology has been utilized by staff to provide a permanent and organisation-wide record of when staff have been underperforming, or when staff have made mistakes, or have worked in the ways that are contrary to established practices inside an organisation. Such visible records articulate the work practices and resulting identity construction of different individuals and occupational groups. Thus, if one reviews a discussion database, these discussion threads can articulate what certain groups will do and will not do, through visible statements of previous identity work. Further, when disputes do occur, these

can be used as a way to reinforce previous practices, thus producing a form of domination. For example, if the dispute between Logistics and Accounting was not resolved through engaging in identity forming activities such as working closely with each other and becoming familiar with the perspectives of the different occupational groups, postings could have been made to discussion forums, or circulated through e-mail carbon copying to politically significant people, pointing out how Accounting's perspective if correct, and their approach should be employed. This would allow them to preserve control over the work and identify formation of the Logistics group. Accounting's argument could include a functional argument that money will be saved if the work is carried out in this manner. These permanent records of the various arguments that surround e-collaborations are also a form of identity work as it represents claims of competence and control over others, through electronic interactions.

Discussion forums may also be utilised by staff who wish to project a successful identity to others in the organisations in the hope that this may enhance their individual and collective chances for career progression. They do so by projecting an identity of competence and creativity to those that may influence their career progression. For example, staff may review and contribute to those discussion databases that they believe senior managers' read, and avoid those that are not (Hayes & Walsham, 2001). Thus they construct their identity in those discussion databases in order to establish a productive identity that influences others. For the staff uninterested in such activity, identity may be shaped in the noncareerist discussion databases they occupy, and a shared understanding of why they avoid these databases may emerge. Thus we suggest that that e-collaboration technology opens up the possibilities and people that an individual can reach to construct and influence identify work, and to the dramaturgical identity claims that staff make, and importantly, do not make. These are central aspects of identity work and research.

CONCLUSION

This article has sought to highlight the much neglected but important concept of identity work in e-collaboration research. In doing this, our aim was to encourage academics and practitioners to reflect upon this theme in their future research and practice. Our article has highlighted that e-collaboration opens up both stabilizing and destabilizing possibilities for identity construction. In terms of stability, the range of identity possibilities could be restricted and configured within specific electronic collaboration technologies such as workflow systems, where a specified series of tasks that cross between different occupational groups stabilizes identity work between individuals across occupational groups. This may allow one group to seek to impose their identity on others. In terms of instability, individuals in various departments and organizations are often thrown together in a new project, using e-mail or discussion databases as the primary communication and coordination tools. This form of identity work requires communication about background, personality, interests, careerism, functional roles, expertise, hierarchical position, and other things in order to figure out one's revised identity in this new context. Indeed we suggest that identity work is likely to be especially noticeable when e-collaboration technologies are introduced or extended so as to require staff to work in an expanded, heterogeneous and often competing set of groups and professions, who have little history of working with each other (Hayes, 2001; Sveningsoon & Alvesson, 2003). Importantly, we argue that identity work is inseparable from the political and normative context of an organisation. Identity work may be bound up with the career or reward structure within an organisation, or importantly with an individual's desire not to be noticed when underperforming (Collinson, 2003). Consequently, the surveillance possibilities of e-collaboration technologies are deeply implicated in shaping the nature and extent of identity work. To conclude, we intend this article to provide the foundations for further conceptual and empirical studies that help us better understand identity work and e-collaboration.

REFERENCES

Alvesson, M., & Willmott, H. (2004). Identity regulation as organizational control producing the appropriate individual. In M. J. Hatch & M. Schulz (Eds.), *Organizational identity* (pp. 436-465). Oxford, UK: Oxford University Press.

Ashforth, B. E. (1998). Becoming: How does the process of identification unfold? In D. A. Whetten & P. C. Godfrey (Eds.), *Identity in organizations: Building theory through conversations*. London: Sage.

Blackler, F., Crump, N., & McDonald, S. (2000). Organizing processes in complex activity networks. *Organization, 7*(2), 277-300.

Brown, J. S., & Duguid, P. (2000). *The social life of information*. Boston: Harvard Business School Press.

Brown, J. S., & Duguid, P. (1998). Organizing knowledge. *California Management Review, 40*(3), 90-111.

Ciborra, C. (1996). Introduction. In C. Ciborra (Ed.), *Groupware and teamwork*. Chichester, NY: Wiley.

Collinson, D. (2003). Identities and insecurities: Selves at work. *Organization, 10*(3), 527-547.

dictionary.reference.com. (2006). Retrieved September 1, 2006, from http://dictionary.reference.com/search?q=identity

Easterby-Smith, M., Crossan, M., & Nicolini, D. (2000). Organizational learning: Debates past, present and future. *Journal of Management Studies, 37*(6), 783-796.

En.wikipedia.org. (2006). Retrieved September 1, 2006, from http://en.wikipedia.org/wiki/Social_capital

Gioia, D. A. (1998). From individual to organizational identity. In D. A. Whetten & P. C. Godfrey (Eds.), *Identity in organizations: Building theory through conversations*. London: Sage.

Goffman, E. (1959). *The presentation of self in everyday life*. Garden City, NY: Doubleday.

Harquail, C. V. (1998). Organizational identification and the "whole person": Integrating affect, behaviour, and cognition. In D. A. Whetten & P. C. Godfrey (Eds.), *Identity in organizations: Building theory through conversations*. London: Sage.

Hayes, N. (2001). Boundless and bounded interactions in the knowledge work process: The role of groupware technologies. *Information and Organization, 11*(2), 79-101.

Hayes, N., & Walsham, G. (2001). Participation in groupware-mediated communities of practice: A socio-political analysis of knowledge working. *Information and Organization, 11*(4), 263-288.

Knights, D., & Willmott, H. (1999). *Management lives: Power and identity in work organizations*. London: Sage.

Ritzer, G. (1992). *Contemporary sociological theory*. New York: McGraw-Hill.

Sveningsoon, S., & Alvesson, M. (2003). Managing managerial identity: Organizational fragmentation, discourse and identity struggle. *Human Relations, 56*(10), 1163-1193.

KEY TERMS

Dramaturgical Identity Work: The ongoing establishement of a groups view of themselves and the working out of disputes and positions between occupational groups are rehearsed within occupational groups prior and during their articulation in electronic or face-to-face format to members of other groups.

E-Collaboration Technologies: Electronic technologies that enable collaboration among individuals engaged in a common task.

Identity: The behaviours and characteristics of people who belong to a specific group. This view sees identity as a cognitive entity possessed by an individual.

Identity Work: Identities depend upon the judgements of colleagues in particular circumstances, which are uncontrollable and unstable. They are not a cognitive possession, but are produced and reproduced in relations between people.

Primary Influences: Primary sources of identity include informal referents such as work relationships and friendship.

Secondary Influences: Secondary sources of identity include formal relations such as those between the different organisational functions which are drawn upon when people are unfamiliar with each other.

The Role of E-Collaboration Systems in Knowledge Management

Sharon A. Cox
Birmingham City University, UK

John S. Perkins
Newman College of Higher Education, UK

INTRODUCTION

Since knowledge retrieval takes place at the interface between social interaction and technology (Gammelgaard & Ritter, 2005) successful systems of e-collaboration intended to manage knowledge involve the effective integration of both their technical and social components. Alongside technical developments, the standardisation of communication protocols has provided the realistic prospect of universal interconnection of businesses. The ubiquity of technology is not, however, reflected in the way that people, using the collaborative infrastructure, make sense of the data that emerges from the collaboration and go on to construct meaning from it. This is mediated not only by technology but by local culture, most explicitly represented by the recurrent activities that represent practice carried out by local communities of workers (Lave & Wenger, 1991; Brown & Duguid, 1996; Wenger, 1998). This interplay of technology, working practice, organisational structure and people traditionally lies at the heart of socio-technical systems (Leavitt, 1965) applied to leverage the skills of knowledge workers. This article examines the role and contribution of e-collaboration systems in inter-organizational knowledge management. The processes underlying this interplay are viewed from three perspectives: communication, collaborative practice and community, opportunities for the future development of e-collaboration systems are then proposed.

BACKGROUND

The role of information systems has been widely debated within knowledge management research. Knowledge is recognised as a critical organizational resource (Alavi & Leidner, 2001). It is claimed that organiza-

tions innovate by drawing on knowledge (Nahapiet, Gratton, & Rocha, 2005) in the "knowledge economy" (Carter & Scarbrough, 2001). The exchange or transfer of knowledge is an essential aspect of knowledge management (Bresman, Birkinshaw, & Nobel, 1999), as reusing knowledge saves time, effort, and money (Bhatt & Emdad, 2001). This is particularly important when the reuse of knowledge prevents "reinventing the wheel" (Hansen, 1999). Organizations can be viewed as distributed knowledge systems (Nahapiet et al., 2005) concerned with how to exploit knowledge already existing in the organization (Kakabadse, Kouzmin, & Kakabadse, 2001). This exploitation is more frequently referred to as knowledge management and the next section identifies some of the major issues surrounding collaboration within this activity.

KNOWLEDGE MANAGEMENT

Typically the features of information systems used for knowledge management activities can be divided into three main categories (Alavi & Leidner, 2001); repositories of stored knowledge collections, search mechanisms to find people with specific features of expertise and virtual spaces or knowledge networks. The first of these three categories lends itself to an ontological approach to the nature of data, information and knowledge (Blackler, 1995). The second category leads to an epistemological approach concerned with what can be known and the nature and residence of skills and knowledge within particular contexts of use. However, it is the third category of virtual spaces and knowledge networks that the rest of this article will address as the most relevant to leveraging of value from knowledge management systems through e-collaboration.

Although technology provides a repository for codified knowledge for people who are separated, socialising is more important than technology (Lagerstrom & Andersson, 2003) and it is the relationships between people that most affect knowledge transfer. Oliver and Kandadi (2006) suggest that organizations need to provide an appropriate communication infrastructure such as knowledge portals to assist in the development of communities of practice. Gammelgaard and Ritter (2005) suggest that electronic communication offers increased opportunity for dialogue and increased frequency of contact and define three categories within which barriers to knowledge transfer may be explored. The first category is knowledge fragmentation, where knowledge is dispersed around the organization and knowledge is inaccessible for a number of reasons. The second category relates to barriers where knowledge is hidden as a result of the quantity of data, information, and knowledge that is held. The third category involves barriers of decontextualisation. In this category, knowledge is located but cannot effectively be retrieved or used due to cultural, technical, or organizational distance between sender and receiver. In the same pattern as for the Alavi and Leidner (2001) model, the first two of these categories relate to the ontological and epistemological issues referred to earlier. The third category presents a different challenge to the development of effective knowledge management systems: the decontextualization of knowledge. A more situated understanding of knowledge and collaboration might address this issue through the analysis of activity resulting from them. One approach to this problem rejects the concept of knowledge altogether and instead proposes the attribute of "knowing" as something that

individuals or organizations *do* rather than contending with knowledge as something they supposedly possess. This approach places recurrent activities going on in a work community at the very centre of an analysis of socially situated knowledge and is used to analyze the dynamics of the systems through which knowing is accomplished.

Socio-historical activity theory (Engestrom, 2001) is another approach increasingly being used to identify knowledge situated within communities of practice. This explores the dynamics between agents, such as the users of collaborative systems, objects of activity, such as trading processes, and the community within which this trading takes place. The fundamental questions of collaboration remain, of how to collaborate, with whom and why Nahapiet et al. (2005) is analyzed in the following sections from the orientations of communication, collaborative practice, and community (Bafoutzou & Mentzas, 2002).

E-COLLABORATION AND COMMUNICATION

Collaboration requires mechanisms for intra- and interorganizational communication. An example of such a structured protocol categorisation is the open systems interconnection (OSI) reference model (Day & Zimmerman, 1983) used to define the requirements for communication across different equipment and applications by different vendors. It divides the communication processes into seven self-contained levels,

Table 1. Layers within information systems

Layer 7	Application layer: Includes the protocols that support user applications and addresses issues of file access and management.
Layer 6	Presentation layer: Deals with data syntax during transfer between two application processes to enable computers using different file formats to communicate.
Layer 5	Session layer: Includes protocols for establishing, maintaining and ending sessions between user applications so that differences between platforms are transparent to the user.
Layer 4	Transport layer: Includes the protocols that are responsible for the reliability of end to end connections.
Layer 3	Network layer: Deals with protocols to establish, maintain and terminate end to end network links routing messages across the network.
Layer 2	Data Link layer: Provides protocols to control logical links.
Layer 1	Physical layer: Includes protocols responsible for establishing, maintaining and ending physical connections (point to point) between computers.

referred to as layers. The upper layers (5-7) focus on application issues and are closest to the end user; the lower layers (1-4) focus on data transport issues in the software.

Information being transferred from a software application in one computer system to a software application in another must pass through the OSI layers as shown in Table 1.

The structure of these seven technical layers can also be used to demonstrate the role of e-collaboration systems from an e-communication perspective.

Moving through levels 1-7 of Table 2 requires incrementally greater integration and commitment to enabling collaboration. The emphasis moves away from the technical aspects of activity carried out to connect people on to activity involving making decisions about with whom to collaborate and how. The higher levels focus on more process related activities of identifying the time to allocate to collaboration, identifying authority to collaborate, and being able to make sense of the communication emerging within the collaborative community.

E-COLLABORATION AND COLLABORATIVE PRACTICE

Collaboration is a more subtle and contextually situated term than co-operation. If cooperation is defined as working together towards the same end, then collaboration is working together with an alternative reference to other contexts of use such as aligning oneself with an enemy occupier. Mason and Lefrere (2003) discuss a six-level interoperability schema to reflect the interdependence between organizational and technical issues in e-collaboration. These levels are:

- **Political:** The extent to which common goals and rules are agreed
- **Jurisdictional:** The ability to map legal interests and define boundaries of responsibility
- **Semantic:** The degree to which a common understanding of key terms is established
- **Cultural:** The establishment of communities of practice within which to create and share knowledge
- **Syntactic:** The agreement of grammars and style sheets used in the communication processes
- **Technical:** The availability of the means to exchange data

Collaboration requires mechanisms to support communication within and between organizations. The lower levels (1-2) of Mason and Lefrere's model (2003) relate to the need for protocols and standards to provide the technical infrastructure for the collaboration. Degrees of commitment in a collaborative relationship are reflected in the degree to which the systems of the collaborating organizations are integrated. The higher levels (5-6) focus on the strategic policy issues relating to formalizing the nature of the collaboration between organizations such as with whom to collaborate, why and how (Barratt, 2004). In social activity theory, the social processes of exchange and combination are criti-

Table 2. Communication layers within e-collaboration

Physical Layer 1	E-collaboration systems provide the physical point to point connection that allows people to communicate across organizational and geographical boundaries.
Logical Layer 2	E-collaboration systems provide a means to structure the data, information or explicit knowledge to be conveyed between two or more people.
Network Layer 3	E-collaboration systems include processes to help knowledge workers within and between organizations to identify the appropriate people with whom to share knowledge.
Transport Layer 4	E-collaboration processes address the issues of reliability of the communication.
Session Layer 5	E-collaboration systems provide the arena for the communication to take place.
Presentation Layer 6	E-collaboration systems address issues of data semantics to ensure that information is accurately interpreted.
Application Layer 7	E-collaboration systems provide access to and management of knowledge within and between organizations.

Table 3. Collaborative practice layers within e-collaboration

OSI Layer	Communication Layers	Collaborative Practice Layers
Application	Access and management of knowledge	Political Jurisdictional
Presentation	Semantic interpretation.	Semantic
Session	Knowledge sharing session.	Cultural
Transport	Reliability of knowledge.	Cultural
Network	Identifying person to share knowledge.	Cultural
Data Link	Structuring knowledge.	Syntactic
Physical	Connecting to knowledge-worker.	Technical

cal for knowledge creation and therefore the quality of relationships between people within the collaborating organizations is a major influence on knowledge management (Nahapiet et al., 2005). The middle levels (3-5) address the cultural and semantic issues that relate to the use of e-collaboration in practice. At the operational level of organizations, knowledge workers also need to consider the same questions, of with whom to collaborate, why and how.

Table 3 correlates the OSI network protocol standard, with the communication and interoperability issues identified in e-collaboration systems.

Factors such as organizational culture, interpersonal trust, and individual capabilities impact the effectiveness of e-collaboration in practice. Electronic means of conducting routine transactions accommodates geographical and cultural diffusion, but more complex knowledge-based working needs a form of community to be in place to allow the development of shared understanding and mutual trust. Social construction of knowledge occurs in communities of practice where knowledge is freely shared through collaborative processes such as conversation and joint work (Wenger, 1998). Communities of practice are self-organizing and cannot be conscripted; they share tacit knowledge through informal interaction (Mason & Lefrere, 2003). These issues of interaction and community are developed further in the next section.

E-COLLABORATION AND COMMUNITY

Hofstede (1991) suggests that understanding differences in thinking is at least as important as understanding technical factors in communication. This is particularly true today in the context of e-collaboration. Differences in the use of language, goals, cognitive views, frames of reference, and organizational pressures all contribute to communication difficulties and lack of trust in collaboration. The literature demonstrates consensus that organizational culture is a core factor in successful knowledge management (Oliver & Kandadi, 2006), however, collaboration within organizations is often critically problematic. A contextual gap exists between the sender and receiver of codified knowledge (Nahapiet & Ghosal, 1998) in the form of age, sex, social position, religion, and political conviction. A shared context and frequent communication between sender and receiver minimizes the contextual distance (Deetz, 1992). The amount of time made available for face-to-face interaction is crucial when knowledge management tasks are complex (Hansen, 1999) or require local adaptation (Nahapiet et al., 2005). This is because the distance between the sender and receiver of knowledge can hinder knowledge transfer due to lack of understanding of the different contexts (Gammelgaard & Ritter, 2005). When storing knowledge it is important to include the contextual background information as the context surrounding the knowledge creation process is not shared, it becomes questionable as to whether retrieval will result in effective use (Alavi & Leidner, 2001).

Trust forms a vital component in bringing together the orientations of communication, collaborative practice, and community within communities of practice and it is in this area that e-collaborative systems might secure most leverage. Features that influence collaboration include trust, norms, identification and obligation

(Nahapiet & Ghoshal, 1998). Processes involving consensus building, consultation, collaboration, and knowledge sharing, are dependent on trust for effectiveness and consistency (Mason & Lefrere, 2003). Where trust exists, there is greater willingness to engage in social exchange (Nahapiet et al., 2005) and overcome suspicion (Todeva & Knoke, 2005). In collaborative partnerships, trust depends on social, political, cultural and economic factors within and between the partners and the practical demonstration of a commitment to the collaboration through continued accountability and transparency (Nahapiet et al., 2005). Collaboration is built upon interpersonal and community relationships and it is trust that differentiates partnerships from traditional relationships (Handy, 1995).

FUTURE TRENDS

The literature on e-business emphasises the role of IT as an enabler and facilitator of collaboration (Shuman & Twombly, 2001). However, Perkins and Cox (2005) emphasize that whilst IT components of e-business are necessary to support collaboration, in themselves they are often insufficient as enablers of collaboration. Culture and behavioural change remain the challenge of e-collaboration systems (Boddy, Cahill, Charles, Fraser-Kraus, & Macbeth, 1998). The introduction of a collaborative system impacts the management and activity of work teams across the partner organizations that need to be addressed in order to sustain the collaboration. Processes, artefacts and working practices are three components of a social practice theory that provides the means to understand human activity systems in terms of what happens in practice (Engestrom, 2001). E-collaboration systems need to employ such approaches to understanding communication but challenges remain in changing culturally embedded work practices and individual capabilities and competencies needed to create a shared collaborative context. The next generation of collaborative systems need to focus on creating this shared context within which knowledge can be meaningfully situated.

CONCLUSION

Those infrastructures to enable e-collaboration in knowledge management that arise from technical is-sues are largely sufficiently stable and established. This successfully addresses the problem of how to communicate across organizations. The question of why and when to collaborate has been widely addressed in the organizational literature at the strategic level; there is acceptance of the generic mutual benefits that can be attained through collaboration within the supply chain. However, organizations comprise groups of individuals and the rationale for individuals to collaborate with colleagues within and across organizations still needs to be addressed. This requires a change of organizational focus and culture that remains a challenge for many organizations. The organizational structures, processes, norms, and values remain the key areas that limit the effectiveness of e-collaboration systems. Social activity theory offers one approach to understanding the communities of practice that may enable a shared context to be developed.

REFERENCES

Alavi, M., & Leidner, D. (2001). Knowledge management and knowledge management systems: Conceptual foundations and research issues. *MIS Quarterly, 25*(1), 107-36.

Bafoutsou, G., & Mentzas, A. (2002). Review and functional classification of collaborative systems. *International Journal of Information Management, 22*(4), 281- 305.

Barratt, M. (2004). Understanding the meaning of collaboration in the supply chain. *Supply Chain Management: An International Journal, 9*(1), 30-42.

Bhatt, G. D., & Emdad, A. F. (2001). An analysis of the virtual supply chain in electronic commerce. *Logistics Information Management, 14*(1/2), 78-84.

Blackler, F. (1995). Knowledge, knowledge work and organizations: An overview and interpretation. *Organization Studies, 16*(6), 1021-1046.

Boddy, D., Cahill, C., Charles, M., Fraser-Kraus, H., & Macbeth, D. K. (1998). Success and failure in implementing supply chain partnering: An empirical study. *European Journal of Purchasing and Supply Management, 4*(2/3), 143-151.

Bresman, H., Birkinshaw, J., Nobel, R. (1999). Knowledge transfer in international acquisitions. *Journal of International Business Studies, 30*, 439-62.

Brown, J. S., & Duguid, P. (1996). Organisational learning and communities of practice: Towards a unified view of working, learning and innovation. In M. D. Cohen & L. S. Sproull (Eds.), *Organisational learning* (pp. 58-82). London: Sage Publications.

Carter, C., & Scarbrough, H. (2001). Towards a second generation of KM? The people management challenge. *Education & Training, 43*(4/5), 215-224.

Day, J. D., & Zimmerman, H. (1983). The OSI reference model. In *Proceedings of the IEEE, 71*(12), 1334-1340.

Deetz, S. A. (1992). *Democracy in an age of corporate colonization: Developments in communication and the policies of everyday life.* State University of Albany: New York Press.

Engestrom, Y. (2001). Expansive learning at work: Toward an activity theoretical reconceptualisation. *Journal of Education and Work, 14*(1), 133-61.

Gammelgaard, J., & Ritter, T. (2005). The knowledge retrieval matrix: Codification and personalisation as separate strategies. *Journal of Knowledge Management, 9*(4), 133-143.

Handy, C. (1995). Trust and the virtual organisation. *Harvard Business Review,* May, 40-49.

Hansen, M. T. (1999). The search transfer problem: The role of weak ties in sharing knowledge across organizational subunits. *Administrative Science Quarterly, 44*(1), 82-111.

Hofstede, G. (1991). *Cultures and organisations: Software of the mind: Intercultural cooperation and its importance for survival.* Maidenhead: McGraw-Hill.

Kakabadse, N. K., Kouzmin, A., & Kakabadse, A. (2001). From tacit knowledge to knowledge management: Leveraging invisible assets. *Knowledge and Process Management, 8*(3), 137-154.

Lagerstrom, K., & Andersson, M. (2003). Creating and sharing knowledge within a transnational team: The development of a global business system. *Journal of World Business, 38*(2), 84-95.

Lave, J., & Wenger, E. (1991). *Situated learning: Legitimate peripheral participation.* Cambridge: Cambridge University Press.

Leavitt, H. J. (1965). Applied organisational change in industry: Structural, technological and humanistic approaches. In J. G. March (Ed.), *Handbook of organisations* (pp. 1144-1170). Chicago: Rand McNally Publishing.

Mason, J., & Lefrere, P. (2003). Trust, collaboration, e-learning, and organisational transformation. *International Journal of Training and Development, 7*(4), 259-270.

Nahapiet, J., & Ghoshal, S. (1998). Social capital, intellectual capital, and the organizational advantage. *Academy of Management Review, 23*(2), 242-266.

Nahapiet, J., Gratton, L., & Rocha, H. O. (2005). Knowledge and relationships: When cooperation is the norm. *European Management Review, 2*(1), 3-14.

Oliver, S., & Kandadi, K. R. (2006). How to develop knowledge culture in organizations? A multiple case study of large distributed organizations. *Journal of Knowledge Management, 10*(4), 6-24.

Perkins, J., & Cox, S. (2005). Towards purposeful collaboration in e-business: A case of industry and academia. In *Proceedings of International Workshop on Requirements Engineering for Information Systems in Digital Economy The Second International Conference on E-Business and Telecommunication Networks* (pp. 51-60).

Shuman, J., & Twombly, J. (with Rottenberg, D.). (2001). *Collaborative communities: Partnering for profit in the networked economy.* Chicago: Dearborn Trade.

Todeva, E., & Knoke, D. (2005). Strategic alliances and models of collaboration. *Management Decision, 43*(1), 123-148.

Wenger, H. (1998). *Communities of practice.* Cambridge: Cambridge University Press.

KEY TERMS

Collaborative Practice: The ways of working which emerge between communities of practice as they work towards the achievement of common goals.

Collaborative Systems: A computer-based system that is accessed and used by more than one organisation to support business transactions in the supply chain. The system allows data to be automatically updated in a partner organisation's systems during the processing of a transaction.

Community of Practice: The emergent process of social learning as a self-organized group of people with shared values, beliefs and goals, work together towards a common aim, sharing knowledge through informal interaction.

E-Collaboration: The use of information technology to establish, facilitate and sustain cooperation between two geographically dispersed parties, who have common goals, to enable them to work together for mutual benefit.

E-Communication: The reciprocal sending and receiving of information using Internet-based technology.

Knowledge Fragmentation: Knowledge is dispersed around an organization and its existence and/or whereabouts is unknown making it inaccessible.

Knowledge Management: Knowledge is contextual information embedded in experience which can be tacit or explicit. Knowledge cannot be managed; only the processes for creating, capturing, codifying, storing, and transferring knowledge in organisations to sustain competitive advantage can be managed.

Social Activity Theory: An approach that explores the dynamic social relationships within communities of practice.

The Role of Individual Trust in E-Collaboration

Terry Nolan
Auckland University of Technology, New Zealand

Linda Macaulay
The University of Manchester, UK

INTRODUCTION

The subject of individual and interpersonal trust within communities has captured the attention of sociologists and psychologists for many decades, having intensified with the advent of virtual or online communities and their potential for increasing social inclusion. E-collaboration, particularly for business purposes, often requires the communication of "rich" information (Daft & Lengel, 1986), of high utility value to its recipients, such that it facilitates "rational action" (Ulrich, 2001).

Communities are identifiable by the levels of trust, reciprocity, dependence, and formality exhibited by their members. The development of e-communities has presented IS developers with a long standing and on-going problem articulated by Kollock and Smith (1996, p. 109) as follows:

At the root of the problem of cooperation is the fact that there is often a tension between individual and collective rationality.

This "tension" has led to confusion amongst researchers and developers, with the result that individual and collective rationalities have often been conflated.

In response to this problem, this article deals explicitly with individual rationality, *distinct from* but *related to* the collective. We adopt for this purpose, Simon's (1957) notion of "bounded rationality" to explain how individuals recognize the cost of gathering and processing information and how its utility contains multiple values. Among the multiple values under consideration, the presence of trust is of primary concern for would-be, online collaborators. Trust is a complex entity, affecting individual and group attitudes and behaviours. Its presence in both techno-scientific and social science literatures on e-collaboration is

recognition of its importance. It is considered to be an essential feature of economy and commerce in reducing complexity by providing "internal security" before taking action (Abdul-Rahman & Hailes, 2000). By deconstructing the elements of individual trust, this article reveals clues to how individuals rationalize participation in e-collaboration.

BACKGROUND

The problem of engendering online trust is conceptualized differently according to two schools of practice: either as an engineering problem or as a social problem. Engineering developments have demonstrated the effectiveness of online community tools for connecting people to one another and helping them to share information. Developments and discussions amongst the technocrats naturally look towards possibilities for making these tools even more powerful. Jordan, Hauser, and Foster (2003), for instance, seek to enhance trust by this method and thereby to "further public discourse" in online communities.

In relation to the broad literature covering trust, developments in engineering at the user interface relate only to one form of online trust relationship: the impersonal institutional phenomenon variously known as "structural" or "system" trust. According to McKnight and Chervany (1996), system trust is not founded on any property or state of the trustee, but rather on the perceived properties of, or reliance on, the system or institution within which that trust exists. The engineers' supposition regarding the relationship between connectivity and trust appears rather tenuous when, according to Kollock (1997), the problems of social interaction and order are often ignored in the software and online industry in their discourse on "social computing." He considers this to be a "thin term" applying more to the user interface design than to actual social interaction between two or more people.

Social science literature focuses on interpersonal relationships. The notion of interpersonal trust can be thought of as one person trusting another specific person(s) in order for meaningful outcomes to occur. For partners in business or retail transactions, it is often defined in terms of trusting beliefs about the other party's predictability, benevolence, honesty, and competence plus a weighting given to events that provide information about the person's motives for being in the relationship (McKnight & Chervany). Identity-based trust, as a subset of interpersonal trust (Lewicki & Bunker, 1996), pervades when individuals comprehend and appreciate the needs of each other: where shared meanings and culture are manifested and there is a commitment to common values, objectives, and a collective identity. If not developed, the lack of identity based trust can be extremely detrimental to group processes and performance.

Castelfranchi and Falcone (1998), with their five-element strategy, addressed a wider agenda, encompassing both the engineering and social paradigms comprising human-computer (or systems) trust, interpersonal trust relationships, dispositional trust, risk, attitude and potential gain. They and others point out the necessity for understanding that virtual communities and their supporting ICTs are embedded in human interpersonal, social, and legal relationships (see also Hartman, 1995; Leiwo & Heikkuri, 1998).

Kollock deals comprehensively with individual perception of risk within a range of community-based contexts, where risk and trust are dynamically related (see also Abdul-Rahman & Hailes, Ibid; Marsh, 1994; Tan & Thoen, 1999). Coetzee and Elof (Ibid, 2005) identify three "important" elements of trust: its dependency on the context (Coetzee & Elof, Ibid 2005); the measurable belief that reflects its strength (Grandison, 2003); its subjective entity that evolves through new experiences and observations (Dimitrakos, 2003). This third property provides the focus for the remainder of this article.

E-collaboration often involves the transfer of rich information when, for example, design blueprints are transmitted or when the complexities of a new piece of legislation are discussed. The community of practice (CoP) (Wenger, 1998) is the most commonly cited medium for the transfer of rich, tacit knowledge. The *process* of e-collaboration is explained partly through the notion of legitimate peripheral participation (LPP) (Lave & Wenger, 1991). LPP is a complex and composite notion, in which the three constituent aspects of legitimation, peripherality, and participation, are indispensable in defining each other and cannot be considered in isolation (Kimble, Hildreth, & Wright, 2001). Whilst LPP explains contributions in terms of social situatedness, social identity and social orientation theories also resonate strongly with this inquiry. Mullins and Hogg (1999) propose that social identification affects both self-conception and intergroup orientation focussing on how the self is defined by group membership.

Of particular relevance is the notion of "social loafing" (Karau & Williams, 1993). They propose that LPP is influenced by a set of individual and group factors that explain why individuals will withhold contributions to a group or community. Social loafing is common where groups undertake "additive tasks," where the group output is greater than the individuals' contributions. This phenomenon can be reduced by ensuring that individual contribution is noted and valued by others and the individuals themselves; enhancing the importance of tasks and information; providing some form of reinforcement (reward or punishment). Small groups are better at providing social cohesiveness, whilst time pressure (a key factor within business communities) is important and can lead to members withholding essential information.

RESEARCH APPROACH

This article is informed by a three-year action research inquiry into the development of an online community comprising over fifty small and medium-sized enterprises (SMEs), information providers, and business experts. The inquiry set out to improve communication quality and information sharing between these actors for the benefit of the SMEs.

By adopting the approach recommended by Kemmis and McTaggart (1990), the enquiry was conducted via a series of research cycles each containing four steps: plan, act, observe, and reflect. To ensure rigor and relevance in the AR process, Davison Martinsons, and Kock (2004), propose the use of "canonical" AR (CAR). Incorporating 31 criteria, embedded within five principles, this inquiry adhered closely to the CAR standard. Qualitative feedback was continually sought to discover why the members, who had voiced great enthusiasm for the online community concept, were reluctant to use it. Whilst technological problems were cited throughout as the underlying barrier to interpersonal communication, interviews uncovered a range of trust-related concerns, attitudes, and behaviors affecting e-collaboration. As part of the AR reflective stage, a series of "mapping" exercises were conducted, aimed at modelling individual's online behaviours. These were compared with qualitative interview data gained in response to questions relating to participants' views and attitudes on e-collaboration. The comparisons revealed five dominant themes, identified as constituents of trust and having direct relevance to e-collaboration. These elements (discussed below) are Utility, Risk, Benefit, Effort, and Power (Expert and Positional). Together they illustrate a limited or bounded process of deconstruction-reconstruction used by individuals to resolve complexity.

DECONSTRUCTING INDIVIDUAL TRUST

Figure 1 illustrates the relationship between the five constituent elements of trust and degrees of participation in e-collaboration. The figure comprises three identical graphs in which sets of opposing elements are paired. In each graph, it can be seen that as one element of individual trust increases, its opponent decreases.

Utility and Risk

The first element of individual trust identified is the information's Utility value (measured by high information quality such that it can be absorbed into immediate practice). Recipients assess Utility through a cognitive process relating to the degree to which information can be applied through action (Ulrich, Ibid). Where Utility value is low, the information, whilst of personal interest, adds little value to an individual's decision making process. This situation is common to Communities of Interest, which describes a group of people connected by a common interest in a specific subject or endeavour (Rheingold, 1993). The level of interest may range from passing to intense, and may, over time, develop into expertise on a subject, but with participation within these communities generally limited to information seeking around pastimes or hobbies they hold only peripheral relevance for businesses.

Risk, the second element, concerns the disclosure of personal or proprietary information, with individuals judging the likelihood of information falling into the wrong hands, or being misused by its recipient. The system's security too comes under scrutiny—the firewalls, methods for authenticating membership and how the system is monitored—each impact upon the perceived risk. Individuals differ with regard to their risk thresholds, which can be low if the value of the transaction is high or high if the individual is a risk seeker (Castelfranchi & Falcone, Ibid; Lewicki & Bunker,

Figure 1. The relationship between individual trust and e-collaboration

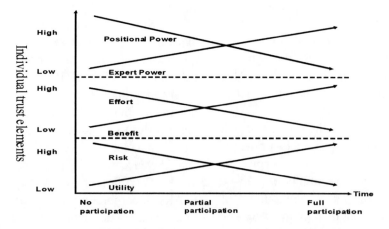

Ibid). Where an individual believes that information is of high Utility value, this factor will outweigh the social, interpersonal element of Risk.

Benefits and Effort

Information givers demand reciprocity or Benefits to accrue in return for giving. Benefits accrue from an overall perception that involvement will provide individual gain. High Utility value information is a benefit but, in addition, "soft" benefits also accrue from full participation within a community, such as the ability to participate in group decisions, the general feeling of belongingness, and the facility to share problems with others. Furthermore, an individual must perceive that the Benefits outweigh the Effort expended in gaining them. Whilst online systems make it easier to share information due to the availability of technologies and the speed of connections, the expenditure of effort is still required in order to form trusting relationships with unknown persons in evaluating the trustworthiness of information and so on.

Expert Power and Positional Power

Power was shown to be a strong and pervasive element within the study, with Expert and Positional Power (also called Legitimate Power) being particularly prevalent (French & Raven, 1959). Expert Power is the power of knowledge—the ability to influence others through their relative expertise in particular areas. This can only be exercised through legitimated inclusion within, and full participation at, the core of a community. CoPs for example, value members according to what they bring as a practitioner in terms of information and their willingness and ability to share it, rather than on any predetermined hierarchical or status value (Wenger, 1998). Therefore, e-collaboration emphasizes Expert Power over the more formalized notion of Positional Power. Those who hold Positional Power—the authority gained from their role in the organization—run the risk of losing that power, which may lead them to attempt to block the community's transition towards e-collaboration. Conversely, those who are likely to gain power through their expertise or who envisage a liberal approach to information sharing are more likely to encourage the transition. The potential for conflict is huge and the likely outcome is that transition to e-collaboration will be blocked. Within the scope of our inquiry, we found that several colocated communities held back from making the transition due to fear held by key people regarding their possible loss of power from such a move.

Thus, as illustrated in Figure 1, the conditions necessary for individuals to fully engage in e-collaboration are seen to be: (1) acceptance of power based upon knowledge or expertise; (2) high perceived levels of benefits accruing from participation; and (3) information of high utility value as an aid to action.

DISCUSSION

Interaction has figured strongly across many research traditions, such as actor network theory, information science, constructivism, and critical theory. This article has examined interaction at a *meta* level—within and between each of the above traditions: the complexities of which are illustrated in Figure 1.

Technology provides connectivity between islands of knowledge and will enhance trust. Yet the relationship between connectivity and trust is relatively weak, leading us to suggest that the presence of engineering-led design features should be considered as a hygiene factor, such that the absence of community tools will hinder trust development, but their presence will not stimulate meaningful discourse per se. The balance therefore lies with social factors.

Both social loafing and legitimate peripheral participation help to explain why individuals will withhold contributions to a group or community. However, our analysis of individual behavior *in relation to* the group or community reveals the presence of a set of subjective and dynamic trust-related elements. These elements are limited in their scope, reflecting individual needs to simplify the complexity of relationships between themselves and the information system, the information and the other community members. An online community can disintegrate at an alarming rate, with the absence of physical cues making continued dialogic communication (or full participation) a critical factor in its continuance. When the numbers of participants, including those lurking at the periphery, fall below a critical mass, even if for only a short period, the trust and confidence of those at the core can be eroded as they begin to doubt the others' motives.

LESSONS

Armed with the understanding that an imprudent reliance upon the techno-engineering paradigm will result in a technically enabled, discourse impoverished membership, IS practitioners and researchers should operate with a light touch to achieve a balance between the technical and social requisites for e-collaboration. Furthermore, an awareness of the dynamics between the trust elements—Utility, Risk, Benefit, Effort, and Power—enables the IS practitioner to diagnose the real-state nature of an online community.

The failure to develop e-collaboration, whilst proving a salutary experience, provides a further lesson for practitioners. Wenger's assertion that communities cannot be determined in the abstract but by the way in which they work, illustrates the existentialist nature of communities. They exist not because they have been created or formed by an outside power, such as an IS developer, but because they embody meaning to their members. Such communities can only be understood from the inside, in terms of the reality created within. A rationalist view of the human participants—as a homogeneous group in pursuit of knowledge, regulated and directed by rational principals—ignores the subjective, dynamic, and limited nature of individual trust.

REFERENCES

Abdul-Rahman, A., & Hailes, S. (2000, January). Supporting trust in virtual communities. *Proceedings of the Hawaii International Conference on System Sciences,* (p. 3). Maui.

Castelfranchi, C., & Falcone, R. (1998). Principles of trust for multi-agent systems: Cognitive anatomy, social importance, and quantification. *Proceedings of the International Conferences on Multi-Agent Systems* (pp. 72-79). AAAI-MIT Press.

Coetzee, M., & Eloff, J. H. P. (2005). Autonomous trust for Web services. *INC 2005 Proceedings, 15*(5), 498-507.

Daft, R. L., & Lengel, R. H. (1986). Organizational information requirements, media richness, and structural design. *Management Science, 32*(5), 554-571.

Davison, R. M., Martinsons, M. G., & Kock, N. (2004). Principles of canonical action research. *Information Systems Journal, 14*(1), 65-82.

Dimitrakos, T. (2003). A service-oriented trust management framework. Cited in Coetzee, M. & Eloff, J.H.P. (2005) Autonomous trust for web services. *INC 2005 Proceedings,* Vol. 15(5) (pp. 498-507).

French, J., & Raven, B. H. (1959). The bases of social power. In D. Cartwright (Ed.), *Studies of social power* (pp.150-167). Ann Arbor, MI: Institute for Social Research.

Grandison, T. W. A. (2003). *Trust management for Internet applications.* Unpublished doctoral dissertation, Imperial College of Science, Technology and Medicine, University of London.

Hartman, A. (1995). Comprehensive information technology security: A new approach to respond to ethical and social issues surrounding information security in the 21st century. *Information security: The next decade: Proceedings of the IFIP TCI 11th International Conference on Information Security* (pp. 590-602). Chapman & Hall: London.

Jordan, K., Hauser, J., & Foster, S. (2003, June). *A report for the link tank.* Paper presented at the Planetwork conference, San Francisco.

Karau, S. J., & Williams, K. D. (1993). Social loafing: A meta-analytic review and theoretical integration. *Journal of Personality & Social Psychology, 65,* 681-706.

Kemmis, S., & McTaggert, R. (1990). *The action research planner.* Geelong: Deakin University Press.

Kimble, C., Hildreth, P., & Wright, P. (2001). Communities of practice: Going virtual. In Y. Malholtra (Ed.), *Knowledge management and business model innovation* (pp. 220-234). Hershey, PA: Idea Group Publishing.

Kollock, P. (1997). Design principles for online communities. In *Internet and society: Harvard conference proceedings.* Cambridge, MA: O'Reilly and Associates (CD-ROM).

Kollock, P., & Smith, M. (1996). Managing the virtual commons: Cooperation and conflict in computer communities. In S. Herring (Ed.), *Computer-mediated communication: Linguistic, social, and cross-cultural perspectives* (pp. 109-128). Amsterdam: John Benjamins.

Lave, J., & Wenger, E. (1991). *Situated learning: Legitimate peripheral participation.* Cambridge: Cambridge University Press.

Leiwo, J., & Heikkuri, S. (1998). An analysis of ethics as foundation of information security in distributed systems. *Thirty-First Annual Hawaii International Conference on System Sciences* (Vol. 6, p. 213).

Lewicki, R. J., & Bunker, B. B. (1996). Developing and maintaining trust in work relationships. In R. M. Kramer & T. R. Tyler (Eds.), *Trust in organisations: Frontiers of theory & research* (pp. 114-139). Thousand Oaks, CA: Sage Publications.

Marsh, S. (1994). *Formalising trust as a computational concept.* Unpublished doctoral dissertation, University of Stirling, UK. Cited in Coetzee, M. & Eloff, J.H.P. (2005). Autonomous trust for web services. *Internet Research, 15*(5), 498-507.

McKnight, D. H., & Chervany, N. L. (1996). *The meaning of trust* (Tech. Rep. 94-04). University of Minnesota, Carlson School of Management.

Mullins, B. A., & Hogg, M. A. (1999). Motivations for group membership: The role of subjective importance and uncertainty reduction. *Basic & Applied Social Psychology, 21*(2), 91.

Rheingold, H. (1993). *The virtual community.* Reading, MA: Addison-Wesley.

Simon, H. (1957). A behavioral model of rational choice. *Models of man.* New York: John Wiley & Sons.

Tan, Y. H., & Thoen, W. (1999). A generic model of trust in electronic commerce. In *Proceedings of the 13th Bled Conference on Electronic Commerce* (Vol. 1, pp. 346-359). Bled, Slovenia.

Ulrich, W. (2001). Critically systemic discourse: A discursive approach to reflective practice. ISD (Part 2). *The Journal of Information Technology Theory & Application, 3*(3), 85-106.

Wenger, E. (1998). *Communities of practice: Learning, meaning, and identity.* Cambridge: Cambridge University Press.

KEY TERMS

Action Research: A methodology that aims to contribute both to the practical concerns of people in an immediate problematic situation and to the goals of social science by joint collaboration within a mutually acceptable ethical framework.

Bounded Rationality: The limits faced by individuals in formulating and solving complex problems and in processing (receiving, storing, retrieving, transmitting) information.

Business Communities: Patterns of formal and informal linkages between individuals, businesses, and other organizations such as government and voluntary agencies.

Community of Practice: A group of people who share an interest in a domain of human endeavor and engage in a process of collective learning that creates bonds between them: a tribe, a garage band, a group of engineers working on similar problems.

Deconstruction: A method used for discovering, recognizing, and understanding the underlying assumptions, ideas, and frameworks that form the basis for thought and belief.

Power: The ability to get someone else to do something you want done and to make things happen the way you want them to.

Rationality: A decision or situation is often called rational if it is in some sense optimal, and individuals are often called rational if they tend to act somehow optimally in pursuit of their goals.

The Role of Leadership in Virtual Teams

Kristi M. Lewis Tyran
Western Washington University, USA

Craig K. Tyran
Western Washington University, USA

INTRODUCTION

As globalization and the prevalence of electronic communication technology has become more widespread, organizations are adapting and changing at a rapid pace. Many organizations are using "virtual teams" of people working across space and time as an organizational structure to enhance organizational flexibility and creativity in this changing environment (Duarte & Snyder, 1999; Townsend, DeMarie, & Hendrickson, 1998). As virtual teams become a more popular organizational tool, many researchers have begun to explore ways in which the performance of such teams may be enhanced (Cohen & Gibson, 2003). One aspect of teamwork that has traditionally had an important impact on team performance is team leadership. Leaders often facilitate effective task performance within a team. By assigning tasks to individuals with the skills, knowledge and abilities to perform best, as well as structuring the team to best accomplish its tasks, a leader can greatly increase the effectiveness and efficiency of a team (Hooijberg, Hunt, & Dodge, 1997; McGrath, 1984; O'Connell, Doverspike, & Cober, 2002). In addition to task-focused behaviors, leaders also motivate, coach and mentor team members toward higher levels of performance (Bass, 1985; Conger & Kanungo, 1998).

The role of leadership for virtual teams and computer-supported teams has not yet been studied extensively (e.g., Kim, 2006; Powell, Piccoli, & Ives, 2004). However, the body of literature in this area has been growing during the past few years. While the early studies suggest that leadership for virtual teams is different from that of traditional face-to-face teams, the contextual factors that may influence virtual team leadership have not been fully explored (Cascio & Shurygailo, 2003; Kayworth & Leidner, 2002; Powell et al., 2004). In this article, we provide a summary of role of leadership in virtual teams and identify some of the ways in which a leader may influence a virtual team.

BACKGROUND

For purposes of this article, a virtual team is defined as a team which has the following attributes: It is a functioning team of people who are interdependent and share responsibility for team outcomes, the members of the team are geographically dispersed, and the team relies on technology-mediated communications instead of face-to-face communications to complete tasks (Cohen & Gibson, 2003). Within the bounds of this definition, virtual teams can vary in numerous ways. For instance, the types of technology-mediated communication media that may be used by a virtual team include a wide variety of media such as telephone, e-mail, videoconference, groupware, or other media. In addition, virtual teams can differ with respect to dimensions such as temporal distribution, diversity of culture, team lifecycle, and member roles (Bell & Kozlowski, 2002). Recent research has focused on a variety of variables that can influence virtual team performance, including sources of structure and social interaction (Avolio, Kahai, & Dodge, 2000), empowerment of the team and face-to-face interaction (Kirkman, Rosen, Gibson, & Tesluk, 2004), interpersonal trust (Paul & McDaniel, 2004), and types of technology (Majchrzak, Malhotra, Stamps, & Lipnack, 2004). Leadership is an aspect of both structure and social interaction within a team, and as such, is a component of virtual teams that has the potential to significantly influence team performance. Leaders help individuals make sense of the world around them, and in a complex, highly technical environment such as a virtual team, leadership may be the key to enhancing the collaborative efforts of team members (Avolio et al., 2000).

Although virtual teams have existed as an organizational structure for decades, recent technological advances have enabled the increased use of virtual teams within organizations. As information technology has expanded, researchers have explored virtual team implementation and performance. A recent literature review of 43 research papers on the topic of virtual teams published between 1991 and 2002 found that researchers have investigated a variety of aspects of virtual teams including inputs (e.g., design of interactions, skills of team), socio-emotional processes (e.g., trust, relationship building), task processes (e.g., communication, coordination), and outcomes (Powell et al., 2004). An example of a study that investigated a virtual team input was research by Piccoli et al. (2004) that demonstrated the importance of team control structures in facilitating virtual team performance. With regard to socio-emotional processes, work by Jarvenpaa, Knoll, and Leidner (1998) found that trust needed for coordination and communication within virtual teams could be enhanced through the use of trust-building intervention. An illustration of virtual team research that addressed task processes was conducted by Yoo and Alavi (2004). This research found that emergent leaders in virtual teams typically sent more electronic messages and longer messages than the followers.

As compared to the traditional face-to-face environment, the "virtual" environment can introduce new types of challenges for those responsible for the management of virtual teams (Pauleen, 2004; Zaccaro & Bader, 2003). One of the key challenges concerns the leadership of virtual teams (Avolio & Kahai, 2003; Cascio & Shurygailo, 2003; Kayworth & Leidner, 2002). Leadership is defined as the process of influencing others toward a goal or objective (Bass, 1985). Although much research has been conducted regarding the role of leadership for traditional teams and organizations, relatively little research has examined the effect of leadership on virtual teams (Kim, 2006). The early studies in this area suggest that leadership for virtual teams is different from that of face-to-face teams. However, many questions remain unanswered (Powell et al., 2004). For example: What are the traits of a good virtual team leader? What can be done to promote efficient and effective group interaction? What role does trust play with respect to the leadership of virtual teams? Answering the foregoing questions can be difficult: As team members work together interdependently, they may require more or less guidance, coaching, monitoring, and structure depending on contextual factors, task progress, and social interaction quality.

LEADERSHIP AND INFLUENCE IN VIRTUAL TEAMS

In their study of "e-leadership" (defined as influencing others across space and time primarily through the use of electronic media), Avolio and his colleagues (2000) have proposed a framework based on adaptive structuration theory (AST). Their framework is relevant to a discussion of virtual team leadership since the framework can be used to enhance understanding of the role of e-leadership in organizations. The framework specifies a number of relevant structural elements of virtual teams, as well as social interaction components (see Figure 1). Within this framework, team structure influences a team's social interaction, which ultimately influences team outcomes. In addition, social interaction and outcomes have a reciprocal effect upon the

Figure 1. Leadership influences on virtual teams

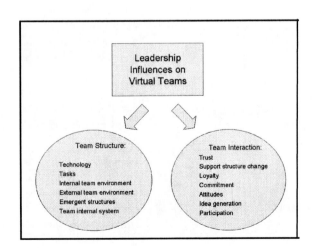

team structural elements. If one posits that a leader can influence the structural and the social interaction aspects of virtual teams, then the framework introduced by Avolio et al. (2000), can serve to identify and organize the different ways in which a leader may influence a virtual team. The following sections consider the structural and the social interaction aspects of virtual teams, respectively.

Leadership and Virtual Team Structure

Avolio and colleagues (2000) suggest that there are many sources of virtual team structure that are relevant for e-leadership. In the virtual team environment, these sources include: (a) the types of information technology used to support the team's work and communication,

(b) the content and complexity of the work tasks, (c) the quality of the internal team environment, (d) the support felt by the external environment, (e) the structures that emerge or evolve as the team makes progress on its tasks, and (f) the team's internal system. In Table 1, we list each of these categories of structures and summarize the ways in which a team leader may potentially influence each type of structure for a virtual team.

A team's "internal system" refers to team aspects such as the team's style of interacting, members' expertise, diversity, and shared mental models. Recent research supports the argument that virtual team leaders may have the greatest potential influence through the team's internal system (Avolio et al., 2000). For example, Beranek, Broder, Reinig, Romano, and Sump (2005) suggest that virtual team project leaders should

Table 1. Leadership influences on sources of structure for a virtual team (Adapted from Avolio et al., 2000)

Sources of Structure	Leader's Potential Influence
Information Technology Used By The Team	• Determine what technology will be used by the team and for what purpose. • Set guidelines and procedures for technology use for team • Model norms for communication frequency, patterns, and channels.
Content and Complexity of the Team's Tasks	• Assign tasks to team members, determining the flow of pooled or sequential interdependence. • Set deadlines for task performance, both final and intermediate. • Determine rewards and consequences for task performance.
Quality of Internal Team Environment	• Establish procedures and policies for internal interaction. • Motivation style influences internal environment through promoting competitive or collaborative functioning within the team.
Support from External Team Environment	• Function as the liaison to the external environment. • Secure needed resources for team functioning. • Act as a gatekeeper for team, reducing unimportant distractions while assuring needed external information reaches team members in a timely and complete fashion.
Types of Emergent Structures	• Can elicit or request emergent structures to assist team functioning. • Respond to emergent structures through encouragement or modification.
Team's Internal System	• Guide and model team interaction style. • Provide guidance in developing shared mental models. • Motivate and inspire team members to identify and commit to the team's goals and vision. • Determine who joins the team, who leaves the team.

pay particular attention to the initial set-up of project teams. In virtual project teams, setting up clearly defined processes, routines, and roles can be crucial to team success (Majchrzak et al., 2004). In their study of virtual team leadership, Kayworth and Leidner (2002) found that the most important factors in leadership effectiveness were all system-oriented. In their study, they found that the key qualities associated with virtual team effectiveness were mentoring capabilities of the leader, perceptions of communication effectiveness, and the ability of the leader to establish role clarity among team members. In another study exploring distributed groups with group support systems, Kim (2006) found that effective virtual team leaders vary their style and behaviors throughout the life of the team. The most effective team leaders focused on clarifying goals and interaction structure earlier in the team's process, and encouraged smooth interaction processes in later stages.

Leadership and Virtual Team Social Interaction

Virtual team leadership may also influence the social interaction of virtual team members. Social interaction within a virtual team is by definition different than that of a collocated team. Face-to-face interaction is limited, at times non-existent, depending on the team's degree of "virtuality" (Cohen & Gibson, 2003). In addition, as proposed in Avolio et al.'s (2000) adapted version of AST, the structure of a virtual team will also influence the team's social interaction in an iterative and uniquely interpreted way. A leader can influence a team's social interaction significantly through a variety of methods, including through the influence of team structure. In collocated teams, a leader can influence social interaction through transformational leadership behaviors and motivational efforts (Bass & Avolio, 1994). In virtual teams, it may be more difficult to influence a team in these ways due to the constraints of computer-mediated communication (Bell & Kozlowski, 2002). However, research clearly indicates that leaders do influence, inspire and motivate their virtual team members (Tyran, Tyran, & Shepherd, 2003). Extending our adaptation of the Avolio et al. (2000) AST model, Table 2 provides an illustration of how leadership can influence the social interaction aspects of a virtual team.

A key factor related to social interaction is trust within a team (Holton, 2001). The importance of trust

in virtual teams has been established by a number of research studies (Cohen & Gibson, 2003; Handy, 1995; Jarvenpaa et al., 1998; Paul & McDaniel, 2004). One way in which a virtual team leader may influence trust may be by providing opportunities for team members to get to know each other through face-to-face meetings. Another way that leaders can help to establish trust in teams is through establishing technological "homes" for the team (Majchrzak et al., 2004). For example, interactive Web pages accessed through company intranets provide an interface for team members to work interdependently and simultaneously on their projects.

Another important aspect of social interaction concerns the group processes used by a team (e.g., participation, task management, conflict management). A team leader can influence social interaction by establishing processes and procedures for team processes. For instance, many virtual team leaders may develop a formal contract outlining team processes for communication, role distribution, and conflict management (Kirkman et al., 2002). Processes can also be developed in an iterative way. As teams go through stages of development, they change and grow in both their performance and their social processes. Leadership can influence team social interaction throughout the team's development. Early during a team's development, the team leader can introduce structures such as a team contract, while at a later time the leader can encourage communication and conflict management between team members (Beranek et al., 2005).

CONCLUSION

Virtual teams have the potential to offer powerful benefits to organizations and change the way that team work is conducted. However, to fully exploit the possibilities afforded by virtual teams, effective leadership will be needed. As illustrated in Tables 1 and 2 of this article, the scope of influence for a virtual team leader is broad, as the team leader may have the potential to influence many aspects associated with the structural and social interaction components of a virtual team. While a review of the information in these tables might suggest that leadership practices for traditional teams and virtual teams are similar (i.e., many of the areas of influence are applicable to both environments), the methods in which an effective leader manages a virtual

Table 2. Leadership influences on virtual team social interaction (Adapted from Avolio et al., 2000)

Social Interaction Aspects	Leader's Potential Influence
Trust Within The Team	• Provide opportunities for team members to demonstrate loyalty, competence, commitment, and other behaviors that establish trust. • Demonstrate trust with team members. • Recognize accomplishments and share information about the team. • Communicate regularly and effectively.
Changes to Team Structure Proposed By Individual Team Members	• Encourages or empowers team members to make changes to the team structure. • Encourages dialogue between team members regarding potential team structure changes. • Channels and develops adaptive structures of all team members through sharing team member mental models.
Team Member Loyalty	• Recognizes and rewards team loyalty. • Enlists team loyalty through communicating the team's vision. • Models team loyalty. • Helps to establish trust in the team and other team members.
Team Member Commitment	• Inspires and motivates commitment to the team and project. • Outlines goals that team members see as relevant to their own goals.
Team Member Attitudes Promoted	• Promotes positive attitudes toward the team and tasks. • Encourages supportive attitudes expressed by team members.
Team Processes for Idea Generation	• Encourages and develops team processes that produce innovative and valued ideas. • Empowers team members to generate their own creative ideas.
Team Member Participation in Decisions, Conflict Management and Task Management	• Includes team members in the development of processes and procedures to address decision-making, conflict management, and task management. • Has a participative leadership style with the team.

team are typically very different from the approaches employed for a traditional team. The research on virtual team leadership indicates that the virtual environment demands a different set of leadership skills than a traditional, face-to-face type of team environment (Powell et al., 2004; Zigurs, 2003). Hence, people who have had success with respect to leading a team in a traditional environment are not necessarily going to be successful with a virtual team unless they can adapt their skills to a virtual type of team environment. For those managers who will be leading others in a virtual team setting, it will be important to learn the new rules of the game. At this point in time, the existing research on virtual team leadership does offer useful insights for practitioners in the field. However, as we look to the future, there are many opportunities to build on the existing body of research.

REFERENCES

Avolio, B.J., & Kahai, S. S. (2003). Adding the E to E-leadership: How it may impact your leadership. *Organizational Dynamics, 31*(4), 325-338.

Avolio, B. J., Kahai, S. S., & Dodge, G. E. (2000). E-Leadership: Implications for theory, research, and practice. *Leadership Quarterly, 11*(4), 615-668.

Bass, B. M. (1985). *Leadership and performance beyond expectation.* New York: Free Press.

Bass, B. M., & Avolio, B. J. (1994). *Improving organizational effectiveness through transformational leadership.* Thousand Oaks, CA: Sage.

Bell, B. S., & Kozlowski, S. W. J. (2002). A typology of virtual teams: Implications for effective leadership. *Group and Organization Management, 27*(1), 14-49.

Beranek, P. M., Broder, J., Reinig, B. A., Romano, N. C., & Sump, S. (2005). Management of virtual project teams: Guidelines for team leaders. *Communications of the Association for Information Systems, 16,* 247-259.

Carte, R.A., Chidambaram, L., & Becker, A. (2006). Emergent leadership in self-managed virtual teams: A longitudinal study of concentrated and shared leadership behaviors. *Group Decision and Negotiation, 15*(4), 323-343.

Cascio, W. F., & Shurygailo, S. (2003). E-Leadership and virtual teams. *Organizational Dynamics, 31*(4), 362-376.

Cohen, S. G., & Gibson, C. B. (2003). In the beginning: Introduction and framework. In C. B. Gibson & S. Cohen (Eds.), *Virtual teams that work: Creating the conditions for virtual team effectiveness* (pp. 1-13). San Francisco: Jossey-Bass.

Conger, J., & Kanungo, R. (1998). *Charismatic leadership in organizations.* Thousand Oaks, CA: Sage Publications.

Duarte, D. L., & Snyder, N. T. (1999). *Mastering virtual teams: Strategies, tools and techniques that succeed.* San Francisco: Jossey-Bass.

Handy, C. (1995). Trust and the virtual organization. *Harvard Business Review, 73*(3), 40-50.

Hooijberg, R., Hunt, J. G., & Dodge, G. E. (1997). Leadership complexity and the development of the Leaderplex model. *Journal of Management, 23*(3), 375-408.

Holton, J. A. (2001). Building trust and collaboration in a virtual team. *Team Performance Management, 7*(3/4), 36-47.

Jarvenpaa, S. L., Knoll, K., & Leidner, D. E. (1998). Is anybody out there? Antecedents of trust in global virtual teams. *Journal of Management Information Systems, 14*(4), 29-64.

Kayworth, T. R., & Leidner, D. E. (2002). Leadership effectiveness in global virtual teams. *Journal of Management Information Systems, 18*(3), 7-40.

Kim, Y. (2006). Supporting distributed groups with group support systems: A study of the effect of group leaders and communication modes on group performance. *Journal of Organizational and End User Computing, 18*(2), 20-37.

Kirkman, B., Rosen, B., Gibson, C., & Tesluk, P. (2004). The impact of team empowerment on team performance: The moderating role of face-to-face interaction. *Academy of Management Journal, 47*(2), 175-192.

Kirkman, B., Rosen, B., Gibson, C., Tesluk, P., & McPherson, S. (2002). Five challenges to virtual team success: Lessons from Sabre, Inc. *Academy of Management Executive, 16*(3), 67-79.

Majchrzak, A., Malhotra, A., Stamps, J., & Lipnack, J. (2004). Can absence make a team grow stronger? *Harvard Business Review, 82*(5), 131-137.

McGrath, J. E. (1984). *Groups: Interaction and performance.* Englewood Cliffs, NJ: Prentice-Hall.

O'Connell, M. S., Doverspike, D., & Cober, A. B. (2002). Leadership and semiautonomous work team performance. *Group & Organization Management, 27*(1), 50-65.

Paul, D. L., & McDaniel, R. R. (2004). A field study of the effects of interpersonal trust on virtual collaborative performance. *Management Information Systems Quarterly, 28*(2), 183-227.

Pauleen, D. J. (2004). An inductively derived model of leader-initiated relationship building with virtual team members. *Journal of Management Information Systems, 20*(3), 227-256.

Piccoli, G., Powell, A., & Ives, B. (2004). Virtual teams: Team control structure, work processes, and team effectiveness. *Information Technology and People, 17*(4), 359-379.

Powell, A., Piccoli, G., & Ives, B. (2004). Virtual teams: A review of current literature and directions for future research. *The DATABASE for Advances in Information Systems, 35*(1), 6-36.

Townsend, A. M., DeMarie, S. M., & Hendrickson, A. R. (1998). Virtual teams: Technology and the workplace of the future. *Academy of Management Executive, 12*(3), 17-29.

Tyran, K. L., Tyran, C. K., & Shepherd, M. (2003). Exploring emerging leadership in virtual teams. In C. B. Gibson & S. Cohen (Eds.), *Virtual teams that work: Creating the conditions for virtual team effectiveness* (pp. 183-195). San Francisco: Jossey-Bass.

Yoo, Y., & Alavi, M. (2004). Emergent leadership in virtual teams: What do emergent leaders do? *Information and Organization, 14*(1), 27-58.

Zaccaro, S. J., & Bader, P. (2003). E-leadership and the challenges of leading e-teams: Minimizing the bad and maximizing the good. *Organizational Dynamics, 31*(4), 377-387.

Zigurs, I. (2003). Leadership in virtual teams: Oxymoron or opportunity? *Organizational Dynamics, 31*(4), 339-351.

KEY TERMS

Adaptive Structuration Theory (AST): A theory used to explain the process in which information technologies are integrated into the workplace. The theory considers change with respect to the structures provided by the technologies, as well as the structures that arise when people use the technologies.

E-Leadership: The process of influencing others through computer-mediated communication.

Leadership: A process used to influence attitudes, behavior, and/or performance with individuals, groups, or organizations.

Team's Internal System: Describes the nature of team member interaction and the member's expertise, diversity, and shared mental models.

Team Leader: A role within a team that involves administering the team, coaching team members, and advising when needed within the team. Often a team leader coordinates tasks and facilitates performance, as well as mediating when conflicts arise.

Team Social Interaction: The manner in which team members work together, including communication style, conflict management approaches, and decision-making techniques.

Trust: A psychological state in which one person has confidence that another party will act in an expected manner.

Virtual Team: A functioning team of people who are geographically dispersed, interdependent and share responsibility for team outcomes, and reliant on technology-mediated communications instead of face-to-face communications to complete tasks.

Scenarios for E-Collaboration are Only Part of the Story

Lydia M. S. Lau
The University of Leeds, UK

INTRODUCTION

Addressing human factors in the development of artifacts has always been crucial for their usability (Norman, 1998). Within the domain of human-computer interaction (HCI), a number of advances have been made since the invention of computer systems, but the paradigm shift to what we are familiar with today started in the early eighties with the invention of GUI, graphical user interface (Pew, 2003). The early work in HCI tended to concentrate on an individual's cognitive system. The range of possible interactions between human and computer then was comparatively simple due to the limited capability of the technologies.

With the emergence of computer-supported cooperative work (CSCW), human factors were taken to a new dimension and opened up a number of different perspectives in terms of human factors challenges—personal, team, and organizational (Dourish, 2001; Olson & Olson, 2003; Orliskowski, 1992). Multiple channels (text, audio, and video) could be used in a variety of ways to facilitate collaborative work.

This article explores the issue of how human factors can be addressed during the development of e-collaboration systems, in particular at the early stage of design since these systems may require new ways of working. It can be difficult to capture requirements for e-collaboration from existing users for two reasons. Firstly, engaging the potential users in requirements capture can be problematic as those users may not perceive themselves as potential users at the outset and hence not putting themselves forward in any consultation exercise. Even with the interested parties (or stakeholders) identified, they may not understand the full potential of how these new tools help people work more effectively with each other. Hence, requirements capture is a challenge for developing e-collaboration systems.

Another area of challenge lies in the evaluation of e-collaboration systems. If requirements cannot be fully articulated at the outset, then evaluation of a prototype may provide useful feedback to the development team for the next iteration. However, evaluation of e-collaboration systems is known to be problematic due to the complexity of influencing factors (Ross, Ramage, & Rogers, 1995). Furthermore, the traditional human factors approach such as usability tests can only examine the hygiene factors and not necessarily the motivators (Herzberg, 1959; Zhang & von Dran, 2000). In the HCI domain, hygiene factors are those that will cause dissatisfaction for the user if they are absent (e.g., readable fonts and good color scheme). Motivator factors are those that will encourage users to continue using the system. For e-collaboration systems, both factors are crucial for their sustainability and any feedback from evaluation on these factors will be valuable.

Scenario-driven techniques have been offered as possible means to engage different stakeholders (Rossen & Carroll, 2003) during a development process. These techniques can be used in a variety of ways and at any stage in the development life cycle (Carroll, 1995; Rossen & Carroll, 2003). In particular, scenarios are used to capture requirements and to provide basis for evaluation. There is a potential to apply the techniques to help meet the two challenges identified above. However, for an inspired user of scenario-driven techniques, there is a danger of stopping at the story-telling stage with the full potential of the techniques not being realized.

This article will share the experience gained from three empirical studies in using scenario-driven techniques at different stages of the development cycle of e-collaboration systems. A discussion on how scenarios can be used in conjunction with the other techniques during the development process will be provided.

BACKGROUND

A scenario is defined as "a concrete description of an activity that the user engages in when performing a specific task, a description sufficiently detailed so the design implications can be inferred and reasoned about" (Carroll, 1995, pp. 3-4). In software engineering, it is commonly used in a story-telling fashion to illustrate how a piece of software is intended to be used. The power of scenarios can be extended by different techniques, such as involving the stakeholders in the writing of the story, hence increasing the credibility and ownership of the story, or using people to role-play the story, hence verifying the logic of the story line. These are collectively labelled as scenario-driven techniques in this article.

Two groups of scenarios are identified in the literature. The first group is the "as-is" scenarios where narrative details are provided on how the operations are currently being performed. These are referred as "problem scenarios" (Carroll, Rosson, Convertino, & Ganoe, 2006; Rosson & Carroll, 2003). The second group is the "to-be" scenarios, which are more visionary and serve as a target for development. These are further split into "activity scenarios," "information scenarios," and "interaction scenarios" (Rosson & Carroll, 2003). The concept of to-be scenarios is particular well-suited for developing e-collaboration systems, as these scenarios can be used as vision statements of how people can collaborate in a different, hopefully better, way.

Scenarios can be used in different stages of a system development life cycle, although they are more commonly used at the requirements stage. It is reported that the design team of the National Digital Library began the requirements with 81 scenarios (Shneiderman & Plaisant, 2005). Apart from using scenarios for capturing requirements, there are also attempts to use them in the requirements analysis stage (Sutcliffe, 1998). Sometimes, scenarios are linked to use cases in the Unified Modeling Language (UML). In that context, each scenario is an instance of a use case (Sutcliffe, Maiden, Minocha, & Manuel, 1998; Preece, Rogers, & Sharp, 2002). A survey on the use of scenarios in system development reveals the following variations (Weidenhaupt et al, 1998):

- Form (text, diagrams, and to a lesser extent, animations)
- Purpose and usage (concretization of abstract models, definition or validation of structural models, facilitation of interdisciplinary development and partial agreement amongst stakeholders, and/or complexity reduction)
- Content (with focus on system context, system interaction, and/or the operations within the system)

- Utilization within the life cycle (as an artifact which evolves throughout the development life cycle, for capturing stakeholders' views, and/or for deriving test cases)

However, most developers in the above survey agreed that scenario creation was a craft and there was very little guidance on the process of scenario construction (Weidenhaupt et al., 1998). There is still a need to collate local innovations of scenario practices to further our understanding on how scenarios can be best exploited (Rosson & Carroll, 2003). Furthermore, a scenario-based method is not a set-book approach and it very often requires the use of other data collection and analysis techniques (Bardram, 2000; Kazman, Carriere, & Woods, 2000; Iacucci & Kuutti, 2002; Haynes, Purao, & Skattebo, 2004).

Following are three examples of scenario construction for e-collaborations. In each case, the context will be explained to illustrate the kind of e-collaboration system involved and at which stage of the development cycle the scenario-based method was applied. A discussion of how other techniques were used in conjunction with the scenario(s) will follow. It will then conclude with the lessons gained from the experience.

CASE ONE: SETTING UP A NEW PRACTICE

- **Context:** This example took place at the concept phase when a new business model was perceived by a legal expert for the provision of mediation services with an innovative e-collaboration system (Barnett & Dew, 2005). The common practice in mediation service required the parties in dispute to prepare the evidence and related paperwork in advance separately, followed by one or more face-to-face hearing sessions. A trained mediator would chair the sessions until an agreed settlement was reached. An example of disputes could be a building project which involved multiple international contractors. The actual mediation process was not entirely prescriptive, with possibilities of break-out sessions in sub-groups during a hearing. This process was known to be time-consuming and getting all parties together at meetings could cause further delays. To test the feasibility of the business model, a proof-of-

S

concept e-collaboration system was required that involved collaborative tools such as a document management system, conferencing system, an events organizer, and communication tools. The challenge was to make the tools work seamlessly together and to adapt the functionalities to the practice of mediation services.

- **Techniques used:** As a first step, the high-level expectations from the legal expert must be translated into requirements specifications for the developers (as the components came from different suppliers). The scenarios-based method was used after the scope of the proof-of-concept prototype was confirmed. The scenario writer produced a typical scenario that captured the challenge explicitly. It covered the whole process of handling a case and the IT functionalities to be used at each step. It was deliberate that the scenario was written in a tabular format (see Figure 1), and not as a prose, to facilitate structured walk-through with the stakeholders. The scenario was split into episodes that would reflect the natural breaks of the process. Any breaks of this kind would imply background coordination or monitoring was required, which could be easily overlooked in requirements capture. The start and end of each episode prompted

questions such as (1) triggers and information needed to be in place at the beginning of each episode, and (2) the remaining house keeping jobs needed at the end of each episode.

- **Participating roles:** Mediator Service Administrator; Mediator; Solicitor A/B (in Party A/B); Solicitor Administrator A/B (in Party A/B); Client A/B (in Party A/B); Expert C/D (not represented by a solicitor):

Episode 1: Assemble material by all parties
Episode 2: Arranging a mediation session
Episode 3: Pre-hearing preparation …cut…
Episode 4: Conduct hearing session …cut…
Episode 5: Follow-up meeting …cut…

The scenario was then used as the focal point for discussion with the legal expert and a couple of potential IT solution providers. The review of every step in the scenario engaged the stakeholders in elaborating details of the collaborative activities (e.g., sequence of events, the form of information, who else might be involved), clarifying scope and noting any other issues.

- **Lessons learned:** The experience showed that a shared understanding of the vision could be

Figure 1. An example of a scenario used at the concept phase

Tasks	IT Support
Task 1a: Solicitor A prepares the opening statement and supporting documents; submits to Mediator Service Administrator and requests for Mediation.	**Document management system**: Administrator can set up an environment for the case for each party to upload documents directly to it. Each party will have control over the security level for each of their documents. (file formats: PDF, Word, JPeg, Tiff)
Task 1b: Solicitor A / Solicitor Administrator A collect further documents from Client A and other sources, then file.	ditto
Task 1c: Solicitor B prepares the opening statement and supporting documents; submits to Mediator Service Administrator and requests for Mediation.	ditto
……cut ………	
Task 1f: Expert D brings in his documents	ditto

Tasks	IT Support
Task 2a: Mediator Service Administrator contacts the mediator and get a few possible sessions, then contact all parties	**Web-based e-mail / group calendar**
…cut…	
Task 2c: Administrator sets up the virtual meeting areas for this case	**multiparty video conferencing system**

achieved by allowing queries of alternatives to be raised and resolved during the walk-through of the scenario. The culture of team work could be written into the scenario to inform the developers how the system might be used in a real situation (e.g., it was quickly established that the role of solicitor administrator was needed to assist the solicitor in episode 1, although solicitors could easily use the system to handle the housekeeping tasks associated with paperwork, it was usually the job for secretaries/administrators).

It was common that a scenario would present a story in which the tasks and the use of IT intertwined. The advantage of separating the tasks from the IT support as presented here enabled alternative IT solutions to be explored, which was found to be useful at the concept phase.

CASE TWO: EVALUATION IN SITU

- **Context:** This example was an evaluation of an experimental e-collaborative system that provided shared workspace and real-time conferencing facilities. The system used a room-based metaphor. Faculty members could each occupy an "office" in the environment. A class of graduate students was split into groups, with each group owning a "group study room" for their tasks over a period of several weeks. This trial focused on the system's support for group work in an online learning situation, in particular, if there were any unforeseen group problem-solving activities unsupported.

- **Techniques used**: A couple of expected usage scenarios were written for the evaluation of the acceptability/usability of the functionalities for some collaborative tasks (see Figure 2 for the first scenario). The students were asked to try out the scenarios in their own time, on campus or from home. They were asked to report on the actual happenings and any deviation from the scenarios in a user log.

The second set of feedback was obtained by a structured questionnaire that asked each student to provide comments on their preferences on the functions provided by the resource room for group working as well as their preferences for the user-interface, and also their reactions to various aspects of the group study room (Lau, Curson, Drew, Dew, & Leigh, 1999). Data collected were in textual format, hence content analysis on those returns was used to categorize the issues emerged.

- **Lessons learned:** Lucy Suchman's situated action theory highlighted the importance of context surrounding the use of systems and that users improvized the use of tools according to the situation at the time when the tools were used (Suchman, 1987). This pilot experiment provided the opportunity for evaluating how different users might collaborate differently in situ during specified collaborative scenarios. Some interesting issues emerged from the evaluation, such as how to prevent a group member unilaterally deleted a shared document in the process of "tidying up the room" (Lau et al., 1999). However, the process of data collection and analysis was qualitative,

Figure 2. An expected usage scenario: a real-time meeting between a student and a professor

A student was stuck when trying to produce a paragraph elaborating on a lecture foil used by the professor. The student needed some clarification from the professor. Using the *system*, the student emailed the professor, requesting a suitable time to run NetMeeting with him.
.. a short time later..
The student received a message from the professor confirming the date and time.
.. at the specified date and time ..
The student entered the reading room and launched a NetMeeting with the professor. Using the "Chat," the student explained the problem and asked for clarification. The student proceeded and showed the professor the lecture foil in question.
.. after some interaction ..
The student had the query answered and finished the meeting happily.

hence there might be an element of subjectivity in the interpretation of the findings.

CASE THREE: BEFORE DEPLOYMENT OF A SYSTEM

- **Context:** This example took place at the completion of an e-collaboration system, getting ready to be rolled out in the workplace as a stable and polished product (Lau et al, 2003). Functionalities were provided which aimed to facilitate academics and industrial partners to work as teams on projects. There were document management facilities, real-time collaboration tools and an expertise matcher, functions which were thought to be useful for the potential users. However, as this was a new venture, some form of realistic use evaluation would be needed to provide an indication of the right level of user expectation to be set during the roll-out and the kind of user training required.
- **Techniques used:** A laboratory based evaluation was set up (Lau, Adams, Dew, & Leigh, 2003). Realistic scenarios were written for (1) team formation by identifying the right members to invite, (2) real-time collaboration with joint authoring and desk-top conferencing, and (3) consultation by an external industrialist, requesting recommendation from the team. Seven different R&D research projects were role-played. Each team consisted of an industrialist and two academics from different fields.

  Apart from designing the scenarios, considerable preparation was required in this evaluation. Firstly, the system was populated with "real" data for these role-played characters. A brief training session was provided so that the participants were familiar with the general functionalities of the system, but they were not trained to use the system for collaborative tasks. At the start of the experiment, documentation was given that explained the context and the role each participant would play. The evaluation session proceeded with the participants going through the scenarios and the tasks. User log was the main data capture device (a tailored template was provided for every role in every scenario). The participants noted down how they completed the tasks and their feelings

at the time. Observations were also conducted by the evaluators.

After the data were collected, every comment was analyzed and issues emerged were classified under their related functionalities. Responses/actions were then decided.

- **Lessons learned:** Experience gained from this case study showed that the outcome of this type of evaluation enabled the deployment team to anticipate training requirements and to manage user expectation appropriately. As the scenarios only specified the goals of the collaborative tasks, participants could explore any route that seemed sensible for them to achieve those goals. This provided useful feedback to the development team if there were any unexpected routes which required better support. Some issues were detected as misuse of the system, which would require better help documentation. It could be argued that this type of laboratory-based evaluation could also be carried out earlier in the development life cycle, for example, after a low-fidelity prototype was produced (Pinelle, Gutwin, & Greenberg, 2003).

SUMMARY OF FINDINGS

In the above three case studies, scenarios were used to unearth some of the human factors that should be addressed during the development of an e-collaboration system. However, even within these three studies, there were variations in their applications. This demonstrated the flexibility and potential uses of scenario-based methods:

At different stages in a development life cycle:

- **Case 1:** Concept stage
- **Case 2:** Prototyping stage
- **Case 3:** Before deployment

With different participants/motivations:

- **Case 1:** With stakeholders (problem owner and IT suppliers)/to share common understanding of the vision;
- **Case 2:** With potential users (teachers and students)/to evaluate the prototype in the context of group learning activities

- **Case 3:** With role-playing participants and the deployment team / to anticipate training requirements and to prioritize further development work.

In conjunction with different techniques:

- **Case 1:** Structured walkthrough
- **Case 2:** User logs and structured questionnaires for data capture; content analysis on the results conducted by a human factors researcher
- **Case 3:** Protocols for a laboratory based evaluation (setting up the "equipments," planned methods and materials in terms of training and documentations, systematic data collection); results analyzed by the deployment team.

CONCLUSION

The experience gained from the case studies confirmed the potential of scenarios for requirements capture. They were particularly powerful in capturing the dynamics of collaborative work and in envisioning new ways of team working.

However, the usefulness of scenarios in evaluations would depend on the design of those scenarios and how they were used with other techniques. When the case studies were conducted, the evaluation and findings were mostly relating to the hygiene factors. The motivator factors were not explicitly articulated and evaluated. In future work, if the motivator factors can be identified at the outset, these should be incorporated into the scenarios for evaluation.

Scenario-driven techniques are still rather peripheral to other activities in systems development. In view of the potential powerful role of scenarios, we should strive for a more rigorous approach to their use, and combine them with more methodical approaches to data collection and analysis. This could become an integral part of systems development for any e-collaboration systems.

REFERENCES

Barnett, J., & Dew, P. (2005). IT enhanced dispute resolution. In J. Zeleznikow & A. R. Lodder (Eds.), *Proceedings of the Second International ODR Workshop* (pp. 1-10). Wolf Legal Publishers.

Bardram, J. E. (2000). Scenario-based design of cooperative systems redesigning a hospital information system in Denmark. *Group Decision and Negotiation, 9,* 237-250.

Carroll, J. M. (Ed.). (1995). *Scenario-based design: Envisioning work and technology in system development.* New York: John Wiley.

Carroll, J. M., Rosson, M. B., Convertino, G., & Ganoe, C. H. (2006). Awareness and teamwork in computer-supported collaborations. *Interacting with Computers, 18*(1), 21-46.

Dourish, P. (2001). Seeking a foundation for context-aware computing. *Human-Computer Interaction, 16*(2), 229-241.

Haynes, S. R., Purao, S., & Skattebo, A. L. (2004). Situating evaluation in scenarios of use. In *Proceedings of CSCW '04* (pp. 92-101). ACM.

Herzberg, F., Mausner, B., & Snyderman, B. B. (1959). *The motivation to work* (2nd ed.). New York: John Wiley & Sons.

Iacucci, G., & Kuutti, K. (2002). Everyday life as a stage in creating and performing scenarios for wireless devices. *Personal and Ubiquitous Computing, 6,* 299-306.

Kazman, R., Carriere, S. J., & Woods, S. G. (2000). Toward a discipline of scenario-based architectural engineering. *Annals of Software Engineering, 9,* 5-33.

Lau, L. M. S., Adams, C. A, Dew, P. M., & Leigh, C. (2003). Use of scenario evaluation in preparation for deployment of a collaborative system for knowledge transfer. In *Proceedings of 12th IEEE International Workshops on Enabling Technologies (WETICE 2003): Infrastructure for Collaborative Enterprises* (pp. 148-152). IEEE Computer Society Press.

Lau, L. M. S., Curson, J., Drew, R., Dew, P. M., & Leigh, C. (1999). Use of virtual science park resource rooms to support group work in a learning environment. In *Proceedings of GROUP '99 International ACM SIG-GROUP Conference on Supporting Group Work* (pp. 209-218). ACM Press.

Norman, D. (1998). *The design of everyday things.* London: MIT.

Olson, G. M., & Olson, J. S. (2003). Groupware and computer-supported cooperative work. In J. A. Jacko & A. Sears (Eds.), *The human-computer interaction handbook* (pp. 583-595). Lawrence Erlbaum Associates.

Orlikowski, W. (1992). The duality of technology: Rethinking the concept of technology in organizations. *Organization Science, 3*(3), 398-427.

Pew, R. W. (2003). Evolution of human-computer interaction: From Memex to BlueTooth and beyond. In J. A. Jacko & A. Sears (Eds.), *The human-computer interaction handbook* (pp. 1-17). Lawrence Erlbaum Associates,.

Pinelle, D., Gutwin, C., & Greenberg, S. (2003). Task analysis for groupware usability evaluation: modeling shared-workspace tasks with the mechanics of collaboration. *ACM Transactions on Computer-Human Interaction, 10*(4), 281-311.

Preece, J., Rogers Y., & Sharp, H. (2002). *Interaction design*. John Wiley & Sons.

Ross, S., Ramage, M., & Rogers, Y. (1995). PETRA: Participatory evaluation through redesign and analysis. *Interacting with Computers, 7*(4), 335-360.

Rosson, M. B., & Carroll, J. M. (2003). Scenario-based design. In J. A. Jacko & A. Sears (Eds.), *The human-computer interaction handbook* (pp. 1032-1050). Lawrence Erlbaum Associates.

Shneiderman, B., & Plaisant, C. (2005). *Designing the user interface* (4th ed.). Pearson.

Suchman, L. (1987). *Plans and situated actions: The problem of human-machine communication.* Cambridge: Cambridge University Press.

Sutcliffe, A. (1998). Scenario-based requirement analysis. *Requirements Engineering Journal, 3*(1), 48-65.

Sutcliffe, A. G., Maiden, N. A. M., Minocha, S., & Manuel, D. (1998). Supporting scenario-based requirements engineering. *IEEE Transactions on Software Engineering, 24*(12), 1072-1088.

Weidenhaupt, K., Pohl, K., Jarke, M. & Haumer, P. (1998) Scenarios in system development: Current practice. *IEEE Software, 15*(2), 34-45.

Zhang, P., & von Dran, G. M. (2000). Satisfiers and dissatisfiers: A two-factor model for Website design and evaluation. *Journal of the American Society for Information Sicence, 51*(14), 1253-1268.

KEY TERMS

As-Is Scenario: A story with narrative details on how the operations are currently being performed.

E-Collaboration System: Computer- and network-based system that supports collaborative activities undertaken by two or more people who are usually in different places.

Hygiene Factors: Factors that will cause dissatisfaction if absent, but their presence does not necessarily lead to satisfaction.

Motivator Factors: Factors that are required to ensure continuing interests in using the system.

Scenario: The use of story-telling techniques to describe the flow of events with a particular focus.

Stakeholder: People or organization who have an interest in the outcome, for examples, owner, customers, suppliers, or employees.

System Development Life Cycle: The process towards delivering a computerized solution from the inception of an idea to the installation of the system (main stages can be requirements capture, analysis, design, implementation, testing, installation, and maintenance).

To-Be Scenario: A story with narrative details on how future operations will be performed. It is more visionary and serves as a target for development.

Usability Test: A methodical way to find out how well a particular artifact is being used by a person and if the purpose is being met.

Setting the Framework of E-Collaboration for E-Science

Andrea Bosin
Università degli Studi di Cagliari, Italy

Nicoletta Dessì
Università degli Studi di Cagliari, Italy

Maria Grazia Fugini
Politecnico di Milano, Italy

Diego Liberati
Italian National Research Council, Italy

Barbara Pes
Università degli Studi di Cagliari, Italy

INTRODUCTION

Collaboration for e-science, namely executing experiments in a cooperative way by sharing data, tools, and expertise towards a common scientific goal, is becoming more and more appealing in a context like the scientific community. In such a context, a critical mass is needed to address very important and complex new questions arising because of the increasing availability of experimental data made possible by continuous technological achievements. An effective collaboration can be set in place by using the increasing empowerment of the Internet to perform such distributed laboratory. By designing a lab environment able to involve accredited actors, public research centers could benefit from technologies and tools, and could become the first promoters and the key players of collaboration initiatives. Moreover, a possible interest by private actors in adhering to such labs should be stimulated in a proactive way by contracting the mutual beneficial and burdens in such a way that both kinds of actors, as well as the civil society (which eventually funds these initiatives) could all take advantage of such arrangements. In addition, privacy concerns that could arise in this type of environment are instead easily granted under appropriate rules that leave public-only scientific data, while keeping both individual and sensitive data and information protected by copyright reserved.

In order to present such initiative as attractive for potential actors, the needed technological tools and applications should be presented as a minimal and non-intrusive tool suite for organizations. To this aim, the technology of Web services (Hey & Fox, 2005) proves very effective, in that the code to be installed at local sites is minimal, easily installable, and easily invoked through a peer-to-peer paradigm. Standard tools coming with a Web service environment should be used, such as XML, UDDI, and WSDL, to exchange data in interoperability environments, to create the services and their public registries, and to communicate in the cooperative environment.

Case studies and best practice in the area of information and communication technology (ICT)-supported collaboration (Tilley et al., 2004) show that overall architectures enabling cooperation should be based on the following logical levels:

1. **Front-end:** Acting as the presentation layer related to the e-experiment
2. **Business logic:** Dealing with collaborative applications
3. **Storage-containing:** A data base of experimental information; an index of experiments, acting as a registry that identifies data and users of the experiment; an *interface* to a set of *informative databases* (for statistical purposes data about experiments, statistics, standard classifications of terms categorized by domains (e.g., medical, biological, remote monitoring), and other reper-

tories and indexes to be consulted in data mining mode (Newman, Ellisman, & Orcutt, 2003).

4. **Interoperability:** For communication towards the centrers adhering to the cooperative environment.

The *management* of experiments between organizations is currently available either in insecure systems, which are appropriate for non-commercial academic collaborations only, or by using proprietary solutions from single vendors that all organisztions participating in the experiment are required to use. This creates a very expensive entry cost for an organization, which is worthwhile only if the collaboration survives for a long period providing substantial benefits. This can be achieved when the existing ICT endowment of each cooperating organization has the potential to interoperate easily.

BACKGROUND

Collaborative environments make use of emerging proposals of networked services that have enabled new types of applications in the field of information systems. These new applications are set in place and executed by several geographically distributed interacting organizations, exchanging data through the network and the Web. Information systems using such paradigms are called *cooperative information systems* (*CoopIS*)—that is, distributed information systems that can be either employed by users of different organizations under a common goal or operate for local activities, hence maintaining their autonomy (CoopIS, 2005).

A core issue of collaborative environments consists of *e-applications* (Mecella & Pernici, 2001), namely orchestrations of e-services provided by different organizations on the Internet. The data exchange and the interleaved execution of processes in such systems bring about issues bound to inter- and intra-organizational structures, to a plurality of actors in the distributed system, and in the heterogeneity of policies existing at the various sites where the portions of the distributed process are executed.

In order to define a framework for cooperative scientific environments, it is necessary to consider their requirements in terms of resource sharing and privacy. In particular, policies have to be defined that constrain the behaviour of system components. Tools for policy

definition and management are becoming available also for dynamic adjustability of applications. Benefits of policy-based approaches include reusability, efficiency, extensibility, and context-sensitivity. Policies are often applied to automate network administration tasks, such as configuration, security, recovery, or quality of service. The adoption of a policy-based approach for controlling the cooperation in an e-science environment is appropriate, since cooperation behaviour can be represented and managed both for policies known at design time, that is, data and resource sharing, or privacy, and for policies that need to be negotiated dynamically among the cooperating processes, such as the quality of services (e.g., performances or delays). It is the responsibility of the experiment manager to specify policies through adequate support tools; for example, using WS Policy with Attachments (Bajàj, Box, Chappell, Curbera, Daniels, Hallam-Baker, et al., 2006), as suggested by standard proposals (W3C, 2004).

The complexity of the *remote components* (i.e., the participative software package at the site of the collaborating actor) that needs to be installed at each adhering to the system, in order to cooperate, has to balance:

- **Performance requirements:** Suggest exporting most of the services to lower the system workload and limit the communications among the systems
- **Privacy requirements:** Prohibit replicating data present on the central system to make them available in the local system
- **Adaptability requirements:** Make the architecture compatible with all the systems used at the various organizations with no substantial modifications
- **Minimal modification requirements:** Impose a limit to the interventions on the applications in the local system
- **Minimal programming efforts:** Required for integrating the existing legacy applications and for further extending the functionalities that have been delivered as the core system portions

Considering also the indications of Web service providers (Tilley et al., 2004), an overall view of the categories of services realizing a collaborative environment can be as follows:

- Management, presentation, administration, data management, and search services are exported at the level of the Web interface and implement functions such as registration, access control, event notification, visualization, and maintenance of information for all users.

- Data access and data integration services are in charge of information management, organization of the data flows (workflow management), of management of user privileges, and security of archives such as:

    ° Information management services, including data search, data aggregation, data mining, and data formatting
    ° Application services, that coordinate the interactions with the system databases

- Collaboration services, implementing the interoperability and communication among all components

MAIN FOCUS

The purpose of the environment for collaboration in e-science is to provide a shared area instrumented with tools where scientists (and possibly also other actors) can make their experiments or share data, and where other scientists can notify their availability to execute a distributed e-experiment. The e-experiment resources (data, tools, and even experts) are the basic data structures of the environment. Data have to be put in common according to both national laws on data privacy and of privacy rules established within the e-science community. Therefore, only some fields of this information can be made public.

On these data, the tools of the environment provide services for collaboration in e-experiment planning, execution, result publications, and system administration functions. Such tools might include Workflow Management Systems, Web applications execution tools, data mining engines, or grid environments, such as Mygrid (http://www.mygrid.org.uk/) (Oinn, Greenwood, Addis, Alpdemir, Ferris, Glover, et al., 2005). Each of the mentioned types of tools proves to provide useful architectural components in order to deploy services for e-science (Hey, 2005):

- Workflows are useful for designing an experiment and for monitoring its execution as various coordinated steps
- Web applications execution tools are helpful for managing the choreography and orchestration of the distributed activities of the experiments in terms of Web services
- Data mining engines are relevant for identifying salient variables, as well as logical and/or mathematical relationships among them
- Grid environments can be a valid support for enhancing the various e-nodes functions in executing specialized tasks to be executed by specific network nodes

In Figure 1, the sequence diagram shows the communication flow needed to set up an e-experiment. The scientists play the role of the *Experiment Designer*, who is in charge of defining the workflow of the experiment activities, no matter where the various parts of the experiment will eventually be executed. They can also play the role of the *Experiment Supervisor*, who checks the feasibility of the experiment from a conceptual point of view (and in subsequent phases not shown here, monitors the results). The *Designer* produces an experiment proposal document (draft of the experiment), composed of various units (algorithms, tools, methods, data). The *Experiment Manager* is in charge of evaluating the proposal in terms of allocation of tasks over the allowable tools of the collaborative network.

After a negotiation phase including both the scientific validity and the physical feasibility (costs, time, nodes, and resources of the collaborative network, and so on) of the experiment, an experiment is generated and is ready for execution.

The designer and the manager actors, mainly involved in the e-experiment design, are described in more detail in the following Figure 2 (designer) and Figure 3 (manager), while in Figure 4 some details about possible tools involved in the execution of the experiment are described.

In particular, the Experiment Designer (Figure 2) produces an experiment proposal in terms of a set of work flows (WF) describing the basic steps of the experiment. The proposal contains the general scientific aim (specified possibly in collaboration with other context experts and scientists) of the experiment, of its methodologies, and of their logical flow declaring also

Figure 1. Sequence diagram of e-collaboration (main actors)

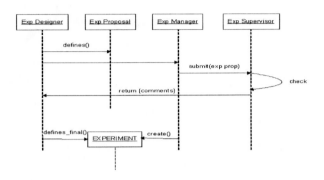

Figure 2. Use case regarding the experiment designer basic tasks

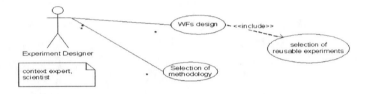

Figure 3. Use case regarding the experiment manager basic tasks

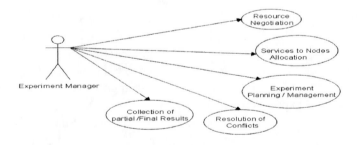

the needed resources (data, tools, experts). In defining the WF, existing portions of previously executed experiments can be reused, thus improving the effectiveness of the environment. A public registry indexing (in the UDDI style) is assumed of reusable experiments is available in the collaborative environment.

In the experiment proposal, the designer can specify also some constraints (e.g., time constraints, costs, data privacy, quality, etc.) that may be defined using a policy document treating non-functional requirements in the WS-Policy attachment style.

The experiment manager (Figure 3), on the basis of the submitted experiment proposal, initiates a negotiation of the network to evaluate the feasibility of the experiment. His evaluation is forwarded to the Supervisor, specifying a detailed proposal of services allocation to the nodes, as well as a proposal of project management. Upon approval, the manager definitely allocates services to nodes and manages the experiment, taking also care of resolution of possible conflicts via negotiation of non-functional parameters contained in the WS-Policy attachments (e.g., temporal or quality

Figure 4. Use case regarding some tools of the collaborative environment

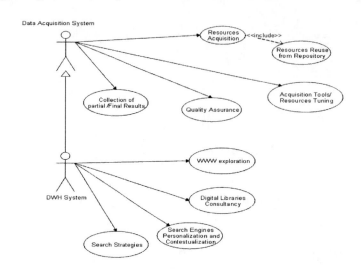

constraints that should be respected by a node, and possibly a task should be reallocated). He is also in charge of collecting the results at each defined milestone of the WF.

Other tasks are possible for the Manager: the privacy authority in the collaborative network, since he a third party with respect to all the involved scientists, or the quality insurance manager, responsible for respect of collaboration and quality contracts in the environment.

As an example, it is shown how tools of the collaborative environment can be modelled as actors (Figure 4). A data acquisition system is modelled as an actor that accesses a common repository of resources (tools, data, publications, and so forth), develops or tunes other specific tools, ensures the quality of acquired resources, and finally exposes the experiment results in public areas.

As a specialization of the data acquisition system, the data warehouse (DWH) system is shown (see Figure 4) acting as an exploration coordinator in the WWW domain, and as an assistant for the users in developing their own strategies for consulting libraries or digital distributed information sources, via search engines personalization, knowledge, and tools contextualization.

FUTURE TRENDS

Issues to be addressed, beyond the core issue of setting in place the proposed collaboration environment, arise

in various areas, such as policy enforcement, trust and security support, collaboration correctness monitoring, quality of service monitoring, transaction logging, and so on. In fact, the future of e-science applications requires the ability to perform long-lived, peer-to-peer collaborations between the participating services, within or across the domains of the organizations. Web service specifications offer a communication bridge between the heterogeneous computational environments used to develop and host applications; such Web service specifications can be enhanced, as in using semantic annotations, to support the declaration of context-related information in order to make this approach a basis for implementation and deployment of the pluggable "experiment handlers" framework.

The approach described in this article has been presented at a conceptual level, and many decisions still need to be made. Standard implementation languages, such as Java, C++, BPEL, can be used to implement the approach and to address the main issues in this area. It is a promising sign that various specification efforts in the area of Web applications take into account collaboration issues and drive the effort to solve them in a standard interoperable manner, since support provided by open standards is crucial for adoption of such solutions. For example, choreography languages (e.g., Web Services—choreography description language [WSCDL, 2004], just to mention one) that describe peer-to-peer collaborations of Web services participants, allow one to define, from a global viewpoint, common and complementary observable behavior of experiment tasks being executed on different nodes, where the order of message exchanges results in the accomplishment of a common "business" goal.

The requirements for distributed experiment management described here can be met by current Web service specifications, but they require secure, stable and interoperating implementations from a variety of information technology vendors. This is a point that encourages ICT vendors to produce secure, stable, and interoperating implementations of the required Web services specifications, which could support truly interoperable collaboration of reusable services (Brown, 2003).

Application scenarios do span over experiment regarding, for instance, brain dynamics investigation, geo-informatics, drug discovery, as well as improving our current bio-informatics application concerning treatment of micro arrays data related to myeloid and lymphoid leukaemia available at the MIT repository (Golub, 1999), processed by both unsupervised Principal Direction Divisive Partitioning (Garatti, Bittanti, Liberati D., & Maffezzoli, 2007) and supervised Adaptive Bayesian Networks (Bosin, Dessì, Liberati, & Pes, 2006).

CONCLUSION

This article has introduced an approach to environments for collaboration in distributed scientific experiments. Concepts are based on the assumption that e-experiments can be mapped to networked e-services that correspond to tasks of the experiment, using both rich service universal descriptions (UDDI or extensions thereof) and end-point specific tools. In both cases, it is possible to configure and execute a distributed experiment and manage its artefacts in a cooperative network.

This possibility is attractive from many points of view, the most important perhaps being clean separation of experiment management, design, and execution, as well as the definition of user roles in the collaborative environment.

The issues debated in this article regard the role of cooperative environments for management of distributed scientific experiments. Such debate regards the definition of models and tools for managing scientific experiments as cooperative services and the grid, pos-

sibly under an evolutionary approach allowing one to adapt the support architecture to accommodate new tools, actors, and classes of experiments (Atkison et al., 2005).

The article has presented basic issues that are quite general and flexible, being adaptable to different contexts and tools. In particular, the nodes of the cooperative environment can be flexibly composed to formalize and execute scientific experiments, making use of Web services, of workflow management tools, and of grid based computation.

Given the challenge of evaluating the effects of applying emerging Web service and grid technology (De Fanti, de Laat, Mambretti, Neggers, & St.Arnaud, 2003; Power, 2005) to the scientific community, our evaluation of the proposal, performed up to now using Web services (Bosin, Dessì, Fugini, Liberati, & Pes, 2006), takes a flexible and multi-faceted approach: It aims at assessing task-user-system functionality and can be extended incrementally according to the continuous evolution of scientific cooperative environments, by incrementally including new classes of experiments.

REFERENCES

Atkison, M., De Roure, D., Dunlop, A., Fox, G., Henderson, P., Hey, P., et al. (2005). Web service grids: An evolutionary approach. *Concurrency and Computation: Practice and Experience, 17*(2-4), 377-389.

Bajaj, S., Box, D., Chappell, D., Curbera, F., Daniels, G., Hallam-Baker, P., et al. (2006). *Web services policy 1.2—Framework (WS-Policy)*. www..w3.org/Submission/WS-Policy/

Bosin, A., Dessì, N., Fugini, M., Liberati, D., & Pes, B. (2006). Applying enterprise models to design cooperative scientific environments. *Lecture Notes in Computer Science, 3812* (pp. 281-292). Springer-Verlag.

Bosin, A., Dessì, N., Liberati, D., & Pes, B. (2006). Learning Bayesian classifiers from gene-expression microarray data. *Lecture Notes in Artificial Intelligence, 3849* (pp. 297-304). Springer-Verlag.

Brown, M. (2003). Blueprint for the future of high-performance networking. *Communications of the ACM, 46*(11), 30-33.

CoopIS. (2005). *Proceedings 6th International Conference on Cooperative Information Systems (CoopIS)*, November, Cyprus, Springer Verlag.

De Fanti, T., de Laat, C., Mambretti, J., Neggers, K., & St. Arnaud, B. (2003). Translight: A global-scale LambdaGrid for e-science. *Communications of the ACM, 46*(11), 34-41.

Garatti, S., Bittanti, S., Liberati, D., & Maffezzoli, P. (2007). An unsupervised clustering approach for leukemia classification based on DNA micro-arrays data. *Intelligent Data Analysis, 11*(2), 175-188.

Golub, T. R., Slonim, D. K., Tamayo, P., Huard, C., Gaasenbeek, M., Mesirov, J. P., et al. (1999). Molecular classification of cancer: Class discovery and class prediction by gene expression monitoring. *Science, 286*, 531-537.

Hey, T., & Fox, F. (2005). Special issue: Grids and Web services for e-science. *Concurrency and Computation: Practice and Experience, 17*(2-4), 317-322.

Mecella, M., & Pernici, B. (2001). Designing wrapper components for e-services in integrating heterogeneous systems. *Very Large Data Base Journal*, Special Issue on e-Services.

Newman, H., Ellisman, M., & Orcutt J. (2003). Data-intensive for e-Science. *Communications of the ACM, 46*(11), 69-75.

Oinn, T., Greenwood, M., Addis, M., Alpdemir, N., Ferris, J., Glover, K., et al. (2005). Taverna: Lessons in creating a workflow environment for the life sciences. *Concurrency and Computation: Practice and Experience, 18*(10), 1067-1100.

Power, D. J., Politou, E. A., Slaymaker, M. A., & Simpson, A. C. (2005). Towards secure grid-enabled healthcare. *Software: Practice and Experience, 35*(9), 857-871.

Tilley, S., Gerdes, J., Hamilton, T., Huang, S., Müller, H., Smith, D., et al. (2004). On the business value and technical challenges of adopting Web services. *Journal of Software Maintenance and Evolution: Research and Practice, 16*(1-2), 31-50.

WSCDL. (2004). *Web Services Choreography Description Language Version 1.0, W3C Working Draft*. Retrieved April 27, from http://www.w3.org/TR/ws-cdl-10/

W3C. (2004, October 12-13). Workshop on Constraints and Capabilities for Web Services, Oracle Conference Center, Redwood Shores, CA, USA

KEY TERMS

Bio-Informatics: The application of the ICT tools to advanced biological problems, like transcriptomics and proteomic, involving huge amounts of data.

Cooperative Information Systems: Independent, federated information systems that can either autonomously execute locally or cooperate for some tasks towards a common organizational goal.

E-Science: Modality of performing experiments in silico in a cooperative way by resorting to information and communication technology (ICT).

Grid Computing: Distributed computation over a grid of nodes dynamically allocated to the process in execution.

Interoperability: Possibility of performing computation in a distributed heterogeneous environment without altering the technological and specification structure at each involved node.

Web Services: Software paradigm enabling peer-to-peer computation in distributed environments based on the concept of "service" as an autonomous piece of code published in the network.

Workflow: Stream of information within the network related to the accomplishment of every single orchestrated task.

Sharing Information Efficiently in Cooperative Multi-Robot Systems

Rui Rocha
University of Coimbra, Portugal

Jorge Dias
University of Coimbra, Portugal

INTRODUCTION

Multi-robot systems (MRS) are sets of intelligent and autonomous mobile robots that are assumed to cooperate in order to carry out collective missions (Arai, Pagello, & Parker, 2002; Cao, Fukunaga, & Kahng, 1997; Rocha, Dias, & Carvalho, 2005). Due to the expendability of individual robots, MRS may substitute humans in risky scenarios (Maimone et al., 1998; Mataric & Sukhatme, 2001; Parker, 1998; Thrun et al. 2003). In other scenarios, they may relieve people from collective tasks that are intrinsically monotonous and repetitive. MRS are the solution to automate missions that are either inherently distributed in time, space, or functionality.

MRS involve the distribution of sensors, computation power and mission-relevant information. This inherent distribution is both an opportunity and a challenge. On one hand, it endows MRS with interesting features, such as space and time distribution, managing complexity through distribution, distribution of risk and increased robustness (Arkin & Balch, 1998). On the other hand, these potential advantages and their utility are to a greater extent dependent on the effective cooperation among robots when performing some collective mission (Rocha, 2005).

Since information is intrinsically distributed, cooperation requires, in turn, efficiently sharing information through communication (Rocha et al., 2005). A method for efficiently sharing information within a MRS is herein presented, which is based on an information utility criterion (Rocha et al., 2005). This concept is illustrated on MRS whose mission is to build cooperatively volumetric maps.

Robotic Mapping

Robotic mapping addresses the problem of acquiring spatial models of physical environments with mobile robots equipped with distance sensors, such as cameras, range finders and sonars. Usually the map is not the goal itself and those mobile robots are used to safely navigate within the environment and perform other useful tasks that require an up-to-date map of the environment (e.g., search and rescue). But mobile robots may also be used for building detailed maps of indoor environments (Martin & Moravec, 1996; Stachniss & Burgard, 2003), being particularly useful on mapping missions of hazardous environments for human beings, such as underground mines (Thrun et al., 2003) or nuclear facilities (Maimone et al., 1998).

As sensors have always limited range, are subject to occlusions and yield noisy measurements, mobile robots have to navigate through the environment and build the map iteratively. Some key challenges in this context are the sensor modeling problem, the representation problem, the registration problem and the exploration problem (Thrun, 2002). This article focuses on efficiently sharing sensory information within a team of mobile robots, so as to build a volumetric map in less time than a single robot.

Sharing Information within Multi-Robot Systems

Most of the work about multi-robot systems (MRS) has been devoted to the definition of different architectures (Gerkey & Mataric, 2002; Mataric et al., 2001; Parker, 1998) that rule the interaction between the behaviors of individual robots. Communication is a central issue of MRS because it determines the possible modes of interaction among robots, as well as the ability of robots to build successfully a world model, which serves as a basis to reason and act coherently towards a global system goal. Communication may appear in three different forms of interaction (Cao et al., 1997): (1) *via environment*, using the environment itself as the communication medium (stigmergy); (2) *via sensing*, when an agent knowingly uses its sensing capabilities to observe and perceive the other robots' actions; and (3) *via communication*, using a communication channel to explicitly exchange messages among robots, thus compensating perception limitations.

This article presents a distributed group architecture which endows robots with a cooperation scheme whereby explicit communication is efficiently used to increase the robot's individual awareness based on a criterion of information utility (Rocha et al., 2005).

PROBABILISTIC VOLUMETRIC MAPS

This section outlines a grid-based probabilistic framework for representing and updating volumetric maps. Further details can be found in Rocha et al. (2005).

Architecture Model

The functional blocks of a mobile robot carrying out a volumetric mapping mission (Rocha et al., 2005) is depicted in Figure 1. The mobile robot's platform is assumed to have a sensor, a localization module and an actuator. The sensor provides new sets of vectors V_{k+1} where obstacles are detected from the current sensor's pose $Y(t)=(\mathbf{x}(t),\mathbf{a}(t))$. The localization module gives the sensor's pose $Y(t)$, including position $\mathbf{x}(t)\in R^3$ and attitude $\mathbf{a}(t)$. The actuator changes the sensor's pose (robot's pose) accordingly with a new selected exploration viewpoint Y^S. New data from the robot's sensor is associated with its current pose, given by the localization module, to form a new batch of measurements $M_{k+1} = (\mathbf{x}_{k+1}, V_{k+1})$. Then, index k is incremented and the new batch of measurements becomes the current batch M_k. The memory of measurements is updated as $\mathrm{M}_k = \mathrm{M}_{k-1} \cup$. The previous map $P(C \mid \mathrm{M}_{k-1})$ is updated upon the new batch of measurements M_k, which yields the current map $P(C \mid \mathrm{M}_k)$. The current map is used to choose a new target pose Y^S which is the reference input to the robot's actuator.

Volumetric Model

Rocha et al. (2005) proposed a grid-based model to represent volumetric maps with an explicit representation of uncertainty through the entropy concept. It is based on coverage maps (Stachniss et al., 2003), which are grid-based probabilistic maps (Moravec et al., 1985) wherein the occupancy of a cell is modeled through a continuous random variable—the cell's coverage. The volumetric model assumes that a 3-D discrete grid Y is defined, which divides the robotic team workspace into equally sized voxels (cubes) with edge $\varepsilon\in R$ and volume ε^3 (Figure 2). The portion of the volume of

Figure 1. Block diagram showing the relation between different parts of the process and the resources of a given mobile robot the fleet

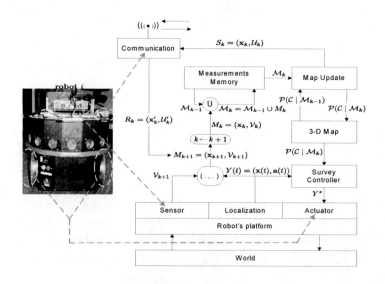

Figure 2. Volumetric discrete grid: (a) the grid divides the workspace into equally sized voxels; (b) the coverage C_l of each voxel $l \in Y$ with edge ε, given the sequence M_k of k batches of measurements, is modeled through a probability density function $p(c_l \mid M_k)$.

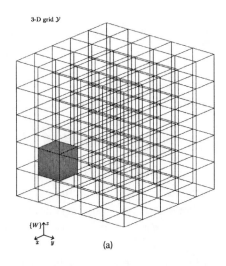
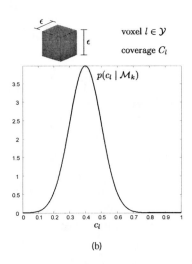

(a) (b)

voxel $l \in Y$ which is covered (occupied) by obstacles is modeled through the continuous random variable C_l taking values c_l in the interval $0 \leq c_l \leq 1$, and having $p(c_l)$ as its probability density function (pdf). The objective when building a map is to obtain for each voxel $l \in Y$ an estimate as accurate as possible about its coverage C_l.

The map's entropy is used to define the mission execution time. After defining a given map's entropy threshold H_{th}; that is, the minimum map's quality that robots must accomplish at the end of the mapping mission, the mission execution time $t_{k_{max}}$ is such that:

$$H(t_{k_{max}}) \leq H_{th} \wedge \forall_k < k_{max}, k \in N_o, H(t_k) > H_{th}. \quad (4)$$

It is associated with the k_{max}-th batch of measurements; that is, the last batch of measurements acquired by the robot with the lowest entropy at the end of the mission.

Figure 4 shows several versions of a map along a mapping mission with two robots.

EFFICIENTLY SHARING INFORMATION

In the absence of a centralized controller, multi-robot systems involve the distribution of control and sensory data gathered from sensors. Therefore, in order to attain a coherent and useful team behavior, a distributed control scheme is required so as to barter efficiently resources and, as well, information (Rocha, 2005). The main advantage of distributing control and data is to easily scale up the robotic systems to an arbitrary number of robots, while maintaining the system's reliability and robustness.

Since information is distributed, efficiently sharing information among robots is crucial. Although there are other forms of conveying information, distributed control relies to a greater extent on explicit communication. Rather than addressing the *communication structure* in multi-robot systems (Balch & Arkin, 1994; Stone & Veloso, 1999; Gerkey et al., 2002; Ulam et al., 2004), it is worth considering the *communication content;* that is, "What is useful (task-relevant) to be communicated?" The question requires in turn assessing information utility (Rocha et al., 2005). The goal behind these questions is to avoid communicating redundant information so as to use efficiently communication.

Information Utility

Whenever a robot gets a new batch of measurements M_k, the associated information utility can be measured in terms of a decrease of the map's entropy $H(C)$. The map's entropy is a measure of the map's uncertainty and its decrease within a period of time is a measure of the information utility of the measurements gathered within the same period of time, in terms of their utility on improving the map's accuracy (Rocha et al., 2005).

Sharing Useful Sensory Information

Although the block diagram depicted in Figure 1 refers to a single robot, it is sufficiently general to represent a multi-robot system. The interaction of a robot with other robots is represented through the communication block and its associated data flow.

Whenever a robot gets a batch of measurements $M_k = (\mathbf{x}_k, V_k)$, it sends to its teammates a subset of measurements $S_k = (\mathbf{x}_k, U_k)$ through the communication module, containing the most useful data $U_k \subseteq V_k$ which has just been acquired from its sensor. This module can also provide the robot with batches of measurements $R_k = (\mathbf{x}'_k, U'_k)$ given by other robots and the map is updated accordingly. Cooperation among robots arise because of this reciprocal interaction (Rocha et al., 2005).

The set:

$$\mathcal{U}_k = \{\vec{\mathbf{u}}_{k,1}, \ldots, \vec{\mathbf{u}}_{k,s_k}\} \subseteq \mathcal{V}_k \qquad (7)$$

contains s_k measurements selected to be communicated. The sensor's position $\mathbf{x}_k$ from where those measurements were gathered is also sent, since it is required for registering those measurements in the local map of other robots.

Whenever a robot receives a batch of u_k communicated measurements $R_k = (\mathbf{x}'_k, U'_k)$, it updates its local map as if measurements U'_k would have been gathered by its own sensor when located at position $\mathbf{x}'_k$.

As communication channels have always limited capacity, a robot acting as information provider should limit the amount of communicated data, *i.e.* select the most useful measurements gathered from its own sensors. Equation (5) provides a way to do this: each robot computes this equation in order to assess the information utility of measurements $\vec{\mathbf{v}}_{k,i} \in V_k$ and classify them by utility.

Let define s_{kmax} as being the maximum number of allowable communicated measurements at a given time instant. Let also define I_{min} as being the minimum allowable information utility for a communicated measurement.

The set (7) is built in such a way that the proposition:

$$(8)$$
$$(s_k \leq s_{k_{max}} \wedge$$
$$s_k < s_{k_{max}} \Rightarrow \forall_{\vec{\mathbf{v}}_{k,z} \in \mathcal{V}_k \setminus \mathcal{U}_k},\ I_{k,z} < I_{min} \wedge$$
$$\forall_{\vec{\mathbf{u}}_{k,j} \in \mathcal{U}_k},\ I_{k,j} \geq I_{min} \wedge \forall_{\vec{\mathbf{v}}_{k,w} \in \mathcal{V}_k \setminus \mathcal{U}_k},\ I_{k,w} \leq I_{k,j})$$

is true. It is true if the batch size is less than the maximum size and it includes the most useful measurements.

RESULTS AND DISCUSSION

The architecture model shown in Figure 1 was implemented in the mobile robots depicted in Figure 3, which use stereo-vision as range sensor. They were used to carry out experiments aiming at studying the influence of the information sharing parameters I_{min} and s_{kmax} on the team's performance, through the comparison of the mission execution time t_{kmax} with different values for those parameters (Rocha et al., 2005). The robots started each experiment with a maximum entropy map and followed the exploration method proposed by Rocha et al. (2005), which is based on the uphill gradient of the map's entropy, in order to explore the environment until the entropy threshold $H_{th} = 500$ was attained.

Table 1 summarizes the obtained results with the team of two robots, which are however extensible and can be generalized to teams having an arbitrary number of robots because the robots' program is intrinsically scalable to any team size. The fourth column shows the ratio between the mission execution time $t_{kmax}(2)$ with two robots and $t_{kmax}(1)$ with one robot. Given that voxels' coverage beliefs were always Gaussians, the values used for I_{min}, {0, 0.00723, 0.01450, 0.07400, 0.15200, 0.32193}, meant an average reduction on the standard deviation of the influenced voxels by a measurement of at least {0%, 0.5%, 1%, 5%, 10%, 20%}, respectively. The fifth column shows the total number of measurements m_T gathered by a robot along the mission, which is given by equation (9).

$$m_T = \sum_{k=1}^{k_{max}} m_k. \qquad (9)$$

The sixth column shows the total number of received measurements from the other robot u_T, which is computed through equation 10.

$$u_T = \sum_{k=1}^{k_{max}} u_k.$$

(10)

In each experiment (row of the table), the results refer to the robot that first attained the entropy threshold H_{th}; that is, the robot having the best map at the end of the mission. Figure 4 presents an example of the maps obtained by the two robots along a 3-D mapping mission. As it can be observed, robot 2 held the best map for the instant times represented in the figure, which means that robot 2 reached H_{th} first. The time $t_k(1)$ that a single robot would need to obtain the represented maps is also shown, so as to better understand the reduction of the mission execution time yielded by a team of cooperative mobile robots.

Advantages Provided by Cooperation

The graph on the left of Figure 5 compares the map's entropy $H(t)$ for the single robot case and for the fastest experiment with two robots (fourth row of Table 1). It shows a non-linear increase of the mission execution time with a decrease of the map's entropy. It also shows that robots' cooperation accelerated the reduction of the map's entropy and led to a reduction of 28% in t_{kmax}. As robots shared useful measurements through communication, each robot was able to integrate in its map a greater number of measurements per time unit and achieved a faster reduction of its map's entropy. The graph on the right of Figure 5 shows that although the two values of m_T were similar, measurements were obtained within time intervals t_{kmax} quite different.

FUTURE TRENDS

Building a volumetric map is essentially an exploration mission. The goal is to completely sense the environment so as to accumulate sufficient sensory evidence and build a consistent spatial model with low uncertainty; that is, a map with low entropy. The previous section showed a performance gain which is far away from a linear improvement with team size. This is mainly due to the lack of coordination of the robots' exploration actions and thus it is worth addressing this problem in the future.

Another important future direction is to demonstrate the application of the information sharing method herein presented to other robotics application domains than robotic mapping and, as well, to domains outside robotics. The essential problem—efficiently sharing information—is indeed relevant to other domains than robotics. For instance, human organizations and human societies involve complex cooperative interactions supported on some flow of information. Redundancy, consistency, and information utility are also crucial issues to these complex societal systems.

CONCLUSION

This article addressed the problem of efficiently sharing information within multi-robot systems in the context of robotic mapping. After briefly presenting a probabilistic framework which allows representing and updating a volumetric map upon range measurements, information utility was formulated through mutual information, an

Figure 3. Mobile robots used in the mapping experiments: (a) Scout mobile robots (top) equipped with stereo-vision sensors (bottom); (b) a stereo image pair (top) and its associated disparity (bottom-left) and depth map (bottom-right)

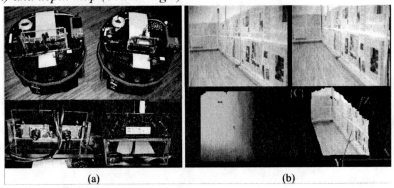

(a)　　　　　　　　　　　　　(b)

Table 1. Results obtained within experiments with two robots and different parameters ruling the information sharing

$s_{k_{max}}$	I_{min}	$t_{k_{max}}$	$\dfrac{t_{k_{max}}(2)}{t_{k_{max}}(1)}$	m_T	u_T	
500	0.01450	8483	0.94	2795351	74729	3 %
1000	0.01450	8387	0.93	2726837	135661	5 %
1750	0.01450	7332	0.81	2447091	184550	8 %
2500	0.01450	6530	0.72	2375273	207636	9 %
5000	0.01450	7955	0.88	2643728	271612	10 %
20000	0	9450	1.04	3192788	1134455	36 %
20000	0.00723	7563	0.84	2453021	457390	19 %
20000	0.01450	6571	0.73	2345844	332270	14 %
20000	0.07400	7007	0.77	2676612	128345	5 %
20000	0.15200	7301	0.81	2595398	59499	2 %
20000	0.32193	7727	0.85	2930155	27323	1 %

Figure 4. Maps' evolution along a volumetric mapping mission with two robots. Each row shows a snapshot of the map of each robot at a different instant time t_k and entropy level $H(C\,|M_k)$. The time $t_k(1)$ that a single robot would need to obtain a map with the same entropy is shown on the bottom-right of the maps of robot 2. The maps' resolution is $\varepsilon = 0.1$ m. The pictures' scale is such that each represented arrow is equivalent to a real length of 1 m. For the presented case, $s_{kmax} = 2500$ and $I_{min} = 0.1520$.

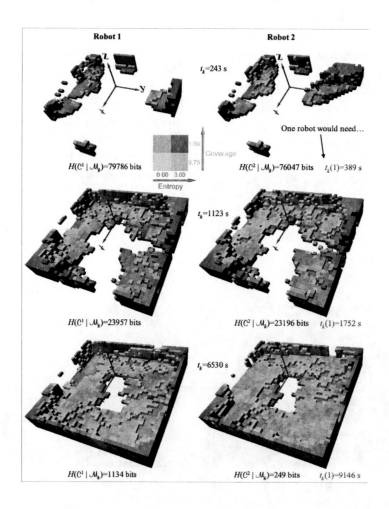

S

Figure 5. Comparison of a volumetric mapping mission using a single robot or two robots: Entropy of the map along the mission (left) and cumulative number of processed measurements along the mission (right). For the presented case, $I_{min} = 0.0145$ and $s_{kmax} = 2500$.

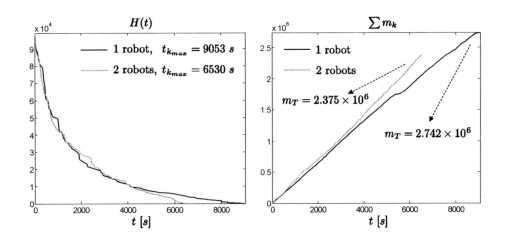

information theoretic concept. Robots use the information utility measure to communicate to other robots only sensory information that is indeed useful.

Results obtained within experiments with mobile robots equipped with stereo-vision demonstrated the validity of the presented formalism. More specifically, it was proven that a team of cooperative mobile robots is able to build a map of the environment in much less time than a single robot, as a result of using judiciously communication resources.

REFERENCES

Arai, T., Pagello, E., & Parker, L. (2002). Special issue on advances in multirobot systems. *IEEE Transactions on Robotics and Automation, 18*(5), 655-864.

Arkin, R., & Balch, T. (1998). Cooperative multiagent robotic systems. In D. Kortenkamp, R. P. Bonasso, & R. Murphy (Eds.), *Artificial intelligence and mobile robots* (pp. 277-298). Cambridge, MA: MIT/AAAI Press.

Balch, T., & Arkin, R. (1994). Communication in reactive multiagent robotic systems. *Autonomous Robots, 1*(1), 27-52.

Burgard, W., Moors, M., Stachniss, C., & Schneider, F. (2005). Coordinated multi-robot exploration. *IEEE Transactions on Robotics, 21*(3), 376-386.

Cao, Y., Fukunaga, A., & Kahng, A. (1997). Cooperative mobile robotics: Antecedents and directions. *Autonomous Robots, 4,* 1-23.

Cover, T., & Thomas, J. (1991). *Elements of information theory.* Wiley.

Gerkey, B., & Matari´c, M. (2002, October). Sold!: Auction methods for multirobot coordination. *IEEE Transactions on Robotics and Automation, 18,* 758-768.

Maimone, M., Matthies, L., Osborn, J., Rollins, E., Teza, J., & Thayer, S. (1998). A photo-realistic 3-D mapping system for extreme nuclear environments: Chernobyl. In *Proceedings of the IEEE/RSJ International Conference on Intelligent Robots and Systems 1998* (pp. 1521-1527). IEEE Press.

Martin, M., & Moravec, H. (1996). *Robot evidence grids* (Tec. Rep. No. CMU-RI-TR-96-06). Pittsburgh, PA: Carnegie Mellon University, Robotics Institute.

Mataric, M. (1992). Minimizing complexity in controlling a mobile robot population. *Proceedings of the IEEE International Conference on Robotics and Automation 1992* (pp. 830-835). IEEE Press.

Mataric, M., & Sukhatme, G. (2001). Task-allocation and coordination of multiple robots to planetary exploration. Proceedings of the 10th International Conference on Advanced Robotics 2001.

Merino, L., Caballero, F., Dios, J. de, & Ollero, A. (2005, April). Cooperative fire detection using unmanned aerial vehicles. *Proceedings of the IEEE International Conference on Robotics and Automation 2005* (pp. 1896-1901). Barcelona, Spain: IEEE Press.

Moravec, H., & Elfes, A. (1985). High resolution maps from wide angle sonar. *Proceedings of IEEE International Conference on Robotics and Automation 1985.* IEEE Press.

Parker, L. (1995). The effect of action recognition and robot awareness in cooperative robotic teams. *Proceedings of the IEEE/RSJ International Workshop on Intelligent Robots and Systems 1995* (pp. 212-219). IEEE Press.

Parker, L. (1998). ALLIANCE: An architecture for fault-tolerant multi-robot cooperation. *IEEE Transactions on Robotics and Automation, 14*(2), 220-240.

Rocha, R., Dias, J., & Carvalho, A. (2005, December). Cooperative multi-robot systems: a study of vision-based 3-D mapping using information theory. *Robotics and Autonomous Systems, 53*(3/4), 282-311.

Stachniss, C., & Burgard, W. (2003). Mapping and exploration with mobile robots using coverage maps. *Proceedings of the IEEE/RSJ International Conference on Intelligent Robots and Systems 2003* (pp. 467-472). IEEE Press.

Stone, P., & Veloso, M. (1999). Task decomposition, dynamic role assignment, and low-bandwidth communication for real-time strategic teamwork. *Artificial Intelligence, 110*(2), 241-273.

Thrun, S. (2002). Robotic mapping: a survey. In G. Lakemeyer & B. Nebel (Eds.), *Exploring artificial intelligence in new millenium* (pp. 1-36). Kaufmann.

Thrun, S., Hahnel, D., Ferguson, D., Montermelo, M., Riebwel, R., Burgard, W., et al. (2003, September). A system for volumetric robotic mapping of underground mines. *Proceedings of IEEE International Conference on Robotics and Automation 2003* (pp. 4270-4275). IEEE Press.

Ulam, P., & Arkin, R. (2004). When good comms go bad: Communications recovery for multi-robot teams. *Proceedings of the IEEE International Conference on Robotics and Automation 2004* (pp. 3727-3734). IEEE Press.

KEY TERMS

Awareness: The extent to which a robot is aware of the state and goals of other robots of a multi-robot system.

Communication: Interaction involving two or more entities that exchange information, especially information that is relevant to the involved entities.

Cooperation: Joint behavior of a group of similar but not necessarily homogeneous entities which interact through different forms of communication, whether implicit or explicit, so as to barter resources and information and carry out collective missions that are not usually feasible by a single entity, or wherein a team may attain better performance.

Distributed Control: A control paradigm for multirobot systems whereby every robot participates in the team's decisions, in the absence of any central controller or hierarchy.

Information Utility: How much a piece of information is relevant to the context it refers to and how much it differs from other similar pieces of information (non-redundancy) and contributes to reduce uncertainty.

Multirobot System: A set of intelligent and autonomous mobile robots that are assumed to cooperate in order to carry out collective missions.

Volumetric Map: A representation model which integrates noisy measurements obtained with range sensors at different instant times or locations, so as to accumulate statistical evidence about the occupancy of a 3-D environment.

Small Business Collaboration Through Electronic Marketplaces

Yin Leng Tan
The University of Manchester, UK

Linda Macaulay
The University of Manchester, UK

INTRODUCTION

It is widely recognized that small businesses with less than 50 employees make significant contributions to the prosperity of local, regional, and national economies. They are a major source of job creation and a driving force of economic growth for developed countries like the USA (Headd, 2005; SBA, 2005), the UK (Dixon, Thompson, & McAllister, 2002; SBS, 2005), Europe (European Commission, 2003), and developing countries such as China (Bo, 2005). The economic potential is further strengthened when firms collaborate with each other; for example, formation of a supply chain, strategic alliances, or sharing of information and resources (Horvath, 2001; O'Donnell, Cilmore, Cummins, & Carson, 2001; MacGregor, 2004; Todeva & Knoke, 2005). Owing to heterogeneous aspects of small businesses, such as firm size and business sector, a single e-business solution is unlikely to be suitable for all firms (Dixon et al., 2002; Taylor & Murphy, 2004a); however, collaboration requires individual firms to adopt standardized, simplified solutions based on open architectures and data design (Horvath, 2001). The purpose of this article is to propose a conceptual e-business framework and a generic e-catalogue, which enables small businesses to collaborate through the creation of an e-marketplace. To assist with the task, analysis of data from 6,000 small businesses situated within a locality of Greater Manchester, England within the context of an e-business portal is incorporated within this study.

BACKGROUND

Small businesses are an important driving force of economic growth and job creation throughout the world. A number of studies (Horvath, 2001; O'Donnell

et al., 2001; MacGregor, 2004; Todeva & Knoke, 2005) show that when firms collaborate or network with each other on a venture, the potential economic and business benefits can be enhanced. The possible network opportunities with other firms include but are not limited to:

1. Collaboration with other businesses to purchase items such as fuel and raw materials, and hence leverage collective buying power in order to negotiate a better deal (Wang & Archer, 2004)
2. Collaboration with other businesses to offer complementary goods in order to increase sales or to enter new markets (Wang & Archer, 2004)
3. Collaboration with other businesses to share information, such as product information, customer demand, transaction information, and inventory information (Ovalle & Marquez, 2003)
4. Liaison with other complementary service businesses to jointly bid for bigger contracts and hence enabling small business to compete with larger counterparts (MacGregor, 2004)
5. Liaison with other similar businesses to jointly bid for a bigger contract than they are able to fulfil by themselves
6. Form collaborative buyer-supplier relationships

Despite government initiatives and support to promote adoption of information collaboration technology (ICT) in small firms, earlier studies show that ICT adoption by small businesses is still very low with a number of barriers to adoption being identified (Dixon et al., 2002; European Commission, 2002; Weiss, 2002; Fillis & Wagner, 2004; Stockdale & Standing, 2004; Taylor & Murphy, 2004a, 2004b; MacGregor & Vrazalic, 2005). Further, small firms are heterogeneous in nature, therefore a single e-business solution is unlikely to be applicable to all firms and treating e-business as a homogeneous concept is probably a mistake (Dixon et al., 2002; Taylor & Murphy, 2004a; Fitzgerald, Papazafeiropoulou, Piris, & Serrano, 2005). In addition, supply chains with buyers and suppliers are not homogeneous (McIvor, Humphreys, & McCurry, 2003). Findings from McIvor et al. suggest that the barriers to the adoption of supply chain systems do not lie primarily with the technology but with the business processes itself. The effective implementation of e-business to support buyer-supplier relationships and to optimize the value chain requires that the e-business application is fully integrated into both the buyer's and the supplier's business architecture and technology infrastructure (McIvor et al., 2003). It is therefore crucial that collaboration technology infrastructure should include the following features: open and low cost connectivity, large and flexible data storage, systems and channel integration, high security, self-service functionalities based on open architectures, and data schemes (Horvath, 2001).

THE E-BUSINESS FRAMEWORK

Doing business is a chain of collaborative processes; a single firm can be a buyer for a business but also a seller for another business, therefore interactions among buyers, suppliers and trading partners are required (Adams, Koushik, Vasudeva, & Galambos, 2003). A retail trade can utilize a one-to-many e-business solution to reach more audiences and maximize profits, whilst at the same time using a many-to-one e-procurement

system to streamline sourcing processes with its trading partners. Firms can also employ a many-to-many e-marketplace to achieve the above. However, in e-marketplace, resources of multiple firms can be pooled together, which cannot be achieved by the one-to-many and many-to-one e-business solutions.

This study draws on the work of Stone (2003), who presents the six "Internet States" of IBM's e-business evolution, and Martin and Matley (2001), who present a DTI e-adoption ladder, which is adapted from the Cisco-led Information Age Partnership study on e-commerce in small business. These two models have been adopted in this study as they highlight the transformational aspects of e-business technology; with a particular focus on the "transact" and "integration" stages.

To further assist with the development of a conceptual e-business framework and a generic e-catalogue, we utilize analysis of data from 6,000 small businesses situated within Tameside, Greater Manchester, England together with the vision of economic growth that could result from greater collaboration, within the context of the Tameside Business Portal.[1] According to Dixon et al. (2002), wholesaling and retailing are the two sectors more likely to adopt ICT than other sectors. In our data, as illustrated in Figure 1, the wholesale/retail trade is in fact the biggest sector within Tameside, with a total of 1,780 (29%) firms. The e-business framework was thus designed to mainly support these two sectors. Nevertheless, these businesses are mixed in nature; therefore, an e-business framework able to support more than one e-business solution is impera-

Figure 1. Business sectors and number of firms in Tameside (Source: TMBC, 2003)

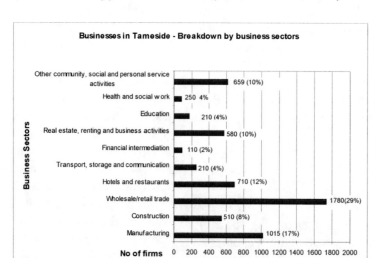

Figure 2. Size of the firms in terms of number of employees in Tameside (Source: TMBC, 2003)

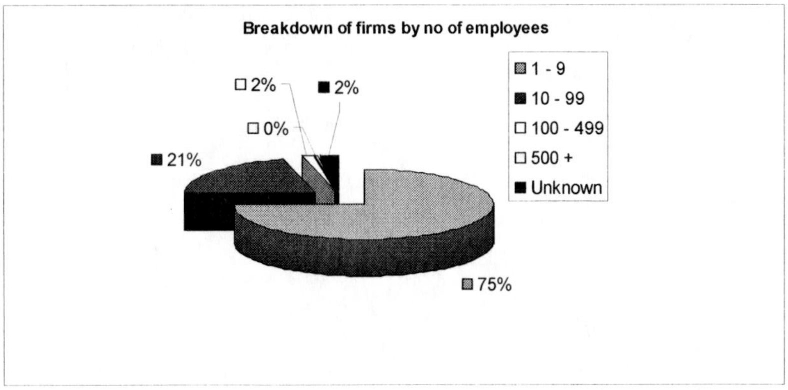

tive. Further, MacGregor (2004) showed that small businesses with fewer than 10 employees were more likely to be part of a formal network. Our data reveals that a substantial number (75%) of firms in the locality are micro enterprises, hiring fewer than 10 people (see Figure 2). Therefore, a framework that would also be able to provide an integrated and collaborative platform for small firms is essential.

Based on the findings from the literature in this domain, IBM's e-business evolution model (Stone, 2003), DTI e-adoption ladder (Martin & Matley, 2001), analysis of the industry sector, Tameside company sizes, and review of the Tameside Business Portal, the proposed e-business framework is as shown in Figure 3. The framework utilizes both the Web portal and e-marketplace models (Piccinelli & Mokrushin, 2001; Dai & Kauffman, 2002; Stockdale & Standing, 2002). Although e-marketplace and Web portal are two different e-business models, they tend to intersect in functionality, and the degree of intersection is increasing (Wang & Archer, 2004). Portal service providers attempt to add transaction facility into their portal applications, whilst e-marketplace service providers offer self-service personalized portals to assist clients administrate their transaction-related content (Wang & Archer, 2004). The framework is drawn up to meet the following criteria:

1.   To support a considerable number of small businesses, who have a narrow product/service range
2.   To support small firms that are mixed in nature, particularly the wholesale and retail trades

3.   To support both business-to-consumer (B2C) and business-to-business (B2B) markets

As shown in Figure 3, the e-business framework comprises an e-catalogue (C), B2C (X), and B2B (Y) applications. Unlike B2B e-business, B2C trading partner relationships are mainly short-term, and transactions are mostly made instantly (spot sourcing); that is, price negotiation (pre-trade) is not required. Besides the pre-trade activity, the other main difference between them is that in B2B, the administration of several catalogues is required (Kim & Lee, 2003; Pham & Fucher, 2004).

The intersection of elements X and Y ($X \cap Y$) is the e-catalogue (C) and shared/common business activities. The shared business processes might include, for example, various business documents and business functionalities that the two e-business solutions have in common. The e-catalogue is the core of the e-business framework and the accomplishment of the framework's objectives depends upon its underlying architecture and data design. It can thus be proposed that the conceptual view of small business e-business framework is $s_e F$:

$$s_e F = P_{portal} \cup E_{e\text{-}marketplaces} \{X, Y\}, \text{ and in which } C_{catalogue} = X \cap Y$$

A number of e-marketplace models and structure have been presented (Kaplan & Sawhney, 2000; Piccinelli & Mokrushin, 2001; Dai & Kauffman, 2002; Rudberg, Klingenberg, & Kronhamn, 2002; Stockdale & Standing, 2002, 2004; Daniel, Hoxmeier, White, & Smart, 2004). Based on the findings, the main focus is on

Figure 3. Conceptual view of the small business e-business framework (Adapted from Tan & Macaulay, 2006)

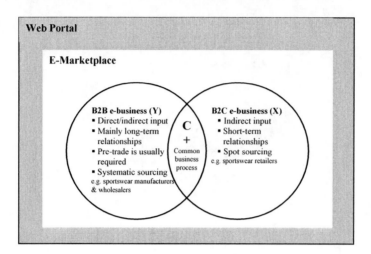

horizontal markets that enable spot and systematic sourcing of both indirect goods (used for the operation of business) and direct goods (used of raw materials for production) that do not require specialized logistics and fulfillment mechanisms.

THE GENERIC E-CATALOGUE FRAMEWORK

As e-catalogue is the core of the proposed e-business framework, to achieve our aims, a generic e-catalogue framework is proposed (as shown in Figure 4). This is carried out based upon findings from the literature (Wong & Keller, 1994; Baron, Shaw, & Bailey, 2000; Fensel et al., 2001; Liu, Lin, Chen, & Huang, 2001; Kim & Lee, 2003) and a review of a number of SME Web sites in the locality. To deal with the problem of heterogeneity of product categories, the United Nations Standard Products and Services Code (UN-SPSC®, 2005) classification schema is also adopted. The proposed e-catalogue framework is divided into a number of levels:

- **Product-level / Goods-level:** Product-level stores detail product item information. It includes standard product information that applies to all items, for example product name and description, as well as product attributes that belong to the similar product family such as colors and sizes.
- **Product Category-level:** Products that have common characteristics are grouped within the same product category, and are not restricted to only one category. A well-defined product classification system is needed and it is crucial for product location and discovery, the UNSPSC® version 7.0901 is, therefore, adopted at this level.
- **Catalogue-level:** To enable a diverse range of firms to participate and collaborate in a multi-vendors catalogue environment, the catalogue-level is proposed. As shown in Figure 4, each firm can have one virtual store, each store has one main catalogue, and under the main catalogue, each firm can have zero to many (0 - *) sub-catalogues. Catalogues can be open (public) or closed (private). A closed catalogue is usually used by B2B businesses and typically includes prenegotiated terms such as product prices, and is only accessible by a specific buyer or a trading partner. However, private catalogues still share the same systems and entries as the public catalogue. By doing so, the streamlining of business processes can be accomplished. Whilst for B2C firms, a sub-catalogue can include either their special offers or winter sale catalogues. Each firm maintains its own set of e-catalogues. The catalogue-level contains information such as catalogue name and status.
- **Resource-level/store-level:** Resource-level holds information of a specific virtual store, such as store contact details.
- **Domain-level:** Domain-level contains the domain that an individual business belongs to (e.g., sportswear and equipment).

Figure 4. Generic five-level e-catalogue framework (Adapted from Tan & Macaulay, 2006)

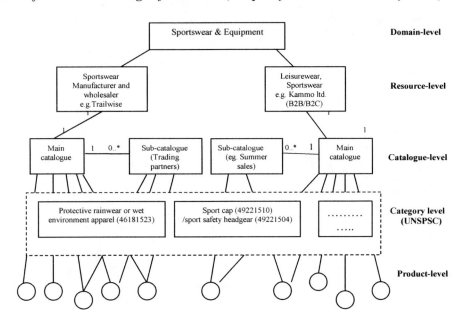

CONCLUSION: THE FRAMEWORK AND E-COLLABORATION

Five different scenarios for collaboration that the framework would support are illustrated in Figure 5 below. Each circle represents a small business (e.g., a1, d2, b4) and the boundary lines show collaborations. A business can belong to more than one collaboration group for example, b1 belongs to group A and group B. The five scenarios are described below.

Scenario A: Strategic Alliances

Strategic alliances involve collaboration among different sellers. Small firms can liaise with each other to offer complementary products so as to boost sales or to enter new markets.

Scenario B: Joint Ventures

Joint ventures involve collaboration between similar or complementary service businesses. Two or more small businesses can be formed to jointly bid for bigger contracts or much larger contracts than they are able to fulfill by themselves, such as responding to an invitation to tender from government agency.

Scenario C: Collaborative Sourcing

Collaborative sourcing involves collaboration among different buyers. Businesses can join together to purchase items such as fuel and hence leverage collective buying power in order to negotiate a better deal.

Scenario D: Sharing of Information, Knowledge, and Resources

Firms can collaborate with other businesses to share information or resources, for example sharing of trade information. They can also organize a joint local or national event, or send a group of representatives to a costly international trade fair with individual firm's product and/or service information.

Scenario E: Collaborative Supply-Chain

Collaborative supply-chain involves collaboration between buyers and suppliers. Businesses can form a collaborative buyer-supplier relationship, thus streamlining their business processes.

The framework utilizes the emerging e-business models of Web portal and e-marketplace, and provides an integrated and collaborative platform for small firms. The Web portal provides a one-stop shop and the e-marketplace offers the aggregation of a wide variety

Figure 5. Possibly collaboration opportunities for small businesses

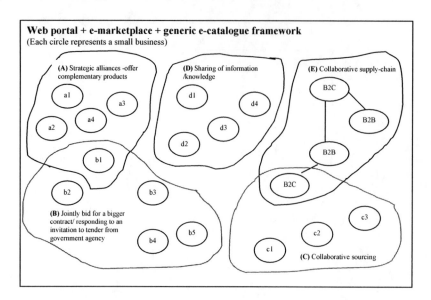

of information such as services and products created by different businesses. With the advent of Extensible Markup Language (XML) standards, the sharing, exchange, and integration of product information among value chains can be accomplished easily and effortlessly (Bagchi & Tulskie, 2000). In addition, through grouping small businesses into one convenient location, these firms would gain a wider audience than they could get on their own, as stated in Terziovski (2003), each business can obtain the advantage of "bigness" or economies of scale while remaining small.

ACKNOWLEDGMENT

An earlier version of this paper, "E-marketplaces: A generic electronic catalogue framework for SMEs," was presented to the 17th International Information Resources Management Association Conference, May 21-24, Washington, D.C., USA.

REFERENCES

Adams, J., Koushik, S., Vasudeva, G., & Galambos, G. (2003). *Patterns for e-business: A strategy for reuse.* Canada: IBM Press.

Baron, J. P., Shaw, M. J., & Bailey, A. D. (2000). Web-based e-catalog systems in B2B procurement. *Communications of the ACM, 43*(5), 93-100.

Bagchi, S., & Tulskie, B. (2000). *E-business models: Integrating learning from strategy development experiences and empirical research.* Retrieved August 24, 2003, from IBM Research, http://www.research.ibm.com/strategy/pub/ebbb.pdf

Bo, J. W. (2005). *Accelerate the reform and innovation of SMEs.* China Economic Times. Retrieved April 5, 2006, from http://theory.people.com.cn/GB/40534/3159065.html

Dai, Q., & Kauffman, R. J. (2002). Business models for Internet-based B2B electronic markets. *International Journal of Electronic Commerce, 6*(4), 41-72.

Daniel, E. M., Hoxmeier, J., White, A., & Smart, A. (2004). A framework for the sustainability of e-marketplaces. *Business Process Management Journal, 10*(3), 277-289

Dixon, T., Thompson, B., & McAllister, P. (2002). *Value of ICT for SMEs in the UK: A critical literature review.* Report for small business service research programme: The College of Estate Management. Retrieved April 25, 2005, from Small Business Service, http://www.sbs.gov.uk/SBS_Gov_files/researchandstats/value_of_ICT_for_SMEs_UK.pdf

European Commission. (2002). *Benchmarking national and regional e-business policies for SMEs.* Final report of the E-business Policy Group of the European Union, Brussels. Retrieved January 4, 2004, from European

Commission, http://europa.eu.int/comm/enterprise/ict/policy/benchmarking/final-report.pdf

European Commission. (2003). *SMEs in Europe 2003*. Retrieved January 11, 2005, from Enterprise and Industry, http://europa.eu.int/comm/enterprise/enterprise_policy/analysis/observatory_en.htm

Fensel, D., Ding, Y., Omelayenko, B., Schulten, E., Botquin, G., Brown, M., et al. (2001). Product data integration in B2B e-commerce. *IEEE Intelligent Systems, 16*(4), 54-59.

Fillis, I., Johannson, U., & Wagner, B. (2004). Factors impacting on e-business adoption and development in the smaller firm. *International Journal of Entrepreneurial Behaviour & Research, 10*(2), 178-191.

Fitzgerald, G., Papazafeiropoulou, A., Piris, L., & Serrano, A. (2005). Organisational perceptions of e-commerce: Re-assessing the benefits. *Electronic Markets, 15*(3), 225-234.

Headd, B. (2005). *Business estimates from the office of advocacy: A discussion of methodology*. Retrieved January 12, 2006, from United States Small Business Administration, http://www.sba.gov/advo/research/rs258tot.pdf

Horvath, L. (2001). Collaboration: The key to value creation in supply chain management. *Supply Chain Management: An International Journal, 6*(5), 205-207.

Kaplan, S., & Sawhney, M. (2000). E-hubs: The new B2B marketplaces. *Harvard Business Review, 78*(3), 97-103.

Kim, G. M., & Lee, G. S. (2003). E-catalog evaluation criteria and their relative importance. *Journal of Computer Information System, 43*(4), 55-62.

Liu, D. R., Lin, Y. J., Chen, C. M., & Huang, Y. M. (2001). Deployment of personalized e-catalogs: An agent-based framework integrated with XML metadata and user models. *Journal of Network and Computer Applications, 24*, 201-228.

MacGregor, R. C. (2004). Factors associated with formal networking in regional small business: Some findings from a study of Swedish SMEs. *Journal of Small Business and Enterprise Development, 11*(1), 60-74.

MacGregor, R. C., & Vrazalic, L. (2005). A basic model of electronic commerce adoption barriers. *Journal of*

Small Business and Enterprise Development, 12(4), 510-527.

Martin, L. M., & Matley, H. (2001). Blanket approaches to promoting ICT in small firms: Some lessons from the DTI ladder adoption model in the UK. *Internet Research: Electronic Networking Applications and Policy, 11*(5), 399-410.

McIvor, R., Humphreys, P., & McCurry, L. (2003). Electronic commerce: Supporting collaboration in the supply chain? *Journal of Materials Processing Technology, 139*, 147-152.

O'Donnel, A., Cilmore, A., Cummins, D., & Carson, D. (2001). The network construct in entrepreneurship research: A review and critique. *Management Decision, 39*(9), 749-760.

Ovalle, O. R., & Marquez, A. C. (2003). The effectiveness of using e-collaboration tools in the supply chain: An assessment study with system dynamics. *Journal of Purchasing & Supply Management, 9*, 151-163.

Piccinelli, G., & Mokrushin, L. (2001). Dynamic service aggregation in electronic marketplaces. *Computer Networks Journal, 37*, 95-109.

Rudberg, M., Klingenberg, N., & Kronhamn, K. (2002) Collaborative supply chain planning using electronic marketplaces. *Integrated Manufacturing Systems, 13*(8), 596-610.

Small Business Services. (2005). *SME statistics UK 2004: Statistical press release*. Retrieved January 1, 2006, from http://www.sbs.gov.uk/analytical/statistics/smestats.php

Stockdale, R., & Standing, C. (2002). A framework for the selection of electronic marketplaces: A content analysis approach. *Internet Research: Electronic Networking Applications and Policy, 12*(3), 221-234.

Stockdale, R., & Standing, C. (2004). Benefits and barriers of electronic marketplace participation: An SME perspective. *The Journal of Enterprise Information Management, 17*(4), 301-311.

Stone, M. (2003). SME e-business and supplier-customer relations. *Journal of Small Business and Enterprise Development, 10*(3), 345-353.

Tan, Y. L., & Macaulay, L. A. (2006). E-marketplace: A generic e-catalogue framework for SMEs. *17th*

International Information Resources Management Association Conference (pp. 954-956). Washington, DC: Idea Group Inc.

Tan, Y. L., Macaulay, L. A. (2007). Adoption of ICTs among small business: Vision versus reality. *International Journal of Business,5*(2), 188-203.

Taylor, M., & Murphy, A. (2004a). SMEs and e-business. *Journal of Small Business and Enterprise Development, 11*(3), 280-289.

Taylor, M., & Murphy, A. (2004b). SMEs and the take-up of e-business. *Urban Geography, 25*(4), 315-331.

Terziovski, M. (2003). The relationship between networking practices and business excellence: A study of small to medium enterprises (SMEs). *Measuring Business Excellence, 7*(2), 78-92.

Todeva, E., & Knoke, D. (2005). Strategic alliances and models of collaboration. *Management Decision, 43*(1), 123-148.

U.S. Small Business Administration. (2005). *Small business profile: United States*. Retrieved January 1, 2006, from http://www.sba.gov/advo/research/profiles/05us.pdf

UNSPCS® (2005). United Nations Standard Products and Services Code. Retrieved December 25, 2005, from http://www.unspsc.org

Wang, S., & Archer, N. (2004). Supporting collaboration in business-to-business electronic marketplaces. *Information Systems and e-Business Management, 2,* 269-286.

Weiss, P. (2002). E-business and small businesses. In B. Stanford-Smith, E. Chiozza, & M. Edin (Eds.), *Challenges and achievements in e-business and e-work* (pp. 173-178). Amsterdam: IOS Press.

Wong, W. T., & Keller, A. M. (1994). Developing an Internet presence with online electronic catalogs. Retrieved November 21, 2006, from Stanford University Info Lab: http://www-db.stanford.edu/pub/keller/1994/cnet-online-cat.pdf

KEY TERMS

Collaborative Sourcing: The collaboration among different buyers. Buyers can join together to purchase items such as fuel and hence leverage collective buying power in order to negotiate a better deal.

Collaborative Supply-Chain: The coordination of all supply activities of a business from its suppliers and partners to its customers, which streamline their business processes.

E-Business Solutions: The use of emerging computer and information technologies, in addition to re-engineered business processes to develop innovative Web applications that support online business activities.

E-Marketplace: A Web-based system that facilitates and encourages buying and selling to induce collaboration among trading partners across a selection of industries.

Spot Sourcing: A trading mechanism commonly used in B2C online shopping. The buyer's goal is to find the required goods at a lowest cost and the purchase of goods is usually made immediately. Buyer-supplier relationship is mainly short-term.

Systematic Sourcing: A trading mechanism mostly used in B2B e-business. Prior to the transactions, buyers and suppliers have established some kind of trading relationships. The procurement of goods is made systematically, which is based on their prenegotiated terms.

Web Portal: A Web-based system that aggregates multiple information sources and applications hence providing a one stop, seamless access for its clients or users.

Speech Act Theory and Communication Modeling

Lewis Hassell
Drexel University, USA

INTRODUCTION

Since the early 1980s there has been an interest in linguistics in general and speech act theory in particular in CSCW, HCI, MIS, and IS modeling in general. The reason for this is simple—computer and information scientists discovered that most work is group work and most group work occurs via language. Winograd and Flores (1986) popularized the use of speech act theory, especially the Searlian variety, for modeling electronic communication and collaboration. However, what one finds if one looks closely is that we have taken the easy road when dealing with language. There are a large variety of speech acts that we ignore when analyzing language, particularly when using speech act theory. Why this is so, the impact on tool-creation, and possible remediation of this problem will be discussed. The importance for such areas as e-collaboration, as well as text mining, computer security, and computing in general will be emphasized.

BACKGROUND

Computer and information science (CIS), being the domain of mathematicians and engineers in the early days of computing, was heavily influenced by logical positivism. As such, the truth or falsity of the propositional content—semantics—was the dominant issue. Syntax, of course, also played an important role. This was for a good reason: These were the only factors a computer could handle. CIS told itself that the programming calculi it used to control machines were, in fact, "languages." Rather than recognizing the metaphorical use of the term, "language," CIS implemented systems under the assumption that human language was as context-free and straightforward as rule-based, mathematical equations.

To support the above conception of language, many researchers used the model of communication developed by Shannon and Weaver (1949), and shown in Figure 1. Since Shannon and Weaver never proposed this as a model of *human* communication, they never explored the complexity for human communication of a number of issues in the model: 1) the nature of "noise," 2) the method of integration with the transmitted/received signal, 3) the social status of source and destination, and the list goes on.

Again, the metaphor (this time of communication), once applied to human communication, took on a life of its own. (For a study of the nature—and insidiousness—of the metaphorical use of language, the reader is directed to classic work of Lakoff and Johnson (2003)).

It was not until the 1980s that some researchers began to look at the broader issues of language, as perhaps the primary avenue of human communication. At this time, a more complete view of language, which included pragmatics, derived from the non-positivist branches of Anglo-American philosophy—for example, speech act theory and the later Wittgenstein—was taken into account, as well as the work of socio- and psycholinguists.

Pragmatics is commonly referred to as the third branch of linguistics (the other two being semantics and syntax). Particularly under the influence of Winograd and Flores (1986), speech act theory—perhaps the most influential branch of pragmatics (Levinson, 1983)—became the dominant branch of pragmatics in e-communication and e-collaboration theory.

SPEECH ACT THEORY

A speech act is a statement, verbal or written, that is designed to accomplish a particular goal. While it may have propositional content, it is not a true or false statement, and is thus not defined by the truth conditions of its propositional content.

Figure 1. Model of communication according to Shannon and Weaver (1949)

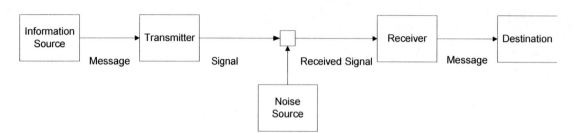

Speech act theory started out (Austin, 1962) as a reaction against the prevailing trend in the philosophy of language during the first half of this century—logical positivism—in which all statements are propositions that were either true or false. Speech act theory contends that besides committing the speaker to the truth or falsity of a statement, there are statements that, by their very utterance, actually constitute performing an action. Because such statements are actions, they cannot be considered to be either true or false but only "felicitous" or "infelicitous." That is, they achieve their stated goal or they somehow "go wrong." John Searle expanded on and codified Austin's work (Searle, 1969). Searle logically enumerated the five possible categories of speech acts. His taxonomy broke speech acts down as follows (Searle, 1979):

1.   ASSERTIVES, which commit the speaker to the truth or falsity of something's being the case. If the propositional content of the statement could be characterized as either "true" or "false," then the statement is classified as an assertive.
      S1: "This Java code does exactly what I wanted the way I wanted it done, and with absolutely zero defects."
2.   DIRECTIVES, which constitute an attempt by the speaker to get the listener to do something.
      S2: "Finish your part of the coding by Friday."
3.   COMMISSIVES, which commit the speaker to do something.
      S3: "I will finish my part of the coding by Friday."
4.   DECLARATIVES, which bring about a correspondence between the propositional content of the statement and reality.
      S4: "I declare this part of the project to be completed."

Obviously, issuing a declarative requires that a speaker/writer have the authority to issue that statement.
5.   EXPRESSIVES, which express the psychological state of the speaker.
      S5: "I feel uncomfortable with that time frame."

Searle also developed the notion of indirect speech acts. Indirect speech acts communicate to the hearer more than the speaker actually says by way of relying on mutually shared background information, both linguistic and nonlinguistic, together with the general powers of rationality and inference on the part of the hearer (Searle, 1979). Indirect expressions (indirection) change the nature of communicative interaction considerably. Let us use statement S1 above. S1, at first blush, is an assertive. It commits the speaker to the truth of something being the case. Now let us suppose a) it was the supervisor who made S1, and b) S1 was made to the programmer responsible for writing the code. S1 could then be (rightly) considered a declarative—that is, the supervisor (who alone has the authority to do so) is indirectly declaring a part of the project completed. The hearer (in this case, the programmer) could also probably assume that his activity in this part of the project is completed. Indeed, if the supervisor were to come back to him at some later date and complain about the Java code, the programmer could rightly object to the supervisor's complaint, using S1 as a source for his objection to the complaint. Indirection is common and natural (usually used for reasons of politeness and decorum—that is, face management), and there is some reason to believe it occurs "automatically" in the brain (Holtgraves, 1998, 1999) and not the result of a deduction, as Grice (1975) and those following him (e.g., van Eemeren & Grootendorst, 2004) seem to assume.

Speech act theory, particularly in Searle's incarnation, is therefore well suited as a foundation for an alternate approach to information systems in general (Lyytinen, 1985; Blair, 1992), and to cooperative systems in particular. Indeed, speech act theory has been used in system modeling by numerous researchers, including (this is only a partial, but representative, list):

- The Language/Action Perspective (LAP) (Winograd, 1987). This was an early model used to analyze and design systems. LAP has actually had a commercial product developed based on it, ActionWorks®, a business process modeler, from Action Technologies (www.actiontech.com).
- SAMPO (Auramäki, Lehtinen, & Lyytinen, 1988) was an early system modeling tool that saw no commercial light.
- DEMO (Dietz, 1994), another system modeling tool, has had some commercial application.
- Change and Woo (1994) used speech act theory to model negotiation protocols for distributed artificial intelligence.
- Denning (2003) used speech acts to model the concept of trust.
- Speech act theory was used to look at information systems security (Dung & Thang, 2004).

One of the virtues claimed for the approaches designed above was voiced by Divitini and Simone (2001)—that of reducing "linguistic opacity." This assumes, of course, that "linguistic opacity" serves no useful role in human communication.

REACTIONS TO SPEECH ACT THEORY

While speech act theory has been used extensively in the field of information systems, it also has its critics. Most of the critics contend not that Searle left certain important aspects out of his analysis or, worse yet, that he was fundamentally wrong. Instead, most of his critics contend that he did not emphasize particular aspects of language enough, or that his taxonomy and/or approach in general are derivative and not fundamental. For example, Bowers and Churcher (1988) argue, "communicative actions should be seen as essentially embedded in dialogical contexts" (p.197). Their contention is that speech act theory ignores social and

historical context. It should be noted that the speech act calculus that Searle and Vanderveken (1985) envisioned includes the element of context as fundamental. In fact, they state, "The single most important question [a theory of illocutionary logic] must answer is simply this: Given that a speaker in a certain context of utterance performs a successful illocutionary act of a certain form, what other illocutions does the performance of that act commit him to?" (p. 6). The computability of this context is, however, another matter.

Levinson (1983) critiques Searle in a way that is especially relevant for this discussion. Searle is committed to the assumption that any indirect speech act can be reduced to its direct form. Levinson calls this the "literal force hypothesis" (LFH). Levinson points out that "most usages [of utterances] are indirect" (p. 264) and that illocutionary force comes solely from sentences in context. The LFH seems to require the hearer (and speaker, for that matter) to make some sort of deduction from the indirect to the direct form of the utterance. As we pointed out above, however, Holtgraves (1998, 1999) has shown that indirection is understood directly, not in a derivative manner. Furthermore, it has been demonstrated in a number of studies that children as young as two years old can respond properly to indirect requests (Papafragou, 2000). It would seem, then, that the exact translatability of an indirect utterance for a direct one is on shaky ground.

SIGNIFICANCE FOR E-COLLABORATION

It would seem obvious that all collaboration—outside dictatorship might not be considered collaboration—requires negotiation of some sort: collaboration of meanings, values, goals, even authority. Thus an adequate—that is, a natural—array of potentialities for appropriate communication becomes especially important in an age of e-collaboration, which we believe is the ultimate message of media naturalness (Kock, 2005). Thus, an attention to the way we have come to be accustomed to communicate is essential. Thus, it is interesting to note that most IS researchers have included only assertives, directives, commissives, and declaratives when modeling information systems. Clearly, however, expressives have an important place in human communication, and thus collaboration. Moreover, it should be noted that IS researchers, outside of

Hassell (Hassell, 2005b, 2005c; Hassell & Christensen, 1996), have paid no attention to indirection. Now it would *seem* obvious that one should make all one's utterances directly. Direct speech acts are more easily understood. They are more clear and distinct. But are they? Certainly, direct—clear and distinct—utterances are more efficient. The question remains, however, are they more *effective*? All too often, effectiveness is sacrificed at the altar of efficiency, particularly when talking about information systems. Great emphasis in IS literature is placed on the cost savings, efficiency, of the computerized data/information/knowledge systems that are implemented. However, in an age of global competition, we should ask if this is what we should be emphasizing.

FUTURE RESEARCH

If, in fact, speech act theory, and in particular, indirection, is unstudied in the communication patterns of e-collaboration, then there are a number of areas that need to be addressed. Certainly, the concept of trust lends itself to such study. Do we trust those who present "clear and distinct" information to their interlocutors more than we do those who use indirection to respect our "face"?

Speech act theory has important applications to the new technology of XML. XML has created great possibilities for data mining. XML, however, provides us with the possibility of mining only semantic or syntactic elements. Adding speech act tagging to XML would add new dimensions to data mining. This, at least on the surface, would be simple, particularly using the work of Ballmer and Brennenstuhl (1981), who classified English verbs for their illocutionary point. When we take indirection into account, however, the situation becomes much more problematic, since there is no convincing method for developing an algorithm for converting direct speech acts to their indirect value.

Perhaps the closest work to what we have been discussing here was done by Brennan and Oheari (1999), not from the point of view of indirection but from the point of view of face management. They noted that there were more hedging and questions in face-to-face interactions than in electronic ones. They speculated that this is to a large extent cost based, since typing is more time consuming than speaking. Controlling

for typing proficiency would be one way to put this observation on a more solid footing.

Research should also be done concerning the cognitive load of direction versus indirection in electronic environments, particularly in collaborative settings. We must do that, however, keeping in mind that not all collaboration is equal. E-collaboration to solve an international governmental dispute is considerably different from using technology for distributed product development.

An examination of effectiveness versus efficiency, and how these influenced by different types of speech acts, would also seem to be an important—indeed, long overdue—subject of inquiry.

CONCLUSION

Many questions remain for the analysis of human communication and its use in collaborative activities. The communication patterns behind e-collaboration have, of course, these same questions. They are complicated, however, by the addition of advanced information and communication technologies (ICTs). The rules (if there are rules) of using this technology are changing by the minute. The world is not standing still. More and more communication and collaboration that relied on old methods of communication are now being done using ICTs. How we use and modify these ICTs, and how we are modified by them, is a story now playing out. What does seem to be certain is that the *application* of these technologies cannot be furthered by using only part of speech act theory. IT theorists call on speech act theory in the name of how we *use* language—that is to say, pragmatics. But by eliminating expressives and indirection, a large portion of speech act theory, what is being used to describe e-collaboration is an impoverished semantic analysis. We must remember that we are describing is not e-communication, which it might be argued, is about the transmission of information. In discussing e-collaboration, we are discussing how we interact and negotiate with one another, which moves us far beyond simple semantics. It is to a renewed discussion of the complete impact of pragmatic analysis on e-collaboration that this paper hopes to offer a beginning.

REFERENCES

Auramäki, E., Lehtinen, E., & Lyytinen, K. (1988). Speech-act-based office modeling approach. *ACM Transactions on Office Information Systems, 6*(2), 126-152.

Austin, J. L. (1962). *How to do things with words.* Cambridge, MA: Harvard University Press.

Ballmer, T., & Brennenstuhl, W. (1981). *Speech act classification: A study in the classification of speech activity verbs.* New York: Springer-Verlag.

Blair, D. C. (1992). Information retrieval and the philosophy of language. *The Computer Journal, 35*(3), 200-207.

Bowers, J., & Churcher, J. (1988). Local and global structuring of computer mediated communication: Developing linguistic perspectives on CSCW in COSMOS. *Office: Technology and People, 4*(3).

Brennan, S. E., & Oheari, J. O. (1999). Why do electronic conversations seem less polite? The costs and benefits of hedging. In *Proceedings of the International Joint Conference on Work Activities, Coordination, and Collaboration* (pp. 227-235). New York: ACM.

Brown, P., & Levinson, S. C. (1987). *Politeness: Some universals in language usage.* Cambridge, MA: Cambridge University Press.

Chang, M. T., & Woo, C. C. (1994). A speech-act based negotiation protocol: Design, implementation and test use. *ACM Transactions on Information Systems, 12*(4), 360-382.

Denning, P. J. (2003). The profession of IT. *Communications of the ACM, 46*(7), 19-23.

Dietz, J. L. G. (1994). Business modeling for business redesign. In *Proceedings of the 27th Hawaii International Conference on System Sciences* (pp. 723-732). Washington, DC: IEEE Computer Society Press.

Divitini, D., & Simone, C. (2001). A computational model of communication for reducing linguistic opacity based on the language-action perspective. *Information and Organization, 11*(2), 157-186.

Dung, P. M., & Thang, P. M. (2004). Stepwise development of security protocols: A speech act-oriented approach. In *Proceedings of the 2004 ACM Workshop on Formal Methods in Security Engineering* (pp. 33-44). Washingtion, DC: ACM Press.

Goffman, I. (1959). *The presentation of self in everyday life.* Garden City, CA: Anchor.

Goffman, I. (1967). *Interaction ritual: Essays in face-to-face behavior.* Garden City, CA: Anchor.

Grice, P. (1975). Conversation and logic. In P. Cole & J. L. Morgan (Eds.), *Syntax & semantics 3: Speech acts.* New York: Academic Press.

Grudin, J. (1994). Computer-supported cooperative work: Its history and participation. *IEEE Computer, 27*(5), 19-26.

Habermas, J. (1979). *Communicative and the evolution of society.* Boston: Beacon Press.

Habermas, J. (1984). *The communicative theory of action: Volume one.* Boston: Beacon Press.

Hassell, L. (1995). *Media, speech act theory, and computer supported cooperative work.* Unpublished doctoral dissertation, Drexel University.

Hassell, L. (2003). Security, trust, and the human factor. *The 1st Security Conference, ISOneWorld 2003.* Las Vegas, NV.

Hassell, L. (2005a). Affect and trust. In P. Herrmann, V. Issarny, & S. Shiu (Eds.), *Trust management: 3rd International Conference, iTrust2005* (pp. 131-145). New York: Springer Verlag.

Hassell, L. (2005b). Language, pragmatics, and knowledge management. *ALOIS 2005,* Limerick, Ireland.

Hassell, L. (2005c). Distribution of assertive versus non-assertive speech acts in email and face-to-face interaction. *ISOneWorld 2005,* Las Vegas, Nevada.

Hassell, L., & Christensen, M. (1996). Indirect speech acts and their use in three channels of communication. In F. Dignum, J. Dietz, E. Verharen, & H. Weigand (Eds.), *Communication modeling: The Language/Action Perspective Proceedings of the First International Workshop on Communication Modeling.*

Holtgraves, T. M. (1998). Interpreting indirect replies. *Cognitive Psychology, 37*(1), 1-27.

Holtgraves, T. M. (1999). Comprehending indirect replies: When and how are their conveyed meanings activated? *Journal of Memory and Language, 41,* 519-540.

Hu, Y.-J. (2001). Some thoughts on agent trust and delegation. In *Proceedings of the Fifth International Conference on Autonomous Agents* (pp. 489-496). Washingtion, DC: ACM Press.

Kock, N. (2005). Media richness or media naturalness? The evolution of our biological communication apparatus and its influence on our behavior toward e-communication tools. *IEEE Transactions on Professional Communication, 48*(2), 117-130.

Lakoff, G., & Johnson, M. (1980). *Metaphors we live by.* Chicago: University of Chicago Press.

Levinson, S. C. (1983). *Pragmatics.* Cambridge, MA: Cambridge University Press.

Lyytinen, K. J. (1985). Implications of theories of language for information systems. *MIS Quarterly, 9*(1), 61-75.

McRoy, S. W., & Hirst, G. (1995). The repair of speech act misunderstandings by abductive inference. *Computational Linguistics, 21*(4), 435-478.

Papafragou, A. (2000). Early communication: Beyond speech-act theory. In S. C. Howell, S. A. Fish, & T. Keith-Lucas (Eds.), *Proceedings from the 24th Annual Boston University Conference on Language Development* (pp. 571-582). Somerville, MA: Cascadilla Press.

Picard, R. (1997). *Affective computing.* Cambridge: MIT Press.

Picard, R. (2003). Affective computing: Challenges. *International Journal of Human-Computer Studies, 59*(1-2), 55-64.

Searle, J. (1969). *Speech acts.* Cambridge, MA: Cambridge University Press.

Searle, J. (1979). *Expression and meaning.* Cambridge, MA: Cambridge University Press.

Searle, J., & Vanderveken, D. (1985). *The foundations of illocutionary logic.* Cambridge, MA: Cambridge University Press.

Solomon, R., & Flores, F. (2002). *Building trust.* New York: Oxford University Press.

Sproull, L. S., & Kiesler, S. (1991). *Connections: New ways of working in the networked organization.* Cambridge, MA: MIT Press.

Twitchell, D. P., Nunamaker, J. F., Jr., & Burgoon, J. K. (2004). Using speech act profiling for deception detection. In *Lecture Notes in Computer Science 3073: Intelligence and Security Informatics: Proceedings of the Second Symposium on Intelligence and Security Informatics* (pp. 403-410). Berlin: Springer-Verlag.

Eemeren, F. H. van, & Grootendorst, R. (2003). *A systematic theory of argumentation: The pragma-dialectical approach.* Cambridge, MA: Cambridge University Press.

Winograd, T. (1980). What does it mean to understand language? *Cognitive Science, 4*(3), 209-242.

Winograd, T. (1987). A language/action perspective on the design of cooperative work. *Human-Computer Interaction, 3*(1), 3-30.

Winograd, T., & Flores, F. (1986). *Understanding computers and cognition.* New York: Addison Wesley.

KEY TERMS

E-Collaboration Technologies: Electronic technologies that enable collaboration among individuals engaged in a common task.

Illocutionary Force: The combination of the illocutionary point of an utterance, and particular presuppositions and attitudes that must accompany that point, including the strength of the illocutionary point, preparatory conditions, propositional content conditions, mode of achievement, sincerity conditions, and strength of sincerity conditions

Mode of Achievement: The means employed by a speaker to accomplish the illocutionary point of an utterance—calling on authority, and so on.

Pragmatics: The study of how language is used in practice, as opposed to the study of signs (semantics) and their interconnection (syntax).

Preparatory Condition: A state of affairs that is presupposed and is a necessary condition for the non-defective employment of the force.

S

Propositional Content Condition: The condition in a commissive act that the commitment to an action must concern a future state of affairs is a propositional content condition.

Sincerity Condition: The psychological state of the speaker.

Strength of Sincerity Conditions: The strength of the psychological state.

The Support of E-Collaboration Technologies for a Blood Bank

P. Sasi Kumar
National Institute of Technology, India

P. Senthil
National Institute of Technology, India

G. Kannan
National Institute of Technology, India

A. Noorul Haq
National Institute of Technology, India

INTRODUCTION

E-collaboration technologies are broadly defined as electronic technologies that enable collaboration among individuals engaged in a common task (Kock, Davison, Ocker, & Wazlawick, 2001; Kock & Davison, 2003; Kock 2004, 2005). The reasons to enter inside the Internet are huge market value and effective data transactions (Perkins, 2000). The developments of electronic collaborations turn out the hard task into a soft one. This technology development allows the whole sectors to leverage the powers of the Internet and communication network to coordinate their efforts and the e-business models have provided the workable infrastructure for group communication and information processing (Jian Cai, 2004). Many published studies have also shown that, besides technologies the social aspects are essential for the success of collaboration (Briggs, 2003; Easley, 2003). The social aspects that lie behind this article are the speedy and effective services provided by the collaboration technologies for the patients. This article mainly speaks on how the deficiency of the blood can be solved by the blood banks. For this purpose a standard model has been created, in which the blood donors can be connected electronically with patients under the network assistance provided by the blood banks and the hospitals.

BACKGROUND

Human blood is the fluid that helps the circulation of oxygen to the cells and carbon dioxide to the heart, to maintain the body temperature, to fight against the foreign organisms, and so on. During the time of emergency the amount of blood decreases inevitably, which can be leveled by injecting the required amount of blood (Encyclopedia of Health Science).

The current setting of the blood bank in various locations of the state is following a conventional procedure. i.e., they can store the donated blood for the extent of one week and the list of donors are also kept in a file for their future reference, but this will not work effectively and there is a chance for the expiry of the valuable blood and a possibility for losing the data.

The new setting of the blood bank is derived from the standard process model, which operates through the help of network and Internet technologies. They typically operate 24 hours a day, 7 days a week, rather than more restricted work hours to search and serve for requirement (Sharp, 1995). For our research, a Web site for the blood bank is designed by Web page designing software, which has the facility for login privacy, donor's registration, and communication with telecom network, mailing, and so forth, as menu driven. It also has the facility to store and access the blood donors list at any time. A password is also provided for security purposes, and can be accessed only by the blood bank user or administrator. All the blood banks situated in different cities of the state are connected with each other by this networking technology. This will help to exchange the details of the blood donors within each blood bank.

LAYOUT OF THE MODEL

The process starts when a patient is in an emergency situation with inadequate blood. At that time, the hospital has to enquire with local blood banks for the availability of particular group of stored blood. In case of unavailability, either the blood bank or the hospital has to search the blood donors, in order to make them available at the hospital for donating blood. The flow of this process can be changed according to the circumstances and requirements through a chain practice of the blood bank user.

THE WEB-BASED BLOOD BANK

To overcome the emergency situation, a standard process model has to be created by the blood bank for the betterment of the patients. Within operations management, the most obvious example model is the European quality model (European foundation for quality management, 2002). The standard process model from the example model is shown in Figure 2 and explained in the following steps, which helps in developing the blood bank Web site.

Incoming Mail and Registration

Initially, the blood bank has to advertise through media or other sources about their Web page. Depending on the advertisement, the blood bank will receive the details of the donor (name, age, gender, blood group, contact number, address, and profession) through mail, which has to be added in the registration space given in the Web site. Or, the blood donors can register themselves by entering the above details in the blood bank Web site. The representative of the blood bank will clarify these details personally. These input details can be stored and retrieved only by the blood bank user/administrator. Editing can also be done only by the user/administrator, in case of improper registration, change of address, phone number, and so forth.

Requesting Mail/Call

The hospital can send mail or call the blood bank to arrange the required amount of particular group of blood. This input will be attended immediately by the local blood bank and can get an early response. In case of unavailability of that particular group of blood, the blood can be arranged by the next nearest blood bank as there is network connection between the blood banks, or the donors may be informed through SMS.

SMS (Short Message Service)

As the blood bank has collaboration with national level mobile network, a request message regarding blood

Figure 1. The structural layout of the model

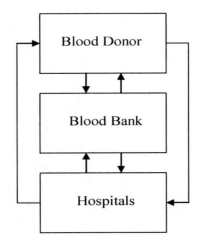

Figure 2. The standard process model

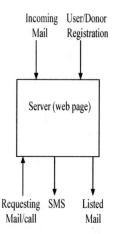

Figure 3. A page of the blood bank Web site

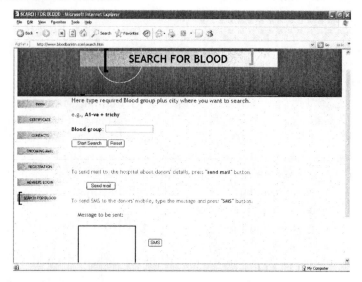

group, hospital name, and contact number can be sent to the list of donors through SMS.

Listed Mail

The list of particular group of blood donors are retrieved and their details are also be given to that particular hospital through mail so that they can contact the donors immediately incase of emergency to save the life of the patients.

A page of the blood bank Web site is displayed below as Figure 3.

RESULTS AND DISCUSSION

The main purpose of this article is to understand the research on design aspects, benefits, and performance of the e-collaboration technologies on the setup of the blood bank. The most important condition for choosing a research strategy is to select a set of methods that fit the type of research questions being asked. In general, research questions can focus on process, and address how decisions are being made by "why" questions (Yin, 1984). The finalized results are organized around pair of questions, at first by focusing on the blood bank's usual process (current setting procedures) and second on the solutions offered by e-collaboration technologies on the new settings to over rule the current one.

- Q.1 What are the current operating procedures and processes followed by the blood bank? What are the key concerns that made the blood bank to change for new setup?
- A.1 The current process is a failure in emergency situations due to the expiry of stored blood and the unavailability of donors. The donor details kept in paper can be damaged by the following key factors like flood and fire. This loss of data takes a long term to be retrieved.
- Q.2 How are the e-collaboration technologies going to give solutions for above concern through this article? Why is e-collaboration needed for blood bank?
- A.2 E-collaboration technologies are used to build the new setup for the blood bank. The process time of searching particular blood donor is reduced by the Web page and retrieval will be fast. The advertisement expense is also much reduced as it can be done easily through e-mail and SMS.

The e-collaboration technologies used in the blood bank will increase the service effectively in saving the life of the patients with elimination of storing problem.

CONCLUSION

This proposed model would help to explain the sequence of implementation and integration of different e-collaboration technologies at the blood bank's properties. It may also be helpful in understanding the benefits of e-collaboration technologies over time. It might also be useful in answering the questions, whether similar research is worthwhile to develop in future. Though this research has some limitations, it will be eliminated by optimization of the Web site and its properties on the basis of suggestion and feedback. The errors and loopholes in the Web page will also be removed. Thus the application of E-collaboration technologies in healthcare department will serve the world usefully and effectively. Apart from this useful e-collaboration technology for blood banks this same technology can also be utilized for the transplant of other vital organs. Therefore this research has much scope for such a study in the future.

REFERENCES

Briggs, R. O. (2003). Collaboration engineering with thinklets to pursue sustained success with group support systems. *Journal of Management Information System, 19*(1), 31-64.

Easley, R. F. (2003). Relating collaborative technology use to teamwork quality and performance: An empirical analysis. *Journal of Management Information Systems, 19*(4), 247-268.

Encyclopedia of health science, from wikipedia- the free encyclopedia.

European Foundation for Quality Management (EFQM). (2003). *The EFQM excellence model.* Available from <http://www.efqm.org/model_awards/model/excellence_model.htm>

Jian, C. (2004). A social interaction analysis methodology for improving e-collaboration over the Internet. *Journal of Management Information Systems, 4*, 85-99.

Kock, N. (2004). The psychobiological: Toward new theory of computer-mediated communication based on Darwinian evolution. *Organization science, 15*(3), 327-348.

Kock, N. (2005). Compensatory adaptation to media obstacles: An experimental study of process redesign dyads. *Information Resources Management Journal, 18*(2), 41-67.

Kock, N., & Davison, R. (2003). Can lean media support knowledge sharing? Investigating a design advantage of process improvement. *IEEE Transactions on Engineering Management, 50*(2), 151-163.

Kock, N., Davison, R., Ocker, R., & Wazlawick, R. (2001). E-collaboration: A look at past research and future challenges. *Journal of systems and Information Technology, 5*(1), 1-9.

Perkins, A., & Perkins, M. (2000). *The Internet bubble.* Seoul, Korea: Gimm-young.

Sharp, J. A., & Cronk, J. (1995). A framework for deciding what to outsource in information technology. *Journal of Information Technology, 10*(6), 259-267

Yin, R. K. (1984). Case study research: Design and methods. Beverly Hills, CA: Sage Publications.

KEY TERMS

Blood Bank: A place where blood is collected from donor, typed, separated into components, stored, and prepared for transfusion to recipients.

E-Collaboration Technologies: Electronic technologies that enable collaboration among individuals engaged in a common task.

E-Mail: A method by which computer users can exchange messages with each other over a network.

Internet: A world wide network of computers that allows the "sharing "or "networking" of information at remote sites from other academic institutions, research institutes, private companies, government agencies, and individuals.

SMS (Short Message Service): A globally accepted wireless service for sending messages between mobile subscribers and external systems such as e-mail, paging, and voice mail.

Supporting Collaborative Processes in Virtual Organizations

Ricardo Mejía
Ecole Supérieure des Technologies Industrielles Avancées, France

Nicolás Peñaranda
Centro de Innovación en Diseño y Tecnología Tecnológico de Monterrey, Mexico

Arturo Molina
Centro de Innovación en Diseño y Tecnología Tecnológico de Monterrey, Mexico

Godfried Augenbroe
Georgia Institute of Technology, USA

INTRODUCTION

The expansion of Internet-based tools has opened new opportunities to improve collaborative work through the development of a new generation of tools designed to support e-work and e-collaboration. The concepts of e-work and e-collaboration have been evolved through the increasing demand of collaborative environments to support distributed networked activities. This trend has triggered the development of new Information Technologies in order to enable the collaboration and interaction of product development teams in distributed environments.

The objective of this article is to describe a technology classification that supports successful collaborative engineering environments (CEE) of virtual organizations (VO). A CEE is then a platform that supports rapid respond to customers, and improves the interaction among members of Virtual Organizations to successful carried out activities of integrated product and process development.

To develop CEE is imperative to integrate the right methodologies and technologies to achieve e-collaboration between VO's members. For this reason an e-collaboration technology classification is proposed to improve and facilitate the understanding, selection and implementation of collaboration technologies in VO.

BACKGROUND

The notion of VO has its origins on the extended enterprise concept (Gott, 1996) where organizations, together with customers and suppliers, are engaged collaboratively in the design, development, production and delivery of their products. Under this approach, the VO concept has evolved as temporary alliances of organizations that come together to share skills or core competencies and resources in order to better respond to business opportunities and produce value-added services and products, and whose cooperation is supported by computer networks (Camarinha-Matos & Afsarmanesh, 2004). Extended enterprises and VOs can be defined in terms of business processes, this article will focused on the "integrated product development" process in order to set the context to define a collaborative engineering environment.

Networked activities demand a huge amount of collaboration and it is here where collaborative environments play a key role when distributed partners exist. A collaborative environment is needed to facilitate team work and, in particular, to enable a group of persons to manipulate shared information objects, and modify them in a coherent manner (Suleiman, Cart, & Ferrié, 1997). Consequently, the concept of e-collaboration is defined by the concept of integrate internal and interorganizational processes supported by Internet-based tools (Gerst, 2003). This is a key concept to accomplish an implementation of collaboration technologies in a VO to, achieve e-work and e-collaboration (Nof, Morel, Monostori, Molina, & Filip, 2005).

CLASSIFICATION OF E-COLLABORATION TECHNOLOGIES FOR PRODUCT DEVELOPMENT

The product development process in VO presents four key phases (Mejía & Molina, 2002): (1) market requirement definition, (2) project planning preparation, (3) project execution, and (4) customer follow-up. The project execution phase is composed of four main activities: (a) ideation, (b) basic development, (c) advanced development, and (d) launching. All these phases and activities use specific tools or applications to assist engineers, designers, marketing people and all the people involved in product life cycle activities to be more effective and efficient. However, successful collaboration environment requires identifying, selecting and implementing ad-hoc technologies in a company. Therefore, it is useful to classify the technologies in a taxonomy to provide a guide to the project team about which technologies and tools are required to launch a collaborative project (Mejía, Canché, Rodríguez, Ahuett, Molina, & Augenbroe, 2004). These e-collaboration technologies are classified as (1) functional tools, (2) information and knowledge management tools, (3) coordination tools, and (4) communication tools.

E-collaboration between VO's members can be accomplished when information/knowledge, coordination and communication tools are integrated on Web platforms. These Web platforms supports access in real time to all the project information and documentation to the product development team; coordinates who, how and when specific tasks have to be undertaken and finally communicates the final objective of the project to all the participants in the team.

The following sections describe in detail the taxonomy of technologies and present a methodology to organize the integration of coordination, collaboration and information/knowledge tools among VO's members.

Functional Tools

These tools support specific execution of a task or a group of tasks allowing the fulfillment of specific objectives within different stages of product development. These kinds of applications are frequently automated or semi-automated engineering methodologies or techniques. Usually a well defined sequential and logical activity performed to accomplish precise objectives, is able to be automated. In addition, such activity can be executed with a computer to support decision making. Two types of functional tools have been identified: functional tools based on information and functional tools based on models.

Functional Tools Based on Information

In this category, the most common tools are the ones used in concurrent engineering. These tools have been commercially developed to generate computational applications facilitating the use of automated methodologies. Some examples can be the QFD (quality function deployment) (Revelle, 1998) which enables the transformation of customers' requirements into actions and design, allowing engineers to focus in content and activities within the QFD instead of documentation issues. For FMEA (failure modes and effect analysis) (Palady, 1995), there are computational tools to process automation, detecting potential failure identification in a product or process design before that these occur. FMEA can be considered as a standardized analytical method to detect and eliminate problems in a systematical and total manner. Finally, IDEF-0 (ICAM definition method zero) (NIST, 1993) is other example, which presents software applications that automate hierarchical deployment, enabling systems definition at any number of detail levels, through the necessary level for required analysis.

These tools can be developed on Web applications, which improve the team coordination and the project performance, using specific functional tools for each activity.

Functional Tools Based on Models

These are engineering software applications based on mathematical, simulation and knowledge based models which support the product development cycle. Applications such as CAD/CAM/CAE are example of these tools. These engineering applications have evolved into knowledge based engineering systems (KBES) which allow the optimization, in time and quality, of engineering processes using rules and expertise embedded in the software application.

One major problem faced by virtual organizations is the need to integrate different software applications. Each application generates different formats, generating integration problems between different VO's members. For this reason, standards of data formats such as EDI-FACT, IGES, SET, STEP, or HTML/XML are widely used. The complete integration of all engineering applications has not been developed, however the use of Web services to achieve interoperability is currently a feasible technical solution (Peñaranda, Galeano, Romero, Mejía, & Molina, 2006).

Tools for Information/Knowledge Management

In the last decades, the continuous introduction of new technologies to improve information/knowledge storage, sharing, processing, and analysis have changed the way engineers execute tasks; from manual processes to computer tools based on information and knowledge management.

The efforts in structuring information to support the product development process have been focused on:

- **Product modeling:** The product model has to represent all the necessary information used through the life cycle of a product such as customer's requirements, design, fabrication, production, assembly, packaging, distribution and collecting/recycle/re-use to support decision making (Krause, Kimura, Kjellberg, & Lu, 1993).
- **Manufacturing modeling:** The manufacturing model must represent the technological and production capabilities. The information entities which define the manufacturing information modeling are the resources, process, and strategies. The manufacturing and process resource describe the technological capabilities. Other issues are new strategies which represent the real capability of the enterprise. The manufacturing strategies represent the information that is specific for each organization. This information helps to identify resources and certain processes which

aid in determining the strategy for the proposed enterprise (Molina & Bell, 1999).

These two information models allow engineers, managers and suppliers to access all the information related with product and/or processes development. The objective is to have access to all process stages of a particular product development process. The integration among VO's members it is possible, because these tools are already available on web platforms.

The evolution of the Information Modeling has lead to a new generation of information and knowledge management technology. The advancement of robust systems has paved the integration and management of different information models, and others strategic enterprise systems. The most known tools in the market are:

- **PDM (product data management):** PDM systems help engineers to manage product information and data for the product life cycle process. These systems keep track of the complete information required for and generated during the product life cycle. They ensure that live data are up-to-date and that old data are archived correctly, provide security measures and allow data to be property tracked and audited. The use of workflow and process management allows an enterprise to define the ways in which changes may be made to product data, providing methods to route data automatically and to audit approval decisions. PDM tools often provide methods by which other software systems may use their resources (Gao & Bowland, 2002).
- **PLM (product life cycle management):** PLM had a heavy emphasis on computer-aided design and product data management. Newer PLM software is designed to help a company deliver a product and continually enhance it by helping to manage and automate materials sourcing, design, engineering change orders and product documentation (such as test results, product packaging, and post-sales data). PLM tools can also help the VO's to handle through a growing number of local, state, federal, and international regulations. PLM applications typically includes a centralized database that stores all product master records, bills of materials, design-history records, packaging and artwork data, and more.

PLM can be integrated with enterprise resource planning (ERP) and supply chain management (SCM) in order to connect the information about the product with human resources, financing and forecasting. This allows a unifying vision and a better understanding of how a simple change in the design can affect the overall system (Bacheldor, 2003). The upgraded and available information for the entire work group it is a requirement, facilitating decision making and to become fast and reliable.

- **ECM (enterprise content management):** An emerging concept, which integrates the management of structured, semistructured, and unstructured information, software code embedded in content presentations, and metadata together in solutions for content production, storage, publication, and utilization in organizations (Päivärinta, 2005). The architecture of ECM is defined by collaboration, workflow, portal, search and retrieval, content management, document management and record management applications (Dillnut, 2006)

Coordination Tools

Engineering design is a process, wherein a requirement or idea is transformed in the necessary information in order to develop a new product or system. Successful management of this process is summarized in the effective handling of three issues (Hales, 1993): (1) activities of the design team (such as task planning, report revision/elaboration, cost estimation, or information recovery); (2) outputs from the design team (is the analysis of measurable variables, being the performance measurements of the design team activities); and (3) influences on the design team (such as markets, customers, management changes, motivation, and technological developments).

In project management and control, the effect of influenced factors must be forecasted, monitored and controlled whenever possible. Thus, it is not only necessary to have planning tools, but also control and monitoring tools. Therefore, this kind of coordination applications are not only meant to be used in specific stages of development, but also in continuous monitoring of activities, execution control and effective planning of activities, resources, time, and results. Business intelligent tools allow keeping track of key

performance indicators in order to analyze and improve process execution.

Various methods and modeling techniques have been proposed over the last decade to improve the enterprise coordination and integration, e.g., ARIS ToolSet, BONAPART, CimTool, FirstSTEP, IDEF methods, IEM, IBM's FlowMark, IMAGIM, METIS, PrimeObject, PROPLAN and BPMN (Business Process Management Notation), to name a few (Molina, Chen, Panetto, Vernadat, & Whitman, 2005). Furthermore, there are business process engines developed that permit run all this business process modeled, current leaders include IBM, IDS Scheer, iGrafx, Mega, Microsoft, Proforma and Telelogic (Blechar, 2006)

Communication Tools

Communication tools enable the interaction between VO's members involved in the different stages of the product life cycle (design, processes and manufacturing systems development). Through the use of these tools, it is possible to support collaboration environments that facilitate remote activities execution.

Individual communication tools enable the interaction between members of a multidisciplinary design team. For this purpose, Web capability to publish information from the concept generation to the product realization and virtual manufacturing, has motivated the adoption of Internet as a collaborative design tool. Currently, it is playing an increasing major role in the systems development for product collaborative development (Wang, Shen, Xie, Neelamkavil, & Pardasani, 2002).

It is important to determine the communication and interaction requirements through the product life cycle. Visualization and virtual environments are critical for the design and development, as well as to the local and remote collaboration, such as face-to-face meeting, telephone, teleconferences and videoconferences. Choosing of a good media depends on the quantity of explicit information that would be used in the activity, and has developed into an interesting media transference richness classification (Van Luxemburg & Ulijn, 2002).

Nowadays, communication tools are becoming more important for the interaction of work teams during the whole product development process. Thus, the selection of the most suitable media depends on the required interaction level for all activities to have an effective and efficient communication and collaboration environment to ensure resources usage optimization.

There are two kinds of synchronous and asynchronous communications systems: (1) traditional and (2) electronical. An extended classification of communications systems are described in the following:

- **Traditional synchronous:** Face to face and telephone
- **Traditional asynchronous:** Mails, documents, notes
- **Electronic synchronous:** Video conference, desktop-conferencing, chatting, whiteboard, shared files
- **Electronic asynchronous:** Instant messaging, e-mail, FTP (File Transfer Protocol), newsgroup, net casting

IMPLEMENTATION OF E-COLLABORATION TECHNOLOGIES IN A VO

The following methodology is proposed to implement a Collaborative Engineering Environment (CEE):

1. **Determine the VO requirements in the product lifecycle development:** The VO has to identify the type of business opportunity that will be dealt including the activities and its requirements. Moreover, it is important to select which activities of the product lifecycle are going to be developed by each VO's member (Product Design, Process Design or Manufacturing system design activities).
2. **Evaluate collaborative technologies:** According to the classification presented in this article, the collaboration tools that are suitable to the VO are evaluated to decide which technologies could support the VO product development process.
3. **Select collaborative technologies:** Based on the evaluation the e-collaboration technologies are selected to design an environment to achieve a successful CEE integration.
4. **Implement and find opportunities for improvement:** The technologies selected are implemented for a VO. And finally reflections about this implementation are taken to find new improvements in the CEE.

An example of a solution for CEE is the European e-HUBs consortium, which was funded by the European Commission's IST program in 2002. This project developed a web hosted platform for planning e-engineering projects. The prototype has been developed on the substrate of an existing web hosted collaboration platform. The platform offers community building, team communication manager and document management. The process modeling module operates on the basis of these functions but has two added functional modules: workflow management and project planning (Mejía, Canché, Rosas, Camacho, Ocampo, & Molina 2005).

CONCLUSION

A complete collaborative engineering environment must incorporate functional, collaborative, coordination and Information/knowledge Management tools. These set of tools allows engineers to be supported in their engineering activities execution. Therefore, a taxonomy of collaboration technologies have been presented to guide developers of collaborative environments to select the best possible combination of tools for virtual organizations.

The functional tools are used for support specific engineering activities, and are selected according to the engineering tasks, while coordination, collaboration and Information management are platforms generalized to the whole engineering lifecycle. Not all the product development activities require communication and functional tools, but all of them require the exchange of information/knowledge and also, control and plan in effective manner activities, resources, time and results. The experiences with e-Hubs platforms have shown the importance of use coordination tools, like workflow process to support product development processes in VO.

Otherwise, the companies can interact with their engineering partners (between employees or company's partners) without the need of the most advanced and complex technologies. However, VOs are associated with strong IT support, because all the interaction between VOs members is achieved using CEE customized to their requirements.

Finally, it is important to develop CEE that supports different cultural behaviors in the organization. All team members must use correctly this collaborative platform using a well defined product development process. It is important to convince team members that using CEE can improve their collaboration and information management process in order to have successful product realization in VOs.

REFERENCES

Bacheldor, B., & Kontzer, T. (2003). PLM—More than just ERP or SCM. *Asia Computer Weekly, July*, p. 1.

Blechar, M., & Sinur, J. (2006). Magic quadrant for bussines process analysis tools. *Gartner Research. Note.* Stamford, CT: Gartner, Inc.

Camarinha-Matos, L. M., & Afsarmanesh, H. (2004). *Collaborative networked organizations: A research agenda for emerging business models.* Springer.

Dillnut, R. (2006). Surviving the information explosion [knowledge management]. *Engineering Management Journal, 16*(1), 39-41.

Gott, B. (1996). *Empowered engineering for the extended enterprise: A management guide.* Cambridge, UK: Cambashi.

Hales, C. (1993). *Managing engineering design.* Longman.

Gerst, M. H. (2003). *The role of standardization in the context of e-collaboration: A snap shot.* Stuttgart, Germany: IEEE.

Gao, J., & Bowland, W. (2002). A product data management integrated product configuration and assembly process planning environment. In *Proceedings of the Institution of Mechanical Engineers* (p. 407). Wilson Applied Science & Technology Abstracts PlusText.

Krause, F.-L., Kimura, F., Kjellberg, T., & Lu, S. C.-Y. (1993). Product modelling. *Annals of the CIRP, 42*(2), 695-706.

Mejía, R., Canché, L., Rodríguez, C., Ahuett, H., Molina, A., & Augenbroe, G. (2004). Designing a HUB to offer e-engineering brokerage services for virtual enterprises. In L.M. Camarinha-Matos (Ed.), *Virtual enterprises and collaborative networks* (pp. 453-460). Kluwer.

Mejía, R., & Molina, A. (2002). Virtual enterprise broker: Process, methods and tools. In L. M. Camarinha-Matos (Ed.), *Collaborative business ecosystems and virtual enterprises* (pp. 81-90). Kluwer.

Mejía, R., Canche, L., Rosas, R., Camacho, R., Ocampo, M. & Molina, A. (2005). Action research as the basis to implement enterprise integration engineering and business process management. In Panetto, H. (Ed.) *Interoperability of Enterprise software and applications*, (pp. 19-30), London: Hermes Science Publisher.

Molina, A., & Bell, R. (1999). A manufacturing model representation of a flexible manufacturing facility. In *Proceedings of the Institution of Mechanical Engineers* (Vol. 213, Part B, pp. 225-246). Journal of Engineering Manufacture.

Molina, A., & Bell, R. (2002). Reference models for the computer aided support of simultaneous engineering. *International Journal of Computer Integrated Manufacturing, 15*(3), 193-213.

Molina, A., Chen, D., Panetto, H., Vernadat, F., & Whitman, L. (2005). Enterprise integration and networking: Issues, trends and vision. In P. Bernus, M. Fox, & J. B. M. Goossenaerts (Eds.), *Knowledge sharing in the integrated enterprise* (pp. 303-313). Springer/Kluwer.

Nof, S. Y., Morel G., Monostori L., Molina A., & Filip, F. (2005). From plant and logistic control to multi-enterprise collaboration. In P. Horacek, M. Simandl, & P. Zitek (Eds), *Selected plenary, milestones and surveys* (pp. 153-166). Prague, Czech Republic: 16th IFAC World Congress.

Palady, P. (1995). *Failure modes & effects analysis.* West Palm Beach, FL: PT Publications.

Päivärinta, T., & Munkvold, B. E. (2005). Enterprise content management: An integrated perspective on information management. In *Proceedings of the 38th Hawaii International Conference on System Sciences* (p. 96). IEEE.

Penaranda, N., Galeano, N., Romero, D., Mejía, R., & Molina, A. (2006). Process improvement in a virtual organization focused on product development using collaborative environments. *12th IFAC Symposium on Information Control Problems in Manufacturing (INCOM 2006)* (pp. 611-616). Saint-Etienne, France: IFAC.

Revelle, J. B., Moran, J. W., & Cox, C. A. (1998). *The QFD handbook.* New York: Wiley.

Suleiman, M., Cart, M., & Ferrié, J. (1997). *Serialization of concurrent operations in a distributed collaborative environment.* Phoenix, AZ: Group 97.

Van Luxemburg, A., & Ulijn, J. (2002). The contribution of different communication media to an effective design process between a company and its customer: communicative and cultural implications of 5 Dutch cases. *IEEE Transactions on Professional Communication, 45*(4), 250-264.

Wang, L., Shen, W., Xie, H., Neelamkavil, J., & Pardasani, A. (2002). Collaborative conceptual design: State of the art and future trends. *Journal of Computer Aided Design, 34*(13), 981-996.

KEY TERMS

Collaborative Engineering Environment (CEE): It is a platform that enables coordination and collaboration among e-engineering groups, supported by methodologies and tools that facilitate product realization process, knowledge sharing and efficient/effective execution of engineering activities.

Communication Tools: These tools enable the interaction between members of the teamwork involved in the different stages of integrated products, processes or manufacturing systems development.

Coordination Tools: There are tools that support the Project management, using project planning and workflow tools.

E-Engineering Groups: There are groups that execute a product development process regardless of their geographic location.

Functional Tools: These tools enable the specific execution of a task or a group of tasks allowing the fulfillment of specific objectives within different stages of product development.

Product Development Process: It is the entire life cycle of product development, from customer requirements to its manufacturing and disposal stages.

Tools for Information/Knowledge Management: These are technologies that permit to structure, integrate, and share the engineering information (product and manufacturing information) among all the members.

Sustainability of E-Collaboration

António Dias de Figueiredo
University of Coimbra, Portugal

INTRODUCTION

In spite of its recognition as a field of research and practice with a lineage of several decades of prolific development (Kock & Nosek, 2005), virtual collaboration is still a domain where mixed results occur and failure crops up without warning (DeSanctis, Poole, Dickson, & Jackson, 1993; Blythin, Hughes, Kristoffersen, Rodden, & Rouncefield, 1997; Kock, 2004; Kock & Nosek, 2005). Even as its theoretical, technical, operational, and conceptual boundaries expand (Kock & Nosek, 2005), we still feel powerless when a promising experience of e-collaboration, which we could swear would last for a long time, suddenly collapses. In this article we discuss some fundamental conditions for sustainable e-collaboration. We start by introducing the concept of value proposal, the common ground of compatible interests required to make collaboration last, and we distill from it what we call the principle of sustainable e-collaboration. We then move to a discussion of the variable levels of collaboration and their relationship to group development, leadership and purpose. Finally, we briefly expound five groups of theories that we view as promising candidates for the future establishment of the theoretical foundations of sustainable e-collaboration. Figure 1 summarizes the key concepts of the article.

VALUE PROPOSAL AND SUSTAINABLE E-COLLABORATION

For any collaboration to be sustainable it must fulfill in permanence the interests and motivations of *all* the parts. Otherwise, sooner or later some of the parts will lose interest, a number of them will break up, and a few may even oppose to the maintenance of the collaboration. This applies to any kind of collaboration, be it within a project involving many collaborators, a business relationship, or the partnership between research student and advisor. It also holds both offline and online.

We use the term *value proposal* to express a common (often tacit) agreement between *all* the parts that keeps them willing to collaborate. The clarification of the value proposal requires that all the *parts*, as well as the *relationships* between them, be identified and the *benefits* of each part plainly recognized. Given the expectable differences of interests between parts, the value proposal tends to emerge from negotiation processes, which may be explicit or implicit. On the other hand, since the interests of the various parts tend to change with time and context, as the collaboration develops, the value proposal needs to keep being negotiated all the time, even if tacitly. A crucial aspect of this negotiation is that each part must be permanently concerned, not just with the satisfaction of her *own* interests, but also with the satisfaction of the interests of all the other parts. Otherwise, the collaboration ceases to be sustainable, and everyone will lose. This primary concern with the interest of all the other parts, which is so often overlooked in most forms of collaboration, justifies the formulation of a principle:

- **Principle of sustainable e-collaboration:** E-collaboration is only sustainable as long as each part feels it is gaining from it and acts so as to grant that all the other parts feel likewise.

The principle of sustainable e-collaboration does not hold, of course, for the cases where e-collaboration is unwilling, which will be discussed later in this article.

SUSTAINABLE E-COLLABORATION AND GROUP DEVELOPMENT

Although the terms "cooperation" and "collaboration" are used interchangeably in everyday language, their frequent application in education, management and politics has led to the refinement of their semantics in different directions. It has also led to their frequent

Figure 1. Key concepts of sustainable e-collaboration

linking to a third term, "coordination." As the three words hold different connotations in the above fields, we attempt here a compromise between these connotations. In *cooperation*, each part recognizes the benefit of working together and is willing to support collective efforts, provided its individual aims and autonomy are not sacrificed. In *coordination*, the recognition of the benefit of working together is not critical, but each part needs to know what, when, and how to do what needs to be done, while accepting the alienation of some of its autonomy in the process. *Collaboration* requires collective commitment to a common mission and a shared effort to get results that would never be achieved by any of the parts in isolation. The distinction between cooperation, coordination, and collaboration has proved to be very useful when studying social groups. However, the insistence on a sharp distinction between the three terms, with no room for integrated visions, often stands in the way of a valuable discussion of subjects such as e-collaboration. Should we be talking about e-cooperation, e-coordination, or e-collaboration, or about them all? Brown and Keast (2003) helped solving this problem by proposing a continuum of patterns of interaction, along an axis of fragmentation/integration, where cooperation stands at the lower extreme, coordination somewhere in the middle, and collaboration at the upper extreme. This

image of a continuum, instead of a break, between collaboration and cooperation lets us see cooperation as a soft kind of collaboration and collaboration as a strong variety of cooperation.

Another advantage of figuring a cooperation/collaboration continuum is that it lets us explore more naturally the theories of group development. These theories make clear that collaboration is exposed to ups and downs, and that some of these ups and downs actually correspond to predictable patterns. Three such theories will be presented here to illustrate how, for a single group, collaboration can change through time. The, earliest, and most influential has been proposed by Bruce Tuckman (1965) who identified four sequential stages: forming, storming, norming, and performing (later expanded with a fifth stage: adjourning). In the forming stage the group members tend to interact superficially and politely while trying to recognize their positions and roles within the group. In the storming stage they begin establishing norms of common behavior, which invariably generate conflict. In the norming stage agreements begin to emerge and the group starts working together as a unit. Finally, in the performing stage, the group becomes more relaxed, flexible, and productive as a collective endeavor.

Another popular model for group development, sequential like Tuckman's, is the punctuated equilibrium framework proposed by Gersick (1988, 1989), based on three important moments along the timeline for the completion of a project—the beginning, the mid-point (which works as a moment of reflection and readjustment), and the end. The three periods are separated by two long periods: one occurring before the mid-point, which establishes, tests out, and refines strategies; and another following the mid-point, which serves to complete the project with a degree of commitment similar to that of Tuckman's performing stage. An alternative view of group development, proposed by McGrath (1991), is the time, interaction and performance (TIP) model, which, unlike the previous two models, is not sequential, assuming the simultaneous occurrence of four modes—inception, problem solving, conflict resolution, and execution—which maintain three distinct functions shared by the group: production, well-being, and member support.

These models, which belong to the face-to-face world, are being gradually tried out in virtual environments. An illustrative study, authored by Johnson, Suriya, Yoon, Berrett, and La Fleur (2002), concluded that Tuckman's model corresponded more closely to their results. This study relates, however, to a specific situation of e-collaboration, leaving much room for future research.

Two issues are often absent from the debates on group development and sustainable collaboration: *leadership* and a *common purpose*. Hersey and Blanchard (1999) have addressed the different kinds of leadership required for the various stages of group development in face-to-face environments. For online group development, however, little knowledge seems to be available. Johnson et al. (2002) suggest that leadership is much more shared in virtual than in face-to-face groups, but Bell and Kozlowski (2002) and Gibson and Cohen (2003) support, on the contrary, that leadership in virtual groups requires much more structure and procedural assistance than leadership in face-to-face groups. The issue of a *common purpose* is also fuzzy for the time being. In the TIP model (McGrath, 1991) the existence of a shared project assumes the identification of an essential common purpose. However, authors devoted to collaboration in communities of practice sometimes argue in favor of modalities of "laisser-faire" where common purposes are not necessarily explicit and leaders may merely play the role of animators (Wenger, McDermott, & Snyder, 2002). This means that the issues of leadership and a common purpose in virtual groups are still open and calling for research.

THEORIZING THE SUSTAINABILITY OF E-COLLABORATION

In spite of the multiple manifestations of the importance of virtual collaboration, namely in the worlds of education and corporate training, management, and politics, where the need for durable periods of collaboration has been acknowledged for many years, relatively little effort seems to have been put so far in building the theoretical foundations of sustainable e-collaboration. A number of theories, all of them from the face-to-face world, seem, however, to carry significant promises.

- **Theories of CoPs.** Brown and Duguid (1991) have used the expression community of practice to describe the work-related social groupings that emerge informally in organizations and enable their members to learn collectively, through the exchange of stories, in ways that go far beyond their formal training and the official directions of the organizations. The expression has also been used by Lave and Wenger (1991) to describe informal networks of people who share common interests, collaborate for extended periods of time, exchange ideas, find solutions, and collectively build knowledge. Wenger (1998) has further explored the concept, in a book where he proposes a social theory of learning centered on a conception of community of practice made up of four key components: practice, meaning, community, and identity. In recent years, the concept of community of practice has started to be explored intensively in connection with virtual environments, in fields as diverse as e-learning (Hung & Der-Thanq, 2001; Stacey, Smith, & Barty, 2004) and knowledge management (Johnson, 2001; Lesser & Storck, 2001). The potential for its exploration in managing the sustainability of e-collaboration is, however, to the best of our knowledge, still untapped. Preliminary attempts we have conducted in this direction suggest that it is very much worth investigating.
- **Activity theory.** Activity theory has been developed in the first half of the 20th century by the Russian psychologists Leont'ev and Rubinstein, within the socio-historical psychology school initiated by Lev Vygotsky. It started to propagate to the West in the mid-1980s, namely through the mediation of authors from the European Nordic

S

countries, who originated the Scandinavian activity theory movement. This movement played an important role in further structuring the theory and in contributing to the creation of the wide international community that currently explores it. The addition to the theory, by Yrjö Engeström (1987), of the concepts of community, rules, and division of labor played an important role in making it useful to analyze, amongst other socio-technical phenomena, the behavior of communities mediated by technologies. In this sense, it may be useful to start theorizing the study of sustainable collaboration within communities, even if, as stated by Bonnie Nardi (1996), an expert in the field, it must be looked more as a descriptive tool than as a predictive theory.

- **Actor-network theory.** Actor-network theory (ANT) is a social theory widely used to help understand the functioning of networks where human and non-human actors interact, form alliances, create relationships of mutual dependence, and resort to artifacts to satisfy their interests (Callon, 1986; Callon & Latour, 1981; Law, 1992). For instance, a community of learners following an online course behaves as an actor-network that includes, as actors, all the learners in the community, its coordinator and e-moderators, the contents of the course, the collaboration mechanisms (e-mail, forums, whiteboard), the digital library (and all its materials), plus the codes of behavior and netiquette supporting the collaboration. Much of the value of ANT results from its ability to help analyze complex socio-technical situations where humans and non-humans interact and become mutually dependent and to support the design of solutions to problems where the introduction of an artifact, a regulation, or a new role for a human actor may redress an otherwise collapsing equilibrium. This particular aptitude of ANT makes it a promising candidate for theorizing sustainable collaboration mediated by technology.

- **Pattern language theory.** Pattern language theory resulted from the adoption, in many disciplines, of the philosophy proposed in *A Pattern Language* (Alexander, Ishikawa, & Silverstein, 1977) and *The Timeless Way of Building* (Alexander, 1979), which views wholeness as a criterion of success for complex systems. Originally proposed as a new approach to the theory and application of

architecture, Alexander's philosophy became an inspiring model for the design and development of socio-technical systems of all kinds. One of its best known consequences outside the field of architecture is the concept of design patterns, in software engineering. Alexander's philosophy supports that everything we build should grow naturally, rather than be strictly planned, and that patterns, or common features and relationships abstracted from previous successful solutions, should be allowed to gain life. To illustrate the concept of pattern with an example from the domain of architecture, we may notice that a window is not, by itself, beautiful or ugly. A window that looks beautiful in a catalog may turn out to be an ugly window when combined with a façade that does not match it, a wall angle that does not favor it, or a beam of sunlight that it is unable to catch. A window is beautiful when it integrates in harmony the reality to which it belongs, which becomes, as a whole, much more beautiful, thanks to that window (Figueiredo & Afonso, 2005). The concerns with wholeness, harmony, and steady growth of Alexander's philosophy suggest its theoretical potential to support strategies for the design of sustainable e-collaboration contexts.

- **Egoistical cooperation theories.** In *The Evolution of Cooperation*, Robert Axelrod (1984) developed a theory about the conditions in which, without the aid of central authority, cooperation can be enforced among individuals who have an incentive to be selfish. The basis of his inspiration, which was explored successfully, for instance, in multiple international attempts to contain conflict during the Cold War, has been the "iterated prisoner's dilemma", a game where two players who are inclined to betray each other end up cooperating. The potential usefulness of this theory is that it may help us explore situations where some of the parts are unwilling to cooperate, letting us push the lack of willingness and mutual commitment from non-cooperation to cooperation. Once there, we may then, using the other theories, be able to go up the cooperation/collaboration continuum to levels where collaboration can flourish and become sustainable.

CONCLUSION

In this presentation of major conditions for sustainable online collaboration we have tried to combine a stable theoretical background with a selection of open issues and proposals that we hope may stimulate and inspire the young researcher. The stable background corresponds to the five theories briefly introduced, the distinction between cooperation/coordination/collaboration, and the literature on face-to-face group development. The open issues pertain to the role, need and nature of leadership and common purpose in sustainable e-collaboration, as well as to the challenges of converting the face-to-face theories of group development to the online world, with any improvements that may be required. The proposals, resulting from work we have in progress in this area, concern the principle of sustainable e-collaboration and the identification of the need for a value proposal accepted by all, as well as the potential of exploring in the future theorization of e-collaboration some of the theories we have pinpointed.

REFERENCES

Alexander, C. (1979). *The timeless way of building.* Oxford: Oxford University Press.

Alexander, C., Ishikawa, S., & Silverstein, M. (1977). *A pattern language.* Oxford: Oxford University Press.

Axelrod, R. (1984). *The evolution of cooperation.* New York: Basic Books.

Bell, B. S., & Kozlowski, S. W. J. (2002). A typology of virtual teams: Implications for effective leadership. *Group and Organization Management, 27,* 14-49.

Blythin, S., Hughes, J. A., Kristoffersen, S., Rodden, T., & Rouncefield, M. (1997). Recognizing "success" and "failure": Evaluating groupware in a commercial context. In *Proceedings of the international ACM SIG-GROUP Conference on Supporting Group Work: The Integration Challenge* (pp. 39-46). ACM Press.

Brown, J. S., & Duguid, P. (1991) Organizational learning and communities of practice: Toward a unified view of working, learning and innovation. *Organization Science, 2*(1), 40-57.

Brown, K., & Keast, R. (2003). Citizen-government engagement: Community connection through networked arrangements. *Asian Journal of Public Administration, 25*(1), 107-132.

Callon, M. (1986). Some elements of a sociology of translation: Domestication of the scallops and the fishermen. In J. Law (Ed.), *Power, action, and belief: A new sociology of knowledge* (pp. 197-225). London: Routledge.

Callon, M., & Latour, B. (1981). Unscrewing the big Leviathan: How actors macro-structure reality and how sociologists help them to do so. In K. D. Knorr-Cetina & A. V. Cicourel (Eds.), *Advances in social theory and methodology: Towards an integration of micro and macro-sociologies* (pp. 277-303). London: Routledge.

DeSanctis, G., Poole, M. S., Dickson, G. W., & Jackson, B. M. (1993). Interpretive analysis of team use of group technologies. *Journal of Organizational Computing, 3*(1), 1-29.

Engeström, Y. (1987). *Learning by expanding: An activity-theoretical approach to developmental research.* Retrieved February 9, 2007 from http://communication.ucsd.edu/MCA/Paper/Engestrom/expanding/toc.htm

Figueiredo, A. D., & Afonso, A. P. (2005). Context and learning: A philosophical approach. In A. D. Figueiredo & A. P. Afonso (Eds.), *Managing learning in virtual settings: The role of context* (pp. 1-22). Hershey, PA: Idea Group.

Gersick, C. J. (1988). Team and transition in work teams: Toward a new model of group development. *Academy of Management Journal, 31,* 9-41.

Gersick, C. J. (1989). Marking time: Predictable transitions in task groups. *Academy of Management Journal, 32,* 274-309.

Gibson, C. B., & Cohen, S. G. (Eds.). (2003). *Virtual teams that work: Creating conditions for virtual team effectiveness.* San Francisco: Jossey-Bass.

Hersey, P., & Blanchard, K. H. (1999) *Leadership and the one-minute manager.* William Morrow.

Hung, D. W. L., & Der-Thanq, C. (2001). Situated cognition, Vygotskian thought and learning from the communities of practice perspective: Implications for the design of Web-based e-learning. *Distance Education, 38*(1), 3-12.

Johnson, C. M. (2001). A survey of current research on online communities of practice. *The Internet and Higher Education, 4,* 45-60.

Johnson, S. D., Suriya, C., Won Yoon, S., Berrett, J. V., & La Fleur, J. (2002). Team development and group processes of virtual learning teams. *Computers and Education, 39,* 370-393.

Kock, N. (2004). The psychobiological model: Toward a new theory of computer-mediated communication based on Darwinian evolution. *Organization Science, 15*(3), 327-348.

Kock, N., & Nosek, J. (2005), Expanding the boundaries of e-collaboration. *IEEE Transactions on Professional Communication, 48*(1), 1-9.

Lave, J., & Wenger, E. (1991). *Situated learning: Legitimate peripheral participation.* Cambridge: Cambridge University Press.

Law, J. (1992). Notes on the theory of the actor-network: Ordering, strategy and heterogeneity. *Systems Practice, 5,* 379-393.

Lesser, E. L., & Storck, J. (2001). Communities of practice and organizational performance. *IBM Systems Journal, 40*(4), 831-841.

McGrath, J. E. (1991). Time, interaction, and performance (TIP): A theory of groups. *Small Group Research, 22*(2), 147-174.

Nardi, B. A. (Ed.). (1996). *Context and consciousness: Activity theory and human computer interaction.* Cambridge, MA: MIT Press.

Stacey, E., Smith, P. J., & Barty, K. (2004). Adult learners in the workplace: Online learning and communities of practice. *Distance Education, 25*(1), 107-123.

Tuckman, B. W. (1965). Developmental sequences in small groups. *Psychological Bulletin, 64*(6), 384-399.

Wenger, E. (1998). *Communities of practice: Learning, meaning, and identity.* Cambridge: Cambridge University Press.

Wenger, E., McDermott, R., & Snyder, W. (2002) *Cultivating communities of practice: Guide to managing knowledge.* Harvard Business School Press.

KEY TERMS

Activity Theory: A theoretical framework, inspired by the Russian socio-historical psychology school, which focuses on the cultural and technical mediation of human activity.

Actor-Network Theory (ANT): Social theory used to help understand the behavior of networks where humans and non-humans interact and support each other.

Collaboration: Act of working together with a collective commitment to a common mission and a shared effort to get results that would never be achieved by any of the parts in isolation.

Community of Practice: Informal network of people who share common interests, exchange ideas, find solutions, and collectively build knowledge.

Cooperation: Act of working together where each part recognizes the benefit of shared action and is willing to support collective efforts, provided its individual aims and autonomy are not sacrificed.

Coordination: Act of working together where each part does not necessarily recognize the benefit of working together but knows (or is told) what, when, and how to do what needs to be done and accepts the alienation of some of its autonomy to accomplish it.

Pattern Language Theory: Theoretical framework maintaining that everything we build should grow naturally and that patterns, or common features and relationships abstracted from previous successful solutions, should be used to support that growth.

Principle of Sustainable E-Collaboration: E-collaboration is only sustainable as long as each part feels it is gaining from it and acts so as to grant that all the other parts feel likewise.

Value Proposal: A common agreement that matches the interests between *all* the parts so that they are willing to keep collaborating.

Task Constraints as Determinants of E-Collaboration Technology Usefulness

Ned Kock
Texas A&M International University, USA

INTRODUCTION

As anyone who looks at the history of research on e-collaboration technologies can attest, much is yet unknown about the impacts of those technologies on people (Kock, 2005; Kock & D'Arcy, 2002; Kock, Davison, Ocker, & Wazlawick, 2001). The development and test of pioneering theoretical models from the 1970s and 1980s, such as the social presence and media richness theories (Daft & Lengel, 1986; Short, Williams, & Christie, 1976), has led to the realization that e-collaboration is a complex phenomenon. This perception of complexity has been met by the development of taxonomies, or classifications, of e-collaboration scenarios.

Since e-collaboration technologies have normally been used to accomplish tasks, hopefully with some advantages over plain face-to-face interaction, taxonomies of both e-collaboration technologies and tasks have emerged (Kock, 2005). The following natural step was the development of theories that proposed that certain types of e-collaboration technologies are better matched with certain types of tasks. Some of those theories hypothesized their e-collaboration technology-task fit links explicitly, which make them easier to test and refine, whereas others have not.

This article provides a brief review of one e-collaboration technology-task fit theory, and argues that it focuses (like most technology-task fit theories) on what can be accomplished through tasks, as opposed to what cannot—that is, the tasks' constraints. The article also argues that task constraints are important explanatory and predictive elements, illustrating that point through an example of a car racing team that employs text-based instant messaging for communication between pilots and support team during races.

BACKGROUND

Zigurs and Buckland's (1998) theory stands out among the task-technology fit theories that can explain and predict human behavior toward e-collaboration tools. The reason is the theory's clarity and parsimony, which are desirable components of any theory that aims to be testable. And, as Popper (1992) pointed out in one of his main contributions to the philosophy of science, a theory that is not testable is not very useful either.

The theory proposed by Zigurs and Buckland (1998) classifies tasks into five main types: simple tasks, problem tasks, decision tasks, judgment tasks, and fuzzy tasks. E-collaboration technologies are differentiated from each other based on three key dimensions, which can be measured in terms of the degree to which each dimension is present in a certain e-collaboration tool. The three dimensions are communication support, process structuring, and information processing. For example, an instant messaging system would provide a higher degree of communication support than a Web-based workflow control system, and a lower degree of process structuring. A group decision support system would generally provide a higher degree of information processing (the compilation, aggregation, presentation, etc., of complex information) than e-mail.

The theory proposed by Zigurs and Buckland (1998) is one of the best developed and, as mentioned before, testable theories of task-technology fit applied to e-collaboration. It highlights e-collaboration technology types and support dimensions that are arguably important in the decision to use this or that type of e-collaboration system (or this or that brand and model of e-collaboration system). The theory places emphasis on what e-collaboration technologies can offer to accomplish certain tasks.

One could argue, however, that the taxonomy of tasks proposed by the theory is missing one key element, which under some circumstances may be the most important in informing decisions to adopt a particular e-collaboration technology. That key element is, essentially, what the process by which the task is accomplished "does not allow"—that is, a task constraint.

GOLDRATT'S THEORY OF CONSTRAINTS

Goldratt's (1999) theory of constraints is perhaps the most popular theoretical model addressing the issue of task constraints, in the sense outlined above. Perhaps its popularity is due to the fact that it was first presented as a best-selling novel titled "The Goal" (Goldratt & Cox, 1986), where a business process improvement consultant helps a manufacturing plant manager deal with a number of professional and personal problems.

The underlying theme of Goldratt's (1999) theory of constraints is that the productivity and quality of the outcomes of a process, by which a task is accomplished (e.g., the process of assembling a car), are strongly determined by the process' constraints. For example, the speed through which cars will be produced by an assembly line is much more strongly defined by the speed of the slowest and more laborious step in the car assembly process than by the faster and simpler steps. In other words, if fitting the windshields is more problematic and takes longer than fitting the doors to the car's main body, then someone looking at improving the process ought to look at the windshield-fitting step more carefully than at the doors-fitting step. This is a very simple idea, but with key implications for decisions related to what e-collaboration tools to use to support one task or another.

A CONSTRAINTS-BASED VIEW OF TECHNOLOGY USEFULNESS

The idea of looking at collaborative tasks from a constraints perspective is not new. For example, Trevino, Daft, and Lengel (1990) already pointed out as part of their symbolic interactionist view of communication media selection and use that a key collaborative task constraint, namely the geographic distribution of the collaborators, strongly influences the decision to which e-collaboration technology to adopt, and how the collaborators view and use the technology.

What is not present in much of the e-collaboration research is a concern with low-level constraints (e.g., task-specific, rather than task-type specific) posed by collaborative tasks. This may be one of the reasons why low-level technology attributes (e.g., system-specific, rather than technology type-specific), are not usually addressed in e-collaboration research. (See, e.g., Markus, 2005, for a more elaborate discussion on this, from a slightly different angle). Low-level collaborative task constraints can influence much more strongly the decision of which e-collaboration technology to use to support the task, as well as the expectations of the technology users and their success in accomplishing the task.

AN ILLUSTRATION: CAR RACING AND INSTANT MESSAGING

Instant messaging is an e-collaboration technology that has been steadily gaining ground in business circles, although its use is still far less widespread than that of e-mail. Instant messaging allows for synchronous communication in a chat-like manner, with a much higher level of interactivity than e-mail (which is primarily used for asynchronous, or time-disconnected, interaction).

Arguably, one of the reasons why instant messaging is not more widely used is that there is another technology that enables synchronous communication and that seems to be better adapted to the design of our biological communication apparatus (Kock, 2004). That other technology is the telephone. We human beings seem to be able to communicate much more easily in an oral fashion than by typing and reading text through computers, which makes text-based instant messaging a somewhat cumbersome alternative to the telephone. Even desktop conferencing using audio only, or audio and video, is likely to be perceived as more natural than instant messaging by the vast majority of us.

But certain task constraints can significantly tip the balance in favor of instant messaging. Take for example the case of the Chip Ganassi Racing team, described by Betts (2004) in a *Computerworld* magazine article. Members of the Chip Ganassi Racing team, which competes in the NASCAR and Indy Racing League, were looking for an alternative to voice communication with the racing car drivers.

Voice communication through radio was problematic not only because it was difficult to find a usable radio channel, but also because of the background noise coming from the driver's car and other cars. These are two key constraints that are inherent in the car-racing task. The solution was instant messaging communication between the crew and the drivers, using an encrypted wireless LAN.

In this example, the task constraints—difficulty finding a usable radio channel, and the background noise coming from the driver's car as well as other cars—were stronger determinants of the choice of e-collaboration technology used than other elements (e.g., perceive communication medium naturalness). Moreover, the task constraints seem to have been more decisive in the choice of technology to be used than the general type of the task.

CONCLUSION

Pioneering theoretical work conducted in the 1970s and 1980s led to key testable theories of e-collaboration. Many empirical tests followed, serving to establish e-collaboration as a distinct field of research. Those theoretical tests led to the generalized perception that e-collaboration phenomena were more complex than originally predicted, and to the search for new theoretical frameworks. Task-technology fit theories emerged, which focused more on what could be accomplished through tasks, that what could not (i.e., task constraints). Here, task constraints are presented as strong determinants of the identification of best fit between e-collaboration technologies and tasks. An illustrative example is provided of a NASCAR and Indy car racing team's use of instant messaging over a wireless network during races.

An exploration is presented here of some ideas that could lead to the development of theories that place emphasis on the surfacing of task constraints as a basis for the selection and use of supporting e-collaboration technologies. This article looks at attempts at task-technology fit theorizing targeted at understanding and predicting e-collaboration phenomena. Those attempts have led to theories that proposed that certain types of e-collaboration technologies are more appropriate for certain types of tasks than other e-collaboration technologies. Among those theoretical efforts, Zigurs and Buckland's (1998) stands out for its resulting theoreti-

cal model's clarity, parsimony, and testability (see also Zigurs, Buckland, Connolly, & Wilson, 1999).

This article calls for a careful look at task constraints when identifying the best fit between an e-collaboration technology and a task. This perspective seems to be missing in Zigurs and Buckland's (1998), as well as other task-technology fit theories addressing e-collaboration phenomena. In fact, rarely one sees theoretical frameworks that address task constraint issues as determinants of e-collaboration technologies selection and use.

The goal of this article is not to develop a new theory of e-collaboration based on task constraints, something that would require significantly more space than available here. The main goal here is to point out that such a theory (or theories) can be developed, and explore some ideas that could lead to the development of such a theory (or theories). As pointed out by Markus (2005, p. 1), "...technologies pose problems for users who want to use them to accomplish particular goals; the solutions users create for those problems during recurrent use may exhibit certain regularities across different contexts." Hypothesizing about such regularities in the context of certain types of task constraints will arguably be the main outcome of related theoretical pursuits.

REFERENCES

Betts, M. (2004). Chat provides competitive edge. *Computerworld, 38*(36), 40.

Daft, R. L., & Lengel, R. H. (1986). Organizational information requirements, media richness and structural design. *Management Science, 32*(5), 554-571.

Goldratt, E. (1999). *Theory of constraints.* New York: North River Press.

Goldratt, E. M., & Cox, J. (1986). *The goal: A process of ongoing improvement.* New York: North River Press.

Kock, N. (2004). The psychobiological model: Towards a new theory of computer-mediated communication based on Darwinian evolution. *Organization Science, 15*(3), 327-348.

Kock, N. (2005). *Business process improvement through e-collaboration: Knowledge sharing through the use of virtual groups.* Hershey, PA: Idea Group Publishing.

Kock, N., & D'Arcy, J. (2002). Resolving the e-collaboration paradox: The competing influences of media naturalness and compensatory adaptation. *Information Management and Consulting, 17*(4), 72-78.

Kock, N., Davison, R., Ocker, R., & Wazlawick, R. (2001). E-collaboration: A look at past research and future challenges. *Journal of Systems and Information Technology, 5*(1), 1-9.

Markus, M. L. (2005). Technology-shaping effects of e-collaboration technologies: Bugs and features. *International Journal of e-Collaboration, 1*(1), 1-23.

Popper, K. R. (1992). *Logic of scientific discovery*. New York: Routledge.

Short, J. A., Williams, E., & Christie, B. (1976). *The social psychology of telecommunications*. London: John Wiley & Sons.

Trevino, L. K., Daft, R. L., & Lengel, R. H. (1990). Understanding manager's media choices: A symbolic interactionist perspective. In J. Fulk & C. Steinfield (Eds.), *Organizations and communication technology* (pp. 71-94). Newbury Park, CA: Sage.

Zigurs, I., & Buckland, B. K. (1998). A theory of task-technology fit and group support systems effectiveness. *MIS Quarterly, 22*(3), 313-334.

Zigurs, I., Buckland, B. K., Connolly, J. R., & Wilson, E. V. (1999). A test of task-technology fit theory for group support systems. *Database for Advances in Information Systems, 30*(3), 34-50.

KEY TERMS

Collaborative Task: Task that is often conducted by a group of people with support of e-collaboration technologies.

E-Collaboration: Collaboration using electronic technologies among different individuals to accomplish a common task.

E-Collaboration Technology: Comprises not only the communication medium created by an e-collaboration technology, but also the technology's features that have been designed to support collaborative work.

Instant Messaging: E-collaboration technology that supports synchronous communication in a chat-like manner, with a much higher level of interactivity than e-mail.

Task-Technology Fit Theories: Theories that include taxonomies of both e-collaboration technologies and tasks, and that propose that certain types of e-collaboration technologies are better matched with certain types of tasks.

Theory of Cconstraints: Theory developed by Goldratt proposing that the productivity and quality of the outcomes of a business process are strongly determined by the process' constraints.

Wireless LAN: Local area network in which computers and other devices exchange data through radio channels.

Technological Challenges in E-Collaboration and E-Business

Fang Zhao
RMIT University, Australia

INTRODUCTION

E-collaboration takes advantage of the current Internet-driven business environment, which integrates the most advanced electronic technologies and the knowledge-based economy. Companies engaging in e-collaboration must participate in external business relationships by using computer interactions (Damanpour, 2001). Implementing e-collaboration strategy can require many sophisticated technologies and systems such as EDI, XML, eCRM. E-collaboration is thus confronted with the great challenge of re-engineering IT strategies and resources. "Nearly 80% of organizations that have rushed to establish Web sites for online retailing have failed to invest in the purchasing and distribution systems that make delivery of their products possible" (Neef, 2001, p.3). System failure has a profound effect on e-collaboration and e-business, both in the short and long-term. The tremendous complexity of information technologies has become a huge hurdle to companies embracing them, affecting their entire management strategy, process, structure, and most importantly, business bottom line results. The main technological issues to be considered are associated with IT infrastructure, and managers' and operatives' knowledge and skills in e-collaboration and e-partnership. The following constitutes some of the key technological issues facing e-collaboration.

- Process and system alignment and integration
- Interoperability of systems
- Accessibility, security and compatibility of inter-organizational information systems
- Traffic in collaborative e-commerce activities
- Sustained IT support and resources
- Transferring and sharing information and data
- Building and sustaining an effective virtual network structure amongst e-partners
- Quality and effectiveness of networking and communications (Zhao, 2004)

This article focuses on the most important technological challenges and issues facing e-collaboration and e-business in the areas of information flow, procurement, logistics, engineering and manufacturing, marketing, customer services, and human resources. It deals specifically with the process and system alignment and integration as well as the issues of interoperability which have become of primary concerns in the practices of e-collaboration

BACKGROUND

Generally speaking, e-collaboration refers to the use of the Internet and/or Internet-based tools among business partners beyond market transactions. The term is often used in the context of supply chain, in particular, in supplier-buyer relationships. E-collaboration is identified as one of the new areas of optimizing the relationship between supplier and OEM via the Internet. It is an Internet-supported, enterprise-spanning cooperation which is viewed as crucial during the development and construction process (the so called e-engineering process) (Kersten, Schroeder, & Schulte-Bisping, 2004). E-collaboration aims to facilitate coordination of various supply chain activities and decision-making processes. It often involves sharing of information and knowledge on which joint supply chain decisions can be made. Information that needs to be shared amongst supply chain partners often include sales data, inventory status, production schedule, promotion plans, demand forecasts, shipment schedule, and new product introduction plans. In addition to information sharing, e-collaboration provides opportunities for collaborative planning and new product development. By resorting to e-collaboration and Web technologies, supply chain partners can exchange product forecasts and replenishment plans and then develop new plans that meet market demand in a timely and effective way. Studies show that this kind of e-collaboration enables the reduction of inventory costs and enhancement of customer service level across the supply chain (Lee & Whang, 2002). New product development is also facilitated by e-collaboration between business partners in which collaborative product development such as product rollover (the transition from one version of a product to its successor) is completed with efficiency and speed powered by Web technologies. Some of the popular e-collaboration methods include virtual workrooms, online visualization of demand forecast, online monitoring of capacity utilization, virtual development platforms and online visualization of business processes (Kersten et al., 2004).

In terms of the e-supply chain collaboration, simplified and standardized solutions based on common technology architecture must be instigated, which may include trading partner processes, multiple levels of connectivity amongst trading partners, internal infrastructure, and system reengineering to ensure e-supply chain interoperability (that is, the ability to be fully compatible and capable of being integrated with each other in e-business), and e-application architecture (Ross, 2003). Given the fact that many companies now operate in more than one electronic supply chain, multiple IT integration becomes paramount to their business operations. Interoperability can be achieved through process standardization and information standards (e.g., EDI and RosettaNet Standards). Studies show that the achievement of multiple IT integration brings significant benefits to the companies that implement it (Davis & Spekman, 2004). A study of implementation of e-SCM solutions shows that often the implementation is "fraught with difficulties, potentially enormous expenses, and significant trauma to even the best of organizations" (Ross, 2003, p.325). There are also issues relating to levels of implementation. As companies vary considerably in terms of the nature of their business, capacities, resources, size, developmental stage, culture, competency of leadership, and so forth, the strategy for e-collaboration infrastructure should be realistic and feasible, and in line with the company's actual need and capacity to embrace e-business. Incremental rather than radical changes are encouraged to implement limited, tactical Web technologies that will enhance existing processes for e-collaboration.

INTEROPERABILITY: ISSUES AND OPTIONS

Interoperability problems are one of the key issues that are paramount to competitiveness and success of corporations in e-collaboration. Interoperability must be viewed as a core business process in managing interorganizational e-collaborations. Unfortunately, the costs of interoperability problems are often underestimated or generally neglected by management. Empirical studies show that interoperability problems can be very costly in terms of financial and time resources and can impact adversely corporate productivity and cause enormous frustration among e-partners. Some large companies have endeavored to alleviate the negative impact of interoperability problems by implementing significant procedures, tools, and infrastructure. However, they are not always successful in resolving the issues (Interoperability best practices, 2004). According to a research presented by David Prawel (2003), president of LongView Advisors Inc. Colorado, at the Time Compression Technologies 2003 Conference, many companies still spend a huge amount of engineering resource on performing the manual tasks associated with sending and receiving data, confirming receipt, tracking contract information, and so forth due to a lack of underlying infrastructure support. Lack of management involvement and assigning responsibility for interoperability to a specific manager or team also contribute to the failure in resolving interoperability problems. Prawel made some key recommendations that may help companies with interoperability problems to improve their interoperability. Interoperability training and consulting, management support, a corporate approach to addressing interoperability issues, developing service relationships with service providers, and investment in infrastructure such as good translation and infrastructure tools are among the key recommendations.

The collaborative technology infrastructure capacities required may vary in different supply chains and e-business contexts, along with the role and size of each e-partner. The following presents fundamental and broad strategies for establishing and maintaining an effective e-infrastructure for e-collaboration in the supply chain.

- Establishing simple and low-cost connectivity to ensure that smaller firms are able to access, and participate fully in, a collaborative infrastructure without having to make a major investment. For example, organizations should have access to networked SCM applications which are browser-based through broadband Internet connections or virtual private networks.
- Establishing and implementing a common data model for data storage across the supply chain, which would be a simpler, faster, and far more efficient than integrating all the various data models.
- Developing high-level self-service technologies that enable supply chain members to not only track orders and obtain logistics and billing information, but also automatically configure products, make payments, and resolve disputes.
- Developing business intelligence technologies to analyze the ongoing flow of information drawn from the entire supply chain, which helps companies make improvements in internal operations and collaborative capabilities on an ongoing process. In fact, applying business intelligence into e-business processes poses enormous opportunity for value creation in the supply chain and enhances SCM practices (Horvath, 2001).

PROCESS AND SYSTEM ALIGNMENT AND INTEGRATION: ISSUES AND OPTIONS

Integration refers to collaborative planning and control, decision integration, information integration, and business process integration between interfirm partners, using information technologies and systems. The technological side of the integration is crucial to e-collaboration. For example, the complexity of integration required by e-marketplaces is one of the big problems that have been attributed to the sharp decline in the number of e-marketplaces. Today, most companies have implemented enterprise resource planning (ERP) systems to automate their back-end planning and scheduling processes and to undertake internal IT integration to meet the needs of multiple vendors and customers for years. B2B software has allowed IT integration across companies with different IT platforms. But there is limited application

of Web technologies to the rest of the procurement process. There is generally a lack of real-time supply and demand information flows amongst supply chain partners, which results in inaccurate planning leading to either inventory shortages or excessive inventories. Therefore, system integration and alignment becomes paramount to e-collaboration, and thus affects directly bottom line results.

E-business provides organizations with opportunities to align their processes for e-collaboration to attain success. However, technology-wise, interoperability requires enhancement of existing systems to transfer them into a cross-firm mode. Electronic supply chain requires integration of software platforms or open systems across the entire network. Integrating processes and systems is paramount to a seamless link with partnering companies. For example, successful implementation of electronic data exchange (EDI) requires realignment of work processes and systems within the network of e-collaboration. However, according to Lowson and Burgess' (2003) study, many organizations, particularly small to medium-sized enterprises (SMEs), have not taken on, or have a limited use of, EDI and other interorganizational systems (IOS) to integrate their supplier processes, operations processes, and sales processes, because they are often not able to undertake the cost of technologies and the management systems integration.

E-collaboration often requires the reengineering of business processes across companies, which is very expensive in terms of time, capital, and human resources. As one former supply chain executive explained, it took major collaboration efforts and 12-18 months to implement business process reengineering between just two trading partners (Davis & Spekman, 2004). In addition, system integration and alignment should take into account the diversity of e-partners. There is hardly a one-size-fits-all solution for all partners. Take Sun Microsystems, for example. Sun has employed three main Web technologies in its e-network: connected ERP systems, B2B e-marketplaces, and Webstores. The company enables its large partners to directly place their orders in its ERP systems. Other partners have the options to choose the e-business application that suits them best. The flexibility of e-business applications that Sun provide facilitates system integration and alignment in an optimal way. In a recent study, de Man and der Zee (2002) suggested that there were a number of technological lessons learned in the process of starting and building Web applications in e-collaboration. These included that systems should never be forced upon partners, and that channel conflicts should be avoided by selecting the right e-business application for each partner and client group. Internal systems should be changed to cater for the requirements of e-network, and finally, the overall process should be guided by the concepts of standardization, harmonization, and simplification.

Establishing e-business process standards is another issue for interfirm B2B integration whose objective is to meet the needs of global supply chains. RosettaNet (2004) has been successful in providing a common language for B2B transactions and in building integrative e-business processes among partners within the global trading network. RosettaNet standards that have been used by Fortune companies worldwide "prescribe how networked applications interoperate to execute collaborative business process."

There are numerous companies that specialize in providing business-to-business integration, synchronization, and collaboration solutions. Global eXchange Services (GXS) is one of leaders in the field. GXS has designed a set of solutions called the "Extended Value Chain" to help streamline cross-enterprise business process. The Extended Value Chain consists of four key layers: transaction, monitoring, synchronization, and collaboration, enabling companies to:

- Transact information with their trading partners by enabling the transmission of information regardless of protocol (e.g., TCP/IP, EDI, XML, etc.)
- Monitor their operations by providing visibility and analytics into the movement of information between enterprise
- Synchronize business processes by enabling their integration, automation and optimization
- Collaborate using solutions that leverage cross-enterprise business processes in real time (Greenfield, 2004)

Cisco System is often cited as a successful example of seamless integration throughout its supply chain operating systems with its partners (Davis & Spekman, 2004). The integration consists of three parts: (1) planning, control, and design integration, (2) information integration, and (3) business process integration. Planning, control, and design integration mainly concerns making collaborative decisions regarding inventory replenishment, and collaborative product development. As the name suggests, information integration refers to the sharing of forecast data, inventory data, customer order, and status information, but it also includes system application integration with trading partners. Business process integration involves allowing partners to access ERP system and MRP processes, automation of routing of EDI data to supplier partners, automation of cross-firm business processes, and real-time flow of customer orders to all partners.

FUTURE TRENDS

As e-commerce and e-business practices will continue to grow, e-collaboration will be more mature (rather than experimental) in nature, in terms of the scope, quality, and credibility of online customer services and products. Participating in e-collaboration will be part of every executive's job in the near future. In terms of supply chain network integration, McCormack et al.'s study shows that most industry supply chains today have not reached the stage at which information and system integration is in place to build a supply chain network (McCormack, Johnson, & Walker, 2003). Full network integration—that is, all key business processes being online and being aligned within the network—will be the next step that organizations need to take to gain competitive advantage over other supply chain networks. E-collaboration in supply chains or virtual supply chains will become a critical part of the future supply chain landscape. Collaboration amongst virtual manufacturers, virtual distributors, virtual retailers, and virtual service providers will dominate the virtual supply chains. E-business infomediaries will leverage the Internet to perform matching of products and buyers or coordinate marketing and transaction processes in e-collaborations.

E-collaboration is, and will continue to be, the key to sustained business success. An e-business strategy will be ineffective without an integrated e-collaboration strategy, because the ability to leverage collaborative relationships becomes essential in today's competitive e-business world. Consumer/purchaser power will dominate the e-business world and propel smaller e-businesses to collaborate to provide customers with an ever-widening array of products and services, real-time and rich information, and speedy and quality transactions. Moreover, e-collaboration helps streamline the product-to-market process through collaborative planning and design, improve efficiency from the channel network by reducing inventories, and ultimately generate profitability (Zhao, 2006).

CONCLUSION

E-collaboration requires interfirm business architecture, including the reengineering of the processes that link companies to their channel trading partners and the development of a collaborative community of trading partners. It also requires closely integrated databases and closely synchronized information flows to eliminate distortions and the "bullwhip" effect in the communication of information between supply chain partners. E-application architecture is imperative to the collaboration and involves "determining individual integration points between the application and data sources, the application and back-end installed software, and between multiple back-end systems" (Hoque, 2001, p.153). This article has demonstrated that information technologies have greatly expanded the way companies do business and partners interact with each other. The value and the prospects that e-collaboration strategy can generate for business are compelling firms to adopt e-collaboration technologies and systems into their business processes. However, technology integration and interoperability issues can be complex. For example, data synchronization using XML can be a formidable task in the transformation process because there are many different data and alert types, and the published XML-based standards do not cover all possible collaboration data. The article highlights many implementation issues regarding technology adoption.

REFERENCES

Damanpour, F. (2001). E-business e-commerce evolution: Perspectives and strategy. *Managerial Finance, 27*(7), 16-32.

Davis, E. W., & Spekman, R. E. (2004). *The extended enterprise: Gaining competitive advantage through collaborative supply chains.* Upper Saddle River, NJ: Prentice Hall.

De Man, A. P., & der Zee, H. V. (2002). *Strategies for e-partnering: moving brick-and-mortar online.* Groningen: Gopher Publishers.

Greenfield, G. (2004). *GXS: Enabling tomorrow's solutions today.* Retrieved July 10, 2004, from www.gxs.com.

Hoque, F. (2001). *E-enterprise: Business models, architecture, and components.* Cambridge, MA: Cambridge University Press.

Horvath, L. (2001). Collaboration: The key to value creation in supply chain management. *Supply Chain Management: An International Journal, 6*(5), 205-207.

Interoperability best practices: The ongoing problems of sharing engineering data. (2004). *Strategic Direction, 20*(5), 31-33.

Kersten, W., Schroeder, A. K., & Schulte-Bisping, A. (2004). Internet-supported sourcing of complex material. *Business Process Management Journal, 10*(1), 101-114.

Lee, H. L., & Whang, S. (2002). Supply chain integration over the Internet. In J. Genunes et al. (Eds.), *Supply chain management: Models, applications, and research directions* (pp. 3-18). Bordrecht: Kluwer Academic Publishers.

Lowson, R. H., & Burgess, N. J. (2003). The building blocks of an operation strategy of e-business. *The TQM Magazine, 15*(3), 152-163.

McCormack, K. P., Johnson, W. C., & Walker, W. (2003). *Supply chain networks and business process orientation: Advanced strategies and best practices.* New York: St. Lucie Press.

Neef, D. (2001). *E-procurement: From strategy to implementation.* Upper Saddle River, NJ: Prentice-Hall.

Prawel, D. (2003). *Interoperability best practices: Advice from the real world.* Paper presented at the TCT 2003 Conference, NEC, UK.

RosettaNet. (2004). *Dynamic trading networks. Operational efficiency. New business opportunities. Investment protection* (p. 6). California: The Author.

Ross, D. F. (2003). *Introduction to e-supply chain management: Engaging technology to build market-*

wining business partnerships. Boca Raton, FL: St. Lucie Press.

Zhao, F. (2004), E-partnerships and virtual organizations: Issues and options. In M. Singh & D. Waddell (Eds.), *E-business: Innovation and change management* (pp.105-119). Hershey, PA: Idea Group Publishing.

Zhao, F. (2006). *Maximize business profits through e-partnerships,* Hershey, PA: Idea Group Publishing.

KEY TERMS

E-Collaboration: E-collaboration refers to the use of the Internet and/or Internet-based tools among business partners beyond market transactions. The term is often used in the context of supply chain, in particular, in supply-buyer relationships.

E-Partnership: Theoretically, e-partnership refers to a business partnership relying on electronic (information) technologies to communicate and interact amongst partners. As e-business has become an integral part of most business practices where consumers, suppliers, buyers are connected by information technologies, the term e-partnership is mostly associated with electronic commerce partnerships, and in a broader sense, electronic business partnerships.

E-SCM (E-Supply Chain Management): E-SCM as the latest advance of SCM has two pillars: the emerging strategic capabilities of SCM and the Web technologies that empower SCM. E-SCM aims to foster agile organizations and supplier-buyer partnerships.

E-Supply Cchain Interoperability: The ability to be fully compatible and capable of being integrated with each other in e-business supply chain.

Informediary: As the name suggests, infomediaries specialize in information management, collecting and storing customer information and controlling the flow of commerce on the Web. Yahoo! is one of the most popular and powerful infomediaries in the world.

Integration: Integration refers to collaborative planning and control, decision integration, information integration and business process integration between interfirm partners, using information technologies and systems.

RosettaNet Standards: RosettaNet standards prescribe how networked applications interoperate to execute collaborative business process. They provide a common language for B2B transactions and assist in building integrative e-business processes among partners. RosettaNet standards consist of a three-level business process architecture for interaction between inter-firm e-partners: (i) partner interface processes, (ii) RosettaNet dictionaries, including the Master Dictionary which contains over 6000 common terms and processes, and grammar that describes how systems communicate, and (iii) RosettaNet implementation framework (RNIF).

Technological Challenges to the Research and Development of Collaborative Working Environments

Karl A. Hribernik
Bremen Institute of Industrial Technology and Applied Work Science (BIBA), Germany

Klaus-Dieter Thoben
Bremen Institute of Industrial Technology and Applied Work Science (BIBA), Germany

Michael Nilsson
Luleå University of Technology, Sweden

INTRODUCTION

Through emerging technological developments, the human being is increasingly becoming liberated in his or her role as a knowledge worker, becoming able to interact seamlessly with colleagues, resources, and information facilitating improved work processes and new value generation opportunities.

In this context, the ability to interact over distance across organizational, geographical, and cultural boundaries, as well as on site, is becoming increasingly important to the overall competitiveness of any organization, as collaborative tools and technologies are rapidly being adopted by both the market and society as soon as they become available. It is evident that the market for more advanced solutions is enormous.

Current developments with regards to e-collaboration technology point toward the integration of previously insular solutions into seamless collaborative working environments (CWE), which represent one such solution to future knowledge society needs (European Commission, 2004). This aim of this article is to identify the current key technological challenges research and development in the field of CWEs is confronted with. The findings are based on a study of twelve collaborative work scenarios from heterogeneous domains, from both commercial and scientific perspectives.

BACKGROUND

Collaborative working environments are defined by Hribernik, Nilsson, Fusco, and Niitamo (2005, p. 10) as "integrated and connected resources providing shared access to contents and allowing distributed actors to seamlessly work together towards common goals." The definition is rooted in the understanding of collaboration as the act or result "of working together to produce a piece of work," according to Sinclair, Fox, and Bullon (1995). It further builds on three key aspects of collaboration (Thoben, Hribernik, Kirisci, & Eschenbaecher, 2003) supported by ICT (information and communication technology). Here, collaboration is viewed first as communication between actors (enterprises, institutions, groups, individuals, machines, etc.). Secondly, it describes cooperation on a technical level, which enables machines and computers to exchange data. The third aspect is coordination, which is described by the exact coordination of communication, cooperation and coordination processes.

CWE represent the next step in the development of collaborative technologies, moving from disparate, insular tools, applications, systems, and services to seamlessly integrated ICT environments for collaborative work from a multidomain perspective. Such CWE should provide full support for optimal interaction within the dimensions of e-collaboration as described in (Thoben et al., 2003): stages (initiation, management, operation, and dissolution), forms (ad-hoc, mediated, planned, and hybrid forms) and assets of collaboration (people, ICT-systems, knowledge, and processes).

In order to achieve an understanding of the type of challenges CWE research and technology development (RTD) needs to address, it is critical to understand requirements as set by real-world and potential future collaborative work scenarios. To gather requirements across a broad spectrum of domains relevant to CWE, a study of twelve scenarios was carried out (Hribernik et al., 2005). The following scenarios were analyzed in the course of the study:

1. **Construction industry:** Large construction projects
2. **Maintenance:** Heavy duty field equipment and aerospace
3. **E-health:** Home care service chain
4. **Engineering:** Virtual enterprise collaboration hub
5. **Engineering:** Professional collaboration in virtual teams and organizations
6. **Humanitarian aid:** Environmental collaboration scenario for humanitarian aid
7. **Rural services:** Collaboration in argriculture and forestry
8. **Urban services:** Collaborative pollution control
9. **Emergency management:** Fire fighting
10. **E-inclusion:** Integrating the disabled into professional life
11. **Collaborative design:** Collaborative work within SME clusters
12. **Knowledge work:** E-professionals in the business and public sectors

From the individual requirements of each of the scenarios, a number of high-level requirements toward CWE can be distilled. These requirements are drawn on to identify major challenges to research and development of CWE, and are described in the following sections.

STUDY APPROACH

The twelve collaborative work scenarios listed above were analyzed in the course of the study. The selection of the scenarios was carried out by an expert group consisting of representatives from major industry players, research funding organizations, innovation agencies, and academic institutions. The scenarios were selected to be representative of sectors in which on the one hand e-collaboration is traditionally successfully employed, and on the other promises to enable new business models and forms of work. The selection encompassed well-documented e-collaboration scenarios in which CWE and related e-collaboration technology

had been implemented in the course of both research and commercial projects. A matrix for the analysis of CWE requirements was developed on the basis of the types (planned, mediated, and ad hoc), stages (initiation, management, operation, and dissolution) and assets (people, process, ICT systems, and knowledge) of each individual process step of the analyzed scenarios. For each of the dimensions, the analysis process was structured according to attributes defined in the CWE taxonomy documented in (Hribernik, Nilsson, Fusco, & Niitamo, 2006). Focus was laid on capturing requirements which identify collaboration functionality lacking or unsatisfactory in the current implementations of the individual e-collaboration scenarios. The requirements analysis matrix was subsequently out filled by actors involved in each of the scenarios described. This was carried out by means of structured questionnaires. Finally, the requirements analysis matrixes were quantified, analyzed, and a set of RTD challenges identified on the basis thereof. The resulting challenges were subsequently verified by means of a systematic benchmarking of 21 national research programmes relevant to CWE.

SCENARIOS STUDIED

The scenario dealing with e-collaboration in the construction industry represents the findings of a number of key German construction industry and academic organizations (Hribernik, Kirisci, & Hünecke, 2004). The scenario highlights the major collaborative processes throughout the life-cycle of a construction project: application for planning, application for building, planning and project management, construction execution, tendering subcontractors, site survey, and finally construction acceptance. The scenario exemplifies the need for hybrid forms of collaboration as well as requirements posed by a wide range of public, civic, and industrial stakeholders.

Customer relationship management (CRM) scenarios can be seen as a "traditional" application area for e-collaboration technology. This field has proven to benefit significantly from the introduction of CWE and mobile technology. The scenario analyzed here describes the introduction of mobile collaborative technology to the assistance/after sales and maintenance processes of a major mechanical engineering company in the field of drilling machinery and structures, down-the-hole accessories and specialized tools for use in oil, gas, and water wells.

A further scenario analyzed in the course of the study was a two-year joint Swedish, Norwegian, and Finnish e-health project focusing on e-collaboration applications for home healthcare. The purpose of the project was to develop a service chain out of the existing home care service structures, which would enable updated information exchange in the care of home-care clients regardless of time and place. It was an aspiration that the integrated systems and work processes would facilitate division of work, availability of nursing instructions and multi-professional collaboration. One of the aims was to decrease double documentation and the transportation between the home of the client and the home care center.

An engineering scenario that dealt with the field of virtual organizations in product development was also analyzed. An essential goal for engineers working in this field is to find suitable ways to deal with time-to-market demands, limited budgets, and increased product complexity, while meeting customer requirements and demands on quality (Ulrich & Eppinger, 1995; Wheelwright & Clark, 1992; and Priest & Sanchez, 2001). With estimates showing as much as 85% of the problems with new products having to do with a poor design process (Ulrich & Eppinger, 1995) product development companies are under constant pressure to improve their design. The globalization of industries implies decreased costs and risks, or shared in the case of collaborative projects (Ulrich & Eppinger, 1995), as well as inevitable changes in traditional organizational structures. Consequently, a growing interest exists in creating and supporting global virtual teams, in which collaboration proceeds across time as well as geographical, cultural, functional, and organizational borders (Baird, Moore, & Jagodzinski, 2000). Virtual Enterprises represent manifestations of such collaborations—they are networks of partners and suppliers that work together to reach common goals. The analyzed scenario deals with the use of CWE for the support of Virtual Enterprises, specifically in the aerospace industry. A further engineering scenario was analyzed in the course of the study. The second scenario dealt specifically with collaborative design within a network of SMEs (small and medium-sized enterprises). This scenario followed the steps of developing and operating a collaborative design team.

In agriculture and forestry, farmers, farming service organizations, forest owners, and forest service organizations require very effective access to services based on Earth observation data. Earth observation data providers and value adding organizations (VAOs) need to work together to communicate, share, and exchange information and notify each other of any recent changes to the coverage and portfolio of related products and services. The related scenarios analyzed illustrate the possibility of delivering services combining information and sub-services from different service providers using distributed data from different data providers, catalogues and archives, multi-modality, and wireless access. In total, three applications e-collaboration were taken into account in the study, regarding forestry, humanitarian aid, and pollution analysis, in rural, remote, and urban areas respectively.

Emergency management scenarios can prove extremely useful for analyzing the most demanding requirements to CWE implementations (Hribernik, Ganzer, & Schmidt, 2003). The implementation studied here exemplifies the use of CWE for emergency management throughout all processes of a fire-fighting scenario. The scenario was split into specific phases: alarm (from the fire alarm until fire engine dispatch), approach (dispatch until arrival on scene), and attack phases (arrival of at the emergency scene until the mission completion). Support for ICT-based collaboration between all actors both in the field and stationary was implemented in this scenario, taking care to enable ad hoc collaboration between the fire fighters and, for example, external experts, members of the public, and other emergency services.

The field of e-inclusion deals ensuring equal benefit from ICT to all members of society. Emphasis is laid on minimizing the digital divide and employing ICT to better integrate into society groups already at a disadvantage. This is especially relevant for disabled and elderly people, with over 10% of the EU population affected by disability (Diamantopoulou, 2002). The scenario analyzed in the study describes CWE requirements for the generation of new opportunities for disabled persons to perform daily activities within organizations or even their own private lives (Interaction Design, 2001). These opportunities are based on collaborative working methods with the related ICT tools and technologies addressing both disabled persons who are interested in employment and to employers consider hiring disabled persons.

A further scenario dealt with the CWE requirements of so-called "E-professionals." The term e-professionals denotes knowledge workers who primarily and intensively use ICT, especially collaboration tools, to perform their daily work. The task of examining this scenario was mainly concerned with the analysis of requirements regarding the processes and enablers for knowledge workers. E-professionals in business and public sectors face technological challenges in view of the unprecedented social, organizational, institutional, and even economic and political challenges in all levels of work and organizations, in both developed and emerging markets. This scenario was designed to capture those requirements as posed by emerging knowledge work challenges for CWE.

STUDY RESULTS: CWE TECHNOLOGICAL CHALLENGES

On the basis of the study of collaborative scenario requirements, the following were identified as technological challenges to the research and development of CWE.

The primary challenge identified was the need for enabling seamless interaction between the actors of collaborations. As described above, CWE need to connect diverse types of actors, encompassing enterprises, institutions, groups, individuals, machines, and so on. Humans involved in collaborative work scenarios display heterogeneous skill profiles, which also need to be addressed by providing suitability integrated synchronous and asynchronous communication channels. Role-based user-profiling, semantic descriptions, and context-sensitive environments represent supporting mechanisms for that interaction.

The second technological challenge identified in the study is the necessity for the provision of end-to-end connectivity. End-to-end connectivity in this context refers to a property of CWE that allows all actors in the environment to connect directly to all other actors, without requiring intermediate elements to relay connection. To achieve this, next generation CWE need to overcome the barriers which currently hinder easy and effective end-to-end connectivity, such as restrictive ICT security policies and other interoperability hindrances. Setup and use of CWE must be simple and manageable for the average user. Secure but flexible and easy-to-use tools and mechanisms for the collaboration using arbitrary media channels need to be developed.

The third most important challenge identified is the requirement for the seamless integration of heterogeneous platforms, tools, applications, and services. Future CWE must allow for the greatest possible flexibility in collaborative work regardless of sector, process, organization, location, task, platform, or application. The development of a flexible, open, architecture based on, for example, grid computing or Web services, allowing flexibility and scalability across these dimensions is necessary.

Furthermore, the study showed that CWE must support the discovery, sharing, creation, and modification of both explicit and tacit knowledge within the collaboration process, implying a seamless integration of both forms of knowledge throughout. It is essential that explicit knowledge used within a CWE be seamlessly accessible and usable for all involved, regardless of the knowledge tools employed in the individuals' enterprises or private work spaces. Ontologies (formal languages for the description of and relationships), semantic descriptions (for example, using markup languages), and folksonomies (collaboratively generated, open-ended labeling systems) need to be applied to all resources to enable seamless integration of all forms of knowledge.

CWE RTD must furthermore take into account hybrid forms of collaboration, and provide flexible integration mechanisms for ICT, organizational and workflow processes, supported for example by a service-based infrastructure.

A further main challenge identified was need to research how to leverage the emerging possibilities regarding the convergence and differentiation of collaboration channels (for example, different types of media and modes of interaction and communication). This refers to the use of multi-modality in CWE. The focus of a worker using CWE should reside on his or her primary tasks. Location- and context-sensitive, role-based differentiation of multiple information presentation and interaction modes (for example, synchronous vs. asynchronous) is required to maintain the highest possible quality of experience for the user in any given situation.

The final technological challenge highlighted in the study is the necessity for the provision of secure and trusted CWE. Suitable security mechanisms must be developed for massively distributed, heterogeneous CWE simultaneously providing ease-of-use (for example, single-sign-on), trusted authentication mechanisms, and rights management for collaborative groups without extensive intervention on the part of system administrators. Without creating CWE with such characteristics that foster trust, users and enterprises will continue to be hesitant about adopting the new technology.

CONCLUSION

Based on the requirements analysis of the twelve collaborative work scenarios listed above, the seven main technological challenges for research and development in the field of CWE as described above were identified. Each individual challenge as such represents fields of RTD toward which a significant amount of research activity is being directed. However, a better coordination of RTD activities across all challenges is currently necessary to move forward in the development of CWE.

This article focused on technological challenges to RTD in the field of CWE. However, innovation is driven by people, not systems—the true benefit of CWE will lie in their ability to facilitate collaboration between individuals. Thus, although CWE RTD should clearly focus on technology, a holistic approach to CWE research and development must also display a significant field of research related to the human being, its ability to innovate and create, to interact with other human beings and artificial entities, with knowledge and information. This implies research in a number of multi-disciplinary fields, which in turn will impose new requirements on the technology.

REFERENCES

Baird, F., Moore C. J., & Jagodzinski, A. P. (2000). An ethnographic study of engineering design teams at Rolls-Royce Aerospace. *Design Studies, 21*(4), 333-355.

Berger, P., & Luckmann, T. (1967). *The social construction of reality*. New York: Doubleday.

Diamantopoulou, A. (2002). *Towards a barrier free Europe for people with disabilities*. Madrid: European Commission.

European Commission. (2004). *Next generation collaborative working environments 2005-2010: First report of the expert group on Collaboration@Work*. Research report. Brussels: European Commission Information Society Directorate-General.

European Commission. (2005a). *Collaboration@Work: The 2005 report on new working environments and practices*. Research report. Luxembourg: Office for Official Publications of the European Communities.

European Commission. (2005b). *A thematic priority for research and development under the specific programme "Integrating and Strengthening the European Research Area" Community Sixth Framework Programme. 2005-06 Work Programme*. Brussels: European Commission.

Grenham, A., le Duc, I., Beco, S., & Fusco, L. (2005). *E-collaboration survey technical note*. Research report. Frascati, Italy: European Space Agency.

Gurvitch, G. (1971). *The social frameworks of knowledge*. Oxford, England: Basil Blackwell.

Holzner, B., & Marx, J. (1979). *The knowledge application: The knowledge system in society*. Boston: Allyn-Bacon.

Hribernik, K. A., Ganzer, M., & Schmidt, T. (2003). Mobile communications for emergency management. In A. Gameiro (Ed.), *Proceedings of the 12th IST Summit on Mobile and Wireless Communications – Enabling a Pervasive Wireless World* (pp. 382-386). Aveiro, Portugal: Instituto Telecomunicacoes.

Hribernik, K. A., Kirisci, P., & Hünecke, H. -H. (2004). Mobile applications for collaboration in the construction industry. In L. Hérault (Ed.), *Proceedings of the 13th IST Mobile & Wireless Communications Summit* (pp. 478-482). Lyon, France: CEA.

Hribernik, K. A., Nilsson, M., Fusco, L., & Niitamo, V. -P. (2005). *BrainBridges IST-015982 D2.2: A set of high-level objectives, definitions and concepts*. Research report. Luleå, Sweden: Centre for Distance-spanning Technology, Luleå University of Technology.

Hribernik, K. A., Nilsson, M., Fusco, L., & Niitamo, V. -P. (2006): *BrainBridges IST-015982 D2.4: A European CWE framework description document including terminology*. Research report. Luleå, Sweden: Centre for Distance-spanning Technology, Luleå University of Technology.

Interaction Design. (2001). *Access to employment for people with disability: Getting to work on the Net* (FlexWork Blueprint No. 003). Welwyn Garden City, UK: Interaction Design, Ltd.

Liebich, T. (2003). *IFC 2X Edition 2, Model Implementation Guide.* International Alliance for Interoperability.

O'Reilly, T. (2006). *What is Web 2.0? Design patterns and business models for the next generation of software.* Sebastopol, CA: O'Reilly Media, Inc.

Priest, J. W., & Sanchez, J. M. (2001). *Product development and design for manufacturing: A collaborative approach to producibility and reliability.* New York: Marcel Dekker, Inc.

Schutz, A. (1962). *Collected papers* (Vol. 1). The Hague, The Netherlands: Nighoff.

Sinclair, J., Fox, G., & Bullon, S. (Eds.). (1995). *Collins Cobuild English dictionary.* London: Harper Collins Publishers.

Thoben, K. -D., Hribernik, K. A., Kirisci, P., & Eschenbaecher, J. (2003). Web services to support collaborative business in manufacturing networks. In F. Weber, K. S. Pawar, & K. -D. Thoben (Eds.), *Enterprise engineering in the networked economy: Proceedings of the 9th International Conference on Concurrent Enterprising* (pp. 453-462). Espoo, Finland.

Thoben, K. -D., Kirisci, P. (2002). Context aware environments for ad hoc collaborative business on construction sites. In *Technology Challenges Workshop.* Stuttgart, Germany.

Thoben, K. -D., Kirisci, P., Hribernik, K., Steinmann, R., Kalbitzer, T., & Eggers, T. (2005). Die Bauorganisation im Wandel: Möglichkeiten der Prozessoptimierung durch den Einsatz mobiler Informations- und Kommunikationstechnologien auf Baustellenumgebungen. *ZWF, 100*(6), 359-364.

Ulrich, K. T., & Eppinger, S. D. (1995). *Product design and development.* Boston: McGraw-Hill.

Wheelwright, S. C., & Clark, K. B. (1992). *Revolutionizing product development: Quantum leaps in speed, efficiency, and quality.* New York: The Free Press.

KEY TERMS

Ad hoc Collaboration: Short-term, on demand, spontaneous and task-specific collaboration.

Collaboration: A mutually influencing activity by one or more persons oriented toward common goals and the solution or completion of a problem or task. This takes place within a mutually agreed and created context (common syntactical space, cooperative setting) in physical co-presence using common resources.

Collaboration Assets: The resources required to carry out collaborations (people, process, ICT systems, and knowledge).

Collaboration Stages: The individual stages of the life-cycle of any given collaboration (initiation, management, operation, and dissolution stages).

Collaborative Working Environments: CWE are defined as integrated and connected resources providing shared access to contents and allowing distributed actors to seamlessly work together towards common goals.

Environment: A collection of hardware, software, network communications and procedures that work together to provide a discrete type of computer service.

Mediated Collaboration: Medium-term collaborations in which a mediator is employed to facilitate collaboration.

Multimodality: Multimodality harmoniously combines the different methods of communication between man and machine.

Planned Collaboration: Long-term collaborations in which collaborative enterprise processes are planned, defined, and implemented.

Telework in the Context of E-Collaboration

Antonio Padilla-Meléndez
University of Málaga, Spain

Ana Rosa Del Aguila-Obra
University of Málaga, Spain

INTRODUCTION

Today, everyone recognizes that we live in the so-called knowledge society. In this society, new possibilities based on and around IT and the Internet arise for human beings. IT technology has also made the organizations where they work change rapidly as well as the wider general business environment. The development of the Internet in the early 1990s was both the catalyst and an example of this phenomenon. This computer network allowed the development of *social networks*, or virtual communities of people who use these networks to communicate and to collaborate. We shall concentrate on the specific changes that have taken place in the workplace because of the introduction and increased usage of IT.

Since the 1970s, a decade in which we can place the beginning of the study of telework or telecommuting (Nilles, 1975), the professional literature has given some insights about the explosion in the number of teleworkers that was going to take place in the next decade. As an example, we can mention the predictions of AT&T in the 1970s. As later with the Internet bubble, nothing happens, the forecasts were not good enough, and the statistics show now the reality is different to the initial predictions: It is difficult to speak about the "real telework." On the other hand, there are some instances, (i.e., in Europe the European Commission), and from some industry e-work and e-collaboration practices, more and more people are starting to speak about a broader concept, especially in the last two years. On this issue, they refer to the situation in which some people have to cooperate, using IT to realize a specific task. However, is e-work (or telework) and e-collaboration the same? To answer this, we have to review historical developments, definitions, clarifications, and analysis of the main reasons of why telework is not yet wide spread or why it can be confused with e-collaboration.

Historical Perspective of Telework

We will discuss shortly the main factors that have affected the development of telework in the last decades. Before the 1990s, only aisle initiatives existed, without using the term *telework*. In the 1970s, the term *telecommuting* is coined, as a result of a research project related to the energy crisis and the big traffic congestion problems in big cities, such as Los Angeles. In the state of California, new laws were developed to force companies to implement telecommuting programs to reduce traffic and pollution. Telecommuting was initiated in big companies that started to measure the results. Conclusions from these first experiences were considered as models for future applications.

In the 1980s, several pilot projects were developed in the United States, where it experienced a greater development than in Europe. In Europe, the implications of telework for rural development, its social implications (and the protection required for teleworkers), and the technological aspects were studied. In the *first part of the 1990s*, companies were looking for new ways of cost reducing, to afford a smaller demand. Telework was designed as a tool for cost reducing. In the *second part of the 1990s*, businesses used teleworking as a way to be closer to clients, having more technological tools to serve client needs in their premises; that is, with mobile teleworkers. Public administrations decidedly promoted telework, more in the European Union, communicating the best practices among the society.

In the first years of the *new century*, we can observe an important change related to the term *telework*; for example the European Commission change it for *e-work* in its well-known annual reports about the area (European Commission, 2005), and more labour and business practices were related to telework and all of them mixed up in new concepts such as e-collaboration.

Summarizing this, we can say that since the 1970s, several lines of research about teleworking have been developed: transportation, management, outsourcing, socio-economic development, and technology. *Teleworking and transportation* was the first line of research, and it studied the way of using telecommuting/teleworking for reducing traffic congestion and noise pollution (Bagley, Mannering, & Mokhtarian, 1994; Handy & Mokhtarian, 1996; Nilles, 1975, 1994).

About *teleworking and management*, teleworking is another aspect of the tendency of business process reengineering driven by IT possibilities in the 1990s and also related to Human Resources Management (Daniels, Lamond, & Standen, 2000).

In the *teleworking and outsourcing* line, they have been studied by big companies that outsource, mainly IT activities, and small businesses that have been created to respond to that demand.

With regard *teleworking and socio-economic development*, and from a political point of view, telework has been proposed and used to help not well-develop areas, either in urban or rural zones. The main objective is to provide to the community with telework facilities, so people living there could work as freelance teleworkers. Finally, about telework and technology, several investigations have been conducted to improve the available technology to telework.

CONCEPT OF TELEWORK

From an etymological point of view, the term *telework* supposes the union between "telou" (*distance*, Greek) and "tripalliare" (*work*, Latin). In summary, this means work at a distance. Nevertheless we use the term *telework*; the first term more accepted internationally and used to talk about this activity was the one of *telecommuting*. This was coined by the North American Jack Nilles (a former physician and NASA researcher, denominated since then as the father of telework) in 1973, who defined it as taking the work to the worker instead of the worker to the work, using telecommunications. Since then, the terms used more frequently have been *telecommuting*, mainly in the United States, and *telework*, mainly in Europe.

After this, other terms have been introduced, such as *flexiplace* (Schiff, 1983). The reason for *telework* being the term most used in Europe is due to the use of it by the European Commission that has promoted several research projects in this field since the 1980s. The confusion about the terms starts when the term is

translated into other languages; for example, in Spanish, *teleworking* and *telecommuting* have the same translation: *teletrabajo*, and there exists several similar expressions. In other languages, the situation is quite the same. Even in English there are several terms similar to teleworking, such as telecommuting.

This is only an example of the lack of consensus on what is the exact meaning of teleworking or telecommuting (Gray, Hodson, & Gordon, 1993) because different forms or organizing the work are included in the concept. Overall, it appears quite easy to understand what telework is about in a colloquial context, but it is more difficult when an empirical study is being conducted and the number of teleworkers has to be considered. Depending on the concept the researches adopt, the results could be different. Korte (1988) considers that it is not appropriate to define telework with only one dimension. He considers telework in the middle of some general changes that are taking place in the organizations, and imply three basic elements: (1) location (the location of the employees change depending on desires of these employees or necessities of the company, so the location in which the work is carried out is independent of the location of the employer); (2) use of IT (telework extensively depends on the use of IT); and (3) the communication link with the employer (it could be completely electronic communication or using other more traditional communication means).

We could propose, as coincident aspects of the different definitions of telework in the literature, the following: (1) Telework is working at a distance—it implies a different location of the worker from the company/client for which her or she is working; (2) telework implies an intensive use of IT; (3) the added value that brings the worker to the company is related with the use of IT.

TYPES OF TELEWORK

From the 1970s, some types of teleworking were defined and related to the place of work: home teleworking, mobile telework, and telecentres or telework centres. The home teleworking was defined as a way of working full time from home instead of working in the office. Mobile telework was referred to as work wherever the employees were needed; for example, in the clients' office or at home. Related with these there are other terms, as *mobile office*, more concentrated in the technological aspect, and *hotelling*, related with office space management (Vischer, 1995). In this practice, the office space is managed as a hotel, where the employees do not have fixed spaces to work, and the space is used in a first-requested first-used basis. Finally, *telecentres* refer to offices with IT equipment, located near residential neighbourhoods, designed to reduce the number of commuting trips and used by one or more employers.

Nowadays, teleworking is understood as an idea of flexible work, optimizing the use of IT, and working where and whenever needed.

Aspects of Telework

Telework can be studied from very diverse perspectives, so its concept is a system of aspects: strategic, organizational, human resources management, economic, social, legal, political, and technological. The strategic aspect is related to the possibilities for a company of getting competitive advantages in applying telework. In this sense, telework supposes a technological and administrative innovation, an opportunity for the companies for being flexible and reduce costs, and this could be a source of competitive advantage. Within the organizational aspect, we refer to the relation of telework with the organizational structure of the company: more flexible, IT-based, and sometimes known as virtual organization. As far as the economic aspect goes, it refers to the valuation of cost savings and benefits that the organization is going to obtain with telework. This it is an important aspect because since the beginning telework has been mentioned in the literature as a way of reduction of the company costs (Gray, Hodson, & Gordon, 1993).

Social aspect is related to the conditions of work of the employees that telework and the potential labour conflicts. At the same time, external social implications when an organization applies telework, like the contribution to the development of economically depressed geographic zones exist. Political aspect refers to the efforts to advance in the knowledge society, where telework is a key element. In this sense, the actions that the European Union comes starting up with respect to telework, with international projects that constitute an important pillar in the situation of telework now. The technological aspect is related to IT developments. Truly, the organizations have at the present time the opportunity to use a technology already available, but previously the implications are due to evaluate and to

know the organizational aspect that it entails. This affirms that now the technological aspect is not the most important aspect of telework. As far as the aspect of the management of human resources, the application of telework in the organizations entails changes for the people, teleworkers, and managers who require a special attention, since their implications are important and numerous. To conclude, we can say that telework means a remote management, and intensive information-based activity, e-collaboration, and an intensive use of technology.

CONCEPT OF E-COLLABORATION

It is not easy to define e-collaboration, as it is a quite new term and is still in its "infancy" (Kock, 2005). Nevertheless, infrastructure for the development of e-collaboration technologies comes for as far as the early 1950s with ENIAC (see Kock & D'Arcy, 2002, for and in-depth description of the evolution of e-collaboration technologies since then). They are also several related concepts that some authors use as the same. An operational definition of e-collaboration is to consider that "as collaboration using electronic technologies among different individuals to accomplish a common task" (Kock & D'Arcy, 2002). Other authors, as Hossain and Wigand (2004) use the term *virtual collaboration* and define it as ICT-enabled collaboration for geographically dispersed groups with no or very little face-to-face communication.

These definitions could be too general (Kock, 2005). In this sense, Kock and Nosek (2005, p. 1) explain two trend that affects how e-collaboration is perceived. The first one is related to the development of an e-collaboration tools industry that has provoked many definitions. This also cause that vendors tend to use more technology-related definitions of e-collaboration, as technological support for electronic meetings over the Internet. The second trend relates that to some IT business-oriented publications that consider e-collaboration technologies as tools to support electronic commerce and supply chain transactions involving two or more organizations.

Some authors considered e-collaboration as an "umbrella" term, which comprises different fields (Kock & D'Arcy, 2002): computer mediated communication (CMC), computer-supportive cooperative work (CSCW), groupware, group support systems, collaboration technologies, and knowledge management. E-collaboration research includes CMC and CSCW research as well as other lines of research (Kock, 2005). Taking into account that telework is not the same that e-collaboration, it seems that "geographically dispersed individuals are more likely than colocated ones to use e-collaboration technologies" (Kock, 2001). According to Kock (2005, p. iv) there are several elements that contribute to define the concept of e-collaboration and which can change its nature: (1) the collaborative task, (2) e-collaboration technology, (3) individuals involved in the collaborative task, (4) mental schemas possessed by the individuals, (5) the physical environment surrounding the individuals, and (6) the social environment surrounding the individuals.

Another line of research related to e-collaboration is Group Decision Support System (GDSS), born in the 1980s, as a new software/technology and a new approach to how people work together in a common project. The groupware refers more to the combination of computers, communications, and decision technologies to support distance-taking decisions. In the IS evolution some of these systems try to help at decision taking, such as the Decision Support Systems (DSS), designed for the individual. However, some advances were made related to increase the possibilities of using DSS for groups or teams. These advances could we summarized in group decision support systems, negotiation support systems, and computer supported cooperative work. GDSS are computer based systems designed to help decision taking in groups. GDSS combines computers, data communications and decision technologies to help solving problems to managers and employees in general (Thierauf, 1989). They try to do this eliminating the communication barriers, using technology. In general, GDSS provides with the same facilities that an individual DSS, but its objective is to reduce some problems associated with conventional decision making in groups and related to (Finlay & Marples, 1992): The lack of organization; dominance of one or some members of the group and the inhibition of the others, and social pressure towards conformity. Some features of GDSS related to the e-collaboration technologies were their communication possibilities (Finlay & Marples, 1992). As the ones specialized in negotiating we have the Negotiation Support Systems (NSS) that are systems designed to resolve opinion or interest conflicts among a group or groups.

CSCW was introduced in the middle eighties (Crowe, 1994) and it is based on systems designed to support communication and cooperative work among a group of people. In the context of CSCW, teamwork is considered as a group of people working together in a common project (Crowe, 1994). A group can work together for a long time and cooperate in a synchronous and asynchronous way. In the meantime, any member can be substituted by other individuals.

In this context, e-work is defined (Nof, 2003) as collaborative, computer-supported activities and communication supported operations in highly distributed organizations of humans and/or robots or autonomous systems. It is interesting that this author includes also the organizations of robots or autonomous systems and he develops a complete field of research around the PRISM Centre (Production, Robotics, and Integration Software for Manufacturing & Management; Purdue University, Indiana, United States).

CONCLUSION

The concept of e-collaboration is broader than the telework one. Talking about history, we can say that if we considered the birth of e-mail as the beginning of e-collaboration, this is older than telework. There are several coincidences between telework and e-collaboration. The more normal types of telework are home-based telework and mobile telework. In both cases, e-collaboration, in the sense of people working together to accomplish task in common is present and needed. The more the teleworkers spend time out of his/her office, the more he/she will need e-collaboration tools to be able to work with supervisors or peers.

The convergence of all technologies towards personal computing and the Internet is provoking a huge increased in the possibilities to telework. If we put together the increasing traffic congestion problems and the petrol crisis, we could predict that more and more people will be a teleworker in the future. The development of appropriate e-collaboration tools will be essential to support this prediction.

REFERENCES

Bagley, M., Mannering, J., & Mokhtarian, P. (1994). *Telecommuting centers and related concepts: A review of practice* (Research report). University of California, Institute of Transportation Studies.

Crowe, M. K. (Ed.). (1994). *Cooperative work with multimedia*. Berlin: Springer-Verlag.

Daniels, K., Lamond, D., & Standen, P. (2001) Teleworking: Frameworks for organizational research. *Journal of Management Studies, 38*(8), 1151-1185.

Daniels, K., Lamond, D., & Standen, P. (2000). *Managing telework: Perspectives from human resource management and work psychology*. London: Thomson International.

European Commission. (2005). *Collaboration@*Work. Information Society and Media. Luxembourg, Luxembourg: Office for Offical Publications of the European Communities.

Finlay, P. N., & Marples, C. (1992). Strategic group decision support systems: A guide for the unwary. *Long Range Planning, 25*(3), 98-107.

Gray, M., Hodson, N., & Gordon, G. (1993). *Telework explained*. New York: Wiley.

Handy, S. L., & Mokhtarian, P. L. (1996). The future of telecommuting. *Futures, 28*(3), 227-240.

Hossain, L., & Wigand, R. T. (2004). ICT enabled virtual collaboration through trust. *Journal of Computer Mediated Communication, 10*(1) Article 8, November 2004.

Kock, N. (2001). The ape that used email: Understanding e-communication behavior through evolution theory. *Communications of the AIS, 5*(3), 1-29.

Kock, N. (2005). What is e-collaboration? *International Journal of e-Collaboration, 1*(1), 1-7.

Kock, N., & D'Arcy, J. (2002). Resolving the e-collaboration paradox: The competing influences of media naturalness and compensatory adaptation [Special issue]. *Information Management and Consulting, 17*(4), 72-78.

Kock, N., & Nosek, J. (2005). Expanding the boundaries of e-collaboration [Special issue]. *IEEE Transactions on Professional Communication, 48*(1), 1-9.

Korte, W. B. (1988). Telework: Potential, inception, operation and likely future situation. In W. B. Korte,

S. Robinson, W. J. Steinle (Eds.), *Telework: Present situation and future development of a new form of work organization* (pp. 159-175). Amsterdam, The Netherlands: North-Holland.

Nilles, J. M. (1975). Telecommunications and organizational decentralization. *IEEE Transactions on Communications, 23*(10), 1142-1147.

Nilles, J. M. (1994). *Making telecommuting happen: A guide for telemanagers and telecommuters*. New York: Van Nostrand Reinhold.

Nof, S. Y. (2003). Design of effective e-work: Review of models, tools, and emerging challenges. *Production Planning & Control, 14*(8), 681-703.

Schiff, F. W. (1983). Flexiplace: An idea whose time has come. *IEEE Transactions on Engineering Management, 30*(1), 26-30.

Thierauf, R. J. (1989). *Group decision support systems for effective decision making: A guide for MIS practicioners and end users*. New York: Quorum-Books.

Vischer, J. C. (1995, Fall). Strategic work-space planning. *Sloan Management Review*, 33-42.

KEY TERMS

Computed mediated communication: A communication method in which the computer is used as the transfer mean, allowing the individual communication without time or geographical restrictions.

E-collaboration technologies: Electronic technologies that enable collaboration among individuals engaged in a common task.

E-collaboration: Collaboration using electronic technologies among different individuals to accomplish a common task.

E-work: Collaborative, computer-supported activities and communication supported operations in highly distributed organizations of humans and/or robots or autonomous systems.

Hotelling: An office space management practice in which the office space is managed as a hotel, where the employees do not have fix spaces to work, and the space is used in a first requested first used basis.

Mobile teleworking: Work wherever the employees are needed, for example in the clients' office, on the move, or at home.

Telecentres: Offices with IT equipment, located near residential neighbourhoods, designed to reduce the number of commuting trips, and used by one or more employers.

Telecommuting: Taking the work to the worker, instead of the worker to the work, using telecommunications.

Telework: Work at a distance. Implies places different from the habitual one to make the labour activity (labour flexibility), a supervision and direction of the employees remote, an intensive labour activity in information and an intensive use of the IT.

Thematic–Based Group Communication

Raymond Pardede
Budapest University of Technology and Economics, Hungary

Gábor Hosszú
Budapest University of Technology and Economics, Hungary

Ferenc Kovács
Budapest University of Technology and Economics, Hungary

INTRODUCTION

These days, communicating, working together, and collaborating without limitation of space and time is a preferable option for some people although they are physically far from each other. For such purposes, the one-to-many communication model is appropriate (Pardede, Szilvássy, Hosszú, & Kovács, 2002). Its well-elaborated but not-widely deployed realization is the IP-multicast, which is a scalable solution since theoretically it improves the efficiency of the use of bandwidth. Nevertheless, the deployment of IP-Multicast is too slow and expensive since it needs upgrading infrastructures (e.g., routers). For solving this problem, the application-level multicast (ALM), with the cost of efficiency, is offered as a solution. The ALM moves the multicast functionality from the network layer to the application layer.

This article reviews the most important fact of the ALM and introduces to the emerging area of the multicasting, namely the multicast over ad hoc networks, including geocast. After these a novel concept of modeling relative density of members called bunched mode and a proposed ALM multicast transport protocol called shortest tunnel first (STF) are described. The bunched mode is based on the thematic multicast concept (TMC), which means that it is a typical multicast scenario where there are many interested hosts in certain institutes and these institutes are relatively far from each other. This situation is called bunched mode, in which the members of a multicast group are locally in the dense mode, and globally their situation similar to the sparse mode because these spots are far from each other. This article also presents a simple chatting program called PardedeCAST as the tools of STF and TMC research.

The article ends with the description of the future trends in the multicast communication and the conclusions of the described technical facts and results.

BACKGROUND

Currently there is an increasing need for scalable and efficient group communication technology. The multicast is theoretically optimal for such purposes. It can be realized in the Datalink-level, IP-level, and Transport/Application-level (Hosszú, 2005). However, the IP-Multicast has a slow deployment; it has been implemented in the most operating systems (OS) and routers, but not widely enabled. Oppositely to the IP-multicast another approach called application-level multicast (ALM) is easy to deploy, but less efficient (Pardede, 2002).

The ALM protocols can be classified into two categories, namely the mesh-based and the tree-based solutions. The mesh-based protocol creates a mesh for the control plane at first with a redundant topology of the connections between members. After creating the mesh, the algorithm starts to construct a multicast tree. Such protocols are the Narada (Chu, Rao, Seshan, & Zhang, 2002), or the Gossamer (Chawathe, 2000).

The opposite of the mesh-based type is the tree-based protocol concept, where the multicast delivery tree is formed first and then each member discovers some others that are not neighbors and creates control links to these hosts. This solution is suitable for data transferring applications, which need high bandwidth, but not efficient for real-time purposes. Such protocols are the Yoid (Francis, 2000) and the host multicast tree protocol (HMTP) from Zhang, Jamin, and Zhang (2002).

Nowadays there is a fundamental change in the computer network technology due to the mobile networks. They are classified into two types, the infrastructure-based networks and the infrastructure-less, so-called ad hoc networks. The first one is based on the various wireless LAN (WLAN) technologies. They need base stations, called hot spots, but their advantage is the direct access to the Internet. Currently the latest computers are generally WLAN-enabled, and so the method to reach of the Internet in under fast change.

The second class of the mobile networks is the ad hoc topologies, which are dynamic structure of mobile devices that spontaneously form the network among them (Ni, Kremer, Stere, & Iftode, 2005). A network is ad hoc if it does not need any infrastructure. Such networks are the Bluetooth (Haartsen, 1998) and Mobile Ad Hoc NETwork (MANET), which comprise a set of wireless devices that can move around freely and communicate in relaying packets on behalf of one another (Mohapatra, Gui, & Li, 2004). The MANET is a self-organizing and self-configuring multihop wireless network, where the network topology changes dynamically due to participants' mobility. Ad hoc networks are getting a promising target platform, with the increasing deployment of smart equipments. Such devices are small wireless equipments with large computing power, and memory, as the PDAs or the smart phones. Furthermore computers can be embedded into cars, ships or fixed architectures as buildings. The topology of their network is dynamically changing as vehicles or persons move. Such networks can become part of larger ad-hoc networks spanning a whole town or even bigger geographical regions.

The ad hoc networks are very attractive for communication in the traffic environment, in convention centers, conferences and electronic classrooms (Hong, Xu, & Gerla, 2002). Nodes in this network model share the same random access wireless channel. They cooperate to engage in multihop forwarding. Each node functions not only as a host but also as a router that maintains routes to and forward data for other nodes in the network that may not be within direct wireless transmission range. Due to the special properties of the mobile host, specific multicast routing protocols have been developed for the multicast over ad hoc environment.

The simplest ad hoc multicast routing methods are flooding and tree-based routing. Flooding is very simple, which offers the lowest control overhead at the expense of generating high data traffic. This situation is similar to the traditional IP-Multicast routing. However, in the wireless ad hoc environment, the *tree-based routing* fundamentally differs from the situation in the wired IP-Multicast, where the tree-based multicast routing algorithms are obviously the most efficient ones, such as in the multicast open shortest path first (MOSPF) routing protocol (Moy, 1994). Although the tree-based routing generates optimally small data traffic on the overlay in the wireless ad hoc network, but the tree maintenance and updates need a lot of control traffic. That is why the both simplest methods are not scalable for large groups.

A more sophisticated ad-hoc multicast routing protocol is the core-assisted mesh protocol (CAMP), which belongs to the mesh-based multicast routing protocols (Garcia-Luna-Aceves & Madruga, 1999). It uses a shared mesh to support multicast routing in a dynamic ad-hoc environment. This method uses cores to limit the control traffic needed to create multicast meshes. Unlike the core-based multicast routing protocol as the traditional Protocol Independent Multicast-Sparse Mode (PIM-SM) multicast routing protocol (Deering, Estrin, Farinacci, Jacobson, Liu, & Wei, 1996), CAMP does not require that all traffic flow through the core nodes. CAMP uses a receiver-initiated method for routers to join a multicast group. If a node wishing to join to the group, it uses a standard procedure to announce its membership. When none of its neighbors is mesh members, the node either sends a join request toward a core or attempt to reach a group member using an expanding-ring search process. Any mesh member can respond to the join request with a join Acknowledgement (ACK) that propagates back to the request originator.

Oppositely to the mesh-based routing protocols, which exploit variable topology, the so-called gossip-based multicast routing protocols exploit randomness in communication and mobility. Such multicast routing protocols apply gossip as a form of randomly controlled flooding to solve the problems of network news dissemination. This method involves member nodes to talk periodically to a random subset of other members. After each round of talk, the gossipers can recover their missed multicast packets from each other (Mohapatra et al., 2004). Oppositely to the deterministic approaches, this probabilistic method will better survive a highly dynamic ad hoc network because it operates independently of network topology and its random nature fit to the typical characteristics of the network.

A type of the gossip-based protocols is the anonymous gossip routing. This method does not require a group member to have any information of the other members of the same group. The procedure has two phases. In the first phase, a host sends multicast packets to the group. While in the second phase, periodic anonymous gossip takes place in the background as each group member recovers any lost data packet from other group members that might have received it.

Due to the increased mobility and changing geographic position of the devices, there are new applications that utilize that such devices or rather a group of such equipments can be identified partly be their position and not their IP addresses as in the past. In such a way the multicast gained a new and increasing importance, where the device (or host, if we use the IP-based terminology) becomes member of a multicast group depending on its actual geographical position. Typically, the scope of the multicasting is limited to this group. Such kind of multicast is called geocast, which is in fact a geographically scoped multicast communication.

In case of geocasting the host that wishes to deliver packets every node in a certain geographical area can use such method. In such case, the position of each node with regard to the specified geocast region implicitly defines group membership. Every node is required to know its own geographical location. For this purpose, they can use the Global Positioning System (GPS). The geocasting routing method does not require any explicit join and leave actions. The members of the group tend to be clustered both geographically and topologically. The geocasting routing exploits the knowledge of location.

The geocasting can be combined with flooding such methods are called forwarding zone methods, which constrain the flooding region. The forwarding zone is a geographic area that extends from the source node to cover the geocast zone. The source node defines a forwarding zone in the header of the geocast data packet. Upon receiving a geocast packet, other machines will forward it only if their location is inside the forwarding zone. The location-based multicast (LBM) is an example for such geocasting-limited flooding (Ko & Vaidya, 2002). With advances in wireless Internet and mobile computing, the different solutions of the LBM are emerging as key value-added services for telecom operators deliver. The LBM enables them to provide personalized location-aware content to subscribers using their wireless network infrastructure. Besides telecom operators, more and more service providers such as public wireless LAN providers, enterprises, and others are developing and deploying location-aware services for users to gain more revenue and productivity (Chen, Chen, Rao, Yu, Li, & Liu, 2004).

Based on the above the traditional, wired-Internet based communication technologies must be generalized and extended to mobile networks, too. The one-to-many communication method named thematic multicast concept (TMC) proposed in this article is usable for not only wired environment, but the emerging wireless networks, too.

THE THEMATIC MULTICAST CONCEPT AND SHORT TUNNEL FIRST

The thematic multicast concept (TMC) is an example of the current research works in the field of ALM; a novel concept of modeling relative density of members called bunched mode. The bunched mode means a typical multicast scenario, where there are many interested hosts in certain institutes but these institutes are relatively far from each other. The members of a multicast group are locally in the dense mode. However, these spots are far from each other, which globally their situation is similar to the sparse mode. A bunch can be a local-area network (LAN) or just an autonomous system (AS). This situation is typical when one collaborative media application has a special topic. That is why this model of communication is called thematic multicast concept.

The shortest tunnel first (STF) is a novel ALM routing protocol, which is optimal for bunched mode delivering. Every group of member hosts in a bunch locally elects their designated member (DM). Every DM after being elected will calculate the shortest unicast IP tunnels among them (see Figure 1).

The DMs also exchange their IP addresses and the information of shortest unicast paths among them. The path is a series of IP links from one host to another. In such a way, all of them know every possible shortest unicast paths and are able to calculate the same topology of the inter-bunch IP tunnels. This mechanism is similar to the *MOSPF* routing in network-level (Moy, 1994). The STF does not require any global rendezvous point for creating the inter-bunch delivery tree, however, suppose that there is only one source per group and constructs unidirectional tree.

Figure 1. The shortest tunnel first (STF) multicast architecture

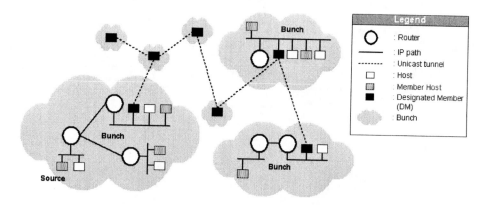

The list of the DMs is maintained by the source application, and the new DM registers here and gets the copy of the current list. Every DM sends periodically IP packets to every other DM and sends a "keep-alive" message to the source if there is one DM that is not available. If a DM is not reported (by other DMs) to the source, it is deleted from the list. The source periodically sends the list to the DMs. If the source is not available, the group state is timed out in the DMs.

The STF, similarly to the MOSPF protocol (Moy, 1994), uses the Dijkstra algorithm (Dijkstra, 1959) for calculating the shortest path, which is composed of IP tunnels in its case. MOSPF uses the underlying open shortest path first (OSPF) unicast routing protocol, in which the routers exchange the link-state information about their neighbors, while in the case of the STF the designated member hosts (DMs) exchange the path information to every other, since there are IP paths to every other DM.

The STF protocol constructs almost similarly optimal tree than the IP-multicast, however it does not require any inter-domain multicast routing mechanism in the routers. It belongs to the mesh-first class. It is optimal for relatively small groups, but due to the TMC method, the topological size of the group does not limit its scalability.

IMPLEMENTATION AND RESULTS

The thematic multicast concept (TMC) for multicasting and the shortest tunnel first (STF) application-level routing protocol is being implemented in the software

PardedeCAST. PardedeCAST is a simple chatting program, which can build a chat room based on one interesting topic (theme). Every interested user can join to one room by contacting the user who first establishes the room.

The current version of the software (version 2.5) is able to create the P2P overlay where the clients use the simple "chain" connection for creating the multicast distribution tree (see Figure 2).

Options are given to the first user of one chat room to select whether this chatting session will use the Dijkstra algorithm or the simple serial algorithm for establishing the multicast distribution tree (see Figure 3).

There are cases, where the users of mobile devices in different ad hoc networks would like to communicate with each other. Such scenario just fits to the base approach of the TMC, in such a way the proposed and partly developed PardedeCast can solve the communication in this specific situation.

The software PardedeCAST version 2.5 has simulator functionality for the developing and testing purposes. It can simulate for example the simple joining process in ALM, and has the ability to handle simple recovery process (see Figure 4) if partition problem occurs (e.g., when one user leaves the chat room). The current version of the software, however, has not yet implemented the Dijkstra algorithm for establishing the multicast distribution tree. The current version of the software also has not yet implemented an algorithm for settling bunches of users for providing the scalability of the system. Those features will be built in the next version of the software.

PardedeCAST is developed based on the object-oriented programming concept, in which several

Figure 2. Simple Serial Algorithm in the software PardedeCAST

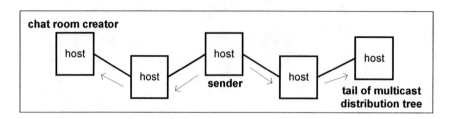

Figure 3. The optional routing protocols in the software PardedeCAST

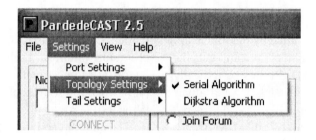

Figure 4. The simple recovery process in the software PardedeCAST

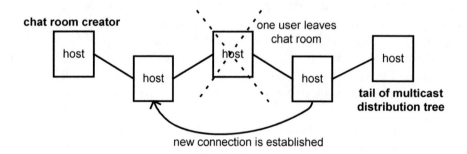

classes build the architecture of this software. In case of the object-oriented programming the programming language model is organized around "objects" rather than "actions" and data rather than logic. PardedeCAST forms the TMC scheme as one package or one library, so this library can be reused for other applications, such as live video streams, online games, video conferencing applications, etc.

CONCLUSION

The article presented the overview of the emerging *application-level multicast* (ALM) protocols, and focused on the novel *thematic multicast concept* (TMC) method. It is intended for describing a typical multicast scenario, where the interested member are located in a restricted number of institutes, but these institutes are far from each other. For modeling the delivery in this situation, the bunched mode delivery is introduced. The novel *shortest tunnel first* (STF) application-level multicast routing protocol is applicable for this *bunched mode* multicast. In each bunch (LAN or AS) designated members are elected. The STF avoids the need for the globally available rendezvous point and in such a way the STF-based multicasting is usable over the whole Internet. The performance of the proposed ALM scheme is demonstrated by the test runs of its implementation named *PardedeCast*. Based on the presented structure the TMC and the STF multicast routing algorithm is applicable and usable for both academics and practitioners.

REFERENCES

Banerjee, S., Bhattacharjee, B., & Kommareddy, C. (2002). Scalable application-layer multicast. *Proceedings of ACM SIGCOMM 2002*, (pp. 205-220). Pittsburgh, PA: Association for Computing Machinery.

Chawathe, Y. (2000). *Scattercast: An architecture for Internet broadcast distribution as an infrastructure service*. Unpublished doctoral dissertation, University of California, Berkeley.

Chen, Y., Chen, X. Y., Rao, F. Y., Yu, X. L., Li, Y., & Liu, D. (2004). LORE: An infrastructure to support location-aware services. *IBM Journal of Research & Development, 48*(5/6), 601-615.

Chu, Y. -H., Rao, S. G., Seshan, S., & Zhang, H. (2002). A case for end system multicast. *IEEE Journal on Selected Areas in Communications, 20*(8), 1456-1469.

Deering, S. E., Estrin, D., Farinacci, D., Jacobson, V., Liu, C. -G., & Wei, L. (1996). The PIM architecture for wide-area multicast routing. *IEEE/ACM Transactions on Networking, 4*(2), 153-162.

Dijkstra, E. W. (1959). A note on two problems in connection with graphs. *Numerische Mathematik, 1*, 269-271.

Francis, P. (2000). *Yoid: Extending the multicast Internet architecture* (Tech. Rep. *ACIRI*). Retrieved August 7, 2006, from http://www.aciri.org/yoid

Garcia-Luna-Aceves, J. J., & Madruga, E. L. (1999). The core-assisted mesh protocol. *IEEE Journal of Selected Areas in Communications, 17*(8), 1380-1394.

Haartsen, J. (1998). Bluetooth: The universal radio interface for ad hoc, wireless connectivity. *Ericsson Review, 3,* 110-117. Retrieved August 7, 2006, from http://www.ericsson.com/review

Hong, X., Xu, K., & Gerla, M. (2002). Scalable routing protocols for mobile ad hoc networks. *IEEE Network, 16*(4), 11-21.

Hosszú, G. (2005). Mediacommunication based on application-layer multicast. In S. Dasgupta (Ed.), *Encyclopedia of virtual communities and technologies* (pp. 302-307). Hershey, PA: Idea Group Reference.

Ko, Y. -B., & Vaidya, N. H. (2002). Flooding-based geocasting protocols for mobile ad hoc networks. *Proceedings of the Mobile Networks and Applications, 7*(6), 471-480.

Mohapatra, P., Gui, C., & Li, J. (2004). Group communications in mobile ad hoc networks. *Computer, 37*(2), 52-59.

Moy, J. (March 1994). *Multicast extensions to OSPF*, Network Working Group RFC 1584. Retrieved July 31, 2006, from http://www.ietf.org/rfc/rfc1584.txt

Ni, Y., Kremer, U., Stere, A., & Iftode, L. (2005). Programming ad-hoc networks of mobile and resource-constrained devices. In *Proceedings of ACM SIGPLAN 2005 Conference on Programming Language Design and Implementation (PLDI)* (pp. 249-260). Chicago: ACM Press.

Pardede, R. E. I. (2002). *Development a classification methodology for multicasting*. Unpublished master's thesis, Budapest University of Technology and Economics, Budapest, Hungary.

Pardede, R. E. I., Szilvássy, G., Hosszú, G., & Kovács, F. (2002). NetSim: Multicast transport mechanism simulator for simplified network model. In *Proceedings of the Meeting Advanced Telecommunication and Informatic Systems and Networks* (pp. 59-60). Budapest: Scientific Association for Infocommunications Hungary, Budapest University of Technology and Economics.

Zhang, B., Jamin, S., & Zhang, L. (2002). Host multicast: A framework for delivering multicast to end users. In *Proceedings of INFOCOM 2002, 21st Annual Joint Conference of the IEEE Computer and Communications Societies, Vol. 3* (pp. 1366-1375). New York: IEEE Computer Society Press.

KEY TERMS

Ad Hoc Network: It is a network which does not need any infrastructure. An example of such a network is the Bluetooth.

Application-Level Multicast (ALM): A novel multicast technology, which does not require any additional protocol in the network routers, since it uses the traditional unicast IP transmission.

ALM Routing Protocol: The members of the hosts construct delivery tree using similar algorithms than the *IP-multicast* routing protocols do.

Application-Level Network (ALN): The applications, which are running in the hosts, can create a virtual network from their logical connections. This is also called *overlay network*. The operations of such software entities are not able to understand without knowing their logical relations.

IP-Multicast: Network-level multicast technology, which uses the special class-D IP-address range. It requires multicast routing protocols in the network routers.

Multicast Island: Multicast capable network, where all routers have multicast routing protocols and the *IP-multicast* is available. One of the main problems of the IP-multicast is to connect the multicast islands into an Internet-wide network.

Multicast Routing Protocol: In order to forward the multicast packets, the routers have to create multicast routing tables using multicast routing protocols.

Peer-to-Peer (P2P): It is a communication way where each node has the same authority and communication capability. They create a virtual network, overlaid on the Internet. Its members organize themselves into a topology for data transmission.

Thematic Multicast Concept (TMC): It means a multicast scenario that is typical in practice, where there are many interested hosts in certain institutes and these institutes are relatively far from each other. The appropriate distribution model takes into account this special topology that is manifested in the name of bunched mode, which is between the sparse- and the dense mode multicast dissemination models.

Thinklets for E-Collaboration

Robert O. Briggs
University of Nebraska at Omaha, USA
University of Alaska Fairbanks, USA

Gert-Jan de Vreede
University of Nebraska at Omaha, USA
Delft University of Technology, The Netherlands

Gwendolyn L. Kolfschoten
Delft University of Technology, The Netherlands

INTRODUCTION

A ThinkLet is a named, scripted collaborative activity that gives rise to a known pattern of collaboration among people working together toward a goal. ThinkLets are design patterns for collaborative work practices (Briggs, Kolfschoten, Vreede, & Dean, in press; Briggs & Vreede, 2001). A thinkLet is the smallest unit of intellectual capital necessary to recreate a known pattern of collaboration. ThinkLets are used by facilitators and collaboration engineers as (1) predictable building blocks for collaboration process design, (2) as transferable knowledge elements to shorten the learning curve of facilitation techniques, and (3) by researchers as parsimonious, consistent templates to compare the effects of various technology-supported collaboration practices. ThinkLets have a rigorous documentation scheme that specifies the information elements needed to adapt the solution it embodies to the problem at hand. This scheme is derived from the design pattern concept of Alexander (1979; Alexander, Ishikawa, Silverstein, Jacobson, Fiksdahl-King, & Angel, 1977). The collection of thinkLets forms a pattern language for creating, documenting, communicating, and learning group process designs. The term thinkLet was coined by David H. Tobey in 2001 when he said "They are like applets…except they are thinkLets."

BACKGROUND

In this section we explain the evolution of the ThinkLet concept. The underlying concept of thinkLets; facilitation techniques, is much older than the term itself. In 1953, for example, A. F. Osborn published a group creativity technique called Brainstorming (Osborn, 1953). His brainstorming method specifies roles and rules for a group to follow in order to generate creative solutions for the problem at hand. Nominal Group Technique, (Delbecq, Ven, & Gustafson, 1975) and Brainwriting, a technique developed in Germany, are two more examples of reusable, repeatable techniques for idea generation.

In 2000, researchers at the University of Arizona and Delft University of Technology in The Netherlands observed that facilitators who used group support systems (GSS) tended to work with a small collection of very effective techniques they had either developed for themselves or learned from mentors. People using these systems routinely reported group project labor savings of 50% and project cycle time reductions of 60% to 90% (Nunamaker, Briggs, Mittleman, Vogel, & Balthazard, 1997; Post, 1993; Vreede, Vogel, Kolfschoten, & Wien, 2003), but people who did not know the techniques were not able to reproduce the successes of others. Researchers set out to collect and document these techniques in sufficient detail that people could reproduce the patterns of collaboration being created by others. They focused on identifying the minimum set of information required to easily transfer techniques from one person to another, and to allow users to produce the pattern of collaboration the technique was meant to invoke in a predictable way, and on a recurring basis.

The first approach to documenting thinkLets captured four elements: (a) the name of the thinkLet; (b) the specific software tools used; (c) the specific configuration of those tools, and (d) a script of everything the moderator and participants were required to do and say in order to complete the activity. The approach worked well for users of the particular software tools, but similar patterns could also be created with other tools, and technology. Therefore a new, technology independent conceptualization was developed. Below, this approach is discussed in more detail.

By 2001, researchers had begun to develop a structured approach to using thinkLets to design collaboration processes for high-value recurring task, and transferring those designs to practitioners to execute for themselves without the ongoing intervention of professional facilitators. The approach came to be called Collaboration Engineering.

APPLICATIONS OF THINKLETS

ThinkLets serve as a pattern language for collaboration process designs (Vreede, Briggs, & Kolfschoten, in press). As such, they have become useful to several populations of users: facilitators, collaboration engineers, practitioners, technology designers, and trainers. For facilitators, thinkLets provide a collection of useful, well-tested, predictable interventions that they can draw upon to conduct processes for groups. Facilitators also use thinkLets as a compact, powerful pattern language for discussing, comparing, and transferring process designs among themselves. A thinkLets-based design created by one facilitator can be readily executed by another who knows the same set of thinkLets.

Collaboration engineers use thinkLets as building-blocks for creating reusable collaboration process designs to be transferred to practitioners to execute for themselves without the ongoing intervention of professional facilitators (Briggs, Vreede, & Nunamaker, 2003; Vreede & Briggs, 2005). Because thinkLets are well tested and fully documented, their likely effects on a group and the levels of skill required to execute them well can be known at design time. Collaboration engineers choose among thinkLets to optimize for ease-of-execution, ease-of-learning, predictability of outcome, and robustness (the degree to which the design can accommodate a wide variety of circumstances, problems, and stakeholder interests) (Kolfschoten, Briggs, Vreede, Jacobs, & Appelman, in press; Vreede et al., in press).

Practitioners can become skilled at executing thinkLet-based collaboration processes after a short training because a component-based learning approach reduces their cognitive load.

Designers of collaboration technology use the collection of thinkLets as a basis for specifying a system's capabilities that are required for a specific (set of) thinkLets. Such capabilities could be an extension to group support capabilities based on the mechanics of collaboration as described in Baker, Greenberg, and Gutwin (2001).

Collaboration engineers, practitioners and facilitators all find that their learners can ramp up to competence far more quickly by learning and practicing a collection of thinkLets than by apprenticing to an experienced professional. In the 1990s the general rule of thumb was that a facilitator who wished to use group support systems required at least a year of apprenticeship before going into the field solo. Trainers using a thinkLets based approach now report that trainees with as little as two days training can conduct simple but successful solo engagements using GSS.

THE STRUCTURE OF THINKLETS

Many books and websites describe useful, well-tested facilitation techniques (FacilitatorU, 2005; Jenkins, 2005). A key distinction between such techniques and thinkLets is in the degree to which they have been formally specified to offer the features and functionalities described above. The current documentation convention (Kolfschoten, Briggs, Vreede, Jacobs, & Appelman, in press; Vreede et al., in press) for a thinkLet includes the following elements:

Identification

Each ThinkLet must have a unique name. These names are typically selected to be catchy and amusing so as to be memorable and easy to teach to others (Buzan, 1974). The name is also selected to invoke a metaphor that reminds the user of the pattern of collaboration the thinkLet will invoke, and visualized with an icon. The names, combined with the metaphor and icon constitute the basis for a shared language. Facilitators can use these names to discuss their collaboration process designs. This makes discussions much more efficient, since facilitators can now discuss the advantages and disadvantages of the use of techniques without explaining these techniques over and over.

Overview

Each thinkLet must have a brief, high-level summary of what happens during the execution of the thinkLet and the patterns of collaboration that emerges when it is used. This helps the user decide whether the thinkLet would be useful without having to read the rest of the details of the thinkLet

Rules

Rules are the core of the thinkLet. The pattern of collaboration that a thinkLet creates is the net effect of the individual actions specified by the rules. The rules are, perhaps, the best basis for comparing, modifying, and adapting thinkLets. Small changes to the rules that guide actions can give rise to very different patterns of collaboration (Santanen, Vreede, & Briggs, 2004). For example, an "add" action guided by a "summarize" constraint gives rise to abstraction, synthesis, and generalization, while an "add" action guided by an "analyze" constraint gives rise to increasingly detailed exposition of the sub- and sub-sub-components of the concepts at hand.

Each thinkLet must specify a set of rules that prescribe the actions that people in different roles must take using the capabilities provided to them under some set of constraints specified in parameters. Consider a group-based selection procedure that allows each participant to tag the three ideas from a brainstorming activity that most closely match the evaluation criteria. The actions required of the participants (role) are *judge* and *choose*. The capabilities necessary for this thinkLet are: a page that displays all brainstorming activities, readable by all participants; a way for each participant to tag up to three ideas. The tagging capability could be afforded by marker pens, stickers, or push-pins, or it could be afforded by an electronic voting system. It is left to the designer of the collaboration process to choose appropriate tools to instantiate the capabilities. The *judge* action is constrained by the evaluation criteria which should be specified in a parameter (e.g., best, cheapest,), the choose action was constrained by a three-item maximum and enabled by a tag and page (capability). In this thinkLet there is only one role specified; all participants perform the same role, and thus contribute according to the same set of rules. ThinkLets can include more roles, for instance in brainstorming a second role can be a devil's advocate (Janis, 1972).

Script

Each thinkLet must provide a script that describes everything a user could do and say to instruct the group in performing their actions based on the rules in the thinkLet. The script makes the thinkLet more readily transferable, because it frames the rules as spoken instructions and guided actions for the user. At design time, collaboration engineers often customize and instantiate the basic script to its specific application.

Selection Guidance

Each thinkLet must explain the pattern of collaboration that will emerge when the thinkLet is executed, and must include guidance about the conditions under which the thinkLet would be useful, and the conditions under which it is known not to be useful.

Insights

Each thinkLet must document insights, tips, and lessons learned from the field to further clarify the way a thinkLet might be used and how it may affect a group. It must also documents known pitfalls that might interfere with its success, and suggests ways to avoid them. This kind of information is useful to first-time users of the thinkLet.

Success Stories

Each thinkLet must recite at least one success story of how a thinkLet was used in a real-life task. Success stories help the user understand how the thinkLet might play out in a group working on a real task. Some documenters of thinkLet also include failure stories to illustrate the consequences of specific execution errors or misapplications of the thinkLet.

THINKLET EXAMPLE

This section presents an example of a fully documented thinkLet (Copyright 2005 Robert O. Briggs and Gert-Jan de Vreede. Used by Permission). Notice that its content includes all elements specified in the previous section.

Identification

- **Name:** DimSum (picture of dim sum restaurant)
- **Methaphor:** A Dim Sum restaurant serves a variety of small treats. Diners choose what ever appeals to them In like manner, in the DimSum thinkLet, team members select the best phrases from sample texts.

Overview

In the DimSum thinkLet, the team works to create a specific, precise, statement, in a way that all understand, and that accommodates the interests of all team members. Each member proposes a candidate wording for the joint statement. Participants then draw the words and phrases they like best from the candidate statements to craft a new joint statement. Periodically all participants propose new candidate statements based on the current draft of the joint statement. The cycle continues until a version emerges that all participants accept. The pattern of collaboration that will emerge is a cycle of generation of concepts, clarification of meaning and consensus building on the final statement.

Rules

- Participants may add written candidate statements for a <topic> to a page
- Participants may read candidate statements proposed by others
- Participant may orally propose words and phrases for a joint statement

Script

- Set up a page that all participants can see and to which all participants can contribute.
- Say: *"Each of you please draft a candidate version of this statement and submit it to the group. In a few minutes, we will review the concepts that emerge."*
- When all concepts are collected, say: *"Review all the statements submitted by your team. What do you see there that you like? Do you see any useful words or phrases we could use in our joint draft?*

- Guide an oral discussion to elicit suggestions of words and phrases that could be incorporated into a joint statement and assemble them into a joint statement. Invite comment on the joint statement as well as comment on the candidate statements.
- If the team arrives at an impasse over the wording of some phrase, collect another round of candidate drafts for just that phrase.
- If the team arrives at an impasse over the wording of the whole statement, ask them to start with the current draft of the joint statement and submit new individual candidate statements.
- Continue the cycle of submitting candidates and converging on a joint statement until a joint statement emerges that all parties accept.

Selection Guidance

With the DimSum thinkLet you can integrate the thoughts of many people into a single statement or a single definition for a key term that all participants can accept. It can be used to overcome an impasse caused by polarized interests about the wording of a joint statement. DimSum is useful in many settings, for example: drafting a mission statement, defining key terms in a project plan, and negotiating the terms of a treaty.

Insights on DimSum

DimSum can significantly speed the negotiating of mutually acceptable wording for joint statements. The magic of DimSum for a harmonious group lies in rapidly developing a variety of approaches to expressing a significant concept, and then drawing from the best words and phrases to arrive at a final draft. In harmonious groups DimSum can cut the time required to draft a joint statement to less than the time required by other means. The quality and clarity of the resulting statement tend to be high.

In a badly conflicted group, like a labor-management contract negotiation team, effective collaboration can be paralyzed by suspicion. Labor may automatically reject any offer from management because if it is offered by management, it must, by definition be bad for labor. Management may likewise reject any offer from labor because it must, by definition, be bad for management if it is proposed by labor. The magic of DimSum for a group in conflict lies in the anonymity of the sample contributions. Because nobody knows for sure which contributions were contributed by which interest group, people can consider the merits of a phrase in light of their own interests, without the certainty that there must be something that will hurt them hidden in it because it was contributed by the other side.

Success Story

Contract negotiations at a major U.S. bus company broke down, and the drivers went on strike. The negotiation team decided to try DimSum. They used two projection screens. On the left screen, they projected one paragraph of the expired contract. On the right screen, they displayed the results of a collaborative comment tool. Any contribution made by any negotiator appeared immediately on the screens of all the other negotiators, and on the right-hand screen. An equal number of negotiators from each side participated in the DimSum activity. Progress was fast on many paragraphs, but it stalled again when the negotiations focused on job security issues. When they were not able to agree on a paragraph, they began using DimSum on sentences. When they could not agree on sentences, they began using DimSum on phrases. When they could not progress on phrases, they used polling techniques on sample phrases. Some phrases received 75% or more approval rates, implying that phrases had support from both sides. The negotiation began to make progress once again. In the end, the negotiators arrived at a contract they all could accept.

CONCLUSION

In e-collaboration settings, especially when people are spread among different locations, process guidance is critical for success. In order to ensure the success of the tools and techniques on the e-collaboration process, their effects should be predictable. The codification of useful facilitation techniques into well-structured, fully documented thinkLets creates reusable modules of facilitation expertise that increase predictability of e-collaboration support and efforts. ThinkLets can be useful in many settings. They can be readily transferred among both expert facilitators and practitioners in the field. Researchers can use the consistent documentation framework of thinkLets to compare different techniques and to find the causes of their added value. This type of

research makes thinkLets even more predictable than best-practices of experts. The predictability of thinkLets and their transferable description allows novices and practitioners to successfully support groups in their collaborative effort, without the need for extensive training and experience. To enable the use of thinkLets among communities of facilitators and practitioners, they should be collected on the web in a library with features to comment on and validate thinkLets and thinkLet combinations.

REFERENCES

Alexander, C. (1979). *The timeless way of building.* New York: Oxford University Press.

Alexander, C., Ishikawa, S., Silverstein, M., Jacobson, M., Fiksdahl-King, I, & Angel, S. (1977). *A pattern language, towns, buildings, construction.* New York: Oxford University Press.

Baker, K., Greenberg, S., & Gutwin, C. (Eds.). (2001). *Heuristic evaluation of groupware based on the mechanics of collaboration.* Heidelberg, Germany: Springer-Verlag.

Briggs, R. O., Kolfschoten, G. L., Vreede, G. J. de, & Dean, D. L. (in press). Defining key concepts for collaboration engineering. In N. Romano (Ed.), *Proceedings of Americas Conference on Information Systems.* Acapulco, Mexico: AIS.

Briggs, R. O., & Vreede, G. J., de. (2001). *ThinkLets, building blocks for concerted collaboration.* Delft, The Netherlands: Delft University of Technology.

Briggs, R. O., Vreede, G. J. de, & Nunamaker, J. F., Jr. (2003). Collaboration engineering with thinklets to pursue sustained success with group support systems. *Journal of Management Information Systems, 19*(4), 31-63.

Buzan, T. (1974). *Use your head.* London: British Broadcasting Organization.

Delbecq, A. L., Ven, A. H. van de, & Gustafson, G. H. (1975). *Group techniques for program planning: A guide to nominal group and delphi processes.* Glenview, IL: Scott, Foresman and Company.

FacilitatorU. (2006, April 26). *Factivities.Com,* Retrieved April 26, 2006, from http://www.factivities.com/exercises.html

Janis, I.L. (1972). *Victims of groupthink: A psychological study of foreign-policy decisions and fiascoes.* Boston: Houghton Mifflin Company.

Jenkins, J. (2006). *IAF mehods database.* Retrieved April 26, 2006, from http://www.iaf-methods.org

Kolfschoten, G. L., Briggs, R. O., Vreede, G. J. de, Jacobs, P. H. M., & Appelman, J.H. (in press). Conceptual foundation of the thinkLet concept for collaboration engineering. *International Journal of Human Computer Science.*

Nunamaker, J. F., Jr., Briggs, R. O., Mittleman, D. D., Vogel, D., & Balthazard, P. A. (1997). Lessons from a dozen years of group support systems research: A discussion of lab and field findings. *Journal of Management Information Systems, 13*(3), 163-207.

Osborn, A. F. (1953). *Applied imagination.* New York: Scribners.

Post, B. Q. (1993). A business case framework for group support technology. *Journal of Management Information Systems, 9*(3), 7-26.

Santanen, E. L., Vreede, G. J. de, & Briggs, R. O. (2004). Causal relationships in creative problem solving: Comparing facilitation interventions for ideation. *Journal of Management Information Systems, 20*(4), 167 -197.

Vreede, G. J. de, & Briggs, R. O. (2005). Collaboration engineering: Designing repeatable processes for high-value collaborative tasks. In R. H. Sprague (Ed.), *Proceedings of the Hawaii International Conference on System Sciences.* Washington, DC: IEEE Computer Society Press.

Vreede, G. J. de, Briggs, R. O., & Kolfschoten, G. L. (in press). Thinklets: A pattern language for facilitated and practitioner-guided collaboration processes. *International Journal of Computer Applications in Technology.*

Vreede, G. J. de, Vogel, D. R. , Kolfschoten, G. L., & Wien, J. S. (2003). Fifteen years of in-situ GSS use: A comparison across time and national boundaries. In R. H. Sprague (Ed.), *Proceedings of the Hawaii International Conference on System Sciences.* Washington, DC: IEEE Computer Society Press.

KEY TERMS

Capability: "The means necessary to contribute, record, read, and manipulate concepts" (Briggs et al., in press).

Collaboration Engineering: "An approach to designing collaborative work practices for high-value recurring tasks, and deploying those designs for practitioners to execute for themselves without ongoing support from professional facilitators" (Briggs et al., in press).

Facilitation Technique: A method, used by facilitators to support a group process.

Group Support Systems (GSS): A class of collaboration software used to move groups through the steps of a process toward their goals.

Pattern of Collaboration: "The nature of a group's collaborative process when observed over a period of time as they move from a starting state to some end state" (Briggs et al., in press).

Rule: "An instruction to perform an action with a certain capability and under one or more specified constraints" (Briggs et al., in press).

ThinkLet: "A named, scripted collaborative activity that gives rise to a known pattern of collaboration among people working together toward a goal" (Briggs et al., in press).

The 3C Collaboration Model

Hugo Fuks
Catholic University of Rio de Janeiro, Brazil

Alberto Raposo
Catholic University of Rio de Janeiro, Brazil

Marco A. Gerosa
University of Vila Velha, Brazil

Mariano Pimental
Federal University of the State of Rio de Janeiro, Brazil

Carlos J. P. Lucena
Catholic University of Rio de Janeiro, Brazil

INTRODUCTION

Computational support for collaboration may be realized through the interplay between communication, coordination, and cooperation tools. Communication is related to the exchange of messages and information among people; coordination is related to the management of people, their activities and resources; and cooperation is the production taking place on a shared workspace. This model, which we call the 3C model, was originally proposed by Ellis, Gibbs, and Rein (1991), with some terminological differences. Cooperation, which Ellis et al. denominates "collaboration," here characterizes the joint operation in a shared workspace.

The 3C model appears frequently in the literature as a means to classify collaborative systems, for example as done by Borghoff and Schlichter (2000). However, a few attempts have been made to use it in the context of groupware implementation. An example is the Clover design model, which defines three classes of functionalities, namely communication, coordination, and production (Laurillau & Nigay, 2002; Calvary, Coutaz, & Nigay, 1997). These three classes of services appear in each functional layer of the model and, during the system design phase, they "must be identified and their access harmoniously combined in the user interface." The Clover model shares the same usefulness of the 3C model in terms of groupware functional specification, because both deal with the three classes of functionalities that a groupware application may support.

Given its complex interactive nature, groupware testing has not yet achieved its maturity. The 3C model may also help evaluators focus their attention on the communication, coordination, and cooperation aspects, guiding the detection of usability problems. A groupware evaluation approach based on a model similar to the 3C one is presented in Neale, Carroll, and Rosson (2004). Differently from the approaches found in the literature, we explore the 3C model as a means to analyze and represent a groupware application domain and also to serve as a basis for groupware development.

The relationship among the 3Cs of the model may be used as a guidance to analyze a groupware application domain. Groupware such as chat, for example, which is a communication tool, requires communication (exchange of messages), coordination (access policies), and cooperation (registration and sharing). Despite their separation for analytic purposes, communication, coordination and cooperation should not be seen in an isolated fashion; there is a constant interplay between them (Pimentel, Fuks, & Lucena, 2004).

For the sake of development, we propose the use of 3C-based components as a means of developing extendable groupware whose assembly is determined by collaboration needs. By conceiving the problem from the viewpoint of the 3C model and using a component structure designed for this model, changes in the collaboration dynamics are mapped onto the computational support. This way, the developer has a workbench with a component-based infrastructure designed specifically for groupware, based on a collaboration model.

INSTANTIATING THE 3C MODEL

Below we present three different groupwork domains that illustrate that the iterative nature of collaboration may be represented as cycles connecting the 3Cs.

We start with the groupwork domain represented in Figure 1. According to this instantiation of the 3C model, while communicating, people negotiate and make decisions. While coordinating themselves, they deal with conflicts and organize their activities in a manner that prevents loss of communication and of cooperation efforts. Cooperation is the joint operation of members of the group in a shared space, seeking to execute tasks, and generate and manipulate cooperation objects. The need for renegotiating and for making decisions about unexpected situations that appear during cooperation may demand a new round of communication, which will require coordination to reorganize the tasks to be executed during cooperation.

Considering media spaces (Mackay, 1999), which are multimedia-enhanced spaces aimed at informal communication among people, the 3C model may be instantiated according to Figure 2a. The media space itself is the shared space. Since it is aimed at informal communication, its main goal is actually to create opportunities for informal meetings, which are coordinated by the standing social protocol, for example, by accessing the availability of remote colleagues. These meetings generate conversation, which may occur using the media provided by the system or any other available means, such as telephones.

Another example is the family calendar (Figure 2b). The main reason for the family calendar is to schedule family activities. Modern family members have a variety of conflicting interests that can render last evening defined schedules ineffective next morning. In order to restore proper family coordination, negotiation among family members is needed. "This process involves seeing what has already been scheduled…and negotiating errand, ride, and other responsibilities are needed" (Elliot & Carpendale, 2005, p.4). The reconciliation obtained after the negotiation round is placed on the shared calendar. But as life never stops, next morning the cycle may start all over again.

These cycles show the iterative nature of collaboration. The participants obtain feedback from their actions and feedthrough from the actions of their companions by means of awareness information related to the interaction among participants (Gerosa, Fuks, & Lucena, 2003). This information mediates each of the 3Cs, which are detailed in the next sessions.

COMMUNICATION

The designer of a communication tool defines the communication elements that will set the communication channel between the interlocutors, taking into consideration the specific usage that is being planned for the tool (time, space, purpose, dynamics, and types of participants) and other factors such as privacy, development and execution restrictions, information overload, and so forth. Then, these elements are mapped onto software components that provide support to the specific needs.

The first communication element that must be considered is the choice of media. They can be textual, spoken, pictorial, or gestured—for example in a video

Figure 1. 3C collaboration model instantiated for group work

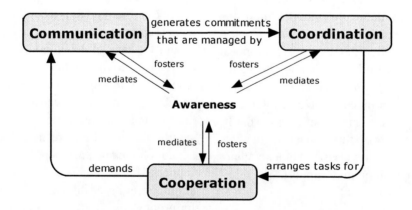

Figure 2. 3C collaboration model instantiated for the media space and for the family calendar domains

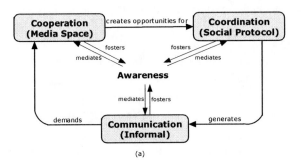

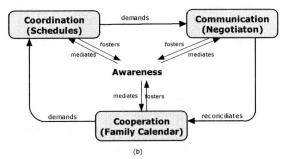

(a)   (b)

or avatar form. The media adopted restrict and influence the vocabulary used in the conversation, which is also influenced by the context.

The *transmission mode* defines whether the information is transmitted in blocks or continuously. In an audio or videoconference and in some chat tools the information is transmitted continuously as it is generated. In asynchronous and in other chat and messenger tools the information is transmitted in blocks: The author edits the message and it is only sent with an explicit command. Asynchronous communication tools are used when one wants to enhance reflection by the interlocutors, since they will have more time before they act, while synchronous tools are used for communication bursts. Restrictions policy is another communication element. For example, it is possible to restrict the text's size, characters allowed, bit rate (for video and audio), allowed vocabulary, and so on. These restrictions are used in some tools to reduce information overload and to save bandwidth.

Meta-information complements the information being transmitted in the body of the message. Common meta-information available in communication tools are the subject of the message, its date, priority, and category.

Conversation structure defines how messages are structured, which can be in a linear, hierarchical, or network form. The conversation structure makes the relations between messages, which are usually implicit within the text, visually explicit. The linear form is used when there are not so many message interconnections, the chronological order of the messages is relevant and fluency in the conversation is desired. Hierarchical structuring is appropriate when the relationships between messages, such as questions and answers, need to be quickly identified. However, as there is no way to link messages from two different branches, the tree can

only grow wide and, thus, the discussion takes place in diverging lines (Stahl, 2001). Network structuring can be used to seek convergence in the discussion.

Conversation paths can be used to restrict the possible directions the conversation can take. Conversation paths formalize the conversation and it is not recommended when fluency is desired.

In order to provide proper support to communication, the designer should also take into account coordination and cooperation elements. Coordination elements deal with access policies to the communication channel, while cooperation elements deal with information rendering and registration.

COORDINATION

Coordination may be viewed as the link connecting the other two Cs in order to enforce the success of collaboration. This is more clearly observed when we analyze the elements that need to be coordinated, namely people, resources, and tasks. The coordination of people, for example, is deeply related to communication and context. The coordination of resources, on the other hand, is related to the shared space (i.e., cooperation). For this reason, coordination aspects also appear in the discussion about the other Cs of the model. In this section, we focus on the coordination of tasks. The management of the tasks being carried out consists in managing interdependencies between tasks that are carried out to achieve a goal (Malone & Crowston, 1990).

Some computer-supported collaborative activities, the so-called *loosely integrated* collaborative activities are deeply associated with social relations and are generally satisfactorily coordinated by the standing social protocol, which is characterized by the absence of any computer-supported coordination mechanism—"a specialized software device, which interacts with a specific software application so as to support articulation work" (Schmidt & Simone, 1996, p. 184)—among the tasks, trusting the users' abilities to mediate interactions. Coordination, in these situations, is contextually established and strongly dependent on mutual awareness. Through awareness information, the participants detect changes in plans and understand how the work of their colleagues is getting along (Dourish & Belloti, 1992). However, there are also the so-called *tightly integrated* collaborative activities, whose tasks are highly interdependent, as the name suggests. They require sophisticated coordination mechanisms in order to be supported by computer systems.

The great challenge of the designer in designing coordination mechanisms in groupware is to achieve flexibility without losing the regulation, which is necessary in some situations in which the social protocol is not enough. The system should not impose rigid work or communication patterns, but rather offer the user the possibility to use, alter, or simply ignore them. Thus, coordination flexibility and accessibility should be pursued by groupware designers. Flexibility is related to the possibility of dynamically allowing redefinition and temporary modifications in the coordination scheme. Accessibility is related to exposing the coordination mechanisms to system users rather than having them deeply embedded in the system's implementation.

Coordination may take place on the temporal and on the object levels (Ellis & Wainer, 1994). On the temporal level, coordination defines the sequence of tasks that makes up an activity. On the object level, coordination describes how to handle the sequential or simultaneous access of multiple participants through the same set of cooperation objects (Raposo & Fuks, 2002).

Communication and coordination, although crucial, are not enough. Given that coordination is required to manage the tasks; according to the 3C model it is also necessary to provide a shared workspace where cooperation will take place (Raposo, Gerosa, & Fuks, 2004).

COOPERATION

Cooperation is the joint operation during a session within a shared workspace. Group members cooperate by producing, manipulating, and organizing informa-tion, and by building and refining cooperation objects, such as documents, spreadsheets, artwork, and so forth. The shared workspace provides a number of tools for managing these artifacts, such as the recording and the recovery of previous versions, access control and permission. By recording the information exchanged, the group is able to count on collective memory, which can be consulted whenever necessary to recover the history of a discussion or the context in which a decision was made.

Production is dependent on how the shared workspace is structured to present the cooperation objects and the interaction that is taking place there. In a face-to-face situation, a large part of how we maintain a sense of who is around and what is going on is related to being able to see and hear events or actions with little conscious effort. On the other hand, in a computer-supported workspace, awareness support is less effective since the means for making information available to sensory organs are limited; however, irrelevant information can be filtered in a way that reduces distractions that usually affect face-to-face collaboration.

Individuals seek the awareness information necessary to create a shared context and to anticipate actions and requirements related to their collaboration goals. Thus, it becomes possible to interpret the intentions of the members of the group in such a way that one can provide assistance in terms of their work whenever it is convenient and needed (Baker, Greenberg, & Gutwin, 2001).

The designer of a digital environment must identify what awareness information is relevant, how it will be obtained, where it is needed and how to display it. Excessive information can cause overload and disrupt the collaboration flow. To avoid disruption, it is necessary to balance the need to supply information with care to avoid distracting the attention required to work. The supply of information in an asynchronous, structured, filtered and summarized form can accomplish this balance (Kraut & Attewell, 1997). The big picture should be supplied and individuals could select which parts of the information they want to work with, leaving further details to be obtained when required. There must also be some form of privacy protection. The shared space must be conceived in a way that group members could seamlessly move from awareness to work.

The register of group interactions is filed, catalogued, categorized and structured within cooperation objects. Ideas, facts, questions, points of view, conversations, discussions, decisions, and so on, are retrievable, providing a history of the collaboration and the context in which learning took place.

DESIGN AND IMPLEMENTATION ISSUES

The 3C collaboration model has been used as a basis for the development of the AulaNet Learning Management System. AulaNet is a freeware web-based environment for teaching and learning. It has been under development since June 1997 by the Software Engineering Laboratory of the Catholic University of Rio de Janeiro (PUC-Rio) (Fuks, Raposo, Gerosa, & Fuks, 2005).

The AulaNet environment's services are subdivided into communication, coordination and cooperation services, as can be seen in Figure 3. The communication services provide tools for forum-style asynchronous text discussion (*conferences*), chat-style synchronous text discussion (*debate*) (Fuks, Pimentel, & Lucena, 2006), instant message exchange between simultaneously connected learners (*instant messaging*), and individual electronic mail with the mediators (*message to participants*) and with the whole class, in a list-server style (*message to the class*).

Coordination services support the management and the enforcement of group activities. In AulaNet, coordination services include tools for notification (*notices*),

evaluation (*tasks* and *exams*) as well as a tool that allows monitoring group participation (*follow-up reports*). Cooperation services in AulaNet include *Lessons* and *Documentation*, a list of course references (*bibliography* and Webliography) and course co-authoring support, both for teachers (teacher co-authoring) and for learners (learner co-authoring).

In developing groupware, the requirements are rarely clear enough to allow for a precise specification of the system's behavior in advance. Groupware development is evolutionary in the sense that it is difficult to predict how a particular group will collaborate and each group has highly distinct characteristics and objectives (Gutwin & Greenberg, 2000). By involving a group, the possibilities of interactions multiply and the demand for synchronism and solving deadlocks increases, posing problems in the construction of suitable interaction mechanisms and conducting tests.

This scenario is suitable for the application of component-based development techniques, which provide the flexibility needed in projects with changing requirements. Groupware services can be seen as groupware components that are plugged and unplugged from the system. The system's architecture comprises component frameworks that define overall invariants and protocols for plugging components.

AulaNet services were developed using a component-framework-based architecture, as can be seen in Figure 4. There is a common structure implemented by the collaboration framework, which defines the skeleton

Figure 3. Classification of AulaNet services based on the 3C model. The 3C triangle appears in Borghoff and Schlichter (2000).

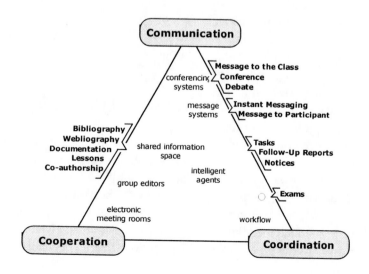

Figure 4. Architecture of a collaboration service

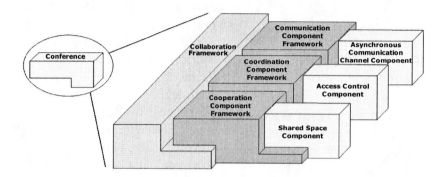

of the services, and plugged to this framework there are the communication, the coordination and the cooperation component frameworks, which support each C of the 3C model. Class frameworks are used to implement components, which are plugged to the corresponding C-framework and implement the specific functionalities of the service (Gerosa et al., 2005).

For example, using the communication class framework, the developer implements components for synchronous and asynchronous communication, message transmission etc. Using coordination class framework, components for task management, participation follow-up, workflow, and so on are implemented. Using cooperation class framework, components for managing the shared space and its awareness elements, version management, among others are implemented.

CONCLUSION

Groupware systems are evolutionary because the composition and the characteristics of workgroups change with time, as well as the tasks that need to be executed. For this reason, even if a groupware designer is able to develop an "optimal" application for a group, it will eventually become inadequate due to new situations and problems that certainly will appear.

Ideally, groupware should be prototyped because collaborative systems are especially prone to failure (Grudin, 1989), hence demanding iterative evaluation during their development. However, given the excessive cost of throwing code away, as demanded by "pure" prototyping (Brooks, 1975), an incremental model can be considered more adequate, leading to the development of more advanced prototypes in the subsequent cycles.

In face of construction and maintenance difficulties, the groupware developer spent more time dealing with technical difficulties than moderating and providing support to the interaction among users. Such problems led to the need to create a quicker and more effective way to develop groupware in which low-level complexities resulting from distributed and multi-user systems were encapsulated into infrastructure components of the architecture. Besides, the concepts of the domain's modeling should permeate all other activities and artifacts of the application's development. This way, the modeling done during the domain analysis could be mapped to implementation, thus increasing productivity in groupware development and maintenance and making the applications more adequate to the users' collaboration needs.

The 3C collaboration model defines three types of services that a groupware may support. The concepts and representation models described in this paper can be used to provide a common language for representing and describing the collaboration aspects of a workgroup and to guide the functional specification and the implementation of computational support for group work.

ACKNOWLEDGMENT

The AulaNet project is partially financed by Fundação Padre Leonel Franca and by the Ministry of Science and Technology through its Program Multi-Agent Systems for Software Engineering Project (ESSMA) grant n° 552068/2002-0. It is also financed by individual grants awarded by the National Research Council to: Carlos José Pereira de Lucena n° 300031/92-0, Hugo

Fuks n° 301917/2005-1, and Marco Aurélio Gerosa n° 140103/02-3. Thanks to Prof. Marcelo Gattass, Head of Tecgraf/PUC-Rio, a group mainly funded by Petrobras, Brazilian Oil & Gas Company.

REFERENCES

Baker, K., Greenberg, S., & Gutwin, C. (2001). Heuristic evaluation of groupware based on the mechanics of collaboration. In M. Little & L. Nigay (Eds.), *Proceedings of 8ᵗʰ IFIP International Conference (EHCI 2001)* (pp. 123-139), Lecture Notes in Computer Science Vol. 2254. Berlin: Springer-Verlag.

Borghoff, U. M., & Schlichter, J. H. (2000). *Computer-supported cooperative work: Introduction to distributed applications*. Berlin: Springer-Verlag.

Brooks, F. P. Jr. (1975). Plan to throw one away. In F. P. Brooks, Jr. (Ed.), *The mythical man-month: Essays on software engineering* (pp. 115-123). Reading, MA: Addison-Wesley.

Calvary, G., Coutaz, J., & Nigay, L. (1997). From single-user architectural design to PAC*: A generic software architectural model for CSCW. In S. Pemberton (Ed.), *Proceedings of the Conference on Human Factors in Computing Systems (CHI'97)* (pp. 242-249). New York: ACM Press.

Dourish, P., & Belloti, V. (1992). Awareness and coordination in shared workspaces. In J. Turner & R. Kraut (Eds.), *Proceedings of the Conference on Computer-Supported Cooperative Work (CSCW)* (pp. 107-114). New York: ACM Press.

Elliot, K., & Carpendale, S. (2005). *Awareness and coordination: A calendar for families* (Technical Report 2005-791-22). Calgary, Canada: University of Calgary, Department of Computer Science.

Ellis, C. A., Gibbs, S.J., & Rein, G. L. (1991). Groupware: Some issues and experiences. *Communications of the ACM, 34*(1), 38-58.

Ellis, C. A., & Wainer, J. (1994). A conceptual model of groupware. In T. Malone (Ed.), *Proceedings of the Conference on Computer-Supported Cooperative Work (CSCW)* (pp. 79-88). New York: ACM Press.

Fuks H., Pimentel, M., & Lucena, C. J. P. (2006). R-U-Typing-2-Me? Evolving a chat tool to increase understanding in learning activities. *International Journal of Computer-Supported Collaborative Learning, 1*(1), 117-142.

Fuks, H., Raposo, A. B., Gerosa, M. A., & Lucena, C. J. P. (2005). Applying the 3C model to groupware development. *International Journal of Cooperative Information Systems (IJCIS), 14*(2-3), 299-238.

Gerosa, M. A., Fuks, H., & Lucena, C. J. P. (2003). Analysis and design of awareness elements in collaboration digital environments: A case study in the AulaNet learning environment. *The Journal of Interactive Learning Research, 14*(3), 315-332.

Gerosa, M. A., Pimentel, M., Fuks, H., & Lucena, C. J. P. (2005). No need to read messages right now: Helping mediators to steer educational forums using statistical and visual information. In T. Koschmann, T. -W. Cha, & D. D. Suthers (Eds.), *Proceedings of Computer Supported Collaborative Learning* (pp. 160-169). Mahwah, NJ: Lawrence Erlbaum Associates.

Grudin, J. (1989). Why groupware applications fail: Problems in design and evaluation. *Office: Technology and People, 4*(3), 245-264.

Gutwin, C., & Greenberg, S. (2000). The mechanics of collaboration: Developing low cost usability evaluation methods for shared workspaces. In R. D. Sriram & N. Shahmehri (Eds.), *Proceedings of the IEEE 9ᵗʰ Workshop on Enabling Technologies: Infrastructure for Collaborative Enterprises, WET ICE* (pp. 98-103). Washington, DC: IEEE Computer Society Press.

Kraut, R. E., & Attewell, P. (1997). *Media use in global corporation: Electronic mail and organisational knowledge, research milestone on the information highway*. Mahwah, NJ: Lawrence Erlbaum Associates.

Laurillau, Y., & Nigay, L. (2002). Clover architecture for groupware. In E. F. Churchill, J. McCarthy, C. Neuwirth, & T. Rodden (Eds.), *Proceedings of the Conference on Computer-Supported Cooperative Work (CSCW)* (pp. 236-245). New York: ACM Press.

Mackay, W. E. (1999). Media spaces: Environments for informal multimedia interaction. In M. Beaudouin-Lafon (Ed.), *Computer supported co-operative work: Trends in software* (Vol. 7, pp. 55-82). Chichester, England: John Wiley & Sons.

Malone, T. W., & Crowston, K. (1990). What is coordination theory and how can it help design cooperative work systems? In F. Halasz (Ed.), *Proceedings of the Conference on Computer-Supported Cooperative Work (CSCW)* (pp. 357-370). New York: ACM Press.

Neale, D. C., Carroll, J. M., & Rosson, M. B. (2004). Evaluating computer-supported cooperative work: Models and frameworks. In J. Herbsleb & G. Olson (Eds.), *Proceedings of the Conference on Computer-Supported Cooperative Work (CSCW)* (pp. 112-121). New York: ACM Press.

Pimentel, M., Fuks, H., & Lucena, C. J. P. (2004). Mediated chat 2.0: Embedding coordination into chat tools. In F. Darses, R. Dieng, C. Simone, & M. Zacklad (Eds.), *Conference Supplement of the 6th International Conference on the Design of Cooperative Systems* (pp. 99-103). Amsterdam: IOS Press.

Raposo, A. B., & Fuks, H. (2002). Defining task interdependencies and coordination mechanisms for collaborative systems. In M. Blay-Fornarino, A. M. Pinna-Dery, K. Schmidt, & P. Zaraté (Eds.), *Cooperative systems design: Frontiers in artificial intelligence and applications* (Vol. 74, pp. 88-103). Amsterdam: IOS Press.

Raposo, A. B., Gerosa, M. A., & Fuks, H. (2004). Combining communication and coordination toward articulation of collaborative activities. In G. J. Vreede, L. A. Guerrero, & G. M. Raventós (Eds.), *Proceedings of the 10th International Workshop on Groupware, CRIWG*. Lecture Notes on Computer Science Vol. 3198 (pp. 121-136). Berlin: Springer-Verlag.

Schmidt, K., & Simone, C. (1996). Coordination mechanisms: Towards a conceptual foundation of CSCW systems design. *Computer Supported Cooperative Work, 5*(2-3), 155-200.

Stahl, G. (2001). WebGuide: Guiding collaborative learning on the Web with perspectives. *Journal of Interactive Media in Education*. Retrieved July 3, 2006, from http://www-jime.open.ac.uk/

KEY TERMS

3C Collaboration Model: A model for the analysis, representation, and development of groupware by means of the interplay between the 3Cs, namely, communication, coordination, and cooperation.

Awareness: The human beings' capability of perceiving the activities of the others and their own activities in the context of collaboration. A groupware generally provides elements and information to enable awareness.

Collaboration: The interplay between communication, coordination, and cooperation.

Communication: Conversation to negotiate and make decisions through an augmentation process.

Component-Based Development Techniques: Techniques that seek to develop modular systems composed of software components that may be adapted and combined as needed, always having reuse and maintenance in mind.

Component Framework: Defines overall invariants and protocols for plugging components.

Cooperation: Joint operation in the shared workspace.

Coordination: The management of people, their activities and resources, in the context of collaboration. In a narrower definition, it consists in managing interdependencies between tasks that are carried out to achieve a goal.

Towards a Collaborative Educational Game Model

Jeane Silva Ferreira Teixeira
Federal Center for Technological Education of Maranhão (CEFET-MA), Brazil
Learning and Interaction Laboratory (LAI) - Aeronautical Institute of Technology (ITA), Brazil

Eveline de Jesus Viana Sá
Federal Center for Technological Education of Maranhão (CEFET-MA), Brazil
Learning and Interaction Laboratory (LAI) - Aeronautical Institute of Technology (ITA), Brazil

Tatiane Macedo Prudêncio Lopes
Learning and Interaction Laboratory (LAI) - Aeronautical Institute of Technology (ITA), Brazil

Inaldo Capistrano Costa
Learning and Interaction Laboratory (LAI) - Aeronautical Institute of Technology (ITA), Brazil

D'Ilton Moreira Silveira
Learning and Interaction Laboratory (LAI) - Aeronautical Institute of Technology (ITA), Brazil

Alessandro Ramos de Oliveira
Learning and Interaction Laboratory (LAI) - Aeronautical Institute of Technology (ITA), Brazil

Clovis Torres Fernandes
Learning and Interaction Laboratory (LAI) - Aeronautical Institute of Technology (ITA), Brazil

José Maria Parente de Oliveira
Learning and Interaction Laboratory (LAI) - Aeronautical Institute of Technology (ITA), Brazil

INTRODUCTION

In a collaborative learning process, concept learning may happen simultaneously with the practice of sociable attitudes and essential values, motivating the learner in the knowledge acquisition process. Among the tools that are likely to help reinforcing the learner motivation, both in and out of classroom settings, are computer games, especially the ones played interactively by two or more players, known as collaborative games,

A collaborative game used as an educational tool in classroom settings or at a distance, integrated to appropriate pedagogic practices, is a collaborative educational game. Aspects as ludic engagement, competitiveness and interactivity inherent to collaborative educational games may deepen children and youngsters' motivation to learn, helping to make the learning process more effective (Prensky, 2001).

This paper presents an e-collaborative educational variation of the widespread computer game Tetris[1], called e-Collaborative Educational Game Tetris or Collaborative JETetris for short, its underlying model and prototype. Simulations done with the JETetris prototype helped to derive a proposal to a Web-based collaborative educational game model—CEG Model, which allows the creation and use of many kinds of games, instead of just JETetris.

In the next section, the background on the use of games in education is presented. Next, Collaborative JETetris, its prototype and the draft of a collaborative educational game model are depicted. Finally, some conclusions and futures perspectives of this work are drawn.

BACKGROUND

In the education setting the applicability of games has been the focus of recent studies. However, the use of games as a tool to assist the teaching of specific contents in the classroom is not a new idea (Hill, Ray, Blair, & Carver, 2003). In the 1970s, for instance, the importance of games in the educational process was discussed along with the use of computers in basic education (Poirot, 1976).

Anderson and Holt (1996) mention the use of games in classroom in the study of instructional contents of social sciences and economics. Almstrum, Ginat, Hazzan, and Morley (2002) encountered the use of games in the teaching of mathematics, and Herr (2002) cites their use to support the teaching of sciences. Navarro and Hoek (2004) present an educational game for teaching the software engineering process, where the learner can assume the role of a project manager of a development team.

The playful characteristics and challenges of games make learners feel attracted and motivated. Coleman, Krembs, Labouseur, and Weir (2005) pointed out that learners have their initial experience with computer games way before they get to know the Internet. They also discuss the need for research related to the design of games, proposing the implementation of courses that include studies on the most diverse contents related to games, including educational games.

Many instructional principles can be related to the process of learning games. Out of them, the following are worth mentioning: *active learning* (Bonwell & Eison, 1991; McKinney 2007), *peer* or *collaborative learning* (Tinto, Goodsell, & Russo, 1993; Wills & Finkel, 1994), *problem-based learning* (Savery & Duffy, 1995) and *game-based learning* (Haas, 1988; Feezel, 1993; Prensky, 2001).

In a collaborative learning context, besides the amusing aspects involved in the educational process, the use of games promotes social interaction encouraging the exchange of knowledge between learners in the group, as well as stimulating their reasoning skills (Elgood, 1990). Playing games allow learners to practice, in classroom settings or at a distance, inestimable values and attitudes that help living in society (Prensky, 2001).

Vartiainen and Ruhomaki (1995) reported their findings about some positive effects of games on groups of participants in the process of collaborative learning, for instance, improvement in communication and cooperation, development of social skills and changes in classroom relationships such as increase of empathy, exchange of information and ability for conflicts solving.

According to Galvão, Martins, and Gomes (2000), collaborative learning through simulation games can promote the development of a number of skills in the learners that facilitates the decision-making and problem solving process as well as their social interaction in different groups.

There are currently a few research works focusing on the potential of games in the process of collaborative learning in the classroom (Galvão et al., 2000). Therefore, the importance of learning and games' model involving such actors as teacher, learner, game and instructional model is subject of reflections in the current literature (Garris, Ahlers, & Driskell, 2002; Jenkins, Klopfer, Squire, & Tan, 2003; Soh, 2004; Teixeira et al., 2005).

COLLABORATIVE JETETRIS

Tetris game (Wikipedia, 2005), besides being an excellent resource for individual amusement, helps the players to evolve both their geometric reasoning and motor coordination skills. Collaborative JETetris is based on Tetris game, allowing the competition between two players in a local area network or through the Internet. In addition, unlike individual Tetris, Collaborative JETetris is to be used in classroom settings or at a distance, allowing the learner to act in a competitive way during the game and in a collaborative way before and after the game, during the definition of specific rules of the game and in the resolution and discussion of questions for enhancing concept learning.

The interactions during a game session always take place in pairs of learners chosen by the teacher. One pair plays against another, where learners of a pair are named "allies" and learners of the other pair are named "opponents."

Although JETetris is also suited for groups of older learners, this project is intended to children and youngsters (6 to 14 years old) of the basic education.

Figure 1 illustrates the flow of learner's actions during a JETetris game session. The teacher's actions, in consonance with the adopted instructional model, allow defining a set of rules that may delimit the scope of the learners' performance. This can avoid contents overloading which the learner might not be prepared for. This can also facilitate the teacher task for defining learning activities intended to work out cognitive and social aspects in classroom settings or at a distance.

The learner's actions comprise three interactive phases: negotiation, competition, and collaboration. The interaction between the pairs of players is done using the chat tool. These interaction phases are described next:

1. **Negotiation:** It consists of the interaction among the pairs of competitive players, in real time, aiming at negotiating the rules of the game, including the game difficulty level. The difficulty levels can be: *beginner, intermediate,* and *advanced.* The difference among these levels is the amount and level of questions to be solved, as well as the bonus value to be earned in case of victory. The allied pair players can play with the respective opponent pair in different difficulty levels, once such differences may benefit the learning process.

2. **Competition:** It is the phase where the learners play Tetris, competing in pairs. Each filled out line is sent to the opponent players, reducing their area of performance and hindering their moves. The winner of the game is the player that fills out and consequently sends a larger number of full lines. As full lines are completed, the corresponding learners will obtain bonus that are accumulated and later exchanged by multiple-choice questions stored in JETetris question repository.

3. **Collaboration:** It consists of the interaction among the allied pairs for exchanging ideas and eliminating doubts during solving the question list obtained in the previous phase. Some interaction among opponent players is allowed in this phase, although such interaction is not the main objective of the phase. After the end of the interactions among the allied or opponent pairs, the result of

the game will be presented to the learners. The teacher may guide the classroom discussions on such results according to the pedagogical planning.

The teacher's actions are oriented by a lesson plan that will contemplate activities to include the use of JETetris considering contents intended to be learned (Figure 2). These actions are modeled by the three following stages:

- **Preparation of the environment based on specific learners characteristics:** elaboration of the list of learners, specification of opponent and allied pairs of players, and selection of educational subjects and corresponding questions according to instructional objectives.
- **Delimitation of Game Rules, considering, at least, the following items:**
  ○ Time spent by collaborative partners
  ○ Time for solving questions
  ○ Game difficulty level
  ○ Number of questions
- **Discussion of results in classroom settings or at a distance**

PROTOTYPE OF COLLABORATIVE JETETRIS

A prototype of the Collaborative JETetris has been implemented in JAVA using RMI to enable the remote communication between players. According to Figure 1, for supporting the Negotiation and Collaboration phases, a chat tool was provided; an adapted TETRIS game was provided for the Competition phase. The

Figure 1. Flow of learner's actions

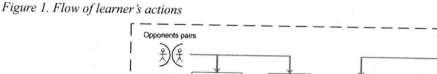
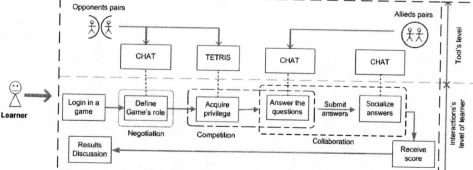

Figure 2. Flow of the teacher's actions

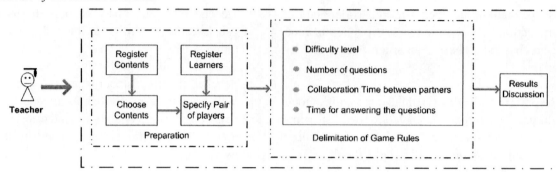

Figure 3. Teacher's actions screen

user interfaces that represent these phases are described next.

Figure 3 illustrates what is available to the teacher for preparing the game environment and for setting up the game rules according to what has been explained in the previous section.

Figure 4 presents two players (Jeane and Vera) chosen by the teacher as a pair. During the Negotiation phase, the competitors are negotiating the rules of the game with an opponent pair (represented by Eveline) in a chat session. In contrast to the dialogue illustrated in the figure, it might not be so easy to come up with a quick consensus in a real life negotiation.

The Competition phase is illustrated in Figure 5, which shows a TETRIS game in execution. Observe that the players should fill in the greatest amount possible of lines so that the corresponding amount of bonus can be used as one of the parameters in the Collaboration phase. This bonus determines the number of ques-

tions to be submitted to the learner in this phase. An example of a group of questions for a certain content area, previously defined by the teacher, is presented in Figure 6.

These three phases have been tested through simulations that accessed a group of questions in the JETetris question repository, related to specific content area, in this case, Geography (Figure 6). The game preparation phase refers to the teacher's actions illustrated in Figure 2.

To carry out the testing tasks, multiple-choice questions have been used. But any kind of question can also be employed, including essay questions.

PROPOSAL OF A COLLABORATIVE EDUCATIONAL GAME MODEL

From the analyses of the testing tasks for the Collaborative JETetris, a Web-based collaborative educational

Figure 4. Negotiation phase supported by the Chat tool

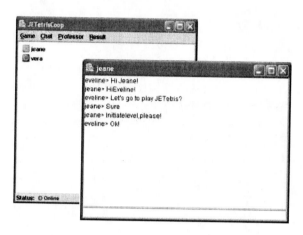

Figure 5. Phase of Competition supported by TETRIS game

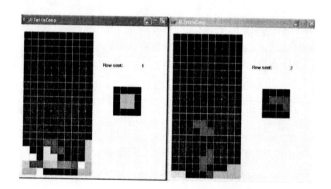

game model (CEG model) has been inferred aiming at conceptually organizing the tools and interactions for all the actors involved: teacher, learner, learning activities, class plans and games.

Figure 7 illustrates the interaction among CEG model and environment's actors.

In the CEG model the teacher will be able to develop a class or lesson plan associated to a suitable game in accordance with the collaborative pedagogic planning, which is guided by the adopted instructional model. The goal of this planning is to allow the teacher to choose the most appropriate game for that class.

With the CEG model it is possible to consider the viability of using other games, instead of only Tetris, in the process of collaborative learning in classroom settings or at a distance. So, depending on the subject to be worked out and the instructional model used, the teacher will be able to choose and use different types of games stored in the game repository.

Purposely resembling the structure of a LMS, learning management system (CMS, 2006), the CEG model is composed by three levels around the game repository: tools' level, learner's level and teacher's level. The *Interaction Tools, Learner's Interaction Phases* and *Teacher's Planning Phases* are respectively inserted in the levels above, as illustrated in Figure 8.

The *Interaction Tools* consist of a set of *communication tools*, conceptually representing any tools that allow the communication between the learners during a game session as a part of the lesson plan.

The *Learner's Interaction Phases* correspond to a set of actions of the learner throughout all learning process. The interaction phases are based on the representation of independent and integrated modules developed exclusively to promote a flexible and stimulating in-

Figure 6. Collaboration phase: Group of questions presented to the learner

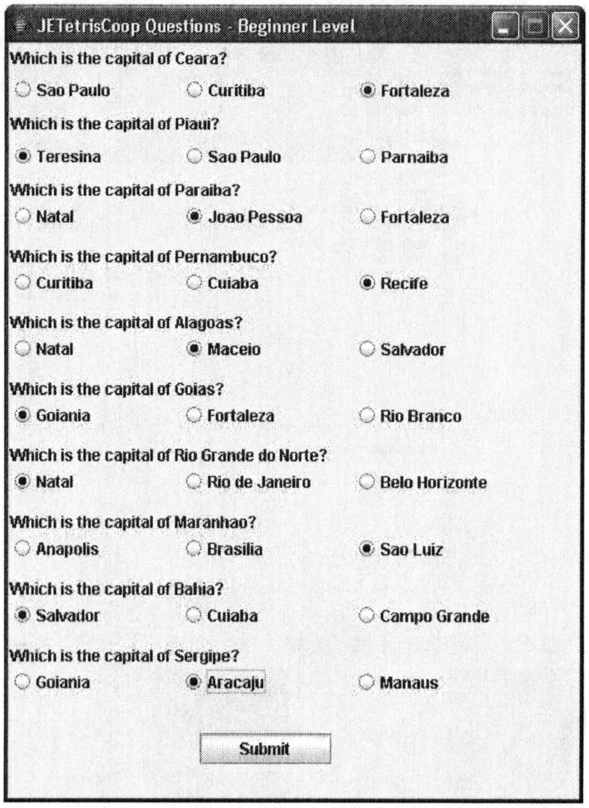

JETetrisCoop Questions - Beginner Level

Which is the capital of Ceara?
○ Sao Paulo   ○ Curitiba   ● Fortaleza

Which is the capital of Piaui?
● Teresina   ○ Sao Paulo   ○ Parnaiba

Which is the capital of Paraiba?
○ Natal   ● Joao Pessoa   ○ Fortaleza

Which is the capital of Pernambuco?
○ Curitiba   ○ Cuiaba   ● Recife

Which is the capital of Alagoas?
○ Natal   ● Maceio   ○ Salvador

Which is the capital of Goias?
● Goiania   ○ Fortaleza   ○ Rio Branco

Which is the capital of Rio Grande do Norte?
● Natal   ○ Rio de Janeiro   ○ Belo Horizonte

Which is the capital of Maranhao?
○ Anapolis   ○ Brasilia   ● Sao Luiz

Which is the capital of Bahia?
● Salvador   ○ Cuiaba   ○ Campo Grande

Which is the capital of Sergipe?
○ Goiania   ● Aracaju   ○ Manaus

[Submit]

Figure 7. Interaction among CEG model and environment's actors

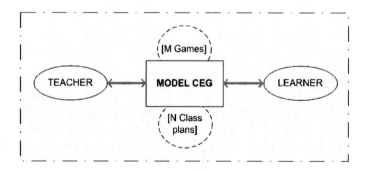

teraction among the learners through the negotiation, competition and collaboration phases, not necessarily in this sequence, as presented in Figure 9.

Although in the testing made with the Collaborative JETetris prototype the learner's interaction phases consisted of the sequence presented in Figure 1, in CEG model such sequence becomes more flexible. So, it is believed that CEG will give a great flexibility to the teacher's actions, enabling the learners to explore each

phase in accordance with the instructional objectives presented in the lesson plan.

The *Teacher's Planning Phases* are composed by three steps: Preparation, Game Selection and Results Discussion (Figure 10). These phases will be supported by the adopted instructional model that can suggest procedures for selecting subjects and activities to be developed by learners in classroom (Ryder, 2005).

The teacher's planning phases are an extension of

Figure 8. Structure of the CEG model

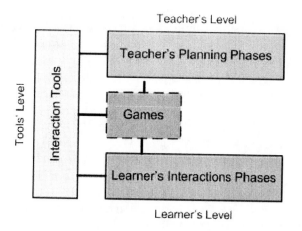

Figure 9. Learner's interaction phases

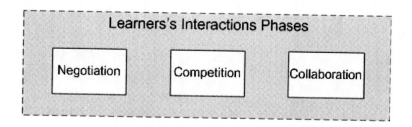

Figure 10. Teacher's planning phases

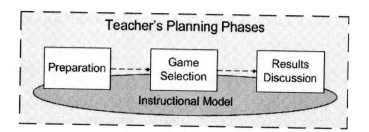

the stages of the teacher's actions presented in Figure 2. For instance, the Game Selection phase allows the teacher to choose the most suitable game to be used to practice certain subjects in classroom settings or at a distance, according to the instructional planning. So, the teacher could plan the use of a great diversity of games, to mediate the interactive process among learners and to follow up all phases of the process, obtaining subsidies for the learners' process evaluation.

The *Games,* as illustrated in Figure 8, represent games instances. Besides its popularity, the Tetris game was chosen in this work because it improves players' motor and reasoning skills. Nevertheless, the choice of which game to use in this process should follow a detailed planning that allows to relate the characteristics of each game to the set of skills they foster. For example, the Memory Game (Giaretta, et al., 1998) can be used for improving concentration and learners memorization capacity, whereas the Gallows Game

(Giaretta et al., 1998) can be used for widening up learners' general vocabulary.

It is important to emphasize that social skills can also be trained in this collaborative setting, independently of the selected underlying domain, once learners will have the opportunity to interact with each other in all phases of the learning process.

CONCLUSION AND FUTURES PERSPECTIVES

The CEG model proposes the use of games integrated with collaboration tasks, in an educational context, that makes possible to take advantage of the collaborative work, considering the ludic engagement and interaction among learners. Since its conception, this model has an e-collaborative aspect (e-collaborative educational game) that supplies user needs in the same way that other models developed for the Web.

An environment supporting the CEG model may provide the teacher with the capability to organize all the subject contents to be learned and reviewed by learners in a ludic perspective, possibly increasing learners' motivation. Considering the case studies carried out and tests based on the Nielsen's usability heuristics (Nielsen, 2007), it was observed that the use of Collaborative JETetris game may promote the practice of social skills negotiation, competition and collaboration among learners, using computers, besides improving the concept learning process.

As future works, the following is planned: Besides discussing the possibilities of structuring specific games to an educational context, the use of games as learning objects is being considered and the development of a game repository is in progress; to specify supporting tools to the phases of interaction (negotiation, competition and collaboration); to specify a hypermedia authoring tool that makes possible the creation of collaborative adaptive hypermedia courses in conjunction with collaborative educational games; to specify a novel ontology of games that allows to choose a suitable game for a specific learning activity given a subject and instructional objectives; and to integrate an instance of CEG model to a open source LMS.

ACKNOWLEDGMENT

All the authors acknowledge FAPEMA (Maranhão State Research Supporting Foundation) and FAPESP—Brazil for partially supporting this research.

REFERENCES

Almstrum, V. L., Ginat, D., Hazzan, O., & Morley, T. (2002, June). Import and export to/from computing science education: The case of mathematics education research. In *Proceedings 7th SIGCSE Conference*, Aarhus, Denmark.

Anderson, L., & Holt, C. A. (1996). Classroom games: Information cascades. *Journal of Economic Perspectives, 10*(4), 187-193.

Bonwell, C. C., & Eison, J. A. (1991). Active learning: Creating excitement in the classroom. *ERIC Clearinghouse on Higher Education*, #ED340272.

CMS. (2006). *Course management systems*. Retrieved July 15, 2006, from http://www.edutools.info/summative/index.jsp?pj=8&i=326

Coleman, R., Krembs, M., Labouseur, A., & Weir, J. (2005, February). Game design & programming concentration within the computer science curriculum. In *Proceedings of the 36th SIGCSE Conference*, St. Louis, USA.

Elgood, C. (1990). *Using management games*. London: Gover.

Feezel, J. D. (1993, April). Preparing teachers through creativity games. In *Proceedings Joint Meeting of the Southern and Central States Communication Associations*, Lexington, KY.

Furtado, A. W., Andrade, G. D., Leitão, A. R., & Andrade Neto, F. C. (2003, July). Cegadef: A collaborative educational game development framework. In S. MacFarlane et al. (Eds.), *Proceedings 2003 Conference on Interaction Design and Children*, Preston, UK.

Galvão, J. R., Martins, P. G., & Gomes, M. R. (2000, December). Modeling reality with simulation games for a cooperative learning. In *Proceedings of the 32nd Conference on Winter Simulation*, Orlando, Florida.

Garris, R., Ahlers, R., & Driskell, J. E. (2002). Games, motivation, and learning: A research and practice model. *Simulation & Gaming, 33*, 441-467.

Giaretta, Letícia L., Alves, Lisandra R., Petry, Tatiana O., & Silveira, M. S. (1998). Chameleon: Tool for building educational computer games. In *Proceedings IV Congresso RIBIE*, Brasília, Brazil. (In Portuguese)

Haas, M. E. (1988, November). Using small group games in social studies. *Proceedings Annual Meeting of the National Council for Social Studies*, Orlando, Florida.

Herr, N. (2002). The sourcebook for teaching science: Strategies, activities, and Internet resources. *Sourcebook*.

Hill, J. M., Ray, C. K., Blair, J. R., & Carver, C. A. (2003, February). Puzzles and games: Addressing different learning styles in teaching operating systems concepts. *Proceedings of the 34th SIGCSE Technical Symposium on Computer Science Education*, Reno, Nevada.

Jenkins, H., Klopfer, E., Squire, K., & Tan, P. (2003). Entering the education arcade. *Computers in Entertainment, 1*(1), 17-28.

Johnson, R. T., & Johnson, D. W. (1986). Action research: Cooperative learning in the science classroom. *Science and Children, 24*, 31-32.

Navarro, E. O., & van der Hoek, A. (2004). SimSE: An educational simulation game for teaching the software engineering process. In *Proceedings of the 9th Annual SIGCSE Conference*, Leeds, UK.

Nielsen, J. (2006). Retrieved August 1, 2007, from http://www.useit.com/papers/heuristic/heuristic_list.html

Poirot, J. L. (1976, February). A course description for teacher education in computer science. In *Proceedings ACM SIGCSE-SIGCUE Joint Symposium*, New York.

Prensky, M. (2001). *Digital game-based learning.* McGraw-Hill.

Ryder, M. *Instructional design models.* Retrieved January 13, 2006, from http://carbon.cudenver.edu/~mryder/itc_data/idmodels.html

Savery, J., & Duffy, T. (1995). Problem-based learning: An instructional model and its constructivist framework.

In B. G. Wilson (Ed.), *Designing construtivist learning environments* (pp. 135-148). Englewood Cliffs, NJ: Educational Technology.

Silberman, M. (1996). *Active learning: 101 strategies to teach any subject.* Allyn & Bacon.

Soh, L. (2004). Using game days to teach a multiagent system class. In *Proceedings of the 35th SIGCSE Conference*, Norfolk, Virginia.

Teixeira, J., SÁ, E., Prudêncio, T., Fernandes, C., Oliveira, J., Costa, I., & Silveira, D. (2005). Collaborative JETetris: Ludicness, competitiveness and collaboration in the learning process. In *Proceedings of the Digital Educational Games* Workshop – SBIE, 2005 Brazilian Symposium on Computers in Education, Juiz de Fora, Brazil. (In Portuguese)

Tinto, V., Goodsell, A., & Russo, P. (1993). Collaborative learning and new college students. *Cooperative Learning and College Teaching, 3*(3), 9-10.

Vartiainen, M., & Ruohomaki, V. (1995). Group training with the teamwork game. *The Simulation and Gaming Yearbook, 3*, 264-270.

Wikipedia. *Tetris.* Retrieved August 19, 2005, from http://en.wikipedia.org/wiki/Tetris

Wills, C. E., & Finkel, D. (1994). Experience with peer learning in an introductory computer science course. *Computer Science Education, 5*(2), 165-187.

KEY TERMS

CEG Model: Collaborative educational game model.

Collaborative Educational Game: A collaborative game used as a collaborative learning tool in classroom settings or at a distance.

Collaborative Game: A computer game to be played interactively by two or more groups of collaborative players.

E-Collaborative Educational Game: A collaborative educational game used as an collaborative learning tool through the Web.

Cooperative Learning: An instructional method in which students at various performance levels work together in small groups toward a common learning goal.

Educational Game: A game used as an educational tool.

Negotiation Phase: Real-time interaction among opponent pairs, for negotiating the rules of the game, including the game difficulty level.

ENDNOTE

[1]   TTC Trademark

Understanding Adverse Effects of E-Commerce

Sushil K. Sharma
Ball State University, USA

Jatinder N.D. Gupta
University of Alabama - Huntsville, USA

INTRODUCTION

The Internet heralded an unprecedented evolution in the transformation of all business and communication. The Internet is growing at an annualized rate of 18% and now has one billion users. Due to this growth, e-commerce will continue to grow in next few years. The United States online population is estimated to be 211 million by 2006 and United States online retail sale are estimated at $112.5 billion for 2006. Jupiter Research predicts that online retail sales are expected to grow from $81 billion in 2005 to $144 billion in 2010 (Jupiter Media Metrix, 2006). E-commerce is defined as buying and selling of information, products, and services via computer networks or internet. Internet and electronic commerce technologies are transforming the entire economy; and changing business models, revenue streams, customer bases, and supply chains. New business models are emerging in every industry of the New Economy. In these emerging models, intangible assets such as relationships, knowledge, people, brands, and systems are taking center stage (Hudson, 2000; Verhoest, Hawkins, & Desruelle, 2003). The relationship and interaction of various stakeholders such as customers, suppliers, strategic partners, agents, or distributors is entirely changed (Sharma & Gupta, 2001, 2003; Sharma, 2005).

The Internet has become an incredibly powerful tool for conducting business electronically. Companies have taken a proactive approach and are moving forward this new business model. E-commerce has been considered a great enabler of organizational change and helping organizations to conduct business with improved efficiencies and productivity (Sen & King, 2003). Meanwhile, it has created many challenges and adverse effects, such as concerns over privacy, consumer protection, and security of credit card purchases, displacement of workers especially low-status ones, and negative impact on quality of work life (Sharma & Gupta, 2001,

2003; Sharma, 2005). This paper describes the various adverse effects that have been created with the advent of internet and e-commerce revolution.

BACKGROUND

The market for e-commerce is growing as more consumers and businesses gain Internet access. Operational benefits of e-commerce include reducing both the time and personnel required to complete business processes and increasing market access and reach, improving internal and market efficiency, and lowering transaction costs. Belief in such benefits has led to the adoption of e-commerce by most of the businesses (Stopford, 2002). Although e-commerce has provided a number of opportunities and benefits to both customers as well as to businesses, loss of privacy, security issues, increasingly sophisticated frauds, and abuse of personal information, have become serious concerns (Fichter, 2003, Siikavirta, Punakivi, Kärkkäinen, & Linnanen, 2003). Many consider that loss of privacy means loss of personal freedom. Consumers are specifically concerned about consumer protection, security of credit card purchases, order fulfillment and delivery. These are undesired adverse effects of e-commerce (Mansell, Schenk, & Steinmueller, 2001). This article describes the various adverse effects or problems that have been created with the advent of Internet and e-commerce revolution (Sharma & Gupta, 2001, 2003 Sharma, 2005).

ADVERSE ECONOMIC AND SOCIAL IMPACTS OF E-COMMERCE

As e-commerce continues to grow at a rapid pace, it could have significant effects on the structure and functioning of economies at the firm, sector and aggregate level. The impacts of these changes are diverse and likely to impinge on prices, the composition of trade, labor markets, and taxation revenues (Sharma & Gupta, 2001, 2003; Sharma, 2005).

Impact on Prices

E-commerce does offer consumers lower prices, greater choices, and increased convenience. On the positive side, e-commerce may increase price transparency and commoditization thus pushing prices down, however, from the business side, lower search costs for buyers may enable them to seek more trusted suppliers and e-commerce might thus effectively shift power from producers to consumers and make it harder for firms to maintain higher prices (Clayton & Waldron, 2003; Smith, Bailey, & Brynjofsson, 2000). Also, companies can direct the market and customize the products as per consumers' preferences and choices and this could lead to more finely differentiated and sophisticated price discrimination for products (Coppel, 2000a, 2000b).

Impact on Competition and Competition Policy

Although theoretically e-commerce offers the ability to reduce barriers to entry and make markets more competitive, in practice, it may kill small players as global players will dominate markets. For example; may perceive that eBay is killing many local stores. The scope for noncompetitive behavior is perhaps strongest among "digital" and knowledge intensive products. Start-up companies may find it difficult to enter due to the large marketing costs needed to develop visibility and a brand name. Such a non-competitive environment may result in "winner-takes-all" scenarios that could hinder innovation and competition (OECD, 2001a, 2001b, 2003a, 2003b; Coppel, 2000a, 2000b; Sharma & Gupta, 2001, 2003; Sharma, 2005).

Tax, Trade, and Regulatory Issues

E-commerce may have a strong impact on taxation and tax policy. Concerns have been expressed that e-commerce could result in the erosion of tax bases. E-commerce may need new regulatory frameworks to handle customs duties and other tax structure across countries since e-commerce transcends national boundaries for buyers and sellers. The laws and regulations

a consumer relies on for protection at home may not apply in the merchant's country (Sharma & Gupta, 2001, 2003; Sharma, 2005; OECD, 2001; Coppel 2000a, 2000b).

Employment and Labor Market Policy

E-commerce is likely to have both direct and indirect impacts on labor markets as well as the composition of employment (OECD, 2001b). E-commerce may generate employment for those with newer IT skills and may kill some traditional skills based jobs. Faster rates of innovation and diffusion may also be associated with more turnovers of jobs. This could put more pressures on employees or workers to continuously upgrade their skills as requirements change frequently (Sharma & Gupta, 2001, 2003; Sharma, 2005).

E-Commerce's Impact on Labor Costs and Employment

E-commerce will have significant effects on the structure and functioning of economies at the firm, sector and aggregate level. Such effects would ultimately be reflected in prices, the composition of trade, labor markets, labor costs and employment. E-commerce is facilitating the shift from the mass labor paradigm to a knowledge-worker paradigm. The shift from mass to knowledge labor has already created a shortage of knowledge workers in several countries where the education system and technology infrastructure has not been strong (Sharma & Gupta, 2001, 2003; Sharma, 2005; OECD, 2001b; Coppel 2000a, 2000b).

Privacy

On one hand, e-commerce provides convenience for buying and selling online, but on the other hand, organizations may be secretly profiling and collecting information about customers. Spamming, which is the practice of sending out unsolicited e-mail, is growing because it costs so little to send out millions of messages or advertisements electronically. Many prominent high-technology companies have already been caught attempting to quietly collect information about their customers via the Internet (Gupta & Sharma, 2001; OECD, 2003a, 2003b). Privacy has become a key issue in the digital age. Technological advances make it easy for companies to obtain personal information and to monitor online activities, thus creating significant potential for abuse. There are three areas of concern in the privacy debate: employers monitoring employee computer and Internet use in the workplace, advertising and market research companies collecting and selling personally identifiable information based on consumers' online activities, and information brokers selling readily available personal information from public-record databases online (Sharma & Gupta, 2001, 2003; Sharma, 2005).

Cyber Slacking

Cyber slacking is the term used to describe employees spending time on the internet for non-work related activities during office time. Cyber slacking is an important social and economic issue. Its effects are seen in debates about computer productivity, Internet censorship, computer monitoring, legal considerations, and managerial challenges. Cyber slacking often causes companies to police Internet use, and has resulted in some publicized firings of involved employees. As more and more companies "crackdown" on appropriate business use of the Internet, more and more people will feel threatened at work and an environment of tension, paranoia, and "shoulder-looking" easily arises (Gupta & Sharma, 2002; Amichai-Hamburger, Wainapel, & Fox, 2002; Sharma & Gupta, 2003).

Social Isolation

The implications of e-commerce are far reaching especially in a social context. While it provides all the comfort of shopping from home, it eliminates old-fashioned human interactions for social needs. Today many organizations allow their employees to work from their homes. There are some positive aspects of telecommuting to organizations and society. But the negative aspect of telecommuting should not be ignored. The stimulation of interacting with colleagues may be lost and the resultant gradual, social isolation may affect opportunities for promotion and the selection of career-advancing assignments (Sharma & Gupta, 2001, 2003; Sharma, 2005; Kling & Lamb, 2000; Coppel 2000a, 2000b; OECD, 2001a, 2001b).

Digital Divide

E-commerce may create a digital divide between those who have access to both technology and skills and those who do not. The digital divide encompasses several dimensions. One dimension is the gap along the gender, race, and social class line, and another is the gap between wealthy and poor countries. Millions of technologically disenfranchised have-nots, who cannot afford the cost of technology or training, are walled off from potentially life-changing tools and knowledge, isolated in the virtual world. Disparities in the location and quality of Internet infrastructure, even the quality of phone lines, have created gaps in access (Sharma & Gupta, 2001, 2003; Sharma, 2005; OECD, 2001a, 2001b)

Security

E-commerce offerings not only provide new opportunities for customers and businesses but also provide new vulnerabilities to companies. Privacy and security concerns in e-commerce have fueled intense pressure from consumers, lawmakers, and regulators to provide foolproof security safeguards and policies to protect their systems and customer privacy (Miyazaki & Fernandez, 2000). A multitude of security technologies are used to prevent electronic snooping but one still hear stories of expert hackers gaining access to sensitive data. Many network vulnerability scanners and intrusion detection systems have been developed and implemented but systems continue to be vulnerable to attacks (Sharma & Gupta, 2001; Coppel 2000a, 2000b, 2003; Sharma, 2005).

FUTURE TRENDS

Despite the privacy and security concerns of consumers and businesses, e-commerce revolution will continue to grow. In the first phase of its existence, it has created a commercial impact and has a minimal impact on improving education, reducing poverty, enhancing health care, supporting community development (Mansell, Schenk, & Steinmueller, 2001). However, in its second phase, it will impact every aspects of day-to-day life in most part of the world. The internal company structures of today will no longer be valid in digital economy of future. Companies have to create more flexible business structures to respond to more dynamic business environment in future. There is a fear that the shift from an industrial/manufacturing economy to a more service and information oriented economy requiring both, highly paid knowledge workers and low paid support workers, may result in a rising wage gap and the social exclusion of a remarkable share of the population (OECD, 2003a, 2003b).

CONCLUSION

E-commerce has provided opportunity for better inter-actions with partners, suppliers and targeted customers for service and relationship. E-commerce provides the customers with choice, information, convenience, time, and savings with improvements that add value to their shopping. However, loss of privacy with e-commerce, security issues, increasingly sophisticated frauds, and abuse of personal information, and the impact on prices have surfaced as the main concerns. Companies increasingly rely on the collection and use of information about consumers for numerous purposes, including targeted advertising and marketing, maximizing the convenience of electronic commerce, and personalizing customer service and support. The collection and use of personally identifiable information has, however, raised significant concerns from lawmakers, regulators, and private litigants, especially when it involves sensitive information such as financial or medical information and information related to children. The e-commerce revolution has been restricted too much of the world's information technology "haves." The "have-nots" lack the technology and training to be part of e-commerce. E-commerce promoters will need to address the question of how to bridge the gap between these "haves" and "have-nots" in order to form a singular society and not a society separated by a digital divide. This article has described some of these adverse affects of e-commerce.

ACKNOWLEDGMENT

Some material in this paper is taken from earlier publication Sharma S.K. and Gupta J.N.D (2003) Adverse Effects of E-Commerce In *The Economic and Social Impact of E-Commerce, Lubbe S. and Heerden J.M.V.,* Idea Group Publishing, USA, pp. 33-49.

REFERENCES

Amichai-Hamburger, Y., Wainapel, G., & Fox, S. (2002). On the Internet no one knows I'm an introvert: Extroversion, neuroticism, and Internet interaction. *Cyber Psychology & Behavior, 5,* 125-128.

Coppel, J. (2000a). E-commerce: Impacts and policy challenges. *Organization for Economic Cooperation and Development Economic Outlook, 67,* 193-213.

Coppel, J. (2000b). *E-commerce: Impacts and policy challenges.* OECD, Economics Department Working Papers (No.252). Paris.

Clayton, T., & Waldron, K. (2003). E-commerce adoption and business impact: A progress report. *Economic Trends ONS, February,* (591), 33-40.

Fichter, K. (2003). E-commerce: Sorting out the environmental consequences. *Journal of Industrial Ecology, 6*(2), 25-41.

Gupta, J. N. D., & Sharma, S. K. (2001). Cyber shopping and privacy. In A. Gangopadhyay (Ed.), *Managing business with electronic commerce: Issues and trends* (pp. 235-249). Hershey, PA: Idea Group Publishing.

Hudson, H. E. (2000). Extending access to the digital economy to rural and developing regions. In E. Brynjolfsson & B. Kahin (Eds.), *Understanding the digital economy: Data, tools, and research.* Cambridge, MA: MIT Press.

Jupiter Media Metrix, Inc. Report. (2006). Retrieved from http://banners.noticiasdot.com/termometro/boletines/autor/boletines-autor-jupiter.htm

Kling, R. (2000). IT and organizational change in digital economies: A socio-technical approach. In E. Brynjolfsson & B. Kahin (Eds.), *Understanding the digital economy: Data, tools, and research* (pp. 261-291). Cambridge, MA: MIT Press.

Mansell, R., Schenk, I., & Steinmueller, W. E. (2001). Net compatible: The economic and social dynamics of electronic commerce. *Communications & Strategies, 38*(2), 241-276.

Miyazaki, A. D., & Fernandez, A. (2000). Internet privacy and security: An examination of online retailer disclosures. *Journal of Public Policy & Marketing, 19*(1), 54-61.

OECD. (2001a). *Understanding the digital divide.* Paris. Retrieved from www.oecd.org//dsti/sti/prod/Digital_divide.pdf

OECD. (2001b). *The economic and social impact of electronic commerce: Preliminary findings and research agenda.*Paris. Retreived from www.oecd.org/dataoecd/3/12/1944883.pdf

OECD. (2003a). *OECD guidelines for protecting consumers from fraudulent and deceptive commercial practices across borders.* Retrieved June 11, 2003 from www.oecd.org/document/50/0,3343,en_2649_34267_2514994_1_1_1_1,00.html

OECD. (2003b). *Consumers in the online marketplace: The OECD guidelines three years later.* Report by the Committee on Consumer Policy on the Guidelines for Consumer Protection in the Context of Electronic Commerce. Retrieved February 3, 2003 from www.nacpec.org/en/links/regfora/oecd.html

Sen, R., & King, R. (2003). Revisit the debate on intermediation, disintermediation and reintermediation due to e-commerce. *Electronic Markets, 13*(2), 153-152.

Sharma, S. K. (2005). Socio-economic impacts and influences of e-commerce in a digital economy. In H. S. Kehal & V. P. Singh (Eds.), *Digital economy: Impacts, influences and challenges* (pp. 1-20). Hershey, PA: Idea Group Publishing.

Sharma, S. K., & Gupta, J. N. D. (2001). E-commerce opportunities and challenges. In M. Singh & T. Thompson (Eds.), *E-commerce diffusion: Strategies and challenges* (pp. 21-42). Heidelberg Press.

Sharma, S. K., & Gupta, J. N. D. (2003). Adverse effects of e-commerce. In S. Lubbe & J. M. V. Heerden (Eds.), *The economic and social impact of e-commerce* (pp. 33-49). Hershey, PA: Idea Group Publishing.

Siikavirta, H., Punakivi, M., Kärkkäinen, M., & Linnanen, L. (2003). Effects of e-commerce on greenhouse gas emissions: A case study of grocery home delivery in Finland. *Journal of Industrial Ecology, 6*(2), 83-97.

Smith, M., Bailey, J., & Brynjofsson, E. (2000). Understanding digital markets. In E. Brynjolfsson & B. Kahin (Eds.), *Understanding the digital economy.* MIT Press.

Stopford, M. (2002). E-commerce implications, opportunities and threats for the sipping business. *International Journal of Transport Management, 1*(1), 55-67.

Verhoest, P. et al. (2003). *Electronic business networks: An assessment of the dynamics of business-to-business electronic commerce in eleven OECD countries: A summary report of the e-Business Impacts Project (EBIS).* Available at http://www.jrc.es/home/publications/publications.html

KEY TERMS

Cyber Slacking: When technologies are implemented in the hope of seeing productivity rise and quality increase but do not see corresponding productivity increases, this activity has been coined as "cyber slacking."

Digital Divide: Digital divide can be defined as the lack of equal access and benefit from computer technologies and the Internet in particular, creating a gap between those who have and those who have not.

E-Commerce: E-commerce is defined as the conduct of buying and selling of products and services by businesses and consumers over the internet. E simply means anything done electronically, usually via the Internet. E-commerce is the means of selling goods on the Internet, using Web pages.

Privacy: Privacy is defined as an individual's right to be left alone, free from

Security: Security refers to the integrity of the data storage, processing and transmitting system and includes concerns about the reliability of hardware and software, the protection against intrusion or infiltration by unauthorized users.

Social Isolation: The separation of individuals or groups resulting in the lack of or minimizing of social contact and/or communication. This separation may be accomplished by physical separation, by social barriers and by psychological mechanisms. In the latter, there may be interaction but no real communication.

Telecommuting: Telecommuting is defined as, the practice of working at home or at a remote site in lieu of working in the office.

Understanding Effective E-Collaboration Through Virtual Distance

Karen Sobel Lojeski
Virtual Distance International, USA

Richard R. Reilly
Stevens Institute of Technology, USA

INTRODUCTION

Virtual distance is a multidimensional perceptual construct resulting from key elements that promote a sense of distance in e-collaborative work environments. Why will virtual distance help to uncover some of the potential downside risks of collaboration using virtual and outsourced resources? Research has shown that the perceived distance between two or more individuals has negative effects on communication and persuasion and promotes a tendency to deceive more than those who do not perceive themselves to be as distant (Bradner & Mark, 2002). Virtual team members and work groups are, by definition, distant from one another, not only in the physical sense but in other ways as well. Socio-emotional factors, for example, can play a role in perceived distance and these factors may contribute to decreased success (Barczak & McDonough, 2003).

The virtual distance model (VDM) was developed after conducting an extensive literature review and combining findings from that effort with executive interview information collected over the course of the first 18 months of this research. The model was tested using a multi-step research method including surveys and follow-up interviews with key executives from a sample of corporations leveraging virtual workspaces.

BACKGROUND

While the notion of distance is, by definition, at the heart of virtual team studies, most of the literature has focused on geographic and temporal factors. Virtual teams (VTs) are those defined as having members that are geographically separate, often with vast distances between one another (Alavi, 1994; Townsend, De-Marie, & Hendrickson, 1998; Majchrzak, Malhotra, Stamps, & Lipnack, 2004). Therefore, the idea that physical distance plays a role in VT behavior is well established. However, research also shows that other variables can contribute to a sense of socio-emotional or psychological distance. Interpersonal, social, organizational, and technical factors also play a role and have important implications for the attitudes and behavior of team members and their ability to succeed (Bradner & Mark, 2002). These factors can include, but are not limited by, building trust and motivating one another, cultural diversity and lack of goal clarity (Barczak & McDonough, 2003). Collaboration, whether it is face-to-face (FtF) or computer mediated, occurs within a much broader context than simply geographic and temporal dispersion. So there is reason to expand the research beyond physical distance constructs. One of the basic assumptions of this thesis was that the use of geographic and temporal distance constructs alone, are not enough to explain performance differences among teams in the 21st century. Instead, it was posited that the construct of distance for VTs be expanded to include socio-emotional distance factors as well.

As Stephen Roach wrote, "it is time to let go of some of our time-honored relationships" (Roach, 2005). While he was referring to macro-economic relationships, the sentiment applies to micro-economic relationships as well, including virtual teams and globally distanced workforces. A paradigm shift in thinking is required to do so and a new, unifying and parsimonious framework is needed to open up the black box that sits between virtual work and performance outcomes; one that reflects the integrative and multi-dimensional nature of the complex interplay of both real and perceived issues at the individual and group level. The development of such a model was the purpose of this thesis and the resulting model has been named, the virtual distance model (VDM).

The model was developed through a review of the major research streams primarily in management and technology, combined with some central tenants of the theories of distance, social science, and psychology. In addition, an initial set of field studies was conducted, in the form of executive interviews, to ground the theoretical discussion in real-world terms as perceived by leaders at major, global organizations.

FACTORS INFLUENCING DISTANCE

Based on a review of management, information systems and psychological literature, a number of socio-emotional distance factors that influence team members were identified. These include spatial, temporal, technical, organizational and social factors that shape the perceptions of individuals engaged in e-collaborative work. In the present investigation these factors were reviewed in terms of how they collectively impacted work related attitudes, behavior and performance. Eleven factors likely to influence the perceptions of distance between team members are discussed in the following sections (see Figure 1).

Geographic Distance (GD)

Research suggests that physical separation or closeness is of great importance to interactions and that the closer one is physically to another, the greater the chance to form social ties (Latane & Herrou, 1996). Physical distance also impacts the tendency to deceive, ability to influence, the likelihood of cooperation (Bradner & Mark, 2002), and has been shown to have some impact on learning behavior (Latane & Bourgeois, 1996; Bulte & Moenaert, 1998; Arbaugh, 2001; Bradner & Mark, 2002; Coppola, Hiltz, & Rotter, 2002).

Temporal Distance (TD)

Differences in time zones between VT members are often cited as one of the factors that plays a role in VT interactions (Jarvenpaa & Leidner, 1998; Montoya-Weiss, Massey et al., 2002; Massey, Montoya-Weiss et al., 2003). It has also been suggested that TD be considered when structuring organizations (Orlikowski & Yates, 2002), globalizing an organization (Boudreau, Loch et al. 1998), assessing team boundary issues (Espinosa, Cummings et al., 2003) and coordinating VTs (Montoya-Weiss et al., 2002)

Relational Distance (RD)

RD refers to the difference between team members' organizational affiliations. For example, an employee is relationally closer to another employee of the same company than to employee from a third party service provider. RD has been shown to play a key role in social cohesion (Moody & White, 2003), information systems networks, as well as leader effectiveness (Klagge, 1997).

Cultural Distance (CD)

Cultural differences have, to date, been a focus of some research in virtual work and innovation, VTs (Jarvenpaa & Leidner, 1999; Dube & Pare, 2001; Massey, Montoya-Weiss et al., 2001), new product teams (Barczak & McDonough, 2003), risk mitigation (Grabowski & Roberts, 1999), virtual societies (Igbaria, 1999), consensus building using group support systems (Mejias, Shepherd et al., 1997), majority influence (Tan, Wei et al., 1998), software development (Tellioglu & Wagner, 1999) and more. CD has also been used to study foreign investment expansion, entry mode choice, and the performance of foreign invested affiliates, among others (Shenkar, 2001). Following the discussion of social network theory and distance related phenomenon, CD has also been used to interpret network ties amongst managers (Stevenson, 2001). Additionally CD is used to explain how international relationships affect responses and behaviors amongst employees (Thomas & Ravlin, 1995).

Social Distance (SD)

SD has been studied in a number of contexts including class or status differences (Akerlof, 1997), feelings of social closeness and distance based on social interactions in social space (Bottero & Prandy, 2003), as a factor in direct and networked exchanges (Buchan, Croson et al., 2002), as a function of management (Fox, 1977), a dimension of the systematic multiple level observation of groups (SYMLOG) management behavior assessment (Jensen, 1993), as a perceived measure contributing to the concept of leader distance (Antonakis & Atwater, 2002), and as a factor in friendship networks (Krackhardt & Kilduff, 1999). Wiesenfeld and colleagues found that virtual work environments may weaken ties that bind organizations and their members together (Wiesenfeld, Raghuram et al., 1999) increasing SD. In another case, centrality, or less distance from the center of the social network, was found to mediate the relationship between social status and virtual R&D groups (Ahuja, Galletta et al., 2003). In a virtual organization with no formalized hierarchy at the outset, the emergence of a hierarchy and SD was found (Ahuja & Carley, 1999).

Relationship History Distance

Relationship history includes both the extent to which members have had a prior relationship or relationships with some of the same people. RH has been shown to be important in mentoring (Siegel, 2000) and trust building (Rousseau, Sitkin et al., 1998). In his study, "When does the medium matter? Knowledge-building experiences and opportunities in decision-making teams," Alge showed that FtF teams exhibited higher levels of openness, trust and information sharing than computer mediated teams that did not have a RH. However when computer mediated teams had prior relationships, many of these issues were eliminated. (Alge, Wiethoff et al., 2003).

Interdependence Distance

Interdependence distance is the degree to which one individual or group perceives that their success is tied to another individual or group member (Thompson, 1967). Thompson claimed distance was a major factor in his classification scheme on Interdependence (Thompson, 1967). Interdependent tasks require more communication (Bishop & Scott, 2000), which should lead to decreased distance between team members. Task interdependence has also been related to both organizational commitment and team commitment and OCB (Pearce & Gregersen, 1991; Bishop & Scott, 2000). In the virtual realm, goals may become less clear amongst players if they are not directly attached to some sort of organizational mandate (Manzevski & Chudoba, 2000). Interdependent goals have also been found to have importance to international teams (Davison, 1994) and embedded, interdependent goal-setting in GSS has been shown to help team cohesion (Huang, Wei et al., 2003).

Face to Face (FtF) Interaction

The notion of social presence has been used in research on virtual work to describe the extent to which team members feel the presence of other group members and the feeling that the group is jointly involved in communicating (Andres & Zmud, 2002; Venkatesh & Johnson, 2002). One end of the continuum of social presence is FtF so frequency of FtF interaction should be related to perceptions of distance. In some cases only email is used and no FtF or phone communications are considered (Jarvenpaa & Leidner, 1999). In other studies, two or more types of mediated communications are investigated. In "Why Distance Matters, Effects on Cooperation, Persuasion and Deception" (Bradner & Mark, 2002), the authors chose instant messaging and video conferencing in an attempt to simulate two ends of the communication spectrum. In other studies it has been found that some FtF meetings blend well with other types of communication mediums. There is some emerging support for the notion that a mix of communication methods improves performance (Aiken & Vanjani, 1997) and produces higher levels of commitment (Alavi, 1994; Alavi, Wheeler et al., 1995).

Team Size

Group or team size has been shown to affect one's sense of belonging (Williams & Wilson, 1997). A sense of belonging is critical to the development of organizational identity, which has been shown to have a direct influence on OCB (Shamir, 1990; Pratt, 1998). Group size in VTs has also been shown to affect team decision making (Baltes, Dickson et al., 2002) and satisfaction (Dennis & Wixom, 2002). Group size in virtual work had impact on group support system processes (Dennis & Wixom, 2002).

Multi-Tasking

Multi-tasking is a term used to describe a person working on more than one task at a time. It can create significant stress on a person if he or she becomes overloaded and it can lead to less efficiency and productivity (Brillhart, 2004). Cognitively distancing oneself from the stress created by multi-tasking and information overload is known as absent presence, "the idea that we may be physically on a street corner, but our distracted minds are not" (Berman, 2003). The absent presence is a form of psychological distance. Some have found that frequent interruptions affect decision-making (Speier, Valacich et al., 1999; Thompson, 2005). During meetings in the new millennium many listen to presentations while also using hand-held PDAs to communicate with others simultaneously. Some experts believe that multi-tasking in this way is detrimental to productivity (Richtel, 2003). While it has been shown that telecommuting can improve satisfaction and work/life balance (Hill, Miller et al., 1998), family and other home-based considerations may represent a form of multi-tasking that creates stresses that are difficult to overcome (Richtel, 2003). The extent to which workers engage in multi-tasking depends, in part, on the organization's desire to increase productivity (Cascio, 1993; Snizek, 1995); another key reason why virtual work is proliferating at "hyperspeed."

Technical Skill

Studies have shown that a member's comfort level with technology plays a role in their interactions with distant team members (Staples, Hulland et al., 1999). Less technically competent team members may be less inclined or able to communicate and form the kinds of relationship that would decrease social distance. The theory of cognitive fit describes the need for matching problem-solving task to problem-solving tools in order to obtain higher levels of performance (Agarwal, Sinha et al., 1996). Major corporations have also found that technical and interpersonal skills are important to the selection of VT members who are most likely to be committed to the project and to each other (Kirkman, Rosen et al., 2002).

RESEARCH RESULTS

The virtual distance model has been applied in two related research studies which offer empirical support for the importance of virtual distance in e-collaboration. A study of 72 collaborative teams showed that a virtual distance index (VDI), based on the eleven factors described above, was directly related to trust and clarity of project vision and goals and indirectly related to organizational citizenship behavior and project success (Reilly, Sobel Lojeski, & Dominick, 2005). A second study (Sobel Lojeski, Reilly, & Dominick, 2006) replicated these findings with 115 collaborative

teams drawn from the financial, pharmaceutical and other industries and also showed that VDI had similar indirect influence on the level of innovation. Figure 2 shows the results of a structural equation model analysis of the relationships between virtual distance, trust, vision/goal clarity, organizational citizenship behavior and innovation and project outcome.

FUTURE TRENDS

Many of the implications of virtual distance have yet to be studied. Some areas that are potentially interesting and important include affective variables, selecting and organizing VTs and managing and leading VTs. For example, how does distance influence the emotional and affective side of work? Do distant employees have more or less satisfaction; are they more or less committed? Recent research has confirmed the increased difficulty of meeting socio-emotional needs of VT members (Chidambaram, 1996; Lurey & Raisinghani, 2001; Maznevski & Chudoba, 2001). In a recent paper, Kock (2004) suggests that human evolution has designed both our brains and bodies for FtF communication. It may be that alternatives to the social interactions of the workplace will have to be found for many virtual workers to meet some of the social and emotional needs required for job and life satisfaction.

Notions of virtual distance may also be applied to selecting and organizing VTs. For example, a critical global project may require understanding and perhaps minimizing the distances between team members by selecting individuals with closely aligned work-related values and organizing the tasks to provide clear opportunities for interdependence and frequent communication. Selecting team members with a history of working together may also be a way to decrease the virtual distance within a team. Lynn and Reilly (2002) found that very high performing teams generally knew one another and had worked on similar projects before. In addition to selecting members with past common experiences, organizations can also plan for the future by providing opportunities for dispersed co-workers to build relationships.

Practitioners may want to start to address CD from a values point of view as well as demographics-based education. Training on cultural demographics helps organizational members who visit different countries on a regular basis. It is also useful for leaders who

Figure 1.

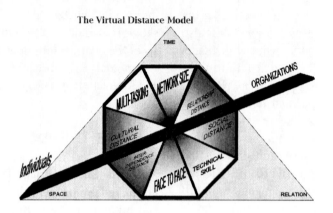

The Virtual Distance Model

Figure 2.

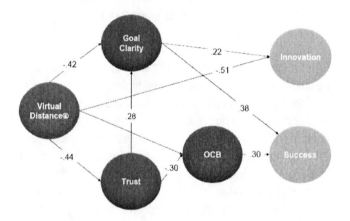

manage multi-cultural team members. However, this type of training does not necessarily improve alignment between cultural value divides—one of the most salient factors contributing to CD and virtual distance. The findings suggest that if cultural values were the focus of more training, team selection processes and mentoring programs, then perhaps higher levels of trust and clarity may emerge, potentially resulting in higher levels of success.

Practitioners also should develop better methods for creating an environment where informal status can be elevated. These practices might include higher levels of recognition for contribution of team members through more frequent and targeted leader messages to all team members. VT reward and recognition programs can also be developed to help bolster team member perceptions of informal status based on rewards for contributions.

Practitioners can help to minimize virtual distance by illuminating social networks within the organizational web. Once visible, they can focus on seeding project teams with a few members that may have worked together in the past or who know some of the same people. This will help to reduce virtual distance and increase trust.

Another area of application of virtual distance notions is in team leadership. Understanding how distant team members are from one another and how they differ on key facets of virtual distance can help project managers to better lead and manage. As virtual work proliferates effective leadership of projects in virtual space will become an essential competitive weapon.

REFERENCES

Akerlof, G. A. (1997). Social distance and social decisions. *Econometrica, 65*(5), 1005.

Ahuja, M. K., & Carley, K. M. (1999). Network structure in virtual organizations. *Organization Science, 10*(6), 741.

Alavi, M. (1994). Computer-mediated collaborative learning: An empirical evaluation. *MIS Quarterly, 18*(2), 159.

Alge, B. J., Wiethoff, C., & Klein, H. J. (2003). When does the medium matter? Knowledge-building experiences and opportunities in decision-making teams. *Organizational Behavior and Human Decision Processes, 91*(1), 26.

Barczak, G., & McDonough, E. F. (2003). Leading global product development teams. *Research Technology Management, 46*(6), 14.

Bradner, E., & Mark, G. (2002). *Why distance matters: Effects on cooperation, persuasion and deception.* Paper presented at the CSCW 2002, New Orleans, LA.

Brillhart, P. E. (2004). Technostress in the workplace managing stress in the electronic workplace. *Journal of American Academy of Business, Cambridge, 5*(1/2), 302.

Coppola, N. W., Hiltz, S. R., & Rotter, N. G. (2002). Becoming a virtual professor: Pedagogical roles and asynchronous learning networks. *Journal of Management Information Systems, 18*(4), 169.

Igbaria, M. (1999). The driving forces in the virtual society. *Association for Computing Machinery. Communications of the ACM, 42*(12), 64.

Jarvenpaa, S. L., Knoll, K., & Leidner, D. E. (1998). Is anybody out there? Antecedents of trust in global virtual teams. *Journal of Management Information Systems, 14*(4), 29.

Latane, B., & Bourgeois, M. J. (1996). Experimental evidence for dynamic social impact: The emergence of subcultures in electronic groups. *Journal of Communication, 46*(4), 35.

Latane, B., & Herrou, T. (1996). Spatial clustering in the conformity game: Dynamic social impact in electronic groups. *Journal of Personality and Social Psychology, 70*(6), 1218.

Majchrzak, A., Malhotra, A., Stamps, J., & Lipnack, J. (2004). Can absence make a team grow stronger? *Harvard Business Review, 82*(5), 131.

Roach, S. (2005). The new macro of globalization. *Global Economic Daily Comment.*

Thompson, J. D. (1967*). Organizations in action.* New York.

Townsend, A. M., DeMarie, S. M., & Hendrickson, A. R. (1998). Virtual teams: Technology and the workplace of the future. *The Academy of Management Executive, 12*(3), 17.

Venkatesh, V., & Johnson, P. (2002). Telecommuting technology implementations: A within-and between-subjects longitudial field study. *Personnel Psychology, 55*(3), 661.

KEY TERMS

Cultural Distance: Cultural distance is a function of differences in values and communication styles that are rooted in culture (demographic OR organizational). Distance is created when individuals or groups perceive that their values and communication styles differ from others.

Interdependence Distance: Interdependence is the degree to which one individual or group perceives that their success is tied to another individual or group member. Interdependence distance is created when individuals or groups do not perceive their goals or tasks as interconnected.

Relational Distance: Relational distance is based on the difference between team members' organizational affiliations. Distance is created when individuals are in different companies.

Relationship History Distance: Relationship history includes both the extent to which members have had prior relationships with one another (strong ties), or relationships with some of the same people (weak ties). Distance is created if the strong and weak ties between individuals are absent.

Social Distance: Social distance is based on perceptions of class or status differences that produce feelings of social closeness or distance based on social interactions. Social distance is created when individuals perceive themselves to be in a different class or status than others. This can be formal status or informal status differences.

Virtual Distance Index: The virtual distance index is an operational measure (quantitative score) of virtual distance. It is a standardized, equally weighted average of all of the virtual distance factors in the model.

Virtual Distance Model: The virtual distance model incorporates a set of eleven physical, social and work-related factors that create a sense of perceived distance in individuals, teams and organizations collaborating in virtual space. Virtual distance is created as distance within each of the eleven factors in the model becomes greater.

Use of E-Collaboration Technologies Among Students of Management

Antonio Padilla-Meléndez
University of Málaga, Spain

Aurora Garrido-Moreno
University of Málaga, Spain

INTRODUCTION

Nowadays technology is seen as an important tool for improving the educational methods at all levels. This has implications for both general and specific education such as management training. Some of the advantages identified in general education include no time limits or geographical barriers and the ability to teach more people without additional cost. In particular, in Europe *digital literacy* is emerging as a new key competency required by workers and citizens for the Knowledge Society (European Commission, 2005). Therefore, the integration of IT-supported learning could help workers acquire the necessary skills and knowledge for their job (European Commission, 2005) and if the student learns the use of technology before starting his/her job, this could be an advantage for both the future professional and the employer. Besides, the use of IT can improve the effectiveness of the learning process.

Referring to specific education, technology is also proving valuable in higher education and in specialised areas such as management training (Ives & Jarvenpaa, 1996). Furthermore, in management education the learner (student) will in their future, work in an environment where technology (in particular information technology (IT) and, indeed, e-collaboration technologies) will have several applications.

It is normally assumed that young people, particularly university students have a developed and sophisticated use of IT and the Internet. In fact, some are included in a new generation of people who are intensive users of technology, particularly for using IT for social contacts, music downloads and so on. Interestingly, almost all students use a lot of mobile phone technology but, normally, this type of technology is not covered in universities teaching. Taking this into account, management teaching should include the use of this type of IT to improve the learning process. But,

to what extent is this true? Are management students advanced users of the Internet? Our own experience is that, students, contrary to the general perception, do not all react positively to an IT based process of teaching-learning management. Some prefer the traditional process.

An understanding of the level of usage of IT by management students would contribute to explain this situation. Several differences exist between the use of IT by university students and the typical user of Internet-based technologies and these are possibly affecting the success of this learning strategy. Consequently, if the students are found not to be sophisticated users, the use of IT for management teaching-learning could be more an obstacle than and advantage. All in all, our objective in this article is to describe the e-collaboration usage by management students and to compare their profile with the profiles of the advanced user of the Internet.

The article is organized as follows. After a literature review, the research design and data analysis are described. The discussion and implications, along with some limitations and preliminary conclusions conclude the article.

BACKGROUND

In general, two different approaches have been developed for using IT in education. The complete substitution of normal learning (based on a physical classroom) by using IT all the time, that is, e-learning, is the first one. This type of learning has been amply criticized in teaching management skills, so a blended learning approach has been proposed. The second use is to employ IT to complement traditional learning methods. In this case, the communication technologies (mainly Internet based) have been used for gaining a bigger interaction of professors with students or even to promote more contacts between them in order to improve its collaboration.

Interestedly, contrary to general idea, statistics about e-learning show a relevant utilization of these tools in past years. According to the results of the Grupo Doxa study, e-learning accounted for 5.2% of total business training in Spain during 2004, rising to 7% when looking at only big companies (Grupo Doxa, 2004). Compared to other European countries, Spain is below the mean but ahead of nations such as Italy or France. The most advanced countries are the UK and Germany (Telefonica, 2004). But, when looking at universities, only 4.1% of their total budget is used for e-learning based training. Moreover, the majority of universities (50%) use *WebCT* as the e-learning platform (Barro, 2004).

In response to the increased importance of IT, several studies analyzing IT use and acceptance, from different theoretical perspectives, have been developed. In this sense, the Technology Acceptance Model (TAM) is very well known (Venkatesh, 2000). In particular, several studies have been done in the context of the application of TAM to management education (Martins & Kellermans, 2004).

As special technologies, we will concentrate in this paper in the e-collaboration technologies. These are technologies that support e-collaboration. An operational definition of electronic collaboration (e-collaboration) is to consider it as collaboration using electronic technologies among different individuals to accomplish a common task (Kock & D'Arcy, 2002; Kock, 2005a). Among these e-collaboration technologies we have several internet-based technologies, such as e-mail, forums, chats, and documents repositories that will be considered in this research. In a historical revision of those technologies, Kock and Nosek (2005) consider the e-mail as the first real e-collaboration technology and today they have evolved towards two different types: browser-based (i.e., run on Web or browser) and nonbrowser-based. The browser-based e-collaboration tools are for example WebEx and eRoom, and e-learning tools as Blackboard and WebCT. As mentioned before this latter one is the most used e-learning tool in Spain. The nonbrower based are either peer-to-peer e-collaboration tools (for example Groove) or client server e-collaboration tools (for example, MSN Messanger, ICQ, AOL Instant Messanger, or Skype). The integra-

tion of IT-supported learning helps students acquire the necessary skills and knowledge for their future job. As these IT are e-collaboration technologies, many positive implications for business process improvement could arise in the future (for an in-depth analysis on e-collaboration technologies effects on business process improvement, see Kock, 2005b).

RESEARCH DESIGN

A Web-based course was used as an environment to study the e-collaboration technologies usage by students. This system has been used for four years for the topic "management control systems" (http://campusvirtual.uma.es/contgest). Over the past 12 months more features have been added, and special support from the IT for education centre at the university has been received. The objective is to adapt the course to the European Space for Higher Education by seeking a greater collaboration between and among students and professors. Consequently, the particular objectives of this virtual classroom have been: (1) put on line the contents of the course; (2) encourage and facilitate students work in groups, using the university's virtual campus e-collaboration tools; and (3) conduct a continuous assessment of student work using this tool.

As mentioned before, apart from the description of the level of use of e-collaboration technologies by students we also compare this usage with the typical use of the advanced Internet user. To do this, we selected a well-known reference such as the results of the study of the AIMC (Association for the Research in Communications Media). This association carries out an annual study from 1996 to the present day about Internet use in Spain. They collect general data on population (computer use and Internet use), and access to the Internet (place of access, frequency of use, most used services, computers utilized, etc.). The methodology in the AIMC questionnaire consists in an auto-administered interview online at a Web site linked to users who visit Spanish Web sites. The interviews were carried out in 2004 (AIMC, 2005). The final size of the sample was 53,647 answers. The most advanced users analyzed in this study will be considered as the advanced users in Spain and we will compare with them our students IT usage.

DATA COLLECTION AND ANALYSIS

The sample units were students in the management control systems course—a subject held in the third year of the administration and management degree (four years) at the Faculty of Economics and Business Studies of a medium-sized university in southern Spain. The selection of these individuals is due to three reasons. They are persons actually using a virtual classroom for their course. They are young students that will be future professionals/managers. They also have very similar age, education, and other characteristics which can mitigate the possible impact of unwanted variables and influences in the analysis. A questionnaire administered in class was used to collect the data. After the literature review, a paper-based questionnaire was designed. The variables used are mainly related to a previous work of Martins and Kellermans (2004) and to the earlier studies mentioned above by AIMC in Spain. This questionnaire was pre-tested by four people, two of them students. After this, some questions were changed to clarify the questionnaire. The questionnaire was administered on March 15, 2005. A total of 225 valid questionnaires were collected. As 597 students were registered in the course, the answer rate is 37.69% when compared to the total registered population. Based on a sample size of 225 valid questionnaires with the population variance unknown and a reliability of 95.5% the maximum error is 5.26%. To analyze the gathered data a one-variable and bi-variable statistical analysis was conducted.

E-COLLABORATION TECHNOLOGIES USAGE

Fully 88% of students have an e-mail account different from (i.e., in addition to) the account provided to them by the university. There is a low usage of the university-provided e-mail account by students. We found 85.8% of students who say they make very little use of their official e-mail address (see Table 1).

Internet access from the IT rooms at the school is also not very high, because half of the population (48.9% of students) report going online from these rooms just a little, very little, or not at all (nothing). Only 4% of the students utilize the Internet from here as well, pointing to disconnect between potential and practice. This problem can be explained partially by the negative perceptions of IT infrastructure at university. Regarding the availability of Internet access points from the university, we discovered that half of the respondents have a negative perception of this item, and a 53.3% of them consider this type of access to be only a little available, very little available or not available at all.

Talking about the general and particular use of IT for learning (see Table 2), there are some important utilization differences that proved very big. These include mobile phones (7.02% general use/4.97% particular use) and SMS use (7.12% general use/5.09% particular use). Another important difference is the instant messaging tool broadly used for general purposes but not for any particular classroom application. The main similarities are the use of an Internet browser, search engines, and

Table 1. Use of technologies at university

	Use of the e-mail account provided by the university	Access to the Internet from the university IT room
	%	%
Nothing	44	9.3
Very little use	26.2	24
Little use	15.6	15.6
Indifferent	1.3	0.9
Something	9.8	30.2
Enough use	2.2	16
Much use	0.9	4
Total	100	100

word processor. In this latter case, respondents tended to use word processing programs more frequently in the learning context than for general purposes.

The Internet is now a familiar entity. Fully 94.2% of respondents have been Internet users for more than a year. So, Internet familiarity is no longer an obstacle for acceptance and usage of new learning technologies. In the AIMC case 34.2% of respondents were Internet users for more than 5 and less than 8 years. In our study the biggest percentage between three and five years (44%).

Regarding the frequency of access to the Internet, we found some differences between the access from the University and from the students' home. 29.3% of students access several times per day from home but from the University the access is less regular, with only 6.7% of respondents that access quite everyday from there. 49.3% access less than one time per week from the University. About the enrolment to any e-learning based training course, only a 3.6% (eight respondents) have done this in the last year. In the AIMC study, an 18.8% of respondents were enrolled in a e-learning

course last year.

In general, there are similarities between the student's online activities and the AIMC ones. The use of Internet mainly involved browser engine or directory searches, music downloads, news reading, and looking for information about cinemas/theatres. In the AIMC study, the activities most cited are: (1) searches (search engines/directories), (2) news reading, (3) information about maps/streets, (4) software downloads, (5) film downloads, (6) music downloads, and (7) information about cinemas/theatres.

With respect to the use of e-mail: in the AIMC study, 60.1% of respondents received between 5 and 50 e-mails per week, and in our study this totalled 95.1% of respondents. Also in the AIMC study, 68.4% sent less than 20 messages per week. In our study this percentage was 96.3%.

DISCUSSIONS AND IMPLICATIONS

After analyzing the empirical data from the study, we can say that students make a little usage of e-collabora-

Table 2. General and particular use of IT

	General Use (1 to 8)		Particular Use (for learning) (1 to 8)	
	Mean	Std dev.	Mean	Std dev.
Word processor	**4.62**	1.746	**5.09**	1.629
Spreadsheet	3.17	1.915	3.78	1.898
Presentations software	1.93	1.406	2.13	1.538
Data management software	1.92	1.721	2.28	1.963
Email software	3.81	2.668	3.38	2.465
Internet navigator	**6.48**	1.626	**6.1**	1.558
Search engines	**6.24**	1.702	**5.82**	1.729
Instant Messaging	**5.55**	2.561	3.97	2.635
Chat	2.45	2.271	1.91	1.855
Web forums	1.98	1.837	1.68	1.381
E-mail lists	1.88	1.713	1.7	1.388
P2P networks	4.24	2.78	2.48	2.218
IP telephony	1.56	1.716	1.41	1.424
Mobile phones	**7.02**	1.456	**4.97**	2.399
SMS	**7.12**	1.47	**5.09**	2.501
Mobile phone to access to Internet	1.51	1.5	1.35	1.335

tion technologies at school, because they have a poor perception about the availability and value of the IT infrastructure at the university. We were surprised by the general negativity expressed towards the incorporation of technology in the curriculum as well as the disconnect between administrator/professor and student perceptions of the new system's importance. This, obviously, will affect the usage of this IT in the learning-teaching process.

In general, the Spanish students surveyed make very little use of e-mail and do not take particular advantage of the university's IT access points even to go out on the Internet. The e-mail account provided by the university may actually be more useful as a way to easily manage one-way sending of official messages to students than to engage in meaningful dialogue. This lack of real communication could contribute to the students' bleak assessment of IT in the learning-teaching process.

When we consider that the majority of respondents are familiar users of Internet (more than a year), this could not be an obstacle for the use of the new system. However, comparing its characteristics as users with the advanced users, we also found that their Internet experience tended to involve relatively simple activities such as searches, news updates or music/film downloads, not directly related with e-collaboration. Furthermore, the use of e-learning tools by the respondents is very limited, with only 3 of each 100 students indicating they were enrolled in an Internet e-learning course during the prior year. Overall, student users were less active and advanced in IT across-the-board than those persons analyzed in the AIMC study.

CONCLUSION

In talking about our preliminary conclusions in Spain we could say that, contrary to the idea that young persons—especially students—are advanced users of Internet-based technologies, we found that they are not particularly prolific. Internet familiarity is no longer an obstacle for acceptance and usage of these technologies for learning. However, management students are not advanced Internet users and this could be a problem in using these technologies for learning purposes. Furthermore, the most popular new technologies for students are not online ones, but rather mobile phones that broadly speaking are e-collaboration tools, but which are rarely used as academic learning tools. Professors

using e-learning systems have to adapt the technology that is used by the system to the one that is more used by their clients (the students).

In seeking to improve the teaching-learning process with e-collaboration technologies we learn that technology for technologies sake is not cost effective, despite the hype over using electronic tools to extend the classroom to new vistas. Rather, users have to clearly understand the benefits of IT from their own perspective, not that of the provider (in this case the University professor). As practical recommendation, we could say that a previous study about IT usage by the students would be positive for getting later better results applying e-collaboration tools in the learning-teaching process.

This study has at least three limitations. First of all, this is non-longitudinal research so we cannot know if the students will change their level of technology acceptance and use after spending more time exposed to the system. Secondly, this study is restricted geographically to one state in Spain and to one university, so this could limit the possibilities of taking a broader view of the results. And, finally, this is, in nature, an exploratory study, so the conclusions are also exploratory.

ACKNOWLEDGMENT

We would like to thank Prof. Del Aguila for her insightful comments about the paper, and Mr. Jan Zadruzynski for their revision of the style and very positive suggestions.

REFERENCES

AIMC. (2005). *Navegantes en la red. 7ª encuesta AIMC a usuarios de Internet*. Madrid, Spain: Sersa. Retreived May 10, 2005, from http://www.aimc.es

Barro, S. (2004). *Las TIC en el sistema universitario español*. Madrid, Spain: Conferencia de Rectores de las Universidades Españolas (CRUE).

European Commission. (2005). *E-learning conference conclusions*. Brussels, Belgium: European Commission.

Grupo Doxa. (2004). *E-learning en las grandes empresas. Panel anual. Resultados año 2004*. Madrid,

Spain: Grupo Doxa.

Ives, B., & Jarvenpaa, S. L. (1996). Will the Internet revolutionize business education and research? *Sloan Management Review, 37*(3), 33-41.

Kock, N. (2005a). What is e-collaboration? *International Journal of e-Collaboration, 1*(1), 1-7.

Kock, N. (2005b). *Business process improvement through e-collaboration: Knowledge sharing through the use of virtual groups.* Hershey, PA: Idea Group.

Kock, N., & D'Arcy, J. (2002). Resolving the e-collaboration paradox: The competing influences of media naturalness and compensatory adaptation. *Information Management and Consulting, 17*(4), 72-78.

Kock, N., & Nosek, J. (2005). Expanding the boundaries of e-collaboration. *IEEE Transactions on Professional Communication, 48*(1), 1-9.

Martins, L. L., & Kellermanns, F. W. (2004). A model of business school students' acceptance of a web-based course management system. *Academy of Management Learning and Education, 3*(1), 7-26.

Telefónica. (2004). *La sociedad de la información en España 2004.* Madrid, Spain: Telefónica.

Venkatesh, V. (2000). Determinants of perceived ease of use: Integrating control, intrinsic motivation, and emotion into the technology acceptance model. *Information Systems Research, 11*(4), 342-365.

KEY TERMS

Digital Literacy: Ability to use digital technology to locate, evaluate, use, and create information.

E-Collaboration Technologies: Electronic technologies that enable collaboration among individuals engaged in a common task.

E-Learning: Electronic learning; the process of learning online, especially via the Internet.

IT-Supported Learning: Use of IT as a support for learning.

Knowledge Society: A society where main of the prosperity and well-being of its people came from the creation, sharing and use of knowledge.

Technology Acceptance Model (TAM): A model developed to study the acceptance of the technology by an individual taking into account, basically, both the perceived easy of use and the usefulness of the technology.

Virtual Classroom: A mode of computer-based education whereby the teacher interacts with students either via video-conferencing, Internet broadcast, or e-mail

A Use-Centered Strategy for Designing E-Collaboration Systems

Daniel H. Schwartz
Air Force Research Laboratory, USA

John M. Flach
Wright State University, USA

W. Todd Nelson
Air Force Research Laboratory, USA

Charlene K. Stokes
Air Force Research Laboratory, USA

INTRODUCTION

The ubiquity of collaboration cannot be overstated. Derived from the Latin *collaborare*, which means "work with" or through, collaboration is the process wherein agents work together through transaction. Collaboration entails the existence of a *team* if a common goal or purpose underlies the transaction. A *virtual* team exists when collaboration takes place (to a varying degree) through technology across time, space, and (often) organizational boundaries; also known as *e*-collaboration. As a general definition, we follow the lead of Kock and colleagues (Kock, Davison, Ocker, & Wazlawick, 2001; Kock & Nosek, 2005), and state that e-collaboration is "collaboration among individuals engaged in a common task using electronic technologies" (Kock et al., 2001, p. 1). This is a very broad definition and includes such historical means of e-collaboration as the U.S. Department of Defense's ARPANET and early group decision support systems (GDSSs) such as Lotus Notes (Kock & Nosek, 2005). Few would argue the contemporary impact computers, the Internet, and network architectures (e.g., local area networks; LANs) have had on collaboration and teams (Schwartz, Divitini, & Brasethvik, 2000). Current instantiations of e-collaborative systems include the Internet (which includes various e-collaborative subsystems such as Internet relay-chat, bulletin boards, and weblogs), videoconferencing, and virtual workstations. The opportunities created by this new wave of e-collaboration and virtual teamwork have, in turn, dramatically transformed military forces (e.g., network-centric warfare; Cebrowski, 1998), business (e.g., B2B collaboration; Rosenberg, 2003),

infrastructure (e.g., traffic flow regulation; Jermann, 2001), and other areas of society (e.g., collaborative music development; Weinberg, 2005).

The fact that e-collaboration wires together so many organizations highlights the idea that there is some advantage in having work virtually distributed across multiple decision makers. Thus, researchers are beginning to frame questions around the nature of e-collaboration and virtual teams. Important dimensions of this phenomenon include the organizational dynamics (e.g., Rochlin, 1997), the technological capabilities (e.g., Iacovou, Benbasat, & Dexter, 1995), and the human factors (e.g., Proctor & Vu, 2005). However, the focus of this article will be on the work domain or problem space as a significant context for understanding the interactions among the lower order dimensions. The central premise is that all work, including teamwork, is situated (e.g., Hutchins, 1995; Suchman, 1987). That is, success depends on adaptation to the demands of the problem (i.e., the work, the situation, the ecology). Therefore, modeling the work domain constraints becomes an essential factor for predicting how the organizational structures, information technologies, and human abilities will interact to determine the overall success and stability of the team. In sum, we would like to make the case for a Cognitive Systems Engineering (CSE) (e.g., Rasmussen, Pejtersen, & Goodstein, 1994; Vicente, 1999) or Ecological (e.g., Flach & Dominguez, 1995; Flach, Hancock, Caird, & Vicente, 1995) approach to questions about e-collaboration or virtual teams.

Below we highlight a military example of the use-focused design and development of an e-collaborative system called Knowledge Web. This is followed by an explication of the role of Cognitive Systems Engineering: a use-centered approach for the design of e-collaborative systems.

KNOWLEDGE WEB

There are several common problems/operational issues the U.S. Navy has faced due to the escalation of e-collaboration. First, when information is most needed, it is rarely in an easily accessible location, in the requisite form, or available at the right time. Often, the data is organized and presented based on where it is coming from rather than where and how it is needed. Also, there exist multiple hardware stovepipes that hinder information access and impede the speed of command. An additional, related problem is that there is great difficulty identifying what information is valuable to whom, and when that information is most crucial. Thus, there is little idea of who uses what information, what form the information should take, why collaboration across organizational departments is rare, and why staff presentations are limited by prescribed slide presentations. When reviewing the ecology through which these problems manifest, a common theme emerges: Naval commands are sharing knowledge in distributed, asynchronous (i.e., temporally staggered) environments with multi-echelon and coalition environments to contend with (Oonk, Rogers,

Moore, & Morrison, 2002). These findings allowed the Navy to propose a concept of operations for command centers that focuses on shared relevant knowledge (i.e., information that is *meaningful* vs. shared data), shared awareness, and speed of command. Thus, the Knowledge Web was born.

Knowledge Web (or K-Web) is an advanced development project that delineates a clear concept of operations for using the Web to improve the effectiveness of command and control (C2) for the U.S. Navy. The goal of K-Web is to increase the speed of command, facilitate collaboration and information sharing, and to store and dynamically present a variety of knowledge in an easily accessible environment—a *Knowledge Web* (Oonk et al., 2002). The K-Web concept utilizes Web-based technologies to share operationally relevant information. In a K-Web, available data are processed, formatted, and stored by "information producers." "Information consumers" use K-Web products to obtain and maintain awareness of current situations. The 'users' of the K-Web, therefore include both information producers and consumers. Figure 1 depicts how, under the K-Web concept, users take advantage of tools developed to facilitate the rapid production of standardized, summarized, knowledge-based information products that are hosted on a Web server. Because the information in the K-Web is viewed through a standard Web Browser, the K-Web facilitates information access and presentation, and information in the K-Web can be integrated with other data as required (e.g., tactical

Figure 1. Knowledge Web concept

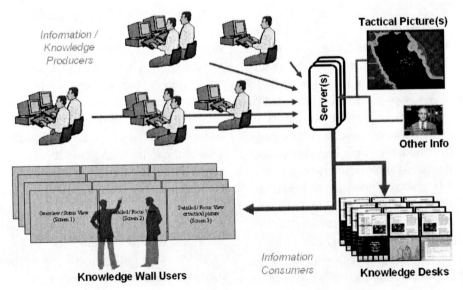

data, satellite imagery, graphical summaries, e-mail; Oonk et al., 2002)

K-Web tools have been developed to facilitate the production, formatting, dissemination, and Web-based presentation of information so that it is easily understood. These tools allow information products to be created and continually updated by information producers. Information products, which are stored in shared locations accessible to information consumers via a standard Web browser, comprise the K-Web. As a result of K-Web deployment, the Space and Warfare systems Center, San Diego, found that "K-Web use was directly correlated to: transformed command operations from 'information sharing' to 'problem solving', improved speed of command, enhanced situation assessment, and increased interaction between functional areas [of the U.S. Navy] " (Oonk, et al, 2002).

DESIGNING E-COLLABORATION SYSTEMS: A COGNITIVE SYSTEMS ENGINEERING PERSPECTIVE

The design of e-collaborative systems in support of virtual teams entails the creation of interface technologies that support work. System design methodologies must fulfill the organization's vision and build work or problem-centered systems that leverage technological advances and support end-user understanding. This is in contrast to more traditional design approaches that focus either on the user (human factors) or the technology (human-computer interaction) for design specifications. The isolated focus of each of these views highlights an artificial separation cut through a functionally whole organization. We argue for an integrated design strategy—that of a transactional, use-centered approach, whereby users, goals, and technology are integrated and the focus is on the *ecology* (i.e., meaning) that emerges from their transaction (Flach, Bennett, Stappers, & Saakes, 2005; Flach & Dominguez, 1995).

Ecology. Each of the areas mentioned earlier (e.g., military forces, business) represents an *ecology* or field through which agents adapt their life processes. Ecology emerges at the intersection of technology, users, and users goals; it is what *matters* (Flach & Dominguez, 1995). Figure 2 depicts an ecology: The dotted boarder around the sphere represents the dynamic, ever changing nature of the ecology whereas the dotted boarder around the agents reminds us that there is no solid boundary between agent and ecology. We employ the

term ecology because of its etymological relation to natural life processes (Ernst Haeckel coined the term, meaning "the economy of nature"; Bramwell, 1996). Within the ecology, we can make a distinction between an *environment* (e.g., e-collaborative environment) and *agent* (e.g., virtual team). The distinction emphasizes the *transaction* that exists across and through the two. We can say that the e-collaborative environment is the medium through which the virtual team adapts its life processes—like a fire burning *through* the medium of wood and oxygen. It would be difficult to imagine fire interacting with wood or burning from its own self-action. Thinking about the fire-wood-oxygen *transaction* as a whole *ecology* facilitates the consideration of emergent and novel behavior that typically evolves through time and is characteristic of fire ecologies (Dewey & Bentley, 1959).

Adaptation through an ecology is afforded when there is adequate feedback (i.e., information) representing the consequences of actions relative to team goals, and the field of possibilities for action (taken to reduce the discrepancy between feedback and goals) is large (Flach et al., 2005). Thus, affording adaptation is another way of saying that there is opportunity for control. When developing e-collaborative systems for virtual teams, the goal of designers is to ensure that adequate control can be maintained. Perturbations to the ecology cannot always be anticipated and the ability to regain control, or adapt to changing situations, should be a primary feature designed into the e-collaborative system.

Designing for opportunity. Traditionally, design processes focused on an a priori determination of the one best way to carry out a particular task or grouping of tasks and a linear support system was developed to assure teams carried out actions according to this predetermined route. Unfortunately, virtual teams face an increasingly complex terrain whereby *many* routes may be sufficient to proceed by in order to maintain control. For example, when designing and developing a business-to-business (B2B) collaborative system for virtual teams engaged in interorganizational commerce, it is important to consider value constraints, information constraints, and constraints on action. Value constraints might include the goal of maximizing supply chain visibility and the objective properties of the supply chain. Information constraints may include the ability of the virtual team to detect opaqueness in the supply chain through an e-collaborative system. Constraints on action represent the ways in which the

Figure 2. Ecology

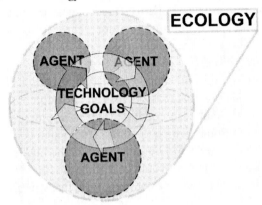

e-collaborative system affords the virtual team a means to keep the supply chain as visible as possible—a way to reduce the discrepancy between the virtual team's goal of visibility and the feedback information about the current visibility of the supply chain.

Constraints determine the functional effectiveness of an organization. The emphasis on "use" reflects our view that virtual teams will naturally adapt to the functional constraints of the ecology if those constraints are visible. Thus the design of collaborative systems cannot be based solely on an analysis of work procedures, structure, and practices in present collaborative systems. In order to understand an existing system or to develop a revolutionary collaborative system, "an analysis is required in terms of organizational goals and constraints, the *potential relationships* among goals, functions and processes, the criteria available for allocation of roles to individual agents, and the coordination needed, i.e., the work organization and management structure" (Rasmussen, 1991). We suggest conducting a work domain analysis and an organizational analysis as an appropriate strategy for modeling and understanding the functional constraints of a virtual team ecology (see Rasmussen, Pejtersen, & Goodstein, 1994; Vicente, 1999).

Work domain analysis. To understand the constraints of the domain in which an e-collaborative system is to be developed to support virtual teams, we feel that it is necessary to conduct a work domain analysis; that is, a functional map of domain constraints. By conducting an analysis of the work domain we can attempt to understand the possibilities and consequences of action within the domain of interest (i.e., the situation constraints; Rasmussen et al., 1994). Situation constraints directly shape belief and action. By "surveying" the field of work, we can gain insight on these possibilities for

action (Bennett & Flach, 1992). Within this framework, designers come to understand the functional purpose of a domain (i.e., why does it exist?), the values and priorities of the system (i.e., the values and priorities or information flows that are the means whereby the system achieves its functional purpose), the general functions (i.e., functions that must accomplished to support the values and priorities), and the physical functions (i.e., the physical interaction/transaction between physical forms within the system—the means whereby the general functions are accomplished). Additionally, each level of abstraction can be decomposed into subcomponents. In other words, a part-whole hierarchy splits a system into its subsystems and then subsystems into components. The goal of decomposition is reducing the complexity of a system by dividing it into smaller functional units.

An abstraction hierarchy can be seen as a way to view a virtual team's ecology through different levels of magnification; for example, microscopic-macroscopic. It is a means-ends description of an ecology whereby higher levels describe the "why" or higher order functionality of the system, and lower levels describe "how" or the means wherein high-level functions are achieved (see Figure 3). The highest level of abstraction is the functional purpose, or the ultimate reason why the ecology exists. The highest level delimits ultimate goals to be achieved and environmental constraints that need to be recognized. For example, the functional purpose of the B2B virtual team mentioned above may be to achieve optimal supply chain management.

The next level of abstraction represents more abstract functions, namely the situation independent values and priorities that must be adhered to in order to accomplish the functional purpose of the virtual team. Priorities and values in the B2B virtual team may include, for example, corporate doctrine. The next level of abstraction delimits the general functions of the virtual team that are the means whereby the measurable priorities and values can be met. In the case of the virtual team, an important general function is e-collaboration. E-collaboration may be a general function that accomplishes the corporate doctrinal notion of B2B synergy. The general functions of a virtual team are supported by specific physical functions (the next level of abstraction). Physical functions are actual specific physical process systems that accomplish a certain role (e.g., Internet relay chat). The lowest level of the abstraction hierarchy is the physical form level. Physical forms are the entities or the most micro-level forms that support

Figure 3. The abstraction hierarchy

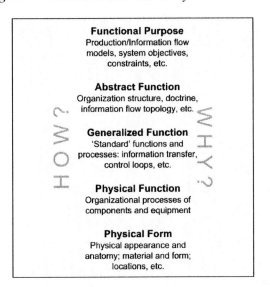

physical functionality (e.g., computers, the location of individual decision making agents).

The abstraction hierarchy also represents a part-whole decomposition of the ecology, whereby functional levels are divided into subsystems and components. This decomposition reduces the complexity of the system by separating functions into individual components. The resulting abstraction-decomposition map provides the foundation for the design of systems. The vertical dimensions represent the dimension of abstraction and the horizontal dimension illustrates varying levels of decomposition (see Figure 4).

The work domain analysis explicitly models work demands and constraints in the work environment as a basis for design. The analysis activity is a means for developing an understanding of the work's demands, and affords the designer the ability to develop and test specific hypotheses about ways to support operators with these demands. The output of a work domain analysis may supply the "design seeds" necessary to instantiate design hypotheses that can be tested through rapid prototyping (Tullis, 1990).

The abstraction hierarchy also provides a means for discriminating between organizational roles and interconnections. Typically, decisions affecting the functional purpose of an organization will be made by higher levels of management. Alternatively, diagnosing system faults at the level of physical function will be exercised by "shop-floor" personnel. Through developing an abstraction hierarchy, the designer may discover unique links between organizational roles that may facilitate the explication of requirements for

a particular e-collaboration system. Consequently, in order to develop a *coordinated* collaborative system for virtual teams, a designer must understand the structure of the work organization—how control is allocated to decision-making agents through social organization.

Organizational analysis. The functional purpose of a virtual team defines the necessary coupling between individual decision-making nodes. Reciprocally, the physical manifestation of coupled decision-making nodes characterizes a virtual team.

The functional purpose describes *why* the virtual team exists, whereas the physical form describes the physical layout of the individual agents. The work domain analysis assists the designer in understanding why a virtual team exists and how, at a broad level, higher order (i.e., abstract) functions are actually accomplished. The organizational analysis helps the designer understand the cooperative decision-making structure that ultimately controls a system (e.g., supply chain management). The requirements for controlling a system depend on particular work situations and may change over time. Ultimately, allocation of roles and responsibilities within a virtual team are contingent upon the work organization architecture established by a particular organization or designer. Often, role allocation reflects the dynamic requirements of the work problem space—there is an attempt to link requirements with resources. Under these circumstances, role allocation will be an informal attempt to maintain control.

Organization structure. There are a variety of control structures that define teams and organizations. For example, there are autocratic coordination structures that are composed of one agent that is responsible for coordinating the activities of all other agents (Rasmussen, 1991). Also, there are hierarchic coordination configurations that are composed of a stratified control structure whereby higher level agents evaluate and plan the activities of lower level agents (Rasmussen, 1991). Other control structures described by Rasmussen (1991) include heterarchic planning (i.e., distributed, stratified), anarchistic planning (i.e., independent, isolated planning), democratic planning (coordination through negotiation), and diplomatic planning (i.e., local negotiation with specific neighbors). The control structure is independent of the characteristics of the work domain yet must be understood when designing an e-collaborative system because it determines whether communicated information is neutral, advice, instructions, or orders. For example, in a democratic-planning architecture, communication between agents consists of

Figure 4. Generic abstraction-decomposition space

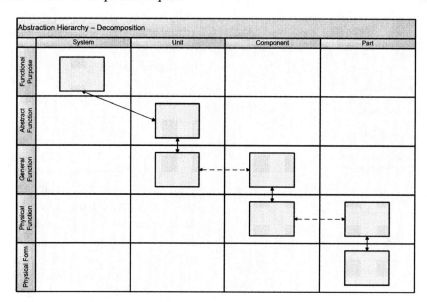

variable amounts of neutral, advice, and instruction information. Therefore, perhaps, a multimodal interface, including voice-communication, videoconferencing, and chat capabilities would be appropriate—depending on the work domain.

CONCLUSION

The idea behind a use-centered approach to the development of e-collaborative systems in support of virtual teams is the focus on work constraints (i.e., *use*) whereby the *ecology*, or agent-technology-goal-system must adapt. Adaptation entails the continuous maintenance of control. Thus, an effective e-collaborative system will afford agents sufficient control of a system in the event of ecological perturbations and uncertainty by behaving like a window through which agents can share relevant knowledge (i.e., information that is *meaningful*). Cognitive systems engineering provides a design methodology for uncovering work domain constraints, including the social-organizational dimensions inherent to a work domain.

Due to the fast pace of change in many systems (e.g., military C2), work analysis will never be complete (Flach, Schwartz, Bennett, & Hughes, 2006). An e-collaborative system will only be as adaptive as how integrated the work analysis is with the system's life cycle. This requires an iterative analysis, design, and development process that promotes the discovery of

new contingencies and problem constraints that can be leveraged by the system to support adaptation.

REFERENCES

Bennett, K. B., & Flach, J. M. (1992). Graphical displays: Implications for divided attention, focused attention, and problem solving. *Human Factors, 34*(5), 513-533.

Cebrowski, A., & Garstka, J. (1998, January). Network-centric warfare: Its origin and future. *Proceedings, 124*(1), 28-35.

Flach, J. M., Bennett, K. B., Stappers, P. J., & Saakes, D. P. (2005). Searching for meaning in complex databases: An ecological perspective. In R. Proctor & K. L. Vu (Eds.), *Handbook of human factors in Web design* (pp. 4-8-423). Mahwah, NJ: Erlbaum.

Flach, J. M., & Dominguez, C. O. (1995, July). Use-centered design. *Ergonomics in Design, 19*(24).

Flach, J. M., Hancock, P. A., Caird, J., & Vicente, K. (1995). *Global perspectives on the ecology of human-machine systems.* Hillsdale, NJ: Erlbaum.

Flach, J. M., Schwartz, D. H., Bennett, A., & Hughes, T. (2006). Integrated constraint evaluation: A framework for continuous work analysis. In A. Bisantz & C. Burns (Eds.), *Applications of cognitive work analysis.*

Hutchins, E. (1995). *Cognition in the Wild.* Cambridge, MA: The MIT Press.

Iacovou, C. L., Benbasat, I., & Dexter, A. S. (1995). Electronic data interchange and small organizations: Adoption and impact of technology. *MIS Quarterly, 19*(4), 465-484.

Jerman, P. (2002). Task and interaction regulation in controlling a traffic simulation. In G. Stahl (Ed.), *Computer support for collaborative learning: Foundations for a CSCL community* (pp. 601-602). Hillsdale, NJ: Erlbaum.

Kock, N., Davison, R., Ocker, R., & Wazlawick, R. (2001). E-collaboration: A look at past research and future challenges. *J. of Systems and Info. Tech., 5*(1), 1-9.

Kock, N., & Nosek, J. (2005). Expanding the boundaries of e-collaboration. *IEEE Transactions on Professional Communication, 48*(1), 1-9.

Oonk, H. M., Rogers, J. H., Moore, R. A., & Morrison, J. G. (2002). *Knowledge web concept and tools: Use, utility, and usability during the Global 2001 War Game* (Tech. Rep. No. 1882). San Diego, CA: The Space and Naval Warfare Systems Center.

Proctor, R. W., & Vu, K.-P. L. (Eds.). (2005). *Handbook of human factors in Web design.* Mahwah, NJ: Erlbaum.

Rasmussen, J. (1991). Modeling distributed decision making. In. J. Rasmussen, B. Brehmer, & J. Leplat (Eds.), *Distributed decision making: Cognitive models for cooperative work* (pp. 111-142). New York: Wiley.

Rasmussen, J., Pejtersen, A., & Goodstein, L. (1994). *Cognitive systems engineering.* New York: Wiley Interscience.

Rochlin, G. I. (1997). *Trapped in the Net: The Unanticipated consequences of computerization.* Princeton NJ: Princeton University Press.

Rosenberg, A. (2003). Focus on collaboration: Collaboration B2B. *Intranet J. of Strategy and Management.*

Schwartz, D. G., Divitini, M., & Brasethvik, T. (2000). *Internet-based organizational memory and knowledge management.* Hershey, PA: Idea Group.

Suchman, L. A. (1987). *Plans and situated actions: The problem of human-machine communications.* Cambridge, England: Cambridge University Press.

Tullis, T. S. (1990). High-fidelity prototyping throughout the design process. In *Proceedings of the Human Factors and Ergonomics Society, 34th Annual Meeting* (pp. 266-275). Santa Monica, CA: HFES.

Vicente, K. J. (1999). *Cognitive work analysis: Towards safe, productive, and healthy computer-based work.* Mahwah, NJ: Erlbaum.

Weinberg, G. (2005). Interconnected musical networks: Toward a theoretical framework. *Computer Music Journal, 29*(2), 23-39.

KEY TERMS

Adaptation: (In reference to e-collaboration and teamwork.) Self-organization around the work domain (i.e., use) constraints.

Cognitive Systems Engineering: A design discipline that uses analyses of work (practice, structure, purposes, and constraints) to inform the design of process and technology for human-system integration. It deals with socio-technical systems, where socio refers to the social processes of communication, cooperation, and competition.

Constraint: A limitation of possibilities.

E-Collaboration: Collaboration among agents engaged in a common task using electronic technologies.

Ecology: (In reference to e-collaboration and teamwork.) A field that emerges at the intersection of technology, users, and goals; it is what matters (i.e., has meaning). Related terms are biological life process (bioprocess), joint cognitive system (JCS), or Umwelt.

Organization: Meaning that emerges from experience with an assemblage or ensemble of interrelated elements.

Virtual Team: A virtual team exists when e-collaboration takes place through a varying degree of technology across time, space, and (often) organizational boundaries.

Work Domain Analysis: A map of the functional structure of an organization or system.

Using IM to Improve E-Collaboration in Organizations

Xin Luo
Virginia State University, USA

Qinyu Liao
University of Texas at Brownsville and Texas Southmost College, USA

INTRODUCTION

Broadly defined as electronic technologies that enable collaboration among individuals engaged in a common task (Kock, 2005b), electronic collaboration (e-collaboration) is now viewed as a new strategic weapon for organizations to fundamentally improve the traditional business relationships and quality of business processes. Since the emergence of innovative information technologies including e-mail, teleconferencing, videoconferencing, and most recently, instant messaging (IM), the importance of e-collaboration has risen as organizations have made the shift from personal computing to interpersonal or collaborative computing that may more effectively and efficiently leverage their business resources for decision-making. Prior literature suggests that the utilization of e-collaboration technologies can avail organizations of facilitating business-to-business interactions and thereby more quickly and easily solving business problems that are in need of integrative operations and smooth information distribution and sharing amid different inter and intra organizational constituents (Johnson & Whang, 2002; Kock, 1999, 2005b; Kock & Hantula, 2005).

As globalization continues to increase organizational connectivity for expanded organizational life, managers have increasingly recognized the strategic value of IM, which can be leveraged to significantly improve e-collaboration among various organizational constituents beyond restricted geographical boundaries. In the past decade, the advent and wide implementation of Internet-based technologies has made it more flexible and less costly for organizations to establish intra and inter business relationships.

IM, which has become ubiquitous among teenagers and recreational users (Grinter & Palen, 2002), is being widely adopted by businesses and implemented in the workplace (Herbsleb, Atkins, Boyer, Handel, & Finholt, 2002; Isaacs, Walendowski, Whittaker, Schiano, & Kamm, 2002; Vos, Hoft, & Poot, 2004). Due to its nature of interactions and outeractions (Nardi, Whittaker, & Bradner, 2000), IM presents a revolution in enterprise communication. This near-synchronous computer-based interactive communication media not only supports informal communication in the workplace where e-mail, phone, and fax are already widely utilized, but also facilitates some of the processes that make information communication possible (Nardi, Whittaker, & Bradner, 2000). In previous research, Rennecker and Godwin (2003), Avrahami and Hudson (2004), Marshak (2004), and IMlogic (2004) have identified such key features of IM as: presence awareness, immediate closed loop communication, multi-party collaboration, anytime, anywhere access, opportunistic interaction, broadcasting of information or questions, negotiation of availability for interaction, within-medium polychromic communication, "pop-up" recipient notification, silent interactivity, and ephemeral transcripts. These unique characteristics make IM "a powerful new tool for business communication" (DeSouza, 2004) by means of revamping employee productivity and efficiency in workplace. As a result, the use of IM has significantly risen. A report from Pew Internet & American Life (2004) reveals that more than four in ten online Americans instant message and, on a typical day, 29% of these send instant messages and one in five IM users send instant messages at work.

While few researches have related IM to business communications and e-collaboration in general, this article attempts to introduce the background and technological innovativeness of IM in organizations, discusses the utilization of IM in organizations for e-collaboration purpose, and provides the future trend of IM development and research in the area of e-collaboration.

BACKGROUND

As "the newest and most popular incarnation of near-synchronous text chat technologies" (Nardi, Whittaker, & Bradner, 2000), IM service can trace its roots back to 1996 when Mirabilis created ICQ (I Seek You) to meet the mushrooming need of a growing Internet community who "was connected but not interconnected." The first Internet-based chat application rapidly ushered in a new category in the virtual world. The public IM arena has flourished with a variety of client chat tools, including MSN Messenger, Yahoo! Messenger, and AOL Instant Messenger (AIM), and so on. Being keenly concerned with time-sensitivity and speed in communications, getting the right information to the right people at the right time, and to be able to make the right decisions, business managers have discovered the strategic value of IM, which aids in interacting with remote individuals, such as employees, suppliers, supporters, contractors, and customers, in a nearly instantaneous fashion. Simply put, the strategic value of IM is able to accelerate decision processes and significantly decrease traditional communication cost because of its real-time nature. The strategic value of IM is the primary force driving companies to leverage IM in workplace in an effort to establish competitive advantage.

IM is now viewed as an enterprise-wide business requirement and thus being adopted by organizations (Marshak, 2004). As such, with the organizational push (Osterman, 2003), employees are zealously embracing this new technology to communicate with each other and with customers and vendors for better productivity and quicker responsiveness. According to IDC's report, approximately 60-70% of all enterprises regard productivity improvements, collaboration, and best practices as the primary business drivers to adopt IM across their employees. Millions of individuals are using IM for business negotiations, real-time reminders, medical emergencies, or any time e-mail isn't fast enough. IBM vice president John Patrick reports that IM has become a mission-critical operation and that IBM employees send over 1 million instant messages each day internally. In fact, Jupiter Media Metrix estimates there was a 34% increase in "at-work users" during 2001. Additionally, IDC reports that the number of IM users will increase by a factor of ten by 2005 (Richardson, 2002).

Industry sources agree that this worldwide community is increasing exponentially, and expansion into wireless access will also expand the user base. Although predictions of IM penetration in workplace vary, all indicators suggest that IM adoption will continue to grow pervasively toward ubiquity in organizations. For instance, IDC estimates that business users will account for nearly half of the 506 million users expected online by 2006 (IMlogic, 2004); Osterman Research Inc. further argues that virtually all enterprises will employ IM and e-mail will incorporate with IM by 2007 (Osterman, 2003).

In an IM environment, a user can type a message and see its text at the bottom of the screen while viewing the exchange of messages with others across the upper majority of the screen. IM systems allow for multiple IM messages to be written and sent simultaneously. According to Park and Sierra (2005), general IM networks are brokered peer-to-peer (P2P) classes. Typically, two or more centralized servers are implemented in the IM service's infrastructure to handle the IM session including authentication, presence tracking, and message routing functions (see Figure 1). As such, server A first authenticates to IM service's Session Management component, which then sends server A's buddies notification of her online presence. Thus, server A receives notification of her buddies' presence. When server A decides to initiate a conversation with a buddy, the Session Management component redirects the message to the Message Routing component, which routes the message to the appropriate recipient, server B in this example. Even if server A and server B are on the same local area network, all messages between them are still routed through the centralized IM message routing component on the public Internet.

TECHNOLOGICAL INNOVATIVENESS

Advances in computer-mediated communication (CMC) have made such media as e-mail and IM common modes of communication for users in the workplace. The popularity of IM use in workplace stems from its technological innovativeness. To some extent, IM is similar to e-mail in terms of text-based communication and to the telephone in terms of interactivity and intrusiveness. Unlike e-mail, IM adds a richer set of features such as real time communication, presence detection, and graphic emotional icons (Chen, Yen, & Huang, 2004), therefore closely resembling face-to-face spoken conversations in which exchanges are

Figure 1.(Adapted from Park and Sierra, 2005)

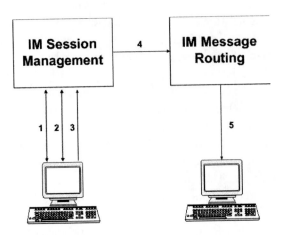

Figure 2. Example of a Yahoo! instant message window (foreground) and a main messenger window (background)

often short, quick, and even incomplete sentences. Recently developed IM products can even allow near-instant video capture of the parties involved in a communication session, providing capabilities similar to video conferencing. Figure 2 and 3 show the MSU and Yahoo messenger dialogues. This section reviews and discusses the most innovative features of IM, whose communicative experience differs from that of other one-to-one workplace communication media.

Presence Awareness

Presence awareness, which provides immediate information about the availability of other users on the dynamic directory, is one of the primary and most

innovative features of IM. This feature allows users to determine whether their coworkers are online or offline. IM users can also set presence messages, such as out for lunch, on the phone, or stepped out, to inform other connected users of the estimated probability of success in making contact with other users. When integrated with other organizational communication technologies, such as mobile telecommunications, IM's presence awareness can be further revamped such that IM system is able to indicate whether the online users logged in IM networks from their desktops or personal digital assistants (PDAs) and even their geographic locations. When working in groups, larger teams seem to have adopted technology to support the coordination of asynchronous work, while smaller teams adopted

Figure 3. Example of a MSN instant video window (background) and a main messenger window (foreground)

technology like IM that primarily supported synchronous collaboration (Bradner, Mark, & Hertel, 2005).

Intrusive Recipient Notification

Similar to telephone, the intrusiveness of IM's recipient notification feature is unique. When an IM message is initialized, a new message window automatically shows on the recipient's computer screen, in most cases, demanding the recipient's attention and response. The intrusion of IM makes users may prefer IM to face-to-face interruptions because they may feel more comfortable deferring electronic messages than people standing in their office and it may be faster for them to reply using IM in short and informal sentences.

Silent Interactivity

Even though IM is somewhat similar to telephone in terms of the negotiation-to-availability procedure, it differs from telephone by means of silent text-based interaction. This silent interactivity allows users to interact each other via the network without talking and being noticed by other people in the same workplace.

Task Dependency

Certain types of e-collaboration technologies are better matched with certain types of tasks. Zigurs and Buckland (1998) classified tasks into five main types: simple tasks, problem tasks, decision tasks, judgment tasks, and fuzzy tasks. An instant messaging system would provide a higher degree of communication support than a Web-based workflow control system and a lower degree of processing structuring. It allows for synchronous communication in a chat-like manner with a much higher level of interactivity than e-mail.

Tasks like communication with car drivers in a car racing would favor instant messaging because of the constraints in radio channel and background noise. In this case, the task constraints are move decisive in the choice of communication technology than medium natureness (Kock, 2006). IM may also be a preferred technology over a face-to-face communication if the response time is more pressing than the richness of the medium.

The building of virtual teams using IM should also be tailored to the type of team and activity to benefit the virtual workspaces (Malhotra, 2005). People adapted their behavior in order to overcome the limitations posed by the new medium, rather than moving to a richer medium such as face-to-face. For example, involved members would prepare longer and more elaborate messages, which partially offset the higher equivocality perceived as inherent in an e-collaboration medium like IM (Kock, 2006).

FUTURE TRENDS

Osterman (2003) indicates that IM will eventually coexist with traditional communication tools like telephone and fax, as IM is poised to become the next necessary item in the corporate communication toolbox. IM with video players allows for the selective incorporation of synchronicity and the ability to convey speech and facial expression to internet-based interactions. Sophisticated text-based chat tools add synchronicity to online business-to-consumer interactions, making easier and more exciting for customers to obtain information about products and services (Kock, 2005a). Text-to-speech voice can significantly increases consumers' cognitive and emotional trust towards the customer service representatives (Qiu & Benbasat, 2005)

Lotus' Sametime uses the Internet to provide instant messaging, electronic voice and video, whiteboard and document sharing, remote computer screen sharing, electronic discussion groups, electronic team rooms and other & audiovisual real-time collaborative tools. Through the internet users can hold conferences and demonstrations, and "chat" with anyone in their offices, or anyone across the country. The optimal goal is to facilitate increased collaboration and more thorough communication.

CONCLUSION

As a hybrid of an asynchronous messaging system such as e-mail and a near-synchronous conversational tool like Web chat, IM is a new yet promising real-time communication method to improve e-collaboration in organizations because it supports impromptu conversations and facilitates remote coordination among different organizational constituents. Users in workplace can utilize IM to instantaneously exchange text messages through a direct linkage. Because of IM's technological innovativeness, such as information immediacy, IM is ideal for spontaneous communications and therefore has become a popular e-collaboration tool in workplace. Supporting near-instantaneous and opportunistic communication makes IM a valuable e-collaboration tool, especially for geographically dispersed constituents. However, the social and ethical issues need to be addressed and proper etiquette followed to protect people's privacy and avoid potential pitfalls. It is believed that IM will eventually become as ubiquitous as e-mail, telephone, and fax in organizations.

REFERENCES

Avrahami, D., & Hudson, S. E. (2004). *Balancing performance and responsiveness using an augmented instant messaging client.* Paper presented at the ACM's CSCW.

Bradner, E., Mark, G., & Hertel, T. D. (2005). Team size and technology fit: Participation, awareness, and rapport in distributed teams. *IEEE Transactions on Professional Communication, 48*(1), 68-77.

Chen, K., Yen, D. C., & Huang, A. H. (2004). Media selection to meet communication contexts: Comparing e-mail and instant messaging in an undergraduate population. *Communication of the AIS, 14,* 387-405.

DeSouza, F. (2004). IM the focus of investigation. *The Information Management Journal, 38*(1), 8.

Folinas, D. (2004). E-volution of a supply chain: Cases and best practices. *Internet Research, 14*(4), 274-283.

Grinter, R. E., & Palen, L. (2002). *Instant messaging in teen life.* Paper presented at the ACM's CSCW.

Gross, E. J., & Gable, S. L. (2001). *Internet use and well-being in adolescence.* Los Angeles, CA: UCLA.

Herbsleb, J. D., Atkins, D. L., Boyer, D. G., Handel, M., & Finholt, T. (2002). *Introducing instant messaging and chat in the workplace.* Paper presented at the CHI.

IMlogic. (2004). *IM has invaded the enterprise: Are you concerned?* Retrieved August 12, 2006 from http://eval.symantec.com/mktginfo/enterprise/white_papers/entwhitepaper_managing_instant_messaging_for_business_advantage_phase_one_assessing_im_usage_03_2006.pdf

Isaacs, E., Walendowski, A., Whittaker, S., Schiano, D. J., & Kamm, C. (2002). *The character, functions, and styles of instant messaging in the workplace.* Paper presented at the ACM's CSCW.

Johnson, M. E., & Whang, S. (2002). E-business and supply chain management: An overview and framework. *Production and Operations Management, 11*(4), 413-422.

Kock, N. (1999). *Process improvement and organizational learning: The role of collaboration technologies.* Hershey, PA: Idea Group Publishing.

Kock, N. (2005a). Media richness or media naturalness? The evolution of our biological communication apparatus and its influence on our behavior toward e-communication tools. *IEEE Transactions on Professional Communication, 48*(2), 117-130.

Kock, N. (2005b). What is e-collaboration? *International Journal of e-Collaboration, 1*(1), 1-7.

Kock, N. (2006). Car racing and instant messaging: Task constraints as determinants of e-collaboration technology usefulness. *International Journal of e-Collaboration, 2*(2), i-v.

Kock, N., & Hantula, D. A. (2005). Do we have e-collaboration genes? *International Journal of e-Collaboration, 1*(2), i-ix.

Malhotra, A. A. (2005). Virtual workspace technologies. *MIT Sloan Management Review, 46*(2), 11-14.

Marshak, D. (2004). *Instant messaging at work*. Retrieved October 14, 2006, from http://interruptions. net/literature/Marshak-PSGP1-22-04CC.pdf

Nardi, B. A., Whittaker, S., & Bradner, E. (2000). *Interaction and outeraction: Instant messaging in action*. Paper presented at the ACM's CSCW.

Osterman. (2003). *Instant messaging: Enterprise market needs and trends*. Retreived August 12, 2006 from www.ostermanresearch.com/or_im03es.pdf

Park, J. S., & Sierra, T. (2005). Security analyses for enterprise instant messaging (EIM) systems. *Information Systems Security, 14*(1), 26.

Pew Internet & American Life Project. (2004). *How Americans use instant messaging*. Retrieved from http://www.pewinternet.org/pdfs/PIP_Instantmessage_Report.pdf

Qiu, L. Y., & Benbasat, I. (2005). Online consumer trust and live help interfaces: The effects of text-to-speech voice and three-dimensional avatars. *International Journal of Human-Computer Interaction, 19*(1), 75-94.

Rennecker, J., & Godwin, L. (2003). *Theorizing the unintended consequences of instant messaging for worker productivity*. Cleveland, OH: Case Western University.

Richardson, R. (2002). Instant messaging goes to work. Retrieved October 14, 2006, from http://www.

callcentermagazine.com/GLOBAL/stg/commweb_ shared/shared/article/showArticle.jhtml?articleId=87 01235&pgno=2

Vos, H. d., Hoft, H. t., & Poot, H. d. (2004). *IM [@ Work] adoption of instant messaging in a knowledge worker organization*. Paper presented at the 37th Hawaii International Conference on System Sciences.

Zigurs, I., & Buckland, B. K. (1998). A theory of task-technology fit and group support systems effectiveness. *MIS Quarterly, 22*(3), 313-334.

KEY TERMS

Computer-Mediated Communication: A cluster of interpersonal communication systems used for conveying written text, generally over the Internet.

E-Collaboration: Electronic technologies that enable collaboration among individuals engaged in a common task.

Globalization: The growing interdependence of countries worldwide through increasing volume and variety of cross-border interactions and transactions in goods and services, free international capital flows, and more rapid and widespread diffusion of technology.

Instant Messaging: An Internet-based application that provides semi-synchronous communication between participants.

Media Richness: A classification of media based on its ability to carry nonverbal cues, provide rapid feedback, convey personality traits, and support the use of natural language.

Synchronized Communication: Communication that enables intensive interaction, immediate clarification, and discussion among all participants at the same time.

Virtual Workspace Technologies: An integrated set of tools that offer a variety of communication support capabilities including a well-organized and searchable common team repository and group discussion forums.

Using the Web for Contract Negotiations

Ben Martz
Northern Kentucky University, USA

Susan Gardner
California State University, USA

INTRODUCTION

Negotiation skills play a critical role for today's knowledge workers. Therefore, the need for university students to develop negotiation and problem-solving skills grows more important each year. Concurrently, the need for students to understand and work with computers continues to grow. This paper presents the exploratory results of using a prototype computer negotiation system developed around a set of real world data. The paper reviews previous research perspectives of negotiation, traditional (face to face) and information system perspectives (electronic). The social information processing theory posits that these characteristics differ between these two groups and that over time the characteristics exhibited by electronic group members should match those exhibited by the traditional group members. The results found differences in the characteristics of satisfaction, trust and tolerance, but did not find a convergence of perceptions between the two groups. The paper concludes by addressing critical success factors for future research in this area.

BACKGROUND

Traditional Perspective: The Face-to-Face Environment

Negotiation is a decision-making process by which "two or more individuals make joint decisions on how to allocate scarce resources" (Thompson, 1998). The competition for scarce resources exists when people perceive each other as wanting the same scarce resources (Thompson & Gonzalez, 1997). The traditional form of labor negotiation is one whereby parties who compete for these scarce resources conduct negotiations face-to-face.

The traditional negotiation process involves both verbal and nonverbal exchanges of information (Thompson, 1998). Kanawattanachai and Yoo (2002) divide the concept of trust in a negotiation into two main attributes: cognition-based trust (CBT) and affect-based trust (ABT). Their results show that the CBT side of trust dominates the ABT in high performing teams. Another critical characteristic is emotion because emotion enables the parties to understand each other (Thompson, 1998). DePaulo (1992) stresses that interpersonal relationships are more successful when people are "sensitive to the emotional, nonverbal cues at the table." According to Ekman (1984), these emotional exchanges include a complex set of facial, vocal and postural cues. Emotion influences the likelihood that negotiators will be able to resolve conflicts with existing resources, pursue cooperative strategies, and consider alternatives made by each other. In so doing, traditional negotiation enables its negotiators to judge accurately one another's interests.

The management of emotions depends upon the negotiator's ability to detect accurately the emotions of others. Ekman (1984) maintains that emotionally skilled negotiators can "detect lies or deceptions in their counterparts." While the least trustworthy sources of lie detection are words and facial expressions, the most reliable sources are body movements. According to Frank (1988), negotiators can judge accurately whether others will cooperate or compete within 30 minutes of interaction. And, when negotiators share similar attitudes and beliefs, when they are physically close to each other during negotiations, they prefer to divide scarce resources equitably.

The way negotiators perceive these interdependent negotiation situations have an important effect on how they will negotiate (Bazerman & Neale, 1992). Lewis and Weigart (1985) point out that trust is a multi dimensional construct with complex interdependencies. Lewicki, Saunders, and Martin (1997) note that although there can be no guarantees that trust leads to collaboration, evidence does suggest that mistrust inhibits collaboration. If negotiators do not trust each other, they act defensively. In so doing, they search for hidden meaning in messages rather than accept information at face value. If parties trust, they most likely communicate their needs accurately. What negotiators say and more importantly, how they say it, affect the conduct of negotiations.

Thus, traditional, face-to-face negotiations provide an environment wherein negotiators can read verbal and nonverbal cues, detect emotions, engage in active listening, and behave in a trusting manner. A negotiation carried out within an electronic environment inherently does not provide access to the nonverbal cues envisioned as vitally important. The purpose of this research is to compare two negotiation environments: the face to face and the electronic.

Information Systems Perspective: The Electronic Environment

The activity of negotiation has a long history, characterized both as an art (Raiffa, 1982) and as a science (Kersten, 1986). Many different models have been developed to attempt to explain and to categorize negotiation. In their work on group support systems, Benbasat and Lim (1993) provide a summary of these models that include game theory, economic, political, and social-psychological. However, whether appearing in a business, economic, or political setting, negotiation strives to have parties come to terms over an issue.

From an information systems perspective, a negotiation system is a subset of the general area of decision support systems. When more than one person, a group, is involved on both sides of the activity, the literature around group decision support systems (GDSSs) becomes relevant. In their work on GDSSs, DeSantis and Gallupe (1985) use a two by two matrix along the two axes of Duration of Decision-Making (limited to ongoing) and Dispersion of Group Members (close proximity to dispersed) created four environments. For this study, the duration remains the same while the dispersion differs.

Jelassi and Foroughi (1989, p. 169) used the DeSantis and Gallupe model to include "behavioral characteristics and cognitive perspectives of negotiators" in their study of negotiation support systems (NSS). They extended the model by proposing, "communication

needs [of a NSS] vary with each bargaining situation." They summarized their exploratory work with a call for a framework that considers both technical and behavioral aspects in NSS designs. This study attempts to answer that call.

Other CMC research is less clear on the success of computer-mediated meetings when compared to face-to-face meetings. Computer-mediated communication meetings have created process losses such as overhead costs (Dennis & Valacich, 1993), stronger identification of non-consensus (Benbasat & Lim 1993), information blocking (Diehl & Stroebe, 1991); information overload (Doyle & Strauss, 1982); and channel conflict (Miranda & Bostrom, 1994). In an effort to reconcile these findings, Walther (1992, p.53) proposes that the CMC channel simply takes "a great deal longer than face-to-face interactions to accomplish more than simple data transfer (does)." Walther believes that the key to understanding these contrary findings is the over-reliance on "one shot, equal time investigations" that offer "no comparison parallel face-to-face interactions." This study does a direct comparison.

Furthermore, Walther (1992) details a social information processing theory to incorporate his ideas and present his propositions. Two fundamental assumptions underpin this theory: (1) the information one receives via nonverbal and verbal-textual channels over the course of interactions with another individual creates the impression of that other individual and (2) in computer-mediated communication, messages take longer to process than do those sent face-to-face. Walther proposes that with all things being equal, the differences between face to face and CMC channels will disappear over time. This current study focuses on this proposal.

Hypothesis: Over time, differences in interpersonal and relational development characteristics between members of computer-supported and face-to-face negotiating groups will disappear.

METHODOLOGY

The authors developed a simple Web-based negotiation system as part of a senior-level, management class in labor negotiation. The prototype system included a proposal-making function where teams could present proposals to their counterparts and also included ways for each team to record relevant data for public viewing (information available to the other team) as well as private viewing (information available only to members of the proposal-making team). The system included a "chat room" where teams conducted virtual bargaining sessions and in which teams more fully explained their proposals and counter proposals. Finally, the prototype allowed for three variations of proposals: "direct negotiations," "conditional," and "non-economic" (discussed in the critical success factors section).

The purpose of the class was to have students experience contract negotiation. Over a five-week period, students "role played" as members of management or labor teams and negotiated a labor agreement. This year's negotiations had special appeal as Lauren Manufacturing, a mid-western rubber extrusion company, agreed to participate in the negotiating experience. The company provided its own managerial, financial, and collective bargaining data for the project and executives of the company served as real-world facilitators to labor and management teams.

Among the many support materials, Lauren Manufacturing provided their most recent financial statements (disguised) and their new, four-year contract with the United Steelworkers Association (USWA). From this information, the instructor selected ten items of bargaining—five involving economic issues and five non-economic issues. Some items were addressed in the current contract; others were not. Student teams had to negotiate over these ten items; bargaining either virtually, using the Web-based system, or traditionally, in face-to-face meetings. Regardless of the venue, each

team had to research extensively the subjects of bargaining in order to provide evidence supporting their proposals and counter proposals.

The instructor divided the class of 31 students into eight teams: four management and four labor. Team members had spent the previous ten weeks together as students in the labor law class and therefore, had some perceptions about their team members and counterparts. She then paired one management team to one labor team for the duration of the project. Two pairs of teams used the computer-support negotiation system (CSN) while the other two pair negotiated face-to-face (FTF). The CSN teams met in a decision support facility (electronic conference room with computers for each member).

Data collection occurred at three separate points in time: Pre-project, during, and post-project. The authors developed five-point, Likert-scale questions and presented them to team members. Based upon the background literature discussed previously, the authors collected trust, commitment and consensus perceptions at all three points in time on five key areas (Table 1). (NOTE: process perceptions at the beginning of a project were not realistic).

DISCUSSION OF RESULTS AND LIMITATIONS

Table 3 presents the raw results categorized by the computer-supported negotiation (CSN) and face-to-face (FTF) columns. In all cases, the face-to-face environment produced scores increasing from T1 (earliest

Table 1. Data collection

Item	Pre-project	During	Post-project
Consensus	Yes	Yes	Yes
Commitment	Yes	Yes	Yes
Trust	Yes	Yes	Yes
Satisfaction	-	Yes	Yes
Pragmatism	Yes	-	Yes
Insight	Yes	-	Yes
Tolerance	Yes	-	Yes
Credibility	Yes	-	Yes
Respect for Process	Yes	-	Yes

recorded measure) to T2 (latest recorded measure). In the CSN environment, seven of the characteristics moved upward while two (consensus and pragmatism) moved downward. In five of the nine characteristics, the T1 raw scores in the FTF environment were lower than the T1 value in CSN environment and the T2 value in the FTF were higher than the T2 value in the CSN environment. This observation implies that the FTF environment contains more volatility than CSN.

A notable exception is the measure of the Commitment characteristic. Here the CSN group dominates the FTF group and the change produced within the CSN group is the largest change (even when standardized for three questions) within the study. In summary, members of CSN groups reported initial higher ratings than the FTF groups on eight of nine measured characteristics

and the CSN ratings moved less than the FTF ratings over time.

The results of a SPSS paired sample T-test (Table 2) show that over the course of the class, three characteristics differ significantly ($p<.05$). However, no characteristics demonstrate significance when segmented by face-to-face and computer-supported as required by the hypothesis. Ultimately, the significance of these results, displayed in Table 3, do not support the basic hypothesis, as there was no statistically difference in the initial values of the characteristics between the two groups.

As this study was exploratory in nature and design, it falls prey to many limitations. First, the implications of the Trust characteristic must be qualified. While the other three constructs: Consensus; Commitment and

Table 2. Overall differences

Consensus	.813
Commitment	.152
Trust	.048
Satisfaction	.007
Pragmatism	.442
Insight	.062
Tolerance	.010
Credibility	.067
Respect for Process	.475

Table 3. Computer supported vs. face to face

Characteristic	Computer-Supported (CSN)		Face-to-Face (FTF)		Significance (1)
	T1	T2	T1	T2	
Consensus	12.63	12.07	11.80	12.67	.233
Commitment	13.19	16.94	11.87	12.93	.654
Trust	12.75	13.27	12.53	13.60	.512
Satisfaction	10.56	11.50	11.47	13.27	.379
Pragmatism	4.56	4.20	4.20	4.27	.267
Insight	4.00	4.27	3.87	4.27	.935
Tolerance	3.88	4.07	3.00	4.07	.106
Credibility	4.25	4.47	4.00	4.40	.870
Process	4.00	4.00	3.87	4.13	.683
(1) Significance is reported based upon differences between T1 and T2 which depend on timing of measures (Table 1)					

Satisfaction had acceptable levels of reliability, .7606, .7190, and .8326 respectively, the low level of reliability, .5229, for the Trust construct is a major limitation. This construct will have to be better analyzed and measured in the future.

As discussed by Walther (1992), time-based studies inherently allow major confounds to impact findings due to lack of imposed control. The five weeks used in the study may still not be sufficient time to observe stabilization of the FTF environment or the accrued communications across the electronic channel.

The FTF and CSN environments could be "functionally" too close as despite best efforts, social channels still existed in the CSN environment (facial expressions, laughter, social presence, etc.). As such, not all information transferred across the negotiation system as needed to fully implement the experiment. In fact, one would expect social information to transfer in other school settings (other classes, email, during group meetings, etc.) and defeat the insulation from a social channel attempted in the study.

In summary, we found no support for the social information processing theory's premise that FTF environments dominate CSN environments on social characteristics. The design attempted to allow for known concerns such as a need for experiments to span greater time periods, to look at technical and behavioral aspects, and to make direct comparisons of face to face and computer supported environments. The results though still seem to suffer from inherent limitations. In the next section we discuss some of these concerns in the context of **critical success factors**; those factors that must be accomplished for the project to be successful.

CRITICAL SUCCESS FACTORS

Critical success factors are those things (actions, assumptions, events, etc.) that must go right for a project to succeed (Rockart, 1986). For our purposes, the critical success factors are identified from the experience of the exploratory research project. This section discusses four factors that would need to be addressed in making future research on this subject more successful.

1. **Computer literacy:** While knowledgeable in labor relations, students participating in the CMC Web-based negotiation project could not be considered experienced computer users. The instructor provided students with one hour of explanation and one day of practice on using the Web-based system. The students' lack of comfort with computers and the negotiation system became immediately evident. As one student noted on the third survey, "I felt that there wasn't enough time to take on a huge assignment like this without more computer training." Therefore one critical success factor for future NSS research is for students to have a high comfort level with computers.

2. **Technology limitations:** Another critical success factor is that the CMC system adequately supports the negotiation process. One student commented, "the computer problems threw a lot of problems into the works. It was a slower process being virtual. We very quickly wished we could talk." Another student agreed, noting that the virtual experience "although more convenient, was not fun and took longer to settle disputes." It is clear that limitations in the technology adversely impacted the experiment; however, the valence of the impact may be attributable more to the next factor.

3. **Multiple negotiation styles:** A third critical success factor is that the Web-based system provides at least three methods resembling that which occurs face-to-face, to propose subjects of bargaining. The first method, denoted as the "direct negotiation method," enabled students to simply propose a specific increase or decrease in a subject of bargaining by merely clicking "increase/decrease," followed by clicking the appropriate subject (e.g., wages, profit-sharing, etc.). Respondents to the proposal either clicked accept or reject to the direct proposal. The second method, denoted as "conditional bargaining," enabled students to propose an increase or decrease on one subject of bargaining, conditioned on an increase or decrease in a different subject of bargaining. Respondents could either accept or reject the conditioned proposal. The third method of bargaining, for so-called "non-economic subjects of bargaining," enabled students to propose items of bargaining for non-economic subjects without initially costing out the consequences of those proposals. Respondents could make a counterproposal. Once an item is agreed to, the paired teams determine its cost and add it to the projected cost of the new contract.

4. **Case scenario flexibility:** A final success factor is that the Web-based system provides the instructor with the ability to monitor the negotiations and to change the case if desired. The instructor is also able to intervene during the negotiations by notifying the teams of issues that may affect their negotiations. For example, the instructor could advise any or all teams that the company's primary customer has cut its orders in half, thereby reducing the amount of resources available for negotiation. Students are then responsible for responding and reacting mid-course to these changes in the negotiation environment.

CONCLUSION

This paper reports on an exploratory project undertaken to develop a prototype negotiation system and study its impact using the foundation of the social information processing theory. As a comparison of computer supported negotiation (CSN) with typical face-to-face (FTF) negotiation, the theory proposes that over time, the ratings of CSN and FTF groups on social relationship development characteristics will converge. While there were significant changes occurring within the group over time with respect to the characteristics, no differences in the characteristics could be attributed to the use of CSN. Ultimately, the experience has been condensed into a set of concerns and critical success factors for future develop and research with computer support negotiation.

REFERENCES

Bazerman, M. H., & Neale, M. A. (1992). *Negotiating rationally*. New York: Free Press.

Benbasat, I., & Lim, L. (1993). The effects of group, task, context and technology variables on the usefulness of group support systems. *Small Group Research, 24*(4), 430-462.

Dennis, A. R., & Valacich, J. (1993). Computer brainstorms. *Journal of Applied Psychology, 78*(4), 531-537.

DePaulo, B. (1992). Nonverbal behavior and self presentation. *Psychological Bulletin, 111*(2), 203-243.

DeSantis, G., & Gallupe, B. (1985). Group decision support systems. *Data Base, 16*(1), 3-10.

Diehl, M., & Stroebe, W. (1991). Productivity loss in idea-generating groups: Tracking down the blocking effect. *Journal of Personality and Social Psychology, 61*(3), 392-403.

Doyle, M., & Strauss, D. (1982). *How to make meetings work*. Jove edition, 1982.

Ekman, P. (1984). The nature and function of the expression of emotion. In K. Scherer & P. Ekman (Eds.), *Approaches to emotion* (pp. 319-344). NJ: Erlbaum.

Frank, R. H. (1988). *Passions within reason: The strategic role of emotions*. New York: Norton.

Jelassi, M.T., & Foroughi, A. (1989). Negotiation support systems: An overview of design issues and existing software. *Decision Support Systems, 5*, 167-181.

Kanawattanachai, P., & Yoo, Y. (2002). *Dynamic nature of trust in virtual teams*. http://sprouts.case.edu/2002/020204.pdf

Kersten, G. D., & Szapiro, T. (1986). Generalized approach to modeling negotiations. *European Journal of Operational Research, 26*, 124-142.

Lewicki, R. J., Saunders, D. M., & Martin, J. W. (1997). *Essentials of negotiation*. Irwin.

Lewis, J. D., & Weigert, A. (1985). Trust as a social reality. *Social Forces, 63*(4), 967-985.

Lim, L., & Benbasat, I. (1993). Negotiation support systems. *Journal of Management Information Systems, 9*(3), 27-44.

Miranda, S. M., & Bostrom, R. P. (1994). The impact of group support systems on group conflict and conflict management. *Journal of Management Information Systems, 10*(3), 63-95.

Raiffa, H. (1982). *The art and science of negotiation*. MA: Harvard University Press.

Rockart, J. F. (1986). A primer on critical success factors. In J. F. Rockart & C. V. Bullen (Eds.), *The rise of managerial computing*. Homewood, IL: Dow Jones-Irwin.

Thompson, L. (1998). *The mind and heart of the negotiator*. NJ: Prentice Hall.

Thompson, L., & Gonzalez, R. (1997). Environmental disputes: Competition for scarce resources and clashing of values. In M. Bazerman, D. Messick, A. Tenbrunsel, & K. Wade-Benzoni (Eds.), *Environment, ethics and behavior* (pp. 75-104). San Francisco: New Lexington Press.

Walther, J. B. (1992). Interpersonal effects in computer-mediated interaction: A relational perspective. *Communication Research, 19*(1), 52-90.

KEY TERMS

Critical Success Factors: Those factors or conditions that must be satisfied or accomplished for a project to be successful.

Group Decision Support Systems (GDSSs): Information systems designed to support groups in the processes involved in making decisions.

Negotiation: Formal process by which two individuals or groups reach consensus or agreement on a topic.

Negotiation Support System: Information system designed to support individuals or groups in the process of negotiation.

Social Information Processing: The theory that when people work together, both verbal and non-verbal channels transfer information to others.

Traditional Negotiation: Formal manual process of negotiation that includes both verbal and non-verbal exchanges of information. Traditional negotiation is positioned "opposite" NSS in that NSSs do not provide for information to transfer over traditional social channels.

Videoconferencing as an E-Collaboration Tool

Michael Chilton
Kansas State University, USA

Roger McHaney
Kansas State University, USA

INTRODUCTION

Videoconferencing (VC) is primarily a synchronous, long distance, e-collaboration tool. Although it offers interpersonal features with some degree of media richness and social presence, it is not a perfect substitute for face-to-face communication. VC can add value in business situations where telephone, text chat, or audio conferencing do not provide adequate secondary communication channels such as nonverbal cues (tone of voice, inflection) and interactions (body language). VC also adds value where it is impossible or undesirable to conduct a personal meeting. Currently there exists an opportunity for organizations and individuals to derive enormous benefit from this medium when used appropriately with necessary tactics and skills, especially when multiple parties are involved in e-collaboration.

VC is a real-time e-collaboration technology phenomenon which enables individuals at different locations to communicate with each other via video monitors and speakers. The participants sit in front of cameras and interact, viewing and hearing each other as if they are in the same room (Barlow, Peter, & Barlow, 2002). VC can be used independently or as an extension to face-to-face meetings with some participants talking in person and others via technology.

To conduct a VC, at least two endpoints and a transmission system are necessary. Endpoints are locations from which video is broadcast or received. Typical endpoints can be a computer-based desktop unit with appropriate software, Webcam, speakers, and perhaps an overhead projector; a conference room unit which is a device that integrates audio/video reception/broadcast for a meeting room setting; or a class room unit that projects the conference to a screen. All endpoints are reliant upon a coder-decoder (codec) to transform video and audio streams into packets that can be transmitted over a network and then returned to viewable and displayable form.

Endpoints are connected to standard networks such those used for telephones, cable television, or the Internet. In order to communicate successfully it is important that the linking network(s) have adequate bandwidth. Typically, dial-up connections will not work. DSL or Cable connections would be better. INTERNET2, the new higher bandwidth version of the Internet, is expected to be much better. Additional technologies that make more effective use of bandwidth can also be used, such as signal compression and conversion.

BACKGROUND

In spite of its current popularity, VC is not new. At best, it is a new version of an old technology. VC traces its origins to 1956 when the first public videoconference was held between AT&T headquarters and their Bell Laboratory in New York City. A few years later at the 1964 World's Fair, Bell Laboratories introduced the Picturephone to the world and offered its VC-like service to businesses in several metropolitan areas (Webster, 1998). As a result, journalists, media philosophers, and marketing teams predicted visual phones in every home in America by the 1970s. In spite of its valuable ability to interconnect with other video phones the technology did not catch on as expected. Instead, high costs, application difficulties, and limited deployment prevented the Picturephone from being a commercial success. The idea was ahead if its time, but the technology was not ready for true e-collaboration (Hansell, 1989).

In the late 1970s this began to change. Improvements in technology once again renewed interest in VC. This time, a codec was developed that enabled a video signal to be digitized and compressed by a factor of 15:1. This allowed transmissions to occur at a much faster rate. In 1982, the availability of newer technologies allowed researchers to develop a codec capable of achieving a compression ratio of 60:1. A few years later, several codec producers improved on this and deployed a compression scheme which allowed video to pass over a single telephone line. While lower in quality, these advances lowered the costs of VC and enabled more usage of the technology (Hansell, 1989). In the past two decades, a steady progression of technical advances has improved VC. Many of the changes involve moving the technology to Internet-based systems.

Phoenix Research projected that within five years, the global percentage of VC users would rise from seven percent to 26% (Bland, 2004). A variety of reasons contribute to this increase. Recent terror attacks and health scares have decreased the desire of people to travel. Additionally, increases in fuel costs have made VC a more cost-effective option.

USING VC APPROPRIATELY

In order to ensure the appropriate hardware and software system is selected for VC use, the user must determine if internal and/or external conferencing will be conducted. Using VC for internal purposes generally means lower image quality is more acceptable whereas external usage generally demands a high quality connection, since the transmission may represent the company's image and potential to the outside world (Bland, 2004). The type of network used plays a major role in video performance quality since it carries data between interacting video systems. Critical factors regarding video quality are picture size and the ability to access the connection consistently (What is Videoconferencing, 2005). Optimal VC quality can only be achieved with a broadband connection and with top quality equipment at both users' ends (Barlow et al., 2002).

Cost considerations must include both startup and usage. For instance, using an ISDN connection (ISDN, 2005) may have a low initial cost, but the ongoing expenses would be much more than using an Internet Protocol (IP) connection (Bland, 2004).

VC-BASED E-COLLABORATION IN BUSINESS

In general, businesses use VC for e-collaboration without having to worry about geographic constraints. Some common instances are international team meetings, board meetings, product demonstrations, and new employee orientation in field offices. Other uses include customer training, shareholder meetings, and multi-site

conferences where participants around the world can interact simultaneously (Barlow et al., 2002). Some organizations even use VC to conduct their potential employee interviews at a distance (Straus, Miles, & Levesque, 2001).

Many businesses report cost savings due to VC usage. Common examples include situations where VC is used to highlight crucial information or make important announcements to multiple recipients at the same time without location constraints (Barlow et al., 2002). Savvas (2004) reported a specific instance where a Pepsi Cola company speaker had a very positive and successful experience conducting simultaneous cross-site training for employees, which under normal circumstances would have taken four weeks.

BENEFITS AND ADVANTAGES FROM VC-BASED E-COLLABORATION

Immense improvements in VC technology, paired with the tendency to communicate with greater frequency via the Internet, make the usage of VC affordable for a wide variety of users around the world (Barlow et al., 2002). Within this context of change, two major benefits become apparent: a first level effect of savings and productivity/efficiency improvements; and a second level effect of improved social interactions (Straus et al., 2001).

In regards to savings, VC is still the primary e-collaboration technology that allows visible face to face communication without requiring the participants to be at the same place (Bland, 2004). Many examples of cost savings due to VC are reported in the literature. For example, many law firms benefit from the existence of this medium by accomplishing successful depositions and meetings with clients or witnesses, as well as conducting regular meetings with remote colleagues through VC (Lore, 2004).

Researchers have reported that a typical break even point for VC usage is reached just 12 months after implementation (Boettger, 2004). PepsiCo covered its VC implementation costs in six months by reducing travel, lodging and related activities (Savvas, 2004). Another financial benefit, especially for smaller businesses, is that companies can rent the VC equipment for one session instead of buying it (Bland, 2004).

In regard to second level advantages, VC enables the user to increase his or her effectiveness in front of an audience that may have previously been only available via telephone (Barlow et al., 2002). Building a social relationship with the audience is dependent on the way individuals express their feelings and reactions through smiling, eye contact, and other nonverbal cues (Straus et al., 2001). Some researchers go further and say that factors like eye contact and physical attractiveness are associated with being extraverted, which in turn might be a signal for predicting managerial job performance (Straus et al., 2001). Since VC's can be recorded they can be viewed repeatedly, which increases the incentive to be fully prepared. This may cause a VC to be conducted more professionally and formally than a regular meeting (Barlow et al., 2002).

In the computer mediated communications literature, VC is considered to be a richer media than is either text-based or audio conferencing (McHaney, Hagmann, Hightower, & Sayeed 2000). Researchers have published findings that indicate it takes much more effort to reach the same level of understanding with lean media than with richer ones. Communication via lean media also has been found to be less efficient (McGrath, 1994). Media richness theory (Daft & Lengel, 1984) describes this phenomenon and suggests that various levels of media provide different capacities for reducing ambiguity and improving understanding in communication. Daft and Lengel (1984) illustrate a media richness hierarchy that uses degrees of: (a) instant feedback; (b) natural speaking language; (c) multiple nonverbal cues; and (d) personal focus to facilitate virtual communication. In related work, Short et al. (1976) propose the idea of social presence theory which states a medium's social effects are principally caused by the degree of social presence it affords its users. This means as a user's awareness of an interacting partner's presence increases, a better person perception will result and communication will be facilitated.

Video conferencing can be used in conjunction with other electronic functions, such as whitespace sharing (analogous to using a chalkboard in a class room), voting and/or polling on specific issues or questions, computer screen sharing/control, Web site sharing and application sharing. Software that provides these additional features (e.g., Lotus™ Sametime® from IBM™) allows a meeting moderator to establish a meeting and include audio and video feeds. Participants are then able to log in and hold a meeting using videoconferencing and utilize these tools. Participants can (1) see and hear other participants, (2) electronically raise their hands to indicate approval or to be recognized by the moderator, (3) share whiteboard, (4) share Web sites, (5) share application programs, and/or (6) share their own computer screens and allow other participants to take control. These functions are especially useful for collaboration and technical support purposes.

RISKS AND CHALLENGES/LIMITATIONS OF VC

In spite of all the advantages, many believe VC has not yet achieved its potential. Reasons for this might include the fact that many corporations still use unsuitable connections to conduct Videoconferencing meetings. Integrated services digital network (ISDN) is a technology that was developed to allow simultaneous transmission of audio, video, and data over the plain old telephone system (POTS). It has not been widely adopted, however, because of the lack of standards and because it is a slower technology than those developed more recently. Because it uses circuit-switching technology, only one connection can be made at a time, thus limiting its effectiveness for use in videoconferencing applications which require several connections at once.

To ensure the best quality of data transfer appropriate bandwidth must be present. Another problem relates to firewalls and network address translation (NAT). Both can create problems because they create barriers preventing video traffic. Complex the dialing schemes should also be considered. No single standard is used by all systems and organizations. This makes connections difficult or in some cases impossible (Meserve, 2004). Another common complaint among VC users is that the technology is not as reliable as the telephone, which nearly always works; it may be disconnected easily or create other technical problems (Barlow et al., 2002). Company networks also play a very critical role by having the capability to deal with video signals (Connolly, 1997).

The challenges mentioned so far have been based on the technological aspects of VC. Many other types of challenges also exist. One of these involves decision making. In organizations using VC, decision making processes become faster. With the ability to address problems immediately comes the greater risk of making the wrong decisions (Barlow et al. 2002). One way to combat this is to plan and organize more carefully in advance. A clear topic must be provided, and major areas of focus within the discussion or mediation should be detailed and finished by a follow-up (Domeisen, 2001). Without these precautions, the videoconference runs a higher risk of failure.

Another challenge of conducting a successful VC is the development of appropriate habits and professionalism. Time and training is required (Barlow et al., 2002). Even in educational environments, research showed participation in a VC environment might lead to more organized and structured interactions. Technical training as well as behavior training are both necessary to reduce potential problems that go along with VC (Bland, 2004).

Given today's level of technology, it is sometime possible to forget that some meeting participants aren't in the room (Lore, 2004). Although VC technology has advanced, it cannot yet substitute for the intangible and nonverbal cues that pass between interacting individuals. Not being able to see some nuance, not being able to shake hands or sense some voice inflections makes it difficult to build trust. Thus, VC still has room to improve to fulfill users' desire to have richer levels of communication (Barlow et al., 2002).

Finally the privacy issues of employees shouldn't be underestimated. It is important to protect user privacy from unauthorized access of the conference. Therefore it must be possible to configure the system to ensure privacy. Considerations such as physical barriers on the camera, alerts to users about who has taken snapshots and other safeguards must be present (Webster, 1998).

Additional limitations of video conferencing include geographic, cultural and process issues (Nunamaker, Briggs, Mittleman, Vogel, & Balthazard, 1997)

VC MARKET LEADERS

The VC industry has been dominated by four primary suppliers: Polycom, Tandberg, Aethra and V-Con. California-based Polycom, founded in the 1990s, is generally considered the leader in developing, manufacturing and marketing VC management software. Polycom is also the leader in market share for voice and video conferencing. Following a model developed by a research firm called Yankee Group, Polycom's goal is have all major decisions made at Headquarters while providing decisions, using VC, to clients, sales teams, and other relevant stakeholders (Rethink It, 2004).

Norwegian vendor Tandberg has two headquarters, one in New York and the other in Norway, with operations in more than 90 countries (Press Release, 2005). Tandberg is the second largest global provider of VC solutions with revenues of $76.5 million in year 2004. Tandberg is involved in developing IP solutions to achieve a broader VC usage in personal and business areas. Through engaging in strategic partnerships it has secured a position as a user of new technologies (Press Release, 2005).

The Italian company Aethra was founded in 1972. It has successfully operated in three major areas: Video-surveillance, Telefinance and Telemedicine (Video Conferencing, 2006). Aethra has a 15% market share in the EMEA (Europe/Middle East/Asia) VC market, and roughly 7% worldwide (Videoconferencing insight newsletter, 2005).

V-Con, former market share leader of personal videoconferencing in 2001, has headquarters in Israel with subsidiaries all over the world. V-Con provides all-around VC technologies with support for interactive communication between remote users of ISDN as well as IP networks (V-Con continues to be recognized, 2001).

FUTURE

A variety of trends show the importance of VC in e-collaboration. These trends also give financial and technological reasons supporting the concept of implementing a VC system in a corporate environment. One major trend is that VC technology and quality continue to improve rapidly while costs decrease. Dual-monitor systems that connect to ten monitor sites and offer touch button capabilities to enable immediate connections to other systems, have increased spontaneity. Desktop systems connected through USB (plug and play) save users time and don't require special expertise. The ability of broadband Internet-connected individual users to purchase a low cost Webcam and microphone and use free software provided by Yahoo! or Microsoft has brought videoconferencing to the masses. Quality has improved in high end systems by offering 3-D images to enhance realism and other aspects of the VC experience.

The second notable trend is that time and costs play a major role in the existence of globally acting companies. Overall, VC can offset the costs of travel by a significant percentage.

The third important trend is that VC can increase personal contact between individuals; instead of just talking on the phone or sending a fax or e-mail, a richer interaction is possible. Involved parties will know with whom they communicate and a more trustworthy relationship can created. By increasing the usage of VC, managers will learn optimize the technology and in many cases it will become a commonplace way to consummate business deals.

The final notable trend is important from an ethical point of view. In today's society, a greater number of companies are concerned about corporate responsibility, especially in regards to the environment. VC allows organizations and their employees to have personal outreach and relationships without contributing to pollution travel by car and/or planes (Barlow et al., 2002).

CONCLUSION

VC is a real time, long distance e-collaboration tool. It enables individuals at different locations to communicate with each other via monitors and microphones. The network connection plays a major role in the quality of video performance. Digital video compression allows VC to become less expensive through lowered transmission costs (Hansell, 1989) and makes it accessible to a broader range of users. VC is used by various sectors, but has achieved most of its success and growth in the mainstream corporate world.

The business world is under pressure. This pressure is not just a financial pressure, but the need for prestige. Growing battles between competitors and their strategies to achieve competitive advantage will lead companies to integrate VC in their organizations. Use of VC and other e-collaboration technologies will increase the effectiveness of the Internet leading the corporate world into sustained growth. Hopefully, at the same time the corporate world will not underestimate the importance and value of interpersonal contact.

ACKNOWLEDGMENT

The authors wish to acknowledge Didem Cuhadaroglu, graduate student at Kansas State University, for her contributions to this article.

Portions of this research were supported by a Kan-Ed Technology and Equipment Grant, 2006.

REFERENCES

Barlow, J., Peter, P., & Barlow, L. (2002). *Smart videoconferencing: New habits for virtual meetings.* San Francisco, CA: Berrett-Koehler Publishers.

Bland, V. (2004). Meeting the smart way: No longer over hyped and under delivered, videoconferencing is being taken seriously by businesses of all sizes. *NZ Business, June,* 1-8.

Boettger, C. (2004). The collaboration revolution. *Successful Meetings, 53*(11), 1-4.

Connolly, A. (1997). Now you see me, now you don't. *Computer Weekly,* Nov 20, 1-5.

Daft, R. L., & Lengel, R. H. (1984). Information richness: A new approach to managerial behavior and organizational design. In L. L. Cummings & B. M. Staw (Eds.), *Research in organizational behavior, 6* (pp. 191-233). Homewood, IL: JAI Press.

Desmond, M. (1995). Videoconferencing coast to coast and face to face. *PC World, 13*(3), 177-186.

Domeisen, N. (2001). Virtual conferences: A new way to network: The Internet offers new opportunities to join in international discussions without the disadvantages of costly and time-consuming travel. *OECD Observer,* Jan, 1-3.

Hansell, K. J. (1989). *The teleconferencing managers guide.* Knowledge Industry, White Plains, NY. "High-quality IP,"

ISDN. (2005). Retrieved January 10, 2005 from PC-Webopedia: http://www.pcwebopedia.com/TERM/I/ISDN.html

Lore, M. (2004). Mediation by videoconferencing—because nothing is lost. *Minnesota Lawyer, November 15,* 1-6.

McGrath, J., & Hollingshead, A. B. (1994). *Groups interaction with technology: Ideas, evidence, issues, and an agenda.* London: Sage.

McHaney, R., Hagmann, C., Hightower, R., & Sayeed, L. (2000). Computer-mediated communications systems. In A. Kent (Ed.), *Encyclopedia of library and information science, vol. 67* (pp. 64-80). New York: Marcel Dekker.

Meserve, J. (2004). Video collaboration still slow to take off. *Network World, 21*(13), 1-5.

Nunamaker, J. F., Jr., Briggs, R. O., Mittleman, D. D., Vogel, D. R., & Balthazard, P. A. (1997). Lessons from a dozen years of group support systems research: A discussion of lab and field findings. *Journal of Management Information Systems, 13*(3), 163-207.

Peterson, C. (2004). Making interactivity count: best practices in video conferencing. *Journal of Interactive Learning Research, 15*(1), 63-75.

Press Release. (2005). Retrieved January 10, 2005, from Tandberg Data: http://www.tandberg.net/press-room/viewPress Release.do'?id=79

Rethink IT. (2004, April) *Video conferencing finds new popularity after a few years in the doldrums.* Retrieved from http://www.rethinkresearch.biz/.

Savvas, A. (2004). Videoconferencing saves PepsiCo executives travel time and costs. *Computer Weekly, December,* 1-2.

Short, J.A., Williams, E., & Christie, B. (1976) *The social psychology of telecommunications.* New York: John Wiley & Sons, Inc.

Straus, S. G., Miles, J. A., & Levesque, L. L. (2001). The effects of videoconference, telephone, and face-to-face media on interviewer and applicant judgments in employment interviews. *Journal of Management, 27*(3), 1-29.

VCON Continues To Be Recognized As Market Share Leader In Personal Videoconferencing. (2001). Retrieved Jan 10, 2005, From VCON: http://www.vcon.com/press room/english/2001/2001.08.27.shtml

Video conferencing. (2006). Press release. Retrieved August 22, 2005, from Aethra: http://www.aethra.com/worldwide/prodselect.asp?M=150

Videoconferencing insight newsletter. (2005). Retrieved January, 10, 2005, From IMP Publications, UK: http://www. videoconferencing.co.uk/

Webster, J. (1998). Desktop videoconferencing: Experiences of complete users, wary users, and no-users. *MIS Quarterly, 22,* 1-20.

What is videoconferencing and what are the benefits. (2005). From Picturephone Direct: http://picturephone.com/products/leam what is vc.htm

KEY TERMS

CODEC: A software coder-decoder that transforms video and audio streams into packets transmitted over a network and then returned to viewable and displayable form at the receiving endpoint.

DSL: Digital Subscriber Line (formerly known as T1), it consists of 24 channels of 64 Kbps bandwidth (+ 8 Kbps control). It is a leased line with 24/7 access.

Endpoints: These are locations from which video-conferences are broadcast or received. Typical endpoints may be a computer-based desktop unit with appropriate software, Webcam, speakers, and overhead projector; a conference room unit which is a device that integrates audio/video reception/broadcast for a meeting room setting; or a class room unit that projects the conference to a screen.

Integrated Digital Services Network (ISDN): A telephone communications technique that allows the simultaneous transmission of audio, video and data over the same telephone network utilizing circuit switching.

INTERNET2: A network proposed and being constructed by consortium of universities, industries and government groups which will implement high speed Internet connections using leading edge technology and innovative applications, with a goal of establishing partnerships to enhance research and the broader Internet community.

Network Address Translation (NAT): A method of converting a private IP address into a public one using port numbers. This is done so that a single public IP address can be utilized at an organization's default gateway to server multiple private IP addresses within the organization's internet. It is often used by a hardware firewall.

Videoconferencing: A real-time e-collaboration technology phenomenon which enables individuals at different locations to communicate with each other via video monitors and speakers.

Virtual Teams Adapt to Simple E-Collaboration Technologies

Dorrie DeLuca
University of Delaware, USA

Susan Gasson
Drexel University, USA

Ned Kock
Texas A&M International University, USA

INTRODUCTION

Increased globalization of enterprises combined with widespread adoption of simple, low cost, asynchronous e-collaboration technologies (e.g., bulletin board, e-mail) provides incentive to attempt increasingly complex problem solving with virtual teams. If complex business process improvement activities could be conducted using asynchronous e-collaboration, the potential to reduce competition for resources by reducing travel time and increasing the communication window to 24/7 improves the ability to simultaneously address the multiple priorities of daily business and business process improvement.

The knowledge that virtual process improvement teams have been successful (DeLuca, Gasson, & Kock, 2006; Kock & DeLuca, 2006; DeLuca & Valacich, 2006; Kock, 2006) and lessons learned from those teams may be what is needed to provide confidence to organizations that virtual process improvement efforts would come to fruition. To manage such initiatives effectively, it is important to understand *how* these virtual teams overcame the difficulties of e-collaboration. Existing theories of information processing in organizations do not scale well to the complex forms of knowledge integration required at the boundary between the diverse teams found in virtual organizations. Thus, we based our investigation on a new theory of communication behavior, compensatory adaptation theory (CAT) (Kock, 2005b) and the relationships suggested by it, explained in the next section. We also operationalize a key construct, compensatory adaptations and present the adaptations made by participants in the study (DeLuca et al., 2006).

BACKGROUND

Empirical research results are inconclusive about the effect of e-collaboration and technologies upon communication (Kock, 2005a, Majchrzak, Rice, Malhotra, King, & Ba, 2000; Miranda & Saunders, 2003; Rice, Kraut, Cool, & Fish, 1994; Riva & Galimberti, 1998). This may be because social norms and availability of media may influence media choice significantly more than media or task characteristics, as previously thought.

Asynchronous, electronic, written communication media are generally familiar, sponsored, and conveniently available, yet not commonly used for complex tasks such as business process improvement because of perceived limitations that must be overcome or compensated for in some way in order to effectively communicate (Daft & Lengel, 1986; Kock, 2005b; Majchrzak et al., 2000; Markus, Majchrzak, & Gasser, 2002).

Earlier theories are based on "richness" of a media (Daft & Lengel, 1986) or "social presence" (Short, Williams, & Christie, 1976) and do not provide for making adaptations to use a media. Research on virtual process improvement shows that team members adapt their communication behaviors to compensate for the deficiencies in the "richness" of the communication channel with which they have chosen to work (Kock & DeLuca, 2006; DeLuca et al., 2006; Kock, 2005b). Compensatory adaptation theory (CAT) (Kock, 2005b), posits the processes shown in Figure 1.

CAT is derived from two principles—the media naturalness principle and the compensatory adaptation principle. The naturalness of a media is proportional to its similarity to face-to-face communications (media naturalness principle). When users of the media perceive a lack of media naturalness, they make adaptations to compensate for the perceived obstacles to communication (compensatory adaptation principle). Media that lack many of the features of face-to-face communications (e.g., immediate feedback, presence of visual, auditory, and social cues) are said to be "lean," like e-mail and bulletin boards. E-collaboration using lean media is referred to as "lean e-collaboration." Based on CAT, the central research question (CQ) is:

CQ: Can process improvement teams using lean e-collaboration be successful and, if so, how do the team members adapt their communication behavior to compensate for perceived shortcomings of the media?

In a study by Graetz, Boyle, Kimble, Thompson, and Garloch (1998), mental demand, temporal demand, effort, and frustration were all more than 50% higher for those using e-collaboration than those using face-to-face (FTF) communications. This supports the assessment of "low" naturalness for e-collaboration and consistent with earlier studies (Daft, Lengel, & Travino, 1987; Rice, 1992; Rice & Shook, 1990). Kock (2004) offers that the human species has been biologically designed for FTF communication. E-collaboration is less "natural" because co-location, synchronicity, body language, facial expressions, and hearing and speech are lacking (Kock, 2005b).

The perceived obstacles posed by e-collaboration must be overcome or compensated for in some way in order to communicate effectively. Studies show (Kock & DeLuca, 2006; DeLuca & Valacich (2006); DeLuca et al., (2006); Kock (2005b); Ocker, Fjermestad, Hiltz, & Johnson, 1998) that virtual team members take additional care in composing messages transmitted via

e-collaboration and this compensation may lead to better quality individual input and successful team completion of the process improvement effort by implementing the improvements. Teams that included use of lean electronic media produced outcomes of the same or higher quality than FTF teams. The concept that the adaptations made affect quality and success is unique to CAT (Kock, 2005b) and provides an explanation for virtual team successes with complex tasks that is not offered by earlier theories.

RESEARCH SITE AND METHOD

This article is derived from a study (DeLuca et al., 2006) that was focused on the communication behavior of four virtual teams from an educational services organization. Virtual teams were studied in their natural environment during one cycle of a larger on-going, traditional (canonical) action research study (see Davison, Martinsons, & Kock, 2004 for principles of canonical action research). All teams chose processes that involved improving the quality of communications among schools, parents, special needs children, and service providers. The teams were cross-functional with from 9 to 11 members. The researcher provided: access to an Internet-based bulletin board and group e-mail to alert virtual team members to check the bulletin board and contribute to it; training on the technology and a typical structured problem-solving process (Kock, 2006); and information on success factors from previous virtual teams and their leadership. The team "outcome" was a re-designed business process.

An analysis of the literature revealed that most discussions related to media naturalness build on some variation of the following three dimensions of media:

Figure 1. Compensatory adaptation process (Adapted from Kock, 2005b)

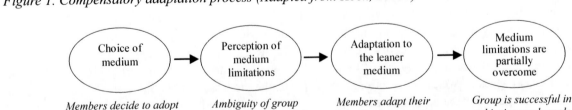

interactivity, channel capacity, and adaptiveness (Kahai & Cooper, 2003; Kock, 2004; Short et al., 1976). The three dimensions provide a typology for operationalization of "compensatory adaptations:"

- **Interactivity** is the potential to obtain immediate feedback from other communicants. Lean e-collaboration media are therefore predicted to be perceived as providing *low* levels of interactivity.
- **Channel capacity** is the ability to transmit a high variety of language and social cues (both verbal and non-verbal). A lack of body language, facial expressions, volume, and tone leads to ineffective communication. Lean media are therefore predicted to be perceived as supporting a *low variety* of language and social cues.
- **Adaptiveness** is the potential to acknowledge, adapt and personalize messages of a particular communicant. The predictions of adaptiveness and personalizability would be *low*, however the written aspect of some lean media may enhance acknowledgement of an individual's contribution by providing a historical record of contributions.

At the end of the business process improvement process, each virtual team member was interviewed to explore perceptions about the relationships among obstacles to communication posed by the media, adaptations to communication behavior, and success of outcomes (DeLuca et al., 2006). Data were collected via open-ended

and neutral questions (Yin, 1994) and a compensatory adaptations scale of Likert statements (DeLuca et al., 2006; Moore & Benbasat, 1991; Nambisan, Agarwal, & Tanniru, 1999). Data were coded by two "raters" into consistent, semantically equivalent phrases and also analyzed adaptations using a chi square summary statistic (Rosenthal & Rosnow, 1991). The responses were integrated using triangulation, "the combination of methodologies in the study of the same phenomenon" (Denzin, 1978, p. 291). The intent of using mixed methods and triangulation is to ensure the variance "measured" is not due to the method (Campbell & Fisk, 1959).

RESULTS

Ninety-five percent of participants reported adaptations made (chi square p < .001 with a large effect size). A positive result is also indicated by the Likert response scale average score of 4.6 of 7, which indicates agreement that adaptations were made to compensate for not being FTF.

In response to the open-ended interview question, participants revealed their perceptions about the relationships among obstacles, the adaptive behaviors engaged in to compensate for the perceived obstacles, and the effect the behaviors had on their communications and thus on the success of the business process redesign.

The perceived obstacles to effective communication using e-collaboration are summarized in Table 1 using

Table 1. Adaptations made to communications overcome perceived obstacles

Perceived Obstacle to Communication Effectiveness	Adaptation to Communication Behavior (written e-collaboration) to Compensate for Perceived Obstacle to Communication
Interactivity – Lack of immediate feedback (and written accountability)	More clear
Interactivity – Lack of immediate feedback	More complete
Channel Capacity – Lack of non-verbal cues (and less personal/social)	More focused
Channel Capacity – Lack of non-verbal cues	More neutral
Channel Capacity – Lack of non-verbal cues	More considerate
Channel Capacity – Lack of language variety	More concrete language
Adaptiveness – Written accountability	More precise
Adaptiveness – Written accountability (size of input box)	More concise
Adaptiveness – Written accountability	More persuasive

the operationalization of adaptations (compensations) into three categories of adaptations. The adaptation(s) made to communication messages to compensate for the perceived obstacle(s) are shown in the right hand column.

E-collaboration provides less immediate feedback (*lack of interactivity*). Virtual team members perceived that asynchronous, electronic, written communication media reduce interactivity when compared to FTF communication. Virtual team members were unable to argue their point as easily as in person, and are unable to quickly identify the need for a clarification. Although the FTF expectation for "immediate feedback" was modified for e-collaboration "feedback within a few days," the frustration was still evident.

The second category of obstacles to communication is the *limited capacity of a channel* to carry cues, indicated by lack of non-verbal cues. To some extent, virtual team members were unable to find symbols to represent language and social cues, such as eye contact, bodily and social cues ranging from intimidation and annoyance to joking and head shake.

Some of the effects of using a written e-collaboration medium were considered beneficial, such as: being able to access and print any virtual team member's message at any time. This allows a member to acknowledge each individual contribution and spend more time on a message, with more considerate language. The size of the input box was perceived by some as a limit on the length of a contribution. The final category of obstacles to communication effectiveness is *lack of adaptiveness*, whereby messages or media are personalized or individuals are acknowledged. Social discussion was seen by many as a waste of time, not focused on the task. E-collaboration provided a (positive) obstacle to socializing and incentive to be more neutral.

CONCLUSION

Four teams in an organization needed to address business process difficulties for their team, but were in a "crunch" time, making a series of face-to-face meetings nearly impossible. Instead of face-to-face communication, the teams used simple asynchronous electronic communication media. All four virtual teams successfully redesigned a business process and successfully implemented all or part of their re-design within six months.

The results cannot be explained by traditional media richness or social presence theories and generally support compensatory adaptation theory. Members of the virtual teams using an asynchronous ECM perceived many obstacles to natural communication when compared to the face-to-face medium. They perceived less interactivity (immediacy of feedback), lower channel capacity (inability to convey non-verbal cues), and less adaptiveness (personalized messages or individual acknowledgements). Some of these were compensated for by the fact that their written messages were posted and printable.

Members of the virtual teams reported making numerous adaptations to their communication behavior in order to effectively collaborate using lean media. *They reported making an effort to change their communications before sending them in order to make them **more focused, clear, precise, neutral, concrete, concise, persuasive, considerate, and complete**, which is the answer to our research question.* They captured language and social cues in writing, requested feedback, reflected more on their own and others' messages, and printed messages for perusal. Somewhat ironically, most of these "adaptations" were also considered "improvements" over the face-to-face environment (Kock & DeLuca, 2006; DeLuca & Valacich, 2006; DeLuca et al., 2006; Robert & Dennis, 2005). These behaviors made the lean media appear to function more richly, and led to better perceived quality.

However, one cannot dismiss the many times participants indicated that they had put forth extra effort to compose messages on a less "natural" medium. It is not clear at what point the extra effort reported to use e-collaboration would be too much of a burden or how long participants could continue the effort on an extended project. It is conceivable that participants would at some point avoid a project knowing the effort required. Studies indicate that largely virtual teams might also choose synchronous media for part of a project. Potential benefits of synchronicity may apply well to: final convergence on a re-design (Dennis & Valacich, 1999; DeLuca & Valacich, 2006); voice mail to support social cohesion; or desire for the lack of a written record for discussion involving political aspects of a project (Gasson & Elrod, 2005).

Improvements achieved when using lean e-collaboration over the FTF environment may be further explained by the media-cognitive-social (MCS) model of creativity as applied in asynchronous creativity theory (ACT) (DeLuca, 2006) and merits further study.

Practitioners are already motivated to attempt virtual teams to save the difficulties and costs of convening face-to-face teams in a global 24/7 enterprise and take advantage of the availability of low-cost Internet-based asynchronous electronic communication media. Yet many organizations have been reluctant to conduct virtual teams for complex tasks, fearing failure. This study shows that adaptations made to use a lean medium may yield effectiveness, efficiency, quality, and success. Practitioners are also cautioned that the price paid in time and effort to make the adaptations listed above may take its toll over time. Yet, for now, the outlook for extending the variety of complex tasks tackled by virtual teams is optimistic.

REFERENCES

Campbell, D. T., & Fiske, D. (1959). Convergent and discriminant validation by the multi-trait, multi-method matrix. *Psychological Bulletin, 56,* 81-105.

Choi, T. Y., & Liker, J. K. (1995). Bringing Japanese continuous improvement approaches to U.S. manufacturing: The roles of process orientation and communications. *Decision Sciences, 26*(5), 589-620.

Daft, R. L., & Lengel, R. H. (1986). Organizational information requirements, media richness and structural design. *Management Science, 32*(5), 554-571.

Daft, R. L., Lengel, R. H., & Trevino, L. K. (1987). Message equivocality, media selection, and manager performance: Implications for information systems. *MIS Quarterly, 11*(3), 355-366.

Davison, R. M., Martinsons, M. G., & Kock, N. (2004). Principles of canonical action research. *Information Systems Journal, 14,* 65-86.

DeLuca, D. C. (2006). Virtual teams rethink creativity: A new theory using asynchronous communications. *Proceedings of the 12th Americas Conference on Information Systems* (pp. 129-138).

DeLuca, D. C., Gasson, S., & Kock, N. (2006). Adaptations that virtual teams make so that complex tasks can be performed using simple e-collaboration technologies. *International Journal of e-Collaboration, 2*(3), 64-85.

DeLuca, D. C., & Valacich, J. S. (2006). Virtual teams in and out of synchronicity. *Information Technology & People, 19*(4), 323-344.

Dennis, A. R., & Valacich, J. S. (1999). Rethinking media richness: Towards a theory of media synchronicity. *Proceedings of the 32nd Annual Hawaii International Conference on System Sciences* (pp. 1-10).

Denzin, N. K. (1978). *The research act* (2nd ed.). New York: McGraw-Hill.

Gasson, S., & Elrod, E. M. (2005). Managing knowledge across the boundaries of a virtual organization. *Proceedings of The International Conference on Knowledge Management (ICKM2005)*, Charlotte, NC.

Graetz, K. A., Boyle, E. S., Kimble, C. E., Thompson, P., & Garloch, J. L. (1998). Information sharing in face-to-face, teleconferencing, and electronic chat groups. *Small Group Research, 29*(6), 714-743.

Kahai, S. S., & Cooper, R. B. (2003). Exploring the core concepts of media richness theory: The impact of cue multiplicity and feedback immediacy on decision quality. *Journal of Management Information Systems,* 263.

Kock, N. (2004). The psychobiological model: Towards a new theory of computer-mediated communication based on Darwinian evolution. *Organization Science: A Journal of the Institute of Management Sciences,* INFORMS: Institute for Operations Research, 327-348.

Kock, N. (2005a). What is e-collaboration? *International Journal of e-Collaboration, 1*(1), i-vii.

Kock, N. (2005b). Compensatory adaptation to media obstacles: An experimental study of process redesign dyads. *Information Resources Management Journal, 18*(2), 41-67.

Kock, N. (2006). *Business process improvement through e-collaboration: Knowledge sharing through the use of virtual groups.* Hershey, PA: Idea Group.

Kock, N., & DeLuca, D. C. (2006). Improving business processes electronically: A positivist action research study of groups in New Zealand and the US. *Proceedings of the 11th Annual Conference of the Center for the Study of Western Hemispheric Trade*, Session VII (pp. 1-35).

Majchrzak, A., Rice, R. E., Malhotra, A., King, N., & Ba, S. (2000). Technology adaptation: The case of a computer-supported inter-organizational virtual team. *MIS Quarterly, 24*(4), 569-600.

Malhotra, Y. (1998). Business process redesign: An overview. *IEEE Engineering Management Review, 26*(3) 27-31.

Markus, M. L., Majchrzak, A., & Gasser, L. (2002). A design theory for systems that support emergent knowledge processes. *MIS Quarterly, 26*(3), 179-212.

Miranda, S. M., & Saunders, C. S. (2003). The social construction of meaning: An alternative perspective on information sharing. *Information Systems Research, 14*(1), 87-106.

Moore, G. C., & Benbasat, I. (1991). Development of an instrument to measure the perceptions of adopting an information technology innovation. *Information Systems Research, 2*(3), 192-222.

Nambisan, S., Agarwal, R., & Tanniru, M. (1999). Organizational mechanisms for enhancing user innovation in information technology. *MIS Quarterly, 23*(3), 365-395.

Ocker, R., Fjermestad, J., Hiltz, S. R., & Johnson, K. A. (1998). Effects of four modes of group communication on the outcomes of software requirements determination. *Journal of Management Information Systems, 15*(1), 99-118.

Rice, R. E. (1992). Task analyzability, use of new media, and effectiveness: A multi-site exploration of media richness. *Organization Science: A Journal of the Institute of Management Sciences*, INFORMS: Institute for Operations Research, 475.

Rice, R. E., Kraut, R. E., Cool, C., & Fish, R. S. (1994). Individual, structural and social influences on use of a new communication medium. *Academy of Management Proceedings*, Academy of Management, 285.

Rice, R. E., & Shook, D. E. (1990). Relationship of job categories and organizational levels to use of communication channels, including electronic mail: A meta-analysis and extension. *Journal of Management Studies, 27*(2), 195-230.

Riva, G., & Galimberti, C. (1998). Computer-mediated communication: Identity and social interaction in an electronic environment. *Genetic, Social and General Psychology Monographs, 124,* 434-464.

Robert, L. P., & Dennis, A. R. (2005). Paradox of richness: A cognitive model of media choice. *IEEE Transactions on Professional Communication, 48*(1), 10-21.

Rosenthal, R., & Rosnow, R. L. (1991). *Essentials of behavioral research: Methods and data analysis.* Boston: McGraw Hill.

Short, J., Williams, E., & Christie, B. (1976). *The social psychology of telecommunications.* London: John Wiley and Sons.

Yin, R. K. (1994). *Case study research.* Newbury Park, CA: Sage.

KEY TERMS

Adaptiveness: Potential to acknowledge, adapt and personalize messages of a particular communicant.

Asynchronous E-Collaboration: Collaboration among individuals engaged in a common task using electronic technologies that allow input at different times.

Asynchronous Creativity Theory (ACT): Theory that considers the effects of communication media capabilities on the traditional social and cognitive factors affecting creativity; the net effect of which is that teams using asynchronous e-collaboration may have and the effect of greater potential for creativity than synchronous face-to-face teams.

Business Process Improvement: Improved use of resources needed to execute a set of interrelated activities performed in an organization with the goal of generating value in connection with a product or service.

Channel Capacity: Ability to transmit a high variety of language and social cues (both verbal and non-verbal).

Compensatory Adaptations: Adaptations made to communicative behavior in order to compensate or overcompensate for the perceived obstacles to communication—operationalized as three constructs—(1)

interactivity; (2) channel capacity; and (3) adaptiveness.

Compensatory Adaptations Principle: Use of e-collaboration media requires that adaptations be made to compensate for the lack of naturalness of the media. The adaptations made to use e-collaboration may generate outcomes of same or better quality than if team members had interacted solely face-to-face.

Compensatory Adaptation Theory: A theory based on the media naturalness principle and compensatory adaptations principle whereby team members compensate or even overcompensate for perceived lack of media naturalness (obstacles) of a communication medium. The outcome of the compensations is improved communication.

Cues: Vocal, non-verbal, paralinguistic, bodily, and social cues.

Interactivity: Potential to obtain immediate feedback from other communicants.

Lean Media: Media that lack many of the features of face-to-face communications (e.g., immediate feedback, presence of visual, auditory, and social cues) like e-mail and bulletin boards.

Lean E-Collaboration: E-collaboration using lean media.

Media-Cognitive-Social Model of Creativity: A model of three influences on creativity—media characteristics (immediacy, cues, parallelism, rehearsability, reprocessability), cognitive influences (attention, diversity, incubation), and social influences (accountability, affiliation, anxiety).

Media Naturalness Principle: The degree of similarity between a given communication medium and the face-to-face medium determines the naturalness of the media and the cognitive effort needed to use the media for communications.

Media Synchronicity Theory: A theory based on media characteristics, task communication function, and team processes. Synchronicity is more important for convergence than for conveyance processes.

Voice–Based Group Support Systems

Milam Aiken
University of Mississippi, USA

INTRODUCTION

One of the primary reasons large meetings utilizing group support systems (GSS) are more efficient and effective than traditional meetings is because the former are based upon typed comments and opinions while the latter are based upon voice input (Nunamaker, Briggs, Mittleman, Vogel, & Balthazard, 1997). Using a keyboard, participants can submit comments to the group anonymously and simultaneously, but in an oral meeting, participants must take turns speaking to avoid confusion. Further, group members in an electronic meeting can skim recorded typed comments easily, while most traditional meetings do not have complete transcripts available as the discussion progresses. Even analyzing new comments is more efficient with a GSS. Most people can read faster than they can listen, and in an oral meeting, the rate of input is limited by the current speaker's voice.

Although many people are now familiar with typing, most are more comfortable with speaking and can generate more words per minute by voice. Are there more efficient means of generating ideas in a GSS session (Briggs, Nunamaker, & Sprague, 1998)? To improve the productivity of electronic meetings, the systems should be made as "typewriter-less" as possible (Gray & Olfman, 1989). Integrating automatic speech recognition (ASR), or simply, speech recognition (SR), with a GSS might improve the rate of comment generation.

Accuracy has been the major barrier to greater SR acceptance, but high accuracy might not be needed in a GSS meeting. For example, if one comment is not understood, there are likely to be other similar, if not redundant, adjoining comments that might be clear or could aid the understanding of the earlier comment. In addition, a participant can submit a new comment asking for clarification from the group.

This paper summarizes research conducted using SR technology during electronic meetings. Results of these voice-based GSS (VGSS) studies show that SR transcription accuracy generally is low due to background noise in these face-to-face meetings. Distributed VGSS meetings are likely to be more efficient and effective.

BACKGROUND

An electronic meeting is one form of e-collaboration that typically involves exchanging comments via a computer-based network (Fjermestad & Hiltz, 2000). The principal reason that these meetings are superior to traditional, oral meetings when sharing ideas among many people is that comments are typed, allowing all comments to be recorded as they are written, anonymous submission of ideas, and simultaneous generation of text.

At any time, GSS group members can read or skim through old comments very quickly. While there is no standard test for measuring reading speed (reading material varies in length, complexity, style, etc.) and tests for reading comprehension are subjective, average readers generally read around 200 words per minute (wpm) with a typical comprehension of 60% (Speed reading test online, n.d.) and average college students can read fiction and non-technical materials between 250 and 350 wpm (Suggestions for improving reading speed, n.d.). Thus, we will assume the average GSS participant can read the public comments at about 300 wpm.

But generating text is much slower than reading. While the maximum typing speed in English has been recorded as high as 212 wpm (Glenday, 2005), most people, of course, type far slower. A student often can type 13 to 41 wpm, a good typist can generate from 61 to 90 wpm, and an excellent typist may produce between 85 to 112 wpm, assuming five characters per word (Cooper, 1983), and typical undergraduate Business students can type 36 "easy" words per minute (commonly occurring words with few syllables) and type 24 "difficult" words per minute (Rebman, Aiken, & Cegielski, 2003). Part of the difference in student typing rates between the two studies can be explained by the far greater use of computers now; many more people are familiar with typing, and many type every day.

However, the rates above were not adjusted for errors. Over an hour-long period, one typist was able to produce 149 wpm, with a 10-word penalty per error (McWhirter & McWhirter, 1973). When simply transcribing text, knowledge workers at IBM were able to generate 32.5

corrected words per minute (cwpm)—time was taken to backspace and type over errors (Karat, Halverson, Horn, & Karat, 1999). In addition, extra time is needed when composing fresh ideas, and the same workers were able to generate only 19 cwpm when thinking of new ideas. Although GSS participants must compose fresh ideas, they do not often take extra time correcting mistakes, as most transcripts have many grammatical (e.g., lack of capitalization and poor punctuation) and spelling errors. For example, in one study (Aiken, Vanjani, Martin, Young, & Govindarajulu, 1994), 30% of all comments typed by undergraduate business students in a GSS meeting had at least one grammatical or spelling error.

Assuming a typical reading speed of 300 wpm and a typing speed of 20 wpm, a participant in an electronic meeting should be able to keep up with 15 other group members typing simultaneously, assuming these other participants are not also spending time reading comments. But GSS participants typically spend 60% of the total meeting time reading others' typed comments (Aiken & Vanjani, 1996). Thus, each minute, 180 words per minute are read and 8 words per minute are typed, on average, by each participant, and each should be able to keep up with the generation of new text until the group reaches a size of 23 people. However, participants do not necessarily need to read everything (especially since a recorded transcript is available for later review), and they could instead decide to skim and selectively read for greater detail, increasing the maximum group size.

One way to address this imbalance is to generate comments faster, perhaps through automatic speech recognition. People speak much faster than they type. The maximum rate of speech may be as high as 637 wpm (Glenday, 2005), but when composing fresh ideas, people typically speak at about 100 to 150 wpm (Lenneberg, 1967). Further, there is no need to worry about spelling or some grammatical errors such as lack of capitalization and punctuation. Assuming a typical speech rate of 120 wpm and a typical typing rate of 20 wpm when composing fresh ideas, it might be possible to generate about six times more text during a GSS meeting with SR, or alternatively, have a meeting last only 1/6 as long.

Figure 1 illustrates the rates of text generation possible in oral, GSS, and SR/GSS meetings. Because group members in a traditional, oral meeting must take turns speaking, the rate of text generation per minute remains constant, such as 120 wpm. However, in a GSS meeting, each group member can type at the same time. Thus, the rate of text generation per minute increases with the group size. Assuming a typical typing rate of 20 wpm, a six-person GSS meeting can generate as much text as a six-person oral meeting. However, integrating SR into a GSS meeting could allow far greater rates of text generation, perhaps as high as 510 wpm for a six-person meeting, assuming each is able to generate 85 wpm using the speech recognition software.

SPEECH RECOGNITION

There are definite advantages to sharing ideas via written text in an electronic meeting, but people often are more comfortable with speaking. Automatic speech recognition has the potential to bridge this gap. The speech recognition process includes several steps (Markowitz, 1996):

1. **Audio input:** The human voice is transmitted through a microphone connected to a microcomputer with a standard sound card.
2. **Acoustic processor:** The acoustic processor converts the captured audio into a series of phonemes.
3. **Word matching:** The software attempts to match the phonemes to the most likely words. First, it uses acoustical analysis to build a list of possible words that contain similar sounds. Then, the software uses contextual information to predict what words should come next, helping the system to distinguish among homonyms, for example.
4. **Decoder:** The decoder selects the most likely word based on the rankings assigned during word matching and assembles the words in the most likely sentence combination. It then transfers the sentence to the word processing application.

To train the software to recognize the speaker's voice, a process known as enrollment is used. During this initial phase, the user reads one or more pre-selected passages of text on the computer screen while the software matches the words with the speaker's distinctive vocal patterns. Although more training usually results in greater accuracy, fairly good results often are achieved within as little as five or 10 minutes.

As one would expect, speech with SR can generate words very quickly, and one study (Rebman & Aiken, 2000) showed that undergraduate Business students can

Figure 1. Words generated per minute in oral, GSS, and SR/GSS meetings

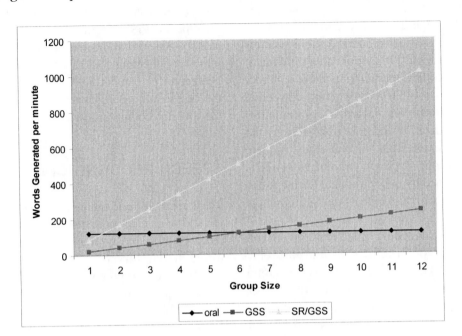

produce 2.36 more words per minute using SR than when typing. This SR rate (85 wpm) was much slower than a typical talking rate, but unfamiliarity with the headset microphone, software, and environment might have reduced the productivity. However, the major barrier to greater SR use has been its relatively poor transcription accuracy.

SR ACCURACY

Several factors can affect SR accuracy, including background noise (e.g., a door slamming, a phone ringing, other people nearby speaking, etc.), quality of the voice (e.g., speaking too soft or too loud), word errors (e.g., slurring words or even mispronouncing words), use of homonyms (e.g., "write" versus "right"), and out-of-vocabulary utterances (e.g., "Kyrgyzstan" or "Djibouti"). As a result, two types of errors can occur:

1. **Rejection errors** occur when the reader has no guess about what the SR-generated text means. Although most systems produce correctly spelled words, the word combination could be completely meaningless, such as "Dog blue apple." In conversation, a listener would typically respond by asking "What?" Repeating the speech slowly and

clearly sometimes helps, resulting in a correct word sequence.

2. **Substitution errors** occur when the reader does not realize a mistake has been made, although correct words have changed. For example, the system might produce "Write now" instead of "right now." This type of error can be especially harmful since the reader believes the text is accurate.

Recent improvements in SR technology have enhanced accuracy almost 15% over previous versions' results. A study using *IBM ViaVoice Pro 8* reported a mean accuracy rate of 93.6% versus 99.6% for human transcription (Al-Aynati & Chorneyko, 2003), and another study of SR using *Dragon Systems Naturally Speaking 7.3* showed accuracies of 98% to 99% at 140 to 160 wpm (Nuance - Dragon NaturallySpeaking 8, n.d.). Studies using earlier versions of the software often achieved accuracies of 80% to 85% (Aiken & Wong, 2001; Pallet, Garofolo, & Fiscus, 2000). In addition, readers often are able to understand the SR-generated text, even with errors. For example, the SR accuracy in one study (Aiken, Wong, & Vanjani, 2002) was only 81.4%, but 95.7% of the text was understood correctly.

SR AND GSS

Although the integration of SR with GSS was first proposed over a decade ago (Aiken, Kim, & Singleton, 1994), the first attempt to study SR within an electronic meeting was not made until seven years later (Aiken, Rebman, & Vanjani, 2001). In this study, 17 undergraduate business students orally discussed the parking problem on campus with a GSS for 10 minutes in groups of six, three, and four while speech was automatically transcribed onto the screen by *Dragon Systems Naturally Speaking Preferred 5*. The students thought it was fairly easy to generate text with their voices, but the accuracy was relatively poor with comment understanding accuracies per group of 80%, 83%, 91%, and 100%. The relatively low accuracies can be attributed to the early version of the SR software and the fact that there was a large amount of background noise. Even though participants were separated into cubicles during the electronic meetings, the noise from several group members talking at once caused some errors in the VGSS transcription.

In the second attempt to study VGSS (Aiken, Rebman, & Paolillo, 2001), eight groups of six undergraduate students each typed comments in an electronic meeting while eight groups of six students each used *Dragon Systems Naturally Speaking Preferred 5* to generate text from speech. Contrary to expectations, twice as many comments were generated using typing than using voice, because students spent much of the time erasing inaccurate, SR-generated text rather than contributing new comments. Much of the inaccurate text came from background noise.

In the final study of VGSS (Rebman et al., 2001), two groups of six undergraduate Business students each typed comments using a GSS, and two groups of six each used a VGSS with *Dragon Systems Naturally Speaking Preferred 5*. SR accuracy was approximately as good as typing (93.9% versus 96.6%), but students involved in the typing meeting generated 69% more comments and 57% more words than those in the SR meeting.

One study (Aiken & Vanjani, 2001) used one speaker who read sample comments from two historical, typed, GSS meeting transcripts with *Dragon Systems Naturally Speaking Preferred 5*. Results showed an accuracy rate of 96%, with 99% of the text understood. Thus, it appears that most of the errors in the prior studies were due to differences in speaker skills with SR and/or background noise.

DISCUSSION

Speech is not always understood, even by human listeners. In addition to contractions, slang, mispronunciations, grammatical errors, and a large vocabulary, people can talk too softly or there can be too much background noise. For example, it is estimated that most human listeners are able to recognize only 96% of words in a freestyle conversation in English (Coursey, 2002). Thus, it is unreasonable to expect automatic speech recognition to consistently achieve 100% accuracy.

What is an acceptable accuracy rate for SR within an electronic meeting? The answer depends upon the importance or seriousness of the discussion, how fast a rough transcription is needed, and other factors. For example, the required accuracy for certification as a court reporter is 96% at 180 words per minute (Automatic speech recognition, 2002). But an informal product focus group needing a printed transcript quickly might accept 85% to 90% accuracy, especially if the understanding accuracy is 95% to 100%. In this case, a VGSS might prove useful.

Results of three experiments utilizing the technology in VGSS meetings showed low effectiveness and efficiency due to subjects' inexperience with the new technique, deficiencies in the software, and perhaps most importantly, background noise. Noise-canceling microphones can almost eliminate constant background sounds but do little to reduce intermittent noise. In addition, even if there is high transcription accuracy with background noise, many group members speaking to their computers at once in close proximity could annoy others in the room.

An alternative to a face-to-face VGSS meeting is placing group members in separate, quiet rooms. However, the group dynamic of such a distributed meeting is changed, and distributed groups could be more difficult to control (Aiken & Vanjani, 1997). In addition, future research should include tests of how VGSS can affect group dynamics. With a greater flow of information, decision making could be made more efficient and group satisfaction with the meeting process could be increased.

CONCLUSION

A Voice-based Group Support System has the potential to increase the efficiency of e-collaboration, but at this point, in face-to-face meetings, background noise appears to present a barrier to acceptable transcription speed and accuracy. Only three studies of VGSS meetings have been conducted, and SR transcription accuracy ranged from 80% to 100% with less text generated. The application of SR to GSS meetings is still in its infancy, however, and future advances in hardware and software could increase the accuracy and subsequent speed of the technology.

REFERENCES

Aiken, M., Kim, D., & Singleton, T. (1994, March). Future developments of group decision support systems. *Proceedings of the 1994 Southeast Decision Sciences Institute Conference*, Williamsburg, VA.

Aiken, M., Rebman, C., & Paolillo, J. (2001, November 17-20). Lessons learned with a voice-based group support system. *Proceedings of the 32nd Annual Meeting of the Decision Sciences Institute*, San Francisco (pp. 22-24).

Aiken, M., Rebman, C., & Vanjani, M. (2001, February 28-March 3). A voice-based group support system. *Southwest Decision Sciences Institute 32nd Annual Conference*, (pp. 15-17). New Orleans, LA.

Aiken, M. & Vanjani, M. (1996). Idea generation with electronic poolwriting and gallery writing. *International Journal of Information and Management Sciences, 7*(2), 1-9.

Aiken, M. & Vanjani, M. (1997). A comparison of synchronous and virtual legislative session groups faced with an idea generation task. *Information & Management, 33*(1), 25-31.

Aiken, M. & Vanjani, M. (2001, November 17-20). Accuracy of typed and spoken group support system transcripts. *Proceedings of the 32nd Annual Meeting of the Decision Sciences Institute*, San Francisco (pp. 25-27).

Aiken, M., Vanjani, M., Martin, J., Young, C., & Govindarajulu, C. (1994). Experiences with a bilingual group decision support system. *International Business Schools Computing Quarterly, 6*(1), 4-9.

Aiken, M., & Wong, Z. (2001, November17-20). The influence of textual complexity on automatic speech recognition accuracy. *Proceedings of the 32nd Annual Meeting of the Decision Sciences Institute*, San Francisco (pp. 28-30).

Aiken, M., Wong, Z., & Vanjani, M. (2002). Speech complexity and automatic recognition accuracy. *Academy of Information and Management Sciences (AIMS), 5*(1-2), 1-12.

Al-Aynati, M., & Chorneyko, K. (2003). Comparison of voice automated transcription and human transcription in generating pathology reports. *Archives of Pathology and Laboratory Medicine, 127*(6), 721-725.

Automatic speech recognition. (2002). Retrieved December 6, 2005, from http://tap.gallaudet.edu/SpeechRecog.htm

Briggs, R., Nunamaker, J., & Sprague, R. (1998). 1001 unanswered research questions in GSS. *Journal of Management Information Systems, 14*(3), 3-21.

Cooper, W. (1983). *Cognitive aspects of skilled typewriting.* New York: Springer Verlag.

Coursey, D. (2002). *Why it's getting easier to talk to your PC.* Retrieved November 11, 2005, from http://reviews-zdnet.com.com/4520-6033_16-4207711.html

Fjermestad, J., & Hiltz, S. (2000). Group support systems: A descriptive evaluation of case and field studies. *Journal of Management Information Systems, 17*(3), 115-159.

Glenday, C. (2005). *Guinness world records 2006.* New York: Guinness, Inc.

Gray, P., & Olfman, L. (1989). The user interface in group decision support systems. *Decision Support Systems, 5,* 119-137.

Karat, C., Halverson, C., Horn, D., & Karat, J. (1999). Patterns of entry and correction in large vocabulary continuous speech recognition systems. In *Proceedings of the SIGCHI conference on human factors in computing systems,* (568-575). Pittsburgh, PA.

Lenneberg, E. (1967). *Biological foundations of language.* New York: Wiley.

Markowitz, J. (1996). *Using speech recognition.* Upper Saddle River, NJ: Prentice Hall.

McWhirter, N., & McWhirter, R. (1973). *Guinness book of world records.* New York: Sterling Publishing.

Nuance - Dragon NaturallySpeaking 8. (n.d). Retrieved December 6, 2005, from http://www.nuance.com/naturallyspeaking/

Nunamaker, J., Briggs, R., Mittleman, D., Vogel, D., & Balthazard, P. (1997). Lessons from a dozen years of group support systems research: A discussion of lab and field findings. *Journal of Management Information Systems, 13*(3), 163-207.

Pallett, D., Garofolo, J., & Fiscus, J. (2000). Measurements in support of research accomplishments. *Communications of the ACM, 43*(2), 75-79.

Rebman, C., Aiken, M., & Cegielski, C. (2003). Speech recognition in the human-computer interface. *Information and Management, 40*(6), 509-519.

Rebman, C., Aiken, M., Reithel, B., & Cegielski, C. (2001). An exploratory study of speech recognition technology and its implications for current electronic meeting support applications. *Proceedings of the Association for Information Systems Americas Conference,* Boston.

Speed reading test online. (n.d.). Retrieved December 6, 2005, from http://www.readingsoft.com/

Suggestions for improving reading speed. (n.d.). Retrieved December 6, 2005, from http://www.ucc.vt.edu/stdysk/suggest.html

KEY TERMS

Automatic Speech Recognition (ASR) or **Speech Recognition (SR):** Transcription of human speech to text by a computer.

Comment: A passage of text typed during an electronic meeting's idea generation phase.

Electronic meeting: A meeting of individuals using Group Support System software to exchange ideas, vote, or collaborate in some other way.

Enrollment: The process of training SR software to recognize an individual's distinctive voice patterns.

Face-to-Face Meeting: A meeting of individuals within the same room, such that each group member is able to see every other group member.

Group Support System (GSS): A computer-based system that automates a meeting.

Voice-Based Group Support System (VGSS): A GSS with SR used to generate text.

Wikis as Tools for Collaboration

Jane Klobas
Bocconi University, Italy
University of Western Australia, Australia

INTRODUCTION

Tim Berners-Lee, the inventor of the World Wide Web, envisioned it as a place where "people can communicate … by sharing their knowledge in a pool … putting their ideas in, as well as taking them out" (Berners-Lee, 1999). For much of its first decade, the Web was, however, primarily a place where the majority of people took ideas out rather than putting them in. This has changed. Many "social software" services now exist on the Web to facilitate social interaction, collaboration and information exchange. This article introduces wikis, jointly edited Web sites and Intranet resources that are accessed through web browsers. After a brief overview of wiki history, we explain wiki technology and philosophy, provide an overview of how wikis are being used for collaboration, and consider some of the issues associated with management of wikis before considering the future of wikis.

In 1995, an American computer programmer, Ward Cunningham, developed some software to help colleagues quickly and easily share computer programming patterns across the Web. He called the software WikiWikiWeb, after the "Wiki Wiki" shuttle bus service at Honolulu International Airport (Cunningham, 2003). As interest in wikis increased, other programmers developed wiki software, most of it (like WikiWikiWeb) open source. Although wiki software was relatively simple by industry standards, some technical knowledge was required to install, maintain and extend the "wiki engines." Contributors needed to learn and use a markup language to edit pages, and even if the markup languages were often simpler than HTML, non-technical users did not find these early wikis compelling.

In the early years of the twenty-first century, a number of developments led to more widespread use of wikis. Wiki technology became simpler to install and use, open source software was improved, and commercial enterprise-grade wiki software was released. The not insignificant issues associated with attracting and managing a community of people who use a wiki to share their knowledge were discussed in forums such as *MeatballWiki* (http://www.usemod.com/cgi-bin/mb.pl?action=browse&id=MeatballWiki&oldid=FrontPage). The public's attention was drawn to wikis following the launch, in January 2001, of the publicly written Web-based encyclopedia, *Wikipedia* (www.wikipedia.org). And wiki hosting services and application service providers (ASPs) were established to enable individuals and organizations to develop wikis without the need to install and maintain wiki software themselves.

By July 2006, nearly 3,000 wikis were indexed at the wiki indexing site www.wikiindex.org, popular wiki hosting services such as *Wikia* (www.wikia.org) and *seedwiki* (www.seedwiki.org) hosted thousands of wikis between them, and *Wikipedia* had more than four and a half million pages in over 100 languages. Moreover, wikis were increasingly being used in less public ways, to support and enable collaboration in institutions ranging from businesses to the public service and not-for-profit organizations.

THE NATURE OF WIKIS

Wiki software allows users to collaboratively edit pages for the Web or intranets. The pages created with wiki software are called "wiki pages" and sites that contain wiki pages are called wiki sites, or simply "wikis."

Technically, wikis consist of four basic elements:

- Content
- A template which defines the layout of the wiki pages
- Wiki engine, the software that handles all the business logic of the wiki
- Wiki page, the page that is created by the wiki engine as it displays the content in a browser

Figure 1. How wikis work (Adapted from Klobas & Marlia, 2006)

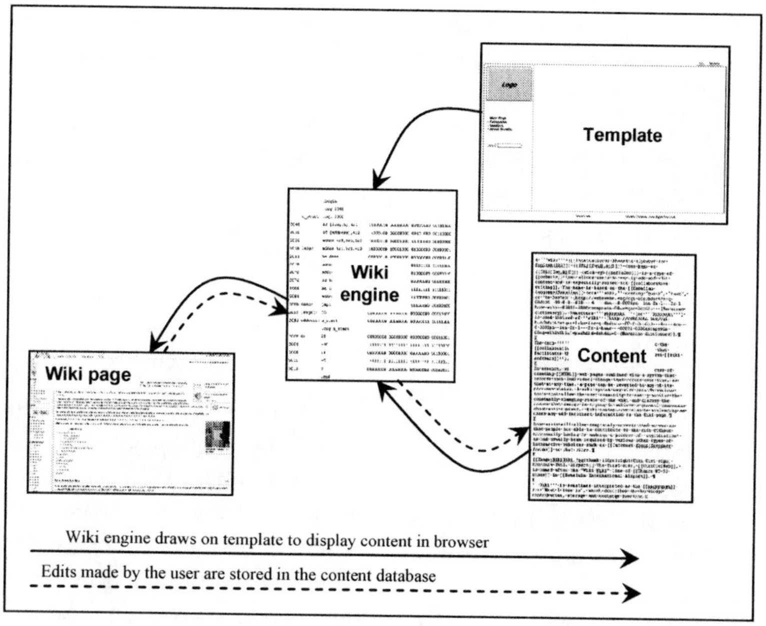

Figure 1 shows how these elements work together.

Wikis consist of pages accessible from a Web browser. They are edited by opening an editing screen in the browser and using either a simple markup language or, increasingly, a rich text editor to edit the text. Links to pages internal or external to the wiki site can be added using simple conventions. These conventions allow a link to be created to a page that does not yet exist; the wiki engine flags such links for future editing. Unless the wiki managers decide otherwise, the content is updated in real time, and once an author saves and closes their changes on the editing screen, the changes are immediately visible online.

Almost all wikis keep track of changes. Older versions of a page can be viewed and, if necessary, restored. Most wikis include a page where recent changes are listed. This feature helps members of the wiki community to keep up to date with changes in content, and can help newcomers get a quick feel for the current concerns of the wiki community. Increasingly, wiki software is integrated with news aggregators like RSS or e-mail notification to alert users to changes without their having to enter the wiki itself.

Another important feature of wikis is the simple permissions structure. Typically, there are three levels of permission: reader, editor, and administrator. Reading permissions may be open to anyone on the World Wide Web or limited to specific, registered individuals. When reading permissions are limited, the wiki is known as a "private wiki."

Various wikis offer support for non-Latin character sets, different media and file types, mathematical notation, style sheets, conflict handling, spam handling, and facilities for merging, exporting and backing up. The different features available in different wiki engines can be seen at the wiki engine comparison site, *WikiMatrix* (www.wikimatrix.org).

Wiki features are based on design principles established by Ward Cunningham. These principles address human as well as technical goals, for example (in the terms used by Wagner, 2004), wikis are:

- **Open:** If any page is found to be incomplete or poorly organized, any reader can edit it as he/she sees fit
- **Organic:** The structure and content of the site evolve over time
- **Universal:** Any writer is automatically an editor and organizer
- **Observable:** Activity within the site can be watched and revised by any visitor to the site
- **Tolerant:** Interpretable (even if undesirable) behavior is preferred to error messages

Wikis usually adopt "soft security," social conventions that assume that most people behave in good faith, establish that users (rather than the software or a system administrator) operate as peer reviewers of content and behavior, allow that people might make mistakes but that mistakes can be corrected, and emphasize the importance of transparency in their management (Meatball, 2006). Together, these technical features and social principles provide a supportive environment for human collaboration.

HOW WIKIS ARE BEING USED FOR COLLABORATION

Wikis are used by groups of people who collaborate to produce information resources that range from meeting agendas to Web sites. They are used in business, government, research, and education.

Public wikis are often built by communities of practice, hobbyists and other interest groups. Most of these wikis are concerned with quite specific topics such as a specific sport, book, author, religion, or philosophy. Often, they are maintained and read by small groups of friends and colleagues. An example is ukcider, a wiki that provides information about real cider (http://ukcider.co.uk/wiki) and supports advocacy for real cider enthusiasts and small producers.

Wikis can be used to create and maintain knowledge repositories for communities of practice (Godwin-Jones, 2003; Roberts, 2005). Members of the community update the repository with information and solutions to problems. As with other types of wiki, the communities served might be public or private. Potential uses of private wikis to increase autonomy among US intelligence workers were described by Andrus (2005).

Committees and working groups, particularly those that work across institutional and/or geographical boundaries, use wikis to develop and maintain agendas and documentation. The DCMI Education Working Group Wiki (http://dublincore.org/educationwiki), for example, contains work plans, FAQs, agendas, notes, documentation, and links to key references for the group.

A common organizational use of wikis is project support (Bean & Hott, 2005). Wikis are used to support the sharing of agendas, ideas, resources, plans and schedules. Documents can be jointly produced within the wiki or made available as attachments. Angeles (2004) describes how wikis have been used by Lucent Technologies for project documentation including preparation of meeting notes, product specification notes, product requirements documents, project deliverables, content audits, technical documentation, and style guides. Other companies that use wikis for project support include Michelin China and the investment bank, DrKW (Paquet, 2006).

Wikis are also being used to write and maintain system documentation. A public wiki has been developed by Mozilla, the provider of the Firefox Web browser, to maintain Mozilla product documentation (http://kb.mozillazine.org). Any member of the public can contribute. Some companies have established private wikis to maintain documentation for their internal information systems (IS). Users contribute to documentation based on their experience with the system.

Organizations and groups can use wikis to produce reports of news or events. Schools, in particular, appear to be experimenting with this type of wiki, in which parents, teachers, administrators, and students can all contribute reports, photographs, audio and video of a school event.

Some groups use wikis to plan conferences or meetings and to continue discussions that were begun at a meeting. Examples include the American Library Association (ALA) 2006 Conference wiki (http://meredith.wolfwater.com/ala2006/) and the Yale Access to Knowledge wiki (http://research.yale.edu/isp/a2k/wiki/). Some experiments in use of wikis during conferences have found them valuable (Boyd, 2004; Suter, Alexander, & Kaplan, 2005). Boyd describes how workshop participants quickly learnt to use wikis to capture "their own streams of consciousness or the comments of others."

Wikis are used in a variety of ways for collaboration in classrooms ranging from primary school to university (Bold, 2006; Ferris & Wilder, 2006; Lamb, 2004; Skiba, 2005). Mitchell (2006) notes that wikis are consistent with modern theories of learning such as "connectivism" (Siemens, 2004), the idea that learning occurs in a complex and changing environment that cannot be controlled. In the classroom, wikis are used in collaborative writing, student projects and group assignments. Collaborative writing tasks can be given to develop writing and collaboration skills or to develop information resources for use by other students.

Wikis are particularly well adapted to shared creation of directories and lists such as bibliographies, staff lists and job vacancies. In schools and universities, teachers who use wikis in this way assign groups of students or whole classes the task of jointly preparing annotated lists of information resources. Although most educational wikis are private, some become public resources. An example is the *Wiki for IS Scholarship* maintained by the Information Systems Department in the Weatherhead School of Management at Case Western Reserve University (http://isworld.student.cwru.edu/tiki/tiki-index.php) which contains summaries of major books and articles in IS as well as reviews of the contributions of major IS scholars. Often, wiki-based directories also provide access to resources. Participants in the Government Open Code Collaborative (http://www.gocc.gov) use a public wiki to list and describe source code that can be shared among its members.

Wiki software is also being used to create Web sites that draw on the contributions of a number of people. One example is the site developed by the Bach-Academie of Montreal to publicize a series of concerts. The musicians themselves added details to the wiki, which was then published to the Web (http://www.bach-academie-de-montreal.com/).

ISSUES IN WIKI MANAGEMENT

The ease with which individuals can contribute to wikis, the open nature of contribution, and the philosophy that the guardians of the content are the people who create the wiki can produce outstanding knowledge resources. On the other hand, these very characteristics can also be associated with problems such as reluctance to contribute or chaos through unstructured contribution. Some management is therefore required by the person or group that wishes to develop and maintain the wiki. The most significant management issues tend to be those of encouraging people to contribute, and managing intellectual property, problem users and spam.

As with any knowledge resource, people need both a reason to contribute to a wiki and a sense that they have the ability to do so. Wikis created for small groups or communities with a specific purpose should quite readily attract users if those users also know how to access and use the wiki. For example, a wiki established for the purpose of jointly preparing an agenda or document, where all members of the group responsible for prepar-

ing the document are motivated to contribute and know how to access the wiki and use the editor, is likely to be used by the intended users for the intended purpose. The more diffuse the purpose and the user group, the more uncertainty there will be about the development and continuation of a wiki. In practice, however we usually find a core group of regular contributors and a peripheral group of occasional contributors and people who read the wiki but do not contribute to it. (This pattern is common in online communities; see, for example, Wenger, McDermott, & Snyder, 2002). The core users can act as role models for other users. Other patterns that have been observed include the champion pattern ("single wiki-nut, encourages coworkers to add, view, improve") and the trellis pattern ("egregiously boring content calls for fixing … My [contribution can be] far more interesting than that!" Confluence, 2006).

Techniques for ensuring that potential contributors feel able to contribute include providing simple, easy to follow documentation for editing pages and assurance that mistakes are permitted and can quickly be remedied. Anti-patterns that discourage use include the gate pattern, "too many procedural barriers to adding content" (Confluence, 2006).

Where the content of a wiki may include information, images and files drawn from other sources, the intellectual property of the original sources must be respected. Standard policies for dealing with material that is the intellectual property of others include: (1) only material that does not violate the copyright of the original creator can be included in the wiki and (2) the sources of all included material should be acknowledged. Equally, a decision has to be made about ownership and conditions for re-use of original material included in a wiki. The fact that the content of any single wiki page is usually produced by multiple authors needs to be acknowledged. A typical statement of the intellectual property in a wiki acknowledges its joint authorship and permits re-use, provided the wiki page is acknowledged as the source of the material. The GNU Free Documentation License used by Wikipedia provides an example (Free Software Foundation, 2002).

The more open a wiki to the public, and the larger the wiki user community, the greater the potential of encountering a problem user. Problem users may post unacknowledged copyrighted material, post material designed to offend or anger readers or contributors, remove material or otherwise "vandalize" the wiki. Some wikis use registration to limit the possibility that a problem user will continue to damage the wiki, but many wiki communities prefer to adopt the soft security principles described earlier in this article. It may be sufficient simply to let someone know (privately, by e-mail) the first time they post unacceptable content, that this type of content is not appropriate for the wiki (Turnbull, 2004). A wiki page that contains problem content can easily be replaced by an earlier, "clean" version. The active members of a wiki community often share the task of watching pages and dealing with problems.

Another potential problem, for public wikis in particular, is spam. Pages can be watched for spam in the same way that they are watched for other problem content, but this can be a time-consuming task. Some wikis use registration processes that require human intervention, such as CAPTCHA (Carnegie Mellon University School of Computer Science, 2000-2005), to prevent mass spam attacks. Spam capturing services are being added to wiki engines to reduce the amount of spam that reaches a wiki page.

FUTURE TRENDS AND CONCLUSION

At the time of writing, wikis were still the domain of early adopters, rather than a part of the mainstream. Despite widespread use of *Wikipedia*, knowledge of what wikis are and how they can be used for collaboration, particularly for private use, is not widespread. As of the end of 2006, only two non-technical books about wikis had been published (Klobas, 2006; Tapscott & Williams, 2006). Nonetheless, in October 2006, Google Inc. bought the wiki software and Web hosting service, JotSpot, an indication that consumer use of wikis is on the rise. The Gartner Group envisages wikis reaching maturity for business use some time before 2010 (Fenn & Linden, 2005). As wiki software matures, we can expect improvements in technical qualities such as ease of implementation and stability, as well as improvements in editing interfaces and graphical quality. These improvements, along with the positive experiences of early adopters, should help gain the interest and confidence of potential users and result in further diffusion of wikis as tools for collaboration.

REFERENCES

Andrus, D. C. (2005). The wiki and the blog: Toward a complex adaptive intelligence community. *Studies in Intelligence, 49*(3). Retrieved October 26, 2006, from http://ssrn.com/abstract=755904

Angeles, M. (2004). *Using a wiki for documentation and collaborative authoring.* Retrieved November 1, 2006, from http://www.llrx.com/features/librarywikis.htm

Bean, L., & Hott, D. D. (2005). Wiki: A speedy new tool to manage projects. *Journal of Corporate Accounting and Finance, 16*(5), 3-8.

Berners-Lee, T. (1999). *Transcript of Tim Berners-Lee's talk to the LCS 35th anniversary celebrations, Cambridge, Massachusetts, 14 April 1999.* Retrieved October 26, 2006, from http://www.w3.org/1999/04/13-tbl.html

Bold, M. (2006). Use of wikis in graduate course work. *Journal of Interactive Learning Research, 17*(1), 5-14.

Boyd, S. (2004, February). Wicked (good) wikis. *Darwin Magazine.* Retrieved November 1, 2006, from http://www.darwinmag.com/read/020104/boyd.html

Carnegie Mellon University School of Computer Science. (2000-2005). *The CAPTCHA Project.* Retrieved November 1, 2006, from http://www.captcha.net/

Confluence. (2006). *Patterns of wiki adoption.* Retrieved November 1, 2006, from http://confluence.atlassian.com/display/PAT/Patterns+of+Wiki+Adoption

Cunningham, W. (2003). *Correspondence on the etymology of wiki.* Retrieved October 25, 2006, from http://c2.com/doc/etymology.html

Fenn, J., & Linden, A. (2005). *Gartner's Hype Cycle special report for 2005.* Stamford, CT: Gartner Group.

Ferris, S. P., & Wilder, H. (2006). Uses and potentials of wikis in the classroom. *Innovate, 2*(5). Retrieved November 1, 2006, from http://www.innovateonline.info/index.php?view=article&id=2258

Free Software Foundation. (2002, November). *GNU free documentation license.* Retrieved November 1, 2006, from http://www.gnu.org/copyleft/fdl.html

Godwin-Jones. (2003). Blogs and wikis: Environments for on-line collaboration. *Language, Learning and Technology, 7*(2), 12-16.

Klobas, J. (2006). *Wikis: Tools for information use and collaboration*. Oxford: Chandos Publishing.

Klobas, J., & Marlia, M. (2006). Creating a wiki: The technology options. In J. Klobas (Ed.), *Wikis: Tools for information work and collaboration* (pp. 149-182). Oxford: Chandos Publishing.

Lamb, B. (2004). Wide open spaces: Wikis, ready or not. *EDUCAUSE Review, 39*(5). Retrieved October 26, 2006, from http://www.educause.edu/ir/library/pdf/erm0452.pdf

Meatball. (2006, September). *Soft Security*. Retrieved October 26, 2006, from http://usemod.com/cgi-bin/mb.pl?SoftSecurity

Mitchell, P. (2006). Wikis in education. In J. Klobas (Ed.), *Wikis: Tools for information work and collaboration* (pp. 119-147). Oxford: Chandos Publishing.

Paquet, S. (2006). Wikis in business. In J. Klobas (Ed.), *Wikis: Tools for information work and collaboration* (pp. 99-117). Oxford: Chandos Publishing.

Roberts, A. (2005). *Introducing a Wiki to a community of practice*. Retrieved January 9, 2006, from http://www.frankieroberto.com/dad/ultrastudents/andyroberts/year2/AEreport/AEtool.html

Siemens, G. (2004). *Connectivism: A learning theory for the digital age*. Retrieved October 26, 2006, from http://www.elearnspace.org/Articles/connectivism.htm

Skiba, D. J. (2005). Do your students wiki? *Nursing Education Perspectives, 26*(2), 120-121.

Suter, V., Alexander, B., & Kaplan, P. (2005). The future of FTF. *EDUCAUSE Review, 40*(1). Retrieved November 1, 2006, from http://www.educause.edu/apps/er/erm2005/erm0514.asp?bhcp=2001

Tapscott, D., & Williams, A. D. (2006). *Wikinomics: How mass collaboration changes everything*. Portfolio Hardcover.

Turnbull, G. (2004). Talking to Ward Cunningham about wikis. *Luvly,* Retrieved November 1, 2006, from http://gorjuss.com/luvly/20040406-wardcunningham.html

Wagner, C. (2004). Wiki: A technology for conversational knowledge management and group collaboration. *Communications of the Association for Information Systems, 13,* 265-289.

Wenger, E., McDermott, R., & Snyder, W. M. (2002). *Cultivating communities of practice*. Boston: Harvard Business School University Press.

KEY TERMS

Editor: The authors of wiki pages may be called "editors" because they have editing permissions.

Soft Security: Social conventions for trust, peer review and correction of errors adopted by contributors to wikis.

Wiki: An information resource that is created by multiple authors who use Web browsers that interacts with wiki software.

Wiki Engine: The software that handles the business logic of a wiki.

Wiki Page: A page of wiki content displayed in a Web browser.

Wiki Site: A set of related wiki pages. When a wiki site can be viewed on the World Wide Web, it is also a Web site.

Wiki Software: The suite of software used to produce and manage a wiki. This software may include, in addition to the wiki engine, add-ons and extensions that extend the feature set and functionality of the wiki.

WikiWikiWeb: The first wiki engine, written by Ward Cunningham.

Workflow Systems in E-Learning Environments

Neide Santos
IME/DICC – Universidade do Estado do Rio de Janeiro, Brazil

Flávia Maria Santoro
DIA – Universidade Federal do Estado do Rio de Janeiro, Brazil

Marcos R. S. Borges
IM/DCC&NCE - Universidade Federal do Estado do Rio de Janeiro, Brazil

INTRODUCTION

In the last few years, the Internet has become the most up-to-date way of structuring educational settings. Different styles of web-based learning technology jump up every day and its acceptance among educational community is unprecedented. With the networking technologies providing infrastructure for new educational formats, the opportunity is ripe to create e-learning community of problem solvers (Singley, Fairweather & Swerling, 1999).

In this new context, Computer-Supported Collaborative Learning (CSCL) emerges as an innovative pedagogical support. Collaborative learning consists of small groups working together to complete academic tasks, which involve a range of objectives, such as searching for facts, applying skills, concepts and principles, problem solving and creative thinking. The teacher acts as the coordinator, setting guidelines, encouraging cooperation, reviewing performance and uniting the groups (Woodbine, 1997). Collaborative learning works positively in promoting inter-group relations, and can help to overcome barriers to friendship, interaction and achievement of academically less able students, and can increase self-esteem.

CSCL studies how computer science can support the learning processes promoted by the Collaborative efforts of students working at a given task. CSCL can supply different kinds of Collaborative learning activities. A current approach used both at school and at organization is project-based learning (PBL). PBL proposes to work with solutions for real world problems, using several educational practices, besides activities in the laboratory (Killpatrick, 1926). The method of projects was reinforced by the ideas of Dewey (1966),

for whom the education depends on action and the knowledge emerges out of situations in which the students have to learn from experiences that make sense and have importance for them. These situations should be proposed within a social context, such as a classroom, a course or a training, where the students collect and manipulate several materials, creating an apprenticeship community that builds knowledge collectively.

In a project, the learning happens by interaction and articulation among different knowledge areas in order to favor the construction of the autonomy and the self-discipline; and develop abilities for the work in team, such as decision making, communication easiness, problem formulation and solving. The definition of the activities in a project is extremely important to the positive interdependence required to stimulate the collaboration. In a collaborative environment, the teacher and the apprentices should have means to define educational processes, to configure different sceneries and projects, and to obtain support, through the available tools, in the accomplishment of their tasks. A project can be described as a process, divided in stages that are related to each other forming a flow of work. Each stage is summed up through the performance of one or more activities, which owns specific objectives and generates some kind of product. These activities should stimulate information sharing and knowledge building.

Many CSCL researches present systems that provide the project-based learning approach. Analyzing these systems, we observed that several support the execution of specific tasks in the context of a project, others support a series of activities, although the definition and the accompaniment of its goals should be done out of the environment, and in other ones, the process is explicitly defined, being, however, fixed. Generally, it was observed that the environments do not provide support to the definition of collaborative processes, neither support the stages for a project development, although the importance of this question is pointed out many times. Planning the interactions and the process in a group project could be a way to stimulate people to collaborate, promoting interdependency and commitment among their work.

In this article, we discuss the importance of an explicit work process along the Collaborative activities, supported by a workflow system. From the belief on the importance of explicit work processes as way to improve collaboration, we develop an infrastructure, called COPLE (Collaborative Project-Based Learning Environment), whose core is COPE, a collaborative process editor. COPE (Collaborative Process Editor) plays the role of an educational workflow system. In the next sections, we discuss workflow systems and its uses in CSCL environments; describe our solution to provide support to the collaborative process and present preliminary results with the COPLE environment. The last section offers the conclusions and points our future work.

PLANNING ACTIVITIES IN E-LEARNING ENVIRONMENTS

Workflow is an information technology which uses electronic systems to manage and monitor business processes. It allows the flow of work between individuals and/or departments to be defined and tracked. The operational level of workflow deal with: how tasks are structured, who performs them, what their relative order is, how they are synchronized, how information flows to support the tasks and how tasks are being tracked. While the concept of workflow is not specific to information technology, support for workflow is an integral part of groupware software.

The structure and the coordination of learning activities can be seen as a process-oriented view. Some authors speak about this kind of process as a learning process. In the domain fields, the description and enactment of coordinated activities is done by workflows system. According to Wang, Haake, Rubart and Tietze (2000), a work procedure in a workflow system is defined by a workflow model composed of a set of discrete work steps with explicit specifications of how a unit of work

flows among the different steps. In general, a workflow coordination model can be defined as a directed graph, (N, L), with a node set N representing individual steps in the procedure and an edge set L representing the coordination structure among the tasks.

Workflow is an important element in e-learning systems because collaboration represents a set of relationships among behaviors and its consequences. A collaborative activity exists in three hierarchical levels: co-ordinate, co-operative, and co-constructive (Engeström, 1987; Bardram, 1998). Co-ordination represents the interaction flow and guarantees that an activity is executed in harmony. Co-operation means that the actors are not simply focused in their own tasks according to pre-defined roles, but also in common and shared goal. Therefore, a higher interaction level exists. Co-construction implies in interactions where the actors establish their concepts continually, building the shared goal collectively. The collaboration assigned to these levels should be supported in the context of the computational environments, besides propitiating the group to evaluate them.

An educational project requires a certain time dedicated to planning and management. For George and Leroux (2001), a project should be structured in time and partitioned in successive stages, forming an action plan. The careful planning of the activities is necessary to provide the project with a temporary structure. The description of human activities as actions performed by the use of operations help understanding the fundamental role that the planning plays in human cognition. Previous experiences advance the possible results of future actions; even so these anticipations should be implemented and adjusted in agreement with the conditions of the real contextualized situation. Therefore, the definition of the activities and their execution flow allows configuring the interaction forms and the products within the development of the project. Besides, it allows the teacher to understand the process and to help the apprentices.

Similar to the approach discussed here is the Instructional Design (ID), if we treat the instruction in a macro-level considering it as a project. ID has been pointed as one of the most important elements in the e-learning process (Winer and Váquez-Abad, 1995). Collaborative e-learning projects aiming at collective knowledge building can make use of design as a cognitive tool. ID can be defined as a cycle of activities, including sequence and structure, the main methods to be used in each activity, and the system control and evaluation (Kemp et al. 1996). Currently, design is seen as an activity shared by the team involved in the technology-mediated learning environment.

Designing is a process of planning and combining learning goals with tasks and procedures that should be performed by the groups of apprentices. Thus, it is also very important to support not only the process of planning but also the execution of the activities, especially if we are interest in promoting collaboration.

The process component alone may not be enough to foster the collaboration. It is also necessary to define the flow of work that makes the collaboration essential for the accomplishment of the project. The flow of work should provide: (i) maintenance of the collaborative posture, with the members of the group conscious of each one's responsibility for the work; (ii) understanding of each stage's goals in the global context; and (iii) motivation for the maximum interaction among the participants. At the operational level, the flow of work definition consists of the preparation, implementation and posterior performance of educational activities by students interacting in groups (Grégoire and Laferrière, 1999).

A typical workflow system helps to define, execute, coordinate and monitor the business processes in an organization. Therefore, the system should contain a representation of the activities structure and work procedures. This representation is usually a sequential or hierarchical decomposition of an activity in tasks, which is built separately from the execution of the activity. Opposite to this classic model, Bardram (1997) affirms that instead of supporting the information routing, the process in workflow systems should mediate the reflection and anticipation of appealing events in the work. Thus, a planning tool should support the construction, execution and monitoring of the collaborative activities along their development.

The use of a model of workflow can be useful to represent the flow of activities. Van der Veen et al. (1998) had carried through an experimental study on the application of workflow systems in the educational context and had identified to the similarities and differences between educational and business-oriented processes. The results had led to the conclusion of that the use of workflow systems as support to project-based learning brings profits in relation to the educational goals. Bardram (1997) concludes on the use of systems of workflow in relation to the activities human

beings: instead of supporting the tracing of information in organizations, as the classic model has supported to the process in workflow systems, the tools would have to mediate the reflection and anticipation of recurrent events in the work processes, thus, a planning tool would have to support the situated planning - construction, alteration, execution and monitoring of the Collaborative activities throughout its development.

The approach presented by Wang, Haake, Rubart and Tietze (2000) can be summarized by: (1) modeling the process (i.e. the topic of learning) using hypermedia; (2) integrating process support (i.e. support for definition and execution of processes) into the hypermedia model; (3) providing Collaborative hypermedia authoring and execution capability (i.e. a shared workspace) for working on the process description and executing it, and (4) providing Collaborative learning support in this shared workspace.

In our work, we used the concepts of collaborative process and workflow, just as they are used by workflow systems, however focused mainly on the planning of the project, the definition of the interactions among the group members, the definition of responsibilities and strategies for the solution of the problems. In the case of the educational processes, there is a commitment to the learning of settled concepts; thus it is necessary to count on the teachers' experience to plan the projects and to think in situations that stimulate the students to work in a collaborative way. It is necessary to lead the students to consider hypotheses of solutions for problems defined in the projects, to discuss them, to ponder them and to generate final products with conscience of the process pursued by them in order to reach their goal.

COPLE AND ITS COLLABORATIVE PROCESS EDITOR

In order to test our idea about the importance of the learning processes design, we implemented a CSCL infrastructure called COPLE - Collaborative Learning Project-Based Environment. (Santoro, Borges and Santos. 2003a). COPLE was supplied with a collaborative text editor (EdiTex) and with a collaborative process editor (COPE) to design the collaboration processes. COPE adopts symbols and conventions that might represent the components of the process: activities, roles, agents, flow, rules, descriptions, and the relationships among them. A process server interprets the model, deploys the execution of the process in the beginning of the project and maintains the relevant information about its activities. In Figure 1, a process for an educational project is presented as defined in COPLE. The process depicts the strategy of the group interaction to reach to the final product with participation and contribution of everybody. The importance of the transitions from individual tasks (I) to group tasks (G) is observed, where the performance of each group member is clear. In one moment, the work is divided in parts; nevertheless each one of them requests the involvement of everyone, because the whole group makes the generation of ideas for posterior elaboration.

A process is a flow of activities. The participants should describe each activity in details in the groupware tool COPE as shown in Figure 2. The information provided is agreed by the members of the group and represents a kind of contract among its members about how they will work together. Understanding individual responsibilities and making participations explicit help people learn how to collaborate. Besides learning some topics or concepts, they learn more about developing a project.

Specific groupware tools such as a Collaborative Text Editor (Santoro, Borges and Santos, 2003b) could support each activity in a project. Students use them to build products, discuss issues related to them and save their findings as they work together to reach their goals. These kinds of tools are also provided within the environment proposed.

After the group members and teachers define a process, they start working accessing individual Work Lists, finding information about the tasks to perform. Educational processes need to count on the teacher experience to plan the projects and to pose situations that stimulate students to work collaboratively. The students must consider hypotheses of solutions for problems defined in the projects, discuss them, and compose final products with conscience of the process.

RESULTS

Four case studies case had been carried out in order to evaluate our approach. Our assumption was that in groups that discuss and define their processes, the collaboration would be improved. In all cases, we proposed that groups develop projects that deal with studying a theme based on a problem stated, discussing questions related to it, and writing a short essay about their findings. We had groups working under an explicit process and others working in an ad hoc manner. We tried to evaluate the collaboration level within the groups during the development of the projects through the examination of the criteria and measures presented in Table 1. Four categories of study were defined: communication, collective knowledge building, coordination and awareness.

Interaction data were collected within the environment and also a questionnaire was applied to the participants. With this instrument we tried to assess the

Figure 1. Educational process design

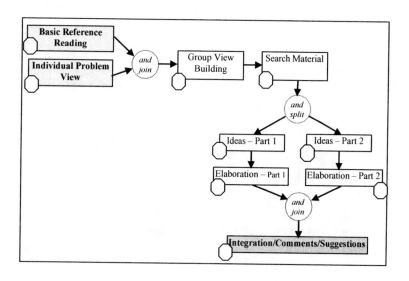

Figure 2. Activity definition in COPE

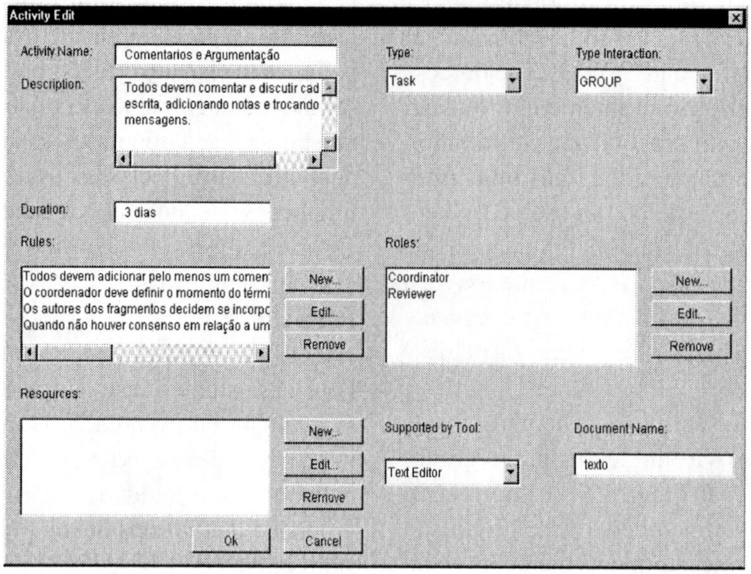

Table 1. Criteria and measures

Criteria	Measure	Type
Communication (interaction and participation)	Number of exchanged messages	Quantitative
	Nature (Quality) of exchanged messages	Qualitative
Collective building (contribution)	Number of contribution in the construction of a collective product	Quantitative
	Quality of contribution in the construction of a collective product	Qualitative
	Construction/Inference on the contribution of other group members	Qualitative
Coordination (engagement and organization)	Engagement in a process definition	Qualitative
	Tasks´ performance	Quantitative
Awareness	Understanding of tasks and the relationship among them	Qualitative

satisfaction and perception of the individuals about the collaboration level. The group composition, the task nature, the collaboration context and the infrastructure for communication are key points for successful collaboration in groupware. In all the case studies, the tasks had been similar; all of them involving the collective production of texts, despite the work processes have been different. The groups had been composed of people with similar academic formation, and still thus, we observed great differences in the dynamic of work. Individual characteristics have strong influence; therefore the environments must both stimulate, and exploit the individualities for the success of the work. We also tested presence/absence of coordinators and teacher support. A summary of the results is presented in Table 2.

Conclusions suggest that mechanisms for the definition and follow-up of processes really help the stimulus to collaboration in work groups. However, interaction based on the new technologies is not familiar for many

Table 2. Summary of results

1st Case Study	Six groups of four persons geographically dispersed: three had explicit processes and the others did not. They were asked to define a coordinator. The teacher defined the work processes and presented to three groups.
	➢ Groups with well-defined processes present best results, relating to interaction and contribution to the final product.
2nd Case Study	One group had to define their work process alone and recall all information about the development of their tasks.
	➢ The group had the feeling that defining a work process and following their initial planning has contributed to improve peers collaboration.
3rd Case Study	Two groups were asked to develop two similar projects. In the first time, they worked in an ad hoc manner. The second one, they used COPLE to plan their process and perform it. The teacher helped students to define their processes.
	➢ The results, mainly of the group 2, showed collaboration improvement from a project ill defined to a other project well defined
4th Case Study	Six groups of four persons working together on a lab: three had explicit processes and the others did not. The teacher helped students to define their processes and suggested forms and rules of interaction among participants in a group.
	➢ Groups that use explicitly defined processes present best results in terms of collaboration than groups that work without or little definition. The first kind of group also present final products with best quality level resulting of good level of contribution.

people. In COPLE, the mechanisms for structuring the knowledge, considered in the model, are not still implemented, such as the typology for chats and messages that aid the communication. The presence of these mechanisms is certainly a factor that could have influenced in the results.

CONCLUSION AND FUTURE WORKS

The findings of the case studies reaffirm the correctness of presented solution approach and provide basis for new research on CSCL area. Now it is important to carry on our research in order to achieve as far as possible a generalization of the results achieved. A characteristic of the formative evaluation, such as the one used in our work, is that less authentic studies, from the point of view of a theory validation, can bring valuable preliminary results. In spite of the limitation in size of the accomplished studies, many problems were discovered and they can be applied in future experiences.

The general conclusion was promising in the direction of the hypothesis proposed, about the relationship between the process design and the level of collaboration. The simple mechanisms implemented for this purpose in COPLE proved to address our main goals and configured real situations of developing a project

and stimulating collaboration. Now, we intend to continue using the environment to analyze its usage in other situations, with new problems being proposed to students and also with different teachers.

The challenge for collaborative learning systems designers is to create computational support that allows the students to follow the necessary paths for the development of the proposed project. To provide a theoretical and practical perspective that relates educational objectives and technical innovations involves the study of the interaction strategies among students' groups and integration of knowledge built collectively. The approach described intends to combine these two dimensions.

REFERENCES

Allen, R. (2006) *Workflow: An introduction.* Workflow Management Coalition. Retrieved February 10, 2006, from http://www.wfmc.org/information/Workflow-An_Introduction.pdf.

Bardram, J. E. (1998). *Designing for the dynamics of collaborative work activities.* Retrieved March 12, 2006, from http://www.daimi.au.dk/~bardram/docs/cscw98-bardram.pdf .

Bardram, J. E. (1997). Plans as situated action: An activity theory approach to workflow systems. J. A. Hughes, W. Prinz, T. Rodden, & K. Schmidt (Eds.), In *Proceedings of the Fifth European Conference on Computer-Supported Cooperative Work* (pp. 17-32). Lancaster, UK: Kluwer Academic Publishers.

Dewey, J. (1966). *Democracy and education.* New York: Free Press.

Dillenbourg, P., Baker, M., Blaye, A. & O´Malley, C.(1996) The evolution of research on collaborative learning. In E. Spada & P. Reiman (Eds.), *Learning in humans and machine: Towards an interdisciplinary learning science* (pp. 189-211). Oxford: Elsevier.

Engeström, Y. (1987). *Learning by expanding.* Helsinki: Orienta-Konsultit.

Gagné, R. M., Briggs, L. & Wager, W. (1992). *Principles of instructional design,* (4th ed.) Fort Worth: Harcourt Brace Jovanovich College Publishers.

George, S., & Leroux, P. (2001). Project-based learning as a basis for a CSCL environment: An example in educational robotics. Retrieved April 2, 2006, from http://www.ll.unimaas.nl/euro-cscl/Papers/57.pdf

Grégoire, R., & Laferrière, T. (1999). *Project-based collaborative learning with network computers- Teachers guide.* Retrieved February 10, 2006, from http://www.tact.fse.ulaval.ca/ang/html/projectg.html.

Kemp, J., Morrison, G., & Ross, S. (1996). *Designing effective instruction.* Upper Saddle River, New Jersey: Prentice-Hall, Inc.

Kilpatrick, W. H. (1926). *Foundations of method: Informal talks on teaching.* New York: Macmillan.

Leontiev, A. N. (1978). *Activity, consciousness, and personality.* Englewood Cliffs, NJ: Prentice-Hall.

Santoro, F., Borges, M., & Santos, N. (2003a). Learning through collaborative projects: The architecture of an environment. *International Journal of Computer Applications in Technology, 16*(2/3), 127-141.

Santoro, F.; Borges, M., & Santos, N. (2003b). Experimental findings with cooperative writing within a project-based environment. In M. Llamas-Nistal, M. J. Fernández-Iglesias, & L. Anido-Rifón (Eds.), *Computers and education: Towards a lifelong learning society* (pp. 179-190). London: Kluwer Academic Publishers.

Tiessen, E.L., & Ward, D.R. (1999). Developing a technology of use for collaborative project-based learning. In *Proceedings of CSCL '99, the International Conference on Computer Support for Collaborative Learning Designing new media for a new millenium : Collaborative technology for learning, education, and training.* Mahwah, N.J.: Lawrence Erlbaum Associates.

Van der Veen, J., Jones, V., & Collis, B. (1998). Workflow applied to projects in higher education. In J. Derrick & E. Najm (Eds.), *Educational case studies in protocols: Proceedings,* (pp. 147-158). Working conference of IFIP WG 6.1. Paris: Ecole Nationale Superieure des Telecommunications & Telecom Paris.

Wang, W., Haake, M., Rubart, J., & Tietze, D. (2000). Hypermedia-based support for cooperative learning of process knowledge. *Journal of Network and Computer Applications, 23*(4), 357-379.

Winer, L. R., & Váquez-Abad, J. (1995). The Present and future of ID practice. *Performance Improvement Quarterly, 8*(3), 55-67.

Woodbine, G. (1997). Can the various forms of collaborative learning techniques be applied effectively in the classroom in content driven accounting courses? In R. Pospisil and L. Willcoxson (Eds.), *Learning through teaching,* (pp. 357-360).

KEY TERMS

Cooperative Learning: Cooperative Learning is accomplished by the division of labor among participants, as an activity where each person is responsible for a portion of the problem solving. The technological support to this kind of activities is called Computer-Supported Cooperative Learning. (Dillenbourg et al., 1996).

Computer-Supported Collaborative Learning (CSCL): Collaborative work involves the mutual engagement of participants in a coordinated effort to solve the problem together. The technological support to this kind of activities is called Computer-Supported Collaborative Learning (Dillenbourg et al., 1996).

W

Groupware: Software designed to help teams or groups of people work together

Instructional Design: Instructional Design is the systematic development of instructional specifications using learning and instructional theory to ensure the quality of instruction. It is the entire process of analysis of learning needs and goals and the development of a delivery system to meet those needs. It includes development of instructional materials and activities; and tryout and evaluation of all instruction and learner activities (Gagné et al., 1992)

Project-Based Learning: Model for classroom activity that shifts away from the classroom practices of short, isolated, teacher-centered lessons and instead emphasizes learning activities that are long-term, in-terdisciplinary, student-centered, and integrated with real world issues and practices

Workflow: The automation of a business process, in whole or part, during which documents, information or tasks are passed from one participant to another for action, according to a set of procedural rules (Allen, 2006).

Workflow Management System: A system that defines, creates and manages the execution of workflows through the use of software, running on one or more workflow engines, which is able to interpret the process definition, interact with workflow participants and, where required, invoke the use of IT tools and applications (Allen, 2006).

Index

Symbols

3C collaboration model 637–644
3C collaboration model, communication 638
3C collaboration model, cooperation 640
3C collaboration model, coordination 639
3C collaboration model, design and implementation issues 641

A

academic blog 2
ACCADEMI@VINCIANA ontology 32
acknowledgement (ACK) 625
adaptability, in business 9
adaptive job performance (AJP) 10
adaptive structuration theory (AST) 193, 541
affect-based trust (ABT) 686
alignment, operational 219
alignment, organizational 219
alignment, strategic 219
Ambassadorial Leadership™ 21–28
ambient intelligence 30, 269
ambient intelligent prototype, and collaboration 29
AmIART prototype 31
AMPLIA 75
anonymity 8
application-level multicast (ALM) 624
application-level networking (ALN) 172
application service providers (ASPs) 355
appropriation 195
artificial partners 493
artificial partners, agent models for cooperation 494
artificial partners, coordinated communication 496
asynchronous creativity 463–471
autonomous system 626
average variance extracted (AVE) 104

awareness net 43

B

bandwidth, availability of 512
Barger, J. 1
blogrolling 2
business-to-business (B2B) 253
business-to-consumer (B2C) 253
business process outsourcing (BPO) 424
business process outsourcing (BPO), cultural misunderstandings 427
business process outsourcing (BPO), risks 425

C

Canada 444–449
Canada, Labrador 445–449
Canada, Newfoundland 445–449
car racing & instant messaging 603
cognition-based trust (CBT) 686
collaboration, definition of 309
collaboration asset 310
collaborative capability 134
collaborative commerce (cCommerce) 309
collaborative delivery of enterprise ICT services 355–362
collaborative delivery of enterprise ICT services, impact on end-user organizations 357
collaborative delivery of enterprise ICT services, impact on supplier organizations 358
collaborative delivery of enterprise ICT services, increasing process orientation 356
collaborative e-learning (CEL) 191
collaborative engineering environment (CEE) 589
collaborative learning 74
collaborative learning project-based environment (COPLE) 720
collaborative learning process editor (COPE) 720

collaborative planning, forecasting, and replenishment (CPFR) 319
collaborative working environments (CWE) 308, 392, 612
communications channels 292–300
communications channels, faculty-student communication 293
communications channels, faculty preferences 292
communications channels, social context of 293
communities of practice (CoPs) 211
compensatory adaptation theory (CAT) 699
computer, definition of 48
computer-mediated communication (CMC) 49, 164, 241, 431, 681
computer-supported collaborative learning (CSCL) 301, 718
computer-supported cooperative work (CSCW) 49, 411, 512
computer-supported negotiation (CSN) 688
computerized post office (CPO) 364
connectivity 8
contract negotiations 686
cooperative information systems (CoopIS) 555
core-assisted mesh protocol (CAMP) 625
creativity, and immediacy 466
creativity, and parallelism 467
creativity, and rehearsability 468
creativity, and reprocessability 468
creativity, and social cues 467
creativity, cognitive influences on 465
creativity, media influences on 465
creativity, social influences on 465
creativity support system (CSS) 248
cultural adroitness 433
cultural awareness 433
cultural sensitivity 433
customer relationship management (CRM) 613

D

data warehouse (DWH) system 558
designated member hosts (DMs) 627
diffusion of innovations (DOI) 412
digital divide 153
direct refinement technologies 227
distributed computer-mediated-communication (DGDSS) 379
DIVA knowledge community 44
dynamic enterprise alignment model 219

E

e-collaboration 512
e-collaboration, adoption by accounting firms 160
e-collaboration, adoption levels and scope 412
e-collaboration, and adaptive workforce 7–13
e-collaboration, and collaborative practice 529
e-collaboration, and communication 528
e-collaboration, and identity work 521–526
e-collaboration, and satisfaction 242

e-collaboration, and the military 673
e-collaboration, and trust 241
e-collaboration, and Weblogs 1–6
e-collaboration, financial audit research 160
e-collaboration, for internationalizing higher education 178–185
e-collaboration, fraud examination research 165
e-collaboration, gender communication issues 259–264
e-collaboration, gender differences and cultural orientation 301–307
e-collaboration, global funding 314–318
e-collaboration, governance mechanisms 319
e-collaboration, governance mechanisms and bargaining power 320
e-collaboration, rural Canadian schools 444–449
e-collaboration, supply chain 280
e-collaboration, the collaborative task 179
e-collaboration, through blogging 198–203
e-collaboration technologies (ECT) 191, 417
e-collaboration technologies (ECT), determinants of use 418
e-collaboration technologies (ECT), for a blood bank 584
e-collaboration technologies (ECT), potential for use 420
e-collaborative knowledge construction 233–239
e-commerce 655
e-commerce, cyber slacking 657
e-commerce, digital divide 657
e-commerce, impact on prices 656
e-commerce, privacy 656
e-commerce, security 657
e-health 153
e-lance networks 324–329
e-procurement for buyer-supplier interaction 14
e-scheduling 253–258
e-scheduling, collaborative production scheduling 254
e-scheduling, office and service tasks 255
electronic brainstorming 330–336
electronic brainstorming, group size effects 330
electronic brainstorming, role 331
electronic data exchange / interchange (EDI) 280, 608
employee-customer interaction 384
employee-employee collaboration 384
expert-centralized knowledge refinement 226
expertise gaps 228

F

face-to-face (FtF) 126, 660
face-to-face (FtF) GDSS groups (FGDSS) 378
face-to-face (FTF) interaction 49, 463, 662

G

geographical information systems (GIS) 133
global exchange services (GXS) 609
globalization, and e-collaboration 8
Goldratt's theory of constraints 603
group decision support system (GDSS) 204, 377
group decision support system (GDSS), communication 378

group decision support system (GDSS), virtual meetings 205
group support system (GSS) 133, 267, 411, 457

H

host multicast tree protocol (HMTP) 624
human-computer interaction (HCI) 412, 481

I

identity 521, 522
identity, and occupational groupings 522
identity work 521, 522, 523
identity work, and e-collaboration 521–526
IM in the workplace (IMW) 104
indirect refinement technologies 227
individual trust 534
individual trust, deconstructing 536
individual trust, expert power and positional power 537
individual trust, utility and risk 536
instant messaging (IM) 383
instant messaging (IM), consequences 102
instant messaging (IM), higher education
instant messaging (IM), support 349–354
intelligent tutoring systems (ITS) 74
intelligent user interfaces (user adaptive interfaces) 30
interaction, face-to-face 463
interaction, synchronous 463
interaction model 398
intercultural communication differences 430
Internet telemedicine to manage health conditions 154
interorganizational systems (IOS) 608
intra-class correlation coefficient (ICC) 452
intrusion detection 173

K

knowledge management (KM) 226, 527
knowledge refinement 226
Knowledge Web (K-Web) 674
knowledge work 9
knowledge worker 9

L

leader-member-exchange (LMX) theory 23
leadership, in virtual teams 343, 540
leadership, transactional 22
leadership, transformational 22
local area network (LAN) 513
Lotus Notes as a knowledge management tool 398

M

machine translation (MT) 457, 458
manufacturing resource planning (MRP II) 253
media-cognitive-social (MCS) model 464
media richness theory (MRT) 378, 499
medical learning 74
MGSS translation 459

mobile collaboration 133
mobile collaboration, human factors 133
multi-robot systems (MRS) 561
multi-robot systems (MRS), efficiently sharing information 563
multi-robot systems (MRS), probabilistic volumetric maps 562
multi-robot systems (MRS), sharing information within 561
multicast open shortest path first (MOSPF) 625
multilevel modeling methods for e-collaboration data 450–456
multilingual collaboration 457
multilingual collaboration, meetings 457
multinational corporation (MNC) 430

N

network intrusion detection systems (NIDS) 174

O

online discussion forum (ODF) 437
online discussion forum (ODF), for collaborative learning 438
online discussion forum (ODF), management and facilitation 439
overlay 172

P

partial least squares (PLS) 103
peer-to-peer (P2P) 681
personal computer 21
personal digital assistant (PDA), use in meetings 136
plain old telephone system (POTS) 695
presence-based real-time communication 487–492
principal investigator 315
production blocking 330
project-based learning (PBL) 718
protection methods 174
protocol independent multicast-sparse mode (PIM-SM) 625
psychological contracts' influence 499

R

real-time communication (RTC) technology 488
real-time communication (RTC) technology, products 490
real-time communication (RTC) technology, usage scenario 489
relational perspective 521
research and technology development (RTD) 612
resource based theory (RBT), to define a firm 218
robotic mapping 561

S

service-oriented architecture (SOA) 393
service-oriented e-collaboration environments 389–397
service-oriented e-collaboration environments, broker pattern 390

service-oriented e-collaboration environments, context modelling 391
service-oriented e-collaboration environments, interaction patterns 390
service-oriented e-collaboration environments, master/slave pattern 391
service-oriented e-collaboration environments, proxy patterns 390
shortest tunnel first (STF) 624, 626
small business collaboration 569
small business collaboration, e-business framework 570
small business collaboration, generic e-catalogue framework 572
small business collaboration, through electronic marketplaces 569–576
small to medium-sized enterprises (SMEs) 535, 608
social capital 521
social facilitation 330
social influence theory (SIT) 378
social loafing 330
social presence theory (SPT) 499
speech act theory 577
speech act theory, language/action perspective (LAP) 579
speech act theory, reactions 579
speech act theory, significance for e-collaboration 579
speech act theory, speech acts 578
synchronous interaction 463
synergy 330

T

technology acceptance model (TAM) 265, 668
technology-mediated learning (or e-learning) 191
technology-supported communities of practice 272–278
telegraph 53
thematic-based group communication 624
thematic multicast concept (TMC) 624, 626
time-interaction-performance (TIP) theory 474
transactional leadership 22
transaction cost economics (TCE) 325
transformational leadership 22

U

ubiquitous communications 30
ubiquitous computing 30
unified communication (UC) 488
U.S. Navy 674

V

value adding organization (VAO) 614
varied discussion forums 370
varied discussion forums, debate discussion 374
varied discussion forums, experience discussion 372
videoconferencing (VC) 693
videoconferencing (VC), market leaders 696
videoconferencing (VC), risks and challenges/limitations 695

Virtual Distance™ 23
virtual distance model (VDM) 660
Virtual Knowledge Park (VKP) 514
virtual organizations (VO) 589
virtual private network (VPN) 172
virtual project management 472
Virtual Science Park (VSP) 514
visibility 524
voice-based group support systems (VGSS) 706–711
voice-based group support systems (VGSS), speech recognition 707
voice-based group support systems (VGSS), sR accuracy 708
voice-based group support systems (VGSS), sR and GSS 709

W

Weblog 1–6
wikis 712
wikis, issues in wiki management 715
wikis, nature of 712
wikis, used for collaboration 714
workflow systems 718

WITHDRAWAL